Algorithms for Intelligent Systems

Series Editors

Jagdish Chand Bansal, Department of Mathematics, South Asian University,
New Delhi, Delhi, India

Kusum Deep, Department of Mathematics, Indian Institute of Technology Roorkee,
Roorkee, Uttarakhand, India

Atulya K. Nagar, School of Mathematics, Computer Science and Engineering,
Liverpool Hope University, Liverpool, UK

This book series publishes research on the analysis and development of algorithms for intelligent systems with their applications to various real world problems. It covers research related to autonomous agents, multi-agent systems, behavioral modeling, reinforcement learning, game theory, mechanism design, machine learning, meta-heuristic search, optimization, planning and scheduling, artificial neural networks, evolutionary computation, swarm intelligence and other algorithms for intelligent systems.

The book series includes recent advancements, modification and applications of the artificial neural networks, evolutionary computation, swarm intelligence, artificial immune systems, fuzzy system, autonomous and multi agent systems, machine learning and other intelligent systems related areas. The material will be beneficial for the graduate students, post-graduate students as well as the researchers who want a broader view of advances in algorithms for intelligent systems. The contents will also be useful to the researchers from other fields who have no knowledge of the power of intelligent systems, e.g. the researchers in the field of bioinformatics, biochemists, mechanical and chemical engineers, economists, musicians and medical practitioners.

The series publishes monographs, edited volumes, advanced textbooks and selected proceedings.

All books published in the series are submitted for consideration in Web of Science.

More information about this series at http://www.springer.com/series/16171

Sandeep Kumar · Sunil Dutt Purohit ·
Saroj Hiranwal · Mukesh Prasad
Editors

Proceedings of International Conference on Communication and Computational Technologies

ICCCT 2021

Editors
Sandeep Kumar
CHRIST (Deemed to be University)
Bangalore, Karnataka, India

Sunil Dutt Purohit
Rajasthan Technical University
Kota, India

Saroj Hiranwal
Rajasthan Institute of Engineering
and Technology
Jaipur, Rajasthan, India

Mukesh Prasad
University of Technology Sydney
Ultimo, NSW, Australia

ISSN 2524-7565   ISSN 2524-7573   (electronic)
Algorithms for Intelligent Systems
ISBN 978-981-16-3248-8   ISBN 978-981-16-3246-4   (eBook)
https://doi.org/10.1007/978-981-16-3246-4

This Springer imprint is published by the registered company Springer Nature Singapore Pte Ltd.
The registered company address is: 152 Beach Road, #21-01/04 Gateway East, Singapore 189721,
Singapore

Preface

This volume contains the papers presented at the 3rd International Conference on Communication and Computational Technologies (ICCCT 2021) jointly organized in virtual format by Rajasthan Institute of Engineering and Technology, Jaipur, and Rajasthan Technical University, Kota, in association with the Soft Computing Research Society during February 27–28, 2021. The International Conference on Communication and Computational Technologies invited ideas, developments, applications, experiences, and evaluations in the field of communication and computing from academicians, research scholars, and scientists. The conference deliberation included topics specified within its scope. The conference offered a platform for bringing forward extensive research and literature across the arena of communication and computing. It provided an overview of the upcoming technologies. ICCCT 2021 provided a platform for leading experts to share their perceptions, provide supervision, and address participant's interrogations and concerns. ICCCT 2021 received 292 research submissions from 35 different countries, viz., Argentina, Australia, Austria, Bangladesh, Brazil, Bulgaria, Burkina Faso, China, Egypt, Ethiopia, Finland, Ghana, Hungary, India, Iraq, Japan, Malaysia, Mexico, Montenegro, Morocco, Nigeria, Oman, Philippines, Portugal, Romania, Russia, Saudi Arabia, Serbia, South Africa, Spain, Sri Lanka, Ukraine, United Arab Emirates, USA, and Zimbabwe. The papers included topics pertaining to varied advanced areas in technology, artificial intelligence, machine learning, and the like. After a rigorous peer review with the help of the program committee members and 92 external reviewers, 75 papers were approved.

ICCCT 2021 is a flagship event of the Soft Computing Research Society, India. The conference was inaugurated by Prof. R. A. Gupta, Hon'ble Vice-Chancellor, Rajasthan Technical University, Kota, along with other eminent dignitaries including Prof. Anoop Singh Poonia, Chairman, RIET, and Prof. Dhirendra Mathur, RTU (ATU) TEQIP-III, Coordinator. The conference witnessed keynote addresses from eminent speakers, namely Prof. Dumitru Baleanu (Cankaya University, Ankara, Turkey), Prof. Nilanjan Dey (JIS University, Kolkata, India), Prof. Nishchal K. Verma (Indian Institute of Technology Kanpur, India), Prof. Aruna Tiwari (Indian Institute of

Technology Indore), Prof. R. K. Pandey (IIT (BHU) Varanasi), and Prof. Balachandran K (CHRIST (Deemed to be University), Bangalore). The organizers wish to thank Mr. Aninda Bose, Senior Editor, Springer Nature, and Mr. Radhakrishnan Madhavamani, Springer Nature, New Delhi, India, for their support and guidance.

Jaipur, India   Sandeep Kumar
Kota, India   Sunil Dutt Purohit
Jaipur, India   Saroj Hiranwal
Ultimo, Australia   Mukesh Prasad

Contents

About the Editors

Sandeep Kumar is currently an Associate Professor at CHRIST (Deemed to be University), Bangalore, India, and a Post-Doctoral research fellow at Imam Muhammad ibn Saud Islamic University, Riyadh, Saudi Arabia. He has worked with ACEIT Jaipur, Jagannath University Jaipur, and Amity University Rajasthan. He received his Ph.D. degree in computer science & engineering from Jagannath University Jaipur in 2015, the Master of Technology degree from RTU Kota in 2011, and the Bachelor of Engineering degree from Engineering College Kota in 2005. He is an associate editor for an SCI journal, "Human-centric Computing and Information Sciences (HCIS)." He has published more than sixty articles in well-known SCI/SCOPUS indexed international journals and conferences and attended several national and international conferences and workshops. He has authored/edited five books in the area of computer science. His research interests include nature-inspired algorithms, swarm intelligence, soft computing, and computational intelligence.

Sunil Dutt Purohit is an Associate professor of Mathematics at Rajasthan Technical University, Kota, India. He did his Master of Science (M.Sc.) in Mathematics and Ph.D. in Mathematics from Jai Narayan Vyas University, Jodhpur, India. He was awarded a University Gold Medal for being a topper in M.Sc. Mathematics and awarded Junior Research Fellowship and Senior Research Fellow of Council of Scientific and Industrial Research. He primarily teaches subjects like integral transforms, complex analysis, numerical analysis, and optimization techniques in graduate and post-graduate level courses in engineering mathematics. His research interest includes special functions, basic hypergeometric series, fractional calculus, geometric function theory, mathematical analysis, and modeling. He has credited more than 160 research articles and four books so far. He has delivered talks at foreign and national institutions. He has also organized many academic events. He is a Life Member of the Indian Mathematical Society (IMS), Indian Science Congress Association (ISCA), Indian Academy of Mathematics (IAM), and Society for Special Functions and their Applications, and Member of American Mathematical Society (AMS) and International Association of Engineers (IAENG). Presently, he is general-secretary of the Rajasthan Ganita Parishad.

Saroj Hiranwal is currently Principal at Rajasthan Institute of Engineering and Technology, Jaipur. Dr. Saroj is a Doctor of Philosophy (CSE) in Computer Science and Engineering; her research interests include high-performance scientific computing, real-time systems, and cloud computing. She has a Bachelor of Engineering and Master and Technology degrees in Information Technology and has 15 years of professional experience teaching technical courses. Her expertise includes operating systems, real-time systems, and object-oriented programming. She has participated and contributed in no. of national and international seminars, workshops, and conferences. Her research work has been published in various national and international journals. She has written books on information protection & security system, and operating systems.

Mukesh Prasad is a Senior Lecturer at the School of Computer Science in the Faculty of Engineering and IT at UTS who has made substantial contributions to machine learning, artificial intelligence, and the internet of things. Mukesh's research interests also include big data, computer vision, brain-computer interface, and evolutionary computation. He is also working in the evolving and increasingly important field of image processing, data analytics, and edge computing, which promise to pave the way for the evolution of new applications and services in the area of healthcare, biomedical, agriculture, smart cities, education, marketing, and finance. His research has appeared in numerous prestigious journals, including IEEE/ACM Transactions and conferences; he has written more than 100 research papers. Mukesh started his academic career as a lecturer with UTS in 2017 and became a core member of the University's world-leading Australian Artificial Intelligence Institute (AAII), which has the vision to develop theoretical foundations and advanced technologies for AI and to drive progress in related areas. His research is backed by industry experience, specifically in Taiwan, where he was the principal engineer (2016–17) at the Taiwan Semiconductor Manufacturing Company (TSMC). There, he developed new algorithms for image processing and pattern recognition using machine learning techniques. He was also a postdoctoral researcher leading a Big Data and computer vision team at National Chiao Tung University, Taiwan (2015). Mukesh received an M.S. degree from the School of Computer and Systems Sciences, Jawaharlal Nehru University, New Delhi, India (2009), and a Ph.D. from the Department of Computer Science, National Chiao Tung University, Taiwan (2015).

Chapter 1
A Novel Type-II Fuzzy based Fruit Image Enhancement Technique Using Gaussian S-shaped and Z-Shaped Membership Functions

Harmandeep Singh Gill⊙ **and Baljit Singh Khehra**⊙

1 Introduction

The process of improving the visual appearance of an image for further study of a particular application is known as image enhancement. It is dependent on the human perception. Image enhancement techniques are classified into spatial domain technique, frequency domain technique and fuzzy domain techniques. In spatial domain methods, pixels of the image were manipulated to enhance the image. Frequency domain methods are based on orthogonal transform of the image performed than the image itself. Sharma et al. [1, 2] deployed firefly algorithm for image enhancement. Nonlinear and knowledge-based systems were used in fuzzy domain methods.

Image enhancement translates the input image into the form that is better in comparison with the acquired image. Generally, enhancement techniques are classified into transform domain and spatial domain. Transformation technique is used to modify the frequency transform of an image, whereas spatial domain transformation technique directly operates on the pixel of an image. Major objective of image enhancement technique is to produce the target image, which is more reliable as compared to the acquire image.

Fruit image enhancement is still an challenging task due to heterogeneous nature of fruits. Due to large variety of fruits, it is tough to identify the type of fruit based on common features. From marketing point of view, it is essential to maintain the quality of fruit. There are various stages in the life of fruit from birth to ripeness. Some fruits are seasonal, and weather also plays a major role during the growth stages of a fruit.

H. S. Gill (✉)
Guru Arjan Dev Khalsa College, Chohla Sahib, Punjab 143408, India

B. S. Khehra
CSE Department, BBSB Engineering College, Fatehgarh Sahib, Punjab, India
e-mail: baljit.singh@bbsbec.ac.in

© The Author(s), under exclusive license to Springer Nature Singapore Pte Ltd. 2021
S. Kumar et al. (eds.), *Proceedings of International Conference on Communication and Computational Technologies*, Algorithms for Intelligent Systems,
https://doi.org/10.1007/978-981-16-3246-4_1

1

To enhance the fruit images, it is mandatory to know about the features of the particular fruit. In proposed work, red apple, orange and watermelon fruits images are enhanced based on shape, size, color and texture features. All these fruits look identical in shape but have different texture, color and intensity features. These fruits images are acquired from the orchards of Agriculture Department, Khalsa college, Amritsar(Punjab)-143005.

Bad weather or environmental conditions also affect the fruit quality. This is the major motivation factor behind this proposed work to enhance the fruit images using Type-II fuzzy-based membership functions. Type-II fuzzy handles the problem of uncertainty and vagueness effectively. Gaussian, S-shaped and Z-shaped fuzzy membership functions are employed in this proposed work for the enhancement of fruit images.

2 Related Work

Histogram equalization is mostly used in image contrast enhancement technique, which is widely used in weather forecasting, medical, agriculture, X-ray, MRI, etc. HE is exploited to increase the visibility of image contents by altering the intensity values at different levels of the image but fails to produce effective results during low contrast-level images. CLAHE is another popular image contrast, operates on small regions in the image called tiles rather than the complete image. CLAHE fails to preserve edges and retains the actual values.

Many algorithms have developed in the recent times for image contrast enhance-ment. Histogram equalization is a famous image contrast enhancement technique, commonly used to enhance medical, agriculture, weather forecasting, X-ray, MRI, etc., till date. Image HE is exploited to increase the visibility of image contents by altering the intensity value. This method is not suitable for low contrast images. Seti-awan et al. [3] enhanced the retinal images by genetic algorithm for both qualitative and quantitative performance using PSNR, AMBE and RMSE. Shi et al. [4] enhanced the satellite image using DWT-SVD by cuckoo search algorithm for improvement of the low contrast images and measured the performance in terms of PSNR, MSE and RMSE.

Transform parameters are determined by different contrast type of input image. Zhang et al. [5] enhanced the fruit images by gray transform and wavelet neural network . In their method, they supposed that HE will improve the contrast of the fruit image but they did not show any results and information which needs further work and research. de Araujo et al. [6] enhanced images based on a new artificial life model, inspired by herbivore organism behavior. This method is used for image enhancement in the spatial domain to find the intensity values and difference between the neighbors; however, this model has some limitations that how we can find the optimized neighbor values to evaluate the best one. The classical image processing techniques process the crisp histograms of the images which do not consider the exactness of the gray values . Also the vagueness and uncertainty are not exactly measured with these existing techniques. Fuzzy logic techniques are applied to overcome these limitations. Evolutionary methods PSO, BFO and ACO were used by Hanmandlu et al. [7] in their different research work on image enhancement, compared with genetic algorithm (GA) and produced results which are better.

The approach used in [7] is extended by enhancing the fruit image by fuzzy logic technique using Batrerial Foraging Optimization (BFO). In this approach, fuzzy technique is used to enhance the image to analyze the image quality and improves the contrast level of the enhanced images by choosing the optimized results produced using BFO. Kuhl [8] enhanced the radiographic, medical, mammography, X-ray angiogram, infrared, microscopic, ultrasound images, used fuzzy techniques based on fuzzy set theory, fuzzy IF-THEN rules, fuzzy operators and fuzzy membership functions, enhanced the medical images using fuzzy set theory by intuitionist fuzzy set to handle the uncertainty in the form of fuzzy membership function and produced results which suggest that the enhanced images are better. Medical images are enhanced using the following formula.

Li et al. [9] deployed fuzzy sure entropy to enhance the low contrast images. In this work, a positive threshold selection is used to tune the image enhancement performance. G.Raju employed fuzzy logic to enhance the color images by considering average intensity values of the image based on M and K parameters by transforming the skewed histogram of the original image into a uniform histogram. They used fuzzy sure entropy to improve the quality of low-quality images. In this work, fuzzy sure entropy is utilized to handle the uncertainty.

Pal and Rosenfeld [10] used fuzzy compactness based on minimization to obtain threshold values for fuzzy and non-fuzzy both fuzzy and ill-defined image. Involved fuzziness in the spatial domain to define in describing the geometry of regions tries to provide more meaningful results than by considering fuzziness in gray level only. Chen and Ludwig [11] used fuzzy clustering to segment the images to overcome the issues of overlapping and poor contrast. Fuzzy clustering is derived from fuzzy C means.

In the fuzzy image processing, before processing an image, the image is transformed from spatial domain to the fuzzy domain using fuzzy membership functions. The membership values are modified to enhance the contrast of the image, and finally de-fuzzification will convert the enhanced image into the spatial domain. Sometimes image quality becomes lower due to camera lens, dim light during capturing or shutter speed. It is mandatory to improve the contrast for further image processing tasks.

3   Problem Formulation

In the suggested work, an image I of size $(A \times B)$ with gray-level values L between $L_{\min}$ and $L_{\max}$ is considered as an array of fuzzy singletons. Each array element describes the rate of illumination, and an image I can be addressed as [12]:

$$I = \mu_A(g_i j)/g_i j, \quad i = 1, 2, \ldots A, j = 1, 2, \ldots B \tag{1}$$

where $\mu_A(g_i j)$ represents the degree of brightness possessed by gray-level intensity g_{ij} corresponding to the pixels (i, j).

Fruit images are converted into fuzzy domain. Fuzzy logic is used to control the uncertainties associated with fruit images. Fuzzy membership functions are most important components to deal with the uncertainty. As proved in the previous work of visibility enhancement of weather, degraded fruit images are using Type-II fuzzy. It has been concluded that Type-II handles uncertainty more robustly than Type-I Fuzzy [13, 14].

Gaussian membership function is utilized to define the presence of gray level. Z and S-membership functions are used to select the dark and bright areas.

3.1 Gaussian Membership Function

Gaussian membership function is used to measure the involvement of gray level in the fruit images. Gaussian membership function is the most commonly used membership function to convert the original image into the fuzzy plane [5]. The standard Gaussian function is defined as follows:

$$g(A, B) = e - \frac{[f(A, B) - L_{\max}]^2}{2b^2} \tag{2}$$

where $f(A, B)$ is the input fruit image, $g(A, B)$ is fuzzy image, $L_{\max}$ is the maximum gray level of input fruit image, and parameter b is bandwidth of the Gaussian function that is calculated from the following equation.

$$b = t_{\mathrm{opt}} - K, L_{\max} - t_{\mathrm{opt}} \tag{3}$$

Selection of Parameter t_{opt} Parameter t_{opt} plays an important role for optimal selection of bandwidth of Gaussian function. Gaussian membership function transformed gray-level values of the fruit image to an interval [0, 1]. Optimal selection of bandwidth of Gaussian function will highlight dark and bright areas. Parameter t_{opt} is an optimal threshold that separates dark region from bright region. The procedure for optimal threshold selection from fruit image is as follows:

Let $f(A, B)$ be the gray-level value of acquired fruit image at the pixel (A, B) and $p_0, p_1, p_2, \ldots p_{L_{\max}}$ be the probability distribution of gray-level values of the fruit image, where p_j is the normalized histogram of fruit image, i.e.,

$$p_j = \frac{h_j}{(A \times B)} \tag{4}$$

where h_j is the gray-level histogram of fruit image. h_j is the number of pixels in the fruit image with gray level $j = 0, 1, 2, \ldots L_{\max}$

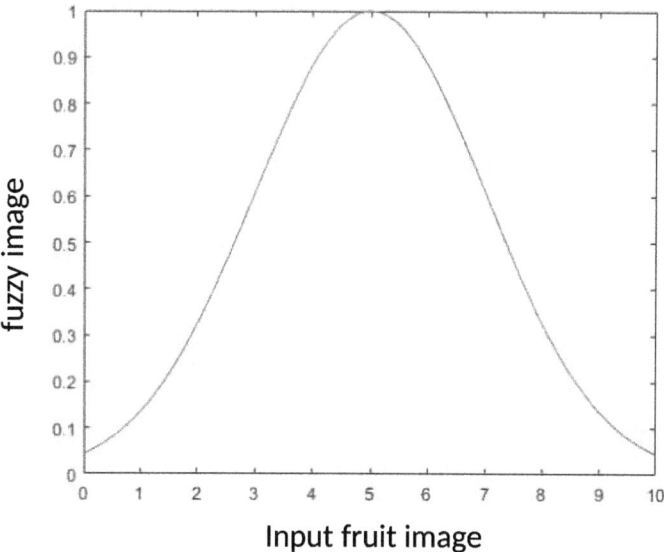

Fig. 1 Graphical representation of Gaussian membership function

To identify dark and bright regions, two fuzzy membership functions (S-function and Z-function) are employed (Fig. 1).

3.2 S-Shaped Membership Function

S-membership function is used to select the dark values from the fruit image and defined as follows:

$$\mu_s(i) = S(i, a, b, c) = \begin{cases} \frac{(j-a)^2}{(c-a)(b-a)} i \leq a, \\ 1 - \frac{(i-c)^2}{(c-a)(c-b)} x < i < y \end{cases} \tag{5}$$

3.3 Z-Shaped Membership Function

Z-membership function is used to select the bright values from the fruit image and defined as follows:

$$\mu_Z(i) = S(i, a, b, c) = \begin{cases} 1 - \frac{(j-a)^2}{(c-a)(b-a)} i \leq a, \\ \frac{(i-c)^2}{(c-a)(c-b)} x < i < y \end{cases} \tag{6}$$

Fig. 2 Graphical representation of S and Z-membership function

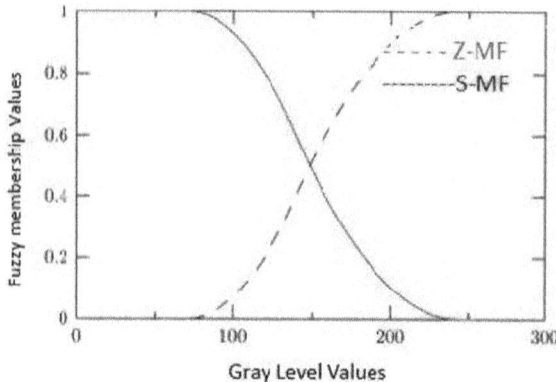

where $0 \le a \le b \le c \le L_{\max}$, a, b and c variables establish the frame of S and Z-membership functions as illustrated in Fig. 2. At intersection point, optimal threshold level (t_{opt}) is achieved for (a, b, c) to maximize $H(a, b, c)$.

In above-specified probability distribution of gray degree rates of fruit image, we can obtain two probability distributions, one for dark region and other for bright.

For dark region, probability is defined as follows:

$$\mu_D(0)p_0, \mu D(1)p1, \ldots, \mu_{D(L_{\max})}p_{L_{\max}} \tag{7}$$

For bright region, probability is defined as follows:

$$\mu_B(0)p_0, \mu B(1)p1, \ldots, \mu_{B(L_{\max})}p_{L_{\max}} \tag{8}$$

Shannon entropy for dark and bright region pixels are defined as follows:

$$H_D(x, y) = -\sum_{j=0}^{L_{\max}} \mu_D(j)p_j log\left(\sum_{j=0}^{L_{\max}} \mu_D(j)p_j\right) \tag{9}$$

$$H_B(x, y) = -\sum_{j=0}^{L_{\max}} \mu_B(j)p_j log\left(\sum_{j=0}^{L_{\max}} \mu_B(j)p_j\right) \tag{10}$$

Now, the sum of dark and bright entropies can be written as follows:

$$H(x, y) = H_D(x, y) + H_B(x, y) \tag{11}$$

The optimal threshold values t_{opt} that divide the fruit image into dark and bright regions can be measured from the optimal values of x and y that can maximize $H(x, y)$ as follows:

$$(x^*, y^*) = \text{Arg-}\max_{x, y = 0} \frac{L_{max}}{}[H(x, y)] \tag{12}$$

4 Experimental Results and Discussions

In this section, the performance of the proposed approach is evaluated through the simulation results using MATLAB 7.7.0 (R2013b) for a set of fruit images captured from the orchards of Agriculture Department, Khalsa College, Amritsar (Punjab).

Results of proposed scheme are compared with HE, CLAHE and MIX-CLAHE for red apple, orange and watermelon fruit images. It has been observed that the proposed scheme results are better. In order to show the efficiency of the suggested algorithm, the results of the suggested and existing enhancement methods are displayed in Fig. 3.

Tables 1, 2 and 3 show the MSE, PSNR, BER and RMSE for the fruit images red apple, orange and watermelon for enhancement techniques HE, CLAHE, MIX-CLAHE and proposed method.

Fig. 3 Results of proposed technique are as follows: **a**, **f** and **k** input fruit images, **b**, **g** and **i** enhanced by HE. **c**, **h** and **m** enhanced by CLAHE, **d**, **i** and **n** enhanced by MIX-CLAHE, **e**, **j** and **o** enhanced by proposed technique

Table 1 Enhancement values for red apple fruit image

Method	HE	CLAHE	Mix-CLAHE	Proposed
MSE	10.73	11.65	16.89	7.55
PSNR	48.99	46.09	45.01	44.08
BER	2.89	4.68	5.79	1.02
RMSE	27.10	26.98	27.99	26.88

Table 2 Enhancement values for orange fruit image

Method	HE	CLAHE	Mix-CLAHE	Proposed
MSE	12.33	11.65	16.08	7.33
PSNR	46.99	46.09	45.01	44.08
BER	3.89	4.08	5.19	1.02
RMSE	27.08	27.98	27.09	26.88

Table 3 Enhancement values for watermelon fruit image

Method	HE	CLAHE	Mix-CLAHE	Proposed
MSE	12.76	9.85	11.08	8.56
PSNR	48.89	46.75	45.10	43.08
BER	1.89	1.08	2.19	1.02
RMSE	26.98	27.56	27.59	26.08

5 Conclusion

In the proposed work, an attempt is made to enhance the fruit images. From experimental results, it is has been observed that Type-II-based fuzzy fruit images enhancement procedure outperforms the existing image enhancement schemes. MSE, BER, PSNR and RMSE parameters are used to evaluate the performance of existing and proposed methods for enhancement of fruit images. Obtained results are quite promising and effective.

References

1. Sharma A, Chaturvedi R, Kumar S, Dwivedi UK (2020) Multi-level image thresholding based on Kapur and Tsallis entropy using firefly algorithm. J Interdiscip Math 23(2):563–571
2. Sharma A, Chaturvedi R, Dwivedi UK, Kumar S, Reddy S (2018) Firefly algorithm based Effective gray scale image segmentation using multilevel thresholding and entropy function. Int J Pure Appl Math 118(5):437–443
3. Setiawan AW, Mengko TR, Santoso OS, Suksmono A (2013) Color retinal image enhancement using CLAHE. In: International conference on ICT for Smart Society. IEEE, pp 1–3
4. Shi Z, Zhu M, Xia Z, Zhao M (2017) Fast single-image dehazing method based on luminance dark prior. Int J Patt Recogn Artif Intell 31(02):1754003
5. Zhang C, Wang X, Zhang H (2006) Contrast enhancement for fruit image by gray transform and wavelet neural network. In: IEEE International conference on networking, sensing and control. IEEE, pp 1064–1069
6. de Araujo AF, Constantinou CET, Jo Manuel RS (2014) New artificial life model for image enhancement. Expert Syst Appl 41(13):5892–5906
7. Hanmandlu M, Verma OP, Kumar NK, Kulkarni M (2009) A novel optimal fuzzy system for color image enhancement using bacterial foraging. IEEE Trans Instrument Measure 8:2867–2879

8. Kuhl C (2007) The current status of breast MR imaging part I. Choice of technique, image interpretation, diagnostic accuracy, and transfer to clinical practice. Radiology 244(2):356–378
9. Li C, Yang Y, Xiao L, Li Y, Zhou Y, Zhao J (2016) A novel image enhancement method using fuzzy sure entropy. Neurocomputing 215:196–211
10. Pal SK, Rosenfeld A (1988) Image enhancement and thresholding by optimization of fuzzy compactness. Pattern Recogn Lett 7(2):77–86
11. Chen M, Ludwig SA (2017) Color image segmentation using fuzzy C-regression model. Adv Fuzzy Syst 2017
12. Khehra BS, Pharwaha APS (2012) Automatic detection of microcalcifications in digitized mammograms using fuzzy 2-partition entropy and mathematical morphology. In: Proceedings of IASTED international conference on signal and image processing (conference proceedings), pp 220–227
13. Singh H, Khehra BS (2018) Visibility enhancement of color images using Type-II fuzzy membership function. Modern Phys Lett B 3(11):18501–18530
14. Gill HS, Khehra BS (2020) Efficient image classification technique for weather degraded fruit images. IET Image Process 14(14):3463–3470

Chapter 2
Lyrics Inducer Using Bidirectional Long Short-Term Memory Networks

Jayashree Domala, Manmohan Dogra, and Anuradha Srinivasaraghavan

Abstract Songs are melodic manifestations that are performed by individuals. These tunes are made altogether by both a lyricist who composes the verses and the artist who sings them. Verse writing in itself is an exceptionally selective and characterized issue. The ever-expanding utilization of innovation and the way that they are effectively accessible to us make human lives comfortable. A lyricist can often have a mind block while considering verses or may even find it difficult to get an idea. The principal reason for this exploration is to enable the lyricist to get a motivation that can assist him in making better verses. To accomplish this, a profound learning method is utilized alongside the idea of natural language processing. Specifically, bidirectional long short-term memory (LSTM) networks are used for lyric generation. The proposed framework can exceptionally create versus relying upon the information seed and the scope of words.

1 Introduction

The verses that are composed by a musician assume a significant function in the arrangement and formation of the song. It gives life to the music and makes the song whole and complete. But the work behind forming the lyrics is often tiresome and time consuming. Additionally, if there is writer's block or no inspiration, then the lyric writing process is delayed further. Machine learning has already done wonders in backing up humans by automating tasks completely or partially. In this case, also machine learning can be used as a tool to help the generation of lyrics. Keeping in mind this as the motivation of the research, a deep neural network-based model is built to generate lyrics of the desired size.

The proposed system aims at generating lyrics according to the seed input which is given by the user. In this way, the lyrics will be generated following the idea that the writer wants. Backed by deep learning and natural language processing (NLP)

J. Domala (✉) · M. Dogra · A. Srinivasaraghavan
Department of Computer Engineering, St. Francis Institute of Technology, Borivali-west, Mumbai, Maharashtra, India

© The Author(s), under exclusive license to Springer Nature Singapore Pte Ltd. 2021   11
S. Kumar et al. (eds.), *Proceedings of International Conference on Communication and Computational Technologies*, Algorithms for Intelligent Systems,
https://doi.org/10.1007/978-981-16-3246-4_2

concepts, this system is trained on a unique dataset of 'romantic' English songs. Long short-term memory networks (LSTM) [1], an uncommon sort of recurrent neural network (RNN), are utilized for the learning process. These LSTM networks are equipped for recalling data for a more drawn out period because of the presence of cell states. The intuition is that the generation of music relies upon the arrangement of input words we feed and subsequently the next word generated must relate to it; therefore, the utilization of the LSTM model is favored over the ordinary RNN model. Before the data can be given to the LSTM networks for learning, it is preprocessed by using NLP techniques. The insight is that since the lyrics consist of the words and strings, we may need to preprocess them using NLP to increase the performance. Specifically, the bidirectional LSTM networks are used since the efficiency is better than the other models it is compared with, namely LSTM and gated recurrent unit (GRU) networks.

2   Related Work

Numerous researchers have worked using a multitude and varied algorithms on the lyrics generation techniques. The authors Pudaruth et al. proposed a framework for automated generation of song lyrics using CFGs. Prior to the execution, an inside and out investigation was done for understanding the necessities of good verses. They completed their assessment by looking over the produced and existing verses present on the web. The outcomes were agreeable, and produced verses were appraised as being existing verses [2]. Another research named Tra-la-Lyrics 2.0 was done by the author Gonçalo Oliveira. The exploration referenced a framework for the automatic generation of song lyrics on a semantic area. It is the improvement of the original Tra-la-Lyrics wherein the content is created which is characterized by at least one seed word. To gage the advancement, the mood, the rhymes, and the semantic intelligence in verses delivered by the first Tra-la-Lyrics were investigated and contrasted with the verses created by the new launch of this framework, named Tra-la-Lyrics 2.0. The assessment demonstrated that in the sections by the new system, words have a higher semantic relationship among them and with the given seeds, while the rhythm is so far composed and rhymes are accessible. A study was led to affirm the aftereffects of the improvement of the verses by Tra-la-Lyrics 2.0 [3]. Furthermore, the author's Malmi et al. proposed DopeLearning. It explains the rap lyrics generation using a computational approach. They build up an expectation model to recognize the following line of the current verses from a lot of applicant next lines. The model is based on two artificial intelligence procedures. They are the RankSVM algorithm and a deep neural network model with a novel structure. Results show that the forecast model can recognize the valid next line among 299 erratically picked lines with a precision of 17%, i.e., more than 50 times more likely than by discretionary. They additionally utilize the forecast model to join lines from existing tunes, creating verses with rhyme and significance. An assessment of the delivered verses shows that as far as quantitative rhyme thickness, the technique beats the best human rappers by 21% [4].

In another paper, Watanabe et al. present novel generation models that catch the point changes between units exceptional to the verses, for example, stanza/theme and line. The research centers around the basic relations in Japanese verses. These changes are displayed by a hidden Markov model (HMM) for representing themes and subject advances. As per the outcomes, the language model is definitely more powerful than the HMM-based model. However, the HMM-based methodology effectively catches the between stanza/ensemble and between line relations. For confirmation, the models are assessed utilizing a log-likelihood of lyrics generation and fill-in-the-spaces type test [5]. The authors Ramakrishnan et al. introduced a model for the programmed generation of Tamil lyrics. A corpus consisting of ten melodies was utilized to prepare the framework to comprehend the syllable examples. Utilizing the prepared model, the syllabic example is speculated for another tune to create an ideal succession of syllables. The obtained sequence is presented to the sentence generation module. This module uses Dijkstra's shortest path algorithm to think of an important expression coordinating the syllabic example [6]. Moreover, the author's Potash et al. exhibited the adequacy of a long short-term memory language model for automatic rap lyrics generation. The model produces sections that are near in style to that of a given rapper. The model characterizes its rhyme plot, line length, and verse length. The examinations show that a long short-term memory language model delivers better 'ghostwritten' verses than a standard model [7]. In the next paper, the authors Son et al. have used deep learning for Korean song lyrics generation. They switched the K-pop verses information and utilized them as learning information. The setting between verses is considered in the proposed song lyrics generation method. Each time the model produces the verse, the model experiences upper randomization. It was affirmed that the verses produced utilizing the opposite information have a more characteristic setting than the verses created utilizing the forward information [8]. The authors Pablo Samuel Castro et al. built up a framework that uses joined scholarly expressive structures and jargon for a verse age. They joined two separately trained language models into a framework that can deliver yield regarding the ideal tune structure while giving wealth and a decent variety of jargon that renders it all the more imaginatively engaging [9]. Additionally, in another research, the authors Fan et al. proposed a hierarchical attention-based sequence-to-sequence model for Chinese lyrics generation. This model advances the subject of significance and consistency of generation by utilizing the encodings of word-level and sentence-level relevant data. For model training, a large Chinese lyrics corpus is likewise utilized. In the end, aftereffects of automatic and human evaluations demonstrate that the model can form total Chinese verses with one joined subject constraint [10]. The authors Manjavacas et al. used hierarchical modeling and conditional templates for the generation of hip hop lyrics. The model created is a straightforward system to separate and apply contingent formats from text pieces. The methodology that is proposed empowers start to finish preparing, focusing on formal properties of text, for example, musicality and rhyme, which are focal attributes of rap messages. A crossover type of hierarchical model is used. This intends to coordinate language modeling at two levels: word and character-level scales [11]. Lastly, in the paper proposed by the authors Jain et al., an automatics lyrics generator is executed for romanized Hindi.

Sr. No	Reference No.	Algorithm Used	Results
1	[2]	Automated generation of song lyrics using CFG's	A majority of 52% of respondents thought that lyrics generated was from a human songwriter
2	[3]	It is the improvement of the original Tra-la-Lyrics on a semantic area	Tra-la-Lyrics 2.0 got the highest scores, not only on the meaning but also sound, but there is still much room for improvement
3	[4]	Rap lyrics generation using a computational approach and expectation model	An accuracy of 17%, i.e., over 50 times more likely than by random, and it was ranked in the top 30 with 53% accuracy
4	[5]	Hidden Markov Model (HMM) based system	This result indicates the superior effectiveness of line generation by the language model than by the contents models
5	[6]	Sentence Generation module using the Dijkstra's shortest path algorithm	Generate a syllable pattern that closely matches the input tune. The identification of strong beats in the melody not considered though.
6	[7]	Long short term memory language model	The correlation between rhyme density and max similarity for the n-gram model is 0.47.
7	[8]	Deep learning based model	Confirmed that the lyrics generated using the reverse data, have a more natural context than the lyrics generated using forward data
8	[9]	Framework which uses joined scholarly expressive structures and jargon	Resulted desired song structure, while providing a richness and diversity of vocabulary.
9	[10]	A hierarchical attention based sequence to sequence model	Model is able to compose complete Chinese lyrics with one united topic constraint with BLEU score 0.288
10	[11]	Hierarchical modeling and conditional templates	Despite advantages of hierarchical modeling, effects of conditional templates did not compound and result is discouraging
11	[12]	Simple techniques to catch rhyming examples previously	54.5% of the paragraphs in N-Gram outputs and 64.5% of the paragraphs in LSTM model outputs were labeled as "Makes Sense"

Fig. 1 Comparison of related work

The proposed model uses simple techniques to catch rhyming examples previously and during the model training process in the Hindi language [12]. There has been quite varied research and implementation in the domain of lyrics generation. The research paper proposes a different method by using the concept of deep learning LSTM networks specifically for romantic English songs. A succinct summary of the related papers is demonstrated in Fig. 1.

3 Implementation

The implementation is demonstrated in steps and shown in Fig. 2.

3.1 Dataset Preparation

The dataset is unique and customized according to the use case. It consists of lyrics of English songs scraped from the web. The total number of words in the dataset is approximately 13k.

3.2 Preprocessing

The dataset is preprocessed to maintain consistency in data which will help further during the training of the model. The preprocessing techniques used are as follows:
Lowercasing The dataset contains both capital and small letters. Since the model will take the same word with a difference in capitalization as distinctive, it will create confusion. For instance, 'The' and 'the' are unique for the model to manage and may not anticipate the similarity on encountering it. Hence, making all the letters lowercase would wipe out this problem and improve the results.
Tokenization The data to be used in the model should be of numeric structure for further analysis and classification. This is accomplished by the process of 'tokenization'. This helps by making the numeric tokens of each word present in the corpus.

3.3 Sequence Creation

The next step is to turn the sentences into lists of values based on the tokens generated by the tokenizer. This is significant as the model will anticipate the next word on feeding the set of input words, we need to train it on a set of words with the next word as output to our model.

3.4 Padding the Vectors

In padding, the list of sentences has been padded out into a matrix of the same length. This is achieved by putting the appropriate number of zeros before the sentence list. The matrix width is kept the same as the width of the longest sentence to make each vector of the same length.

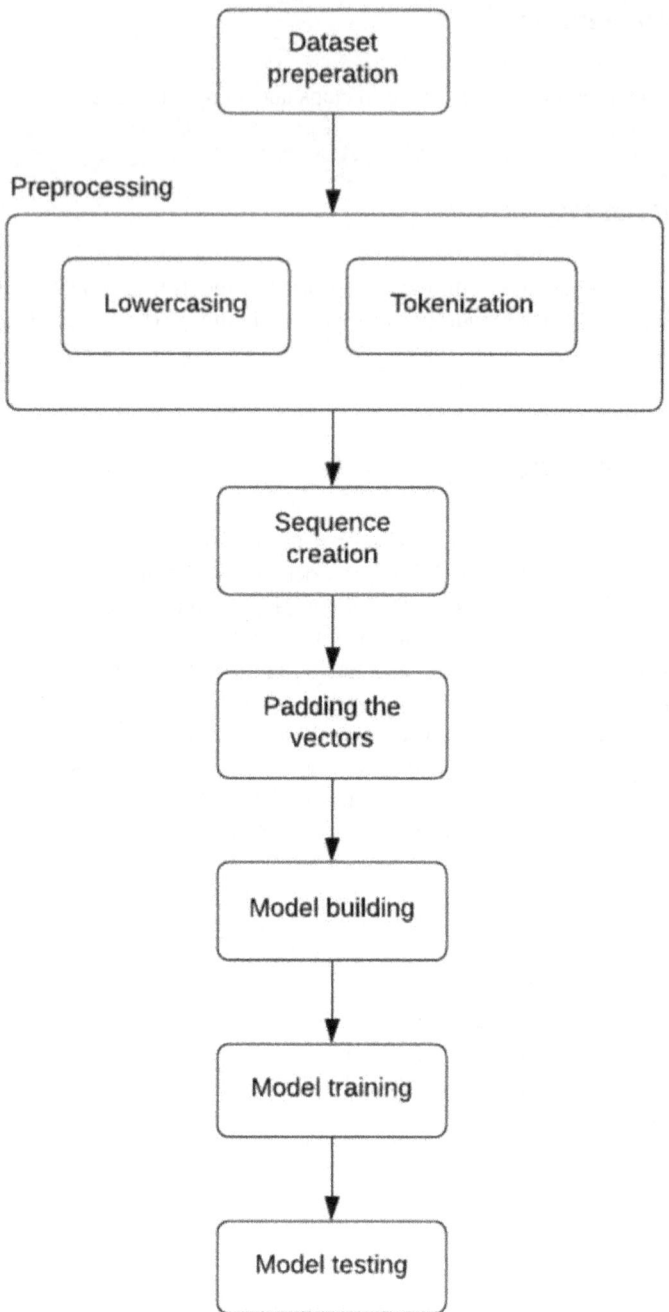

Fig. 2 Flowchart of the implementation

3.5 Model Building

Now, since all the required preprocessing is done, the next step is to build the model for training. The layers that are used for model building are as follows:

Embedding Layer It is the first hidden layer of the neural network. The intuition is that words and related words are bunched as vectors in a multi-dimensional space. The input and output dimension along with the length are specified. For this layer, the yield is a 2D vector with one embedding for each word in the input sequence of words.

LSTM Layer LSTM layer and bidirectional LSTM layer are used. The bidirectional layer helps in providing additional context to the network and results in a faster learning process. The bidirectional LSTM trains two LSTMs on the input sequence. The first one is the input sequence the way it is, and the second on a switched duplicate of the input sequence.

Dropout Layer This layer is introduced to reduce overfitting. At each update of the training phase, it works by arbitrarily setting the outgoing edges of hidden units to 0.

Dense Layer This layer is fully connected. All the neurons in a layer are associated with those in the following layer. It is followed by the activation function which in this case is 'softmax'.

3.6 Model Training

The model is compiled using the loss function 'categorical cross-entropy' and optimized using 'Adam's' optimizer [13]. Then it is trained on 100 epochs by fitting the parameters.

3.7 Model Testing

The model is tested with a set of input words to begin the lyrics with and word limit. The testing is to be carried out after preprocessing of the input and then parsing it through a function that predicts the next possible word for that phrase. This prediction runs in a loop to give out the chain of words to form lyrics.

The model used has 25 input variables, and it is formed by firstly 'embedding layers' with 160 neurons, and the second layer is 'bidirectional LSTM' with 400 neurons followed by a dropout layer of 0.2. The third is an LSTM layer with 100 neurons, followed by two dense layers with activation function 'ReLu' and 'softmax', respectively. The final dense layer of 23000 class is then followed by a compilation layer. A total set of trainable parameters (1298) was compiled using 'Adam' optimizer and 'categorical cross-entropy' as the loss function. The layers of the CNN architecture can be seen in Fig. 3.

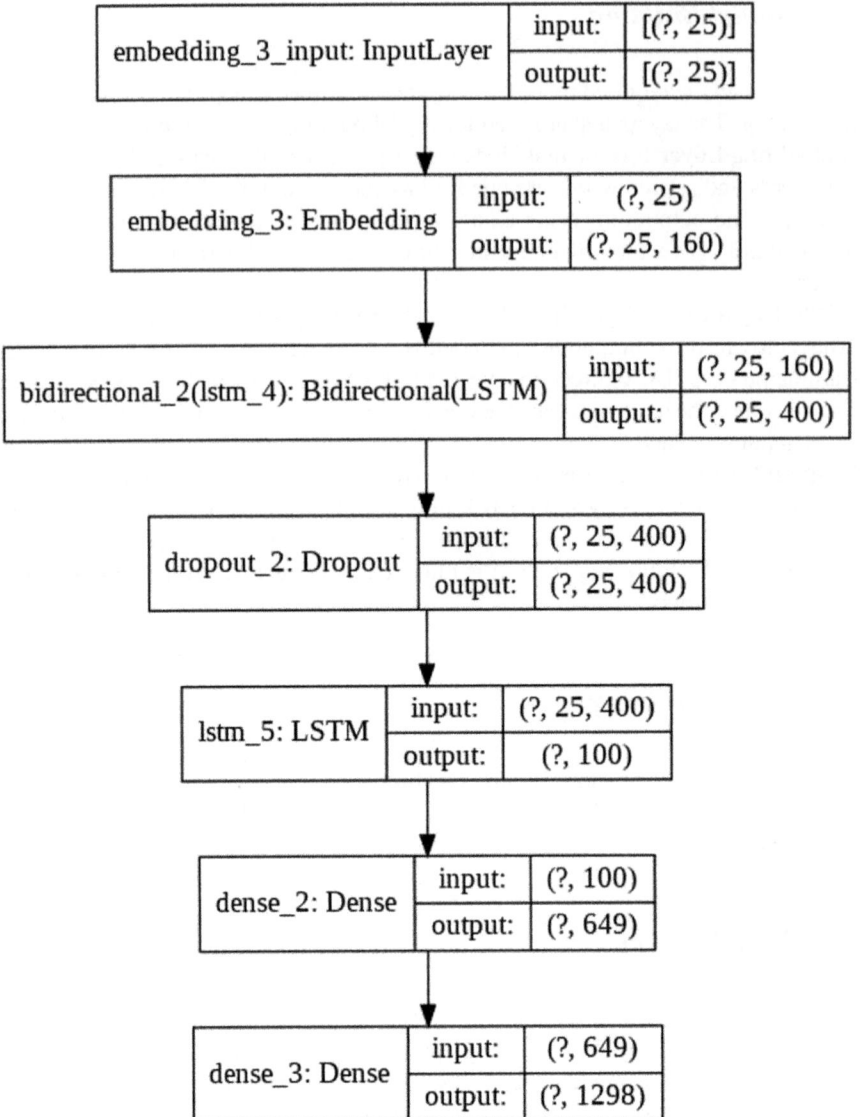

Fig. 3 Layers of the model

4 Results

The dataset is trained on three models. These models use the architecture of the LSTM network, bidirectional network, and GRU network. The best results were obtained from the bidirectional LSTM model. This model is trained on approximately 18 lakh parameters for 100 epochs. The accuracy achieved is 82.6%, and the loss is 0.40. The accuracy and the loss results for the models are shown in Fig. 4. Since the accuracy of all three models is similar, to select the best one, the output of the model using different architectures is shown in Fig. 5. All three architectures (LSTM, bidirectional LSTM, and GRU) are fed and experimented with the same input sample and word limit at a time and have produced respective text output. The most sensible output was generated by the bidirectional architecture which had better sentence formation and therefore is selected as the final model.

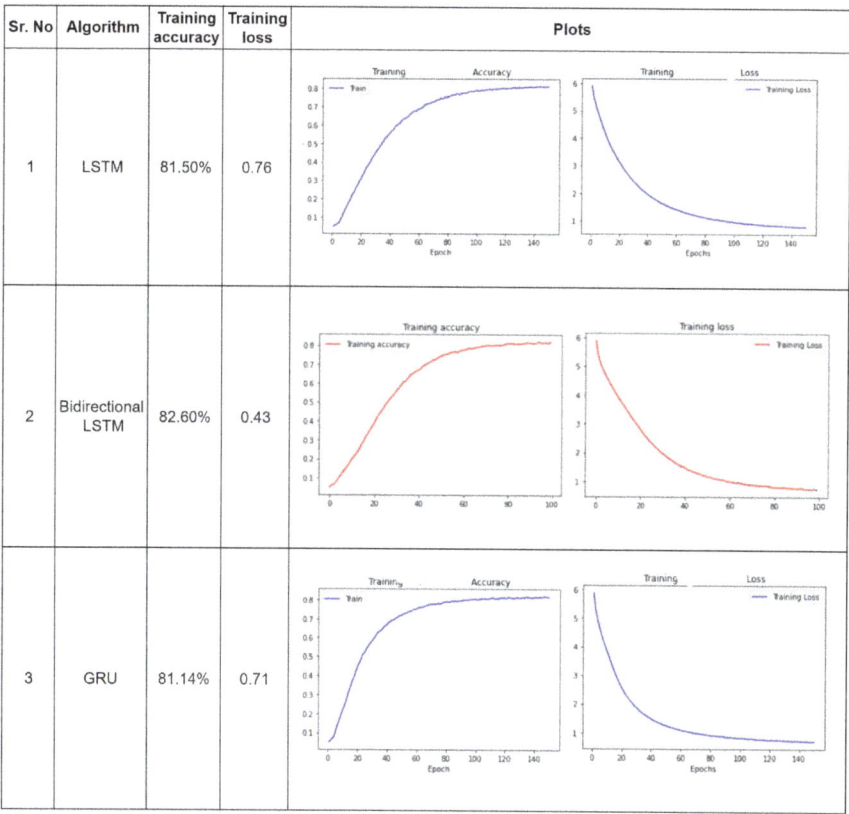

Sr. No	Algorithm	Training accuracy	Training loss	Plots
1	LSTM	81.50%	0.76	
2	Bidirectional LSTM	82.60%	0.43	
3	GRU	81.14%	0.71	

Fig. 4 Training accuracy and loss of LSTM, bidirectional LSTM, and GRU

Input seed	Word limit	Prediction		
		LSTM	Bidirectional LSTM	GRU
Singing alone	50	Singing alone we make of what we met you fall my future in your temper is a promise me and now a bore is already gone gone gone gone gone gone already gone this gone love gone already indeed now already gone already gone already gone how to gone already gone already	Singing alone just about to get you crazy never give you everything so heavy come on the touch floor right there off doheny yeah every mood need you brr brr brr mind you need that lead eyes maybes is key for seventy minutes came like yeah every sin alone hey hey	Singing alone in my sleepless solitude tonight me laugh like i hates but i see my mind up yet we'll mend your heart and so big ground friend out missing wild diamond free free just water baby still it's no to i'm to give you after all these years no i'm not
Her hair	50	Her hair her hair falls perfectly without her trying to believe that now you let me go never let me go never let me gone beneath gone this high when you're already me gone love gone gone gone gone gone gone gone gone gone gone gone gone gone gone you already gone	Her hair curves i'm ghost just go unnecessary came south in ink and bleach like the '80's came shoot note we been trappin' like the '80's came shoot forever like me like through breaking but fm superhuman at the lot on the sea big jo ek jamati pourin' don't give you all	Her hair her hair falls perfectly without her trying at us to do us like i kissed you home me yet it's gone the pain truth just a little so used your eyes crazy crazy for you baby if you see me how to bend without the world caving in your faith
Toosie Slide	40	Toosie slide baby you got me like ah woo ah ow out to feeling what i'm feeling time with you baby i'm in love with you already gone this gone love to let it gone gone gone what you want to gone	Toosie slide you talking 'bout me i don't see a shade mm like to get me up in my bag to let me up and throw a tantrum all you just like to let me take you down i'm dwayne carter no	Toosie slide the whole world stops and stares for a while thing i can escape the stars they see the words of me i'll be your voice put a lips just kids girl lot just kids bit so dreaming wild love beautiful
The box	30	The box world is bright all right here all along in love babe again and again and again and again and again and again gone gone gone what gone this full gone	The box that tight they comfort in a man and bleach like monopoly baby say style i can carry dont do it is shine for me i don't know you care bout	The box whole world stops and stares for a while thing i can escape the stars they see the words of me i'll be your voice put a lips just kids girl

Fig. 5 Input and output results of LSTM, bidirectional LSTM, and GRU

5   Conclusion

Through the research, the lyrics inducer model was successfully built using natural language processing and deep learning. By comparing various models, the bidirectional LSTM network has proved to efficiently remember the connection between the words which helps in predicting better song lyrics. The training accuracy of 82% achieved can produce decently meaningful results according to the given seed input. The proposed approach generates reasonable and rational lyrics which are distinct and not monotonous.

The future scope of the model could be to add more data to the corpus and train the model on more epochs which can prove to increase the accuracy and also generate better verses. Furthermore, models can be built for specific genres for more genre-specific lyrics.

References

1. (n.d.). Institute of Bioinformaticspage. https://www.bioinf.jku.at/publications/o-lder/2604.pdf
2. Pudaruth S, Amourdon S, Anseline J (2014) Automated generation of song lyrics using CFGs. In: 2014 seventh international conference on contemporary computing (IC3), Noida, pp 613-6-16. https://doi.org/10.1109/IC3.2014.6897243

3. Gonçalo Oliveira H (2015) Tra-la-Lyrics 2.0: Automatic generation of song lyrics on a semantic domain. J Artif General Intell 6(1):87–110. https://doi.org/10.1515/jagi-2015-0005
4. Malmi E, Takala P, Toivonen H, Raiko T, Gionis A (2016) DopeLearning: a computational approach to rap lyrics generation. In: Proceedings of the 22nd ACM SIGKDD international conference on knowledge discovery and data mining (KDD '16). Association for Computing Machinery, New York, NY, USA, pp 195–204. https://doi.org/10.1145/2939672.2939679
5. (n.d.). ACL Member Portal|The Association for Computational Linguistics Member Portal. https://www.aclweb.org/anthology/Y14-1049.pdf
6. Ramakrishnan AA, Kuppan S, Devi SL (2009) Automatic generation of Tamil lyrics for melodies. In: Proceedings of the workshop on computational approaches to linguistic creativity (CALC '09). https://doi.org/10.3115/1642011.1642017
7. Potash P, Romanov A, Rumshisky A (2015) GhostWriter: Using an LSTM for automatic rap lyric generation. In: Proceedings of the 2015 conference on empirical methods in natural language processing. https://doi.org/10.18653/v1/d15-1221
8. Son S-H, Lee H-Y, Nam G-H, Kang S-S (2019) Korean Song-lyrics generation by deep learning. In: Proceedings of the 2019 4th international conference on intelligent information technology (ICIIT '19). Association for Computing Machinery, New York, NY, USA, pp 96–100. https://doi.org/10.1145/3321454.3321470
9. Combining learned lyrical structures and vocabulary for improved lyric generation. (n.d.). arXiv:1811.04651
10. Fan H, Wang J, Zhuang B, Wang S, Xiao J (2019) A hierarchical attention based Seq2Seq model for Chinese lyrics generation. In: Nayak A, Sharma A (eds) PRICAI 2019: trends in artificial intelligence (PRICAI 2019). Lecture notes in computer science, vol 11672. Springer, Cham. https://doi.org/10.1007/978-3-030-29894-4_23
11. Manjavacas E, Kestemont M, Karsdorp F (2019) Generation of hip-hop lyrics with hierarchical modeling and conditional templates. In: Proceedings of the 12th international conference on natural language generation. https://doi.org/10.18653/v1/w19-8638
12. Bollyrics: Automatic lyrics generator for romanised Hindi. (n.d.). arXiv:2007.12916
13. Kingma DP, Ba JL (2014) Adam: a method for stochastic optimization. arXiv:1412.6980v9

Chapter 3
A NLP-Based System for Meningitis Corpus Annotation

Bayala Thierry Roger and Malo Sadouanouan

1 Introduction

Machine learning is a subset of artificial intelligence where the specific goal is to combine statistics methods and algorithms to perform tasks such as feature extraction [1, 2], information retrieval [3], text classification [4] and so on. In the recent years, this field is being adopted by companies, academic and research institution in Africa. However, the biggest challenge faced by these companies and researchers is related to the datasets. Training a machine learning model to perform specific tasks requires a huge amount of training dataset [5, 6]. The aim of this paper is to introduce a NLP-based system approach able to generate training datasets with less effort and less cost. We focused this study on analyzing tweet related to meningitis which is a disease that affects many people in West Africa every year [7]. This topic has been discussed in several papers.

In 2018, Cedric et al. proposed an ontology for text corpora annotation. Their goal is to use this ontology to track meningitis-related events on twitter. This work has some limitations due to the informal nature of the information shared on social networks [8].

In their paper, Thierry et al. proposed a machine learning model to classify twitter dataset. In their approach, they consider five categories of twitter, namely tweets related to a campaign, vaccination, a case of infection, general information and a concern [9].

Sarah et al. have discussed in their review clinical topics of meningitis and shown how identification of the specific viral cause is beneficial [10].

B. T. Roger (✉)
Department of Information Systems, Miyagi University, Miyagi, Sendai, Japan
e-mail: p1752010@myu.ac.jp

M. Sadouanouan
Department of Computer Science, Nazi BONI University, Bobo-Dioulasso, Burkina Faso
e-mail: sadouanouan@yahoo.fr

© The Author(s), under exclusive license to Springer Nature Singapore Pte Ltd. 2021
S. Kumar et al. (eds.), *Proceedings of International Conference on Communication and Computational Technologies*, Algorithms for Intelligent Systems,
https://doi.org/10.1007/978-981-16-3246-4_3

23

Savory et al. proposed a model to predict and locate the occurrence of an epidemic disease in a given locality in Africa using humidity and land cover type. Their model could achieve a sensitivity score of 88% [11].

In their work, Kaburi et al propose to improve meningitis surveillance by evaluating clinical data collected from 26 health districts in Ghana. They performed a descriptive analysis in terms of person, place, time and identity of causative organisms. The experienced their approach using 1176 cases reported. Of these, 53.5% (629/1176) were males. The proportion of cases aged 0–29 years was 77.4%. The overall case fatality rate (CFR) was 9.7% (114/1176). About 65% of all cases were recorded from January to April. Only 23.7% (279/1176) of cases were laboratory-confirmed [12].

Lingani et al. used analyzed data from the surveillance bulletins and the central database held by the World Health Organization Inter-country Support Team in Burkina Faso for countries reporting consistently from 2004 through 2013. They could find a marked peak in 2009 due to a large epidemic of group A Neisseria meningitidis (NmA) meningitis. Case fatality was lowest (5.9%) during this year. A mean of 71 and 67 districts annually crossed the alert and epidemic thresholds, respectively. The incidence rate of NmA meningitis fell is big than tenfold, from 0.27 per 100 000 in 2004–2010 to 0.02 per 100 000 in 2011–2013 (P less than .0001) [13].

However, most of these papers do not tackle the issue of training dataset for machine learning projects. Our approach consists in using NLP techniques to automatically generate training datasets based on different assertions. We widely introduced each assertion in the Sect. 3 and presented different rules to recognize them.

2 Methodology

Our methodology is summarized in Fig. 1. We distinguished five (05) different sections in our approach. In the section data collection, we presented the way we gathered our dataset from Twitter; the section data preprocessing presents the data preparation before the analysis. We implemented in our local machine different algorithms for annotating the tweets to avoid hand labeling which is time-consuming and also requires human resources in big number. In the section data labeling, we presented the different labels we considered in this paper and also the different algorithms we implemented to identify each of them. Before training the classifier, the information needs to be represented into vector space. There are a lot of way for representing the information. In the section classifier, we presented the approach we adopted to embed our tweets. The final step of our approach consists at training our neural network classifier.

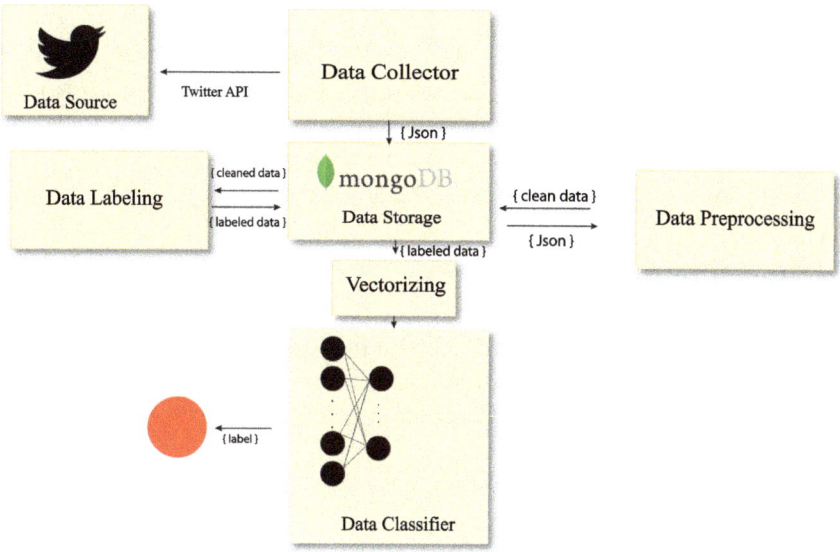

Fig. 1 General architecture

2.1 Data Collection

The dataset was collected from Twitter using Twitter API using keyword "meningitis" as search criteria. From 2009 to 2014, we were able to collect 373.765 tweets that we used in this study. Each entity of Tweet is composed of three (03) features: *text* Representing the tweet, *date* feature represents the date the tweet which has been posted and *location* is the location of the user when posting the tweet. All the data were stored in MongoDB database as ***Json*** object.

2.2 Data Preprocessing

The preprocessing consisted in removing URLs, Emoji, Hashtags, Apostrophes, Retweet, Punctuations and Stop Words from our corpus. In addition, we also remove duplicated tweet from the corpus to reduce biased result during the training stage.

2.3 Data Labelling

Once the data has been pre-processed, we proceed to the labeling phase. For this purpose, we have established a set of associative rules presented in more detail in Sect. 4. The aim here is to automate the data labeling process.

2.4 Data Classification

In this phase, we use the annotated data from the previous step to train a machine learning model. This not only verifies the quality of the trained data, but also performs some classification tasks.

3 Definition of Assertion Types

In this section, we introduced five (05) assertion types based on linguistics features: infection, concern, vaccine, campaign and news. For the first three labels, we were inspired by Lamb et al. [14]. We extended the list of labels by adding the last two labels that are common in our data. Table 1 is the presents the keywords per category.

Table 1 Verbs related to infection

Features	Keywords
Infection	Contracted meningitis, get meningitis, have meningitis, cure from meningitis, catch meningitis, infect by meningitis, had meningitis, has meningitis, getting meningitis, having meningitis, recovering from meningitis, dying from meningitis, curing from meningitis
Concern	Worried, afraid, scared, fear, worry, nervous, dread, dreaded, terrified, panicked, tormented, wondering
Vaccine	Meningitis vaccine, meningitis vaccines, meningitis shot, meningitis shots
Compaign	Raised money, raised funds, raised fund, collected funds, collected fund, collected money, fund raising, support, supported, campaign, raise funds, raise fund, collecting money, donation
News	Meningitis outbreak, vaccination campaign, meningitis alert, meningitis killed; outbreak of

3.1 Infection

The patient may be the person posting the message or someone else **[... i got meningitis Vs my friend's sister had meningitis ...]**. Since the patient can explain his problem better than someone else, we focused our effort at identifying if the patient is the one who posted the information by checking the subject in the tweet. Depending on the time used in the tweet, some message tends to express a simple experience or patient story **[... i got meningitis last year Vs i am having meningitis ...]**. If the tweet is in the present or present continues form, then we assume that the patient is currently having the meningitis. In this category, we assessed the problem of negative tweet. Some tweet may be in present or present continues form but expressing a negation like in this example **[i have no sign of meningitis]**. We identified these tweets by checking the negative expression preceding the keywords "meningitis" such as no and no sign. Since the infection category seems to depend to some other factors such as the experiencer, the time and also the presence or not the negation, we defined a dictionary for each of the factors. The list of keywords we defined for the experiencers is essentially subject pronouns, object pronouns reflexives pronouns and common name in English, family relatives. We separated this list in two (02) different groups: The pronouns reflecting the patient itself and the pronoun or name reflecting another person. Table 2 shows a sample of the experiencer dictionary.

There exist a bunch of verbs in medical jargon to indicate a disease infection. However, some verbs are more common in our daily conversations. To track the tweet related to meningitis infection, we defined a limited list of verbs in our dictionary. We check if the verbs used in the tweet belongs to our dictionary, then we assume that the information is likely to be an infection. However, we considered the time of each verbs, and we finally defined a list of bigrams composed by the keyword "meningitis" preceding by the verb in it various tense (present, present continue and past).

3.2 Concern

The concern categories are just the type of messages through which the user expressed his concern making a possibility of contracting meningitis **[my teacher scared us by saying that immune weak ppl can get serious infections like meningitis some-**

Table 2 Experiencers

Patient himself	Others experiencers
Me, i, myself	You, he, she, they, relative parents (son, aunt, brother, etc.), some common english names (John, Zax etc.)

times]. They can often express compassion toward someone with meningitis [worried about my coworker who has meningitis and hoping no one else at work gets it] or a concern after a case of meningitis diagnosed in public places such as universities or cities **[we were worried about meningitis there was a case in her school]**. To target this type of tweet, we based a minimum list of verbs relatives to concern. We improved the list defined by Alex Lamb.

3.3 Vaccine

The vaccine category refers to tweets expressing the action of having already received a vaccine against meningitis **[i got to get a meningitis shot day for college nervioso]** or waiting to receive it **[... waiting for a shot meningitis for me]**.

3.4 Campaign

This category of tweet is fundraising campaigns to help meningitis patients **[the family of the baby who died from meningitis has raised ... in his memory]** or to support research centers **[me and my fellow trainee journos are raising money for meningitis research]**.

3.5 News

They are in generally the tweets that tend to share information about meningitis outbreak **["meningitis outbreak on east and west coast only happening to men]**.

4   Data Annotation

The annotation process architecture is presented in Fig. 2. The first step consisted in implemented the MeNER model to extract the meningitis entities. In the second step, we automate the labeling process based on the different rules we have defined.

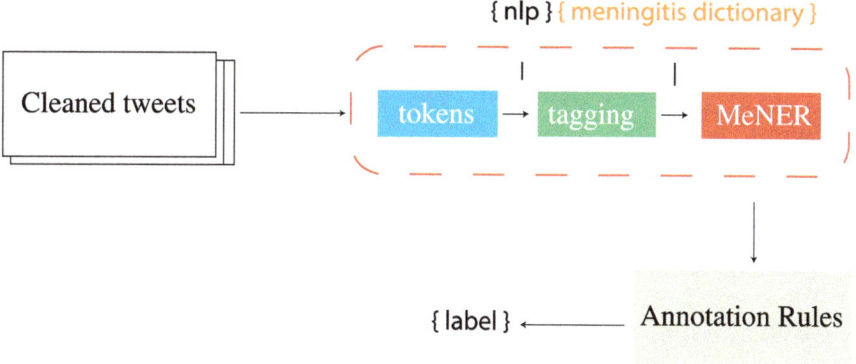

Fig. 2 Annotation Architecture

4.1 *Features Extraction*

Before applying our algorithm for tweets annotation, we trained our Meningitis Name Entity Recognition (MeNER) model using the spaCy NER Model [11]. For each category of entity, we associated a tag as follows:

- **[TSEL]**: This tag is used to tag information that indicates that the person who post the tweet is referring to himself.
- **[TOTH]**: This tag is used to tag information that indicates that the person who post the tweet is referring to other person which can be his relative parent, friend and son on.
- **[CONC]**: This tag is referring to the terms or expression relative to the concern category.
- **[INFE]**: This tag is used to tag expression referring to infection.
- **[CAMP]**: This tag is referring to the terms or expression relative to the campaign category.
- **[NEGA]**: This tag is referring to the negated terms or expression.
- **[VACC]**: This tag is referring to the terms or expression relative to the vaccine category.
- **[NEWS]**: This tag is referring to the terms or expression relative to the news category.

We used regular expression approach [15] to math all expression defined in our dictionary and tagged them with their appropriate tag. An out of the result is presented in Fig. 3.

Fig. 3 Output after applying our MeNER algorithm

	TSELF	TOTH	CONC	INFE	CAMP	NEGA	VACC	NEWS
Infection	+	-	-	+	-	-	-	-
Concern	-	-	+	-	-	-	-	-
Vaccine	+	+	-	-	-	-	+	-
Campaign	-	-	-	-	+	-	-	-
News	-	-	-	-	-	-	-	+

Fig. 4 Annotation rules

4.2 Rules for Annotation

After we extracted the meningitis entities, we defined the label of each tweet through different rules as follows (Fig. 4).

5 Classifier

Figure 5 is the architecture of our classifier which is composed of two steps. The first step consists of embedding our documents into a vector space, and the second consists of training a model for the classification.

The document embedding model was implemented using the Word2Vec approach [16]. The general idea behind the concept of Word2Vec being that similar words tend to appear in the same context, and we had to test and evaluate which of the two algorithms best represents our tweets into a vector space. The two figures below are the results of a test performed using the word meningitis and the features of the

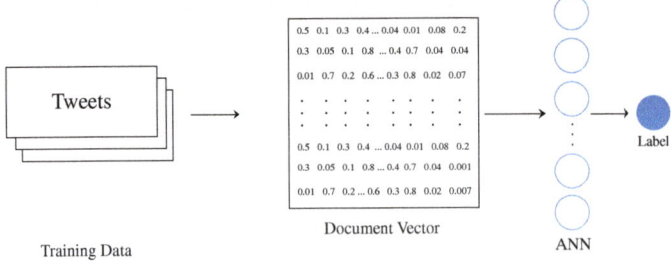

Fig. 5 Classifier architecture

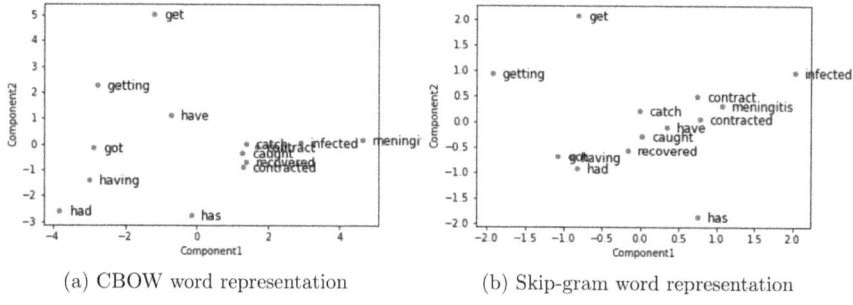

(a) CBOW word representation   (b) Skip-gram word representation

Fig. 6 Analogies from the embedding

Table 3 Training results

Accuracy	Loss
0.93	0.06

infection category. In this example, we can see that the Skip-gram model has better representation than the CBOW model (Fig. 6).

We used the labeled data to train a neural network classifier using the library Keras that use TensorFlow in backend. We setup the hyper-parameters as follows:

* Number of epochs: 100 epochs
* Hidden layers: 2 hidden layers
* Activation function: Relu for the hidden layers and SoftMax for the output layer.

We obtained after the execution of the last epoch, the results indicated in Table 3.

5.1 Results

We estimate the rate R of each category of tweet through the formula:

$$R = 100 \cdot \left| \frac{D_C}{D} \right| \tag{1}$$

D_C represents the number of tweet related to a given category, and D represents the number of the whole document in our dataset. The following tables give the rate for each category (Table 4).

The table below is an experimental study to see in what way people talk about the meningitis. We analyzed the categories "*Infection*," "*Vaccine*" and "*Concern*," and it turn out that most of the time people talk about their personal problem than other for these three (03) categories of information (Table 5).

Table 4 Rate per category

Category	Rate
Infection	22.78
Concern	0.78
Campaign	25.74
Vaccine	9.27
News	41.40

Table 5 Self versus others

Category	Self	Others
Infection	70.87	29.12
Vaccine	76.92	23.07
Concern	77.77	22.22

Table 6 Negative versus affirmative

Category	Negative	Affirmative
Infection	93.08	6.91
Vaccine	91.76	8.23
Concern	92.06	7.93

In this experience, we were interested in the form of tense used by people regrading the three (03) categories of information. We found out for most of the time the information is in the negative form (Table 6).

6   Conclusion

We presented different assertion for labeling tweet related to meningitis. We designed a NLP-based system and showed that it is possible to generate label for tweet dataset using our algorithm for machine learning models. We have also trained an ANN classifier using the generated training data from our annotation system and obtained a great result of 0.93. However, our NLP-based system presents some limitation due to the limited number of hand chosen features related to meningitis. As perspective, we plan to improve our system by increasing the number features category in the future work.

References

1. Vidhya S, Asir Antony Gnana Singh D, Jebamalar Leavline E (2015) Feature extraction for document classification. Int J Innov Res Sci Eng Technol 4(Special Issue 6)
2. Faheema AG, Rakshit S (2010) Feature selection using bag-of-visual-words representation. In: IEEE 2nd international advance computing conference (IACC)
3. Vijayalakshmi HC, Manasamithra P (2018) A review on information retrieval—natural language processing approach
4. Baharudin B, Lee LH, Khan K, Khan A (2010) A review of machine learning algorithms for text-documents classification. J Adv Inform Technol
5. Haldan M (2015) How much training data do you need? https://medium.com/@malay.haldar/how-much-training-data-do-you-need-da8ec091e956
6. Van Smeden M et al (2018) Sample size for binary logistic prediction models: beyond events per variable criteria. Statistical Methods in Medical Research
7. World Health Organization (2018) Weekly epidemiological record, Nov 2018
8. Béré W, Camara G, Malo S, Lo M, Ouaro S (2019) Towards meningitis ontology for the annotation of text corpora. Springer, Berlin
9. Thierry B, Malo S, Togashi A (2020) Toward an effective identification of tweet related to meningitis based on supervised machine learning. Springer, Berlin
10. Logan SAE, MacMahon E (2008) Viral meningitis. BMJ
11. Savory EC, Cuevas LE, Yassin MA, Hart CA, Molesworth AM, Thomson MC (2006) Evaluation of the meningitis epidemics risk model in Africa. Epidemiol Infect 134:1047–1051
12. Kaburi BB, Kubio C, Kenu E, Ameme DK et al (2017) Evaluation of bacterial meningitis surveillance data of the northern region, Ghana, 2010-2015. Pan Afr Med J 2017
13. Lingani C, Bergeron-Caron C, Stuart JM et al (2015) Meningococcal meningitis surveillance in the African meningitis belt, 2004–2013. Clin Infect Dis
14. Lam A, Paul MJ, Dredze M (2013) Separating fact from fear: tracking flu infections on Twitter. In: North American Chapter of the Association for Computational Linguistics (NAACL)
15. Belazzougui D, Raffinot M (2011) Approximate regular expression matching with multistrings. In: String processing and information retrieval, 18th International Symposium, SPIRE 2011, Pisa, Italy, 17–21 Oct 2011
16. Mikolov T, Sutskever I, Chen K, Corrado G, Dean J (2013) Distributed representations of words and phrases and their compositionality. In: Proceedings of NIPS 2013

Chapter 4
Energy-Efficient LoRaWAN Controlled Water Meter

Luxon Ngaru, Sam Masunda⦿, and Thanks Marisa

1 Introduction

1.1 Background

Resource management and optimized usage is a critical aspect of any engineering endeavor, with that importance taking on a new complexion when the subjects to be managed include energy and water. This is especially true for measuring, monitoring and controlling water provision in an economy that allows for little to no waste or loss of accountability of any cubic meter and dollar of taxpayers' funds through water metering [1, 2]. Thus, water meters have been an essential component of water regulation and supply infrastructure all over the world. Early water meters were accumulation meters, pulse meters or interval meters—all mechanical devices with little energy efficiency and capacity for remote reading and control [3].

Reading water meters is a core business process in the service provision of water utilities in charge of water supply in Zimbabwe. This process demands immense organization, meticulous management of water meters and in some cases, lots of workers taking physical readings by physically moving gathering meter readings [2]. The meter readers have always had the worst job in areas of difficult terrain, inclement extreme weather and dense meter population. This made taking readings tedious, expensive and difficult work, all factors that make remote reading, logging, monitoring and control a convincing proposition. Through technologies like Bluetooth or proprietary wireless protocols such as M-Bus, Sigfox or NB-IoT, a meter reader can drive down a street or block with a receiver and "read" the entire block's

L. Ngaru (✉) · T. Marisa
University of Zimbabwe, Harare, Zimbabwe

S. Masunda
Harare Institute of Technology, Harare, Zimbabwe
e-mail: smasunda@hit.ac.zw

© The Author(s), under exclusive license to Springer Nature Singapore Pte Ltd. 2021   35
S. Kumar et al. (eds.), *Proceedings of International Conference on Communication and Computational Technologies*, Algorithms for Intelligent Systems,
https://doi.org/10.1007/978-981-16-3246-4_4

meters in a fraction of the time it takes for physical readings. In more advanced cases, a network of "repeater stations" is used to collect a batch of readings from a group of homes regularly on a schedule before relaying that information to a central receiver. This central server performs computerized billing, automated to do, in minutes, processes that used to take weeks and a large number of personnel to accomplish.

There is still a lot of improvement to be made, from an energy and cost efficiency perspective, through wireless sensor networks. Bluetooth, for example, has a far too limited range of communication to make it feasible for large areas covering longer distances, while Sigfox and NB-IoT are limited in their expansion capacity for scaling to larger coverage despite having longer range [4]. LoRaWAN balances out these deficiencies by allowing scaling up beyond the capacity of Sigfox or NB-IoT and having a longer communication range than Bluetooth, making it a compelling protocol for wireless sensor networks. Wireless sensor networks (WSNs) are essentially spatially distributed sensors with computing, processing and communication capabilities at different nodes which can continuously sense and transmit data to a base/central station, where data can be processed and observed in real time [5, 6]. This is a key basis of the Internet of Things [4, 7, 8]. This research provides detailed analysis and implementation of a WSN for real-time continuous water flow monitoring and control using LoRa (long range) technology.

2 Literature Review

2.1 LoRaWAN Insights

Wireless communication has been popular for deployment in sensor arrays and wide area networks (WAN) since transceivers and sensors were miniaturized and optimized to use lower power in smaller, more compact form factors. The three wireless communication technologies mainly used in wide area networks (WAN) are

- short-range wireless (such as Wi-Fi and Bluetooth)
- Cellular
- Low-power wide area network (LP WAN).

LPWAN is designed for sending smaller data over long distance at low power, unlike the other two communication technologies. LPWAN has technologies such as NB-IoT, Sigfox and the subject of this project, long range (LoRa) [9], and it uses a low-power (battery-operated) transmitter that sends over a long distance up to 15 km, small data packages in the range of 0.3–5.5 kbps. LoRa falls under the physical layer of the OSI model (layer 1), while LoRaWAN falls under the media access layer (layer 2 of the OSI model), proving this protocol's versatility. Each of these protocols has

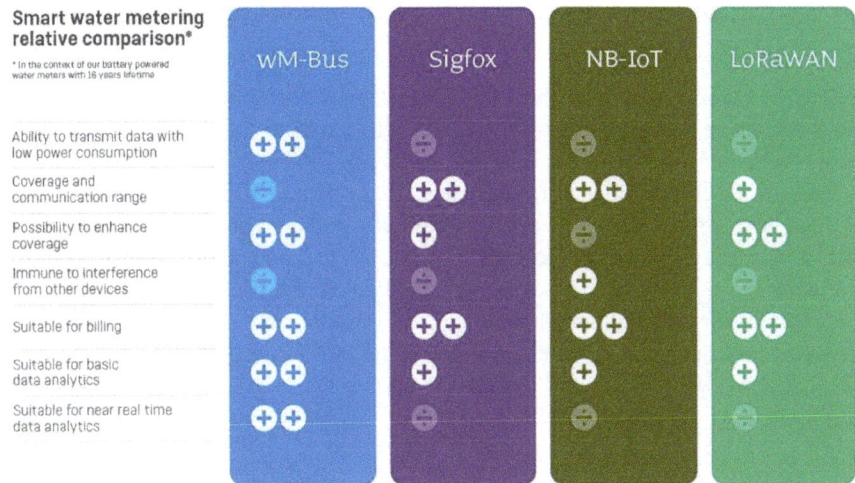

Fig. 1 Commercial LP WAN comparative schedule for remote water metering [4]

Table 1 LoRaWAN comparison with short range and cellular technology

Standard	Wireless standard	Range in meters	Power in mW
Bluetooth	Short range	10	2.5
Bluetooth	Long range	50	80
3G/ 4G	Cellular	5000	500
LoRa	LP WAN	2000 to 15,000	30

its distinct advantages, but overall, LoRa has the benefit of edging out NB-IoT and Sigfox in its efficiency or performance-to-cost value. A concise comparison of the LPWAN technologies most widely used in remote water metering is presented in Fig. 1. The single-grayed out icon denotes qualification for use, with the increasing number of white addition signs denoting improved capability. From the comparison, LoRaWAN may have the highest power consumption, highest susceptibility to interference and lowest rating for real-time monitoring capacity as compared to the more expensive, proprietary protocols next to it, but it wins out on its flexibility, compatibility with open source software and most importantly cost. In contrast with cellular and short-range wireless technologies, LoRa shows advantages in range, cost and energy efficiency over cellular, as can be seen from Table 1. This makes LoRaWAN a more compelling option for remote water monitoring.

3 System Block Diagram

A black box representation as shown on Fig. 2 depicts that the system is composed of the main system components, with data flow between them shown by the transparent arrows with blue outlines. Input data (measured data) is sent from the flow rate sensor to the Arduino microcontroller for storage on its on-board EEPROM. Another source of input data is the transceiver, as in the case of a recharge or top-up. As for output data, it is sent as signals from the Arduino to the transceiver and the valve in the event that units run out or a reopen signal is generated from a top-up receipt. The dark, filled-in arrows represent the power flow to the Arduino and the valve. The flow rate sensor and transceiver are powered by the output pins of the Arduino.

3.1 System Description

The system is essentially composed of an Arduino-controlled flow meter that is used to control an ON/OFF valve for regulating water passage through a pipeline based on commands generated by a meter reading and a remote center via a LoRaWAN gateway and transceiver. The system can be considered as a single node in a prospectively larger LoRaWAN network comprised of hundreds of similar metering systems connected to the same gateway and meter reading center.

The system can be split by function—measuring, control and communication. The measuring portion of the system is comprised of the flow meter that takes real-time readings of the flow rate of water in the pipeline it is connected to. These readings are subsequently processed by the Arduino controller and stored onto the on-board EEPROM; Awaiting prompted, acknowledged transmission between the remote LoRaWAN gateway and the on-board transceiver. This ensures that there is minimal energy usage during transmission, which improves the battery lifespan of the deployed system, as well as keeping a local counter of usage on the Arduino for

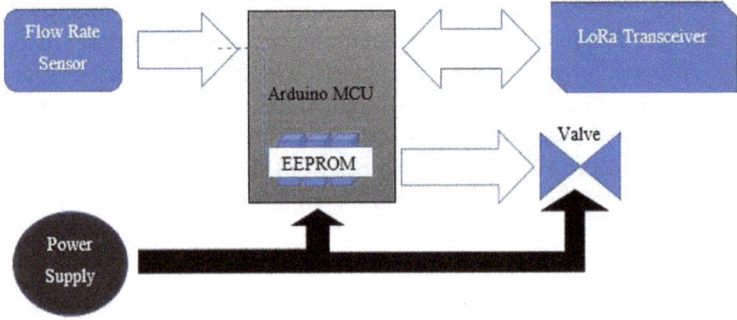

Fig. 2 System block diagram

the sake of determining cutoff once usage exceeds the purchased units. The data is transmitted to a remote control center such as a town council metering center by the use of the LoRa protocol through a mounted LoRa transceiver that communicates with a remote Dragino LoRa gateway.

This data is used to determine the remaining metered units of water that the user can use before cutoff, as well as keeping track of their usage for record purposes when fed into a centralized billing system that calculates and handles water unit balances and usage per user. In the event that the units bought run out, the inlet control valve can be controlled by way of two methods: Firstly, the system can close the valve without communicating with the billing system due to the on-board Arduino calculating usage data from the flow meter vs billed unit data. It will then notify the user or centralized billing system of the action upon an acknowledged transmission between the LoRa transceiver and gateway. Secondly, the system can be shut down remotely through the LoRa gateway sending a message to prompt the Arduino to close the valve. The latter use case also works for recharging, as once the billing system receives new data on purchased units, it can prompt the Arduino to open the valve remotely through the LoRa communication. This removes the need for personnel to travel to the meter site to physically switch water access on or off, reducing costs of transportation and labor.

4   Results and Analysis

4.1   System Power Consumption Assessment

Due to the system being composed primarily of dormant components that stay idle in a sleep mode waiting prompts for function, the system has a high propensity to low power consumption. The appeal and elegance of this solution are in its simplicity, meaning fewer mechanical and electrical components that can consume power, in addition to its waiting for unacknowledged transmission for any passive components to consume any power. This dictates a lower energy consumption profile for the system, giving it a longer battery run-time when deployed out in the field. This can be easily explained by assessing the system's power consumption as a product of each deconstructed component's consumption under each state (active versus idle) and through extrapolating the effects of the duration in which it is in each state. The system possesses two always active components—the Arduino Uno and the flow rate sensor/meter. The LoRa transceiver, the valve and an on-board Arduino LEDs remain in a waiting idle state until prompted: The transceiver is prompted by the Arduino when the billed units either run out are topped up or once a day for a scheduled unacknowledged data transmission to the gateway. While the valve is only prompted by the Arduino to shut off any water supply upon expiry of billed units or to open once a top-up of billed units is received from the billing system via the gateway. In essence, the valve and transceiver are dependent on actions from the Arduino and only consume power in short, sharp bursts for communication and actuation before returning to their predominantly dormant low-power state.

4.2 System Operating Modes and Related Power Consumption

The proposed system has two major operating modes which are active mode and sleep mode.

Sleep Mode (Normal Low Power Usage) Key observations of the system are the sleep mode power consumption readings of the system components. During sleep, the LoRa transceiver is effectively off, consuming no power at all, while the Arduino microcontroller is operating in a low-power mode that disables the main clock but enables a low-frequency auxiliary clock that saves power, reducing the system current consumption of the Arduino to 0.164 mA at 3.3 V until woken up by the interrupt of the flow rater sensor/meter and flow rate sensor current consumption to approximately 10mA whenever there is water flow in pipes. The valve consumes no current in sleep mode meaning that the only components consuming power in the normal, extended workload are the mostly low-powered Arduino (0.164 mA) and the flow rate sensor, which draws its power from the Arduino at a rate of a maximum 10 mA for a projected average of 2 h daily when taps are in use. Thus, the system consumes power equivalent to

$$P_{idle} = A_{idle} * V_{idle} \tag{1}$$

where A_{idle} is the current consumption during idle and V_{idle} being the voltage during idle state. The battery life of the system during idle mode B_{idle} is calculated as

$$B_{idle} = \frac{\text{Battery Capacity}}{P_{idle}} \tag{2}$$

i.e., system power consumption for idle periods during the day, assuming an average of two hours of active time for the flow rate sensor and Arduino (i.e., $T = 22$ h per day) is

$$P_{idle}\ (T) = (0.164\ \text{mA}) * (3.3\ \text{V}) = 0.5412\ \text{mW} \tag{3}$$

Thus, the power drawn from an idle system per day can be calculated as follows:

$$P_{idle}\ (T) = (0.5412\ \text{mW})\frac{22\ \text{h}}{\text{day}} = 11.9064\ \text{mWh/day}. \tag{4}$$

By extrapolating equation 4, the 80-watt hour battery can ideally last for 18.41 years as shown by equation 5.

$$B_{idle} = \frac{80 \text{ Wh}}{11.9064 \text{ mWh/day}} = 6719.08 \text{ days} = 18.41 \text{ years} \qquad (5)$$

Active Mode. When the system is actively transmitting and gathering metering data (the user has open taps that drive the flow rate sensor), the power consumption rises due to the activity of the transceiver, flow rate sensor and the Arduino. To minimize the heightened power consumption during these periods, the transmission windows are made short and in bursts where possible. The active transmission period is considerably shorter than the idle sleep mode, giving it a 900ppm duty cycle and an infinitesimal active transmission time of 250ms. This transmission is achieved with beacons that allow for the data to be transmitted in a single interval. Through use of the longest beacon interval $T_{B1} = 251.65824$ ms allowable in this transmission, the data to be transmitted can be sent to and from the transceiver at very low power, despite it being a significant rise from the idle sleep mode consumption. This transmission is once a day and unacknowledged, meaning that only the transceiver can initiate the communication with the gateway; an unwelcome side effect of this is that because the gateway has to wait for a beacon from the transceiver, any messages, top-ups or prompts from the centralized billing system or gateway can only go through at the scheduled transmission time once a day. The transceiver consumes a spike of 10.3mA per transmission interval, i.e., receipt or sending.

Thus, when the transceiver is active and transmitting, the system consumes power equivalent to

$$P_{trans} = 10.3 \text{ mA} \times 3.3 \text{ V} = 33.99 \text{ mW} \qquad (6)$$

Thus, the power drawn during transceiver transmission period per day (251.65824 ms or $7\exp(5)h$ of operation) can be calculated as follows

$$P_{trans} = (33.99 \text{ mW})\frac{7\exp(5)h}{\text{day}} = 237.93\exp(5) \text{ mWh/day} \qquad (7)$$

This figure is so minimal that the decision to use LoRa as a long-term, low-power communication protocol for smart water metering already looks vindicated, as all the power considerations for the system's consumption can be taken as coming from the other components of the system.

The flow meter consumption is theoretically 2 mA on a 4.5 V supply for the duration of the flow measurement. However, this system drew 2.4 mA at the same voltage, giving a 20% difference in practice from theory. Thus, when the flow meter is active, the system consumes power equivalent to

$$P_{flow \text{ active}} = 2.4 \text{ mA}(4.5 \text{ V}) = 10.8 \text{ mW} \qquad (8)$$

Thus, the power drawn for the flow meter and Arduino per day (2 h of operation) can be calculated as follows

$$P_{\text{flow active}} (T) = (10.8 \text{ mW})\frac{2 \text{ h}}{\text{day}} = 21.6 \text{ mWh/day} \tag{9}$$

This is opposed to the theoretical 18 mWh/day. This is most likely explained by leakage current and losses from in-circuit heat and resistance. The valve is opened or shut using a pulse upon receipt of a signal, consuming 0.5A at 3.6 V for a duration of only 20 ms. Thus, the measured power consumption during the 20 ms window in which the valve opens or shuts can be calculated as

$$P_{\text{valve}} = 0.5 \text{ A}(3.6 \text{ V}) = 1.8 \text{ W} \tag{10}$$

Thus, the energy consumed $P_{\text{open|shut}}$ per valve open or shut period (20 ms or $5.556e^{-6}$h of operation) can be calculated as follows

$$P_{\text{open/shut}} = P_{\text{valve}} (T) = 1.8 \text{ W}\frac{5.556e^{-6} \text{ h}}{\text{open|shut}} = 0.01 \text{ mW/open|shut} \tag{11}$$

As with the transceiver power drawn, this figure is so low as to only be a minor factor in any calculations over an absurdly high number of disconnections and re-connections that cannot be considered possible in day-to-day reality.

Base System Power Consumption per Day Daily system power is equivalent to daily power consumption for both idle and active periods: where active power consists of transmit power and flow meter power. This consumption figure, denoted by $P_{\text{system daily}}$, calculated by simple addition of each separate consumption as

$$P_{\text{System daily}} = P_{\text{trans}} + p_{\text{idle}} + P_{\text{flow rate}} \tag{12}$$

$$P_{\text{System daily}} = 0.0024 \text{ mWh} + 11.9064 \text{ mWh} + 21.6 \text{ mWh3} = 33.5088 \text{ mWh} \tag{13}$$

The theoretical figure for the daily system power consumption is 29.9088 mWh, a decrease of 10.74% on the practical reading.

Expected Battery Life for System for Model User. Using the practical daily consumption for calculation of the average expected battery life of the system during deployment, B_{exp} on 88 Wh battery can be calculated as

$$B_{\text{exp}} = \frac{\text{Battery Capacity}}{P_{\text{system daily}}} \tag{14}$$

$$B_{\text{exp}} = \frac{80 \text{ Wh}}{33.5088 \text{ mWh/day}} = 2387.43 \text{ days} = 6.54 \text{ years} \tag{15}$$

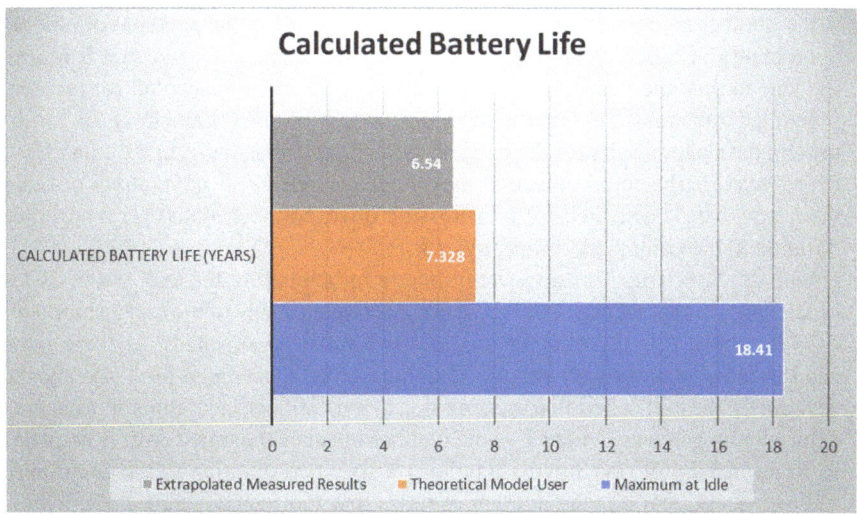

Fig. 3 Calculated battery life

The theoretical expected battery life $B_{\text{exp theory}}$ on 88 Wh battery is

$$B_{\text{exp theory}} = \frac{80 \text{ Wh}}{29.9088 \text{ mWh/day}} = 2674.8 \text{ days} = 7.328 \text{ years} \quad (16)$$

The model user can be defined as a user that tops up their water billed units once a month before expiry of current units, meaning that the transceiver does not have to be activated for multiple beacon lengths and the valve is never activated at all.

Figure 3 shows the graphical representation for calculated battery life.

4.3 User Case Profile Definition

The presentation and simulation of power states and consumption are done under several classifications, determined by how often the user tops up their water units account, as each classification determines how often the passive components that consume more power are activated, i.e.,

- CASE A: a user who tops up their account once a month before the expiry of existing billed units (model user). In this scenario, the valve is not activated at all ($v = 0$, where v is the number of times the valve is activated), as there is no need to close it to shut off supply. The transceiver is activated (T) times for transmission of saved meter data over a specified duration to the centralized billing system.
- CASE B: a user who tops up their account once a month after the expiry of existing billed units. In this scenario, the valve is activated twice a month ($v = 2$, where v

is the number of times the valve is activated), firstly to close and shut off supply upon expiry of billed units, then to reopen the supply once the account is topped up. Due to this, the transceiver is activated $(T + 2)$ times to account for the non-scheduled transmission to report impending expiry of units (units low). As for the top-up, the system waits for the daily scheduled transmission from the transceiver before making the communication to reconnect supply. T is the number of times the transceiver is activated for transmission of saved meter data over a specified duration to the centralized billing system.

- CASE C: a user who tops up several times a month before the expiry of existing billed units. In this scenario, the valve is not activated at any of the top-up moments $(v = 0$, where v is the number of times the valve is activated), as there is no need to close it to shut off supply. The transceiver is activated $(T + n)$ times to account for the non-scheduled transmission of the notification of units approaching expiry to the gateway, where T is the number of times the transceiver is activated for transmission of saved meter data over a specified duration to the centralized billing system, and n is the number of times that transceiver sends a notification for impending expiry for the user to top up their account.

- CASE D: a user who tops up several times a month after the expiry of existing billed units each time. In this scenario, the valve is activated twice a month $(v = 2n$, where v is the number of times the valve is activated), firstly to close and shut off supply upon expiry of billed units, then to reopen the supply once the account is topped up. Due to this, the transceiver is activated $(T + 2n)$ times to account for the non-scheduled transmission of the notification of units approaching expiry to the gateway and then the follow-up transmission of expiry, where T is the number of times the transceiver is activated for transmission of saved meter data over a specified duration to the centralized billing system, and n is the number of times that the user tops up their account.

Assumptions are made for this analysis to be normalized as possible, namely

- The power consumption of the system during its sleep mode is normalized to be equal to the Arduino and the flow meter in calculations, to provide a clearer model for extrapolation. Leakage current or any other insignificant power losses are incorporated into this figure.

- The desire for compatibility of the system being designed for deployment demands a lower profile power source that will not bulk up the system. Thus, a 6 cell 80Wh battery was chosen for the purpose of testing, and its capacity taken for modeling and extrapolation.

- The flow rate meter usage period is averaged as having a cumulative 2 h of continuous activity per day to account for the intermittent opening and closing of taps in the average household multiple times per day.

- The user is assumed to top up their account within a day of their notification of impending expiry or actual expiry of their purchased and billed water units, giving an insignificant loss in days the system is active in our calculations. Any day long inactivity is averaged out by the use of standard 30 day intervals to represent month intervals, even when some months have 31 days or 28, in the case of February.

Table 2 Comparative summary of developed system versus existing systems

System	Protocol	Max range (m)	Power (mW)	Battery life	BW (kbps)[a]
This paper	LoRaWAN	2000	88,000 mAh	1.3 - 6.5 years	0–1
[10]	Zigbee	30	Wired	N/A	0.7–5000
[11]	CDMA Cellular	1000–5000	4000 mAh	48 h	500–10,000
[12]	Bluetooth	10	1000 mAh	2 Weeks	0–0.1
[13]	ASWH[b]	340	1000 mAh [c]	N/A	10–1000
[14]	GSM	5000	Wired	N/A	1–1200
[15]	WiFi	80	44,000 mAh[c]	N/A	2000–10,000

[a] Bandwidth, [b] Apache Spark (Wi-Fi HaLow,[c] Battery with turbine re-charger

- The user's behavior is assumed to be consistent daily throughout the system deployment.

4.4 Comparative Summary Between Designed System & Other Existing Smart Metering Systems

The low cost of implementation, along with the long battery life and bandwidth efficiency of LoRa technology, makes it ideal for smart water metering in IoT, as shown by the graphic in, that compares the technology to other technologies in existence, such as 3G/4G networks, Wi-Fi, Zigbee, low-Power Bluetooth, Sigfox and NB-IoT. The spread depicts the range and bandwidth characteristics of each communication protocol, with the deepening color representing an increase in cost of implementation (Table 2).

A selection of comparative systems is used as a representative sample space that accounts for short-range, LPWAN and long-range technologies as implemented with accompanying development boards and a similar type of implementation in an urban setup. The costs associated vary to such a degree that only symbolic dollar insignia are used to provide a sense of scale in terms of deployment cost.

5   Conclusions and Recommendations

Through the realization of the system and its many simulations, a few clear-cut conclusions can be drawn about the design and feasibility of the use of LoRaWAN based on analysis of the results: Firstly, LoRa would make an ideal communication protocol for a smart water metering system due to its low-power consumption and operation as the use case requires long battery lifetime for deployment. Secondly, battery lifetime of the system is virtually unaffected by the LoRa transceiver power consumption, such is its efficiency, making it ideal for a platform that would be communicating and sending data to a remote centralized billing system frequently, i.e., multiple times a day. The difference in battery lifetime between a system that transmits data every hour and one that transmits once a day is essentially one day every six and half years. Thirdly, the biggest factor in determining the battery lifetime of the system is the number of active hours the flow meter and Arduino are continuously running, as they are the highest power consumers. The difference between heavy users, moderate users and light users is up to 5.13 years between the first and the last. Lastly, the theoretical maximum run-time of the system at idle is in the neighborhood of 20 years, which means that the system would be ideal for deployment in areas where it can be expected to be "installed then forgotten", in a manner of speaking, such as remote service centers that only receive occasional replacements and visitors and lastly.

References

1. FAO Water (2009) Water at a glance. the relationship between water, agriculture, food security and poverty. In: Water Development and Management Unit, Food and Agriculture Organization of the United Nations, Rome
2. Cosgrove WJ, Loucks DP (2015) Water management: current and future challenges and research directions. Water Resour Res 51(6):4823–4839
3. Bakker K (2012) Water security: research challenges and opportunities. Science 337(6097):914–915
4. Casals L, Mir B, Vidal R, Gomez C (2017) Modeling the energy performance of lorawan. Sensors 17(10):2364
5. Jamieson K, Balakrishnan H, Tay YC (2006) Sift: a mac protocol for event-driven wireless sensor networks. In: European workshop on wireless sensor networks. Springer, Berlin, pp 260–275
6. Lambebo A, Haghani S (2014) A wireless sensor network for environmental monitoring of greenhouse gases. In: Proceedings of the ASEE (2014) Zone I Conference. University of Bridgeport, Bridgeport, CT, p 2014
7. Kaur K, Kaur P, Sharanjit Singh E (2014) Wireless sensor network: architecture, design issues and applications. Int J Sci Eng Res (IJSER) 2
8. Zennaro M (2017) Introduction to the internet of things. In: Telecommunication and ICT4D Lab, The Abdus Salam International Centre for Theoretical Physics Trieste, Italy, pp 1–48
9. Workgroup LATM (2015) A technical overview of Lora and Lorawan. In Technical Report, LoRa Alliance
10. Zhang B, Liu J (2010) A kind of design schema of wireless smart water meter reading system based on zigbee technology. In: 2010 international conference on E-Product E-Service and E-Entertainment, pp 1–4. IEEE
11. Cao L, Jiang W, Zhang Z (2008) Networked wireless meter reading system based on zigbee technology. In: 2008 Chinese control and decision conference, pp 3455–3460. IEEE
12. Mehta S, Saraff N, Sanjay SS, Pandey S (2018) Automated agricultural monitoring and controlling system using hc-05 bt module. Int Res J Eng Technol (IRJET) 5(05)

13. Frank Domoney W, Ramli N, Alarefi S, Walker SD (2015) Smart city solutions to water management using self-powered, low-cost, water sensors and apache spark data aggregation. In: 2015 3rd international renewable and sustainable energy conference (IRSEC), pp 1–4. IEEE
14. Kashid PV, Gaikwad N, Deshmukh P, Marchande P (2019) Smart water monitoring system for real-time water quality and usage monitoring
15. Li XJ, Chong PHJ (2019) Design and implementation of a self-powered smart water meter. Sensors 19(19):4177

Chapter 5
Surface Material Classification Using Acceleration Signal

Naveeja Sajeevan, M. Arathi Nair, R. Aravind Sekhar, and K. G. Sreeni

1 Introduction

The way we understand the world around us is greatly influenced by tactile perception. Objects differ from each other in their texture, color and roughness. As a result, there arises a difference in the way we handle these objects. We tend to regulate our responses according to the feedback signals obtained from surfaces. Today's technology has enabled us to capture these feedback signals and process them to have a deeper understanding of its material properties. Wide range of applications has come up that make use of these feedback signals, which include product design, education, entertainment, health care, telecommunication, manufacturing and marketing.

Systems that create virtual environments by engaging solely the visual and auditory senses of the user are restricted in their capability to interact with the objects. But by including tactile knowledge, the system will be able to convey a sense of touch and feel of objects, thus providing a more immersive experience to the user. The user can press, grasp, squeeze, or stroke the virtual objects, can explore object properties such as surface texture, shape, softness and can manipulate tools such as a pen or a jack-hammer. Thus, the incorporation of even a simple haptic interface with auditory and acceleration feedback along with visual display can produce an outsized improvement within the virtual setting.

The rest of the paper is organized as follows. Section 2 discuss the existing works in the field. Section 3 explains the dataset details. Section 4 elaborates the surface classification experiment and Sect. 5 presents the result. Finally, Sect. 6 concludes the paper.

N. Sajeevan (✉) · M. Arathi Nair · R. Aravind Sekhar · K. G. Sreeni
College of Engineering, Trivandrum, Kerala, India

© The Author(s), under exclusive license to Springer Nature Singapore Pte Ltd. 2021
S. Kumar et al. (eds.), *Proceedings of International Conference on Communication and Computational Technologies*, Algorithms for Intelligent Systems,
https://doi.org/10.1007/978-981-16-3246-4_5

2 Existing Works

Image classification tasks have become very popular, and it is the primary domain in which deep neural networks play its role. This led to drastic developments in the field of image classification, detection and recognition [1]. Converting the haptic signals into its corresponding image representation makes the surface material classification task into an image classification task as done in [2].

Two surfaces that appear visually the same can have completely different haptic properties, and vice versa. In such instances, better classification performance can be achieved by combining both haptic and visual information. Thus, [2] introduces FusionNet which takes in both acceleration and image data as input and classifies surfaces with outstanding accuracy. The paper [3] addresses fusion of weakly paired sound and acceleration signal measurements from TUM dataset. In [4] Gao et al. proposed a method of classifying surfaces with haptic adjectives (e.g., hard or soft, smooth or textured) by employing both visual and haptic data. It was done on Penn Haptic Adjective Corpus-2(PHAC-2) dataset. Before convolutional neural networks came into play, certain discriminative features were extracted from the signals and fed into classic machine learning models [5]. In [6] Strese et al. first introduced LMT haptic texture database with both freehand and controlled recordings of 43 classes of surfaces. Several additions were made to this dataset and now it contains 184 classes. The paper [7] presents an approach for tool-mediated surface material classification which is immune to varying scan-time parameters on 69 textures. The classification system uses features such as hardness, roughness, and friction and features adapted from speech recognition such as modified cepstral coefficients. Combined with the Naive Bayes Classifier, the method proposed in [7] leads to a classification accuracy of 95%. Similarly, [8] creates a robotic system capable of touching everyday objects and describing them with haptic adjectives. In the paper [9], Bednarek et al. presented a material classification method using force signals and achieved 100% accuracy on data recorded from a robotic manipulator whereas it achieved 97.96% accuracy for the data from a one-legged machine. In a recent work, Strese et al. [10] presented a methodology for acquisition and parametrization of object material properties. A novel set of mathematical features were introduced and the materials were classified using machine learning techniques. For converting three-dimensional acceleration signal into one dimensional signal for generating spectrogram, the DFT321 technique [11] is used in this project. The neural network model used here is ResNet50. It is one of the most powerful deep neural networks which has achieved close-to-human performance results in the ImageNet classification challenge 2015 [12].

3 Dataset Details

Surface material classification is done on LMT haptic texture database [6] which is freely accessible at https://zeus.lmt.ei.tum.de/downloads/texture/. The dataset consists of feedback signals from 184 different texture materials. The signals include

(a) AluminiumMesh   (b) CrushedRock   (c) GlitterPaper

(d) Marble   (e) RedVelvet   (f) VelcroHooks

Fig. 1 Images of selected classes from the dataset. *Data courtesy* https://zeus.lmt.ei.tum.de/downloads/texture/

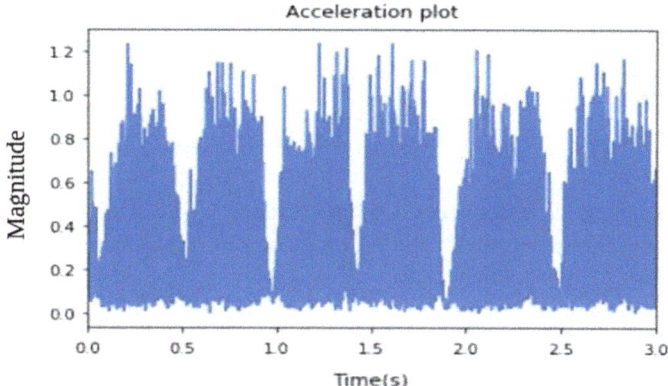

Fig. 2 Magnitude v/s time plot of acceleration signal (Marble)

acceleration scan components, friction scans, image scans, metal detection scans, reflectance scans and sound scans recorded during controlled and well-defined texture scans, as well as uncontrolled human free hand texture explorations. In this project, we have chosen acceleration signals from six different classes. Each class consists of 10 sampled haptic traces of three-dimensional acceleration signal. Figure 1 shows some typical images of selected classes from the dataset.

Acceleration signal is given as separate text files for each coordinate. Each file consists of 48,000 samples, sampled at 10 kHz. Magnitude-time plot of acceleration signal for Marble is given in Fig. 2.

4   Surface Material Classification

For the purpose of classification, we have chosen acceleration data of six different texture surfaces from the TUM dataset. Two approaches are followed for preprocessing these signals and they are compared based on the accuracy of results produced during classification. The first approach converts the acceleration signal into spectrogram while the second approach converts the acceleration signal directly into an image. These approaches are discussed in detail in the following subsections.

4.1   Converting Acceleration Signal into Spectrogram

The three-dimensional acceleration signal is first converted into one-dimensional signal using the DFT321 technique [11]. Here, the acceleration signal consists of components in x, y and z directions. DFT321 uses frequency domain processing to convert a three dimensional signal into a one-dimensional signal without compromising the spectral power. Let 'A_x', 'A_y' and 'A_z' represents the components of acceleration signal in x, y and z directions. Toward calculating the DFT321 signal, DFT of the individual components 'A_x', 'A_y' and 'A_z' needs to be calculated. Let '$A_x(f)$', '$A_y(f)$' and '$A_z(f)$' denotes the DFT of the signals 'A_x', 'A_y' and 'A_z', respectively. Determination of DFT321 is done as per Eq. 1.

$$|A(f)| = \sqrt{|A_x(f)|^2 + |A_y(f)|^2 + |A_z(f)|^2} \tag{1}$$

where $|A_f|$ magnitude in frequency domain of the new DFT 321 signal corresponding to acceleration. The determination of the phase of the corresponding DFT321 signal denoted by '$\theta(f)$' is shown in Eq. 2.

$$\theta(f) = \tan^{-1} \left\{ \frac{I_m(A_x(f) + A_y(f) + A_z(f))}{R_e(A_x(f) + A_y(f) + A_z(f))} \right\} \tag{2}$$

Here, 'I_m' represents the imaginary part and 'R_e' denotes the real part, respectively. The new time domain signal can be obtained by applying an inverse DFT on the magnitude and phase given by Eqs. 1 and 2.

Hence obtained one-dimensional signals are then converted into spectrogram. The Hanning window length is given as 256 and the range of overlapping is given as 128 datapoints. The spectrogram is plotted as a colormap. Fourier frequencies are calculated using sampling frequency of 44,100 samples per unit. Here, we have selected only the components with frequency below 2 kHz as most of the energy lies in the low-frequency region. The spectrogram obtained is augmented by image data augmentation techniques-like horizontal flip and random noise addition to image array. Thus dataset is increased from 10 to 1210 samples for each class. Spectrograms are further resized to suit the training network (224 × 224 in this case). The spectrogram of acceleration signal of marble before resizing is shown in Fig. 3.

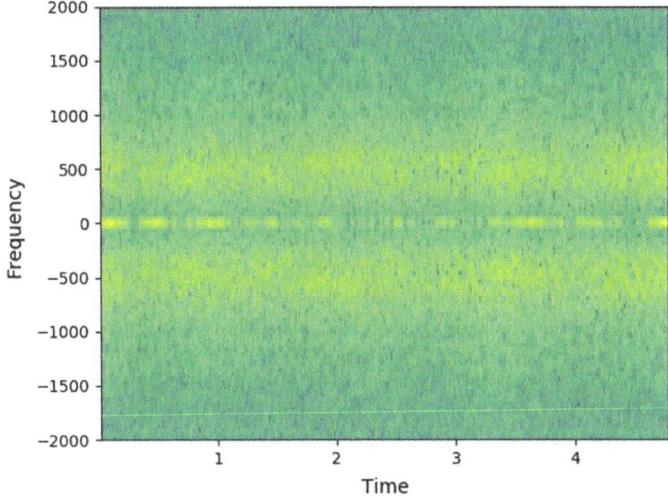

Fig. 3 Spectrogram of sample acceleration signal of Marble

4.2 Direct Conversion of Acceleration Signal into Image

This section explains a method for converting the coordinates of three-dimensional acceleration signal into an RGB image. The three axes coordinates are made positive by shifting the signal by addition of minimum value. Positive values hence obtained are range shifted to [0, 255] so that they denote pixel intensity values. An RGB image of size 224 × 224 is chosen and its pixel values are replaced by the above-obtained values. Since there are only 48,000 values in the acceleration data, the remaining 4176 pixel values are taken as zeros. The obtained image is augmented by image data augmentation techniques-like horizontal flip and random noise addition to image array inorder to increase the number of images in the dataset from 10 to 1210 image samples for each class. Figure 4 shows the images thus constructed for the six classes. The black intensity values at the bottom of the image represent the 4176 pixel values which were assigned zero value.

4.3 Training Details

In both approaches discussed in Sects. 4.1 and 4.2 acceleration signal gets converted into its image representation. Thus, surface material classification task turns into an image classification task. For an easy comparison, the same neural network model is chosen for both the approaches. Our implementation is based on Keras deep learning library. The neural network used is ResNet50. The model is compiled with Adam optimizer. Training is done on google colaboratory with GPU runtime. ResNet is

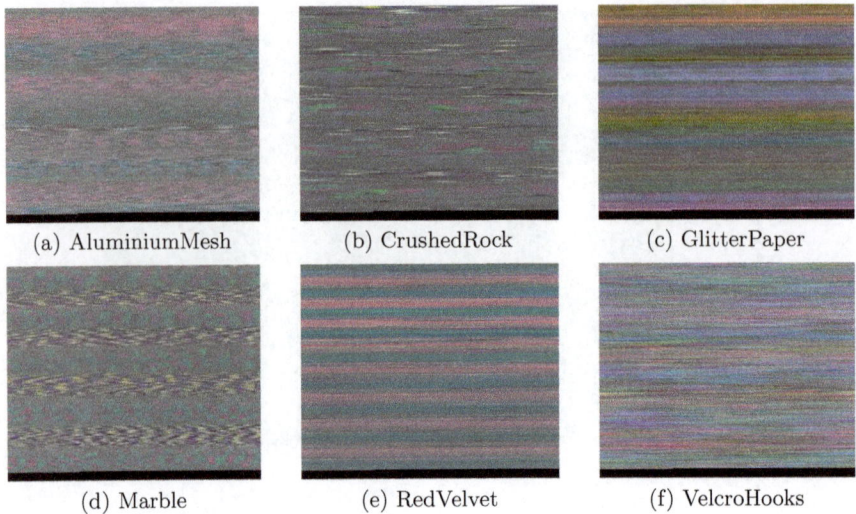

Fig. 4 Some typical images generated from acceleration data as discussed in Sect. 4.2

Table 1 Parameters used for training

Data	Learning rate	Batch size	Validation split	Epochs
Spectrogram	0.0001	10	0.4	35
Image converted	0.0001	10	0.4	20

initialized with ImageNet weights (Transfer learning). Only the last fully connected layer is trained. Experiments have been done with different learning rates for training the model. The model was seen to attain its best accuracy with the learning rate of 0.0001. Spectrogram data took 35 epochs to converge while data converted into image converged faster with just 20 epochs. The parameters for training are given in Table. 1.

5 Results and Discussion

In this paper, two approaches have been followed for preprocessing acceleration signal. These approaches are compared based on the classification results. Acceleration data converted into spectrogram and that directly converted into image were trained and tested individually and the results are compared in this section.

Table 2 Peformance comparison of the proposed methods

Data	Acceleration-spectrogram	Acceleration-image
Test accuracy (%)	83.333	96.66
Test loss	1.1202	0.0757

```
[[10  0  0  0  0  0]          [[10  0  0  0  0  0]
 [ 0 10  0  0  0  0]           [ 0 10  0  0  0  0]
 [ 0  0  9  0  1  0]           [ 1  0  9  0  0  0]
 [ 0  0  0 10  0  0]           [ 0  0  0 10  0  0]
 [ 0  0  1  0  9  0]           [ 0  0  0  0 10  0]
 [ 0  0  0  0  7  3]]          [ 1  0  0  0  0  9]]
```

(a) Acceleration-spectrogram   (b) Acceleration-image

Fig. 5 Comparison of confusion matrices of two classification methods

The test result is given in Table 2. The test accuracy of acceleration converted into spectrogram is found to be 83.33% and that of acceleration converted into image is 96.66%. This shows that acceleration into image is a better approach than acceleration into spectrogram approach. To evaluate the classifier performance, metrics used are confusion matrix, classification report, and receiver operating characteristics (ROC) curve.

5.1 Confusion Matrix

The confusion matrix for spectrogram based approach and image based approach are shown in Fig. 5. The classes are ordered in the same way they appear in Fig. 1. From Fig. 5a, it can be seen that VelcroHooks is misclassified as RedVelvet in 70% of the cases which agrees with the lower classification accuracy of acceleration-spectrogram method as compared to acceleration-image method.

5.2 Classification Report

The classification report shows a per-class basis representation of the main classification metrics. Classification report is given in Fig. 6.

In this case, the test data consists of 10 samples for each class. So the macro average and weighted average are the same. For acceleration-image method the F1 score for all the classes is 0.95 and above (Fig. 6b) while acceleration-spectrogram method has F1 score as low as 0.46 for VelcroHooks (Fig. 6a).

	Precision	Recall	F1-score	Support
Aluminium Mesh	1.00	1.00	1.00	10
Crushed Rock	1.00	1.00	1.00	10
Glitter Paper	0.90	0.90	0.90	10
Marble	1.00	1.00	1.00	10
Red Velvet	0.53	0.90	0.67	10
Velcro	1.00	0.30	0.46	10
Accuracy			0.85	60
Macro Average	0.90	0.85	0.84	60
Weighted Average	0.90	0.85	0.84	60

(a) Acceleration-spectrogram

	Precision	Recall	F1-score	Support
Aluminium Mesh	0.83	1.00	1.00	10
Crushed Rock	1.00	1.00	1.00	10
Glitter Paper	1.00	0.90	0.95	10
Marble	1.00	1.00	1.00	10
Red Velvet	1.00	1.00	1.00	10
Velcro	1.00	0.90	0.95	10
Accuracy			0.97	60
Macro Average	0.97	0.97	0.97	60
Weighted Average	0.97	0.97	0.97	60

(b) Acceleration-image

Fig. 6 Comparison of classification reports of proposed methods

5.3 ROC Curve

ROC curve plots true positive rate against false positive rate. Larger area under the curve (AUC) shows better classification accuracy. ROC curves for both approaches are shown in Fig. 7. It can be seen that AUC for all classes in Fig. 7b is nearly equal to one showing better classification accuracy than Fig. 7a.

6   Conclusions

This paper compares two techniques for preprocessing the acceleration signal towards the purpose of surface material classification. Experiments on TUM haptic texture database indicate that the proposed conversion of acceleration signal directly into image gives a classification accuracy of 96.66%, while the acceleration converted into spectrogram yields an accuracy of 83.33%. Also the number of training

Fig. 7 Comparison of ROC curves of proposed methods

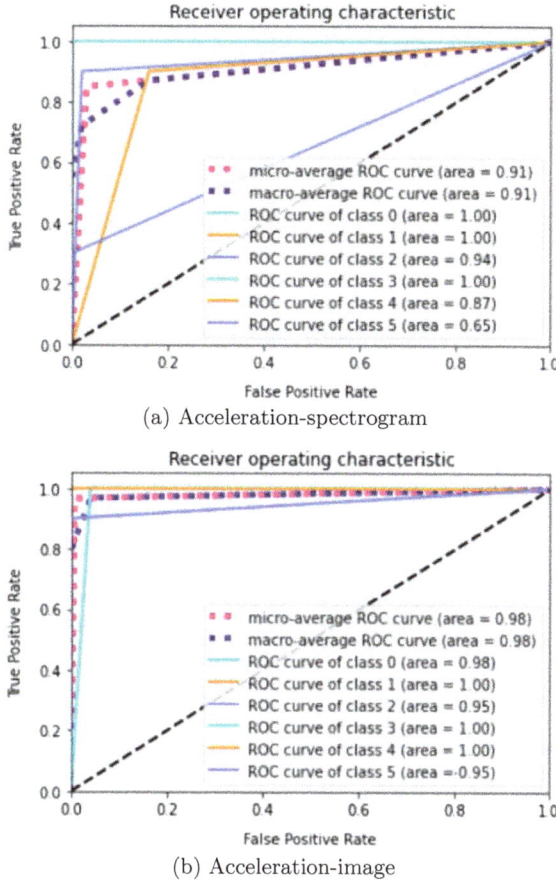

(a) Acceleration-spectrogram

(b) Acceleration-image

epochs required was less in the case of image based approach which can reduce the computation time to a great extent. As a future work the method aims to be extended over a larger dataset.

References

1. Krizhevsky A, Sutskever I, Hinton GE (2015), ImageNet classification with deep convolutional neural networks. In: Proceedings of advances in neural information processing systems (NIPS), pp 1097–1105
2. Zheng H, Fang L, Ji M, Strese M, Ozer Y, Steinbach E (2016) Deep learning for surface material classification using haptic and visual information. IEEE Trans Multimedia 18:12
3. Liu H, Sun F, Fang B, Shan L (2018) Multimodal measurements fusion for surface material categorization. IEEE Trans Instrum Measure 67(2):246–256

4. Gao Y, Hendricks LA, Kuchenbecker KJ, Darrell T (2015) Deep learning for tactile understanding from visual and haptic data. In: IEEE International conference on robotics and automation (ICRA), pp 536–543
5. Strese M, Schuwer C, Lepure A, Steinbach E (2017) Multimodal feature based surface material classification. IEEE Trans Haptics 10(2):1–15
6. Strese M, Lee JY, Schuwerk C, Han Q, Kim HG, Steinbach E (2014) A haptic texture database for tool mediated texture recognition and classification. In: Proceedings of IEEE international symposium on haptic audio-visual environments games, pp 118–123
7. Strese M, Schuwerk C, Steinbach E (2015) Surface classification using acceleration signals recorded during human free hand movement. In: Proceedings of IEEE world haptics conference, pp 214–219
8. Chu V, McMahon I, Riano L, McDonald CG, He Q, Perez-Tejada JM, Arrigo M, Darrell T, Kuchenbecker KJ (2014) Robotic learning of haptic adjectives through physical interaction. Haptics Group, GRASP Laboratory. https://doi.org/10.1016/j.robot.2014.09.021
9. Bednarek J, Bednarek M, Kicki P, Walas K (2019) Robotic touch: classification of materials for manipulation and walking. In: 2019 2nd IEEE international conference on Soft Robotics (RoboSoft), pp 527–533
10. Strese M, Brudermueller L, Kirsch J, Steinbach E (2020) Haptic material analysis and classification inspired by human exploratory procedures. IEEE Trans Haptics 13(2):404–424
11. Landin N, Romano JM, McMahan W, Kuchenbecker KJ (2010) Dimensional reduction of high frequency accelerations for haptic rendering. In: Generating and perceiving tangible sensations. Springer, Berlin, pp 79–86
12. He K, Zhang X, Ren S, Sun J (2016) Deep residual learning for image recognition. In: IEEE conference on computer vision and pattern recognition (CVPR), pp. 770–778. https://doi.org/10.1109/CVPR.2016.90

Chapter 6
Genetic Algorithm with Approximation Algorithm Based Initial Population for the Set Covering Problem

Hajar Razip and Nordin Zakaria

1 Introduction

The two main classes of computational complexity are P and NP [23]. P (polynomial time), encompasses problems solvable via some deterministic polynomial-time algorithms. NP (nondeterministic polynomial time) include problems that have solutions which can be verified efficiently. The hardest NP problems are considered "NP-complete"—any NP problem can be reduced to it. Combinatorial optimization problems belong to a subset of NP-complete (COP) [19]. In an optimization problem, we seek to minimize or maximize an objective function by building an optimal configuration for a set of variables subject to some constraints. A COP requires that these configurations be discrete. Generally, continuous optimization problems do have solutions in P, linear programming (LP) being the standard one. In the case of COP, the two main tracks have been metaheuristics and approximation algorithms (AA).

Metaheuristics are essentially algorithms that attempt at improving a candidate solution iteratively guided by a measure of quality [10]. Genetic algorithm (GA) is a popular example, developed by John Holland and his team in the 1970s [11]. GA is prevalent in a wide range of applications including data analysis and prediction, genomics, automatic programming, evolutionary neural networks, structural engineering, and many more [8, 16]. As a nature-inspired metaheuristic, GA adapts some key characteristics of biological evolution; the key processes are selection, crossover, and mutation. A GA's input is an initial population (IP) filled with feasible solutions to a problem instance.

An approximation algorithm (AA) is a heuristic with guaranteed bounds on the value of the solution it provides [29], and this is an advantage over metaheuristics.

H. Razip (✉) · N. Zakaria
HPC3, Universiti Teknologi Petronas, Seri Iskandar, Malaysia

© The Author(s), under exclusive license to Springer Nature Singapore Pte Ltd. 2021 59
S. Kumar et al. (eds.), *Proceedings of International Conference on Communication and Computational Technologies*, Algorithms for Intelligent Systems,
https://doi.org/10.1007/978-981-16-3246-4_6

An η-approximation algorithm, for example, is capable of generating a solution that is a factor η of the optimal solution (OPT).

It has been widely accepted that the initial population (IP) of a GA affects its overall performance significantly [5, 8, 14, 17, 20, 26, 30]. A good IP can prevent premature convergence, enhance convergence speed and improve the chances of finding good solutions by moving the search towards more promising regions in the search space. Despite this, compared to other GA operators, literature focusing on the IP of GA is relatively scarce. Furthermore, IP generation (IPG) methods are limited to randomization and heuristics, neither of which guarantees the quality of the generated solutions the way AA does. An implementation of AA at the IPG for the Set Covering Problem (SCP) has been presented in [21] and to the best of our knowledge, we have yet to find other work that explores such an approach.

Hence, this paper builds upon the work in [21]; we present a number of new approaches, implementing some of the other well-known AAs for the SCP. The rest of the paper is organized as follows; Sect. 2 lays out the state-of-the-art of the set covering problem (SCP), focusing on current AA solutions and IP generation methods. Section 3 presents the materials and methods, detailing the problem data sets, experimental setup, and algorithms. Section 4 discusses results and Sect. 5 concludes the paper. Suggestions for future work follows in Sect. 6.

2 The Set Covering Problem

The Set Covering Problem (SCP) is a problem of covering a ground set of elements $\mathcal{E} = \{e_1, e_2, \ldots, e_m\}$ with weighted subsets of those elements in $\mathcal{S} = \{s_1, s_2, \ldots, s_n\}$, while minimizing the total weights. This is a minimization problem that is generally formulated as in Eq. 1 [4]. Hence, the SCP is a problem of covering the rows $i = \{1, \ldots, m\}$ of a zero-one $m \times n$ matrix (a_{ij}) with a subset of the columns $j = \{1, \ldots, n\}$ at minimal cost. Each a_{ij} corresponds to whether the row i is covered by a column j and vice versa. The integrality constraint ensures that x_j (which corresponds to each column in the solution) set to either 0 or 1, 1 for inclusion and 0 otherwise. The equation $\sum_{j=1}^{n} a_{ij} x_j \geq 1$ ensures that each row is covered by at least one column.

$$
\begin{aligned}
\text{minimize} \quad & \sum_{j=1}^{n} w_j x_j \\
\text{subject to} \quad & \sum_{j=1}^{n} a_{ij} x_j \geq 1, \qquad i = 1, \ldots, m \\
& x_j \in \{0, 1\}, \ j = 1, \ldots, n
\end{aligned}
\tag{1}
$$

The problem has a wide-range of applications—facilities location [25], information retrieval [28], airline crew scheduling [3, 22], bus driver scheduling, tanker routing, switching theory [2], simplification of boolean expressions [6], wireless sensor networks [31], error detection circuit [15] and more.

A number of heuristics and exact approaches to solve the SCP have been proposed in literature. An overview of the state-of-the-art is treated succinctly in [7].

2.1   Approximation Algorithms for the SCP

A number of the well-known AA methods for the SCP has been described by Williamson and Shmoys in [29]. In this section, we briefly describe the ones we have implemented in our novel IPG designs. Many of them depend on the LP relaxation solution Z_{LP} as a lower bound to the approximation factor (i.e. an η-approximation algorithm will calculate a solution within $\eta \cdot Z_{LP}$) where an LP relaxation is a method where the integrality constraints on the problem are relaxed and the problem is solved using LP.

Deterministic Rounding Algorithm (AAD) Given the relaxation solution x^*, the corresponding variable in the approximation solution x is rounded off to 1 if $x_j^* \geq 1/f$, where f is the maximum number of sets in which any elements appear. This gives us an f-approximation algorithm.

Greedy Algorithm (G) To date, the greedy algorithm is claimed to be the best AA for the SCP. A greedy solution is built by iteratively adding the next column that minimizes the ratio between its weight and all the containing uncovered elements to the solution while arbitrarily breaking ties. This is an H_m-approximation algorithm where H_m is the mth harmonic number.

Randomized Rounding Algorithm I (AAR) The LP relaxation solution is treated as a set of probabilities that a certain column is included in a solution and the inclusion of a column depends on a random event. To ensure that the generated solution is a feasible one, the random event is repeated $2 \ln m$ times. This gives a $2 \ln m$-approximation algorithm.

Randomized Rounding Algorithm II (AAR2) Another implementation of a randomized rounding algorithm is presented in [27] where, instead of calculating the frequency of random events per column, a number of passes are run until the solution is feasible. This is a $2(\ln m + 3)$-approximation algorithm. Nonetheless, both AAR and AAR2 are $O(\ln m)$ in the worst case.

2.2   State-of-the-art in IP Generation for the Multicost SCP Problem

A population in a GA consists of a number of *chromosomes* that encodes feasible solutions to the problem instance it is optimizing. A "gene" in a chromosome corresponds to a variable of the objective function. In the case of the SCP, a gene corresponds to a column. The binary representation is the most popular representation scheme for the SCP—a gene that encodes '1' indicates that a column is included in a solution and '0' otherwise. The chromosomes are also referred to as *individuals* or *members* of a population.

A solution string can hence be generated by a random sequence of 1s and 0s. However, this method, while good for diversity, is usually unreliable in maintaining feasibility. A repair mechanism is typically required. Another popular generation method is by heuristically building a solution string. A typical pattern has been identified—a solution is usually built by iteratively adding covering columns based on some predefined rules. Various column picking mechanisms have been introduced in the literature. Forms of greedy heuristics are popular, where the next best choice is based on a carefully calculated score. Local optimization steps typically follow.

Some of the earliest GA implementation for the SCP can be found in [2, 4, 12]. In [12], a feasibility operator was proposed that first calculates the cost of an infeasible chromosome, track excluded columns and uncovered rows, then associate the columns with a cost-ratio (cost/number of uncovered rows). Subsequent lowest-cost-ratio columns are included until the solution is feasible. In [4], a similar feasibility operator was proposed but with an added local optimization step that, in order of decreasing costs, removes redundant columns [4]. Aside from that, the initial string was randomly generated by assigning each row a random column from the first 5 covering columns to maintain the population size. A local optimization step follows with an algorithm similar to its feasibility operator, but the redundant columns were chosen arbitrarily.

Iwamura, Horiike, and Shibahara (2003) proposed a more deterministic approach to IP generation by considering domain-specific knowledge [13]. In decreasing order of cost-effectiveness, the columns are iterated and after each iteration, a feasible solution is generated. A new solution started from the iterating column, and uncovered rows are covered by adding new covering columns thereafter following rules based on cost-effectiveness. This algorithm was more effective for high-density problems with run times proportional to the problem size. However, it was unable to provide an IP good enough to reach the optimal solution (OPT) in the tested data sets, unlike Beasley and Chu's GA [4]. Iwamura et al.'s approach have a number of disadvantages. Most notably, the way that all columns were considered in solution generation meant that even the most expensive one has a sure chance to be included. In contrast, the design of the previous two approaches made sure to discard the more expensive columns. Furthermore, since Iwamura et al.'s method is deterministic, there are a limited number of configurations that it can provide (the number of columns, less in case of redundancies). If none of those configurations supply enough building blocks for the OPT, and mutation fails to introduce them, then the OPT will never be found.

In Aickelin's indirect GA for the SCP, the rows are iterated in an "optimum" order [1]. The GA part of their algorithm generates an optimal permutation and later decoded into a feasible solution, followed by a post-optimization step. While GA was used as a parameter optimizer instead of directly providing the optimizing function, Aickelin's decoder function was worth considering as an inspiration for future solution generation approaches. At each row iteration, a column is added in the solution based on a weighted column score strategy. Subsequently considered rows are the ones that had not been covered yet, a step that would save computation time. As a GA IP generation method, using a GA as permutation generator might not be the best approach due to the time it requires. Otherwise, it should be run in parallel. Alternatively, the order of row iterations can be decided randomly (which would remove the "optimally ordered iteration" aspect of the algorithm).

Nonetheless, the SCP comes in various flavors. Hence, many approaches have been presented in literature, carefully designed to maintain feasibility based on the problem constraints. The IP generation approaches mentioned in this section cover only the multi-cost SCP, and their steps can be generalized as having a constructive heuristic part with occasional pre-processing steps, followed by a local optimization step.

3   Materials and Methods

3.1   Problem Data Sets

The tested data sets are obtained from OR-Library.[1] Table 1 lists the details of the data sets. Density is the proportion of 1s in the problem matrix.

3.2   Experiment 1: Sampling Pool Quality

Experiment 1 is conducted to test the new IPG method designs (described in Sect. 3.4) and compare their quality alongside a benchmark population generated by a traditional method. Each method is used to generate a sampling pool of size $N \leq 1000$ (a method is run 1000 times, and the resulting solutions may or may not duplicate, generating a sampling pool of size ≤ 1000) and the generated individuals will be used as GA input. The GA performances will be tested in experiment 2.

3.2.1   Quality Measures

The quality is measured in terms of redundancy rate ($\mathscr{R}$), diversity ($\mathscr{D}$), and closeness to the OPT.

The redundancy rate ($\mathscr{R}$) is the rate at which a solution is duplicated in a sampling pool. We calculate $\mathscr{R}$ using the formula in Eq. 2; the redundancy rate is the number of duplicates divided by the sum of the number of unique individuals and the number of duplicates i.e. the portion size of duplicates among the entire sampling

---

[1] The OR-Library is an open internet repository hosting a collection of data sets for various Operation Research (OR) problems, accessible at http://people.brunel.ac.uk/ mastjjb/jeb/info.html.

Table 1 Description of problem instances

Problem set	m	n	Density (%)	Number of problems
4	200	1000	2	10
5	200	2000	2	10
6	200	1000	5	5
A	300	3000	2	5
B	300	3000	5	5
C	400	4000	2	5
D	400	4000	5	5
E	500	5000	2	5
F	500	5000	5	5
G	1000	10,000	2	5
H	1000	10,000	5	5

pool. A redundancy rate $\mathcal{R}_M = 0.999$, where M is an IPG method, indicates that all individuals in a sampling pool are a duplicate of the first.

$$\mathcal{R} = \frac{\# \text{ duplicates}}{\# \text{ duplicates} + \# \text{ unique}} \tag{2}$$

The Moment of Inertia (MoI) based metric proposed by Morrison and De Jong is used to evaluate diversity [18]. The score measured by MoI is identical to the sum of pairwise Hamming distances (HD) averaged by the population size with a more efficient calculation time. The diversity measure that we used is a modified MoI that gives the sum of pairwise Hamming distances averaged by the number of calculations instead of population size (Eq. 3). This will result in a score that is within the range $[0, l]$ where l is the chromosome length.

$$\mathcal{D} = \frac{2 \times (\sum_{j=1}^{l} \sum_{i=1}^{N} (a_{ij} - \frac{\sum_{i=1}^{N} a_{ij}}{N})^2)}{N - 1} \tag{3}$$

Closeness to the OPT is measured in terms of average HD between every individual in a sampling pool to the OPT solution (Γ_μ), the standard deviation of fitness error of the sampling pool individuals to the OPT ($\mathcal{S}$) and the difference of fitness values between a sampling pool's best individual and the OPT ($\mathcal{S}_{\text{best}}$). Equation 4 describes $\mathcal{S}$, where OBJ_i is the objective function value of individual i, and $(\frac{\text{OBJ}_i}{\text{OPT}} - 1.0)^2$ calculates the fitness error between individual i and the OPT. A fitness error > 0 indicates that individual i has an objective function value greater than OPT.

$$\mathcal{S} = \sqrt{\frac{\sum_{i=1}^{N} (\frac{\text{OBJ}_i}{\text{OPT}} - 1.0)^2}{N - 1}} \tag{4}$$

Table 2 List of GAs

GA	IP source
GA_R	P_R
GA_{AADLO}	P_{AADLO}
GA_{AAR}	P_{AAR}
GA_{AARLO}	P_{AARLO}
GA_{AAR2}	P_{AAR2}
GA_{AAR2LO}	P_{AAR2LO}

Aside from the three main quality measures, we also performed PCA[2] to visualize the distribution of our sampling pool individuals. Furthermore, the sampling pools were evaluated for the fitness-distance correlation (FDC) [9] score to test whether the objective function values correlate with the distances of the solution configurations to the OPT i.e. the closer an individual to the OPT in terms of HD, the lesser the objective function value difference. This has previously been performed in [9] on some of the selected problem data sets, but with a different distance measure based on the number of distinct columns (we used HD instead). Furthermore, the correlation between fitness and distance was also evaluated using Spearman's correlation test between $\mathscr{S}$ and Γ_{μ}.

3.3   Experiment 2: GA Performance

We adapted the GA design from the one proposed in [4] as our base GA. It uses the binary representation for the chromosomes, the binary tournament selection for parent selection, fusion crossover for generating offsprings, a variable mutation rate with parameters same as the original implementation, and the steady-state replacement model for population replacement. The GA terminates under three conditions; after 100,000 unique solutions are generated or if the best feasible solution (BFS) remains unchanged after 10,000 iterations or the BFS found matches OPT.

Each type of GA compared is based on the source of its IP, which is a sampling pool generated in experiment 1. Each GA_M, where M represents an IPG method listed in Sect. 3.4, starts with an IP of size 100 whose members were picked randomly from a sampling pool P_M. Table 2 lists the GAs and their source of IP.

---

[2] The principal component analysis is a dimensionality reduction method used to enable visualization of high dimensional data in lower dimensional space [24].

3.3.1 Performance Measures

Each GA is run 10 times and the GA performances were compared pairwise in terms of average number of iterations elapsed ($\mathcal{T}$) before the BFS is found, the best-performing trial in terms of least number of iterations ($\mathcal{T}_{\text{best}}$), the frequency of successes in finding the BFS across 10 trials and the average of solution values.

3.4 Algorithms

Our AA-based IP generation methods followed the convention of using a constructive heuristic, and on some of them, implementing a local optimization step. The AA methods described earlier took up the constructive heuristic part. The local optimization step, aside from improving the solutions, provided diversity to the population. Hence, including the traditional method for the SCP as benchmark, the random heuristic as described in [4], seven IP generation methods were tested;

1. **Random heuristics (R)**: A random heuristic typically used for building initial solutions for the set covering problem in a genetic algorithm. Used as benchmark against other sampling methods
2. **AADLO**: The deterministic rounding f-approximation algorithm (AAD), followed by a local optimization step.
3. **GLO**: The greedy H_m-approximation algorithm (G), followed by a local optimization step.
4. **AAR**: The randomized $2(\ln m)$-approximation algorithm.
5. **AARLO**: AAR, followed by a local optimization step.
6. **AAR2**: The randomized $2(\ln m + 3)$-approximation algorithm.
7. **AAR2LO**: AAR2, followed by a local optimization step.

We called AAD, G, AAR, and AAR2 the base methods that took up the constructive heuristic part of the IPG design. The idea is that the initial solution from these base methods will not be locally optimal, hence containing redundant columns that can be arbitrarily dropped to obtain new solutions that are locally optimal.

Next, the algorithms are described for each method. The following denotations apply in this section only;

I = set of all row indices
J = set of all column indices
β_j = the set of indices of rows covered by column j, $j \in J$
S = the set of column indices in a solution
U = the set of indices of uncovered rows
c_i = the number of columns that cover row $i \in I$ in S
w_j = cost of column j.

Algorithm 1 describes the local optimization algorithm (LO). First, a copy of S is created in T (1). S initially is a feasible solution. While T is not empty (2), a random column k from T is selected and set aside from T (3). If there are other columns covering any of the rows covered by column k, then column k is discarded from the solution (5–7). This effectively removes any redundant columns in the solution.

Algorithm 1 Local Optimization (LO)

Input: feasible solution S
Output: locally optimized S
1: $T \leftarrow S$
2: **while** $T \neq \emptyset$ **do**
3:    $k \leftarrow random(T)$
4:    $T \leftarrow T - k$
5:    **if** $c_i \geq 2, \forall i \in \beta_k$ **then**
6:        $S \leftarrow S - k$
7:        $c_i \leftarrow c_i - 1, \forall i \in \beta_k$
8:    **end if**
9: **end while**
10: **return** S

AADLO is described in Algorithm 2. The AAD part is covered in steps 1–6 and in step 7, the generated solution S is input into the LO algorithm. The algorithm starts off with the LP relaxation solution x^* and an empty solution S (1). For every x_j, if x_j is more than $1/f$, where f is the maximum number of columns covered by any row, then column j is included in S (2–4).

Algorithm 2 Deterministic Rounding Algorithm (AAD) and AAD + LO (AADLO)

Input: $x*$
Output: feasible S
1: $S \leftarrow \emptyset$
2: **for** $j \in J$ **do**
3:    **if** $x*_j \geq 1/f$ **then**
4:        $S \leftarrow S \cup j$
5:    **end if**
6: **end for**
7: $S \leftarrow LO(S)$
8: **return** S

GLO is described in Algorithm 3. The greedy heuristic is described in steps 1–8 and step 9 inputs the resulting feasible solution S into LO. The algorithm starts with an empty solution S (1). Let $\hat{\beta}_j$ be the set of rows in column j uncovered by the solution (2). While there are still uncovered rows (3), assign to l the column j that minimizes the ratio between the cost w_j and the number of rows uncovered by j i.e. choose the column that covers the most uncovered rows at minimum cost (4). Include column l in the solution (5) and update $\hat{\beta}_j$ (6) and the list of uncovered rows (7).

Algorithm 3 Greedy approximation algorithm + LO (GLO)

Output: feasible S
1: $S \leftarrow \emptyset$
2: $\hat{\beta}_j \leftarrow \beta_j, \forall j$
3: **while** $U \neq \emptyset$ **do**
4:    $l \leftarrow arg\ min_{j:\hat{\beta}_j \neq \emptyset} \frac{w_j}{|\hat{\beta}_j|}$
5:    $S \leftarrow S \cup l$
6:    $\hat{\beta}_j \leftarrow \hat{\beta}_j - \beta_l, \forall j$
7:    $U \leftarrow U - \beta_l$
8: **end while**
9: $S \leftarrow LO(S)$
10: **return** S

AARLO is described in Algorithm 4. Like AAD, the input is the LP relaxation solution x^*. AAR takes up steps 1 to 11, and the resulting feasible solution S is input into LO at step 12. Starting with an empty solution S (1), the frequency of random events d is calculated (2). For every x_j^*, a random number r within the range [0.0, 1.0] is generated d times (3–8). If in any of those trial, r turned out to be less than x_j^*, then column j is included in the solution.

Algorithm 4 Randomized Rounding Algorithm (AAR) and AAR + LO (AARLO)

Input: $x*$
Output: feasible S
1: $S \leftarrow \emptyset$
2: $d \leftarrow \lfloor 2 \ln |I| \rfloor$
3: **for** $j \in J$ **do**
4:    **for** $k \in \{1, ..., d\}$ **do**
5:       $r \leftarrow random([0.0, 1.0])$
6:       **if** $r \leq x_j^*$ **then**
7:          $S \leftarrow S \cup j$
8:          Goto step 10
9:       **end if**
10:    **end for**
11: **end for**
12: $S \leftarrow LO(S)$
13: **return** S

AAR2LO is described in Algorithm 5. AAR2 takes up steps 1 to 9 and the resulting feasible solution S is input to LO at step 10. The algorithm is similar to AARLO. However, instead of calculating the frequency of random events, we run the random event once for every column and repeat the process until all rows are covered.

Algorithm 5 Randomized Rounding Algorithm II + LO (AAR2LO)

Input: $x*$
Output: feasible S
1: $S \leftarrow \emptyset$
2: **while** $\exists i : i \notin \cup_{j \in S} \beta_j$ **do**
3:    **for** $j \in J$ **do**
4:        $r \leftarrow random([0.0, 1.0])$
5:        **if** $r \leq x_j^*$ **then**
6:            $S \leftarrow S \cup j$
7:        **end if**
8:    **end for**
9: **end while**
10: $S \leftarrow LO(S)$
11: **return** S

---

4   Results and Discussion

All computational experiments were conducted on a machine Intel(R) Xeon(R) CPU E3-1231 v3 @ 3.40 GHz. The codes were written in Python 2.7.13 and we utilized the IBM(R) ILOG(R) CPLEX(R) Interactive Optimizer 12.7.0.0 (CPLEX) for optimization functions. CPLEX's MIP[3] function was used to obtain the OPT for problem sets 4–6, and A–G. For problem set H which could not be solved the same way, R-GA was run 10 times for each instance and the BFS found across all trials were used in place of OPT.

Significant differences between groups were established using the Friedman test, followed by the Wilcoxon signed rank test on every pairwise comparison. Significance α level is set at 0.05 before Benferroni corrections.

4.1   Sampling Pool Quality

4.1.1   Redundancy Rate, $\mathscr{R}$

A number of observations were obtained on redundancy rates $\mathscr{R}$. For all instances, $\mathscr{R}_R$ is 0 and $\mathscr{R}_{GLO}$ is always greater than 0.9, resulting in the exclusion of P_{GLO} as GA input. Larger sized instances in problem sets A–H generally has 0 redundancy rates for any sampling pool while that of many of the smaller-sized instances in problem sets 4–6 obtained greater than 0.9, many of which scored $\mathscr{R} = 0.999$ i.e. all generated individuals were the same.

We compared six of the $\mathscr{R}$ groups ($\mathscr{R}_R$ excluded due to its all-zero scores) for instances whose at least half of their $\mathscr{R}$ scores scored greater than zero. This is observed for most of the smaller-sized instances in sets 4–6, excluding the rest of

---

[3] Mixed-Integer Programming is an exact method to solve combinatorial optimization problems.

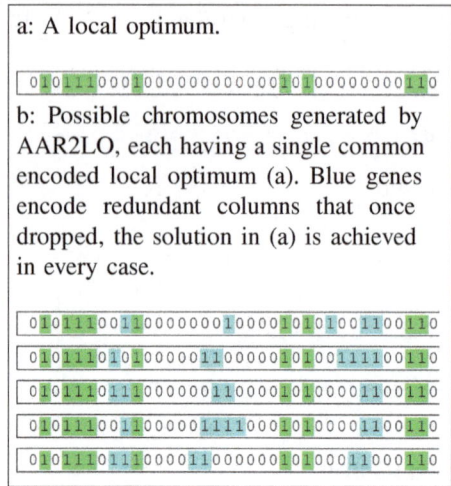

Fig. 1 Chromosome generation using AAR2LO

the larger-sized problems. Comparison would be meaningless for the latter sets since $\mathscr{R}_M$ were mostly zero. Significant difference were observed ($X^2 = 21.756$, df $= 5$, $p < 0.001$). Post-hoc analysis was performed with $\alpha = 0.05/15 = 0.003$.

Our observations showed that $\mathscr{R}_{AAR2}$ is significantly lower than most groups. The group with the highest redundancy rates was $\mathscr{R}_{GLO}$ for achieving 0.9 for all instances, but the scores were not significantly different from other groups except $\mathscr{R}_{AAR2}$ ($p < 0.001$).

Our hypothesis was that a greater randomization measure will improve population diversity. Hence, methods with LO-counterparts such as AAR and AAR2 should generate sampling pools with greater redundancy rates than their LO-counterparts. It is not always the case for the set of instances compared here. While $\mathscr{R}_{AARLO}$ was indeed lesser than $\mathscr{R}_{AAR}$ ($p < 0.001$), it is not the case for the AAR2 pairs. $\mathscr{R}_{AAR2LO}$ was generally greater than $\mathscr{R}_{AAR2}$ ($p = 0.001$). This occurrence is explained in Fig. 1. Instead of deriving new locally optimal solutions by applying LO, the initially generated solutions may share the exact same local optimal configurations which, upon removal of redundant columns, the algorithm would arrive to.

Furthermore, the ability of LO to derive new solutions is also limited depending on the frequency of redundant columns in an initial solution. Figure 2 further details this situation. Using AADLO and GLO as an example, the former method has a base method whose solution guarantee is greater than that of the latter. Hence, the former solution would have more redundant (a) columns compared to the latter (b). While various local optimums can be derived from (a), the same local optimums can not be derived from (b) as they are not encoded in it, hence lesser possibility of generating new solutions.

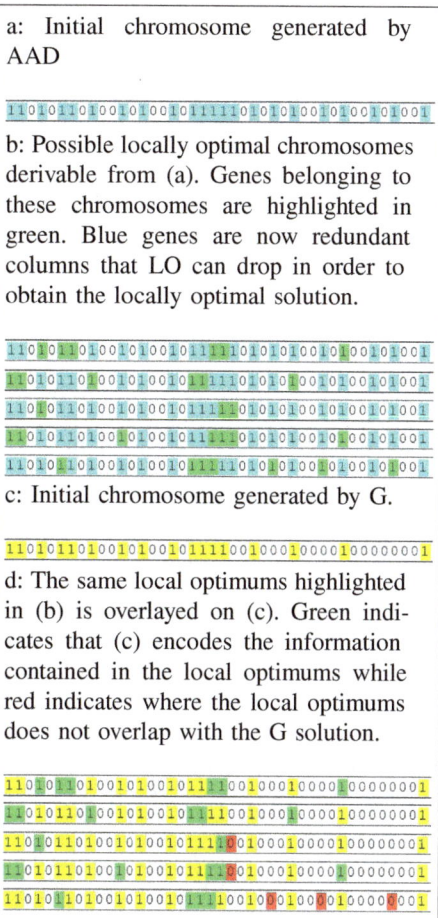

Fig. 2 Chromosome generation using AAD versus greedy

4.1.2 Diversity, $\mathcal{D}$

The measure of diversity $\mathcal{D}$ is indeed related to redundancy rate but describes variation within a sampling pool to a greater extent. While $\mathcal{R}$ indicates whether two solution strings are different, $\mathcal{D}$ indicates how different they are. For example, the strings 00100 and 11011 have a redundancy rate of 0, which indicates that both strings are different. The same goes for 00100 and 00111. The former pair, however, has a diversity score of 5 while the latter scored 2. Clearly, the first pair are more different to each other compared to the second in terms of solution configuration. In our case, if $\mathcal{R}_M = 0.999$, clearly $\mathcal{D}_M$ would be 0, as all individuals were the same. Due to the effect of redundancy rate on diversity in our observations, in this section, only instances which all $\mathcal{R}_M$ (excluding $\mathcal{R}_{GLO}$) scored 0 are included in the comparisons.

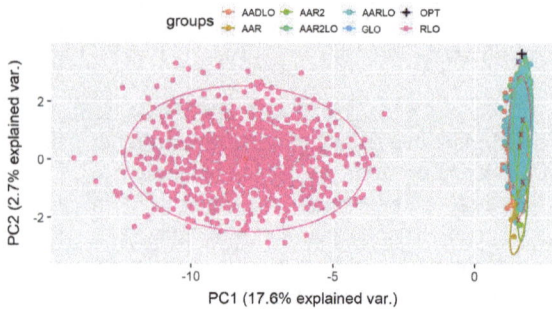

Fig. 3 PCA plot for problem instance C.1

This comprised of the bigger-sized problem instances in contrast to the ones tested for the previous criteria.

A significant difference was observed between the groups ($X^2 = 145.035$, df $= 5$, $p < 0.001$). Post-hoc analysis is performed with significance level set at $\alpha = 0.05/15 = 0.003$.

Of all groups, $\mathscr{D}_R$ scored greater, which was expected of the benchmark sampling pools. $\mathscr{D}_{\text{AADLO}}$ was only lesser than $\mathscr{D}_R$ ($p < 0.001$) but greater than others ($p < 0.001$). $\mathscr{D}_{\text{AAR}}$ has the lowest diversity scores compared to others ($p < 0.001$), but its LO-counterpart, $\mathscr{D}_{\text{AARLO}}$ achieved scores greater than any of the AA-based groups ($p < 0.001$). A similar observation was made on the pair $\mathscr{D}_{\text{AAR2}}$ and $\mathscr{D}_{\text{AAR2LO}}$ where the latter was generally greater than the former ($p < 0.001$). This was unlike the observation on $\mathscr{R}$ – LO was apparently successful in improving the diversity in the larger-sized problems.

The difference in diversity between the groups can be clearly distinguished via an observation of the PCA plots. Figure 3 is an example, generated for problem instance C.1. The points corresponding to the AA-based sampling pools are visibly clustering closer to one another, indicating more similar configurations, while that corresponding to the benchmark sampling pool P_R is more spread out.

4.1.3 Closeness to OPT

Closeness is measured in terms of average HD of individuals to OPT (Γ_μ), the standard deviation of fitness error to OPT ($\mathscr{S}$), and the difference of solution values between the best individual in a sampling pool and the OPT ($\mathscr{S}_{\text{best}}$). Comparisons were made between all groups, including that corresponding to GLO.

Γ_μ for all groups were observed to significantly different from each other ($X^2 = 160.347$, df $= 6$, $p < 0.001$) and post-hoc analysis were performed with $\alpha = 0.05/21 = 0.002$. $\Gamma_\mu(R)$ was significantly greater than any other groups ($p < 0.001$), signifying that P_R for most instances were generally further from the OPT compared to any other AA-based sampling pools. Based on this measure of closeness and the number of groups one group has managed to outperform, for most instances and among the AA-based groups, P_{GLO} were closest to OPT, and P_{AADLO} the furthest. In between, P_{AAR} were generally the furthest, followed by P_{AAR2}, P_{AARLO} and P_{AAR2LO}. Compared to their non-LO-counterparts, P_{AARLO} and P_{AAR2LO} were closer to OPT.

The standard deviation of fitness error to OPT ($\mathscr{S}$) groups observed a significant difference between each other ($X^2 = 179.481$, df $= 6$, $p < 0.001$). It is ideal for the fitness values to correlate positively with HD to OPT. However, the post-hoc analysis on $\mathscr{S}$ did not seem to completely reflect the observation on Γ_μ i.e. where $\mathscr{S}_A > \mathscr{S}_B$, it is not always the case that $\Gamma_\mu(A) > \Gamma_\mu(B)$. The pairs in which this contradiction was found where both comparisons are statistically significant are; (R, AAR), (R, AAR2) and (AADLO, AAR2), which is 3/21 comparisons. The majority, however, mirrors the direction of their counterparts.

The FDC measure is evaluated on all instances using P_R and it was observed that $r > 0$ for everyone, with only 4 instances scoring low ($r < 0.035$).[4] On the other hand, the Spearman correlation ρ between $\mathscr{S}$ and Γ_μ scored mostly positive from moderate strength ($\rho = 0.372$) to very strong ($\rho = 0.943$). This indicated that $\mathscr{S}$ and Γ_μ were indeed dependent—$\mathscr{S}_A > \mathscr{S}_B$ indicates that $\Gamma_\mu(A) > \Gamma_\mu(B)$. Hence, we can rely on a method that guarantees the solution value to be close to OPT to also guarantee that the solution configuration is close to OPT e.g. a solution with a value closer to OPT would have lower hamming distance to OPT.

$\mathscr{S}_{best}$ showed a significant difference between the groups ($X^2 = 162.8799$, df $= 6$, $p < 0.001$). Post-hoc comparisons reflect the observation on $\mathscr{S}$, where P_R generally contained the furthest best solution from OPT, and P_{GLO} the closest best solution.

The distances between the sampling pools can be observed on the PCA plots. In Fig. 3, the AA-based sampling pools can be observed to be closer to the OPT point compared to P_R. Similar observations can be seen for most instances.

Furthermore, we look at the problem instances that scored high redundancy rates in most groups $\mathscr{R}_M > 0.9$ and consequently very low diversities. There were six instances in set 4 and three in set 5 with redundancy rates $\mathscr{R}_M = 0.999$ for any sampling pools but P_{GLO} and P_R. For all of them, $\mathscr{S}_{\mathscr{M}} = 0$, which indicated that all AA-based IPG method (except GLO) solutions equals OPT. Clearly, these instances were outliers, therefore not included in previous comparisons. Despite generating solutions closer to OPT than most methods, none of GLO's solutions equals OPT. Most of them have seemed to be stuck in local optima a with $2\% \leq \mathscr{S} \leq 21\%$. Due to the high redundancy rates, P_{GLO} was ineligible as GA inputs in the current design. Instead, to further optimize the solutions, other heuristics can be applied on top of it. However, that is beyond the scope of this research.

---

[4] The ideal score is $r = 1.0$ which indicates positive correlation between fitness values and distances in terms of average HD.

4.2 GA Performance

4.2.1 Average GA Iterations ($\mathcal{T}$) and GA Iterations of Best Performing Trial ($\mathcal{T}_{best}$)

Table 3 presents the percentage of instances under GA_R that achieved greater scores under $\mathcal{T}$ and $\mathcal{T}_{best}$ than AA-based GAs.

On average, more than half of the instances under almost all AA-based GAs achieved lesser number of average iterations compared to the benchmark GA_R. $\mathcal{T}(GA_R)$ is greater than $\mathcal{T}(GA_{AADLO})$ in 56% of instances, $\mathcal{T}(GA_{AAR2})$ in 67% of instances, $\mathcal{T}(GA_{AARLO})$ in 63% of instances and $\mathcal{T}(GA_{AAR2LO})$ in 69% instances. Only $\mathcal{T}(GA_{AAR})$, when compared to $\mathcal{T}(GA_R)$ contained less than half (43%) of the instances that has lesser number of average GA iterations.

The observation on best performing GA trial is similar. $\mathcal{T}_{best}(GA_R)$ is greater than $\mathcal{T}_{best}(GA_{AAR2})$ in 69% of instances, $\mathcal{T}_{best}(GA_{AARLO})$ in 71% of instances and $\mathcal{T}_{best}(GA_{AAR2LO})$ in 79% instances. $\mathcal{T}_{best}(GA_{AADLO})$ is similar to $\mathcal{T}_{best}(GA_R)$ with 50% of instances scoring less, and only $\mathcal{T}_{best}(GA_{AAR})$ contained less than half (37%) of the instances for which best GA trials scoring lesser number of iterations.

Our hypothesis is that AA-based GAs will perform better than the benchmark GA_R. This is true in most cases. Based on these observations, more instances performed better on average under the AA-based GAs compared to GA_R. It can be said that an AA-based IP can oftentimes be more beneficial in speeding up convergence.

Earlier, it was observed that the standard deviation of fitness errors from the OPT ($\mathcal{S}$) for AA-based sampling pools were significantly lower compared to P_R i.e. quality for AA-based sampling pools is better than R. Hence, we expect that the GAs associated with the better quality sampling pool i.e. the AA-based ones will perform better. To test this, we have performed the Spearman's ρ measure to test the correlation between the paired differences $\mathcal{S}_{RM}$ ($\mathcal{S}_R - \mathcal{S}_M$, where M represents any AA methods) for sampling pool quality and a GA performance measure $T(RM)$ ($T(R) - T(M)$), where T is either $\mathcal{T}$ (average number of GA iterations) or $\mathcal{T}_{best}$ (best GA trial in terms of number of iterations). We wanted to test for any positive correlations between them e.g. if $\mathcal{S}_R > \mathcal{S}_M (\mathcal{S}_{RM} > 0)$, then $T(R) > T(M)(T(RM) > 0)$ i.e. if the sampling pool is closer to OPT, then the GA performance is better. $\mathcal{S}_{RM}$ was observed to be more strongly correlated to $\mathcal{T}_{best}(RM)$ ($\rho > 0.4$) than $\mathcal{T}$, which correlation strengths ranged from negligible ($\rho = 0.078$) to moderately positive ($0.358 < \rho < 0.470$).

Table 3 The percentage (%) of instances under GA_R that has achieved greater scores than AA-based GAs

		GA$_{AADLO}$	GA$_{AAR}$	GA$_{AARLO}$	GA$_{AAR2}$	GA$_{AAR2LO}$
$\mathcal{T}$	GA$_R$	56	43	63	67	69
$\mathcal{T}_{best}$	GA$_R$	50	37	71	69	79

The observation on $\mathscr{S}_{RM}$ versus $T(RM)$ suggests that GA performance do rely on the population distance to OPT; sampling from an AA-based sampling pool with known lower average distance to OPT may reduce the number of GA iterations compared to sampling from one generated using the benchmark method.

4.2.2 Solution Quality

A GA is solved when the best solution found (BFS) equals OPT. The GA performances, in terms of solution quality, were evaluated based on the frequency of successes on each instance and the means of solution values (BFS$_\mu$) across 10 trials.

Whether AA-based GAs were superior to GA$_R$, or otherwise, was difficult to conclude. Some instances were unsolvable by any GAs while many observed high rates of success under any GA. Some instances were difficult to solve by some of the GAs while for most, some GAs performed better than others.

No significant differences were found between solution value means (BFS$_\mu$). However, a Spearman ρ test did show a positive correlation between the paired differences of BFS$_\mu$ (BFS$_\mu(RM)$) and that of the closeness of sampling pools to OPT, again represented by $\mathscr{S}$ ($\mathscr{S}(RM)$). A moderately positive correlation was observed ($3.20 < \rho < 3.59$). This showed that BFS$_\mu$ decreased with sampling pool closeness to OPT. A manual count (summarised in Table 4) showed that while almost half of all instances showed no difference in terms of BFS$_\mu$, among those that did ($\mathcal{N}$), more instances under AA-based GAs have lower BFS$_\mu$ and has closer-to-OPT AA-based sampling pools compared to their R counterparts than not, combined e.g. 19 instances fall under this count for GA$_{AADLO}$, which is more than the combined counts of 2 and 7 of the other observations (both lower BFS$_\mu$ and $\mathscr{S}$ and both going opposite directions). This suggested that in many cases, AA-based IPs can indeed be a better alternative to random ones.

From the building block hypothesis [11], we know that a potential substring of the final solution is embedded in each chromosome of a GA population. If parts of the OPT did not exist from the beginning and later operations in the GA (e.g. mutation) failed to introduce them, the OPT may never be found. In our experiment,

Table 4 The counts of instances under AA-based GAs, compared to GA$_R$, that; achieved no difference in mean of solution values BFS$_\mu$ (no difference), differed in BFS$_\mu$ ($\mathcal{N}$), and among those that differ, both paired differences BFS$_\mu(RM)$ and $\mathscr{S}_{RM}$ are positive (+), negative (−) and of opposing signs (<>)

	GA$_{AADLO}$	GA$_{AAR}$	GA$_{AARLO}$	GA$_{AAR2}$	GA$_{AAR2LO}$
No difference	24	20	24	25	20
$\mathcal{N}$	28	26	28	29	32
+	19	4	21	13	22
−	2	12	1	6	1
<>	7	10	6	10	9

where the GA has failed to find the OPT, either the sampling method has failed to generate chromosomes that contained the necessary building block or later operations has failed to coax them into existence. Hence, we believe that an IPG method's ability to produce such chromosomes were highly instance dependent as one GA that performed well on one instance does not always repeat it on another instance e.g. GA_{AAR} was outperformed by GA_{AADLO} on instance 5.2 but not on instance A.3.

5   Conclusion

This paper explored the incorporation of approximation algorithms (AA) in the initial population generation (IPG) of GA. Six AA-based IPG methods were designed and tested on the set covering problem (SCP) using some of the well-known AA for the SCP and compared to a benchmark random heuristic method. The design followed a typical pattern for the SCP, where AA took up the constructive heuristic part, and a randomized local optimization method was implemented on some of them. The resulting AA-based sampling pools were observed to be of better quality compared to the benchmark in terms of closeness to OPT, but not diversity. Subsequently, only five were eligible as GA inputs as one of the AA-based IPG method generated solutions that are insufficiently unique to make up a population. The GA performances varied among different AA-based IPs. Compared to the benchmark IP, more of the AA-based IP GA performed better in terms of number of GA iterations. GA performances were also observed to correlate positively with closeness to OPT. While it can not be concluded that AA-based IPs will be better than random-based ones, our results have shown that in many cases, AA-based IPs can be a better alternative. However, we can conclude that the effectiveness of AA-based IP on GA performance relies more heavily on problem instance structure i.e. AA-based IP might be better on certain types of instances but not others.

6   Suggestions for Future Work

The only characteristics identified of the tested problem data sets were density and dimension. This provided little knowledge as to how instance structure affects the performance of AA-based IPG methods. An exploration into this topic would provide interesting insights into what kind of problem instances would be more suited for AA-based IPG generation.

One of the sampling pools was excluded in the GA performance testing due to insufficient unique solutions. However, the available chromosomes, which were evidently of good quality due to its closeness to the optimal compared to randomly generated ones, can be used in a GA by way of population seeding i.e. mixing them with randomly generated solutions. This is a popular method in GA at the IP stage and has been shown to be successful in a number of literature.

The instances tested in this research were limited to the popular OR Library data sets. Needless to say, a set of 65 instances were insufficient to reflect real-world behaviors. The research presented would benefit greatly from testing with a greater variety of instances which can be computer-generated or extracted from real-world problem instances.

Aside from the multi-cost SCP, a wide range of other AAs are available for various well-established combinatorial optimization problems. As our research has shown, having some problem knowledge, AAs can be easily incorporated into the IPG design of GA or any population-based metaheuristics. Hence, this is a promising area to explore for future research.

References

1. Aickelin U (2002) An indirect genetic algorithm for set covering problems. J Oper Res Soc 53(10):1118–1126. https://doi.org/10.1057/palgrave.jors.2601317
2. Al-Sultan KS, Hussain MF, Nizami JS (1996) A genetic algorithm for the set covering problem. J Oper Res Soc 47(5):702–709. https://doi.org/10.1057/jors.1996.82
3. Baker EK (1979) Efficient heuristic solutions to an airline crew scheduling problem. AIIE Trans 11(2):79–85. https://doi.org/10.1080/05695557908974446
4. Beasley J, Chu PC (1996) A genetic algorithm for the set covering problem. Eur J Oper Res 94(2):392–404. https://doi.org/10.1016/0377-2217(95)00159-X
5. Beg AH, Islam MZ (2015) Clustering by genetic algorithm-high quality chromosome selection for initial population. In: Proceedings of the 2015 10th IEEE conference on industrial electronics and applications, ICIEA 2015, pp 129–134. https://doi.org/10.1109/ICIEA.2015.7334097
6. Breuer MA (1970) Simplification of the covering problem with application to boolean expressions. J ACM 17(1):166–181. https://doi.org/10.1145/321556.321572
7. Caprara A, Toth P, Fischetti M (2000) Algorithms for the set covering problem. Ann Oper Res 98(1):353–371. https://doi.org/10.1023/A:1019225027893
8. Diaz-Gomez P, Hougen D (2007) Initial population for genetic algorithms: a metric approach. In: Proceedings of the 2007 international conference on genetic and evolutionary methods, pp 43–49
9. Finger M, Stützle T, Lourenco H (2002) Exploiting fitness distance correlation of set covering problems. In: Lecture notes in computer science (including subseries Lecture notes in artificial intelligence and lecture notes in bioinformatics), vol 2279. LNCS, pp 61–71. https://doi.org/10.1007/3-540-46004-7_7
10. Gendreau M, Potvin JY (2010) Handbook of metaheuristics, 2nd edn. Springer, Incorporated
11. Goldberg DE (1989) Genetic algorithms in search, optimization and machine learning, 1st edn. Addison-Wesley Longman Publishing Co., Inc., Boston, MA, USA
12. Huang WC, Kao CY, Horng JT (1994) A genetic algorithm approach for set covering problems. In: Proceedings of the first IEEE conference on evolutionary computation. IEEE world congress on computational intelligence, no 1, pp 569–574. https://doi.org/10.1109/ICEC.1994.349997
13. Iwamura K, Horiike M, Sibahara T (2003) Input data dependency of a genetic algorithm to solve the set covering problem. Tsinghua Sci Technol 8(1):14–18
14. Khaji E, Mohammadi AS, Students G (2014) A heuristic method to generate better initial population for evolutionary methods
15. Konjevod G (1994) Solving a set covering problem with genetic algorithms. Tech. rep., University of Zagreb, Faculty of Mathematics and Natural Sciences, Department of Applied Mathematics and Computer Science

16. Li L, Huang Z, Liu F (2009) A heuristic particle swarm optimization method for truss structures with discrete variables. Comput Struct 87(7):435–443. https://doi.org/10.1016/j.compstruc.2009.01.004
17. Maaranen H, Miettinen K, Mäkelä MM (2004) Quasi-random initial population for genetic algorithms. Comput Math Appl 47(12):1885–1895. https://doi.org/10.1016/j.cainwa.2003.07.011
18. Morrison RW, De Jong KA (2002) Measurement of population diversity. In: Collet P, Fonlupt C, Hao JK, Lutton E, Schoenauer M (eds) Artificial evolution: 5th international conference, Evolution Artificielle, EA 2001 Le Creusot, France, 29–31 Oct 2001 selected papers. Springer, Berlin, Heidelberg, pp 31–41. https://doi.org/10.1007/3-540-46033-0_3
19. Papadimitriou CH, Steiglitz K (1998) Combinatorial optimization: algorithms and complexity. Courier Corporation
20. Raja PV, Bhaskaran VM (2013) Improving the performance of genetic algorithm by reducing the population size. Int J Emerg Technol Adv Eng 3(8):86–91
21. Razip H, Zakaria MN (2017) Combining approximation algorithm with genetic algorithm at the initial population for np-complete problem. In: 2017 IEEE 15th student conference on research and development (SCOReD), pp 98–103. https://doi.org/10.1109/SCORED.2017.8305413
22. Rubin J (1973) Technique for the solution of massive set covering problems, with application to airline crew scheduling. Transp Sci 7(1):34–48. https://www.scopus.com/inward/record.uri?eid=2-s2.0-0015585112&partnerID=40&md5=fa6a29621381d1227fafc071f242528e
23. Sipser M (1996) Introduction to the theory of computation, 1st edn. International Thomson Publishing
24. Smith LI (2002) A tutorial on principal components analysis introduction. Statistics 51:52. https://doi.org/10.1080/03610928808829796
25. Toregas C, Swain R, ReVelle C, Bergman L (1971) The location of emergency service facilities. Oper Res 19(6):1363–1373. http://econpapers.repec.org/RePEc:inm:oropre:v:19:y:1971:i:6:p:1363-1373
26. Toğan V, Daloğlu AT (2008) An improved genetic algorithm with initial population strategy and self-adaptive member grouping. Comput Struct 86(11–12):1204–1218. https://doi.org/10.1016/j.compstruc.2007.11.006
27. Trevisan L (2011) 8.1 A linear programming relaxation of set cover. In: Combinatorial optimization: exact and approximate algorithms, chap 8
28. Wang X, Ju S (2009) A set-covering-based approach for overlapping resource selection in distributed information retrieval. In: 2009 WRI world congress on computer science and information engineering, vol 4, pp 272–276. https://doi.org/10.1109/CSIE.2009.702
29. Williamson DP, Shmoys DB (2011) The design of approximation algorithms, 1st edn. Cambridge University Press, New York, NY, USA. https://doi.org/10.1017/CBO9780511921735
30. Zhang G, Gao L, Shi Y (2011) An effective genetic algorithm for the flexible job-shop scheduling problem. Expert Syst Appl 38(4):3563–3573
31. Zhang XY, Zhang J, Gong YJ, Zhan ZH, Chen WN, Li Y (2016) Kuhn-Munkres parallel genetic algorithm for the set cover problem and its application to large-scale wireless sensor networks. IEEE Trans Evol Comput 20(5):695–710. https://doi.org/10.1109/TEVC.2015.2511142

Chapter 7
Modified Generalised Quadrature Spatial Modulation Performance over Weibull Fading Channel

Kiran Gunde and Anuradha Sundru

1 Introduction

Energy-efficient open-loop transmission technique for multiple-input multiple-output (MIMO) wireless communication system is called as spatial modulation (SM) [1–3]. It was developed to increase the transmission data rate of the system without increasing bandwidth. SM adds an extra spatial dimension to the conventional MIMO system to increase the data rate. SM uses a single radio frequency (RF) chain and activates only one antenna to transmit the data at every symbol period. Hence, it decreases the system hardware complexity. In SM, data bits are split into two mapping blocks; the first block is antenna mapping block, and it maps the data bits to the index of the transmitting antenna; the remaining block is symbol mapping block, and it maps the data bits to QAM modulation symbol. With this structure, SM increases the transmission data rate over conventional MIMO systems without changing the system bandwidth. The performance of MIMO and SM systems was studied over Weibull fading channel [4–6], respectively. In [4], it is shown that the average channel capacity increases by increasing the number of transmitting antennas. In [5, 6], the theoretical expression for SM symbol error rate is derived and compared the performances of various techniques of SM systems for different fading environments, respectively.

The data rate of SM is $R_{SM} = \log_2(N_T) + \log_2(M)$ bits per channel use (bpcu), where N_T and M denotes the number of transmit antennas and QAM symbol, respectively. The first $\log_2(N_T)$ bits selects the transmitting antenna index using antenna mapping block, and the remaining $\log_2(M)$ bits select the QAM symbol using symbol mapping block. At every symbol time period, SM activates only a single antenna to transmit the data bits. SM is free from inter channel interference (ICI) because of single antenna transmission and also avoids inter symbol interference (ISI) due to one symbol transmits at every channel use. The limitations of SM are that the number

K. Gunde (✉) · A. Sundru
National Institute of Technology, Warangal, India
e-mail: anuradha@nitw.ac.in

© The Author(s), under exclusive license to Springer Nature Singapore Pte Ltd. 2021
S. Kumar et al. (eds.), *Proceedings of International Conference on Communication and Computational Technologies*, Algorithms for Intelligent Systems,
https://doi.org/10.1007/978-981-16-3246-4_7

79

of RF chains and antennas for the data transmission is restricted to one and integer power of 2, respectively. By relaxing these two limitations, generalised SM (GSM) [7] and multiple active SM (MA-SM) [8] techniques were introduced to increase the SM rate by providing multiple RF chains. GSM and MA-SM schemes activate multiple antennas to transmit the same modulation symbol and different modulation symbols, respectively. GSM is free from ISI, due to the same symbol transmission on multiple antennas. However, the data rate of GSM is low when compared to MA-SM.[1]

Quadrature SM (QSM) [9] is a simplified technique of SM. In QSM, the complex information symbol is separated as real coefficient ($\Re\{\cdot\}$) and imaginary coefficient ($\Im\{\cdot\}$) symbols. Furthermore, these symbols are modulated on in-phase (I) and quadrature (Q) elements, respectively. Hence, it improves the data rate of SM over spatial dimension. At one time instant, it conveys data through either one or two transmit antennas. QSM is also avoiding ICI because of the orthogonality of I and Q elements. In [10], the performance of QSM system is studied with imperfect channel estimation. Recently, generalised QSM (GQSM) [11] technique with antenna grouping was introduced to enhance the QSM data rate by allowing multiple RF chains. In GQSM, transmitting antennas are separated as groups in such a way that each group works according to QSM transmission principle. More recently, modified GQSM (mGQSM) [12] scheme is introduced to enhance the data rate of GQSM by one bpcu. In mGQSM, the grouping of transmitting antennas is relaxed. Furthermore, the codebook size increases with respect to the transmitting antennas. Coded mGQSM system performance is presented with convolutional encoding and also presented system performance under imperfect channel conditions. The reduced codebook mGQSM (RC-mGQSM) is a reduced complexity scheme for mGQSM system. In this paper, we present the uncoded mGQSM performance over Weibull fading channel with deep fade and non-fading environment scenarios.

The remaining paper is ordered as follows. Section 2 presents the mGQSM system implementation and Weibull fading channel, Sect. 3 describes the performance of mGQSM system under imperfect channel conditions, Sect. 4 discusses the simulation results, and Sect. 5 presents the conclusion.

2 System and Channel Models

The modified GQSM (mGQSM) technique is developed to improve the data rate of variant SM schemes. For the same system configuration, mGQSM scheme improves additional one bpcu data rate over GQSM by relaxing antenna grouping. Considering multiple RF chains, the data symbols are transmitted on any antenna according to antenna activation pattern (AAP) procedure given in [12]. With the constraint

---

[1] Notation: $\lfloor \cdot \rfloor$ denotes floor operation, $\{\cdot\}$ denotes the fractional part, $\binom{\cdot}{\cdot}$ denotes binomial coefficient, bold uppercase, and lowercase letters denote matrices and column vectors, respectively. $\|\cdot\|_0$ denotes norm zero vector, $|\cdot|$ denotes cardinality of a set.

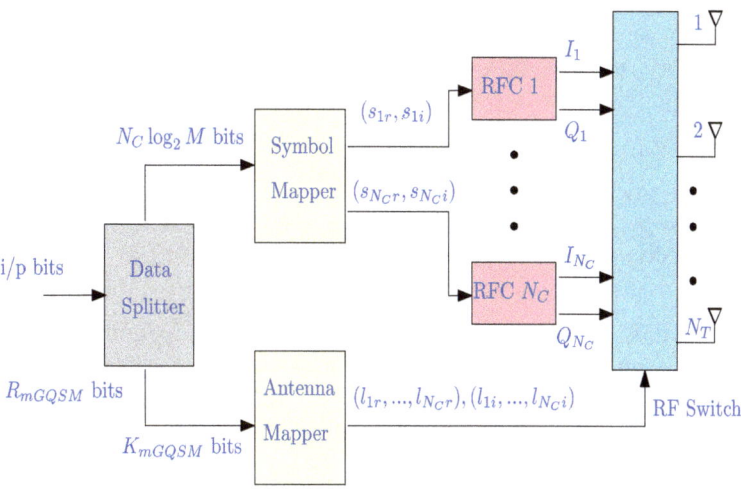

Fig. 1 mGQSM transmitter system diagram [12]

$\{\log_2 \binom{N_T}{N_C}\} \geq 0.5$, where N_C denotes the number of RF chains, $1 \leq N_C \leq \lfloor \frac{N_T}{2} \rfloor$, it improves an extra one bpcu data rate of GQSM by relaxing the antenna grouping. The AAP in mGQSM is being doubled when compared with GQSM, which leads to increase in codebook size. In mGQSM, the data rate increases by increasing the codebook size, and it is given by [12],

$$R_{\mathrm{mGQSM}} = \left\lfloor 2\log_2 \binom{N_T}{N_C} \right\rfloor + N_C \log_2 M \text{ bpcu} \tag{1}$$

The mGQSM transmitter system diagram with N_T transmitting antennas is shown in Fig. 1. The data splitter divides the input bits, R_{mGQSM} bits into two blocks and processes through antenna mapper and symbol mapper. The first block of bits, $K_{\mathrm{mGQSM}} = \left\lfloor 2\log_2 \binom{N_T}{N_C} \right\rfloor$, maps the AAP with the help of antenna mapper, and remaining block of bits, $N_C \log_2 M$, map the QAM symbols with the help of a symbol mapper. These symbols are separated as real coefficient which is modulated on in-phase component and imaginary coefficient which is modulated on quadrature component and transmitted. At every time instant, the transmitting antennas in mGQSM system vary from N_C to $2N_C$. RC-mGQSM scheme is developed to decrease the complexity of mGQSM system by reducing the codebook size. The transmitting antennas in RC-mGQM are less when compared to the mGQSM. Hence, the rate decreases with reducing the codebook size.

The codebook of mGQSM is defined as $\mathcal{C}_{\mathrm{mGQSM}} = \{\mathbf{s}|s_j = s_{jr} + js_{ji}; s_{jr} \in \Re\{\mathbb{A}_M\}, s_{ji} \in \Im\{\mathbb{A}_M\}, N_C \leq \|\mathbf{s}\|_0 \leq 2N_C, \mathcal{I}(\mathbf{s}) \in \mathbb{S}\}$, where $\mathbb{A}_M$ denotes the modulation symbol set, $j = 1, 2, \ldots, N_C, \mathbb{S}$ denotes the AAP $\ni, \mathbf{p} \in \mathbb{S}$ and $N_C \leq \|\mathbf{p}\|_0 \leq 2N_C, p_j \in \{0, 1\}$, and $\mathcal{I}(\mathbf{s})$ denotes the function that generates AAP for $\mathbf{s}$.

The mGQSM system can select any $2^{\left\lfloor 2\log_2 \binom{N_T}{N_C} \right\rfloor}$ AAPs out of $\binom{N_T}{N_C}^2$ possible number of AAPs to transmitting the information symbols.

For example, let $N_T = 4$ for the mGQSM, then we have $N_C = 2$ according to AAP selection given in [12]. The total number of AAPs are $\binom{N_T}{N_C}^2 = \binom{4}{2}^2 = 36$ out of which any $2^{\left\lfloor 2\log_2 \binom{N_T}{N_C} \right\rfloor} = 32$ AAPs are selected for the information symbols transmission. Let 4-QAM modulation for data transmission, therefore the data rate of mGQSM system is 9 bpcu. Consider the data bits [0 1 0 0 1 0 0 1 0] at one time instant. The first block of bits $K_{\text{mGQSM}} = \left\lfloor 2\log_2 \binom{N_T}{N_C} \right\rfloor = 5$ bits, [0 1 0 0 1] select the AAP $(l_{1r}, l_{2r}) = (2, 3)$ for real, and $(l_{1i}, l_{2i}) = (1, 3)$ for imaginary coefficients transmission. The remaining block of bits $N_C \log_2 M = 4$ bits, [0 0 1 0] select two data symbols $s_1 = -1 + j1$ and $s_2 = +1 + j1$. Moreover, these symbols separated as real ($s_{1r} = -1, s_{2r} = +1$) and imaginary ($s_{1i} = +1, s_{2i} = +1$) coefficients. The antennas $l_{1r} = 2$ and $l_{2r} = 3$ transmit the symbols $s_{1r} = -1$ and $s_{2r} = +1$, respectively. Hence, the signal vector for the real coefficient is $\mathbf{s}_r = [0 - 1 + 1 \ 0]^T$. Similarly, the antennas $l_{1i} = 1$ and $l_{2i} = 3$ transmit the symbols $s_{1i} = +1$ and $s_{2i} = +1$, respectively. Hence, the signal vector for the imaginary coefficient is $\mathbf{s}_i = [+1 \ 0 +1 \quad 0]^T$. Therefore, the final signal vector for the transmission is $\mathbf{s} = \mathbf{s}_r + j\mathbf{s}_i = [+1 - 1 + 1 + j1 \ 0]^T$.

2.1 System Model

The data bits are processed through the system with N_R receiving antennas and N_T transmitting antennas with $N_R \times N_T$ dimension. Let the noise $\mathbf{n}$ with $N_R \times 1$ dimension and its elements are assumed as iid random variables and follow Gaussian distribution with mean 0 and variance N_0, i.e. $\mathcal{CN}(0, N_0)$. Consider the Weibull channel $\mathbf{H}$ with $N_R \times N_T$ dimension. The characteristics of $\mathbf{H}$ are discussed in the following subsection.

Let $\mathbf{h}_{l_{jr}}$ and $\mathbf{h}_{l_{ji}}$ are the l_{jr}^{th} and l_{ji}^{th} columns of $\mathbf{H}$, respectively, $\mathbf{h}_{l_{jr}} = [h_{1,l_{jr}}, \ldots, h_{N_R,l_{jr}}]^T$ and $\mathbf{h}_{l_{ji}} = [h_{1,l_{ji}}, \ldots, h_{N_R,l_{ji}}]^T$, where $j = 1, 2, \ldots, N_C$. The signal vector at the receiver is given as

$$\mathbf{y} = \sqrt{E_s}\mathbf{Hs} + \mathbf{n}$$

$$= \sqrt{E_s}\sum_{j=1}^{N_C}(\mathbf{h}_{l_{jr}}s_{jr} + j\mathbf{h}_{l_{ji}}s_{ji}) + \mathbf{n} \tag{2}$$

where E_s denotes the transmitted symbol energy, $l_{jr}, l_{ji} = 1, 2, \ldots, N_t, j = 1, 2, \ldots, N_C$, and the transmitted signal $\mathbf{s}$ is selected from the codebook, i.e. $\mathbf{s} \in \mathcal{C}_{\text{mGQSM}}$.

2.2 Weibull Fading

Consider the frequency flat Weibull fading channel. It is characterised by the PDF, and it is given as

$$f(x; \lambda, \beta) = \frac{\beta}{\lambda} \left(\frac{x}{\lambda}\right)^{\beta-1} e^{-\left(\frac{x}{\lambda}\right)^\beta} \tag{3}$$

where β denotes the shape parameter, and λ denotes the scale parameter. The complex envelope for the Weibull fading channel is given by [4],

$$h_{N_R, N_T} = (U + jV)^{2/\beta} \tag{4}$$

where U and V denote the Gaussian I and Q elements of the multipath components, respectively. The Weibull distribution exhibits good fit to practical channel measurements for indoor and outdoor environment scenarios. The fading severity is controlled by the parameter β. For the simulations, we consider deep fade environment when $\beta = 0.5$ and the non-fading environment when $\beta = 5$. The negative exponential and Rayleigh distributions are the special cases of Weibull fading when $\beta = 1$ and $\beta = 2$, respectively.

At the receiver, a maximum likelihood (ML) decoding algorithm is applied to jointly estimate transmit symbols and antenna indices. It is given as

$$\widehat{\mathbf{s}} = \underset{\mathbf{s} \in \mathcal{C}_{\text{mGQSM}}}{\arg\min} \|\mathbf{y} - \sqrt{E_s} \mathbf{H} \mathbf{s}\|^2. \tag{5}$$

$$[\widehat{l}_{jr}, \widehat{l}_{ji}, \widehat{s}_{jr}, \widehat{s}_{ji}]$$

$$= \underset{l_{jr}, l_{ji}, s_{jr}, s_{ji}}{\arg\min} \|\mathbf{y} - \sqrt{E_s} \sum_{j=1}^{N_c} (\mathbf{h}_{l_{jr}} s_{jr} + \mathbf{j}\mathbf{h}_{l_{ji}} s_{ji})\|^2 \tag{6}$$

where $\widehat{l}_{jr}$ and $\widehat{l}_{ji}$ denote the estimated antenna indices corresponding to $\widehat{s}_{jr}$ and $\widehat{s}_{ji}$. $\widehat{s}_{jr}$ is real coefficient, and $\widehat{s}_{ji}$ is imaginary coefficient of the detected symbols for $j = 1, 2, \ldots, N_C$.

3 MGQSM System Performance: Imperfect Channel

This section discusses the uncoded mGQSM system performance with imperfect channel conditions. In addition to Weibull channel $\mathbf{H}$, we use $N_R \times N_T$ dimension error channel $\delta\mathbf{H}$ for the imperfect channel. The entries of $\delta\mathbf{H}$ are modelled as i.i.d. $\mathcal{CN}(0, \sigma_h)$, where σ_h denotes the error channel variance. We consider Weibull non-fading environment ($\beta = 5$) scenario. At the receiver, the ML decoding algorithm makes a decision with respect to the error channel matrix, $\delta\mathbf{H}$.

ML decoding algorithm for the signal vector **s** is given as

$$\hat{\mathbf{s}} = \underset{\mathbf{s} \in \mathcal{C}_{\text{mGQSM}}}{\arg \min} \|\mathbf{y} - \sqrt{E_s}(\mathbf{H} + \delta\mathbf{H})\mathbf{s}\|^2 \tag{7}$$

where $\mathcal{C}_{\text{mGQSM}}$ denotes codebook for mGQSM system, and $\hat{\mathbf{s}}$ denotes estimated signal vector for the input vector **s**.

4 Simulation Results

In this section, we present uncoded mGQSM system performance and compared it with the QSM and SM system performances. Consider frequency flat Weibull fading channel with two different fading environment scenarios which depend on the Weibull shape parameter β, deep fade environment with $\beta = 0.5$ and the non-fading environment with $\beta = 5$. The entries of noise vector are followed as i.i.d. $\mathcal{CN}(0, N_o)$.

Signal-to-noise ratio (SNR) is defined as $\frac{E_s}{N_o}$, where E_s is the symbol energy, and N_o is a variance of the noise. We compute the bit error rate (BER) for different SNR values.

We use 10^5 data symbols to calculate BER with different SNR values. For the simulations, we consider $N_T = 4$, $N_R = 4$ system model for all the systems and $N_C = 2$ for mGQSM and RC-mGQSM systems.

The error performance comparison of mGQSM, QSM and SM systems with Weibull deep fade ($\beta = 0.5$) and the non-fading ($\beta = 5$) environment scenarios is shown in Figs. 2 and 3, respectively. We compare the error performance of systems with the same transmission data rate of 9 bpcu. In Fig. 2, at BER of 10^{-2}, we noticed mGQSM system performance with 4-QAM ($M = 4$) scheme gain in SNR of ~ 4 dB over QSM with 32-QAM ($M = 32$) and ~ 8 dB over SM with 128-QAM ($M = 128$). In Fig. 3, at BER of 10^{-3}, we noticed mGQSM system performance with 4-QAM ($M = 4$) scheme gain in SNR of ~ 2 dB over QSM with 32-QAM ($M = 32$) and ~ 4 dB over SM with 128-QAM ($M = 128$) modulation.

The error performance comparison of RC-mGQSM, QSM and SM systems with Weibull deep fade ($\beta = 0.5$) and non-fading ($\beta = 5$) environment scenarios is shown in Figs. 4 and 5, respectively. We compare the error performance of systems with the same transmission data rate of 8 bpcu. In Fig. 4, at BER of 10^{-2}, we noticed RC-mGQSM system performance with 4-QAM ($M = 4$) scheme gain in SNR of ~ 2 dB over QSM with 16-QAM ($M = 16$) and ~ 6 dB over SM with 64-QAM ($M = 64$). In Fig. 5, at BER of 10^{-3}, we noticed RC-mGQSM system performance with 4-QAM ($M = 4$) scheme gain in SNR of ~ 3 dB over QSM with 16-QAM ($M = 16$) and ~ 0.8 dB over SM with 64-QAM ($M = 64$) modulation.

The mGQSM performance under imperfect channel conditions is shown in Fig. 6. We consider a non-fading environment ($\beta = 5$) scenario and $N_R \times N_T$ dimension error channel $\delta\mathbf{H}$, with the variance σ_h varies from 0.01 to 0.05. For 4-QAM ($M = 4$)

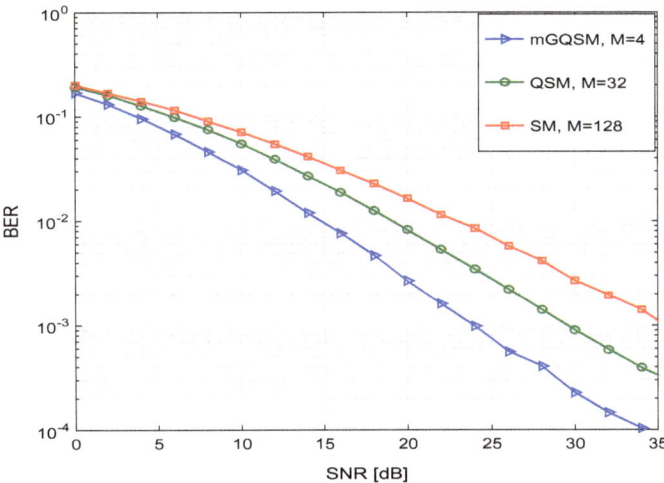

Fig. 2 BER versus SNR for mGQSM, QSM and SM with $N_T = 4$, $N_R = 4$ system model over Weibull fading environment with $\beta = 0.5$ and the rate, $R_{\mathrm{mGQSM}} = R_{\mathrm{QSM}} = R_{\mathrm{SM}} = 9$ bpcu

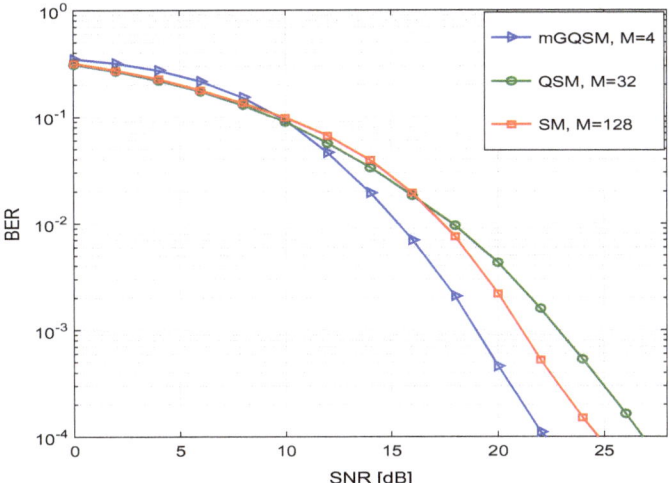

Fig. 3 BER versus SNR for mGQSM, QSM and SM with $N_T = 4$, $N_R = 4$ system model over Weibull fading environment with $\beta = 5$ and the rate, $R_{\mathrm{mGQSM}} = R_{\mathrm{QSM}} = R_{\mathrm{SM}} = 9$ bpcu

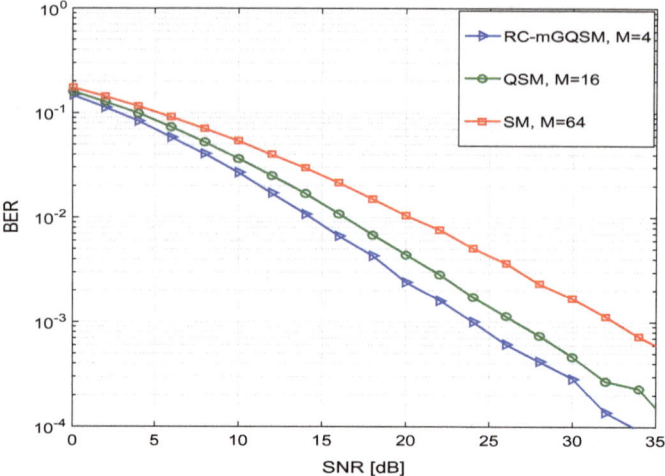

Fig. 4 BER versus SNR for RC-mGQSM, QSM and SM with $N_T = 4$, $N_R = 4$ system model over Weibull fading environment with $\beta = 0.5$ and the rate, $R_{\text{RC-mGQSM}} = R_{\text{QSM}} = R_{\text{SM}} = 8$ bpcu

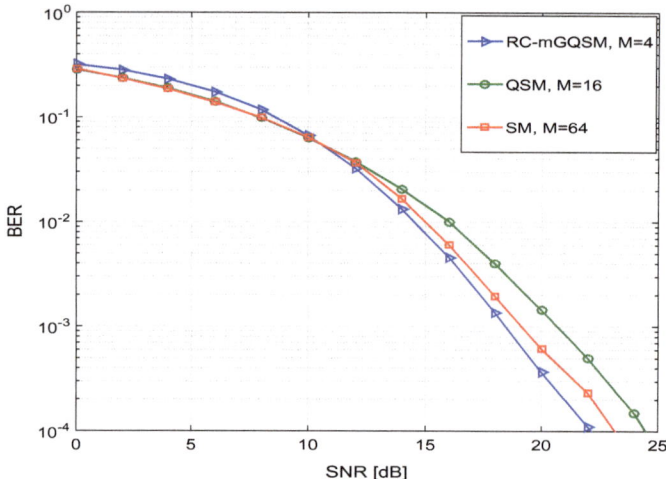

Fig. 5 BER versus SNR for RC-mGQSM, QSM and SM with $N_T = 4$, $N_R = 4$ system model over Weibull fading environment with $\beta = 5$ and the rate, $R_{\text{RC-mGQSM}} = R_{\text{QSM}} = R_{\text{SM}} = 8$ bpcu

scheme, we compared the mGQSM system performance with different error channel variance performances. At the BER level of 10^{-2}, we noticed that mGQSM system performance with error channel variance 0.01 and 0.02 is loses ~ 2 dB and ~ 6 dB SNR values, respectively. Hence, the system performance degrades by adding an error channel to the perfect channel.

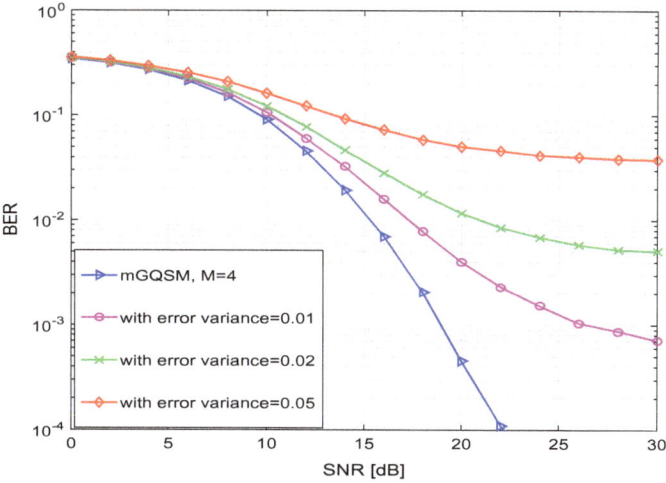

Fig. 6 BER versus SNR for mGQSM with $N_T = 4$, $N_R = 4$ system model over Weibull fading environment with $\beta = 5$ and the error variances are equal to $0.01, 0.02, 0.05$ and $R_{mGQSM} = 9$ bpcu

5 Conclusion

This paper presented the uncoded mGQSM error performance over Weibull fading channel with deep fade and non-fading environment scenarios. We compared the mGQSM and RC-mGQSM system performance with QSM and SM system performances. We described the mGQSM system model with a suitable example. BER performance of RC-mGQSM scheme is compared with QSM and SM scheme over Weibull fading. Using the ML decoding algorithm, we studied the system error performances via simulations. We also presented mGQSM system performance under imperfect channel conditions.

References

1. Mesleh RY, Haas H, Sinanovic S, Ahn CW, Yun S (2008) Spatial modulation. IEEE Trans Veh Technol 57:2228–2241
2. Jeganathan J, Ghrayeb A, Szczecinski L (2008) Spatial modulation: optimal detection and performance analysis. IEEE Commun Lett 12(8):545–547
3. Renzo MD, Haas H, Ghrayeb A, Sugiura S, Hanzo L (2014) Spatial modulation for generalized MIMO: challenges, opportunities, and implementation. Proc IEEE 102(1):56–103
4. Lupupa M, Dlodlo ME (2009) Performance of MIMO system in Weibull fading channel–channel capacity analysis. IEEE EUROCON 2009:1735–1740
5. Alshamali A, Aloqlah M (2013) Performance analysis of spatial modulation over Weibull fading channels. WSEAS Trans Commun 12:604–607

6. Goutham Simha GD, Koila S, Neha N, Sripati U (2015) Performance of spatial-modulation and spatial-multiplexing systems over Weibull fading channel. In: 2015 international conference on computing and network communications (CoCoNet), pp 389–394

7. Younis A, Serafimovski N, Mesleh R, Haas H (2010) Generalised spatial modulation. In: 2010 conference record of the forty fourth ASILOMAR conference on signals, systems and computers, pp 1498–1502

8. Wang J, Jia S, Song J (2012) Generalised spatial modulation system with multiple active transmit antennas and low complexity detection scheme. IEEE Trans Wirel Commun 11:1605–1615

9. Mesleh R, Ikki SS, Aggoune HM (2015) Quadrature spatial modulation. IEEE Trans Veh Technol 64:2738–2742

10. Afana A, Atawi I, Ikki S, Mesleh R (2015) Energy efficient quadrature spatial modulation MIMO cognitive radio systems with imperfect channel estimation. In: 2015 IEEE international conference on ubiquitous wireless broadband (ICUWB), pp 1–5

11. Castillo-Soria FR, Cortez-González J, Ramirez-Gutierrez R, Maciel-Barboza FM, Soriano-Equigua L (2017) Generalized quadrature spatial modulation scheme using antenna grouping. ETRI J 39(5)

12. Gunde K, Hari KVS (2019) Modified generalised quadrature spatial modulation. In: 2019 national conference on communications (NCC), pp 1–5

Chapter 8
CESumm: Semantic Graph-Based Approach for Extractive Text Summarization

S. Gokul Amuthan and S. Chitrakala

1 Introduction

The rapid growth of textual data being recorded and made available online has been growing even since the Internet era and searching for a particular topic, indexes back humongous amount of content. The process of manually reading, comprehending and generating summary from all the data requires immense task force making it a time consuming, demanding and expensive task [1]. And there is no guarantee that the document read, comprehend and summarized by the user which is useful as the user skims through the text to get important information in the shortest possible time. The solution to this problem is text summarization (TS) system which automatically extracts the key essence of the document provided. The document could be of any domain or any language and can vary in magnitude [2]. The goal of any text summarization system is to create a machine-generated summary that is shorter in length compared to input source document, but also retains the key information in the source document and does not include redundant data while preserving the readability of the summary being generated. The text summarization system can be broadly classified into following three categories based on input, output and approach [3] used—(1) Based on the input size: Single-document text summarization system attempts to create a summary from a single input document, whereas multi-document text summarization system generates a summary from group or cluster of related document. (2) Based on the output type: Basic text summarization system creates summary by the considered the key essence of the document (there can be more than one key essence). Query-based text summarization system formulates summary based on the input query given by the user. (3) Based on approach: Extractive text summarization system focuses on generating summary by scoring each sentence and retrieving

S. Gokul Amuthan (✉) · S. Chitrakala
College of Engineering Guindy, Department of Computer Science and Engineering,
Anna University, Chennai, Tamil Nadu, India
URL: http://cs.annauniv.edu

the top-ranked sentences. Abstractive text summarization system attempts to create a summary that is entirely different from the input text and also retains the key aspect of summary, whereas hybrid text summarization system attempts to utilize both extractive and abstractive strategy in generating the summary.

This paper presents a strategy for extractive text summarization system from single document that utilizes an unsupervised and domain-independent framework which does not require training process also thereby aiming to improve the readability quality of summary by preserving the cause and effect relationship in the extracted text. The proposed system overcomes the disadvantage of previous graph-based models that fail to consider the significance of noun words in the source text. The proposed system aims to generate extractive text summary for single document by constructing text graph and semantic link network, which fixes the issue of treating all words in sentence contributing equally for sentence score, by constructing text graph based on nouns; to resolve issue of domain dependency by using unsupervised graph-based approach and to avoid biasing the dominant topic in generated summary. The application of text summarization is vast and proven useful in areas such as outline generation for biographies, highlighting central theme of text in documents such as legal and medical and compressing text content without losing the context. This work contributes a strategy to preserve the semantic cause and effect relation in the extracted summary.

2 Related Works

The field of text summarization has been and still is extensively researched and particularly in the field of extractive summarization. A wide range of approaches has been proposed in each work that extracts summary from text faster while preserving the readability and key meaning in the source text. One of the most straightforward approach is to use statistical feature score-based approach [4] which assigns feature vector value and calculates score for each sentence using this feature; despite being simple and of low processing capacity, the methods fail to include important topic in source text and may include redundant information. Another approach is clustering and optimization technique where the foremost objective is to improve topic coverage and diversity in sentence selected in summary, but still fails to remove redundancy problem completely. An alternative study proposes an innovative approach of utilizing a bio-inspired optimizer which evolves the candidate sentence selected for final summary aiming to improve the quality [5]. The problem with this approach is convergence of summary and summary produced depends on candidate population. Yet another approach to summarization proposed is a text mining policy and entropy function for generation of summary but still lacks to take in consideration of semantic relation in the text to generate summary [6].

With recent interest and trend in machine learning techniques, the summarization task is reduced to problem for generating summary which is viewed as a supervised classification problem that aims to classify sentence whether belong in final summary or not [7]. But this method fails when it comes to producing quality summary with limited dataset. Also with progress in deep learning, newer models such as RNN and LSTM are proposed which views summarization as sequence generation problem [8]. The major disadvantage is not only constraint on huge dataset availability but also the portability of the document to another domain requires retraining the model which is inefficient. Another system aims to bio-mimicrise the human brain model of generating summary which uses hierarchical cognitive model that retrieves summary based on set of causal and intentional relations present in the source text [9]. The drawback of this approach is that the quality of summary depends on construction of efficient knowledge base which requires mining for huge knowledge on document domain making technique domain dependent.

Another approach to extractive summary generation is the construction of graph representation of the text where each node represents the sentence in the text, and edge represents the similarity between two sentences. The sentence is then ordered according to their similarity score, and the highest score is selected for final summary. The pitfall to this approach is that it does not consider the importance of each word inside a sentence and fails to establish a relation of word-topic relevance. Addressing this issue, yet these graph-based approaches still face issue of maintaining semantic coherence in the text [10]. Other graph-based approaches [11] use semantic relation in the text to construct graph but fails to preserve the word-topic relevance in extracted summary and often the summary generated tends to be large. Other research on preserving semantic relation on output summary by applying latent semantic approach which transforms input document to a representation that captures the semantics, and the quality of summary is directly dependent on the quality of representation [12] being used.

Hence, briefing the issue of previous related works, the following issues are to be addressed in proposed system.

- Not robust to domain variation (adaptation problem)
- Issue in generating concise summary
- Content redundancy issue
- Lack of readability in generated summaries
- Bias to topics with high score
- Limited availability of source data
- Issue in selecting sentence for generated summaries.

3 CESumm: System Architecture and Methodology

The proposed system CESumm aims at generating extractive summary from single document which retains the semantic cause and effect relation in the final summary and resolves the problem discussed in the previous section. The extracted summary generated can also be controlled with the number of words parameter. The proposed system architecture is depicted in Fig. 1.

3.1 Document Pre-processing

The document pre-processing task involves sentence and word level processing such as segmentation, reduction procedure such as removing hyphens, unwanted quotations, citations, tokenization and lemmatization. This helps to extract meaningful word units which help to assign relevant score the document representation. Word frequency count and the bigram frequency are computed for the pre-processed text.

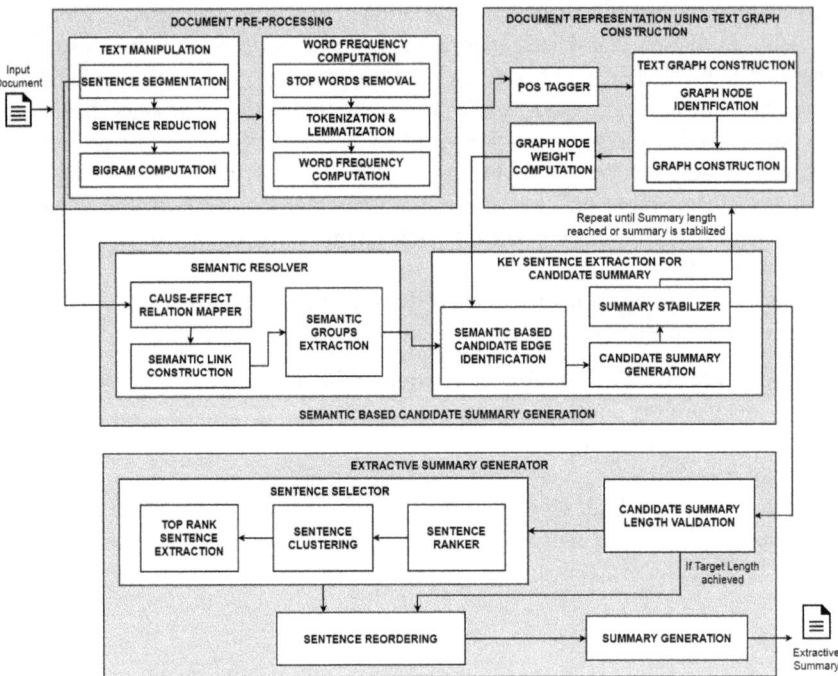

Fig. 1 System architecture—CESumm

3.2 Document Representation Using Text Graph Construction

The pre-processed document is converted into a graph representation that uses noun in the text for nodes and the rest of non-noun words, if exists are maintained at edges. Each sentence starts with dummy start node and ends with dummy end node. The graph edge also maintains the order of the edge from the start node, which later helps in construction of candidate summary. Each node in the graph is assigned with a weighed based on word frequency and bigram frequency computed earlier. The node weight is calculate as given in Eq. (1). The algorithm for construction of text graph is depicted in Fig. 2

$$\text{Node(Label).weight} = \text{WordFreq(Label)} + \text{BigramFreq(Label)} * \text{AbsWt} \qquad (1)$$

where AbsWt is the absolute difference in the average word frequency and median word frequency.

Fig. 2 Algorithm—text graph construction

Algorithm: Text Graph Construction

Input: Preprocessed document d
Output: Text Graph → TG
src_node = 'S', edge_ord = 0
for each sentence s **in** document d **do**
    word_tags = pos_tag(s)
    for each word w, tag t **in** word_tags **do**
        if t **is** NOUN **then**
            dest_node = lemmatize(w)
            n_wt = wordFreq(w) +
                        BigramFreq(w) * AW
            TG.add_node(dest_node, w, n_wt)
            TG.add_edge(src_node, dest_node,
                        edge_label, edge_ord)
            edge_ord++
            src_node = dest_node
            dest_node = edge_label = ''
        else
            edge_label += ' ' + w
        end if
    end for
    TG.add_edge(src_node, 'E', edge_label,
                    edge_ord)
end for

3.3 Semantic Based Candidate Summary Generation

The graph constructed is traversed by considering only the nodes that have weight above average of all the adjacent nodes, and a version of candidate summary is generated from traversing the edges. The candidate summary version is stored as edge list which contains all edges that are visited. This candidate version is iteratively constructed by pruning of unnecessary low weighted nodes in the graph using current version of candidate summary, and constructing a newer version of candidate summary until the candidate summary is stabilized. A candidate summary is stabilized when the expected word length of the summary is met or the size of summary cannot be further reduced.

While generation of candidate summary, the edges selected are tested whether they are part of any semantic groups as the algorithm defined in Fig. 3. The semantic group construction can be done by using a naïve Bayes classifier or a simple pattern matching algorithm. The consideration of cause and effect within the same sentence is not needed to be maintained as they will be added as whole in the extracted summary. The relation that span across two sentences must be maintained as a group. This proposed work uses a Bayesian classifier to identify the cause and effect across sentence. The algorithm can be easily modified to adapt any other algorithm that can be used to extract the cause–effect relation.

3.4 Extractive Summary Generator

The stabilized candidate summary is chosen as extractive summary if the summary length is matching the expected summary length specified by user then its chosen as final summary else the stabilized summary is run through a sentence selector algorithm which clusters most similar sentences based on ranking criteria and then

Fig. 3 Algorithm—semantic group construction

Algorithm: Semantic Group Construction

Input: Preprocessed document d
Output: Semantic Groups → SG
SG = Group()
for each sentence s **in** document d **do**
    sent_num = sentNum(d, s)
    prev = sent_num − 1
    next = sent_num + 1
    cause, effect = Model.Match(d, prev, next)
    if cause **or** effect is **not** null **then**
        SG.addGroup(cause, effect)
    end if
end for

select the top-ranked sentence from each cluster. This also ensures that the cause–effect semantic group is preserved. The sentence selector performs a simple clustering algorithm such as k-means for building sentence that matches the expected final summary length. While selecting sentence from each cluster, the algorithm also makes sure that the entire semantic group is chosen for the final summary.

The clustering algorithm in sentence selector works on basis of set of ranking criterions define in Eqs. (2)–(4).

$$SentenceOrderRank(S_i) = \frac{|CS| - i}{|CS|} \tag{2}$$

$$WeightRank(S_i) = \frac{\sum_{w \in S_i} Node(w).weight/|S_i|}{Maximum\ Sentence\ Weight} \tag{3}$$

$$BigramWeightRank(S_i) = \frac{\sum_{w \in S_i} BigramFreq(w)}{Maximum\ Bigram\ Weight} \tag{4}$$

where $|CS|$ is the number of sentences in candidate summary, S_i is the statement "i" in the candidate summary, and $|S_i|$ is number of words in statement "i."

The statements selected for final summary are reordered according to their appearance in the original text so that the flow is maintained. The algorithm for summary generation is shown in Fig. 4.

Fig. 4 Algorithm—extractive summary generator

Algorithm: Extractive Summary Generator

Input: Candidate Summary CS
       Semantic Group SG.
Output: Extractive Summary → ES
if |CS| > expected_length **then**
    ranks = sent_rank(CS)
    cluster = kMeans(ranks, k=5)
    for each topic **in** cluster **do**
        CS = Add topic from each cluster until
        expected length is met
    end for
end if
ES = reorder_sentence(CS)

4 Results and Discussion

4.1 Evaluation Dataset

The work proposed is evaluated using the standard datasets for single document summarization DUC2001 [13] and DUC2002 [14] which contains English news documents. The DUC2001 dataset contains 30 clusters of 580 documents, and DUC2002
dataset contains 59 clusters of 524 documents after removing duplicates.

4.2 Evaluation Metrics

The most commonly used evaluation tool for summary is ROUGE [15] which evaluates the effectiveness and performance of the summary generated by the machine
with respect to the human-generated summary. The measures used for evaluation
in the literature are ROUGE-1, ROUGE-2, ROUGE-L and ROUGE-SU4 which is
defined by Eqs. (5)–(7).

$$\text{ROUGE}_N = \frac{\sum_{s \in X} \sum_{\text{n-gram} \in Y} \text{CountMatch}(\text{gram}_s)}{\sum_{s \in X} \sum_{\text{n-gram} \in Y} \text{Count}(\text{gram}_n)} \tag{5}$$

$$\text{ROUGE}_{\text{LCS}} = \frac{\text{LCS}(X, Y)}{m} \tag{6}$$

$$\text{ROUGE}_{\text{SU4}} = \frac{\text{Skip4Gram}(X, Y)}{mC_4} \tag{7}$$

where X is reference summary of length "m" words, and Y is system-generated summary. N refers to n-gram chosen for comparing the reference human summary and
machine-generated summary. (Here, $N = 1, 2$) LCS refers to the longest common
subsequence, and Skip4 represents the skip bigram constructed by skipping at most
four words.

4.3 Evaluation Results

The CESumm is compared against the baseline system in Tables 1 and 2 for various
ROUGE metrics used in baseline. There is improvement in ROUGE-1 and ROUGE-
LCS score implying that more key phrases are covered as part of the summary
generated by the proposed system. The decrease in other metric because in human-
generated summary the cause-effects statements are often condensed as single unit

Table 1 Evaluation result for DUC2001

Metric	Baseline	CESumm
ROUGE-1	0.5137	0.5590
ROUGE-2	0.2716	0.2347
ROUGE-L	0.4736	0.4947
ROUGE-SU4	0.2565	0.2191

Table 2 Evaluation result for DUC2002

Metric	Baseline	CESumm
ROUGE-1	0.5379	0.6076
ROUGE-2	0.2858	0.2666
ROUGE-L	0.4979	0.5505
ROUGE-SU4	0.2765	0.2551

Table 3 Evaluation result of CESumm varying word length for documents in DUC2001 and DUC2002

Word length	DUC 2001		DUC 2002	
	Best model	Average model	Best model	Average model
50	0.5714	0.2912	0.6058	0.3268
100	0.7629	0.4588	0.7576	0.4865
200	0.9901	0.6268	0.9802	0.6737
400	0.9978	0.7599	0.9904	0.8061

as possible with all key phrases, but in the proposed system, the key phrases are scattered across multiple sentence making bigram appear less.

The CESumm system is evaluated with various final summary length and corresponding human summary of same length. The best and average model of the each word length parameter are shown in Table 3, where the CESumm tends to perform well compared to human gold summary when the number of word expected in the extract from the source document is more, which is usually set for document of larger number of words ranging from 5000 to 20,000 words.

The histogram plot in Fig. 5 depicts the evaluation of CESumm system on DUC2001 dataset by setting system-generated summary of word length 100 with a threshold ($\pm$ 15 words). The best model of the proposed system achieves best for certain document reaching a best ROUGE score of nearing 0.8–0.9. Similar phenomenon can be observed for DUC2002 dataset for constraint of word length of 100 words with a threshold ($\pm$ 15 words), which is illustrated in Fig. 6.

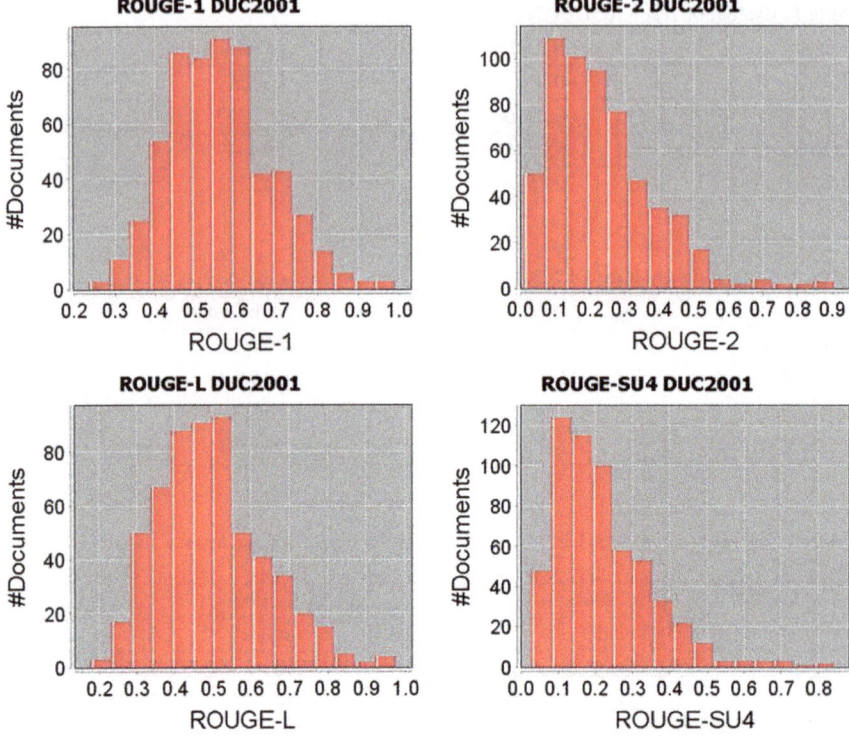

Fig. 5 ROUGE metric—DUC2001 document set

5 Conclusion

Text summarization is very essential in current day increase in amount of information and saves a lot of time in summarizing automatically compared to manual summarization which user can quickly understand the gist of the document. The proposed work picks more key phrases than the baseline and also improves the readability by retaining cause and effect statement in extracted summary and eliminating dangling causes or effects in the final summary. On evaluating system with standard evaluation tool ROUGE, its proven to outperform compared to other state-of-the-art techniques by retaining more key phrases in the generated summary. The system removes the problem of dangling cause–effect relation in the extracted summary by using a simple Bayesian classifier that identifies the cause–effect relation in the document, which can easily be modified to adapt another model to identify cause and effect relation which is a potential open work to find if any other cause and effect extraction and encapsulation mechanism are better than proposed approach which could improve

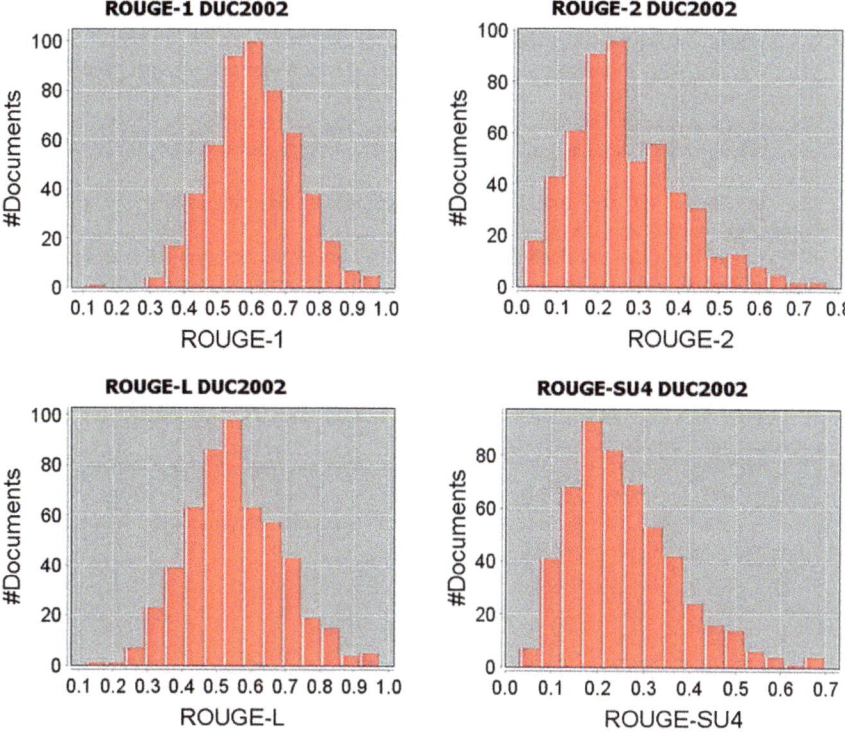

Fig. 6 ROUGE metric—DUC2002 document set

the bigrams generated in the final summary. The experiments show that CESumm on evaluating with standard evaluation dataset of DUC2001 and DUC2002 is found to be 8–8.7% better than the existing state-of-the-art baseline systems.

References

1. Gupta V, Lehal GS (2010) A survey of text summarization extractive techniques. J Emerg Technol Web Intell 2(3):258–268
2. Afsharizadeh M et al (2018) Query-oriented text summarization using sentence extraction technique. In: 2018 4th international conference on web research (ICWR). Tehran, pp 128–132
3. Gambhir M, Gupta V (2017) Recent automatic text summarization techniques: a survey. Artif Intell Rev 47:1–66
4. Afsharizadeh M et al Query-oriented text summarization using sentence extraction technique. In: 2018 4th international conference on web research (ICWR). Tehran, pp 128–132
5. Saini N et al (2019) Extractive single document summarization using multi-objective optimization: exploring self-organized differential evolution, grey wolf optimizer and water cycle algorithm. Knowl-Based Syst 164(15):45–67

6. Hark C et al (2020) Karcı summarization: a simple and effective approach for automatic text summarization using Karcı entropy. Inf Process Manag 57(3)
7. Mao X et al (2019) Extractive summarization using supervised and unsupervised learning. Expert Syst Appl 133:1
8. Zhou Q, Yang N et al (2020) A joint sentence scoring and selection framework for neural extractive document summarization. IEEE/ACM Trans Audio Speech Lang Process 28:671–681
9. Marx R, Chitra A (2019) Extractive document summarization using an adaptive, knowledge based cognitive model. Cogn Syst Res 56:56–71
10. El-Kassas WS et al (2020) EdgeSumm: graph-based framework for automatic text summarization. Inf Process Manag 57(6)
11. Cao M et al (2020) Grouping sentences as better language unit for extractive text summarization. Future Gener Comput Syst J 109
12. Chen K-Y et al (2018) An information distillation framework for extractive summarization. IEEE/ACM Trans Audio Speech Lang Process 26(1):161–170
13. DUC2001: document understanding conference (2001). https://www-nlpir.nist.gov/projects/duc/data/2001_data.html. Accessed 11 Nov 2020
14. DUC2002: document understanding conference (2002). https://www-nlpir.nist.gov/projects/duc/data/2002_data.html. Accessed 11 Nov 2020
15. Lin C-Y (2004) ROUGE: a package for automatic evaluation of summaries. Association for Computational Linguistics, pp 74–81

Chapter 9
Detection of Macular Diseases from Optical Coherence Tomography Images: Ensemble Learning Approach Using VGG-16 and Inception-V3

L. R. Ashok, V. Latha, and K. G. Sreeni

1 Introduction

Diabetic Macular Endema (DME) affects about one-third of the patients having Diabetic Mellitus (DM) which is a sight-threatening condition having a wide social and economic impact [1]. Diabetic Mellitus has spread over the globe, and the number of patients reported with DM is increasing day by day. All of them are in the high-risk category for developing DME which will eventually lead to vision loss unless it is diagnosed and treated well in advance [2].

Choroidal Neovascularization (CNV) is a serious complication of age related changes in the eye, which is an abnormal Choroidal capillary proliferation at Bruch's membrane [3].

Drusen is another condition of the eye which manifests due to the accumulation of yellow substances beneath Retinal Pigment Epithelium(RPE) [4]. Even though Drusen occurs due to normal ageing process, it is also an important sign associated with retinal disease for which proper treatment is required [5].

Optical coherence tomography (OCT) is a non-invasive high resolution optical imaging technology based on interference between the signal from an object under investigation and a local reference signal. OCT can produce images of the cross-section of an object, i.e. a two-dimensional image in the space [6]. By proper investigation using OCT, it is possible to diagnosis various macular conditions in the early stage itself, and many further complications can be avoided.

Figure 1 shows OCT images of Normal, DME, CNV, and Drusen categories. Traditionally clinicians and ophthalmologists have to go through each OCT frame for the diagnosis of various pathologies. It is both time-consuming and error-prone. With the advent of state-of-the-art machine learning and deep learning algorithms, it

L. R. Ashok (✉) · V. Latha · K. G. Sreeni
Department of Electronics and Communication, College of Engineering,
Trivandrum, Kerala, India
e-mail: vlatha@cet.ac.in

© The Author(s), under exclusive license to Springer Nature Singapore Pte Ltd. 2021   101
S. Kumar et al. (eds.), *Proceedings of International Conference on Communication and Computational Technologies*, Algorithms for Intelligent Systems,
https://doi.org/10.1007/978-981-16-3246-4_9

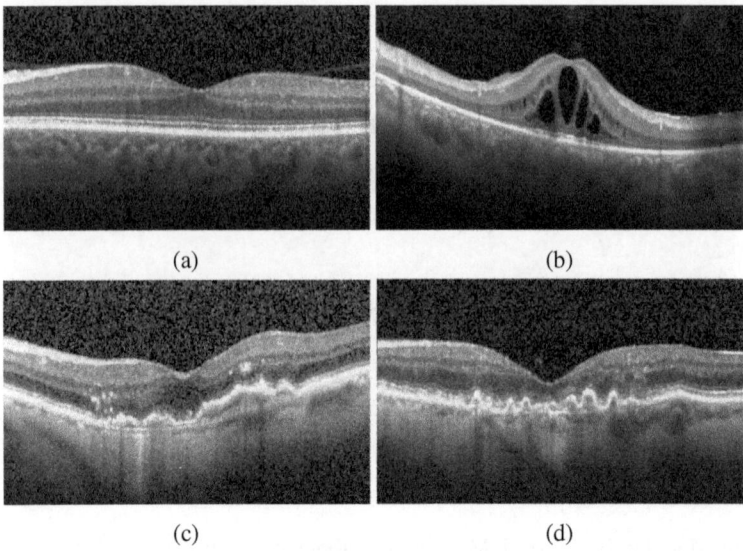

Fig. 1 OCT images of eye with various medical conditions. **a** Normal, **b** DME, **c** CNV, **d** Drusen

is now possible to design compute algorithms for assisting in diagnosis and automated diagnosis of various pathologies. In this proposed approach, a probabilistic prediction of the category of disease (i.e. Normal, DME, CNV, Drusen) will be made on the OCT images. We can choose highest probable prediction for the classification and detection of the macular diseases.

The remainder of the paper is as organised as follows: Sect. 2 gives brief introduction of various researches in the classification of OCT images. Section 3 discusses about the proposed methodology. Dataset, training, environment used for training and evaluation matrices are discussed in Sect. 4. Section 5 describes classification performances of the proposed methods and compare them with the existing methods. Finally, conclusion and future work are given in Sect. 6.

2 Literature Review

For identification and classification of macular diseases, a number of algorithms have been proposed. References [7–12] are machine learning based approaches, in which hand crafted features extracted from the images are used for classification.

Yu-Ying et al. proposed a machine learning approach to detect macular edema, macular hole, and age-related macular degeneration from OCT images in [7]. It is based on global image descriptors formed from a multi-scale spatial pyramid with the help of 2 class support vector machine(SVM). In [9], Sreenivasan et al. used SVM for classification, in which he used feature descriptors as multi-scale histograms of oriented gradient to classify OCT into dry age-related macular degeneration, diabetic macular edema and Normal. Yu Wang et al. proposed sequential minimal optimization (SMO)-based model selection from the linear configuration pattern-based features to classify OCT images. Venhuizen et al. in [10] proposed a machine learning method to identify high-risk age-related macular degeneration and stages it into early, intermediate or advanced AMD without requiring retinal segmentation using receiver operator characteristic analysis and Cohen's k statistics. Yankui Sun et al. proposed a frame work using dictionary learning for classification and detection of AMD and DME from OCT images in [11]. In this approach, the global representations of images are obtained using sparse coding and a spatial pyramid, and then a multi-class linear SVM is utilised for classification. In [12], Mousavi et al. proposed a dictionary learning based classification method on Histogram of Oriented Gradients (HOG) feature descriptors and successfully classified DME, AMD, and Normal OCT images.

Now, research has been shifted from traditional machine learning architectures to deep learning architectures which employ multiple hidden layers for feature extraction and classification. [13–19] use deep learning-based approach for classification. The deep learning models have shown extremely good classification accuracy in biomedical image classification than the traditional methods.

Karri et al. in [13] fine tuned pre-trained convolutional neural network (CNN) and GoogLeNet to provide accurate prediction of OCT disease classes. This method can effectively learn to identify pathologies than the classical learning on limited dataset. Reza Rasti et al. introduced Wavelet-based CNN (WCNN) for classification of macular diseases over 3D volumes of OCT scans in [14]. It is a fully automatic classification algorithm which operates on 3D OCT scans and classifies it into AMD, DME, and healthy macula. A multi-scale convolutional neural network is used by Rasti et al. in [15] for classification of OCT images. A Multi-Scale Convolutional Mixture of Expert (MCME) ensemble model is proposed to distinguish between healthy macula from AMD and DME affected macula. Leyuan Fang et al. proposed a method in [16] to mimic human attention system for classification in which they proposed a novel Lesion-Aware Convolutional Neural Network (LACNN) method for more accurate classification. The attention map generated by LACNN is utilised by the classifier, which results in improved OCT classification. In [18], an iterative fusion strategy is proposed by Leyuan Fang et al. for automatic classification of retinal OCT. L.Huang et al. suggested a novel Layer Guided Convolutional Neural Network (LGCNN) to identify normal retina from common types of macular pathologies such as DME, Drusen, and CNV in [17]. Here, layer segmentation is performed first, and then LGCNN is used to extract lesion related region which will provide improved OCT classification. In [19], Das et al. used Generative Adversarial Network (GAN) for semi-supervised learning and classified OCT images using limited amount of dataset.

The proposed method is inspired from the classification performance of VGG-16 [20] and Inception-V3 [21]. Ensemble learning-based approach is used here by combining the classification capabilities of these two models.

3 Design and Algorithm

In this method, two well-known CNN architectures Inception-V3 [21] and VGG-16 [20] are jointly used for classification. Convolution layers of Inception-V3 and VGG-16 are pre-trained on imagenet dataset [22] followed by fully connected layers and a classifier.

In this method, feature extraction capabilities of these CNNs are combined using ensemble learning to get a classification with a better accuracy. Following two schemes are used for combining VGG-16 and Inception-V3.

1. Arithmetic Mean (AM) Model : Two separate deep learning Sub-models are trained independently and results are combined.
2. Arithmetic Sum (AS) Model : Single model is created by combining two deep learning Sub-models and the resulting model is trained.

Figure 2 shows the structure of fully connected layers. It preforms global average pooling [23] of each feature vector received from preceding CNN layers and flatten it into a single dimensional vector of features and then process it through three layers of hidden neural network layers having Rectified Linear Unit (ReLU) [24] activation function with size 2048, 1024, and 512, respectively. Dropout [25] is added to avoid over-fitting. The output of fully connected layers is given to a softmax classifier layer in AM model and is summed in AS model.

3.1 Arithmetic Mean (AM) Model

Here, two independent deep learning models, Sub-model1 & Sub-model2, are created as shown in Fig. 3.

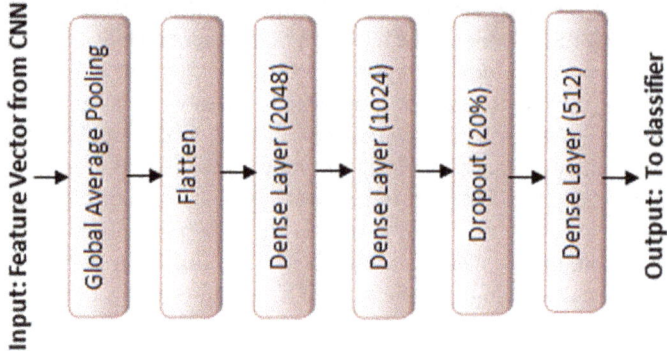

Fig. 2 Block diagram of fully connected layer: It contains three number of hidden neural network layers which take feature vector extracted from CNN layers as input. The output of this layer is used for classification

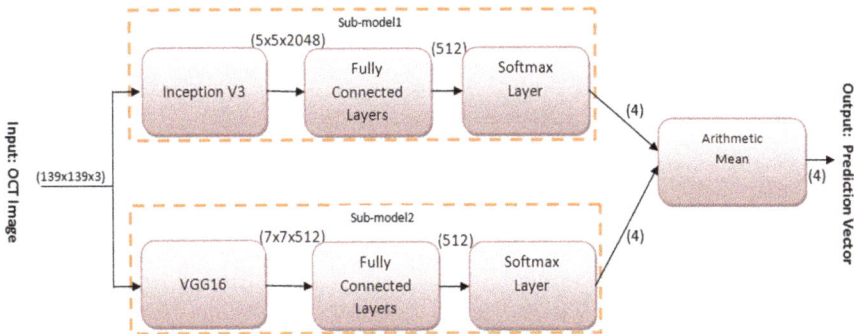

Fig. 3 Block diagram of AM model: Arithmetic mean of prediction vector from Sub-model1 and Sub-model2 is taken for final prediction. Output size of each block is shown in the bracket

Sub-model1 contains convolution layers of Inception-V3 followed by fully connected layers with a softmax output layer. Sub-model2 contains convolution layers of VGG-16 followed by fully connected layers with a softmax output layer.

$$\text{Sub-model1}\,Pr_{\text{class}}(I; \theta_{sm1}) = \text{softmax}_{\text{class}}(f_{\text{Inception}}(I; \theta_{sm1}))$$

$$\text{Sub-model2}\,Pr_{\text{class}}(I; \theta_{sm2}) = softmax_{class}(f_{VGG16}(I; \theta_{sm2}))$$

$$\text{AMPr}_{\text{class}}(I; \theta_{AM}) = \frac{1}{2}(\text{Sub-model1}\,Pr_{\text{class}}(I; \theta_{sm1}) + \text{Sub-model2}\,Pr_{\text{class}}(I; \theta_{sm2}))$$

where Sub-model1 Pr_{class} and Sub-model2 Pr_{class} denote the prediction vectors of Sub-models over the input image I, which are independently trained neural networks. Each contains softmax activation layer which gives a probabilistic prediction vector for each OCT class. $f_{\text{Inception}}$ and f_{VGG16} contain convolution layers of Inception-V3 and VGG-16 pretrained on imagenet dataset [22] and fully connected layers. The parameter θ_{sm1} & θ_{sm2} are learned during the training process. θ_{AM} is the joint parameter of θ_{sm1} & θ_{sm2}. AMPr$_{class}$ is the mean of prediction vector of Sub-model1 Pr_{class} & Sub-model2 Pr_{class}.

Where, class={'Normal', 'DME', 'CNV', 'Drusen'}

$$\sum_{\text{class}} \text{Sub-model1}\,Pr_{\text{class}}(I; \theta_{sm1}) = 1$$

$$\sum_{textclass} \text{Sub-model2}\,Pr_{\text{class}}(I; \theta_{sm2}) = 1$$

$$\sum_{\text{class}} \text{AMPr}_{\text{class}}(I; \theta_{AM}) = 1$$

The class corresponding to maximum value of prediction vector in $\text{AMPr}_{\text{class}}(I; \theta_{AM})$ gives the class prediction

$$\text{Class Prediction} = \underset{\text{class}}{\arg\max} \, \text{AMPr}_{\text{class}}(I; \theta_{AM})$$

3.2 Arithmetic Sum(AS) Model

In this approach, single model is created by combining convolutional layers of VGG-16 and Inception-V3 followed by fully connected layers. Arithmetic sum of last dense layer from both fully connected layers taken and a dense layer with softmax activation function is used for output prediction as shown in Fig. 4.

$$f_{\text{sum}}(I; \theta_{\text{AS}}) = f_{\text{VGG16}}(I; \theta_{\text{VGG}}) + f_{\text{Inception}}(I; \theta_{\text{Inception}})$$

$$\text{ASPr}_{\text{class}}(I; \theta_{\text{AS}}) = \text{softmax}_{\text{class}}(f_{\text{sum}}(I; \theta_{\text{AS}}))$$

where $f_{\text{Inception}}$ and f_{VGG16} contain convolution layers of VGG-16 and Inception-V3 pretrained on imagenet dataset [22] with fully connected layers which are parameterized by θ_{VGG} and $\theta_{\text{Inception}}$. f_{sum} denotes linear addition of final dense layers which is connected to a softmax layer ($\text{softmax}_{\text{class}}$) which gives probability of prediction $ASPr_{class}$ of each class of images I. θ_{AS} is the combined parameter vector.

$$\sum_{\text{class}} \text{ASPr}_{\text{class}}(I; \theta_{\text{AS}}) = 1$$

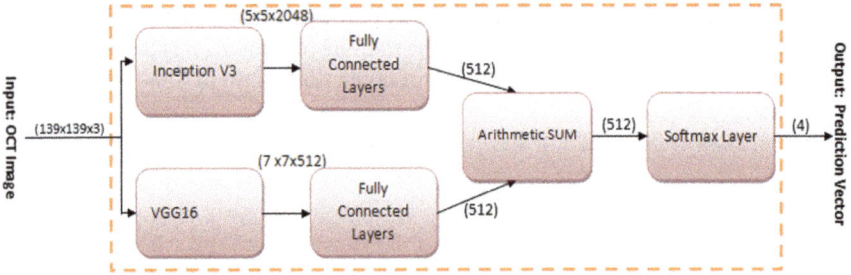

Fig. 4 Block diagram of AS model: AS model is created by summing last layer in fully connected layers of VGG-16 & Inception-V3 based Sub-models. Softmax layer at the output gives prediction vector. Output size of each block is shown in the bracket

The class corresponding to maximum value of prediction vector in $\text{ASPr}_{\text{class}}$ gives the class prediction

$$\text{Class Prediction} = \arg\max_{\text{class}} \text{ASPr}_{\text{class}}(I; \theta_{AS})$$

4 Experimental Setup

In order to optimise the parameters of AM and AS models which are designed as explained in Sects. 3.1 and 3.2, proper training has to be carried out using suitable training dataset in an environment having sufficient computational power.

4.1 Dataset

In the above proposed model, publicly available labelled Kaggle dataset of OCT images [26] is used for training. This model contains 84,495 OCT scanned images in the Normal, DME, CNV, and Drusen categories which are carefully labelled by trained graders. 242 Number of images in each category are used for testing and rest of the data is split into 80:20 for training and validation. In order to compare the results with other existing methods, UCSD V3 dataset of OCT images [27] are used.

4.2 Training

Training of AM and AS model are carried out for 100 epohs, and the progress of training accuracy, validation accuracy, training loss, and validation loss are closely monitored for each step. The parameters that provide highest training accuracy are saved and used for classification.

1. AM Model Training: As explained in 3.1 Sub-model1 h VGG-16 convolutional layers and Sub-model2 have Inception-V3 convolutional layers which are pre-trained on imagenet dataset [22] followed by fully connected layers and a softmax layer. Sub-model1 & Sub-model2 are trained independently using stochastic gradient descent optimisation algorithm using Kaggle dataset [26]. Plot of accuracy and normalised loss curve vs no of epochs is shown in the Fig. 5a, b. Inception-V3 Model trained for 100 epochs (see Fig. 5a). Highest training accuracy achieved is 99.46% on 56th epoch and corresponding loss is 0.0151 and validation accuracy is 93.63% and validation loss is 0.3283. VGG-16 Model is trained for 100 epochs (see Fig. 5b). Highest training accuracy achieved is 100% on 36th epoch and corresponding loss is 0.0833. Validation accuracy and validation loss are 90.03 and 0.2993%, respectively.

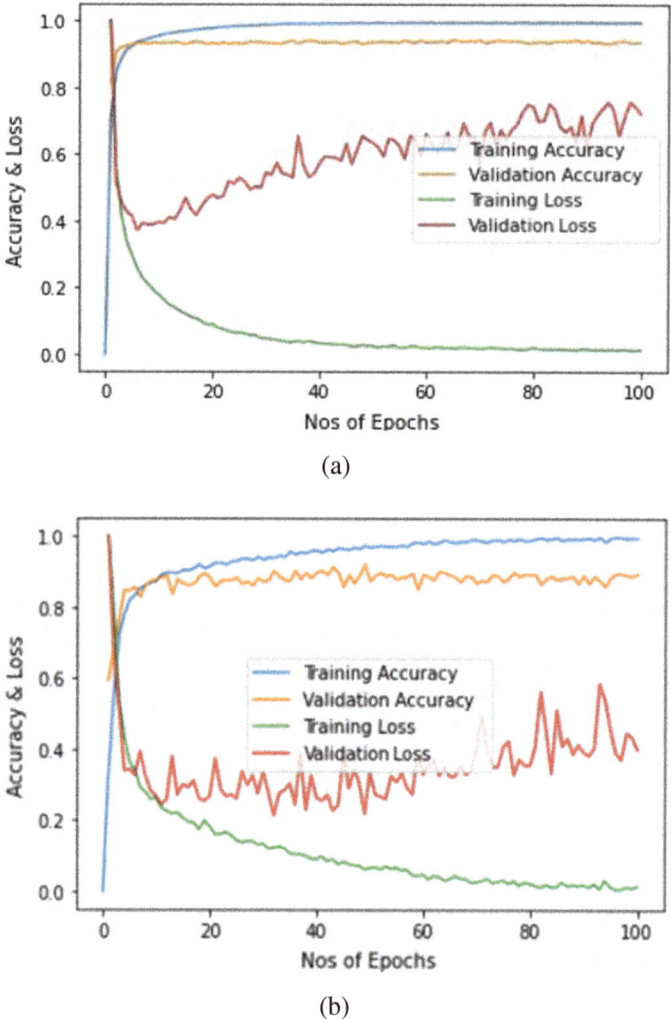

(a)

(b)

Fig. 5 Accuracy and normalised loss curves vs number of epochs during training of Sub-models of AM **a** Inception-V3-based Sub-model1, **b** VGG-16-based Sub-model2

2. AS Model Training: The AS model is created as explained in the Sect. 3.2 contains pretrained convolution layers of VGG-16 and Inception-V3 on imagenet dataset [22]. It is trained with stochastic gradient decent optimisation algorithm using the Kaggle dataset [26]. Plot of accuracy and normalised loss curve vs no of epochs is shown in the Fig. 6. Highest training accuracy achieved is 99.62% on 82th epoch and corresponding loss is 0.0103. Validation accuracy and validation loss are 95.89 and 0.2019%, respectively.

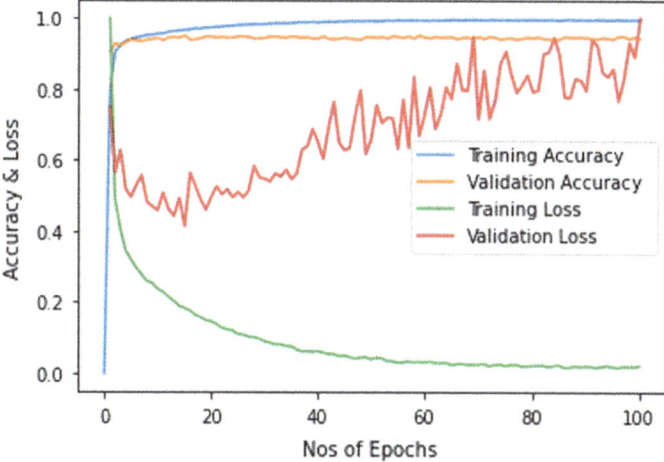

Fig. 6 Accuracy and normalised loss curves vs number of epochs during training of AS model

4.3 *Environment*

Jupyter notebook interface provided by Google colab with GPU support is used for training. Keras functional API is used for designing the neural network layers on TensorFlow framework.

4.4 *Evaluation Matrices*

In order to evaluate the classification performance of the model, following matrices are used [28].

1. Sensitivity (SE): It shows the number of positive samples correctly classified to total number of positive samples in a class.

$$SE = \frac{TP}{TP + FN}$$

2. Precision (PR): It shows the number of positive samples correctly classified to total number of positively classified samples in a class.

$$PR = \frac{TP}{TP + FP}$$

3. Specificity (SP): It shows the number of negative samples correctly classified to total number of negative samples in a class.

$$SP = \frac{TN}{TN + FP}$$

4. Accuracy (ACC): It shows the number of samples correctly classified to total number of samples in a class.

$$ACC = \frac{TP + TN}{TP + TN + FP + FN}$$

5. Overall Accuracy (OA), Overall Sensitivity (OS) and Overall Precision (OP): OA, OS & OP are ACC, SE, PR respectively with TP, TN, FP, FN summed over all classes.

where,
True Positive (TP): Number of positive samples correctly classified.
True Negative (TN): Number of negative samples correctly classified.
False Positive (FP): Number of negative samples incorrectly classified.
False Negative (FN): Number of positive samples incorrectly classified.

5 Results and Discussions

The models are evaluated by predicting OCT classes of test data provided by Kaggle dataset [26] which are not part of the training dataset and comparing it with the ground truth class of the image. Classifications of models are compared with the ground truth and the resulting confusion matrix are shown in Table 1. The rows of the matrix denote predicted class and columns denote ground truth class. Table 1a, b shows the confusion matrices of two AM Sub-models while Table 1c shows confusion matrix of AM model. Table 1d shows confusion matrix generated by the AS model. Sub-Model1 (Table 1a) correctly classifies Normal and DME categories, but make six incorrect classifications on CNV and two incorrect classifications on Drusen categories. Sub-Model2 (Table 1b) incorrectly classifies 2 normal, 5 DME, 4 CNV, and 4 Drusen categories. AM model (Table 1c) incorrectly classifies 3 Normal, 2 DME, 1 CNV, and 1 Drusen. AS model (Table 1d) correctly classifies Normal and DME categories but makes 4 incorrect classification on CNV and 1 incorrect classification on Drusen categories.

Table 2 shows the prediction vector generated for OCT images that are wrongly classified by Sub-model1 & Sub-model2. Both AM and AS models correctly classify first and second images even though one Sub-model can not classify them correctly, using the higher prediction probability of the other Sub-model. This is the motivation for using combined models AM and AS for more accurate prediction. However in certain images, AM and/or AS model can not classify correctly even though one of the Sub-model make correct classification.

Table 1 Confusion matrices of prediction by models over the test dataset provided in [26]. (a) Inception-V3-based Sub-model1, (b) VGG-16-based Sub-model2, (c) AM model, (d) AS model

	Normal	DME	CNV	Drusen
(a)				
Normal	**241**	0	0	1
DME	0	**242**	0	0
CNV	0	0	**241**	1
Drusen	0	0	6	**236**
(b)				
Normal	**236**	2	0	4
DME	2	**240**	0	0
CNV	0	3	**239**	0
Drusen	0	0	4	**238**
(c)				
Normal	**241**	1	0	0
DME	0	**241**	0	1
CNV	3	0	**239**	0
Drusen	0	1	1	**240**
(d)				
Normal	**241**	0	0	1
DME	0	**242**	0	0
CNV	0	0	**242**	0
Drusen	0	0	4	**238**

Table 3 shows comparison of classification performance matrices as specified in Sect. 4.4 for Sub-model1, Sub-model2, AM model, and AS model for each of the macular classes on the dataset [26]. The highest parameters obtained are shown in bold letters. Sub-model1 has 99.59% overall accuracy; Sub-model2 has given overall accuracy of 99.23%. The AM and AS models have given overall accuracy 99.64% and 99.74%, respectively. Table 4 shows comparison of classification performance matrices as specified in Sect. 4.4 with other existing classification methods on dataset [27]. The highest parameters obtained are shown in bold letters. AM model has 96.60% overall accuracy. AS model has given overall accuracy of 97.25% which is the highest value obtained in this dataset.

It can be observed from Tables 3 and 4 that AS model achieved better performance in most of the parameters in dataset [26, 27].

Table 2 Prediction vector generated for OCT images that were wrongly classified by Sub-models

SI No.	Image	Ground truth	Model	Prediction vector				Class prediction
				Normal	DME	CNV	Drusen	
1		CNV	Sub-model1	0.000	**0.583**	0.417	0.000	DME
			Sub-model2	0.000	0.055	**0.945**	0.000	CNV
			AM	0.000	0.319	**0.681**	0.000	CNV
			AS	0.000	0.010	**0.990**	0.000	CNV
2		Drusen	Sub-model1	0.000	0.000	0.005	**0.995**	Drusen
			Sub-model2	0.000	0.002	**0.617**	0.382	CNV
			AM	0.000	0.001	0.311	**0.689**	Drusen
			AS	0.000	0.000	0.124	**0.875**	Drusen
3		CNV	Sub-model1	0.000	0.047	**0.953**	0.000	CNV
			Sub-model2	0.000	**0.947**	0.053	0.000	DME
			AM	0.000	0.497	**0.503**	0.000	CNV
			AS	0.000	**0.606**	0.393	0.001	DME
4		Normal	Sub-model1	0.027	**0.973**	0.000	0.000	DME
			Sub-model2	**0.970**	0.029	0.000	0.000	Normal
			AM	0.498	**0.501**	0.000	0.000	DME
			AS	**0.825**	0.087	0.034	0.054	Normal

(1) CNV is wrongly classified by Sub-Model1, (2) Drusen is wrongly classified by Sub-Model2, (3) CNV is wrongly classified as CNV by Sub-model1 & AS model, (4) Normal is wrongly classified by Sub-Model2 and AM model

Table 3 Performance of Sub-model1, Sub-model2, AM model, and AS model on sensitivity (SE %), precision (PR %), specificity (SP %), and accuracy (AC%) on each of the four classes and overall accuracy (OA %), overall sensitivity (OS %), and overall precision (OP %) on test dataset provided in [26] are shown

Model	OCT class	SE (%)	PR (%)	SP (%)	ACC (%)	OA (%)	OS (%)	OP (%)
Sub-model1	CNV	97.57	99.59	99.86	99.28	99.59	99.17	99.17
	Drusen	99.16	97.52	99.18	99.17			
	DME	**100.00**	**100.00**	**100.00**	**100.00**			
	Normal	**100.00**	99.59	99.86	99.90			
Sub-Model2	CNV	98.35	98.76	99.59	99.28	99.23	98.45	98.45
	Drusen	98.35	98.35	99.45	99.17			
	DME	97.96	99.17	99.72	99.28			
	Normal	99.16	97.52	99.18	99.17			
AM model	CNV	**99.58**	98.76	99.59	**99.59**	99.64	99.28	99.28
	Drusen	**99.59**	**99.17**	**99.72**	**99.69**			
	DME	99.18	99.59	99.86	99.69			
	Normal	98.77	**99.59**	**99.86**	99.59			
AS Model	CNV	98.37	**100.00**	**100.00**	**99.59**	**99.74**	**99.48**	**99.48**
	Drusen	99.58	98.35	99.45	99.48			
	DME	**100.00**	**100.00**	**100.00**	**100.00**			
	Normal	**100.00**	**99.59**	**99.86**	**99.90**			

6 Conclusion

Fast and accurate diagnostic tool for various macular diseases can save from many sight-threatening conditions. CAD approach using deep learning is a promising method which can assist ophthalmologists by classifying OCT scan images from Normal to various disease categories. In this proposed approach, OCT images are classified into Normal, CNV, Drusen, and DME for which the feature extraction capabilities VGG-16 and Inception-V3 are utilised. Both of these models are capable of predicting the class of the images independently. Using ensemble learning, it is possible to increase the classification accuracy by combining both of these models. Two such models, i.e. arithmetic mean (AM) model and arithmetic sum (AS), model are designed, trained, and the results are compared with other models. Former approach has an overall accuracy of 99.64% in Kaggle dataset [26] and 96.60% in UCSD V3 dataset [27] while later approach has an overall accuracy 99.74% in Kaggle dataset and 97.25%in UCSD V3 dataset. It is observed that the AS model achieves better

Table 4 Performance comparison of proposed methods with existing methods on sensitivity (SE %), precision (PR %), specificity (SP%), and accuracy (AC %) on each of the four classes and overall accuracy (OA %), overall sensitivity (OS %), and overall precision (OP %) in UCSD V3 dataset [27]

Model	OCT class	SE (%)	PR (%)	SP (%)	ACC (%)	OA (%)	OS (%)	OP (%)
HOG-SVM [9]	CNV	74.60	89.90	90.30	81.80	70.90	59.60	51.00
	Drusen	38.90	10.70	90.50	89.10			
	DME	55.10	15.20	88.00	86.80			
	Normal	70.00	87.90	93.70	84.30			
Transfer learning [29]	CNV	90.50	81.50	86.30	87.90	80.30	73.20	75.20
	Drusen	46.10	56.00	95.20	88.60			
	DME	69.00	69.40	95.20	91.50			
	Normal	87.20	90.00	95.30	92.60			
VGG16 [30]	CNV	89.50	89.50	91.60	90.60	81.20	73.80	70.20
	Drusen	49.00	37.30	93.00	89.50			
	DME	77.10	61.60	94.20	92.30			
	Normal	79.60	92.50	96.30	90.10			
IFCNN [18]	CNV	94.80	87.90	90.90	92.40	87.30	82.50	84.70
	Drusen	64.40	76.80	97.30	93.00			
	DME	79.20	81.90	97.20	94.40			
	Normal	91.50	92.20	96.40	93.00			
LGCNN [17]	CNV	93.30	91.50	93.30	94.40	88.40	84.60	82.90
	Drusen	71.00	65.20	96.00	93.30			
	DME	85.70	79.40	96.80	95.40			
	Normal	88.50	95.50	97.90	94.60			
AM model	CNV	80.32	**99.60**	**99.86**	93.80	96.60	93.20	93.20
	Drusen	98.47	77.20	92.91	94.00			
	DME	**98.79**	97.60	99.20	**99.10**			
	Normal	**99.60**	98.40	99.47	**99.50**			
AS model	CNV	**99.03**	82.00	94.33	**95.30**	**97.25**	**94.50**	**94.50**
	Drusen	**98.79**	**98.00**	**99.34**	**99.20**			
	DME	96.48	**98.80**	**99.60**	98.80			
	Normal	85.81	**99.20**	**99.72**	95.70			

classification performance in classifying 'Normal', 'Drusen', 'CNV', 'DME' than the existing methods.

In AM model, arithmetic mean of Sub-model1 and Sub-model2 is used where equal weights for prediction vectors of both models are given. It may be possible to increase the classification accuracy by taking weighted average of prediction vectors where weights are learned from the dataset. Other ensemble learning methods such as bagging, boosting along with AM and AS models can also be attempted. Here, only four classes of OCT images are considered; the model can be trained to predict more classes in future.

References

1. Maniadakis N, Konstantakopoulou E (2019) Cost effectiveness of treatments for diabetic retinopathy: a systematic literature review. PharmacoEconomics 37(8):995–1010. https://doi.org/10.1007/s40273-019-00800-w
2. Flikier S, Wu A, Wu L (2019) Revisiting pars plana vitrectomy in the primary treatment of diabetic macular edema in the era of pharmacological treatment. Taiwan J Ophthalmol 9(4):224. https://doi.org/10.4103/tjo.tjo_61_19
3. Freund KB, Yannuzzi LA, Sorenson JA (1993) Age-related macular degeneration and choroidal neovascularization. Am J Ophthalmol 115(6):786–791
4. Khan KN et al (2016) Differentiating drusen: Drusen and drusen-like appearances associated with ageing, age-related macular degeneration, inherited eye disease and other pathological processes. Progress Retinal Eye Res 53:70–106. https://doi.org/10.1016/j.preteyeres.2016.04.008
5. Ardeljan D, Chan CC (2013) Aging is not a disease: distinguishing age-related macular degeneration from aging. Prog Retin Eye Res 37:68–89. https://doi.org/10.1016/j.preteyeres.2013.07.003
6. Podoleanu AGh (2012) Optical coherence tomography. J Microscopy 247(3):209–219. https://doi.org/10.1111/j.1365-2818.2012.03619.x
7. Liu Y-Y, Chen M, Ishikawa H, Wollstein G, Schuman JS, Rehg JM (2011) Automated macular pathology diagnosis in retinal OCT images using multi-scale spatial pyramid and local binary patterns in texture and shape encoding. Medical Image Anal 15(5):748–759. ISSN 1361-8415. https://doi.org/10.1016/j.media.2011.06.005
8. Wang Y et al (2016) Machine learning based detection of age-related macular degeneration (AMD) and diabetic macular edema (DME) from optical coherence tomography (OCT) images. Biomed Opt Exp 7:4928–4940
9. Srinivasan PP et al (2014) Fully automated detection of diabetic macular edema and dry age-related macular degeneration from optical coherence tomography images. Biomed Opt Exp 5:3568–3577
10. Venhuizen FG et al (2017) Automated staging of age-related macular degeneration using optical coherence tomography. Invest Ophthalmol Vis Sci 58(4):2318. https://doi.org/10.1167/iovs.16-20541
11. Sun Y, Li S, Sun Z (2017) Fully automated macular pathology detection in retina optical coherence tomography images using sparse coding and dictionary learning. J Biomed Opt 22(1), Art. no. 16012
12. Mousavi E, Kafieh R, Rabbani H (2020) Classification of dry age-related macular degeneration and diabetic macular oedema from optical coherence tomography images using dictionary learning. IET Image Process 14(8):1571–1579. https://doi.org/10.1049/iet-ipr.2018.6186

13. Karri SP, Chakraborty D, Chatterjee J. Transfer learning based classification of optical coherence tomography images with diabetic macular edema and dry age-related macular degeneration. Biomed Opt Express 8(2):579–592. https://doi.org/10.1364/BOE.8.000579

14. Rasti R, Mehridehnavi A, Rabbani H, Hajizadeh F (2017) Wavelet-based convolutional mixture of experts model: an application to automatic diagnosis of abnormal macula in retinal optical coherence tomography images. In: 10th Iranian conference on machine vision and image processing (MVIP). IEEE, pp 192–196

15. Rasti R, Rabbani H, Mehridehnavi A, Hajizadeh F (2018) Macular OCT classification using a multi-scale convolutional neural network ensemble. IEEE Trans Med Imag 37(4):1024–1034

16. Fang L, Wang C, Li S, Rabbani H, Chen X, Liu Z (2019) Attention to lesion: lesion-aware convolutional neural network for retinal optical coherence tomography image classification. IEEE Trans Med Imaging 38(8):1959–1970. https://doi.org/10.1109/TMI.2019.2898414

17. Huang L, He X, Fang L, Rabbani H, Chen X (2019) Automatic classification of retinal optical coherence tomography images with layer guided convolutional neural network. IEEE Signal Process Lett 26(7):1026–1030. https://doi.org/10.1109/LSP.2019.2917779

18. Fang L, Jin Y, Huang L, Guo S, Zhao G, Chen X (2019) Iterative fusion convolutional neural networks for classification of optical coherence tomography images. J Vis Commun Image Represent 59:327–333

19. Das V, Dandapat S, Bora PK (2020) A data-efficient approach for automated classification of OCT images using generative adversarial network. IEEE Sens Lett 4(1):1–4. https://doi.org/10.1109/LSENS.2019.2963712

20. Simonyan K, Zisserman A (2014) Very deep convolutional networks for large-scale image recognition. CoRR, vol. abs/1409.1556

21. Szegedy C, Vanhoucke V, Ioffe S, Shlens J, Wojna Z (2015) Rethinking the inception architecture for computer vision. arXiv:1512.00567 [cs], Dec 2015 [Online]. Available: http://arxiv.org/abs/1512.00567

22. Deng W, Dong R, Socher L, Li KL, Li F-F (2009) ImageNet: a large-scale hierarchical image database. In: 2009 IEEE conference on computer vision and pattern recognition, June 2009, pp 248–255. https://doi.org/10.1109/CVPR.2009.5206848

23. Lin M, Chen Q, Yan S. Network in network. arXiv:1312.4400

24. Nair V, Hinton GE (2010) Rectified linear units improve restricted Boltzmann machines. In: ICML, 2010

25. Srivastava N, Hinton G, Krizhevsky A, Sutskever I, Salakhutdinov R (2014) Dropout: a simple way to prevent neural networks from overfitting. J Mach Learn Res 15(1):1929–1958

26. Mooney P. Retinal OCT Images (optical coherence tomography). https://kaggle.com/paultimothymooney/kermany2018

27. Kermany D, Zhang K, Goldbaum M (2018) Large dataset of labeled optical coherence tomography (OCT) and chest X-Ray images. Mendeley Data, V3. https://doi.org/10.17632/rscbjbr9sj.3

28. Tharwat A (2020) Classification assessment methods. ACI, vol. ahead-of-print, no. ahead-of-print, Aug 2020. https://doi.org/10.1016/j.aci.2018.08.003

29. Kermany DS et al (2018) Identifying medical diagnoses and treatable diseases by image based deep learning. Cell 172(5):1122–1131

30. Simonyan K, Zisserman A (2015) Very deep convolutional networks for large-scale image recognition [Online]. Available: arXiv:1409.1556

Chapter 10
Optimization Methods for Energy Management in a Microgrid System Considering Wind Uncertainty Data

Yahia Amoura, Ana I. Pereira, and José Lima

1 Introduction

Renewable energy sources (RES) are widely used to address the increasing energy demand, minimize environmental pollutants and provide sustainable development and socioeconomic benefits [1]. In counterpart, renewable energy sources suffer from several obstacles, mainly their intermittent nature, which makes difficult to precisely predict their production [2]. However, to address this problem, an aggregation of (RES) at a local level as a hybrid energy system (HES) gives rise to the microgrid (MG) concept. Achieving a secure electrical equilibrium between the offer and the requirement can be difficult when using a large renewable energy system, this is why an energy management strategy is necessary in the case of a microgrid [3].

Quoc-Tuan et al. [4] have classified energy management systems in microgrids into four categories according to the type of backup system employed, regarding non-renewable energy generators, energy storage system, management of the demand side noted DSM, and other hybrid energy systems.

Y. Amoura · A. I. Pereira · J. Lima
Research Centre in Digitalization and Intelligent Robotics (CeDRI), Instituto Politécnico de Bragança, Bragança, Portugal
e-mail: apereira@ipb.pt

J. Lima
e-mail: jllima@ipb.pt

A. I. Pereira
ALGORITMI Center, University of Minho, Braga, Portugal

J. Lima
INESC TEC—INESC Technology and Science, Porto, Portugal

Y. Amoura (✉)
Higher School in Applied Sciences of Tlemcen, Tlemcen, Algeria

S. Kumar et al. (eds.), *Proceedings of International Conference on Communication and Computational Technologies*, Algorithms for Intelligent Systems,
https://doi.org/10.1007/978-981-16-3246-4_10

Microgrids are defined as low-voltage (LV) distribution networks including a set of decentralized power systems called distributed generators (DGs), storage devices and controllable loads operating in islanded mode, else interconnected within the principal distribution network as a controlled entity [5], usually based on a central controller that enables the optimization of their functioning during an interconnected operation by reducing the production of distributed generators and electricity trade with the rural grid.

The microgrid control operation contains three main levels, the first level characterizes the micro-sources controller (MSC) which uses the local information to maintain the voltage and frequency during process. Besides the two other levels that concern the microgrid system controllers (MGSC) and the distribution management system (DMS) which are intended to maximize the microgrid value and the optimal management of its operation considering electricity bids in order to quantify the electrical flow that the MG should extract from the supply network [6].

Deployment of these systems offers several earnings for either consumer and the electricity distributor. For the user entity, the MG can lead to quality improvement of the network and reduce the operation cost. On the other hand, the local production help the distributor to reduce energy circulation on transmission lines, minimizing losses and extra energy costs [7], as well as contributing on the reduction of greenhouse gas emissions. Microgrids are designed to enhance the sustainability, cost-effectiveness, and clean production of electricity and its provisioning to ensure consumer satisfaction. The inclusion of RES in a local production system has made it possible to generate, distribute, and follow electrical energy of the microgird in order to obtain the optimal combination [8]. Hence, several research works have been developed in the area of microgrid energy management. The authors of [9] developed optimal energy management of microgrid system considering it as being as optimal scheduling of power flow, in [10] authors treat the energy management issues by the mean of an economic objective function using a matrix real-coded genetic algorithm (MRC-GA). The linear programming (LP) algorithm was used in [11] to manage the microgrid for the purpose of minimizing the daily operating cost. In [12], Kerboua et al. proposed evolutionary PSO-based algorithm for the energy management strategies in smart cities using load scheduling. In [13], a genetic algorithm (GA) was used for an enhanced EMS model intended determine the most effective strategies for reducing energy bids and exhaust emissions. Other authors have considered the energy management in microgrid as a multi-objective optimization problem considering both economic and environmental aspects, and in [14], a multi-bacterial foraging optimization (MBFO) was proposed for the optimal energy dispatch of a microgrid system. In [15], a multi-objective particle swarm optimization was proposed (MOPSO) for management and optimal distribution of energy resources, for the same purpose a non-dominated sorting genetic algorithm (NSGA) was adopted on [16].

Further to its remarkable development in the field of renewable energies, according to the Portuguese Renewable Energy Association (PREA), in 2019, the wind power production in Portugal was estimated at 5429 MW. In fact, this represents an encouraging statistic to increase wind production capacity in the country. As a matter of fact, wind generation in microgrid systems represents an important resource, but its widely fluctuating effect makes it scheduling with other distributed energy resources more difficult. However, a wind forecasting model allowing the prediction of the power capacity regarding the local wind farm in the microgrid is important to improve the system reliability, and to do this, several models have been proposed in the literature. Liang et al. proposed in [17] a wind velocity prediction model based on the preceding wind velocity measurements using the artificial neural networks based on back-propagation algorithm as learning process for short-term wind speed prediction. In [18], the authors established the development of a forecasting device for wind energy using an artificial neural network and the integration of the wind prediction results within a unit commitment (UC) scheduling considering forecasting error using a confidence interval probabilistic concept. In [19], a prediction model was proposed using a hybrid Kalman filter with an artificial neural network (KF-ANN) based on the linear autoregressive integrated moving average (ARIMA). In [20], the authors proposed several prediction models based on ANN which conjointly employs several local metrics including wind speed, temperature and pressure values, temperature and pressure values, the results allowed to analyze and compare the impact of providing several metrics instead of wind speed only.

This paper proposes optimization strategies for an optima management in a microgrid system while analyzing the impact of wind intermittency. To perform the hourly forecast of wind energy production during the day, a multilayer neural network algorithm is proposed, the performances of the model are evaluated according to the mean squared error (MSE) value. On the other hand, energy management is formulated as a uni-objective optimization problem. To allocate the power set-points for the optimal scheduling of microgrid generators, five optimization methods are proposed and compared: linear programming (LP) based on simplex method, two particle swarm optimization (PSO) algorithms, genetic algorithm (GA), and a hybrid approach (LP-PSO). Finally, two management scenarios are proposed to illustrate the economic and environmental impact of energy exchange between the microgrid and the main grid.

The remaining parts of the paper are organized as follows: Sect. 2 describes the wind forecasting model. In Sect. 3, the architecture, as well as the operation of the microgrid, is presented. The storage system has been modeled in Sect. 4. The operation of the energy management system, the optimization problem, and these constraints is explained in Sect. 5. In Sect. 6, we present and discuss results obtained under the computational simulations. Section 7 concludes the study and proposes guidelines for future works.

2   Wind Forecasting Model

Wind energy is one of the most energy-efficient ways to produce electrical power in a microgrid. The wind farms require a continuous and sufficient wind speed for proper electricity production [21]. However, to increase the microgrid optimality, a wind

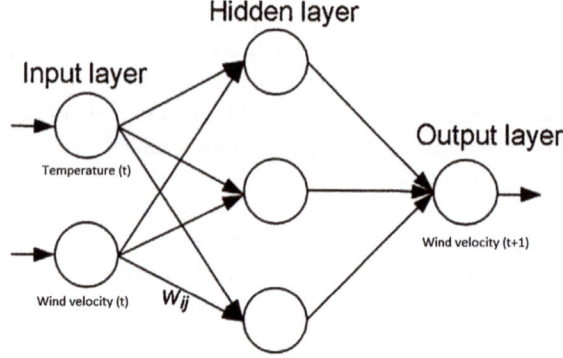

Fig. 1 Structure of the ANN model

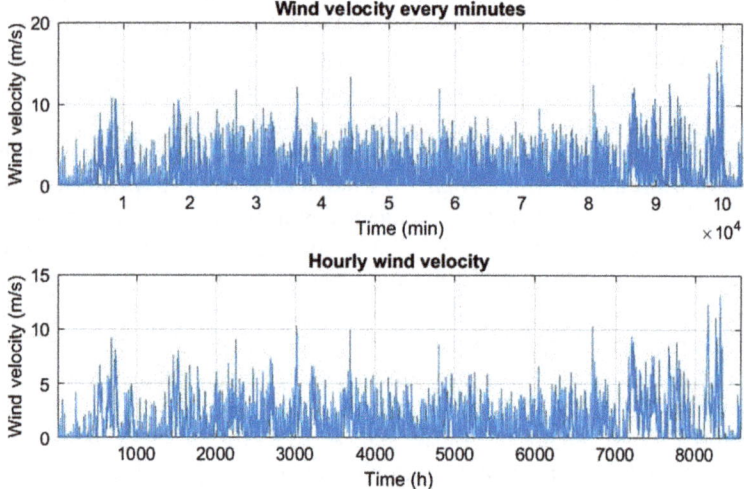

Fig. 2 Measured wind speed data in five minutes' interval in Polytechnic Institute of Bragança

speed forecasting model based on ANN neural network is employed in this paper. The wind speed is precisely forecasted by ANN based on various weather observations. The ANN model uses the previously recorded wind speed and temperature together weighted by W_{ij} to predict the future value of wind speed as shown in Fig. 1.

The real data are obtained by using a datalogger system which can register the measurement of the sensor at 5 min intervals. The data are measured by the meteorological station of the laboratory at the Polytechnic Institute of Bragança (latitude: 41° 47′ 52.5876″ N—longitude: 6° 45′ 55.692″ W) from January 1, 2019, to December 31, 2019. Figures 2 and 3 show the data of wind and temperature.

Wind speed data of five-minute intervals between January 1, 2019, and December 31, 2019, are obtained as an input representing 103 104 samples of which 90% are used for training, 5% for testing, and 5% for validation. The ANN structure has two layers.

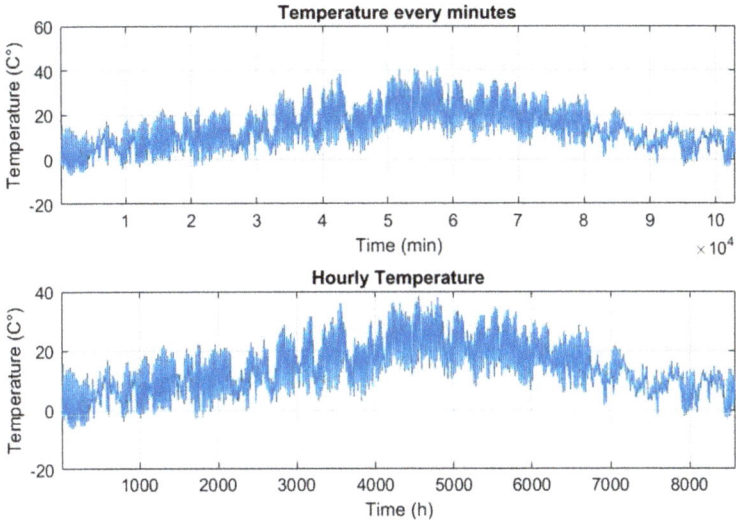

Fig. 3 Measured temperature data in five minutes' interval in Polytechnic Institute of Bragança

Feed-forward is addressed as a network type. The activation function is choose in a sigmoid. Because the Levenberg–Marquardt algorithm has fast convergence, this latter is adopted by the learning process for all ANN structure. The evaluation of model performances is made using the mean square error (MSE) value as:

$$\text{MSE} = \frac{1}{n} \sum_{i=1}^{n} (x_i - \bar{x}_i)^2 \qquad (1)$$

where n is the number of periods of time, x_i is the desired neural network output value associated to the wind velocity, and $\bar{x}_i$ is the estimated value obtain by neural network associated to the wind velocity.

3 Microgrid Architecture

The chosen microgrid consists of two renewable sources photovoltaic (PV) and wind turbine (WT), a conventional source micro-turbine (MT) and an energy storage system (ESS) in addition to the load. The latter are interconnected via two busses (DC and AC) through the bidirectional inverter. The MG system is connected to the main grid. The exchange of energy between the microgrid and the main grid is mutual in a way that the main grid supplies (sells) energy when its unit price is cheap and absorbs (buys) surplus energy from renewable generators.

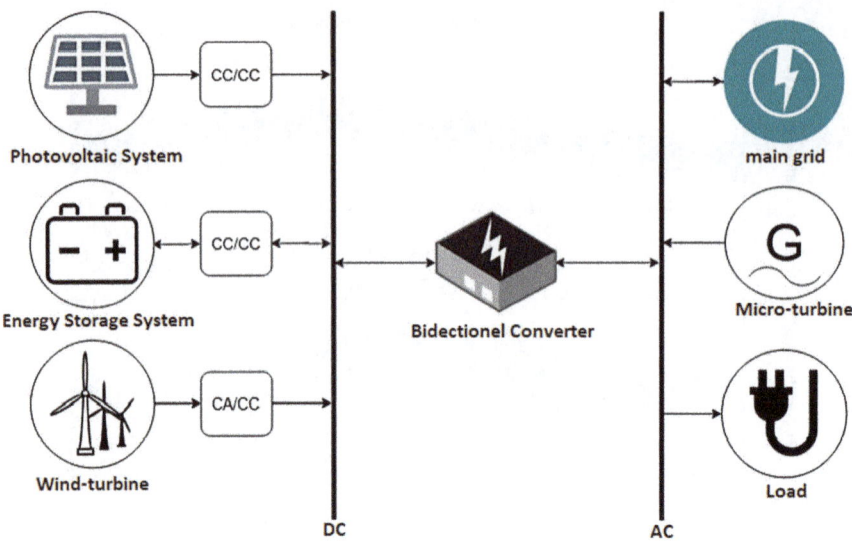

Fig. 4 Microgrid architecture

Table 1 Maximum and minimum limits for microgrid production units

MG system	Min power (kW)	Max power (kW)
P_{gr}	0	90
P_{WT}	0	20
P_{PV}	0	25
P_{MT}	6	30
P_{ESS}	−25	30

The real-time energy management of different elements of the microgrid is mainly based on the unit cost of energy per kWh by satisfying the load balance constraint while minimizing the cost. Figure 4 shows the microgrid architecture adopted in this study.

The power limits of the microgrid generators are presented in Table 1.

The daily photovoltaic and wind production power profiles are shown in Fig. 5 [14].

The average daily consumption for the community of the microgrid is shown in Fig. 6.

4 Energy Storage System Modeling

The development of microgrids with an energy storage system (ESS) has been a subject of considerable research in recent years [22]. To ensure reliable, resilient, and cost-effective operation of the microgrid, the ESS must have a proper model

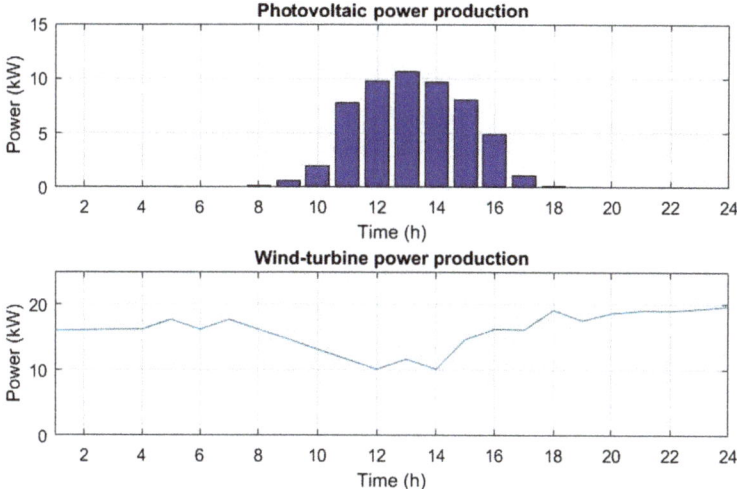

Fig. 5 Photovoltaic and wind energy production profile

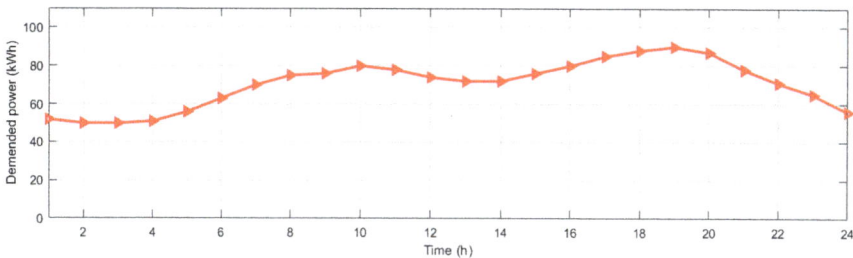

Fig. 6 Daily profile of the microgrid demand

with a correct type choice. Several types of energy storage systems can be used in a microgrid system, and each storage type has different characteristics, including response times, storage capacities and peak current capacities, which are addressed at different applications and different time scales [23].

In the literature, electrochemical batteries have shown the best performance in microgrid systems as well as their ability to store electrical energy for a long period of time [24]. Within this context, an ESS composed of electrochemical batteries is introduced in this study, and a complete mathematical model is used to imitate the states of charge and discharge of the ESS.

Several factors are necessary to describe the battery behavior, such as capacity and charge/discharge rate [25]. To increase the lifespan of the battery energy system (BES), deep discharges must be avoided, considering that $E(t)$ represents the battery stored energy at time t, the energy flows entering (Charging mode) or exiting (discharging mode) from the battery at each time step t are computed as follows:

$$\begin{cases} E(t+1) = E(t) - \Delta_t P_c(t)\eta_c, & \text{charging mode,} \\ E(t+1) = E(t) - \frac{\Delta_t P_d(t)}{\eta_d}, & \text{discharging mode,} \end{cases} \quad (2)$$

where $P_c(t)$ and $P_d(t)$ are the charging and discharging powers of the battery at time t; Δ_t is the interval of time considered, and finally, η_c and η_d are the charging and discharging efficiency.

For the reliable operation, battery must remain within the limits of its capacity and its charging/ discharging is limited by a maximum rate that must not be exceeded

$$E^{\min}(t) \leq E(t) \leq E^{\max}(t) \quad (3)$$

$$\begin{cases} P_c(t)\eta_c \leq P_c^{\max} & \text{charging mode,} \quad P_c(t) < 0 \\ \frac{P_d(t)}{\eta_d} \leq P_d^{\max} & \text{discharging mode,} \quad P_d(t) > 0 \end{cases} \quad (4)$$

where $E^{\min}(t)$ and $E^{\max}(t)$ are the minimum and maximum energy levels of the battery, respectively, and $P_c^{\max}$ and $P_d^{\max}$ are the maximum rates of charge/discharge of the battery that must be respected in each operation.

5 Energy Management System Operation

In this section, the optimization model of the energy management system adopted for the proposed microgrid will be presented. The state variables to be optimized in this case are the output powers of the different generators, the storage system and the main grid. The goal is to determine the power set-points of all microgrid generators by formulating the management problem as an objective function to be optimized. Indeed, five optimization methods are proposed in this study including linear programming (LP) based on the simplex method, two particle swarm optimization (PSO) algorithms, a genetic algorithm (GA), and a hybrid (LP-PSO) algorithm. Besides, greenhouse gas emissions (GHG) released during an operational day will be evaluated through an environmental function.

The optimization model used in the energy management system is shown in Fig. 7.

The purpose of the microgrid operator is to manage the system in order to find the optimal daily profiles for each source of the microgrid that will allow us to obtain the lowest possible daily energy price, and the management will be based mainly on three essential factors:

1. The nominal hourly power $P_x(t)$ available in each source x (renewable or conventional) in each hour t.
2. The hourly energy unit price $B_x(t)$ for each generator of the microgrid system.
3. The state of charge $\text{SOC}(t)$ of the energy storage system.

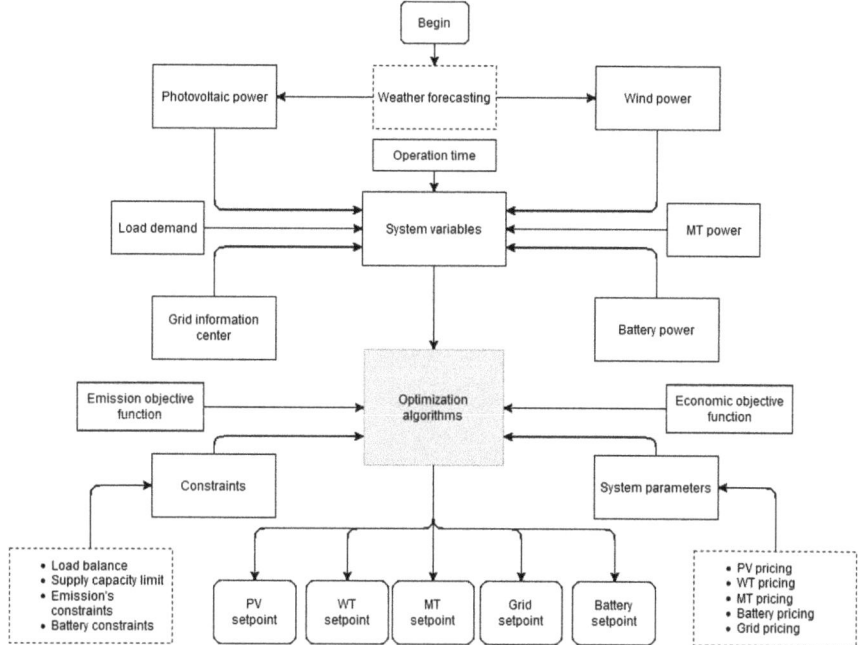

Fig. 7 Microgrid optimization model

The energy management system (EMS) problem intent to find the optimal set-points of the distributed generators, the storage system, and the amount of energy exchanged with the power grid taking into account the economic and environmental constraints.

5.1 Problem Formulation

Energy management in the microgrid system is formulated as an optimization problem based on economic and environmental objective functions as described as follows.

Energy Price Evaluation. The choice of the cost function is the most relevant issue for the optimization problem. It depends on several parameters mainly the type of architecture of the microgrid. Several functions have already been used; in [14], the cost of exploitation from the distributed resources and the storage system was considered constant during the day, and the buying/selling price of the main network was different. In [26–28], the cost of the distributed resources and the storage system were considered dynamic throughout the day, and also the cost of selling / buying energy supplied by the grid or injected varies during the day. In this case, the main

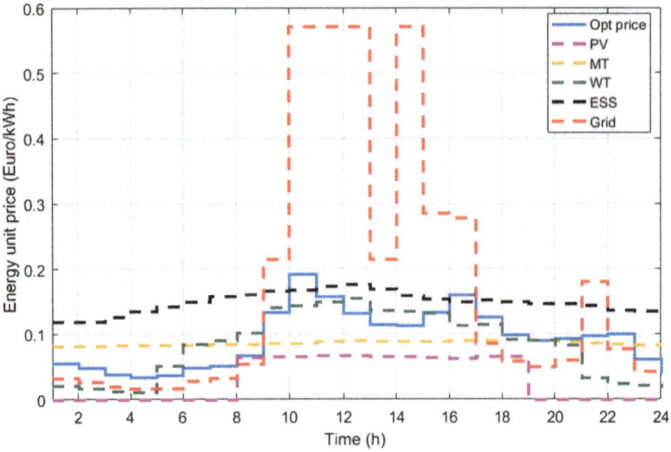

Fig. 8 Unit energy prices of the MG generators and the main grid

objective of the cost function is to satisfy the demand of load during the day in a most economical way. So, in each hour t the cost function $(C(t))$ can be calculated as:

$$C(t) = \sum_{i=1}^{N_g} P_{DGi}(t) B_{DGi}(t) + \sum_{j=1}^{N_s} P_{SDj}(t) B_{SDj}(t) + P_g(t) B_g(t) \qquad (5)$$

where N_g and N_s are the total number of generators and storage devices, respectively. The $B_{DGi}(t)$ and $B_{SDj}(t)$ represent the bids of ith DG unit and jth storage device at hour t. $P_g(t)$ is the active power which is bought (sold) from (to) the utility at hour t and $B_g(t)$ is the bid of utility at hour t as shown in Fig. 8

Emissions Evaluation. In addition to the operating cost, the aspect of greenhouse gas emissions is also taken into consideration. The emission objective function consists of the atmospheric pollutants such as nitrogen oxides NO_X, sulfur dioxide SO_2, and carbon dioxide CO_2. The mathematical formulation of total pollutant emission in kg can be expressed as:

$$EM(t) = \sum_{i=1}^{N_g} P_{DGi}(t) EF_{DGi}(t) + P_g(t) EF_g(t) \qquad (6)$$

where $EF_{DGi}(t)$ and $EF_g(t)$ are GHG emission factors which described the amount of pollutants emission in kg/MWh for each generator and main grid at hour t, respectively. Table 2 presents the emission factors for non-renewable sources as defined in [14].

Table 2 Emission factors

EF	Micro-turbine (kg/MWh)	Grid (kg/MWh)
CO_2	724	922
NO_X	0.2	2.295
SO_2	0.00136	3.583

The energy management optimization problem can be defined as follows:

$$\min_{(P_{DGi}, P_{SDj}, P_g)} \sum_{i=1}^{N_g} P_{DGi}(t) B_{DGi}(t) + \sum_{j=1}^{N_s} P_{SDj}(t) B_{SDj}(t) + P_g(t) B_g(t) \qquad (7)$$

$$\text{s.t.} \quad \sum_{i=1}^{N_g} P_{DGi}(t) + \sum_{j=1}^{N_s} P_{SDj}(t) + P_g(t) = P_L(t) \qquad (8)$$

$$P_{DGi}^{\min}(t) \le P_{DGi}(t) \le P_{DGi}^{\max}(t) \quad \text{for } i = 1, \ldots, N_g, \qquad (9)$$

$$P_{SDj}^{\min}(t) \le P_{SDj}(t) \le P_{SDj}^{\max}(t) \quad \text{for } j = 1, \ldots, N_s, \qquad (10)$$

$$P_g^{\min}(t) \le P_g(t) \le P_g^{\max}(t) \qquad (11)$$

where the total price is calculated by $CT = \sum_{t=1}^{T} \min C(t)$ and the total quantity of emissions in kg can be determined by $EM = \sum_{t=1}^{T} EM(t)$. Equation (8) represents the total power generation needs to satisfy the total demand. Equations (9)–(11) are the simple bounds associated with the decision variables.

5.2 Management Operation

Several management systems have been presented in the literature, and Sedighizadeh et al. [29] have proposed a multi-objective operational strategy of a microgrid for a residential application. In this context, the economic and environmental aspects have been formulated as a multi-objective problem with nonlinear constraints. For this purpose, the terms of operating cost, maintenance cost, start-up cost, and the cost of CO_2, SO_2, NO_X emissions are taken into account. In this study, the management is developed as a uni-objective optimization problem whose main goal is to optimize the economic aspect. However, the environmental aspect will be evaluated but will not be taken into account in the optimization process. Therefore, the aim is to select the cheapest power in a given hour and to allocate it to the load, ensuring the energy balance required by the consumer while obtaining the cheapest possible daily energy bill. During this process, the storage system is managed in detail as follows:

- **In case of** $(E(t) = E^{max})$: The storage system will be considered as the main source with the four other sources (photovoltaic, wind, micro-turbine, and grid), and its energy supply will be operated according to the quantity of energy requested and its unit energy price per hour. It should be noted that the discharge rate is limited by a maximum quantity that must not be exceeded according to the constraints presented before.
- **In case of** $(E(t) = E^{min})$: The storage system will require a certain amount of energy for the charging process from the cheapest sources in the microgrid at a given time. In this situation, the storage system will be considered as a load by the microgrid. If all unit energy prices of the different generators are considerably high, and the load is satisfied, the charging process of the storage system will not happen at this time and will wait until the energy prices are sufficiently low.
- **In case of** $(E^{min} < E(t) < E^{max})$: Depending on the energy unit price of the storage system, two cases can occur:

  1. In the event that the price of the energy delivered by the storage system is the most expensive and the energy demanded by microgrid consumers can be largely satisfied by other sources, the storage system will continue to be charged, and its energy will not participate in supplying the load. But, if the energy supplied by the various generators is insufficient, the energy from the storage system will be used as a compensating energy source to satisfy the energy balance constraint.
  2. Otherwise, if the price of the energy delivered by the storage system is cheaper compared to other sources, the storage system will participate in supplying the load and provides maximum energy equal to the limit of its discharging power rate.

The objective of the management system presented in this paper is to reduce the energy bill over a 24-h day. The target point in this study case is the determination of the power set-points calculated by the five optimization methods. The remaining renewable energy not used to power the microgrid consumers and to charge the battery storage system will be sent to the main grid. In fact, we present the two scenarios proposed for this purpose:

- **Scenario 01**: The energy surplus from the different RES of the microgrid is used to cover the energy needs of the storage system while preserving the economic aspect by choosing the times when the price is the cheapest. However, if the batteries become fully charged, the energy surplus will be considered as energy loses. During this management, we will take into account the optimal price retained from the optimization as well as the rate of GHG emissions resulting from the energy operations performed by the microgrid.
- **Scenario 02**: The energy from the different renewable energy resources is used to cover the energy needs of the storage system in order to charge it while preserving the economic aspect by choosing the times when the energy prices are relatively low. However, if the batteries prove to be fully charged, the energy surplus from

renewable sources in this case will be distributed and sold to the grid with the same purchase energy prices. During this management, we will evaluate the optimal price retained from the optimization procedures as well as the rate of GHG emissions resulting from the energy operations achieved by the microgrid. In addition, the power of the renewable energy generators (photovoltaic and wind) in this case is fully exploited, in order to highlight the impact of the energy injection to the main grid and its economic-environmental consequences.

6 Results and Discussions

6.1 Wind Forecasting Results

The multilayer neural network proposed is being trained with predefined function "nntool" in MATLAB. The feed-forward network based on a back-propagation algorithm ensures the weights adjustment which is identified in the offline training. Table 3 gives the characteristics of the network.

Figure 9 represents the mean squared error, and the best MSE obtained is 0.48 in the ninth epoch.

To evaluate the reliability of the prediction model proposed in this paper, Fig. 10 shows a comparison of the results obtained by the wind forecasting model based on the artificial neural network and the real wind speed results. According to this latter, the prediction speed follows the real speed; on the other hand, some deviation occurs between values due to the stochastic character of the problem under study which has already been deduced from the MSE value.

Table 3 ANN characteristics

Parameter	Characteristic
Back-propagation strategy	Levenberg–Marquardt
Network type	Feed-forward
Number of hidden neurons	10
Performance evaluation	Mean squared error
Number of samples	103,104
Training samples	90 %
Testing samples	5 %
Validation samples	5 %

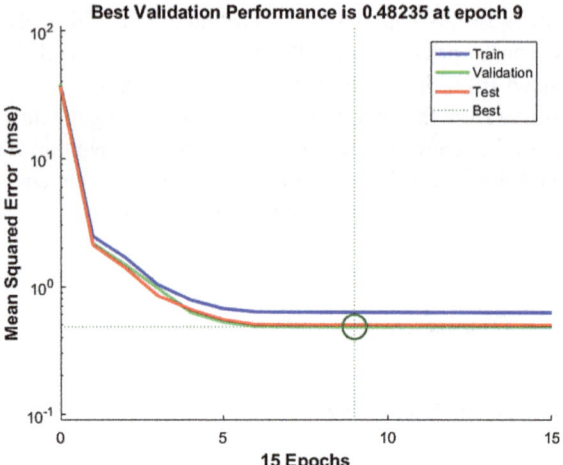

Fig. 9 Mean squared error of the network

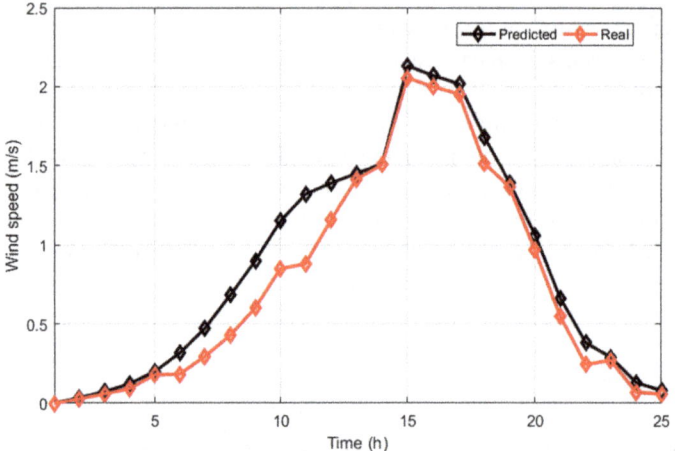

Fig. 10 Hourly comparison of predicted wind velocity with real data

6.2 Optimization Methods Comparison

The objective of the energy management system is to reduce the energy microgrid consumer bill over a 24-h day. The target point in this section is the determination of the power set-points calculated by the energy management system based on optimization algorithms.

The optimization problem is represented by a linear objective function and constraints; for its treatment, five optimization methods was applied, namely the linear programming LP based on the simplex method, two variants of particle swarm opti-

Table 4 Algorithmic performances

Results	LP	PSO1	PSO2	GA	LP-PSO
Total price (Euro)	143.0492	144.9574	143.0492	143.0492	143.0492
Total emissions (kg)	1353.7329	1351.4000	1353.7329	1353.7329	1353.7329
Simulation time (s)	3.120	0.040	0.031	5.620	0.038

mization PSO algorithm with different starting conditions, the first noted PSO1, whose particle starting point represents a random value that translates between the problem bounds while in the second one noted PSO2 a new approach of particle initialization has been proposed by fixing the particle starting point using an upper bounds vector of the problem. The fourth method used for the treatment of the problem is a hybrid LP-PSO, and it is an innovative optimization strategy whose goal is to improve the performance of the PSO for optimal treatment of the management problem characterized by a linear optimization function. The approach adopted in this method lies on the use of linear programming as a technique for generating the initial starting points of the swarm particles, and the PSO continues with those particles the search for the optimum to deliver the optimal set-points to ensure the minimization of the energy price evaluation function. And finally, a genetic algorithm GA was used in the management system to be compared to the four methods explained above. The performances of each method are presented in Table 4.

Taking into account the available power shown in Fig. 5 as well as the microgrid energy unit prices shown in Fig. 8, the EMS allows to have the optimal set-points of the distributed generators and the storage system through one of the optimization algorithms LP, PSO1, PSO2, GA, and LP-PSO as shown in Table 4.

According to the results presented in Table 4, the operating cost for LP, PSO2, GA, and the hybrid LP-PSO is 143.0492 Euro, and on the other side for PSO1 was 144.9574 Euro.

A comparison is made between the performances of the optimization methods used to solve the energy management problem. According to Tables 5 and 6, it is noted that in the five programs the cheapest source at a given hour has the most important set-point without exceeding the power limits as shown in Table 1.

The optimal power set-points of the LP, PSO2, LP-PSO, and GA showed in Table 5 converged to the global optimum, opposite to PSO1 where the set-points showed in Table 6 do not represent the global optimum (convergence to local optimality); due to the nature of the optimization problem, this convergence with a certain error influenced the total daily energy price.

The linear nature of the optimization problem judges the reliability of linear programming LP based on the simplex method. The adoption of the LP-PSO method also delivered optimal results but in terms of convergence rapidity, it was not the best method; in fact, the PSO2 has demonstrated the best performances compared to the

Table 5 Optimal power set-points using LP and PSO2 and GA and LP-PSO method

Time (h)	PV (kWh)	WT (kWh)	MT (kWh)	Battery (kWh)	GRID (kWh)	Load (kWh)
01:00	0	16.0133	6	−33.3333	63.32	52
02:00	0	16.08	6	−33.3333	61.2533	50
03:00	0	16.16	6	−33.3333	61.1733	50
04:00	0	16.1733	6	−33.3333	62.16	50
05:00	0	0	6	−8.8889	58.8889	51
06:00	0	0	6	0	57	63
07:00	0	0	6	0	64	70
08:00	0	0	6	0	69	75
09:00	0.59	14.7333	30	22.5	8.1767	76
10:00	1.9800	13.16	30	22.5	12.36	80
11:00	7.7500	11.6667	30	22.5	6.0833	78
12:00	9.8	10.1468	30	22.5	1.5532	74
13:00	10.65	11.6667	30	19.6833	0	72
14:00	9.7	10.146	30	22.1540	0	72
15:00	8.12	14.6467	30	12.1627	11.0706	76
16:00	4.9500	16.2133	30	0	28.8367	80
17:00	1.1	0	27.2333	−33.3333	90	85
18:00	0.1	1.2333	30	−33.3333	90	88
19:00	0	3.3333	30	−33.3333	90	90
20:00	0	18.6493	11.6840	−33.3333	90	87
21:00	0	19.04	30	22.5	6.46	78
22:00	0	19.03	6	−33.3333	79.3033	71
23:00	0	19.3330	6	−33.3333	73.0003	65
24:00	0	19.6900	6	−5.5556	35.8656	56

four other optimization methods in terms of accuracy and rapidity convergence of optimal power set-points.

The performances of the genetic algorithm GA have given good results in terms of precision, knowing that the stopping condition is taken similar to that of the PSO which is the reaching of the global optimum and with the same starting condition; the path to the optimum by the genetic algorithm is much slower than that of the PSO, this is judged by the very large search space generated by the GA mechanism following their genetic operators like crossover and mutation, in PSO particles update themselves with the internal velocity. Besides, the information-sharing mechanism in PSO is significantly different than the genetic algorithm.

Table 6 Optimal power set-point using PSO1 method

Time (h)	PV (kWh)	WT (kWh)	MT (kWh)	Battery (kWh)	GRID (kWh)	Load (kWh)
01:00	0	16.0133	6	−33.3333	63.32	52
02:00	0	16.08	6	−33.3333	61.25333	50
03:00	0	16.16	6	−33.3333	61.1733	50
04:00	0	16.1733	6	−33.3333	62.16	50
05:00	0	0	6	−8.8889	58.8889	51
06:00	0	0	6	0	57	63
07:00	0	0	6	0	64	70
08:00	0	0	6	0	69	75
09:00	0.3576	14.6848	29.9823	22.4610	8.5142	76
10:00	1.9148	12.5053	29.9690	22.4348	13.1761	80
11:00	7.7358	11.64	29.9995	22.4732	6.1515	78
12:00	9.7986	10.0716	29.9870	22.4306	1.7122	74
13:00	10.6289	11.6662	29.9588	22.4320	3.3142	72
14:00	8.0985	9.8576	29.9923	22.4730	7.5786	72
15:00	7.9557	14.5146	29.9509	9.1309	14.4480	76
16:00	4.9077	16.2132	29.9986	0.1489	34.7316	80
17:00	1.0035	3.3301	29.9999	−33.3333	89.9999	85
18:00	0.0358	2.5170	28.7815	−33.3333	89.9990	88
19:00	0	3.3652	29.9682	−33.3333	90	90
20:00	0	18.6418	11.6917	−33.3333	89.9999	87
21:00	0	18.9552	29.9523	22.4275	6.6650	78
22:00	0	19.03	6	−33.3333	79.3033	71
23:00	0	19.3330	6	−33.3333	73.0003	65
24:00	0	19.6900	6	−5.4466	35.7566	56

6.3 Comparison of the Two Scenarios

Following the algorithmic performances illustrated by the PSO2, this latter will be used as an optimization tool in the energy management system (EMS) for the two scenarios described above. The rest of the paper presents the economic and environmental results of the two proposed scenarios.

Scenario 1. The results obtained in Fig. 11 are the optimal power set-points for the different energy sources of the microgrid, and the sum of these values in a given hour t is equal to the power value of the load for the same hour t. The cheapest source in a given hour has the highest set-point without exceeding these power limits. The second cheapest source is added to it and so on until the power balance constraint is verified. In this way, the operating cost is minimized and the emissions are evaluated.

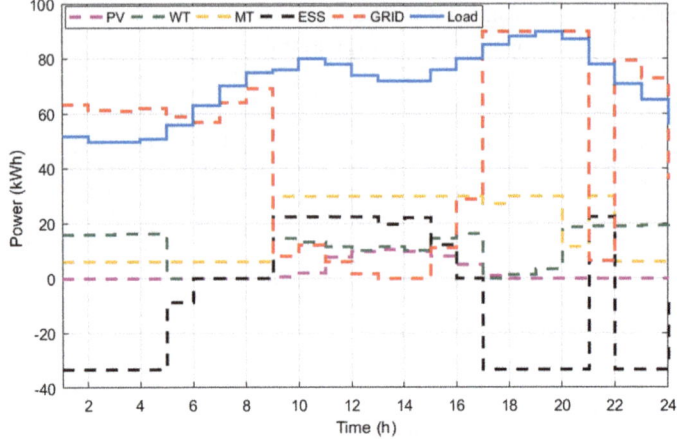

Fig. 11 Optimal power set-points obtained in the first scenario

The charging of the storage system is ensured during the part of the day when consumption is low and characterized by low energy unit costs. Otherwise, the battery provides energy to compensate the deficit during the day. In this study case, the energy from the grid is supplied unidirectionally; i.e., the energy is only sold from the grid and delivered to the microgrid, reverse operation is not allowed. For maintenance and safety reasons, the micro-turbine is present all day long either by its minimum power of 6 kW or by its delivered power to compensate the energy deficit that should be supplied to the microgrid consumers.

Figure 12 presents the hourly unit prices of the optimal energy flows from the various sources and the optimal price obtained by the energy management system in function of the operating hours during the day.

It is remarkable that the photovoltaic source is fully exploited during the day because of its low price compared to the other four sources and the wind source is widely exploited during the night because of its low price as well. However, during peak hours, the grid price is very high; in this case, the use of the storage system allows to compensate the energy deficit and reduce the dependence on the main grid. This demonstrates the importance of the battery during the day when the grid price is high. The storage system itself follows its charging process during the night when the consumption of the microgrid is smaller and the energy unit price is low. Figure 13 shows the daily energy exchange of the batteries with the microgrid. It is well-observed that when batteries demand energy, the SOC increases, and when they supply energy, the SOC decreases.

The emissions quantity is directly related to the two sources: the main grid and the micro-turbine, which are responsible for greenhouse gas emissions. According to Fig. 14, it is clear that emissions are higher during the night due to the reduced unit prices of the grid, and thus, the primary operation of the microgrid is to supply consumer and take advantage to charge the ESS system.

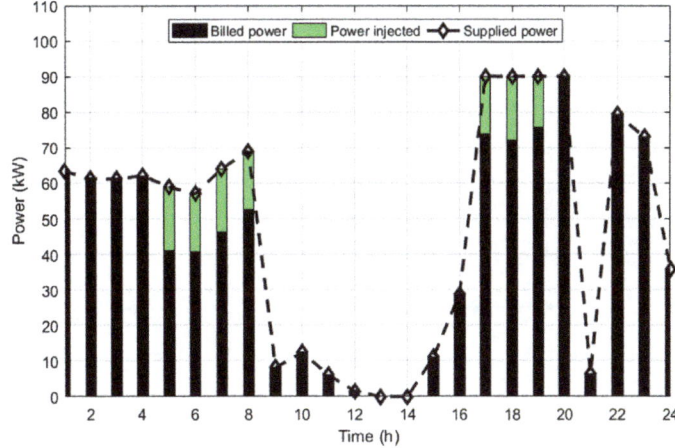

Fig. 12 Unit prices of the powers resulting from the optimal management and the optimal billing prices for the first scenario

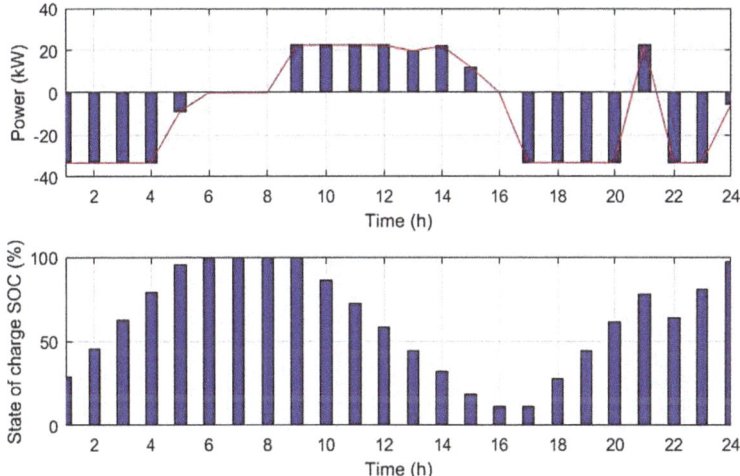

Fig. 13 Energy exchange of the batteries with the microgrid during the day

Scenario 2. According to the results obtained in Table 7 and Fig. 15, it can be seen that the renewable sources are fully exploited, and no loss of power is caused. The excess energy, after satisfying the local needs of the microgrid, allowed the successful charging of the storage system in such a way that at the end of the day the battery was fully charged. In addition, an amount of 116,0529 kW was also delivered to the main grid, which reduced the total daily energy bill of the microgrid to 137.6627 Euro, and reduced GHG emissions to 1246.1 kg.

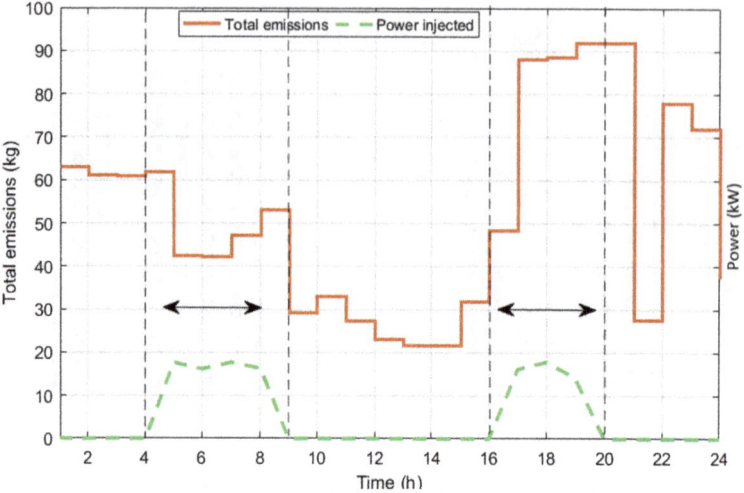

Fig. 14 Total daily emissions due to the use of fossil sources in the microgrid without injection

Table 7 Results of both scenarios

Scenarios	Scenario 01	Scenario 02
Total price (Euro)	143.0492	137.6627
Total Emissions (kg)	1353.7329	1246.1000

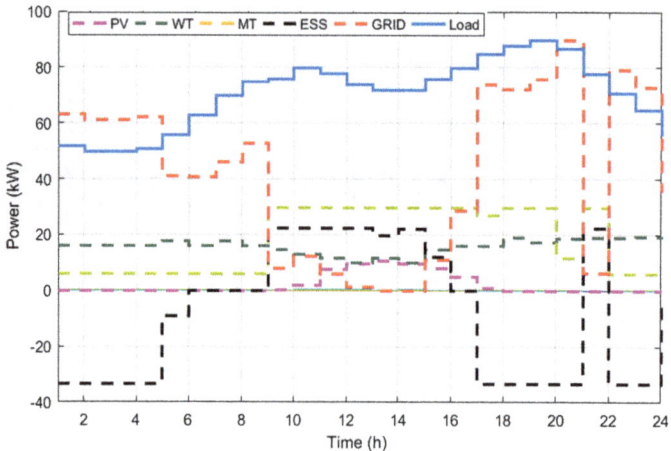

Fig. 15 Optimal power set-points obtained in the second scenario

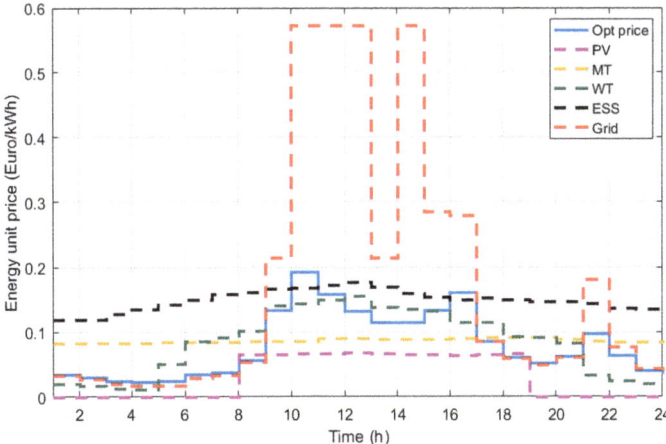

Fig. 16 Unit prices of the powers resulting from the optimal management and the optimal billing prices

It is remarkable that the photovoltaic and wind energy sources are fully exploited during the day, in order to take advantage of the benefits of injecting green energy into the main grid, thus reducing the energy bill and the rate of GHG emissions. The power management of the other sources seems identical to the first scenario, except for the main grid that is modified due to its price, which remains very high during the day compared to the micro-turbine and the storage system.

The grid provides precise power to meet the load requirements. However, this energy is not fully counted in the energy bill, and the power injected during a given hour is subtracted and compensates for the energy that is supposed to be supplied by the network. In this way, during consumption billing, only the power paid will be considered. Figure 16 shows the optimal daily energy price of the microgrid obtained from its sources management .

Figure 17 shows the power supplied by the grid, the power injected, as well as the power taken into account in the billing.

The quantity of emissions is directly related to the two sources: the main grid and the micro-turbine, which are responsible for greenhouse gas emissions. According to Fig. 18, it is clear that emissions are higher during the night due to the reduced unit prices of the grid and thus the primary operation of the microgrid, taking advantage of this to recharge the storage system. The emission rate, in this case, remains lower than the first scenario.

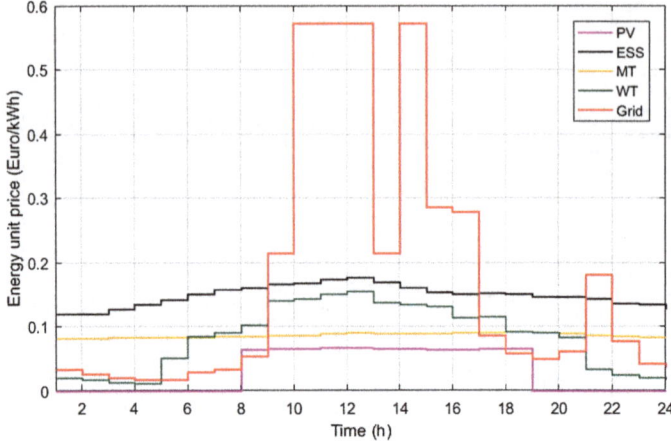

Fig. 17 Unit prices of the powers resulting from the optimal management and the optimal billing prices for the second scenario

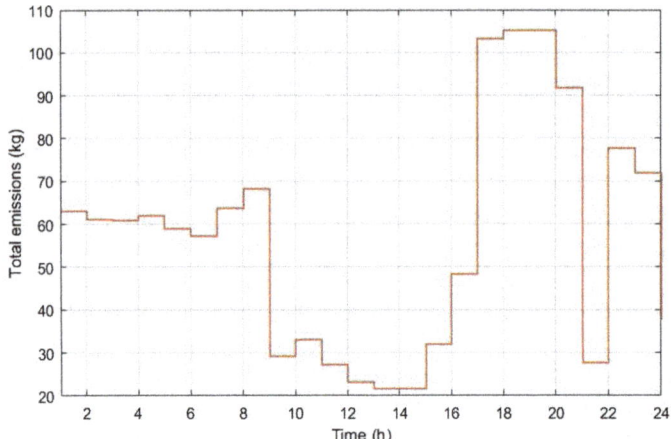

Fig. 18 Total daily emissions due to the use of fossil fuel sources in the microgrid with injection

7   Conclusion and Future Works

This study investigated the problem of energy management in microgrid systems by considering the impact of the wind speed intermittent aspect on wind turbine power production. For that matter, a prediction model based on the artificial intelligence of neural network (ANN) has been developed to ensure a forecast of the wind velocity parameter, and the performance of the model was evaluated by the mean squared error (MSE) value. On the other hand, this work showed a comparison between several optimization methods used by the energy management system (EMS) proposed for the optimal dispatch of energy inside a microgrid, ensuring a reduced energy cost. In particular, five optimization approaches were proposed, including two versions of particle swarm optimization (PSO) algorithm, a genetic algorithm (GA), linear programming (LP) based on the simplex method, and finally a hybrid approach (LP-PSO), all programmed in the MATLAB software. However, the proposed PSO has shown a high level of performance. Two scenarios were adopted to assess the technical-economic and environmental impact of bidirectional interconnection between the microgrid and the main grid. In fact, the low energy price and the reduced rate of emissions have made it possible to present one of the important advantages that a microgrid could bring in the reduction of the energetic cost as well as in the contribution to the reduction of the greenhouse gases (GHG) emissions responsible for the global warming. Differently to the uni-objective approach that gave an optimal point, a multi-objective optimization approach will be developed as future work on which an energy management system is dedicated to ensuring the optimal scheduling of the distributed generators and the energy storage system accompanied by a moderate exchange between the MG and the main grid while considering the simultaneous optimization of both economic and environmental criteria. The results will deliver a set of optimal solutions (Pareto front), that will represent scenarios, in which the best trade-off between price and emission is selected by the microgrid operator to give the optimal scheduling.

References

1. Zia MF, Elbouchikhi E, Benbouzid M (2018) Microgrids energy management systems: a critical review on methods, solutions, and prospects. Appl Energy 222:1033–1055. https://doi.org/10.1016/j.apenergy.2018.04.103
2. Vlachogiannis J (2005) Probabilistic constrained load flow considering integration of wind power generation and electric vehicles. IEEE Trans Power Syst 24:1808–1817. https://doi.org/10.1109/tpwrs.2009.2030420
3. Hua H, Qin Y, Hao C, Cao J (2019) Optimal energy management strategies for energy internet via deep reinforcement learning approach. Appl Energy 239:598–609. https://doi.org/10.1016/j.apenergy.2019.01.145
4. Quoc-Tuan A, Tung LN (2016) Optimal energy management strategies of microgrids. In: IEEE symposium series on computational intelligence (2016). https://doi.org/10.1109/ssci.2016.7849851
5. Zhang X, Sharma R, He Y (2012) Optimal energy management of a rural microgrid system using multi-objective optimization. IEEE PES Innov Smart Grid Technol (ISGT) 1–8. https://doi.org/10.1109/ISGT.2012.6175655
6. Aghajani G, Ghadimi N (2018) Multi-objective energy management in a microgrid. Energy Rep 218–225. https://doi.org/10.1016/j.egyr.2017.10.002
7. Tsikalakis AG, Hatziargyriou ND (2011) Centralized control for optimizing microgrids operation. IEEE Power Energy Soc Gen Meet 1–8. https://doi.org/10.1109/pes.2011.6039737
8. Shi W, Xie X, Chu C-C, Gadh R (2014) Distributed optimal energy management in microgrids. IEEE Trans Smart Grid 1137–1146. https://doi.org/10.1109/tsg.2014.2373150
9. Ouammi A, Achour Y, Zejli D, Dagdougui H (2019) Supervisory model predictive control for optimal energy management of networked smart greenhouses integrated microgrid. IEEE Trans Autom Sci Eng 117–128. https://doi.org/10.1109/tase.2019.2910756

10. Zhang F, Cho K, Choi J, Lee Y-M, Lee KY (2011) A study on wind speed prediction using artificial neural network at Jeju Island in Korea II. In: IEEE 54 the international midwest symposium on circuits and systems(MWSCAS), pp 1–4 (2011). https://doi.org/10.1109/mwscas. 2011.6026399

11. Dolara A, Grimaccia F, Magistrati G, Marchegiani G (2017) Optimization models for islanded microgrids: a comparative analysis between linear programming and mixed integer programming. Energies 241. https://doi.org/10.3390/en10020241

12. Shayeghi H, Shahryari E, Moradzadeh M, Siano P (2019) A survey on microgrid energy management considering flexible energy sources. Energies 1–26. https://doi.org/10.3390/ en12112156

13. Adnan R, Ruslan FA, Zain ZM (2013) New artificial neural network and extended Kalman filter hybrid model of flood prediction system. In: IEEE 9th international colloquium on signal processing and its applications 252–257. https://doi.org/10.1109/cspa.2013.6530051

14. Motevasel M, Seif AR (2014) Expert energy management of a micro-grid considering wind energy uncertainty. Energy Convers Manag 58–72: https://doi.org/10.1016/j.enconman.2014. 03.022

15. Filik UB, Filik T (2017) Wind speed prediction using artificial neural networks based on multiple local measurements in Eskisehir. Energy Proc 264–269. https://doi.org/10.1016/j. egypro.2016.12.147

16. Kumar Y, Ringenberg J, Depuru SS, Devabhaktuni VK, Lee JW, Nikolaidis E, Andersen B, Afjeh A (2016) Wind energy trends and enabling technologies. Renew Sustain Energy Rev 209–224. https://doi.org/10.1016/j.rser.2015.07.200

17. Faisal M, Hannan MA, Ker PJ, Hussain A, Mansor MB, Blaabjerg F (2018) Review of energy storage system technologies in microgrid applications. IEEE Issues Challeng 35143–35164. https://doi.org/10.1109/access.2018.2841407

18. Tan X, Li Q, Wang H (2013) Advances and trends of energy storage technology in microgrid. Int J Electr Power Energy Syst 179–191. https://doi.org/10.1016/j.ijepes.2012.07.015

19. Chemali E, Preindl M, Malysz P, Emadi A (2016) Electrochemical and electrostatic energy storage and management systems for electric drive vehicles State of the art review and future trends. IEEE J Emerg Sel Top Power Electron 1117–1134. https://doi.org/10.1109/jestpe.2016. 2566583

20. Tian R, Park S-H, King PJ, Cunningham G, Coelho J, Nicolosi V, Coleman JN (2019) Quantifying the factors limiting rate performance in battery electrodes. Nat Commun 1–11. https:// doi.org/10.1038/s41467-019-09792-9

21. Zhang D, Evangelisti S, Lettieri P, Papageorgiou LG (2016) Economic and environmental scheduling of smart homes with microgrid Distribured energy resources operation and electrical tasks. Energy Convers Manag 113–124. https://doi.org/10.1016/j.enconman.2015.11.056

22. Elsied M, Oukaour A, Gualous H, Hassan R (2014) Energy management and optimization of the multisource system based on the genetic algorithm. In: IEEE 23rd international symposium on industrial electronics (ISIE). https://doi.org/10.1109/isie.2014.6865020

23. Min W, Sheng C, Ming-Tang T (2015) Energy management strategy for microgrids by using enhanced bee colony optimization. Energies 9:1–16. https://doi.org/10.3390/en9010005

24. Dukpa A, Duggal I, Venkatesh B, Chang L (2010) Optimal participation and risk mitigation of wind generators in an electricity market. IET Renew Power Gen 165–175. https://doi.org/10. 1049/iet-rpg.2009.0016

25. Hartono B, Setiabudy R (2013) Review of microgrid technology. In: IEEE international conference on QiR, pp 127–132. https://doi.org/10.1109/qir.2013.6632550

26. Abdelfettah K, Boukli-Hacene F, Mourad KA (2020) Particle swarm optimization for microgrid power management and load scheduling. Int J Energy Econ Policy 71–80. https://doi.org/ 10.32479/ijeep.8568

27. Chakraborty S, Weiss MD, Simoes MG (2007) Distributed intelligent energy management system for a single-phase high-frequency AC microgrid. IEEE Trans Ind Electron 97–109. https://doi.org/10.1109/TIE.2006.888766

28. Waseem M, Wang J, Xiong L, Huang S (2020) Architecture of a microgrid and optimal energy management system. Multi Agent Syst Strateg Appl. https://doi.org/10.5772/intechopen.88472
29. Sedighizadeh M, Esmaili M, Mohammadkhani N (2018) Stochastic multi-objective energy management in residential microgrids with combined cooling, heating, and power units considering battery energy storage systems and plug-in hybrid electric vehicles. J Clean Product 301–317. https://doi.org/10.1016/j.jclepro.2018.05.103

Chapter 11
Certain Expansion Formulae for Incomplete I-Functions and $\bar{I}$-Functions Involving Bessel Function

Kamlesh Jangid, Mudit Mathur, Sunil Dutt Purohit, and Daya Lal Suthar

1 Introduction and Definitions

Rathie [6] defined I-function in the prescribed manner:

$$
\begin{aligned}
I_{p,q}^{m,n}[z] = I_{p,\,q}^{m,\,n}\left[z \,\middle|\, \begin{array}{l} (e_1, \nu_1; \chi_1), \ldots, (e_p, \nu_p; \chi_p) \\ (f_1, \omega_1; \varpi_1), \ldots, (f_p, \omega_p; \varpi_p) \end{array}\right] \\
= \frac{1}{2\pi i} \int_L \psi(s)\, z^s \,\mathrm{d}s,
\end{aligned}
\tag{1}
$$

where

$$
\psi(s) = \frac{\prod_{j=1}^{m}[\Gamma(f_j - \omega_j s)]^{\varpi_j} \prod_{j=1}^{n}[\Gamma(1 - e_j + \nu_j s)]^{\chi_j}}{\prod_{j=m+1}^{q}[\Gamma(1 - f_j + \omega_j s)]^{\varpi_j} \prod_{j=n+1}^{p}[\Gamma(e_j - \nu_j s)]^{\chi_j}}
\tag{2}
$$

and $m, n, p, q \in N_0$ with $0 \leq n \leq p$, $0 \leq m \leq q$, $\nu_j, \chi_j (j = 1, \ldots, p)$, $\omega_j, \varpi_j (j = 1, \ldots, q) \in R^+$, $e_j, f_j \in C$. The suitable conditions for the L contour convergence depicted in (1) and portrayals just as the I-function information can be found in [6].

Next, we characterized the recognizable incomplete gamma functions; lower gamma function $\gamma(s, x)$ and upper gamma function $\Gamma(s, x)$, separately as (see [1]):

K. Jangid · S. D. Purohit (✉)
Department of HEAS (Mathematics), Rajasthan Technical University Kota, Kota, India

M. Mathur
RTU Campus, B-9, Staff Colony, Kota, India

D. L. Suthar
Department of Mathematics, Wollo University, P.O. Box 1145 Dessie, Ethiopia

© The Author(s), under exclusive license to Springer Nature Singapore Pte Ltd. 2021 143
S. Kumar et al. (eds.), *Proceedings of International Conference on Communication and Computational Technologies*, Algorithms for Intelligent Systems,
https://doi.org/10.1007/978-981-16-3246-4_11

$$\gamma(s, x) := \int_0^x y^{s-1} e^{-y} \, dy \quad (\Re(s) > 0; \ x \geq 0) \tag{3}$$

and

$$\Gamma(s, x) := \int_x^\infty y^{s-1} e^{-y} \, dy \quad (x \geq 0; \ \Re(s) > 0 \ \text{if } x = 0). \tag{4}$$

These functions satisfies the associated decomposition relationship:

$$\gamma(s, x) + \Gamma(s, x) =: \Gamma(s) \ (\Re(s) > 0). \tag{5}$$

Jangid et al. [3] presented a group of the incomplete I-functions $^\gamma I_{p,q}^{m,n}[z]$ and $^\Gamma I_{p,q}^{m,n}[z]$ which prompts a characteristic speculation and disintegration equation for I-function:

$$
\begin{aligned}
^\gamma I_{p,q}^{m,n}[z] &= {}^\gamma I_{p,q}^{m,n} \left[z \ \middle| \ \begin{array}{l} (e_1, v_1; \chi_1 : y), (e_2, v_2; \chi_2), \ldots, (e_p, v_p; \chi_p) \\ (f_1, \omega_1; \varpi_1), (f_2, \omega_2; \varpi_2), \ldots, (f_q, \omega_q; \varpi_q) \end{array} \right] \\
&= \frac{1}{2\pi i} \int_L \phi(s, y) \, z^s \, ds
\end{aligned}
\tag{6}
$$

and

$$
\begin{aligned}
^\Gamma I_{p,q}^{m,n}[z] &= {}^\Gamma I_{p,q}^{m,n} \left[z \ \middle| \ \begin{array}{l} (e_1, v_1; \chi_1 : y), (e_2, v_2; \chi_2), \ldots, (e_p, v_p; \chi_p) \\ (f_1, \omega_1; \varpi_1), (f_2, \omega_2; \varpi_2), \ldots, (f_q, \omega_q; \varpi_q) \end{array} \right] \\
&= \frac{1}{2\pi i} \int_L \Phi(s, y) \, z^s \, ds
\end{aligned}
\tag{7}
$$

for all $z \neq 0$, where

$$\phi(s, y) = \frac{[\gamma(1 - e_1 + v_1 s, y)]^{\chi_1} \prod_{j=1}^m [\Gamma(f_j - \omega_j s)]^{\varpi_j} \prod_{j=2}^n [\Gamma(1 - e_j + v_j s)]^{\chi_j}}{\prod_{j=m+1}^q [\Gamma(1 - f_j + \omega_j s)]^{\varpi_j} \prod_{j=n+1}^p [\Gamma(e_j - v_j s)]^{\chi_j}} \tag{8}$$

and

$$\Phi(s, y) = \frac{[\Gamma(1 - e_1 + v_1 s, y)]^{\chi_1} \prod_{j=1}^m [\Gamma(f_j - \omega_j s)]^{\varpi_j} \prod_{j=2}^n [\Gamma(1 - e_j + v_j s)]^{\chi_j}}{\prod_{j=m+1}^q [\Gamma(1 - f_j + \omega_j s)]^{\varpi_j} \prod_{j=n+1}^p [\Gamma(e_j - v_j s)]^{\chi_j}}. \tag{9}$$

The definitions (6) and (7) at once yield the following division relation for $\chi_1 = 1$:

$$^\gamma I_{p,q}^{m,n}[z] + {}^\Gamma I_{p,q}^{m,n}[z] = I_{p,q}^{m,n}[z] \tag{10}$$

for the familiar I-function.

Above incomplete I-functions defined in (6) and (7) exist for all $y \geq 0$, within the parameters specified by Rathie [6]:

$$\Omega > 0, \ |\arg(z)| < \Omega \frac{\pi}{2}, \tag{11}$$

where

$$\Omega = \sum_{j=1}^{m} \varpi_j \omega_j - \sum_{j=m+1}^{q} \varpi_j \omega_j + \sum_{j=1}^{n} \chi_j \nu_j - \sum_{j=n+1}^{p} \chi_j \nu_j. \tag{12}$$

If we give the specific values to the parameters such as $\varpi_j = 1$ $(j = 1, \ldots, m)$ in (6) and (7), then we obtain the following incomplete $\overline{I}$-functions:

$$^{\gamma}\overline{I}_{p,q}^{m,n}(z) = {}^{\gamma}I_{p,q}^{m,n}\left[z \ \middle| \ \begin{matrix} (e_1, \nu_1; \chi_1, y), (e_j, \nu_j; \chi_j)_{2,p} \\ (f_j, \omega_j; 1)_{1,m}, (f_j, \omega_j, \varpi_j)_{m+1,q} \end{matrix} \right]$$

$$= \frac{1}{2\pi i} \int_L \overline{\phi}(s, y) \, z^s \, ds \tag{13}$$

and

$$^{\Gamma}\overline{I}_{p,q}^{m,n}(z) = {}^{\Gamma}I_{p,q}^{m,n}\left[z \ \middle| \ \begin{matrix} (e_1, \nu_1; \chi_1, y), (e_j, \nu_j; \chi_j)_{2,p} \\ (f_j, \omega_j; 1)_{1,m}, (f_j, \omega_j, \varpi_j)_{m+1,q} \end{matrix} \right]$$

$$= \frac{1}{2\pi i} \int_L \overline{\Phi}(s, y) \, z^s \, ds, \tag{14}$$

where

$$\overline{\phi}(s, y) = \frac{[\gamma(1 - e_1 + \nu_1 s, y)]^{\chi_1} \prod_{j=1}^{m} \Gamma(f_j - \omega_j s) \prod_{j=2}^{n} [\Gamma(1 - e_j + \nu_j s)]^{\chi_j}}{\prod_{j=m+1}^{q} [\Gamma(1 - f_j + \omega_j s)]^{\varpi_j} \prod_{j=n+1}^{p} [\Gamma(e_j - \nu_j s)]^{\chi_j}} \tag{15}$$

and

$$\overline{\Phi}(s, y) = \frac{[\Gamma(1 - e_1 + \nu_1 s, y)]^{\chi_1} \prod_{j=1}^{m} \Gamma(f_j - \omega_j s) \prod_{j=2}^{n} [\Gamma(1 - e_j + \nu_j s)]^{\chi_j}}{\prod_{j=m+1}^{q} [\Gamma(1 - f_j + \omega_j s)]^{\varpi_j} \prod_{j=n+1}^{p} [\Gamma(e_j - \nu_j s)]^{\chi_j}}. \tag{16}$$

2 The Integrals

This section derives improper integrals containing the Bessel function, incomplete I-functions, and incomplete $\overline{I}$-functions. These integrals will be used to prove the expansions for incomplete I-functions and incomplete $\overline{I}$-functions in Sect. 3.

Throughout the paper existence conditions of the incomplete I-functions and $\bar{I}$-functions are assumed to be true unless otherwise stated.

Theorem 1 *Let $\Re(\wp) > 0$, $\Omega > 0$, $h > 0$, $|arg(z)| < \frac{\pi\,\Omega}{2}$, $\Re\left(u + v + h\frac{f_i}{\omega_i}\right) > -1$, $\Re(u) < -\frac{1}{2}$, then the below integral stands for $y \geq 0$*

$$\int_0^\infty e^{ix} x^u J_v(x) \, {}^\Gamma I_{p,\,q}^{m,\,n}\left[z\,x^h \,\middle|\, \begin{matrix} (e_1, \nu_1; \chi_1 : y), (e_j, \nu_j; \chi_j)_{2,p} \\ (f_j, \omega_j; \varpi_j)_{1,q} \end{matrix}\right] dx$$

$$= \frac{e^{\frac{1}{2}(u+v+1)i\pi}}{2^{u+1}\,\Gamma(\frac{1}{2})} \, {}^\Gamma I_{p+2,\,q+1}^{m+1,\,n+1}\left[z\left(\frac{e^{\frac{i\pi}{2}}}{2}\right)^h \,\middle|\, \begin{matrix} (e_1, \nu_1; \chi_1 : y), (-u - v, h; 1), \\ (-u - \frac{1}{2}, h; 1), \end{matrix}\right.$$

$$\left. \begin{matrix} , (e_j, \nu_j; \chi_j)_{2,p}, (v - u, h; 1) \\ , (f_j, \omega_j; \varpi_j)_{1,q} \end{matrix}\right]. \quad (17)$$

Proof We start with the left-hand side to prove the result. Using definition (7), we obtain

$$\text{LHS} = \frac{1}{2\pi i} \int_0^\infty e^{ix} x^u J_v(x) \int_L \Phi(\wp, y) \, (z\,x^h)^\wp \, d\wp \, dx, \quad (18)$$

change the order of the integrations, we obtain

$$\text{LHS} = \frac{1}{2\pi i} \int_L \Phi(\wp, y) \, z^\wp \int_0^\infty e^{ix} x^{u+h\wp} J_v(x) \, dx \, d\wp. \quad (19)$$

To calculate the above integral make use of [4, p. 106, (1)], viz

$$\int_0^\infty e^{ix} \, x^u \, J_v(x) \, dx = \frac{e^{\frac{1}{2}(u+v+1)i\pi} \Gamma(u + v + 1)\Gamma(-u - \frac{1}{2})}{2^{u+1}\Gamma(\frac{1}{2})\Gamma(v - u)}, \quad (20)$$

$$\Re(u + v) < -1, \quad \Re(u) < -\frac{1}{2}$$

We obtain,

$$\text{LHS}$$

$$= \frac{1}{2\pi i} \int_L \Phi(\wp, y) \frac{e^{\frac{1}{2}(u+v+1+h\wp)i\pi} \Gamma(u + v + 1 + h\wp)\Gamma(-u - \frac{1}{2} - h\wp)}{2^{u+1+h\wp}\Gamma(\frac{1}{2})\Gamma(v - u - h\wp)} z^\wp \, d\wp, \quad (21)$$

using (7), we get the desired RHS of (17).

Theorem 2 *Let* $\Re(\wp) > 0$, $\Omega > 0$, $h > 0$, $|arg(z)| < \frac{\pi\,\Omega}{2}$, $\Re\left(u + v + h\frac{f_i}{\omega_i}\right) > -1$, $\Re(u) < -\frac{1}{2}$, *then the below integral stands for* $y \geq 0$

$$
\int_0^\infty e^{ix} x^u J_v(x) \, {}^Y I_{p,\,q}^{m,\,n}\left[z\,x^h \,\middle|\, \begin{array}{c} (e_1, v_1; \chi_1 : y), (e_j, v_j; \chi_j)_{2,p} \\ (f_j, \omega_j; \varpi_j)_{1,q} \end{array}\right] dx
$$

$$
= \frac{e^{\frac{1}{2}(u+v+1)i\pi}}{2^{u+1}\,\Gamma(\frac{1}{2})} \, {}^Y I_{p+2,\,q+1}^{m+1,\,n+1}\left[z\left(\frac{e^{\frac{i\pi}{2}}}{2}\right)^h \,\middle|\, \begin{array}{c} (e_1, v_1; \chi_1 : y), (-u - v, h; 1), \\ (-u - \frac{1}{2}, h; 1), \end{array}\right.
$$

$$
\left. \begin{array}{c} , (e_j, v_j; \chi_j)_{2,p}, (v - u, h; 1) \\ , (f_j, \omega_j; \varpi_j)_{1,q} \end{array}\right]. \quad (22)
$$

Proof Same line of proof as in Theorem 1.

Theorem 3 *Let* $\Re(\wp) > 0$, $\overline{\Omega} > 0$, $h > 0$, $|arg(z)| < \frac{\pi\overline{\Omega}}{2}$, $\Re\left(u + v + h\frac{f_i}{\omega_i}\right) > -1$, $\Re(u) < -\frac{1}{2}$, *then the below integral stands for* $y \geq 0$

$$
\int_0^\infty e^{ix} x^u J_v(x) \, {}^\Gamma \overline{I}_{p,\,q}^{m,\,n}\left[z\,x^h \,\middle|\, \begin{array}{c} (e_1, v_1; \chi_1 : y), (e_j, v_j; \chi_j)_{2,p} \\ (f_j, \omega_j; 1)_{1,m}, (f_j, \omega_j; \varpi_j)_{m+1,q} \end{array}\right] dx
$$

$$
= \frac{e^{\frac{1}{2}(u+v+1)i\pi}}{2^{u+1}\,\Gamma(\frac{1}{2})} \, {}^\Gamma \overline{I}_{p+2,\,q+1}^{m+1,\,n+1}\left[z\left(\frac{e^{\frac{i\pi}{2}}}{2}\right)^h \,\middle|\, \begin{array}{c} (e_1, v_1; \chi_1 : y), (-u - v, h; 1), \\ (-u - \frac{1}{2}, h; 1), (f_j, \omega_j; 1)_{1,m}, \end{array}\right.
$$

$$
\left. \begin{array}{c} , (e_j, v_j; \chi_j)_{2,p}, (v - u, h; 1) \\ , (f_j, \omega_j; \varpi_j)_{m+1,q} \end{array}\right]. \quad (23)
$$

Proof We start with the left-hand side to prove the result. Using definition (14), we obtain

$$
\text{LHS} = \frac{1}{2\pi i} \int_0^\infty e^{ix} x^u J_v(x) \int_{\mathcal{L}} \overline{\Phi}(\wp, y) \, (z\,x^h)^\wp \, d\wp \, dx, \quad (24)
$$

change the order of the integrations, we have

$$
\text{LHS} = \frac{1}{2\pi i} \int_{\mathcal{L}} \overline{\Phi}(\wp, y) \, z^\wp \int_0^\infty e^{ix} \, x^{u+h\wp} \, J_v(x) \, dx \, d\wp. \quad (25)
$$

To calculate the above integral use the formula (20), we obtain

LHS

$$= \frac{1}{2\pi i} \int_{\mathcal{L}} \overline{\Phi}(\wp, y) \frac{e^{\frac{1}{2}(u+v+1+h\wp)i\pi} \Gamma(u+v+1+h\wp)\Gamma(-u-\frac{1}{2}-h\wp)}{2^{u+1+h\wp}\Gamma(\frac{1}{2})\Gamma(v-u-h\wp)} z^{\wp} \, d\wp,$$

$$(26)$$

making use of (14), we get the desired RHS of (23).

Theorem 4 *Let* $\Re(\wp) > 0$, $\overline{\Omega} > 0$, $h > 0$, $|arg(z)| < \frac{\pi\overline{\Omega}}{2}$, $\Re\left(u + v + h\frac{f_i}{\omega_i}\right) > -1$, $\Re(u) < -\frac{1}{2}$, *then the below integral stands for* $y \geq 0$

$$\int_0^{\infty} e^{ix} x^u J_v(x) \, {}^{\gamma}\overline{T}_{p,\,q}^{m,\,n}\left[z\,x^h \left| \begin{array}{l} (e_1, v_1; \chi_1 : y), (e_j, v_j; \chi_j)_{2,p} \\ (f_j, \omega_j; 1)_{1,m}, (f_j, \omega_j; \varpi_j)_{m+1,q} \end{array} \right. \right] dx$$

$$= \frac{e^{\frac{1}{2}(u+v+1)i\pi}}{2^{u+1}\,\Gamma(\frac{1}{2})} \, {}^{\gamma}\overline{T}_{p+2,\,q+1}^{m+1,\,n+1}\left[z\left(\frac{e^{\frac{i\pi}{2}}}{2}\right)^h \left| \begin{array}{l} (e_1, v_1; \chi_1 : y), (-u-v, h; 1), \\ (-u-\frac{1}{2}, h; 1), (f_j, \omega_j; 1)_{1,m}, \end{array} \right. \right.$$

$$\left. \begin{array}{r} , (e_j, v_j; \chi_j)_{2,p}, (v-u, h; 1) \\ (f_j, \omega_j; \varpi_j)_{m+1,q} \end{array} \right]. \quad (27)$$

Proof Same line of proof as in Theorem 3.

Remark 1 Alternatively, if we set $\varpi_j = 1$ $(j = 1, \ldots, m)$ into Theorems 1 and 2, then we obtained the outcomes of Theorems 3 and 4, respectively.

3 Expansion Formulas

The expansions for incomplete I-functions, and incomplete $\overline{I}$-functions containing Bessel function are derived in this section using the integrals presented in Sect. 2 and the orthogonal properties of Bessel functions.

Theorem 5 *Let* $h > 0$, $\Omega > 0$, $|arg(z)| < \frac{\Omega\pi}{2}$, $r = \delta + 2\wp + 1$, $\Re\left(u + v + 2\,h\frac{f_i}{\omega_i}\right) > 0$ $(i = 1, \ldots, m)$, $\Re(u) < \frac{1}{2}$, *then the following expansion formula holds true for* $y \geq 0$

$$e^{ix} x^u \ {}^{\Gamma} I_{p,\,q}^{m,\,n} \left[z\,x^h \, \middle| \, \begin{matrix} (e_1, \nu_1; \chi_1 : y), (e_j, \nu_j; \chi_j)_{2,p} \\ (f_j, \omega_j; \varpi_j)_{1,q} \end{matrix} \right]$$

$$= \frac{1}{2^{u-1}\Gamma(\frac{1}{2})} \sum_{\wp=0}^{\infty} r\, e^{\frac{1}{2}(u+r)i\pi}\, J_r(x) \times$$

$${}^{\Gamma} I_{p+2,\,q+1}^{m+1,\,n+1} \left[z \left(\frac{e^{i\pi/2}}{2} \right)^h \, \middle| \, \begin{matrix} (e_1, \nu_1; \chi_1 : y), (-u - r + 1, h; 1), \\ (-u + \frac{1}{2}, h; 1), \end{matrix} \right.$$

$$\left. \begin{matrix} , (e_j, \nu_j; \chi_j)_{2,p}, (r - u + 1, h; 1) \\ , (f_j, \omega_j; \varpi_j)_{1,q} \end{matrix} \right]. \quad (28)$$

Proof To prove the result (28), let

$$f(x) = e^{ix} x^u \ {}^{\Gamma} I_{p,\,q}^{m,\,n} \left[z\,x^h \, \middle| \, \begin{matrix} (e_1, \nu_1; \chi_1 : y), (e_j, \nu_j; \chi_j)_{2,p} \\ (f_j, \omega_j; \varpi_j)_{1,q} \end{matrix} \right] = \sum_{\wp=0}^{\infty} C_\wp\, J_{\delta+2\wp+1}, \quad (29)$$

here $f(x)$ is continuous and bounded in the interval $(0, \infty)$, for $u \geq 0$. Hence assertion (28) is valid.

On multiply Eq. (29) with $x^{-1} J_{\delta+2t+1}(x)$ and integrate it w.r.t. x from 0 to ∞, we have

$$\int_0^{\infty} e^{ix} x^{u-1} J_{\delta+2t+1}(x) \ {}^{\Gamma} I_{p,\,q}^{m,\,n} \left[z\,x^h \, \middle| \, \begin{matrix} (e_1, \nu_1; \chi_1 : y), (e_j, \nu_j; \chi_j)_{2,p} \\ (f_j, \omega_j; \varpi_j)_{1,q} \end{matrix} \right] dx$$

$$= \sum_{\wp=0}^{\infty} C_\wp \int_0^{\infty} x^{-1} J_{\delta+2t+1}(x) J_{\delta+2\wp+1}(x)\, dx.$$

Now using assertion (17) and the orthogonal property for the Bessel functions [4, p. 291, (6)], we obtain

$$C_t = \frac{\nu\, e^{\frac{1}{2}(u+v)i\pi}}{2^{u-1}\Gamma(\frac{1}{2})} \ {}^{\Gamma} I_{p+2,\,q+1}^{m+1,\,n+1} \left[z \left(\frac{e^{i\pi/2}}{2} \right)^h \, \middle| \, \begin{matrix} (e_1, \nu_1; \chi_1 : y), (-u - r + 1, h; 1), \\ (-u + \frac{1}{2}, h; 1), \end{matrix} \right.$$

$$\left. \begin{matrix} , (e_j, \nu_j; \chi_j)_{2,p}, (r - u + 1, h; 1) \\ , (f_j, \omega_j; \varpi_j)_{1,q} \end{matrix} \right], \quad (30)$$

where, $v = \delta + 2t + 1$.

Equations (29) and (30), give the required result (23).

Theorem 6 Let $h > 0$, $\Omega > 0$, $|\arg(z)| < \frac{\Omega\pi}{2}$, $r = \delta + 2\wp + 1$, $\Re\left(u + v + 2h\frac{f_i}{\omega_i}\right)$ > 0 $(i = 1, \ldots, m)$, $\Re(u) < \frac{1}{2}$, then the following expansion formula holds true for $y \geq 0$

$$e^{ix} x^u \, {}^{\gamma} I_{p,\,q}^{m,\,n} \left[z \, x^h \, \middle| \, \begin{array}{l} (e_1, \nu_1; \chi_1 : y), (e_j, \nu_j; \chi_j)_{2,p} \\ (f_j, \omega_j; \varpi_j)_{1,q} \end{array} \right]$$

$$= \frac{1}{2^{u-1} \Gamma(\frac{1}{2})} \sum_{\wp=0}^{\infty} r \, e^{\frac{1}{2}(u+r)i\pi} \, J_r(x) \times$$

$${}^{\gamma} I_{p+2,\,q+1}^{m+1,\,n+1} \left[z \left(\frac{e^{i\pi/2}}{2} \right)^h \, \middle| \, \begin{array}{l} (e_1, \nu_1; \chi_1 : y), (-u - r + 1, h; 1), \\ (-u + \frac{1}{2}, h; 1), \\ , (e_j, \nu_j; \chi_j)_{2,p}, (r - u + 1, h; 1) \\ , (f_j, \omega_j; \varpi_j)_{1,q} \end{array} \right]. \quad (31)$$

Proof Same line of proof as in Theorem 5.

Theorem 7 *Let* $h > 0, \overline{\Omega} > 0, |arg(z)| < \frac{\pi \overline{\Omega}}{2}, r = \delta + 2\wp + 1, \Re \left(u + v + 2 \, h \frac{f_i}{\omega_i} \right)$
$> 0 \ (i = 1, \ldots, m), \Re(u) < \frac{1}{2},$ *then the following expansion formula holds true for*
$y \geq 0$

$$e^{ix} x^u \, \Gamma \overline{I}_{p,\,q}^{m,\,n} \left[z \, x^h \, \middle| \, \begin{array}{l} (e_1, \nu_1; \chi_1 : y), (e_j, \nu_j; \chi_j)_{2,p} \\ (f_j, \omega_j; 1)_{1,m}, (f_j, \omega_j; \varpi_j)_{m+1,q} \end{array} \right]$$

$$= \frac{1}{2^{u-1} \Gamma(\frac{1}{2})} \sum_{\wp=0}^{\infty} r \, e^{\frac{1}{2}(u+r)i\pi} \, J_r(x) \times$$

$$\Gamma \overline{I}_{p+2,\,q+1}^{m+1,\,n+1} \left[z \left(\frac{e^{\frac{i\pi}{2}}}{2} \right)^h \, \middle| \, \begin{array}{l} (e_1, \nu_1; \chi_1 : y), (-u - r, h; 1), \\ (-u - \frac{1}{2}, h; 1), (f_j, \omega_j; 1)_{1,m}, \\ , (e_j, \nu_j; \chi_j)_{2,p}, (r - u, h; 1) \\ , (f_j, \omega_j; \varpi_j)_{m+1,q} \end{array} \right]. \quad (32)$$

Proof To prove the result (32), let

$$g(x) = e^{ix} x^u \, \Gamma \overline{I}_{p,\,q}^{m,\,n} \left[z \, x^h \, \middle| \, \begin{array}{l} (e_1, \nu_1; \chi_1 : y), (e_j, \nu_j; \chi_j)_{2,p} \\ (f_j, \omega_j; 1)_{1,m}, (f_j, \omega_j; \varpi_j)_{m+1,q} \end{array} \right] = \sum_{\wp=0}^{\infty} C_\wp \, J_{\delta+2\wp+1}, \quad (33)$$

here $g(x)$ is continuous and bounded in the interval $(0, \infty)$, for $u \geq 0$. Hence assertion
(32) is valid.

On multiply above Eq. (33) with $x^{-1} J_{\delta+2t+1}(x)$ and integrate it w.r.t. x from 0 to
∞, we obtain

$$\int_0^\infty e^{ix} x^{u-1} J_{\delta+2t+1}(x) \, {}^\Gamma\overline{I}_{p,q}^{m,n} \left[z \, x^h \; \middle| \; \begin{matrix} (e_1, v_1; \chi_1 : y), (e_j, v_j; \chi_j)_{2,p} \\ (f_j, \omega_j; 1)_{1,m}, (f_j, \omega_j; \varpi_j)_{m+1,q} \end{matrix} \right] dx$$

$$= \sum_{\wp=0}^\infty C_\wp \int_0^\infty x^{-1} J_{\delta+2t+1}(x) J_{\delta+2\wp+1}(x) \, dx.$$

Now using assertion (23) and the orthogonal property for the Bessel functions [4, p. 291, (6)], we get

$$C_t = \frac{v \, e^{\frac{1}{2}(u+v)i\pi}}{2^{u-1}\Gamma(\frac{1}{2})} \; {}^\Gamma\overline{I}_{p+2,\,q+1}^{m+1,\,n+1} \left[z \left(\frac{e^{\frac{i\pi}{2}}}{2} \right)^h \; \middle| \; \begin{matrix} (e_1, v_1; \chi_1 : y), (-u - v + 1, h; 1), \\ (-u + \frac{1}{2}, h; 1), (f_j, \omega_j; 1)_{1,m}, \end{matrix} \right.$$

$$\left. \begin{matrix} , (e_j, v_j; \chi_j)_{2,p}, (v - u + 1, h; 1) \\ , (f_j, \omega_j; \varpi_j)_{m+1,q} \end{matrix} \right],$$

(34)

where, $v = \delta + 2t + 1$.

Equations (33) and (34), give the required result (32).

Theorem 8 *Let* $h > 0, \overline{\Omega} > 0, |arg(z)| < \frac{\pi\overline{\Omega}}{2}, r = \delta + 2\wp + 1, \Re\left(u + v + 2\,h\frac{f_i}{\omega_i}\right) > 0$ $(i = 1, \ldots, m), \Re(u) < \frac{1}{2},$ *then the following expansion formula holds true for* $y \geq 0$

$$e^{ix} x^u \, \gamma\overline{I}_{p,q}^{m,n} \left[z \, x^h \; \middle| \; \begin{matrix} (e_1, v_1; \chi_1 : y), (e_j, v_j; \chi_j)_{2,p} \\ (f_j, \omega_j; 1)_{1,m}, (f_j, \omega_j; \varpi_j)_{m+1,q} \end{matrix} \right]$$

$$= \frac{1}{2^{u-1}\Gamma(\frac{1}{2})} \sum_{\wp=0}^\infty r \, e^{\frac{1}{2}(u+r)i\pi} \, J_r(x) \times$$

$$\gamma\overline{I}_{p+2,\,q+1}^{m+1,\,n+1} \left[z \left(\frac{e^{\frac{i\pi}{2}}}{2} \right)^h \; \middle| \; \begin{matrix} (e_1, v_1; \chi_1 : y), (-u - r, h; 1), \\ (-u - \frac{1}{2}, h; 1), (f_j, \omega_j; 1)_{1,m}, \end{matrix} \right.$$

$$\left. \begin{matrix} , (e_j, v_j; \chi_j)_{2,p}, (r - u, h; 1) \\ , (f_j, \omega_j; \varpi_j)_{m+1,q} \end{matrix} \right].$$

(35)

Proof Same line of proof as in Theorem 7.

Remark 2 Alternatively, if we set $\varpi_j = 1$ $(j = 1, \ldots, m)$ into Theorems 5 and 6, then we obtained the results of Theorems 7 and 8, respectively.

4 Special Cases and Conclusion

In this paper, we described the integral formulas and expansion formulas of incomplete I-functions and incomplete $\bar{I}$-functions involving the Bessel function. The presented results are of a general character, and their special cases are captured throughout the literature. If we allocate unique values to the incomplete I and $\bar{I}$-function parameters, the primary results are reduced to the incomplete H and incomplete $\bar{H}$-functions recently obtained by Meena et al. [5] results. In addition, since the incomplete I and $\bar{I}$-functions generalize the incomplete $\Gamma^{m,\,n}p,\ q$-function, the incomplete Meijer $^{(\Gamma)}G$-function, the incomplete Fox-Wright $_p\Psi_q^{(\Gamma)}$-function, and the incomplete generalized hypergeometric $_p\Gamma_q$ function (see [2, 7]), therefore, the expansion formulas investigated here can be used to obtain the number of new results involving verity of special functions.

References

1. Chaudhry MA, Zubair SM (2001) On a class of incomplete gamma functions with applications. Chapman and Hall (CRC Press Company), Boca Raton, London, New York & Washington, DC
2. Fox C (1961) The G and H-functions as symmetrical Fourier kernels. Trans Amer Math Soc 98:395–429
3. Jangid K, Bhatter S, Meena S, Baleanu D, Qurashi MA, Purohit SD (2020) Some fractional calculus findings associated with the incomplete I-functions. Adv Differ Equ 2020:265
4. Luke YL (1962) Integrals of Bessel functions. MacGraw-Hill, New York
5. Meena S, Bhatter S, Jangid K, Purohit SD (2020) Some expansion formulas for incomplete H and $\bar{H}$-functions involving Bessel functions. Adv Differ Equ 2020:562
6. Rathie AK (1997) A new generalization of generalized Hypergeometric functions. Le Math LII 297–310
7. Srivastava HM, Saxena RK, Parmar RK (2018) Some families of the incomplete H-functions and the incomplete $\bar{H}$-functions and associated integral transforms and operators of fractional calculus with applications. Russ J Math Phys 25:116–138

Chapter 12
Adoption of Microservice Architecture in the Processing and Extraction of Text from Images

Werliney Calixto Conceicao, Fabio Gomes Rocha, and Guillermo Rodríguez

1 Introduction

The extraction of information from documents can become a daunting and burdensome issue [1]. This happens because, depending on these documents' origin and precedence, many of them will be damaged. Documents that are mismanaged tend to lose their information. The damage caused over time is a problem for organizations that will need their information in the future. Several techniques and technologies have emerged over the years that facilitate the extraction and conservation of information from these documents to deal with this problem. One of them is the digitization of these documents, which will prevent deterioration and, consequently, information loss. Moreover, for the extraction of information from these scanned documents, Optical Character Recognition (OCR) can be used, which according to Akhil, is the mechanical or electronic conversion of typed, handwritten, or printed images into machine-coded text. It is the easiest method of scanning printed and handwritten texts so that they can be easily searched, stored more compactly, displayed and edited online, and used in various other processing tasks such as language translation and text mining [2].

The use of OCR's is widely used for document information extraction, but according to Kaur and Jindal [3]. If the scanned documents have noise due to damage or wear of time, it may be necessary for the images to go through a preprocessing

W. C. Conceicao · F. G. Rocha (✉)
Universidade Tiradentes and ITP, Aracaju, Sergipe, Brazil
e-mail: fabio.gomes@souunit.com.br

W. C. Conceicao
e-mail: werliney.calixto@souunit.com.br

G. Rodríguez
ISISTAN(UNICEN-CONICET) Research Institute, Tandil, Argentina
e-mail: guillermo.rodriguez@isistan.unicen.edu.ar

© The Author(s), under exclusive license to Springer Nature Singapore Pte Ltd. 2021   153
S. Kumar et al. (eds.), *Proceedings of International Conference on Communication and Computational Technologies*, Algorithms for Intelligent Systems,
https://doi.org/10.1007/978-981-16-3246-4_12

to remove the noise [3] to reduce the problems concerning data extraction. However, when we address features such as image preprocessing and image information extraction, if they are applied to monolithic systems, which are systems developed in a unified, compact, and with little flexibility, it can result in a high workload, resulting in resource consumption too. Taking this into consideration, there are approaches to software development that, if well applied, can contribute greatly to the extraction of information from texts, which is the case of microservice architecture. Microservice architecture is an organizational approach to software development where the system functionalities are divided into small and independent services that communicate using well-defined APIs. This approach has several positive points that allow greater flexibility, agility, and better control of resource consumption in systems that aim to extract information from images [4].

Through the use of software development techniques and approaches and technologies such as OCR, the extraction of information from scanned documents can become faster and more effective. This allows companies to speed up the productivity in processing high amounts of images to extract important data. In light of the above, this work introduces two microservices whose goal is the preprocessing and extraction of text from images that come from scanned documents. The workflow of these microservices is divided into 3 stages. First, a scanned image is collected from a document. Second, the image is introduced into the preprocessing microservice to remove noise. Third, this image that has been preprocessed is finally introduced in the text extraction microservice.

To evaluate our proposal, the "Sergipe Military Fire Department" documents were used as a case study. These documents present historical and informative reports of the organization and the firemen, such as the organization's first members and their origins. The rest of the article is organized as follows. Section 2 describes what preprocessing is and the preprocessing flow that was performed on the scanned images. Section 3 describes what data extraction is, what an OCR is, and the benefits of using it. Section 4 explains what microservices are and the benefits of using them, and comparisons with other architectures in the computing landscape. Finally, Sect. 5 concludes the article and also describes the plans for the future of the project.

2 Pre-processing

Buades et al. [5] point out that image preprocessing is the step that makes up a set of low abstraction operations. The goal is to improve image data that suppress unwanted distortions or enhance some image features that are important for further processing. These operations compose studies, applications, and algorithms applied to the image to be preprocessed. As previously mentioned, this research's focus is on scanned images, which, unlike analog images, there is a wider range of operations for the application of preprocessing and processing.

The preprocessing techniques aim to make the image better for the step that follows it, which is image processing. There is a wide approach to preprocessing

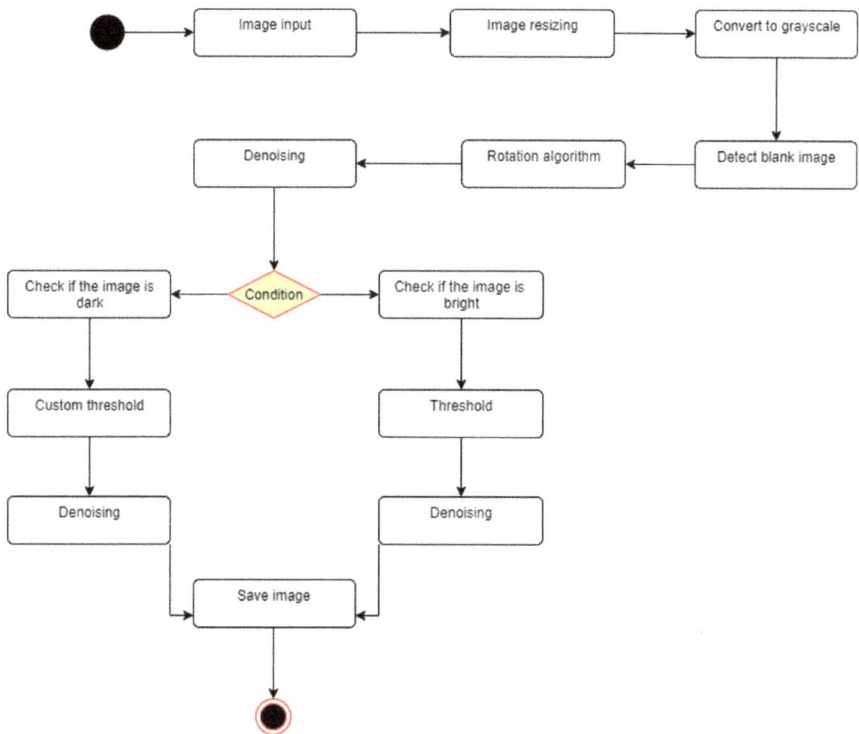

Fig. 1 Processing flowchart, adapted from [6]

methods. A certain method will only be applied if it proves to be effective for the image to be preprocessed, because as said before, one of the stages of preprocessing is the study. This stage consists of analyzing which methods will be necessary and which are really effective for image improvement.

According to Rocha and Rodriguez [6], an image can present several problems, such as noise, low contrast, inadequate inclination, and elements that make it difficult to recognize its characteristics. The work presents a stream of digitized image processing from image acquisition to the final stage of processing. Considering this, Fig. 1 shows the flow applied to the image processing of this project.

First of all, it is necessary to obtain the image that will be applied to the preprocessing. In the **image reading** step, the image of the document that has been scanned is entered into the preprocessing service, where it then goes through the resizing step. The **image resizing** is applied when the image is too large to compromise the extraction of its information, then a reduction in its size is made. The **conversion of the image to grayscale**, is an essential step in image processing. At this stage, the scanned color image, which has 3 color channels and, consequently, more information displayed on the image that can compromise data extraction, is converted to a single grayscale channel image, reducing the amount of information on the image.

According to Rocha and Rodriguez, the grayscale image is generated after a conversion performed from the weighted sum of the color channels, considering the human eye's ability to absorb the light emitted by each color rocha2020spedu.

The **white image detection** was made to make the image processing flow more productive. At this stage, it is checked if the image is white. If the condition is true, it will not have data to be extracted, and then this image is removed.

The **image rotation** is a step that is also important to take into account when working with OCRs. This step is responsible for receiving the image and adjusting its tilt angle so that images that, for some reason, are outside the ideal 90° angle do not hinder the extraction of the data. This is because OCR has more difficulties extracting data from an image tilted than images at 90°, the extraction time is longer on tilted images, and sometimes words come out distorted.

The **noise suppression** is one of the most important of the whole image processing flow. As said before, documents have a natural tendency to age. Consequently, they are damaged by several possible motives, losing information, or leaving them compromised with time. This is one of the reasons for the appearance of noise. Considering this, as the name itself says, the goal of this stage is the removal of noise from scanned images. These noises hinder the extraction of OCR data because it becomes more complex to identify and distinguish words from those images with distortions. For this step's execution, the "fastNlMeansDenoising" functionality was used of the OpenCV library,[1] a feature that uses the Non-local Means Denoising algorithm according to Buades et al. [7]. According to Buades, Coll, and Morel, this algorithm is based on a simple principle: replace the color of a pixel with the average colors of similar pixels. It averages all pixels of the image, weighted by the similarity of these pixels with the target pixel. This results in better filtering of the image and little loss of characteristics [7].

After the noise removal step, there is a condition that will be checked to proceed with image processing. As shown in the flowchart, this condition will check whether the image is clear or dark, and once checked. Different algorithms will be applied to each condition. The step that verifies if the **image is clear** will be the starting point that will direct the flow to the application of algorithms that deal with this type of problem. It is essential so that we can allow the improvement of data extraction. Once it is confirmed that the image is clear for OCR, the step that deals with this problem, which is applying **threshold (binarization)**, will start. According to Rocha and Rodriguez, binarization is a technique that allows the segmentation of objects in the background. This process transforms the image in gray tones into a binary image. It contains only 2 colors—black (represented by 0) and white (represented by 1), removing the other colors. The binarization of the images intended to improve the extraction of data. This happens because the Tesseract OCR, due to binarization, is easier to extract data from images since the binarization allows greater highlighting of words due to the difference in colors of the image's background (which turns white) and words (which turns black). In this step's execution, the functionality "Imgproc.threshold" of the OpenCV library was used. This functionality receives one of the parameters, the type

---

[1] https://opencv.org/.

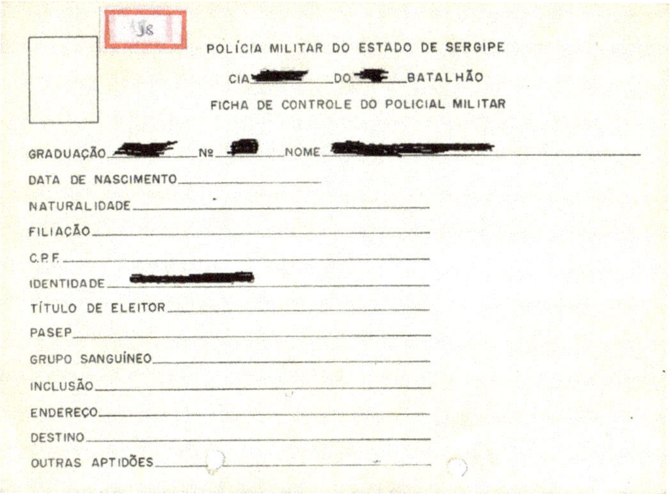

Fig. 2 Image before preprocessing

of threshold, which is one of the thresholding operations to be applied. In this case, 2 types of thresholding have been applied: the common binarization explained above; and the "otsu binarization", which is a threshold based on a single limit [8]. Right after the application of the threshold, there is also another step for noise removal. This step was applied again so that noises that could appear after the threshold was applied, or if they had not been removed previously, could be removed.

In Figs. 2 and 3, the before and after of an image considered clear, which was processed and applied the threshold.

Verifying if the **image is dark**, is essential when we are working with OCR's because it can reduce the noise and improve the information extraction, so this will be the starting point that will direct the flow to the application of algorithms that will allow better extraction of data. This happens because the Tesseract OCR has problems extracting data from images that it considers to be dark. After all, the distinction of words and recognition of characteristics are impaired due to the image's color. This results in the return of distorted, incomplete words or even an extraction completely without text. Once it has been confirmed that the image is dark, the stage dealing with this type of problem, which is the application of **threshold modified**, will begin. It is given the name "modified threshold" because to solve the problem of the dark image, a different approach has been taken from the light image approach. The approach applied was creating a method that, in order to lighten the image, takes as a parameter the image, and the desired threshold value. The method will leave all pixels of the image white above the received threshold value. This value goes from 0 to 255, allowing flexibility in which the user can pass the value to determine how much an image needs to be whitened. Right after the modified threshold, noise removal is applied again. Immediately after the modified threshold, noise removal is

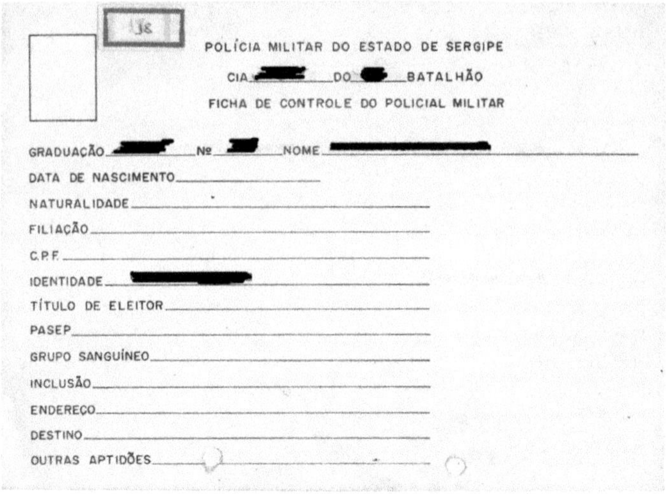

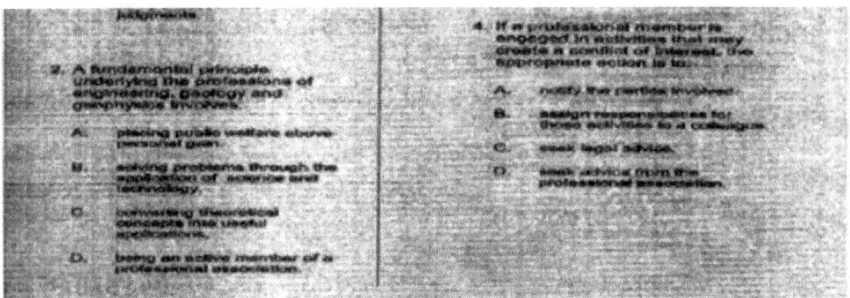

Fig. 3 Image after preprocessing

Fig. 4 Dark image before preprocessing

applied again. To ensure that noises that could appear after the modified threshold has been applied, or if they had not been removed previously, can be removed. Figs. 4 and 5, represent the before and after of an image that is considered dark, which was processed and applied the modified threshold. Noise reduction is noted after application of the threshold.

3 Data Extraction

Data extraction is the process of collecting and retrieving various types of data from a variety of sources, where the data from these sources can usually be disorganized, damaged, or completely unstructured. Data extraction makes it possible to consoli-

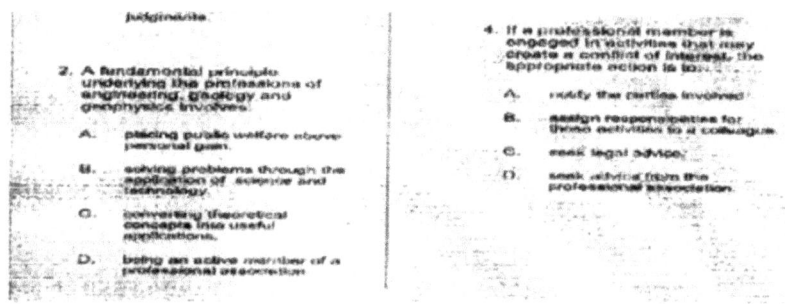

Fig. 5 Dark image after preprocessing

date, process, organize, and refine data from these various sources, allowing this data to be later saved for future work.

Data can be extracted from various sources, from Web sites that are accessed daily, to documents and scanned images that have important information in the context introduced. With the advance of technology and the fact that data is increasingly concentrated in digital sources, information extraction is becoming increasingly important and crucial for companies in this digital age.

According to Jung, Kim, and Jain, text data present in images contains useful information for automatic annotation, indexing, and image structuring [9]. The extraction of this information involves detection, location, tracking, enhancement, and text recognition from a given image. However, variations in text for size, style, orientation, alignment, and low image contrast make automatic data extraction extremely challenging.

Besides, Kaur and Jindal indicate that for text recognition and detection to be possible, preprocessing images is necessary, removing the noise [3] considering historical documents are sources of complex data because due to their inevitable aging or mismanagement, the information contained begins to compromise.

Taking this into account, as said before, in this project, developed data sources are the scanned images that came from scanned documents, where the goal is the extraction of texts. These documents are mostly old or damaged, so their images needed to go through the preprocessing process to extract their texts to be effective.

The text extraction service of the images of this project was developed with the use of Tesseract OCR. That according to [2], the Tesseract Ocr is an open-source optical character recognition (OCR) mechanism. Tesseract started as a Ph.D. research project in HP's laboratories in Bristol. In 1995, it was perfected with greater accuracy, and in 2005 it became open source [2].

According to [2], optical character recognition is the mechanical or electronic conversion of typed, handwritten, or printed images into machine-encoded text. It is the easiest method of scanning printed and handwritten text to be easily searched, stored more compactly, displayed and edited online, and used in various other processing tasks such as language translation and text mining [2].

To complement the importance of using an OCR in this project, Rajeswari and Magapu state that in the digital age, the conversion of printed documents into electronic format has become a necessity for the availability of information [10]. Moreover, to enable the localization of documents, their metadata must be extracted through optical character recognition (OCR) tools [10].

Besides, for data extraction through OCR, Vasilopoulos, and Kavallieratou state that it is necessary to employ methods that combine document layout analysis with text detection [11]. According to Gomes and Rodriguez, the first stage of the process occurs with the document's digitalization. This can be done through photographs or digitization of sources. However, the scanning process aims to turn the printed or manuscript document into a digital document, but this does not automatically allow the document to be searched. In this way, the researcher will remain dependent on the transcription, cataloging, and indexing of documents [6].

Taking all this into consideration, an explanation of how OCR works would be an important thing to mention, as it would give a better understanding of the microservice that performs the extraction of data from scanned images. Furthermore, it is also worth mentioning the benefits of using it and how these benefits impact the productivity of the produced tasks.

3.1 How an OCR Works

The workflow of an OCR is relatively simple. Figure 6 this flow is displayed, which starts from the document that has been scanned to the final step, which is the extraction of the texts.

Due to the great variety of fonts for writing, and the different writing methods, image data extraction is not such an easy job to do. Nevertheless, before an OCR can be chosen, the image must be preprocessed. As explained above, it is responsible for removing the noise and leaving the image ready to have its data extracted more effectively. With the image preprocessing and ready to be used in OCR, the texts' steps of recognition and extraction begin. There are two different techniques (or algorithms) in optical character recognition: pattern recognition and resource extraction, and it is worth examining each technique in a little more detail. Starting with pattern

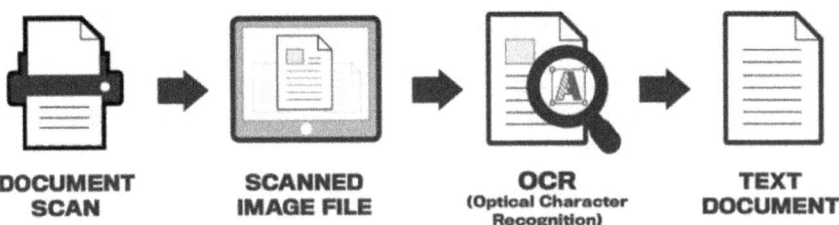

DOCUMENT   SCANNED   OCR   TEXT
SCAN   IMAGE FILE   (Optical Character Recognition)   DOCUMENT

Fig. 6 Flow of an OCR [12]

Fig. 7   Capital letter A [13]

recognition, using this technique, the computer tries to recognize the entire character and tries to make the combination with the character matrix stored in the software. As a result, this technique is also known as pattern recognition or matrix recognition. This disadvantage depends on the input characters and the stored characters being of the same font and the same scale. Moreover, the resource extraction step is a much more sophisticated way of identifying characters. It breaks down characters into "resources" such as lines, closed loops, line directions, and intersections. Let's take the letter "A" in Fig. 7, for example.

If the computer identifies that the character has two angular lines at the top, and a horizontal line joins both lines in the middle, it will identify that character as an "A". By using rules like these, the program can identify most of the capital A's, regardless of the font in which it is written. After all, the identification of the scanned image is made. In the end, it receives the extracted text.

3.2   Benefits of OCR

The use of an OCR for data extraction in a context with many documents and images can bring several benefits. Starting with **escalability**, where software such as OCR is key to helping organizations collect data in scale. Without a tool like this, users would have to analyze the sources to collect this information manually. Regardless of how much data an organization has, its ability to leverage the collected data is limited by manual processing. By automating extraction, organizations increase the amount of data that can be deployed for specific use cases. There is also a further increase in the **efficiency** of extraction, as the automation that OCR delivers in data extraction contributes to greater efficiency, especially when considering the time involved in data collection. Doing the data extraction manually is tiring and also unproductive. Organizations that use data extraction tools substantially reduce data-driven processes, having more time to develop other tasks. With the use of OCR, a **higher accuracy rate** is also noted, as automated data entry tools such as OCR data entry result in errors and reduced inaccuracies, resulting in efficient data entry. Besides, problems such as data loss can also be successfully solved by OCR data entry. There is no labor involved, problems such as accidentally typing in erroneous information or otherwise eliminated. The flexibility that OCR provides to organizations is an important positive point to mention. Since scanned documents need to be edited at times, especially when some information needs to be updated. OCR converts data into any preferred format, such as Word, which can be easily edited. This can be of great help when there is content that needs to be constantly updated or corrected. Also, OCR allows scanned documents to have their texts searched. This helps professionals quickly search for numbers, addresses, names, and various other parameters that distinguish the document being searched. It also brings a greater modernity texture to the context being applied, as OCR processing improves the user experience by eliminating manual data from companies and organizations. Accessing a digital archive is much faster than locating a paper document in a stack of other unrelated documents. Moreover, we also know that storing documents is a complicated task, and you can have several files stacked in your storage space. The important uses of OCR lie in its ability to organize this storage space. If they are not properly organized on drives, important folders may be accidentally lost or deleted. OCR helps you store.

3.3   Data Extraction Workflow

The data extraction service works in conjunction with the image preprocessing service. Figure 8, shows the extraction flow that is performed in the project.

First of all, to run the application correctly, it is necessary to go first through the Image Processing module, which is the module that has the preprocessing service, where the images will have their noises removed and will be prepared to have their texts extracted. After that, the image that has been preprocessed will be saved in a folder of the user's choice. It is from the saved image that the data extraction service and all its related processes will start. With the image being introduced in Tesseract OCR, the algorithms explained above will be applied. The texts that were in a digital image and that could not be changed or deleted will be extracted by OCR as normal texts so that later they can be worked on according to need.

In Fig. 9, an image is shown that came from a scanned document, which I chose to demonstrate the texts' extraction. The choice was the larger amount of texts that facilitated the project's demonstration so that confidential information of the firemen would not be displayed. In Fig. 10 is displayed the text extracted from the image.

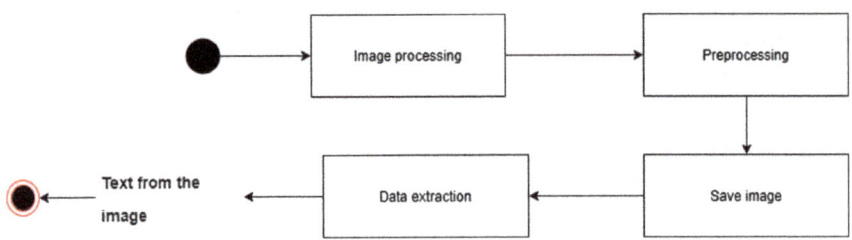

Fig. 8 Data extraction flow

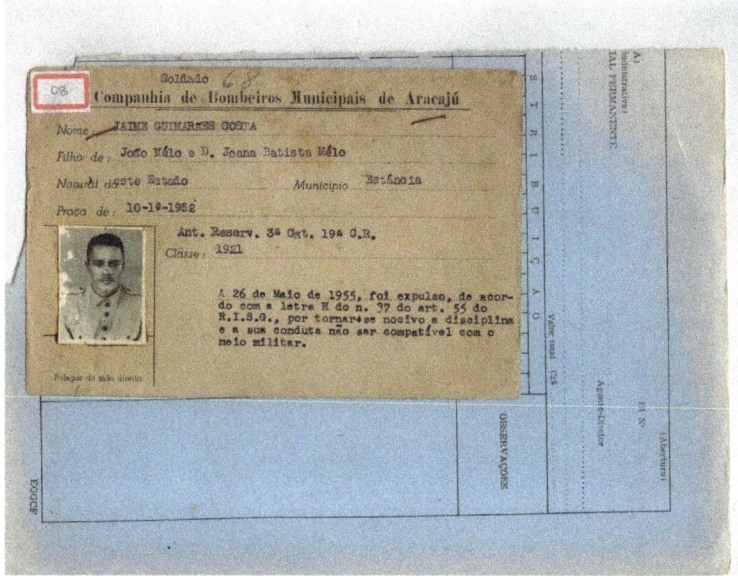

Fig. 9 Image that will have your texts extracted

4   Microservice Architecture

Jamshidi et al. point out that microservices is an architectural approach that has emerged from service-oriented architecture, emphasizing self-management and lightness as means of improving software agility, scalability, and autonomy [14]. The architectures of microservices facilitate and speed up software development, increase business productivity, enable innovation, and accelerate the introduction of new features in the market [14].

Because of these positive points, the architecture of microservices is a subject that has been much talked about and has gained much strength in recent years. For this reason, it is increasingly being applied by companies and software developers who seek a more modern approach.

As we are talking about microservices, which is a relatively new architecture compared with others, it is important to explain the differences between existing architectures, like service-oriented architecture (SOA), that confuse people with its similarities. The differences it has with Monolithic architecture are the oldest and opposite approach of microservice architecture. This explanation will explain how micro-services work and show the challenges that software development had, which motivated the emergence of microservices architecture.

REGINALDO SANTOS MOURA - MAJOR QOBM

RESP/ P/ SUBCOMANDO DO CBM/SE.

CONTINUAÇÃO DO BGO Nº 061 DE 04/04/2000

- Ajudante — Sd. 4404 CLÉBSON 2º SGI/GI

Motorista — Sd. 0350 FONSECA 2º SGI/GI

Í GUARNIÇÃO DO CCI/065

- Chefe da 1º Linha — Sd. 3470 NOBERTO 2º SGI/GI

- Ajudante — Sd. 3567 CLEVERTON 2º SGI/GI

Motorista — Sd. 0350 FONSECA 2º SGI/GI

- GUARNIÇÃO DO CCI/025 ,

Chefe da 1º Linha — Sd. 4631 CARVALHO 2º SGI/GI

2º SGI/GI

Chefe da 2º Linha - 2º SGI/GI

2º SGI/GI

Motorista — Sd. 3333 CARLOS - 2º SGI/GI

(QUARTEL DO SGBS)

1- (Guarnição Comando,

Chefe da Guarnição de Guarda-Vidas — 1º Sgt. JECONIAS SGBS

Auxiliar - Cbs. 0145 ALMERINDO, 0175 J. CARLOS SGBS

Sds. 4444 JOSIVAL (PLANTÃO), 4477 GILDÁSIO, 4624 ADEGILDO, 4982 AGUIAR, 5219 SGBS

e RAMOS |

o Motoristas — Sd. 29061 ELIELSON (AA-2) SGBS

' GUARNIÇÃO DO SIET/UTI)

. Socorristas de Dia — CB 0312 SALMERON, Sd. 5224 MATEUS

Motorista de Dia — Sd. 2124 GERALDO

GUARNIÇÃO DA UTE / AA-3

, Socorristas de Dia — CB 0327 SANTOS FILHO, Sd. 0306 MARQUES

Fig. 10 Text extracted by OCR

4.1   Microservices Versus Service-Oriented Architecture

Given the similarities of microservice architecture and service-oriented architecture (SOA), people often think that microservices and SOA are the same things. The service-oriented architecture (SOA) is a type of software design that makes the components reusable using service interfaces with a common communication language in a network. The main characteristic that differentiates the two is the scope: SOA is an architecture approach adopted by the company, while microservices are an implementation strategy of each application's development team. The communication between the components is also different. SOA uses ESB (which is a type of integration infrastructure [15]), while the microservices communicate with each other in a stateless way through language-independent APIs. Because of this aspect of APIs in microservices, development teams can also choose the tools they prefer to work with. Thus, microservices are more tolerant and flexible.

Also, according to Jamshidi, Pahl, Mendonça, Lewis, and Tilkov, SOA is often associated with web services, tools, and formats such as SOAP (which is a protocol for information exchange [16]) [14]. In contrast, microservices usually depend on REST and HTTP, or other formats made for web development [14].

4.2   Microservices Versus Monolithic

When we talk about Microservices, monolithic architecture is usually the first subject compared with microservices architecture. This happens because these two are opposite versions of architecture, with different philosophies and objectives. Hence, it is important to show each one's differences to understand the evolution of software development from an older approach of architecture, which is monolithic, to a more modern approach that is the microservices.

Monolithic architecture has been the standard way of architecting applications for years. Monolithic applications use a single code base, making deployment and development easier without adding any additional complexity as long as the application size remains relatively small. Because of these mentioned qualities, it is usually a good choice at the beginning of the project development. In monolithic, the application is divided into layers, being them: the UI layer, which is the part responsible for the user interface; the service/business rules layer, which contains the logic of the application; and the data access layer, which as the code base is unique, a monolithic application usually uses a single database to manage all the data. Figure 11 shows an example of monolithic architecture.

Fig. 11 Monolithic
architecture

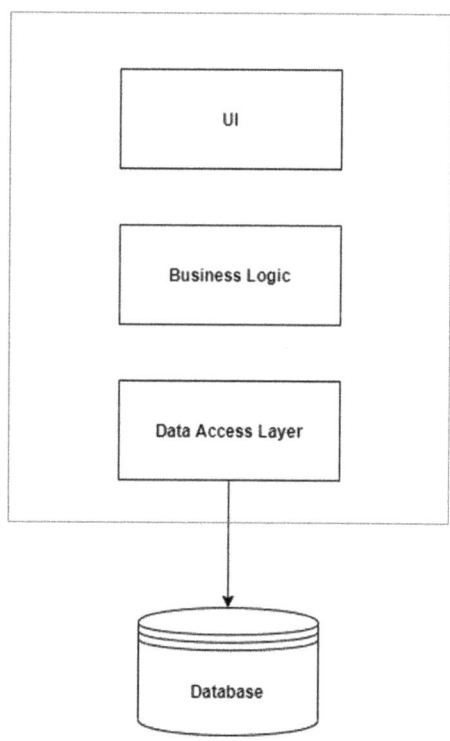

However, when working with a monolithic architecture, some problems and challenges must be considered. These problems usually arise due to the system's growth over time, which makes the application consume more resources and be more complex. It is from these obstacles in a monolithic application that barriers such as lack of flexibility, which forces developers to get stuck to the technology originally chosen for the system, even if in some situations it is not the best choice; The complexity that developers have in doing maintenance due to the infinity of the size of the code that grows over time; Delay of acculturation, since a new employee will have difficulties to understand better how the application works; Difficulty to put changes in the application in production, because any change, even if it is in only one line of code, will require the system to be rebooted, decreasing the productivity of the development team; And one of the main obstacles among all the others is the limited scalability, which requires that the whole system be replicated, even if only part of its functionality is needed in the new instance, causing higher costs than expected.

In microservices, the difference in how an application's functionalities are approached compared to monolithic is quite different. As [17], the architecture of microservices consists of multiple small standalone services deployed and developed separately. It enables more refined scalability and allows development cycles to be faster by decreasing the amount of required regression testing because each service can be deployed and updated separately from each other. After all, one of its main goals was to bring solutions to the problems monolithic applications had when the code base got too large. Figure 12 shows an example of a microservice architecture.

Fig. 12 Microservice architecture

The purpose of this project, as previously mentioned, is the development of services for the extraction of data from scanned images, in which these services need to be flexible and simple to the point of being used by others when necessary, without having to download or compile an entire project with a multitude of code to use one functionality. Moreover, from these needs, and understanding the advantages and disadvantages of monolithic architecture and the benefits of microservice architecture, we concluded that the microservice approach would be the best option for the development of this project.

4.3   Benefits of Using Microservices

There were several benefits to its application when it adopted the architecture of microservices instead of monolithic. Starting with **autonomy**, where the microservice architecture services are developed, deployed, operated, and scaled without affecting other services' functioning. This allows services to share no code or implementation. All communications between the services of the different modules occur through APIs. The services are **specialized** to perform a task. Each of them is designed to have a set of resources dedicated to solving a specific problem. If a service starts to get bigger and increases its complexity over time, developers can divide it into smaller services. The [agility] is also a notorious benefit when adopted the architecture of microservices. Microservices promote an organization where services have small and independent teams that are responsible for them. The teams act within a small context, where they have the autonomy to work more independently and faster. This results in the acceleration of development cycles, giving more productivity in the delivery of functionalities. The increase in productivity also happens due to the high rate of **reusable code** in microservice architectures. Since the division is made into small, well-defined modules, this allows teams to use functions for various purposes. A service created to perform a function can be reused to develop another resource. This allows some code snippets to be reused, increasing the productivity of development cycles. The **flexibility in technologies** in microservices is one of the highlights compared to monolithic. Microservice architectures do not follow a generalist approach, where only one language is addressed. Teams are free to choose the best tool to solve specific problems. This allows teams that create microservices to choose the best tool or language for each task. There is also a great **resilience to failures** in microservice applications. The independence the services have from each other increases the application's resistance to failures. In a monolithic architecture, the failure of a single component can cause the entire application's failure. With microservices, applications only deal with the service that is failing without interrupting the whole application. The [easy to deploy] is a positive point because the microservices allow continuous integration and delivery, which facilitates the testing of new features and their reversal if something does not work properly. The low cost of failure allows experimentation, making it easier to update the code and speed up, introducing new features to the market. There is also probably the greatest positive point of the microservice architecture compared to monolithic, which is **escalability**. Microservices allow each service to be scaled independently when the demand for that service increases with the application's use. This gives teams the flexibility to correctly scale infrastructure needs, accurately measuring the cost of a resource, and maintaining availability when a service experiences a peak demand.

5 Case Study

The operation of a microservice architecture is relatively simple to understand. Figure 13 shows how the services work from a deployment diagram.

In the case of the developed application, **Spring Boot in version 2.4.0** was used, a framework that uses the programming language **Java**, and is used to develop REST applications and microservices. The deployment diagram shows the deployment of the developed application. Everything starts from the client, who on his computer, will make the functionality request, and this request will be redirected to a **Api Gateway**. To make the **Api Gateway** functionality, the **Eureka Server** technology was used, which is a technology that contains the information about all the client services of the application. Each microservice will be registered on the Eureka server, and the Eureka server will know all the client applications running on each port and IP address. With Eureka, each client can act simultaneously as a server to replicate its status to a connected pair. In other words, a client retrieves a list of all connected pairs from a service record and makes all additional requests to any other services through a **load balancing algorithm**. It also allows a better **escalability** of the application as

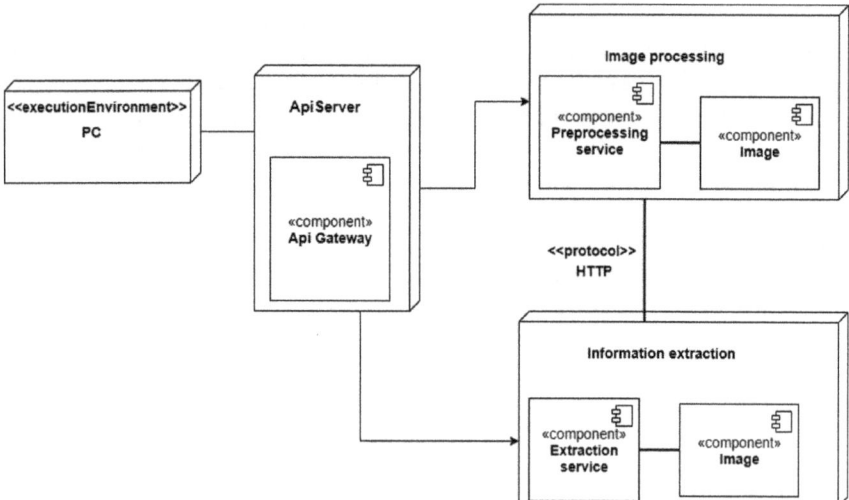

Fig. 13 Application deployment diagram

the application, and the number of microservices grows. With this API Gateway, the **communication** with the application modules will be made. In the case of the project, it will have the image processing module. This module will have the microservice responsible for preprocessing the images so that they are noiseless. Moreover, the application will have another module for the extraction of data. For example, the user would need to communicate first with the image processing module to preprocess the image to extract the information. It is also possible to do **communication** between services the HTTP protocol since in Spring Boot there is a functionality called **Rest Template**, which allows communication between different REST services bypassing the HTTP URL of the service you want to communicate with.

5.1   *Analysis of Results*

To statistically analyze the results of the project, the accuracy rate of the information extraction was calculated. This precision was calculated based on the joint work of the 2 services since the extraction will be evaluated in preprocessed images, so the calculation of precision was made based on Feng and Manmatha [18]. The results can be seen in the Table 1, being analyzed the number of characters and words extracted correctly in the OCR output compared with the original characters and words of the image that was introduced.

It can be noted that there is a higher accuracy rate in character recognition because for OCR, it is easier to recognize those characters that, regardless of the font, do not have their formats drastically changed, such as parentheses, hyphens, and accents. In general, due to the comparisons that OCR makes with the texts of images and the texts in its database, it is easier to recognize singular texts as vowels and consonants than to recognize the complete words formed from these vowels consonants.

In the adoption of the microservice architecture in the project, one of the main positive points is the application's resilience. Since when a service stays off the air for some reason, the other service independent of it will continue in execution. It is also worth mentioning that it is possible to expand the computational resources of only one service, reducing the application's total cost. As negative points, one of the biggest challenges is the planning of the distribution of the services and their integration.

Table 1 Accuracy rate

Num. of samples in the image	Num. of samples in the OCR	Accuracy rate
12,650 characters	12,112 characters	95.74
2230 words	2120 words	95.06

6 Conclusions

In this article, we introduced the concept of microservices and the development of two microservices to work together in the extraction of texts from scanned images of documents. We presented the differences between the architecture of microservices and service-oriented architecture (SOA), the differences with the monolithic architecture, and how these differences provided a better understanding of the architecture of microservices. We also showed the benefits of using the microservices approach and how they encouraged me to develop this architecture approach. The first microservice is responsible for the preprocessing of the images, which is applied to a set of operations that approach studies, applications, and algorithms where the objective is to improve image resources important for further processing. The second microservice is responsible for data extraction, where the Tesseract OCR was implanted. OCR is the optical character recognizer responsible for obtaining the scanned images, extracting the unalterable texts from them, and transforming them into normal texts that can be worked according to the needs. In the end, the results of the development of this project were displayed, showing the extraction of the texts from a scanned image that was preprocessed. For the future, the project's improvement will be continued, so that new techniques of development of microservices that can deliver better results. Furthermore, better strategies for extracting texts using Tesseract OCR will be explored.

Acknowledgements FAPITEC partially supported this work.

References

1. Mattia Z, Komminist W (2017) Extracting information from newspaper archives in Africa. IBM J Res Dev 61(6):12–1
2. Akhil S (2016) An overview of tesseract OCR engine. In: A seminar report. Department of Computer Science and Engineering National Institute of Technology, Calicut Monsoon
3. Kaur RP, Jindal MK (2019) Headline and column segmentation in printed Gurumukhi script newspapers. In: Smart innovations in communication and computational sciences. Springer, pp 59–67
4. Baresi L, Garriga M (2020) Microservices: the evolution and extinction of web services? In: Microservices. Springer, pp 3–28
5. Buades A, Coll B, Morel JM (2005, June) A non-local algorithm for image denoising. In: 2005 IEEE computer society conference on computer vision and pattern recognition (CVPR'05), IEEE, vol 2, pp 60–65
6. Rocha FG, Rodriguez G (2020) SPEdu: a toolbox for processing digitized historical documents. In: Mexican international conference on artificial intelligence. Springer, pp 363–375
7. Antoni B, Bartomeu C, Jean-Michel M (2011) Non-local means denoising. Image Process OnLine 1:208–212
8. Nobuyuki O (1979) A threshold selection method from gray-level histograms. IEEE Trans Syst Man Cybernet 9(1):62–66
9. Jung K, Kim KI, Jain AK (2004) Text information extraction in images and video: a survey. Pattern Recogn 37(5):977–997

10. Rajeswari S, Magapu SB (2018) Development and customization of in-house developed OCR and its evaluation. The Electronic Library, Emerald
11. Nikos V, Ergina K (2017) Complex layout analysis based on contour classification and morphological operations. Eng Appl Artif Intell 65:220–229
12. llango R (2020) Using NLP (bert) to improve OCR accuracy. Disponível em. https://medium.com/states-title/using-nlp-bert-to-improve-ocr-accuracy-385c98ae174c. Acessado em 09 Dec 2020 às 18:05
13. Victoria (2020) How does OCE work? a short explanation. Disponível em https://www.scan2cad.com/tips/how-does-ocr-work/. Acessado em 09 Dec 2020 às 18:10
14. Jamshidi P, Pahl C, Mendonça NC, Lewis J, Tilkov S (2018) Microservices: the journey so far and challenges ahead. IEEE Softw 35(3):24–35
15. Falko M (2007) Enterprise service bus. In: Free and open source software conference, vol 2, pp 1–6
16. Box D, Ehnebuske D, Kakivaya G, Layman A, Mendelsohn N, Nielsen HF, Thatte S, Winer D (2000) Simple object access protocol (SOAP) 1.1. W3C. USA
17. Kalske M et al (2018) Transforming monolithic architecture towards microservice architecture. M.Sc. Thesis University of Helsinki, Department of Computer Science, Finland
18. Manmatha R, Feng S (2006) A hierarchical, hmm-based automatic evaluation of ocr accuracy for a digital library of books. In: Proceedings of the 6th ACM/IEEE-CS joint conference on digital libraries (JCDL'06). IEEE, pp 109–118

Chapter 13
A Hybridization Technique for Improving a Scheduling Heuristic in Mass Production

Tibor Dulai, György Dósa, Ágnes Werner-Stark, Gyula Ábrahám, and Zsuzsanna Nagy

1 Introduction

Scheduling is an intensively researched field of operations research. If a production process is scheduled to be efficient, it may result in huge benefit for the company. Hybridization techniques seem to be effective in solving different combinatorial optimization problems [3]. The considered scheduling problem—that was introduced by Auer et al. [1]—does not fit completely into any of the job-shop scheduling problem extension that can be found in the literature. It is a multi-stage scheduling problem for mass production (its production processes have to be executed several times) with multi-purpose parallel machines, where each process has parallel branches. In contrary to the job-shop scheduling with processing alternatives (AJSP) presented by Kis [4], the operations of the same process in the parallel branches can be executed parallel. The problem has some common characteristics with hybrid flow shop (HFS) problems [5], too, but in contrary to HFS problems, here, the allocated machine sequence for the sequence of the operations of the processes may differ.

There are two kinds of products that have to be produced in big number during the mass productions. Their production processes are similar, both of them have parallel branches and an assembly operation. One of the two processes has an extra operation related to the other one. Both production processes are well defined by precedence constraints between their operations. Some operations can be performed by more than one machines; in these cases, the operation times may differ. This feature allows us to classify the problem as flexible job-shop problem (FJSP), although here successive activities of a job can be processed on the same machine.

All of these properties give the significant complexity of the problem. Based on Chen et al. [6], the considered problem is a variant of $R_m|\text{prec}|C_{\max}$ problems. In a $R_m|\text{prec}|C_{\max}$, there are m unrelated machines, precedence constraints are defined

---

T. Dulai (✉) · G. Dósa · Á. Werner-Stark · G. Ábrahám · Z. Nagy
University of Pannonia, Veszprém 8200, Hungary
e-mail: dulai.tibor@virt.uni-pannon.hu

© The Author(s), under exclusive license to Springer Nature Singapore Pte Ltd. 2021   173
S. Kumar et al. (eds.), *Proceedings of International Conference on Communication and Computational Technologies*, Algorithms for Intelligent Systems,
https://doi.org/10.1007/978-981-16-3246-4_13

between the operations of the problem, and the goal is to find a solution with minimal makespan. If the problem is classified either into the group of $R_m |\text{prec}| C_{\max}$ problems or is handled as an FJSP problem, some characteristics were lost. We have found no paper in the scientific literature that considers the same problem type.

2 The Problem to Solve

The considered problem includes eight machines of four machine types and two hardly different production processes. They are shown in Fig. 1.

Because of mass production, both processes have to be executed with big cardinality. The desired number of products is $n(p_1)$ and $n(p_2)$. In the figure, the possible resource allocations are highlighted by different patterns, assigned with the operations. There are operations that can be performed by more than one machine, and there are machines that are able to execute more than one operation type. The opera-

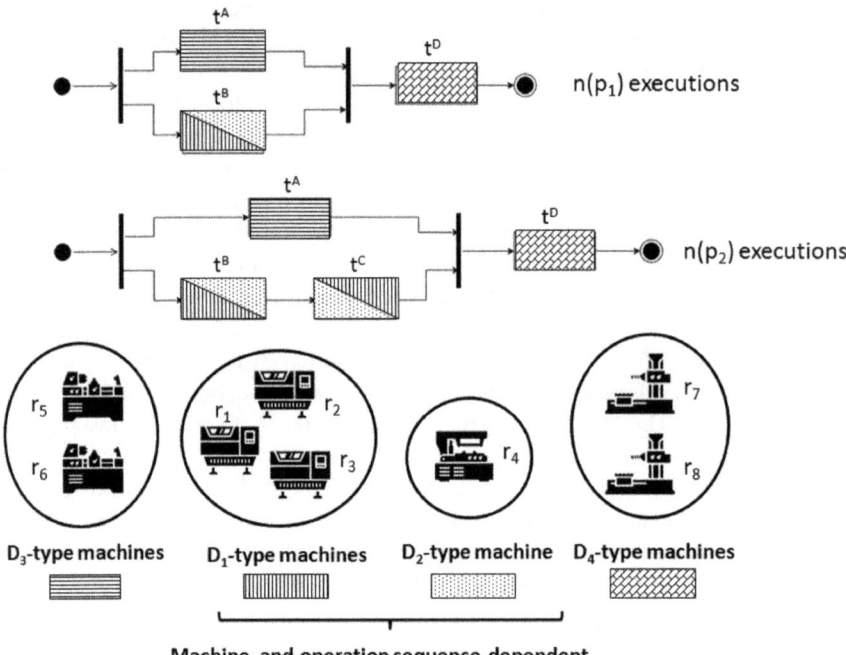

Fig. 1 Production processes, the resource set and the resource allocation possibilities of the scheduling problem

tion time is influenced by the resource-operation pairing. When a machine executes more than one operation type in a sequence, between them setup time has to be applied. The goal is to find the solution that has minimal makespan.

3   The Improvement Phase of the Heuristic Scheduler of Auer et al. [1]

The main idea of the heuristic scheduler of Auer et al. [1] was to divide the initial problem into three less complex subproblems, as Fig. 2 shows. Two of the subproblems (1 and 3) were solved optimally by simple greedy methods. For the remaining subproblem (2), a series of "tricky" steps were applied, including simplifications, determination of subcases, calculation of a preemptive optimal solution, rounding and improvement. For the improvement phase, two methods were applied: local search and Tabu search.

3.1   Local Search

Local search applies an elementary operation in a loop's each step. The two elementary operations are as follows: MOVE and SWAP. MOVE selects randomly a critical operation from a machine and moves to another machine that is capable to perform that. Critical operation is an operation that is an element of the critical path. The critical path is the shortest path from the latest operation to the earliest operation in the graph of the schedule as it is defined by Hurink et al. [2]. SWAP picks a random

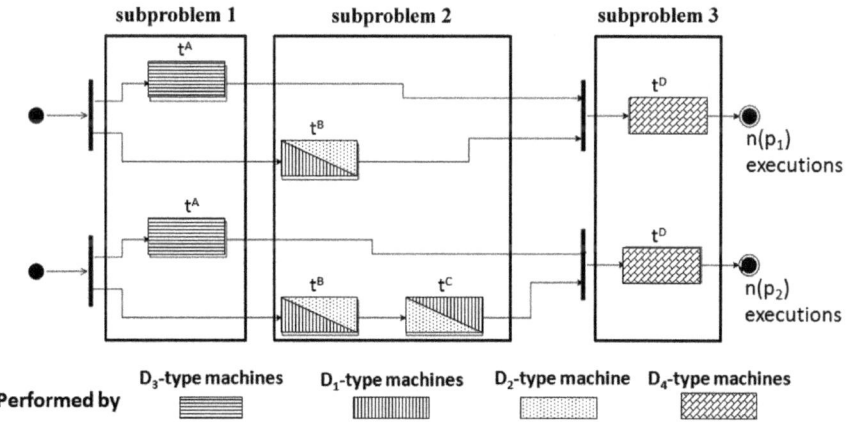

Fig. 2 Subproblems of the initial problem are handled in separate subproblems

critical operation and changes it with another random operation that has different operation type and is originally assigned to a different machine. In each step, the result of the elementary operation is accepted only if its makespan is smaller than the makespan of the solution in the beginning of the iteration.

3.2 Tabu Search

The Tabu search algorithm used in Auer et al. [1] was based on the algorithm of Hurink et al. [2]. It generates neighbours by picking a random operation from a block of a critical path and moves it either to another machine or to the beginning/end of the block to a feasible solution. A block was defined as a successive maximum—at least two-sized—sequence of operations of a critical path whose operations are assigned to the same machine. It was also highlighted by Auer et al., that although the Tabu search of Hurink et al. was implemented in their improvement phase, the considered problem differs from the scheduling problem of Hurink et al.

4 Four Variants of Tabu Search to Improve the Result of the Heuristic Algorithm

Auer et al. reached a small improvement by both LS and TS. Here, we implemented four variants of Tabu search to investigate whether we could get higher improvements.

The Tabu search algorithm that was implemented in [1] resolves some prior simplifications of the heuristic algorithm. The heuristic algorithm assumed that there is no machine where a t^C-type operation precedes a t^B-type operation, and setup time can be applied at most once on each machine (it is guaranteed by the sequence of the operations). Contrary to these simplifications, after applying Tabu search, there can be cases where a t^C-type operation is before a t^B-type operation in the operation sequence of a machine, and machines also may exist with more than one setup time. To maintain the original simplifications of the heuristic algorithm, a variant of the Tabu search was implemented, where the presented simplifications exist.

The Tabu search algorithm of Hurink et al. lets to select operation only from a block of a critical path to create a neighbour. Due to the specificities of the problem considered in [1], the move of any element of a critical path (even if it is not part of a block) may improve the schedule. An example for that is illustrated in Fig. 3. Noticing this property, another variant of Tabu search was implemented where any operation of a critical path can be selected to reallocate in the process of neighbour creation.

Altogether, it results in four variants of the Tabu search:

- maintaining the operation sequence constraints, any element of a critical path can be selected to reallocate during the neighbour creation process,

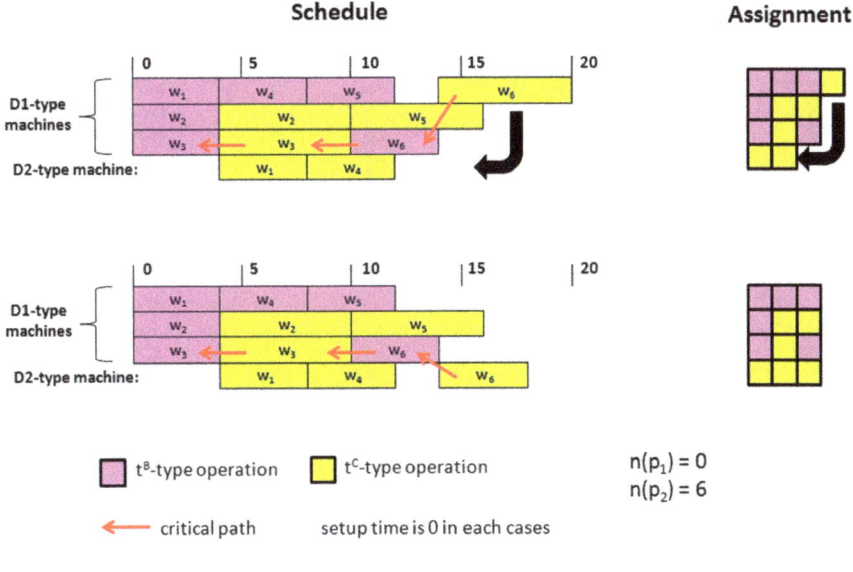

Fig. 3 Improvement reached by moving a non-block element from a critical path

- maintaining the operation sequence constraints, only elements of blocks can be selected to reallocate during the neighbour creation process,
- resolving the operation sequence constraints, any element of a critical path can be selected to reallocate during the neighbour creation process and
- resolving the operation sequence constraints, only elements of blocks can be selected to reallocate during the neighbour creation process.

5 Comparison of the Different Improvement Techniques

The local search of the heuristic algorithm of [1] was changed to the four variants of Tabu search, one-after-one, and the makespan of the whole schedule obtained by the application of each variant was compared with the makespan of the whole schedule obtained by the algorithm that was improved by LS. During this comparison, we varied the size of the Tabu list. The applied values of the size of the Tabu list were 10, 30, 50, 70 and 100. For the comparison, we used 1000 different random test inputs, where the number of the products ($n(p_1)$ and $n(p_2)$), the operation times and the setup times were chosen randomly from a predefined interval.

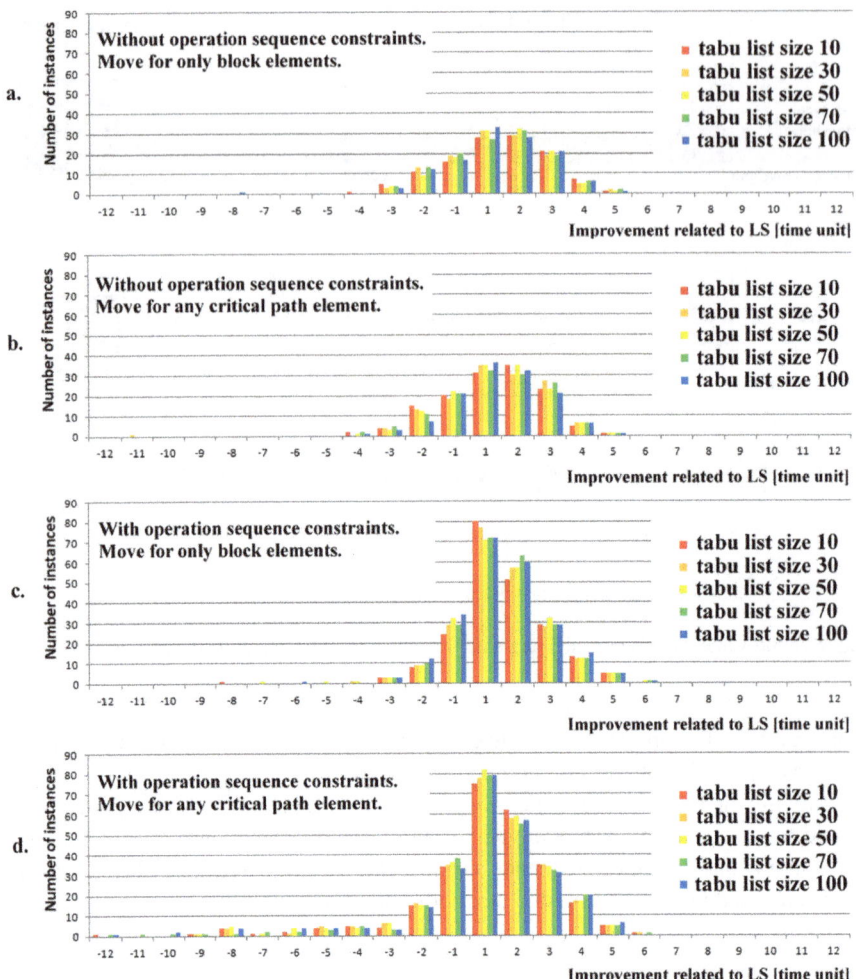

Fig. 4 Comparison of the results obtained by the application of the four variants of TS with the results obtained by the application of LS

The values of the improvement are highlighted in Fig. 4, where the horizontal axis is the difference between the makespan obtained by the application of LS for the improvement and the makespan obtained by the application of TS for the improvement, and the vertical axis shows the number of the cases of the 1000 test inputs, for what the difference was obtained.

The comparison shows that the application of TS instead of the LS is more efficient. In most of the cases, they result in the same makespan, however, for the remaining inputs TS results in shorter makespans in more cases than the local search. As it can be seen, the best results were obtained when the TS was applied in the improvement phase of the heuristic scheduler that maintains the operation sequence-related constraints and moves only the elements of the blocks of a critical path during the neighbour creation process. Moreover, the best results belong to the case when the Tabu list size was set to 70.

In case of capacity restrictions and a long Tabu list, usually instead of the whole solutions, only their typical properties are stored in the Tabu list. In that case, aspiration criteria (see [7]) are often applied to accept special neighbours that are elements of the Tabu list. As additional parameters, the application of aspiration criteria and some other parameters to control these special solutions' selection also could be used in the algorithm. Although in our case, Tabu list size 70 gave the best results; thus, the store of the whole solutions in the Tabu list does not require huge capacity.

6 Conclusion

This paper gives a hybridization technique, where four different variants of Tabu search algorithm were used instead of the algorithms that were used by Auer et al. [1] for the improvement phase of their heuristic scheduler. These four variants were tested on 1000 random test instances, and the most effective one was determined. We have experienced that the best results are obtained when the constraints introduced by the original heuristic algorithm are kept, only block elements are allowed to move in the neighbour generation phase of the TS, and the size of the Tabu list is 70.

Acknowledgements Supported by the ÚNKP-20-4 New National Excellence Program of the Ministry for Innovation and Technology from the source of the National Research, Development and Innovation Fund. We acknowledge the financial support of Széchenyi 2020 under the EFOP-3.6.1-16-2016-00015.

References

1. Auer P, Dósa G, Dulai T, Fügenschuh A, Näser P, Ortner R, Werner-Stark Á (2020) A new heuristic and an exact approach for a production planning problem. Cent Eur J Oper Res 1–35. https://doi.org/10.1007/s10100-020-00689-3
2. Hurink J, Jurisch B, Thole M (1994) Tabu search for the jobshop scheduling problem with multi-purpose machines. Oper Res Spektrum 15:205–215
3. Muthuraman S Venkatesan VP (2017) A comprehensive study on hybrid meta-heuristic approaches used for solving combinatorial optimization problems. In: 2017 world congress on computing and communication technologies (WCCCT), pp 185–190
4. Kis T (2003) Job-shop scheduling with processing alternatives. Eur J Oper Res 151:307–332
5. Ruiz R, Rodríguez JAV (2010) The hybrid flow shop scheduling problem. Eur J Oper Res 205:1–18

6. Chen B, Potts CN, Woeginger GJ (1998) A review of machine scheduling: complexity, algorithms and approximability. In: Handbook of combinatorial optimization, vol 3. Kluwer Academic Publishers
7. Dell'Amico M, Trubian M (1993) Applying Tabu search to the job-shop scheduling problem. Ann Oper Res 41:231–252

Chapter 14
Harmonic Estimator Using Design Atom Search Optimization Algorithm

Aishwarya Mehta, Jitesh Jangid, Akash Saxena, Shalini Shekhawat, and Rajesh Kumar

1 Introduction

In the present scenario of power system network, the widespread use of nonlinear and power electronics devices is increasing. These devices deteriorate power quality which mainly depends on the voltage, current, and frequency components. Power quality is a vital problem in front of researchers and electrical engineers because power quality reduces the reliability of the system and increases power losses. Nonlinear and power electronics devices generate adverse effects on the power system network by generating harmonics. In the power system, harmonics, [1] are the distorted signal of the current or voltage waveform. This harmonics signal also impacts the quality of power. Harmonics are mainly divided into two categories: First one is even harmonics which are even multiple of the fundamental frequency. The second one is odd harmonics that are an odd multiple of the fundamental frequency component. For accurate estimation of phase and amplitude components of power signal, harmonic estimation is necessary. Accurate estimation of harmonics is helpful in design mitigation filters that help to remove the spikes of frequency and improve the power quality of the power system network.

In past decades, many researchers try to get accurate estimation of phase and amplitude components by different conventional and metaheuristic techniques. In this consequence of estimation of harmonic components, mostly researchers used fast Fourier transform (FFT) [2] but during the picket fence effect, FFT fails to provide an accurate estimation. Phase-locked loop (PLL) [3], Kalman filter (KF)

A. Mehta · J. Jangid · A. Saxena (✉)
Department of Electrical Engineering, SKIT, Jaipur, India

S. Shekhawat
Department of Mathematics, SKIT, Jaipur, India

R. Kumar
Department of Electrical Engineering, MNIT, Jaipur, India

© The Author(s), under exclusive license to Springer Nature Singapore Pte Ltd. 2021   181
S. Kumar et al. (eds.), *Proceedings of International Conference on Communication and Computational Technologies*, Algorithms for Intelligent Systems,
https://doi.org/10.1007/978-981-16-3246-4_14

[4], and adaptive notch filtering (ANF) [5] also used to design mitigation filters. For KF state, matrix identification and statistical information are necessary requirements which make a disturbance in the estimation of harmonic components. Later, artificial neural network (ANN) [6] is also used that uses directional links. These directional links increase when the complexity or dimension of the signal increases which is the cause of ANN failure. FFT, KF, ANF, and PLL are known as conventional methods. Conventional methods are failed when the number of dimension of electrical signal increases, and these methods stuck into local optima and do not find global optima. To overcome this problem, researchers adopt the newly developed metaheuristic techniques which have the ability to escape from the local optima and search global optima without consuming more time. These metaheuristic techniques are classified into five categories

- Environment-Based Techniques: These techniques are influenced by the surrounding and the environment. Grey wolf optimization (GWO) [7], firefly algorithm (FA) [8], ant colony optimization algorithm (ACO) [9], etc., are examples of these techniques.
- Evolutionary-Based Techniques: These techniques follow Darwin's theory of evolution. According to the evolution theory, every individual has different abilities and tries to give better results in their next generation by mutation process. Examples of these techniques are differential evolution (DE) [10], genetic algorithm (GA) [11], etc.
- Chemistry- and Physics-Based Techniques: Chemical reactions and physics-based principles are followed by these techniques. These principles help to find the optimal solution to the research problem. Chemistry-based algorithms are chemical reaction optimization (CRO) [12], while physics-based algorithms are Big Bang–Big Crunch algorithm (BB-BC) [13], space gravitational algorithm (SGA) [14], central force optimization (CFO) [15], magnetic optimization algorithm (MOA) [16], etc.
- Human Nature-Based Techniques: These techniques are inspired by human nature. Teaching-based optimization algorithm (TLBO) [17], harmony search (HS) [18], exchange market algorithm (EMA) [19] are examples of these techniques.
- Hybrid Techniques: Hybrid techniques are the combination of two or more algorithms which helps to improve the result of the problem. Examples of these techniques are grey wolf optimization (GWO) with particle swarm optimization (PSO) [20], ant colony optimization (ACO) with PSO [21] and adaptive neuro-fuzzy inference system (ANFIS), artificial bee colony (ABC) algorithm [22], etc.

The free lunch theorem is stated that in nature an algorithm has no capability to solve all the research problems. This theorem motivates researchers to adopt and develop more algorithms. In this consequence, recently developed atom search algorithm is introduced in this paper to solve the harmonic estimation problem. This ASO algorithm developed by Zhao et al. in [23] and applied in many research problems like dispersion coefficient estimation [23] and hydro-geologic parameter estimation

[24]. ASO algorithm has more ability to solve the problem without trapping into local optimas and find global optima in less time. These qualities of ASO motivate us to solve the harmonic problem with this algorithm.

1.1 Objectives

Following the literature review on the developments of recent metaheuristic in the problem of harmonic estimation, the following research goals are formed

- To evaluate components of the harmonic estimation problem by atom search optimization algorithm (ASO).
- To design a mathematical model of the harmonic estimator and test the reliability, effectiveness of ASO for estimation.

1.2 Paper Structure

This paper is segmented into numerous parts. The literature on harmonic estimation problems and introductory details of the problem are presented in Sect. 1. Problem formulation of harmonics is described in Sect. 2. Applied algorithm ASO briefly discussed in Sect. 3. Section 4 is used to check the applicability of ASO on the harmonic estimator problem, and finally, in Sect. 5, we concluded the paper.

2 Problem Formulation

Basically, voltage or current waveform of power harmonics evaluation can be defined as

$$A(t) = \sum_{n=1}^{N} Y_n(w_n t + \gamma_n) + Y_{dc} \exp\left(-\theta_{dc}t\right) + \sigma(t) \tag{1}$$

Here, N is used to express the order of harmonics and known as the angular frequency of the n component. The fundamental frequency is denoted by f while it represents the additive white Gaussian noise (AWGN). To denote decaying term …is used. Unknown parameters of the waveform are for nth harmonics. Now, it is necessary to the sampled original signal with sampling period which is represented as

$$A(m) = \sum_{n=1}^{N} Y_n \sin\left(w_n m T_x + \gamma_n\right) + Y_{dc} \exp(-\theta_{dc}m T_x) + \sigma(m) \tag{2}$$

Then, in Eq. 2, we applied Taylor series expansion which is represented as below

$$A(m) = \sum_{n=1}^{N} Y_n \sin(w_n m T_x + \gamma_n) + Y_{dc} - Y_{dc}\theta_{dc}m T_x + \sigma(m) \tag{3}$$

After that, a generalized and concise formula is obtained which is helpful to estimate the amplitude and phase component of overall harmonics. In addition, the nonlinearity of the proposed sinusoidal model is increased by the phase component, and to reduce this problem, it is necessary to generate a more simplified formulation which is used to implement optimization methods. Now, Eq. 3 is simplified using sine and cosine terms which are written as

$$A(m) = \sum_{n=1}^{N} [Y_n \sin w_n m T_x \cos \gamma_n + Y_n \cos(w_n m T_x) \sin \gamma_m]$$
$$+ Y_{dc} - Y_{dc}\theta_{dc}m T_x + \sigma(m) \tag{4}$$

Later, the electrical signal of the non-parametric form is transformed into parametric form as below

$$A(m) = H(m)\gamma(m) \tag{5}$$

Here, $H(m)$ is defined as

$$H(m) = [\sin(w_1 m T_x)\cos(w_1 m T_x)\ldots\sin(w_n m T_x)\cos(w_n m T_x)1 - p T_x]^T \tag{6}$$

Unknown parameters of the whole vector are expressed as

$$\gamma(m) = [\gamma_{1m}\gamma_{2n}\cdots\gamma_{(2n-1)}\gamma_{2mn}\gamma_{(2n+1)m}\gamma_{(2n+2)m}]^T \tag{7}$$

$$\gamma = [Y_1 \cos(\gamma_1)Y_1 \sin \gamma_1]\ldots Y_n \cos(\gamma_n)Y_n \sin \gamma_n \ldots Y_{dc}Y_{dc}\theta_{dc}]^T \tag{8}$$

C is the objective function of the harmonic estimation problem, which helps to optimize unknown parameters (phase and amplitude) and can be estimated as

$$C = min\left(\sum_{m=1}^{M} E_m^2(m)\right) = min\left(\sum_{m=1}^{M}(A_m - A_{m_{est}})^2\right) = MSE(A_m - A_{m_{est}}) \tag{9}$$

Here, A_m refers t original harmonic signal, and $A_{m_{est}}$ denotes the evaluated estimated output signal.

3   Overview of Atom Search Optimization (ASO)

Acceleration of the atom can be estimated as

$$a_x = \frac{F_x + C_x}{M_x} \tag{10}$$

Here, F_x represents interaction force of atom, and C_x is constraint force which is implied on the xth atom having a mass M_x.

3.1   Force of Attraction

Lennard potential [25] (a mathematical representation of interaction force between atoms) between ith and jth atom is expressed as

$$Z(r_{xy}) = 4P\left[\left(\frac{\Delta}{r_{xy}}\right)^{12} - \left(\frac{\Delta}{r_{xy}}\right)^{6}\right] \tag{11}$$

where P refers to potential well depth, and Δ represents a bound distance where the inter-particle potential is zero. $\left(\dfrac{\Delta}{r_{xy}}\right)^{12}$ and $\left(\dfrac{\Delta}{r_{xy}}\right)^{6}$ are repulsion and attraction between atoms.

r_{xy} is Eulicidian distance between j_x and j_y which is estimated as

$$r_{xy} = \| j_y - j_x \| = \sqrt{(j_{x1} - j_{y1})^2 + (j_{x2} - j_{y2})^2 + \cdots + (j_{xn} - j_{yn})^2} \tag{12}$$

The interaction force operation yth to xth atom in a time interval of t for d dimension is represented as

$$F_{xy}^{D}(t) = -\nabla Z(r_{xy}) = \frac{24P(t)}{\Delta(t)}\left[2\left(\frac{\Delta(t)}{r_{xy}(t)}\right)^{13} - \left(\frac{\Delta(t)}{r_{xy}(t)}\right)^{7}\right]\frac{r_{xy}}{r_{xy}^{D}} \tag{13}$$

$$F_{xy}^{D}(t) = -\nabla Z(r_{xy}) = \frac{24P(t)}{\Delta(t)}\left[2\left(\frac{\Delta(t)}{r_{xy}(t)}\right)^{13} - \left(\frac{\Delta(t)}{r_{xy}(t)}\right)^{7}\right]\frac{r_{xy}}{R_{xy}^{d}} \tag{14}$$

In this phenomena, atom maintains a distance from other atoms that changes during the attraction or repulsion between atoms.

$$F_{xy}'(t) = \frac{24P(t)}{\Delta(t)}\left[2\left(\frac{\Delta(t)}{r_{xy}(t)}\right)^{13} - \left(\frac{\Delta(t)}{r_{xy}(t)}\right)^{7}\right] \tag{15}$$

$$F'_{xy}(t) = -\gamma(t)\left[2(b_{xy}(t))^{13} - (b_{xy}(t))^7\right] \tag{16}$$

Here, $\gamma(t)$ is known as depth function which helps to regulate attraction or repulsive area. It is measured as

$$\gamma(t) = \beta\left(1 - \frac{t-1}{T}\right)^3 e^{\frac{-20t}{T}} \tag{17}$$

where β represents depth weight, and T shows maximum number of iterations.

$$b_{xy}(t) = \begin{cases} b_{min} \; if \; \frac{r_{xy}(t)}{\Delta(t)} < b_{min} \\ \frac{r_{xy}(t)}{\Delta(t)} \; if \; b_{min} \le \frac{r_{xy}(t)}{\Delta(t)} \le b_{max} \\ b_{max} \; if \; \frac{r_{xy}(t)}{\Delta(t)} > b_{max} \end{cases} \tag{18}$$

Here, b_{max} and b_{min} are maximum and minimum boundaries of b_{xy}, respectively. $\Delta(t)$ is known as length scale which is estimated as

$$\Delta(t) = \left\| j_{xy}(t), \frac{\sum_{y \in Hbest} j_{xy}(t)}{H(t)} \right\| \tag{19}$$

Hbest represents best function fitness value for a subset of H atoms.

$$\begin{cases} b_{min} = q_o + q(t) \\ b_{max} = p \end{cases} \tag{20}$$

where p is used to represent upper limit, and q is known as drift function which is used to drift from exploration to exploitation, and it is calculated as

$$q(t) = 0.1 * \sin\left(\frac{\pi}{2} * \frac{t}{T}\right) \tag{21}$$

Now, the total working force is calculated on xth atom. On this xth atom, force is acting from all other atoms which is the weighted sum of force on atoms in the D dimension.

$$F_x^D(t) = \sum_{y \in Hbest} \text{random}_y F_{xy}^D(t) \tag{22}$$

where random_y is a random number which lies between 0 to 1.

The geometric constraints in molecular dynamics plays an important role in the movement of atoms. For simplicity, an atom is considered to have a covalent bond with the best atom in ASO, so each of them is operated by the best atom by a constraint force.

$$Q_x^D = \lambda(t)(j_{best}^D(t) - j_x^D(t)) \tag{23}$$

where λ represents Lagrangian multiplier which is expressed as

$$\lambda(t) = \rho e^{\frac{-20t}{T}} \tag{24}$$

3.2   Atomic Motion

Acceleration of xth atom for D dimension at t time considering both geometric constraints and interaction force is estimated as

$$
a_x^D(t) = \frac{F_x^D(t) + G_x^D(t)}{M_x} \alpha \left(1 - \frac{-t-1}{T} \right)^3 e^{\frac{-20t}{T}}
$$

$$
\times \sum_y \frac{random_y \left[2\left(b_{xy}(t)\right)^{13} - \left(b_{xy}(t)\right)^7 \right]}{M_x(t)} \frac{\left(j_x^D(t) - j_x^D(t)\right)}{j_x(t), j_y(t)} +
$$

$$
\rho e^{-\frac{20t}{T}} \frac{j_{best}^D(t) - j_x^D(t)}{M_x(t)} \tag{25}
$$

For xth atom at ith iteration, mass is $m_x(t)$ which is determined as

$$m_x(t) = e - \frac{\text{Fit}_x(t) - \text{Fit}_{best}(t)}{\text{Fit}_{worst}(t) - Fit_{best}(t)} \tag{26}$$

$$M_x(t) = \frac{m_x(t)}{\sum_{y=1}^{N} m_y(t)} \tag{27}$$

where

$$\text{Fit}_{best}(t) = \min_{x=1,2,\dots n} \text{Fit}_x(t) \text{ and } \text{Fit}_{worst}(t) = \max_{i=1,2,\dots,n} \text{Fit}_x(t).$$

Velocity and position of xth atom at the time $(t+1)$ are represented as

$$j_x^D(t+1) = j_x^D(t) + u_x^D(t+1) \tag{28}$$

$$u_x^D(t+1) = random_x^D u_x^D(t) + a_x^D(t) \tag{29}$$

Exploration ability needs improvement in order to implement it to the optimization problem. Therefore, in the beginning, like its neighbors, each atom requires a large number of atoms with better fitness values, and therefore, the value of H must be high. To improve exploitation in ending of algorithm, number of neighbors K must be decreased. So, H is calculated as

$$H(t) = N - (N-2)\sqrt{\frac{t}{T}} \qquad (30)$$

Algorithm 1: Atom Search Optimization (ASO)

1: **Begin**
2: Initialize the Atoms (A_T)
3: **For** each atom A_{Ti} do
4: Fitness Calculation of Atom Fitness;
5:     **if** $Fitness_i < Fitness_{Best}$ then
6:         $Fitness_{Best} = Fitness_i$
7:         $A_{T\,Best} = A_{Ti}$
8:     **End If**
9:     **end**
10: Massing calculation using equations (25) and (26)
11: Using equation (29) to evaluate the neighbors H
12: Calculate Fi (interaction force) and Qi (restriction force) using equations (22) and (23), respectively
13: Acceleration Calculation using Eq (25)
14: **End**
15: Update Best solution Obtain so far $A_{T\,Best}$

---

4 Case Studies for Harmonic Problem Estimator

From the brief research study of the harmonic estimator problem and the consistent reporting of metaheuristic approaches in the design of harmonic estimators, it is clear that this problem still has exciting opportunities for analysis. For this assessment, we will assess the ASO algorithm on the two problems which are taken from the literature.

4.1 Case Study of Design Problem-1

Design problem-1 is made up of fundamental, 3rd, 5th, 7th and 11th harmonics. Here, fundamental harmonics of 50 Hz is used. The main motive of this design problem is to assess the amplitude and phase components with their respective harmonics. Wave equation of this design problem-1 is expressed as

$$P_1(m) = \frac{3}{2} \times \sin(2\pi f^{\text{fundamental}} m + 80) + \frac{2}{4} \times \sin(2\pi f^3 m + 60)$$
$$+ \frac{2}{10} \times \sin(2\pi f^5 m + 45) + \frac{3}{20}$$
$$\times \sin(2\pi f^7 m + 36) + \frac{2}{20} \times \sin(2\pi f^{11} m + 30)$$
$$+ (\frac{1}{2}\exp -5m) + \gamma_n$$

Assessment results of design problem-1 are depicted in Tables 1, 2 and 3. Where Table 1 represents the estimated values of phase and amplitude components of harmonic estimator design problem-1. These amplitude and phase components of ASO are compared with particle swarm optimization with passive congregation least square (PSOPC-LS) [26, 27], modified artificial bee colony algorithm (MABC) [28],GA-LS [26, 27], F-BFO [26, 27], and BFO [26, 27]. From these results, it is evident that ASO provides an exact estimation of amplitude and phase components in comparison with other algorithms.

Table 2 is representing the % error analysis of problem-1 for phase and amplitude components of the harmonic wave. % Error analysis is used to check which algorithm provides an optimal solution to the problem. From Table 2 results, it indicate that ASO provides minimum values of % error of phase and amplitude components in compare to BFO-RLS, PSOPC-LS, GA-LS, MABC, which represents that harmonic problem can be solved by ASO successfully.

Table 1 Algorithms comparison on problem-1

Algorithms	Parameters	Fun	3rd	5th	7th	11th
Actual	F (Hz)	50	150	250	350	550
	A (V)	1.5	0.5	0.2	0.15	0.1
	P (deg.)	80	60	45	36	30
PSOPC-LS [26, 27]	A (V)	1.4828	0.4886	0.1822	0.1561	0.0948
	P (deg.)	80.5423	62.2445	46.6343	34.6214	27.3154
MABC [28]	A (V)	1.5006	0.4997	0.1995	0.1498	0.1003
	P (deg.)	80.0187	60.0098	45.0905	36.0894	29.5511
GA-LS [26, 27]	A (V)	1.48	0.485	0.18	0.158	0.0937
	P (deg.)	80.61	62.4	47.03	34.354	26.7
F-BFO [26, 27]	A (V)	1.488	0.5103	0.198	0.1545	0.1028
	P (deg.)	80.42	58.1	45.75	34.73	29.358
BFO [26, 27]	A (V)	1.48	0.5108	0.1945	0.1556	0.1034
	P (deg.)	80.4732	57.9005	45.8235	34.5606	29.358
Proposed	A (V)	**1.4997**	**0.5000**	**0.1998**	**0.1500**	**0.0999**
	P (deg.)	**79.9947**	**59.9947**	**44.9272**	**35.5330**	**29.9430**

Table 2 % error analysis of harmonics estimator design problem-1

Algorithms	Parameters	Order of harmonics				
		Fun	3rd	5th	7th	11th
BFO [26, 27]	% Phase error	4.340E−01	5.660E-01	1.550E+00	3.310E+00	2.130E-01
	% Amp. error	3.840E−01	2.860E−01	9.020E−01	1.760E+00	1.750E+00
PSOPC-LS [26, 27]	% Phase error	6.750E−01	3.670E+00	3.560E+00	3.820E+00	8.970E+00
	% Amp. error	1.200E+00	2.400E+00	9.000E+00	4.060E+00	5.200E+00
GA-LS [26, 27]	% Phase error	7.630E−01	4.000E+00	4.510E+00	4.570E+00	1.100E+01
	% Amp. error	1.330E+00	3.000E+00	1.000E+01	5.330E+00	6.300E+00
MABC [28]	% Phase error	2.340E−02	1.640E−02	2.010E−01	2.490E−01	1.500E+00
	% Amp. error	4.120E−02	5.530E−02	2.420E−01	1.550E−01	3.370E−01
Proposed	% Phase error	**6.619E−03**	**8.821E−03**	**1.619E−01**	**1.297E+00**	**1.900E−01**
	% Amp. error	**3.365E−04**	**2.233E−04**	**3.927E−03**	**4.298E−04**	**3.444E−03**

Table 3 represents the phase and amplitude values in the presence of noise signals. In this, we take into account noise signals which are below

1. 10 dB
2. 20 dB
3. 30 dB
4. 40 dB

In this, these noise signals are mixed with the original signal and check the effect of the noisy signal on the original signal. From these Table 3 results, it is evident that the noisy signal disrupts the phase and amplitude components of the harmonic wave signal. From the result assessment, it is concluded that ASO provides better results in comparison with other algorithms. Figure 1 shows the estimated wave, which is plotted with the analysis of the original design problem of case 1 and the estimated wave of ASO. From this plot, it is seen that the estimated and original design problem is similar to each other.

Table 3 Problem-1 analysis with noise signals

Signal's	Parameters	Fun	3rd	5th	7th	11th
Actual signal parameter's	F (Hz)	50	150	250	350	550
	A (V)	1.5	0.5	0.2	0.15	0.1
	P (deg.)	80	60	45	36	30
Wave without noise	A (V)	**1.5000**	**0.5000**	**0.2000**	**0.1500**	**0.1000**
	P (deg.)	**79.9947**	**59.9947**	**44.9272**	**35.5330**	**29.9430**
Original with 10 dB noise	A (V)	1.4954	0.4762	0.1948	0.1607	0.1223
	P (deg.)	79.9949	59.9946	44.9154	35.4433	29.9391
Original with 20 dB noise	A (V)	1.4979	0.4905	0.1955	0.1592	0.1144
	P (deg.)	79.9954	59.9948	44.9046	35.3332	29.9464
Original with 30 dB Noise	A (V)	1.5003	0.5004	0.1960	0.1537	0.1036
	P (deg.)	79.9950	59.9955	44.8951	35.4776	29.9634
Original with 40 dB noise	A (V)	1.5001	0.4999	0.1985	0.1513	0.1005
	P (deg.)	79.9950	59.9952	44.8957	35.4991	29.9442

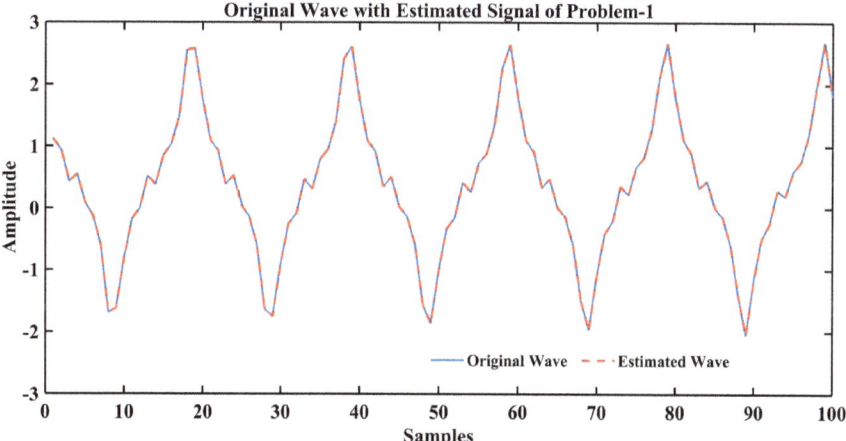

Fig. 1 Estimated wave of problem-1

4.2   Case Study of Design Problem-2

After inspecting design problem-1, it is concluded that ASO provides an exact estimation of phase and amplitude components with their respective harmonics. Now, we take the second design problem which aims to estimate the components of harmonics in the presence of inter and sub-harmonics [29]. Inter-harmonics [30] are non-integer multiples of fundamental which is greater than the fundamental frequency such 180, 230 Hz. While sub-harmonics are also non-integer multiple of harmonics which is less than fundamental frequency like 20 Hz. These inter- and sub-harmonics have an adverse effect on the electrical power system network such as thermal heating, interference in power line communication.

Design problem-2 is made up of sub, fundamental, 3rd, inter-1, inter-2, 5th, 7th, and 11th harmonics. Here, fundamental harmonics 50 Hz is used. The main motive of this design problem is to assess the amplitude and phase components with their respective harmonics. Wave equation of this design problem-2 is expressed as

$$
\begin{aligned}
P_2(m) = {} & \frac{101}{200} \times \sin(2\pi f^{\text{sub}} m + 75) + \frac{3}{2} \times \sin(2\pi f^1 m + 80) + \frac{1}{2} \times \sin(2\pi f^3 m + 60) \\
& + \frac{1}{4} \times \sin(2\pi f^{\text{inter1}} m + 80) + \frac{7}{20} \times \sin(2\pi f^{\text{inter2}} m + 20) + \frac{1}{5} \times \sin(2\pi f^5 m + 45) \\
& + \frac{3}{20} \times \sin(2\pi f^7 m + 36) + \frac{1}{10} \times \sin(2\pi f^{11} m + 30) + (\frac{1}{2}\exp{-5m}) + \gamma_n
\end{aligned}
$$

The estimated results of problem-2 are recorded in Tables 4, 5, and 6. Here, Table 4 represents the estimated values of phase and amplitude components of harmonic estimator design problem-2. These amplitude and phase components of ASO are compared with PSOPC-LS, MABC, GA-LS, F-BFO, and BFO. From these results, it is evident that ASO provides an exact estimation of amplitude and phase components in the presence of sub- and inter-harmonics.

Table 5 represents the % error analysis of problem-2 for phase and amplitude components of the harmonic wave. % error analysis is used to check which algorithm provides an optimal solution to the problem in the presence of sub- and inter-harmonics. From Table 5 results, it indicate that ASO provides minimum values of % error of phase and amplitude components compared to BFO-RLS, PSOPC-LS, GA-LS, MABC, which represents that harmonic problem can be solved by ASO successfully.

Table 6 represents the phase and amplitude values in the presence of noise signals. In this, we take into account noise signals like 10, 20, 30 and 40 dB in the presence of sub- and inter-harmonics. In this, these noise signals are mixed with the original signal and check the effect of the noisy signal on the original signal. From these Table 6 results, it is evident that the noisy signal disrupts the phase and amplitude components of the harmonic wave signal. From the result assessment, it is concluded that ASO provides better results in comparison with other algorithms. Figure 2 shows

Table 4 Results of harmonics estimator design problem-2

Algorithm comparison on problem-2

Algorithms	Parameters	Sub	Fun	3rd	Inter-1	Inter-2	5th	7th	11th
Actual	F (Hz)	20	50	150	180	230	250	350	550
	A (V)	0.505	1.5	0.5	0.25	0.35	0.2	0.15	0.1
	P (deg.)	75	80	60	65	20	45	36	30
PSOPC-LS [26, 27]	A (V)	0.53	1.5049	0.281	0.24	0.377	0.211	0.165	0.11
	P (deg.)	73.51	79.45	58.12	63.28	18.23	48.1	37.109	31.87
MABC [28]	A (V)	0.5052	1.5008	0.4998	0.25	0.35	0.1997	0.15	0.1
	P (deg.)	74.9539	79.98	60.1254	64.9374	20.04	45.1636	35.9987	29.958
GA-LS [26, 27]	A (V)	0.532	1.5083	0.472	0.238	0.381	0.251	0.172	0.117
	P (deg.)	73.02	79.23	57.55	62.41	17.64	48.33	38.78	32.56
F-BFO [26, 27]	A (V)	0.521	1.489	0.489	0.261	0.371	0.28	0.1468	0.1019
	P (deg.)	74.61	79.86	61.16	64.33	19.729	47.22	36.658	30.52
BFO [26, 27]	A (V)	0.525	1.4788	0.4877	0.2664	0.3729	0.2052	0.1464	0.1016
	P (deg.)	74.48	79.8361	61.2316	63.991	19.6887	47.698	36.7372	29.3928
Proposed algo	A (V)	0.5049	1.4999	0.5000	0.2500	0.3501	0.2001	0.1500	0.1000
	P (deg.)	74.9948	79.9947	59.9946	64.9948	19.9954	44.9958	35.9961	29.9948

Table 5 % error analysis of harmonics estimator design problem-2

Algorithms	Parameters	Order of harmonics							
		Sub	Fun	3rd	Inter-1	Inter-2	5th	7th	11th
PSOPC-LS [26, 27]	% Phase error	1.99E+00	6.87E-01	3.13E+00	2.65E+00	8.85E+00	6.88E+00	3.08E+00	6.23E+00
	% Amp. error	4.95E+00	3.26E-01	3.80E+00	4.00E+00	7.70E+00	5.50E+00	1.00E+01	1.10E+01
MABC [28]	% Phase error	6.15E-02	2.50E-02	2.09E-01	9.64E-02	2.00E-01	3.64E-01	3.60E-03	1.40E-01
	% Amp. error	4.68E-02	5.30E-02	4.41E-02	1.33E-02	1.10E-03	1.39E-01	3.19E-02	1.84E-02
GA-LS [26, 27]	% Phase error	2.64E+00	9.62E-01	4.08E+00	5.52E+00	1.18E+01	7.40E+00	7.72E+00	8.53E+00
	% Amp. error	5.35E+00	5.53E-01	5.60E+00	4.80E+00	8.85E+00	7.50E+00	1.47E+01	1.70E+01
F-BFO [26, 27]	% Phase error	5.20E-01	1.75E-01	1.93E+00	1.03E+00	1.39E+00	4.93E+00	1.83E+00	1.73E+00
	% Amp. error	3.25E+00	7.33E-01	4.89E-01	4.40E+00	6.00E+00	4.00E+00	2.13E+00	1.90E+00
BFO [26, 27]	% Phase error	2.53E-01	1.06E-01	1.54E+00	5.30E-01	6.61E-01	2.84E+00	1.24E+00	2.14E-01
	% Amp. error	1.19E+00	1.95E-01	1.59E+00	3.24E+00	3.97E+00	4.54E-01	1.41E+00	1.48E+00
Proposed	% Phase error	**6.98E-03**	**6.65E-03**	**9.01E-03**	**8.07E-03**	**2.32E-02**	**9.38E-03**	**1.07E-02**	**1.75E-02**
	% Amp. error	**5.05E-03**	**1.09E-03**	**6.90E-04**	**7.06E-03**	**1.09E-02**	**1.14E-02**	**2.28E-03**	**8.00E-03**

Table 6 Problem-2 analysis with noise signals

Signals	Parameters	Sub	Fun	3rd	Inter-1	Inter-2	5th	7th	11th
Actual signal parameters	F (Hz)	20	50	150	180	230	250	350	550
	A (V)	0.505	1.5	0.5	0.25	0.35	0.2	0.15	0.1
	P (deg.)	75	80	60	65	20	45	36	30
Wave without noise	A (V)	**0.5050**	**1.5000**	**0.5000**	**0.2500**	**0.3500**	**0.2000**	**0.1500**	**0.1000**
	P (deg.)	**74.9948**	**79.9947**	**59.9946**	**64.9948**	**19.9954**	**44.9958**	**35.9961**	**29.9948**
Original with 10 dB noise	A (V)	0.4789	1.4682	0.4989	0.2503	0.3508	0.1982	0.1187	0.0761
	P (deg.)	74.9947	79.9947	59.9951	64.9957	19.9945	44.9951	35.9955	29.9961
Original with 20 dB noise	A (V)	0.4904	1.4854	0.5008	0.2489	0.3506	0.2005	0.1430	0.0903
	P (deg.)	74.9946	79.9950	59.9943	64.9949	19.9953	44.9949	35.9958	29.9959
Original with 30 dB noise	A (V)	0.5008	1.4942	0.4999	0.2503	0.3500	0.1999	0.1486	0.1006
	P (deg.)	74.9946	79.9964	59.9948	64.9951	19.9952	44.9961	35.9957	29.9953
Original with 40 dB Noise	A (V)	0.5034	1.4983	0.5003	0.2499	0.3499	0.1999	0.1503	0.1000
	P (deg.)	74.9944	79.9950	59.9951	64.9956	19.9952	44.9957	35.9947	29.9957

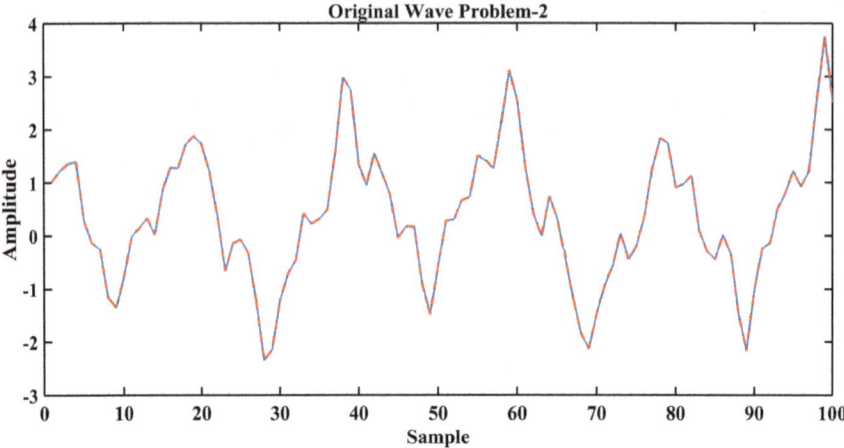

Fig. 2 Estimated wave of problem-2

the estimated wave, which is plotted with the analysis of the original design problem of case 2 and the estimated wave of ASO. From this plot, it is seen that the estimated and original design problem is similar to each other.

5 Conclusion

Exact estimation of phase and amplitude parameters is necessary to develop harmonic mitigation filters, so harmonic estimation is important. Mitigation filters help to reduce the losses in the power system network and provides a reliable power supply to consumers. In this work, ASO is used to design harmonic mitigation filters and also prove the superiority of ASO harmonic mitigation filter which is proved by comparing other algorithms that are proposed in the literature.

We notice that the harmonic estimator based on ASO demonstrates acceptable results. We have measured the phase and amplitude of waves with their respective measurement errors in these case studies. We get that ASO provides more accurate results in comparison with PSOPC-LS, MABC, GA-LS, F-BFO, and BFO. It can be observed from this application that ASO is a powerful algorithm and can be an appropriate choice to design harmonic estimators.

References

1. Subjak JS, Mcquilkin JS (1990) Harmonics-causes, effects, measurements, and analysis: an update. IEEE Trans Ind Appl 26(6):1034–1042
2. Nussbaumer HJ (1981) The fast fourier transform. In: Fast fourier transform and convolution algorithms. Springer, pp 80–111
3. Benhabib MC, Saadate S (2005) A new robust experimentally validated phase locked loop for power electronic control. EPE J 15(3):36–48
4. Abdelsalam A, Eldesouky AA, Sallam AA (2012) Classification of power system disturbances using linear kalman filter and fuzzy-expert system. Int J Electric Power Energy Syst 43(1):688–695
5. Rao DB, Kung S-Y (1984) Adaptive notch filtering for the retrieval of sinusoids in noise. IEEE Trans Acoustics Speech Signal Process 32(4):791–802
6. Hassoun MH et al (1995) Fundamentals of artificial neural networks. MIT Press, Cambridge
7. Mirjalili S, Mirjalili SM, Lewis A (2014) Grey wolf optimizer. Adv Eng Software 69:46–61
8. Yang X-S et al (2008) Firefly algorithm. Nature-Inspired Metaheuristic Algorithms 20:79–90
9. Dorigo M, Di Caro G (1999) Ant colony optimization: a new meta-heuristic. In: Proceedings of the 1999 congress on evolutionary computation-CEC99 (Cat. No. 99TH8406), vol 2. IEEE, pp 1470–1477
10. Price KV (2013) Differential evolution. In: Handbook of optimization. Springer, pp 187–214
11. Whitley D (1994) A genetic algorithm tutorial. Statistics Comput 4(2):65–85
12. Albert Lam YS, Victor Li OK (2012) Chemical reaction optimization: a tutorial. Memetic Comput 4(1):3–17
13. Osman KE, Ibrahim E (2006) A new optimization method: big bang–big crunch. Adv Eng Software 37(2):106–111
14. Hsiao Y-T, Chuang C-L, Jiang J-A, Chien C-C (2005) A novel optimization algorithm: space gravitational optimization. In: 2005 IEEE international conference on systems, man and cybernetics, vol 3, pp 2323–2328. IEEE
15. Formato RA (2007) Central force optimization. Prog Electromagn Res 77:425–491
16. Tayarani-N MH, Akbarzadeh-T MR (2008) Magnetic optimization algorithms a new synthesis. In: 2008 IEEE Congress on Evolutionary Computation (IEEE World Congress on Computational Intelligence), pp 2659–2664. IEEE
17. Venkata Rao R, Savsani VJ, Vakharia DP (2011) Teaching–learning-based optimization: a novel method for constrained mechanical design optimization problems. Computer-Aided Design 43(3):303–315
18. Geem ZW, Kim JH, Loganathan GV (2001) A new heuristic optimization algorithm: harmony search. Simulation 76(2):60–68
19. Ghorbani N, Babaei E (2014) Exchange market algorithm. Appl Soft Comput 19:177–187
20. Singh N, Singh SB (2017) A novel hybrid gwo-sca approach for optimization problems. Eng Sci Technol Int J 20(6):1586–1601
21. Shelokar PS, Siarry P, Jayaraman VK, Kulkarni BD (2007) Particle swarm and ant colony algorithms hybridized for improved continuous optimization. Appl Math Comput 188(1):129–142
22. Padmanaban S, Priyadarshi N, Bhaskar MS, Holm-Nielsen JB, Ramachandaramurthy VK, Hossain E (2019) A hybrid ANFIS-ABC based MPPT controller for pv system with anti-islanding grid protection: experimental realization. IEEE Access 7:103377–103389
23. Zhao W, Wang L, Zhang Z (2019) A novel atom search optimization for dispersion coefficient estimation in groundwater. Future Generation Comput Syst 91:601–610
24. Zhao W, Wang L, Zhang Z (2019) Atom search optimization and its application to solve a hydrogeologic parameter estimation problem. Knowledge-Based Syst 163:283–304
25. Stone A (2013) The theory of intermolecular forces. OUP, Oxford
26. Singh SK, Sinha N, Goswami AK, Sinha N (2016) Robust estimation of power system harmonics using a hybrid firefly based recursive least square algorithm. Int J Electric Power Energy Syst 80:287–296

27. Ray PK, Subudhi B (2012) BFO optimized rls algorithm for power system harmonics estimation. Appl Soft Comput 12(8):1965–1977
28. Kabalci Y, Kockanat S, Kabalci E (2018) A modified ABC algorithm approach for power system harmonic estimation problems. Electric Power Syst Res 154:160–173
29. Testa A, Langella R (2005) Power system subharmonics. In: IEEE Power Engineering Society General Meeting, pp 2237–2242. IEEE
30. Yacamini R (1996) Power system harmonics. iv. interharmonics. Power Eng J 10(4):185–193

Chapter 15
Robust Control and Synchronization of Fractional-Order Complex Chaotic Systems with Hidden Attractor

Ahmad Taher Azar, Fernando E. Serrano, Nashwa Ahmad Kamal, Tulasichandra Sekhar Gorripotu, Ramana Pilla, Sandeep Kumar, Ibraheem Kasim Ibraheem, and Amjad J. Humaidi

1 Introduction

Hidden attractors are found in some chaotic systems when the domain of attraction is not the equilibrium point. Fractional-order chaotic systems have become a vital topic to be studied. The fractional-order dynamic model provides a more realistic mathematical model than integer order systems [1, 2]. It is important to remark that

A. T. Azar (✉)
College of Computer and Information Sciences, Prince Sultan University, Riyadh, Saudi Arabia

Faculty of computers and Artificial Intelligence, Benha University, Benha, Egypt
e-mail: aazar@psu.edu.sa; ahmad.azar@fci.bu.edu.eg; ahmad_t_azar@ieee.org

F. E. Serrano
Research collaborator at Robotics and Internet-of-Things Lab (RIOTU), International Group of Control Systems (IGCS), Prince Sultan University, Riyadh, Saudi Arabia
e-mail: serranofer@eclipso.eu

N. A. Kamal
Faculty of Engineering, Cairo University, Cairo, Egypt

T. S. Gorripotu
Department of Electrical & Electronics Engineering, Sri Sivani College of Engineering, Srikakulam, Andhra Pradesh 532402, India

R. Pilla
Department of Electrical & Electronics Engineering, GMR Institute of Technology, Rajam Srikakulam, Andhra Pradesh 532127, India
e-mail: ramana.pilla@gmrit.edu.in

S. Kumar
CHRIST (Deemed to be University), Bengaluru, Karnataka 560074, India

I. K. Ibraheem
Department of Electrical Engineering, College of Engineering, University of Baghdad, Al-Jadriyah, Baghdad 10001, Iraq
e-mail: ibraheemki@coeng.uobaghdad.edu.iq

© The Author(s), under exclusive license to Springer Nature Singapore Pte Ltd. 2021
S. Kumar et al. (eds.), *Proceedings of International Conference on Communication and Computational Technologies*, Algorithms for Intelligent Systems,
https://doi.org/10.1007/978-981-16-3246-4_15

fractional-order complex; chaotic systems offer a novel insight into the chaos phenomenon study. The real and imaginary dynamics are even more complex compared to real chaotic systems. Besides, it is important to remark that hidden attractors are found in this kind of system's dynamics as similar to its real counterpart.

In [3–5], fractional-order chaotic systems with hidden attractors are presented and proved that hidden attractors and hidden oscillations exist with fractional-order dynamics. Then, in [6], a complete analysis of chaotic systems with hidden attractors is shown, evincing the fundamentals of this kind of phenomenon. It is also essential to explain the attractors and domain of attractions' mathematical background, so in papers like [7], crucial results for this study are provided; compact sets are essential to determine the region of attraction of a chaotic system.

The control for fractional-order chaotic systems has been extensively studied in papers such as [8] in which a controller for a fractional-order chaotic chemical reactor system is shown. In [9], a terminal sliding mode controller for synchronization purposes of a fractional-order uncertain chaotic system is designed and provided. Then, in [10], a controller is designed for the stabilization of the fractional-order chaotic ecological system. Then, in [11], a fractional-order sliding mode controller is shown to stabilize a chaotic uncertain system.

The synchronization of chaotic systems is an important issue considering the performance of coupled chaotic systems [1, 12–18]. Some examples of such systems can be found in [19]. The synchronization of a simple fractional-order chaotic system with a hidden attractor is shown. Then, in [20], the synchronization of a 5D chaotic system representing a homopolar disk dynamo is established. Also, in [21, 22], the synchronization of a fractional-order complex chaotic system with terminal sliding mode control is presented.

It is essential to mention some studies related to robust control for chaotic systems due that they are the fundamental basis for the results displayed in this study. For example, in [23], a $H - \infty$ controller for a nonlinear uncertain chaotic system is shown. Then, in [24], a neural fuzzy controller for stabilizing a chaotic, insecure system is presented. Another example can be found in [25], in which a robust backstepping controller for a chaotic system is presented. Finally, in [26], a powerful fuzzy logic controller for an unknown system is presented.

A robust controller and synchronizer for fractional-order complex chaotic systems with hidden attractors are shown in this paper. The controller and synchronizer techniques consist of dividing the complex, chaotic system into real and imaginary parts. Using a robust fractional-order Lyapunov function, these strategies are obtained. The imaginary part of the fractional-order complex chaotic systems is used by selecting a compact set that contains the appropriate region of attraction.

In Sect. 2, the problem formulation is established. Section 3 presents the design of robust control and synchronization of fractional-order complex chaotic system with the hidden attractor. In Sect. 4, a couple of experiments performed to prove the

A. J. Humaidi
Department of Control and Systems Engineering, University of Technology, Baghdad 10001, Iraq
e-mail: 601116@uotechnology.edu.iq

efficiency of proposed approach. Finally, Sects. 5 and 6 discussed about results and concluded this work, respectively.

2   Problem Formulation

Consider the following fractional-order complex chaotic system:

$$D^\alpha Z = f(Z) = f_R(Z) + f_{IM}(Z)j \tag{1}$$

with $Z \in \mathbb{C}^n$, $Z = [Z_1, Z_2, \ldots, Z_n]^T$, α is the fractional order and the complex vector field $f(Z) \in \mathbb{C}^n$ with:

$$f_R(Z) = [f_{R1}(Z), f_{R2}(Z), \ldots, f_{Rn}(Z)]^T$$
$$f_{IM}(Z) = [f_{IM1}(Z), f_{IM2}(Z), \ldots, f_{IMn}(Z)]^T \tag{2}$$

Now, for the robust controller and synchronizer design, consider the following system:

$$D^\alpha Z = f_R(Z) + f_{IM}(Z)j + U_R + U_{IM}j \tag{3}$$

with

$$U_R = [U_{1R}(Z), U_{2R}(Z), \ldots, U_{nR}(Z)]^T$$
$$U_{IM} = [U_{11M}(Z), U_{21M}(Z), \ldots, U_{nIM}(Z)]^T \tag{4}$$

One of the main contributions of this study is that the domain of attraction of the hidden attractor is defined by the following compact set with accumulation points ξ_i and $\zeta_i \; \forall \, i = 1 \ldots n$ [7]:

$$A_{ij} = \left\{ Z_{ij} \in \mathbb{C} : \{\Re(Z_{ij})\}_{j=0}^{\infty} \to \xi_i \right\}$$
$$B_{ij} = \left\{ Z_{ij} \in \mathbb{C} : \{\Im(Z_{ij})\}_{j=0}^{\infty} \to \zeta_i \right\} \tag{5}$$

So, in order that the fractional-order chaotic system reaches the domain of attraction for the real and imaginary part, the latest must be obtained by the following null space:

$$M = \left\{ f_{IM}(Z) \in \mathbb{R}^n : \mathcal{T}(f_{IM}(Z)) = 0 \right\} \tag{6}$$

where $\mathcal{T}(f_{IM}(Z)) = f_R(Z) >< f_{IM}(Z)$ (outer product).

3 Robust Control and Synchronization of Fractional Order Complex Chaotic System with Hidden Attractor

For the theoretical derivations, consider the real part of system (3) [3] as $f_{R1}(Z) = a(Z_{2R} - Z_{1R})$, $f_{R2}(Z) = bZ_{1R}Z_{3R}$ and $f_{R3}(Z) = c(1 - Z_{1R}Z_{2R})$ with $a = 6$, $b = 4$, $c = 5$ and $\alpha = 0.98$ with initial conditions $Z_R(0) = [0.1, 0, 0.1]^T$. The complex part is found as explained in the previous section and the respective phase portraits are shown in Figs. 1 and 2 for the real and imaginary part, respectively.

3.1 Robust Controller Design

For the robust controller design, consider the following theorem:

Theorem 15.1 *The following robust control laws for the complex and real parts of the fractional-order chaotic system are obtained:*

$$U_{iR} = -f_{Ri}(z) - \frac{Z_{Ri}^2}{sign(e_{Ri})}$$

$$U_{iIM} = -f_{IMi}(z) - \frac{Z_{IMi}^2}{sign(e_{IMi})} \tag{7}$$

$\forall i = 1, \ldots, n$ *by selecting appropriate robust control Lyapunov functions.*

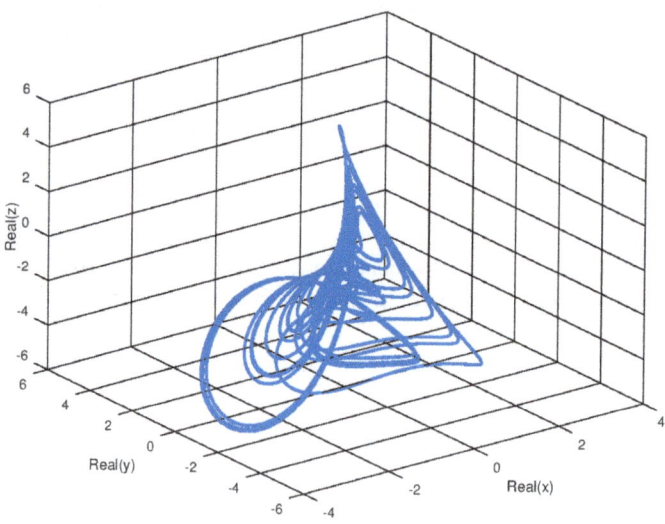

Fig. 1 Phase portrait of the real part of the studied system

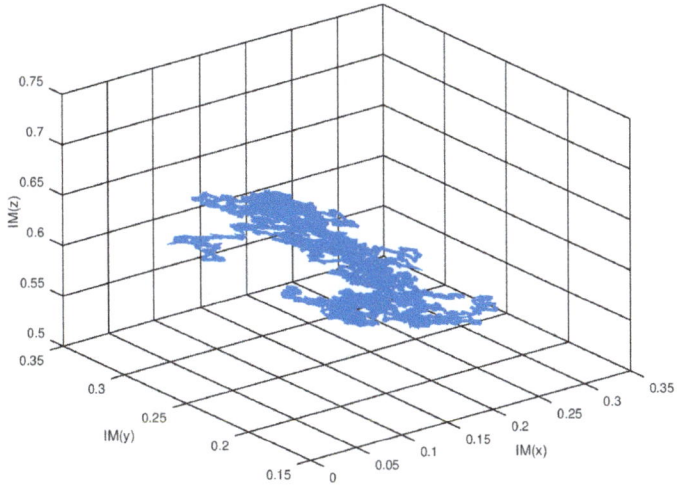

Fig. 2 Phase portrait of the imaginary part of the studied system.

Proof Consider the following variable $e_{Ri} = D^{\alpha-1}Z_{Ri}$ and the following Lyapunov function and its derivative:

$$V = \sum_{i=1}^{n} |e_{Ri}|$$

$$\dot{V} = \sum_{i=1}^{n} sign(e_{Ri})\dot{e}_{Ri} \tag{8}$$

with $\alpha_{vi} = Z_{Ri}^2$ and the following robust control Lyapunov function:

$$W = \dot{V} + \sum_{i=1}^{n} \alpha_{vi}$$

$$W = \sum_{i=1}^{n} sign(e_{Ri})\dot{e}_{Ri} + \sum_{i=1}^{n} Z_{Ri}^2 \tag{9}$$

Now, by substituting U_{iR} as appears in (7), the robust control Lyapunov function becomes in $W \leq 0$. The control law U_{iIM}, as shown in (7), is obtained in a similar way with $e_{IMi} = D^{\alpha-1}Z_{IMi}$ completing the proof of this theorem.

3.2 Robust Synchronization Design

For the robust synchronization, consider the following theorem:

Theorem 15.2 *The following robust synchronization laws for the complex and real parts of the fractional-order chaotic system are obtained taking into consideration the drive system reference variable for the real and imaginary parts $Z_{Ri_{REF}}$ and $Z_{IMi_{REF}}$ and the variables of the response system Z_{Ri} and Z_{IMi}:*

$$U_{iR} = D^\alpha Z_{Ri_{REF}} - f_{Ri}(z) + \frac{e_{Ri}^2}{sign(e_{Ri})}$$

$$U_{iIM} = D^\alpha Z_{IMi_{REF}} - f_{IMi}(z) + \frac{e_{IMi}^2}{sign(e_{IMi})} \qquad (10)$$

$\forall i = 1, \ldots, n$ *by selecting appropriate robust control Lyapunov functions.*

Proof Consider the following variable $e_{Ri} = D^{\alpha-1}(Z_{Ri_{REF}} - Z_{Ri})$ and the following Lyapunov function with its derivative:

$$V = \sum_{i=1}^{n} |e_{Ri}|$$

$$\dot{V} = \sum_{i=1}^{n} sign(e_{Ri})\dot{e}_{Ri} \qquad (11)$$

with the variable $\alpha_{vi} = e_{Ri}^2$ and the following robust control Lyapunov function:

$$W = \dot{V} + \sum_{i=1}^{n} \alpha_{vi}$$

$$W = \sum_{i=1}^{n} sign(e_{Ri})\dot{e}_{Ri} + \sum_{i=1}^{n} e_{Ri}^2 \qquad (12)$$

Now, by substituting U_{iR} (10) into (12), it makes $W \leq 0$ completing the proof. To find U_{iIM} (10), the following variable $e_{IMi} = D^{\alpha-1}(Z_{IMi_{REF}} - Z_{IMi})$ is necessary.

4 Numerical Experiment

These experiments are done by considering the real part of the fractional-order chaotic system with hidden attractor as appears in Sect. 2. The first experiment consists in the stabilization of the chaotic system, and the second one consists in the synchronization of two identical chaotic systems.

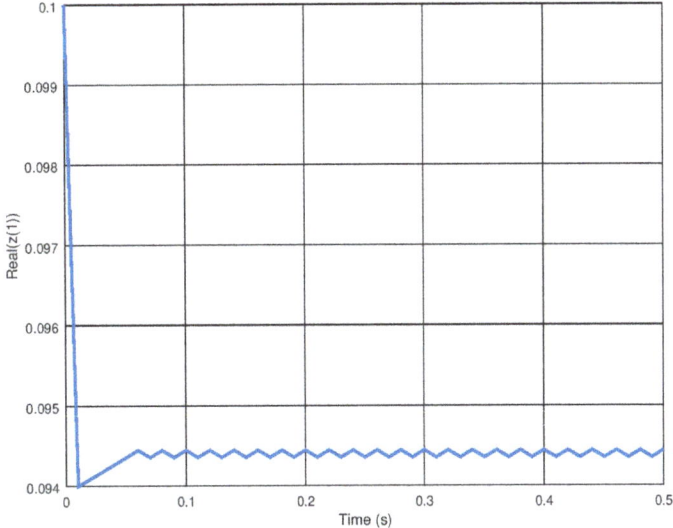

Fig. 3 Stabilized variable Z_{1R} obtained with the controller

4.1 Experiment 1

In Figs. 3 and 4, the evolution in time of the stabilized variables Z_{1R} and Z_{11M} with the proposed robust controller is shown. It is observed how the variables reach stability faster and accurately, proving the proposed controller's effectiveness, considering the complexity of the problem.

Meanwhile, in Figs. 5 and 6, the control input of the variables U_R real part and U_{IM} of the imaginary parts are shown, observing that the control inputs provide a low control effort, something that is very important taking into consideration that in real physical systems saturation must be avoided.

4.2 Experiment 2

Finally, in Figs. 7 and 8, the variable Z_{1R} and the synchronization error are shown, and as it can be noticed, the synchronization error is considerably smaller and remains zero most of the time.

5 Discussion

As corroborated in this paper's theoretical and experimental results, the robust controller and synchronizer for fractional-order complex chaotic systems with hidden attractors are achieved satisfactorily. Considering the dynamic model's complexity,

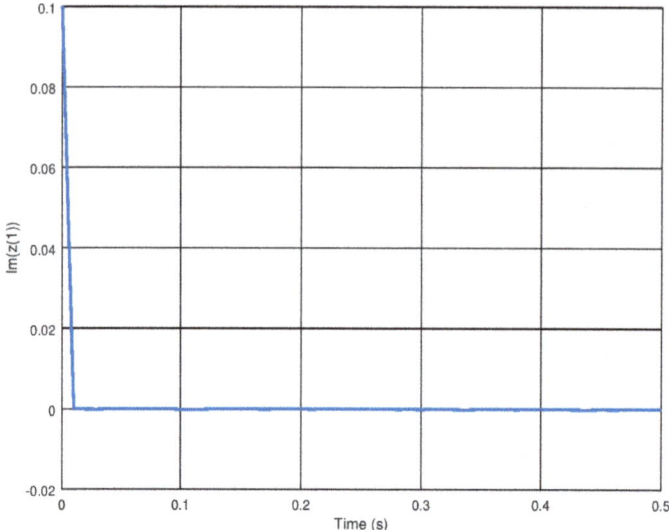

Fig. 4 Stabilized variable Z_{1IM} obtained with the controller

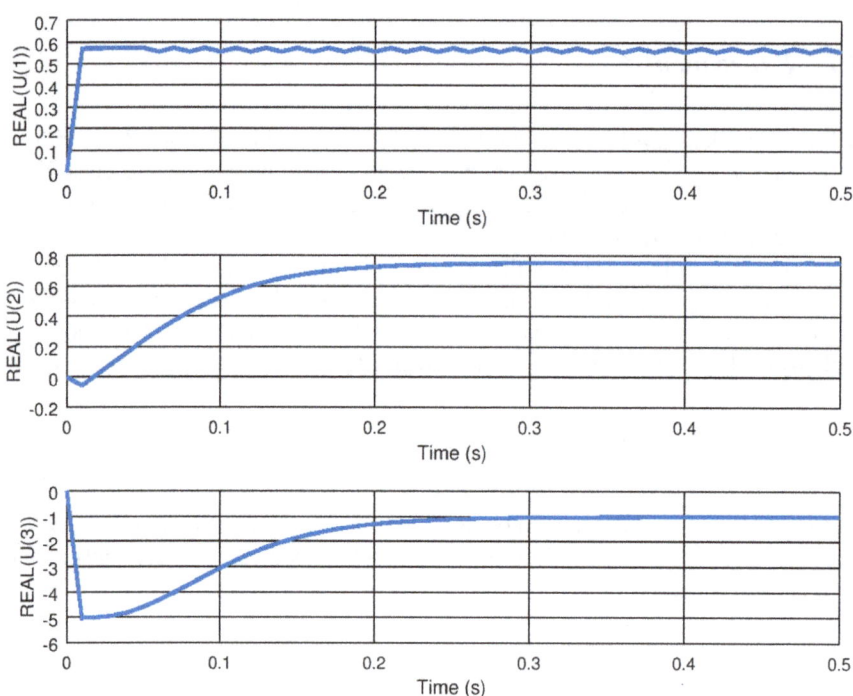

Fig. 5 Input variables for the real part obtained with the designed controller

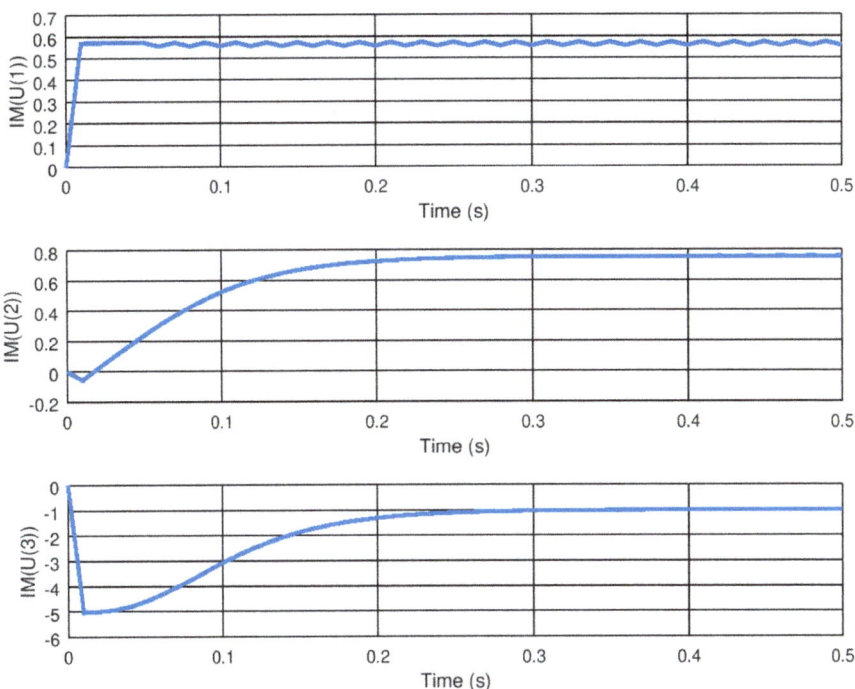

Fig. 6 Input variables for the imaginary part obtained with the designed controller

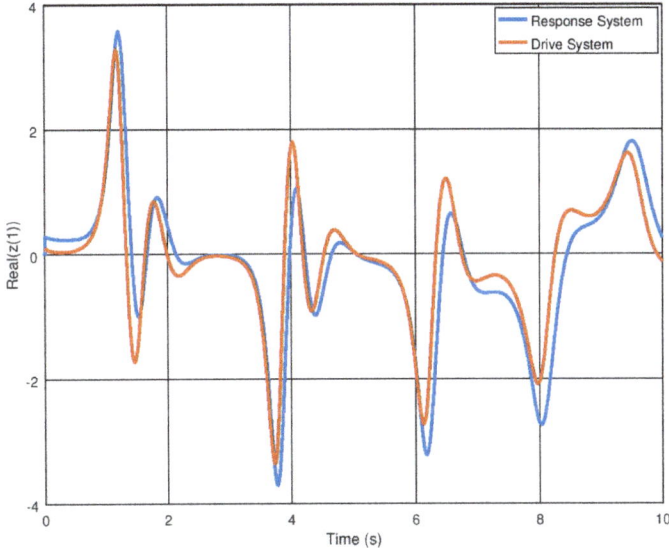

Fig. 7 Synchronized variable Z_{1R}

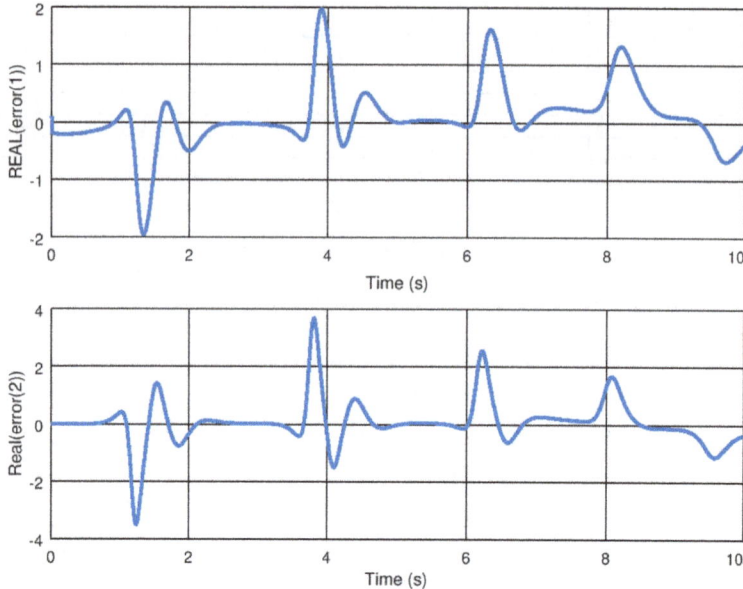

Fig. 8 Synchronized error for the variable Z_{1R}

these strategies provide a robust control technique that makes it ideal for overcoming the complex dynamics and other external conditions that can affect this kind of system for stabilization and control purposes. It is a better alternative than soft control techniques. Chaos suppression is done faster and accurately until the desired final value is reached. In the synchronization case, the response system follows the drive system variables, considering reference variables, accurately. One of this study's main contribution is that the topological conditions for the domain of attraction of the imaginary part of the system are established.

6 Conclusion

In this paper, the design of a robust controller and synchronizer for fractional-order complex chaotic systems with hidden attractors is evinced. The robust controller is designed by dividing the system into a real and complex part, and by using a robust control Lyapunov function, the control law for each part of the system is designed. A robust control Lyapunov function is implemented, and this time, the drive system's output is considered a reference for the response systems. Two numerical examples are provided that validate the theoretical results obtained in this study.

Acknowledgements Special acknowledgement to Robotics and Internet-of-Things Lab (RIOTU), Prince Sultan University, Riyadh, Saudi Arabia. We would like to show our gratitude to Prince Sultan University, Riyadh, Saudi Arabia.

References

1. Ouannas A, Azar AT, Ziar T, Radwan AG (2017) Generalized synchronization of different dimensional integer-order and fractional order chaotic systems. In: Azar AT, Vaidyanathan S, Ouannas A (eds) Fractional order control and synchronization of chaotic systems, studies in computational intelligence, vol 688. Springer International Publishing, Cham, pp 671–697
2. Ouannas A, Azar AT, Ziar T, Vaidyanathan S (2017) Fractional inverse generalized chaos synchronization between different dimensional systems. In: Azar AT, Vaidyanathan S, Ouannas A (eds) Fractional order control and synchronization of chaotic systems, studies in computational intelligence, vol 688. Springer International Publishing, Cham, pp 525–551
3. Cui L, Lu M, Ou Q, Duan H, Luo W (2020) Analysis and circuit implementation of fractional order multi-wing hidden attractors. Chaos Solitons Fractals 138(109):894
4. Wang Z, Volos C, Kingni ST, Azar AT, Pham VT (2017) Four-wing attractors in a novel chaotic system with hyperbolic sine nonlinearity. Optik—Int J Light Electron Optics 131:1071–1078
5. Doungmo Goufo EF (2019) On chaotic models with hidden attractors in fractional calculus above power law. Chaos Solitons Fractals 127:24–30
6. Dudkowski D, Jafari S, Kapitaniak T, Kuznetsov NV, Leonov GA, Prasad A (2016) Hidden attractors in dynamical systems. Phys Rep 637:1–50
7. Maslanka L (2020) On a typical compact set as the attractor of generalized iterated function systems of infinite order. J Math Anal Appl 484(2):123,740
8. Yadav VK, Das S, Bhadauria BS, Singh AK, Srivastava M (2017) Stability analysis, chaos control of a fractional order chaotic chemical reactor system and its function projective synchronization with parametric uncertainties. Chin J Phys 55(3):594–605
9. Modiri A, Mobayen S (2020) Adaptive terminal sliding mode control scheme for synchronization of fractional-order uncertain chaotic systems. ISA Trans 105:33–50
10. Mahmoud EE, Trikha P, Jahanzaib LS, Almaghrabi OA (2020) Dynamical analysis and chaos control of the fractional chaotic ecological model. Chaos Solitons Fractals 141(110):348
11. Rashidnejad Z, Karimaghaee P (2020) Synchronization of a class of uncertain chaotic systems utilizing a new finite-time fractional adaptive sliding mode control. Chaos Solitons Fractals: X 5(100):042
12. Vaidyanathan S, Azar AT (2016) Dynamic analysis, adaptive feedback control and synchronization of an eight-term 3-D novel chaotic system with three quadratic nonlinearities. In: Advances in chaos theory and intelligent control. Springer, Berlin, pp 155–178
13. Vaidyanathan S, Azar AT (2016) A novel 4-D four-wing chaotic system with four quadratic nonlinearities and its synchronization via adaptive control method. In: Advances in chaos theory and intelligent control. Springer, Berlin pp 203–224
14. Vaidyanathan S, Azar AT (2016) Adaptive control and synchronization of Halvorsen circulant chaotic systems. In: Advances in chaos theory and intelligent control. Springer, Berlin, pp 225–247
15. Vaidyanathan S, Azar AT (2016) Generalized projective synchronization of a novel hyperchaotic four-wing system via adaptive control method. Adv Chaos Theory Intell Control. Springer, Berlin, Germany, pp 275–290
16. Vaidyanathan S, Azar AT (2016) Qualitative study and adaptive control of a novel 4-d hyperchaotic system with three quadratic nonlinearities. In: Azar AT, Vaidyanathan S (eds) Advances in chaos theory and intelligent control. Springer International Publishing, Cham, pp 179–202
17. Ouannas A, Azar AT, Ziar T (2017) On inverse full state hybrid function projective synchronization for continuous-time chaotic dynamical systems with arbitrary dimensions. In: Differential equations and dynamical systems. https://doi.org/10.1007/s12591-017-0362-x

18. Vaidyanathan S, Azar AT, Ouannas A (2017) Hyperchaos and adaptive control of a novel hyperchaotic system with two quadratic nonlinearities. In: Azar AT, Vaidyanathan S, Ouannas A (eds) Fractional order control and synchronization of chaotic systems, studies in computational intelligence, vol 688. Springer International Publishing, Cham, pp 773–803

19. Wang M, Liao X, Deng Y, Li Z, Su Y, Zeng Y (2020) Dynamics, synchronization and circuit implementation of a simple fractional-order chaotic system with hidden attractors. Chaos Solitons Fractals 130(109):406

20. Mahmoud EE, Abualnaja KM (2020) Control and synchronization of the hyperchaotic attractor for a 5-d self-exciting homopolar disc dynamo. Alexandria Eng J

21. Azar AT, Serrano FE, Vaidyanathan S (2018) Sliding mode stabilization and synchronization of fractional order complex chaotic and hyperchaotic systems. In: Azar AT, Radwan AG, Vaidyanathan S (eds) Math Tech Fractional Order Syst. Elsevier, Advances in Nonlinear Dynamics and Chaos (ANDC), pp 283–317

22. Vaidyanathan S, Sambas A, Azar AT, Rana K, Kumar V (2021) A new 5-d hyperchaotic four-wing system with multistability and hidden attractor, its backstepping control, and circuit simulation. In: Vaidyanathan S, Azar AT (eds) Backstepping Control Nonlinear Dyn Syst. Academic Press, Advances in Nonlinear Dynamics and Chaos (ANDC), pp 115–138

23. Peng YF (2009) Robust intelligent backstepping tracking control for uncertain non-linear chaotic systems using h-infinity control technique. Chaos Solitons Fractals 41(4):2081–2096

24. Chen CS, Chen HH (2009) Robust adaptive neural-fuzzy-network control for the synchronization of uncertain chaotic systems. Nonlinear Anal Real World Appl 10(3):1466–1479

25. He N, Gao Q, Jiang C (2011) Robust adaptive control for a class of chaotic system using backstepping. Procedia Eng 15:1229–1233

26. Poursamad A, Davaie-Markazi AH (2009) Robust adaptive fuzzy control of unknown chaotic systems. Appl Soft Comput 9(3):970–976

Chapter 16
Deep Learning-Based Approach for Sentiment Classification of Hotel Reviews

Sarah Anis, Sally Saad, and Mostafa Aref

1 Introduction

Customer reviews on social media often reflect joy, dissatisfaction, frustration, happiness, and different sentiments. Sentiment analysis is an automated process of transforming those sentiments expressed in text reviews into meaningful information. In the field of tourism, most of the travel-related sites allow users to describe their experience by writing reviews and rating hotels they have visited. These reviews actually influence other people either positively or negatively. Taking advantage of these huge volumes of subjective information could help tourism associations know their limitations and improve them as well as their strength points. Sentiment polarity classification is the task of assigning a polarity to a text review whether it is positive or negative. Some reviews can be misleading sometimes that it does not include opinion or any subjective information. These kind of reviews are called objective reviews [1], which should be removed before training our system to avoid confusing the classifier with misleading reviews. The process of identifying subjective and objective reviews is often called sentiment detection. Sentiment detection is a prior step to sentiment classification which is used to enhance the performance of the classifier. There are several levels for sentiment analysis which are word, sentence, paragraph, and document levels. In document level, the whole document is assigned a single sentiment, while in paragraph-level sentiment analysis, the sentiment of each paragraph is determined separately. Other approaches consider applying sentiment analysis on each sentence or even each word individually. It is more challenging to precisely extract polarity in sentence level since sentences contain fewer number of words compared with paragraphs and documents. Deep learning has demonstrated superior performance on a wide variety of classification tasks. The proposed model

S. Anis (✉) · S. Saad · M. Aref
Faculty of Computer and Information Sciences, Ain-Shams University, Cairo, Egypt

© The Author(s), under exclusive license to Springer Nature Singapore Pte Ltd. 2021   211
S. Kumar et al. (eds.), *Proceedings of International Conference on Communication and Computational Technologies*, Algorithms for Intelligent Systems,
https://doi.org/10.1007/978-981-16-3246-4_16

presents a deep learning-based approach that uses word embedding and gated recurrent unit, to create an improved sentiment classifier. The structure of this paper includes a rundown of recent work in the field of sentiment analysis in Sect. 2. The system framework is presented in Sect. 3. Section 4 portrays the Implementation of the system. Section 5 presents the evaluation of methods and results. Lastly, Sect. 6 presents our Conclusions.

2 Related Work

User reviews found on tourism-based websites are increasing tremendously [2]. These amount of information shed light on the importance of sentiment analysis in tourism industry. Sentiment analysis process include several steps which initially starts with cleaning text. In this step, text is prepared for sentiment classification process by excluding any unnecessary or irrelevant content in text [3]. One of the challenges in sentiment analysis is identifying subjective reviews. Some reviews are objective which mean they don't contain any sentiment; these reviews may mislead our classification process. So as an important prior step to sentiment classification, reviews are classified into subjective and objective categories. This process is called subjectivity or sentiment detection. After this step, subjective sentences are retained and sentences with objective expressions are discarded. Recent work showed that the use of fuzzy logic techniques has proved to be powerful in sentence-level sentiment detection as opinions are fuzzy in nature [4, 5].

Sentiment analysis of subjective sentences is one of the most dynamic research domains. The classification process of sentiment analysis determines the overall polarity of opinion behind text by learning the relationship between inputs and outputs through available data which is called training data [6, 7]. After the training phase, we test and validate our classification model with new unlabeled data, also known as testing data. The process of extracting relevant features that demonstrate a specific sentiment polarity is an essential step in building our classification model. Yue et al. [6] discussed the effective features for sentiment classification and their impact on the performance of the overall classification process.

Deep learning has arisen as a dynamic machine learning technique that is widely used in sentiment analysis [8]. Most deep learning models use their first layer as an embedding layer for feature extraction from input data. Dang et al. [9] showed that it is better to use word embedding with deep learning than term frequency–inverse document frequency (TF-IDF) in the process of sentiment analysis. They also compared between different deep learning models and their performance on several datasets. Habimana et al. [10] also discussed the importance of using word embedding with deep learning approaches which gives promising results for sentiment analysis. Word2vec is one of the popular word embedding techniques used with deep learning [11]. Guggilla et al. [12] presented a deep neural network for claim classification from online user comments that is based on LSTM and CNN. They used word2vec and linguistic embedding with their deep learning model. Their work showed significant improvement over other methods. Zulqarnain et al. [13] also investigated the impact of word embedding namely word2vec in improving the sentiment classification process. They also proposed a variant of gated recurrent unit (GRU) that is used to preprocess data by means of encoding, and their model outperformed the accuracy of three different recurrent approaches. Clearly, word embedding plays a significant role in deep learning-based methods for sentiment classification. It additionally demonstrated that word embedding can be utilized as feature extraction method for other tasks in natural language processing.

Deep learning is prosperous in the new decade due to many significant reasons that include the improvements made for capacities of chip processing (GPU units) and the tremendous upgrades in machine learning methods [14, 15]. Yuebing et al. [16] integrated bidirectional GRU along with other methods in their deep learning approach in character-level text classification. Their experimental results achieved sufficient performance at fast convergence speed. Khedkar et al. [17] compared between traditional machine learning classifiers and different deep learning-based classifiers for sentiment classification of online hotel reviews. They classify reviews into two main categories which are praises and complaints. Their evaluation results showed that the deep learning-based approaches performed better than other existing machine learning methods for sentiment analysis.

3 Proposed System

The process of sentiment analysis of user reviews from tourism-related websites is divided into several stages. Figure 1 depicts our proposed system architecture of sentiment analysis of hotel reviews. Starting from how the dataset was used in the domain of tourism, second is the step that includes cleaning and preparing the data for the classification process. Next step involves extracting features from the preprocessed data. Fourth step is the sentiment detection. Fifth step is where we apply our approach for sentiment classification of hotel reviews. Sixth is evaluating the performance of our system.

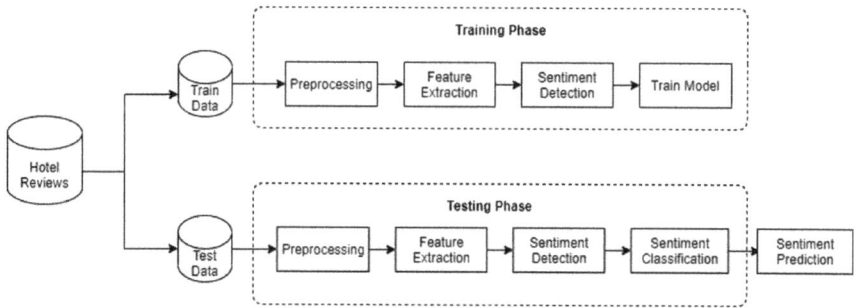

Fig. 1 The proposed system architecture of sentiment analysis

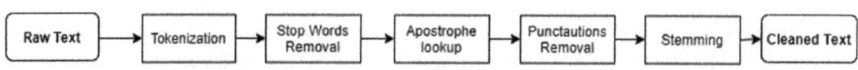

Fig. 2 Text preprocessing pipeline

Initially, the input for the system contains labeled hotel reviews. Dataset contains information about each review like ID, text review, device used to post the review, and polarity of review. The input is divided into training and testing subsets. Data that will be used to build our classification model is called training data, while the testing data is used for validating the performance of our system. Preprocessing step is used to remove any irrelevant content in text that may mislead our classifier. Figure 2 describes the pipeline of text preprocessing.

For feature extraction, word embedding technique is employed to transform textual words into numeric vectors to be easily interpreted by the classifier [17]. Word2vec [11] is the word embedding method that will be used for feature extraction. Word2vec captures deep semantic and syntactic features between words and compute them into an output numeric vector for each word [18, 19].

In sentiment detection, we have used the fuzzy C-means clustering algorithm to classify data into subjective and objective. Subjective reviews are the ones that actually contains sentiment or opinionated text, while objective reviews have only facts. This step aims to filter out any factual information and retain only subjective reviews. The fuzzy C-means clustering algorithm uses soft clustering for classification, so instead of assigning each data point to a single cluster, a point can have a degree of membership to multiple number of clusters. The higher the degree of membership, the closer the point is to the center of that cluster. Sentiment detection is a prior step before sentiment classification that ensures that objective data is discarded and only subjective reviews are passed on to the classifier.

For sentiment classification, a deep learning model is introduced to classify hotel reviews. Our deep learning model is composed of three layers. The embedding layer which is the first hidden layer is where we used our pretrained word2vec model. Gated recurrent unit (GRU) is the second hidden layer. It is a simplified variation of long short-term memory (LSTM) [20], but more powerful as it is faster, uses less parameters, and it can also capture long-term dependencies between the words [13]. The output layer is a dense neural network that has one node and uses the sigmoid activation function.

Finally, the output of the system is the sentiment polarity of each review. The evaluation of our model is measured in terms of accuracy and *F*-score.

4 Implementation

Dataset is composed of 38k hotel reviews from the Kaggle website (https://www. kaggle.com/harmanpreet93/hotelreviews/activity). The dataset contains 26k positive reviews and 12k negative reviews. We have used a random subset of 75% of the

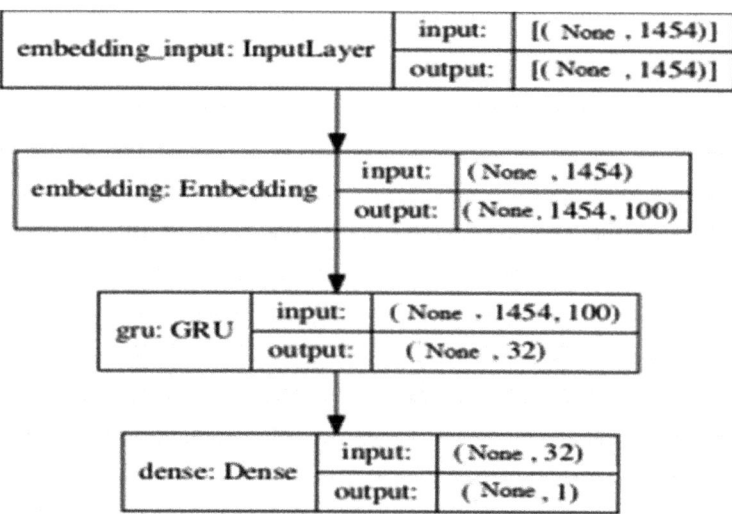

Fig. 3 Deep learning model architecture for sentiment analysis

dataset for training, and the other 25% were used in testing. We have compared our model performance with different traditional machine learning techniques used in [21] with the same hotel review dataset we have used. All the experiments were executed in Python using Keras library [22] for developing our deep learning model. Keras is a powerful Python library because of its capability of running on top of TensorFlow [23] or Theano [24]. Initially, the input training data are utilized to build our model. To evaluate our models, we've used the testing dataset to test the system. The preprocessing step is done for both training and testing sets. The next step is feature extraction. The output feature vector is then normalized. Then, sentiment detection is used to filter out any objective reviews and pass only subjective reviews to the classifier. The best parameters were chosen using manual testing.

In our deep learning model, the input for the embedding layer is the maximum length of word sequences in our dataset. Its output dimension is the size of the output feature vector of each word in a sequence. The second hidden layer is GRU with 32 units' dimensionality of the output space and 0.2 as a dropout rate. Figure 3 shows the architecture of our deep learning model based on Keras visualization.

5   Results

The evaluation of the traditional machine learning techniques from [21] compared with our deep learning model is presented in Fig. 4. The results obtained show that our deep learning model outperformed the traditional methods in terms of accuracy and *F*-score. The deep learning model achieved 89% accuracy and 92% *F*-score.

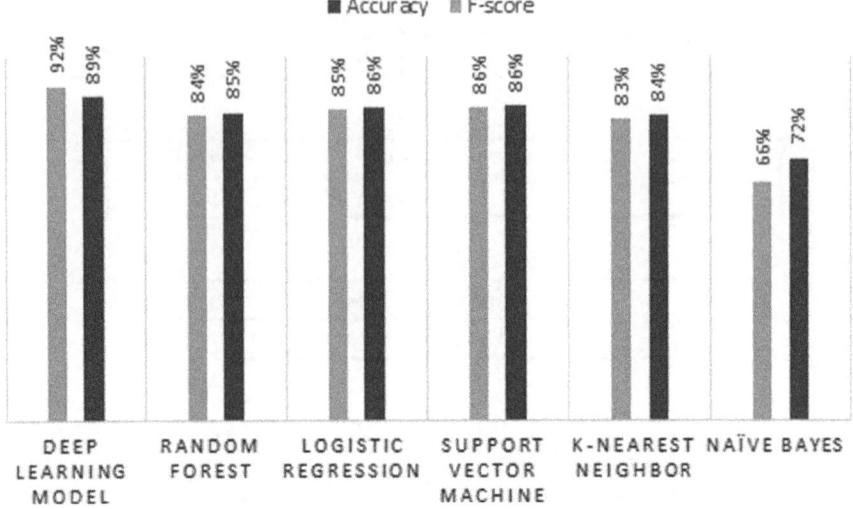

Fig.4 Results of our deep learning model compared to other machine learning methods

On the other side, the highest performance of the machine learning methods was obtained by support vector machine of 86% accuracy and 86% *F*-score.

6 Conclusion

Sentiment analysis of text reviews is of great consideration nowadays, for its ability to automatically analyze people's opinions. As people recently depend on online reviews to get information and make decisions based on these reviews, the demand for sentiment analysis has been extremely increased. Travelers express their opinions regarding products and services online daily on tourism-related websites. Deep learning has showed promising results in sentiment classification of hotel reviews. In this research, a deep learning model based on gated recurrent unit and dense layers was proposed for sentiment analysis in tourism. Different traditional machine learning techniques were investigated and compared with our proposed deep learning model. Results showed that our proposed deep learning approach performed better than other investigated machine learning techniques. Our deep learning approach achieved 89% accuracy and 92% *F*-score. On the other side, the highest accuracy of the machine learning methods was 86%, and an 86% *F*-score was obtained by support vector machine.

References

1. Feldman R (2013) Techniques and applications for sentiment analysis. Commun ACM 56(4):82–89
2. Alaei A, Becken S, Stantic B (2017) Sentiment analysis in tourism: Capitalising on big data. J Travel Res 58(9):175–191
3. Haddi E, Liu X, Shi Y (2013) The role of text pre-processing in sentiment analysis. Procedia Comput Sci In: Shi Y, Xi Y, Wolcott P, Tian Y, Li J, Berg D, Chen Z, Herrera-Viedma E, Kou G, Lee H, Peng Y, Yu L (eds) First international conference on information technology and quantitative management, vol 17. Elsevier, Suzhou, China, pp 26–32
4. AL-Maimani M, Salim N, Al-Naamany AM (2014) Semantic and fuzzy aspects of opinion mining. J Theor Appl Inf Technol 63(2)
5. Rustamov S, Clements MA (2013) Sentence-level subjectivity detection using neuro-fuzzy models, In: Proceedings of the 4th workshop on computational approaches to subjectivity sentiment and social media analysis, pp 108–114
6. Yue L, Chen W, Li X, Zuo W, Yin M (2018) A survey of sentiment analysis in social media. Knowl Inf Syst 1–47
7. Nakov P (2016) Developing a successful SemEval task in sentiment analysis of twitter and other social media texts. Lang Resour Eval 50(1):35–65. Available: https://doi.org/10.1007/s10579-015-9328-1
8. Zhang L, Wang S, Liu B (2018) Deep learning for sentiment analysis: a survey. ArXiv e-prints
9. Dang NC, Moreno-García, MN, De la Prieta F (2020) Sentiment analysis based on deep learning: a comparative study. Electron 9:483
10. Habimana O, Li Y, Li R, GU X, YU G (2020) Sentiment analysis using deep learning approaches: an overview. Sci China Inf Sci 63:111102
11. Ray P, Chakrabarti A (2019) A mixed approach of deep learning method and rule-based method to improve aspect level sentiment analysis. Appl Comput Inform. https://doi.org/10.1016/j.aci.2019.02.002
12. Guggilla C, Miller T, Gurevych I (2016) CNN-and LSTM-based claim classification in online user comments. In: Proceedings of the international conference on computational linguistics
13. Zulqarnain M, Ishak SA, Ghazali R, Nawi NM, Aamir M, Hassim YMM (2020) An improved deep learning approach based on variant two-state gated recurrent unit and word embeddings for sentiment classification. Int J Adv Comput Sci Appl 11:594–603
14. Ain QT, Ali M, Riaz A, Noureen A, Kamran M, Hayat B, Rehman A (2017) Sentiment analysis using deep learning techniques: a review. Int J Adv Comput Sci Appl 8(6):424
15. Guo Y, Liu Y, Oerlemans A, Lao S, Wu S, Lew MS (2016) Deep learning for visual understanding: a review. Neurocomputing 187:2748
16. Yuebing Z, Zhifei Z, Duoqian M, Jiaqi W (2019) Three-way enhanced convolutional neural networks for sentence-level sentiment classification. Inf Sci 477:55–64, ISSN 0020-0255. https://doi.org/10.1016/j.ins.2018.10.030
17. Sujata K, Subhash S (2020) Deep learning and ensemble approach for praise or complaint classification. Procedia Comput Sci 167:449–458, ISSN: 1877-0509. https://doi.org/10.1016/j.procs.2020.03.254
18. Zhang D, Xu H, Su Z, Xu Y (2015) Chinese comments sentiment classification based on word2vec and SVMperf. Expert Syst Appl 42(4):1857–1863
19. Mikolov T, Yih W, Zweig G (2013) Linguistic regularities in continuous space word representations. In: Proceedings of the 2013 conference of the north american chapter of the association for computational linguistics: human language technologies, Georgia, pp 746–751
20. Sutskever I, Vinyals O, Le QV (2014) Sequence to sequence learning with neural networks. In: Proceedings of advances in neural information processing systems 27:3104–3112
21. Anis S, Saad S, Aref M Sentiment analysis of hotel reviews using machine learning techniques. In: Hassanien AE, Slowik A, Snášel V, El-Deeb H, Tolba FM (eds) Proceedings of the international conference on advanced intelligent systems and informatics 2020. AISI 2020. Advances

in intelligent systems and computing, vol 1261. Springer, Cham. https://doi.org/10.1007/978-3-030-58669-0_21

22. Chollet F (2015) Keras. https://github.com/fchollet/keras
23. Dong H, Supratak A, Mai L, Liu F, Oehmichen A, Yu S, Guo Y. 2017 (2017) TensorLayer: a versatile library for efficient deep learning development. In: Proceedings of the 25th ACM international conference on Multimedia (MM 17). Association for Computing Machinery, New York, NY, USA, 1201–1204. https://doi.org/10.1145/3123266.3129391
24. Bergstra J, Breuleux O et al. (2010) Theano: a CPU and GPU math compiler in python technical report

Chapter 17
Routing Protocol Based on Probability Along with Genetic Algorithm for Opportunistic Network

Shivani Sharma and Sandhya Avasthi

1 Introduction

Opportunistic network is a diversification of MANETs (mobile ad hoc networks), where in implementations are unconfined to transmit and proceed with one another via wireless interface [1, 2]. Each node is treated as individualistic router, therefore putting an end to the requirement of central unit for supervising network unit and its operations. However, multiple resemblance is shared by mobile ad hoc network and opportunistic networks. Auxiliary constraints are faced by opportunistic network [3, 4]. Such networks possess unstable connection from source to target node. They require anticipating on interconnection opportunities present among the nodes. Assembling of network topology is inconsistent often because such network position of the node is unspecified. Therefore, the paradigm of store and carry forward is implemented to achieve the transmission of information from origin to target node. Every node ferry and stash message is stored in buffer and promotes that message whenever there is interaction via an advanced carrier [5, 6].

Researchers are seeking to create a classic approach of routing packets through past research in the last few years. Most of the routing techniques are based on three main paradigms.

(1) As in HiBop routing protocol, where context information is stored via node for the recognition of best relevant upcoming hop. For the forecast of upcoming hope, usage of related information may change as per various implementations. Bygone data of the node can be calculated and the behavior of the node in operating system. The prediction and interaction of node is the same as in geographic and energy aware routing (GAER) [7, 8].

S. Sharma (✉)
GGSIPU, Dwarka, Delhi, India

S. Avasthi
Krishna Engineering College, Ghaziabad, India

© The Author(s), under exclusive license to Springer Nature Singapore Pte Ltd. 2021   219
S. Kumar et al. (eds.), *Proceedings of International Conference on Communication and Computational Technologies*, Algorithms for Intelligent Systems,
https://doi.org/10.1007/978-981-16-3246-4_17

(2)   The estimation of distance of the node from target node and later, forwarding
      the message from transferring one to transitional node, considering more prox-
      imity, will be achieved the same as in enhanced data rate (EDR) protocol of
      routing [9].
(3)   The possibility of interconnection among transitional node and target node is
      calculated the same as in PRoPHET protocol of routing [10].

An advanced routing protocol that can incorporate the contextual information
and advantages of probabilistic routing and genetic algorithm is proposed here [11–
14]. It utilizes the preference of above-specified conventions for the enhancement
of routing from origin to target node of a message. The vision of the nature of the
adjacent node is performed by putting into action the genetic algorithm over the stored
situation. Genetic search is performed in comparatively skilled manner by proposing
mutation and two-point crossover operators. Goyal et al. [15] developed new routing
mechanism using dragon fly algorithm [16, 17] for efficient routing. The efficiency for
the route anticipated via operating genetic algorithm is calculated by the introduced
protocol using latest fitness function. By utilizing the approach for probabilistic
routing, the threshold value is set effectively to perform legislatively, transfer of
information from origin/active node to the nearest node. Hence, geographic and
energy routing protocol of routing lags the gateway access protocol of routing.

2   Opportunistic Networks and Proposed Protocol

The nodes which consist of operating status of Oppnets are borne by human beings,
and human beings also pursue familiar in their act, which is built on gateway access
protocol (GAP). Human beings have a propensity of traveling to a few places often.
For example, accommodation to headquarter places. Such spots are likely to possess
nodes, and these recurrent places are called home locations for the node, and the
anticipating protocol elucidates the three home locations being each one of the nodes.
Based on actions, nodes are categorized into node groups (NG) (Fig. 1).

The node having alike node groups possess identical home locations. IDs are
the parameter used for categorizing all the node groups. In gateway access protocol

ID	A	B	C	D	D	D	D	D	D	D	D
A	1123, 123	223, 923	1465, 13	A	A	A	B	C	B	D	E
B	1253, 464	323, 1233	323, 145	B	B	D	F	B	A	S	G

Fig. 1 These groups are interpreted as per the resemblance in movement specimen. (A—familiar
location 1, B—familiar location 2, C—familiar location 3, D—nearest group 0, ID—group ID)

(GAP), each node stores the context information. Therefore, efficiency of adjacent nodes can be calculated effectively by using prior-obtained nodes' information.

2.1   Gateway Access Routing Protocol (GARP)

The gateway access routing protocol comprises of three phases. First phase is known as predictive phase. Second phase is called evaluation phase. Third phase is known as transfer phase. Basically, as the name implies, predictive phase anticipate the route the message would opt for when transmitted to the nearest node to upcoming three hop counts. Now, here comes evaluation phase. The anticipated route is calculated with the help of fitness function based on location. Last one, the prominent judgment of transmission of message to the nearest node lies within transfer phase. It is dependent upon dynamic threshold with the usage of postulates for probabilistic routing. The protocol is described through three phases as follows.

2.1.1   Predictive Phase

This protocol pertains genetic algorithm upon the related information available in the particular table. Firstly, N arbitrary chromosomes are denoted via length as nine of the binary strings are established. The hop count in the anticipated route and total of node group hoard via node in the statistics table decide the chromosome's length. In the anticipated protocol, there are three bits which are vital for the depiction of each hop count and node group in the envisage path, which also produce equivalent of three. Therefore, nine is chromosome length. There are three actions: selections. Crossover along with mutation for genetic algorithm is subjected after the chromosomes depict encoded information for the anticipated route. The very first operation is selection operation, and it is followed by sorting of chromosomes based on their respective fitness value along with formation of twined chromosome's set. The effectiveness of the anticipated path is elucidated by the chromosome's fitness assess. The expounded description for the fitness value is mentioned in later phase. The twined chromosomes are, then, followed by crossover potency. Thus, opting in such a manner for chromosome that it is replaced generate two new offspring chromosomes. The gateway access protocol has crossover of two points. (a) Start along with and (b) end point, which are chosen arbitrarily, and string is replaced among these points. But it is not the case in geographic and energy routing protocol because it just has access to crossover operation for single point. The geographic and energy routing protocol does not scrutinize the impact of mutation operative. Thus, resulting in the shortfall in the investigation potential of genetic algorithm. This is the second limitation of the geographic and energy routing protocol. The mutation operators in gateway access protocol chooses a bit among the offspring chromosomes randomly and reverse the value and assuring variety amidst population. The probability of mutation is denoted by m. The value of m is set small, i.e., 4%, otherwise it will suffer from arbitrary search. For the three age-groups, the algorithm mentioned in upper portion is executed followed by opting for the chromosome which is the fittest being the envisage route. The initial step is to disassociate the chromosome in three domains of same length. The statistic table consists of position of node group of anticipated initial hop in decimal view of forth part. It consists of position of node group of envisaged second hop in decimal view for second portion. It consists of position of node group of envisaged third hop in decimal view for third portion.

2.1.2 Evaluation Phase

The twice, the number of chromosomes which are produced after the first phase are deciphered for generating envisaged route. However, the effectiveness of the route is evaluated for the transmission of message to the target. For this purpose, fitness function is required to be accessed. The mathematical declarations for the fitness function as in gateway access protocol (Eq. 1):

$$Fitness = \alpha \times \text{fitness of group (data group 3)}$$
$$+ \beta \times \text{fitness of group (data group 2)}$$
$$+ \gamma \times \text{fitness of group(data group 3)} \tag{1}$$

The very first decoded group is higher weighted as constant along with. The last decoded group has lowest weight. It calculates the fitness function for every entity of the data group (DG) when we pay attention to the replica area for two-dimensional system of coordinate. Represented by Fig. 2a and b, two of the feasible situations are represented during replica for gateway access protocol. Now, data group (1), data group (2), and group (3) are familiar positions for data group. Figure 2a represents the phase just when one familiar position of target node is present in the exterior of triangle. Though, Fig. 2b represents the situation of familiar positions of target node

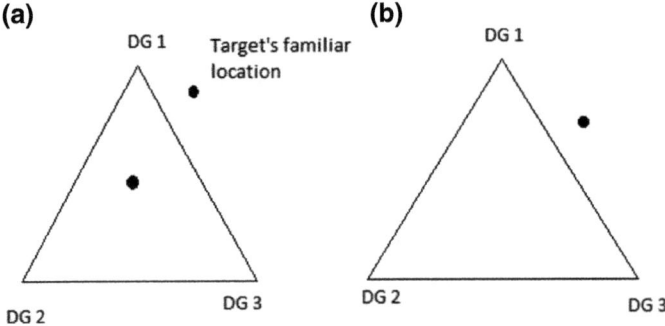

Fig. 2 **a** Target's familiar location inside triangle; **b** target's familiar location outside triangle

which is present in the interior of triangle. The chances for an encounter among the deciphered node group as well as in the targeted node are low. Rather, in Fig. 2b as the target node is present in the interior of triangle, the chance of encounter among node group along with target node group is quite high. It is so because throughout the activity of data group's node, to the familiar position, target node has effectively more possibility of covering its route. Hence, providing route to victorious transmission of message. Thus, researchers summarized that effectiveness of node is linked to the location of target's familiar position, in relation with the familiar position of data group. If few or every of the familiar position of the target node is present in the interior of triangle, in that case, it represents the higher chances of successful transmission of message, thus providing the effectiveness of the nearest node. But, when all the positions are present in the exterior of triangle, it represents the ineffectiveness of the nearest node. The group fitness is used to evaluate the fitness for data group. It calculates the ineffectiveness for deciphered node group by evaluating the distance among familiar positions of deciphered node group, which are present in the exterior of the triangle along with familiar positions of target node. Its mathematical equivalent function is depicted in Eq. 2.

$$\text{GrpFitness}(\text{DG}_i) = \sum_i \frac{\sum_{j=1}^{3} \text{distance}(H_j, \text{dest}H)}{\text{Normalized factor } Z} \tag{2}$$

Here, the distance(x,y) is called the Euclidean distance, calculated between the points x and y. Data group evaluates the distance, if and only if the target's familiar location is in the exterior of triangle. To provide the most effective carrier node, smallest value of fitness, the factor named as normalized factor has been defined. This factor is represented by Z. The function of fitness is equivalent to the reciprocal of effectiveness of anticipated path.

2.1.3 Transfer Phase

In the earlier discussed phases, fitness of the nearest node is examined based on how much effective the anticipated route is. The effectiveness of the nearest node is dependent on related information hoarded by intermediate node. However, a situation can pop up, like the given nearest node can be appropriate carrier of message depending upon the evaluated value of fitness. While in real-life scenario, the chances of meeting target node are quite less. Delivery predictability is used by gateway access protocol for managing such type of situations as mentioned by PRoPHET protocol of routing to adjust the value for threshold which acts as a deciding factor of level of acceptance for the fitness. Though, delivery predictability of gateway access protocol is introduced by NG rather of specific node. As soon as the node related to node group (a) comes in touch with the nearest node related to node group (b), the chances of encounter among node group (a) along with node group (b). The mathematical equation is expressed as follows.>

Aging factor confirms that no two nodes should be in contact for a longer period because in this case, they are quite less likely to have contact in future ever. It is expressed as:

$$P(A, B) = P(A, B)\text{old} \times f^k$$

k = total units of time proceeded since the previous time, this was upgraded.

Transitivity in the delivery predictability means that if node U is in contact with node V often and node V is in contact with node W often, which means that, node A will act as good transmitter of message having target as W. Its mathematical expression is as:

$$P(A, C) = P(A, C)\text{old} + (1 - P(A, C)\text{old}) \times P(A, B) \times P(B, C) \times \delta$$

δ = constant of scaling. A, B, C = node groups (NG).

The chances of being in contact of target node groups along with the nearest node group are known by delivery predictability. For the effective transmission of the message in nearest node, the value of delivery predictability must be low, and for node group to act as correct transmitter, the fitness function's value must be ideal. While for the effective transmission of the message in the nearest node, the value of delivery predictability must be high, for NG to act as best transmitter, the fitness function's value is not quite ideal. Therefore, there is restriction in the value of acceptable fitness along with the minimum energy of fitness function in delivery predictability. Hence, fitness for the anticipated route is lower than delivery predictability. Here, C acts as a constant, and it is inferred that the nearest node acts a great transmitter as well, and the message can be carried to node.

$$\text{GrpFitness}(DG_i) = \sum_i \frac{\sum_{j=1}^{3} \text{distance}(H_j, \text{dest}H)}{\text{Normalized factor } Z}$$

3 Simulation Results and Discussions

The presented protocol is replicated and contrasted with rest of the protocols of opportunistic routing. Opportunistic network environment simulator has been used for carrying out every simulation. It can be observed that the model of movement outlined for gateway access protocol is a replica of geographic and energy protocol. Thus, providing an ease at comparative study. Node group consist of levant number as twenty. Each node group is consisting of five nodes. However, for each of the message, the welshing TTL (time-to-live) has been set as three hundred. For each node, welshing buffer size has been set up as ten megabytes. The counterfeit is performed after every twelve hours. The size of buffer should be from 500 to 1MB

Fig. 3 Values for specifications introduced in GAP protocol

Specifications	Value
Normalized factor, Z	50
P_{init}	0.75
C	100
A	4
B	2
Υ	1
Δ	0.25

along with the interval of message generation at 25 to 35 s. For the communication purpose, speed of the transmission is around 250 kbps, and its range lies within 10 m (Fig. 3).

The production for gateway access protocol is calculated against Spray and Wait, PRoPHET, and geographic and energy routing protocol by changing the values for nodes between 20 and 160 in the intervals of 20 and TTL between 100 and 300 min in the intervals of 50 min. For distinguishing the display for gateway access protocol from rest of the protocols of routing, ratio of overhead, and average latency along with total count of delivered messages.

The variation of the value of node is depicted in Figs. 4, 5, 6, and 7. It can be easily concluded that increase in the count of nodes is directly proportional to the total count of delivered messages as shown in Fig. 2.

of delivered messages as shown in Fig. 2. The reason behind this is that as the count of nodes gets increased, the feasible opportunities for appropriate upcoming node also increase. While it can also be inferred that GAP protocol outperforms than rest of the protocols because this protocol enhances the message route comparatively more effectively than rest of the protocols of routing as shown in Figs. 2 and 3. Similarly, the more the count of ratio of overhead, the more the count of nodes. Ever since after the enhancement of message route, messages are being delivered effectively to their target which significantly produces the fact that the average latency for presented protocol is preferable than rest of the routing protocol. It is evident that

Fig. 4 Count of delivered messages to that of the count of nodes

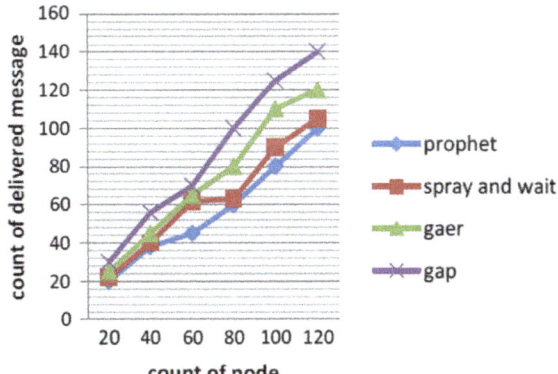

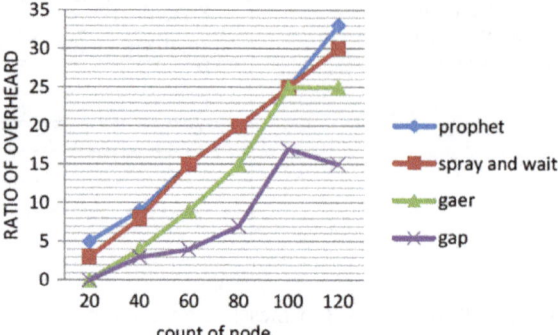

Fig. 5 Ratio of overhead to that of count of nodes

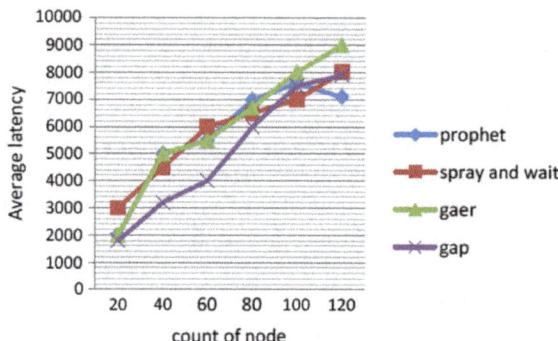

Fig. 6 Average latency for time to that of count of node

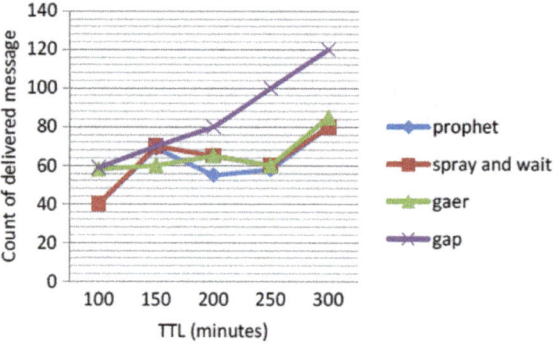

Fig. 7 Count of delivered messages to that of time to live

count of messages transmitted via GAP protocol ensures higher chances than rest of the protocols because of finer decision making during the message forwarding as shown in Fig. 4. The execution while varying the values in GAP protocol is depicted from Figs. 4, 5 and 6. Even, count of delivered messages rises as TTL also uplifts since messages are allotted quite additional TTL for holding up for upcoming best suited node. The messages are allocated faster to their targets in the proposed protocol because of very low latency in comparison to that of other protocols. The presented

Fig. 8 Ratio of overhead to the time to live

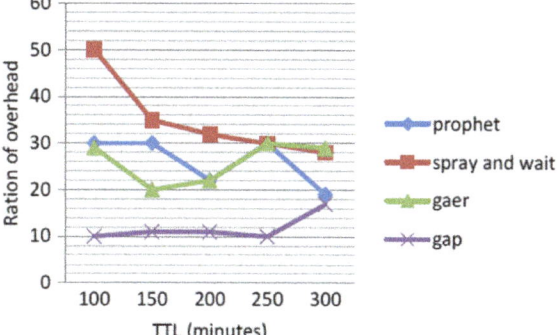

protocol enhances the message route which is resultant toward the less requirement for repetitive copies of the messages leading to the fortunate delivery completion of message. Thus, ratio of overhead for presented protocol outperforms rest other protocols (Fig. 8).

4 Conclusion and Future Proposals

Gateway access protocol, a protocol of foster-related routing, has been introduced. It consists of three insignificant strategies of routing. It enhances the message route to the target effectively. Firstly, in the prediction phase, it anticipates the path the message would opt for when transmitted to the nearest node. Second phase is the evaluation phase. It evaluates the fitness value for computation of efficiency for the anticipated route. Lastly, there comes the transfer phase. This phase determines whether the message will be transferred or not, with the help of probabilistic routing. The presented protocol resulted as much better than rest of the protocols in the aspects of ratio of overhead, time of average latency, and count of delivered messages. There can be better decision making during the transfer phase for future work. Multiple latest protocols provide understanding capability till few extents. During the last phase, for the arrangement of the effective value of threshold, cognitive capability can be utilized.

References

1. Lilien L, Gupta A, Yang Z (2007) Opportunistic networks for emergency applications and their standard implementation framework. In: 2007 IEEE international performance, computing, and communications conference. IEEE, pp 588–593
2. Toh C. K (2001) Ad hoc mobile wireless networks: protocols and systems, 1st edn. Prentice Hall PTR, Englewood Cliffs

3. Lilien L, Kamal ZH, Bhuse V, Gupta A (2007) The concept of opportunistic networks and their research challenges in privacy and security. In: Mobile and wireless network security and privacy. Springer, Boston, MA, pp 85–117

4. Dhurandher SK, Sharma DK, Woungang I, Gupta R, Garg S (2014) GAER: genetic algorithm-based energy-efficient routing protocol for infrastructure-less opportunistic networks. J Supercomput 69(3):1183–1214

5. Huang CM, Lan KC, Tsai CZ (2008) A survey of opportunistic networks. In: 22nd International Conference on Advanced Information Networking and Applications-Workshops (AINA). IEEE, pp 1672–1677

6. Lindgren A, Doria A, Schelen O (2003) Probabilistic routing in intermittently connected networks. ACM SIGMOBILE Mobile Comput Commun Rev 7(3):19–20

7. Dhurandher SK, Borah S, Woungang I, Sharma DK (2016) EDR: An encounter and distance based routing protocol for opportunistic networks. In: Proceedings of 30th International Conference on Advanced Information Networking and Applications-Workshops (AINA). IEEE

8. Boldrini C, Conti M, Jacopini J, Passarella A (2007) Hibop: a history-based routing protocol for opportunistic networks. In: 2007 IEEE international symposium on a world of wireless, mobile, and multimedia networks. IEEE, pp 1–12

9. Dhurandher SK, Sharma DK, Woungang I, Bhati S (2013) HBPR: History based prediction for routing in infrastructure-less opportunistic networks. In: Proceedings. of 27th International Conference on Advanced Information Networking and Applications-Workshops (AINA), IEEE. Spain, pp 931–936

10. Keranen A (2008) Opportunistic network environment simulator. Special assign–ment report. Helsinki University of Technology, Department of Communication Networking

11. Avasthi S, Chauhan R, Acharjya DP (2021) Techniques, applications, and issues in mining large-scale text databases. In: Advances in information communication technology and computing. Springer, Singapore pp 385–396

12. Avasthi S, Chauhan R, Acharjya DP (2021) Processing large text corpus using N-gram language modeling and smoothing. In: Proceedings of the second international conference on information management and machine intelligence. Springer, Singapore, pp 21–32

13. Nautiyal N, Malik S, Avasthi S, Tyagi E (2021) A dynamic gesture recognition system for mute person. In: Proceedings of the second international conference on information management and machine intelligence. Springer, Singapore, pp 33–39

14. Gupta A, Bansal A, Naryani D, Sharma DK (2017) CRPO: Cognitive routing protocol for opportunistic networks. In: Proceedings of International conference on high performance compilation, computing and communications (HP3C), Kuala Lumpur, Malaysia

15. Goyal M, Goyal D, Kumar S (2020) Dragon-AODV: efficient Ad Hoc on-demand distance vector routing protocol using dragon fly algorithm. Soft computing: theories and applications. Springer, Singapore, pp 181–191

16. Goyal M, Kumar S, Sharma VK, Goyal D (2020) Modified dragon-Aodv for efficient secure routing. In advances in computing and intelligent systems. Springer, Singapore, pp 539–546

17. Goyal A, Sharma VK, Kumar S (2020) Development of hybrid Ad Hoc on demand distance vector routing protocol in mobile Ad hoc network. Int J Emerg Technol 135–9

Chapter 18
A Hybrid Recommendation System for E-commerce

Shefali Gupta and Meenu Dave

1 Introduction

Every year, e-commerce websites introduce a vast amount of items on their websites. This huge amount of information urges the need to develop a system that filters out irrelevant information and provides recommendations that fit user's interest [1]. Over the past decade, many recommendation systems have been developed in various areas. These systems aim at providing personalized recommendation to the users by understanding the taste of each user and finding items that might be desirable by them [2–5]. A personalized recommendation system uses various methods such as content-based filtering, collaborative filtering, hybrid filtering, and knowledge-based filtering [6].

Content-based filtering looks at features of items and users' taste to recommend similar items to the user [7]. On the other hand, collaborative filtering uses user's behavior to create a neighborhood of similar users and, then, provides recommendations [8]. Both these filtering techniques have advantages as well as disadvantages with cold start, diversity, scalability, and data sparsity being some of the challenges faced by these techniques [9]. To bring synergy between them, hybrid algorithms are developed that uses combinations of various recommendation techniques. Numerous researchers have combined collaborative and content-based filtering together to levy the advantages of both.

Burke [10] divided hybrid techniques into seven different classes:

1. Weighted: It linearly combines recommendations obtained from different recommendation techniques
2. Switching: This technique switches from one technique to another based on certain defined criteria

S. Gupta (✉) · M. Dave
Jagannath University, Jaipur, India

© The Author(s), under exclusive license to Springer Nature Singapore Pte Ltd. 2021
S. Kumar et al. (eds.), *Proceedings of International Conference on Communication and Computational Technologies*, Algorithms for Intelligent Systems,
https://doi.org/10.1007/978-981-16-3246-4_18

3. Mixed: This technique provides recommendations from several techniques at the same time
4. Feature combination: This technique combines features from different recommendation techniques and input it in a single recommendation technique
5. Cascade: Output from one technique is used as an input to other technique to improve the results
6. Feature augmentation: Output from one technique is injected as an input feature to another technique
7. Meta-level: Model learned on one recommendation technique is used on another technique.

In particular, the main contribution of this paper is as follows:

a. A hybrid algorithm that combines content-based and collaborative-based filtering using the cascade method has been proposed
b. This algorithm has been implemented on mobile data that consists of mobile ratings datasets and mobile attributes datasets
c. The performance of the above algorithm is measured using recall parameter, and comparative analysis is presented with other recommendation techniques.

The rest of the paper is arranged as follows. Section 2 provides background knowledge about various hybrid algorithms developed so far. The proposed hybrid algorithm is explained in detail in Sect. 3. In Sect. 4, experimental results obtained from above-developed hybrid algorithm are displayed along with a comparative analysis of results obtained from other recommendation techniques. Finally, in Sect. 5, the conclusion of the paper and future research work are presented.

2 Background

In order to take advantage of recommendation system techniques, various algorithms have been proposed by researchers that use a mixture of collaborative filtering or content-based filtering, or any other recommendation techniques. Jain et al. [11] developed a hybrid algorithm that calculates the similarity of users among others grouped around various genres. This algorithm uses the user's preference of movies in terms of genres as a deciding factor for providing recommendations of movies to the user.

Doke and Joshi [12] combined user- and item-based collaborative filtering model for providing suggestions to the user. Li et al. [13] developed a hybrid algorithm for Q&A documents, in which they combined collaborative filtering, content-based filtering, and a complementarity-based recommendation method to find documents that match user needs. Paul and Das [14] used item-based and user-based filtering algorithm along with factor-based hybrid models to build an effective recommendation system for an artist.

Stanford and Carnegie Mellon [15] suggested the concept of a personalized recommendation system. Netflix and Amazon were the first e-commerce websites that used personalized recommendation system for recommending movies or items to shoppers based on the similarities in the products they purchase. However, currently existing recommender system faces some weaknesses such as cold-start, data sparsity, and overspecialization in its recommendation strategy.

Researchers like Liu et al. [16] introduced a hybrid algorithm for the news recommendation system. They improved the correlation coefficient formula by adding news as an important parameter while calculating the similarity of users to provide better recommendations to users. Yang and Yan [17] proposed a framework called MACBR, multi-agents collaboration case-based reasoning, which aims to reduce information overload by recommending personalized e-learning resources to the students with different characteristics such as prior knowledge, academic background, learning styles, learning goals, and objectives.

These researches show that a combination of two or more recommendation techniques results in more robust model as limitations of one technique is overcome by strengths of other technique.

3   Proposed Hybrid Filtering Algorithm

In this section, the proposed hybrid algorithm has been discussed which is developed by combining content-based filtering and collaborative filtering using the cascade method. The cascade method helps in refining the recommendations obtained by applying the first technique.

Figure 1 depicts the flow of content-based filtering. Content-based filtering takes user profile and item attributes as input and suggest items to the user based on their interest. For example, if a user likes movies starring Robert Downey Jr., the system tracks its choice and recommends movies starring Robert Downey Jr.

Figure 2 depicts the working of collaborative filtering. Collaborative filtering takes user as well as community data as input and recommends items to the user based on items liked by similar users [18]. For example, if user A and user B have the same

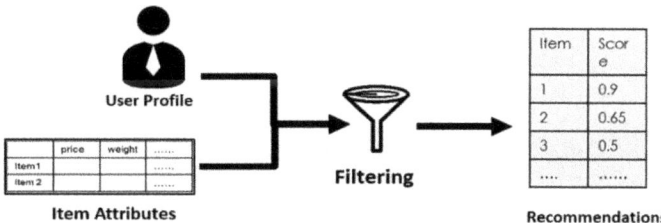

Fig. 1 Content-based filtering

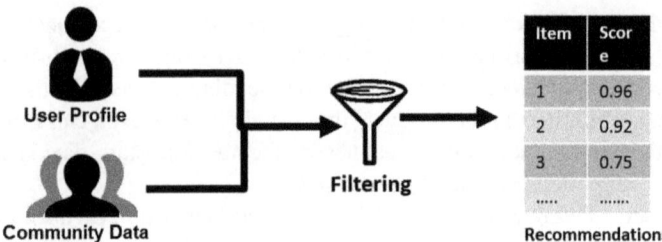

Fig. 2 Collaborative filtering

preferences in movies and user B has liked some movies different than user A, the system will recommend other movies liked by user B to user A.

The proposed hybrid algorithm combines content-based filtering and collaborative filtering. Figure 3 depicts the general procedures of hybrid algorithm which will be described in detail as follows:

Fig. 3 Proposed hybrid filtering algorithm

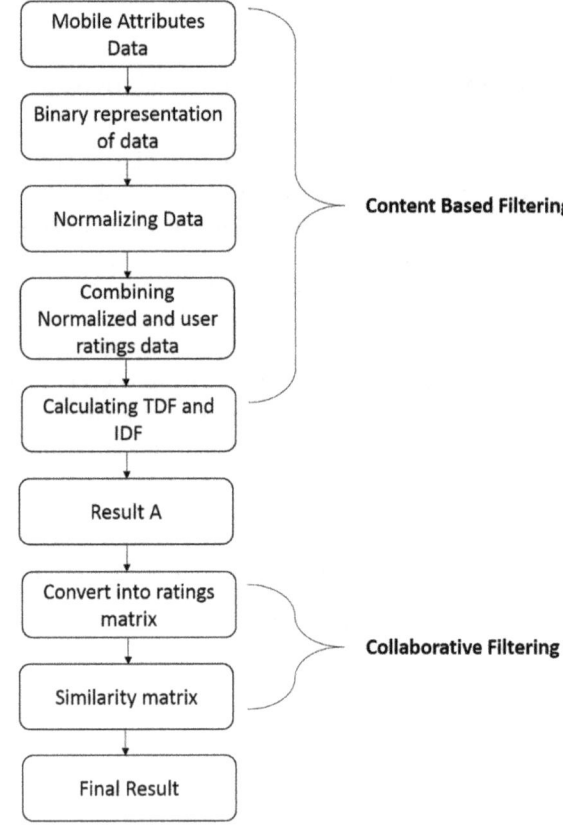

Items	Brand	SIM
Item9	Apple	Nano
Item10	Apple	Nano
Item11	Apple	Nano
Item12	Sony	Mini
Item13	Apple	Nano

Items	Brand.Apple	Brand.Sony	SIM.Micro	SIM.Nano	SIM.Mini
Item9	1	0	0	1	0
Item10	1	0	0	1	0
Item11	1	0	0	1	0
Item12	0	1	0	0	1
Item13	1	0	0	1	0

Fig. 4 Binary representation of data

- **Step 1**: Content-based filtering works by recommending items to the user based on its profile. In order to do it, item attribute data is first converted into binary representation as shown in Fig. 4.
- **Step 2**: Each data vector is then normalized before any similarity calculations are to be done so that the ratings lie between 0 and 1 instead of getting 0 or 1. This is done by calculating the magnitude of all the items by taking the square root of the sum of the squares of all the item attributes for each data vector Eqs. 1 and 2.

$$\text{Magnitude} = \sqrt{x^2 + y^2 + z^2 + \cdots} \tag{1}$$

A new data vector is then created by dividing the items attribute value with the magnitude value obtained in (1).

$$\text{Data vector} = \frac{x}{\text{magnitude}}, \frac{y}{\text{magnitude}}, \ldots \tag{2}$$

- **Step 3**: User taste matrix is obtained by taking the product of ratings for each item with the data vector obtained in (2).

- **Step 4**: In order to calculate similarity between items, TF (term frequency) and IDF (inverse term frequency) are used [12] Eqs. 3 and 4.

$$\text{TF} = \frac{\text{Frequency of the word in the document}}{\text{total number of words in document}} \tag{3}$$

$$\text{IDF} = \log 10 \left(\frac{\text{Total number of document}}{\text{Number of documents having the term}} \right) \tag{4}$$

For example, in case if the word apple appears ten times in the dataset with 100 total words, it will represent a term frequency of ten and an inverse document frequency of 1.

These measures help in evaluating the importance of a word in the document. As content-based filtering depends largely on contextual information about items, this helps in providing more information about the item in order to distinguish between them.

- **Step 5**: Final scores are obtained by taking the sum product of the user taste matrix and IDF matrix calculated above.
- **Step 6**: Collaborative filtering algorithm takes final scores matrix as input and convert it into ratings matrix for applying similarity measure [19].
- **Step 7:** Similarity measure is used to build a user–user similarity matrix through which neighbors for the current users can be computed.
- **Step 8**: Similarity matrix thus obtained is then merged with the actual user ratings data by applying various aggregation functions, resulting in final dataset of items that is recommended to the user.

4 Experimental Evaluation

The evaluation of results of any model is very important; therefore, recommendation system needs these metrics to compare the accuracy of the model. There are several types of evaluation metrics used for comparing models such as MAE, MAPE precision, and recall. [20]. In this section, recall metric will be used for evaluating the performance of the model [21] Eq. 5.

Recall also known as sensitivity is defined as:

$$Recall = \frac{true\,positive}{true\,positive + false\,negative} \tag{5}$$

For this paper, the dataset has been sourced from Kaggle [22] and represents multiple cell phones and their reviews on Amazon. This dataset can be leveraged to investigate a wide range of recommendation algorithms as it includes contextual features about mobile phones as well as ratings given by different users. The complete dataset is divided into two parts: train dataset (80%) and test dataset (20%). Recall measures how accurately the predictions made from train data match with test data. The following chart compares various recommendation techniques against the proposed hybrid model.

As shown in Fig. 5, the predictive accuracy of item-based collaborative filtering is the lowest. The user-based CF is slightly better than the item-based algorithm on account of the more accurate neighbor selection. Content-based filtering performs better than collaborative filtering with a 72% recall measure.

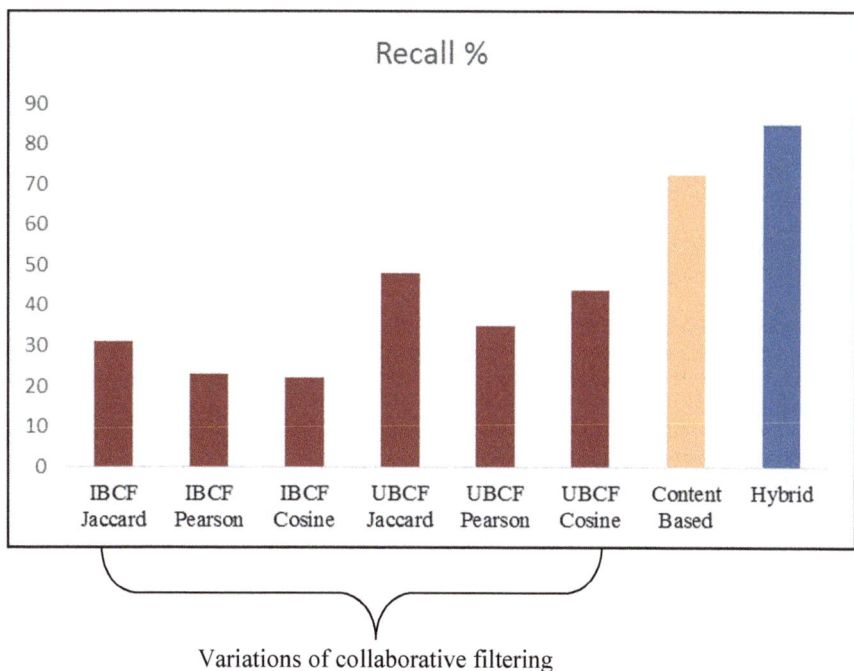

Fig. 5 Comparison of recall among different recommendation techniques

The proposed hybrid algorithm works best compared to all recommendation techniques mentioned here. Therefore, the proposed hybrid model is proved to improve the quality and accuracy of recommendation.

5 Conclusion and Future Work

The main contribution of this paper was to describe the framework of the proposed hybrid algorithm. This method combines content-based and collaborative filtering to leverage the advantages of both techniques. The experiment result on the mobile data set confirms that the proposed model outperforms other recommendation techniques.

In the future work, the aim is to do the following things:

1. Above-mentioned algorithm will be applied on various other datasets in order to verify the feasibility.
2. Modifying it to be able to solve the cold start problem.

References

1. Zhang T, Li W, Wang L, Yang J (2019) Social recommendation algorithm based on stochastic gradient matrix decomposition in social network. J Ambient Intel Human Comput
2. Celdrán AHF (2016) Design of a recommender system based on users' behavior and collaborative location and tracking. J Comput Sci 12:83–94
3. Batmaz, Z (2019) A review on deep learning for recommender systems: challenges and remedies. Artif Intel Rev 52(1):1–37
4. Ricci F, Rokach L, Shapira B (2015) Recommender systems: introduction and challenges. In: Ricci F, Rokach L, Shapira B (eds) Recommender systems handbook. Springer US, Boston, MA, pp 1–34
5. Campos PG, Díez F, Cantador I (2014) Time-aware recommender systems: a comprehensive survey and analysis of existing evaluation protocols, user model. UserAdap Inter 24:67–119
6. Çano E (2017) Hybrid recommender systems: a systematic literature review. Intel Data Anal 21(6):1487–1524
7. Son J, Kim S (2017) Content-based filtering for recommendation systems using multiattribute networks. Expert Syst Appl 89:404–412
8. Sarwar B, Karypis G, Konstan J, Riedl J (2001) Item-based collaborative filtering recommendation algorithms. In: Proceedings of the 10th international conference on World Wide Web. ACM, pp 285–295
9. Jain K, Kumar V, Kumar P, Choudhury T (2018) Movie recommendation system: hybrid information filtering system
10. Burke R (2002) Hybrid recommender systems: survey and experiments, User Model. User Adapt Interact 12:331–370
11. Aggarwal CC (2016) An introduction to recommender systems. Springer International Publishing, pp 1–28
12. Doke N, Joshi D (2020) Song recommendation system using hybrid approach. In: Bhalla S, Kwan P, Bedekar M, Phalnikar R, Sirsikar S (eds) Proceeding of international conference on computational science and applications, algorithms for intelligent systems. Springer, Singapore
13. Li M, Li Y, Lou W, Chen L (2019) A hybrid recommendation system for Q&A document. In: Expert systems with applications, vol 144
14. Paul S, Das D (2020) User-item-based hybrid recommendation system by employing mahout framework
15. Chen C-M (2008) Intelligent web-based learning system with personalized learning path guidance. Comput Educ 51:787–814
16. Liu S, Dong Y, Chai J (2016) Research of personalized news recommendation system based on hybrid collaborative filtering algorithm. In: 2016 2nd IEEE international conference on computer and communications (ICCC), Chengdu, pp 865–869
17. Yang L, Yan Z (2011) Personalized recommendation for learning resources based-on case reasoning agents. In: 2011 international conference on electrical and control engineering (ICECE), pp 6689–6692
18. Logesh R, Subramaniyaswamy V, Malathi D, Sivaramakrishnan N, Vijayakumar V (2018) Enhancing recommendation stability of collaborative filtering recommender system through bio-inspired clustering ensemble method. Neural Comput Appl
19. Yang X, Guo Y, Liu Y, Steck H (2014) A survey of collaborative filtering based social recommender systems. Comput Commun 41:1–10
20. Isinkaye F, Folajimi Y, Ojokoh B (2015) Recommendation systems: principles, methods and evaluation. Egypt Inform J 16(3):261–273
21. García-Sánchez F, Colomo-Palacios R, Valencia-García R (2020) A social-semantic recommender system for advertisements. Inf Process Manag 57(2)
22. Amazon Cell Phones Reviews Dataset. https://www.kaggle.com/grikomsn/amazon-cell-phones-reviews, accessed on May 2020

Chapter 19
The Convexity of Fuzzy Sets and Augmented Extension Principle

D. D. Gadjiev, G. V. Tokmazov, and M. M. Abdurazakov

1 Definitions and Main Principles of the Fuzzy Sets

We need to recollect some concepts and definitions occurring in the articles [1–5] in order to proceed with the sequel.

Definition 1.1 Let X be denoted a non-empty set, and $L = [0, 1]$. A function φ from $x \in F \subset X$ to the interval L is a fuzzy set in X. For $\forall x \in X$, $\varphi(x) \in L$ is the grade of membership of x in φ [6–14]:

$$\varphi_F(x) = 1 \quad x \in F$$

$$\varphi_F(x) = 0 \, , x \notin F$$

A fuzzy set F is defined by the set of ordered pair: $\{(x, \varphi(x)), x \in X\}$.

Definition 1.3 $\varphi_F(x): X \to \Psi[0,1]$, where $\Psi[0, 1]$ is a group of the functions on the closed intervals. This type of the fuzzy set is called the closed-interval fuzzy set [6].

Definition 1.4 The fuzzy sets are equivalent if and only if:

$$\forall x \in X \, , \varphi_{F_1}(x) = \varphi_{F_2}(x), F_1 \equiv F_2 \, .$$

D. D. Gadjiev (✉)
Department of Mathematics Lee Campus, Florida South Western College, 8099 College Parkway, Fort Myers, FL 33919, USA
e-mail: dgadjiev@fsw.edu

G. V. Tokmazov
Admiral Ushakov State Maritime University, Lenin Avenue, 93, Novorossiysk 353918, Russia

M. M. Abdurazakov
Russian Academy of Education, Salaryevskaya, 14-3-8, Moscow 108811, Russia

© The Author(s), under exclusive license to Springer Nature Singapore Pte Ltd. 2021
S. Kumar et al. (eds.), *Proceedings of International Conference on Communication and Computational Technologies*, Algorithms for Intelligent Systems,
https://doi.org/10.1007/978-981-16-3246-4_19

Definition 1.5 The support plane of the fuzzy set is the set with the non-zero membership: $\text{supp}\, F = \{x \in X, \varphi_F(x) > 0\}$.

Definition 1.6 The point in $x \in X$, where $\varphi_F(x) = \frac{1}{2}$ is the crossover point of $\varphi(x)$ [6].

Definition 1.7 A fuzzy set of $\varphi(x)$ defined in $x \in X$ is called a singleton if and only if $\varphi(x) = s, 0 < s \leq 1$, x is its support and $\varphi(y) = 0$.

A fuzzy set $\varphi(x)$ in X is called the fuzzy point if and only if $\varphi(x) = s, 0 < s < 1, \varphi(y) = 0$.

Definition 1.8 The height of a fuzzy set is defined as it is:

$$H = sup_{x \in X} \varphi_F(x) .$$

Definition 1.9 *of λ-cut fuzzy set*

The λ-cut is comprising of the elements of a fuzzy set at least at the degree of λ:

$$F^\lambda = \{x \in X, \varphi_F(x) \geq \lambda\}.$$

$\mathcal{F}^\lambda$ is a strict $\lambda-$ cut fuzzy set, if $\varphi_F(x) > \lambda$.

We can determine the cuts of the fuzzy sets as it is [1–6]:

Definition 1.10 For the cut l fuzzy set there is

$$F_\lambda(x) = \lambda F^\lambda(x) \tag{1}$$

Based on (1) we want to describe the fuzzy set as the union of the fuzzy subsets:

$$F = \cup_{[0,1]} F \tag{2}$$

The (2) is the illustration of the union of the fuzzy subsets, which is also considered as the decomposed λ-cut fuzzy sets.

There is the following theorem of the decomposed cut of the fuzzy sets:

Theorem 1 $\forall F \in F(X)$ *is the union of fuzzy sets λ-cut*

$$F(X) = \cup_{\lambda \in [0,1]} F_\lambda \tag{3}$$

where $F_\lambda(x) = \lambda F^\lambda(x), F^\lambda(x)$ is the fuzzy power set.

Proof Since $F_\lambda(x) = \lambda F^\lambda(x)$, where $\lambda \in [0, 1]$.

Then,

$$F_{\lambda_1}(x) = {^{\lambda_1}}/_{x_1} + {^{\lambda_1}}/_{x_2} + \cdots + {^{\lambda_1}}/_{x_n},$$
$$F_{\lambda_2}(x) = 0 + {^{\lambda_2}}/_{x_2} + \cdots + {^{\lambda_2}}/_{x_n},$$
$$\cdots$$
$$F_{\lambda_n}(x) = 0 + 0 + \cdots + 0 + {^{\lambda_n}}/_{x_n},$$

where 0 is the empty set $\emptyset$: $\forall x \in X$, $\varphi_\emptyset(x) = 0$, $\lambda_n = 1$,
$\lambda_1, \lambda_2, \ldots, \lambda_n$ are the λ-cut sets, where $\lambda \in [0, 1]$.
Thereafter, $F_{\lambda_1} \cup F_{\lambda_2} \cup \ldots \cup F_{\lambda_n} = \cup_{\lambda \in [0,1]} F_\lambda = F(X)$.

Corollary 1 *The cardinality of fuzzy set of $\mathcal{F}$ is the following:*

$$|F| = \Sigma_{x \in F} \, \varphi_F(x).$$

Corollary 2 *For the fuzzy sets $G, J, G \subseteq J$ if $\forall x \in X, \varphi_G(x) \leq \varphi_J(x)$.*

2 The Convex Fuzzy Sets

Definition 2.1 A fuzzy set is convex if:

$$\varphi_F(vx_1 + (1-v)x_2) > \max(\varphi_F(x_1), \varphi_{\mathcal{F}}(x_2)),$$

$$x_1, x_2 \in \mathbb{R}, v \in [0,1] \tag{4}$$

Since the convexity of a fuzzy set is given, then we define the quasi-convexity.

Definition 2.1.1 Let F be non-empty fuzzy convex set in $\mathbb{R}$ and let $\varphi_F(x): X \to [0,1] \in \mathbb{R}$..
The fuzzy set is a quasi-convex if for each element $x_1, x_2 \in F$ there is the following inequality holds:

$$\varphi_F(vx_1 + (1-v)x_2) \geq max(\varphi_F(x_1), \varphi_F(x_2)) \tag{5}$$

There is a theorem presented to show that the fuzzy sets can be determined by the convexity of its λ cut sets.

Theorem 2 Let $F = (x, \varphi(x)) \neq \emptyset$ be non-empty convex fuzzy set and $\varphi_x: X \to [0,1] \in \mathbb{R}$
The fuzzy set is quasi-convex if

$$F_\lambda(x) = \lambda F^\lambda(x) = \cup_{[0,1]} F_\lambda = \{x \in X, \varphi_F(x) \geq \lambda\}$$

for each $\lambda \in [0, 1]$.

Proof If $\hat{x}$ is a point of the support plane. Let us suppose that there exists some neighborhood $O_\varepsilon(\hat{x})$, such that $\varphi_F(\hat{x}) \geq \varphi_F(\bar{x}), \bar{x} \in F \cap O_\varepsilon(\hat{x}), F = \cup_{[0,1]} F_\lambda$.

Further, let us assume that there exists another point $\bar{x} \in F, \bar{x} \neq \hat{x}$, and $\varphi_F(\bar{x}) \geq \varphi_F(\hat{x})$. The strong quasi-convexity brings us to the inequality:
$$\varphi_F(\upsilon \bar{x} + (1 - \upsilon)\hat{x}) > \max(\varphi_F(\bar{x}), \varphi_F(\hat{x})) = \varphi_F(\bar{x}) \text{for all } \upsilon \in [0, 1].$$
However, $\upsilon \bar{x} + (1 - \upsilon)\hat{x} \in F, \bar{x} \in F \cap O_\varepsilon(\hat{x})$. However, this result contradicts the local convexity at $\bar{x}$.

There is a corollary:

Corollary 3 *If all cuts of the fuzzy set are convex, then the fuzzy set is quasi-convex.*
Extension principle, originally defined by Zadeh, is a tool perform fuzzification on the mathematical applied problems.

Definition 2.2 Let $X \neq \emptyset, Y \neq 0$ are non-empty sets and f-mapping from X into Y and $\varphi(x) \subset X$.
Then, the support is $\delta(y) = \sup \varphi(x), f^{-1}(y) \neq \varphi(x)$. Otherwise, $\delta(y) = 0$.

Definition 2.3 Let f-mapping from X into Y and $\delta(y) \subset Y$. the inverse $f^{-1}(\delta(y)) = \delta(f(x)), \forall x \in X$.

Extension Principle Let us consider mapping from X into $Y : f : X \rightarrow Y, x \in X$, $y \in Y$.
If $\varphi(x) \subset \mathcal{X}$, then a fuzzy subset of $\mathcal{X}$ has the image by the fuzzy function of φ of the fuzzy variable: $Y = \varphi_F(x)$.

Case 1 If φ is one-to-one mapping, then $\varphi(y) = \varphi_F(\varphi^{-1}(y)) = \varphi_F(x)$.

Case 2 If φ is not one-to-one mapping, then there is the following condition applies:
$\varphi(y) = \sup_{x|y=\varphi(x)} \varphi_F(x)$.

There is presented theorem. According to this theorem, the function is evaluated at two points, when the fuzzy function is not one-to one.

Theorem 3 Let $\neq \emptyset$ is a non-empty quasi-convex fuzzy set on $[x_1, x_2]$.

Suppose there are $\eta, \mu \in [x_1, x_2], \eta < \mu$.

Part 1 If $\varphi_F(\eta) > \varphi_F(\mu)$, then there $\theta \in [x_1, x_2]$ such that $\varphi_F(\theta) \geq \varphi_F(\mu)$ for $\theta \in [x_1, \eta)$ Then $\varphi_F(\theta) = \sup \varphi_F(x)$ at $\theta \in [x_1, \eta)$.

Part 2 If $\varphi_F(\theta) \leq \varphi_F(\mu)$, then $\varphi_F(\theta) \geq \varphi_F(\eta)$ for $\theta \in (\mu, x_2]$.
Then $\varphi_F(\theta) = \sup \varphi_F(x)$ for $\theta \in (\mu, x_2]$.

Proof Suppose $\varphi_F(\eta) > \varphi_F(\mu)$ and $\theta \in [x_1, \eta)$.. By contradiction, let

$$\varphi_F(\theta) < \varphi_F(\mu).$$

Since η is the point between θ, μ and fuzzy function is strictly quasi-convex, we have then $\varphi_F(\eta) < \max\{\varphi_F(\theta), \varphi_F(\mu)\} = \varphi_F(\mu)$. However, this result contradicts to $\varphi_F(\eta) > \varphi_F(\mu)$. Therefore, $\varphi_F(\theta) = \sup \varphi_F(x)$.

The part 2 can be proven by using similar strategy of the contradiction.

There is a theorem of the strictly quasi-convex fuzzy function over the given interval $[x_1, x_2]$.

and the mid-point of this interval $\eta = \frac{1}{2}(x_2 - x_1)$ is the combination of the other two points.

Theorem 4 $F \neq \emptyset$ *is a non-empty fuzzy set on* $[x_1, x_2]$.

Let $\eta = \frac{1}{2}(x_2 - x_1)$ and $\xi > 0$, where ξ is sufficiently small. Next, let $[x_1, \eta - \xi), \mu \in (\eta + \xi, x_2]$, such that $\lambda < \eta < \mu$.

Part 1 If $\varphi_F(\eta) > \varphi_F(\mu)$, then $\varphi_F(z) \geq \varphi_F(\mu)$ for $z \in [x_1, \eta)$.
   If $\varphi_F(\eta) \leq \varphi_F(\mu)$, then $\varphi_F(z) \geq \varphi_F(\eta)$ for $z \in (\mu, x_2]$.

Part 2 If $\varphi_F(\eta) > \varphi_F(\lambda)$, then $\varphi_F(z) \geq \varphi_F(\lambda)$ for $z \in (\eta, x_2]$.
   If $\varphi_F(\eta) \leq \varphi_F(\lambda)$, then $\varphi_{\mathcal{F}}(\eta)$ for $z \in [x_1, \lambda)$.

Proof

Part 1 For $\varphi_F(\eta) > \varphi_F(\mu)$ and $z \in [x_1, \eta)$ let us assume $\varphi_F(z) < \varphi_{\mathcal{F}}(\mu)$.

Since η is a of z, μ, then using the property of the strictly convex fuzzy sets, we have that $\varphi_F(\eta) < \max\{\varphi_F(z), \varphi_F(\mu)\} = \varphi_F(\mu)$. This results bring us to contradiction since $\varphi_F(\eta) > \varphi_F(\mu)$.

Therefore, $\varphi_F(z) \geq \varphi_F(\mu)$

Part 2 Let $\varphi_F(\eta) \leq \varphi_F(\mu)$ and $z \in (\mu, x_2]$. Using the contradiction strategy let us assume $\varphi_F(z) < \varphi_F(\eta)$., Because μ is a convex combination of $z\eta$ we have that $\varphi_F(\mu) < \max\{\varphi_F(z), \varphi_F((\eta)\} = \varphi_F(\mu)$, which contradicts to $\vartheta_{\mathcal{F}}(\eta) < \vartheta_{\mathcal{F}}(\mu)$. Therefore, $\vartheta_{\mathcal{F}}(z) \geq \vartheta_{\mathcal{F}}(\eta)$.

3 Modeling with Fuzzy Sets by Using Analytical Forms

The fuzzy sets can be described in the different ways such as the modeling presented here by (1) through to (6), correspondingly [3–5].

a. Discrete representation of the fuzzy sets: $\varphi_F(x_n)/x_n$.
b. Analytical representation, when the fuzzy sets are described as the parametric or piecewise functions.

(1) The fuzzy function in a triangular graphical representation:

$$\varphi_F(x) = 0, x < x_1, \varphi_F(x) = (x - x_1)/(x_2 - x_1), x_1 \leq x \leq x_2,$$

$$\varphi_F(x) = \frac{(x_3 - x)}{(x_3 - x_2)}, x_2 \leq x \leq x_3,$$

$$\varphi_F(x) = 1, x_n < x$$

(2)   The Fuzzy function as Γ(gamma)-function type of the membership function:

$$\varphi_F(x) = 0, x < x_1,$$

$$\varphi_F(x) = \frac{x - x_1}{x_2 - x_1}, x_1 \leq x \leq x_2,$$

$$\varphi_{\mathcal{F}}(x) = 1 \, for \, x_n < x$$

(3)   Fuzzy function represented by smoothed alternative Γ−function:

$$\varphi_F(x) = 0, 0 \leq x \leq x_1,$$

$$\varphi_F(x) = 1 - e^{-k(x-x_1)^2}, x_1 > x, k > 0.$$

(4)   Fuzzy function is represented by S-function introduced by Zadeh:

$$\varphi_F(x) = 0, x < x_1,$$

$$\varphi_F(x) = 2\left(\frac{x - x_1}{x_2 - x_1}\right)^2,$$

$$x_1 \leq x \leq x_2$$

$$\varphi_F(x) = 1 - 2\{(x - x_3)/(x_3 - x_1)\}^2,$$

$$x_2 \leq x \leq x_3,$$

$$\varphi_F(x) = 1, x_2 = \frac{x_1 + x_3}{2}, x > x_4$$

(5)   Fuzzy functions represented by a generalized trapezoidal function:

$$\varphi_F(x) = 0, \, x < x_1,$$

$$\varphi_F(x) = \frac{b_2(x - x_1)}{x_2 - x_1}, \, x_1 \le x \le x_2,$$

$$\varphi_F(x) = \frac{(b_3 - b_2)(x - x_2)}{x_3 - x_2} + b_2, \, x_2 \le x \le x_3,$$

$$\varphi_F(x) = b_3 - b_4, \, x_3 \le x \le x_4,$$

$$\varphi_F(x) = \frac{(b_4 - b_5)(x_5 - x)}{x_5 - x_4} + b_5, \, x_4 \le x \le x_5,$$

$$\varphi_F(x) = \frac{b_5(x_6 - x)}{x_6 - x_5}, \, x_5 \le x \le x_6,$$

$$\varphi_F(x) = 0, \, x_6 < x.$$

(6)   Fuzzy function represented by the smoothed concave function:

$$\varphi_F(x) = 1 - e^{-k)(x-x_1)^2},$$

$$k > 1.$$

The establishment and modeling with fuzzy functions by utilizing the modeling with functions 1–6 are an important process to present the fundamental approach to the various modeling techniques in the theory and applications of fuzzy systems.

Utilizing modeling types from 1 to 6 is highly suggested, especially for the problems, where the parameters of the related fuzzy function are estimated in terms to be aligned with the parameters of the fuzzy function modeling 1–6.

Modeling with the chosen parameters approximated to match to the parameters of the membership function by modeling with functions 1–6 is considered as the modeling of real phenomena problems [4–6].

4   Conclusion

The fuzzy modeling presents the fuzzy parameters by the fuzzy functions and their graphs. The fuzzifying mathematical application and problems in fuzzy set applications occur in the form of the modeling of linguistic problems. Further, the linguistic problems can be represented in the form of fuzzy sets, which enable us to represent the fuzzy functions to be described analytically and/or graphically.

The fuzzifying criteria can be achieved by the application of the extension principle. The extension principle provides us with the opportunity to apply modeling with fuzzy functions. Furthermore, such modeling acre practicable to be applied to the wide variety of the functions, which are not one-to-one. The mathematical and theoretical developments were given by the introduced theorems.

Theorem 1 is proved to show that the convex fuzzy cut sets are represented the convex power fuzzy set.

Theorem 1 represents the inclusion of the fuzzy sets.

Theorem 2 is proven to show the quasi-convex property of the fuzzy set.

Theorem 3 is proven to apply the extension principle to the fuzzy functions, which are not one-to-one. Furthermore, the fuzzy functions are identified as the pairs of elements of the fuzzy support plane.

Theorem 4 provides the opportunity to define quasi-convexity of the fuzzy sets.

The problems involving the modeling with fuzzy membership functions refer to the problems of knowledge engineering. The membership functions can be classified as it is:

(a)   construction of the membership function based on the heuristic models,
(b)   fuzzy function is modeling the problems arisen from human concept;
(c)   the establishment of the fuzzy function is based on demand for neural networks.

References

1. Gadjiev D, Rustanov (2020) A fuzzy topology and fuzzy geometry of the topological concepts 2020. In: IOP conference series. https://doi.org/10.1088/1757-899x/1001/1/012071
2. Gadjiev D, Rustanov A (2021) New methods of finding support planes of cut-level fuzzy power sets and geometry of the convex power fuzzy sets. IOP Conf Ser Mater Sci Eng 1001(1): https://doi.org/10.1088/1757-899x/1001/1/012070
3. Gadjiev D, Kochetkov I, Rustanov A (2021) The convex fuzzy sets and their properties with the application to the modeling with fuzzy convex membership function. Adv Intel Syst Comput 1259:276–284. https://doi.org/10.1007/978-3-030-57453-6_24
4. Gadjiev DD (2020) The soft programming and fuzzy logic mathematics modeling concepts with the application of the mathematical analysis. Chebyshevskii Sbornik 1(73):1
5. Gadjiev DD, Kochetkov I, Rustanov AR (2019) Aggregation of the fuzzy logic sets in terms of the functions of the triangular norm and triangular co-norm. IOP Conf Series 403: https://doi.org/10.1088/1755-1315/403/1/012187
6. Rafik AA, Fazlollahi B, Aliev RR (2004) Soft computing and its applications in business and economics. Springer, Berlin
7. Klir GJ, Yuan B (1995) Fuzzy sets and fuzzy logic. Theory and applications. Prentice Hall, NJ
8. Klir GJ, Yuan B (1997) Fuzzy sets theory. Foundations and applications. Prentice Hall, NJ
9. Aliev RA, Aliev FT, Babaev MD (1991) Fuzzy process control and knowledge engineering. Verlag, TUV Rheinland, Koln
10. Aliev RA, Aliev RR (2001) Soft computing and its applications. World Scientific

11. Zadeh LA (1971) Fuzzy orderings. J Inf Sci 3:117–200
12. Zadeh LA (1978) Fuzzy sets as a basis for a theory of possibility. J Fuzzy Sets Syst 1:3–22
13. Zadeh LA (1965) Fuzzy sets. J Inf Controls 8:338–353
14. Zadeh LA (1973) Outline of a new approach to the analysis of complex systems and decision processes. J IEEE Trans Syst Man Cybern 3:28–144

Chapter 20
Nonlinear Technique-Based ECG Signal Analysis for Improved Healthcare Systems

Varun Gupta, Monika Mittal, Vikas Mittal, Nitin Kumar Saxena, and Yatender Chaturvedi

1 Introduction

To reduce death rate due to different heart diseases, it is important to know correct status of the heart [1]. If any disease is found, then classification of the disease becomes important for timely diagnosis of the patient heart. Due to unhealthy routine and busy lifestyle, heart diseases are responsible for majority of deaths around the world. Due to heart diseases, 17.9 million people die each year [2]. To detect heart disease efficiently, the whole cardiac activity is monitored based on three main waves, viz. P-wave, QRS complex wave, and T-wave [3]. These three waves are combined as P-QRS-T, known as one cardiac cycle, and bunch of these cardiac cycles are called electrocardiogram (ECG) signal. For clarification of any heart-related issue, ECG signal is investigated [4]. Among these three waves of ECG signal, features of QRS complex wave are most important especially frequency response which is 10–25 Hz. ECG signal is also useful for various applications such as emotion detection, biometric pattern recognition, and blood pressure signal analysis. Unfortunately, recorded ECG signal has different types of noncardiac components apart from cardiac components (clinical attributes). In noncardiac components, respiration, cough, electromagnetic interference, unstable connection of the electrode from the body, power line interference, and body movements are common [5]. These components make ECG signal chaotic in nature which cannot be analyzed using

V. Gupta (✉) · N. K. Saxena
KIET Group of Institutions, Delhi-NCR, Ghaziabad, UP, India
e-mail: varun.gupta@kiet.edu

M. Mittal · V. Mittal
National Institute of Technology (NIT), Kurukshetra, Haryana, India

Y. Chaturvedi
Sunder Deep Group of Institutions, Ghaziabad, UP, India
e-mail: yatendra.chaturvedi@kiet.edu

© The Author(s), under exclusive license to Springer Nature Singapore Pte Ltd. 2021
S. Kumar et al. (eds.), *Proceedings of International Conference on Communication and Computational Technologies*, Algorithms for Intelligent Systems,
https://doi.org/10.1007/978-981-16-3246-4_20

Table 1 Different related works with involved techniques

References	Involved techniques	Performance
[1]	Higher-order statistics (HOS) and entropy-based feature selection methods	Accuracy = 99.83%
[5]	Hurst nonlinear feature and HOS	Accuracy = 92.87%
[6]	Empirical mode decomposition (EMD)	Accuracy = 94.92%
[7]	Short-time Fourier transform (STFT)	Standard deviation = 1.77 (cross-validation)
[8]	Hilbert transform (HT)	Recall = 99.88%

naked eyes. It makes the requirement of nonlinear signal processing technique to point out the abnormal fluctuations in ECG signal. To achieve this, necessary algorithms (mathematical tools) should be selected from different domains. Therefore, in this paper, nonlinear signal processing technique, i.e., chaos analysis is used which makes the detection system fast and accurate. For preprocessing and optimization of ECG datasets, discrete wavelet transform (DWT) and African buffalo optimization (ABO) techniques are used, respectively.

The rest of the paper is structured as follows; Sect. 2 covers the related work on ECG datasets, Sect. 3 shows materials and methods, Sect. 4 showcases important simulated results and discussion over those, and finally, Sect. 5 covers conclusion of the paper.

2 Related Work

In this section, different related works are presented in Table 1 with corresponding detection parameter and its value.

3 Materials and Methods

In ECG, a change in heart rate (HR) comes because of systolic and diastolic activity within the heart. Mainly, alteration in HR is resulted by the His-Purkinje system with specialized cells of the sinoatrial (SA) node and atrioventricular (AV) node. It leads to spectrum of diseases known as arrhythmias [6, 9–13]. Figure 1 shows proposed methodology. It covers different stages, viz. data acquisition, ECG datasets, preprocessing, feature extraction, optimization, performance evaluating parameters, and classification of subjects based on HR.

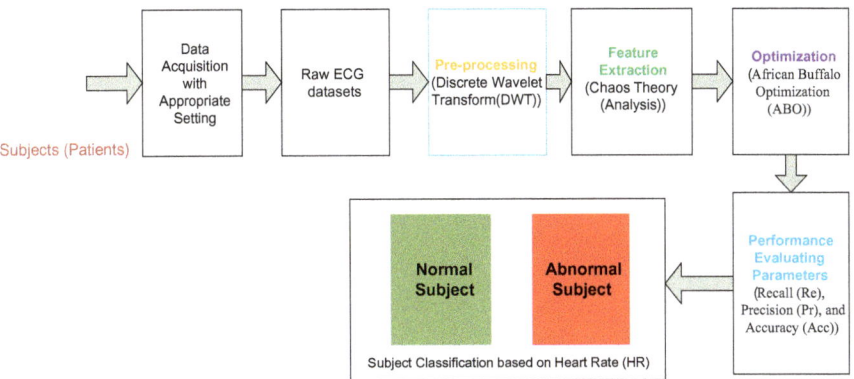

Fig. 1 Proposed methodology

3.1 ECG Datasets

In this paper, real-time datasets are recorded at room temperature with prior consent of associated institute research council. Total thirty-one datasets are recorded, and among these, eight datasets are analyzed (R01 DB-R08 DB). Because most of the researchers used standard database, thus datasets of MIT-BIH Arrhythmia database [14–18] no. 102, 104, 106, 107, 108, 112, 114, 115, 116, 118, 119, 202, 208, 209, 210, 212, 213, 214, 217, 219, 221, 230, and 232 are included in this paper.

3.2 Preprocessing

The recorded datasets are corrupted due to different artifacts and distortions including respiration, cough, electromagnetic interference, unstable connection of the electrode from the body, interferences of electrical equipment [19], power line interference (PLI) [20], body movements, and base line wanders (BLWs). These artifacts and distortions reduce signal-to-noise ratio (SNR) which may hide important clinical information of the recorded ECG dataset [21]. In past literature, various techniques have been used such as digital filters, modeling techniques [22], adaptive filtering [23]; singular value decomposition [24]; independent component analysis, S-transform [25], nonlinear filter bank [21]; fast Fourier transform and adaptive nonlinear noise estimator [26]; Empirical Mode Decomposition [27]; neural networks and wavelet transform. For removing these artifacts and distortions and to preserve original pathological characteristics, Daubechies 8 (Db8) wavelet is selected at decomposition level of 5 after wide iterations. The obtained result using Db8 is presented in Fig. 2. Wavelet transform is used because it effectively handles the non-stationary signals with better resolution in time–frequency domain than existing techniques [28].

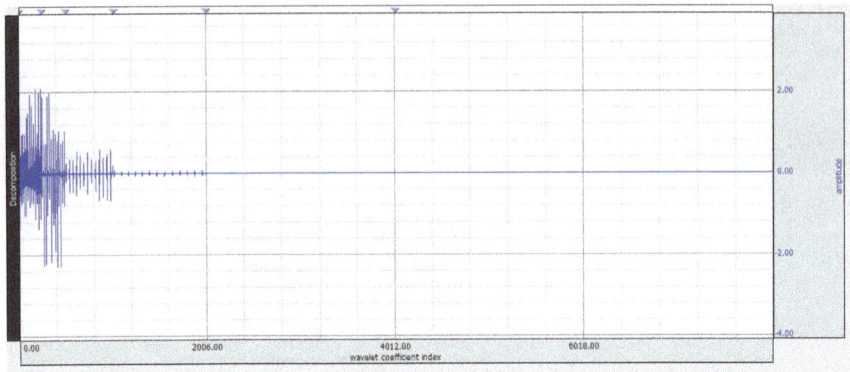

Fig. 2 DWT operation by selecting Daubechies 8 at decomposition level of 5 (vertical axis and horizontal axis show amplitude in mV and wavelet coefficient index, respectively)

3.3 Feature Extraction

The strength of dynamic analysis is to quantify the regularity and complexity involved in heart abnormalities using heart rate variability [29]. For performing dynamic analysis, different key tools are phase plane plot, Poincare map, fractal dimension, return map, and Lyapunov exponent. In this paper, different trajectories are plotted in phase plane plot. To quantify it, approximate entropy (ApEn) is used to estimate the randomness in the dataset [30, 31].

$$\text{ApEn}(i, k) = \lim_{M \to \infty} \left[\theta^i(k) - \theta^{i+1}(k) \right] \tag{1}$$

where

$$\theta^i(k) = \frac{1}{M - i + 1} \sum_{m=1}^{M-i+1} \log C_m^i(k) \tag{2}$$

where M denotes length of the dataset, i denotes non-negative integer ($i \leq M$), and k denotes positive real number.

3.4 Optimization

For optimization of feature extracted datasets, African buffalo optimization (ABO) technique [32, 33] is used in this paper, illustrated below.

African Buffalo Optimization (ABO)
This optimization theory was motivated by African buffalos due to their extensive migrant lifestyle which has outstanding memory, intelligence, and communication

(MIC). Basically, waaa calls of buffalo k show unsafe (mathematically w.k), and maaa calls of buffalo show safe (mathematically m.k). Mathematically, the movement of buffalos is given by

$$m.k + 1 = m.k + lp1(bgmax - w.k) + lp2(bpmax - w.k) \tag{3}$$

where lp1 and lp2 are learning factors.
The location of buffalo k is expressed as

$$w.k + 1 = \frac{w.k + m.k}{\pm 0.5} \tag{4}$$

The basic steps involved in ABO are as follows: (i) objective function, (ii) randomly place buffalos, (iii) fitness values (bgmax and bpmax), (iv) modify the location of buffalo, (v) check best fitness value, i.e., bgmax, (vi) If no, initialize the buffalos again, and (vii) output best solution.

3.5 Subject Classification

For subject classification, heart rate (HR) [34–45] is estimated. Based on HR, different heart diseases (arrhythmias) are investigated.

4 Results and Discussion

For graphical representation, ECG signal of 8 s is shown in Fig. 3a of real-time database (R02 DB). Figure 3b shows chaos analysis of same dataset resulting into sinus arrhythmia at coordinates (−0.77857, 0.15703). It reveals all nonlinear states within the signal and concludes the complete dynamics of cardiac rhythm and conduction disturbances. Further, using ApEn, randomness is estimated. Low value of ApEn indicates rhythmic signal (periodic), and high value of ApEn indicates random signal (nonlinear segments). Using phase space mapping (attractor plots), clear visualization of QRS complex waves can be obtained [46].

The proposed technique detected R-peaks of 53,218 in actual R-peaks of 53,294, TP of 53,212, FN of 61, FP of 37, and output SNR of 17.42. Further, the proposed technique is evaluated using the statistical parameters such as recall (Re), precision (Pr), and accuracy (Acc) [34], securing Re of 99.89%, Pr of 99.93%, and Acc of 99.82%.

As normal HR lies between 60 and 100 beats/min, arrhythmia (heart disease) detects when HR goes outside this range. In case of atrial fibrillation (AFi), HR varies chaotically in higher range, i.e., 100–175 beats/minute which may increase chances of heart strokes and heart failure [47]. In AFi patient, ventricle beats so

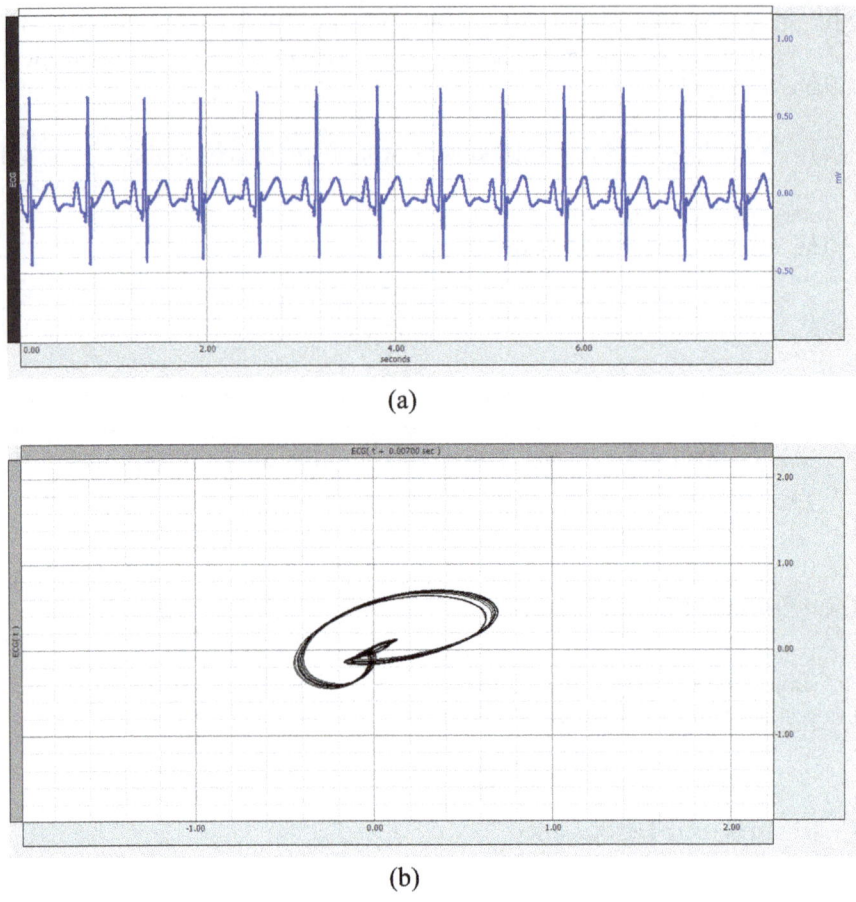

(a)

(b)

Fig. 3 **a** Real-time ECG dataset (vertical axis and horizontal axis show amplitude in mV and time in sec, respectively) and **b** chaos analysis (vertical axis and horizontal axis show amplitude of ECG signal in mV of signal delayed by T msec and original signal, respectively)

rapidly as compared to the atria and atria beat irregularly than the ventricles [48]. The estimated HR in AFi was 157 beats/min. The main symptoms of AFi are weakness, serious heart injury, and shortness of breath. During atrial flutter (AF), atria are more settled with low randomness than AFi. In AF patient, chances of stroke and health failure are more [1]. In another dataset, supraventricular tachycardia (SVT) is investigated which occurs if faulty electrical connections in the heart trigger and sustain an abnormal rhythm. In SVT patient, the HR is very high.

5 Conclusion

The concept of chaos analysis using ApEn with DWT and ABO clearly demonstrates the clear dynamics and HR in both databases, viz. real time and standard. Using proposed technique, different arrhythmias are classified successfully. The proposed technique finds its major applications in the detection of heart disease mostly to urban areas due to its robustness, good recall rate and precision rate. The obtained results clearly reveal that the nonlinear analysis of ECG signal captures all nonlinear segments that cannot be seen in manual analysis of ECG signals. In future, the proposed work can be integrated with sophisticated classification technique to enhance the detection accuracy.

References

1. Chashmi AJ, Amirani MC (2019) An efficient and automatic ECG arrhythmia diagnosis system using DWT and HOS features and entropy-based feature selection procedure. J Electr Bioimp 10:47–54
2. Cardiovascular Diseases. https://www.who.int/health-topics/cardiovascular-diseases#tab=tab_1. Accessed on 21 Nov 2020
3. Gupta V, Mittal M (2019) A novel method of cardiac arrhythmia detection in electrocardiogram signal. Int J Med Eng Inf 12(5):489–499
4. Jangra M et al (2020) ECG arrhythmia classification using modified visual geometry group network (mVGGNet). J Intel Fuzzy Syst 38:3151–3165
5. Selvaraj J, Murugappan M, Wan K, Yaacob S (2013) Classification of emotional states from electrocardiogram signals: a non-linear approach based on hurst. BioMed Eng OnLine 12:44. http://www.biomedical-engineering-online.com/content/12/1/44
6. Maji C, Sengupta P, Batabyal A, Chaudhuri H (2020) Nonlinear and statistical analysis of ECG signals from arrhythmia affected cardiac system through the EMD process. Electr Eng Syst Sci-Sig Proc, 1–24
7. Dasgupta H (2016) Human age recognition by electrocardiogram signal based on artificial neural network. Sens Imaging 17(4):1–15
8. Gupta V, Mittal M (2018) R-peak based arrhythmia detection using Hilbert transform and principal component analysis. In: 3rd International conference proceedings on innovative applications of computational intelligence on power, energy and controls with their impact on humanity, Ghaziabad, India. IEEE, pp 116–119
9. Sheetal A et al (2019) QRS detection of ECG signal using hybrid derivative and MaMeMi filter by effectively eliminating the baseline wander. Analog Integr Circ Sig Process 98(1):1–9
10. Gupta V, Mittal M (2020) Efficient R-peak detection in electrocardiogram signal based on features extracted using Hilbert transform and burg method. J Inst Eng India Ser B. https://doi.org/10.1007/s40031-020-00423-2
11. Kora P (2017) ECG based myocardial infarction detection using Hybrid firefly algorithm. Comput Methods Programs Biomed. https://doi.org/10.1016/j.cmpb.2017.09.015
12. Draghici AE, Taylor JA (2016) The physiological basis and measurement of heart rate variability in humans. J Physiol Anthropol 35(1):1–8
13. John RM, Kumar S (2016) Sinus node and atrial arrhythmias. Contem Rev in Cardiovasc Med 133:1892–1900
14. Sharma LD, Sunkaria RK (2020) Myocardial infarction detection and localization using optimal features based lead specific approach. IRBM 41:58–70

15. Halder B (2019) Classification of complete myocardial infarction using rule-based rough set method and rough set explorer system. IETE J Res. https://doi.org/10.1080/03772063.2019. 1588175
16. Gupta V et al (2020) R-peak detection based chaos analysis of ECG signal. Analog Integr Circ Sig Process 102:479–490
17. Gupta V et al (2019) R-Peak detection using chaos analysis in standard and real time ECG databases. IRBM 40(6):341–354
18. Xingyuan W, Juan M (2009) Wavelet-based hybrid ECG compression technique. Analog Integr Circ Sig Process 59(3):301–308
19. Bahoura M, Ezzaidi H (2010) FPGA-implementation of wavelet-based denoising technique to remove power-line interference from ECG signal. In: Proceedings of the 10th IEEE international conference on information technology application biomedicine (ITAB), Greece Corfu, New Jersey. IEEE, pp 1–4
20. Rahman MZU, Shaik RA, Reddy DVRK (2010) Baseline wander and power line interference elimination from cardiac signals using error nonlinearity LMS algorithm. In: International conference on systems in medicine and biology (ICSMB), Kharagpur, India. IEEE, pp 217–220
21. Łęski JM, Henzel N (2005) ECG baseline wander and powerline interference reduction using nonlinear filter bank. Signal Process 85(4):781–793
22. Acharya UR et al (2008) Automatic identification of cardiac health using modeling techniques: a comparative study. J Inform Sci 178:4571–4582
23. AlMahamdy M, Riley HB (2014) Performance study of different denoising methods for ECG signals. Procedia Comput Sci 37:325–332
24. Bandarabadi AAJGM, Karami-Mollaei MR (2010) ECG denoising using singular value decomposition. Aust J Basic Appl Sci 4(7):2109–2113
25. Das M, Ari S (2013) Analysis of ECG signal denoising method based on s-transform. IRBM 34(6):362–370
26. Shirbani F, Setarehdan SK (2013) ECG power line interference removal using combination of FFT and adaptive non-linear noise estimator. In: Proceedings of the 21st Iranian conference on electrical engineering (ICEE), Mashhad, Iran. IEEE, pp 1–5
27. Agrawal S, Gupta A (2013) Fractal and EMD based removal of baseline wander and powerline interference from ECG signals. Comput Biol Med 43(11):1889–1899
28 Aouinet A, Adnane C (2014) Electrocardiogram denoised signal by discrete wavelet transform and continuous wavelet transform. Akram Aouinet & Cherif Adnane. J Signal Process Int J (SPIJ) 8:1–9
29. Krstacic G, Krstacic A, Smalcelj A, Milicic D, Jembrek-Gostovic M (2007) The chaos theory and nonlinear dynamics in heart rate variability analysis: does it work in short-time series in patients with coronary heart disease? Ann Noninvasive Electrocardiol 12(2):130–136
30. Pincus SM, Goldberger AL (1994) Physiological time-series analysis: what does regularity quantify? Am J Physiol Heart Circul Physiol 266:1643–1656
31. Delgado-Bonal A, Marshak A (2019) Approximate entropy and sample entropy: a comprehensive tutorial. entropy 21. https://doi.org/10.3390/e21060541
32. Odili JB, Kahar MZM (2016) African buffalo optimization. Int J Softw Engi Comput Syst 2:28–50
33. Odili JB, Kahar MNM, Anwar S, Ali M (2017) Tutorials on African buffalo optimization for solving the travelling salesman problem. Int J Softw Eng Comput Syst 3:120–128
34. Gupta V et al (2020) Performance evaluation of various pre-processing techniques for R-peak detection in ECG signal. IETE J Res. https://doi.org/10.1080/03772063.2020.1756473
35. Gupta V, Mittal M (2018) Dimension reduction and classification in ECG signal interpretation using FA & PCA: a comparison. Jangjeon Math Soc 21(4):765–777
36. Gupta V, Mittal M (2016) Respiratory signal analysis using PCA, FFT and ARTFA. In: Proceedings of the international conference on electrical power and energy systems (ICEPES), MANIT Bhopal, India, pp 221–225
37. Gupta V, Mittal M (2018) ECG (Electrocardiogram) signals interpretation using Chaos Theory. J Adv Res Dyn Cont Syst (JARDCS) 10(2):2392–2397

38. Gupta V, Mittal M (2015) Principal component analysis & factor analysis as an enhanced tool of pattern recognition. Int J Elec Electr Eng Telecoms 1(2):73–78
39. Gupta V et al (2011) Principal component and independent component calculation of ECG signal in different posture. AIP Conf Proc 1414:102–108
40. Gupta V et al (2019) Auto-regressive time frequency analysis (ARTFA) of electrocardiogram (ECG) signal. Int J Appl Eng Res 13(6):133–138
41. Gupta V, Mittal M (2021) R-peak detection for improved analysis in health informatics. Int J Med Eng Inf https://www.inderscience.com/info/ingeneral/forthcoming.php?jcode=ijmei
42. Gupta V, Mittal M (2019) QRS complex detection using STFT, chaos analysis, and PCA in standard and real-time ECG databases. J Inst Eng (India): Ser B 100 489–497
43. Gupta V, Mittal M (2018) KNN and PCA classifier with autoregressive modelling during different ECG signal interpretation. Procedia Comput Sci 125:18–24
44. Gupta V, Mittal M, Mittal V (2020) Chaos theory: an emerging tool for arrhythmia detection. Sens Imaging 21(10):1–22
45. Gupta V, Mittal M (2019) R-Peak detection in ECG signal using yule–walker and principal component analysis. IETE J Res. https://doi.org/10.1080/03772063.2019.1575292
46. Albert DE (1991) Chaos and the ECG: fact and fiction. J Electrocardiol 24:102–106
47. Atrial Fibrillation, https://www.mayoclinic.org/diseases-conditions/atrial-fibrillation/symptoms-causes/syc-20350624. Last accessed 12 Nov 2020
48. Atrial Fibrillation, https://www.heart.org/en/health-topics/atrial-fibrillation/what-is-atrial-fibrillation-afib-or-af. Last accessed 13 Nov 2020

Chapter 21
Recommendation System for Adoption of ICT Usage and Its Impact on Academic Performance of Students in India

J. S. Shyam Mohan, Vedantham Hanumath Sreeman,
Vanam Venkata Chakradhar, Harsha Surya Abhishek Kota,
Challa Nagendra Panini, Narasimha Krishna Amruth Vemuganti,
Naga Venkata Kuladeep, Vankadara Raghuram Nadipalli,
Surekuchi Satya Swaroop, and M. U. M. Subramanyam

1 Introduction

With the incremental growth of information and communication technology, learning has become very easy in today's environment. Learning can be effectively handled and delivered by successfully using information and communication technology tools across the globe. Government and other public and private sectors have started investing huge amounts of money for adopting ICT [1, 2]. Most of the colleges and Universities are in various stages of adopting ICT tools for their academic usage, research activities and etc. Even some Universities have started to transfer credits into their regular curriculum. Adoption of ICT in the proposed work is comprehended as a way of automating the educational process by automizing it into a learning model called as learning management system (LMS) for hosting all the e-learning materials in a repository for future use. Majority of the colleges and Universities in India have been using blackboard as their regular teaching method for decades. With the adoption of ICT tools in India, it has promoted education to reach even to the remote places where there is no awareness of education at all. The data collected from the institutions is shown in Table 1. All the institutions mentioned in Table 1 are at different locations. Previous studies reveal that the viability of ICT usage improves

J. S. Shyam Mohan · V. H. Sreeman · V. V. Chakradhar · H. S. A. Kota · N. K. A. Vemuganti (✉) ·
N. V. Kuladeep · V. R. Nadipalli · S. S. Swaroop · M. U. M. Subramanyam
Department of CSE, Sri Chandrasekharendra Saraswathi Viswa Mahavidyalaya, Kanchipuram,
Tamil Nadu, India

C. N. Panini
Department of IT, Shri Vishnu Engineering College for Women, Kovvada, Andhra Pradesh, India

S. Kumar et al. (eds.), *Proceedings of International Conference on Communication
and Computational Technologies*, Algorithms for Intelligent Systems,
https://doi.org/10.1007/978-981-16-3246-4_21

Table 1 Data collected from the institutions

S. No.	Name of the institution	Place	State
1	Institution A	Chennai	Tamil Nadu
2	Institution B	Tiruchirappalli	Tamil Nadu
3	Institution C	Chennai	Tamil Nadu
4	Institution D	Kanchipuram	Tamil Nadu
5	Institution E	Kadapa	Andhra Pradesh
6	Institution F	Chitoor	Andhra Pradesh

students' academic performance; however, these studies failed to prove precisely the exact effect of ICT on students' performance [3].

These studies have posed some challenges [4]:

1. Determine the exact performance of students is difficult as there is no standard mechanism of transferring credits to student's GPA.
2. Technological changes making it difficult to determine the impact of ICT.

Reference [5] provided methodology for adoption of ICT in institutions can improve the standards of the institution. Many studies have focused only cognitive results that were proved to be successful in adopting ICT toward development. The proposed model is implemented for the institutions that have started using ICT tools. The proposed model identified some factors that may contribute to increase the performance of students and how these factors are affected by ICT. The objectives of the proposed work are given below:

a. Institution accessibility for adoption of ICT.
b. To check the relationship between ICT usage and the performance of the students
c. To assess the impact of ICT in institutions and students performance.

2 The Research Study Model (RSM)

The research model is stated by taking the considering the following aspects [6, 7]:

1. Adoption of ICT for the institutions shown in Table 1 has provided benefits in enhancing the students' academic performance. The evaluation and overall credit transfer benefit has been much better than existing methods. Research study model 1 (RSM1) is framed as follows:
   RSM1: Shows the significant relationship of ICT adoption with student's academic performance.
2. Faculties are able to design their own schedule and methods using ICT. All the institutions have started implementing ICT as a part of their academic curriculum that has provided students to have easy access at their own convenience. Therefore, RSM2 is stated as follows:

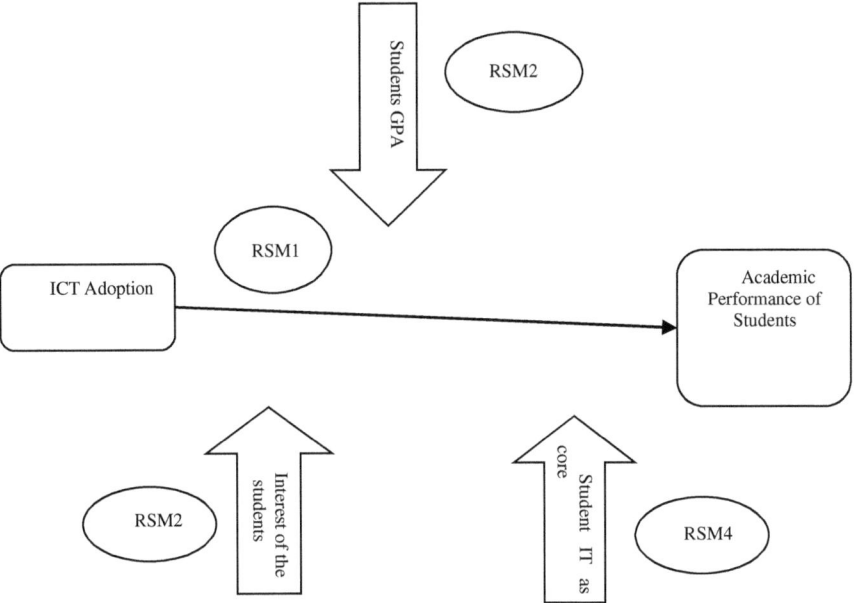

Fig. 1 The research study model

RSM2: Adoption of ICT provided a good and healthy relationship between students GPA and improved the students' academic performance (Fig. 1).

3. Some of the students do not have interest in using ICT mode for their regular academic activities. They follow the traditional way of securing the GPA. However, the institutions have not mandatory for all the students to use ICT. Hence, the number of ICT usage and access to LMS varies across institutions. This has led to creating awareness among students to use ICT for credit transfer to their GPA. Hence, RSM3 is stated as follows:
   RSM3: Students' interests depend upon the adoption of ICT for their academic curriculum.

4. Many students in institutions choose IT as a major and show interest to learn new technologies. Therefore, institutions show major responsibility to support students to fulfill their career aspects also. Hence, RSM4 is stated as follows:
   RSM4: Relation between ICT and student academic performance is based on students' interests.

3 Methodology

For data collection, traditional techniques and tools such as questionnaires were used. Another method used for data collection is by downloading the data from

NPTEL nodal coordinators across institutions to know the number of students who have enrolled for the courses and the number of students who have passed in the certification course [8, 9]. All the data collected are used for performing statistical analysis.

3.1 Data Sampling

The data collected from few of the institutions had adopted nearly 80% of ICT in their regular academic curriculum. Other institutions have adopted nearly 50–60% of ICT usage as a part of their academic curriculum. Sample size of 3000 respondents, nearly 300 students from each institution is taken for analysis. The respondents were categorized into two: Faculty members and students. There were some challenges faced like the institutions were geographically at longer distances [10].

3.2 Data Collections Tools Used

The questionnaire was collected from students using Google forms. The students had access to ICT tools. Some of the data was collected from friends those who were in the institutions [11].

3.3 Data Analysis

Data are checked for any bias, inconsistencies, etc. after data collection. For measuring the ICT usage, SPSS structural equation modeling (SEM) was used. Variations in moderating the variables were analyzed using the same. The difficulties faced by using ordinary least squares method were overcome by using SPSS–AMOS. SEM specifies the causal relationships among variables. Confirmatory factor analysis (CFA) is commonly used measurement model used for measuring latent constructs along with correlations [12]. AMOS specifies models and allows users to create path diagrams by simply writing equations or commands. The responses given by students in Google forms were nearly 70%. Nearly 900 candidates have responded to the questionnaires in Google form. Rest of the details received was excluded because the details were only not completely answered and hence they were discarded for the analysis [13]. Among the recorded responses, there are 90 faculties and the rest were students. Majority of the candidates belong to IT background. Over 57% of the candidates have GPA between 7.0 and 8.0 on a 10–point scale. The values are only indicative and vary from institution to institution [14]. Few of the students were in their second and first year of their studies. Some of the candidates have exposure to the latest technology. In Cronbach's alpha α, if the reliability indicator is 0.84, the model is stated as well-fitted, with good internal consistency, the mean score of the candidates is 3.68 (Standardized Alpha 0.84) states that the ICT adoption affects the students' academic performance. Reliability indicators are satisfactory and cannot be changed as shown in Table 4 [15, 16]. Incremental model fit is used for the proposed model in order to measure the fit [17]. Some of the models used in incremental model fit are shown in Table 3. In the proposed model, we consider two variables: Independent variable as ICT adoption and dependent variable as students' academic performance. The variables vary with the moderating values, students GPA, interest of students and students choosing IT as core [18] (Table 2).

The relationship between independent and dependent variables gets affected with moderating values [20, 21]. The moderating values remain unaffected and specify that the relationship between independent and dependent variables differ during its interaction. For qualitative research, independent variable is considered as X and for dependent variable Y is considered. In the proposed model, for qualitative studies,

Table 2 Details about the candidates who participated in the study

Institution A	Institution B	Institution C	Institution D	Institution E	Institution F	Total
142	75	263	221	96	103	900
16%	9%	29%	25%	10%	11%	100%

Table 3 Independent variable ICT adoption–statistics (Cronbach's alpha 0.84) [19]

Items	Cronbach's alpha	Std. alpha
Flexibility of ICT usage in institution	0.84	3.68
Whether your institution supports credit transfer of ICT certification?	0.81	3.26
Accessibility of ICT for employability	0.86	3.12
Average score	0.83	3.35

Table 4 Incremental model fit [27]

S. No.	Incremental model	Acceptance values	Model fit
1	Normed Fitted Index (NFI)	0.96	Acceptable
2	Relative Non-Centrality Index (RNI)	0.90	Good
3	Tucker-Lewis Index (TLI)	0.93	Acceptable
4	Comparative Fit Index (CFI)	–	–
5	Goodness-of-Fit-Index (GFI)	0.95	Acceptable
6	Adjusted GFI (AGFI)	–	–
7	Standardized Root Mean Square Residual (SRMR)	0.05	Acceptable
8	Root Mean Square Error of Approximation (RMSEA)	0.03	Acceptable

there are three variables: Students GPA, interest of students and students choosing IT as core besides X and Y [22].

In Figs. 2 and 3, all the independent and dependent variables are linear. If all the variables are linear or moving in the same direction, the model must be selected as default model. If there are any variations, the effectiveness of the assumed model will stop [23]. Both the figures represent the standardized regression weight (0.62) as

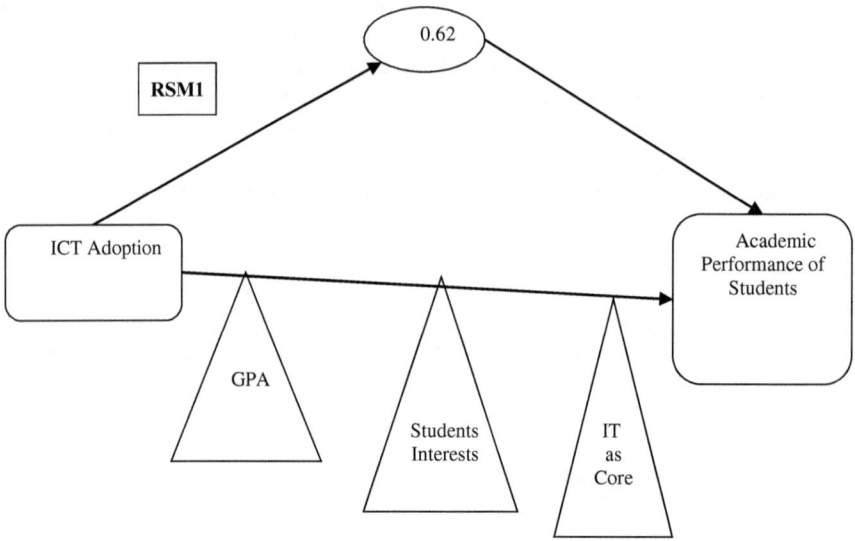

Fig. 2 The relationship between independent and dependent variables without the moderators

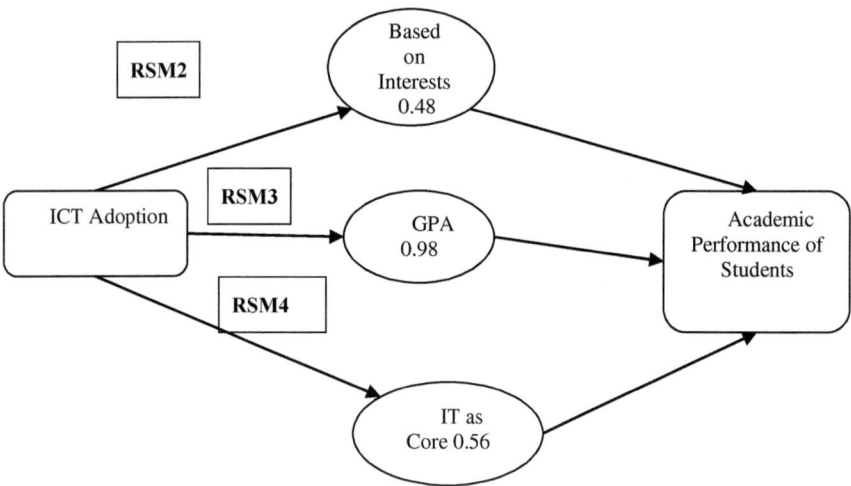

Fig. 3 Relationship between independent and dependent variables with moderators effect

Table 5 Acceptance of model (values ≥ 0.90)

Nature of the model	Acceptance values
Good fit	≥ 0.95
Absolute fit	GFI + AGFI + SRMR + RMSEA
Good model–Data fit	RNI ≥ 0.95, SRMR ≤ 0.08 and RMSEA ≤ 0.06

Table 6 Regression weight and research model

	Moderator	Regression weight	Research model
Students academic performance	All	0.62	Supported
	GPA	0.98	Supported
	Based on Interests	0.48	Not supported
	IT as core	0.56	Supported

calculated in the relationship of independent and dependent variables [24] (Tables 5 and 6).

Table 4 also shows the incremental model fit for adoption of ICT and performance of students is consistent [25, 26]. All the proposed models in this paper have been tested successfully and found supported excluding the fourth model. From the studies stated above, it is clear of the following points:

1. The overall impact of adoption of ICT on academic performance is good (RSM1).
2. In some cases, some students did not showed interest in ICT and relied upon the traditional methods.
3. Majority of the students showed interest in transferring credits to GPA for improving their academic performance.

Online LMS has facilitated the institutions to rely upon adoption of ICT [28]. The statistical results reveal that over 50 percent of the candidates consider ICT useful, few (6%) considered ICT for enhancing their skill sets. Nearly, 70% of the candidates admitted that they use their smartphone's for completing their assignments etc. [29–32]. Recently, few.

+ 1—Value is maximum—Same direction.

− 1—Value is minimum—Opposite direction.

0.5 To − 0.5—Values Are not Supported.

0.56, 0.98 and 0.62—High association.

4 Conclusion

The findings of the proposed model proved to be useful to the institutions for framing ICT policies for higher education. Students can be aware of the ICT usage can be helpful for improving their academic performance and enhancing their technical skills. Researchers interested to carry on their research in ICT can find this article useful. Information communication technology (ICT) is the latest filed in the education that provided a platform and opportunities to some extent of knowledge. From the findings, it is evident that a majority of the candidates find ICT as useful for improving their skills. Furthermore, ICT requires candidates to remain brighter to adapt to new technology. Another interesting finding of this study is that most of the students spend two to three hours daily on ICT. Even institutions have also started to give credits transferred directly to GPA. The proposed model can be effectively implemented if deployed using blockchain technology.

References

1. Fu JS (2013) ICT in education: a critical literature review and its implications. Int J Educ Dev using Inf Commun Technol (IJEDICT) 9(1):112
2. Ali F, Zhou Y, Hussain K, Nair PK, Ragavan NA (2016) Does higher education service quality effect student satisfaction, image and loyalty? A study of international students in Malaysian public universities. Qual Assur Educ 24(1):70–94
3. Rose A, Kadvekar S (2015) ICT (information and communication technologies) adoption model for educational institutions. J Commer Manage Thought 6(3):558
4. Ellis V, Loveless A (2013) ICT, pedagogy and the curriculum: subject to change. Routledge, London
5. Chan D, Bernal A, Camacho A (2013) Integration of ICT in higher education: experiences and best practices in the case of the University of Baja California. In: Proceedings of the Edulearn13, Barcelona, Spain, pp 1040–1049
6. Sari A, Mahmutoglu H (2013) Potential issues and impacts of ICT applications through learning process in higher education. Procedia Soc Behav Sci 89:585–592
7. Iniesta-Bonillo MA, Sanchez-Fernandez R, Schlesinger W (2013) Investigating factors that influence on ICT usage in higher education: a descriptive analysis. Int Rev Public Nonprofit Mark 10(2):163–174
8. Castillo-Merino D, Serradell-Lopez E (2014) An analysis of the determinants of students' performance in e-learning. Comput Hum Behav 30:476–484
9. Attuquayefio SN, Addo H (2014) Using the UTAUT model to analyze students' ICT adoption. Int J Educ Dev Inf Commun Technol (IJEDICT) 10(3):75
10. Voogt J, Knezek G, Cox M, Knezek D, ten Brummelhuis A (2013) Under which conditions does ICT have a positive effect on teaching and learning? A call to action. J Comput Assist Learn 29(1):4–14
11. Croteau AM, Venkatesh V, Beaudry A, Rabah J (2015) The role of information and communication technologies in University students' learning experience: the instructors' perspective. In Proceedings of the 48th Hawaii international conference on system sciences (HICSS'2015), Kauai, HI, USA. IEEE, pp 111–120
12. Cruz-Jesus F, Vicente MR, Bacao F, Oliveira T (2016) The education-related digital divide: an analysis for the EU-28. Comput Hum Behav 56:72–82

13. Kreijns K, Vermeulen M, Kirschner PA, Buuren HV, Acker FV (2013) Adopting the integrative model of behaviour prediction to explain teachers' willingness to use ICT: a perspective for research on teachers' ICT usage in pedagogical practices. Technol Pedagog Educ 22(1):55–71

14. Sabi HM, Uzoka FME, Langmia K, Njeh FN (2016) Conceptualizing a model for adoption of cloud computing in education. Int J Inf Manage 36(2):183–191

15. Sanchez RA, Cortijo V, Javed U (2014) Students' perceptions of Facebook for academic purposes. Comput Educ 70:138–149

16. Solar M, Sabattin J, Parada V (2013) A maturity model for assessing the use of ICT in school education. Educ Technol Soc 16(1):206–218

17. Gallego JM, Gutierrez LH, Lee SH (2014) A firm-level analysis of ICT adoption in an emerging economy: evidence from the Colombian manufacturing industries. Ind Corp Change 24(1):191–221

18. Babaheidari SM, Svensson L (2014) Managing the digitalization of schools: an exploratory study of school principals' and IT managers' perceptions about ICT adoption and usefulness. In: Proceedings of the E-Learn: world conference on e-learning in corporate, government, healthcare, and higher education, New Orleans, LA, USA, vol 2014 no 1, pp 106–113

19. Lin CY, Huang CK, Chen CH (2014) Barriers to the adoption of ICT in teaching Chinese as a foreign language in US universities. ReCALL 26(1):100–116

20. Wastiau P, Blamire R, Kearney C, Quittre V, Van de Gaer E, Monseur C (2013) The use of ICT in education: a survey of schools in Europe. Eur J Educ 48(1):11–27

21. Venkatesh V, Croteau AM, Rabah J (2014) Perceptions of effectiveness of instructional uses of technology in higher education in an era of Web 2.0. In Proceedings of the 47th Hawaii international conference on system sciences (HICSS'2014), Washington, DC, USA. IEEE, pp 110–119

22. Macharia JK, Pelser TG (2014) Key factors that influence the diffusion and infusion of information and communication technologies in Kenyan higher education. Stud High Educ 39(4):695–709

23. Basri W, Suliman M (2012) Factors affecting information communication technology acceptance in public organizations in Saudi Arabia. Int J Comput Sci Inf Secur 10(2):118–139

24. Venkatesh V, Sue B, Bala H (2013) Bridging the qualitative–quantitative divide: guidelines for conducting mixed methods research in information systems. MIS Q 37(1):21–54

25. Awang Z (2012) Structural equation modeling using AMOS. Penerbit Universiti Teknologi MARA, Shah Alam, Selangor Darul Ehsan, Malaysia

26. Hox JJ, Bechger TM (1998) An introduction to structural equation modelling. Family Sci Rev 11:354–373

27. Lei P-W, Wu Q (2007) Introduction to structural equation modeling: issues and practical considerations. Educ Meas Issues Pract 26(3):33–43

28. Hu L-T, Bentler P (1999) Cutoff criteria for fit indexes in covariance structure analysis: conventional criteria versus new alternatives. Struct Equ Model 6:1–55

29. Hilbert M, Lopez P (2011) The world's technological capacity to store, communicate, and compute information. Sci 332(6025):60–65

30. Basri WSh et al (2018) ICT adoption impact on students' academic performance: evidence from Saudi universities. Educ Res Int 2018, Article ID 1240197, 9p. https://doi.org/10.1155/2018/1240197

31. Cloke C, Sharif S (2006) Why use information and communications technology? Some theoretical and practical issues. J Inf Technol Teach Educ 10(1&2). ISSN: 0962-029X (Print) (Online) Journal homepage: https://www.tandfonline.com/loi/rtpe19

32. Shyam Mohan JS, Challa NP, Chakravarthy VVK, Kumar GPS, Rao RS, Raju PVR (2021) Recent trends and challenges in Blockchain technology. In: Sekhar GC, Behera HS, Nayak J, Naik B, Pelusi D (eds) Intelligent computing in control and communication. Lecture notes in electrical engineering, vol 702. Springer, Singapore. https://doi.org/10.1007/978-981-15-8439-8_19

Chapter 22
Performance Comparison of Dispersion Compensating Techniques for Long Distance Optical Communication System

Rajkumar Gupta and M. L. Meena

1 Introduction

With the increase in high-data rates and low cost, the demand of optical fiber technology is increased day-by-day. Wavelength division multiplexed system provides high-data rates at very low cost. But there are some hurdles when transmitting the signal from transmitter to receiver. These are dispersion and attenuation. To overcome the problem of dispersion d compensation techniques are used. There are two important techniques to reduce the dispersion effects in optical fiber are dispersion compensation fiber (DCF) and fiber bragg grating (FBG). Different chirped apodized FBG are used like linear chirped, square root chirped and cube root chirped. Uniform FBG is also used to overcome the problem of dispersion. A single wavelength signal is reflected in uniform FBG. To overcome the problem of attenuation some amplifiers are used in optical fiber. There are different types of amplifiers used in optical fiber. These are Raman amplifier, erbium doped fiber amplifier (EDFA). In this paper, EDFA is used to reduce the attenuation [1–4].

Chakkour et al. [5, 6] reduced the dispersion effects by using FBG with EDFA technique. Initially, this technique is performed on single channel, but authors used this technique on four channel system. Dar bashir ashif et al. [1] used various dispersion techniques like DCF, DCF + FBG and FBG with different chirp function. He found that DCF + FBG gave better result but at high cost. And FBG with linear chirp function also gave good result compare to DCF at low cost. Mohammed et al. [2] used various dispersion compensation methods and various chirping functions. He found among all chirping function linear chirping function is preferable and among all the techniques Dispersion compensation fiber with linearly chirped tanh FBG gave better result. Meena et al. [7] in the paper, DWDM network with linear chirped

R. Gupta (✉) · M. L. Meena
Department of Electronics Engineering, Rajasthan Technical University, Kota, India

© The Author(s), under exclusive license to Springer Nature Singapore Pte Ltd. 2021   267
S. Kumar et al. (eds.), *Proceedings of International Conference on Communication and Computational Technologies*, Algorithms for Intelligent Systems,
https://doi.org/10.1007/978-981-16-3246-4_22

fiber bragg gating is designed and performance is evaluated in the terms of bit error rate (BER), Q-factor.

2 Dispersion Compensation Unit

2.1 *Dispersion Compensation Fiber (DCF)*

In long distance optical fiber transmission system, dispersion problem occurs due to the nonlinearity of fiber. So, dispersion compensation fiber plays a significant role for designing the long distance transmission system. In this technique, a special single mode fiber is designed to reverse the harmful effect of dispersion and upgrade the quality of optical fiber transmission system. Consequently, the DCF has a higher negative dispersion coefficient in the range of −70 to −90 ps/nm/km and a single mode fiber (SMF) has a positive dispersion coefficient around 17 ps/nm/km. A DCF is connected to SMF to reduce the dispersion effect as shown in Fig. 1 [7, 8].

For the perfect dispersion compensation given Eqs. (1–3) should be satisfy: [9].

$$D_{SMF}L_{SMF} + D_{DCF}L_{DCF} = 0 \tag{1}$$

$$L_{DCF} = -L_{SMF}\left(\frac{D_{SMF}}{D_{DCF}}\right) \tag{2}$$

$$S_{DCF}L_{DCF} + S_{SMF}L_{SMF} = 0 \tag{3}$$

where
D_{SMF} = Dispersion of single mode fiber.
L_{SMF} = Length of single mode fiber.
D_{DCF} = Dispersion of dispersion compensation fiber.
L_{DCF} = Length of dispersion compensation fiber.
S_{SMF} = Dispersion slope of SMF.

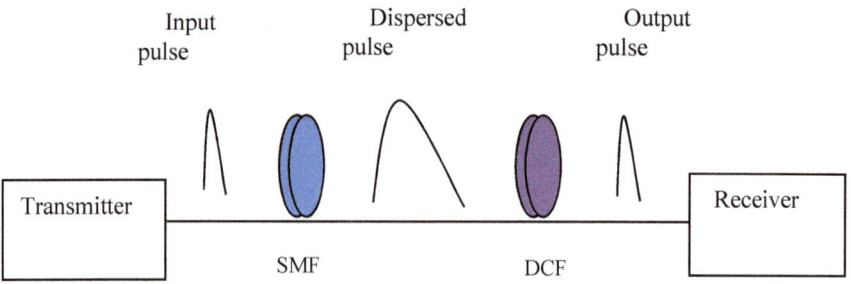

Fig. 1 Basic principle of DCF technique

Table 1 Parameter of DCF

Parameter	Value
Length (in km)	21
Dispersion (ps/km/nm) at 1550 nm	−80
Dispersion slope (ps/nm^2/km) at 1550 nm	−0.3
Attenuation (dB/km)	0.5

S_{DCF} = Dispersion slope of DCF.

Parameter of DCF for this paper is given in Table 1.

2.2 Fiber Bragg Grating

FBG works as a filter. It is designed in a short portion of a optical fiber that contains periodic dissimilarity in a refractive index of the optical fiber core by which a particular wavelength signal can be reflected and all other wavelength signals can be transmitted [7, 9].

The reflected wavelength is called Bragg's wavelength that is given by Eqs. (4 and 5).

$$\lambda_{\text{Bragg}} = 2\eta_{\text{eff}}\Lambda \tag{4}$$

where

$$\lambda_{\text{Bragg}} = \text{The braggs wavelength}$$

$$\eta_{\text{eff}} = \text{Effective refractive index}$$

Λ = Grating period of the fiber.

A uniform FBG has uniform refractive index along the optical fiber length shown as in Fig. 2, and if there are a linear dissimilarity in the grating period that is called non uniform or chirped fiber bragg grating as illustrated in Fig. 3.

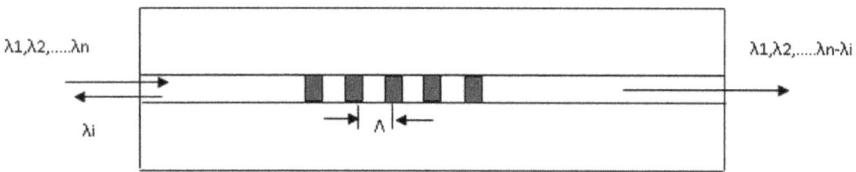

Fig. 2 Schematic diagram of a uniform fiber Bragg grating

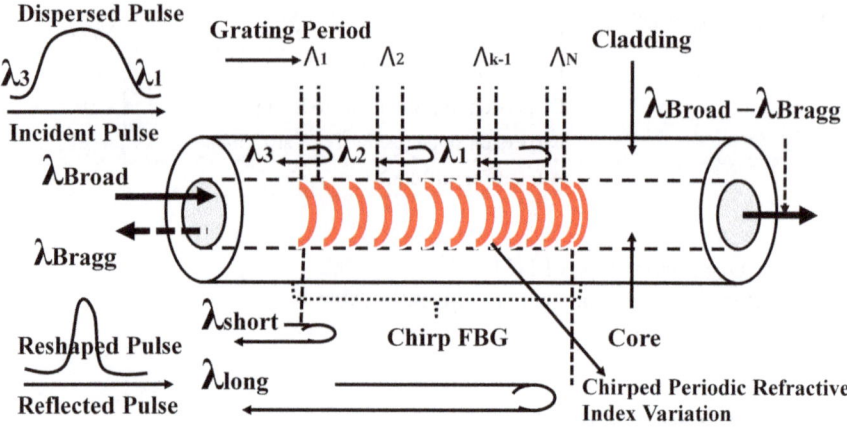

Fig. 3 Schematic diagram of non-uniform chirped FBG [9]

2.3   Linear Chirped Fiber Bragg Grating

A single wavelength signal is reflected by uniform FBG, but to reflect all wavelength signal we require to be chirping the grating period along the FBG length. The chirped fiber Bragg's grating (CFBG) contains a non-uniform grating period. It introduces divergent time-delay for every wavelength. In CFBG, speedy wavelengths return from the shortest grating period with taking the longest time and slow wavelengths reflect from the longest grating period taking the shortest time as shown in Fig. 3. The range of all reflected wavelengths is given by [10, 11]:

$$\Delta\lambda_{\text{chirp}} = \lambda_{\text{long}} - \lambda_{\text{short}} = 2\eta_{\text{eff}}.(\Lambda_{\text{long}} - \Lambda_{\text{short}}) = 2\eta_{\text{eff}}.\Delta\Lambda_{\text{chirp}} \qquad (5)$$

where Λ_{long}, Λ_{short} are the longest and the shortest grating period and λ_{long}, λ_{short} are the longest and the shortest wavelength, consequently.

The optimized parameters of FBG for all three chirping techniques for this paper are given in Table 2. In this table, the parameter of FBG for the joint technique (DCF + FBG) is also given [12, 13].

Table 2 Parameters of FBG

Parameters	LcAFBG	ScAFBG	CcAFBG	FBG in joint technique
Length (in mm)	49	47	22	74
Effective refractive index	1.95	1.95	1.95	1.45
Apodization function	Tanh	Tanh	Tanh	Uniform
Tanh parameter	4	4	4	-
Chirp parameter	0.0001	0.0001	0.0001	0.0001
AC modulation	0.0001	0.0001	0.0001	0.0001

In combine technique, parameter of DCF is also changed. The length of DCF is now 19 km and reduced by 2 km with respect to only DCF technique and dispersion value is −89.47 ps/km/nm.

3   Proposed Simulation Model

The block diagram of simulated model is presented in Fig. 4.It consist of some important parts like transmitter, optical fiber, dispersion compensation techniques, optical amplifier and optical receiver.

3.1   Transmitter

The transmitter comprises of a pseudo-random bit sequence generator that has the bit rate of 10 Gbps. A transmitter also has a non-return to zero pulse generator which is used to generate NRZ pulse and output of NRZ pulse generator is connected to Mach-Zender modulator that has extinction ration of 30 dB [14].

A continuous wave (CW) laser source which is operating at frequency of 193.1 THz is connected to Mach-Zender modulator. After the modulation, an optical pulse is transmitted over the fiber length of 100 km. The block diagram of transmitter is illustrated in Fig. 5.

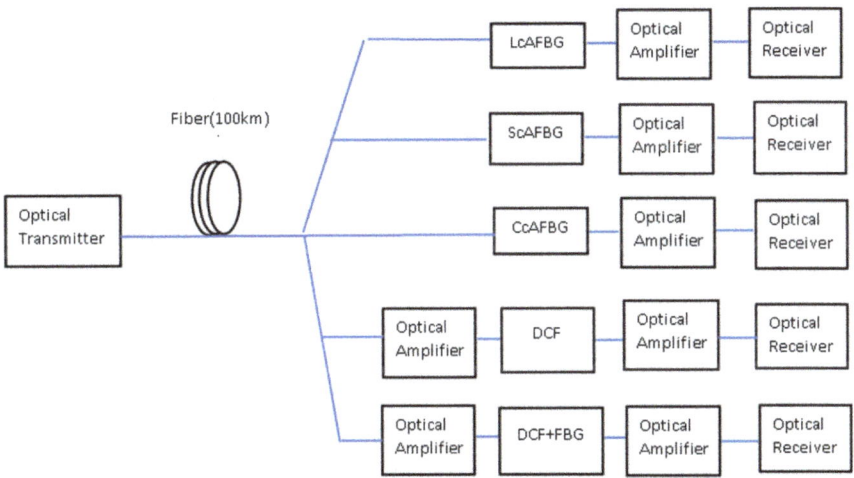

Fig. 4 The block diagram for the comparison of different dispersion compensation techniques [1]

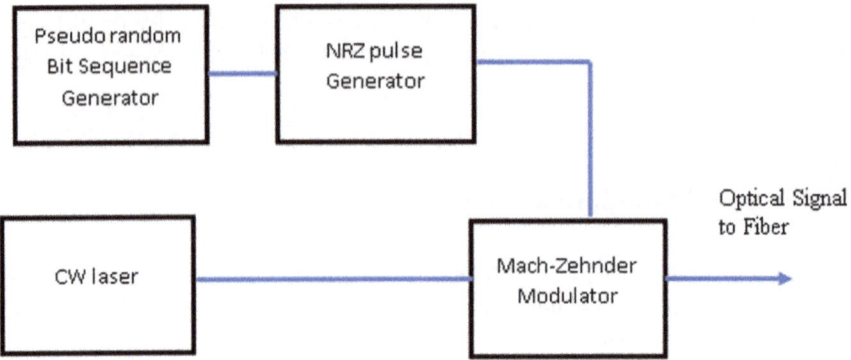

Fig. 5 The block diagram of transmitter [1]

Table 3 Parameters of SMF

Parameter	Value
Length (in km)	100
Dispersion (ps/km/nm) at 1550 nm	17
Dispersion slope (ps/nm^2/km) at 1550 nm	0.058
Attenuation (dB/km)	0.2
Effective area (μm^2)	80

3.2 Optical Fiber

In this paper, a single mode fiber (SMF) length of 100 km is used. The parameters of SMF are described in Table 3.

3.3 Optical Amplifier

In this paper, erbium doped fiber amplifier (EDFA) with 20 dB gain is used for amplification. A preamplifier also used for DCF system to overcome the losses [15].

3.4 Optical Receiver

At the receiver side, amplifier output is connected to photodetector. In this paper, PIN diode is used as a detector that convert optical signal into electrical signal [16].Then, output of PIN detector is, followed by Bessel filter and regenerator. Then output of regenerator is connected to eye diagram analyzer to show the result in terms of quality factor. The block diagram of optical receiver is illustrated in Fig. 6.

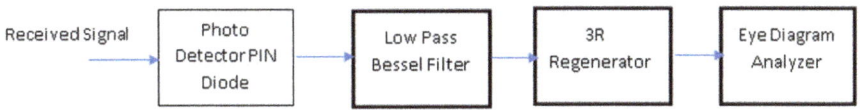

Fig. 6 The block diagram of receiver [1]

4 Results

Simulated results using different chirp FBG techniques are shown in Table 4. This table shows the variation in quality factor with respect to different grating length. From this table, it is clear that Q-factor is maximum around 37.78 with 49 mm grating length for LcFBG with 1.95 refractive index and it is maximum around 20.13 with 47 mm grating length for ScFBG with 1.95 refractive index and for CcFBG, Q-factor is maximum around 21.23 with 22 mm grating length and 1.95 refractive index. Quality factor is maximum around 47.39 at 74 mm grating length for 1.45 refractive index by using DCF + FBG technique.

Figure 7 shows the variation of the Q-factor with grating length for different refractive index by using different chirp FBG techniques in the form of graph.

It is clear from these graphs that for LcFBG, maximum Q-factor is around 37.78 at 49 mm grating length with 1.95 refractive index and for ScFBG, maximum Q-factor is around at 20.13 with 47 mm grating length with 1.95 refractive index and for CcFBG, maximum Q-factor is around 21.23 at 22 mm grating length with 1.95 refractive index and for DCF + FBG technique, Quality factor is maximum around 47.39 at 74mm grating length with 1.45 refractive index.

Quality factors and BER using different dispersion compensation techniques have been calculated with the help of eye diagram analyser as shown in Fig. 8 and simulated results analysis are shown in Table 5. The results have been simulated on Opti system.

5 Conclusion

Quality factors and BER using different dispersion compensation techniques have been calculated that is shown in Table 5. It is found that only DCF technique with 21 km length gives quality factor of 45.15 and combination of DCF and FBG technique gives quality factor of 47.39 with 19 km DCF length. By using combine technique length of DCF can also be reduced by 2 km. Performance of different chirp FBG techniques have also been calculated using different grating length and refractive index in terms of quality factor and BER that has shown in Table 4 and found that LcFBG gives better performance compare to other chirp FBG techniques. Overall combine DCF and FBG technique give best performance compare to all other techniques in terms of quality factor and BER.

Table 4 Quality factors and BER using different chirp FBG techniques

Techniques	Refractive index	Grating length (mm)	Q-Factor	BER
LcFBG	1.95	47	34.48	$5.19e^{-26}$
		48	36.94	$2.59e^{-29}$
		49	37.78	$6.77e^{-31}$
		50	35.65	$6.91e^{-27}$
		51	32.65	$2.18e^{-23}$
	1.45	47	24.02	$5.06e^{-12}$
		48	23.79	$1.27e^{-12}$
		49	23.20	$1.49e^{-11}$
		50	22.31	$8.64e^{-11}$
		51	21.32	$2.27e^{-10}$
ScFBG	1.95	45	18.87	$9.77e^{-08}$
		46	19.85	$5.35e^{-08}$
		47	20.13	$1.90e^{-09}$
		48	19.96	$5.27e^{-08}$
		49	19.41	$2.89e^{-08}$
	1.45	45	14.98	$4.32e^{-05}$
		46	15.01	$3.12e^{-05}$
		47	14.54	$2.99e^{-04}$
		48	13.97	$1.06e^{-04}$
		49	13.38	$3.81e^{-04}$
CcFBG	1.95	20	20.12	$2.04e^{-09}$
		21	20.81	$1.58e^{-09}$
		22	21.23	$2.23e^{-10}$
		23	21.13	$1.80e^{-09}$
		24	20.74	$7.23e^{-09}$
	1.45	20	13.93	$1.85e^{-04}$
		21	15.35	$1.58e^{-05}$
		22	16.61	$2.59e^{-06}$
		23	17.71	$1.57e^{-07}$
		24	18.71	$1.74e^{-07}$
DCF+FBG	1.95	72	3.68	$8.16e^{-005}$
		73	3.86	$3.98e^{-005}$
		74	3.70	$7.39e^{-005}$
		75	3.66	$8.75e^{-005}$
		76	3.76	$6.57e^{-005}$
	1.45	72	40.31	0

(continued)

Table 4 (continued)

Techniques	Refractive index	Grating length (mm)	Q-Factor	BER
		73	41.15	0
		74	47.39	0
		75	42.48	0
		76	43.72	0

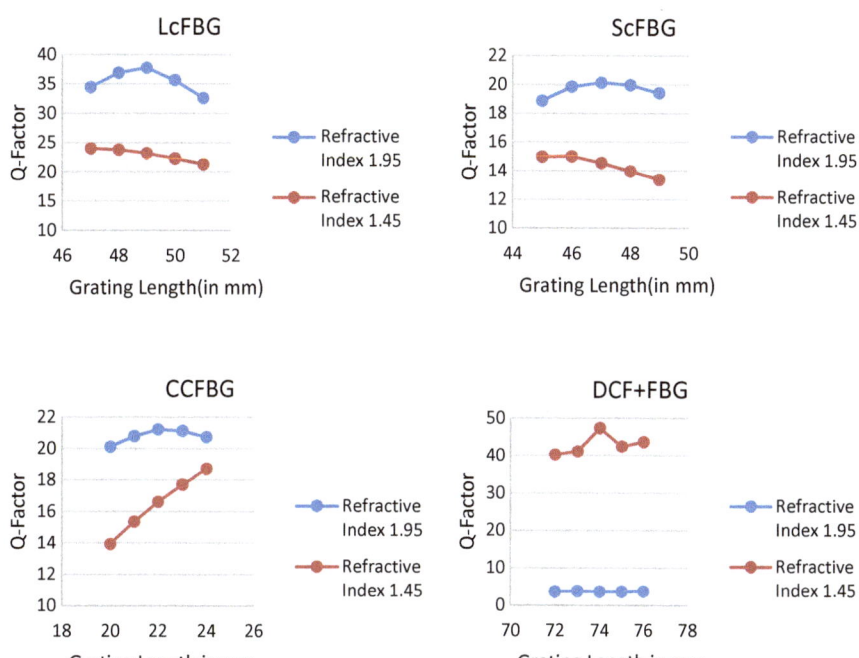

Fig. 7 Graphs between quality factor and grating length with dissimilar chirp FBG techniques

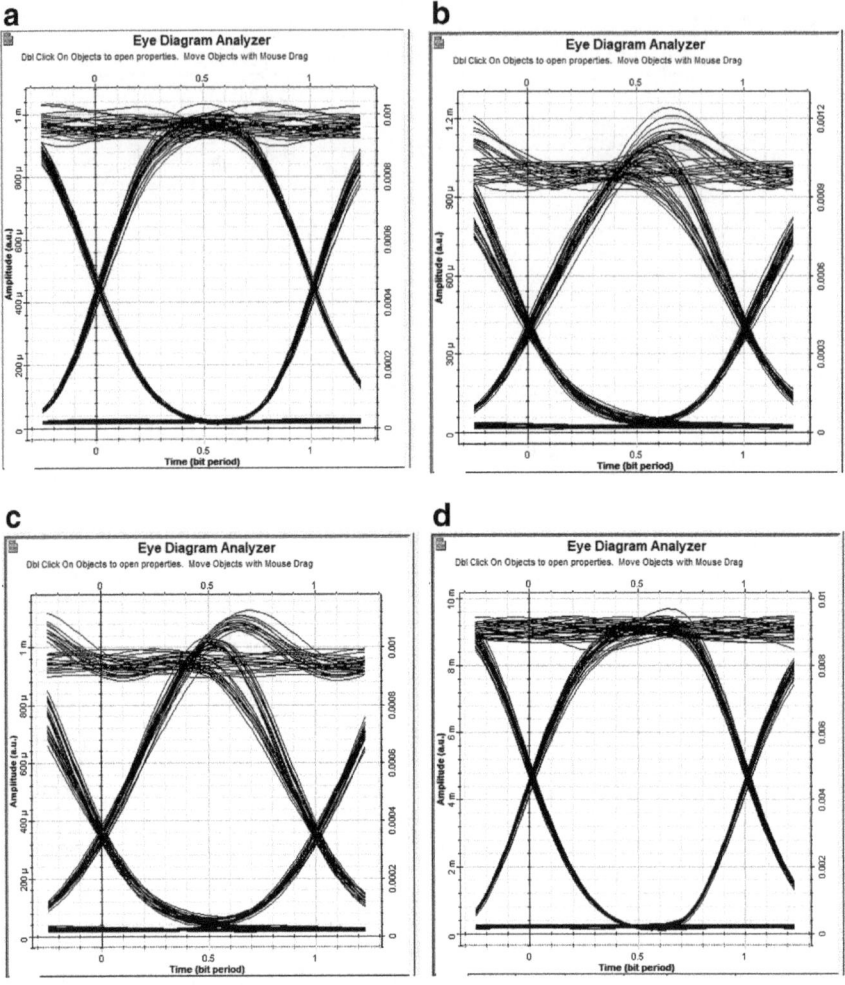

Fig. 8 Eye diagram using different dispersion compensation techniques **a** LcFBG. **b** ScFBG. **c** CcFBG. **d** DCF. **e** DCF + DBG

e

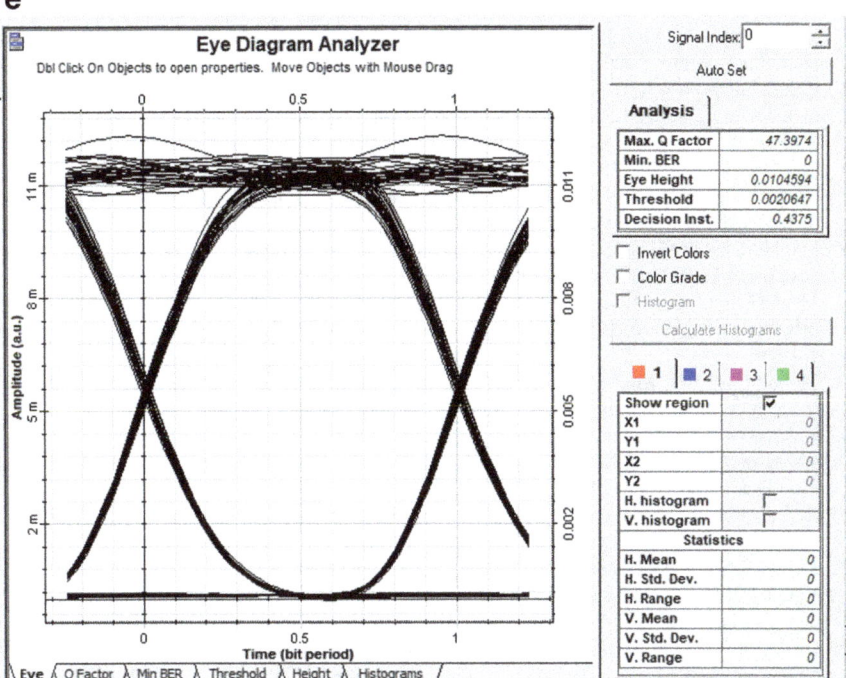

Fig. 8 (continued)

Table 5 Simulated result analysis

Parameters	LcAFBG	ScAFBG	CcAFBG	DCF	DCF + FBG
Quality factor	37.78	20.13	21.23	45.15	47.39
BER	$6.77e^{-31}$	$1.90e^{-09}$	$2.23e^{-10}$	0	0

References

1. Dar AB, Jha RK (2017) Design and comparative performance analysis of different chirping profiles of tanh apodized fiber Bragg grating and comparison with the dispersion compensation fiber for long-haul transmission system. J Mod Opt 64(6):555–566
2. Mohammed NA, Solaiman M, Aly MH (2014) Design and performance evaluation of a dispersion compensation unit using several chirping functions in a tanh apodized FBG and comparison with dispersion compensation fiber. Appl Opt 53(29):H239–H247
3. HU BN, Jing W, Wei W, Zhao RM (2010) Analysis on dispersion compensation with DCF based on optisystem. 2nd International conference on industrial and information systems, pp 40–43
4. Choi BH, Park HH, Chu MJ (2003) New pump wavelength of 1540-nm band for long-wavelength-band erbium-doped fiber amplifier (L-Band EDFA) IEEE. J Quantum Electron 39

5. Chakkour M, Hajaji A, Aghzout O (2015) Design and study of EDFA-WDM optical transmission system using FBG at 10 Gbits/s chromatic despersion compensation effects. Mediterranean conference on information and communication technologies
6. Chakkour M et al. (2017) Chromatic dispersion compensation effect performance enhancements using FBG and EDFA-wavelength division multiplexing optical transmission system. Int J Opt
7. Meena D, Meena ML (2019) Design and analysis of novel dispersion compensating model with chirp fiber bragg grating for long-haul transmission system. Optical and wireless technologies, lecture notes in electrical engineering book series, vol 546
8. Singh P, Chahar R (2014) Performance analysis of dispersion compensation in long haul optical fiber using DCF. Int J Eng Sci 3:18–22
9. Meena ML, Gupta RK (2019) Design and comparative performance evaluation of chirped FBG dispersion compensation with DCF technique for DWDM optical transmission systems. Optik 188:212–224
10. Mohammadi SO, Mozzaffari S, Shahidi M (2011) Simulation of a transmission system to compensate dispersion in an optical fiber by chirp gratings. Int J Phys Sci 63:7354–7360
11. Kashyap R (1999) Fiber Bragg grating. Academic Press, San Diego, USA, pp 310–316
12. Bhardwaj A, Soni G (2015) Performance analysis of 20 Gbps optical transmission system using fiber Bragg grating. Int J Sci Res Publ 5:2250–3153
13. Sayed AF, Barakat TM, Ali IA (2017) A novel dispersion compensation model using an efficient CFBG reflectors for WDM optical networks. Int J Microw Opt Technol 12:3
14. Kaler RS, Sharma AK, Kamal TS (2002) Comparison of pre-, post- and symmetrical-dispersion compensation schemes for 10Gb/s NRZ links using standard and dispersion compensated fibers. Opti Commun 209:107–123
15. Ismail M, Othman M (2013) EDFA-WDM Optical Network Design System. Procedia Eng 53:294–302
16. Agrawal G (2002) Fiber-optic communication systems. 3rd edn. Wiley-India edition

Chapter 23
Priority-Based Shortest Job First Broker Policy for Cloud Computing Environments

Nitin Kumar Mishra, Puneet Himthani, and Ghanshyam Prasad Dubey

1 Introduction

Cloud computing is a model for providing Resources and Services to Users in a convenient and on-demand manner. It manages resources and services as a shared pool to perform provisioning and release without any manual intervention. It is a distributed model based on the Internet; the Internet is essential for Cloud computing [1]. Another important aspect of Cloud is that it follows the "Pay per Use" Model; the User will have to pay only for the consumed, utilized, or subscribed resources [2]. Cloud computing is based on the principle of Virtualization. Virtualization improves the throughput of the system by increasing the utilization of resources to a greater extent depending upon these resources' computing capabilities and configuration [3].

The major characteristics of Cloud computing, as specified by NIST [1], include On-Demand Self Service, Broad Network Access [4], Resource Pooling, Rapid Elasticity and Measured Service. Cloud computing models are mainly classified as Service Models and Deployment Models. Cloud Service Models are commonly defined as X as a Service (XaaS), where X represents the type of the Cloud's service. Standard Cloud Service Models are primarily classified as Software as a Service (SaaS), Platform as a Service (PaaS), and Infrastructure as a Service (IaaS) [5]. Cloud Deployment Models are classified as Private Cloud, Community Cloud, Public Cloud, and Hybrid Cloud [6].

N. K. Mishra
ABV—IIITM, Gwalior, Madhya Pradesh, India

P. Himthani (✉)
Department of CSE, TIEIT, Bhopal, Madhya Pradesh, India

G. P. Dubey
Department of CSE, SISTec, Bhopal, Madhya Pradesh, India

© The Author(s), under exclusive license to Springer Nature Singapore Pte Ltd. 2021   279
S. Kumar et al. (eds.), *Proceedings of International Conference on Communication and Computational Technologies*, Algorithms for Intelligent Systems,
https://doi.org/10.1007/978-981-16-3246-4_23

1.1 Broker Policy

Data Center Broker plays a vital role in Cloud Environment as it is responsible for Scheduling and Load Balancing. It acts as an interface (Fig. 1) between the User and the Cloud environment [7]. Data Center Broker hides the process of creating VMs, submitting Cloudlets to VMs and destroying VMs from User. The Data Center processes all the Requests of the User for necessary action [8]. It is the central entity in a Cloud environment comprising Physical Hosts or Resources with Computational Power and capabilities. Virtualization of these Resources or Hosts allows the execution of multiple Virtual Machines over a single Host [9].

Data Center Broker is responsible for assigning a VM to a Host (Fig. 1) based on VM Scheduling policy. VM Scheduling is classified as Space-Shared and Time-Shared [10]. The former allows exclusive access of a VM to a Host, leading to low Utilization of Resources. The latter allows parallel access of multiple VMs to Resources of a Host depending upon the Configuration of Physical Host, leading to improved Resource Utilization [11].

Cloudlets are basic execution units in Cloud environments. Another important function performed by the Data Center Broker is Cloudlets' assignment to VM (Fig. 1) based on Cloudlet Scheduling policy [12]. It is also classified as Space-Shared and Time-Shared. The former allows exclusive access of Cloudlet to VM, while the latter allows VM to execute multiple Cloudlets simultaneously depending upon the Computation Power and Configuration of VMs and Physical Hosts [13].

Cloudlets are assigned to VM for execution in First Come First Serve (FCFS) order by the Data Center Broker [14]. Data Center Broker is also responsible for deciding

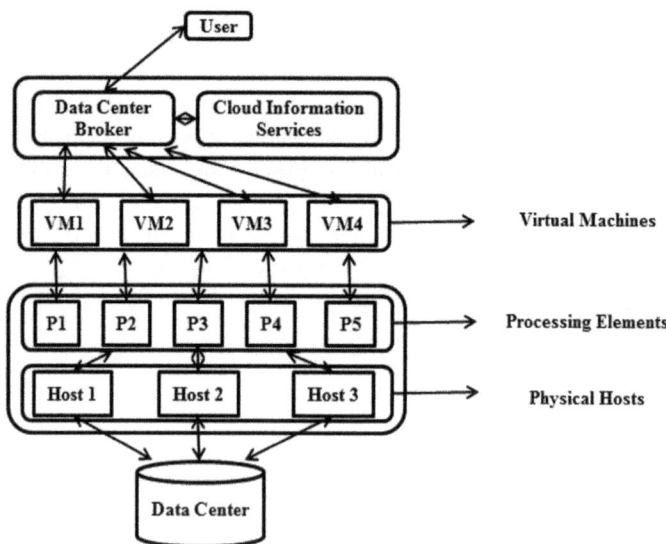

Fig. 1 Data Center Broker in cloud environment

the order of executing the Cloudlets. Real-Time scenarios may be deterministic or uncertain [15]. Sometimes, it happens that certain tasks are required to be performed before others; no matter when they are generated. In such a case, we have to complete these tasks before others. This is because these tasks are of higher importance than others. This justifies their priority over others [16].

1.2 Objective

Scheduling of Cloudlets in a suitable order is necessary for the proper utilization of resources. The FCFS approach schedules the Cloudlets for execution according to the order of their arrival. SJF schedules the Cloudlets for execution in the increasing order of their length or execution time. However, neither FCFS nor SJF supports the concept of priority. Real-time systems exhibit probabilistic nature and uncertainty and priority play an important role in such systems for execution of tasks. This paper proposes a priority-based SJF policy, called Priority_DCB for Cloud environments, that schedules the Cloudlets for execution in the non-decreasing order of their priority.

The proposed approach is very simple and easy to understand and implement. The major objective of this work include is to implement a Broker Policy for Cloud environments based on priority, called *Priority_DCB*, applicable in Real-Time Systems. The proposed policy will reduce the Turn-around Time and Waiting Time of Cloudlets, improves the resource utilization and execute the essential tasks without Starvation and at the earliest.

1.3 Organization of Paper

The remaining part of the paper is organized as: Sect. 2 provides a brief overview of the related work and Sect. 3 provides an introduction about the proposed Broker Policy. Section 4 provides an introduction to Simulation environment and parameters and Sect. 5 provides the results of the proposed Algorithm. Section 6 provides the Conclusion; followed by the References.

2 Related Work

R. Buyya, R. Ranjan and R. N. Calheiros in 2009 suggested the First Come First Serve (FCFS) Broker Policy; which is also the default Data Center Broker Policy supported by CloudSim [3]. A. Wadhonkar and D. Theng in 2016 proposed a Task Scheduling Algorithm as Broker Policy based on the two basic parameters of Cloudlet: Task Length and Deadline. All the Cloudlets to be executed are scheduled in the ascending order according to the sum of Normalized Task Length and Normalized Deadline.

However, their approach does not consider "priority" as one of the parameters for specifying the order of execution of Cloudlets, thereby, making it infeasible and inapplicable in real-time Cloud systems [16].

D. Saxena, R. K. Chauhan and R. Kait in 2016 proposed Dynamic Fair Priority Optimization Task Scheduling Algorithm as Broker Policy in Cloud computing environment based on three core concepts: Constraint-based Grouping of Tasks, Weighted Fair Priority Queue, and Greedy Resource Allocation. The major drawback of their approach is that it is impractical for Multi-Facet environments where the requests for resources are uncertain or probabilistic, rather than static or fixed. Another issue in their approach is that it doesn't incorporate the concept of "priority" for deciding the order of execution of Cloudlets, hence will not be effective in real-time Cloud systems [17].

M. L. Chiang, H. C. Hseih, W. C. Tsai and M. C. Ke in 2017 proposed Improved Task Scheduling and Load Balancing Algorithm for heterogeneous Cloud computing network. They proposed the Advanced MAX Suffrage Algorithm to improve the Resource Utilization. The major bottleneck of their approach is that it does not consider "priority" as one of the parameters for deciding the order of execution of Cloudlets [18].

S. Elmougy, S. Sarhan and M. Joundy in 2017 proposed a Hybrid Task Scheduling Technique as Broker Policy based on Shortest Job First and Round Robin with Dynamic Variable Quantum Time. Their approach divides the tasks into two categories as Short Tasks and Long Tasks. For this, they are calculating the Median of execution time for all Cloudlets. Their approach allows for the execution of Cloudlets in a Pre-Emptive manner, thereby avoiding the Starvation; which is one of the common drawbacks of the Non-Pre-Emptive approaches. The other drawback of the proposed approach is that the Response Time is on the higher side as compared to Round Robin. Secondly, their approach does not support the concept of "priority", which is typically common to real-time systems [19].

B. P. Rimal and M. Maier in 2016 proposed workflow scheduling for multi-tenant Cloud computing environments. They proposed a workflow scheduling architecture and algorithm for multi-tenant Cloud environments. According to their approach, a workflow representing a set of Cloudlets is assigned to the VM for execution. Once the scheduler assigns a workflow to a VM for execution, there are three possible orders for Cloudlets of that workflow to be executed, viz. FCFS, Deadline First/Easy Backfilling Scheduling and Minimal Completion Time approach. The drawback of their proposed model is that it leads to the problem of Starvation, as in all three cases of scheduling, the only consideration is to reduce the completion time to as low as possible. Secondly, none of the three proposed techniques incorporates the "priority" of a Cloudlet as a parameter for deciding the order of execution, thereby makes these approaches quite ineffective in real-time systems, where "priority" is a common, yet typical parameter, that plays an important role in decision making [20].

Wu and Wang in 2018 proposed a distributed algorithm for energy efficient scheduling in cloud-based on multi-modal estimation technique. They proposed a DAG-based optimal scheduling algorithm using optimized multi-objective approach. However, their approach does not consider the uncertainties and abnormalities that arise in the system in real-time. Secondly, "priority" of a Cloudlet is not considered as a parameter for deciding the order of execution. Secondly, this approach does not ensure the proper utilization of resources [21].

Panda and Jana in 2015 proposed efficient task scheduling algorithm for heterogeneous multi-cloud environments. They proposed three approaches for scheduling of Cloudlets, as MCC, MEMAX and CMMN, where MCC is a single-phase algorithm while remaining two algorithms are having two phases of operation. MCC follows earliest completion time approach. MEMAX computes Median in first phase and based on that it schedules Cloudlets based on earliest completion time approach. CMMN computes normalized value for each Cloudlet in first phase and based on that normalized value, it schedules them for execution based on earliest completion time approach. But, they do not incorporate the "priority" of a Cloudlet as a parameter for deciding their order of execution; thereby making them highly ineffective for being deployed in real-time environments, which is highly uncertain and probabilistic and considers "priority" is one of the most important parameters for decision making and deciding the order of execution [22].

Researchers, Scholars and Intellects across the world had already done a lot of work in developing an efficient Broker policy for Cloud environments and still a lot of work is already being going on. Some approaches are also based on the concepts of Optimization theory, Machine Learning, Operational Research, etc. But, only a handful of these approaches inculcate the concept of priority, which is typical in Real-Time scenarios. In this regard, this paper proposes a simple and effective priority-based SJF Broker policy, termed as Priority_DCB, that schedules the execution of Cloudlets in Cloud environment according to the non-decreasing order of their priority, yet ensures the efficient resource utilization and performance of the Cloud environment.

3   Proposed Priority-Based SJF Broker Policy

This paper proposes a priority-based Shortest Job First Data Center Broker Policy "*Priority_DCB*" for Cloud environments. Data Center Broker Policy specifies the order for execution of Cloudlets over the Virtual Machines in the Cloud computing environment. By default, execution of Cloudlets over Virtual Machines follows First Come First Serve (FCFS) Broker Policy. However, Real-Time environments are uncertain, where some specific tasks are required to be performed immediately. It is evident that such tasks are of higher importance than others, having a higher priority associated with them. Further, it does so without considering the impact on the system due to the out of ordered execution. This type of execution of tasks is called priority scheduling.

By default, a Cloudlet is characterized by its length, which specifies its execution time. Here, each Cloudlet submitted for execution will have a priority associated with it along with the length. Now, the Data Center Broker will rearrange all the Submitted Cloudlets according to their priorities from highest to lowest for execution. By this,

Fig. 2 Flow chart of
proposed priority-based Data
Center Broker policy

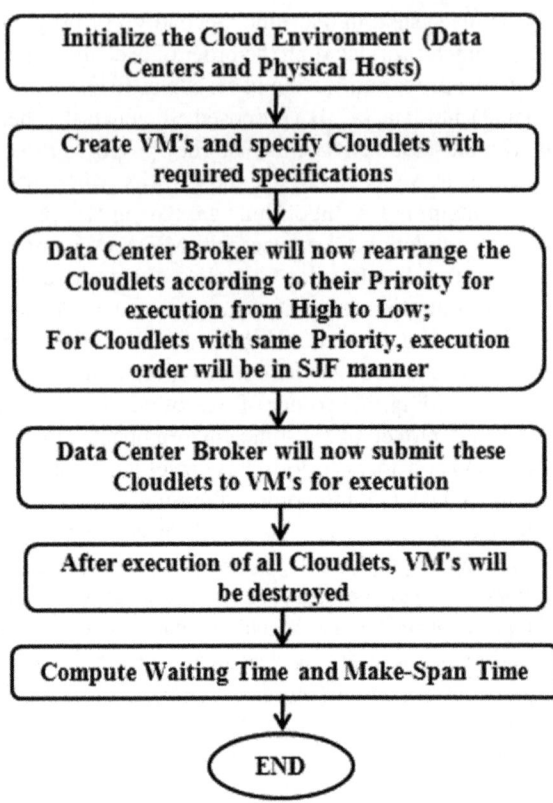

Initialize the Cloud Environment (Data
Centers and Physical Hosts)

Create VM's and specify Cloudlets with
required specifications

Data Center Broker will now rearrange the
Cloudlets according to their Priroity for
execution from High to Low;
For Cloudlets with same Priority, execution
order will be in SJF manner

Data Center Broker will now submit these
Cloudlets to VM's for execution

After execution of all Cloudlets, VM's will
be destroyed

Compute Waiting Time and Make-Span Time

END

a Cloudlet with higher priority will be assigned to VM for execution before others.
If two or more Cloudlets have the same priority, they are arranged in the increasing
order of their execution time (Cloudlet Length). Fig. 2 specifies the flow chart of the
proposed algorithm.

4 Simulation Environment

The proposed approach has been simulated on CloudSim; the most popular and
widely used Cloud Simulating Framework, developed by Dr. Rajkumar Buyya and
his team at the University of Melbourne, Australia [23]. The major components of
CloudSim include Cloudlets, Cloudlet Scheduler (Time-Shared and Space-Shared),
Data Center, Data Center Broker, Host, Log (Output Stream), Pe (Processing Element
or simply a Processor or a Core of Processor), Vm, Vm Scheduler (Time-Shared and
Space-Shared), PeList, HostList, CloudletList, etc. [24].

4.1 Performance Evaluation Parameters

Waiting Time and Turn-around Time are two parameters taken into consideration to evaluate the performance of the proposed priority-based Broker Policy. These two are the main parameters that researchers working on this topic evaluate to justify the performance of their proposed approach [25]. Waiting Time is the total time duration for which the Cloudlet waits from arrival to completion of execution. In the case of Non-Pre-Emptive Scheduling approaches, Waiting Time is equal to the Response Time [26]. Turn-around Time is the total time for which the Cloudlet is in the system from arrival to completion. The lower the value of Turn-around Time; the better the Broker Policy will be [27].

Let, Cl_i be a Cloudlet with S_i being its Submission Time, F_i being the Finish Time and L_i being the Length of Cloudlet, then the value of Turn-around, and Waiting Time for Cl_i will be evaluated, as:

$$cl^{\text{turn-around}} = F_i - S_i$$

$$cl_i^{\text{waiting}} = cl_i^{\text{turn-around}} - L_i$$

Suppose, there are n Cloudlets, then the values of average Turn-around Time (AVG–TAT) and average Waiting Time (AVG–WT) will be computed, as:

$$\text{AVG} - \text{TAT} = \frac{1}{n} \sum_{i=1}^{n} cl_i^{\text{turn-around}}$$

$$\text{AVG} - WT = \frac{1}{n} \sum_{i=1}^{n} cl_i^{\text{waiting}}$$

4.2 Simulation Scenario

Table 1 specifies the scenario of the Cloud environment where the performance of the proposed algorithm is evaluated and compared with other state-of-the-art approaches. All simulations of the compared approaches are carried out in this scenario. For Cloudlet, a new parameter "priority" is defined based on which the order for execution of Cloudlets will be decided by the Data Center Broker (*Priority_DCB*). By default, there is no specification like "priority" exists for the Cloudlet in CloudSim or Cloud environment.

Table 1 Simulation scenario specifications

S. No.	CloudSim component	Specification	Value
1	DATA CENTER (1 Data Center)	Architecture	X86
		OS	LINUX
		VM Manager	XEN
2	HOST (4 Quad Core Hosts in each Data Center)	MIPS	5000
		PE	4
		RAM	2560 MB
		Storage	1,000,000 MB
		Bandwidth	10,000
		VM Scheduler	Space-Shared, Time-Shared
3	VIRTUAL MACHINE (Number of VMs will be specified by User)	MIPS	1000
		PE	1
		RAM	512 MB
		Storage	10,000 MB
		Bandwidth	1000
		VM Manager	XEN
		Cloudlet Scheduler	Space-Shared
4	CLOUDLET (Number of Cloudlets will be specified by User)	Length	User-defined (random)
		File Size	300
		Output Size	300
		PE	1
		Priority	User-defined (1–5) [5—Highest and 1—Lowest]

5 Results

Five scenarios are considered for evaluating the performance of the proposed approach and comparing it with other state-of-the-art approaches. Case 1 considers five Cloudlets are submitted for execution on a single VM with lengths 7, 5, 3, 2 and 8 respectively with 7 having the highest priority and 8 having the lowest priority. Case 2 considers 100 Cloudlets with lengths in the range 1–20 are submitted for execution over 15 VMs. Table 2 provides a comparative analysis of the proposed approach with other approaches for Case 1 and Case 2.

Case 3 assumes that there are 15 VMs over which the number of Cloudlets is to be executed in different Simulations with 50, 100, 150, 200, and 250 Cloudlets. In all Simulations, VM Scheduler and Cloudlet Scheduler are Space-Shared. Each Simulation is executed at least 10 times to achieve more precise results. Case 4 assumes that 100 Cloudlets are required to be executed over a varying number of VMs from 5, 10, 15, 20, and 25 VMs. Both VM and Cloudlet Schedulers are Space-Shared.

Table 2 Comparison of FCFS and SJF with proposed algorithm for Case 1 and Case 2

Algorithm	CASE 1		CASE 2	
	Turn-around	Waiting	Turn-around	Waiting
FCFS [3]	15.47	10.47	41.82	30.78
SJF [16]	11.8	6.8	32.64	21.32
Priority SJF	14.93	9.93	39.09	28.38

Case 5 assumes that 100 Cloudlets are required to be executed over a varying number of VMs from 15, 20, and 25 VMs. In this case, VM Scheduler is Time-Shared and Cloudlet Scheduler is Space-Shared. Tables 3, 4 and 5 provide a comparative analysis of the proposed approach with other approaches for Case 3, Case 4 and Case 5 respectively.

Table 3 Comparison of FCFS and SJF with proposed algorithm for Case 3

Cloudlets	50		100		150		200		250	
Algorithm	TAT	WT	TAT	WT	TAT	WT	TAT	WT	TAT	WT
FCFS [3]	25	13.7	41.8	30.8	61.4	50.3	80	69	101	89.7
SJF [16]	19.4	8.1	32.6	21.3	44.2	33.3	57.4	46.4	67.4	56.7
Priority SJF	22.1	11.4	39.1	28.4	55.5	44.7	74.7	63.8	92	80.8

Table 4 Comparison of FCFS and SJF with proposed algorithm for Case 4

VMs	5		10		15		20		25	
Algorithm	TAT	WT	TAT	WT	TAT	WT	TAT	WT	TAT	WT
FCFS [3]	114.6	103.3	57.2	46.6	41.8	30.8	41	29.8	41	29.8
SJF [16]	90.5	78.8	42.2	31.5	32.6	21.3	28	17.4	28	17.4
Priority SJF	117.3	106	58.9	47.9	39.1	28.4	38.4	27	38.4	27

Table 5 Comparison of FCFS and SJF with proposed algorithm for Case 5

VMs	15		20		25	
Algorithm	TAT	WT	TAT	WT	TAT	WT
FCFS [3]	42.46	31.49	33.72	22.47	33.72	22.47
SJF [14, 16]	28.87	18.45	24.33	13.6	24.33	13.6
Priority SJF	39.55	28.64	32.16	21.02	32.16	21.02

5.1 Discussion

We start the discussion by observing the results of Case 1, Case 2, and Case 3. We notice that FCFS broker achieves the worst results and proposed *Priority_DCB* approach performs better than the FCFS Broker. However, the performance of *Priority_DCB* is inferior to SJF because SJF is the most Optimal Non-Pre-Emptive Scheduling technique.

An interesting observation from the results of Case 4 and Case 5 is that there is saturation in the Turn-around Time and Waiting Time, once the number of VMs exceeds 15 and 20 respectively, which is mainly due to the fact that it is not possible to create more than 16 VMs in Space-Shared manner and more than 20 VMs in Time-Shared manner on the available infrastructure and resources.

We then notice that if the priority of processes with smaller lengths is higher than the priority of processes with long lengths, then *Priority_DCB* exhibit almost similar behavior to SJF. For scenarios where priority of processes with long length is higher than others, *Priority_DCB* behaves in similar manner as FCFS. In other cases, *Priority_DCB* performs better than FCFS; while its performance is inferior to SJF in all such cases.

These observations verify the effectiveness of the proposed method over the other state-of-the-art Data Center Broker Policies in terms of Cloudlets' turnaround time and waiting time.

6 Conclusion

Cloud computing is being widely used as a model for providing Services, due to its ease and automated process. Utilization of Resources is of utmost importance in Cloud environments, as the optimal utilization of resources will lead to the generation of high revenues, as it follows the pay-per-use model. The proposed Broker Policy will execute the Cloudlets over VMs based on their priority. Results show that the proposed priority SJF approach performs better than FCFS. Secondly, whenever priority is associated with a task, it doesn't matter how long it will take to complete, how much overheads it will incur, how much it will affect the performance of the system, etc.; the only aspect that matters is to perform that task before others.

The Turn-around Time and Waiting Time in *Priority_DCB* reduces by 5–10% when resources are less. When deployed on large Cloud environments, the *Priority_DCB* may lead to a reduction in Turn-around Time and Waiting Time by around 15–20% at least.

The proposed approach is not as optimal as traditional SJF, but it carries out the operation for which it has been implemented, i.e., scheduling the Cloudlets for execution according to their priority. The future enhancements in this work are to optimize this approach using some optimization techniques like Ant Colony Optimization, Particle Swarm Optimization, Genetic Algorithms, Machine Learning techniques, etc.

References

1. Mell P, Grance T (2011) The NIST definition of cloud computing. Recommendations of the National Institute of Standards and Technology
2. Mittal S, Katal A (2016) An optimized task scheduling algorithm in cloud computing. In: Proceedings of 6th international conference on advance computing. IEEE, pp 197–202
3. Buyya R, Ranjan R, Calheiros RN (2009) Modeling and simulation of scalable cloud computing environments and the cloud sim toolkit: challenges and opportunities. In: Proceedings of international conference on high-performance computing and simulation. IEEE
4. Yang H, Tate M (2012) A descriptive literature review and classification of cloud computing research. In: Communications of the association for information systems (CAIS), vol 31, Issue 1, No 2, pp 36–60
5. Ahmed M, Chowdhury ASMR, Ahmed M, Rafee MMH (2012) An advanced survey on cloud computing and state-of-the-art research issues. Int J Comput Sci 9(1):201–207
6. Sriran I, Hosseini AK (2010) Research agenda in cloud technologies (arXiv:1001.3259)
7. Li F, Liao TW, Zhang L (2019) Two level multi task scheduling in a cloud manufacturing environment. J Rob Comput Integr Manuf 56:127–139
8. Zheng PY, Zhou MC (2017) Dynamic cloud task scheduling based on a two stage strategy. IEEE Trans Autom Sci Eng
9. Chatterjee T, Ojha VK, Adhikari M, Banerjee S, Biswas U, Snasel V (2014) Design and implementation of an improved data center broker policy to improve the QoS of a cloud. In: Proceedings of 5th international conference on innovations in bio-inspired computing and applications, advances in intelligent systems and computing. Springer, pp 281–290
10. Manasrah AA, Smadi T, Momani AA (2016) A variable service broker routing policy for data center selection in cloud analyst. J King Saud Univ Comput Inf Sci
11. Liu L, Qiu Z (2016) A survey on virtual machine scheduling in cloud computing. In: Proceedings of 2nd international conference on computer and communications. IEEE, pp 2717–2721
12. Anuragi R, Pandey M (2019) Review paper on cloudlet allocation policy. Emerg Trends Data Min Inf Secur Adv Intel Syst Comput 319–327
13. Banerjee S, Adhikari M, Kar S, Biswas U (2015) Development and analysis of a new cloudlet allocation strategy for QoS improvement in Cloud. Arab J Sci Eng 1409–1425
14. Kumar M, Sharma SC (2017) Dynamic load balancing algorithm for balancing the workload among virtual machine in cloud computing. In: Proceedings of 7th international conference on advances in computing and communications. Elsevier, pp 322–329
15. Singh S, Chana I (2016) A survey on resource scheduling in cloud computing: issues and challenges. J Grid Comput 217–264
16. Wadhonkar A, Theng D (2016) A task scheduling algorithm based on task length and deadline in cloud computing. Int J Sci Eng Res 7(4):1905–1909
17. Saxena D, Chauhan RK, Kait R (2016) Dynamic fair priority optimization task scheduling algorithm in cloud computing: concepts and implementations. Int J Comput Netw Inf Secur 2: 41–48

18. Chiang ML, Hsieh HC, Tsai WC, Ke MC (2017) An improved task scheduling and load balancing algorithm under the heterogeneous cloud computing network. In: Proceedings of 8th international conference on awareness science and technology. IEEE, pp 290–295

19. Elmougy S, Sarhan S, Joundy M (2017) A novel hybrid of shortest job first and round robin with dynamic variable quantum time task scheduling technique. J Cloud Comput Adv Syst Appl 6:1–12

20. Rimal BP, Maier M (2015) Workflow scheduling in multi-tenant cloud computing environments. IEEE Trans Parallel Distrib Syst

21. Wu CG, Wang L (2018) A multi-model estimation of distributed algorithm for energy efficient scheduling under cloud computing system. J Parallel Distrib Comput

22. Panda SK, Jana PK (2015) Efficient task scheduling algorithms for heterogeneous multi cloud environments. J Super Comput 71:1505–1533

23. Calheiros RN, Ranjan R, Rose CAFD, Buyya R (2009) Cloud sim: a novel framework for modeling and simulation of cloud computing infrastructures and services. arXiv:0903.2525

24. Banerjee S, Chowdhary A, Mukherjee S, Biswas U (2018) An approach towards development of a new cloudlet allocation policy with dynamic time quantum. J Autom Control Comput Sci 52(3):208–219

25. Kapur R (2015) A cost-effective approach for resource scheduling in cloud computing. In: Proceedings of international conference on computer, communication and control. IEEE

26. Kapur R (2015) A workload balanced approach for resource scheduling in cloud computing. In: Proceedings of 8th international conference on contemporary computing. IEEE

27. Banerjee S, Adhikari M, Biswas U (2017) Design and analysis of an efficient QoS improvement policy in cloud computing. J Serv Oriented Comput Appl 11:65–73

Chapter 24
Multi-cipher Encrypter Using Symmetric Key Algorithms

Vaibhav Tripathi, Akriti Yadav, Rithwik Chithreddy, and N. Subhashini

1 Introduction

Encryption is considered as an efficient means to maintain data confidentiality. Despite, the best efforts to develop a fully secure algorithm for encryption, no such cryptography algorithm has been devised which can guarantee complete confidentiality of data. In the context of cryptography, encryption serves as a mechanism to ensure confidentiality and to maintain secrecy [1]. Since data may be visible on the internet, sensitive information such as passwords, bank account details and personal communication information may be exposed to potential interceptors live on the network. To protect this information, encryption algorithms convert plaintext into cipher-text to transform the original data to a non-readable format accessible only to authorized parties who can decrypt that back to a readable format. Simple symmetric key text encryption algorithms have become practically inefficient as they easily succumb to known plaintext attacks [2]. Although these algorithms may have lost their individual strength against attackers, by combining algorithms and applying multiple encryption layers, the strength of encryption can be enhanced by a large degree [3].

In this paper, we have proposed one such approach to build an easy to implement yet secure multi-cipher encrypter which utilizes simple symmetric key algorithms such as Base Change Cipher, Hill Cipher and Caesar Cipher along with AES encryption to encrypt text data with relatively higher resistance against attacking softwares. By combining these techniques in any permutation and a wise selection of keys according to the choice of the user, a unique encrypted file can be generated with a unique decrypting method only known to the user. The paper further discusses the scope of possible combination with other encryption methods so as to obtain a

V. Tripathi · A. Yadav · R. Chithreddy · N. Subhashini (✉)
Vellore Institute of Technology, Chennai 600127, India
e-mail: subhashini.n@vit.ac.in

© The Author(s), under exclusive license to Springer Nature Singapore Pte Ltd. 2021
S. Kumar et al. (eds.), *Proceedings of International Conference on Communication and Computational Technologies*, Algorithms for Intelligent Systems,
https://doi.org/10.1007/978-981-16-3246-4_24

highly secure file, while maintaining an efficient yet user-friendly interface. Thus, the authors propose a way to utilize even those ciphering attacks which are initially weak against the modern plain text attacks into an encrypter which can generate strongly encrypted texts.

This paper will walk through the literature survey and their findings in Sect. 2 which were utilized to develop the proposed encryption model. Section 3 explains about the proposed encryption model and the key operations involved in developing the model. Section 4 talks about the implementation of the model and its algorithmic overview which can be utilized to develop the proposed model and understand its working. An analysis of the encrypter is carried out and the results obtained are compared to the existing base encryption techniques in Sect. 5. The performance and strength of the proposed encrypter against brute force attacks are then evaluated and compared to that of the custom cipher based encryption algorithms. Finally, Sect. 6 draws a conclusion and proposes the future work which can be carried out for developing even stronger multiple encryption-based models to enhance cipher strength and protect confidential data against attacks.

2 Literature Survey

Multiple encryption is one of the popular techniques used for encrypting simple alphanumeric text data and offers strength of much higher magnitude as compared to traditional encryption algorithms [4]. By referring to the research work carried out on these algorithms one can understand and compare the strength of these algorithms to evaluate the proposed multiple encryption model. Applying multiple encryption to text files enhance their strength against attacks by a great degree as compared to the traditional ciphering algorithms. The proposed encryption model utilizes symmetric key encryption algorithms such as Hill cipher, Base change cipher and Caesar cipher for implementing multiple encryption to the text file along with an AES encryption as the top level of encryption to the ciphered text.

Hill cipher [5] is a polygraphic substitution cipher based on linear algebra. To encrypt a message, each block of n letters (considered as an n-component vector) is multiplied by an invertible $n \times n$ matrix, against modulus L where L denotes the length of character set containing all possible characters in the text. The matrix used for encryption is the cipher key, taken as input from the user of length n^2 formed into invertible $n \times n$ matrices which is modulo with L. To decrypt the message, each block is multiplied by the inverse of the matrix used for encryption. The resultant integers are the indices of the original characters in the character set which are obtained and concatenated to form the original text. This ciphering based on linear algebra therefore finds application in modern encryption for data security [6].

Universally, the decimal number system (base 10) system is followed for representing numerical data. By converting the number from decimal base to any other natural base (except 1), the numerical representation is changed and a new numerical representation is obtained which enables one to safeguard the original numerical data, especially account details, phone numbers, etc. For decryption, the number from the known changed base is simply converted into a decimal number system. Both encryption and decryption are carried out by standard number system conversion method, by dividing and taking mod of the number NUM to new base N to obtain the new base changed number.

Caesar cipher [7] is one of the simplest yet widely used encryption techniques. It is a substitution cipher in which all the characters are shifted by an integer N to obtain a new string with the shifted characters, earning this algorithm the name of Shift cipher. Encryption of a letter x by a shift N can be described mathematically by Eq. 1.

$$E_N(x) = (x + N)\%26 \tag{1}$$

Decryption is performed similarly using Eq. 2.

$$D_N(x) = (x - N)\%26 \tag{2}$$

Here, 26 is taken for English alphabet length. Alternatively, for any other character set, modulus with L can be taken where L is the length of the character set.

By utilizing these algorithms and incorporating them in a multiple encryption model with any number of layers of encryption and a variety of keys according to the users choice, the proposed model can multiply their individual strengths to create an encrypted text file of higher strength and security against attacks [8]. Applying a single AES encryption layer will further guarantee complete security of data as AES algorithms are one of the toughest to break and there have been no evidence till date of it being decoded and exploited by an attacker [9].

3 Proposed Multi-cipher Encrypter

The proposed ciphering encryption utilizes multiple symmetric key algorithms to create an encrypted file by applying multiple layers of cipher to the input text file. The number of layers, their order and their respective keys are determined by the user. While the base algorithms discussed in the paper are Hill Cipher, Base Change Cipher and Caesar Cipher, any number of algorithms can be added to the model according to the user's needs along with a top layer of AES encryption. By selecting any number of encryption layers, encryption method and its keys in any permutation, the users can define a unique encrypter every time with exponential increase in strength. This will ensure data confidentiality even after an attacker gains access to confidential files on a system.

Figure 1 illustrates the encryption and decryption carried out in symmetric key multiple encryption which is implemented in the proposed multi-cipher encrypter.

Multiple encryption uses a collection of keys which comprise a number of encryption keys as, $K_{AES}, K_1, K_2, K_3, \dots K_{N-1}, K_N$ each of lengths and values according

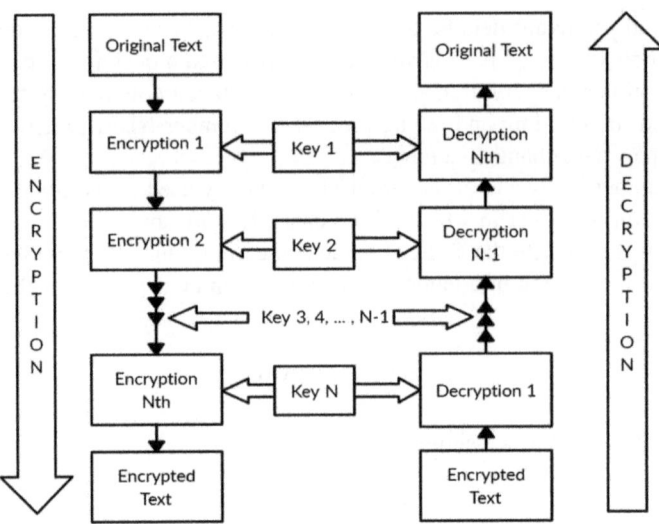

Fig. 1 Symmetric key encryption and decryption

to the ciphering algorithm and the user's choice. When applied on a message M, the encryption operation of multiple encryption can be described as shown in Eq. 3:

$$\text{Ciphertext} = \text{Encrypt}_{\text{AES}}(K_{\text{AES}}, \text{Encrypt}_N(K_N, \text{Encrypt}_{N-1} \dots \text{Encrypt}_2$$
$$\times (K_2, \text{Encrypt}_1(K_1, M)))) \tag{3}$$

Here, N denotes the number of encryption layers, whereas M denotes the original message. These N number of layers can have any combination of ciphering techniques. It is recommended to use at least 3 or more than 3 encryption algorithms in a non-repetitive and random order with unique keys for each level so as to strengthen the encryption and achieve greater level of security. After every encryption, the keys are added to a key vector. Finally, an AES encryption is applied to the encrypted file as a top layer of encryption and its key is added to the key vector, which is used for decrypting the text file by the authorized recipient. Decryption operation is performed in the reverse order by using the keys stored in key vector in the order given below as Eq. 4:

$$\text{Plaintext} = \text{Decrypt}_N(K_1, \text{Decrypt}_{N-1} \dots \text{Decrypt}_3(K_{N-2}, \text{Decrypt}_2$$
$$\times (K_{N-1}, \text{Decrypt}_1(K_{N-1}, \text{Decrypt}_1(K_N, \text{Decrypt}_{\text{AES}}(K_{\text{AES,C}}))))) \tag{4}$$

Here, the AES decryption is applied to the encrypted cipher text C using AES key and then decrypted by the keys in the key vector starting from K_N, then further result is decrypted with key K_{N-1} and this process is continued till final decryption by key K_1 takes place.

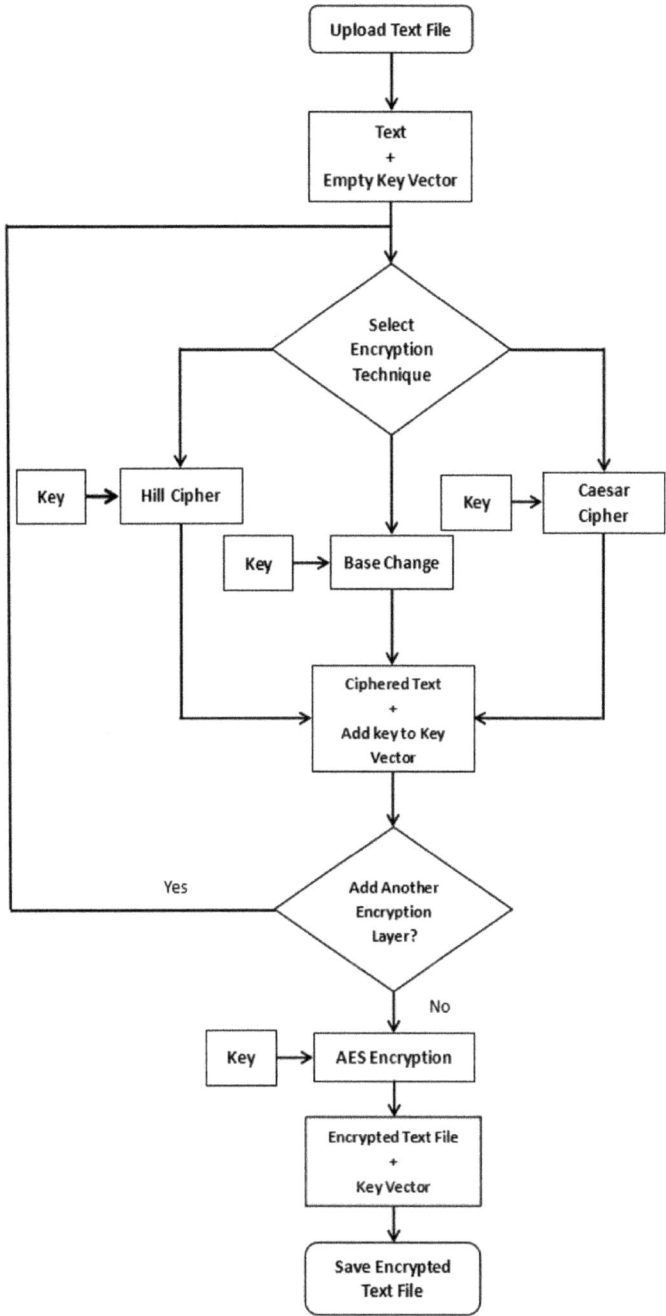

Fig. 2 Proposed multiple encryption algorithm

The flow diagram of the proposed multiple encryption model is shown in Fig. 2. The encrypter will take a text file as input from the user and obtain the text data to be encrypted. A key vector is initialized for storing the keys. After the first encryption, the key is stored, and the encrypted file is generated. As evident from the flow diagram, whenever the user selects the option to add another layer, the program allows a choice of encrypting method and key selection to add another encryption to the encrypted text. AES encryption is applied to the encrypted text to generate the final encrypted text file which is returned to the user along with the key vector.

4 Implementation

The proposed multi-cipher encrypter takes a text file input from the user and obtains the text message M from the file. An empty key vector KV is initialized to store the keys and method index used for encryption which will be used at the end for decryption. For every encryption, the key is stored in this key vector. The user now gets a choice of encryption technique for the first layer of encryption. The techniques used are Hill cipher (9 character key/3X3 matrix method, 16 character key/4X4 matrix method, 25 character key/5X5 matrix method), Base change and Caesar cipher. After selecting the encryption method, the user enters the key for encryption. This key used for encryption is stored in the key vector, and the encrypted message E_1 is generated. The user now gets the choice for adding another layer of encryption. If the user selects *Yes*, the user now gets to select any encryption method except the one used for previous encryption to maintain the security. Furthermore, the new key entered by the user must be unique, i.e., it must not exist in the key vector KV. For every ith encryption the program flow takes E_i as the message and encrypts it. This process is repeated n times for applying n layers of encryption to the message. The user selects *No* when asked to add another layer of encryption to generate the encrypted message E_n with N layers of encryption. Now an AES encryption is applied and the key is added to key vector KV. The encrypted text file is generated and returned along with the key vector KV. This text file can be now transmitted securely and only personnel with access to KV can decrypt the text file to maintain data confidentiality. For decryption, the encrypted text file and the key vector are utilized. The AES decryption is applied and the key vector pops keys from the end one-by-one. Using these keys, the text file is decrypted sequentially. Finally after N decryptions are carried out, the original text is retrieved and saved.

The Caesar cipher encryption algorithm used in the proposed encrypter takes an integer key k for encryption. This keys denotes the shift of alphabetical characters during the encryption. Each character from the message is extracted and converted to its integral/ASCII value. The key k is added to this value and the new integer formed is converted back to character and added to the encrypted message. This process is repeated till every character in the message gets shifted by k and added to the encrypted text string. Then, the encrypted string is returned to the user.

In base change encryption, a numeric key is *newBase* is input by the user. The program iterates through the message and finds the numbers present in the message. After a number is extracted, it is converted to base specified by key *newBase*. This is achieved by dividing the base-10 number by the newBase and storing the remainder till the number becomes 0. The stored reminder list is then reversed and added in this new sequence to the message instead of the original base-10 number. Thus, base change of the number is achieved.

For Hill cipher encryption using $n \times n$ key matrix method, a key of length n^2 is taken by the user and each character is given an integral value equal to its position in the character set S and then split into a $n \times n$ key matrix K. Now from the message text, groups of n character are taken one by one to form a $n \times 1$ character matrix M. These matrices are multiplied and then modulo with the length of character set S to form the encrypted character matrix, *EncryptedMatrix = ([K] X [M]) % len(S)*. The resultant *EncryptedMatrix* is of dimension $n \times 1$ and contains index of encrypted characters in character set S. Using the indices in the *EncryptedMatrix*, the characters are taken from S and added to a new string for storing encrypted message. This process repeats till all character are encrypted and then the encrypted string is returned to the user.

5 Results and Discussion

For decryption using brute force approach, the following combinations are required to break the encryption.

1. Caesar cipher $n_1 = 25$ for English alphabet character set.
2. Base change $n_2 = 98$ for base range 2 to 100 (except 10), i.e., [2,100]-{10} numerical base range.
3. Hill cipher $n_3 = 72^9$ combinations for 3×3 matrix method and character set containing 72 characters (a set containing 72 characters is taken as it contains alphabetical, numeric, space, line breaks as well as the commonly used symbols)

Although these are the mathematical combinations required to obtain the key, a ciphertext attack will require much less time for breaking such single-layered encryption techniques [10]. When these methods are combined for multiple encryption along with AES encryption, the resultant combinations required to break the algorithms becomes $(25 \times 98 \times 72^9) \times 3!$ which is of the order 10^{20} and an AES-128 or AES-256 encryption which is virtually unbreakable. Thus, the proposed multi-cipher encrypter increases the strength of encryption by a large degree[3]. The model offers 3 variations of hill cipher using 3×3, 4×4 and 5×5 matrix methods which require even higher combinations for brute force attacks of magnitude 72^9, 72^{16} and 72^{25}. The character set length can be modified according to the character set used in the file. By adding just 2 more layers, the encryption strength increases 100 times. Adding more layers will result in increased strength of encryption. Furthermore, adding a top layer of AES encryption ensures maximum safety to the text information [11].

It may seem like adding encryption layers indefinitely will increase the strength exponentially, but there are a few limitations which should be considered as well [12]. Firstly, key selection must be non-repetitive and taken randomly to avoid any pattern in keys which can be utilized by attackers. Secondly, as the number of encryption layers increases, the time needed for decryption also increases. Hence an optimum trade-off between the number of encryption layers and program performance must be made depending on the system and data size. Thirdly, adding layers indefinitely does not guarantee exponential strength as after a certain threshold some repetitions or patterns are bound to occur which will make those additional layers of negligible significance. Hence, for proper encryption which can be secure against modern attacks [13], the users must use an optimum number of layers along with appropriate keys so as to ensure fast yet secure encryption and decryption [14].

6 Conclusion

The proposed model helps implement a multi-cipher encryption-based model using which the user can define a custom encryption method. On evaluating this encrypter it is found that encryption strength is greatly increased by adding multiple layers of encryption. The use of modern encryption algorithms such as AES algorithms will help obtain encrypted files safe from attackers and hence maintain data confidentiality. This model thus serves as a naive implementation of multiple encryption which can be modified by adding numerous other encryption methods according to need to develop even strong encryption algorithms using the base ciphering algorithms in the future.

References

1. Kumari S (2017) A research paper on cryptography encryption and compression techniques. Int J Eng Comput Sci 6(4) Art no 4
2. Shallal Q, Bokhari M (2016) A review on symmetric key encryption techniques in cryptography. Int J Comput Appl 43
3. Gupta H (2010) Multiphase encryption: a new concept in modern cryptography
4. Haboush A (2018) Multi-level encryption framework. Int J Adv Comput Sci Appl 9. https://doi.org/10.14569/IJACSA.2018.090422
5. Vijayaraghavan N, Narasimhan S, Baskar M (2018) A study on the analysis of Hill's Cipher in cryptography. Int J Math Trends Technol IJMTT
6. Siahaan MDL, Siahaan APU (2018) Application of Hill Cipher algorithm in securing text messages
7. Balogun A, Sadiku P, Mojeed H, Hameed R (2017) Multiple Ceaser Cipher encryption algorithm. Abacus 44
8. Dai Y, Lee J, Mennink B, Steinberger J (2014) The security of multiple encryption in the ideal cipher model. Advances in Cryptology—CRYPTO 2014, Berlin, Heidelberg, pp 20–38. https://doi.org/10.1007/978-3-662-44371-2_2
9. Abdullah A (2017) Advanced encryption standard (AES) algorithm to encrypt and decrypt data

10. Biryukov A, Kushilevitz E (2001) From differential cryptanalysis to ciphertext-only attacks. Lect Notes Comput Sci. https://doi.org/10.1007/BFb0055721
11. Al Hasib A, Haque A (2008) A comparative study of the performance and security issues of AES and RSA cryptography. International Conference on Convergence Information Technology, vol 2, pp 505–510. https://doi.org/10.1109/ICCIT.2008.179.
12. Merkle R, Hellman M (1981) On the security of multiple encryption. Commun ACM 24:465–467. https://doi.org/10.1145/358699.358718
13. Kagita MK, Thilakarathne N, Gadekallu TR, Maddikunta PKR, Singh S (2020) A review on Cyber Crimes on the internet of things arXiv:2009.05708 [cs]
14. Nadeem A, Javed M (2005) A performance comparison of data encryption algorithms. 84–89. https://doi.org/10.1109/ICICT.2005.1598556

Chapter 25
Optimal Combined-Coordination of Overcurrent and Distance Relays Using Jaya Algorithm

Saptarshi Roy, P. Suresh Babu, and N. V. Phanendra Babu

1 Introduction

Relay coordination is a planning study that would be done before the operation of power system is commenced. The main aim of discussed problem is to obtain the performance characteristics of relays so efficiently that it reduces the interruptions due to short circuits as minimum as most possible [1]. During the coordination of relays, all possible disturbances will be considered so that the characteristics thus obtained from relay coordination are able enough to protect the element. During coordination, the characteristics of each relay will be designed such that it must have at least one relay backs it up. For this to happen, their characteristics at each value of fault current will be provided with a definite time interval named as coordination time interval (CTI). Here, the first relay seen by the fault current is called the primary relay, and the next immediate relay that backs it up is called the backup relay. With minimum operating time, the primary relay will always be there as the first line of defense. During relaying, the backup protection is effective only if the primary relay removes less number of power system elements than that of backup protection. And, the backup relay should respond only when the primary relay suffers unsuccessful operation.

In early stage, the relay coordination is done for only overcurrent relays. To perform this, several authors have applied several optimization tools. Initially, they have used deterministic techniques such as linear programming (LP) and simplex

S. Roy
Mirmadan Mohanlal Government Polytechnic, Plassey, West Bengal 741156, India

P. Suresh Babu
National Institute of Technology, Warangal, Telengana 506004, India

N. V. Phanendra Babu (✉)
Chaitanya Bharathi Institute of Technology, Gandipet, Hyderabad 500075, India
e-mail: phanendrababu_eee@cbit.ac.in

methods. In [2], an LP-based optimal coordination is obtained without modifying the relays' settings that already exist. In [3], another sequence of coordination is recommended using simplex method. Using heuristic techniques such as Genetic Algorithm (GA) and particle swarm optimization (PSO), the coordination problems are answered in [4, 5], for the internal fault coordination using overcurrent and IDMT relays. In later times, meta-heuristic techniques such as cuckoo search [6] to minimize total operating time, fire-fly algorithm [7], harmony search algorithm [8], and water cycle algorithm [9] are employed toward obtaining the optimal sequence of overcurrent relays. Then the research has turned toward applying hybrid algorithms to the problem of relay coordination [10].

Now a day, the trend is to coordinate overcurrent relay in combination with distance relay. This strategy is especially used for the protection of EHV lines. Here, the methodology will be improved by reducing the CTI. In this combination of overcurrent and distance relays, each of the relay must coordinate with other irrespective of their operational characteristics to ensure the backup protection [11, 12]. Otherwise, the inconvenience to the customer, and the damage to the equipment become inconceivable. Sometimes, these situations get even worsened by causing cascading outages in the system which may drive the complete system to be collapsed [13, 14]. This, unnecessarily, demands the protection engineers to design for wide-area measurement systems [15, 16], which incur additional cost. So, the coordination of overcurrent and distance relays has become a serious issue before/during the planning/operation of EHV transmission network. Initially, for obtaining the optimal coordination among overcurrent relays with distance relays, a discrimination-time-based approach using a Genetic Algorithm (GA) is suggested [17]. And, then an approach of modeling objective function for this optimization problem is presented [18]. In later times, an optimal coordination problem using heuristic algorithms is proposed [19–21].

In this paper, the distance and overcurrent relay coordination using the Jaya algorithm is presented. Several recent meta-heuristic algorithms came in research during last few years in different domains as discussed in the references [22–25], etc. But for this work we have chosen Jaya algorithm for the following reason:

Jaya algorithm is a recent meta-heuristic algorithm which is any algorithm-specific parameterless and hence capable of further optimize the relay operation time for smooth operation of the power system. As a result, Jaya algorithm is very useful corresponding to the aspect of our problem discussed. So, Jaya algorithm has been chosen for this specific problem. Later in this paper, it is shown that the choice of this technique is very apt with the help of different results and case studies.

Interestingly, it uses several (8types) of intelligent overcurrent relay characteristics during the selection of overcurrent relay for coordination. It obtains optimal Time Multiplier Settings (TMS) and operating times for various intelligent relay characteristics. The main motto of this research is to optimize the relay operation time subjected to the constraints. The contribution of this work includes the use of intelligent characteristics of overcurrent relay using Jaya algorithm to optimize relay operation time and observation of the results with and without optimization for the improvement shown by the algorithm. This novel method is found to be better than any other methods suggested w.r.t several aspects. After discussion, the proposed algorithm is implemented on standard IEEE test systems (IEEE-9 and 14 test systems). In this way, it is concluded that Jaya algorithm is no doubt very good and better than other contemporary techniques.

2 Modeling of Coordination Problem

Distance and overcurrent relays are combined together as a unit to achieve better protection in power transmission system. For solution of the problem and for proper coordination, the fitness function is chosen as Eqs. 1–4.

$$\text{Fitness function} = \min\left(\alpha \sum_{i=1}^{n} t_i + \beta \sum_{i=1}^{n} |T_{\text{DIOC}i} - |T_{\text{DIOC}i}||\right.$$
$$\left. + \lambda \sum_{i=1}^{n} |T_{\text{OCD}Ii} - |T_{\text{OCD}Ii}|| + \delta \sum_{i=1}^{n} |T_{\text{OC}i} - |T_{\text{OC}i}||\right) \quad (1)$$

where

$$T_{\text{OC}i} = T_{\text{oc backup}\,i} - T_{\text{oc main}\,i} - \text{CTI}' \quad (2)$$

$$T_{\text{DIOC}i} = T_{\text{oc}\,i} - T_{z2\,i} - \text{CTI}' \quad (3)$$

$$T_{\text{OCDI}i} = T_{z2\,i} - T_{\text{oc}\,i} - \text{CTI}' \quad (4)$$

T_{oc} is the required time of overcurrent relay operation and T_{z2} is the 2nd zone operation time of the distance relay α, β, λ, δ are constants. The constants are introduced to reduce the chances of mis-coordination. In case of mis-coordinations, the penalty factors increase the value of the fitness function. Since the discussed problem is a minimization problem, such solutions fail to sustain due to increased value of fitness function for the penalty factors. But in case of proper coordination, 2nd, 3rd and 4th term of fitness function vanishes. Only relay operation time is optimized in case of proper coordination. The discussed problem is also required to satisfy several constraints. These constraints are useful regarding maintaining primary and backup relay operation time intervals, pickup currents, TSM, operation timings, etc. Without satisfying all the constraints, the optimum solution cannot be found. Therefore, the different constraints important to find optimum solution for this problem is discussed below.

2.1 Constraints

Several constraints are required to be satisfied to find optimal solution. The various constraints for this particular problem are described as Eqs. 5 and 6.

2.1.1 Coordination Constraints

$$T_{z2\,\text{backup}} - T_{\text{oc\,main}} \geq \text{CTI}' \tag{5}$$

$$T_{\text{oc\,backup}} - T_{z2\text{main}} \geq \text{CTI}' \tag{6}$$

In this particular problem, CTI is taken as 0.25 s.

2.1.2 Relay Characteristics

The overcurrent relay characteristics usually have following characteristics Eqs. 7 and 8:

$$t = \text{TSM}\left(\frac{K}{M^{\alpha} - 1} + L\right) \tag{7}$$

t indicates relay operation time.

TSM indicates time setting multiplier.

K, L and α are constants. Those values are different for different intelligent overcurrent relay characteristics.

M is the ratio between short circuit current I_{sc} and pickup current I_{p}.

8 Types of Intelligent Characteristics Available for Digital Overcurrent Relay is Taken from Ref. [26].

2.1.3 Pickup Current Constraints

The pickup current constraints can be expressed as below [27]:

$$Ip_{\text{min}} \leq Ip \leq Ip_{\text{max}} \tag{8}$$

The limiting values depend on manufacturer specifications.

2.1.4 TSM Constraints

Mathematically TSM constraints can be expressed as follows Eq. 9:

$$TSM_{min} \leq TSM \leq TSM_{max} \tag{9}$$

2.1.5 Constrains on Relay Operating Time

Time of operation of relay (t_{op}) constraint is shown as Eq. 10.

$$t_{op\,min} \leq t_{op} \leq t_{op\,max} \tag{10}$$

Least operation time of relay is 0.1 s and mostly depends on the requirement of the particular power system.

3 Jaya Algorithm

Jaya is introduced very recently. It is a meta-heuristic technique that can be used to solve constraints and unconstraint problems. The tuning of control parameters plays an important role as they influence the performance of the meta-heuristic algorithms. If the tuning of those parameters is improper, then either the computational time increases or results in a local optimal solution [28]. Toward answering these anomalies, R. V. Rao et al. have suggested an algorithm called Jaya algorithm. It is parameter independent. When compared this algorithm with TLBO which has two phases (one for teaching and another for learning), t the Jaya algorithm has only one phase and relatively simpler to apply for solving problems. The working of the technique also is dissimilar than TLBO.

For the objective function $f(x)$, there will be p number of design variables, and q number of candidate solution exists. Here, p and q are natural numbers of any value (i.e., $k = 1, 2, …, q$). If $f(x)_{best}$ and $f(x)_{worst}$ are the best and worst solutions obtained, the current value of the kth candidate $X_{j,k,i}$ where i represents iteration and j represents variable, can be updated using:

$$X^1_{j,k,i} = X_{j,k,i} + r_{1,j,i}\left(X_{j,best,i} - |X_{j,k,i}|\right) - r_{2,j,i}\left(X_{j,worst,i} - |X_{j,k,i}|\right) \tag{11}$$

$X_{j,best,i}$ = The value of the variable j for the best candidate.
$X_{j,worst,i}$ = The value of the variable j for the worst candidate.
$X^1_{j,k,i}$ = The updated value of $X_{j,k,i}$.
$r_{1,j,i}, r_{2,j,i}$ are two random numbers whose values are between 0 and 1.

The second term of the above expression shows the tendency of the solution toward best value of the solution. And, from the third term, it is clear that the solution will be moved away from the worst. Unless the better solution is assured, the $X_{j,k,i}$ will not be accepted. The accepted functions will be carried to the next iteration as inputs.

The flow chart of the algorithm is shown in Fig. 1. The algorithm always tends toward best value and avoids worst solution, thus proceed. With the use of this algorithm many problems are properly solved and get optimized solutions which are very useful. The advantage of this technique also bestows some good results in our case also. The results with a comparison of the discussed problem are shown in the upcoming sections of this paper.

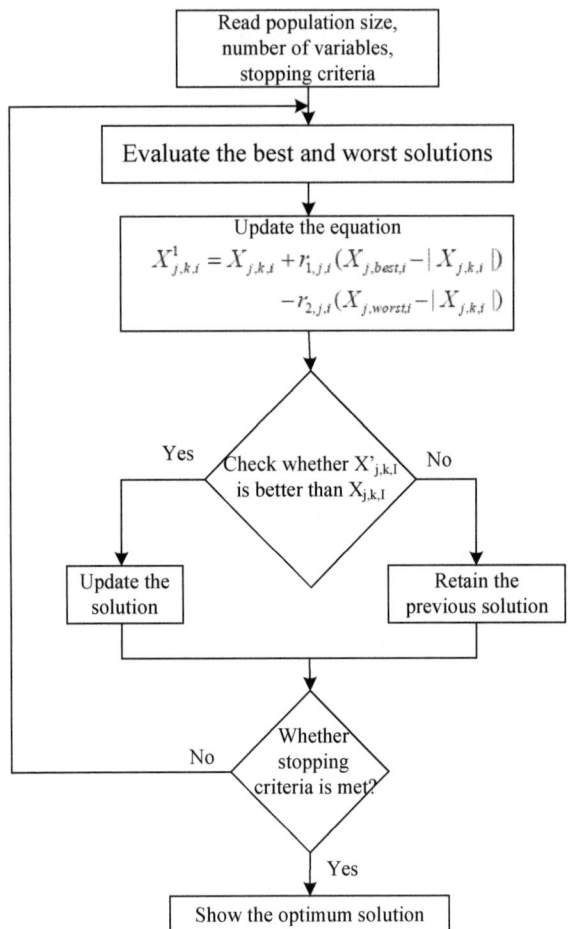

Fig. 1 Flow chart of Jaya algorithm

4 Test Results

To show the efficacy of the method, the method is tested on several standard test systems. The results and discussions are presented in this section. The detailed results are shown below.

4.1 WSCC-9 Bus Test System Results

The relay locations for the WSCC-9 bus system are as shown in Fig. 2. The directional mho relays are utilized here. Each feeder is looked after by two relays (one relay is a combination of one overcurrent and one distance relay) except the transmission lines with transformer. The lines are having transformer with different kind of protections like differential or other protection. Therefore, distance and overcurrent relay combinations are not used here. The relay arrangements for this problem are shown in Fig. 2. Relay positions are marked with a box in the figure.

The main and backup relay pairs for the above 9-bus system are shown in Table 1. Relay backup pair is formed by formation of loop technique after assigning the direction of the relays. The load current is obtained through power world simulation in normal condition of the power system. Pickup current is obtained by multiplying load current with a factor 1.25 in approximately integer form. The short circuit current is obtained by creating an LLLG fault at 15% length from the near end bus of the transmission line, except the lines having a transformer for all test systems.

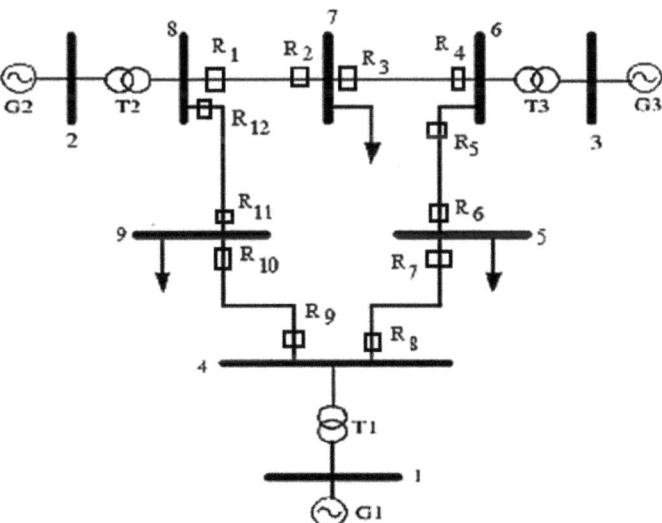

Fig. 2 Relay locations of WSCC-9 bus system

Table 1 Main and backup relay pairs

Main relays	Backup relays
R_3	R_1
R_5	R_3
R_7	R_5
R_9	R_7
R_{11}	R_9
R_1	R_{11}
R_{12}	R_2
R_{10}	R_{12}
R_8	R_{10}
R_6	R_8
R_4	R_6
R_2	R_4

Suppose for line 1–2; we have created fault at 15% length from bus 1 in line 1–2. Table 2 shows the output table for WSCC-9 bus system without optimization results. It shows, the relay is selecting only overcurrent relay characteristics number 2, without the use of optimization. Table 3 shows the output table for WSCC-9 bus system with suitable second zone operation time and TSM values with Jaya algorithm optimization. With the use of optimization technique, the system is capable of selecting different intelligent overcurrent relay characteristics. As a result, the optimum time for relay operation minimizes compare to without optimization case.

Table 2 Output table: without optimization: WSCC-9 bus system

Relay (R_i)	Second zone operation time (T_{z2}) (s)	No. of selected characteristic from Table 1
R_1	0.5215	2
R_2	0.7260	2
R_3	0.5491	2
R_4	0.5673	2
R_5	3.078	2
R_6	3.439	2
R_7	0.5057	2
R_8	0.6022	2
R_9	0.4958	2
R_{10}	0.4848	2
R_{11}	3.949	2
R_{12}	0.5743	2
Average	1.2911	–

Table 3 Output results with Jaya algorithm optimization: WSCC-9 bus system

Relay (r_i)	Second zone operation time (T_{z2}) (s)	TSM	No. of selected characteristic from Table 1
1	0.3734	0.1624	6
2	0.4766	0.0621	1
3	0.5946	0.28	3
4	0.5588	0.225	2
5	0.7974	0.05717	7
6	0.5601	0.2312	1
7	0.2768	0.2575	7
8	0.6662	0.3812	2
9	0.393	0.2745	6
10	0.2653	0.3166	6
11	0.9877	0.07	1
12	0.4	0.0873	7
Average value	0.5291	0.2004	–

One thing that need to mention here, we considered LLLG fault to obtain short circuit data as it is a balanced fault. The balanced fault is most severe fault, and moreover all the three-phase currents in this case is same, which is helpful in calculating relay settings. In case if any unbalanced fault is considered to obtain short circuit data, the relay setting calculations may become more complex, as the short circuit current will be different in all three phases. As a result, three different phases may have three different TSM values and may obtain three different intelligent overcurrent relay characteristics which may increase complexity.

The output results are shown in Table 3.

Figure 3 shows a comparison of 2nd zone operation time with and without optimization. The obtained results clearly show that with the use of optimization the relay operation time drastically reduced which is the main objective of this problem. The next section of this paper shows the brief results of IEEE 14 bus test system with this discussed technique.

4.2 IEEE-14 Bus Test System Brief Results

The discussed method is applied to IEEE 14 bus test system also. The detailed results are obtained. The consolidated results are shown here.

Figure 4 shows the relay arrangements for IEEE 14 bus test system. Figure 5 shows comparison of second zone operation time with and without optimization for IEEE 14 bus test system.

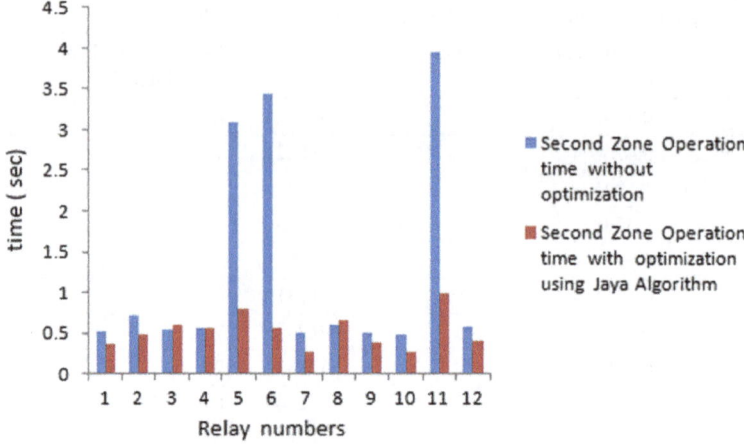

Fig. 3 Comparison of 2nd zone operation time with and without optimization WSCC-9 bus system

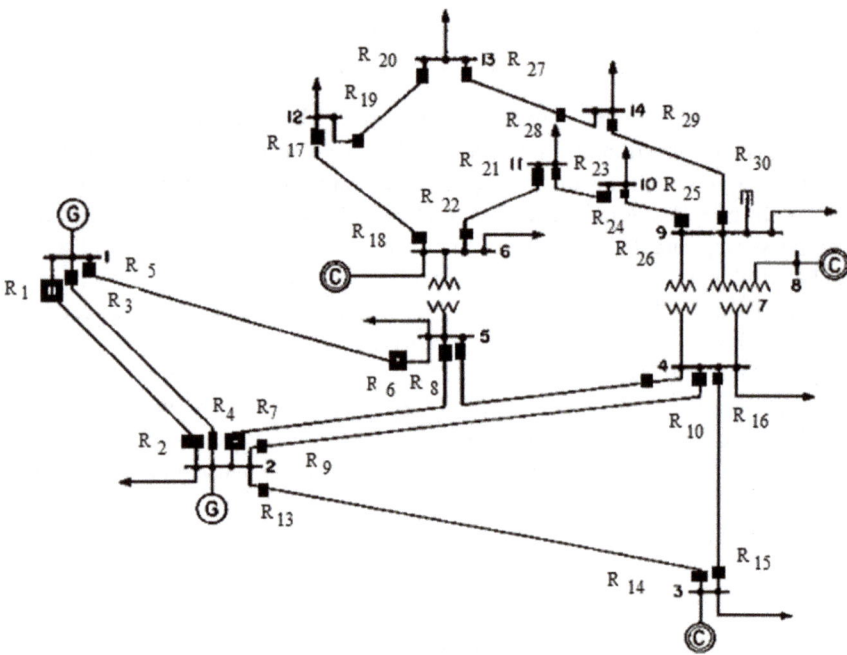

Fig. 4 Relay arrangements for IEEE 14 bus test system

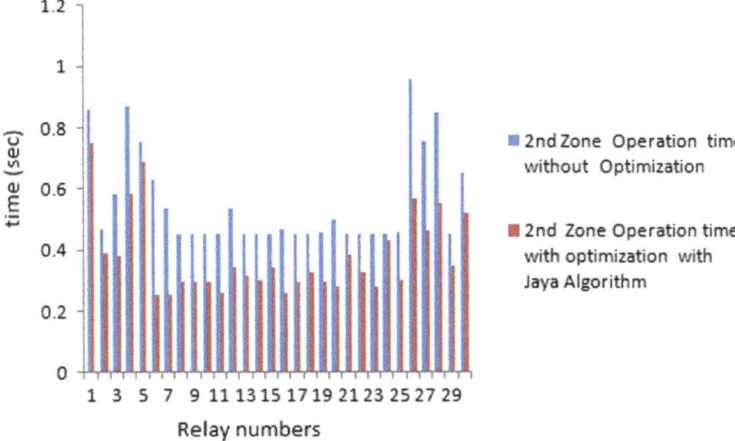

Fig. 5 Comparison of 2nd zone operation time with and without optimization IEEE 14 bus test system

4.3 Comparative Study with Other Optimization Techniques

This section discusses the comparison of the performance of proposed method with its contemporary algorithms.

Figures 6, 7, and 8 show the comparison of 2nd zone operation time, Comparison of TSMs and comparison of fitness functions for WSCC-9 bus system with different algorithms used for optimization. Table 4 shows a comparative study of the Jaya algorithm, GA, PSO and TLBO w.r.t three attributes, namely converging iteration numbers, average time/iteration taken, and total converging time for WSCC-9 bus

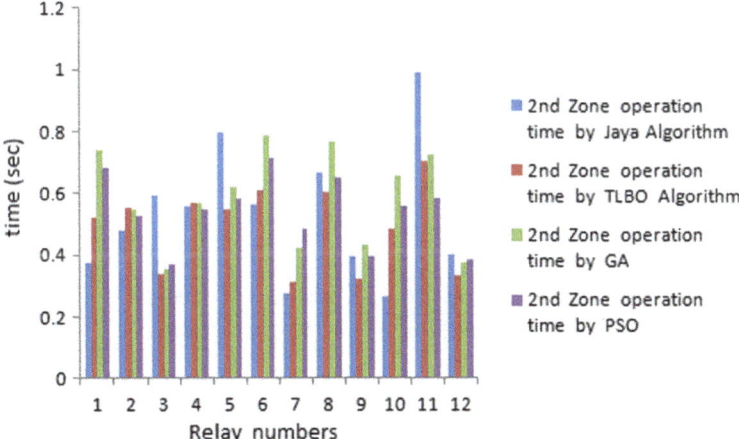

Fig. 6 Comparison of 2nd zone operation time for IEEE-9 system

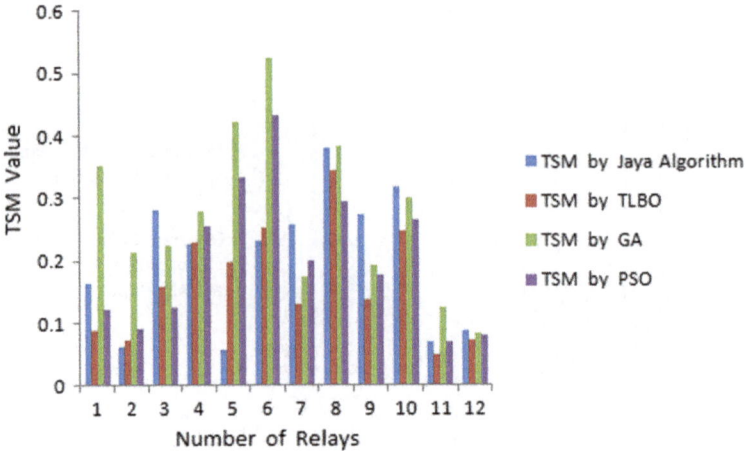

Fig. 7 Comparison of TSMs for IEEE-9 system

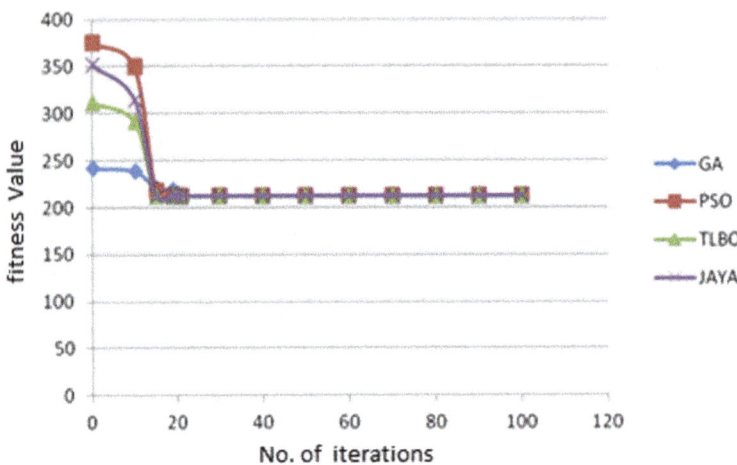

Fig. 8 Comparison of fitness versus iteration graph for WSCC-9 bus system with different algorithms

Table 4 Comparative study of Jaya, GA, PSO, TLBO (WSCC-9 bus system)

Parameters	GA	PSO	TLBO	Jaya
Converging iteration numbers	21	19	15	15
Avg. time/iteration (s)	0.00626	0.00833	0.00404	0.00273
Total converging time (s)	0.13138	0.15822	0.0606	0.04102

system. It is seen from the table that the Jaya algorithm is fastest w.r.t process time to converge hence best among the four w.r.t process speed. The number of iteration may be more sometimes w.r.t other algorithms to converge but one iteration time may be different in the case of different algorithms. Therefore, if process quickness is considered, the iteration may be often deceptive. So, the process fastness should be considered w.r.t total process time always, and in this respect, Jaya algorithm is best among the four. The cause of this is, not the use of any algorithm-specific parameters or phases. Jaya algorithm has only one phase, thus it is even faster than TLBO.

Figures 9, 10 and 11 shows the comparison of 2nd zone operation time, comparison of TSMs and comparison of fitness functions for IEEE 14 bus system with different contemporary algorithm used for optimization. Table 5 shows a comparative

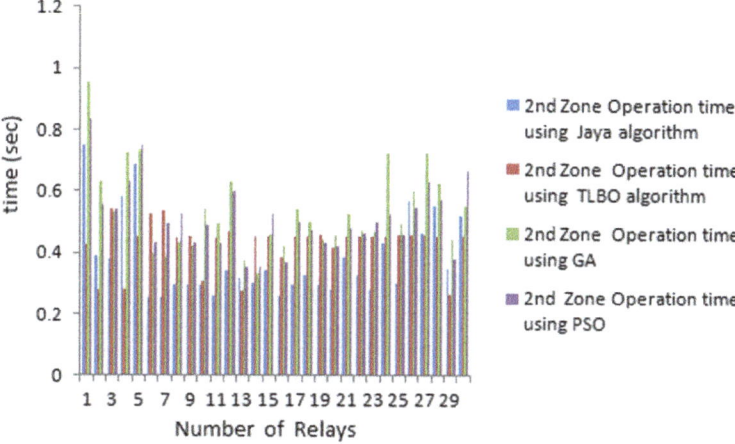

Fig. 9 Comparison of 2nd Zone operation time with different algorithms for IEEE 14 bus system

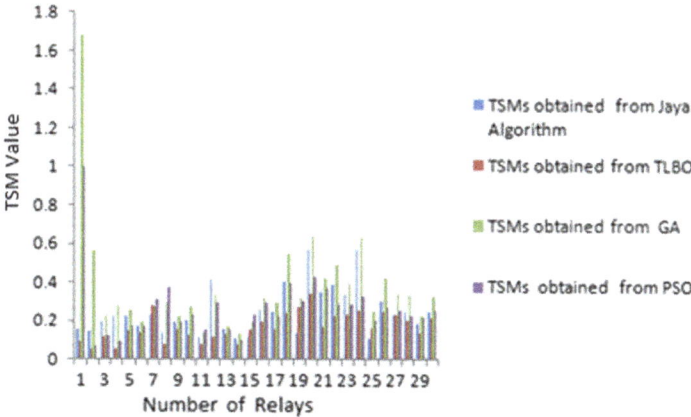

Fig. 10 Comparison of TSMs with different algorithms for IEEE 14 bus system

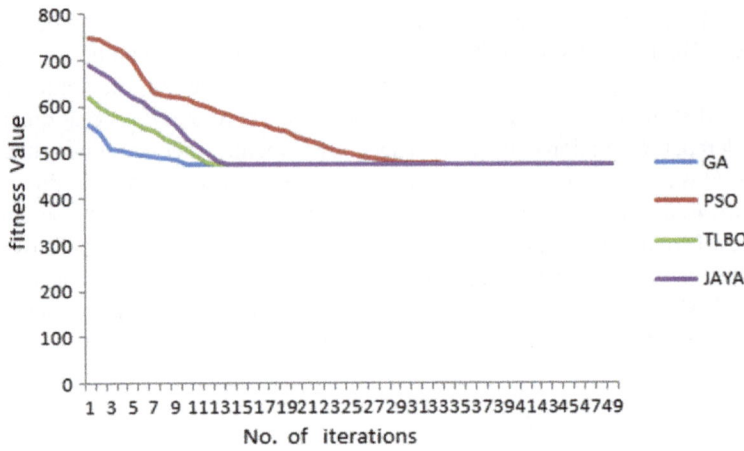

Fig. 11 Comparison of fitness versus iteration graph for IEEE 14 bus system with different algorithms

Table 5 Comparative study of Jaya, GA, PSO, TLBO (IEEE-14 bus system)

Parameters	GA	PSO	TLBO	Jaya
Converging iteration numbers	10	34	12	14
Avg. time/iteration (s)	0.014	0.0054	0.0078	0.0026
Total converging time (s)	0.14	0.1836	0.0936	0.0364

study with other contemporary meta-heuristic algorithms w.r.t converging iteration numbers, avg. time/iteration and total converging time for IEEE 14 bus test system. Similar to 9 bus system, here also it is found that Jaya algorithm is the fastest among all other contemporary algorithms w.r.t total converging time.

5 Conclusions

The main motto of this research is to find the optimal coordination among the over-current and the distance relays. The coordination of the relays is the most important aspect subject to the satisfaction of all the constraints. The coordination problem deals with a large number of constraints. Jaya algorithm is utilized to deal with the coordination problem that has more number of constraints. The comparison of results with no-optimization case, optimization with GA, PSO and TLBO says that the remarkable reduction in the operation time is achieved using optimization technique Jaya. It has given the planning engineer the provision for selecting different intelligent overcurrent relay characteristics. Different other parameters like Time Setting Multipliers (TSM) are also calculated. The process is successfully working on IEEE-9 and IEEE 14 bus systems. And it shows that the proposed methodology is proven to be successful and worthy.

References

1. Inc Institute of Electrical and Electronics Engineers (1986) IEEE recommended practice for protection and coordination of industrial and commercial power systems. Institute of Electrical and Electronics Engineers, Incorporated
2. Urdaneta AJ, Restrepo H, Marquez S, Sanchez J (1996) Coordination of directional overcurrent relay timing using linear programming. IEEE Trans Power Deliv 11(1):122–129
3. Askarian HA, Keyhani R (1995) Optimal co-ordination of overcurrent relays in power system by dual simplex method. In: Proceedings of 1995 AUPEC Conference, Perth, Australia, vol 3, pp 440–445
4. Uthitsunthorn D, Kulworawanichpong T (2010) Optimal overcurrent relay coordination using genetic algorithms. In: 2010 international conference on advances in energy engineering
5. Kheshti M, Tekpeti BS, Kang X (2016) The optimal coordination of over-current relay protection in radial network based on particle swarm optimization. In: 2016 IEEE PES Asia-Pacific power and energy engineering conference (APPEEC). IEEE, pp 604–608
6. Gokhale SS, Kale VS (2015) Time over-current relay coordination using the Levy flight Cuckoo search algorithm. In: TENCON 2015–2015 IEEE region 10 conference. IEEE, pp 1–6
7. Gokhale SS, Kale VS (2014) Application of the firefly algorithm to optimal over-current relay coordination. In: 2014 international conference on optimization of electrical and electronic equipment (OPTIM). IEEE, pp 150–154
8. Rajput VN, Pandya KS (2017) Coordination of directional overcurrent relays in the interconnected power systems using effective tuning of harmony search algorithm. Sustain Comput Inf Syst 15:1–15
9. Kudkelwar S, Sarkar D (2019) Online implementation of time augmentation of over current relay coordination using water cycle algorithm. SN Appl Sci 1(12):1628
10. Atha R, Santra T, Karmakar A (2019) Directional overcurrent relay coordination using hybrid BBO-PSO technique. In: International conference on innovation in modern science and technology. Springer, Berlin, pp 364–372
11. Kumar A, Babu PS, Babu NP, Roy S (2017) A back-up protection of teed-transmission line using taylor-kalman-fourier filter. In: 2017 third international conference on sensing, signal processing and security (ICSSS), IEEE, pp 237–241
12. Roy S, Babu PS, Babu NP, Kumar A (2017) An efficient fault locating technique with backup protection scheme using wide area measurement for power system with simultaneous faults. Int J Electr Eng Inf 9(1):100
13. Babu NP, Babu PS, SivaSarma DVSS (2015) A wide-area prospective on power system protection: a state-of-art. In: 2015 international conference on energy, power and environment: towards sustainable growth (ICEPE), pp 1–6
14. Babu NP, Babu PS, Sarma DS (2017) A new power system restoration technique based on WAMS partitioning. Eng Technol Appl Sci Res 7(4):1811–1819
15. Babu NP, Babu PS, Sarma DS (2015) A reliable wide-area measurement system using hybrid genetic particle swarm optimization (HGPSO). Int Rev Electr Eng 10(6):747–763
16. Phanendrababu NV (2021) Optimal selection of phasor measurement units. In: Wide area power systems stability, protection, and security. Springer, Berlin, pp 127–166
17. Abyaneh HA, Kamangar SSH, Razavi F, Chabanloo RM (2008) A new genetic algorithm method for optimal coordination of overcurrent relays in a mixed protection scheme with distance relays. In: 2008 43rd international universities power engineering conference. IEEE, pp 1–5

18. Chabanloo RM, Abyaneh HA, Kamangar SSH, Razavi F (2008) A new genetic algorithm method for optimal coordination of overcurrent and distance relays considering various characteristics for overcurrent relays. In: 2008 IEEE 2nd international power and energy conference. IEEE, pp 569–573
19. Damchi Y, Sadeh J, Mashhadi HR (2016) Optimal coordination of distance and overcurrent relays considering a non-standard tripping characteristic for distance relays. IET Gener Transm Distrib 10(6):1448–1457
20. Roy S, Babu PS, Babu NP (2019) Optimal combined overcurrent and distance relay coordination using TLBO algorithm. In: Soft computing for problem solving. Springer, Singapore, pp 121–135
21. Rivas AEL, Pareja LAG, Abrão T (2019) Coordination of distance and directional overcurrent relays using an extended continuous domain ACO algorithm and an hybrid ACO algorithm. Electr Power Syst Res 170:259–272
22. Reddy TG, Bhattacharya TS, Maddikunta PDR, Hakak S, Khan WZ, Bashir AK, Jolferi L, Tariq U (2020) Antlion re-sampling based deep neural network model for classification of imbalanced multimodal stroke dataset. In: Multimedia tools and applications. Springer, Berlin
23. Reddy TG et al (2020) A novel PCA–whale optimization-based deep neural network model for classification of tomato plant diseases using GPU. J Real-time Image Process
24. Reddy PK et al (2020) Green communication in IoT networks using a hybrid optimization algorithm. Comput Commun 159:97–107
25. Priya SR et al (2020) An effective feature engineering for DNN using hybrid PCA-GWO for intrusion detection in IoMT architecture. Comput Commun 160:139–149
26. Chabanloo RM, Abyanch HA, Kamangar SSH, Razavi F (2011) Optimal combined overcurrent and distance relays coordination incorporating intelligent overcurrent relay characteristics selection. IEEE Trans Power Deliv 26(3):1381–1391
27. Saha D, Dutta A, Saha Roy BK, Das P (2016) Optimal coordination of DOCR in interconnected power systems. In: IEEE 2nd international conference on control, instrumentation, energy and communication, Kolkata, India
28. Rao RV (2016) Jaya: a simple and new optimization algorithm for solving constrained and unconstrained optimization problems. Int J Industr Eng Comput 7:19

Chapter 26
Creativity in Machines: Music Composition Using Artificial Intelligence that Passes Lovelace 2.0 Test

Shagaf Hasnain, Palash Goyal, and Rishav Kumar

1 Introduction

Music generation using AI is being researched for a long time, and a lot of developments have been done in this area. But these developments are limited to the generation of simple musical notes. The music generated lacks the mood detection that is the emotional meaning which can be easily detected in a human-composed music. With this paper, we are proposing a system based on the framework PALASH 1.0 which we suggested in our earlier paper [1], and it was an approach to enhance the creativity of machines based on the earlier work of Turing's and Riedl [2, 3]. This framework uses deep learning techniques to generate music based on different moods/genres.

From the last few centuries, researchers have been using mathematical techniques to generate music [4, 5]. Music is a sequence of elements (or sound), and this was first mentioned by Iannis Xenakis in the early 1950s. His music, popularly known as 'Stochastic Music' [6, 7], was composed using the concepts of statistics and probability. Later the 'IILIAC Computer' which is the best known work of Hiller and Isaacson's [8] that generates music using the 'generate and test' approach. The generated notes were tested first by heuristic compositional rules of classical harmony and only the notes that pass the test are kept.

'CONCERT' is one of the earliest designed generative models which was architectured to compose simple melodies [9]. The model though had some limitations as it was unable to capture the structure of music that was used globally. Later, it was observed that sequential modeling techniques like Markov chains or recurrent

S. Hasnain
Gazelle Information Technologies, Dwarka, Delhi, India

P. Goyal (✉) · R. Kumar
Mount Carmel School, Dwarka, Delhi, India
e-mail: Palashgoyal1608@mountcarmeldelhi.com

© The Author(s), under exclusive license to Springer Nature Singapore Pte Ltd. 2021
S. Kumar et al. (eds.), *Proceedings of International Conference on Communication and Computational Technologies*, Algorithms for Intelligent Systems,
https://doi.org/10.1007/978-981-16-3246-4_26

neural networks are the most prevailing methods to be used to create models that can learn probable transitions of notes in the given class of music [10, 11]. In the last few years, researchers have proposed a lot of new deep neural network models for the generation of music [12–17]. These AI models assign a certain probability to every piece of music and captures the uniformities in a class of music whether in terms of genres, style, category, etc. Music consists of emotions (or moods) that can be impacted by attributes like tempo, timbre, harmony, loudness, etc.

In our approach, we are taking some existing music data to train our model using these existing data. The model will understand and learn the patterns in music. It will get trained on different moods and compositions of music to be able to understand and differentiate between the different classes of music. Once the model gets trained, it should be able to generate a new sequence of music. It will not just copy-paste the sequence from the training data instead it will understand the patterns and the different genres of music from the training datasets to generate new music.

We will pass our model generated output through Lovelace 2.0 [3] test to see if the model generated quality music that passes the test.

2   Methodology

First, we must represent the music in the form of a sequence of events as we are using RNN which takes input in form of sequences. Representation of music can be done in three forms:

- Sheet Music: A Pictorial representation of music is known as sheet music in which a sequence of musical notes is represented.
- ABC Notation: In the ABC notation, there are two parts, the first part represents the metadata, and the second part represents the tune which is a sequence of characters representing musical notes
- MIDI: MIDI represents a series of messages such as 'note on,' 'note off,' 'pitch bend,' etc., which is implied by MIDI instruments to generate music.

Here, we are representing our music in form of ABC notation. These musical notes will then be segmented into different classes of music such as happy, sad, and soul.

3   Data Analysis

We have taken different classes of musical data as a source to train our model. We then represent these in form ABC notation. In Fig. 1, some sample musical data represented in form of ABC notation can be seen. In the first part, it provides metadata to understand the tunes such as (X:), the title (T:), the time signature (M:), the default note length (L:), the type of tune (R:) and the key (K:). In the second part,

```
S:EF
M:4/4
K:A
M:6/8
P:A
f|"A"ecc c2f|"A"ecc c2f|"A"ecc c2f|"Bm"BcB "E7"B2f|
"A"ecc c2f|"A"ecc c2c/2d/2|"D"efe "E7"dcB| [1"A"Ace a2:|
 [2"A"Ace ag=g||\
K:D
P:B
"D"f2f Fdd|"D"AFA f2e/2f/2|"G"g2g ecd|"Em"efd "A7"cBA|
"D"f^ef dcd|"D"AFA f=ef|"G"gfg "A7"ABc |1"D"d3 d2e:|2"D"d3 d2||
```

Fig. 1 Sample training data

the tune is given which is a sequence of characters where each character represents some musical note Fig. 1.

To train our model, we will create batches of data and then feed these batch of sequences into our model. Each of these unique characters is assigned some numerical index value. These unique characters will be store based on their emotional moods. A dictionary is created that stores these unique indices and moods as the value and the key is the identified unique characters Fig. 2.

Then, we will feed these batches into our RNN models. Here, we are using many-to-many RNN, which gives output equal to the number of inputs. It takes both current and the previous output as input. So, for the first iteration, we will feed zero/dummy input as the previous output along with our input data.

As our model gets trained on these data, we will give some random character to the model from the unique characters that have been identified during training. The model will generate the sequence of characters automatically based on its learning from the training phase Fig. 3.

	Batch-1	Batch-2	...	Batch-150	Batch-151
0	0...63	64...127	...	9536...9599	9600...9663
1	9701...9764	9765...9829	...	19237...19300	19301...19364
.	...	...	...	...	...
.	...	...	...	...	...
.	...	...	...	...	...
14	135814...135877	135878...135941	...	145350...145413	145414...145477
15	145515...145578	145579...145642	...	155051...155114	155115...155178

Fig. 2 Batches of data

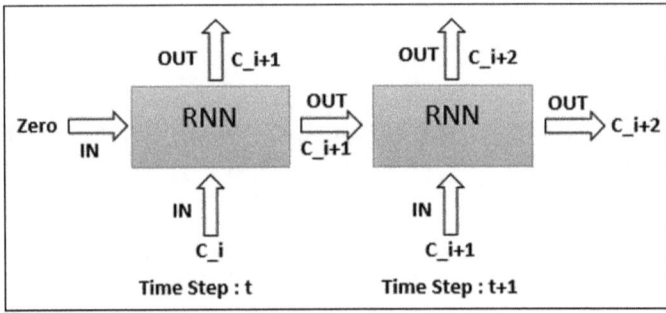

Fig. 3 Many-to-many RNN

4 Techniques Involved

Char-RNN is a type of RNN with character-based learning that predicts the next character given a sequence of characters [18, 19]. Using char-RNN has two benefits, first, the form of the text representation of music has no constraints, and second, it has fewer number of states, i.e., a decreased vocabulary which is a drawback of word-based learning methods. Here, we have given a sequence of such characters as an input to train our model. Suppose we have a sequence of music as [d, a, e, o, b, a, p, …]. Now we will give 'd' as input to the model and expect 'a' as the output, then we will give 'a' as input and expect 'e' as an output, then again, we give 'o' as input and expect 'o' as the output, and so on. We train our model in such a way that it will output the next character in the sequence. It will learn these whole sequences and identify the patterns and will be able to generate a new sequence on its own.

Despite the widespread use of RNNs, it has certain limitations like vanishing gradients and long-term dependency issues. These issues were resolved by LSTM which uses addition operations and allows the gradient to flow by a separate path [20]. In this approach, we are using three RNN layers each having 256 LSTM units. At each time step, the output generated from all the LSTM units will be given as input to the next layers, and the same output will be again given as an input to the same LSTM unit. After these three layers of RNNs, we added a 'Time Distributed' dense layer with Softmax activation in it as shown in Fig. 4.

Softmax is a type of logistic regression that normalizes an input value into a vector of values following a probability distribution whose total sums up to 1. Decimal probabilities are assigned to each class in a multi-class problem. It is implemented using a neural network just before the output layer. The number of nodes must be same in both the Softmax layer and the output layer.

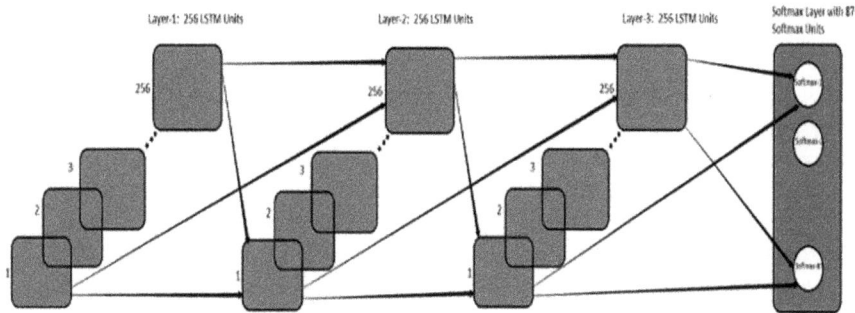

Fig. 4 Model architecture

5 Result

At first, the model was trained with a learning accuracy of 82% which is not very good for melodious music generation. To increase the accuracy, we transferred the learning of our previous model and again trained our model again with two extra layers of LSTM units. Now, the learning accuracy of our model has increased to 91%.

After we have trained our model and found the most effective weights, now our model will be predicting and generating music. For prediction, we will provide input from any of the 87 unique characters identified during the training process and the emotional mood (happy, sad, joy, etc.). The model will generate 87 probable values as output through the Softmax layer. From these returned values, the model will again choose the next character probabilistically, and finally, the chosen character will again be fed back to the model and so on. This process continues and keeps on concatenating the output characters to generate music of the given length as shown in Fig. 5.

```
MUSIC SEQUENCE GENERATED:

"(37)"E2E D2)|"Am"E2c "G7"B=GB|"Am"D2c cBc|
"D"d2A ABd|"G"g2d e2G|"Am"B2A "D7"A2G|"G"G3 -G2:|
P:B
|:A|"G"BGD "D7"G2A|"G"BGB dBd|"C"efg "B7"b2g|"Em"gfe "Am"dBG|
"D"DEF AGF|"G"GBd g2a|"G"g2d "D7"c2B|"G"G3 -G2:|
P:B
d|"G"dBd gfg|"C"e^de g2e|"F"dBA "D7"ABA|"G"G3 G2:|
```

Fig. 5 Generated musical sequence represented in ABC notations

6 Conclusion

In this paper, we proposed a system that is able to generate music by giving a set of characters and an emotional mood. The generated output is more creative than the earlier models [11, 21, 22], in terms of generating mood-based music. We passed the generated output through Lovelace 2.0 test [3] to check the accuracy of our system.

We have four input parameters for our model the epoch weight, the initial character, the length of the music sequence, and the mood based on which the model will generate the musical sequence. Below is the table that shows the Lovelace 2.0 test [3] output Table 1.

7 Ideas for Future Improvements

Although we can generate music using the creativity of the machine, there is a huge scope of improvement possible in the future.

Further enhancement of creativity and consciousness in machines can be done by training the model on metadata and data which is as per the human behavioural parameters. These systems can be helpful in various areas like education, law and order, and healthcare.

Table 1 Results of Lovelace 2.0 for generated outputs

Input	Generated output	Passed	Not passed
50, 45, 350	K:C l:c2e geelg2f e2cl"G"d2d d2Bl"Am"ABA "D7"B2Al "G"G3 "G"gfgl"D"f2e "D/f + "def "G"g2fl"Em"e2d "D"d2cl"Em"B2E "A7"EFEl "D"D3 "A7"F2El"G"D2D d2Bl"Am"A2G "D"FEDl"G"G3 G3l "G"ded dcBl"D7"A2D DEFl"G"G3 "D7"FGAl"G"B2G d2Bl "G"ded Bddl"G"B2B B^{A}ABl"G"d2d "G7"e2dl"C"c2c "G"B2Bl\ "Am"c2B "D7"A2Gl "G"B2G GABl"C"c2c edcl"G"B2G "D"AAGEl"G"G3-A2:l\	✓	✗
30, 78, 250	M::/8 K:D l:dl"D"d2f "A7"edcl"G"d2d d = cBl"Am"ABA "D7"ABcl"G"d2G "D7"AGFl"G"G3 G2:l P:B g/2g/2l"C"agg g2el"C"c3-d2dl(A7"e3 g2el"D7"d3 d3l"G7"g3 g3l"C"e2d cdel"G"d2B d2Ac\ "G"D2b "C7"g2el"G"d3 d2el"C"f2f "G"dcBl"D7"c2d e2dl"G"B2G BAGl "C"c3 ccB	✓	✗

(continued)

Table 1 (continued)

Input	Generated output	Passed	Not passed
80, 86, 500	"Em"e3 "Bm"d3|\ "C"cde "G"dcB|"D7"A3 "G"G3||	✓	✗
60, 59, 400	M:6/8 K:Em E|"Am"A2B c2d|eef e2d|"G"G2B "G7"GAB|"C"c6|G2A A3| "F"c3 AFA|"G7"G2A B2d|"C"e2c GAB|"G"def g2B| "Am"efg "D"f2d|"Am"cBA "E7"BAG|"Am"cAA A2B|"Am"cAA ABc|"E7"e3-e2d|"G"dBG GBd|"G"gdB "D7"AGF|"G"GGB d2:: g|"F"a2f "C"g2e|"G"dBG "C"e3|"G"dBG "D"A2B|"Em"GEE "Em"GEE| "Bm"DFD FED|"E"B,2B B2A|"Em"G2A BA:|	✗	✓
40, 56, 200	FC|"Dm"F2F "A7"GFG|"D"A2A "G"BAG|"C"G2A "D"B2A| "Am"A2G "D7"DEF|"G"GAB "D"D2A|"G"B2G BAG|"A7"AGA "D"AGF|"G"G3-G2:| P:B |:b|"G"dAcd def|"G"g2d cB'2 f|"C"e2c e2c|"G"B2B BAG|"G"GAG g2f| "C"g2G g2G|	✓	✗

Acknowledgements We would like to thank Mr. Kirit, the CEO of Gazelle Information Technologies Pvt. Ltd., for his expert advice and a supply of required resources for the implementation of this project.

References

1. Kumar L, Goyal P, Kumar R (2020) Creativity in machines: music composition using artificial intelligence. Asian J Conv Technol ISSN NO: 2350-1146 I.F-5.11
2. Turing AM (1950) Computing machinery and intelligence. Mind, New Series 59(236):433–460
3. Riedl MO (2014) The lovelace 2.0 test of artificial creativity and intelligence. arXiv:1410.6142v3
4. Kirchmeyer H (1968) On the historical constitution of a rationalistic music, vol 8. Die Reihe, pp 11–24
5. Hiller L, Isaacson LM (1959) Experimental music: composition with an electronic computer. McGraw-Hill, New York
6. Roberts GE (2012) Composing with numbers: Iannis Xenakis and his stochastic music. Math/Music: Aesthetic LinksMontserrat Seminar Spring
7. The origins of stochastic music, Translated by Hopkins GW from Xenakis's paper 'Les Musiques formelles' in Revue Musicale No. 253/254
8. Hiller L, Isaacson L (1993) Musical composition with a high-speed digital computer (1958). In: Reprinted in Schwanauer SM, and Levitt DA (eds) Machine models of music, pp 9–21. The MIT Press, Cambridge Mass
9. Schulze W, Van Der Merwe A (2011) Music generation with Markov models. IEEE Multim 18(3):78–85. ISSN 1070986X. https://doi.org/10.1109/MMUL.2010.44
10. Franklin JA (2006) Recurrent neural networks for music computation. INFORMS J Comput 18(3):321–338
11. Pachet F, Roy P (2011) Markov constraints: steerable generation of markov sequences. Constraints 16(2):148–172

12. Bretan M, Weinberg G, Heck L (2016) A unit selection methodology for music generation using deep neural networks. arXiv preprint arXiv:1612.03789
13. Yang L-C, Chou S-Y, Yang Y-H (2017) MIDINET: a convolutional generative adversarial network for symbolic-domain music generation. arXiv:1703.108472v2
14. Conklin D (2003) Music generation from statistical models. In: Proceedings of the AISB 2003 symposium on artificial intelligence and creativity in the arts and sciences, Aberystwyth, Wales, pp 30–35
15. Choi K, Fazekas G, Sandler M (2018) Text-based LSTM networks for automatic music composition. arXiv:1604.05358v1
16. De Lopez R (2006) Mantaras making music with AI: some examples. In: Proceedings of the 2006 conference on Rob Milne: a tribute to a pioneering AI scientist, Entrepreneur and Mountaineer
17. Dong Y et al (2020) Research on how human intelligence, consciousness and cognitive computing affect the development of artificial intelligence. Hindwai vall 2020
18. Pratheek I, Paulose J (2019) Prediction of answer keywords using char-RNN. Int J Electr Comput Eng (IJECE) 9(3)
19. Sutskever I, Martens J, Hinton GE (2011) Generating text with recurrent neural networks. In: Proceedings of the 28th international conference on machine learning (ICML-11), pp 1017–1024
20. Hochreiter S, Bengio Y, Frasconi P, Schmidhuber J (2001) Gradient flow in recurrent nets: the difficulty of learning long-term dependencies
21. Coca AE, Romero RAF, Zhao L (2011) Generation of composed musical structures through recurrent neural networks based on chaotic inspiration. In: The 2011 international joint conference on neural networks (IJCNN). IEEE, pp 3220–3226
22. Eck D, Schmidhuber J (2002) Arst look at music composition using LSTM recurrent neural networks. Istituto Dalle Molle Di Studi Sull Intelligenza Articiale

Chapter 27
Imbalanced Dataset Visual Recognition by Inductive Transfer Learning

Raji S. Pillai and K. Sreekumar

1 Introduction

Human Brains are exceptionally good in transferring knowledge from one context to another context. Most Machine Learning algorithms are inspired by the architecture of human brain. But the limitation regarding the machine architectures is that it requires enormous amount of training examples to train a novel architecture from scratch and often newly trained architectures fail to apply the learned knowledge to a test data which differ from the train data. This performance deviation is mainly due to the dissimilarity in domain and task. Considering the case of image recognition where so many external factors like lighting, background clutter, and view of angle can cause the classification or recognition task difficult.

To exploit previously at hand data, effectively for latest tasks with scarce data, architectures or knowledge learned from one domain tasks can be effectively utilized by transferring knowledge from previous domain to the latest domain. In most of the image classification algorithms features considered are texture, shape or color channel or its combinations. So they are problem specific and they lack generalization capability.

R. S. Pillai (✉)
Department of Computer Applications, St Teresa's College (Autonomous), Ernakulam, Kerala, India

K. Sreekumar
Department of Computer Science and IT, Amrita Viswa Vidyapeetham, Kochi Campus, Kochi, India

S. Kumar et al. (eds.), *Proceedings of International Conference on Communication and Computational Technologies*, Algorithms for Intelligent Systems,
https://doi.org/10.1007/978-981-16-3246-4_27

325

2 Literature Review

With the extreme advancement in the area of deep learning, many algorithms were developed for image categorization. Most of the algorithm uses visual cues obtained from local parts using detection or segmentation method for categorization [1]. In 2017 Xia and Cui proposed an image classification model based on Inception—v3 Tensor flow model. Their model reported 95% classification accuracy on Oxford 17 flower dataset and 94% on Oxford 102 flower dataset [2]. In 2019 Çibuk et al. proposed a model based on VGG16 and AlexNet. The feature extracted using these pre-trained DCNN models is concatenated and best features are selected by the application of mRmr algorithm. The selected features are then fed to an SVM classifier (Cibuk, UmitBudak, YanhuiGuo, M. Cevdet Ince, 2019). The Transfer Learning methodology is widely used in medical image processing also. In 2020 Saranya et al. implemented VGG, ResNet and Xception with SVM classifier for pneumonia detection using X-ray images [3]. In 2017, Gavai et al. developed a model on MobileNets on TensorFlow platform to retrain the flower category datasets, where the accuracy increased by minimizing the classification time [4]. In 2018 Wu et al. developed a flower classification model using convolution neural network and transfer learning. They compared it with different deep neural network models and reported that transfer learning can viably maintain a strategic distance from local optimal problems and over-fitting problems [5]. In 2017 Li et al. employed the features learned in the convolutional sparse auto-encoder to initialize convolutional neural networks for classification [6]. In 2020, Pillai et al. employed densenet for classifying black and white images [7].

3 Data Sets

3.1 Datasets

3.1.1 Caltech 101

Caltech 101 dataset is a digital dataset compiled by Li et al. [8] which consists of 9146 images belonging to 101 different object categories and a background clutter category. Each category contains around 45–400 images. The images were scaled approximately to 300×200 pixels size. The significance of this database is its large inter-class variability (Fig. 1).

3.1.2 Oxford Flower 102

The Oxford Flowers 102 dataset introduced by Nilsback and Zisserman in 2008 includes digital images of 102 flower categories. Each flower species includes images

Fig. 1 Some images from Caltech-101 dataset

in a range of 40 and 258 images. The images in the dataset vary considerably in quality, scale, clutter, resolution, lighting, etc. In addition, the intra-class and inter-class similarities are very large. The total no of images in the dataset is 8189. The images were rescaled approximately to 300×500 to 500×700 pixels size [9] (Fig. 2).

Fig. 2 Some images from Oxford Flower 102 dataset

4 Methodology

Deep Learning is the automation of machine learning process where machines automatically select the features to be extracted from the fed image.

4.1 Transfer Learning

Deep Learning algorithms perform remarkably well in classifying images. But the problem with this type of architecture is that it requires a large amount of data, computational power and also it is time-consuming. If a network is trained with sufficient data and if it is a proven architecture, the same architecture training can be used for another application. This can be achieved by using a technique called Transfer Learning. "Transfer learning is the improvement of learning in a new task through the transfer of knowledge from a related task that has already been learned Residual Neural Network [10]. Residual Neural Network is a novel deep learning architecture widely used for image classification [11]. The heuristics of Convolution Neural Network going deeper raised the problem of vanishing gradient descent. Since the gradient is back-propagated to prior layers, repeated multiplication forge gradient immensely small. Due to this, as network goes deeper, after a certain level its presentation gets immersed or begins lessening.

In residual neural network architecture the authors explicate the layers by learning residual functions of the input layers, instead of learning unreferenced functions. Deep Residual networks also have Convolution, subsampling and fully connected layers piled up. The extra connection that makes a residual network is the identity connection established in between the layers (Fig. 3).

The identity connections do not hold any parameters and it just passes the output from one layer to the next layer ahead. If the dimension of input function x and residual function $F(x)$ is different, then a linear projection W is multiplied with the identity mapping which expands the shortcut channels to match with the residual channels. This process combines the input x and residual function $F(x)$ and pass it

Fig. 3 Residual learning block

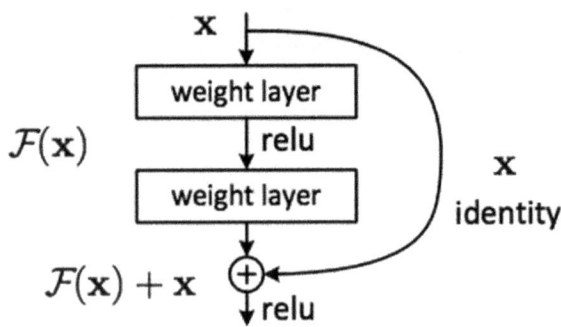

to the next layer Eq. 1.

$$Y = \mathcal{F}(x, \{W_i\}) + W_s x \tag{1}$$

The linear projection term W_s is implemented by 1×1 convolution which adds additional parameter to the architecture.

5 Experiment

Our proposed architecture is based on ResNet50 model of TensorFlow platform. The experimental datasets include Oxford 102 Flower dataset and Caltech101 dataset. In this architecture we implement higher level representation approach which initializes the weights from a pre-trained model and then the target task parameters are fine-tuned using the target training data. We have used two categories of imbalanced experimental datasets with homogeneous and heterogeneous feature and label spaces.

5.1 Residual Neural Network as Fixed Feature Extractor

In proposed architecture Residual neural network is being used as the feature extractor. The identity connection or residual block of ResNet50 deep neural network is being exploited here.

Figure 4 depicts the process flow. Raw input images of the dataset are fed into the ResNet model, the parameters of the previous layers are maintained and the last layers are removed, then the new layers are added, trained with the target datasets. The number of output nodes is set to 102 for flower dataset and 101 for Caltech 101 dataset. In order to augment the training dataset, we synthetically created training images by applying data transformations like flipping, rotation and cropping. The mean values and standard deviation calculated normalizes the input tensor image. The optimizer we used is Adagrad with a learning rate started at 0.01. We have run 70 epochs of training for adjusting the classifier weights and network parameters.

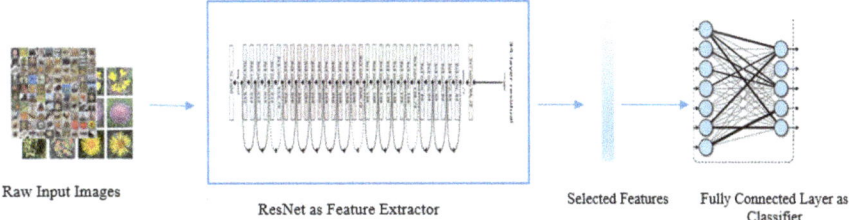

Raw Input Images   ResNet as Feature Extractor   Selected Features   Fully Connected Layer as Classifier

Fig. 4 Proposed methodology

5.2 Experiment's Result

During the initial training phase the learning rate started at 0.01 and later moved up to `0.00001000` delivering an accuracy of 97.56% for Oxford 102 flower dataset with a test loss of 0.058919.

Test Loss: 0.058919   Test accuracy: 97.56%

And for Caltech 101 dataset, the accuracy of 94.58% with a test loss of 0.174624

Test Loss: 0.174624   Test accuracy: 94.58%

Figures 5 and 6 graphically represent the correctly predicted and missed out samples representation of each dataset. Out of 409 samples, only a few were missed out. The missed out samples were depicted away from the prediction line.

Validation accuracy of the dataset images, which is the exact percentage level of effectively classified images is depicted in Figs. 7 and 8.

The parameter adaptation during the whole training process is depicted in Figs. 9 and 10.

Classification Performance Comparison of Different Algorithms on Oxford 102 Flower Dataset is depicted in Table 1.

Classification Performance Comparison of Different Algorithms on Caltech 101 Dataset is depicted in Table 2.

The accuracy analysis of existing models in literature is depicted in Tables 1 and 2. For Oxford 102 Flower dataset the accuracy we obtained is 97.56% and for Caltech-101 dataset the accuracy we obtained is 94.58%. The results clearly depict

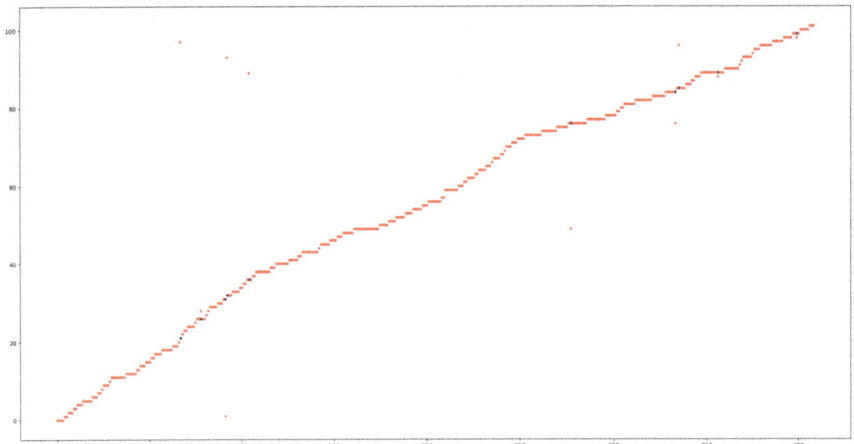

Fig. 5 Oxford 102 flower predicted samples line

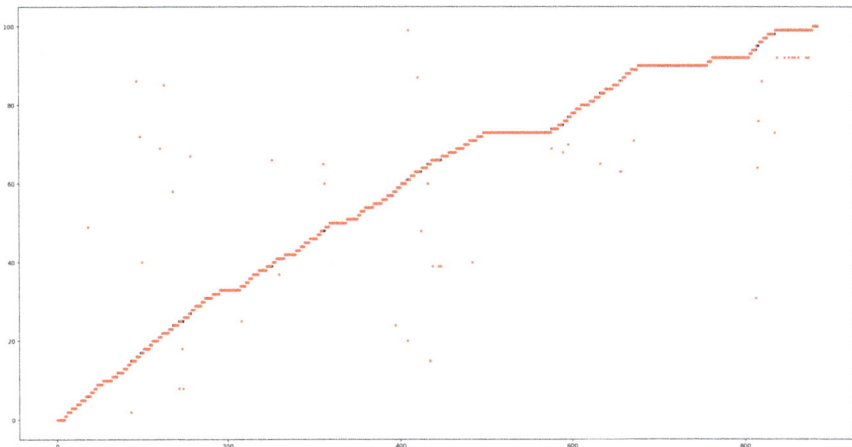

Fig. 6 Caltech 101 predicted samples line

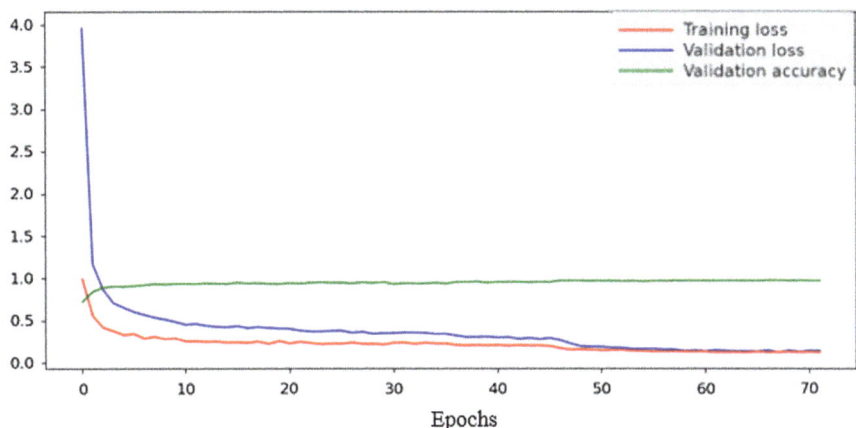

Fig. 7 Accuracy analysis of Oxford 102 Flower dataset

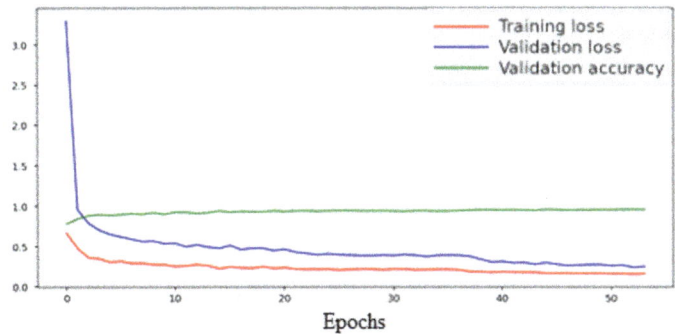

Fig. 8 Accuracy analysis of Caltech-101 dataset

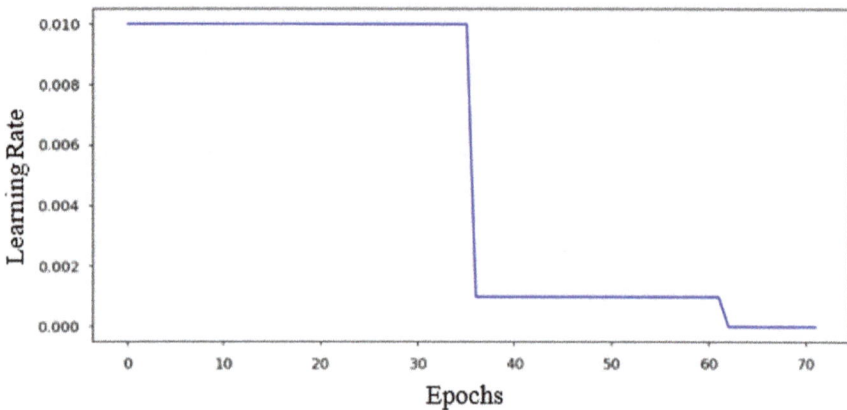

Fig. 9 Training phase parameters evolution of Oxford 102 flower

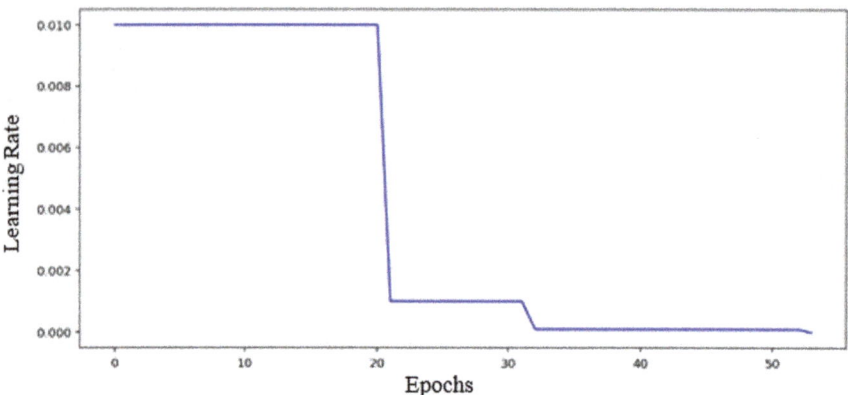

Fig. 10 Training phase parameters evolution of Caltech-101 dataset

Table 1 Performance comparison of literatures of Oxford 102 flower dataset

Method	Classification accuracy (%)
Flower classification via convolutional neural network [12]	84.02
MobileNets for Flower classification using TensorFlow [4]	85
Inception V3 flower classification model [2]	95
Convolution neural network-based transfer learning for classification of flowers [5]	95.29
Our proposed model	97.56

Table 2 Performance comparison of literatures of Oxford 102 flower dataset

Method	Classification accuracy (%)
CSAE LLC-SVM [6]	71.4
M-SVM [13]	81.3
Cubic SVM [14]	90
Our proposed model	94.58

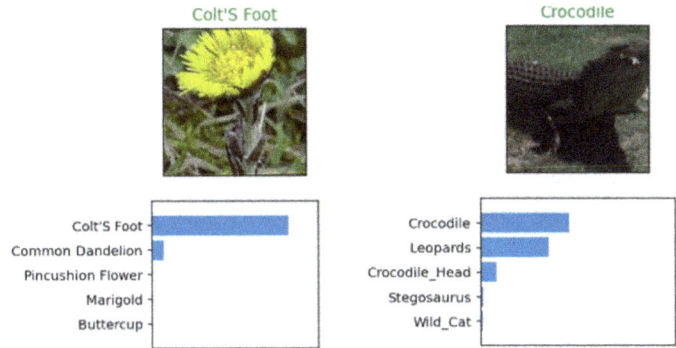

Fig. 11 Prediction for an image in Oxford 102 flower dataset and Caltech-101 dataset

that by using Residual Network as feature extractor, the model outperforms existing state-of-the-art deep neural network models.

During the validation phase an image is fed into the network and it will result in the most probable label name for that particular image (Fig. 11).

6   Conclusion

In this conferred work we proposed a model dependent on ResNet50 for visual classification of two imbalanced benchmarked dataset with variable attributes. Since the datasets are of homogeneous feature spaces we implemented inductive transfer learning methodology with supervised feature construction for feature representation. The results we obtained precisely indicate that our proposed model exhibits better accuracy compared to other state-of-the-art algorithms. In future, we are planning to apply our proposed transfer learning technology for solving challenging applications like video classification, social network analysis, etc.

References

1. Cai D, Chen K, Qian Y, Kämäräinen J-K (2019) Convolutional low-resolution fine-grained classification. Pattern Recogn Lett 119:166–171
2. Xia X, Xu C, Nan B (2017) Inception-v3 for flower classification. In: 2nd international conference on image, vision and computing (ICIVC), Chengdu
3. Saranya SS, Shubham J, Gangesh B (2020) Pneumonia detection using transfer learning. Int J Adv Sci and Technol, 29(3):986–994
4. Gavai NR, Jakhade YA, Tribhuvan SA, Bhattad R (2017) MobileNets for flower classification using TensorFlow. In: 2017 international conference on big data, IoT and data science (BID), Pune
5. Wu Y, Qin X, Pan Y, Yuan C (2018) Convolution neural network based transfer learning for classification of flowers. In: 2018 IEEE 3rd international conference on signal and image processing (ICSIP), Shenzhen
6. Luo W, Li J, Yang J, Xu W, Zhang J (2018) Convolutional sparse autoencoders for image classification. In: IEEE transactions on neural networks and learning systems
7. Pillai RS, SreeKumar K (2020) Classification of fashion images using transfer learning. Adv Intel Syst Comput 1176
8. Li F-F, Fergus R, Peron P (2004) Learning generative visual models from few training examples: an incremental bayesian approach tested on 101 object categories. In: Conference on computer vision and pattern recognition workshop, Washington
9. Nilsback M-E, Zisserman A (2008) Automated flower classification over a large number of classes. In: Indian conference on computer vision, graphics and image processing, Bhubaneswar
10. Olivas ES, Martin Guerrero JD, Martinez Sober M, Benedito JRM, Lopez AJS (2009) Transfer learning. In: Handbook of research on machine learning applications and trends: algorithms, methods and techniques, Hershey, Information Science Reference
11. He K, Zhang X, Ren S, Sun J (2016) Deep residual learning for image recognition. In: IEEE conference on computer vision and pattern recognition (CVPR), Las Vegas
12. Liu Y, Tang F, Zhou D, Meng Y, Dong W (2016) Flower classification via convolutional neural network. In: IEEE international conference on functional-structural plant growth modeling, simulation, visualization and applications (FSPMA), Qingdao
13. Bosch AZ, Munoz X (2007) Image classification using random forests and ferns. In: IEEE 11th international conference on computer vision, Rio de Janeiro
14. Loussaief S, Abdelkrim A (2016) Machine learning framework for image classification. In: 7th international conference on sciences of electronics, technologies of information and telecommunications (SETIT), Hammamet

Chapter 28
Cloud-Based Adaptive Exon Prediction Using Normalized Logarithmic Algorithms

Md. Zıa Ur Rahman, **Chaluvadi Prem Vijay Krishna**, **Sala Surekha**, and **Putluri Srinivasareddy**

1 Introduction

Genomics is an immense field in which areas that code for proteins is identified using smart AEP-based system presented here. Exon areas have a role to play in the assessment of diseases and drug design. Intergenic and genic sections are included in DNA sequence [1]. The formation of primary protein segments is studied to support both exon sections tertiary also secondary structure. After determining this for overall exon segments, any irregularities are identified and rehabilitate diseases [2, 3]. Living things all are divide into eukaryotes and prokaryotes. Protein coding regions are still referred to as exons in part of eukaryotes, whereas non-protein coding sections are referred as introns. Just 3% of the eukaryotic human gene sequence has coding areas and rest remain non-coding areas. Consequently, it is an important task to detect coded regions in the DNA data [4, 5]. Therefore, in literature, several methodologies [6–10] to recognition of exon rely on various techniques of digital signal processing.

Adaptive techniques using AEP-based smart communication system in a number of iterations may process more lengthy sequences. Our current work presents new ENLMLS algorithms for adaptive exon predictor (AEP). ENLMLS with its signed variants is taken into consideration for enhancing the performance of AEP than LMS. LMS drawbacks are resolved by ENLMLS algorithm, thereby increases speed and ability of exon tracking. During the prediction of exon regions, there is a reduction in excess mean square error (EMSE) [11]. Sign-based algorithms also reduce the sign function by quantity of multiplication calculations [12]. Several errors also do not meet the monitoring requirements due to the data-independent stationary step

Md. Z. Ur Rahman · C. P. V. Krishna · S. Surekha (✉) · P. Srinivasareddy
Department of ECE, Koneru Lakshmaiah Education Foundation, Green Fields, Vaddeswaram, Guntur, AP 522002, India

C. P. V. Krishna
e-mail: 170040132@kluniversity.in

S. Kumar et al. (eds.), *Proceedings of International Conference on Communication and Computational Technologies*, Algorithms for Intelligent Systems, https://doi.org/10.1007/978-981-16-3246-4_28

size techniques [13]. Lower EMSE and larger step size are necessary for the best convergence rate. Disadvantages of LMS are overcome by use of ENLMLS-based techniques.

The error term obtained from the iterations shows instances of forbidden step size in the variable step size adaptive algorithms [14–18]. These techniques give better performance than that of the least mean squares (LMS) algorithm. We combine ENLMLS technique with three sign variants to minimize computational complexity. The new variants of implemented AEPs include error normalized least mean logarithmic squares (ENLMLS), error normalized sign regressor ENSRLMLS (ENSRLMLS), error normalized sign ENLMLS (ENSLMLS), as well as error normalized sign LMLS (ENSSLMLS) algorithms. The performance of developed AEPs is evaluated by using standard genomic data sequence which is taken from the NCBI gene database [19]. The parameters considered for the performance assessment of several proposed AEPs are convergence features, computational complexity, sensitivity (Sn), precision (Pr), and specificity (Sp). In [20–22], several techniques for the prediction of exon regions are discussed [23–25], various adaptive filtering techniques are studied. In the following sections, we discussed about adaptive filtering methods and performance efficiency of several proposed AEPs.

2   Adaptive Algorithms for Exon Prediction

Let us consider mapped digital sequence is $M(n)$, DNA data sequence is $x(n)$, gene sequence is $d(n)$ that follows TBP, the output of adaptation process is $y(n)$ and error signal is $e(n)$, feedback signal obtained in the feedback loop used for updating the filter coefficients. In LMS algorithm, filter length is considered as 'T.' Based upon step size 'P' in present weight coefficient of the adaptation algorithm, the subsequent weight coefficient is estimated. In the proposed algorithm at the present instant, the current weight coefficient is $h(n)$, and input mapped digital sequence is $M(n)$. In [12], the mathematical model of LMS algorithm is described. Figure 1 describes the proposed block diagram representation of AEP. Weight updating equation of LMS adaptation process is given by

$$h(n + 1) = h(n) + Px(n)e(n) \tag{1}$$

In applications of nano-bioinformatics, the adaptive algorithms should have least computational complexity for the prediction of exon regions. For this purpose, the clipping of gene information sequence, or a feedback signal or both is viable. The suitable techniques used for this purpose are described in [18]. For the proposed technique, three clipped variants are used to minimize the computation burdens. The three clipped variants are error clipped, data clipped, and both data and error clipped.

The expression for clipping function is given by

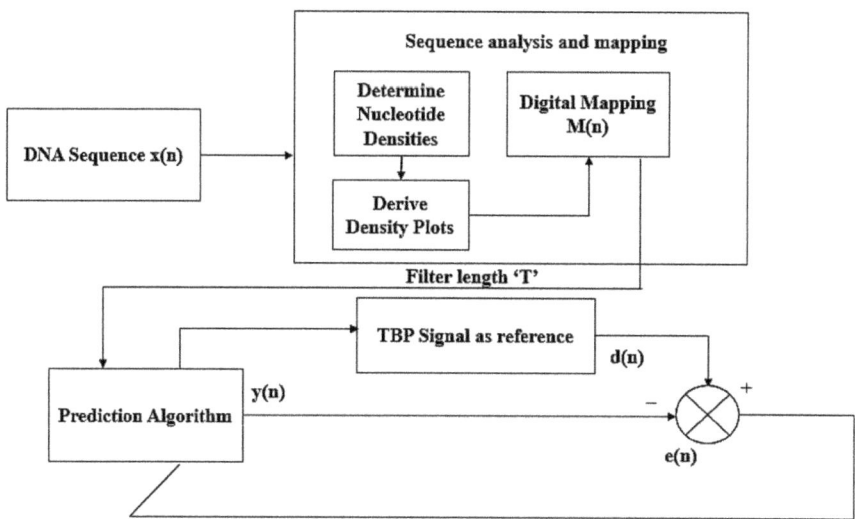

Fig. 1   Block diagram of proposed exon predictors

$$C\{x(n)\} = \left\{ \begin{array}{l} 1 : x(n) > 0 \\ 0 : x(n) = 0 \\ -1 : x(n) < 0 \end{array} \right\} \tag{2}$$

Clipping function is used to reduce the computation complexity. Generally, without clipping variants, the conventional LMS has greater computational complexity. The weight updating expression of data clipped LMS (DCLMS) is obtained from LMS weight updating expression by modifying the input data vector as $C\{x(n)\}$. Mean values of $C[x(n)]$ is replaced for $x(n)$. The clipping function C is applied to $x(n)$ on element-by-element basis.

The weight updating expression of DCLMS algorithm is represented as

$$h(n + 1) = h(n) + PC\{x(n)\}e(n) \tag{3}$$

Weight updating expression of error clipped LMS (ECLMS) is obtained from the LMS weight updating expression by modifying the error vector as $C\{e(n)\}$.

The weight updating expression of ECLMS algorithm expressed as

$$h(n + 1) = h(n) + Px(n)C\{e(n)\} \tag{4}$$

The weight updating expression of both data error clipped LMS (DECLMS) is obtained from the LMS weight updating expression by modifying the both input data vector as $C\{x(n)\}$ and error vector as $C\{e(n)\}$.

The weight updating expression of DECLMS algorithm is represented as

$$h(n + 1) = h(n) + PC\{x(n)\}C\{e(n)\} \tag{5}$$

Normalized LMLS algorithm is considered as a unique LMS algorithm application that takes into consideration signal level variation at the filter output also chooses a logarithmic normalized cost function which leads to a faster converging as well as stable adaptation algorithm. ENLMLS algorithm overwhelms LMS limitations and increases the convergence speed and exon prediction ability. Here, we have used ENLMLS and its adaptive algorithm based on SRA to enhance AEP efficiency. The ENLMLS algorithm overcomes the LMS disadvantages and increases the ability of exon identification and quicker convergence when error is high. This also reduces the surplus EMSE in the exon identification process. These ENLMLS adaptive algorithms are used for developing AEPs in order to cope with computing difficulty of an AEP in practical applications. In addition to weight update equation of ENLMS algorithm, sign function is used to reduce complexity. Three sign versions ENSRLMLS, ENSLMLS, and ENSSLMLS algorithms are derived with help of sign function to ENLMLS algorithm. Then weight updated equations for ENSRLMLS, ENSLMLS, and ENSSLMLS techniques become

$$h(n + 1) = h(n) + \frac{P'}{\varepsilon + (||e(n)||)^2} C[x(n)]e(n) \left[\frac{\alpha(e(n))^2}{1 + \alpha(e(n))^2} \right] \tag{6}$$

$$h(n + 1) = h(n) + \frac{P'}{\varepsilon + (||e(n)||)^2} x(n)C\left[e(n) \left[\frac{\alpha(e(n))^2}{1 + \alpha(e(n))^2} \right] \right] \tag{7}$$

$$h(n + 1) = h(n) + \frac{P'}{\varepsilon + (||\mathbf{x}(n)||)^2} C[x(n)]C\left[e(n) \left[\frac{\alpha(e(n))^2}{1 + \alpha(e(n))^2} \right] \right] \tag{8}$$

Finally, implemented four different AEPs based on these algorithms and compared their performance with LMS adaptive algorithm. The parameters measurement such as precision, specificity, and sensitivity proved that ENSRLMLS is nearly below the non-sign regressor technique. Hence, ENSRLMLS gives better performance than the signed variant; in the computation complexity, scenario also exhibits better performance among several proposed techniques selected for the real-time application.

3   Computational Complexities and Convergence Issues

The number of multiplications required for the computations is taken as the one of the performance metrics to compare the complexity of the implemented techniques. Our aim is not only on the particular computational complexity of the computations, but also on the process of evaluation of various ENLMLS-based adaptive techniques. By using the sign function in the weight updating expression of the techniques, the

Table 1 Computations required for LMS and various ENLMSS-based AEPs

S. No.	Algorithm	Multiplications	Additions
1	LMS	$T + 1$	$T + 1$
2	ENLMLS	$2T + 7$	$2T + 2$
3	ENSRLMLS	$2T$	$2T + 2$
4	ENSLMLS	$2T + 5$	$2T + 2$
5	ENSSLMLS	$T + 2$	$T + 2$

computations do not require complex multiplications in the exon prediction applications. In the LMS case for the computation of weight updating expression, it requires $T + 1$ multiplication and 1 addition. However, in the ENSRLMLS technique, the weight updating expression requires $2T$ multiplications only. For other two, sign variants of ENLMLS technique require $2T + 1$ multiplication for the computation of weight updating expression. Hence, the ENSRLMLS technique requires a smaller number of multiplications compared to other variants that gives ENSRLMLS-based AEPs exhibits less computation complexity compared to other variant AEPs. Table 1 describes the computational complexities of LMS and three sign variants of ENLMLS algorithm.

Figure 2 represents the convergence performance of the proposed ENLMLS and its three sign variants signed variants. From the figure, it is observed that ENLMLS-based AEP converges faster than the LMS-based AEP. Therefore, the ENSRLMLS adaptive algorithm is considered better, based on computing difficulty as well as convergence efficiency in contrast to LMS and its other signed algorithms, among the algorithms considered for AEP implementation. It was obvious that ENSRLMLS

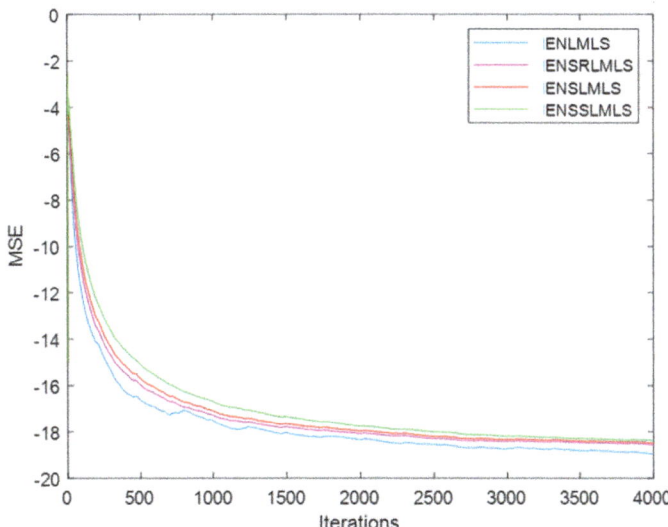

Fig. 2 Convergence curve of ENLMLS algorithm and its sign variants

converges quicker compared to ENSLMLS and ENSSLMLS-based AEPs from convergence features.

4 Results and Discussion

In this section, we discuss the comparison of performance analysis of several AEPs. In Fig. 1, AEP system block diagram representation is shown. Several AEPs are developed based on ENMLS technique and its three sign models. We also implemented an AEP-based LMS for comparison purpose. 10 genomic data sequences are taken from NCBI data for analysis purpose [19]. The performance of the several implemented AEPs is evaluated by taking the parameters such as precision (Pr), sensitivity (Sn), and specificity (Sp). In [13], the mathematical expressions and description about these parameters are mentioned. The results of these parameters for various techniques are represented in Table 3. As part of the determination of exon segments by using DSP methods, there are few measures based on changes in the threshold level in the output spectrum used for comparison. Nucleotides amount situated like introns in exon locating phase is considered as true negative (TN), whereas exon areas exactly identified is stated for instance true positive (TP). Also, the entire amount of exon areas positioned as intron areas is indicated to be false negative (FN), comparison with the exact amount of anticipated introns areas of exons to be false positive (FP). 10 gene datasets of NCBI are taken to evaluate the efficiency of various techniques. The accession for these sequences remains X59065. 1, E15270. 1, U01317. 1, X77471. 1, AF009962, X92412. 1, AB035346. 2, AJ223321. 1, AJ225085. 1, and X51502. 1, respectively, as shown in Table 2.

Specificity (Sp) is the number of exons exactly exist in the part of exon, whereas sensitivity (Sn) is the number of exons that remains effectively forecasted. The results related to exon prediction of AF009962 sequence using various ENLMLS techniques

Table 2 Gene datasets from NCBI gene databank

Seq. No.	Accession No.	Sequence definition
1	E15270.1	Human gene for osteoclastogenesis inhibitory factor (OCIF) gene
2	X77471.1	Homo sapiens human tyrosine aminotransferase (TAT) gene
3	AB035346.2	Homo sapiens T-cell leukemia/lymphoma 6 (TCL6) gene
4	AJ225085.1	Homo sapiens Fanconi anemia group A (FAA) gene
5	AF009962	Homo sapiens CC-chemokine receptor (CCR-5) gene
6	X59065.1	Homo sapiens human acidic fibroblast growth factor (FGF) gene
7	AJ223321.1	Homo sapiens transcriptional repressor (RP58) gene
8	X92412.1	Homo sapiens titin (TTN) gene
9	U01317.1	Human beta globin sequence on chromosome 11
10	X51502.1	Homo sapiens gene for prolactin-inducible protein (GPIPI)

Table 3 Performance measures of various ENLMLS and its signed-based AEPs with respect to Sn, Sp, and Pr calculations

Algorithm	Metric	Gene sequence serial number									
		1	2	3	4	5	6	7	8	9	10
LMS	Sn	0.6286	0.6384	0.6457	0.6273	0.6481	0.6162	0.6193	0.6241	0.6268	0.6202
	Sp	0.6435	0.6628	0.6587	0.6405	0.6518	0.6324	0.6529	0.6289	0.6452	0.5965
	Pr	0.5922	0.5894	0.5934	0.5858	0.5904	0.5786	0.5896	0.5856	0.5814	0.5761
ENLMLS	Sn	0.8016	0.7782	0.8032	0.8024	0.8006	0.8014	0.7996	0.8088	0.7985	0.3062
	Sp	0.8006	0.7893	0.8012	0.7994	0.8023	0.7963	0.8084	0.7988	0.8036	0.7981
	Pr	0.8028	0.8003	0.7996	0.8013	0.8008	0.8022	0.7985	0.8035	0.7983	0.3014
ENSRLMLS	Sn	0.7899	0.7563	0.7898	0.7873	0.7786	0.7864	0.7782	0.7893	0.7785	0.7877
	Sp	0.7787	0.7674	0.7783	0.7783	0.7894	0.7782	0.7895	0.7789	0.7394	0.7754
	Pr	0.7784	0.7792	0.7871	0.7894	0.7798	0.7892	0.7782	0.7854	0.7782	0.7857
ENSLMLS	Sn	0.7780	0.7456	0.7679	0.7653	0.7668	0.7673	0.7568	0.7659	0.7572	0.7644
	Sp	0.7673	0.7457	0.7678	0.7664	0.7645	0.7561	0.7693	0.7565	0.7578	0.7582
	Pr	0.7568	0.7671	0.7652	0.7673	0.7679	0.7649	0.7587	0.7692	0.7633	0.7676
ENSSLMLS	Sn	0.7561	0.7129	0.7458	0.7422	0.7455	0.7478	0.7349	0.7468	0.7349	0.7454
	Sp	0.7348	0.7235	0.7896	0.7349	0.7468	0.7452	0.7471	0.7343	0.7387	0.7336
	Pr	0.7459	0.7342	0.7483	0.7474	0.7457	0.7476	0.7356	0.7484	0.7385	0.7466

is depicted in Fig. 5. The threshold levels are taken from 0.4 to 0.9 at regular intervals of 0.05. The efficiency of metrics Pr, Sn, and Sp is evaluated by using these values. The exon prediction is perfect at value of threshold equal to 0.8. Consequently, the measures for performance at threshold value 0.8 are depicted in Table 3. Regions having more A + T percentage of nucleotides of DNA sequence usually show intergenic sequence components, while low A + T and greater G + C nucleotides show potential genes. Mostly, high CG dinucleotide content is often found ahead for a gene. Functions of statistics for a gene sequence remain beneficial for determining whether the input gene sequence has protein coding segments.

The processing steps for AEP steps are described below

(a) Gene data sequences of the input are analyzed which are taken from NCBI database with help of density plots shown in Fig. 5 for evaluating existence of gene locations dependent upon nucleotide density base pairs for G + C also A + T dimers. Following the assessment, this sequence is then converted into digital notation after analysis using the digital mapping technique, while input of AEP remains to be binary information from Fig. 1.

(b) This ensuing sequence is then given as an input to developed AEP following assessment. For proposed ENLMLS dependent AEPs, as a reference signal TBP obedient biological sequence is given.

(c) For updating filter coefficients, derived $e(n)$ feedback signal from Fig. 1 has been used.

(d) Once this signal becomes minimal, the genes from DNA sequences are accurately located with plot of PSD.

(e) Plots of desired exon areas are shown in PSD. Moreover Sp, Pr, and Sn are calculated and compared.

The performance metrics values of AF009962 sequence for the sign variants of ENLMLS technique are measured by using MATLAB and are given in Table 3. From this table, it is observed that performance of AEP based on ENLMLS technique is just inferior than the performance of AEP implemented with ENSRLMLS technique-based AEP, which has a smaller number of iterations because of the low computation complexity and greater exon prediction capability by exactly locating at 3934–4581 samples with high resolution, and at that samples, a spiky peak is observed in PSD plot. Of all these algorithms, ENSRLMLS-based AEP is effective in terms of accurate exon prediction when compared to LMS, ENLMLS and its other signed variants with specificity Sp, 0.7890 (78.90%), sensitivity Sn 0.7789 (77.89%), also precision, Pr 0.7806 (78.06%), respectively. At 0.8 threshold value, the exon prediction appears to be better for ENSRLMLS-based AEP. The PSD plots of ENLMLS algorithm and its three sign models are depicted in Fig. 3b, c and d, respectively. Finally, all proposed ENLMLS-based AEPs are more effective to discover exon areas in genomic sequences compared with the prevailing LMS technique.

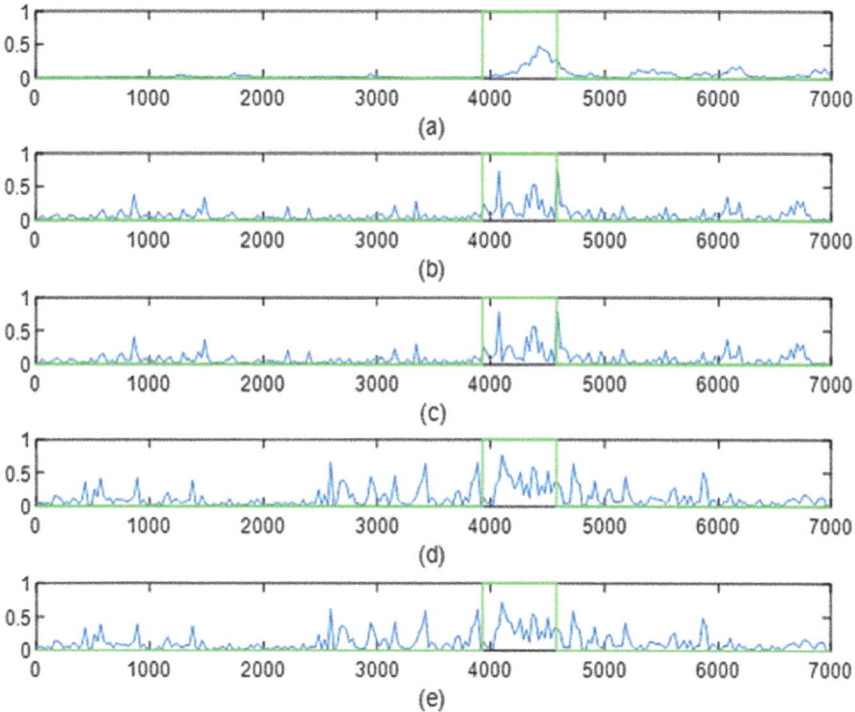

Fig. 3 Exon prediction in sample AF009962 **a** AEP of LMS, **b** AEP of ENLMLS, **c** AEP of ENSRLMLS, **d** AEP of ENSLMLS, **e** AEP of ENSSLMLS (*Relative Base Location on x-axis and Power Spectrum on y-axis*)

5 Conclusion

The detection concepts of prediction of exon in the gene sequence are described in this work. Here we presented a new approach for adaptive exon prediction using smart communication-based system. The AEPs developed with ENLMLS adaptation techniques are used in the processing of several DNA sequences effectively. The parameter measurements for exon prediction are mentioned in Table 3, and power spectral density plots of various developed techniques are shown in Fig. 3. The developed AEPs exactly predicted exon regions at 3934–4581 samples with more resolution in the described plot. The AEPs implemented with ENSRLMLS technique exhibit best performance in terms of computation complexity, and performance evaluation parameters for the AF009962 gene sequence at threshold value of 0.8 are nearly below the values of AEP implemented with ENLMLS technique. Though, it is preferably chosen because it exhibits less computation complexity also exhibits fast convergence speed in the prediction process of exon regions. The AEPs implemented with ENSRLMLS technique are advised for exon prediction and also used in processors, which are used for real-time nano-bioinformatics applications.

References

1. Ning LW, Lin Ding H, Huang J, Rao N, Guo FB (2014) Predicting bacterial essential genes using only sequence composition information. Genet Mol Res 13:4564–4572
2. Li M, Li Q, Ganegoda GU, Wang JX, Wu FX, Pan Y (2014) Prioritization of orphan disease-causing genes using topological feature and go similarity between proteins in interaction networks. Sci China Life Sci 57:1064–1071
3. Inbamalar TM, Sivakumar R (2013) Study of DNA sequence analysis using DSP techniques. J Autom Control Eng 1:336–342
4. Maji S, Garg D (2013) Progress in gene prediction: principles and challenges. Curr Bioinform 8:226–243
5. Srinivasareddy P, Ur Rahman MZ (2016) New adaptive exon predictors for identifying protein coding regions in DNA sequence. ARPN J Theor Appl Sci 11:13540–13549
6. Saberkari H, Shamsi M, Hamed H, Sedaaghi MH (2013) A novel fast algorithm for exon prediction in eukaryotes genes using linear predictive coding model and Goertzel algorithm based on the Z-curve. Int J Comput Appl 67:25–38
7. Wazim Ismail M, Ye Y, Haixu T (2014) Gene finding in metatranscriptomic sequences. BMC Bioinform 15:01–08
8. Ghorbani M, Hamed K (2015) Progress in gene prediction: principles and challenges. Bioinformatics approaches for gene finding, vol 4, pp 12–15
9. Devendra Kumar S, Rajiv S, Narayan Sharma S (2013) An adaptive window length strategy for eukaryotic CDS prediction. Trans Comput Biol Bioinform 10:1241–1252
10. Azuma Y, Onami S (2014) Automatic cell identification in the unique system of invariant embryogenesis in *Caenorhabditis elegans*. Biomed Eng Lett 4:328–337
11. Liu G, Luan Y (2014) Identification of protein coding regions in the eukaryotic DNA sequences based on Marple algorithm and wavelet packets transform. Abstr Appl Anal 2014:1–14
12. Simon Haykin O (2014) Adaptive filter theory, 5th edn. Pearson Education Ltd., Harlow, UK, pp 320–380
13. Saberkari H, Shamsi M, Hamed H, Sedaaghi MH (2013) A fast algorithm for exonic regions prediction in DNA sequences. J Med Signals Sens 3:139–149
14. Nagesh M, Prasad SVAV, Rahman MZ (2016) Efficient cardiac signal enhancement techniques based on variable step size and data normalized hybrid signed adaptive algorithms. Int Rev Comput Soft 11:1–13
15. Sayin MO, Vanli ND, Kozat SS (2014) A novel family of adaptive filtering algorithms based on the logarithmic cost. IEEE Trans Signal Process 62(17):4411–4424
16. Gogineni VC, Mula S (2017) A family of constrained adaptive filtering algorithms based on logarithmic cost. IEEE Trans Signal Process 1–14
17. Mula S, Gogineni VC, Dhar AS (2017) Algorithm and architecture design of adaptive filters with error non-linearities. IEEE Trans VLSI Syst 25(9):2588–2601
18. Paula Diniz SR (2013) Adaptive filtering, algorithms and practical implementation, 4th edn. Springer, New York
19. National Center for Biotechnology Information [Online]. Available at: www.ncbi.nlm.nih.gov/. Accessed 25 Jan 2019
20. Srinivasareddy P, Ur Rahman MZ, Chandra Sekhar A, Nagireddy P (2019) New exon prediction techniques using adaptive signal processing algorithms for genomic analysis. IEEE Access 7:80800–80812
21. Putluri SR, Ur Rahman MZ (2018) Identification of protein coding region in DNA sequence using novel adaptive exon predictor. J Sci Ind Res 77:1–5
22. Srinivasareddy P, Ur Rahman MZ, Fathima SY (2018) Cloud based adaptive exon prediction for DNA analysis. IET Healthc Technol 5(1):1–6
23. Sulthana A, Ur Rahman MZ (2018) Efficient adaptive noise cancellation techniques in an IOT Enabled Telecardiology System. Int J Eng Technol (UAE) 7(2):74–78

24. Gayathri NB, Thumbur G, Rajesh Kumar P, Rahman MZU, Reddy PV, Lay-Ekuakille A (2019) Efficient and secure pairing-free certificateless aggregate signature scheme for healthcare wireless medical sensor networks. IEEE Internet Things J 6(5):9064–9075
25. Salman MN, Trinatha RP, Ur Rahman MZ (2017) Adaptive noise cancellers for cardiac signal enhancement for IOT based health care systems. J Theor Appl Inf Technol 95(10):2206–2213

Chapter 29
Energy Detection in Bio-telemetry Networks Using Block Based Adaptive Algorithm for Patient Care Systems

Sala Surekha⊙ and **Md. Zia Ur Rahman**⊙

1 Introduction

Because of advances in wireless technologies, explored various technological advancements in the medical area is increases and this type of research called as wireless body area networks or medical wireless body area networks. Advantages of medical wireless body area networks includes patient's recovery by continuous observing, expenditure of health care decreases, monitoring of patients in remote areas, patient's mobility is increased, spreading disease was reduced because of placing sensors on different parts of body [1–4]. Medical devices are very delicate to electromagnetic interference due to wireless transmissions. Interferences may result in malfunctioning of medical devices which leads to harmful for patient's healthcare condition. To avoid this interference, we are using wireless communication particularly in clinics or hospitals. So implemented recent cognitive radio technique, it is based on software defined radio, it was used to improve wireless communication efficiency by reducing interference and by increasing radio spectrum utilization. Basically, cognitive radio receiver detects, then acquired operating environment status so that we conclude and adjusted parameters accordingly of wireless transmission. This cognitive radio concept is used in wireless applications with strict limits on EMI [5, 6] related to various health devices and quality of service requirements for several healthcare applications considering into account. By using wireless body area network sensors, we can know information of patient like blood pressure, temperature and heart rate.

By using the idea of cognitive radio, in this paper it mainly deals with problems of wireless communication system for mobile applications. Firstly, a brief summary

S. Surekha (✉) · Md. Z. Ur Rahman
Department of ECE, Koneru Lakshmaiah Education Foundation, K L University, Green Fields, Vaddeswaram, Guntur 522002, India

© The Author(s), under exclusive license to Springer Nature Singapore Pte Ltd. 2021 347
S. Kumar et al. (eds.), *Proceedings of International Conference on Communication and Computational Technologies*, Algorithms for Intelligent Systems,
https://doi.org/10.1007/978-981-16-3246-4_29

of needs and challenges for wireless communications in mobile applications [7–9] is discussed. To avoid harmful EMI causes to healthcare devices, designed a cognitive radio system which are used to consider to protect users and also to get best quality of service (QoS) requirements for different healthcare applications like to know hospital and telemedicine information. Telemedicine applications provide medical services for faraway patients, while hospital information systems give information about storage, recovery and processing of medical records. Telemedicine applications have higher priority compared to hospital information applications, so telemedicine applications are taken as primary users and hospital information applications are considered as secondary users. To know information about interference occurred due to medical devices, channel access mechanism for cognitive radio systems is designed, so that it determines transmission parameters of wireless medical devices safely and it satisfy with EMI limitations. Basically, all healthcare devices in medical environment are sensitive to interference, so we have to satisfy electromagnetic compatibility of all medical applications related to wireless communications [10, 11]. Let us consider one example, to avoid harmful interference we are limiting transmit power of healthcare devices. Delay and loss are two main communication QoS performance measurements of health care applications. For example, heart attack patient has to monitor continuously, if there is delay in data transmission, then patients did not receive treatment timely and may lead to serious condition of patient. So, we are using various wireless communication technologies like IEEE 802.11 for health care applications, and IEEE 802.15.4a is used for ultra-wide band based personal area networks for mobile applications. Wireless communications improve the mobility of health care applications. Bio signal data is continuously monitored when patient was away from hospital. For achieving better performance in wireless communications, mobility management is necessary. In horizontal unlicensed band spectrum sharing [12, 13], cognitive radio techniques are used to provide efficient coexistence of different users to improve utilization of spectrum radio. For avoiding interferences occurred with medical devices, we have to choose transmission parameters very carefully. Generally, for healthcare monitoring, considered medical telemetry applications and gathering information of a patient from hospital applications. Here, medical telemetry applications are considered as primary user, and gathering data of patient form hospitals are considered as secondary user. Two radio interfaces are introduced in cognitive radio controller, i.e., control channel and data channel, in these primary users have more priority for accessing data channel and also secondary user is uses this data channel when if primary user was not occupied. In this paper, spectrum sensing using energy detection for providing spectrum utilization efficiently by licensed users and unlicensed users and block based normalized algorithm for spectrum sensing to avoid interferences was explained in Sect. 2 and results are explained in Sect. 3, respectively.

2 Spectrum Sensing

Wireless communication technologies are used in healthcare applications to provide patient information properly. Cognitive radio method is used because it is efficiently used by co-existed users so that to improve frequency spectrum utilization. In this frequency spectrum, concept is mostly used because it searches for unused frequency bands continuously then shares to vacant users without causing interference to primary users. Vacancy bands are used by both primary users and secondary users. Energy detection using spectrum sensing is commonly used method because it does not need any licensed user data. Presence or absent of users are detected by this energy detection method. For spectrum sensing problems, hypothesis test formulation is considered as shown in Eq. (1). Secondary user signals are sensed, and it is indicated by $s(t)$. Hypothesis testing is used for identifying channel gain and it is also clearly defined hypothesis of not receiving any signal from licensed users for target of frequency band, $g(t)$ is additive noise. H_0 and H_1 are the basic two hypotheses considered to identify primary user presence.

H_0: Primary user is absent.

H_1: Primary user is present at known value of 't'.

Cognitive radio user detects primary user at bandwidth 'w' and time 't', then H_1 was observed at a spectrum "s" and compares with threshold value.

$$H_0 : S(t) = g(t)$$

$$H_1 : S(t) = p(t) + g(t) \tag{1}$$

where $t = 0, 1, 2, ..., T$,

$S(t)$: received signal of secondary user (SU),

$P(t)$: Primary users (PU) signal,

$g(t)$: AWGN channel noise variance.

Performance of spectrum sensing was measured with below parameters,

Probability of detection (P_{ode}): It indicates primary user is available or not, and it value at its maximum, it also helps to avoid primary user from interferences.

Probability of false alarm (P_{ofa}): It actually shows as primary user is present but in reality, primary user is absent. To increase spectrum utilization factor value, false alarm probability value should be minimum.

Probability of detection specifies that a channel is occupied with H_1, then it provides interference level provided to primary user. False alarm probability occurs when it shows H_1 actually right decision of hypotheses is H_0. When false alarm probability occurs, secondary user should not provide free spectrum because it can misuse free channel. So P_{ofa} should be kept as small for avoiding underutilization of transmission opportunities. False alarm probability influences the spectrum sensing performance.

In spectrum sensing, energy detection is preferred method because of its low computational complexity and their cost of implementation was low. By using FFT, spectrum sensing detects licensed user signals then it will change from time variance to frequency domain. Then, power spectrum density of every energy signal is calculated and detection parameters are calculated by using fast fourier transform (FFT) as a series by using Eq. (2) as

$$D(k) = \frac{1}{T} \sum_{t=1}^{T} [s(t)p(t)] > \text{ or } < \gamma \tag{2}$$

where $D(k)$ is the probability of detection,
T is the number of samples,
γ is the detection threshold value.
Probability of detection based on threshold γ was calculated as

$$P_{\text{ode}} = P_r\left(D(k) > \frac{\gamma}{H_1}\right) \tag{3}$$

Probability of false alarm was calculated as

$$P_{\text{ofa}} = P_r\left(D(k) > \frac{\gamma}{H_0}\right) \tag{4}$$

Missed detection probability was calculated as

$$P_{\text{md}} = 1 - P_{\text{ode}} = P_r\left(1 - D(k) > \frac{\gamma}{H_1}\right) \tag{5}$$

Signal power amount and their corresponding amount of variations in noise conflicts taken as signal to noise ratio band. Energy detection for proposed algorithm worked on basis of SNR along with channel estimation. Signal to noise ratio was based on Gaussian distribution for AWGN channel, it was expressed as

$$\text{SNR} = \frac{\alpha}{\sigma_t^2} \tag{6}$$

where σ is standard power for autocorrelation.
Depends on low SNR values, threshold value for primary user is selected by the following probability detection and probability of false alarm Eqs. (7) and (8) as follows

$$P_{\text{ofa}} = \mathbb{Q}\left(\frac{\gamma - \omega_0}{\sigma_0}\right) \tag{7}$$

$$P_{\text{ode}} = \mathbb{Q}\left(\frac{\gamma - \omega_1}{\sigma_1}\right) \tag{8}$$

Here $\mathbb{Q}$ is complementary error function.

$$\mathbb{Q}(k) = \frac{1}{2\gamma} \int\limits_{-\infty}^{\infty} S_{tt}(w)d_w \tag{9}$$

where $F - 1(S_{tt}) = $ autocorrelation function $= R_{tt}(w)$

$$\gamma = \sigma_j \mathbb{Q}(P_{\text{ofa}}) + w_{j+1} \tag{10}$$

So that obtained threshold value requires improvement for decreasing SNR value by interpreting probabilities depending on availability of number of users and environmental conditions.

Now let us consider length of adaptive filter as 'N', for $i(t)$ input vector, $g(t)$ is generated system output signal, then the equation becomes [14, 15].

$$g(t) = i(t)^T b(t) = b(t)^T i(t) \tag{11}$$

$[i(t), i(t-1), i(t-2), \ldots, i(t-N+1)]$ are taps inputs forms the elements by N by 1 tap input vector $i(t)$, tap weights are also represented as $b(t) = [b_0, b_1, \ldots, b_{N-1}]^T$ by the L by 1 element formed the weight vector b(t), then weight update equation of LMS is given as

$$b(t + 1) = b(t) + \beta i(t)e(t) \tag{12}$$

where β is step size, for stationary LMS algorithm converges mean in between $0 < \beta < \frac{2}{\delta_{\max}}$.

For normalized LMS weight update equation is given as

$$b(t + 1) = b(t) + \frac{\beta}{\varepsilon + ||i(t)||^2} i(t)e(t) \tag{13}$$

Here β is normalized step size with $0 < \beta < 2$.

Also, various sign based NLMS adaptive filter structures are represented for low-computational complexity, good filtering capability and fast convergence rate. Sign regressor versions of NLMS algorithm are named as normalized sign regressor LMS (NSRLMS), normalized sign LMS (NSLMS), normalized sign sign LMS (NSSLMS) algorithms and they represented by Eqs. (14)–(16)

$$b(t+1) = b(t) + \frac{\beta}{\varepsilon + ||i(t)||^2} \text{sign}\{i(t)\}e(t) \tag{14}$$

$$b(t+1) = b(t) + \frac{\beta}{\varepsilon + ||i(t)||^2} i(t)\text{sign}\{e(t)\} \tag{15}$$

$$b(t+1) = b(t) + \frac{\beta}{\varepsilon + ||i(t)||^2} \text{sign}\{i(t)\}\text{sign}\{e(t)\} \tag{16}$$

In the contest of adaptive spectrum sensing, LMS algorithm is widely used. But it has limitations of low-convergence rate and signal to noise ratio is low. Hence to reduce computational complexity block processing of normalized algorithms are adopted. In block-based algorithm approach, input data is partitioned into blocks, then the maximum magnitude in each block used for computing variable step size parameters and its flow chart is shown in Fig. 1. By taking consideration of NLMS algorithm, and weight update equation of block based normalized LMS (BBNLMS) algorithm for $\varepsilon = 0$ and $i_{Nj} \neq 0$ is represented as

$$b(t+1) = b(t) + \frac{\beta}{i_{Nj}^2} i(t)e(t) \tag{17}$$

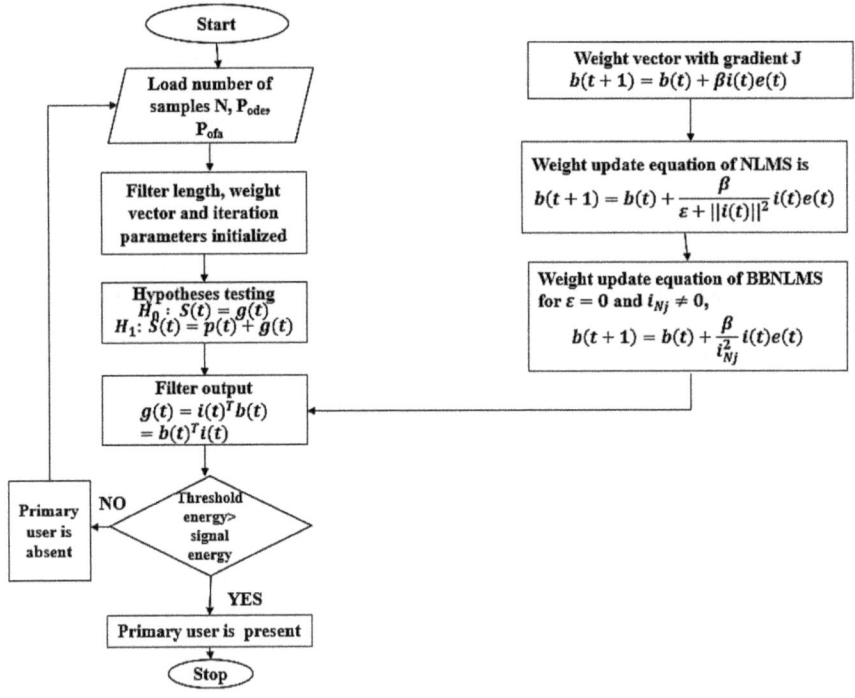

Fig. 1 Flow chart of spectrum sensing using block based normalized LMS

Sign regressor versions of BBNLMS algorithm are named as block based normalized sign regressor LMS (BBNSRLMS), block based normalized sign LMS (BBNSLMS), block based normalized sign sign LMS (BBNSSLMS) algorithms and they represented as

$$b(t + 1) = b(t) + \frac{\beta}{i_{Nj}^2} \text{sign}\{i(t)e(t)\} \tag{18}$$

$$b(t + 1) = b(t) + \frac{\beta}{i_{Nj}^2} i(t) \text{sign}\{e(t)\} \tag{19}$$

and

$$b(t + 1) = b(t) + \frac{\beta}{i_{Nj}^2} \text{sign}\{i(t)\} \text{sign}\{e(t)\} \tag{20}$$

where, $i_{Nj} = \max\{|i_h|, h \in Z_j'\}$, $Z_i' = \{jN, jN + 1, \ldots, jN + N - 1\}$, $j \in Z$, and for $i_{Nj} = 0$ and $\varepsilon = 0$ then the weight update equation becomes $b(t + 1) = b(t)$.

3 Results and Discussion

Energy detection capability for secondary users in spectrum sensing are evaluated in this section. All simulation results are done using MATLAB for proposed mathematical modeling. For validating the performance of proposed block based normalized least mean square (BBNLMS) algorithm for spectrum sensing, relation between number of samples and signal to noise ratio is studied. In noise uncertainty cases, assumed noise has AWGN with zero mean and variance σ_t^2, with range of noise uncertainty factor range $0 < \alpha < 1.05$, SNR in between range of -15 and 0 dB for 1500 samples with probability of false alarm range $0 < P_{ofa} < 0.1$, then varied number of samples. From theoretical studies, as noise uncertainty is increases, detection performance duration or number of samples increases, whereas with energy detection noise uncertainty did not change well. It means that even with increase in number of samples also, noise uncertainty problem was not solved. Then, Monte Carlo method is used for simulations it is stochastic method. Spectrum sensing performance is identified by depicting receiver operating characteristics. Probability of detection versus false alarm probability parameter characteristics is taken into consideration. While plotting ROC curves, one parameter is varying another parameter is fixed, then it will be studied various parameters of interest. In single user non-fading channel detection, SNR effect on detection performance using energy detector is operated. Probability of false alarm set as 0.01, for -10 dB number of Monte Carlo points are considered as $N = 1000$, performance is checked for more than one SNR value. Increased probability of false alarm effect on detection performance is identified.

For probability of false alarm, with varying number of samples, detection probability does not change so much. Primary user signal is transmitted based on binary phase shift keying, it has cyclic frequencies with carrier frequency and bit duration. For evaluating, the performance of proposed block based NLMS algorithm, primary user signal is transmitted in the sample range 0–4000. Then, primary user signal is turned on particular intervals and it is turned-off in another time interval. Detecting the presence or absence of primary user signal done using energy detection and its flowchart is shown in Fig. 1. Throughout simulations, step size is fixed as 0.1 and five taps are used for block based NLMS algorithm, $g(t)$ is an additive white Gaussian noise with zero mean, SNR value is set as -10 dB. Two cyclic frequencies are used in BPSK signal, for estimated weights 1000 iterations are used in proposed block based NLMS algorithm for spectrum sensing using energy detection. Probability of detection is compared with proposed BBNLMS algorithm for spectrum sensing in three different cases for five different filter taps. Spectrum sensing is basically a combination of energy detection with stochastic resonance and then adaptive based spectrum sensing using energy detection is presented in this paper. Performance is at high SNR values for three different nodes with basic spectrum sensing method. By the proposed block based NLMS algorithm outperforms the other algorithm in low SNR region also. Performance of proposed algorithm is more noticed when in adaptive filer number of taps are increased in this spectrum sensing, then it decreases error in the estimated weights of block based NLMS algorithm.

Performance of proposed spectrum sensing using block based NLMS algorithm compared with probability of detection and it is shown in Fig. 2. For enhancing primary user signals using energy detection in adaptive based spectrum sensing spatial and coherence time signals are considered. For each time, threshold level is

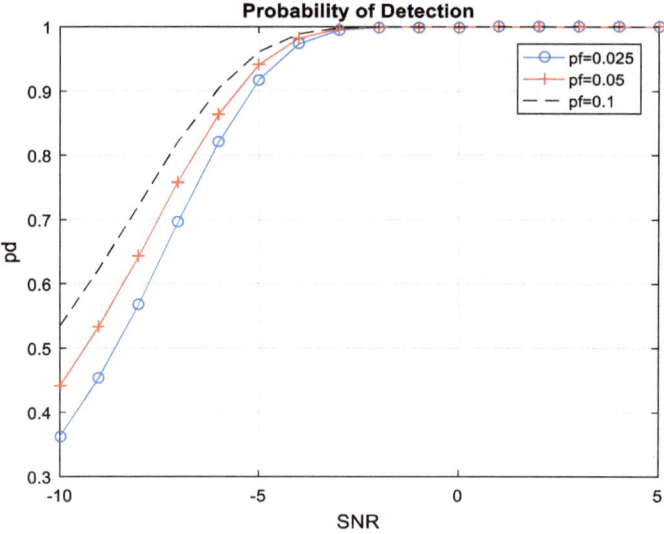

Fig. 2 Probability of detection for different P_{ofa} values

compared with norms, where as in energy detector threshold level is compared with $D(k) = \frac{1}{T} \sum_{t=1}^{T} [s(t)p(t)]$, where T is the number of samples. Here, we assumed primary user signal as BPSK signal with a carrier frequency, FIR filters are also used with 11 taps with frequency shifts. Relation between threshold value and detection gives as a constant false alarm rate for value of 0.1. Proposed spectrum sensing using block based NLMS performs well even for small number of samples. In practical communication systems, OFDM signals are used, investigated detection performance using OFDM signal. Simulations are done with parameters like varying SNR value, number of samples and threshold. For spectrum sensing probability of detection is considered as function of SNR. If SNR value increases, probability of detection also increases. Values of SNR with greater than 0 dB has probability of detection with greater than 90%. For high SNR values in which the signal with higher noises achieves 100% detection, for number of samples were low also get high probability of detection. But for energy detection, false alarm is very high then the detector does not distinguish noise and signal clearly.

Autocorrelation based spectrum sensing has lower probability detection when compare to other spectrum sensing methods, but if there is lack of false alarm it does not allow performance of spectrum sensing. For obtaining better detection, each spectrum sensing method have a particular behavior for certain environmental conditions. Spectrum sensing using energy detection gives good better performance under high SNR values, it works on high-powered signals, large number of samples and high threshold values. However, this spectrum sensing method is not reliable because of distinguish problems between noise and signals. So, to avoid this noisy signal problems adaptive filter-based algorithm is proposed for energy detection. After initialization of samples for proposed block based NLMS algorithm for spectrum sensing, for first half samples it converges to hypotheses test 1 and for next half samples converged for hypotheses test 2, hence identified the presence of primary user or secondary user in a cognitive radio network. Computational complexity of proposed BBNLMS algorithm is low when compared to LMS and NLMS and its convergence curve is shown in Fig. 3. For number of iterations and mean square error, convergence curve is simulated. This cognitive radio network concept is used in medical telemetry system who are far away from hospital. In remote health care areas, this cognitive radio concept is widely used by allocating spectrum frequencies to users with help of sensors, then acquired patient data.

4 Conclusion

In this paper, block based normalized LMS (BBNLMS) algorithm for spectrum sensing is proposed. Mathematical modeling for spectrum sensing using energy detection for noise uncertainty is studied. Performance of noise uncertainty is taken into consideration and their effect on parameters like number of samples, probability detection and false alarm probability is studied. By using this parameters, energy detector performance in detection of unused spectrums is evaluated. With

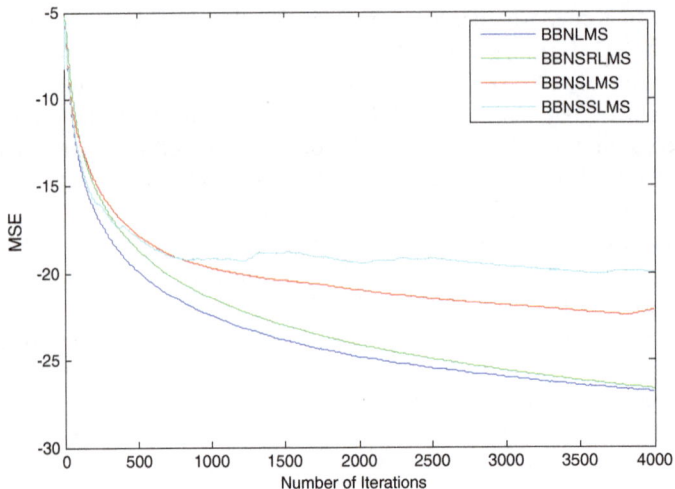

Fig. 3 Convergence behavior for block based normalized algorithms in spectrum sensing

this proposed method of BBNLMS in spectrum sensing it has low-computational complexity because of sign function is used in normalized weight update equation and also it uses very a smaller number of samples for sensing when compared to spectrum sensing alone. Spectrum sensing using energy detection is preferred when compared to other sensing methods because it does not require any prior information about primary user signal but is does not distinguish signal and noise. By the proposed algorithm noise uncertainty problems are decreased and performance is good at low SNR values. By using threshold values, it gives better performance when compared to static threshold value. Cognitive radio concept used in medical telemetry for providing treatment to patients on time by allocating spectrums to users.

References

1. Sodagari S, Bozorgchami B, Aghvami H (2018) Technologies and challenges for cognitive radio enabled medical wireless body area networks. IEEE Access 6:29567–29586
2. Phunchongharn P, Hossain E, Niyato D, Camorlinga S (2010) A cognitive radio system for e-health applications in a hospital environment. IEEE Wirel Commun 17(1):20–28
3. Doost-Mohammady R, Chowdhury KR (2012) Transforming healthcare and medical telemetry through cognitive radio networks. IEEE Wirel Commun 19:67–73
4. Chavez-Santiago R et al (2012) Cognitive radio for medical body area networks using ultra wideband. IEEE Wirel Commun 19(4):74–81
5. Ur Rahman MZ, Akanksha S, Krishna Kalyan RP, Nayeem S (2020) Energy detection in medical telemetry systems using logarithmic adaptive algorithm. Int J Eng Trends Technol 68(9):1–8
6. Patel A, Ali Khan MZ, Merchant SN, Hanzo L (2017) The achievable rate of interweave cognitive radio in the face of sensing errors. IEEE J Mag 5:8579–8605

7. Bhatia M, Kumar K (2018) Network selection in cognitive radio enabled Wireless Body Area Networks. Digit Commun Netw
8. Ur Rahman MZ, Vishnu Vardhan B, Jenith L, Rakesh Reddy V (2020) Spectrum sensing using NMLMF algorithm in cognitive radio networks for health care monitoring applications. Int J Adv Trends Comput Sci Eng 9(5):1–7
9. Ali A, Hamouda W (2017) Advances on spectrum sensing for cognitive radio networks: theory and applications. IEEE Commun Surv Tutor 19(2):1277–1304
10. Sulthana A, Rahman MZU, Mirza SS (2018) An efficient Kalman noise canceller for cardiac signal analysis in modern telecardiology systems. IEEE Access 6:34616–34630
11. Surekha S, Ur Rahman MZ (2020) Artifact elimination in thoracic electrical bioimpedance signals using new normalized adaptive filters. Int J Emerg Trends Eng Res 8(9):1–8
12. Yasmin Fathima S, Ur Rahman MZ, Krishna KM, Bhanu S, Shahsavar MS (2017) Side lobe suppression in NC-OFDM systems using variable cancellation basis function. IEEE Access 5:9415–9421
13. Surekha S, Ur Rahman MZ, Lay-Ekuakille A, Pietrosanto A, Ugwiri MA (2020) Energy detection for spectrum sensing in medical telemetry networks using modified NLMS algorithm. In: 2020 IEEE international instrumentation and measurement technology conference (I2MTC), Dubrovnik, Croatia, 2020, pp 1–5
14. Boroujeny BF (1998) Adaptive filters—theory and applications. Wiley, Chichester, UK
15. Haykin S (2013) Adaptive filter theory. Prentice-Hall, Eaglewood Cliffs, NJ

Chapter 30
Framework and Model for Surveillance of COVID-19 Pandemic

Shreekanth M. Prabhu⬤ **and Natarajan Subramanyam**⬤

1 Introduction

Medical experts and public policy experts are increasingly called upon to advise governments on administrative decisions to better manage pandemics through critical interventions such as lockdowns that have huge social implications. Traditionally, data from surveillance of infectious diseases informs appropriate policy options to the authorities concerned. However, in the case of pandemics such as the recent COVID-19 which has spread world over with a high degree of intensity, the exercise of conducting the surveillance and arriving at the right recommendations that can apply to varied geospatial, spatial-social, and social-temporal contexts can be daunting. Hence, we need appropriate technology-based frameworks that can focus on what is truly critical and models that can decipher insights amidst continuing flux in the state of the pandemic within and across contexts.

The problem of surveillance is all the more acute in the case of COVID-19 as many infected people remain asymptomatic for long periods. In Table 1 [1], we can see that COVID-19 has the longest incubation period compared to other viruses. Thus, a person can be infectious even before the symptoms start. General information on COVID-19 is presented lucidly by Britt [2]. Further, it is hard to diagnose the onset of disease just based on symptoms.

To make the matters worse, the real-time reverse transcription-polymerase chain reaction (RT-PCR) test used to detect COVID-19 infection can result in a large number of false negatives. Sethuraman et al. [3] emphasize the importance of timing

S. M. Prabhu (✉)
CMR Institute of Technology, Bengaluru 560037, India

N. Subramanyam
PES University, Bangalore 560085, India

© The Author(s), under exclusive license to Springer Nature Singapore Pte Ltd. 2021   359
S. Kumar et al. (eds.), *Proceedings of International Conference on Communication and Computational Technologies*, Algorithms for Intelligent Systems,
https://doi.org/10.1007/978-981-16-3246-4_30

Table 1 Comparing COVID-19 symptoms with infection due to other viruses [1]

	COLD	FLU	NOROVIRUS	COVID-19
Incubation period	1–3 days	1–4 days	A few hours	2–14 days
Symptom onset	Gradual	Abrupt	Abrupt	Gradual
Typical illness duration	7–10 days	3–7 days	1–2 days	Undetermined
Symptoms				
Sore throat	**Common**	Some times	*Rare*	Some times
Sneezing	**Common**	Some times	*Rare*	*Rare*
Stuffy, runny nose	**Common**	Some times	*Rare*	*Some times*
Cough, chest discomfort	Some times	**Common**	*Rare*	**Common**
Fatigue, weakness	Some times	**Common**	Some times	Some times
Fever	*Rare*	**Common**	Some times	**Common**
Aches	*Rare*	**Common**	Some times	Some times
Chills	*Rare*	**Common**	Some times	Some times
Headache	*Rare*	**Common**	Some times	Some times
Shortness of breath	*Rare*	*Rare*	*Rare*	**Common**
Nausea	*Rare*	*Rare*	**Common**	*Rare*
Vomiting	*Rare*	*Rare*	**Common**	*Rare*
Diarrhea	*Rare*	*Rare*	**Common**	*Rare*
Stomach pain	Rare	Rare	**Common**	*Rare*

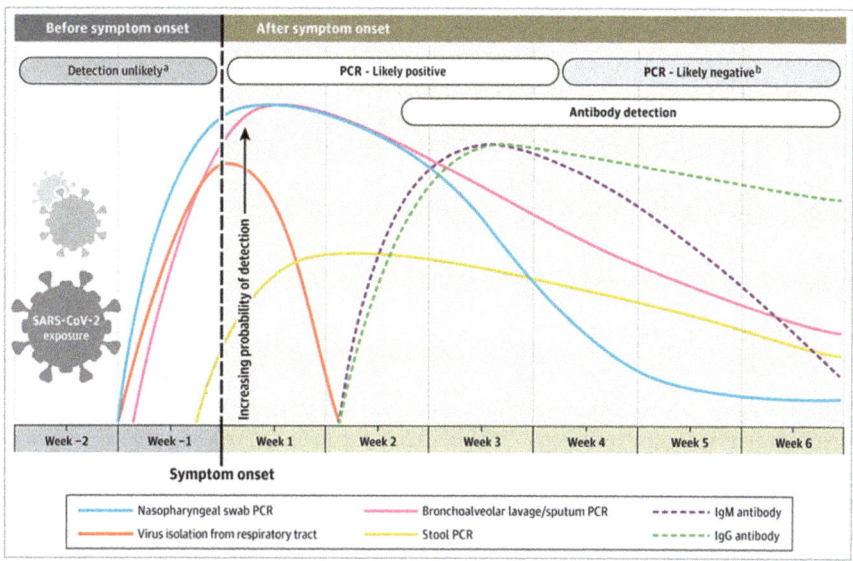

Fig. 1 Estimated variation over time in diagnostic tests for detection of SARS-CoV-2 infection relative to symptom onset [3]

to get the correct results in tests that are used to detect SARS-CoV-2 RNA as illustrated in Fig. 1. Thus, there is a very real possibility that a sizable number of COVID-19 cases are not detected on testing. See Fig. 1.

Considering the limitations cited above, it is important that we also look at non-pharmaceutical interventions to tackle COVID-19.

Ferguson et al. [4] from Imperial College London classified possible non-pharmaceutic interventions to tackle COVID-19 into mitigation and suppression. Under mitigation, the focus is on slowing the epidemic spread, and under suppression, the focus is on reversing the epidemic growth. Mitigation included detection, isolation, and quarantine of suspect cases. Suppression required lockdown of large regions to save for bare essentials. As mitigation required enormous healthcare infrastructure, suppression was the only viable option provided by the governments to have the will to implement severe social restrictions.

In India, government has used both mitigation and suppression to tackle COVID-19, at times switching back and forth between them. Despite all the measures taken over the last 9 months or so, a policy framework that can guide the government on how to calibrate the administrative measures at regional/locality/community levels seems to be still lacking. Further, the exercise of surveillance is highly fine-grained and intrusive and instruments available to the authorities are overly blunt and affect large sections of the populace. There is a need to flip this paradigm. The objective of this paper is to address both these issues.

In our paper, we make use of the system dynamics framework as an overarching framework to analyze the pandemic. In conjunction with the framework, we make use of an Improvised Hidden Markov Model to infer the latent spread of the pandemic at regional levels.

Rest of the paper is structured as follows: Section 2 covers "Literature Survey" where we cover surveillance practices. Section 3 covers the "Proposed Methodology" for evolving a framework and model for the surveillance of COVID-19 pandemic. Section 4 presents "Results and Discussions" on the application of the proposed methodology to Indian context. Section 5 covers "Conclusions".

2   Literature Survey

In this section, first, we cover literature related to surveillance framework in general, followed. In their paper published in 2002, McNabb et al. [5] describe the contours of a conceptual framework for public health surveillance and actions health authorities can take. As shown in Fig. 2, the framework comprises six core activities and four support activities as well as two types of responses, namely epidemic type and routine management type. They proposed that surveillance can be done in a unified manner instead of being focused on specific diseases. Further, their framework envisaged concurrent systems operating on the same data, from local level to national level.

One important takeaway from their framework is that it reflected the changing definition of surveillance from one which was limited to collecting data and communicating to the stakeholders till 1960s, to a newer definition which covered the "ongoing systematic collection, analysis, and interpretation of outcome-specific data for use in the planning, implementation, and evaluation of public health practice"

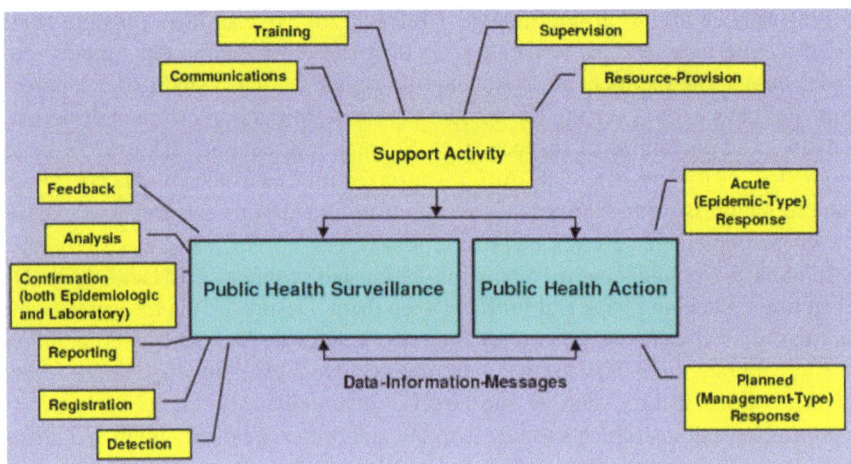

Fig. 2   Conceptual framework of public health surveillance and action [5]

[6]. The importance of surveillance in planning cannot be starker than in COVID-19 pandemic. There is a wealth of information on surveillance of communicable diseases in the guide produced by the World Health Organization [7].

Next, we cover the literature related to surveillance taxonomy and shortcomings of the current surveillance framework with a focus on COVID-19. This is followed by a survey of literature on past pandemics, the frameworks, and models used.

Very recently, Ibrahim [8] did a study on the types, challenges, and implications of epidemiological surveillance for controlling the COVID-19 pandemic. Table 2 lists and describes different types of global surveillance done under the directions of WHO.

More broadly, surveillance can be categorized as passive and active. When there is no outbreak of an infectious disease, passive modes are adequate. Here healthcare facilities routinely and continually inform public health authorities on the incidence of infectious diseases of interest as and when patients visit healthcare facilities. Active surveillance in contrast involves reaching out to the society at large looking for potential suspects who may have already encountered the disease.

Peixoto et al. [9] have raised apprehension on uncertainty and under-ascertainment of COVID-19 surveillance. In their words, "it is the role of the scientific community in public health and epidemiology not only to communicate what is certain and based in robust established science, but also of what is uncertain when it may be of relevance for decision making and can be further researched, discussed, and shared".

The Center for Infectious Diseases Research and Policy (CIDRAP), University of Minnesota [10], has published their view-point on COVID-19 surveillance detailing the state of the field and the mammoth task involved. Some of the challenges they have outlined are limited COVID-19 pandemic testing, incomplete reporting of critical information, the omission of mildly symptomatic and asymptomatic persons from surveillance, inconsistent data collection and reporting, non-specific case definitions, clustering of cases detected in the outbreaks, and difficulties in assessing the exposures in communities. Further, there are issues such as lack of timeliness and regularity of reporting and lack of integrated reporting infrastructure. There are also challenges such as how to utilize serological surveillance, which is critical to better understand how many infections have occurred at different points in time, in different locations, and within different populations in the country. In particular, inferences drawn from serological surveillance studies must recognize the challenge of data interpretation due to the potential for false positives. Another issue is that ongoing surveillance of other diseases suffers due to the pandemic as people are given a choice to avoid accessing healthcare facilities. There is also a proliferation of tools for digital surveillance. These tools, however, need to be used carefully being cognizant of privacy concerns.

In summary, surveillance of COVID-19 can be a daunting challenge from a data management view-point. On one hand, the data involved is huge and requires an enormous cost to manage. On the other hand, even with the best effort, the data collected is most likely incomplete. On top of that, certain data cannot be directly obtained but only inferred indirectly or with a time lag.

Table 2 Types of surveillance of COVID-19

S. No.	Surveillance type	Method	Drawbacks
1	Comprehensive routine	Test all suspected cases and track confirmed cases among a group of 100,000 people. Gives accurate assessment of regional spread	Depends on the wide availability of testing facilities, and the cost of testing can become high
2	Case-based routine surveillance	Reporting probable and confirmed cases as and when they are identified with strict time limits. Easy to administer	If we limit only to this, we cannot prevent new cases. We also need to check if contacts of confirmed cases are also infected
3	Aggregated routine surveillance reporting	Here all countries are required to report weekly newly confirmed cases, deaths due to COVID-19, hospitalizations, and discharges along with age-groups they belong to	If applied alone, is the loss of diagnosis of many cases
4	Active surveillance	Fast recognition of the infected persons and their prompt isolation by mechanisms such as contact tracing and having containment zones to prevent the spread of diseases beyond known clusters	Requires training of health workers and cooperation of people
5	Syndromic (clinical) surveillance	It means surveillance of health data about the clinical manifestations that have an important impact on health. Each case is described and documented in detail and categorized, then utilized to inform public policy [13]	Need good facilities, processes, and personnel to undertake this exercise thoroughly
6	Sentinel surveillance	Select hospitals/facilities are recruited to conduct surveillance that have good laboratory facilities and qualified staff as well as have high possibility of seeing the cases. Enables assessing magnitude of COVID-19 spread in a certain population	The drawback is that it is based only on the cases in sentinel healthcare facilities

(continued)

Table 2 (continued)

S. No.	Surveillance type	Method	Drawbacks
7	Sentinel syndromic surveillance	Here select facilities are used for syndromic surveillance	The drawback is that it is based only on the cases in sentinel healthcare facilities
8	Virological (serological) surveillance	Makes use of antigen and antibody tests. Such methods are less costly but less accurate	Need to prioritize and strategize who should go through this kind of testing
9	Virological sentinel surveillance	The surveillance is restricted to sentinel centers	
10	Population serological surveillance	Anybody coming for a blood test is asked to give an additional sample for COVID-19 tests	Involves huge costs when applied on a large scale
11	Hospital-based surveillance for SARI	These countries need to monitor the percentage of confirmed COVID-19 from all SARI	Case detection may be late as it depends only on the surveillance of symptomatic patients
12	Routine surveillance of nosocomial outbreaks and outbreaks in long-term care facilities	Here outbreaks originating from hospitals or healthcare facilities are tracked	
13	Enhanced surveillance of hospitalized cases	Here surveillance is applied to all hospitalized cases. Such surveillance is a very important to detect new cases	The case detection methods are restricted to symptomatic patients
14	Mortality surveillance	Here mortality is tracked not only in hospitals but also in long-term care facilities	The population tracked may not be representative of the larger population
15	Healthcare surveillance	Here surveillance of healthcare facilities is done. Useful to get help from all quarters to upgrade the facilities	
16	Wildlife surveillance	Surveillance of people who are in touch with wild-life and become carriers of infection	
17	Participatory surveillance	Using digital means, information can be collected from people. Enables follow-up with people	At times some cases may go out of the radar

(continued)

Table 2 (continued)

S. No.	Surveillance type	Method	Drawbacks
18	Electronic reporting system	Using web-based tools for reporting disease and epidemics in a timely manner	
19	Digital surveillance	Many countries utilize digital surveillance including apps, location data, and electronic tags/	There may be privacy issues in addition to the cost of running the surveillance over long periods
20	Event-based surveillance (EBS)	EBS captures information from channels such as online content, radio broadcasts, and print media, to complement conventional public health surveillance efforts	Requires a lot of human resources that can intelligently sift information for insights

2.1 Learning from the Past Pandemics

The three pandemics in the twentieth century killed between 50 and 100 million people. 2009 H1N1 "swine flu" exposed vulnerabilities that we still have to influenza epidemics. Hutton [11] has done a review of operations research tools and techniques used for influenza pandemic planning from a public health view-point. In his study, he covers a variety of models such as system dynamics which is helpful to study problems as varied as social distancing, vaccine-delivery mechanisms [12], antiviral treatments [13], optimal use of healthcare resources, and portfolio analysis of interventions. Larson [14] showed an approach that required limited disruption by targeting social distancing to correct sub-populations. Nigmatulina and Larson [15] extended that work to multiple interconnected communities spread over different towns in a region. They found that travel restrictions generally are not effective. These analyses [16] even while using simple spreadsheet-based models provide powerful policy insights.

Rath et al. [17] made use of the hidden Markov model to segment the spread of influenza as non-epidemic and epidemic phases, modeled as hidden states. Rath et al. in their paper have used exponential distribution for the non-epidemic phase and Gaussian distribution for the epidemic phase. This is a refinement over the work done by Le LeStrat and Carrat [18] who made use of only Gaussian distributions. Recently, Singh et al. [19] developed an SVM-based model for COVID-19 prediction, Bhatnagar et al. [20] anticipated a model for descriptive analysis, while Kumari et al. [21] introduced a machine learning model for analyzing spread, recovery, and death caused by COVID-19.

Even though there is a flurry of literature on COVID-19 pandemic, the majority of papers are focused on prediction and prognosis. The literature on surveillance is relatively scant. There is a need for research to understand how pandemic spreads in an end-to-end manner, in particular before it reaches the syndromic stage.

3 Proposed Methodology

We propose the use of system dynamics as an overarching framework to understand the spread of COVID-19 pandemic across geospatial, social-temporal, and syndromic contexts. Also, we propose the use of the hidden Markov model to infer the latent status of COVID-19 pandemic using hidden Markov model, with hospital reportage as visible observations, with required improvisations and refinements in the model.

At a high level, the state changes people go through when there is an outbreak in their region are depicted in Fig. 3. Every person in a new neighborhood starts off being a **prospect,** and he/she may get classified as a **suspect** in the event they develop symptoms or were in touch with other suspects. Suspects move into **confirmed** state when they are tested positive; a confirmed case may result in **recovery** or **mortality** over time. Some recovered may get reinfected and face death due to post-recovery.

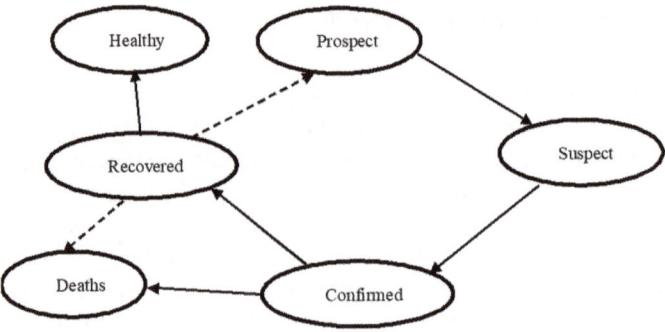

Fig. 3 People state life cycle

3.1 Representing Pandemic Context Using System Dynamic

System dynamic framework is extremely useful to represent causal relationships and flow dynamics in the context of COVID-19 pandemic. The flows are across geospatial, social, syndromic, and clinical contexts.

Firstly, we look at the flow between **geospatial** and **social** contexts. We start with a finite population group living in a given geospatial context, as a fundamental unit. The initial condition of this group is that none is infected with COVID-19 pandemic, and none is in contact with outsiders, so the whole population can be considered as **prospects**. Then, we have a set of people who would like to move out of the context who are modeled as **emigrants**. These emigrants after some delay migrate back into the context, as **immigrants**. Immigrants lead to an increase in **social contacts,** possibly infectious. Figure 4 represents the **geospatial-social flows**. Here, we can easily track emigrants, immigrants, and contacts with newly arrived people and treat them as predictor variables which can help us infer say overall migration and enhanced social contact.

Secondly, we look at the flow between **social** and **syndromic** contexts. It is much easier to track symptomatic people who are called **suspects** in WHO terminology. Once they get tested and correctly diagnosed, they are added to the list of known

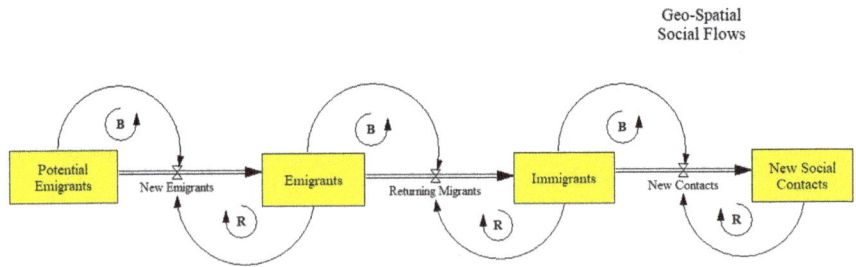

Fig. 4 Stock and flow diagrams of migrations and contacts

confirmed cases. Here again, those whose tests turn negative have to be treated as **probable suspects** for some time, till their symptoms subside. Figure 5 illustrates the **social-syndromic flows**.

Thirdly, we look at the **clinical** context. Here some may be hospitalized, whereas others may be getting treatment or under observation without getting admitted to the hospitals. Here some of the confirmed cases turn to active cases. They, in turn, may move from regular care, severe care, and critical care in the clinical context. However, we have chosen to represent the flows at a coarser granularity. Figure 6 represents **clinical flows.**

As we can see among the three different contexts, the information in the clinical context has the highest degree of completeness as well as authenticity. However, the phenomena in geospatial, social, and syndromic contexts happen in a latent manner. Further, it may be difficult to completely capture all the flows. This information on social phenomena has a far more critical bearing on containment and cessation of the pandemic. To address this information gap, we propose the use of the hidden Markov model, which is detailed hereafter.

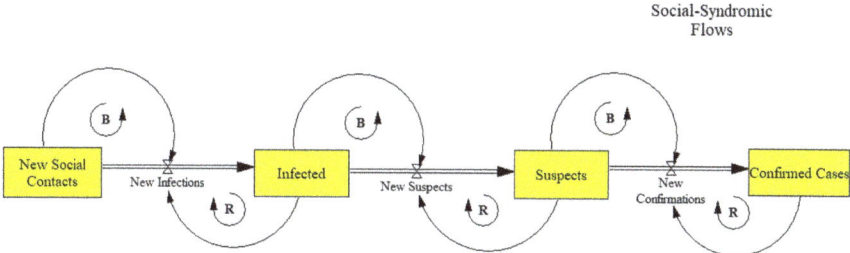

Fig. 5 Stock and flow diagrams of suspects and confirmations

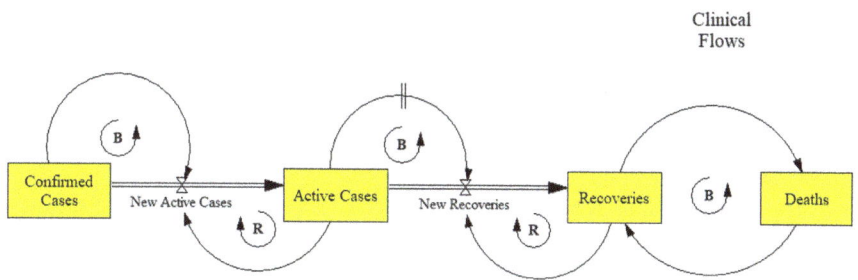

Fig. 6 Stock and flow diagrams of recoveries and deaths

Table 3 State of the regions

Region state	Description
Healthy	The region has very few new migrants. Overall social contact in the population group is stable and routine
Infected	The region may have sizable flows of migrants, and the social contacts in the population groups are intense, with a high degree of churn
Symptomatic	The region has already seen many suspects with a history of contact with a larger population group
Detected	The region possibly has contacts who, in turn, have had social contact with suspects, but they are yet to be treated as suspects

Table 4 Inferring hidden state of regions from hospital reportage

Active	Recovered	Deaths	Inferred state
Low	High	Medium	Healthy
High	Low	Medium	Infected
High	Medium	Low	Symptomatic
Medium	High	Low	Detected

3.2 Inference of Latent State in the Regions Using Hidden Markov Model

To start with, we assume that each population group/region is in one of the following states (Table 3).

To infer the state in geospatial-social and in particular social-syndromic context, we make use of information in the clinical context and workback. We use the following criteria which are summarized in Table 4.

(1) If the region were to be healthy (or reverting to being healthy), then recoveries should exceed active cases.

(2) If the region were to be infected, there may be more active cases, and deaths may exceed recoveries on a given day as the infection is spreading surreptitiously.

(3) If the region were to be symptomatic, then active cases should be high and recoveries should be more than deaths.

(4) If the region were to be in a detected state (with easy availability of testing), then recoveries should be a lot higher than new cases on any given day. Since people are majorly asymptomatic, some may recover without being categorized as active.

(5) Further, if there is a severe shortfall of capacity at the hospital end, it may lead to catastrophic states.

This scenario where observation sequences are known along with possible state values and possible observations are known lends itself to be modeled as a learning problem in HMM. In learning problem, given some training observation sequence V and general structure of HMM (number of hidden and visible states), we need to determine HMM parameters $\mathbf{M} = (\mathbf{A}, \mathbf{B}, \pi)$ that best fit the training data where $\mathbf{A}$ is state-transition probability matrix, $\mathbf{B}$ is visible symbol probability matrix, and π is the probability distribution of hidden states at the initial stage.

Generally, the learning problems of the hidden Markov model are solved using the Baum–Welch forward–backward EM algorithm. In this case, such an approach is unwieldy. The kind of examples in literature that use the EM algorithm and problem at are hand is very different. For example, what should we use as observation and what should be termed as a hidden state in the problem are not that apparent. Any approach to use the EM algorithm will lead to a formulation that is overly fine-grained and possibly leading to results that are far less interpretable and communicable to public health authorities. Considering this, we propose the following refinement and improvisation to the hidden Markov model, detailed below:

(1) Choose a vector of observation sequences catering to a different region. Each region on a given day can generate a, d, r, and a~-symbols in different numbers that account for the change of observed states of patients on a given day.

(2) Based on the visible symbols generated above, we infer state symbols for that day according to Table 4.

(3) For each region and each day, generate the π, the probability distribution of hidden states at that period.

(4) For each day and each hidden state, compute the visible symbol distribution matrix and compute/update the visible symbol probability distribution matrix.

(5) Compare the state inferred for the previous day and the current day for different regions, and populate the state-transition distribution matrix. Update the state-transition probability matrix with incremental state transitions for every day. This starts getting updated from day 2.

(6) Iterate between steps 1 and 4 till symbols are exhausted.

Thus, at the end of steps 1–6, we have a hidden Markov model comprising of $\mathbf{A}$, $\mathbf{B}$, and π. The model can then be used for evaluation problem and decoding problem. The state distributions that were arrived at by using this model on a given day can be used by governments to design interventions in a calibrated manner. We can further refine the model by using numeric values for probability for $\mathbf{B}$ from the model in step 2 and arrive at more refined values of $\mathbf{A}$.

Once we have a determination of state as well as typical transition to other states, we need to circle back to the system dynamics framework to arrive at the recommendations to the medical experts/authorities. Table 5 illustrates our approach which draws on the system dynamics representation of the surveillance problem. All the cell values are net additions.

The scenarios where death rates are increasingly much faster than the increase in active rates or recoveries are deemed as catastrophic. This typically requires action at the hospital end.

Table 5 Designing interventions based on inferred states

S. No.	Predominant state	Recommendations
1	Repeatedly getting into an infected state	Need to rigorously track and control inward migration into affected regions. Do a data audit on migrations
2	Prolonged stay in symptomatic state	Need to expedite isolation and treatment of suspects. Perform a process audit about typical delays in these processes
3	Detected not reverting to healthy and staying there	Have a more rigorous process for establishing contact chains of active cases in the affected region
4	Switching back and forth between infected and symptomatic states	Need to have a watch on social contacts and do an analysis of how infection may have spread using surveys and digital surveillance
5	Switching back and forth between symptomatic and detected	Have a widespread testing regimen with a judicious combination of antigen/antibody testing and RT-PCR testing

4 Results and Discussions

The approach we used was to collect data daily which was published by the Ministry of Health and Family Welfare, Government of India. This not only gave counts of confirmed, active, recoveries, and deaths at national and regional levels. This reportage was generally accompanied by reportage in media which covered impressions from the field as to which regions were doing well and which regions were faring badly.

Our basic premise in this paper is that every region goes through the following state sequence broadly.

Healthy -> Infected -> Symptomatic -> Detected -> Healthy.

This is in contrast to the application of the hidden Markov model to influenza by Rath et al. which had only epidemic and non-epidemic phases.

This, however, may happen only after encountering multiple periods of volatility. Another assumption we make is that propensity to repeatedly return to say the infected or symptomatic state may be addressed by specific interventions by the regional authorities. In our approach, we also indicate a "local" state change which may call for a temporary measure in detecting migrants and contexts to be done intensely for short while to other "macro or global" state changes as well, where issues are more enduring. In such cases, the actions required may be more pervasive and strategic.

Table 6 gives details on how the data was collected. Then we analyzed the data at the national level and regional level over time between April and August 2021. Figures 7 and 8 depict state changes at the national level and regional levels. For both the national and regional levels, we suggest appropriate recommendations along with the rationale. Figure 5 shows the number of regions in healthy, infected, symptomatic,

Table 6 Dataset specifications

How data were acquired	The data was acquired from the official website of the Ministry of Health and Family Welfare, Government of India (https://www.mohfw.gov.in/)
Parameters for data collection	Data consists of the reported cumulative confirmed, active, recovered, and death cases on a given day and previous day(s)/past day(s). We look at state-wise data on the Ministry of Health and Family Welfare site
Description of data collection	Data was obtained daily as of 8 a.m. IST from the Ministry of Health and Family Welfare website. Data was collected between April 29 and May 20, June 10, July 11, and July 12, 2020, as well as August 23 and August 24, 2020
Data source location	https://www.researchgate.net/publication/344044688_COVID-19_Surveillance_usingIHMM_Data_files It is in the form of three supplementary files containing (i) excel tabulations that cover hidden state inferences on daily basis, (ii) excel tabulations containing calculations to refine the hidden Markov model over time, and (iii) excel tabulations used to generate the outputs

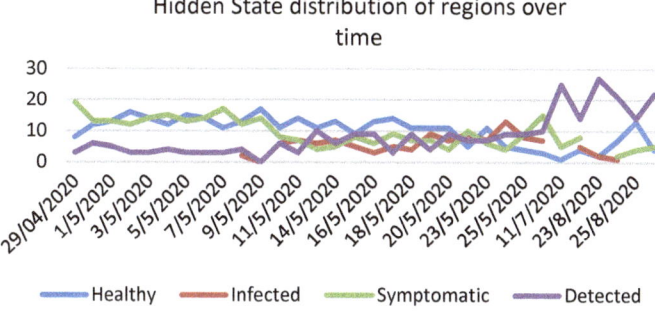

Fig. 7 Hidden states over time in India

and detected states over the observed duration. Figure 7 can provide inputs/insights to the government to shape the national strategy regarding imposing/relaxing social restrictions. It can be observed that the number of regions in the detected state is steadily increasing. Figure 8 depicts the hidden states in six different regions over time along with analysis. Here, we cover the six most affected states in India as of September 1, 2020. We make recommendations based on system dynamics representation of COVID-19 pandemic surveillance which is illustrated in Figs. 3, 4, and 5 and insights summarized in Table 5. Region-wise recommendations follow the figure.

Maharashtra is predominantly in a **symptomatic state**. Need to expedite isolation and treatment of suspects. Perform a process audit to address delays.

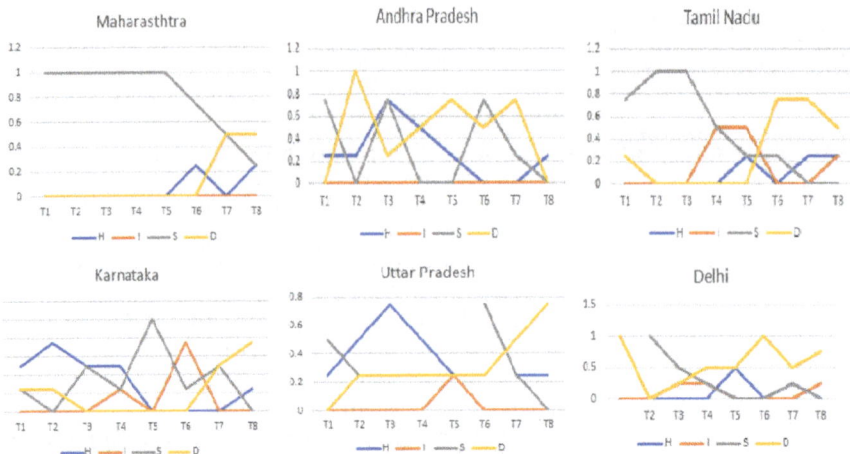

Fig. 8 Hidden states in different regions of India

Andhra Pradesh is predominantly in **detected state**. Have a more rigorous process on establishing contact chains of active cases in the affected region.

In **Tamil Nadu**, symptomatic cases are coming down. A broader testing regimen with a judicious combination of antigen/antibody testing and RT-PCR testing can be useful.

Karnataka has sporadic spread of infections. Need to control movements between zones.

Uttar Pradesh needs to perform intense testing of migrants and their contacts.

Delhi needs to have a stronger watch on social contacts using digital surveillance tools to assess possibilities of relapse.

In this section, we have detailed how governments can track hospital reportage in each region/geospatial context, infer the hidden state for that region, and take recommended actions corresponding to the hidden state, based on routine behaviors and social processes operating within those regions. The governments, however, need to take a lot more actions comprehensively to ensure containment. An important enabler for containment of pandemic is to manage flows across geospatial contexts. In certain cases, it is hard to even track migrations as many may choose to cross the boundary by feet or using inner roads that are not widely used. Equally important is being able to continually track the whereabouts of new migrants if and when they are found to be possible suspects. A lot of infections happen if people are in-transit, participate in congregations and crowded events, with invitees from outside regions. These need in situ interventions with a good degree of follow-up.

The ability to differentiate different regions/geospatial contexts is very useful for governments to calibrate their non-pharmaceutical interventions. In place of whole-sale lockdowns, the approach can be a lot more nuanced that varies from district to district. It is also needed to study the social contexts. It may be better not to prescribe stay-at-home in neighborhoods where too many people cohabit in the same household. Further, how each community gears itself to isolate the infected while continuing to operate with the right regimen that allows movements within and outside in a controlled guided manner is a matter of study by itself. Overall, the objective of our research was to enable governments to use non-pharmaceutical interventions in a targeted manner to enable containment and cessation of pandemic spread in as many regions as possible without overly inconveniencing the public.

5   Conclusions

Surveillance of COVID-19 pandemic is a mammoth exercise that is supposed to guide action from public health authorities. Surveillance research is as important if not more compared to the prediction and prognosis of the COVID-19 pandemic. In this paper, we have proposed the use of system dynamic framework to study the spread of pandemic across geospatial, social-temporal, syndromic, and clinical contexts. Among these, we have recognized that the process of spread of infectious disease in geospatial and social contexts happens as a latent phenomenon, whereas in syndromic and clinical contexts, it happens in a rather overt manner. Hence, we have made use of periodic hospital reportage that is publicly available to infer latent states in geospatial contexts, i.e., states and union territories of India with the help of an improvised and refined hidden Markov model. This approach has enabled us to arrive at region-wise recommendations that are found to be in broad agreement with reportage from news media about the specific regions. More work is needed to better understand flows within the social contexts which play an important role in arresting or expediting the spread of the pandemic.

References

1. Britt RR (2020) How to tell if you have Flu, Coronavirus or something else. Medium.com, 4–9 March 2020. https://elemental.medium.com/how-to-tell-if-you-have-the-flu-coronavirus-or-something-else-30c1c82cc50f
2. Britt RR (2020) The latest Coronavirus Q&A: everything you need to know now. Medum.com, 7 March 2020
3. Sethuraman N, Jeremiah SS, Ryo A (2020) Interpreting diagnostic tests for SARS-CoV-2. JAMA 200101 (ahead of print)
4. Ferguson NM et al (2020) Impact of non-pharmaceutical interventions (NPIs) to reduce COVID-19 mortality and healthcare demand, Author's website, Imperial College, London
5. McNabb SJN, Chunong S, Ryan M, Wuhib T, Nsubuga P, Alemu W, Carande-Ku-lis V, Rodier G (2002) Conceptual Framework of public health surveillance and action and its application in health sector reform. BMC Public Health 2:2
6. Thacker SB, Berkelman RL (1988) Public health surveillance in the United States. Epidemic Rev 10(164):190
7. World Health Organization (2006) Communicable disease surveillance and response systems. Guide to monitoring and evaluation, WHO/CDS/EPR/LYO/2006.2

8. Ibrahim NK (2020) Epidemiologic surveillance for controlling Covid-19 pandemic: types, challenges, and implications. J Infect Public Health. https://doi.org/10.1016/j.jiph.2020.07.019
9. Ricoca Peixoto V, Nunes C, Abrantes A (2020) Epidemic surveillance of Covid-19: considering uncertainty and under-ascertainment. Port J Public Health 38:23–29. https://doi.org/10.1159/000507587
10. Ulrich A et al (2020) COVID-19: the CIDRAP viewpoint, July 9, 2020. Part 5: SARS-CoV-2 infection and COVID-19 surveillance: a national framework
11. Ferguson NM, Cummings DA, Cauchemez S, Fraser C, Riley S, Meeyai A, Iamsirithaworn S, Burke DS (2005) Strategies for containing an emerging influenza pandemic in Southeast Asia. Nature 437:209–214
12. Hutton D (2013) Review of operations research tools, and techniques used for influenza pandemic planning, Chapter 11. In: Operations research and health care policy, pp 225–247. https://doi.org/10.1007/978-1-4614-6507-2_11
13. Khazeni N, Hutton DW, Garber AM, Hupert N, Owens DK (2009) Effectiveness and cost-effectiveness of vaccination against pandemic influenza (H1N1) 2009. Ann Intern Med 151:829–839
14. Larson RC (2007) Simple models of influenza progression within a heterogeneous population. Oper Res 55:399–412
15. Nigmatulina KR, Larson RC (2007) Stopping pandemic flu: government and community interventions in a multi-community model. Massachusetts Institute of Technology Engineering Systems Division Working Paper Series, No. ESD-WP-2007-28
16. Lee VJ, Chen MI (2007) Effectiveness of neuraminidase inhibitors for preventing staff absenteeism during pandemic influenza. Emerg Infect Dis 13:449–457
17. Rath TM, Carreras M, Sebastiani P (2003) Automated detection of influenza epidemics with hidden Markov models. In: Berthold MR, Lenz HJ, Bradley E, Kruse R, Borgelt C (eds) Advances in intelligent data analysis V. IDA 2003. Lecture notes in computer science, vol 2810. Springer, Berlin, Heidelberg
18. LeStrat Y, Carrat F (1999) Monitoring epidemiologic surveillance data using hidden Markov models. Stat Med 18(24):3463–3478
19. Singh V, Poonia RC, Kumar S, Dass P, Agarwal P, Bhatnagar V, Raja L (2020) Prediction of COVID-19 corona virus pandemic based on time series data using Support Vector Machine. J Discrete Math Sci Cryptogr 23(8):1583–1597
20. Bhatnagar V, Poonia RC, Nagar P, Kumar S, Singh V, Raja L, Dass P (2020) Descriptive analysis of COVID-19 patients in the context of India. J Interdiscip Math 24(3):489–504
21. Kumari R, Kumar S, Poonia RC, Singh V, Raja L, Bhatnagar V, Agarwal P (2021) Analysis and predictions of spread, recovery, and death caused by COVID-19 in India. Big Data Min Anal 4(2):65–75

Chapter 31
A Survey on IoT Applications in Health Care and Challenges

Soumeya Mahamat Yassin, Dalal Batran, Asmaa Al Harbi,
and Mohammad Zubair Khan

1 Introduction

In recent years, the Internet of things (IoT) has gained popularity as a means of
connecting objects and things such as computers, appliances, sensors, information services, and applications [1]. It enables these things to communicate with
one another through the Internet. Communication, recognition, utilities, computing,
sensing, and semantics are the fundamental components of the Internet of things. The
communication aspect connects various objects in order to provide specific intelligent
services. The identity factor associates the requested services with the identification
element. The service factor is divided into four categories: ubiquitous or all-around
services, information-gathering services, identity-related services, and cooperative
services. The processing of IoT units constitutes the computation components. The
most widely used computing components are microcontrollers and microproces-
sors. The sensing factor collects data from various network objects and sends it to
a database or the cloud. The semantic element is used to extract knowledge from
different devices.

The Internet of things (IoT) was first proposed in the 1980s. Simultaneously, the
electromagnetic telegraph was invented in Russia by Baron Schilling in 1832. This
uses encoded electrical impulses to transmit the message's text from point to point and
then develops later. In 1833, Wilhelm Weber and Carl Friedrich Gauss devised a code
that enabled them to communicate with devices located 1200 m within Göttingen,
Germany. Tim Berners-Lee created the first Internet webpage in 1991 [2].

S. M. Yassin (✉) · D. Batran · A. Al Harbi · M. Z. Khan
Department of Computer Science, College of Computer Science and Engineering, Taibah
University, Madinah, Saudi Arabia

M. Z. Khan
e-mail: mkhanb@taibahu.edu.sa

© The Author(s), under exclusive license to Springer Nature Singapore Pte Ltd. 2021
S. Kumar et al. (eds.), *Proceedings of International Conference on Communication
and Computational Technologies*, Algorithms for Intelligent Systems,
https://doi.org/10.1007/978-981-16-3246-4_31

However, during a presentation at Procter and Gamble in 1999, Kevin Ashton coined the phrase "Internet of things." P&G had already adopted the new definition of radio-frequency identification (RFID) in their products. The Internet was a new and popular interest at the time, so his idea seemed rational. The title of the presentation, the Internet of things, has been made public. However, the Internet of things (IoT) did not receive much attention in the following ten years [2, 3].

Many businesses are pouring a lot of money into the beginning of 2020. According to Statista Research Department estimates, there are numerous items related to the Internet of things, as the Internet of things rule of things may reach more than 40 billion connected devices worldwide. People wish to link the entire globe [4].

Many researchers have attempted to establish an accurate description of the Internet of things. Since no one person owns or regulates the Internet of things, there can be no official description. Still, all meanings literally flow into one concept, which we'd like to explain with the following text:

> Internet of Things is an evolving concept of the Internet so that all things in our life can connect to the Internet or to each other to send and receive data to perform specific functions through the network. [5]

We can define it in a non-theoretical definition. Simply the Internet of things is the world that we are beginning to live in some of its aspects now that some of the things we use can connect to the Internet, for example, watches, televisions, wristbands, glasses, and others. But what does this world hide from us other than what has appeared so far, and what is meant by "things" in the phrase Internet of things? [6, 7].

Everything could be everything comes under the concept of the Internet of Things like clothes, furniture, home devices, kitchen appliances, body parts, buildings, streets, light bulbs, and even animals. Anything that does a process and has an Internet connection is something in the IoT world. For example, many cow farms around the world are beginning to link their cows' bodies to the Internet in order to track their health, fertility, and the percentage of certain hormones in their bodies, which indicates the best time to milk them and helps to improve the production process.

This paper presents the concept of IoT architecture. Then it reviews the different services and applications of IoT. It is discussing various issues in the healthcare services of the IoT. Finally, it also explained the most important challenges facing the Internet of things.

1.1 Motivation

The Internet of things systems have become a major role in running many areas in our lives. We were motivated to explain the Internet of things, and we identified the Internet of things in health care because of its special importance in now with the conditions of the Corona pandemic. We motivate to do this research paper to present healthcare applications, service and challenges that face this area.

2 Preview Internet of Things

2.1 IoT Architecture

As we mentioned earlier, the Internet of things is a collection of physical objects, devices, buildings, communication protocols, and sensors that collect data, exchange, store, analyze, and process data. Because of the wide range of IoT objects, there is no consensus on IoT architecture. As different researchers have proposed several architectures, some have based, for example, on RFID, industry, service-oriented architecture, supply chain management, wireless sensor network, logistics, smart cities, connected living, health care, big data, security, social computing, and cloud computing [8]. Figure 1 shows IoT architectures.

We will present the basic architecture, three layers of IoT, the perception layer, the network layer, and the application layer [9, 10].

The perception layer includes the physical intelligent devices, which is a sample of data witch gathering from its environment and processes the data to get useful information to send it to the network layer through the access devices.

The network layer oversees data transfer from the perception layer to the upper layers via the Internet. It also uses protocols to process, gather data, and protect data from any attacks [11].

The application layer includes providing services that merge and store the source of information from the network layer and manage information and make decisions

Fig. 1 Common IoT architecture. **a** Three-layer. **b** Five-layer

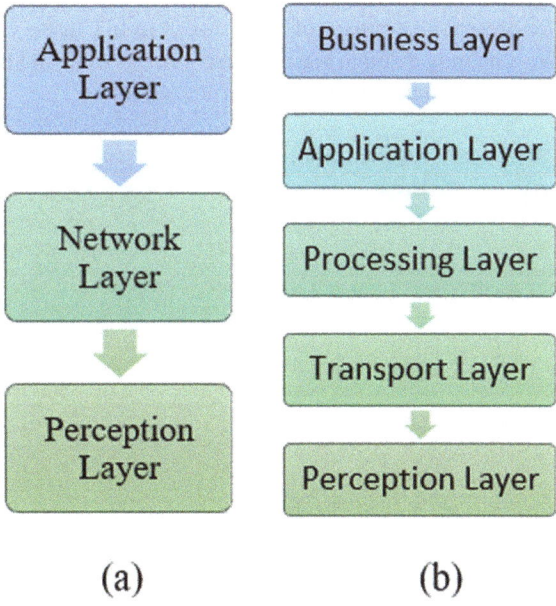

(a)   (b)

in a data centre. The applications combine all the necessary functions for the lower layers and provide a specific service as needed of an organization or an environment.

In addition, the most of the research mentioned the five layers of the Internet of things, which can explain of the layers from the previous Fig. 1b in a simplified way as follows:

The transport layer is in charge of transmitting data from the processing layer to the perception layer through the Internet using Bluetooth, wireless networks, RFID, NFC, 5G, and so on. The modeling of the Internet of things can address the communication between smart objects through RFID technology. Each marker has a unique ID as the RFID reader transmits the signal from the surroundings to determine the object's activity through its unique key and the information in real time to transfer it to the system [12].

The process layer is also referred to as the middleware layer. It can do any process on data like analyzing, processes, managing, and storing huge amounts of data from another lower layer, providing services, and using a database or cloud computing and decision-making [13].

The business layer that manages the Internet of things system and is considered the highest layer in the structure and needs a high level of requirements to build a successful and sustainable design, especially for the business field and includes applications, user privacy, etc. [14].

3   Related Work

We present a summary of recent work in IoT in the healthcare field in this section. We are going through various papers in this field and briefly explaining each paper's method, domain, and what the article addresses.

References	Research method/techniques	Domain	Description
[1]	Survey	IoT in health care	This paper illustrates or describes the wireless body area network (WBAN) focused IoT healthcare system and examines network architecture, applications, and topology in IoT healthcare solutions Furthermore, investigate the privacy and security features, such as privacy, resources, authentication, control, and so on

(continued)

(continued)

References	Research method/techniques	Domain	Description
[15]	Review	IoT in health care	The paper reviews existing literature on IoT in health care, as well as its supporting technologies, applications, and crucial challenges
[16]	Comprehensive review	IoT	This paper reviews the concept of the Internet of things and how to connect through the Internet. It also presented a detailed review of the most important applications in the Internet of things and the challenges and issues you may face from the researcher's .perspective. And giving new researchers in this field a future scene to improve the Internet of things system, with innovative and unique ideas
[17]	Brief study	IoT	In this paper, a number of research papers focus on IoT applications. As a result, the smart city has the highest proportion of application approaches but the lowest proportion of industrial applications

(continued)

(continued)

References	Research method/techniques	Domain	Description
[18]	Conference paper	HCI and IoT	The paper introduced a smart IoT-based monitoring system to monitor the elderly. The home healthcare architecture and proposed monitoring or ICE (IoT cares for elderly) relies on the Intel Edison platform. And the addition of sensors built into the system to measure vital human signs, sleep patterns, and movement. Output readings will be sent to ICE's main central system to detect any anomalies or errors, give health advice, and even call for help if the elderly are in an emergency. Besides, output readings are also sent to cloud storage, where they can provide real-time information to close family members and caregivers by developing the ICE web and phone application. The system allows the elderly to live a comfortable and healthy life
[19]	Survey	IoT in health care	A survey paper in IoT-based reviews network architectures (platforms) and healthcare technologies, IoT-based in industrial healthcare solutions and applications. Also, it analyzes Internet of things security and privacy features. Further, suggested a smart, collaborative security model reduce security danger
[20]	Survey	IoT	This paper presents different problems and challenges in the smart healthcare system (SHCS) and proposed solutions to overcome them

4 IoT Healthcare Services and Applications

IoT was used in health areas to improve people's quality and nature, manage real-time diseases, improve patient outcomes, and improve user knowledge. Concerning IoT healthcare services and applications, the area can include: care for pediatric, managing of private fitness and health, elderly patients, the management of chronic diseases, among others. In order to provide the best understanding of the subject, we divide the discussion into two categories in this paper: applications and services [21], as shown in Fig. 2.

A. **Healthcare Services**

a. **Ambient-Assisted Living (AAL)**

It supplies health help to the independent living of disabled people, the aged people. It also tries to help the human servant with any problems that arise. In [21], this paper is proposed an IoT AAL architecture for the management of insulin therapy and blood glucose level. The proposed approach was to observe the sick persons at their residence; give a personal or unique card heath depending on the radio-frequency identification and web diabetes management portal. It was proposed in [22] that a combination of KIT technology and closed-loop healthcare systems provide a safe association between people to objects, things to things, or people to people.

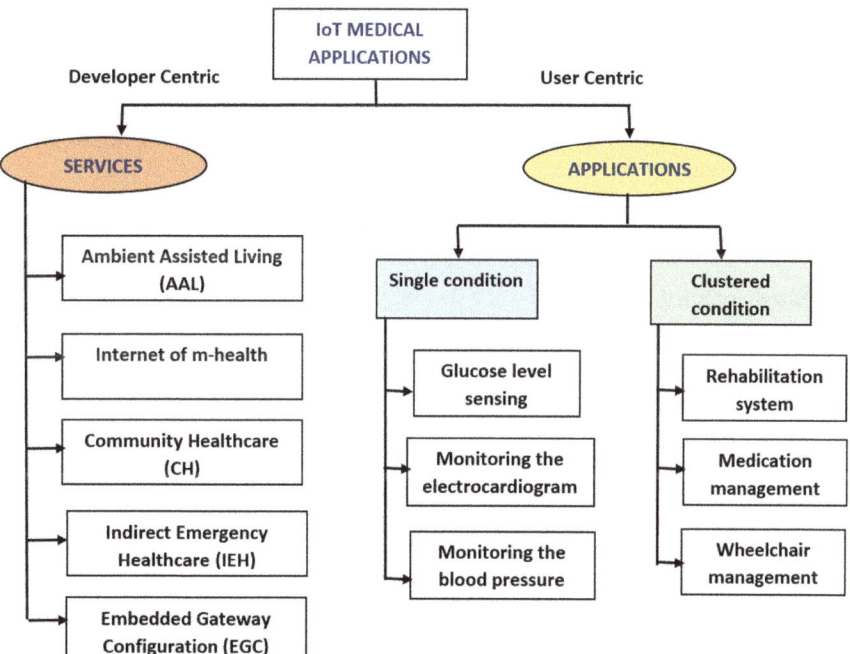

Fig. 2 IoT in medical applications

b. **Internet of m-health**

Utilizing communication technologies and health sensors makes m-health supply healthcare services. [23] introduced a new application concept for m-IoT sensing used for noninvasive blood glucose (BG) level and management in heterogeneous and diverse domain or setting, implemented the method by using the 6LoWPAN and TelsoB nodes sensor for system output verification. In [24], the idea and concept of the 4G health suggested addressing the problems and future implementation concerns in m-health.

c. **Community Health Care (CH)**

Observing the CH brings out an IoT network covering a region around the local community such as a hospital, housing strict or location, and countryside. The network structure is generated by linking many smaller IoT networks. A cooperative IoT network for improved health monitoring and control in the countryside or rural healthcare monitoring has been suggested [25].

This paper proposed the community-medical-network (CMN) for observing local medicine and health system based on GSM/3G/Internet infrastructure and wireless body area sensor network in [26]. The proposed CMN decreases the time and cost of diagnosing and treating the disease.

d. **Indirect Emergency Health Care (IEH)**

In [27], a theory-based immune health observing and risk assessment model was proposed, which analyzes risks and tracks health in eastern locations and provides highly detailed outputs on the health hazards of the one environmental aspect of earthen locations. [28] is proposed a smart community security system.

e. **Embedded Gateway Configuration (EGC)**

An architectural service links the network to the Internet as well as other health devices. Embedded services are used by healthcare systems in the paper [29]. Rasid et al. [30] proposed an IoT-based platform for healthcare applications, that is, stable, accessible, and resilient.

B. **Applications**

There are two types of IoT applications: clustered condition and single condition.

i. **Single condition**

The single condition applications, it considered for a specific disease.

a. **Glucose level sensing**

Diabetes mellitus is a health case that any human who has high glucose levels for a long time. An interactive healthcare system for mobile was proposed in [28] to link diabetic illness and IoT technology in a two-way manner.

b. **Monitoring the electrocardiogram (ECG)**

It is used to check the electrical track of a heart rate activity. In IoT-based ECG monitoring, the sensors are located in the place shown in Fig. 3 provides steady data about the rhythm and the heart level or rate. In [20], a smart home-based platform the iHome Health-IoT is proposed for enhancing interchangeability and connectivity. The wearable biomedical sensor appliance or device joined to inkjet printing technology.

Fig. 3 ECG [31]

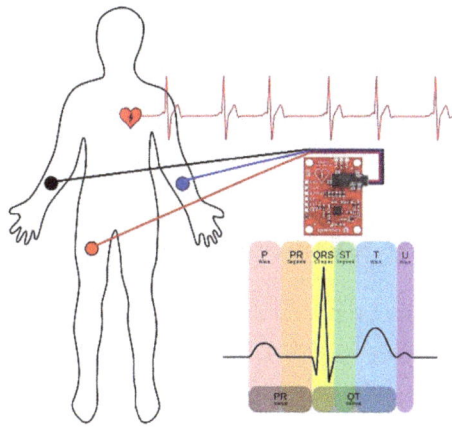

c. **The blood pressure Monitoring (BP)**

(BP) The force uses the heart to pump the blood in the human body. In [25], the cooperative IoT method was proposed for controlling and monitoring the variables of health like (BP), blood sugar, and hemoglobin (HB). In [32], an intelligent health service (HIS) is empowered to provide an overall evaluation by collaborating to share risk factors for tracking blood pressure, obesity, and diabetes between personal health devices.

ii. **Clustered condition**

The term "clustered condition applications" refers to the ability to treat many diseases at the same time.

a. **Rehabilitation system**

The recovery method increases the quality of life, while IoT aims to solve issues related to the unavailability of health specialists or experts and the aging population. The body sensor network (BSN) is suggested in this paper [33] to enhance therapeutic operation. In [34], an ontology-based automating design methodology (ADM) is proposed for providing intelligent recovery systems.

b. **Medication management**

It is one of the major concerns in public health, as are the many financial issues or burdens that the drug imposes. The Internet of things ensures a novel solution to this dilemma [35]. In paper [36], a preventive and pervasive drug control approach addressing the medication noncompliance problem is proposed to resolve the issues associated with it.

c. **Wheelchair management**

An intelligent wheelchair is specially developed for disabled persons. The wheelchair movements are controlled by using the computerized system. [37] was suggested a wireless body sensor network (WBSN) for monitoring a wireless heart level, control actuators, ECG sensors, and detection of the pressure.

5 IoT Challenges

Daily, many people died due to their health carelessness because they do not have time for their health monitoring and management. Therefore, the latest research and implementation have been made to use IoT technology in the healthcare industry. IoT-based healthcare systems are developed for normal users, patients, and doctors. These devices properly measure the human body's parameters, such as electrocardiogram (ECG) records, temperature, and heartbeat. Several IoT-based healthcare products and applications are working in daily life, but many open issues and challenges still require proper research and improvements very carefully. This segment addresses some of the most popular yet crucial problems in IoT healthcare systems.

5.1 Scalability

In any device, scalability is important. Especially in the healthcare system, which requires many changes and has space to insert new functionalities. Scalability refers to a device's ability to adapt to changes in the environment and meet potential specifications. Since IoT-based healthcare systems are made up of several sensors and other smart devices that exchange data over the Internet, they tend to be less scalable. However, the majority of linked systems and sensors do not perform consistently. This lack of uniformity between medical devices decreases the healthcare system's scalability. To function properly for an IoT-based healthcare system, all these devices should be managed, maintained, controlled, follow protocols, and power conventions. There is huge work available, but current approaches to handling this challenge are ineffective and fail to scale many IoT objects. The most important benefit of scalability is system performance, where the system operates gracefully and function properly without any delay and makes effective use of the available resources [1]. To make it more efficient, proper functioning with high performance, it is important to improve healthcare devices' scalability.

5.2 IoT Healthcare Security

Healthcare devices and software deal with sensitive data and personal healthcare records, but instruments in the IoT healthcare systems use the Internet to share information. IoT-based healthcare systems and their applications are developed based on private information acquisition and sharing over the Internet [15]. Many patients and the doctor's massive stored data can be easily attacked and violated by many patients. On the other side, hackers can generate fake IDs to buy drugs and medications. Security is an open challenge in various Internet-based applications. These concerns persist, however, since the IoT architecture in health care isn't well-established, and it lacks knowledge about how to maintain data privacy and protection. As a result, data privacy and security are significant challenges in IoT-based healthcare systems and applications.

5.3 Low Power in IoT Healthcare Device

Almost all the IoT-based sensors and devices are resource constraints such as small size, limited battery, and little processing power. Their resource limitation can be considering their advantages but, in many applications, these are deal with as disadvantages. Similarly, in an IoT-based healthcare system, these resource constraints, especially low power, are considered a challenge because these devices save energy when running these devices on the power-saving mode. Even so, when the device isn't performing well, most sensors switch off automatically.

Despite being switched on in power-saving mode, the instrument has a power shortage due to several duty requirements. Long-term monitoring applications have a low power consumption and require minimal input from the system wearer [1]. If the device's power level drops below a certain threshold, the machine automatically shuts it down and takes the patient offline user of low power IoT systems are critical for lowering the risk of patients being disconnected.

5.4 Network Architecture

No single IoT architecture is agreed upon universally; therefore, many architectures with different properties are proposed for other applications. The most common IoT architectures are three-layer and five-layer IoT architecture. But the main concern in IoT-based healthcare systems is that the IoT network architecture should give authorization and authentication for the users in the healthcare system. Researchers have suggested various types of network architectures. Both architectures, three-layer, and five-layer, for IoT-based healthcare system, have their cons. The three-layer architecture cannot satisfy healthcare applications' requirements, and there are low storage and low energy capabilities issues in the latest five-layer IoT architecture. Low-cost, lightweight health-care devices can only save a few bytes of data. The IoT network architecture in health care remains a source of concern because it does not meet the power, storage, security, and privacy requirements of IoT-based healthcare systems [20].

5.5   Cost Analysis of the Internet of Things Healthcare System

Healthcare equipment has always been expensive, but today's IoT-based healthcare system dominates them. There is no comparative cost work between past and today's devices in literature. As a result, cost analysis is still a work in progress in the IoT-based healthcare system [20]. The high cost of monitoring equipment in an IoT-based healthcare system is a significant issue even in developing countries. The Internet of things has not yet made health care more affordable to the average person. As it is a smart era and everyone wants to use IoT-based healthcare devices, healthcare devices' high cost is a worrying sign for several healthcare institutes and individual users.

5.6   The Application Development Process

Smartphone application developers face security, privacy, database, and integrity issues. It is possible to control, track, monitor, and maintain multiple devices via a single application. It is necessary to take care of patient and users' data by properly ensuring that data is shared with others. Security and privacy are major concerns for protecting patient or consumer information from leakage. On a minute-by-minute basis, the mobile application gathers patient health-related data, making it difficult for developers to properly handle the big data and derive useful insights from it. These mobile healthcare applications are often suggested to patients who want to keep track of their everyday routines; the most pressing requirement is that the application be driven by artificial intelligence, which will provide real-time activity feedback to patients and send data to relevant doctors.

5.7   The Quality of Services

IoT-based systems are now primarily used for everyday and real-time applications. The accuracy and timeliness of IoT data used for decision support are inextricably related to service quality (QoS). Since real-time applications cannot tolerate delays, it entails collecting, transmitting, processing, analyzing, and using data created by healthcare sensors in a timely manner. In certain situations, IoT devices fail to deliver needed data on time, resulting in low performance and unnecessary long delays for the system and application. The IoT-based healthcare framework, where applications are built to adopt QoS, faces a major challenge. IoT devices often produce massive quantities of real-time data, resulting in a massive data crisis. Big data presents a major challenge when it comes to integrating and matching such data with historical patient data in order to obtain realistic diagnostic feedback in a timely manner (while retaining QoS) [15]. Medical wearable systems need a strict QoS guarantee to operate properly since they perform real-time and life-critical applications. This creates a significant gap in heterogeneous data collection, real-time patient tracking, and automated decision support based on QoS. As a result, it is critical to overcome these obstacles in order to achieve real time, high-performance systems, and applications.

5.8 Continuous Monitoring for Medical Purposes

Most diseases are long term in nature, necessitating long-term care of many patients, mostly the elderly and those suffering from chronic diseases. Constant data gathering, sharing with doctors, data monitoring by patients and doctors, and logging are very important in this context. Smart devices and sensors must submit data to the appropriate doctor on time in order to provide proper and consistent monitoring. [20] Formalized paraphrase However, it has some disadvantages that make continuous monitoring difficult. If the patient crash devices or its battery dies, it is not possible to achieve remote data sharing. There continuous monitoring would also be adversely affected by how it mainly depends.

6 Conclusion

To summarize, it became clear why the Internet of things is necessary. Information technology has become important in our lives and has entered into many areas such as infrastructure, health care, retail, personal use, manufacturing, etc. The Internet of things was created to collect data and information from end-users and applications in our lives. Therefore, many major companies have focused on developing and improving the Internet of things for users and many fields. IoT health care is one of the most important technical areas of great importance. In this paper, we have discussed services, applications, and IoT challenges in health care.

References

1. Dhanvijay MM, Patil SC (2019) Internet of Things: a survey of enabling technologies in healthcare and its applications. Comput Netw 153:113–131
2. Lueth K (2014) Why it is called internet of things: definition, history, disambiguation [Online]. Iot-analytics.com. Available at https://iot-analytics.com/internet-of-things-definition
3. Postscapes (2019) Internet of Things (I.O.T.) History|Postscapes [Online]. Available at https://www.postscapes.com/iot-history/

4. O'Dea S (2019) Global 5G I.O.T. Endpoint installed base by segment 2023|Statista [Online] Statista. Available at https://www.statista.com/statistics/1061195/5g-iot-endpoint-installedbase-by-segment-worldwide
5. Ranger S (2020, 3 Feb) https://www.zdnet.com/article/what-isthe-internet-of-things-everything-you-need-to-know-about-the-iot-rightnow/. Retrieved from ZDNet
6. Betters E (2018, 20 Feb) https://www.pocketlint.com/apps/news/126559-internet-of-things-explained-your-completeguide-to-understanding-iot. Retrieved from pocket-lint
7. Burgess BM (2018, 16 Feb Friday) https://www.wired.co.uk/article/internet-of-things-what-is-explained-iot. Retrieved from WIRED https://www.wired.co.uk/article/internet-of-thingswhat-is-explained-iot
8. Teksun (2017, 2 Sep) https://medium.com/@TeksunGroup/importance-of-internet-of-things-iotin-our-live-b71e53d50a44. Retrieved from medium
9. Al-Fuqaha A, Guizani M, Mohammadi M, Aledhari M, Ayyash M (2015) Internet of things: a survey on enabling technologies, protocols, and applications. IEEE Commun Surveys Tutor 17(4):2347–2376
10. Duan R, Chen X, Xing T (2011, Oct) A QoS architecture for I.O.T. In: 2011 international conference on Internet of Things and 4th international conference on cyber, physical and social computing IEEE, pp 717–720
11. Sharma R, Pandey N, Khatri SK (2017, Sept) Analysis of IoT security at network layer. In: 2017 6th international conference on reliability, Infocom technologies and optimization (trends and future directions) (I.C.R.I.T.O.). IEEE, pp 585–590
12.] Jabraeil Jamali MA, Bahrami B, Heidari A, Allahverdizadeh P, Norouzi F (2020) IoT architecture. In: Towards the Internet of Things. E.A.I./Springer innovations in communication and computing. Springer, Cham
13. Pallavi S, Sarangi SR (2017) Internet of Things: architectures protocols and applications. J Electr Comput Eng 2017:1–0147
14. Calihman A (2019) IoT architectures—common approaches and ways to design IoT at scale. Retrieved 11 Nov 2020, from https://www.netburner.com/learn/architectural-frameworks-in-the-iot-civilization/
15. Hussain F (2018) The application of the Internet of Things in healthcare. Int J Comput Appl 180(18):19–23
16. Khanna A, Kaur S (2020) Internet of Things (IoT), applications and challenges: a comprehensive review. Wirel Pers Commun 114:1687–1762. https://doi.org/10.1007/s11277-020-07446-4
17. Sathiyanathan N, Selvakumar S, Selvaprasanth P (2020) A brief study on IoT applications. Int J Trend Sci Res Dev (IJTSRD) 4(2). Available online www.ijtsrd.com e-ISSN: 2456-6470
18. Hu BDC, Fahmi H, Yuhao L, Kiong CC, Harun A (2018, Aug) Internet of Things (IOT) monitoring system for elderly. In: 2018 international conference on intelligent and advanced system (ICIAS). IEEE, pp 1–6
19. Islam SR, Kwak D, Kabir MH, Hossain M, Kwak KS (2015) The internet of things for health care: a comprehensive survey. IEEE Access 3:678–708
20. Yang G, Xie L, Mantysalo M, Zhou X, Pang Z, Da Xu L et al A health-IoT platform based on the integration of intelligent packaging, unobtrusive bio-sensor, and intelligent medicine box. IEEE Trans Ind Inform 10
21. Riazul Islam S, Kwak D, Humaun Kabir M, Hossain M, Kwak K-S (2015) The internet of things for health care: a comprehensive survey. IEEE Access 3:678–708
22. Dohr A, Modre-Opsrian R, Drobics M, Hayn D, Schreier G (2010) The internet of things for ambient assisted living. In: 2010 Seventh international conference on information technology, pp 804–809
23. Istepanian RSH, Hu S, Philip NY, Sungoor A (2011) The potential of Internet of m-health Things "mIoT" for noninvasive glucose level sensing. In: 2011 annual international conference of the IEEE engineering in medicine and biology society, pp 5264–5266
24. Istepanaian RSH, Zhang YT (2012) Guest editorial introduction to the special section: 4G health—the longterm evolution of m-Health. IEEE Trans Inf Technol Biomed 16:1–5

25. Rohokale VM, Prasad NR (2011) A cooperative Internet of Things (IoT) for rural healthcare monitoring and control. In: Proceedings of international conference on wireless communication vehicle technology information theory aerospace electronics system technology (wireless VITAE), pp 1–6
26. Lei Y, Chungui L, Sen T (2011) Community medical network (C.M.N.): architecture and implementation. In: 2011 global mobile congress (GMC), pp 1–6
27. Xiao Y, Chen X, Wang L, Li W, Liu B, Fang D (2013) An immune theory based health monitoring and risk evaluation of earthen sites with Internet of Things. In: IEEE international conference on and IEEE cyber, physical and social computing green computing and communications (GreenCom)
28. Chang SH, Chiang RD, Wu SJ, Chang WT (2016) A context-aware, interactive M-health system for diabetics. IT Prof 18:14–22
29. Liu J, Yang L (2011) Application of Internet of Things in the community security management. In: Third international conference on computational intelligence, communication systems and networks (CICSyN), pp 314–318
30. Rasid MFA, Musa WMW, Kadir NAA, Noor AM, Touati F, Mehmood W et al (2014) Embedded gateway services for Internet of Things applications in ubiquitous healthcare. In: 2nd 298 international conference on information and communication technology (ICoICT), pp 145–148
31. https://csdt.org/culture/performingarts/sensing-infosheet.html
32. Lee BM, Ouyang J Intelligent healthcare service by using collaborations between IoT personal health devices. Int J Bio-Sci Bio-Technol 6
33. Tan B, Tian O (2014) Short paper: using B.S.N. for tele-health application in upper limb rehabilitation. In: IEEE world forum on Internet of Things (WF-IoT), pp 169–170
34. Fan YJ, Yin YH, Xu LD, Zeng Y, Wu F (2014) IoT-based smart rehabilitation system. IEEE Trans Ind Inf 10:1568–1577
35. Riazul Islam SM, Kwak D, Kabir MH, Hossain M, Kwak KS (2015) The Internet of Things for health care: a comprehensive survey. IEEE Access
36. Pang Z, Tian J, Chen Q (2014) Intelligent packaging and intelligent medicine box for medication management towards the Internet-of-Things. In: 16th international conference on advanced communication technology (ICACT), pp 352–360
37. Yang L, Ge Y, Li W, Rao W, Shen W (2014) A home mobile healthcare system for wheelchair users. In: IEEE 18th international conference on computer supported cooperative work in design (CSCWD), pp 609–614

Chapter 32
Machine Learning Algorithms for Predication of Traffic Control Systems on VANET's

G. Bindu and R. A. Karthika

1 Introduction

The intelligent Transport System (ITS) targets towards the delivery of worthy traffic administration with the provisions of automated user alerts on traffic condition based on situations in particular surrounding or environment [1]. The quantity of vehicles in urban areas are ceaselessly expanding with financial and social turn of events. The incessant event of traffic clog in cities display an impact on economy with present condition. Because of the constrained road structure of enormous urban communities and limitations to transportation foundation development from financial elements, to manage the traffic and control measures in a sensible and compelling manner to enhance the effectiveness of existing transportation agencies, and oblige the developing traffic in huge urban areas have become critical exploration substance for neutralizing urban traffic blockage [2]. Traffic control is one of the most significant specialized intends to direct traffic stream in order to reduce the clog, and even to diminish pollution. The recent growth and advancement has consistently been joined by the improvement of data innovation, PC innovation with networks supports in VANET, and a framework science with the assistance of OBU and Road Side Units (RSU) [3] For instance, the adaptability of nodes communicates to each vehicle from Vehicle to Infrastructure (V2I), V2V and RSU a greater network is formed with the dynamics of VANET for the road safety and diver efficiency. The self-versatile control framework can alter the sign planning boundaries continuously as indicated by the main focus of the administrator, for example, the base deferral of the convergence and the appearance attributes of the traffic stream at the crossing point will be organized with the V2V and V2I communication for an incited control, the

G. Bindu (✉) · R. A. Karthika
Department of Computer Science Engineering, School of Engineering, Vels Institute of Science Technology and Advanced Studies (VISTAS), Chennai, India
e-mail: bindu.se@velsuniv.ac.in

© The Author(s), under exclusive license to Springer Nature Singapore Pte Ltd. 2021
S. Kumar et al. (eds.), *Proceedings of International Conference on Communication and Computational Technologies*, Algorithms for Intelligent Systems,
https://doi.org/10.1007/978-981-16-3246-4_32

self-versatile regulatory framework can utilize the general traffic limit of the street arrangements and successfully improve the effectiveness of vehicles in a street to organize traffic [4]. The traffic information gathered by traffic control framework using a traffic detection indicator and other existing sensors are constrained with limited data. Thus there is a great option to organize the activity of urban traffic by the participation of machine learning algorithm to control traffic signal control and driving practices in VANET's [5]. The vehicular sensors are additionally getting more astute with time, bringing about a capacity of the vehicles to more readily evaluate the earth. This headway has prompted the chance of acknowledging self-ruling driving that depends on mirroring human driving conduct while relieving human mistakes, there are plenty of utilizations have been created beginning from dynamic and normal street condition with a upgrade in traffic alerts, going from independent vehicles to smart transportation through communication between vehicles to road side units (V2R) and V2V [6].

Figure 1, displays the classification of machine learning algorithms and their association in demonstrating the analytical data in a supervised and unsupervised methods. These machine learning algorithms are unique with its features on classifying and making regression, in further the deep neural networks of machine learning

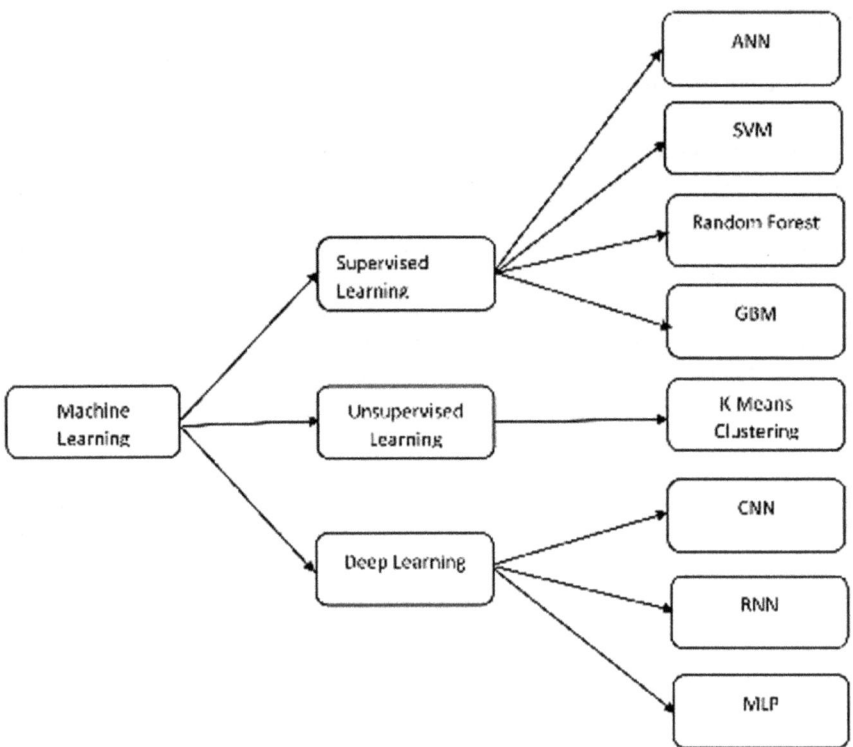

Fig. 1 A machine learning model

assist in classifying complex data structures and provide authenticable information at VANET's.

2 Related Works

The significant analysis on traffic prediction have been initiated in the course of recent years and distinctive directing conventions are proposed to address the difficulties in VANET [7]. In this segment, a portion of these as of late created directing conventions alongside their confinements are examined with machine learning algorithms [8]. Utilizing a trusted steering conventional algorithm, an improved algorithm is proposed for the traffic correspondence thinking about the location, speed, course, and safety. This conventional algorithm endeavors to diminish the connected mishaps, traffic jamming using decision tree algorithm [9]. In any case the high traffic density which causes channel clog and neglects the course support process for various reasons. A machine learning algorithm is introduced to direct and predicts the neighbor vehicles expected position. Nevertheless, the quality and outer strings are not considered in this algorithm for estimating delay during traffic correspondence and not ready to anticipate future connection disappointments during correspondence [10]. The high-density urban areas have been utilized to take care of the blockage issue. Be that as it may, security issues have not been proposed in this study. There is an autonomic processing design that copies the inspirations and responses of body's sensory system to accomplish the self-administration of a neural network [11]. This neural network is made out of six capacities that structure a round technique. Six capacities have observing, examination, arranging, execution, detecting, and affecting. This paper coordinates the autonomic figuring engineering with a VANET-based route framework [12]. Vehicles gather messages around as onlookers, and the indicators illuminate while they start a movement. The Traffic Information Centre (TIC) controlling the first four capacities takes after a cerebrum that plans courses relying upon traffic data [13]. The vehicles get a proposed course that coordinates the driver by means of alerts [14]. Thus, a Global Positioning System (GPS) supports in knowing the direction and position of vehicle in order to finding routes for destination. In urban it could be shaky of GPS signals hindered by towers or structures; in any case, vehicles despite everything have precision area in light of sign sharing. Drives utilize the route framework to get some place efficiently or to maintain a strategic distance from hindrances on streets cautiously even it can give the fuel-efficient courses satisfied for their clients [15]. The propelled route framework additionally called the circulated route framework, claims vehicular interpersonal organizations where they share traffic data. In the disseminated route framework, vehicles are separated into a few groups and the significant traffic data is amassed into one way. The machine learning algorithms proving a greater efficiency in processing a raw data to an optimized data [16].

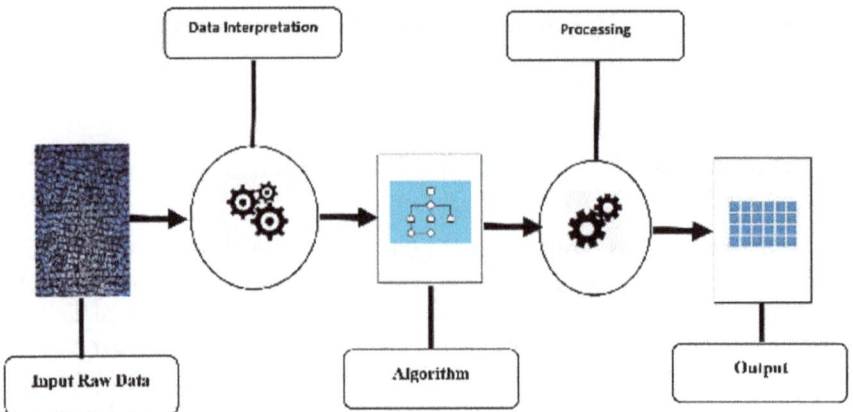

Fig. 2 Traffic prediction using machine learning

3 Traffic Prediction with Machine Learning

There are different types of machine learning algorithms ant comes under Artificial Intelligence (AI). The machine learning algorithms is divided into three major kind that is supervised learning, unsupervised learning and deep learning [17]. The various types of machine learning procedures have its own features and that can be identified with the outputs obtained from traffic at urban areas during peak hours by training and testing these data. In light of the reasonable information, the data is trained and processed for testing. At that point in the testing stage the forecast is made. The supervised machine learning is administered by taking in gains from a lot of marked information mostly by regression and classification of data. In the event that the preparation information just incorporates distinct qualities, at that point it is an issue related to regression and the yield of the prepared model, as it is an order which is additionally discrete [18]. Then again on the off chance that the preparation information contains consistent qualities, at that point it is the relapse issue and the yield of the prepared model will be a forecast.

In Fig. 2, the data processing to algorithm is depicted with data input in raw for further it is interpreted to train and tested in algorithm, as the data from a remote source it is processed to various algorithms according to the purpose get the highest accuracy on output (Fig. 3).

4 Supervised Learning

There are some general supervised machine learning algorithms (SMLA) are Random Forest (DT), Support Vector Machine (SVM), Decision Tree (DT) and Gradient Boosting Machine (GBM). The regression output obtained is a persistent worthy

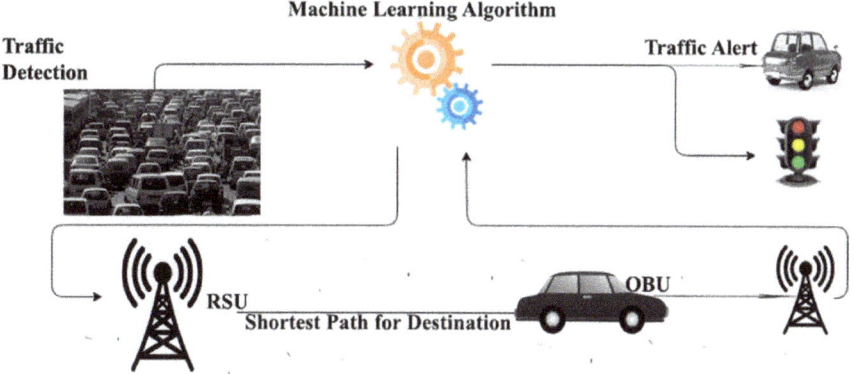

Fig. 3 Traffic control using machine learning

data may provide the traffic information based of an electric vehicle battery, level of traffic blockage at different crossing points, and sticking estimates. In vehicular informal organizations, the results can be utilized to foresee boundaries, for example, arrange throughput of dynamic driver data frameworks, obstruction discovery, and anticipating complex traffic types by means of available unlabeled information [19]. This plan attempts to locate an effective portrayal of the unlabeled information. For instance, the highlights of an information can be caught by some shrouded factors that can be spoken to by the Naive Bayesian learning methods. Bunching is a type of solo discovering that gatherings tests with comparable highlights. The info highlights of every information point can be its total portrayal or a relative comparability level with other information focuses. In the remote systems worldview, the bunch arrangement for the progressive conventions is vital as far as vitality for the board, where every part simply needs to speak with the group head before speaking with the individuals from different groups. Some customary grouping algorithms, range bunching, and hierarchical bunching. Measurement decrease is another subclass of unaided ML conspire [20].

The fundamental thought behind measurement decrease is to down-example the information from a higher measurement to a lower measurement with no huge information disaster. The application of SMLA for most applications require measurement decrease because of various reasons. Revile of information dimensionality is the primary explanation. In grouping, arrangement and improvement, the general model complexity increments significantly with the expansion in include measurements [21]. The subsequent explanation is the obstacle in the learning procedure. In a large portion of the cases the highlights of the information tests are connected in certain angles, however in the event that the component esteem is influenced by commotion or impedance, at that point the individual result of the relationship will be defiled and the learning procedure will be influenced. Such sort of measurement decrease in the vehicular informal communities is the development which prompts a vehicular bunch [22]. The bunch head gathers and transmits the data to the vehicle nodes in order to lessen the correspondence cost. The scourge of dimensionality can be diminished by the dimension decrease strategies. Measurement decrease techniques are assembled in two classes.

5 Unsupervised Learning

In this method of machine learning mainly depends on unlabeled information. This method attempts to locate an effective portrayal of the unlabeled information. For instance, the highlights of an information can be caught by some hidden layers that can be spoken to by the Bayesian learning methods. Bunching is a type of unaided discovering that gatherings tests with comparative highlights. The information highlights of every information point can be its outright portrayal or a relative comparability level with other information focuses on the cluster change for the various leveled conventions that is vital for learning methods, as far as vitality is administered with every part of data need is communicated with the facts of other data with is relationship. Some customary grouping calculations is K means clustering and hierarchical clustering are used in prediction. The measurement decrease is another subclass of unaided data, whereas the principle thought behind measurement decrease is to minimize the sample information from a higher capacity to a lower quantity with no huge missing data. The submission of these methods requires data measurement as it can be decreased because of various reasons [24]. The subsequent explanation is the obstacle in the learning procedure. In the greater part of the cases the highlights of the information tests are connected in certain angles, however in the event that the component esteem is influenced by clamor or obstruction, at that point the separate result of the relationship will be ruined and the learning procedure will be influenced. Such sort of measurement decreases in the vehicular informal communities and in arrangement which prompts a vehicular group. The bunch head gathers and transmits the data to the different nodes to decrease the correspondence cost. The scourge of dimensionality can be diminished by the data measurement decrease techniques.

6 Deep Learning (DL)

The deep learning is firmly identified as a part of Machine Learning Algorithms these are used to estimate a more complex system of neurons in different layers. It expects to extricate information from the information portrayals that can be produced from the other algorithms. The system comprises of various layers like input, hidden and output for every neuron has a distinct change and work with the nonlinear function to transmit the data, like ReLU, the mounting of information is extremely vital as this can seriously influence the forecast or arrangement of a VANET's. Since the quantity of layers found as hidden will expands, the ability of the system to adapt the changes and likewise it develops a network. Nevertheless, after a specific point, any expansion in the layers gives no improvement in the prediction of VANET's [25]. The preparation of a more profound system is additionally testing since it requires broad computational assets, and the inclinations of the systems may detonate or disappear. The positioning of devices with nodes are work with more profound systems that has raised the significance of edge registering innovation. Vehicles progressing can get advantage from portable edge registering servers with the possibility of neurons, layers in neural network for instance Multi-Layered Perceptron (MLP) [26].

7   Traffic Prediction to Elude Traffic Jamming

There is a framework which permits vehicles to speak with traffic lights commonly known as Green Light Optimal Speed Advisory (GLOSA). This is initiated to educate drivers about the speed they should drive as they approach intersections to evade the red lights [27]. The practice of V2V communication will be in this way to keep the drivers from fast driving so as to attain a gradual traffic movement without congestions. Thus, this also improve the air quality by diminishing the general cruel quickening to reach their destinations or slowing down vehicles unnecessarily as the vehicles connect with RSU and infrastructure simultaneously it communicates between V2V with the influence OBU deployed in vehicles of VANET's. The purpose of GLOSA application is to assist in providing information related to the utilization of fuel to reach their targets and also gives the alerts associated with time taken at traffic signal. The traffic movements estimation in reality has as of late, pulled in the VANET research network, because of its more noteworthy impact in the Intelligent Transport System (ITS). The unwavering quality and security of VANET is critical to ensure the fruitful organization of VANET's applications. The unpredictable condition of VANET has made the system abnormality identification all the more testing because of portability, structures, and fleeting connections by utilizing CNN this condition will be resolved, as their fundamental investigations have demonstrated that ML based strategy is viable when contrasted with the conventional inconsistency location strategies [28].

8   Traffic Movement Control in VANET

The traditional and active traffic light control has been utilized in programming the signal lights at intersection by a Fuzzy Logic method, because of restrictions of computational assets and recreation devices but in the case of VANETs the vehicles can communicate with various sources with a machine learning algorithm which provide the instructions depends on position and speed of vehicles, with the activity engaged using CNN [29]. The state of vehicle is also communicated by the Queue length and the activity utilizes at various stages, the distinction between streams in dual segments at the rush hour gridlock headings also predicted by utilizing Stacked

Auto-Encoders (SAE) for knowing the vehicle directions and the traffic conditions. Since the vehicle state is picked as the situation of activity that is made out of 2 stages, which incorporates the time taken at the stop and the adjustment for deferral time, the time utilized the position and speed as states, the activity is made out of 4 stages and the change in combined staying time, utilizing CNN experience an improved method to manage traffic signals and V2V communication, as the control of traffic crossing point situation contains different stages, which speaks to a high-measurement activity space [30]. The work likewise ensures that the traffic signal time easily changes between two neighboring activities. The deep neural network has presented an option that exploits the constant GPS information and figures out how to control the traffic lights at a disengaged traffic interval with the Recurrent Neural Network (RNN) to associate between intersections, its presentation with the standard traffic convergence (Fig. 4).

The data about time productive traffic stream is significant preceding the organization of different ITS applications. These applications can effectively utilize this data for the errands, for example, traffic blockage amendment, better usage of PEV charge, limiting fuel utilization and improving area-based administrations. The traffic stream information can be procured from various sources, for example, traffic cameras, swarm-based data administrations, vehicles and so on. This information can be taken for progressive or in disconnected vehicle environment e.g., hourly, week after week, month to month or yearly. This information can be extremely valuable for the pre-expressions of different ITS applications by utilizing AI methods. The traffic expectation strategies are partitioned into classes, to attain high exactness of traffic stream expectations is a major test for traditional traffic stream assessment

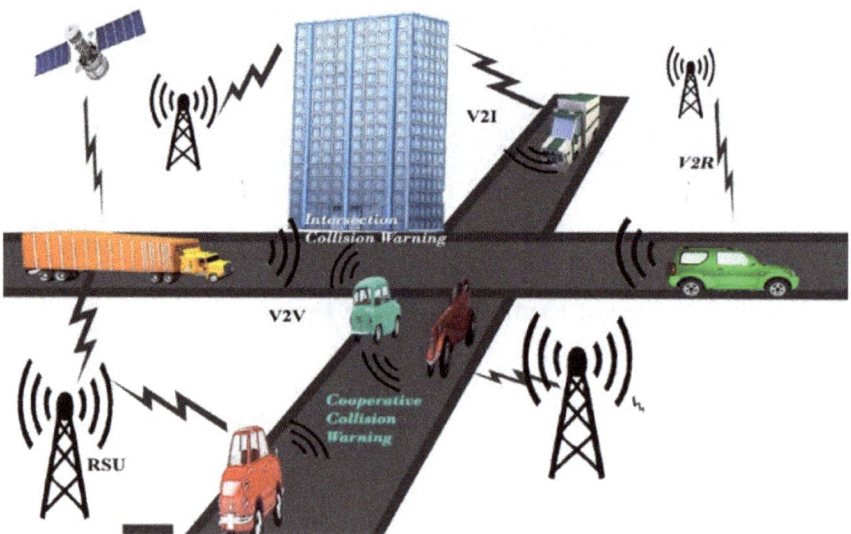

Fig. 4 VANET traffic control system

strategies. Some examination on traffic stream forecast shows magnificent execution improvement over regular methodologies [31]. A novel stacked auto encoder-based traffic stream forecast strategy is presented to encode the traffic data as this auto encoders are utilized as building squares to speak to highlights of traffic stream that are utilized for the Deep Learning and forecasts with improved precision.

9   Vehicle to Vehicle Correspondence at VANET

The vast majority of the recently examined research concentrated on the expectation of traffic stream for the interstates, where the traffic stream is generally smooth. In any case, in the urban city situation the traffic lights would have more prominent effect on the traffic stream because of speed varieties of vehicles that anticipates on traffic [32]. The V2V correspondence mainly based on the spectrum availability during peak hours through RSU and Road side Infrastructure (RSI) [33]. The traffic stream expectation is a difficult undertaking without joining the publicly supported data, because of the mind-boggling nature of connection between the group and the vehicles. The traffic and system clog are basic at the cross segments in the urban condition. A machine learning algorithm can resolve the issues related to spectrum availability, which depends on k nearest neighbor, utilizes a focal controlled way to deal with the vehicle correspondence and the vehicles halting at the red light at the convergences can utilize data from RSI and transmit the controls to remote channel between areas of traffic jam. The transmission information is bunched into various gatherings by utilizing k-nearest neighbor as this approach dependent on the data transmitted to vehicle to everything (V2X) thus the separation of message types and headings of message sending to vehicles is autonomous based on this algorithm and the corresponding boundaries, for example, transmission rate, trans-mission power will be established. The channel must be made accessible before transmission to maintain a strategic distance from the vehicles. In V2X situations, the profoundly unique correspondences, complex relationship among different transmis-sion modes challenging for the restricted range resources, time shifting information rates, and vehicle-portability, make it trying to ideally dispense the accessible range is established through Cognitive Radio (CR) at VANET's [34].

10   Conclusion

The machine learning algorithms has a major role in predicting the traffic based on supervised, unsupervised, deep learning methods from the data transmitted and obtained from Vehicles. This study has showcased various machine learning methods of AI, as these algorithms are driven from the data of VANET's applications based on the communication from V2V, V2I, V2R and V2X. The determination by classifying the data could be achieved by supervised algorithms that help each other to draw out an ideal arrangement which would not cause or produce issues in the area they are not proposed for. The data transmitted from vehicle using various devices are accessed using machine learning algorithm especially with CNN and the KNN which has the ability to elude collision between vehicles. Thus, the machine learning algorithm has it vital role in traffic prediction at VANETs.

References

1. Hasan MK, Sarker O (2020) Routing protocol selection for intelligent transport system (ITS) of VANET in high mobility areas of Bangladesh. In: Uddin M, Bansal J (eds) Proceedings of international joint conference on computational intelligence. Algorithms for intelligent systems. Springer, Singapore
2. Elhoseny M, Shankar K (2020) Energy efficient optimal routing for communication in VANETs via clustering model. In: Elhoseny M, Hassanien A (eds) Emerging technologies for connected internet of vehicles and intelligent transportation system networks. Studies in systems, decision and control, vol 242. Springer, Cham
3. Dutta AK, Elhoseny M, Dahiya V, Shankar K (2020) An efficient hierarchical clustering protocol for multihop Internet of vehicles communication. Future Internet of Veh 31(5):e3690
4. Gao H, Liu C, Li Y, Yang X V2VR: reliable hybrid-network-oriented V2V data transmission and routing considering RSUs and connectivity probability. IEEE Trans Intell Transp Syst. https://doi.org/10.1109/TITS.2020.2983835
5. Kamble SJ, Kounte MR (2020) Machine learning approach on traffic congestion monitoring system in internet of vehicles. Proc Comput Sci 171:2235–2241. https://doi.org/10.1016/j.procs.2020.04.241
6. Abassi R (2019) VANET security and forensics: challenges and opportunities. Wire Forensic Sci 1(2):e1324. https://doi.org/10.1002/wfs2.1324
7. Sun P, AlJeri N, Boukerche A DACON: a novel traffic prediction and data-highway-assisted content delivery protocol for intelligent vehicular networks. IEEE Trans Sustain Comput. https://doi.org/10.1109/TSUSC.2020.2971628
8. Wu J, Fang M, Li H, Li X (2020) RSU-assisted traffic-aware routing based on reinforcement learning for urban vanets. IEEE Access 8:5733–5748. https://doi.org/10.1109/ACCESS.2020.2963850
9. Balta M, Özçelik I (2020) A 3-stage fuzzy-decision tree model for traffic signal optimization in urban city via a SDN based VANET architecture. Future Gener Comput Syst 104:142–158, https://doi.org/10.1016/j.future.2019.10.020
10. Srivastava A, Prakash A, Tripathi R (2020) Location based routing protocols in VANET: issues and existing solutions. Veh Commun 23:100231. https://doi.org/10.1016/j.vehcom.2020.100231
11. Aqib M, Mehmood R, Alzahrani A, Katib I (2020) A smart disaster management system for future cities using deep learning, GPUs, and in-memory computing. In: Mehmood R, See S, Katib I, Chlamtac I (eds) Smart infrastructure and applications. EAI/Springer innovations in communication and computing. Springer, Cham
12. Behbahani H, Amiri AM, Nadimi N, Ragland DR (2020) Increasing the efficiency of vehicle ad-hoc network to enhance the safety status of highways by artificial neural network and fuzzy inference system. J Transp Safety Secur 12(4):501–521. https://doi.org/10.1080/19439962.2018.1501785
13. Abbas G, Abbas ZH, Haider S, Baker T, Boudjit S, Muhammad F (2020) PDMAC: a priority-based enhanced TDMA protocol for warning message dissemination in VANETs. Sensors 20(1):45

14. Haider S, Abas G, HaqAbbas Z, Boudjit S, Halim Z (2020) P-DACCA: a probabilistic direction-aware cooperative collision avoidance scheme for VANETs. Futur Gener Comput Syst 103:1–17. https://doi.org/10.1016/j.future.2019.09.054

15. Beheshti S, Adabi S, Rezaee A (2020) Location-aware distributed clustering with eliminating GPS in vehicular ad-hoc networks. Preprints, 2020010367. https://doi.org/10.20944/preprints 202001.0367.v1

16. Patil S, Ragha L (2020) Deployment and decentralized identity management for VANETs. In: 2020 3rd international conference on emerging technologies in computer engineering: machine learning and Internet of Things (ICETCE), Jaipur, India, pp 202–209. https://doi.org/10.1109/ICETCE48199.2020.9091766

17. Tong W, Hussain A, Bo WX, Maharjan S (2019) Artificial intelligence for vehicle-to-everything: a survey. Emerg Technol Veh Everything (V2x) 7:10821–10843

18. Tang Y, Cheng N, Wu W, Shen X Delay-minimization routing for heterogeneous VANETs with machine learning based mobility prediction. IEEE Trans Vehic Technol 99:1. https://doi.org/10.1109/TVT.2019.2899627

19. Manikandan S, Chinnadurai M, Vianny DM, Sivabalaselvamani D (2020) Real time traffic flow prediction and intelligent traffic control from remote location for large-scale heterogeneous networking using tensor flow. Int J Futur Gener Commun Netw 13(1):1006–1012

20. Ata A, Khan MA, Abbas S, Khan MS, Ahmad G Adaptive IoT empowered smart road traffic congestion control system using supervised machine learning algorithm. Comput J. https://doi.org/10.1093/comjnl/bxz129

21. Zhang T, Zhu Q (2020) Differentially private collaborative intrusion detection systems for vanets. Comput Sci Cryptogr Secur, preprint, May 2020

22. Patel SR, Ajmeri M (2020) Machine learning approach for transportation services using vehicular ad hoc network. Int J Trend Innov Res (IJTIIR) 2(2)

23. Hossain MA et al (2020) Faster convergence of Q-learning in cognitive radio-VANET scenario. In: Zakaria Z, Ahmad R (eds) Advances in electronics engineering. Lecture notes in electrical engineering, vol 619. Springer, Singapore

24. Ramalingama M, Thangarajan R (2020) Mutated k-means algorithm for dynamic clustering to perform effective and intelligent broadcasting in medical surveillance using selective reliable broadcast protocol in VANET. Comput Commun 150(15):563–568

25. Ndikumana A, Tran NH, Kim DH, Kim KT, Hong CS Deep learning based caching for self-driving cars in multi-access edge computing. IEEE Trans Intell Transp Syst. https://doi.org/10.1109/TITS.2020.2976572

26. Guo Z, Zhang Y, Lv J, Liu Y, Liu Y An online learning collaborative method for traffic forecasting and routing optimization. IEEE Trans Intell Transp Syst. https://doi.org/10.1109/TITS.2020.2986158

27. Bisht N, Shet RA (2020) Platoon-based cooperative intersection management strategies. In: 2020 IEEE 91st vehicular technology conference (VTC2020-Spring), Antwerp, Belgium, 2020, pp 1–6.https://doi.org/10.1109/VTC2020-Spring48590.2020.9129543

28. Khan S, Alam M, Fränzle M, Müllner N, Chen Y (2018) A traffic aware segment-based routing protocol for VANETs in urban scenarios. Comput Electr Eng 68:447–462

29. Ravish R, Shenoy DP, Rangaswamy S (2020) Sensor-based traffic control system. In: Mandal J, Mukhopadhyay S (eds) Proceedings of the global AI congress 2019. Advances in intelligent systems and computing, vol 1112. Springer, Singapore

30. Kulanthaiyappan S, Settu S, Chellaih C (2020) Internet of Vehicle: effects of target tracking cluster routing in vehicle network. In: 2020 6th international conference on advanced computing and communication systems (ICACCS), Coimbatore, India, pp 951–956. https://doi.org/10.1109/ICACCS48705.2020.9074454

31. Xhafa F, Barolli L (2020) The convergence of evolutionary intelligence and networking. Evol Intel 13:69–70. https://doi.org/10.1007/s12065-020-00368-x

32. Mekonnen WG, Hailu TA, Tamene M, Karthika P (2020) A dynamic efficient protocol secure for privacy-preserving communication-based VANET. In: Das A, Nayak J, Naik B, Dutta S, Pelusi D (eds) Computational intelligence in pattern recognition. Advances in intelligent systems and computing, vol 1120. Springer, Singapore

33. Martuscelli G, Boukerche A, Foschini L, Bellavista P (2016) V2V protocols for traffic congestion discovery along routes of interest in VANETs: a quantitative study 16(17)
34. Ibrahim BF, Toycan M, Mawlood HA (2020) A comprehensive survey on VANET broadcast protocols. In: 2020 international conference on computation, automation and knowledge management (ICCAKM), Dubai, United Arab Emirates, pp 298–302. https://doi.org/10.1109/ICCAKM46823.2020.9051462

Chapter 33
Dimensionality Reduction and Feature Extraction Using Image Processing Techniques for Classification of Single Partial Discharge Patterns

B. Vigneshwaran, M. Subashini, A. Sobiya, H. Rubla, G. Vigneshwari, and K. Kumar

1 Introduction

Insulation coordination on power equipment needs more concentration in the power grid due to transmission and distribution advancements. Due to this phenomenon, electric field intensity is gradually increased in that partial bridges as compared to the rest of the system. PD phenomenon is characteristically self-quenching and stochastic in which their properties are variable in nature [1]. In practical situations, PD can occur both in single and multiple sources in nature. In multiple sources, they overlap in nature and it is very hard to recognize the type of discharges occurs in the power apparatus. The most important adjustable parameters which illustrate the occurrence of PD are the phase relationship of phase angle (φ), volume of discharge (q) and the discharge count (n) can be represented as a PRPD signature [2].

The PRPD patterns find its application for recognizing the patterns since each electrical discharge pulse reproduces the characteristics of the discharge phenomenon and establishes the strong relationship between the type of the signatures and the type of the fault [3]. Hence, the proposed work deals with the above-said signatures.

The main motive for extracting the features and dimensionality reduction from PRPD pattern was to achieve the appropriate input features from the huge dataset in order to represent the type of discharges that take place with less computation time and less storage space. Owing to prosperous impact, extraction of high-quality features from the signature and identify the exact fault is emerged as a profoundly studied topic in the past decades [4]. Classification of PD signal from PRPD patterns can be categorized into various dissimilar types. Some researchers were taking an

B. Vigneshwaran (✉) · M. Subashini · A. Sobiya · H. Rubla · G. Vigneshwari · K. Kumar
Department of Electrical and Electronics Engineering, National Engineering College, Kovilpatti, Tamil Nadu, India

© The Author(s), under exclusive license to Springer Nature Singapore Pte Ltd. 2021   405
S. Kumar et al. (eds.), *Proceedings of International Conference on Communication and Computational Technologies*, Algorithms for Intelligent Systems,
https://doi.org/10.1007/978-981-16-3246-4_33

effort to categorize the dimension of cavities in dielectrics [5], though others were concentrated in the time-resolved signatures in the power apparatus [6]. However, in some research articles mention that fault occurs in the system can be differentiated using PRPD signatures.

Several feature extraction techniques like statistical parameters [7], image processing techniques [8], frequency and time features [9], chaos theory [10], Weibull features [11], and auto-correlation techniques [12] were commonly used by various researches for PD recognition. In general, four different techniques have been acknowledged to carry pattern recognition tasks. They are distance classifiers [13], statistical approach [14], syntactic approach [15], and artificial intelligence classifiers.

Recently, numerous works have been carried out by various researchers in the area of intelligent classifiers. The most advantages of the classifiers are fast response to the query data and high accuracy of recognition rate once the model is well-trained. Some examples of ANNs are [16] fuzzy inference system [17], linear discriminant analysis [18], classification techniques [19], evolutionary algorithms [20], and machine learning techniques [21].

In this research methodology, a novel approach has been projected for dimensionality reduction and feature extraction from PRPD patterns for recognizing PD sources. Investigational samples are synthetically created with surface discharge, corona discharge, and cavity discharge in air and oil medium, and the equivalent waveforms are gathered in laboratory. Initially, PRPD patterns are evaluated from the PD signal and then the PD images are converted into grayscale images. After that, the noise in the PRPD patterns is eliminated by detecting edges using canny edge detection filters [22], because it has the competence to categorize the anemic boundaries correctly. At last, the edge detected PRPD patterns are subjected to dimensionality reduction and feature extraction using SF and RICA techniques. These elements are supplied as input vectors for multi-class support vector machine (MCSVM) classifier. The effectiveness of the proposed algorithm is determined by varying the number of features to be extracted as well as optimal hyperplane by linear and nonlinear SVMs with different kernel functions.

The paper is structured as follows. Section 2 demonstrates the experimental arrangement and laboratory setup. Section 3 explains the feature extraction techniques and classifiers, followed by the results in Sect. 4. At last, the conclusion is done in Sect. 5.

2 PD Measurement and Detection

2.1 Laboratory Models for PD Pattern Classification

In this work, 14 dissimilar types of PD configuration like corona discharge, surface discharge, cavity discharge both in air and in oil medium, and their combinations

Table 1 Design of single discharges

Configuration	Various discharges
C1	Corona in air
C2	Corona in oil
C3	Surface in air
C4	Surface in oil
C5	Cavity in air
C6	Cavity in oil

are considered for single and multiple PD sources. The substance used between the electrode configurations C3–C6 for single PD sources is dielectric material and in C5 and C6 cavity position is not accurately determined, but the dimension of the void is approximately 0.5 cm.

The point electrode with radius of 50 μm on the tip and plane electrode has been used as a HV side electrode and LV side electrode with a gap spacing of 1 cm in air medium. The inception voltage for this setup is approximately 13 kV under air medium. In configuration C2, the setup is same as C1 but the medium used here is oil and the inception voltage for this setup is approximately 18 kV. The configuration C3 is same as C1, by implementing plane electrode for HV side and the inception voltage is found to be 17 kV for air medium and 23 kV for C4. For configuration C5 and configuration C6, a hole of diameter 0.5 cm created in the polycarbonate material artificially of thickness 2.5 cm. The corona free electrode profile is used, and the PD inception voltage is around 15.5 kV for air and 17 kV for oil medium. In this proposed work, the configuration is shown in Table 1.

2.2 PD Laboratory Test Setup

The experimental setup of PD measurement was shown in Fig. 1. In common, at the time of measuring the PD signatures from the faulty environment, it is important to mention that practical voltage is little bit higher than the inception voltages for all setup. Changeable voltage sources up to 20 kV are given across different electrode configurations.

Figure 2 represents the PRPD signature for C1 and C3 configuration before and after canny detected are shown. The most important problem arises during PD evaluation is noise (white noise, random noise, and discrete spectral interference). In this work, canny edge detection filter is used to denoise the PD signal. The canny technique is at variance from the other edge detection technique since it makes use of two different thresholds.

Fig. 1 Different electrode setup configuration—experimental setup

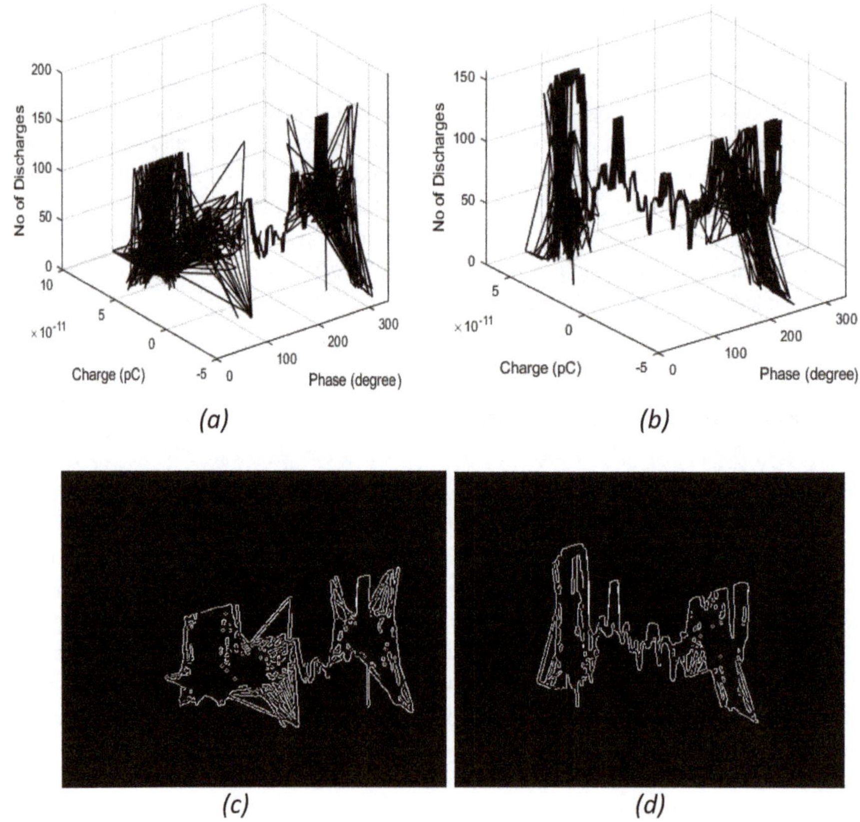

(a)

(b)

(c)

(d)

Fig. 2 3D PRPD patterns of various electrode setup **a** and **b** 3D PRPD pattern for C1 and C3, **c** and **d** 3D PRPD pattern after canny edge detection for C1 and C3

3   Feature Extraction Algorithm

Feature extraction is a processing of extracting new output features vectors from input features vectors. Mostly supervised learning algorithm is used for feature extraction. The aim of supervised machine learning algorithm deals with recognized group of input patterns and renowned group of output patterns. Based on the above algorithm, the model is trained for predictions for the response to query data. In this proposed work, SF and RICA with optimizing nonlinear objective functions are used for dimensionality reduction and feature extraction from PRPD patterns. By means of these features can show the way to improved classification of single PD accuracy.

3.1   Sparse Filtering

The SF algorithm consists of input data (X) matrix of order $n \times p$ where n denotes the particular label and one column of p represents the one feature or predictors. Initially, the number of features q to be extracted from the input matrix should be determined. The models used in this proposed work with different number of features with description are shown in Table 2. In general for any type of feature representations, the value of q can be less or greater than the number of predictor variables. The SF algorithm generates an initial random W weight matrix of order $p \times q$ [23].

Transformation of X into q number of output features by weight transformation W. The SF algorithm stops its iteration once the norm is less than step tolerance or norm of the gradient is less than gradient tolerance times a scalar ∂, which is shown in Eq. (1).

$$\partial = \text{maximum}(1, \text{minimum}(|k|, \|h_\infty\|)) \tag{1}$$

The objective function of SF algorithm creates a new set of features from input data with equal weight W [24]. $h\infty$ is the infinity norm of the initial gradient. For improving the classification accuracy rate of PRPD patterns, the iteration limit should be 100. Over iteration limit causes generalization problems for new features. The

Table 2 Proposed model with descriptions

Model	Description
M1	100 transform features with SF
M2	100 original features with SF
M3	200 transform features with SF
M4	450 transform features with SF
M5	200 transform features with RICA
M6	450 transform features with RICA

following steps are involved in computation of objective function using SF algorithm. The objective function is mainly depends on X and W. Initially compute the $n \times q$ matrix by multiplying the input data and weight matrix. F matrix is obtained by applying approximate absolute value function (u) to each element of $n \times q$ matrix. The value of φu is shown in Eq. (2).

$$Q(u) = \sqrt{u^2 + 10^{-8}} \tag{2}$$

where Q is a soft nonnegative balanced purpose. Next to that normalize the row and column of F by the approximation L^2 norm. The normalization matrix is defined in Eqs. (3) and (4).

$$\|F(j)\| = \sqrt{\sum_{i=1}^{n} (F(i, j))^2 + 10^{-8}} \tag{3}$$

$$\tilde{F}(i, j) = F(i, j)/\|F(j)\| \tag{4}$$

At last SF find the weight which minimizes the objective function. Thus, the requested number of features q that SF computes.

3.2 Reconstruction Independent Component Analysis

The RICA algorithm is mainly performed in the principle of minimizing an objective function. The independent component analysis (ICA) model is given in Eq. (5).

$$x = \mu + As \tag{5}$$

where x is a column of length p, μ is a constant value of column vector of length p, s is a column vector of length q and A is a mixing matrix of size $p \times q$. In this proposed work, RICA model is used to compute the matrix A.

The RICA algorithm initiates with X that has n rows and p columns. The data matrix denotes in Eq. (6).

$$X = \begin{bmatrix} x_1^T \\ x_2^T \\ \vdots \\ x_n^T \end{bmatrix} \tag{6}$$

The weight matrix W is comprised of articles wi of size $q \times 1$ and denoted in Eq. (7).

$$W = [w_1, w_2, \ldots w_q] \tag{7}$$

In RICA algorithm, objective function uses a contrast function and it is denoted in Eq. (8).

$$g = \frac{1}{2} \log(\cosh(2x)) \tag{8}$$

The objective function in terms of the $p \times q$ matrix W is given in Equation

$$h = \frac{\lambda}{n} \sum_{i=1}^{n} \left\| W W^T x_i - x_i \right\|_2^2 + \frac{1}{n} \sum_{i=1}^{n} \sum_{j=1}^{q} \sigma_j g\left(w_j^T x_i\right) \tag{9}$$

In general, RICA supports only the standardization transform. It makes the predictors have zero mean and unit variance. The resulting minimal matrix W provides the transformation from input data X to output features XW.

3.3 Classifiers

In this work, MCSVM classifiers with optimized hyperplane are used for PD source discrimination. To evaluate the effectiveness of the PD recognition, the proposed classifier is compared with linear SVM. These classifiers are trained and then used to classify different configurations.

3.3.1 Support Vector Machine and Multi-Class Support Vector Machine

A number of research works on insulator performance under contaminated conditions have been carried out based on theoretical or experimental models in order to obtain an optimized test results with minimum time interval [25]. But most of the proposed methods are hard to apply in practice and time-consuming and do not dependable to environmental variations. Similarly, it is difficult to forecast the conditions of the insulator under polluted conditions to check the reliability of its operation in service. Hence, artificial intelligence techniques have been proposed in this paper to carry out classification for polluted insulators by determining the input parameters such as fractal dimension and lacunarity.

In recent times, support vector machine (SVM) has been extensively used in machine learning field due to its significant performance when compared to other computing techniques [26]. The hyperplane in SVM is intentionally used to separate different classes of data into a separate group. With the help of linear SVM, datasets are easily separated using hyperplane by a straight line. To make the non-separable data into separable one, nonlinear SVM is implemented where the nonlinear data are mapped to higher-dimensional space to classify the results. Moreover, linear SVMs are speedy to train and execute, but they be likely to underperform on complex datasets. Nonlinear SVMs can be more reliable in performance across diverse problems and are the most preferred option in many complex applications. The RBF kernel is defined as follows with $\sigma = 0.5$ in Eq. (10).

$$K\left(x_i, x_j\right) = e^{-\frac{\|x_i - x_j\|^2}{2\sigma^2}} \tag{10}$$

where σ is the width of the RBF function.

In this research work, the four different kernel functions provided by SVM classifier are considered for the prediction of dry band location on insulator surface under polluted conditions. The four different kernel functions considered in this work are radial basis function, sigmoidal function, polynomial, and linear function, and their mathematical expression is given by Eqs. (11) and (12).

$$\text{Radial basis function: } K(x_i, y_i) = \left(-\gamma \|X_i - X_j\|\right), \gamma > 0, \tag{11}$$

$$\text{Polynomial: } K(x_i, y_i) = (\gamma X_i^T X_j + r)^d, \gamma > 0, \tag{12}$$

In general, SVM is a binary classifier which deals with two different classes at instant. In this paper, single PD discharge sources can be classified at a distant with six different configurations. Thus, MCSVM is used to overcome the above issues by applying one versus one (OVO) and one versus all (OVA) approach.

4 Result and Discussion

Based on the database obtained, totally 900 input PRPD images are considered with 150 images for each configuration with 784 feature vectors. Out of 150 samples, for training 100 samples were taken and 50 samples are used for testing each category. It is clearly inferred that throughout detailed evaluation and analysis in the change of number of features (q) beyond certain limit to be extracted, the number of misclassifications has been marginally increased in MCSVM. Figure 3 has minimum objective function which settled quickly with less function evaluations as compared to other model.

The SF and RICA feature extraction method has been elaborately briefed in Sects. 3.1 and 3.2 which has been executed for both single-source datasets. The single-source PD dataset encompasses six types of categories of PD specifically coronal discharge, surface discharge, and cavity discharge for both air and oil medium. It is relevant to observe that pattern recognition of the second sub-group of dataset has become a theme of attention to investigators newly due to the characteristic nature

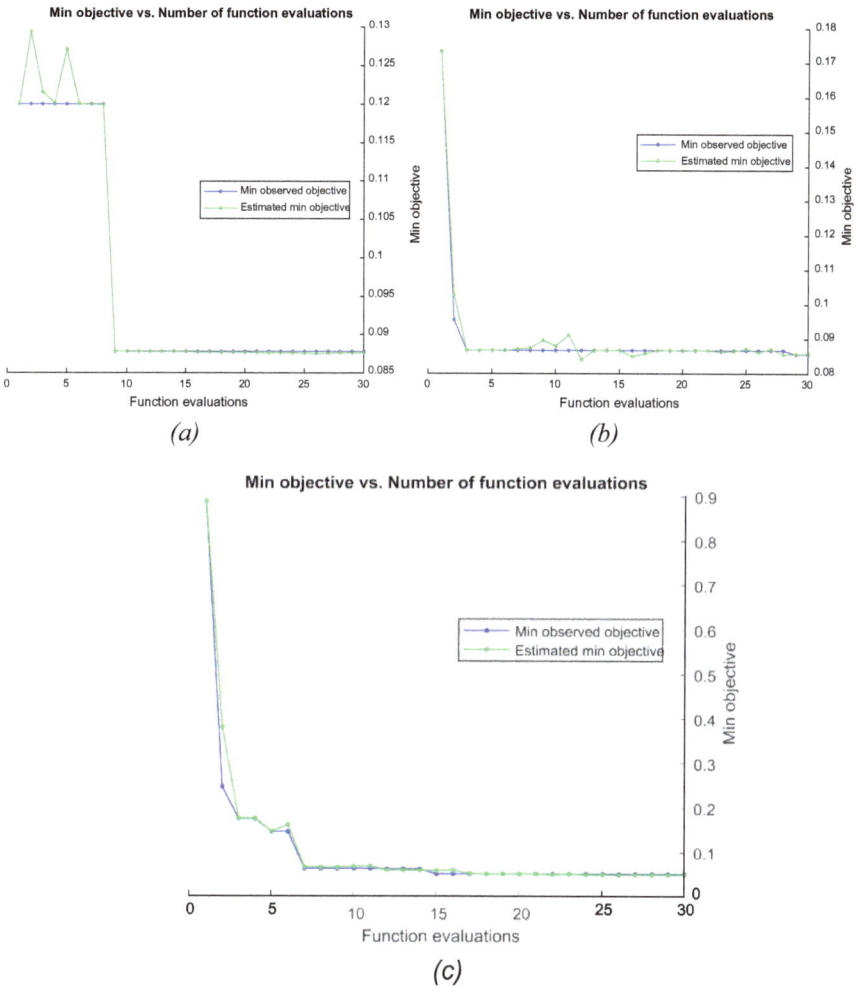

Fig. 3 Function evaluations plot for various models **a, b,** and **c** for Model 4 with linear, polynomial, and RBF SVM

of the complex patterns and difficulties in discriminating the nature of overlap in the PD pulses.

Error correcting output code is used for automatic hyper-parameters optimization. Expected improvement plus function is used for reproducibility. In this work, the eligible parameters for ECOC are OVA and OVO coding. In OVA, one class is termed as positive and remaining all class is termed as negative. In OVO, one class is labeled as positive and one class is labeled as negative and the rest of class is ignored at one particular instant. In this proposed work, the two optimization parameters are box constraint and kernel scale. The limit of box constraint is between positive values log-scaled in the range [1e − 3, 1e3], and kernel scale is between log-scaled in the range [1e − 3, 1e3]. Two different kernel functions are used for the classification of single PD patterns, namely "RBF" and "polynomial." The main advantages of ECOC are reducing the classification problem error with more than two classes of binary classifiers.

It is obvious from Table 3 and the following investigation that the SF with 450 transformed feature extraction scheme not only reduces the dimension of the input vector, but also the objective function evaluation time is also reduced for fast response with excellent classification rate at optimal hyperplane. Further, from Table 3 it is clearly understood that box constraint, coding, and kernel scale value for both observed and estimation are same with high accuracy. The RBF kernel is commonly more flexible than the linear SVM or polynomial kernel SVM in that you can model whole functions with its function space.

Table 3 Observations on the classification capability of radial basis SVM for different number of features

q	M1	M2	M3	M4	M5	M6
Filters	SF				RICA	
Max objective evaluations	30	30	30	30	30	30
Total function evaluations	30	30	30	30	30	30
Total elapsed time	92.7538	257.7097	94.2149	97.7423	111.10	201.338
Total objective function evaluation time	55.0229	222.218	63.9778	60.132	65.636	164.3569
Coding	OVA	OVA	OVA	OVA	OVO	OVA
Box constraint	30.203	989.07	233.51	990.23	991.15	971.66
Kernel scale	0.60683	996.01	0.83419	1.1493	197.12	36.663
Observed objective function value	0.084444	0.068889	0.06555	0.05111	0.0844	0.08111
Estimated objective function value	0.086996	0.068928	0.0659	0.04970	0.0846	0.08250
Function evaluation time	1.2397	8.3233	1.3898	3.7921	1.9995	3.9894
Coding	OVA	OVA	OVA	OVA	OVO	OVA
Box constraint	98.228	614.66	233.51	990.23	978.14	947.57
Kernel scale	0.57914	996.9	0.83425	1.1493	137.38	32.307
Estimated objective function value	0.086996	0.068928	0.0659	0.049704	0.0846 2	0.08250
Estimated function evaluation time	1.4912	8.2509	1.3255	4.0254	1.9546	4.4666
Loss	0.0867	0.0999	0.0711	0.0550	0.0733	0.0844

5 Conclusion

A new PD recognition algorithm using dimensionality reduction and feature extraction SF and MCSVM is presented in this paper. The proposed recognition algorithm is examined with artificial models which are created experimentally in HV laboratory. A grayscale PRPD image pattern based on canny edge detection can provide the weak and strong edges clearly which is developed for the classification of single PD. Canny edge detected cannot be affected by the noise immersed in the PD signal. The SF and RICA algorithms combined with dimensionality reduction are applied to grayscale image for extracting appropriate features. Extracted features using SF and RICA are used as input vectors to the MCSVM with linear, RBF, and polynomial kernel function. Finally, 450 transformed features with SF using RBF kernel MCSVM show better results. In the past methods (used for single PD classification) presented in the literature, there are following shortcomings like combination of single, two, and three different PD source classifications in a single algorithm and dimensionality reduction of PRPD patterns using image processing techniques to enrich the system performance with less storage capacity. The optimal hyperplane with variable parameters like coding (OVO or OVA), box constraints, and kernel scale and RBF MCSVM having 450 transformed feature vectors extracted by SF shows better results

References

1. Abdel-Galil TK, Hegazy YG, Salama MMA, Bartnikas R (2004) Partial dis- charge pulse pattern recognition using hidden Markov models. IEEE Trans Dielectr Electr Insul 11(4):715–723
2. Sahoo NC, Salama MMA, Bartinkas R (2005) Trends in partial discharge pattern classification: a survey. IEEE Trans Dielectr Electr Insul 12(2):248–264
3. Satish L, Zaengl WS (1995) Can fractal features be used for recognizing 3-D partial discharge patterns? IEEE Trans Dielectr Electr Insul 2(3):352–359
4. Hui M, Chan JC, Saha TK, Ekanayake C (2013) Pattern recognition techniques and their applications for automatic classification of artificial partial discharge sources. IEEE Trans Dielectr Electr Insul 20:468–478
5. Mazroua AA, Salama MMA, Bartnikas R (1993) PD pattern recognition with neural networks using the multilayer perceptron technique. IEEE Trans Electr Insul 28:1082–1089
6. Su M-S, Chen J-F, Lin Y-H (2013) Phase determination of partial discharge source in three-phase transmission lines using discrete wavelet transform and probabilistic neural networks. Int J Electr Power Energy Syst 51(10):27–34
7. Gulski E (1993) Computer-aided measurement of partial discharges in HV equipment. IEEE Trans Electr Insul 28:969–983
8. Basharan V, Siluvairaj WIM, Velayutham MR Recognition of multiple partial discharge patterns by multi-class SVM using fractal image processing technique. IET Sci Measur
9. Contin A, Cavallini A, Montanari GC, Pasini G, Puletti F (2002) Digitaldetection and fuzzy classification of partial discharge signals. IEEE Trans Dielectr Electr Insul 9:335–348
10. Xiaoxing Z, Song X, Na S, Ju T, Wei L (2014) GIS partial discharge pattern recognition based on the chaos theory. IEEE Trans Dielectr Electr Insul 21:783–790

11. Yu H, Song YH (2003) Using improved self-organizing map for partial discharge diagnosis of large turbogenerators. IEEE Trans Energy Convers 18:392–399
12. Rana KPS, Singh R, Sayann KS (2009) Auto-correlation based intelligent technique for complex waveform presentation and measurement. J Instrum 4:P05007
13. Salama MA, Bartnikas R (2002) Determination of neural-network topology for partial discharge pulse pattern recognition. IEEE Trans Neural Netw 13:446–456
14. Bishop CM (1995) Neural network for pattern recognition. Clarendon Press, Oxford, UK
15. Majidi M, Fadali MS, Etezadi-Amoli M, Oskuoee M (2015) Partial discharge pattern recognition via sparse representation and ANN. IEEE Trans Dielectr Electr Insul 22:1061–1070
16. Gulski E, Krivda A (1993) Neural networks as a tool for recognition of partial discharges. IEEE Trans Electr Insul 28:984–1001
17. Cavallini A, Montanari GC, Contin A, Pulletti F (2003) A new approach to the diagnosis of solid insulation systems based on PD signal inference. IEEE Electr Insul Mag 19:23–30
18. Satish L, Gururaj BI (1993) Use of hidden Markov models for partial discharge pattern classification. IEEE Trans Electr Insul 28:172–182
19. Robles G, Parrado-Hernández E, Ardila-Rey J, Martínez-Tarifa JM (2016) Multiple partial discharge source discrimination with multiclass support vector machines. Expert Syst Appl 55:417–428
20. Zhang L, Wang J (2015) Optimizing parameters of support vector machines using team-search-based particle swarm optimization. Eng Comput 32(5):1194–1213. https://doi.org/10.1108/EC-12-2013-0310
21. Lai K, Phung B, Blackburn T (2010) Application of data mining on partial discharge part I: Predictive modelling classification. IEEE Trans Dielectr Electr Insul 17:846–854
22. Das S (2016) Comparison of various edge detection technique. Int J Sig Process Image Process Pattern Recogn 9(2):143–158. http://dx.doi.org/10.14257/ijsip.2016.9.2.13
23. Nocedal J, Wright SJ (2006) Numerical optimization. Springer Series in Operations Research, 2nd edn. Springer Verlag
24. Ngiam J, Chen Z, Bhaskar SA, Koh PW, Ng AY (2011) Sparse filtering. Adv Neural Inf Process Syst 24:1125–1133
25. Vapnik V (1995) The nature of statistical learning theory. Springer, Red Bank, NJ
26. Mitiche I, Morison G, Nesbitt A, Hughes-Narborough M, Stewart BG, Boreham P (2018) Classification of EMI discharge sources using time–frequency features and multi-class support vector machine. Electr Power Syst Res 163:261–269

Chapter 34
A Survey of Data Storing and Processing Techniques for IoT in Healthcare Systems

Anwar D. Alhejaili(ID)**, Marwa Alsheraimi**(ID)**, Nojoud Alrubaiqi**(ID)**, and Mohammad Zubair Khan**(ID)

1 Introduction

The origin of the concept of IoT goes back more than 15 years, and there is still no stander definition for IoT. In [1] it can define the IoT according to The International Telecommunication Union (ITU) as the following: "a global infrastructure for the Information Society, enabling advanced services by interconnecting (physical and virtual) things based on, existing and evolving, interoperable information and communication technologies" [1]. The IoT can also be defined as a network through which electronic devices, motors, sensors, and programs communicate with each other [2].

IoT has incredible growth recently in many different fields. These fields can be seen in the following: smart environment, industrial applications, domestic applications, security and emergencies, logistics, and transport [3].

IoT in the healthcare industry is considered the most noticeable and dominant field. The IoT healthcare system can be defined as a global infrastructure in the healthcare sector that provides diversified and advanced services by connecting sensors, patients, doctors, etc., and collects data via ICT.

A. D. Alhejaili (✉) · M. Alsheraimi · N. Alrubaiqi · M. Z. Khan
Department of Computer Science, College of Computer Science, Taibah University, Madinah, Saudi Arabia
e-mail: TU4160117@taibahu.edu.sa

M. Alsheraimi
e-mail: TU4160118@taibahu.edu.sa

N. Alrubaiqi
e-mail: Tu4160161@taibahu.edu.sa

M. Z. Khan
e-mail: mkhanb@taibahu.edu.sa

IoT healthcare systems' main contribution is supporting conventional healthcare by replacing paper and local records with electronic records that can be accessed from everywhere. There is no doubt that this technology's emergence was crucial to the digital transformation of modern health care, as it is possible through networks and sensors to build a large volume of data, which helps to acquire knowledge. One of IoTs most important goals in health care is to help IoT users know their health status continuously and with less effort and cost [4]. Also, it saves more time for the patient who needs constant monitoring. Moreover, it allows doctors to expand their services to geographical locations that were previously difficult to reach.

Moreover, the IoT in health care has led to creating devices that monitor patients remotely, with patient participation and interaction, with the health practitioner. These devices include many major aspects of healthcare, such as providing information about commitment to taking medicines, adherence to treatment, and using the medical device for some health problems such as breathing, as the patient participates in the treatment plan.

In addition to that, IoT health care provide various types of services for patients at any time and from anywhere, such as ambient assisted living (AAL), adverse drug reaction (ADR), wearable device access (WDA), and community healthcare (CH).

There are various techniques used to store and support data processing in IoT healthcare systems. These techniques are cloud computing-based, database-based, and AI-based. This survey paper will discuss the previously mentioned techniques and point out the fundamental issues in IoT healthcare systems.

2   Cloud Computing-Based IoT Healthcare Systems

One of the most popular techniques used in IoT healthcare is cloud computing. NIST presented a definition of the Cloud in September 2011:

> [A]s a model for allowing universal, convenient, on-demand network access to a common pool of configurable computing resources (e.g., networks, servers, storage, software, and services) that can be easily provisioned and released with minimal management effort or interference with service providers. [5]

Could computing can facilitate communication, information exchange, and continuous patient monitoring. Using cloud computing leads to an increased number of healthcare system users. Cloud computing is based on the Internet to provide distributed shared resources and infrastructure for processing. Also, it allows using the service and application remotely without the need for installation. According to [6], there are many reasons mentioned in the following lead humanity to use cloud computing at any distributed system: The mobilization, which refers to the availability of services anywhere in the World even without infrastructure, contributes to saving time. Also, the device becomes more independent, has lower cost and maintenance and provides productivity and scalability. Regarding the capacity of service provided by the Cloud, there are some authors and researchers who were proposed using the Cloud to store patient information in the healthcare system.

The following section will introduce some research papers that using cloud computing to store patient's data:

- The authors of [7] suggest a comprehensive healthcare system for depressed patients based on cloud computing. It is considered as a complete system because it allows all healthcare participants to exchange and coordinate data on one platform. In other words, it will enable all healthcare entities such as patients, doctors, hospitals, laboratories, pharmacists, and nurses to collaborate and exchange data easily, as described in Fig. 1. The data can be collected in various ways, such as wearable monitoring devices, placed strategically on the human body (sensors/RFID tags), or jewelry. All these devices are wireless body area sensor network, which means the following: these devices are capable of sampling, sensing, and processing data. They may also help determine the user's position, distinguishing between the states of the user (walking, sitting, and running) and estimating the form and the level of the user's physical activity. All patient data will be stored on a cloud EHR application as a patient's medical profile. Figure 1 presents a scenario that illustrates all participants' roles in the system.

Unfortunately, researchers and users discover many problems related to cloud computing in the healthcare system. Cisco tried to solve some of the cloud problems

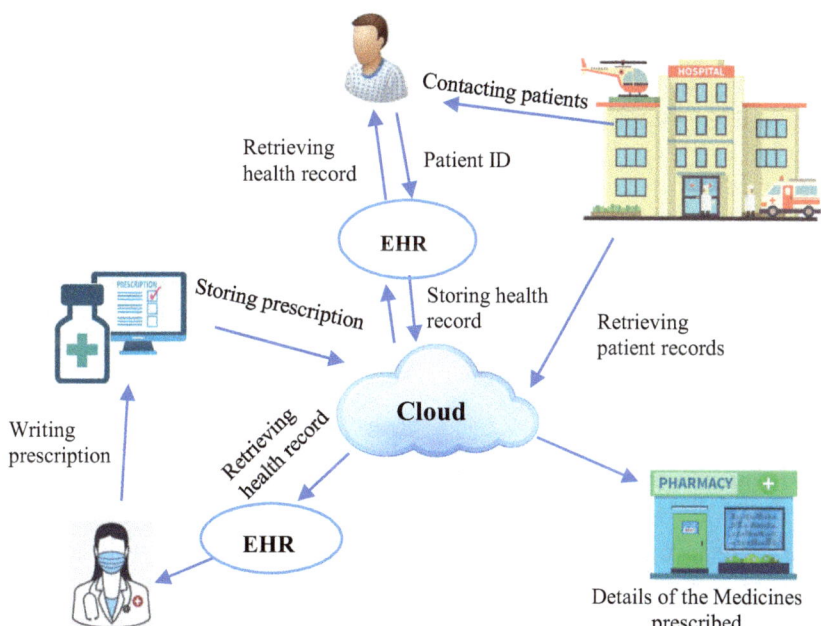

Fig. 1 Roles of participants in the system

through the invention of new technology, known as fog computing, in 2012, according to [8]. Fog computing aims to improve the speed of response and analysis of the data that is close to the source.

The next section will present some studies based on using cloud and fog computing together in healthcare systems:

- Nandyala and Kim in [9] suggest a system for ubiquitous healthcare monitoring in real time based on cloud and fog. They mention a remarkable benefit of using fog computing with the cloud. The benefit can be seen through the low latency, improving reliability and scalability, providing flexible processing, and faster responding. The authors suggest four tiers in the system as mentioned in the following from bottom to top: smart devices or things network tier, fog tier, core tier, and cloud tier. The first tier is where the patient's data is collected. Patient's information is collected either by sensors or wearable devices. In this tire, the decision is made, and the interaction is designed as (M2M) tier. This tire filters the locally consumed data and sends the remainder to the upper levels. The second tier is the fog tier, known as a multiservice edge. Fog tier provides local analysis and processing based on defining optimal health patterns. The controller on this tier also alerts medical staff and family members via mobile devices.
- Furthermore, the fog tier supports wired and wireless connectivity and various protocols such as IEEE 802.11, Zigbee, 3G, and 4G to easily communicate with potentially different endpoint devices. The core network is the third tier of the system. It provides paths for communicating between different sub-networks as well as transmitting data. The core network looks like the traditional network architecture, but the only difference is containing a traffic profile for the provided paths. The last tier is the cloud or data center tier. It offers hosting applications and maintaining the architecture of IoT as well. Also, it helps store important data for the long term, which might be retrieved later to detect chronic diseases and patient history.
- The authors of [10] propose an M-healthcare system based on fog and cloud computing. The proposed system's main contribution is providing a service in emergencies to the patient whenever the patient's location is changed. This service offers the nearest healthcare center regarding the patient's current location. As a result, the proposed system reduces the latency by $\sim 28\%$ and the device energy consumption by $\sim 27\%$ than other cloud healthcare systems.

3   Database-Based IoT Healthcare Systems

There is a remarkable feature of using a database to store medical data in healthcare IoT-based systems. This feature eliminates the traditional way of manually keeping a patient's medical records. The conventional way might cause problems such as writing a patient's medical data falsely, which leads to a wrong diagnosis and an incorrect treatment decision accordingly. In some cases, the patient's medical records might get lost, requiring a new diagnosis. Consequently, the patient may lately receive medical treatment.

IoT-based healthcare systems will be mentioned in the following save medical data in either the Firebase database or SQL database. Firebase database is a "NoSQL cloud-based database that syncs data across all clients in real time and provides offline functionality" [11]. In contrast, SQL database "is a standard language for dealing with Relational Database" [12] and allows accessing and manipulating data successfully.

The following section will introduce some research papers that are using the database to store patient's data:

- The authors of [13] mention home-based ECG monitor device. This device provides an effective way of connecting patients with their medical professionals regardless of their location to monitor their heart rate conditions. This device controls patients' health efficiently because it allows an advanced detection and prevention of any degradation. The device contains hardware of ECG acquisition and an Android-based mobile application. Figure 2 displays the system design.

The ECG acquisition in-home hardware is named Mythro. Mythro catches a flow of a single signal of ECG and sends it by Bluetooth wireless communication to a developed application selected for an Android-based smartphone. The mobile application receives the ECG data, which can be showed in real time and automatically stored in the smartphone as a text file. The ECG signal processing built in the mobile application analyses the signal according to the heart rate. By using the Firebase database, a Google technology, patients can upload ECG data, and doctors can access ECG data for medical analysis and treatment decisions. Besides, doctors can reach the patient's location in an emergency case based on the mobile application's GPS service.

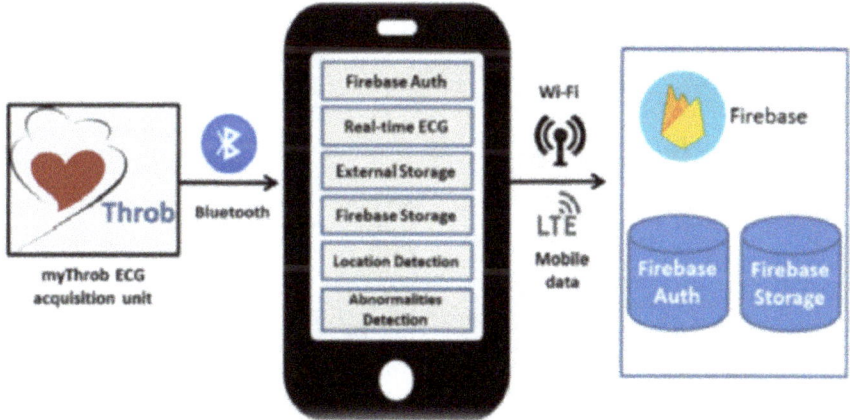

Fig. 2 System design [13]

- The authors of [14] present a CardioSys device that is IoT-based and aims to have an early diagnosis of hypertension. This device consists of two parts: CardioSys hardware and CardioSys software. Figure 3 shows the CardioSys diagram.

The CardioSys hardware, which consists of several sensors, is responsible for measuring the following in a patient: cardiology system, particularly systolic heart condition and a diastole heart condition, heart rate, and temperature. The measurements of the above mentioned are processed and sent by a microcontroller to CardioSys software. Moreover, these measurements help in predicting any hypertension occurrence. CardioSys software is an android application. This application

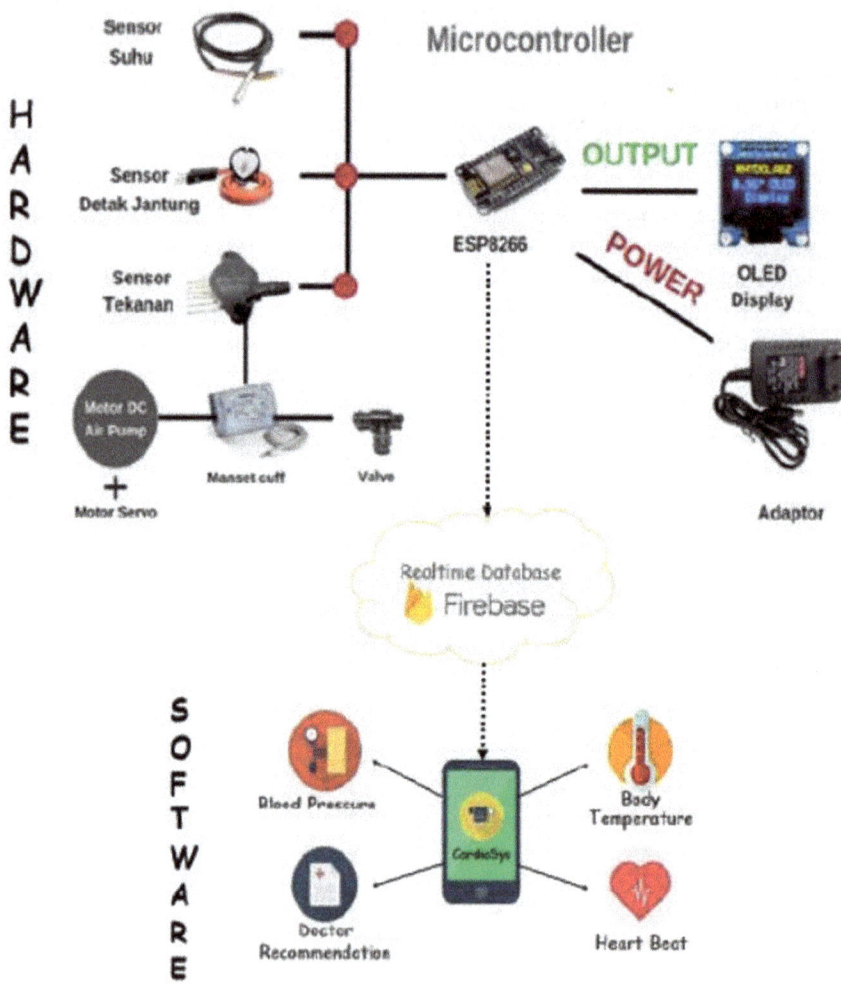

Fig. 3 CardioSys diagram [14]

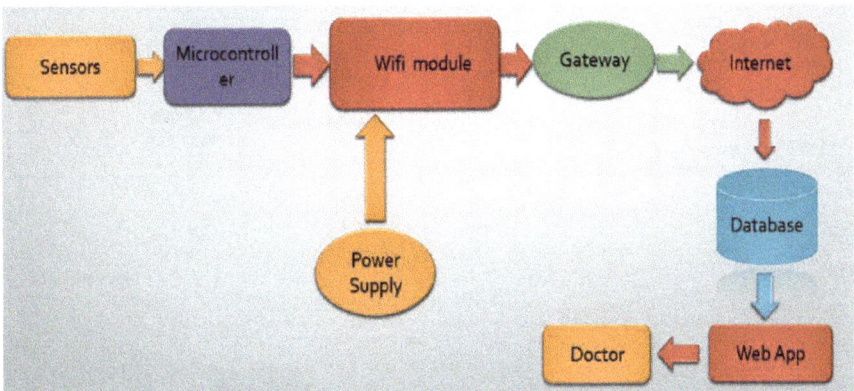

Fig. 4 Proposed model [15]

can save the results of the measurements in the Firebase database automatically and monitor them simply as well as periodically by patients, medical staff, and hospitals.

- The authors of [15] propose a smart healthcare system based on IoT. This system determines patients' health conditions by using several IoT sensors such as pulse sensors, blood pressure sensors, heart rate sensors, and temperature sensors. These sensors precisely gather patients' medical data and send them to the database—the database stores all patients' medical records. Consequently, by visiting only a Web site or URL, doctors can preserve their patients' records and provide a better treatment method. Figure 4 demonstrates the proposed model.

The sensors used in the system are a temperature sensor that measures the temperature of the room and patients' body temperature and an oximeter sensor that measures the oxygen level in blood and either heartbeat or pulse rate. Microcontrollers are an Arduino UNO and a NodeMCU that provides the wireless connection. The NodeMCU will be connected to the Firebase database. Therefore, whatever devices or sensors are connected to NodeMCU will send patients' data to the database. Therefore, doctors can check them and determine the best way of treatment.

- The authors of [16] propose a glucose monitoring system with a prognosis algorithm. The system is an IoT architecture that uses a prognosis algorithm to avoid any potential complications which elderly patients might encounter. The architecture consists of three layers: perception layer, network layer, and application layer. Figure 5 displays these layers.

Perception layer: It is accountable for gaining the patient's glucose levels using a noninvasive wireless sensor. This sensor is linked to NodeMCU, which is an open-source platform. NodeMCU contains the following: a 10-bit resolution A/D converter and a microcontroller. The reading is converted to digital values through the use of the A/D converter. Network layer: It has a wireless router that connects to NodeMCU. Hence, glucose values will be sent to the server. Application layer: It has a Web server.

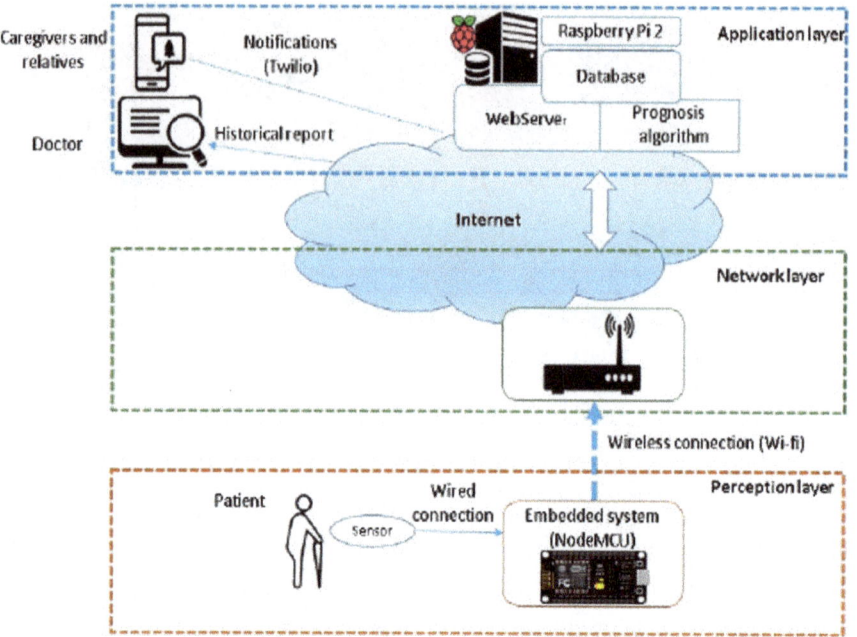

Fig. 5 Architecture of the system [16]

The server stores a prognosis algorithm. This algorithm can determine the patient's state and the need to send a warning through SMS to either doctors or the patient's relative. The Web server also depends on Raspberry Pi and a database. The database is where the value of glucose levels for different patients is stored. This database is developed in SQL and can be accessed either remotely or locally. Therefore, doctors can check out patient's records via query language.

4 Artificial Intelligence (AI)

Artificial intelligence (AI) is a simulation of human intelligence through technologies in computers and machines. These technologies have been developed to perform many tasks that require human intervention in all fields. Among those areas is the health field. Artificial intelligence has successfully analyzed a large and different set of health data, including clinical, behavioral, patient data, and all vital medical data.

Artificial intelligence methodologies contributed to speech recognition, natural language processing, machine learning, and deep learning, facilitating the digitization of health data. Learning levels in artificial intelligence vary between machine learning and more complex deep learning, allowing learning from data in the medical field and using more complex structures to build [17].

In addition to that, AI enables the patient's records and his disease history to discover the disease, gives predictions of possible future disease, suggests treatment, and appropriates drugs, for each patient [18].

Both artificial intelligence and the IoT are developing to meet human needs. One of the powerful innovations that carry multiple features, making them more used in the medical field, is the rechargeable devices, which provide better interaction, lightweight, can be carried anywhere, and clarity in the communication process. These devices are considered to be more comfortable in collecting data and facilitating monitoring of patients' health through the support of artificial intelligence, which relies on arm badges, or ECG trackers, to generate, organize, and manage data.

The following study will review the use of artificial intelligence techniques and IoT in diagnosing heart diseases.

- The authors of [19], developed a model with high efficiency in diagnosing heart disease. This system is based on using the Internet of things, computing techniques, managing data, and machine learning algorithms to monitor and predict heart disease. The UCI Repository dataset and healthcare sensors were used to predict who might suffer from heart disease by using classification algorithms, sort patient data, and then train this data so that it is ready to be tested to determine the presence of the disease. The proposed model consists of five main parts: Internet of things sensors in medical devices, a cloud database, a patient database, a cardiac disease data set, a system for prediction, and based on machine learning, and Fig. 6 shows the proposed model. In this model, artificial intelligence and machine learning algorithms played a fundamental role in decision-making in dealing with the vast amount of data. The following classifiers (J48, logistic regression, multilayer, and support vector machine) were compared to a dataset, when the system was applied. In the results, the classifiers differed in accuracy. The classifier J48 is the most accurate algorithm for IoT-based cardiac disease prediction model.

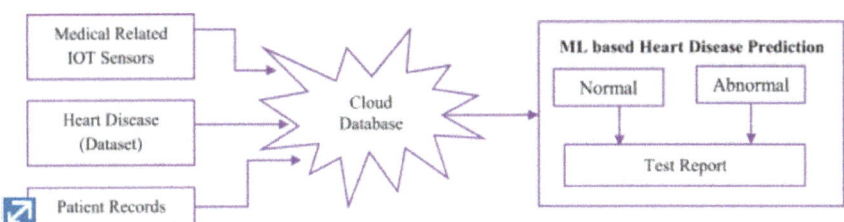

Fig. 6 Block diagram of the diagnosing heart disease model [19]

5 Machine Learning

Artificial intelligence techniques are very effective in healthcare. Accordingly, AI performs large data processing through different algorithms such as machine learning, which performs prediction and analysis of data, with the lowest possible error rate. Compared to the medical practitioner's capabilities in analyzing this data, it has very high accuracy.

Machine learning technology has excelled in analyzing the samples that are taken from the patient to diagnose the disease in laboratories. It can also process large numbers of data collected from various smart Internet devices to predict the result in a specific period.

Many companies concerned with technology development have invested in health care to develop healthcare systems based on machine learning, improving algorithms, and retraining them. The above-mentioned contributes to diagnosing complex diseases through three-dimensional images and training the machine to recognize lung cancer through a set of samples and large 3D images. Internet of interconnected medical devices for predictive analysis, preventive analysis, and surgical procedures are now being applied to a very broad field [20].

The next section will present some suggested systems using machine learning, which supports the interaction between the medical practitioner and the patient and improves patient health by developing strategies, which depend on deep learning.

- The authors of [21] designed a technology platform for monitoring patients with chronic obstructive pulmonary disease (COPD). The proposed system contributes to forming a new model and a new way of dealing with infected patients due to its use of artificial intelligence algorithms in patient monitoring, aiming to support interaction between health professionals and patients. It supports the process of partnership between the two parties. The data that is monitored by the patient is integrated. The doctor is provided daily with a dynamic picture of the disease and its impact on the patient's life. Through this monitoring, the doctor can diagnose disease development and give timely advice and solutions. The system depends on two important frameworks, and the first is the system methodology. Patient monitoring devices rely on the various types of available communication protocols to collect data, as volume and processing capacity must be required for the correct application of artificial intelligence algorithms. As shown in Fig. 7, the platform consists of techniques and algorithms. These devices will be related to the patient first by wearing monitoring devices such as smartwatches or smartphones and installing additional sensors to detect environmental changes such as air quality and their impact on sleep. It will also monitor physiological information through custom AI algorithms such as heart rate, physical activity, lung function, and patient adherence to taking medications through smart inhalers. All this information is also collected to develop predictive models that outline correct treatment plans. The second framework is the personal records of the patient, as the systems approach to patient monitoring relies on two aspects, the first aspect

Fig. 7 Concept of monitoring of patients platform [21]

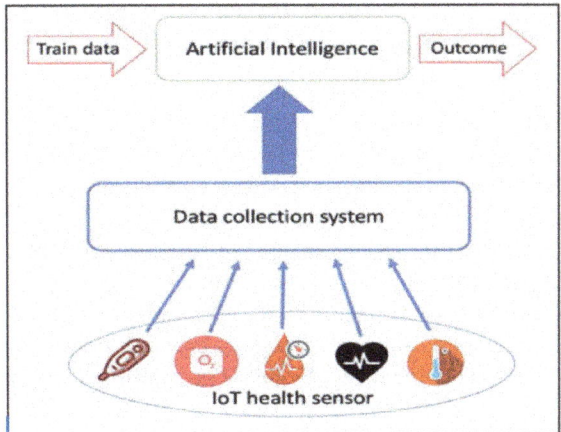

is the handling of the decision support and management system, which organizes all heterogeneous internal and external data, and methods of accessing them and the other part is the aspect that relies on artificial intelligence approaches in describing and measuring vital medical images and applying analysis techniques for three-dimensional graphics. According to the Semantic Web model, we also need to encode this extracted information and index data and resources, making this system available for use and application even on diseases of the muscular system.

- The authors of [22] developed a platform that supports Alzheimer's patients and helps them perceive their faces due to poor visual perception or poor memory. This is done by providing an assistant smart device, which can be wearable, a device developed with the technology of the Internet of things and deep learning. The developed system is applying deep learning technology and the use of this technology in facial recognition. People who have a close relationship with the patient, such as friends and family, also include pets. The face data for these people is generated through the platform, where faces are recognized and re-identified in real-time situations and different situations. Deep learning algorithms can create other representations of the face, extracting each image's unique, and distinctive features to redefine it. The proposed system consists of three main components: first, the facial recognition system, to recognize the face through feature extraction. Secondly, a patient database, through which it is possible to record the basic truth of the facial features of people related to the relationship and friendship of the patient, employs supervised learning technology. Third, the congruence scale is a measure used to compare the characteristics of the current faces, of any person related to the patient, with the features of the faces extracted from the images captured in real time. The experiments' results show that this system effectively recognizes faces, as it helped Alzheimer's patients identify friends and determine their identities. It is a scalable system does not require many computational resources and can be used in real time.

5.1 Deep Learning-Based IoT Healthcare Systems to Combat Coronavirus (COVID-19)

Deep learning is used tremendously in the medical field because it can predict, recognize images, and solve classification problems. One of the characteristics of deep learning is that it can adapt to multiple data types. Deep learning helps with an accurate diagnosis due to its ability to quickly analyze complex medical images and the enormous computational capacity to process a large amount of data. This speed in diagnosing diseases helps doctors in the healthcare field detect diseases early, instead of manual work, which made this technique the best way for doctors to diagnose COVID-19 disease, which has spread recently.

Artificial intelligence and deep learning techniques have contributed to the processing of medical images in COVID-19 in many ways, such as tracking the virus's spread, predicting the virus's spread, diagnosing, and treating the virus, immunization, and drug discovery. The spread of COVID-19 disease was monitored by developing applications that use deep learning and the Internet of things through computer vision concepts and image processing approaches, for example: through deep learning techniques and the Internet of things, the temperature of individuals is measured for continuous thermal examination, as high temperature is one of the most important symptoms of infection with the virus. Respirators, protective equipment, and automatic disinfectants are also used to treat the injured and hospital specialists [22].

This section reviews some cases of Internet of things and deep learning use in COVID-19 in terms of medical image processing:

- The authors of [23] developed assistive diagnostic systems to detect the disease COVID-19 as quickly as possible to contain the disease to limit its spread as cases of COVID-19 have spread frighteningly and terrifyingly and threatened the economy and public health. In cooperation with the USA, China conducted this study to apply deep learning techniques in artificial intelligence with radiographic procedures to monitor the disease's spread and detect it faster. The study revealed that deep learning technology, the Internet of things, and artificial intelligence-based tomography images have very high accuracy in detecting positive cases of COVID-19. The data used in the study conducted in China were collected, and reviews were also conducted on cases outside China of COVID-19 cases. The study relied on the use of two-dimensional and three-dimensional models of deep learning, and their integration with artificial intelligence models, and the inclusion of complete clinical knowledge. Numerous experiments have been conducted to uncover deep learning CT features related to COVID-19 and detect the disease and monitor its progression using a 3D volume display. The results were 99% accurate in classifying positive or negative cases.
- The authors of [24], from the University of Waterloo in Canada, have designed a program called COVID-Net, which uses deep learning and the X-ray method, to increase the polymerase chain reaction (PCR) of the SWAB tests performed on COVID-19 patients. This increase, which depends on advanced computational

techniques, contributed to improved accuracy, which led to a decrease in the examination time. One of the positive results of this program is, it contains a larger number of patients, and chest radiographs of 3000 patients were collected to develop the examination tool from a large set of data. COVID-Net was developed in the next stage by the developer to design COVID-RiskNet, which diagnoses dangerous conditions in an infected patient so that treatment priorities are set. Through this performance, doctors can separate severe patients from relatively well-off patients to give them the right to health care according to their critical health conditions.

- The authors of [25], in Toronto company—BlueDot developed a platform that uses machine learning technology and big data to predict the outbreak of the infectious disease COVID-19. This platform has saved millions of lives, especially workers in the private and government policymakers, as their mitigation plans have been implemented. The platform has achieved great success in alerting departments to several pneumonia patients around the Wuhan market in China. The platform collects various data on 150 types of diseases around the World, and the database is updated every quarter of an hour around the clock. The data repository contains information on traveler history, the center for disease control, population and animal data, and local knowledge that journalists collect on diseases every day. Analysts compile this data, and it is collected manually. After the classification process is searched by keywords, applying machine learning and natural language processing to train the model. Highly filtered results are generated, and in the final stage, they are analyzed by experts to take necessary actions.

6 The Motivation of Our Work

IoTs emerge in healthcare systems extremely and positively shifts the healthcare systems. The most significant contribution of IoT in healthcare systems is continuously monitoring patients' conditions at any time and from anywhere. This assists medical staff to take a crucial decision as needed. The contribution mentioned above cannot exist without the employment of different techniques to store and process patients' collected data. These techniques can be cloud computing-based, database-based, and AI-based. This survey paper aims to study the methods mentioned above integrated in IoT healthcare systems. The paper gathers different proposed systems of the techniques mentioned above from multiple reviewed studies.

Besides the proposed systems, the paper also presents the most important challenges and issues that may face integrating these techniques with IoT healthcare systems. The paper is very useful for new researchers to recognize the current research in the IoT healthcare systems and give them overall knowledge about these techniques and how they can be proposed. Table 1 shows the different contributions of other existing surveys.

Table 1 Different contributions of other existing surveys

References	Year	Contributions
[26]	2020	The paper reviews the multiple benefits of the IoT in health care and the most prominent challenges such as issues of security and privacy
[27]	2020	A literary survey was presented for healthcare applications, and for the various technologies used in developing IoT systems such as device platforms such as Arduino, Raspberry Pi, Intel's Galileo, and BeagleBone
[28]	2020	The survey paper clarifies the contribution of IoT to healthcare systems by providing different proposed systems as well as presenting their challenges and limitations

7 Review of Recent Trends Using IoT in Health care

The Internet of things in health care has become popular research in recent years. It highlights the development of electronic devices and many wearable devices, which can be used for various patient monitoring systems, assisting in improving treatment, and processing patient data in health care. These devices measure the patient's health parameters and send them for further processing, where the acquired data are analyzed. The analysis provides patients or their relatives with the required medical support or predictions based on the received data. Cloud computing and database technologies, deep learning, and machine learning play a prominent role in consecutive data processing, analysis, and prediction.

In Table 2, we present the latest studies and research for healthcare systems based on the Internet of things, various sensors used to measure health standards, and the various deep learning and machine learning approaches provided to diagnose various diseases.

8 Issues of IoT Healthcare Systems

Most of the IoT issues in healthcare systems are privacy and security issues. Privacy issues are related to preserving confidential information pertaining to the patient, medical staff, research centers, hospital work environment, and many others.

Security issues are concerned about sensors. The sensors are often the ones that process sensitive data in patient monitoring and control health information. Hence, sensors should ensure that user data is protected from cyber-attacks.

Different techniques are used to overcome IoTs privacy and security issues in healthcare systems. The following illustrates the proposed solutions for these issues while using AI-based and cloud-based healthcare systems.

Table 2 Review of recent work in the area of IoT in health care

References	Key idea	Results
[7]	Cloud-based	The framework for all entities that may participate in health care. It allows all healthcare sector participants to communicate and collaborate through a safe environment
[9]	Cloud and fog-based	The proposed architecture is for healthcare monitoring in smart homes and hospitals based on cloud and fog computing. Also, it illustrates the key features that provided by fog computing
[10]	Cloud, fog, and edge-based	Healthcare framework supports mobility. The proposed framework was better than the existing health care based on the cloud by $\sim 28\%$ less delay and $\sim 27\%$ less in the device's energy consumption
[13]	Database-based: Firebase	The integration of Google technology in home-based ECG monitor device allows the improvement of patients' heart care through the interaction between patients as well as doctors despite their different locations
[14]	Database-based: Firebase	The combination of the CardioSys hardware and CardioSys software works satisfactorily. So, patients can be periodically diagnosed
[15]	Database-based: Firebase	Smart healthcare system IoT-based assists doctors by knowing patients' health state and analyzing their collected data in real time stored in Firebase
[16]	Database-based: SQL	A glucose monitoring system with a prognosis algorithm helps elderly patients who cannot come to their hospital appointments to be daily monitored
[19]	AI-based: machine learning algorithms	An effective cardiology diagnostic model was developed for monitoring and predicting it. The model was based on the cloud and the Internet of things. Classification algorithms were used to classify patient data to identify heart disease

(continued)

Table 2 (continued)

References	Key idea	Results
[21]	AI-based: machine learning algorithms	A new framework has been created to integrate artificial intelligence technology and the Internet of things to collect and process heterogeneous data from environmental and wearable sensors to monitor patients' health
[23]	AI: deep learning	The system that was developed to monitor the spread of the disease by applying deep learning techniques and radiographic techniques contributed greatly to detecting the disease COVID-19 with high accuracy of 99% to classify cases
[24]	AI: deep learning	The COVID-Net software, which was adopted using the X-ray method and deep learning, reduced the examination time for patients and contributed to the rapid diagnosis of serious cases. This greatly helped set treatment priorities
[29]	AI: deep learning	A platform has been developed to help the Alzheimer's community using face perception through the visual understanding of facial images related to family, friends, and colleagues to help with memory problems Try facial reminder glasses to help by using deep learning to recognize and identify people in a patient's life to enhance social skills

8.1 A Using Artificial Intelligence

In smart healthcare, emergency services, and healthcare analytics systems that rely on artificial intelligence have an essential requirement toward security to create a smart hospital environment, sensors, and laboratory as well [16].

- The authors of [21] proposed solutions to ensure user data privacy and patient data in healthcare, from cyber-attacks, which aim to steal medical data by providing different techniques to anonymize the data using artificial intelligence algorithms, in the following machine learning methods. Anonymity is hidden by following the classic and well-known (unsupervised) clustering approach. For data privacy is the anonymity of type, the k-mean algorithm can be used as this algorithm can predict identity revocation and extract data through geolocation while on the move. One of the most prominent features of the k-mean algorithm is that it maintains privacy when data features differ. Another approach, which is the k-means-based clustering approach, can be adapted to distinguish between different sensitive data and anonymous data, using anonymized algorithms.

- The authors of [30] proposed an algorithm to protect the data generated from machine learning, which is trained to predict the solution of specific problems. The reason of that is the datasets can be possibly manipulating to produce specific results. However, the proposed algorithm is based on using blockchain which is great for security to the private cloud that stores its data.

- The authors of [31] proposed a deep neural network (DNN), a machine learning-based. DNN depends on a hybrid PCA-GWO and is a basic technique for developing a classification model. This technique is used to create a very useful intrusion detection system (IDS) in the Internet of medical things (IoMT) environment. IoMT is a special type of IoT where the medical devices such as medical sensors connect to exchange hypersensitive data. Using DNN in IoMT, a high percentage of unpredicted cyber-attacks can be classified and expected. The proposed DNN model achieved a better performance than the already existing machine learning approaches by 15% of rising accuracy and 32% of reducing in the time complexity.

8.2   Using Cloud

- The authors of [32] propose security architecture for a smart medical system based on the cloud to face the different types of attacks such as impersonation, MITM, anonymity, known-key security, and stolen verifier attack. One of the basic concerns issues in smart medical system (SMS) based on cloud is patient data security and privacy. The authors' secure authentication framework is made by using the elliptic-curve cryptography (ECC) algorithm. The main participants in this system are patients, doctors, healthcare centers, and the cloud. Moreover, there are six phases that the proposed architecture consists of. These phases are mentioned in the following: patient's registration, healthcare center uploading, patient's data uploading, treatment phase, patient's checkup, and emergency phase.

- The authors on [33] propose a BAMHealthCloud system used for healthcare management. BAMHealthCloud is related to biometric signature-based authentication. The authors use a training model on the signature to ensure secure electronic patient records access. As a result, the authors achieve speedup by 9 times and an error rate of 0.12.

- The authors on [34] proposed a CBMH system which is a platform to support patients in an emergency remotely based on the cloud. The system's main contribution is providing a framework to enhance authenticity for secret session key among communication.

9 Conclusion

The following techniques as cloud computing, database, and AI are widely used today in healthcare systems. These techniques help improve healthcare systems efficiently to assist any healthcare team members, such as patients, doctors, and medical staff. However, due to the widespread and development of technology, some fundamental issues exist in the IoT healthcare system. Some of these issues are privacy and security. These two issues should be considered to perform more reliable and credible healthcare systems.

References

1. Wortmann F, Flüchter K (2015) Internet of Things: technology and value added. Bus Inf Syst Eng 57(3):221–224. https://doi.org/10.1007/s12599-015-0383-3
2. Smart and secure IoT and AI integration framework, vol 01, no 03, pp 172–179 (2019)
3. Suresh P, Daniel JV, Parthasarathy V, Aswathy RH (2014) A state of the art review on the Internet of Things (IoT) history, technology and fields of deployment. In: 2014 international conference on science engineering management research ICSEMR 2014. https://doi.org/10.1109/ICSEMR.2014.7043637
4. Sundaravadivel P, Kougianos E, Mohanty SP, Ganapathiraju MK (2018) Everything you wanted to know about smart health care: evaluating the different technologies and components of the internet of things for better health. IEEE Consum Electron Mag 7(1):18–28. https://doi.org/10.1109/MCE.2017.2755378
5. Ray PP (2016) A survey of IoT cloud platforms. Futur Comput Inform J 1(1–2):35–46. https://doi.org/10.1016/j.fcij.2017.02.001
6. Arasaratnam O (2011) Introduction to cloud computing. Audit Cloud Comput A Secur Priv Guid:1–13. https://doi.org/10.1002/9781118269091.ch1
7. Tyagi S, Agarwal A, Maheshwari P (2016) A conceptual framework for IoT-based healthcare system using cloud computing. In: Proceedings of 2016 6th international conference on cloud system big data engineering confluence, pp 503–507. https://doi.org/10.1109/CONFLUENCE.2016.7508172
8. Puliafito C, Mingozzi E, Anastasi G (2017) Fog computing for the internet of mobile things: issues and challenges. In: 2017 IEEE international conference on smart computing SMARTCOMP 2017. https://doi.org/10.1109/SMARTCOMP.2017.7947010
9. Nandyala CS, Kim HK (2016) From cloud to fog and IoT-based real-time U-healthcare monitoring for smart homes and hospitals. Int J Smart Home 10(2):187–196. https://doi.org/10.14257/ijsh.2016.10.2.18
10. Mukherjee A, Ghosh S, Behere A, Ghosh SK, Buyya R (2020) Internet of Health Things (IoHT) for personalized health care using integrated edge-fog-cloud network. J Ambient Intell Humaniz Comput 0123456789. https://doi.org/10.1007/s12652-020-02113-9
11. Moroney L (2017) The firebase realtime database. In: The definitive guide to firebase. Apress, Berkeley, CA, pp 51–71
12. Nordeen A (2020) Learn SQL in 24 Hours. Guru99
13. Goh V, Ng K, Teoh J, Lim J, Hau Y (2019) Home-based electrocardiogram monitoring device with google technology and Bluetooth wireless communication. Int Sem Inf Commun Technol (ISICT 2019) 1(1):30–35
14. Negara R, Tulloh R, Alfasyanah A, Umbarawati A (2019) Cardiosys: human cardiology measurement system based on internet of things. J Theor Appl Inf Technol 97(24):3836–3849
15. Hasnat M, Al Mamun S, Hossain F, Hossain S (2019) IoT based smart healthcare

16. Valenzuela F, García A, Ruiz E, Vázquez M, Cortez J, Espinoza A (2020) An IoT-based glucose monitoring algorithm to prevent diabetes complications. Appl Sci 10(3):1–12. https://doi.org/10.3390/app10030921

17. Wang F, Preininger A (2019) AI in health: state of the art, challenges, and future directions. Yearbook Med Inform 28(1):16

18. Mohanta B, Das P, Patnaik S (2019) Healthcare 5.0: a paradigm shift in digital healthcare system using artificial intelligence, IOT and 5G Communication. In: 2019 international conference on applied machine learning (ICAML), Bhubaneswar, India, pp 191–196. https://doi.org/10.1109/ICAML48257.2019.00044

19. Ganesan M, Sivakumar N (2019) IoT based heart disease prediction and diagnosis model for healthcare using machine learning models. In: 2019 IEEE international conference on system, computation, automation and networking (ICSCAN), Pondicherry, India, pp 1–5. https://doi.org/10.1109/ICSCAN.2019.8878850

20. Ahamed F, Farid F (2018) Applying Internet of Things and machine-learning for personalized healthcare: issues and challenges. In: 2018 international conference on machine learning and data engineering (iCMLDE), Sydney, Australia, pp 19–21. https://doi.org/10.1109/iCMLDE.2018.00014

21. Mongelli M et al (2020) Challenges and opportunities of IoT and AI in pneumology. In: 2020 23rd Euromicro conference on digital system design (DSD), Kranj, Slovenia, pp 285–292. https://doi.org/10.1109/DSD51259.2020.00054

22. Bhattacharya S, Maddikunta PKR, Pham QV, Gadekallu TR, Chowdhary CL, Alazab M, Piran MJ (2020) Deep learning and medical image processing for coronavirus (COVID-19) pandemic: a survey. Sustain Cities Soc 65:102589

23. Gozes O, Frid-Adar M, Greenspan H, Browning PD, Zhang H, Ji W, Bernheim A, Siegel E (2020) Rapid ai development cycle for the coronavirus (covid-19) pandemic: initial results for automated detection and patient monitoring using deep learning CT image analysis. arXiv:2003.05037

24. Ozturk T, Talo M, Yildirim EA, Baloglu UB, Yildirim O, Acharya UR (2020) Automated detection of COVID-19 cases using deep neural networks with X-ray images. Comput Biol Med 121:103792

25. Simonite T (2020) Chinese hospitals deploy AI to help diagnose Covid-19 [Online] Wired 2020. Available at https://www.wired.com/story/chinese-hospitals-deploy-ai-help-diagnose-covid-19/. Accessed 23 Jan 2021

26. Thilakarathne NN, Kagita MK, Gadekallu TR (2020) The role of the internet of things in health care: a systematic and comprehensive study. Int J Eng Manage Res 10(4):145–159

27. Bhatia H, Panda SN, Nagpal D (2020) Internet of Things and its applications in healthcare—a survey. In: 2020 8th international conference on reliability, Infocom technologies and optimization (trends and future directions) (ICRITO), Noida, India, pp 305–310. https://doi.org/10.1109/ICRITO48877.2020.9197816

28. Lakshmi PS (2020) Survey on IoT healthcare 6(5):1–7

29. Roopaei M, Rad P, Prevost JJ (2018) A wearable IoT with complex artificial perception embedding for Alzheimer patients. In: 2018 world automation congress (WAC), Stevenson, WA, pp 1–6.https://doi.org/10.23919/WAC.2018.8430403

30. Gadekallu TR et al (2020) Blockchain based attack detection on machine learning algorithms for IoT based E-Health applications. arXiv:2011.01457

31. Swarna Priya RM et al (2020) An effective feature engineering for DNN using hybrid PCA-GWO for intrusion detection in IoMT architecture. Comput Commun 160:139–149. https://doi.org/10.1016/j.comcom.2020.05.048

32. Chen C (2020) CSEF : cloud-based secure and efficient framework for smart medical system using ECC, vol 8. https://doi.org/10.1109/ACCESS.2020.3001152

33. Shakil KA, Zareen FJ, Alam M, Jabin S (2020) BAMHealthCloud: a biometric authentication and data management system for healthcare data in cloud. J King Saud Univ Comput Inf Sci 32(1):57–64. https://doi.org/10.1016/j.jksuci.2017.07.001
34. Deebak BD, Al-Turjman F (2020) Smart mutual authentication protocol for cloud based medical healthcare systems using internet of medical things. IEEE J Sel Areas Commun

Chapter 35
Quantity Discount Algorithm in an E-Commerce Environment

Olli-Pekka Hilmola

1 Introduction

Ordering decision of manufacturers, wholesalers, retailers and consumers has been central question for last century in purchasing and logistics. How many units, and when to order—these have been in constant debate and base of development of analytical models. Economic order quantity (EOQ) models were firstly developed in the early parts of twentieth century to answer these questions [1, 2]. EOQ research continued after these for the specific applications in purchasing and warehouses (like min–max ordering and inventory control) as well as to production lot sizing [3]. Later on, applications expanded to specific areas such as purchasing quantity discounts [4]. Even ordinary consumer has faced in retail outlets offers such like, *"select three items with the price of two"* or *"two items with price of X."* Research has shown that these sorts of discounts are workable and increasing sales in some product groups, like functional products [5]. It is rather evident that sales will increase with this sort of policies; however, supply chain is under stress and will end to excessive inventory holding and capacity due to distorted signals to upstream from demand changes (even if overall demand in long-term would not change at all; [6, 7]). Quantity discount is not only privilege of consumers, but it has been often used in business-to-business environment to stimulate demand and enable higher production quantities at factories and within distribution. Even if higher order quantities stimulated with lower prices would distort real demand at final markets, it will enable many scale and synergy factors [8] within production supply chains (e.g., higher capacity utilization, lower amount of transportation needed, lower packaging amounts and lower handling

O.-P. Hilmola (✉)
LUT University, Kouvola Unit, Tykkitie 1, FI-45100 Kouvola, Finland
e-mail: olli-pekka.hilmola@taltech.ee; olli-pekka.hilmola@lut.fi

Tallinn University of Technology (Taltech), Estonian Maritime Academy, Kopli 101, 11712 Tallinn, Estonia

© The Author(s), under exclusive license to Springer Nature Singapore Pte Ltd. 2021   437
S. Kumar et al. (eds.), *Proceedings of International Conference on Communication and Computational Technologies*, Algorithms for Intelligent Systems,
https://doi.org/10.1007/978-981-16-3246-4_35

amounts). Most often large order quantity is associated in literature with the same amount of supply quantity. However, applications in industry are such that transfer or transportation quantity could be much smaller and entire lot being transported smoothly within continuous fashion throughout the manufacturing supply chain [9, 10]. In addition, suppliers might provide discounts based on longer time horizon total amount of purchases, not depending on single purchase transaction [8].

In e-commerce, the role of product price discounts is even more significant than in traditional retailing or wholesale trade. Mostly research has concentrated on consumer side of e-commerce, and the role of discounts is in the business success [11, 12]. However, situation remains the same, if the examination is moved to business-to-business e-commerce, like the platform of Alibaba. As Lv et al. [12] have concluded through their empirical work that low-reputation supplies in e-commerce are forced to use price discounts, in order to attract customers. Price discounts also differ between environments as single purchased unit and its lower price attracts consumers, where in business context it is the old quantity-based discount what matters. This is also the main motivation for this research work: To examine price discount use in business-to-business environment by e-commerce platform, and to introduce algorithm to utilize these discounts in economically most favorable form by purchasing party. Based on previous research out of traditional wholesale/retail, price discounts were argued to be beneficial in wholesaling or retailing environment, but not necessarily in manufacturing [13]. Intention in this research is just objectively analyze e-commerce offers and provide algorithm for the use of analyzing benefits of price discounts. Therefore, results should be more on the side of wholesaling or retail and are possibly in favor of utilizing attractive prices, even if they result on large inventory holdings. It should be noted that not necessarily one actor has needed sales volume to acquire highest volume classes; however, purchasing collaboration is then one alternative with the same branch companies [14].

This research work is used quantitative and analytical research method. Examples in this study are taken from real life and out of e-commerce platform of Alibaba. All information has been gathered solely by this author, and it mostly concerns low priced consumer sales items (sold in bulky terms through business-to-business principles in e-commerce platform). Analytical model of EOQ with price discounts is as first taken and introduced using previous literature and research findings. This formula is then fitted in an e-commerce environment, where there exists typically number of price discount steps, but also freight and customs payments in the liability of ordering party. Analysis is done concerning one illustrative product (facemask), however, research problem area is further elaborated through the analysis of arbitrarily selected 50 items of Alibaba e-commerce platform.

This chapter is structured as follows: In Sect. 2, quantity discount algorithm using economic order quantity framework is introduced. In the following Sect. 3, selected 50 e-commerce platform products (business-to-business, wholesale) are analyzed through quantity discount perspective concerning provided discounts and required volumes in each discount step. Application of quantity discount algorithm follows in Sect. 4, where facemask e-commerce offer (wholesale) is analyzed in details.

Chapter ends in Sect. 5, where work is concluded and future research avenues are being proposed.

2 Quantity Discount Algorithm and Economic Order Quantity

Economic order quantity formula expanded for quantity discount situations consists three parts to be considered for overall minimization. First component is the order cost in the planning period, second is the inventory holding cost and the third one is purchase cost. Last cost item is altered by possible discounts (as also in somewhat inventory holding cost). Formula is as follows [e.g., 8, 13]:

$$\min \frac{D}{Q}C_o + \frac{1}{2}QC_i + QP_q \tag{1}$$

$$P_q = 0 < Q \leq n\vee \tag{2}$$

$$n + 1 \leq Q \leq m\vee \tag{3}$$

$$m + 1 \leq Q \ldots \tag{4}$$

$$n + 1 < m \tag{5}$$

where

D	Demand in period (year).
Q	Order lot size.
C_o	Cost of ordering (fixed).
Ci	Inventory holding costs of item.
P_q	Price paid from product-based order quantity, Q (volume).
n, m	Quantity limits n and m for discounted price range.

As the effects of quantity discounts are taken into consideration, it will lead to situation, where lowest possible cost could be found in one of the following two: (1) economic order quantity of each price class, or (2) if calculated EOQ is outside of volume limits, the closest volume point of EOQ in particular price class (it could be lowest possible volume, but also highest as it depends, what is the level of EOQ in this particular situation in price class). EOQ needs to be calculated for each quantity discount area separately as inventory costs change due to lower item acquisition price. Typically, quantity discount possibility will lead to larger lot sizes used or even order all the need of one period (like year) at once. This could yield considerable savings for ordering organization, however, it should be remembered that inventory holding risks also do increase (e.g., spoilage, quality issues, used technology or design becoming obsolete, theft, market changes, etc.). If purchased item takes a lot of space or is weighty, then own warehouse space could be limited. In that case, further analysis needs to take into account rental market of warehouses with varying sizes and prices. This will also lead to further modification of algorithm [see, 15].

In traditional and e-commerce, bulky offers of products could lead to situations, where "cost of ordering (fixed)" is actually zero. This could be the case in traditional sales that supplier promises to deliver these products and take care of all the practicalities, if customer is just willing to make bulky order (e.g., rest of their excessive inventory holding, see example, [16]). In e-commerce, this could happen, if e-commerce platform is having country-level customs warehouses, and it is in the need to reduce inventory holdings of some products quickly in some particular region. Typically, this cost benefit is not taken into consideration, but is possible in real life.

It should be noted that above presented quantity discount formula just takes into account cycle stock caused by an order (for the sake of simplicity, it is assumed that consumption of item is stable, and average inventory amount in units is $Q/2$). This will result in situation, where 50% out of time, company is not able to meet customer requests [17], in other words, facing stock-outs. Therefore, safety stocks are typically used. These could be built either making order earlier than intended (in future, when company starts to run out of items) or alternatively order excessive amount in the first order [18]. Again, bulkier orders of price discounted items give an opportunity here to serve customers in better service level. As inventory levels are high, then safeties are excessively high. This is one practical aspect, what is typically forgotten in research, but is apparent in real life. Large inventory holdings also provide opportunities for unexpected bulky sales.

3 Price Discounts of Selected Items

The purposes of this research work were sample of fifty different products selected from Alibaba [19]. These products were typical consumer items such as wristwatches, shoes, shirts, mobile phone accessories, and kitchenware. Average starting price of these products was 16.6 USD and median 6.6 USD per unit. As average is so much above median, it indicates that some more expensive products were in this analyzed group (as average is skewed). Highest priced product was "small kitchen" with price of 395 USD per one unit, and following this as the second dearest was "stainless steel cookware set," having price of 40 USD per unit. Lowest starting price of items was on "cell phone grip," 0.24 USD per unit. This sort rather high difference in prices resulted on standard deviation of 55.3 USD.

Selected quantity discount products were typically having three different levels of prices depending on the volume (41 observations out of 50–82%, see Table 1), and quite many still having four levels even (21 observation out of 50–42%, see Table 1). As could be noted from Table 1, these lower prices do not come with

Table 1 Minimum quantity development in different price levels as first price level is 100 ($n =$ 50)

	Level 1	Level 2	Level 3	Level 4
Average index	100	377.9	749.3	2357.9
Median index	100	3000.0	5000.0	10,000.0
Min index	100	1000.0	5000.0	10,000.0
Max index	100	333.3	333.3	1666.7
Observations	50	50	41	21

small volume additions. On the average proceeding from level 1 prices to level 2 will require increase of 3.78 times, and further to level 3 volumes need to increase from base by 7.5 times. Highest discount level 4 could be reached on the average by 23.58 times higher volumes. However, these are average volume increase needs. In median terms increases needed are much higher as from level 1 to level 2 will require 30 times more volume, while from base to level 3 it is 50 times and level 4 in turn 100 times higher volumes.

Most important quantity discount level change is in the analyzed group of products proceeding from base level 1 to level 2 (Table 2)—this will grant on the average 15% lower prices (in median terms decline is somewhat lower). As taking this first volume level increase, there are some products, where price decrease is very significant, 73%; however, in other extreme, it is just one percent. As proceeding further to levels 3 and 4, prices will decrease on the average 10 and 11% further from previous price level. Median values follow this development but are somewhat lower. There are some products, which show also very steep declines in prices as proceeding to level 3 (40%) or to level 4 (34%). In other extreme, in some products decrease is just 1–2%. Situation in level 2 is further elaborated in Figs. 1 and 2. The most common frequencies in price discount terms are classes − 1 to − 11% and − 11 to − 22% (Fig. 1). It is rather uncommon to have higher price reductions as other discount classes account 18% out of all observations. Two products out of fifty products belong to steepest price discount class. Highest discount is that of "Thermal Travel Mug" (73% discount as volumes increase to level 2, however, this requires at least order of 10,000 units), where the second highest discount is for electronic wristwatch (63% discount as volumes increase to at least 10).

Further examination of minimum volume level to reach first quantity discount reveals that most of the products fall in the first class (45 out of 50, or 90%), which

Table 2 Quantity discount percent in different discount levels ($n = 50$)

	Level 2 (%)	Level 3 (%)	Level 4 (%)
Average	− 15	− 10	− 11
Median	− 11	− 8	− 9
Min	− 73	− 40	− 34
Max	− 1	− 1	− 2

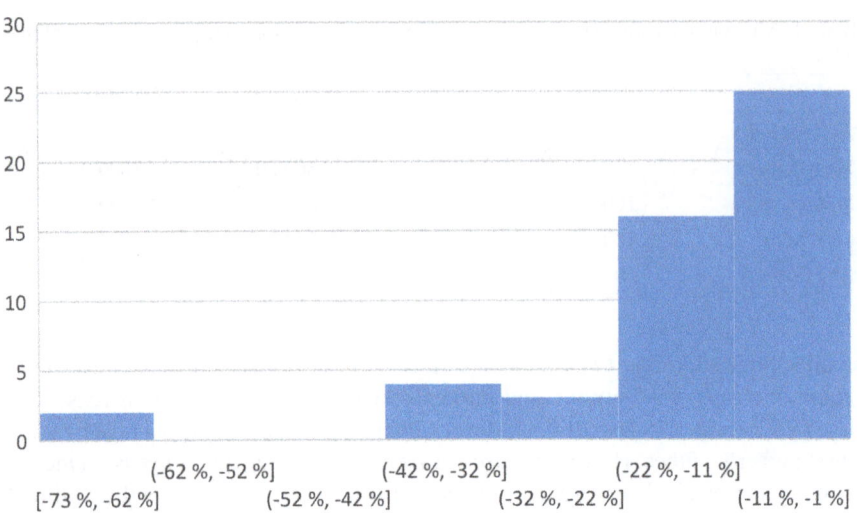

Fig. 1 Quantity discounts in percent as examining first discount level 2 from selected products of Alibaba ($n = 50$)

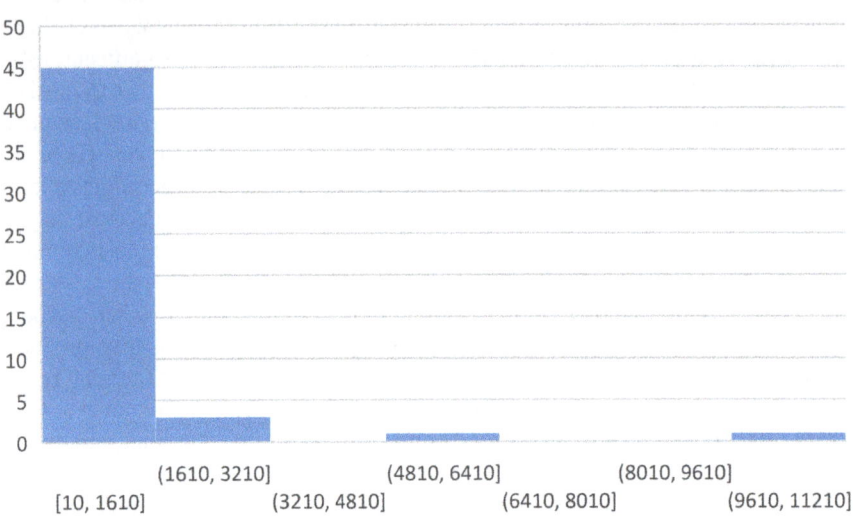

Fig. 2 Minimum amount of quantity ordered to reach first discount level 2 in selected products of Alibaba ($n = 50$)

starts from quantity of 10 and ends to 1610. Three products (6%) belong to following quantity class from 1610 to 3210. Two other higher volume classes have one product in each one. This in overall reveals that reaching first quantity discount level requires some volumes, but not necessarily very high ones. It is at the reach of many companies regardless of size.

4 Price Discount Algorithm for Lowest Total Costs: Illustrative Example

To illustrate analysis and possible use of quantity discounts, in the following is analyzed quantity discount of regular facemask. This data is taken from one real supplier existing in Alibaba e-commerce platform [20]. Facemask is offered with three levels of prices, and lowest volume (2000–9999) will result in price of 0.09 USD per unit (Table 3). Prices shall decline to 0.06 USD per unit, if order is above 10,000 units, but below 999,999. Orders having size of one million units or more are having price of 0.03 USD per unit. Although customs globally, and particularly in European Union with China, have declined, this commodity group is still having import duty of 4.3%. As products enter EU area in import operations, it is import duty calculated based on paid purchase costs, e.g., in China, and then adding freight costs into this sum. This sum together is the basis for customs payment. As facemasks

Table 3 Illustrative example from facemask quantity discounts offered and lowest total costs order lot size

Demand in period (annum)		1,000,000		
Setup cost (USD)		$500.00		
Inventory holding cost, p.a		40%		
Customs import duty		4.3%		
Price of mask (incl. customs)	$0.094	$0.063	$0.031	
Economic Order Quantity	163,195	199,872	282,662	
Is feasible?	No	Yes	No	
Units ordered	2000–9999	10,000–999,999	≥ 1 mill	
Original price per unit:	$0.09	$0.06	$0.03	
Price of mask (incl. customs)	$0.094	$0.063	$0.031	
Lower limit	2000	10,000	1,000,000	
Upper limit	9999	999,999		
Lower limit feasible?	No	No	Yes	
Upper limit feasible?	Yes	No	No	

	Lowest volume	EOQ	Highest volume	Cost difference to	
				Lowest volume	EOQ
Order costs	$50,005.00	$2501.60	$500.00	$49,505.00	$2001.60
Inventory holding costs	$469.30	$6253.99	$15,645.00	$– 15,175.70	$– 9391.01
Purchase costs	$93,870.00	$62.580.00	$31,290.00	$62,580.00	$31,290.00
Total costs	$144,344.30	$71,335.60	$47,435.00	$96,909.30	$23,900.60
Per mask	$0.144	0.071	$0.047	0.097	$0.024

are extremely light weighted and take small space, therefore, freight payments do not make significant effect on Table 3 example, and they have been neglected in the calculations (in some items freight costs are very important, and will change significantly with volume, making algorithm development even more complex, [21]). However, customs duty level is taken into consideration in Table 3 prices.

As analyzing through Table 3 regarding price discount possibilities, it is important to start analysis by calculating economic order quantity (EOQ) within each price class. Single EOQ calculation is not feasible as price change has its effect on inventory holding costs (lower prices result on lower inventory holding costs), and EOQ needs to be calculated for each price class separately. As could be noted, only second price class is giving EOQ, which is within the volume area of this class. Lowest price class has too high EOQ as compared to its maximum amount, while highest class EOQ is too low as thinking about order of one million units. Another area of minimum total costs is the lowest level of highest volume class (lowest volume of second lowest class is not having minimum costs as EOQ is higher than its minimum), together with highest purchasing amount of lowest volume class (this due to the reason that highest volume is nearest to first volume class EOQ).

In the last part of Table 3 is examined, these three different options, highest volume of the first price class, EOQ of the second price class and lowest volume of the third price class. The beginning of Table 3 is given parameter values for this economic examination. Setup costs include typically all administration and management costs of making an order from suppliers (and making payment), following its progress and taking care of entrance to own country (through customs, making payments of customs and value added tax as well as checking the level of quality and quantity of the shipment). Setup cost is fixed sum per order, and in illustrative example, EOQ will result on somewhat higher costs than highest volume class order (as only one order is being made in the entire period, e.g., one year). Of course, clearly highest cost is in the first price class option (due to numerous orders being made). Inventory holding costs are set in this illustrative case as 40% p.a. Inventory holding cost is of course dearest in the highest volume order as average inventory is 0.5 mill. units (vs. 99,936 units in EOQ or 4999.5 units in first price class), even if price of product is much lower (making economic holding cheaper). Third component in total cost calculation is actual purchase cost. This is of course in favor of highest volume class and increases total costs in two other options rather significantly.

Based on cost analysis of Table 3, it is apparent that aiming for highest volume class in facemasks is worthwhile to pursue as in total costs it will lead to 23,900.6 USD as saved in annum as compared to EOQ option. If comparison is made to first price class option, then it is saving 96,909.3 USD per annum. This result needs of course in the real life be analyzed thoroughly as it will increase the risk of holding obsolete or excessive inventory (and not only that but also additional warehouse space is possibly needed), however, in other regards effects are only positive. If all costs are calculated per mask, it could be seen that highest volume class order shall lead to 4.7 USD cents per mask, which is 2.4 USD cents less than EOQ and 9.4 USDD cents less than first price class. Lower total purchase cost gives room to defend own sales within final markets as prices could be lowered in situation of lower or sluggish demand.

5 Conclusions

Two decades ago, as digitalization and Internet became available for larger public, company and public applications grew in the short amount of time. From that time onwards, digitalization research also started concerning supply chains, and it has argued that shorter lead times, lower lot sizes and higher clock speed are the future [22]. In one way, they are and were, however, as emerging markets started to take foothold in manufacturing, it reversed a little bit these plans. Suddenly, digitalization made all markets closer together, and even consumers were able to order long-distance items to their homes. This meant that price competition increased even further, and also all sized companies got a change to expand their sourcing channels to emerging markets. In addition, this all without having intermediaries or country-level offices in distant countries. As was shown in previous sections and research completed, this of course is coming in companies with some restrictions. Very competitive prices are available in distant markets, however, they often require higher volumes. As quantity discounts contain typically three to four levels, analysis needs to be completed with algorithms. Even if total costs are minimized with this approach, decision-makers need to consider all risks involved. Like having excessive inventories at hand as well as having possible obsolete problem due to poor quality or inappropriate materials used. However, as shown in this research, rewards are also handsome, and high volume orders could provide really good opportunity for profitable wholesale or retail trade in home markets. As was shown in this research, larger inventory has also some positive sides, like the ability of company to serve its customers better (even sudden demand spikes) and also being able to avoid excessive administration work due to numerous small orders.

As a further research in this area, it would be interesting to analyze higher number of quantity discount items from the perspective of business buyer, like was done in illustrative example. This would give further insights, what should be done to algorithm to make it as automated as possible, and that no human interaction would be needed to gain results. Another interesting further research area would be to expand examination only from one e-commerce platform to other competing companies in China, and other emerging markets, like India and Vietnam.

References

1. Harris FW (1913) How many parts to make at once. Factory Mag Manage 10(2):135–136
2. Wilson RH (1934) A scientific routine for stock control. Harv Bus Rev 13(1):116–128

3. Tersine RJ (1985) Production-operations-management: concepts, structure, and analysis, 1st edn. Elsevier-North Holland Publishing, New York, US
4. Pereira V, Costa HG (2015) A literature review on lot size with quantity models: 1995–2013. J Model Manage 10(3):341–359
5. Drechsler S, Leeflang PSH, Bijmolt THA, Natter M (2017) Multi-unit price promotions and their impact on purchase decisions and sales. Eur J Mark 51(5/6):1049–1074
6. Tan A, Hilmola O-P, Binh DH (2016) Matching volatile demand with transportation services in Vietnam: a case study with Gemadept. Asia Pacific J Marketing Logist 28(1):160–174
7. Zotteri G (2013) An empirical investigation on causes and effects of the Bullwhip-effect: evidence from the personal care sector. Int J Prod Econ 143(2):489–498
8. Benton WC, Park S (1996) A classification of literature on determining the lot size under quantity discounts. Eur J Oper Res 92(2):219–238
9. Inkaya T, Akansel M (2017) Coordinated scheduling of the transfer lots in an assembly-type supply chain: a genetic algorithm approach. J Intell Manuf 28:1005–1015
10. Koskinen P, Hilmola O-P (2008) Supply chain challenges of North-European paper industry. Ind Manage Data Syst 108(2):208–227
11. Jiang Y, Shang J, Liu Y, May J (2015) Redesigning promotion strategy for e-commerce competitiveness through pricing and recommendation. Int J Prod Econ 167:257–270
12. Lv J, Wang Z, Huang Y, Wang T, Wang Y (2020) How can e-commerce businesses implement discount strategies through social media? Sustainability 12:7459
13. Aghazadeh S-M (2001) A comparison of just-in-time inventory and the quantity discount model in retail outlets. Logist Inf Manage 14(3):201–207
14. Zeng RL, Qu H, Wang L (2018) Optimizing the new coordinated replenishment and delivery model considering quantity discount and resource constraints. Comput Ind Eng 116:82–96
15. Rajabi AM (2016) A revision on cost elements of the EOQ model. Stud Bus Econ 11(1):5–14
16. Drury C (1996) Management and cost accounting. Pearson, London, UK, p 717
17. King PL (2011) Crack the code: understanding safety stock and mastering its equations. APICS Mag:33–35
18. Simchi-Levi D, Kaminsky P, Simchi-Levi E (2003) Designing and managing the supply chain, 2nd edn. McGraw-Hill, Irwin, Boston, pp 58–61
19. Alibaba homepage. https://www.alibaba.com/, data gathering of items. Last accessed 2019/10/21
20. Alibaba homepage, medical mask. https://www.alibaba.com/product-detail/Wholesale-has-cer tification-3d-face-mask_1600054702711.html. Last accessed 2020/12/5
21. Russell RM, Krajewski LJ Optimal purchase and transportation costs lot sizing for a single item. Decis Sci 22(4):940–952
22. Kayikci Y (2018) Sustainability impact of digitization in logistics. Proc Manuf 21:782–789

Chapter 36
Application of Watermarking Along with Reversible Image Compression in Secure Medical Image Transmission Through a Network

Smita Khond, Prashant Sharma, and Bellamkonda Vijayakumar

1 Introduction

Information can be downloaded effectively without any proprietor authorization, as today's generation is based on E-era that includes E-learning, E-health, E-banking, and E-commerce. Different issues can be resulted due to these circumstances, including proprietor verification and copyright security. In such type of situations, the advanced information with validation and assurance is needed before exchanging the data over an open-source transmission medium. Different strategies of information concealing are proposed by analysts to overcome these issues: watermarking, stenography, and cryptography [1, 2]. For validating and ensuring the information, the watermarking technique is used primarily. This technique overcomes steganography constraints through the integration of a watermark image into host content that helps in not allowing the determination of hidden watermark by the primary client. Three components have included the framework of watermarking, such as a watermark extractor, and a wireless or wired correspondence channel, as per the literature review [3–5]. For producing a watermarked image, a watermark is embedded into host images by the watermark embedder. In contrast, a watermark extractor is used to extract the test image's watermark that would be the watermarked image without or with attacks. The digital image watermarking technique requires some major needs which are mentioned here as follows:

S. Khond (✉)
Pacific Academy of Higher Education and Research University, Udaipur, Rajasthan, India

P. Sharma
Computer Science and Engineering Department, Pacific Academy of Higher Education and Research University, Udaipur, Rajasthan, India

B. Vijayakumar
Vidya Jyothi Institute of Technology (VJIT), Hyderabad, Telangana, India

© The Author(s), under exclusive license to Springer Nature Singapore Pte Ltd. 2021   447
S. Kumar et al. (eds.), *Proceedings of International Conference on Communication and Computational Technologies*, Algorithms for Intelligent Systems,
https://doi.org/10.1007/978-981-16-3246-4_36

a.   Robustness: To protect from any manipulations, the scheme of watermarking should preserve owners' information without compromising on robustness.

b.   Imperceptibility: The visual quality of host data should not be impacted much after inserting the watermark into host data. Here, the watermark must be invisible.

c.   Embedding capacity: The hiding watermarks with large size should be allowed by the watermarking scheme.

Based on three processing domains like hybrid domain, transform domain, and spatial domain, watermarking schemes are improved [6]. The easier implementation could be possible with the spatial domain schemes, but less imperceptibility is provided due to directly modifying host image pixels. The complexity is more in transform domain watermarking but excels in robustness when compared to the spatial domain watermarking. Before embedding the watermark, the conversion of the host image is made into the frequency domain in all the transform domain schemes. These schemes are developed based on different image transforms like discrete wavelet transform (DWT), discrete cosine transform (DCT) [7, 8], and discrete Fourier transform (DFT). Then, it is transformed inversely. Most of the researches are based on new transforms with the implementation of watermarking techniques such as finite ridgelet transform (FRT) [9], non-sub-sampled contourlet transform (NSCT) [10], and fast discrete curvelet transform (FDCuT) [11]. In most of the watermarking schemes, the cover image's mid-frequency DCT weights are utilized extensively for embedding the watermark image out of all these transforms. These weights can provide more robustness than the other weight values against watermarking attacks.

For providing the security for medical images, the significant challenges are included in designing watermarking schemes such as

(a)   It should provide more robustness against watermarking attacks.

(b)   It should be non-blind or blind.

(c)   It must be provided with the information about hidden or embedded patient or owner identification into the cover medical image.

The medical image transfers from one hospital or remote healthcare center to another hospital; the watermarking schemes have been exploited in telemedicine applications to provide medical image security. Based on singular value decomposition (SVD), finite ridgelet transform (FRT), and Arnold scrambling-based encryption, a new watermarking scheme is designed in this paper. The robustness is achieved with the Arnold scrambling to avoid the watermarking attacks and is incorporated more owner or patient identification.

The remaining paper is described as follows. In Sect. 1, the recent research published works relevant to the proposed method with a summary is demonstrated. In Sect. 2, the proposed work's main contribution is illustrated. The proposed process of embedding and extraction is discussed in Sect. 3. A brief discussion of the proposed work with experimental results is presented in Sect. 4. The conclusion points are discussed in Sect. 5.

2 Related Work

This section presented the essential researches about medical image security. The watermarking schemes were reported by many researchers recently to secure the medical image. Arunkumar et al. [12] have proposed a method of non-blind and hybrid domain medical image watermarking by using different transforms like singular value decomposition (SVD), redundant integer wavelet transform (RIWT), and discrete cosine transform (DCT). A patient's information is embedded in the medical image's hybrid coefficients in this scheme based on additive watermarking. The robustness and good imperceptibility have been achieved with this scheme based on the experimental results. In [13], a watermarking scheme-based encryption is proposed using the substitution of IWT and the least significant bit (LSB). In [14], a model is proposed for medical image security in telemedicine applications.

In [15], a scheme of multilayer watermarking is demonstrated with the implementation of SVD, DCT, DWT, and chaotic encryption. After completion of embedding watermark information into the watermarked medical image, its encoding is performed through chaotic encryption. In [16], a method of quartic digital signature-based medical image authentication is considered. In [17], a robust watermarking scheme is presented by using DWT and DCT. Under the scheme, the medical image is categorized into two parts, like the region of non-interest (RONI) and the region of interest (ROI). After that, the embedding of watermark into the medical image's RONI hybrid coefficients is processed.

In [17], a scheme of robust and multiple layer watermarking is proposed with the use of SVD, redundant discrete wavelet transform (RDWT), non-sub-sampled contourlet (NSCT), and chaotic encryption. In telemedicine applications, this scheme is used by inserting three encrypted watermarks into the medical image for the purpose of security. In [18], a fragile watermarking scheme was introduced based on NSCT and compressive sensing (CS)-based encryption on authenticating the medical image. In [19], a watermarking scheme was considered using CS-based encryption and DCT for tamper detection in the medical image. Gupta et al. [20, 21] developed new approaches for identification age, gender, and race. Yadav et al. [22] processed video images for same purpose. Sharma et al. [23, 24] deployed firefly for image thresholding and segmentation. Chaturvedi et al. [25] improved security with the help of watermarking.

3 Embedding Procedure

For the gray-level portraits, the principle is to be discussed significantly, but protracting is made to color portraits virtually [15]. Based on a bit rate of 8, the gray-level picture is presented and is represented. Based on the computation of the picture element, the portrait histogram is determined using a gray-level value j for $j \in \{0, 1, 2, …, 254, 255\}$. To represent the picture elements, count with a j value, h_i

is utilized by assuming that I have N different characters of the picture element. The observation of N non-empty bins is done in h_I, including selecting the two highest bins and representing complimentary high and low values I_R and I_S cooperatively. In h_I with the charge I, the picture element is tallied and data sinking is accomplished using below element:

$$i' = \begin{cases} i-1 \text{ for } i < I_s \\ I_s - b_k \text{ for } i = I_s \\ i \text{ for } I_s < i < I_r \\ I_r + b_k \text{ for } i = I_r \\ i+1 \text{ for } i > I_r \end{cases}$$

i' the value of modified picture element
b_k hidden information about k-th message.

For each picture element, the above equation is implemented, and its value is counted in h_I. The integration of binary elements is processed as $h_I(I_S) + h_I(I_R)$. In the altered histogram, $N + 2$ bins are present, and the bounding values are not available in I. Between two summits, the available bins remain unaltered. In contrast, external ones fluctuate outward that categorize summits into two opposite bins, such as I_S and $I_S - 1$ and $I_R - 1$ and I_R.

To extract the embedded information, the implementation of summit values I_S and I_R is required. From the histogram ascertaining, the refusing of 16 picture elements in I is one way of maintaining them. The confidential information is included in the least critical bits. For making the replacement of the 16 refused pictured elements' least significant bits, the peak values are utilized based on the bitwise functioning after exploiting the expression for each picture element, which is performed in h_I to retrieve the embedded data. The recollection of summit values and calculation of the marked image i' histogram value through the elimination of 16 picture elements are to be done to retrieve the embedded information. The below-mentioned operation is processed for any picture element, and it is counted in the histogram with the values of $I_S - 1$, I_S, I_R, and $I_R - 1$.

$$b'_k = \begin{cases} 1 \text{ for } i' = I_s - 1 \\ 0 \text{ for } i' = I_s \\ 0 \text{ for } i' = I_r \\ 1 \text{ for } i' = I_r + 1 \end{cases}$$

$b'_k = k$-th parallel worth which is taken from the checked picture. To recover the authentic portrait, the operation on each picture element is determined, as shown below. From the derived binary values, the least significant bits of 16 picture elements are generated, not included in this original image. By scripting the forbidden picture elements, they can be re-established to recover the primary portrait.

$$i = \begin{cases} i' + 1 \text{ for } i' < I_s - 1 \\ I_s \text{ for } i' = I_s - 1 \text{ or } i' = I_s \\ I_r \text{ for } i' = I_r \text{ or } i' = I_r + 1 \\ i' - 1 \text{ for } i' > I_r + 1 \end{cases}$$

A. Preprocess for the entire recovery

It is required to modify the pixel values of 0 and 255 since the data range of unit 8 is not supporting the numbers less than 0 and more than 255. From 0 to 1 and 255 to 254, all pixel values are to be set out, respectively. In decryption of information, errors have been resulted using this technique. By ignoring the pixel values of 0 and 255, the algorithm is modified in this paper. Based on the histogram, the data hiding is changed on the third and fourth highest peaks. It will lead to the encrypted cover image with the exact recovery.

LZW Compression

The LZW algorithm comes from an improvement proposed by Terry Welch (1984) to the algorithms proposed by Abrahm Lempel and Jacob Ziv. It is a compression method without losses, since the encrypted data can be perfectly rebuilt on the receiver.

The advantage of LZW lies in its dynamism, since it performs the encoding of the data and the generation of new entries, creating a semi-adaptive dictionary. It does not require to be known by the receiver. The repeating patterns form a table to code, where several concatenated symbols can be grouped for each entry. This arrangement called a dictionary or pattern table. It is necessary to assign the corresponding code. With the above, you can assign code to a greater quantity of symbols that would normally be designated. It is, that is, two symbols could be given the same code in length that would normally be assigned to only one, to three or four or two, among other houses that may appear. With this reduction, we have the corresponding compression; the final code length leads to a reduction in the quantity of initial symbols. The following pseudo-code shows the process of data encoding.

As indicated by Algorithm 1, the corresponding inputs to the dictionary are generally formed by the arrangement ω_k, which will grow in length when compared to all previous entries and is no different to any of them. The decompression is a bit more complicated, since the dictionary is reformed and is not previously known to the receiver from of the entries formed by ω_k. It should be noted that it knows only the initial alphabet of symbols. Variables CODE, OLDcode, and FINsim are the helpers for forming the dictionary again. It must be foreseen if the entrance assigned to CODE has not yet been formed in the table of patterns, for which the previous entries will be auxiliary in the formation of said entry.

Algorithm 1 Compression LZW

1. $w = [], k = [];$
2. Initialize the table or dictionary that will contain the patterns
3. Read the first symbol → assign it to ω

4.   Repeat
5.   Read the following symbol k
6.   if there is no next symbol k to read in the plot then
7.   Assign code to $\omega \rightarrow$ output;
8.   LEAVE;
9.   End yes
10.   if ω_k exists in the dictionary: then eleven:
11. $\omega_k = \omega$, repeat
12.   otherwise, if ω_k does not exist in the dictionary so
13.   Assign code to $\omega \rightarrow$ output;
14.   Add ω_k to the dictionary;
15. $k \rightarrow \omega$;
16.   End yes
17.   Until there are no symbol left to code

Algorithm 2 LZW Decompression

1.   First input code $\rightarrow$ CODE $\rightarrow$ OLDcode
2.   Having CODE = code (k), $k \rightarrow$ output;
3. $k \rightarrow$ FINsim
4.   Next input code $\rightarrow$ CODE $\rightarrow$ INcode
5.   if there is no new code then
6.   LEAVE;
7.   if CODE is not defined in the dictionary: then
8.   FINsim $\rightarrow$ exit;
9.   OLDcode $\rightarrow$ CODE;
10.   code (OLDcode, FINchar) $\rightarrow$ INcode;
11.   End yes
12.   end yes
13.   if CODE = CODE (ω_k) then
14.   stack $k \rightarrow$ stack
15.   CODE (ω) = CODE;
16.   Go to the next symbol
17.   end yes
18.   if CODE = code (k) then
19. $k \rightarrow$ output;
20. $k \rightarrow$ FINchar;
21.   end yes
22.   while the stack is not empty do
23.   Top of the stack $\rightarrow$ exit;
24.   Remove stacked item
25.   OLDcode;
26. $K \rightarrow$ dictionary

27.   INcode → OLDcode
28.   Go to next code
29.   end while

The vector of codes given has values from 0 and will represent the indexes of the dictionary, which does not know the receiver as a whole, only the initial values which correspond to the four QPSK symbols.

Compressed Data Encryption
Bit shuffling is employed in the proposed method to encrypt the compressed data. The compressed image data needs to be securely transmitted in the Internet. The bit shuffling is very fast and effective and was used to scramble the pixels at a pixel level.

4 Experimental Results

Table 1 shows the input images used for analyzing the proposed method along with the compression sizes of the data.

5 Conclusion

In this paper, a novel secure image compression technique is presented for telemedicine applications. The algorithm designed uses reversible watermarking technique to hide the patient-specific information. The watermarked image is then compressed using LZW lossless image compression technique. The compressed data is then scrambled using bit scrambling. This secure data is then transmitted over the Internet for telemedicine applications. In the present COVID era, the proposed algorithm plays a major role providing security to the medical image data sent over the Internet.

Table 1 Input images used for analyzing the proposed method

Input images	Watermaking data	Uncompressed image size	Compressed data size	Compressed data size after encryption
 a. Input image 1	Pradeep Raj, 46, 485968, Finger fracture	196608	8314	8314
 b. Input image 2	Swapna, 31, 478596, Abdomen scan	196608	13543	13543
 c. Input image 3	Surajkumar, 55, 465875, Brain MRI	196608	11310	11310
 d. Input image 4	Pankajkumar, 43, 468751, Chest X-ray	216480	14995	14995

References

1. Borra S, Lakshmi HR (2015) Visual cryptography based lossless watermarking for sensitive images. In: International conference on swarm, evolutionary, and memetic computing. Springer, Cham, pp 29–39
2. Thanki RM, Kothari AM (2017) Digital watermarking: technical art of hiding a message. In: Intelligent analysis of multimedia information. IGI Global, pp 431–466
3. Ashour AS, Dey N (2017) Security of multimedia contents: a brief. In: Intelligent techniques in signal processing for multimedia security. Springer, Cham, pp 3–14
4. Banerjee S, Chakraborty S, Dey N, Pal AK, Ray R (2015) High payload watermarking using a residue number system. Int J Image Graph Sig Process 3:1–8
5. Surekha B, Swamy GN (2011) A spatial domain public image watermarking. Int J Secur Appl 5(1):1–12
6. Singh AK, Kumar B, Singh G, Mohan A (eds) (2017) Medical image watermarking: techniques and applications. Springer, Berlin
7. Leng L, Zhang J, Khan MK, Chen X, Alghathbar K (2010) Dynamic weighted discrimination power analysis: a novel approach for face and palm print recognition in DCT domain. Int J Phys Sci 5(17):2543–2554
8. Leng L, LiM KC, Bi X (2017) Dual-source discrimination power analysis for multi-instance contactless palm print recognition. Multimed Tools Appl 76(1):333–354

9. Thanki R, Borra S (2018) A color image steganography in hybrid FRT–DWT domain. J Inf Secur Appl 40:92–102

10. Singh S, Singh R, Singh AK, Siddiqui TJ (2018) SVD-DCT based medical image watermarking in NSCT domain. In: Quantum computing: an environment for intelligent large-scale real application. Springer, Cham, pp 467–488

11. Thanki R, Borra S, Dwivedi V, Borisagar K (2017) An efficient medical image watermarking scheme based on FDCuT–DCT. Eng Sci Technol Int J 20(4):1366–1379

12. Arunkumar S, Subramaniyaswamy V, Vijayakumar V, Chilamkurti N, Logesh R (2019) SVD-based robust image steganographic scheme using RIWT and DCT for secure transmission of medical images. Measurement 139:426–437

13. Priya S, Santhi B (2019) A novel visual medical image encryption for secure transmission of authenticated watermarked medical images. Mob Netw Appl 1–8

14. Pirbhulal S, Samuel OW, Wu W, Sangaiah AK, Li G (2019) A joint resource-aware and medical data security framework for wearable healthcare systems. Future Gener Comput Syst 95:382–391

15. Thakur S, Singh AK, Ghrera SP, Elhoseny M (2019) Multi-layer security of medical data through watermarking and chaotic encryption for tele-health applications. Multimedia Tools Appl 78(3):3457–3470

16. Babu S, Kapinaiah V (2019) Medical image authentication using quartic digital signature algorithm. Int J Intell Inf Syst 7(4):38–41

17. Thakur S, Singh AK, Ghrera SP, Mohan A (2018) Chaotic based secure watermarking approach for medical images. Multimedia Tools Appl 1–14

18. Thanki R, Borra S (2018) Fragile watermarking for copyright authentication and tamper detection of medical images using compressive sensing (CS) based encryption and contourlet domain processing. Multimedia Tools Appl 1–20

19. Borra S, Thanki R (2019) Crypto-watermarking scheme for tamper detection of medical images. Comput Methods Biomech Biomed Eng Imaging Vis 1–11

20. Gupta R, Kumar S, Yadav P, Shrivastava S (2018, July) Identification of age, gender, & race SMT (scare, marks, tattoos) from unconstrained facial images using statistical techniques. In: 2018 International conference on smart computing and electronic enterprise (ICSCEE). IEEE, pp 1–8

21. Gupta R, Yadav P, Kumar S (2017) Race identification from facial images using statistical techniques. J Stat Manag Syst 20(4):723–730

22. Yadav P, Gupta R, Kumar S (2019) Video image retrieval method using dither-based block truncation code with hybrid features of color and shape. In: Engineering vibration, communication and information processing. Springer, Singapore, pp 339–348

23. Sharma A, Chaturvedi R, Kumar S, Dwivedi UK (2020) Multi-level image thresholding based on Kapur and Tsallis entropy using firefly algorithm. J Interdisc Math 23(2):563–571

24. Sharma A, Chaturvedi R, Dwivedi UK, Kumar S, Reddy S (2018) Firefly algorithm based effective gray scale image segmentation using multilevel thresholding and entropy function. Int J Pure Appl Math 118(5):437–443

25. Chaturvedi R, Sharma A, Dwivedi U, Kumar S, Praveen A (2016) Security enhanced image watermarking using mid-band DCT coefficient in YCbCr space. Int J Control Theory Appl 9(23):277–284

Chapter 37
Microstrip Line Fed Rectangular Split Resonator Antenna for Millimeter Wave Applications

S. Murugan and E. Kusuma Kumari

1 Introduction

In recent years, split ring resonator (SRR)-based antennas are developed for multi-frequency and enhancing the bandwidth. There are two basic types, namely rectangular split ring resonator (RSRR) and circular split ring resonator (CSRR). The double split ring resonators are popular among SRR. They are having meta-material properties such as negative permittivity at particular frequency/band, while properly designed. This meta-material-inspired antenna enhances bandwidth and gain. There are number of literature available based on this split ring resonator antenna. A wideband compact microstrip antenna is a useful research area. In near future, most of the countries shall switch from present 4G to 5G communication. The frequency band below 6 GHz is already used in some countries. However, most of 5G frequency bands are ranging 24–40 GHz, i.e., in millimeter wave frequency bands. Though millimeter waves offer wide bandwidth and channel capacity, the attenuation of millimeter wave in atmosphere is very high compared to present microwave communication. It restricts the use of millimeter wave technology useful for indoor/short distance applications and not suitable for long-distance communication. With these background and motivation, a compact split ring resonator antenna is proposed in this paper. A few of the close literatures relevant to this paper is discussed below.

In [1], a complementary folded triangular SRR is introduced on the dumbbell-shaped radiating element in order to get multiple resonant frequencies 1.8 and 2.4 GHz, besides the actual resonant frequency of 5 GHz due to radiating element. The RSR and CSR were designed to achieve meta-material properties and used as superstrates for the radiating element. The mathematical equation for the design of RSR and CSR is found in [2]. In [3], a split ring-based microstrip antenna operating at 12.2 GHz is discussed. Four complementary circular split ring resonators

S. Murugan (✉) · E. K. Kumari
ECE Department, Sri Vasavi Engineering College, Tadepalligudem, Andhra Pradesh 534101, India

© The Author(s), under exclusive license to Springer Nature Singapore Pte Ltd. 2021
S. Kumar et al. (eds.), *Proceedings of International Conference on Communication and Computational Technologies*, Algorithms for Intelligent Systems,
https://doi.org/10.1007/978-981-16-3246-4_37

are etched on the rectangular patch antenna to enhance the bandwidth. The substrate length and width are 10 mm × 7.7 mm, and thickness is 0.8 mm. The patch length and width are 10 and 7.7 mm. However, there is no comparison of performance with and without split ring resonator. In this work, it is compared with other split ring resonator-based antennas operating at lower frequencies below 12 GHz. The dimensions of the microstrip antenna indirectly proportional to frequency of operation, hence, lower the dimensions of microstrip antenna are for higher resonant frequency antennas In [4], a dual-band patch antenna based on CSRR was proposed, which reduces the dimensions of antenna as well as more gain achieved but the bandwidth of antenna below 2% for −5 dB.

In [5], the design of split ring resonators inspired inset feed rectangular patch antenna is discussed. Multiple split ring resonators placed at appropriate places increase the bandwidth, and a new resonant frequency is achieved. In [6], the gain and bandwidth enhancement of microstrip patch antenna array are obtained by loading the RSRR. The gain improvement of 1.5 dB and bandwidth of 185 MHz at resonant frequency of 5.8 GHz are obtained. A circular patch loaded with slots and split ring resonators is discussed. Multiple resonances are obtained at 2.4, 3.5 and 5.8 GHz [7]. An inset fed rectangular patch antenna with complementary split ring resonator etched on ground plane was fabricated and tested. The radiation efficiency of antenna is below 60% at all resonant frequencies [8]. In this paper, a wideband microstrip line fed rectangular split ring resonator [9, 10] is designed and simulated. The paper is organized as follows: Sect. 2 discusses the design methodology, and Sect. 3 discusses the simulation results, followed by conclusion of the paper.

2 Design of Rectangular Split Ring Resonator

The design of rectangular split resonator is explained in [2]. The design equations are as follows:

$$f_{r1} = \frac{c}{2L_1 \sqrt{\varepsilon_r}} \tag{1}$$

$$L_1 = 4l_1 - s - 4w_m \tag{2}$$

$$f_{r2} = \frac{c}{2L_2 \sqrt{\varepsilon_r}} \tag{3}$$

$$L_2 = 4l_2 - s - 4w_m \tag{4}$$

where l_1 and l_2 represent external and internal side length, respectively, and w_m is the width of each external ring with external and internal resonant frequencies f_{r1} and f_{r2}, respectively. In order to verify if the designed split ring resonator represents a meta-material ($\varepsilon < 0$, $\mu < 0$) and ε, μ value can be estimated from refractive index (*n*) and impedance (*z*) values.

$$\varepsilon = \frac{n}{z} \tag{5}$$

$$\mu = n.z \tag{6}$$

where n and z are related to reflection coefficients S_{11} and S_{21}. The initial design of rectangular split ring resonator is carried out with frequencies of 1.8 GHz and 2.4 GHz as f_{r1} and f_{r2}, respectively. By using above Eqs. (1)–(4), the values of L_1 and L_2 are found as 39.7 mm and 29.7 mm. From the values of L_1 and L_2, and assuming $w_m = 1.5$ mm and $s = 1.5$ mm, the side lengths of outer and inner ring values of l_1 and l_2 are found to be 11.8 mm and 9.3 mm. With these initial values, the rectangular split resonator is designed using HFSS. FR4 material with 20 mm × 20 mm cross section and thickness 1.6 mm is used as substrate. A microstrip line feed having width 1.5 mm is used to excite the resonator. The design structure of antenna is shown in Fig. 1a, b.

The final dimensions of the proposed antenna are shown in Table 1.

3 Results and Discussion

The S_{11} (dB) versus frequency (GHz) plot for microstrip line fed RSRR is shown for frequency range between 1 and 80 GHz (Fig. 2). The return loss is minimum of 32 dB at 41.5 GHz. The S_{11} (dB) value is below −15 dB between 14 and 80 GHz. The impedance bandwidth of antenna is covering unlicensed ISM bands 24 GHz, 60 GHz and 77–80 GHz and also 37–40 GHz which can be used for millimeter wave 5G applications.

The radiation efficiency of the antenna is 54.33%, at mid-band frequency of 40.1 GHz as shown in Fig. 3. The radiation pattern of the antenna is obtained at $\varphi = 0°$ (co-polarization) and $\varphi = 90°$ cross-polarization planes, as shown in Fig. 4. The antenna is radiating perpendicular to the boresight axis. It has a maximum gain of 9.3 dB at $\theta = -90°$, $\varphi = 0°$ and 6.34 dB at $\theta = 90°$. In case of $\varphi = 90°$, i.e., cross-polarization, the maximum gain of 5.25 dB is obtained at $\theta = 90°$ and 2.48 dB at $\theta = -90°$. It covers the entire upper hemisphere and maximum radiation perpendicular to axis of antenna. Figure 5 shows the 3D polar plot of antenna. It shows that the maximum radiation is present in the co-polarization direction, i.e., $\varphi = 0°$, and it is minimum along $\varphi = 90°$.

Figure 6 shows the axial ratio (dB) plot versus frequency. The axial ratio (AR) is an important parameter for getting the information about polarization of the antenna. When the axial ratio is less than 3 dB, it is radiating circular polarization (CP), and the value of AR is 3–6 for elliptically polarized wave. For a linearly polarized antenna, the value is above 10. From Fig. 6, the antenna radiates CP wave at 26, 30, 35, 52, 69 and 77 GHz. However, the antenna is mostly radiating linearly polarized wave.

a

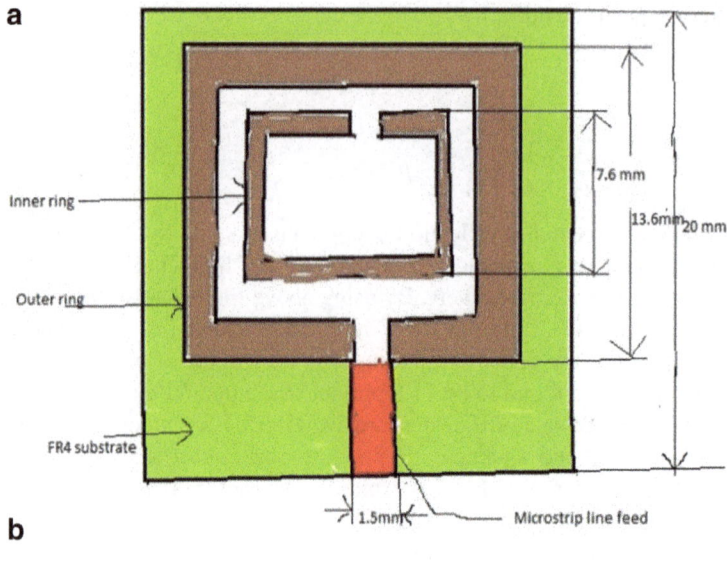

b

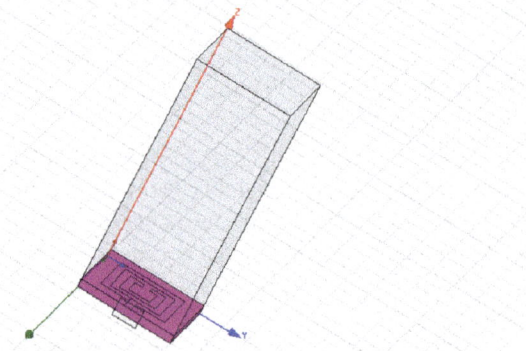

Fig. 1 a Top view of proposed antenna. **b** Simulation model rectangular split resonator with MS line feeding

Table 1 Proposed antenna dimensions

Parameter	Dimension (mm)/Value
Substrate permittivity	1.6
Substrate size	20
Circumference length of outer ring (L_1)	39.7
Circumference length of inner ring (L_2)	29.7
Outer ring side length (l_1)	11.8
Inner ring side length (l_2)	9.3
Microstrip line feed width	1.5

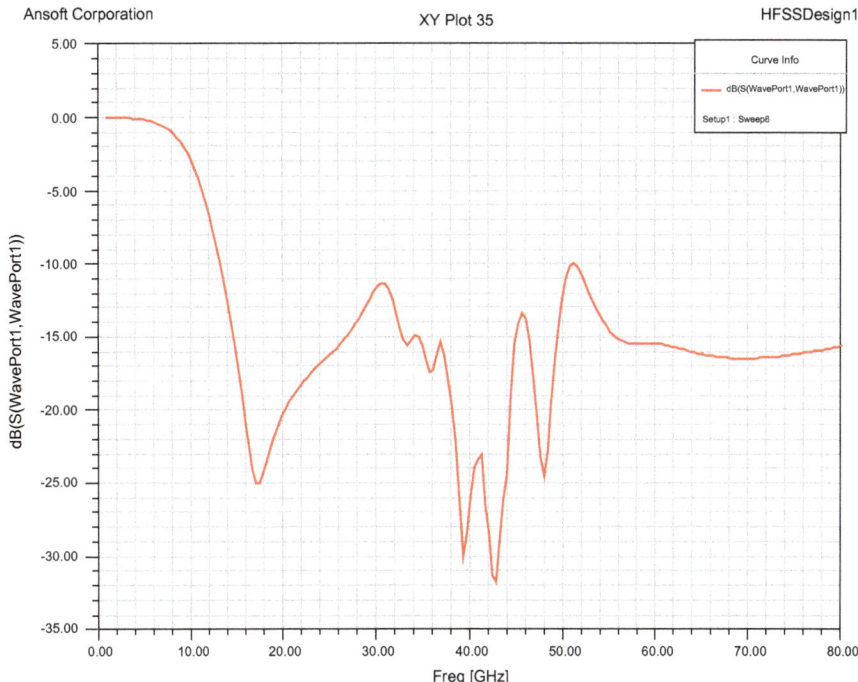

Fig. 2 S_{11} (dB) versus frequency (GHz)

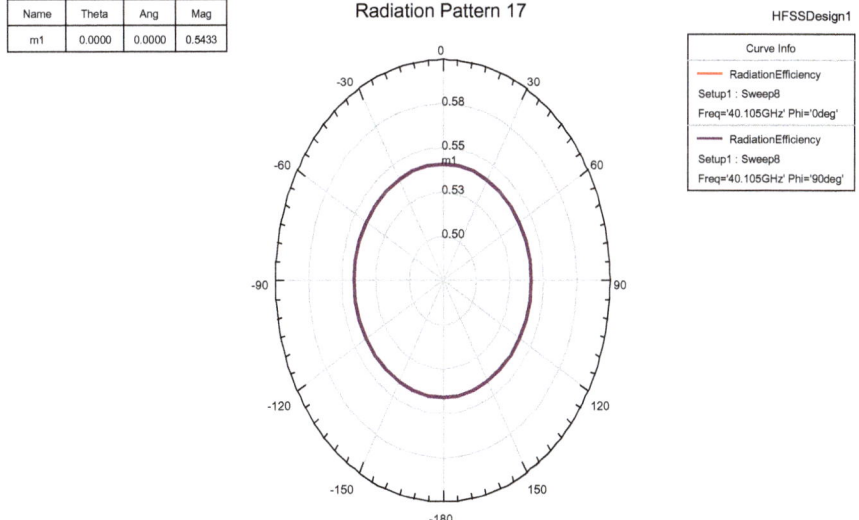

Fig. 3 Radiation efficiency at 40 GHz

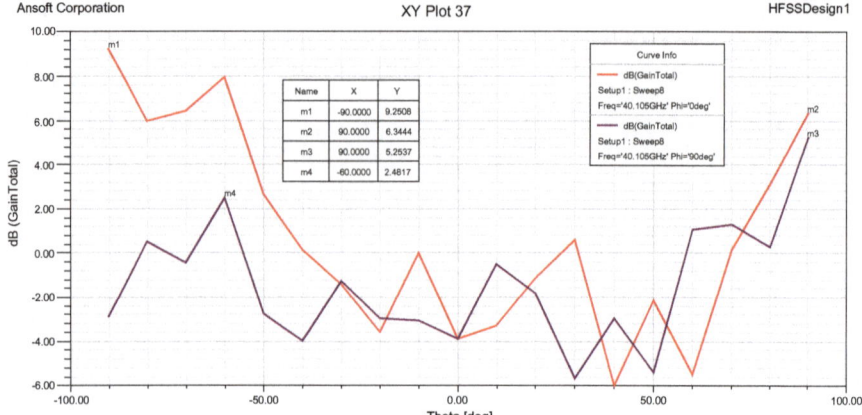

Fig. 4 Gain at $\varphi = 0°$ and $\varphi = 90°$ at 40 GHz

Fig. 5 3D polar plot at
40.5 GHz

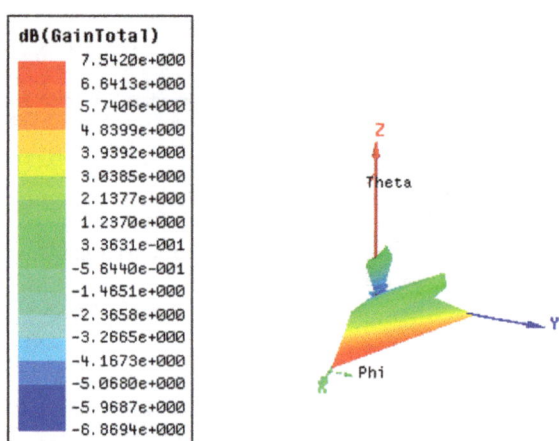

Figure 7 shows the impedance plot for proposed antenna. The antenna impedance is well matched with 50 Ω in the wide frequency range from 15 to 80 GHz. The antenna impedance is varying within VSWR = 1 circle (Table 2).

The proposed antenna is compared with [11], which is a planar inverted F antenna. The radiating plate is at the height of 3 mm from the dielectric substrate having thickness 1 mm. The thickness of the antenna is higher compared to proposed antenna. In [12], gain improvement is obtained by placing a split ring resonator metaplate above the radiating element at height of 8 mm, which increases the overall volume of the antenna.

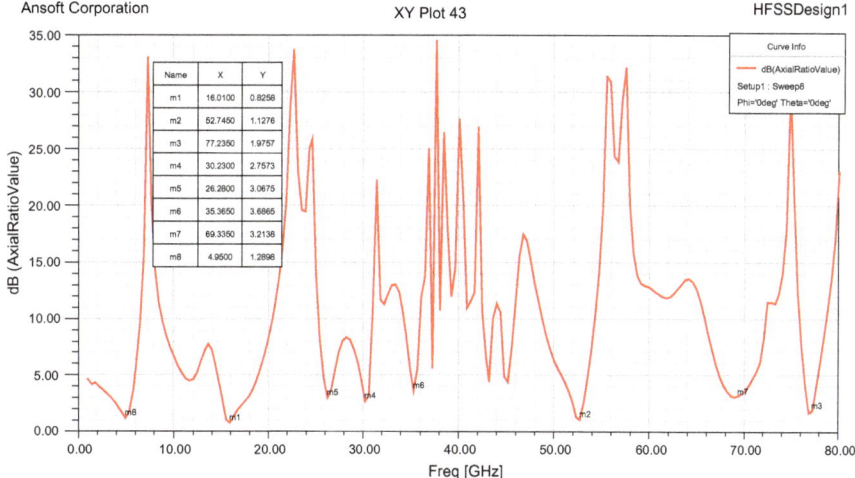

Fig. 6 Axial ratio versus frequency

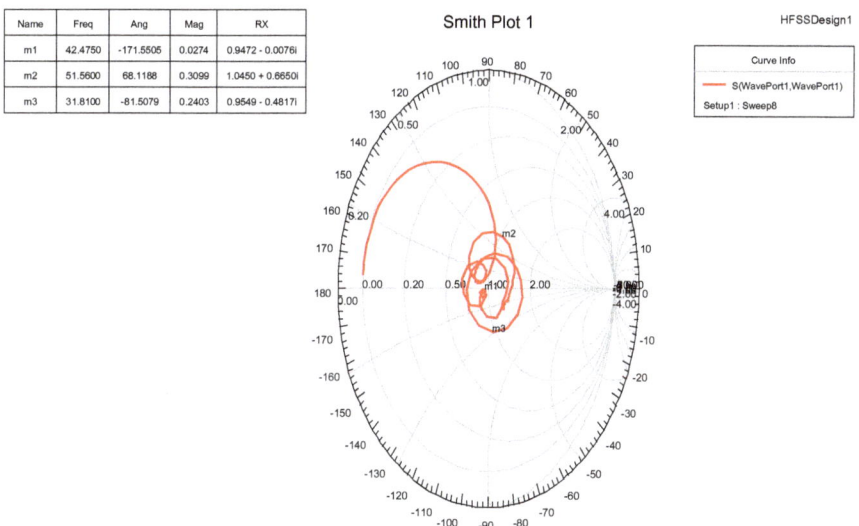

Fig. 7 Smith plot for impedance

4 Conclusion

Hence, a compact split ring resonator fed by microstrip line is designed and simulated. The size of antenna is $20 \times 20 \times 1.6 \text{ mm}^3$. A wide impedance bandwidth from 14.32 to 80 GHz is obtained, covering many 5G cellular frequency bands. The antenna radiates with maximum gain of 9 dB in the end-fire direction. It radiates CP wave

Table 2 Comparison of proposed antenna with existing literature

Ref. paper no	Resonant frequency (GHz)	Size (mm^2)	Gain (dBi)	Range of operating bandwidth (GHz)
[11]	6 10 15	18 × 12	3.4, 4.9, 5.8	4.8–7, 9.5–10.8, 14–15
[12]	27	18 × 22	11.94	26.58–29.31
Proposed work	40.5	20 × 20	9	14–80

at few selected frequencies. The proposed antenna may be integrated with other components of transceiver to fabricate system on chip.

References

1. Rajalakshmi P, Gunavathi N (2019) Compact complementary folded triangle split ring resonator tri-band mobile handsets for voice and Wi-Fi applications. Progr Electromagn Res C 91:254–263
2. Xavier GVR, Serres AJR, Costa EGD, Olivera ACD, Noberga LAMM, De Souza VC (2019) Design and application of a metamaterial superstrate on a bio-inspired antenna for partial discharge detection through dielectric windows. Sens J 19:4255
3. Robinson S (2019) Design and analysis of split ring resonator based microstrip patch antenna for X band applications. ICTACT J Microelectron 4(4):687–692
4. Ortiz N, Falcone F, Sorolla M (2011) Gain improvement of dual band antenna based on complementary split ring resonator. ISRN Commun Netw 2012
5. Nambiyappan Selvi T, Pandeeswari R, Selvan PT (2018) An inset fed rectangular microstrip patch antenna with multiple split ring resonator loading for WLAN and RFID applications. Progr Electromagn Res C 81:41–52
6. Arora C, Pattnaik SS, Baral RN (2015) SRR inspired microstrip patch antenna array. Progr Electromagn Res C 58:89–96
7. Zhang Y, Hang W, Yu C, Kuai Z-Q, Don Y-D, Zhou JY (2008) Planar ultrawideband antennas with multiple notched bands based on etched slots on the patch and/or split ring resonator on the feed line. IEEE Trans Antennas Propag 56(9):3062–3067
8. Ali W, Hamad E, Bassiuny M, Hamdallah M (2017) Complementary split ring resonators based triple band microstrip antenna for WLAN/WiMax applications. Radio Eng 26(1)
9. Zahetar S, Yalcinkaya AD, Torun H (2015) Rectangular split ring resonators with single split and dual split under different excitations at microwave frequencies. AIP Adv 5:117220
10. Saha C, Siddiqui JY (2009) Estimation of the resonance frequency of conventional and rotational circular split ring resonators. In: 2009 Applied electromagnetics conference (AEMC), Kolkata, pp 1–3. https://doi.org/10.1109/AEMC.2009.5430631
11. Ishfaq MK, Rahman TA, Chattah HT, Rehman MU (2017) Multiband split ring resonator based planar inverted F antenna for 5G applications. Int J Antennas Propag 2017:1–7. https://doi.org/10.1155/2017/5148083
12. Jeong MJ, Hussain N, Park JW, Park SG, Rhee SY, Kim N (2019) Millimeter wave microstrip patch antenna using vertically coupled split ring metaplate for gain enhancement. Microw Opt Technol Lett 1–6

Chapter 38
An Assessment of Noise Pollution at Some Critical Locations of Udaipur City by the Use of Sound Level Metre

Parth Samdani and Bhopal Singh Singhvi

1 Introduction

Noise pollution is considered as a big problem that affects human life in urban as well as rural areas across the globe [1]. Noise is an irregular, unwanted, unpleasant and irritating sound with no musical quality. Noise can be defined as a wrong sound which is produced at a wrong time and at a wrong place. Lots of unwanted sounds are dumped continuously into the environment, creating noise pollution through human activities. Many studies have revealed that the impacts of noise pollution could be divided clearly into four types, namely physical, psychological, physiological and performance effects. Physical means hearing impairment, etc. Physiological means elevated blood pressure and irregular heartbeats, etc. Psychological means irritation and disturbance in sleeping, etc., and performance effects mean reduction in output of work, misunderstanding in hearing, etc. [2]. According to International Organization for Standardization, 1990 data, it is clear that at the sound pressure level of 80–100 decibels, physical ears feel irritation. The maximum range of frequencies that can be heard by the healthiest human ears ranges from 16 to 20 Hz on the lower end to about 20,000 Hz on the higher end.

2 Sources of Noise Pollution

Noise may arise through human-induced activities or from natural sources. The human-induced activities are heavy machineries used in the industries, mining operations, construction sector machineries, vehicles, aeroplanes, trains, television, loud

P. Samdani (✉) · B. S. Singhvi
Department of Civil Engineering, College of Technology and Engineering, MPUAT, Udaipur, India

S. Kumar et al. (eds.), *Proceedings of International Conference on Communication and Computational Technologies*, Algorithms for Intelligent Systems, https://doi.org/10.1007/978-981-16-3246-4_38
465

speakers, dish washers, etc. The natural sources are earthquakes, volcanic eruptions, cyclone and thunder, etc. Study on noise pollution at some places of residential, commercial, industrial and silence zone within Jagiroad town in Assam revealed that the maximum noise level is found in commercial area due to high traffic congestion, outdated vehicles, heavy vehicular movement, narrow poorly managed roads, unplanned urban area and commercial zones [3]. Evaluation of present noise level in Jabalpur city reveals that the silence and residential zones of the Jabalpur city are highly polluted from the noise pollution. Heavy traffic flow, unplanned urbanization and vehicular traffic are the main reasons which cause noise pollution in the Jabalpur city [4]. Noise pollution analysis during Chhath Puja in Gorakhpur city showed that the noise levels are created by congestion and heavy traffic, generators and public on the streets and roads. It is also observed that, on public holidays at many sites, a considerable increase in noise is observed during the whole day [5]. The human-induced noise can be categorized into transport noise, occupational noise and neighbourhood noise.

2.1 Transport Noise

(a) Road Traffic Noise: A large number of heavy vehicles on the roads and the high speed of them along with the use of pressure horns are basic causes of this type of noise. An evaluation of indoor and outdoor noise level in various areas of Gorakhpur City revealed that outdoor noise levels are influenced by congestion, large traffic volume and big crowd on roads [6].

(b) Rail Traffic Noise: The introduction of diesel and electrical engines instead of stream engines, improved coaches and welded tracks has contributed in minimizing the noise level.

(c) Aircraft Noise: A high intensity noise is produced during the process of landing and take-off of the flights. Jet engines create a severe noise due to the fact that faster and larger aeroplanes create more noise than the slower and smaller ones.

2.2 Occupational Noise

Occupational noise is normally produced by mills and factories, mining operations, motor body building and many other processes like blasting operations and shipbuilding, etc. Vacuum cleaners, dish washers, mixer grinders, televisions and washing machines, etc., which are used for domestic work, also produce occupational noise.

(a) Industrial Activities: Most of the industries contain heavy machinery which produces sound of high intensity. The noise is produced by stacking or loading of the goods, friction of the machines, etc. Usually, the industrial noise level ranges from 90 to 130 dB (Table 1).

Table 1 Occupational noise levels

S. No.	Domestic/Industrial source	Noise level (dBA)
1.	Vacuum cleaner	70
2.	Dish washer	76
3.	Washing machine	82
4.	Waste grinder	83
5.	Food blender	90
6.	Milling machines	90
7.	Steel plate riveting	130
8.	Newspaper press	101
9.	Farms tractor	103
10.	Circular saw	110
11.	Textile loom	112
12.	Boiler marker's shop	120
13.	Pneumatic metal chipper	122

Table 2 Noise emissions from opencast machinery

S. No.	Noise source	Noise level (dBA)
1.	Front-end loader	85–105
2.	Shovel	90–100
3.	Dragline	85–105
4	Pneumatic drill	105–115
5	Dumper	80–100
6	Dozer	90–106

(b)   Mining Activities: Noise emissions from operation of heavy duty underground and opencast mining, crushing, screening, loading and unloading activities, etc., are shown in Table 2.

2.3   Neighbourhood Noise

The machines which are used for building, drainage and road construction work and demolition work are the sources of neighbourhood noise. Dance in late evenings, use of public address system at public functions and music systems in parties also cause noise pollution to the neighbours. An assessment of environmental noise pollution in Bikaner city at residential zone, commercial zone, industrial zone and even in the silence zone, e.g. PBM Hospital area, have shown that the noise levels in all these zones are higher than the prescribed permissible noise level [7]. The prescribed permissible sound levels for cities by Central Pollution Board of India are shown in Table 3.

Table 3 Permissible sound levels for cities

S. No.	Area/Zone	Day (dBA) 6.00 a.m. to 10.00 p.m.	Night (dBA) 10:00 p.m. to 6:00 a.m.
1.	Silence	50	40
2.	Residential	55	45
3.	Commercial	65	55
4.	Industrial	75	70

3 Effects of Noise Pollution

- Noise levels which are above 80 decibels may cause harmful effect to the ear. Ears when exposed to high noise, i.e. Udaipur city is known as the city more than 100 decibels for a long time period, causes a major hearing problem and may cause a permanent damage to ears.
- Increase in number of vehicles, construction activities, cleaning of land due to deforestation, domestic release of pollutants like methane and chlorofluorocarbons may lead to increase in levels of various pollutants. Increased levels of these pollutants have a direct impact on health of animals and humans [8].
- High intensity noise increases heart rates and thereby increasing the blood pressure. It may lead to heart attack.
- The ability of a person to read, learn and to understand may reduce if regularly exposed to high noise. Excess noise affects humans causing disturbances in daily activities like work efficiency, sleep and level of judgement [9].
- High noise level interrupts a person's sleep which generally results in the loss of their energy level and efficiency of working.
- The noise levels more than 50–60 decibels make it very difficult for two peoples to talk with each other and thereby lead to confusion.
- Exposure to high noise may cause a rise in stress of a person, resulting in wild behaviour. A constant noise can cause headaches, irritates people and thereby leads to anxiety disorder.
- High intensity sound induces hearing loss and fear and decreases a cow's capacity of milk production. It adversely affects the behaviour and pattern of breeding and feeding of some animals.

4 Study Area

Udaipur city is known as the city of lakes and is the district headquarters of the Udaipur district located in the southern part of Rajasthan state, near the Gujarat border. It was founded by Maharana Udai Singh Ji of Mewar in 1559. Three representative locations of the Udaipur city have been chosen, representing silence zone (MB), commercial zone (SP) and residential zone (AM) as per Fig. 1 (Table 4).

Fig. 1   Three critical places of Udaipur city

Table 4   Locations of the area and type of zone

S. No.	Name of location	Type of zone
1.	M.B. Hospital	Silence zone
2.	Suraj Pole	Commercial zone
3.	Amba Mata	Residential zone

5   Methodology

The objective of the study was to investigate the concentration of the pollutants from various sources like transport, industry and neighbours over the ambient noise quality of the Udaipur city. As such, Udaipur is a big city, and it was not possible to record the percentage of these main pollutants in all areas, so the research was restricted to some critical points only. A survey to decide the three locations which could be considered as silence, commercial and residential zones was performed. We, in consultation with the Regional Office, Rajasthan State Pollution Control Board (RSPCB), Udaipur [10], worked on three critical places of Udaipur city and took the observations of noise quality on two different dates, i.e. 24.10.2016 (normal day) and 30.10.2016 (Diwali day). The sites were so selected that the silence area,

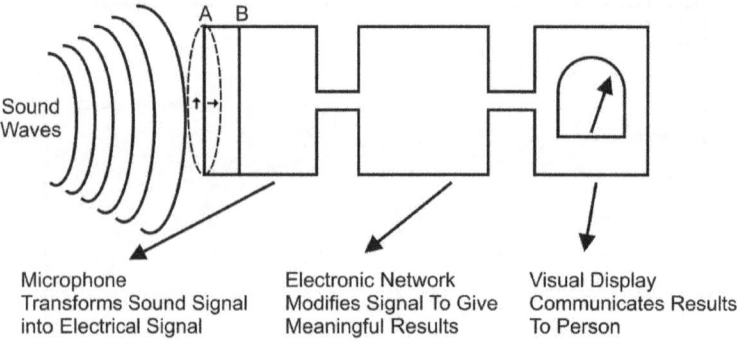

Microphone   Electronic Network   Visual Display
Transforms Sound Signal   Modifies Signal To Give   Communicates Results
into Electrical Signal   Meaningful Results   To Person

Fig. 2 Internal system of sound level metre

commercial area and residential area are covered under the study. A sound level
metre was used for noise measurements. The sound level metre also known as a
sound pressure level is a handy instrument and is used for the measurement of any
sound which travels through air. It consists of a digital display unit, an electronic
circuit and a microphone (Fig. 2) which converts the sound signals into equivalent
electrical signals. The signals were passed through the weighing network of the sound
level metre, and the sound pressure level in decibels was displayed on the screen.

6 Results and Discussion

The survey results for identification of various sources of noise pollution and the
readings of noise quality control on two different dates, i.e. 24.10.2016 (normal
day) and 30.10.2016 (Diwali day) at silence zone (M.B. Hospital), commercial zone
(Suraj Pole) and residential zone (Amba Mata) were observed and analysed.

6.1 Sources of Noise Pollutions and Their Percentage Share

A survey (in the month of March 2017) in which 327 respondents were asked a
question about the sources of noise pollution was performed. The results are tabulated
in Table 5, and a pie chart showing their percentage share is drawn as shown in Fig. 3.
The highest noise pollution (40.3%) was from road traffic.

Table 5 Sources of noise pollution and their share percentage

S. No.	Sources of noise	Percentage share (%)
1.	Road traffic	41.3
2.	Railway	2.1
3.	Aeroplanes	1.5
4.	Neighbourhood	17.4
5.	Construction work	16.7
6.	Commercial activity	8.5
7.	Schools/educational institute	6.8
8.	Sports	3.4
9.	Others	2.3

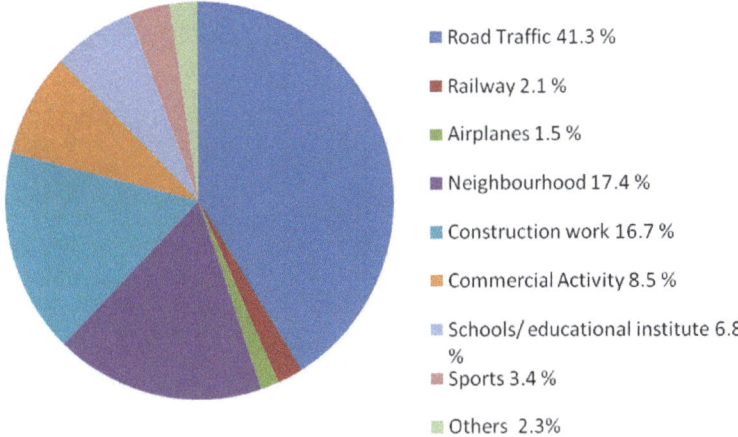

Fig. 3 Sources of noise pollution and their share percentage

6.2 Noise Levels in Silence Zone

M.B. Hospital was selected for the observation of sound levels as this location is declared as a silence zone. The observations were tabulated in Table 6, and corresponding graphs were plotted for normal day and Diwali day as shown in Fig. 5. It was observed that minimum noise pollution was at M.B. Hospital (silence zone) on normal day. An interesting point was observed that at M.B. Hospital (silence zone), the noise pollution level at 12 p.m. was much lower on the Diwali day (60.7) than the normal day (66.6). It may be explained that movement of the people and patients to visit doctors and wards in the night at 12 p.m. may be self-avoided due to holiness of the Diwali day (Fig. 4).

Table 6 Noise levels at monitoring location: M.B. Hospital (latitude and longitude: 24° 35′ 17″ N, 73° 41′ 29″ E)

Time	Normal day (dB)	Diwali day (dB)
7 p.m.	52.4	71.7
8 p.m.	58.9	75.9
9 p.m.	68.5	75.5
10 p.m.	60.7	69.7
11 p.m.	63.5	67.1
12 p.m.	66.6	60.7

Fig. 4 Monitoring location: M.B. Hospital

Fig. 5 Noise levels at monitoring location: M.B. Hospital

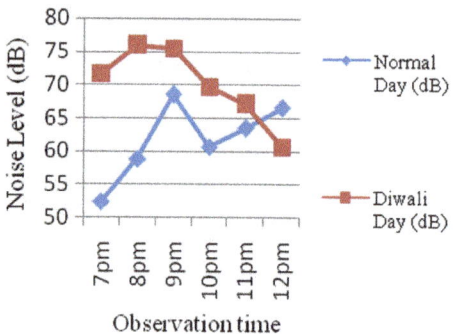

6.3   *Noise Levels in Commercial Zone*

Suraj Pole Chouraha was selected for the observation of sound levels, and this location is considered as a commercial zone. The observations were tabulated in Table 7, and corresponding graphs were plotted for normal day and Diwali day as shown in Fig. 7. The maximum noise pollution was at Suraj Pole (commercial zone) on Diwali day. Out of the three zones, i.e. silence, commercial and residential, on an average, commercial zone, i.e. Suraj Pole has the highest noise levels on both the days, i.e. 74.13 dB on normal day and 85.6 dB on Diwali day (Fig. 6).

Table 7   Noise levels at monitoring location: Suraj Pole (latitude and longitude: 24° 34′ 46″ N, 73° 41′ 46″ E)

Time	Normal day (dB)	Diwali day (dB)
7 p.m	77.5	90.2
8 p.m.	74.8	85.9
9 p.m.	72.5	85.7
10 p.m.	74.1	83.3
11 p.m.	64.6	80.2
12 p.m	63.3	88.3

Fig. 6   Monitoring location: Suraj Pole Chouraha

Fig. 7 Noise levels at
monitoring location: Suraj
Pole Chouraha

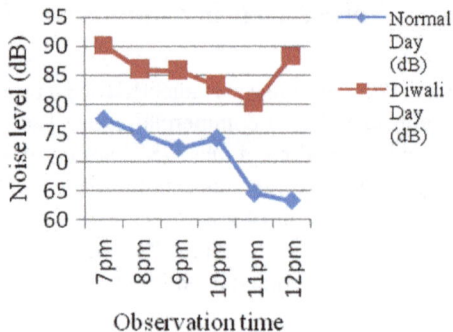

Observation time

Table 8 Noise levels at
monitoring location: Amba
Mata (latitude and longitude:
24° 35′ 08″ N, 73° 40′ 30″ E)

Time	Normal day (dB)	Diwali day (dB)
7 p.m.	74.7	65.0
8 p.m.	71.9	67.0
9 p.m.	71.4	77.1
10 p.m.	70.1	71.2
11 p.m.	73.2	74.6
12 p.m.	67.7	81.2

6.4 Noise Levels in Residential Zone

Amba Mata was selected for the observation of sound levels, and this location is
considered as a residential zone. The observations were tabulated in Table 8, and
corresponding graphs were plotted for normal day and Diwali day as shown in Fig. 9
(Fig. 8).

7 Suggestions to Control Noise Pollution

The health of public is a very important matter for us. Therefore, the control over
noise pollution is the prime necessity of the day. The control at source, control in
the transmission path and the control using preventive equipment are some of the
important techniques which can be used as remedial measures for the control of
noise pollution [11]. The techniques and measures employed for controlling the
noise pollution are as follows:

- The selective and judicious use of domestic appliances such as television sets,
  radios, tape recorders, washing machines, mixers and grinders can minimize the
  noise level.
- Speaking at low voices during communication can reduce the noise level up to a
  huge extent.

Fig. 8 Monitoring location: Amba Mata

Fig. 9 Noise levels at monitoring location: Amba Mata

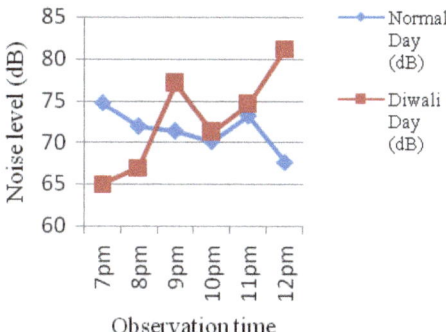

- The proper design of foundations and use of suitable packing materials reduce the noise due to vibrations.
- Servicing and tuning of vehicles at regular intervals will reduce the noise levels.
- Noise level may be reduced by restricting the use of public address system in the residential zone except for important events.
- Proper servicing of vehicles and machineries will not only minimize the noise pollution but also improves their life.
- Plantation of trees and shrubs, especially evergreen species to work as planted buffers are very effective in controlling the noise pollution.
- The proper use of noise absorbing construction material for walls, windows, doors and ceilings, if incorporated in the design of the buildings, will be highly useful in reducing the noise levels. It is the responsibility of the engineers and builders, who are responsible for the construction of high-rise buildings [12].

- The cities should be developed in a planned manner such as green belt development around the residential areas and separation of industry and transport areas into different zones.
- The schedule of the workers and the rotation of the job between the workers and employees should be set in such a manner that their over exposure to the high noise may be avoided.

8 Conclusions

Average of the recording period (6 p.m. to 12 p.m.) showed that at all the three locations, i.e. M.B. Hospital, Amba Mata and Suraj Pole: noise pollution was much higher on Diwali day (70.1, 72.68 and 85.6) than normal day (61.76, 71.5 and 71.13). The safe intensity level of sound, as prescribed by World Health Organization, is 45 dB. At every monitoring location of the Udaipur city, the intensity level of sound was more than the prescribed. At all the three locations, i.e. M.B. hospital, Amba Mata and Suraj Pole, average noise pollution were much higher on Diwali day (70.1, 72.68 and 85.6) than the normal day (61.76, 71.5 and 71.13). Awareness should be made among the people about the harmful effects of playing with fire crackers and using high sound producing music and public address system on the social events. Implementation of new stringent laws, such as motor vehicle act, introduction of new regulation to lower speed limits of vehicles, promotion of education and research, creating awareness among the public about the harmful effects of noise pollution should be encouraged. Various techniques such as limiting vehicle speeds, limiting heavy duty vehicles, improving tyre design, use of noise barriers and use of silencers in the vehicles for road traffic noise should be adopted. The cities should be developed in a planned manner, making green belt around the residential areas, around airport and on the sides of the roads.

References

1. Kumar A, Mishra RK, Kumar A (2015) Noise pollution analysis in different mega cities of India during Deepawali festival. J Environ Res Dev 9(04):1075–1080
2. Olayinka OS (2013) Effective noise control measures and sustainable development in Nigeria. World J Environ Eng 1(1):5–15
3. Sharma C, Kalita DJ, Bordaloi R (2015) A study on noise pollution in some places of industrial, commercial, residential and silence zone within Jagiroad town, Assam. Int J Sci Technol Manag 4(01):1574–1580
4. Baghel PS, Bhatia RK, Rahi DC (2016) Evaluation of present scenario of ambient noise level in residential zone and silence zone of Jabalpur city. Int Res J Eng Technol (IRJET) 03(10):410–416
5. Baniya SK, Mishra AK (2016) Noise pollution analysis during Chhath Puja in Gorakhpur City. Int J Res Appl Sci Eng Technol (IJRASET) 4(XII):445–450
6. Pritam U, Pandey G, Singh SP (2014) Assessment of outdoor and indoor noise pollution in commercial areas of Gorakhpur city. Int J Eng Res Technol (IJERT) 3(12):777–783

7. Charan PD (2017) An assessment of environmental noise pollution in Bikaner city of western Rajasthan, India. Int J Life Sci Technol 10(3):33–37
8. Gulabchandani L, Sethi T (2020) A study of ambient air quality of Ajmer city, Rajasthan, India. J Emerg Technol Innov Res 7(3):1060–1064
9. Saraswat P (2018) An assessment of environmental noise pollution of Jodhpur, Rajasthan, India. MATTER Int J Sci Technol 4(2):207–219
10. Observation report by "The Regional Office, Rajasthan State Pollution Control Board (RSPCB)", Udaipur (2016)
11. Bande SU, Nawathe MP (2013) Management of traffic noise on express highway—an ergonomic approach. Int J Eng Sci Res Technol 2(11):3235–3239
12. Singh B, Choudhary MP (2017) Evaluation of noise levels and ascertaining noise indices in an urbanizing city of Kota, Rajasthan, India. Nat Environ Pollut Technol 16(4):935–938

Chapter 39
A New Z-Source Inverter Topology with Less Number of Passive Elements

Tushar Tyagi, Amit Kumar Singh, and Rintu Khanna

1 Introduction

Globally, the conversion of renewable energy, electric transport, and many other industrial applications involving power converters/inverters are major research efforts. Therefore there is a wide scope for developing effective and reliable power converters that are economically viable and technologically feasible [1].

For short-circuit safety, a dead time is needed in the traditional inverter between the switches connected in the same leg, which results in distortion of the waveform of the ac output voltage. If the DC source voltage is weak to provide the desired output voltage, it also requires an intermediate power conversion stage (DC–DC) which reduces efficiency. Issues stated above motivated the researchers to work on a single-stage power converter by overcoming the issues as stated [1].

The Z-Source first proposed by Peng [1] in 2002. It is an X shape impedance network having two capacitors and two inductors. This network is capable of providing a single-stage power conversion process with buck-boost inverter and converter. The general topology of this Z source inverter (ZSI) is shown in Fig. 1. In ZSIs, switches in the very same leg can be flipped on concurrently without compromising inverter's reliability, thus no dead time is needed which further improves the waveform shape of the output voltage and reduces the distortion.

In this paper, a new topology of the Z-Source inverter is proposed. The proposed topology has less number of passive elements as compared to the conventional Z Source inverter topology. The detailed analysis of the different operating modes are also presented here. A simple modulation technique to generate the control signals

T. Tyagi (✉) · A. K. Singh
Lovely Professional University, Phagwara, India
e-mail: tushar.20586@lpu.co.in

R. Khanna
Punjab Engineering College (Deemed to be University), Chandigarh, India

© The Author(s), under exclusive license to Springer Nature Singapore Pte Ltd. 2021   479
S. Kumar et al. (eds.), *Proceedings of International Conference on Communication and Computational Technologies*, Algorithms for Intelligent Systems,
https://doi.org/10.1007/978-981-16-3246-4_39

Fig. 1 Z source topology

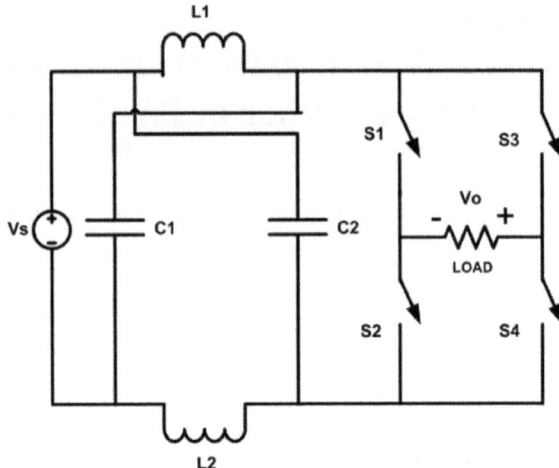

for the devices is used. The proposed topology is simulated in MATLAB and various performance characteristics like effect of boost factor on voltage and THD have been analyzed. In the end, comprehensive comparative analysis has been carried out based on parameters such as number of passive devices, switching frequency, boost factor and losses, to prove the superiority of the proposed topology.

2 State of the Art in Z-Source Inverter

There has been a significant advance in the field of Z-source inverter in the last decade or so. Some of the researchers focused on the modulation strategies, applications, and modeling [2, 3] while some of them work on the development of new topologies. Most of the work is done for three-phase ZSI, while for low power or small renewable energy system a lot of work has to be done for single-phase ZSIs. This article presents a new topology for renewable energy applications using a single-step half-bridge Z-source inverter. A brief introduction of other single-phase topologies developed in recent years is also discussed here. In [4] a topology is proposed that demonstrates similar properties to those of ZSI with a reduced number of passive components but due to a rise in the number of semiconductor devices, the Switched Boost Inverter (SBI) needs a better safety circuit compared to ZSIs. SBI's average dc link voltage is just $(1 - D)$ times that of ZSI. Therefore, in order to get the same ac output voltage, SBI must run at a higher modulation index as opposed to ZSI.

Topologies proposed in [5] have a high boost factor in comparison with conventional types with less number of passive components and has high efficiency in comparison to conventional ZSI. An improved topology is proposed in [6, 7] for the half-bridge switched boost inverter with low capacitor tension. The proposed half-bridge quasi-switched boost inverter has fewer passive components as compared to the half-bridge Z-source inverter with two Z-networks. Furthermore, the proposed topology produces high voltage gain as opposed to traditional topologies and has stable operation.

A new CUK converter-based three-switch single-phase Z-source inverter (ZSI) is proposed in [8, 9]. This topology has the ability to buck-boost and the dual-grounding properties. Additionally, in the proposed inverter the voltage gain is greater than in the single-phase quasi-Z-source and semi-Z-source inverters. This topology however has a high amount of passive elements which would be a major drawback to this topology.

In [10] a dual input quasi Z-Source inverter is proposed for PV applications. In [11], a novel switching technique is proposed for a Z-source inverter, which has unequal short-term intervals to the current ripple in its impedance source inductors. This proposed method could also reduce power ripple of the inductor by 26.9% compared to a conventional method which has an equal short-through interval, without increasing the number of inverter switches. In [12] a new sinusoidal modulation strategy is proposed for the reduction of the losses. A prototype is made for the validation of successful implementation of sinusoidal modulation strategy. In [13] different single-stage boost inverter topologies for Nano-grid application are discussed. A new, switched boost network (SBN)-based structure for Z-source inverters is developed in [14]. The mentioned inverter is labeled as dual switched boost (DSBI) inverter. DSBI raises the SBI boost factor by using two boost networks switched in cascades. Among the most significant features of DSBI is that they can generate the proposed structure, even if the SBN is available as a block with two inputs and two output terminals. A detailed overview of various Z-source inverter topologies based on the switched impedance network is provided in [15].

3 Proposed Topology

Topology proposed in this paper has two inductors (L_1 and L_2), two switches, two sources, and one capacitor (C). It has less number of passive and active elements as compared to the conventional Z source inverter topology as shown in Fig. 1. The topology circuit diagram is described in Fig. 2. Other advantage with this topology is that the current drawn from the source is continuous.

3.1 Operating Modes

There are four operating modes of the proposed topology. Different switching states along with the detailed analysis of the topology are explained below.

Mode-1 Both the switches (S_1 and S_2) are closed in this mode, and the inverter is in shoot-through configuration. The inductor charges in this mode. A care should be taken while choosing the value of the capacitor as the value of the capacitor will

Fig. 2 Proposed topology

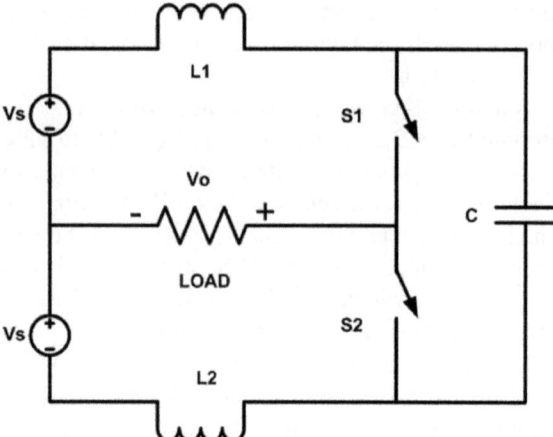

directly govern the inrush current at the time of the starting. Inductor current increases and limited by the value of the inductor. Equivalent circuit is shown in Fig. 3.

Mode-2: **Non-shoot Through Mode** After the mode-1, this mode is called non-shoot through mode. Switch S_1 and S_2 are ON and OFF respectively. Boosting of output voltage happened when the inductor discharges and the inductor voltage is added to the source voltage. The detailed circuit and its equivalent circuit is presented in Fig. 4 and Fig. 5 respectively.

Voltage and current relations.

$$I_{L_1} = I_R + I_C$$
$$V_S = V_{L_1} + V_O$$

Fig. 3 Shoot through state

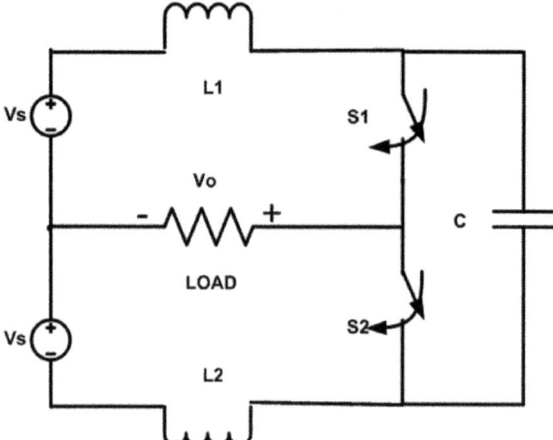

Fig. 4 Non-shoot through state

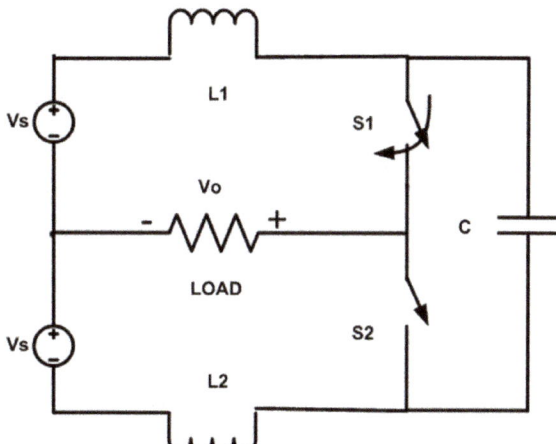

Fig. 5 Equivalent circuit in Mode-2

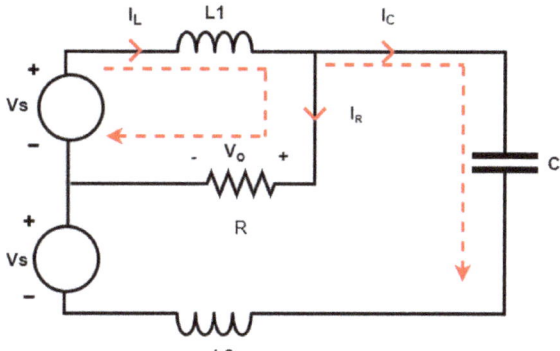

Mode-3 In this operating mode both the switches in the same leg are closed and the inverter will be in the shoot-through state. Inductor current increases and is limited by the value of the inductor. The circuit diagram for the same is given in Fig. 3.

Mode-4 In mode-4 S_1 is OFF and S_2 is ON. Boosting of output voltage happened when the inductor discharges and the inductor voltage is added in the source voltage similar to Mode-2. The detailed circuit and its equivalent circuit is presented in Fig. 6 and Fig. 7 respectively.

Voltage and current relations.

$$I_{L_2} = I_R + I_C$$
$$V_S = V_{L_2} + V_O$$

Fig. 6 Circuit diagram of
Mode-4

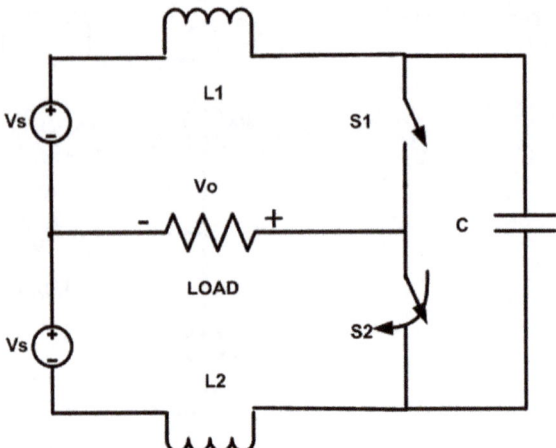

Fig. 7 Equivalent circuit in
Mode-4

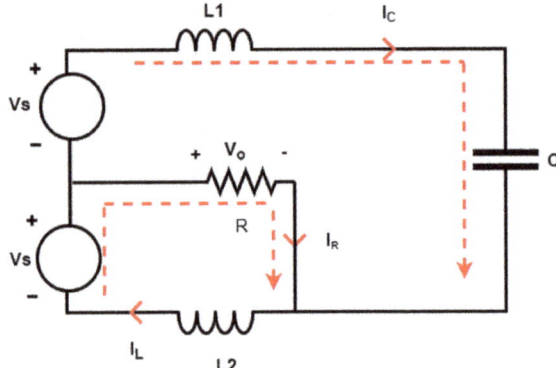

3.2 Boost Factor Calculation

Boost factor is calculated by making inductor voltage average to zero. By applying
KVL in the equivalent circuit in mode-1. It is clear that,

$$V_S = V_{L_1} \tag{1}$$

Applying KVL in the loop in Mode-2. It is clear that

$$V_{L_1} = V_S - V_O \tag{2}$$

Now applying Voltage-second balance in inductor.

$$\frac{D_{ST} \times V_S}{2} + \frac{(1 - D_{ST}) \times (V_S - V_O)}{2} = 0 \tag{3}$$

$$V_O = \frac{V_S}{1 - D_{ST}} \tag{4}$$

$$B = \frac{1}{1 - D_{ST}} \tag{5}$$

It is clear from Eq. 4 $V_O > V_S$ and hence the boosting of voltage is happens in single-stage conversion.

4 Performance Characteristics

In this section different performance characteristics such as variation of output voltage with boost factor and shoot through duty ratio, and Total Harmonic Distortion (THD) variation with the boost factor for the proposed topology are discussed.

4.1 Effect of Shoot Through Duty Ratio with Output Voltage

It is clear from Eq. 4 by increasing the shoot through duty ration the average output voltage will increase. The variation is plotted in Fig. 8.

The shoot through can be increased up to a certain level, further increment may lead to the unstable system.

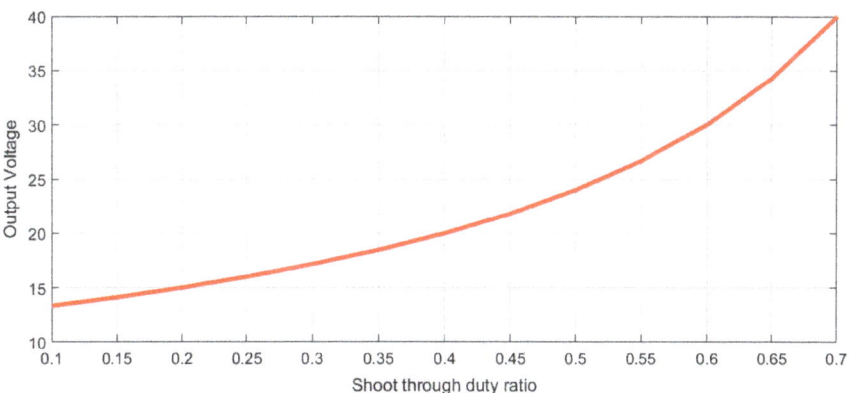

Fig. 8 Output voltage variation with shoot through

4.2 *Effect of Boost Factor on Output Voltage*

It is observed from Fig. 9 by increasing the boost factor the output voltage increases however the waveform deviates from the pure sinusoidal voltage. Since the shoot through time is very low hence it will not affect the performance of the inverter much. Figure 10 shows the variation of voltage with $D_{ST} = 0.2$ and $D_{ST} = 0.3$ respectively.

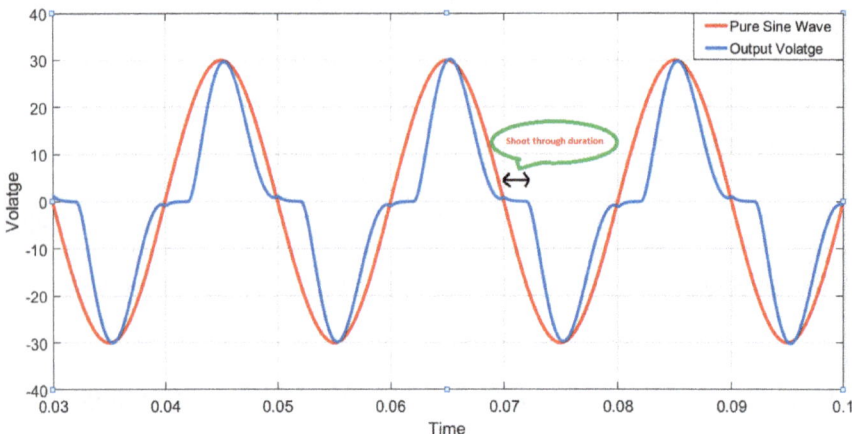

Fig. 9 Comparison of output voltage with pure sinusoidal waveform

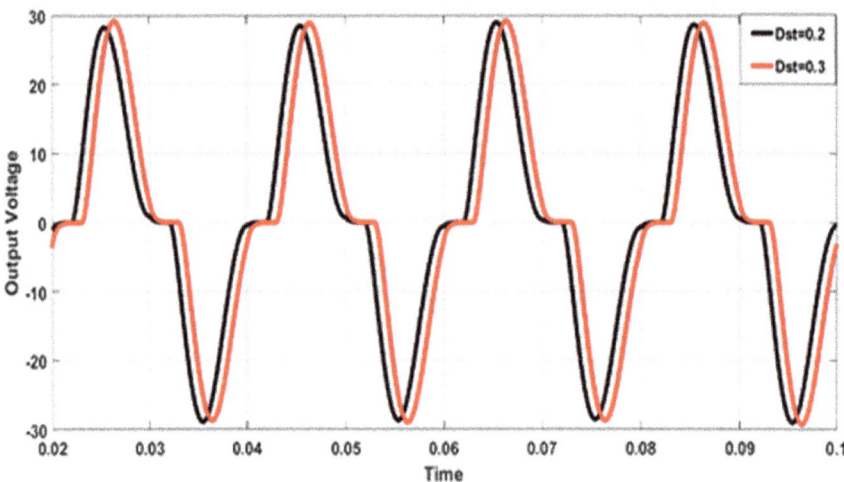

Fig. 10 Variation in output voltage with shoot through duty ratio

Table 1 Variation of THD with boost factor and shoot through duty ratio

D_{ST}	0.1	0.2	0.3	0.4	0.5
B	1.11	1.25	1.43	1.67	2.00
THD (%)	26.7	35.81	41.76	42.56	51.85

4.3 Boost Factor and THD Variation

Total harmonic distortion of the output voltage increases with the increment in the shoot through duty ration as shown in Table 1. Since the output voltage waveform deviates from the pure sinusoidal waveform THD also increases. Selection of shoot through duty ratio depends on the application whether the power quality is important or the voltage boosting. A trade-off will be there between THD and shoot through duty ratio.

5 Results and Discussion

5.1 Modulation Technique

Control signals for the switch are generated as per the logical diagram shown in Figure 11. Topology proposed here has only two switches so only two control signals are required.

Triangular signal is compared with two constant signals V_{ST1} and V_{ST2}. Magnitude of the two signals is decided directly by the value of D_{ST}. If $V_{ST1} = D_{ST}$ then $V_{ST2} = 1 - D_{ST}$. A signal S is generated by comparing sinusoidal signal to zero. While S' represents the NOT of S. Output frequency of inverter is 50 Hz. The advantage of the proposed inverter is that the switching happens at very low frequency which reduces the switching losses and increases the efficiency.

5.2 Simulation Results

Proposed topology is implemented in MATLAB (Simulink) to check the performance and characteristics under different loading conditions. Emphasis is given to check the capacitor current variation with shoot through duty ratio.

Output voltage is shown in Fig. 12 and its variation with shoot through duty ratio. Variation of capacitor current is also shown in Fig. 13. It is observed from the simulation that the start-up current of capacitor is low as D_{ST} increases, however the RMS value of output voltage decreases.

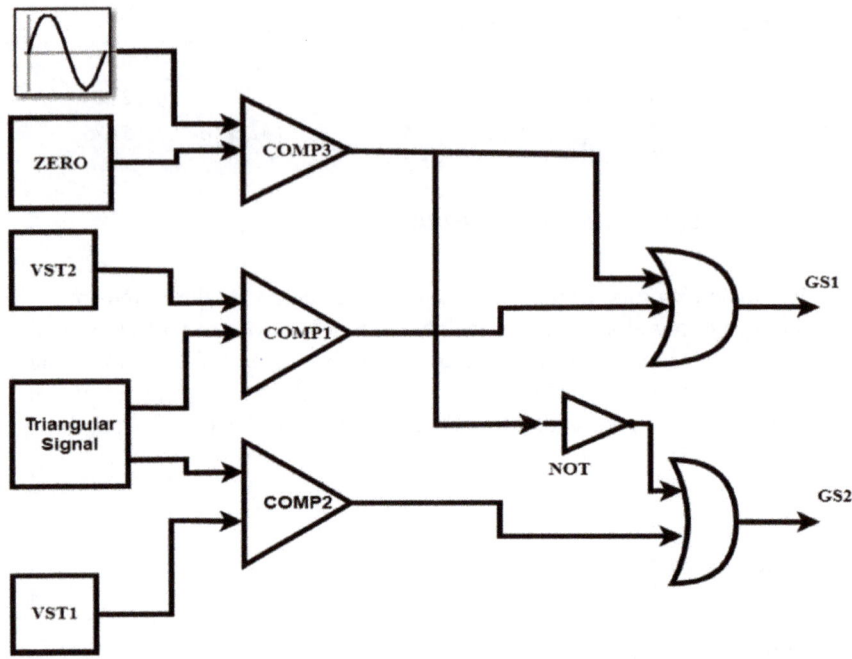

Fig. 11 Control signal for switches

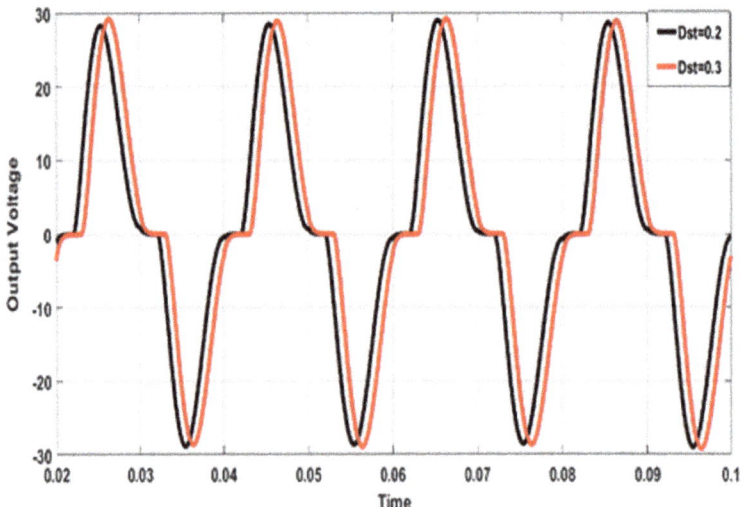

Fig. 12 Output voltage waveform of the inverter

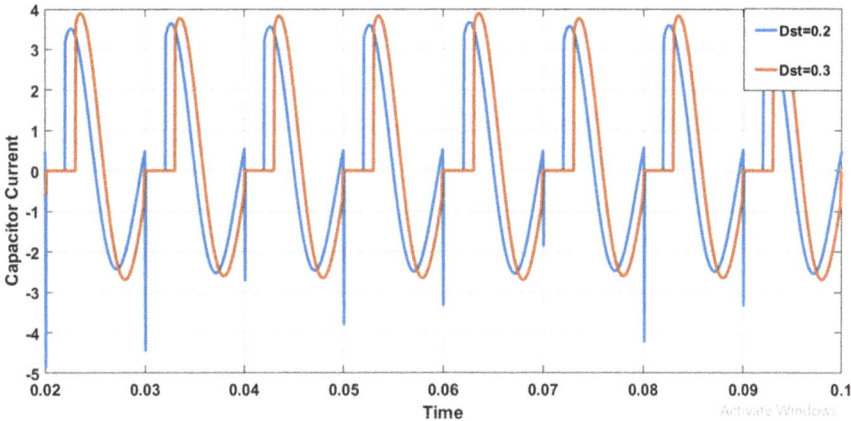

Fig. 13 Capacitor current comparison with different shoot through duty ratio

5.3   Comparative Analysis

Proposed topology is compared with the recently reported topologies and the comparison in terms of the number of components is presented in Table 2. In the original Z-Source inverter topology there are two inductors and two capacitors but in the proposed topology one less capacitor is utilized. Also the value of the capacitor is less so does the size will reduce of inverter. Another benefit of topology would be that the switching frequency is much lower than the traditional frequency, resulting in the good efficiency of the proposed topology for the inverter.

6   Conclusion and Future Scope

A new topology of Z source inverter with fewer passive elements is presented in this paper. Switch frequency and output frequency are maintained at 50 Hz. Higher switching frequency results in fewer inverter losses and greater inverter efficiency. Due to less number of elements the overall size of the inverter reduces. The modulation technique is simple and easy to implement. Proposed topology is compared with the recently published topology based on various parameters like number of passive elements, active elements, and switching frequency.

In this paper only the simulation of the proposed topology is carried out, in future for practical analysis and application the hardware of the topology will be developed and efficiency of the inverter in real-time with different types of loads can be computed.

Table 2 Comparison with other topology

Topology	Count of inductor	Value of inductors (in mH)	Count of capacitor	Value of capacitors (in μF)	Count of diodes	Count of switches	Switching frequency (Hz)	Switching losses
ZSI conventional [1]	2	5	2	200	5	4	10,000	Medium
SBI [4]	1	5	1	100	1	5	10,000	Medium
HB-SBI [5]	1	4	2	25	6	4	10,000	Medium
HB-QSBI [6]	2	2.4	2	47	6	4	30,000	High
Proposed topology	1	9	1	120	0	2	50	Low

References

1. Peng FZ (2003) Z-source inverter. IEEE Trans Ind Appl 39:504–510. https://doi.org/10.1109/TIA.2003.808920. Author F, Author S (2016) Title of a proceedings paper. In: Editor F, Editor S (eds) Conference. LNCS, vol 9999. Springer, Heidelberg, pp 1–13
2. Peng FZ, Shen M, Qian Z (2005) Maximum boost control of the Z-source inverter. IEEE Trans Power Electron 20:833–838. https://doi.org/10.1109/TPEL.2005.850927
3. Shen M, Wang J, Joseph A et al (2006) Constant boost control of the Z-source inverter to minimize current ripple and voltage stress. IEEE Trans Ind Appl 42:770–778. https://doi.org/10.1109/TIA.2006.872927
4. Ravindranath A, Mishra SK, Joshi A (2013) Analysis and PWM control of switched boost inverter. IEEE Trans Ind Electron 60:5593–5602. https://doi.org/10.1109/TIE.2012.2230595
5. Asl ES, Babaei E, Sabahi M et al (2018) New half-bridge and full-bridge topologies for a switched-boost inverter with continuous input current. IEEE Trans Ind Electron 65:3188–3197. https://doi.org/10.1109/TIE.2017.2752118
6. Asl ES, Babaei E, Sabahi M (2017) High voltage gain half-bridge quasi-switched boost inverter with reduced voltage stress on capacitors. IET Power Electron 10:1095–1108. https://doi.org/10.1049/iet-pel.2016.0291
7. Nozadian MHB, Babaei E, Hosseini SH, Asl ES (2019) Switched Z-source networks: a review. IET Power Electron 12(7):1616–1633
8. Wang B, Tang W (2018) A new CUK-based z-source inverter. Electronics 7(11):313. https://doi.org/10.3390/electronics7110313
9. Miao S, Gao J (2019) A family of inverting buck-boost converters with extended conversion ratios. IEEE Access 7:130197–130205
10. Lashab A, Sera D, Martins J, Guerrero JM (2020) Dual-input quasi-Z-source PV inverter: dynamic modeling, design, and control. IEEE Trans Ind Electron 67:6483–6493. https://doi.org/10.1109/TIE.2019.2935927
11. Iijima R, Isobe T, Tadano H (2019) Optimized short-through time distribution for inductor current ripple reduction in Z-source inverters using space-vector modulation. IEEE Trans Ind Appl 55:2922–2930. https://doi.org/10.1109/TIA.2019.2898848
12. Abdelhakim A, Blaabjerg F, Mattavelli P (2019) Single-phase quasi-Z-source inverters: switching loss reduction using a quasi-sinusoidal modulation strategy. In: Conference proceedings—IEEE Applied power electronics conference and exposition (APEC), pp 1918–1925
13. Sriramalakshmi P, Sreedevi VT (2018) Single-stage boost inverter topologies for nanogrid applications. In: Lecture notes in electrical engineering. Springer, Singapore, pp 215–226
14. Nozadian MHB, Babaei E, Ranjbarzad V (2018) Steady state analysis of dual switched boost inverter. In: Proceedings—2018 IEEE 12th International conference on compatibility, power electronics and power engineering, CPE-POWERENG 2018. IEEE, pp 1–6
15. Subhani N, Kannan R, Mahmud A, Blaabjerg F (2021) Z-source inverter topologies with switched Z-impedance networks: a review. IET Power Electron

Chapter 40
Forecasting the Price of Potato Using Time Series ARIMA Model

C. J. Jamuna⬤, Chetana Patil⬤, and R. Ashok Kumar⬤

1 Introduction

Potato scientifically known as Solanum tuberosum, belongs to the family of Solanaceae. It is one of the important crops grown in India from past 300 years with high nutritional value and the staple food in many parts of India. It has achieved importance all over the world due to its consumption because it contains 20% of carbohydrates, 1% crude fiber, 2% proteins, and only 0.3% of fat thereby enriched with Vitamin C and B1 minerals. Almost all the states in India are involved in its production due to which its claimed that "Potato is Poor Man's Friend". The production rate has achieved greater heights of about 51 Million metric ton during the year 2018, which accounted 6% increase compared to 2017. This encountered major part of the economy was contributed by the states such as Northern state of Uttar Pradesh around 30%, West Bengal 26% and Bihar of about 12% [1].

Some of the top-rated production of potatoes in India are from, Karnataka, Punjab, Uttar Pradesh, Madhya Pradesh, Rajasthan, Gujarat, Maharashtra, Himachal Pradesh, Madhya Pradesh, West Bengal, Bihar, and Assam. Even though India is contributing to greater world's economy in terms of potatoes its yield for a hector is comparatively low to that of the developed countries. Hence it is challenging to farmers not only to have huge production but also to have high-quality potato growth. Sometimes even these farmers are prone to problems such as, drastic change in the climatic factors,

C. J. Jamuna (✉) · C. Patil · R. A. Kumar
Department of Information Science and Engineering, BMS College of Engineering, Bangalore, India
e-mail: jamunacj.scn19@bmsce.ac.in

C. Patil
e-mail: chetanapatil.scn19@bmsce.ac.in

R. A. Kumar
e-mail: ashokkumar.ise@bmsce.ac.in

© The Author(s), under exclusive license to Springer Nature Singapore Pte Ltd. 2021
S. Kumar et al. (eds.), *Proceedings of International Conference on Communication and Computational Technologies*, Algorithms for Intelligent Systems,
https://doi.org/10.1007/978-981-16-3246-4_40

middleman intervention, low and unsatisfied marketing strategies, rates and their facilities, and even lack of knowledge about the future price. All these factors not only degrade the production but also hinders the farmers growth intern resulting in decrease of Indian economy. Hence the paper proposes the usage of ARIMA Model to forecast the future market price of potatoes over one particular major market in 5 different states such as Karnataka, Gujarat, Madhya Pradesh, Maharashtra, and Rajasthan for leading 10 years from now.

As the name suggests Autoregressive Integrated Moving Average (ARIMA) model was introduced by Box and Jenkins in 1970, comprises of time series data that performs identifying, estimating, and diagnosing activities on the models. The combined effect of past values and past errors in a linear fashion is the main criteria to design the future values of the variable in an ARIMA Model. This paper utilizes ARIMA Model, the most popular and widely used forecasting model on time series data that contains the past 10 years average model price of potatoes to forecast the future 10 years price value.

ARIMA is one such model that gives more accuracy when compared to other modelling. Using other methods could yield to lower estimation of the future price and could possibly increase the error rate when we try to predict the price for long term [2]. But in this project, the future 10 years price is been easily estimated for potato crop with 95% confidence interval. Sometimes using the basic machine learning algorithms could not give us control to feed the input dataset to the algorithm. Also, it will not allow us to go through the steps that the algorithm follow that includes the equations and the relationship that exists among the coefficients, rather would directly yield us the final result [3]. Henceforth here in our project we are trying to use the ARIMA Model to forecast the potato price for future 10 years from now.

2 Objectives of the Study

The main aim of this paper is portrayed with the help of implementation of time series using ARIMA model with the following objectives:

- Forecast the future price of potato by depicting the modal price
- Guides farmers on how much potato crop to be cultivated for the next harvesting
- Directly benefits farmers to earn more profit by predicting market price
- Enhance production rate significantly by forecasting market price.

3 Review of Literature

The main interest in choosing potato to forecast its price is because Central Potato Research Institute (CPRI) has around 521 ha of farmland which is been circulated more than 15 units across India. India is the main nation in Asia except for Japan, that possess a settled seed creation program. CPRI yields around 2500–2700 metric huge amounts of hybridizer seed every year. The reproducer seed provided is sufficient to cook over 55% of all the ensured seed necessity of the nation at the pace of 25% yearly seed substitution rate. Income of around 4.5–5.0 crore rupees is created every year by offer of hybridizer seed over Revolving Fund plan of action. CPRI has greater than 3900 acquisition of developed and wild potatoes germplasm, the greatest group of South-West Asia. Refined reproducing lines and world-class hereditary stocks had been created and enrolled for different qualities, for example, earliness, protection from bug and frost resilience [4].

Chandran and Pandey [5] presented an idea of Potato Price Forecasting using seasonal ARIMA technique in which Potato discount prices of Delhi advertise were examined. In view of the Schwarz Bayes Criterion (SBC) and Akaike Information Criteria (AIC), the forecasts dependent on this model were near the examined values. Periodical evidence ratios determined normally indicates, for most part of the cost is low from December to May and it elevates from June, and arrives at the greatest in October. Since Potato costs vary by seasons because of the varieties in producing and wholesale appearance, planning and predicting the monthly value-based conduction throughout the years is of much significant. The perfect apt model (AIC and SBC) was ARIMA (1, 1, 1) × (1, 0, 0) with intensity of 12. The graph forecasted by the author showed that the wholesale potato cost of Delhi market had huge variation during 1995–2000 compared to previous years, 1989–1994, and was stabilized from 2000.

Kumar et al. [6] presented a paper on Price Estimating System for Crops during Seeding where in which they tested by experimenting the previous data of cost of harvests for estimating in India. Utilizing South-India as a sector for dates and their cost before planting of seeds, by producing a framework which can estimate the cost of yields on yearly just as month to month premise. Further inferred that there is potential to use previous data to estimate with ARIMA model to anticipate future cost of given yield based on time sequence by utilizing rolling mean, standard deviation, dickey fuller test, halfway auto-correlation function to survey areas of helpful probabilistic expectation. This framework demonstrated a possibility to anticipate all things considered 95% trust in its forecasts and will be expanded further in getting more data on regular routine. Author has forecasted price of potato in a decreasing trend between 2017 and 2018.

Darekar and Amarender Reddy [7] applied ARIMA modelling for the time-series data to forecast and analyze the price of maize in major states of India. Forecasted prices by author on maize have an increasing trend between 2006 and 2018. According to forecasted price the market value is, Rs. 1200–1600 per quintal in kharif harvesting season. Price from September to December 2017–18 are high in Madhya Pradesh, Andhra Pradesh, and Rajasthan. Prices are low in Karnataka and Bihar, respectively are represented by the author through the graph.

Darekar and Amarender Reddy [8] both these authors again forecasted price of cotton in major states of India (0, 1, 0) was the best fitted ARIMA model with a constant trend as forecasted by the author. According to forecast the prices of cotton is high in Karnataka, Andhra Pradesh, and Maharashtra, i.e., 4500, 4450, and 4400 per quintal respectively and prices are low Gujarat and Haryana i.e., 4350 and 4300 per quintal respectively, which are as represented by the author below.

The above two reviews made on different crops by the authors, Ashwini Darekar and A. Amarender Reddy, helped us to estimate the model price or the average price of any crop over a year for every month to forecast the potato price for leading 10 years in this paper.

4 Analysis and Design Methods

4.1 Time Series

Our paper deals with the Time Series data which is nothing but taking ordered price of potatoes on an equal monthly time interval to analyse its future trend using Time Series Analysis [9]. The main aim of the paper is to forecast the price of potatoes which is univariate (single variable i.e. price of potatoes) in nature over an independent variable time for future years [10].

4.2 Data Collection

The dataset which is available on AGMARKNET website [11] was collected and the average modal price of potatoes on monthly basis was estimated as required for time series analysis [12]. However, the dataset collected for the price estimation included the top potato grown states in India namely, Karnataka, Gujarat, Madhya Pradesh, Maharashtra, as well as Rajasthan that had the minimum, maximum, and the modal price of Potatoes on a daily basis.

4.3 Components of Time Series

There are four different categories which affects the observation values of time series.

The first two components on which time series can be applied are Trend and Seasonality. Trend is the tendency of a data to either increase or decrease over a long period that need not be always in the same direction but it can be upward, downward, or in a stable condition [13]. The graph having time series value around the straight line is called as linear trend or else a non-linear trend. In this paper some of the time series were having a linear upward trend. Whereas, Seasonality the rhythmic values of time series operating at a regular interval over a span of time (not less than a year) [13]. However, we try to obtain seasonality (length of the season) from autocorrelation plot that will have recurrent pattern for a period of 12 months (a year) where the data recorded will be an hour, day, week, or a month [9].

The other two components which the time series can't handle or applied are Irregularity (Random or Residual) and Cyclic. Irregularities are the variations in the observation values or the errors present in time series that are uncontrollable, unpredictable, and erratic in nature [13]. Though such observations have randomness in them few of the correlation components useful for time series are hidden in them, which can be correlated using Auto-Correlation Function [14]. However, Cyclic is the variation in time series having a complete period of oscillations that are non-periodic but, are operated by themselves for more than a year [13].

4.4 Stationarity

Before we could apply ARIMA Model some of the transformations must be applied to the time series data to remove non-stationary behavior present in them [9]. Stationarity being one of the major characteristics of time-series maintains consistency in the time series analysis. Three main criteria required to make time series data stationary are, having a constant mean and variance, which is dependent on time, and covariance which is independent of time.

In order to analyze the stationarity of time series we conducted two tests;

Rolling Statistics
To support this, we plotted the graph between moving average or moving standard deviation with time to check stationarity [15]. It's mainly a visual technique where, if mean or standard deviation changes over time then the time series is considered to be non-stationary [14].

Augmented Dickey-Fuller Test (ADCF)
Another important test conducted for analyzing stationarity is, a statistical test on time series known as "Dickey-Fuller Test" [16], that checks for the unit root, $\beta = 1$ (beta is the first lag on 'Y') present in the Null Hypothesis, H_0 [17]. If the root is present, then ($p > 0$) null hypothesis is not rejected and the time series is non-stationary. But, if the value of ($p = 0$) means, the null hypothesis is rejected and the time series is stationary [9].

Further, we also test the time series using "Augmented Dickey-Fuller Test" which is evolved from the Dickey-Fuller equation to include the regressive process having higher order (more differencing terms to bring thoroughness to the test) [16]. Even here, the test checks for the unit root, $\beta = 1$ present in the Null Hypothesis, H_0 [17]. But, the p-value is changed to 0.05 i.e. if the root is present, then ($p > 0.05$) null hypothesis is not rejected and the time series is non-stationary. Whereas, if the value of ($p \leq 0.05$) means, the null hypothesis is rejected and the time series is stationary [16].

Along with this, two other parameters are considered in both the tests namely, the Test-Statistics and the Critical Value. If the test-statistics are significantly lesser than the critical-values then, we reject null hypothesis and the time series data has stationarity [14]. Else, null hypothesis can never be rejected and the time series is non-stationary in nature.

4.5 Autoregressive Integrated Moving Average (ARIMA) Model

ARIMA Method can also be called as Box-Jenkins Method [18], which makes the complex models in time series by using the combination of simpler models [9]. In our paper, we try to forecast the price of potatoes (univariate time series data) for future 10 years from now using the ARIMA Model because, it generates a good result on a large sample size particularly, when applied for short time prediction and also for predicting time series having highest seasonality [10].

White Noise 'ε' is a series that is independent and purely random in nature having finite mean and variance. If its normally distributed with mean(μ) $= 0$ and are correlated with constant variance, then its termed to be as "Gaussian White Noise".

In order to use this ARIMA Model for modelling our system, we have to identify few of the key parameters which plays an important role in forecasting the price (univariate time series) of potatoes that should not have any trend or seasonality in them.

Autoregressive (AR) It is the determination of the current values (present month price of potatoes) by the Weighted Average of the previous values (previous month price of potatoes) along with some lag [9]. This lag is termed to be as "Lag Order" and is denoted by the parameter 'p' [19], that represents the maximum number of lags [20]. For example, if $p = 0$ then there is no autocorrelation present in the series and if $p = 1$ then an auto correlation exists with one lag and so on [18].

This process can be expressed as

$$y_t = c + \phi_1 y_{t-1} + \phi_2 y_{t-2} + \cdots + \phi_p y_{t-p} + \varepsilon_t$$

where 'y_t' is the current value, 'y_{t-p}' is the previous value, 'c' is the constant, 'ϕ_p' is the coefficient and 'ε_t' is the white noise [21]. For the effective working of our model, the correlation between the past and the current values are maintained to be high.

Note: The p-value in ADCF test and the parameter 'p' in ARIMA Model are different from one another.

Integrated (*I*) It helps to convert a non-stationary time series into stationary time series by differencing the raw observations (time series data). This differencing is termed to be as "Degree of Differencing" and is denoted by the parameter '*d*' [19], that represents the count of differences made on raw observation [20]. For example, if $d = 0$ then the series is stationary and if $d = 1$ then one differencing is done in order to make the series stationary and so on [18].

Moving Average (MA) It uses the dependent parameter present between the observations and the residual error [19], that was applied to the lagged observation from a moving average model i.e. the biggest lag (window) present in the autocorrelation plot [9]. This parameter is termed to be as "Order of Moving Average" and is denoted by the parameter '*q*' [19], that represents the moving average window size [20]. For example, if $q = 1$ then there is an error term having autocorrelation with one lag [18].

This process can be expressed as

$$y_t = c + \varepsilon_t + \theta_1 \varepsilon_{t-1} + \theta_2 \varepsilon_{t-2} + \cdots + \theta_p \varepsilon_{t-p}$$

where 'y_t' is the current value, 'ε_t' is the previous white noise value, 'c' is the constant, 'θ_q' is the coefficient and 'ε_{t-q}' is the white noise [21]. Here the present values of the time series (present month price of potatoes) are the linear combination of the past errors (errors present in the AR) [21].

4.6 Auto-correlation

Considering too many features of correlation during our modelling causes Multi-linearity Issues [21]. Hence, we try to filter out and retain only relevant features for modelling our system through autocorrelation. Autocorrelation has the similarities as a time lag function between the observations where the sinusoidal function helps to find the seasonality of time series through its period [9]. Therefore, for determining the autocorrelation between the series and the error term we use ACF and PACF plots [18].

Auto Correlation Function (Complete Auto Correlation) provides auto-correlation values (the relation between the present and past monthly price of potato) of time series with its lagged values by considering trend, seasonality, cyclic and residual components [21]. It is used to find autocorrelation between time-stamp t and $t - k$, where $k = 1, 2, 3, \ldots, t - 1$ [14] i.e. the value of 'q' for the ARIMA can be found [15]. On the other hand, Partial Auto Correlation Function removes already obtained correlation feature and finds correlation for the residuals along with the next lag value [21]. This allows the usage of next lag feature for modelling the hidden information within the residuals [21] i.e. the value of 'p' for the ARIMA can be found [15].

4.7   Akaike Information Criteria (AIC)

Keeping all these parameters and tests in mind we use the best diagnostic tool to quantify the goodness of fit in a statistical model using Akaike Information Criteria. It helps to choose the predictors for regression and determines order of an ARIMA Model. It is expressed as

$$\text{AIC} = -2\log(L) + 2(p + q + k + 1)$$

where L is the likelihood of the model, $(k = 1)$ if $c \neq 0$ and $(k = 0)$ if $c = 0$ and the last term is the number of parameters in the model [22].

In our paper to choose appropriate model, lowest absolute value of AIC is preferred that could be either positive or negative. The correlated AIC expression goes as follows [22]

$$\text{AICc} = \text{AIC} + \frac{2(p + q + k + 1)(p + q + k + 2)}{T - p - q - k - 2}$$

5   Results and Discussions

Series of steps were followed in our paper to forecast the price of potatoes in each state. First of all, cleaning of raw data was done by taking average price of potatoes for every month in a year on a single major market in a state that includes, past 10 years of dataset ranging from 2010 to 2019. The plot of the datasets of different states had varying trends of time series.

Since modelling a system in time series is only possible if the data is stationary (as mentioned in Sect. 4.4 of analysis and design methods), we calculated the rolling mean and rolling standard deviation for the following datasets. Under this test, all the datasets were non-stationary due to variation in the values of mean and standard deviation, which was visualized by plotting a graph between those two parameters.

Further, Dickey-Fuller test was conducted on the original sets of data to check the stationarity (as mentioned in Sect. 4.4 of analysis and design methods). As we already know that, if the value of p is less than the unit root and if the test statistics value is greater than the critical values of 1, 5, and 10%, the data is said to have stationarity component [14]. Keeping this in mind we could observe that except for the states of Maharashtra and Karnataka other states like Madhya Pradesh, Rajasthan, and Gujarat datasets were stationary as shown in Figs. 1 and 2 respectively.

As logarithmic transformation helps us to straighten the inflationary growth pattern to fit into a linear model and also linearizes the relationships among the variables in the regressing model, we modeled the system for the other two states by taking the logarithm of the time series datasets and plotted the graph to see their

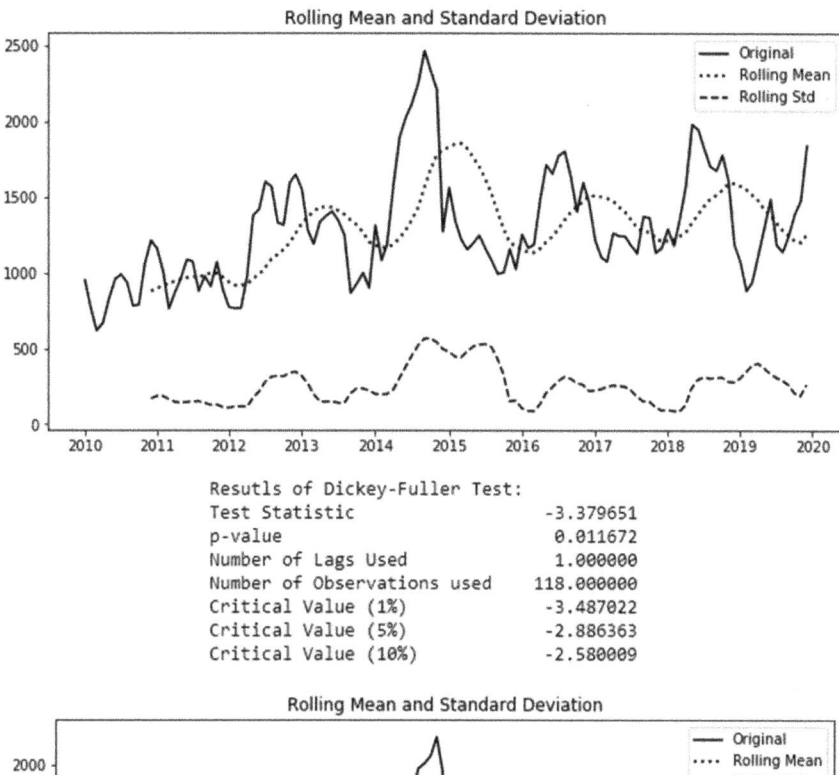

Resutls of Dickey-Fuller Test:
Test Statistic   -3.379651
p-value   0.011672
Number of Lags Used   1.000000
Number of Observations used 118.000000
Critical Value (1%)   -3.487022
Critical Value (5%)   -2.886363
Critical Value (10%)   -2.580009

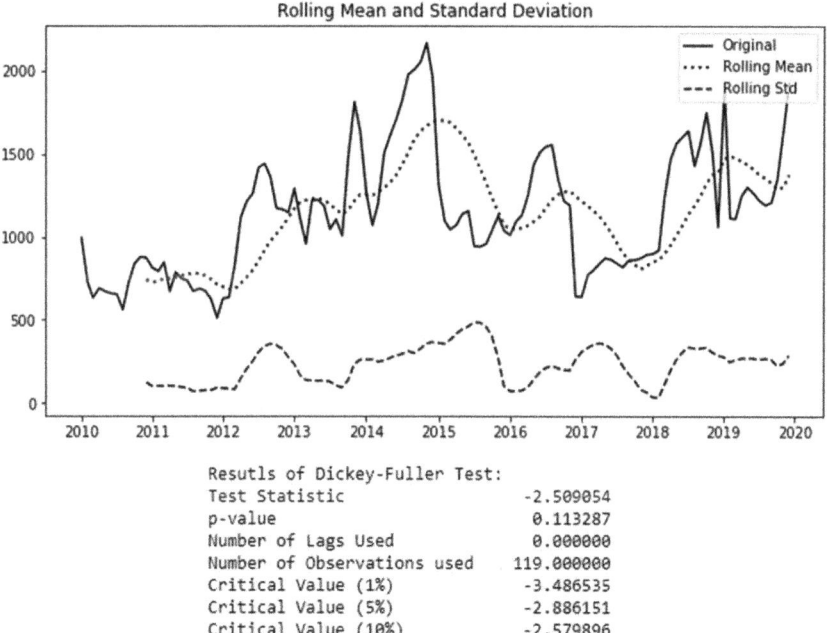

Resutls of Dickey-Fuller Test:
Test Statistic   -2.509054
p-value   0.113287
Number of Lags Used   0.000000
Number of Observations used 119.000000
Critical Value (1%)   -3.486535
Critical Value (5%)   -2.886151
Critical Value (10%)   -2.579896

Fig. 1 Results of ADCF and rolling mean and standard deviation of Karnataka and Maharashtra respectively

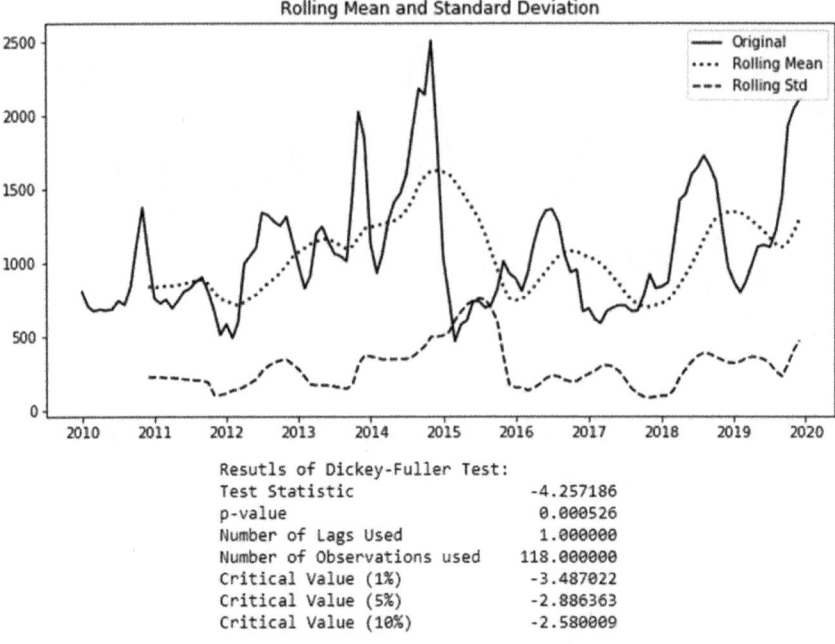

Resutls of Dickey-Fuller Test:
Test Statistic   -4.257186
p-value   0.000526
Number of Lags Used   1.000000
Number of Observations used   118.000000
Critical Value (1%)   -3.487022
Critical Value (5%)   -2.886363
Critical Value (10%)   -2.580009

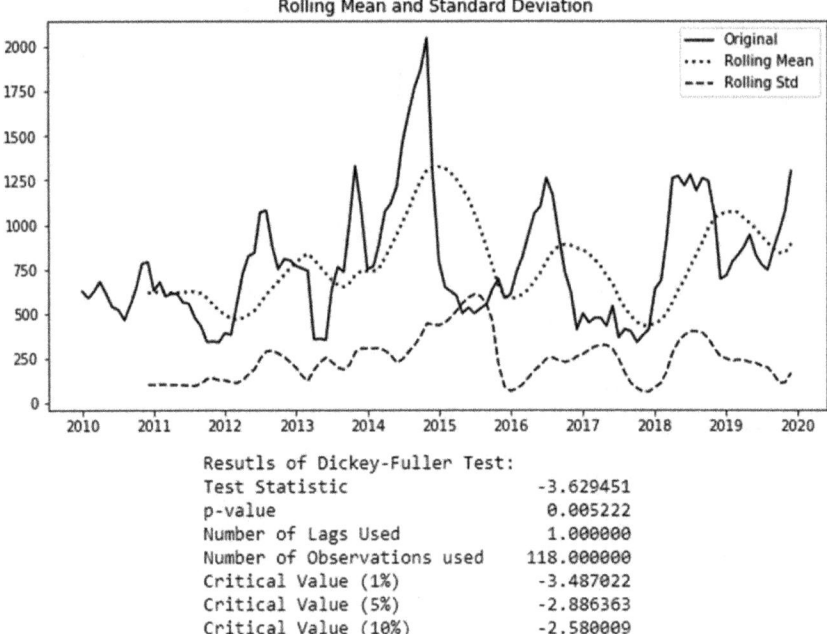

Resutls of Dickey-Fuller Test:
Test Statistic   -3.629451
p-value   0.005222
Number of Lags Used   1.000000
Number of Observations used   118.000000
Critical Value (1%)   -3.487022
Critical Value (5%)   -2.886363
Critical Value (10%)   -2.580009

Fig. 2 Results of ADCF and rolling mean and standard deviation of Gujarat, MP, and Rajasthan, respectively

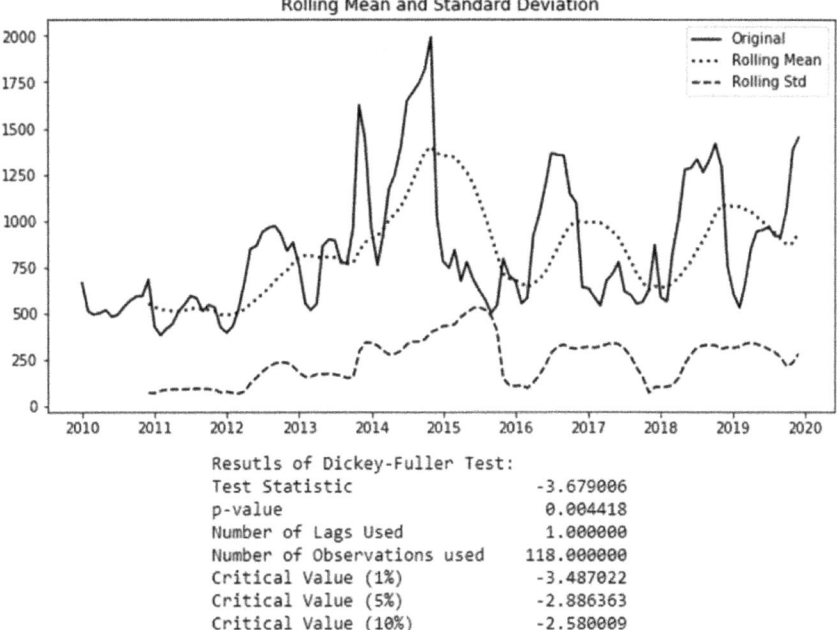

Fig. 2 (continued)

trends. We observed that the time series data had a moving average component in it which made it non-stationary. Hence, we took the difference between the moving average and the logarithmic value and calculated the Augmented Dickey-Fuller test (as mentioned in Sect. 4.4 of analysis and design methods) for the same [14] and observer that, the values of p and the test statistics were significant that rejected the null hypothesis and made the system stationary. In spite of performing all these tests if the dataset doesn't become stationary then we can calculate the weighted average. After looking at its trend we can subtract the weighted average from the logarithmic value [14]. This will further make the system more stationary.

The above-mentioned procedures of moving average and weighted average can be followed to the datasets of other states to make them still more stationary. Performing these stationarity tests depends on the p-value and the test statistics parameters. It is not recommended to decrease the already reduced value of these parameters, as it could lead to increase in its value.

Successfully, after making the model stationary we try to obtain the value of the first ordered difference, 'd' of an ARIMA Model (as mentioned in Sect. 4.5 of analysis and design methods) by shifting the value of time series with some lag [18]. The value of 'd' for all our time series datasets is considered to be one ($d = 1$) because differencing a series more than twice makes it unreliable [18]. We found that the values of datasets of all the states were stationary i.e. ($p > 0.05$) and the

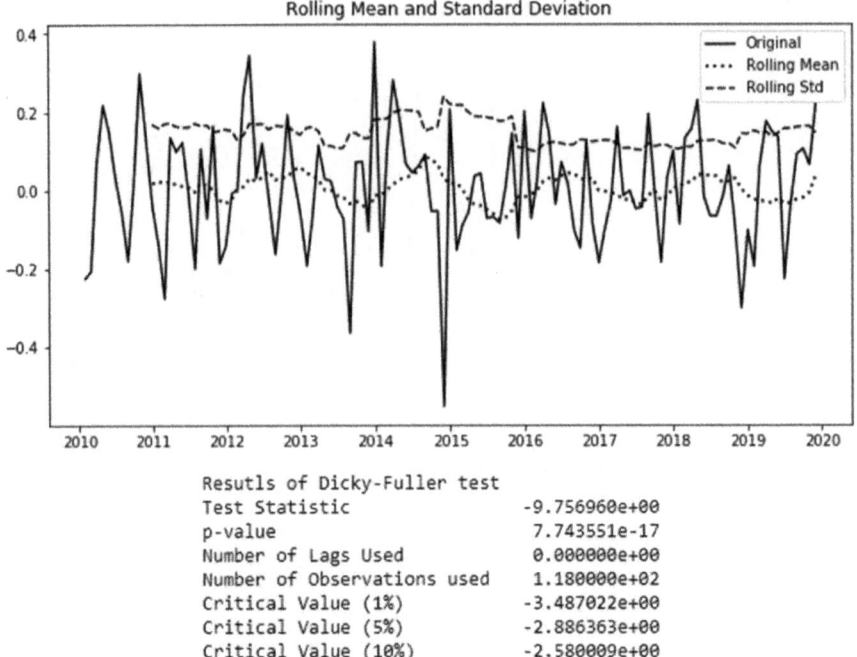

Fig. 3 Results of ADCF and rolling mean and standard deviation of Karnataka

test statistics value rejected the null hypothesis (as mentioned in Sect. 4.4 of analysis and design methods) as shown in Figs. 3, 4, 5, 6 and 7.

Now we try to plot all the components of time series (as mentioned in Sect. 4.3 of analysis and design methods). This is because as moving average depends on residual component of time series (as mentioned in Sect. 4.5 of analysis and design methods), making it stationary smoothens our job to forecast the future values [9]. Since all the datasets passed the ADCF test thereby indicating that the residual component of time series is stationary in nature.

Now that we have determined value of 'd' the other two values required for ARIMA Model is 'p' and 'q' which is estimated using PACF and ACF plots respectively (as mentioned in Sect. 4.6 of analysis and design methods). To determine 'p' (AR term) in PACF, Ordinary Least Square (OLS) is adopted. As it estimates the coefficients of AR to be fitted in a linear regression model through a single iterated autocorrelation [23]. Here the value that cuts off the origin (precisely the critical level) for the first time is taken into consideration in both plots. All our datasets don't exceed more than 3 for 'p' as well as 'q' value.

After detecting the values for ARIMA(p, d, q) we tried to plot the model and found that (2, 1, 1) was the best fit for Karnataka Fig. 8, (2, 1, 3) was the best fit for Gujarat Fig. 9, (2, 1, 3) was best fit for Madhya Pradesh Fig. 10, (2, 1, 2) was the best fit for Maharashtra Fig. 11 and (2, 1, 2) was the best fit for Rajasthan Fig. 12. We

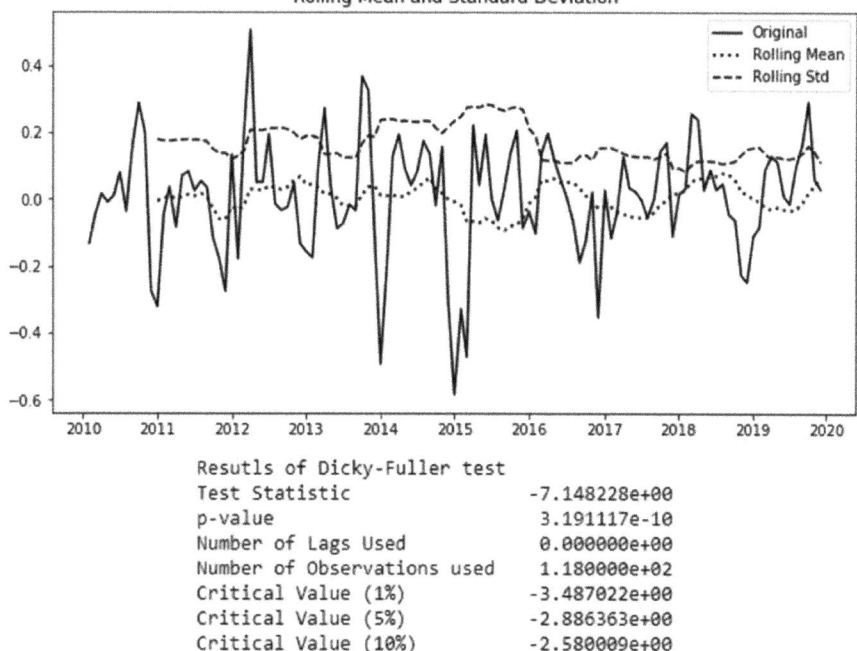

Fig. 4 Results of ADCF and rolling mean and standard deviation of Gujarat

even recorded the least value of AIC as possible for all the models (as mentioned in Sect. 4.7 of analysis and design methods), that were all negative as shown in Table 1, that shows less information loss yielding a better model [24].

Now that we have fitted all these values in an ARIMA Model, we reversed the process once (since we have used only the first-order difference) to verify if it depicts the original curve or not. For that, we shifted the fitted value one time and placed the first item from the log scaled value at index 0 and then performed the cumulative sum for retrieving the logarithmic scaled value [14]. Lastly, we took the exponential of the values this gave almost a good fitting plot with respect to the original time series value for all the states price dataset as shown in Figs. 13, 14, 15, 16 and 17.

Finally, as mentioned in our objective we forecast the price of potatoes for future 10 years from now, i.e., 2020–2029 for all the states, and could observe variations in the price with 95% significance level as shown in Figs. 18, 19, 20, 21 and 22. All these values are in exponential form as we have taken the logarithmic value of the original dataset.

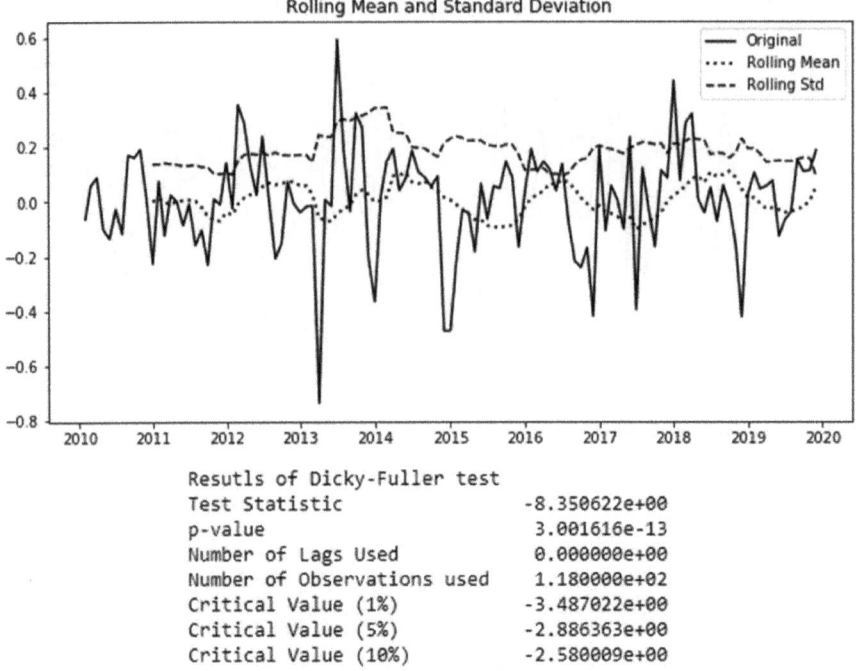

```
                    Resutls of Dicky-Fuller test
                    Test Statistic                    -8.350622e+00
                    p-value                            3.001616e-13
                    Number of Lags Used                0.000000e+00
                    Number of Observations used        1.180000e+02
                    Critical Value (1%)               -3.487022e+00
                    Critical Value (5%)               -2.886363e+00
                    Critical Value (10%)              -2.580009e+00
```

Fig. 5 Results of ADCF and rolling mean and standard deviation of MP

6 Conclusion

This paper forecasted the future prices of potatoes using ARIMA Model and thereby it helps farmers to know how much to invest in the production of potatoes for the upcoming 10 years as mentioned in our objective. It also provides better feedback to the farmers for growing the potato with 95% confidence interval. The prediction system intimates that there is higher possibility of getting profit on potato production in Karnataka, whereas there is a stabilized condition in other states on a moderated investment. If production rate is made high it shows that there could be a possibility of getting more profit, if all the agroclimatic factors supports a healthy environment for the crop growth. But under certain worst circumstances this decision could also incur loss to the farmers of all the other states.

The price forecasting made in this paper provides a greater support to the farmers by avoiding middleman interruption, as the farmer would already know the price that could be fixed according to the market price forecasted here. Nevertheless, the model is very successful in forecasting the future price, as ARIMA model requires only some of the past time series observations on the univariate time variable. As this is a mathematical model-based analysis, extensions on the model can be conducted in future where there may be possibility that the forecasted value could vary from the actual price by considering the tentative developments made in the market. Some

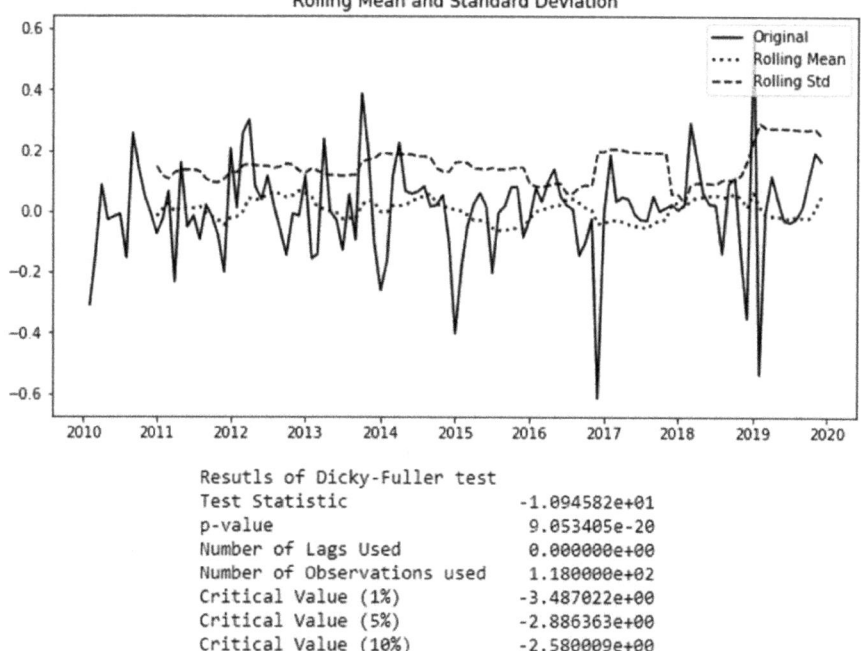

Resutls of Dicky-Fuller test
Test Statistic   -1.094582e+01
p-value   9.053405e-20
Number of Lags Used   0.000000e+00
Number of Observations used   1.180000e+02
Critical Value (1%)   -3.487022e+00
Critical Value (5%)   -2.886363e+00
Critical Value (10%)   -2.580009e+00

Fig. 6 Results of ADCF and rolling mean and standard deviation of Maharashtra

of the reasons could be changes in International price, the market value, and even an import and export that takes place within and across the countries.

Potato is one such crop that is sensitive to the above-mentioned impacts. Since its price keeps on fluctuating continuously due to natural calamities, we can forecast the price accurately by adopting Seasonal ARIMA (SARIMA) modelling or Neural Network Auto Regressive (NNAR) modelling [25], Support Vector Regression (SVR), AdaBoost or Bayesian networks [26] could be used as a future enhancement for the same project. By adopting these advanced techniques, the price forecasting could be made more accurate by considering even more attributes for forecasting potato prices.

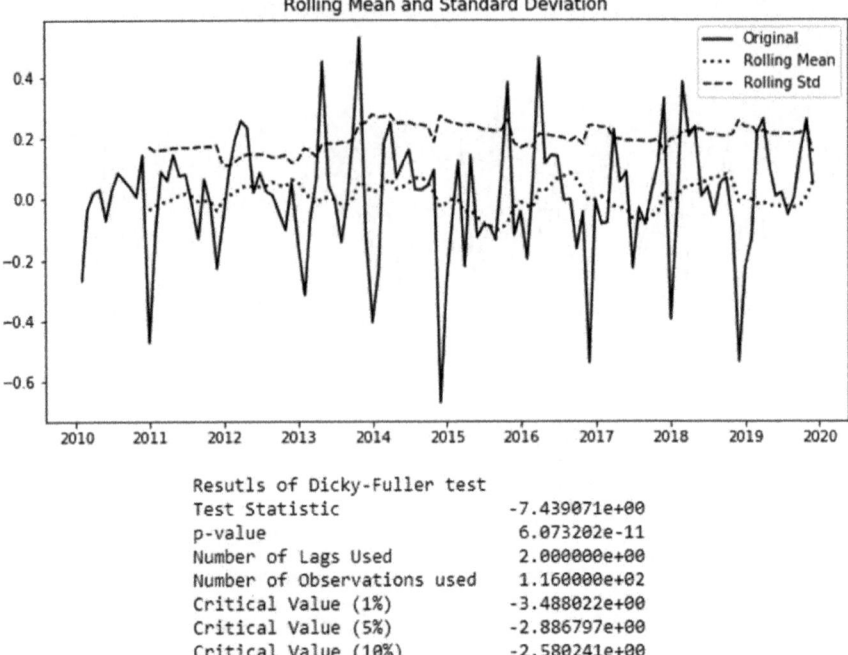

```
Resutls of Dicky-Fuller test
Test Statistic                   -7.439071e+00
p-value                           6.073202e-11
Number of Lags Used               2.000000e+00
Number of Observations used       1.160000e+02
Critical Value (1%)              -3.488022e+00
Critical Value (5%)              -2.886797e+00
Critical Value (10%)             -2.580241e+00
```

Fig. 7 Results of ADCF and rolling mean and standard deviation of Rajasthan

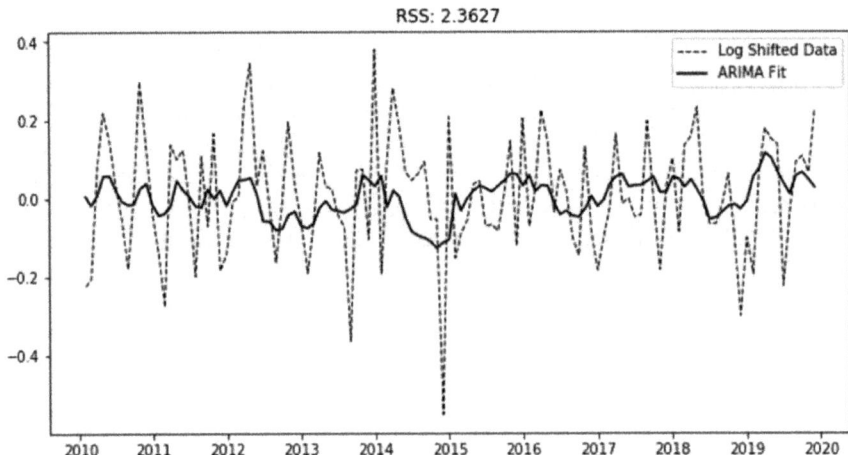

Fig. 8 Best fit ARIMA model at (2, 1, 1) for Karnataka

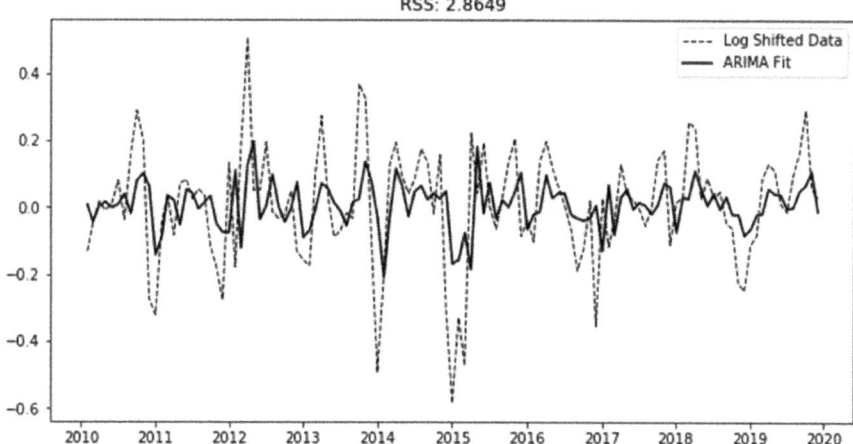

Fig. 9   Best fit ARIMA model at (2, 1, 3) for Gujarat

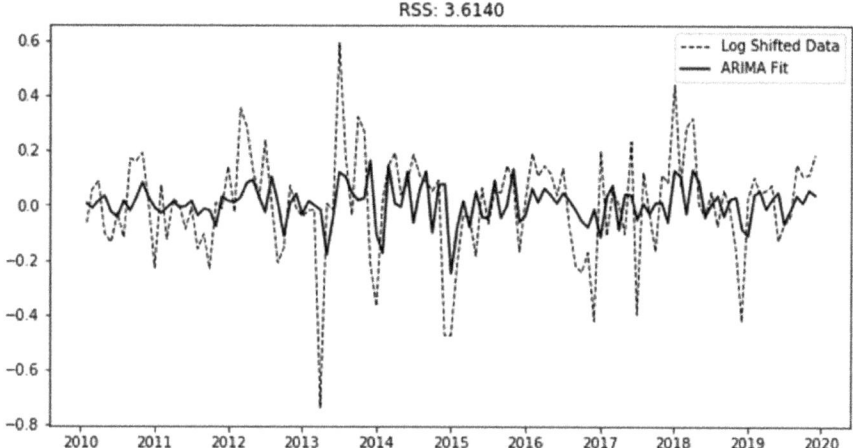

Fig. 10   Best fit ARIMA model at (2, 1, 3) for MP

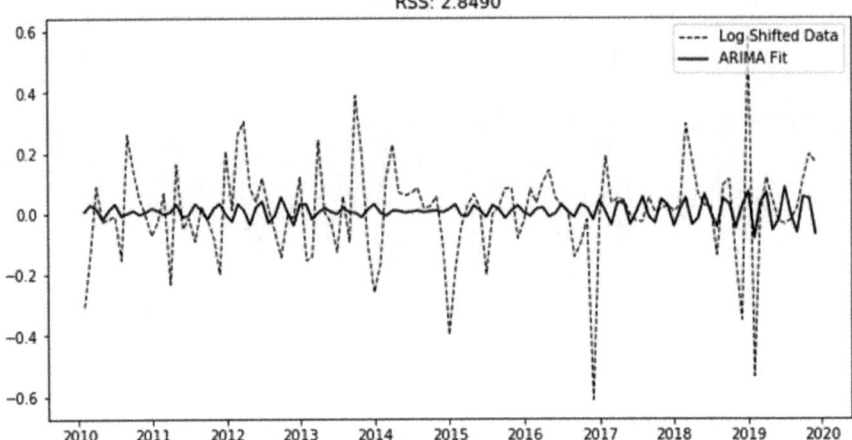

Fig. 11 Best fit ARIMA model at (2, 1, 2) for Maharashtra

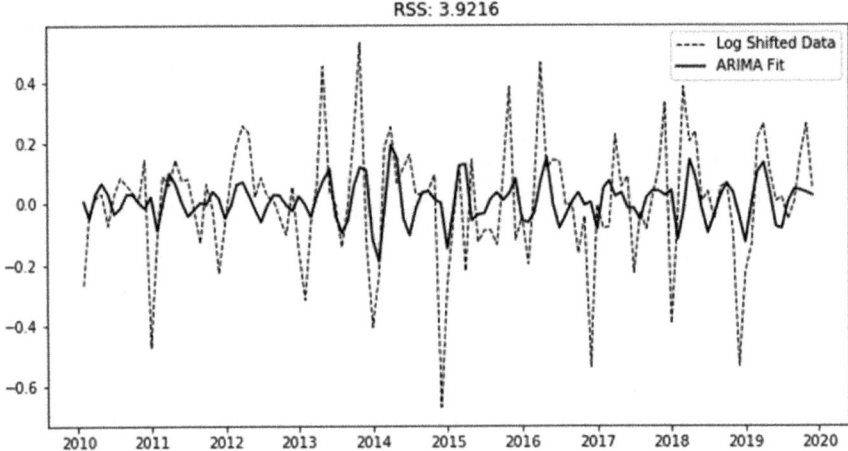

Fig. 12 Best fit ARIMA model at (2, 1, 2) for Rajasthan

Table 1 Depicts the AIC, RSS, and likelihood values of different state potatoes

State	ARIMA	RSS	P > |z|						AIC	Log likelihood
			ar.L1. D.Price	ar.L2. D.Price	ma.L1. D.Price	ma.L2. D.Price	ma.L3. D.Price			
Karnataka	(2, 1, 1)	2.3627	0	0.028	0	–	–		−118.859	64.429
Gujarat	(2, 1, 3)	2.8649	0	0	0	0	0		−91.75	52.875
Madhya Pradesh	(2, 1, 3)	3.614	0	0	0.023	0	0		−63.57	38.737
Maharashtra	(2, 1, 2)	2.849	0	0	0	0	–		−94.224	53.112
Rajasthan	(2, 1, 2)	3.9216	0	0	0.017	0.006	–		−56.342	34.171

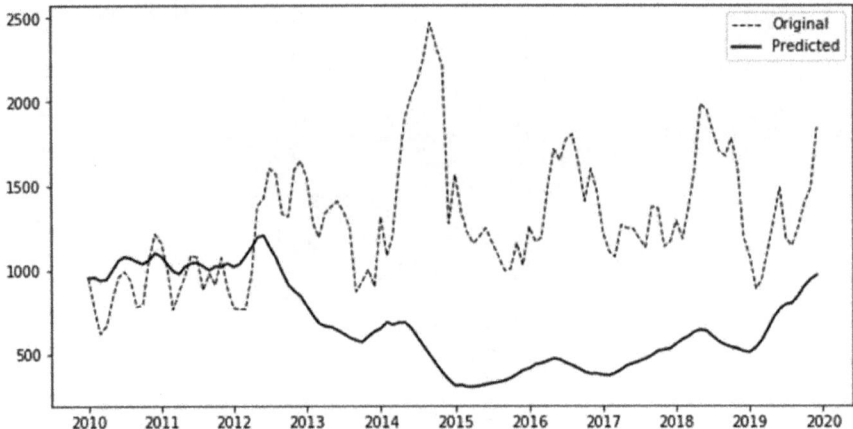

Fig. 13 Best fitted plot for Karnataka

Fig. 14 Best fitted plot for Gujarat

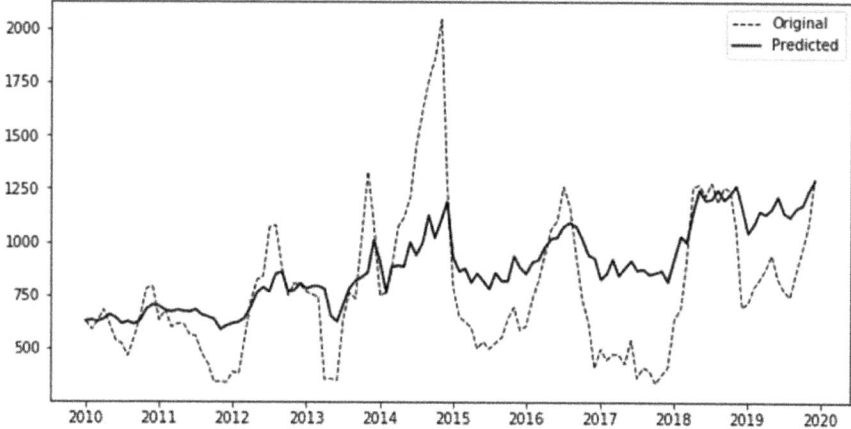

Fig. 15 Best fitted plot for MP

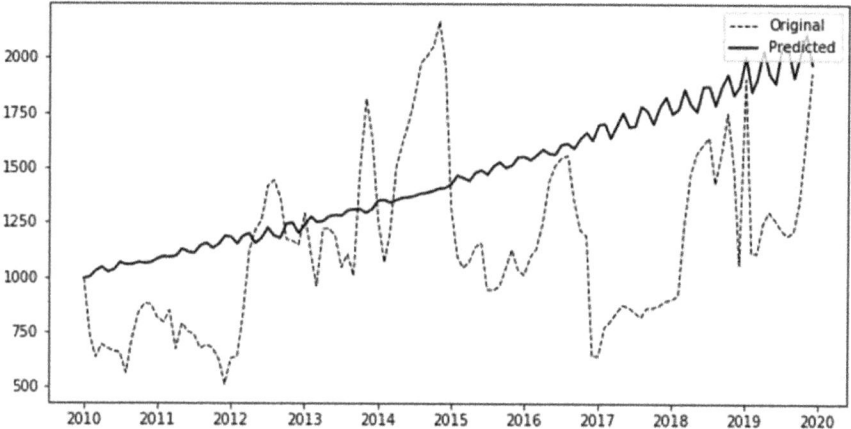

Fig. 16 Best fitted plot for Maharashtra

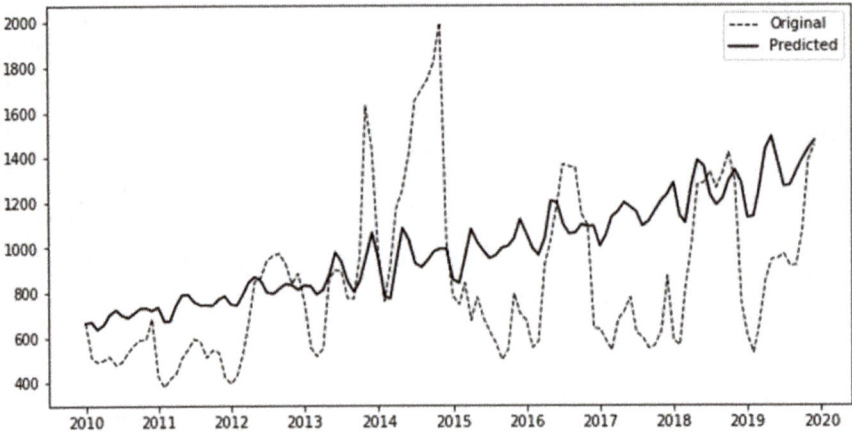

Fig. 17 Best fitted plot for Rajasthan

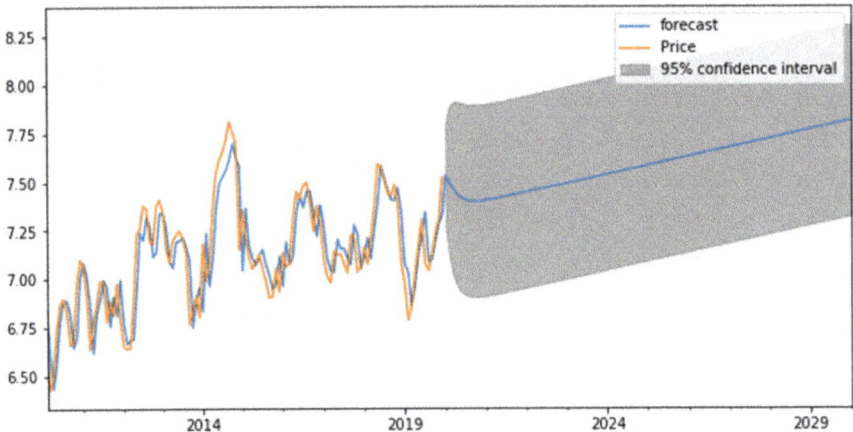

Fig. 18 Forecasted prices of potato for Karnataka

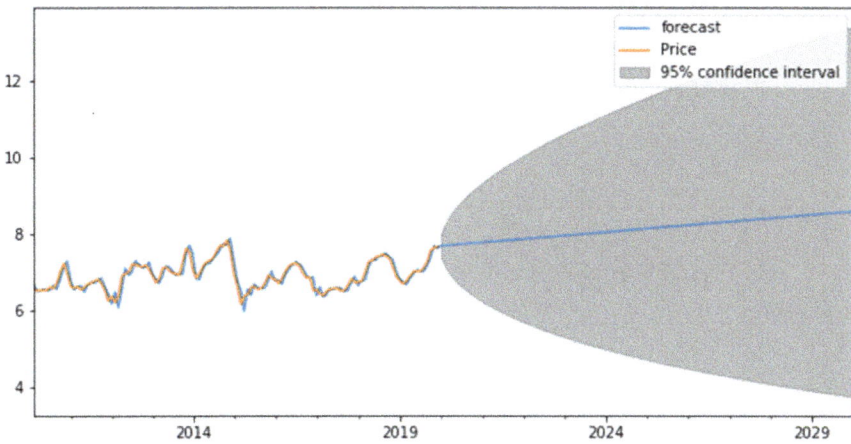

Fig. 19 Forecasted prices of potato for Gujarat

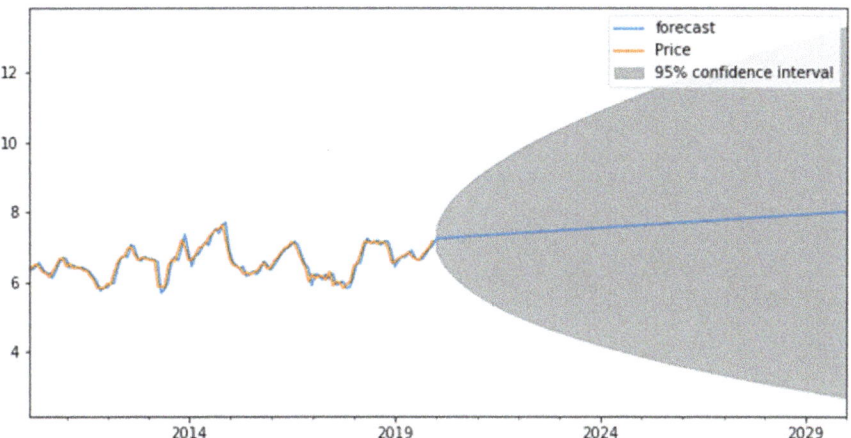

Fig. 20 Forecasted prices of potato for MP

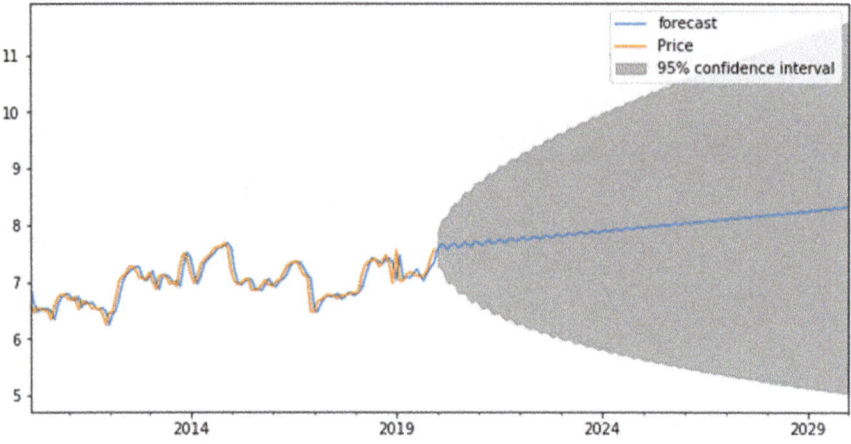

Fig. 21 Forecasted prices of potato for Maharashtra

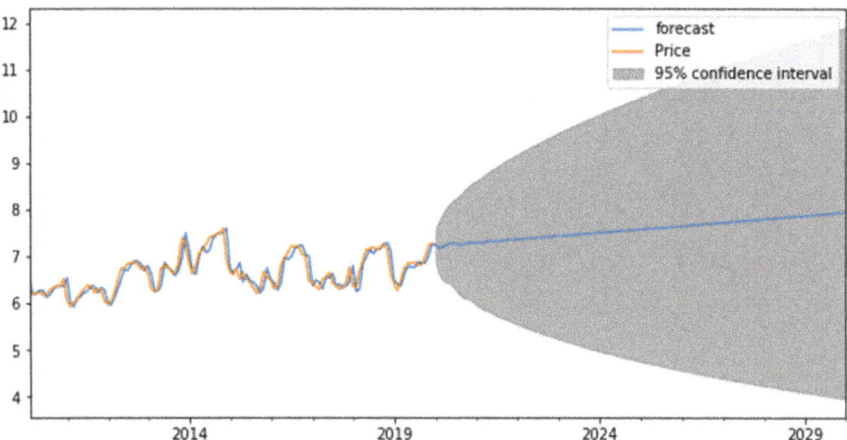

Fig. 22 Forecasted prices of potato for Rajasthan

References

1. Jaganmohan M (2019) Statista. Production of potato in India 2014–2018, 23rd Sept 2019. https://www.statista.com/statistics/1038959/india-production-of-potato/
2. Zhang Y, Na S (2018) A novel agricultural commodity price forecasting model based on fuzzy information, granulation and MEA-SVM model. Math Probl Eng 2018:10. Article ID 2540681. https://doi.org/10.1155/201+*98/2540681
3. Li G, Xu S, Li Z (2010) Short-term price forecasting for agro-products using artificial neural networks. Agric Agric Sci Procedia 1:278–287. ISSN 2210-7843. https://doi.org/10.1016/j.aas pro.2010.09.035

4. Singh B, Rana R (2014). History of potato and its emerging problems in India. https://www. researchgate.net/publication/267654356_History_of_Potato_and_its_Emerging_Problems_ in_India

5. Chandran KP, Pandey NK (2007) Potato price forecasting using seasonal ARIMA approach. Potato J 34(1–2):137–138

6. Kumar S, Chaddha H, Kumar H, Singh RK (2019) Price forecasting system for crops at the time of sowing. Int Res J Eng Technol (IRJET) 6(4). https://www.irjet.net/archives/V6/i4/IRJET-V6I4384.pdf

7. Darekar A, Amarender Reddy A (2017) Price forecasting of maize in major states. Maize J Res Pap 6. https://www.researchgate.net/publication/325755970_Price_forecasting_of_maize_in_ major_states

8. Darekar A, Amarender Reddy A (2017) Cotton price forecasting in major producing states. Econ Affairs 62(1–6):10. 5958/0976-4666, 00047. https://www.researchgate.net/publication/ 320482197_Cotton_Price_Forecasting_in_Major_Producing_States

9. Peixeiro M (2019) Medium towards data science. The complete guide to time series analysis and forecasting, 7th Aug 2019. https://towardsdatascience.com/the-complete-guide-to-time-series-analysis-and-forecasting-70d476bfe775

10. Mahapatra SK, Dash A (2019) Forecasting area and production of green gram in Odisha using ARIMA model. Int J Chem Stud. IJCS 7(3):3899-3904. P-ISSN: 2349–8528, E-ISSN: 2321-4902. URL: http://www.chemijournal.com/archives/2019/vol7issue3/PartBL/7-2-148-779.pdf

11. AgMarknet, Directorate of Marketing & Inspection (DMI), Ministry of Agriculture and Farmers Welfare, Government of India. https://agmarknet.gov.in/

12. Darekar A, Reddy AA (2017) Price forecasting of maize in major states. Maize J Res Pap 6. https://www.researchgate.net/publication/325755970_Price_forecasting_of_maize_in_ major_states

13. Toppr. Components of time series. https://www.toppr.com/guides/business-mathematics-and-statistics/time-series-analysis/components-of-time-series/

14. Sayef (2019) Sayef's tech blog. Time series analysis in python—introduction to ARIMA, 12th June 2019. http://sayef.tech/post/time-series-analysis-in-python-inroduction-to-arima/

15. Chatterjee A (2018) Kaggle. Time series for beginners with ARIMA, 10th July 2018. https://www.kaggle.com/freespirit08/time-series-for-beginners-with-arima

16. Prabhakaran S. Machine learning plus. Augmented dickey fuller test (ADF test)—must read guide. https://www.machinelearningplus.com/time-series/augmented-dickey-fuller-test/

17. Unit root and ARIMA models. http://www.fsb.miamioh.edu/lij14/672_2014_s6.pdf

18. Statistics Solutions. Time series analysis. https://www.statisticssolutions.com/time-series-ana lysis/

19. Brownlee J (2017) Machine learning mastery. How to create an ARIMA model for time series forecasting in python, 9th Jan 2017. https://machinelearningmastery.com/arima-for-time-ser ies-forecasting-with-python/

20. Ippolito PP (2019) Medium towards data science. Stock market analysis using ARIMA, 21st May 2019. https://towardsdatascience.com/stock-market-analysis-using-arima-8731ded2447a

21. Salvi J (2019) Medium towards data science. Significance of ACF and PACF plots in time series analysis, 27th Mar 2019. https://towardsdatascience.com/significance-of-acf-and-pacf-plots-in-time-series-analysis-2fa11a5d10a8

22. Hyndman RJ, Athanasopoulos G. Otexts. Forecasting: principle and practice. https://otexts. com/fpp2/arima-estimation.html

23. Darekar A, Reddy AA (2017) Cotton price forecasting in major producing states. Econ Affairs 62:1–6. https://doi.org/10.5958/0976-4666.2017.00047.X. https://www.researchgate. net/publication/320482197_Cotton_Price_Forecasting_in_Major_Producing_States

24. Reddit. If my AIC and BIC are negative, does that mean that more negative values indicate a better fit or the number closer to 0?. https://www.reddit.com/r/AskStatistics/comments/5yd t2c/if_my_aic_and_bic_are_negative_does_that_mean/

25. Aashiqur Reza DSA, Debnath T (2020) Towards data science. An approach to make comparison of ARIMA and NNAR models for forecasting price of commodities, 26th Jan 2020. https://towardsdatascience.com/an-approach-to-make-comparison-of-arima-and-nnar-models-for-forecasting-price-of-commodities-f80491aeb400

26. Zhang D, Chen S, Liwen L, Xia Q (2020) Forecasting agricultural commodity prices using model selection framework with time series features and forecast horizons. IEEE Access 8:28197–28209. https://doi.org/10.1109/ACCESS.2020.2971591. https://ieeexplore.ieee.org/abstract/document/8981960

Chapter 41
Factors Influencing Employees Turnover and Measuring Its Impact in Pharmaceutical Industry: An Analytical Analysis with SPSS Method

Geeta Kumari and Krishna Murari Pandey

1 Introduction

Turnover can be described as the act of changing or replacing old employees with new ones. Separations between the company or any organizations and employees may include dismissal, retirement, death, inter-agency transfers, and resignations. The turnover of an organization is measured in turnover rate which is represented in percentages. Turnover rate is the percentage of persons leaving the organization in a certain fiscal year to the average total persons working in the organization for that same fiscal year.

On the off chance that a business is esteemed to have a high turnover rate contrasted with its rivals, this implies the employees of this organization have a normal span of work lower than that of different organizations in the same area. High turnover may influence the profitability of an organization if talented labourers leave regularly and the working populace contains a high level of beginners who are completely new to the job and inexperienced. Organizations regularly follow the inward developments of offices, divisions, or other segment gatherings, for example the turnover pace of ladies comparative with men. Most organizations permit administrators to terminate employees whenever, out of the blue, or for reasons unknown by any means, regardless of whether the employee is on favourable terms. Likewise, organizations all the more precisely track the wilful turnover rate by introducing studies to employees who find employment elsewhere, hence distinguishing the particular explanations

G. Kumari
Department of Management, Eternal University, Rajgarh, Himachal Pradesh 173101, India

K. M. Pandey (✉)
Department of Mechanical Engineering, National Institute of Technology, Silchar, Assam 788010, India

© The Author(s), under exclusive license to Springer Nature Singapore Pte Ltd. 2021   519
S. Kumar et al. (eds.), *Proceedings of International Conference on Communication and Computational Technologies*, Algorithms for Intelligent Systems,
https://doi.org/10.1007/978-981-16-3246-4_41

behind their choice to leave. Numerous industries have found that turnover is incredibly decreased when issues influencing employees are settled quickly and expertly. Organizations are attempting to diminish turnover rates by offering advantages, for example paid leave if u got sick, even if u want to take some time off then also u get paid and adaptable working hours.

A significant worry for most organizations is that staff turnover turns out to be a costly affair, particularly in some of the lower-paying employments, where we observe that the turnover rate is the most noteworthy or highest in many cases. Numerous components or factors are responsible for determining the rate of turnover of any organization and it emerges out of both the organization that gives employment and from the employees. Pay rates, business benefits, employee participation, and occupation execution are altogether factors that assume a critical responsibility in staff turnover. With regard to HR, turnover can be called the rate at which a business procures or do recruitment and loses or fires employees. Straightforward approaches to describe it are "how employees will, in general, remain associated with the company" or "the pace of traffic entering through the spinning entrance door". Turnover is estimated for singular organizations and their industry in general. On the off chance that a business is regarded to have a high turnover rate contrasted with its rivals, this implies that the employees of this organization have a normal length of work and is lower than that of different organizations in a similar division. High turnover can influence a company's efficiency if experienced specialists leave frequently and the labourer populace incorporates a high level of amateur specialists.

2 Literature Review

Le et al. [1] in their review investigated the influence of Western work-life balance on the work-life balance of employees in Asian corporate jobs and industries. They concluded that there are various factors such as culture, economy, and many institutional factors of Western countries that affected the perception of Asian employees and their views on work-life balance. Another important factor that plays a huge role in reducing employee's job stress and better job satisfaction and more commitment towards the organization is the vision of the leader. It has been found that job stress increases with hierarchical levels of different leaders each having a different vision as compared to a single leader and one vision of the organization [2]. Pfister et al. [3] investigated the role of appreciation on employee's behaviour and their satisfaction towards the job over time. Appreciation of employees can help them validate their judgments about their performance and gives them a feeling of success, which leads to better job satisfaction and they don't feel any kind of resentment towards the organization. The effects of job insecurity on employees' job performance over some time were more dynamically studied by Debus et al. [4]. They suggested that cognitive appraisal theory should be implemented within the organization to provide a sense of job security to the employees; the organization should look into various factors affecting the employee's performance such as job stress, any personal events that may be bothering the person. By providing them enough time and believing in their potential can help in increasing job security that ultimately leads to better job performance.

Parker et al. [5] studied the performance-based pay model also known as the stress appraisal model. Performance-based pay leads to more strain among employees in the organization and it also increases job stress and job insecurity. The performance-based pay model leads to controlled motivations for the employees and shows fatigue and hindrance in employees of the organization whereas autonomous forms of motivation are more supported by the employees of the organization. Bullying is such a negative term, but what if bullying leads to produce positive results in favour of the victim. Majeed and Naseer [6] hypothesize eustress mechanism based on cognitive appraisal theory that leads to improvement in job performances, success in career, creativity in the work field, etc. for the people having high psychological values and strong individuals who were once bullied. Yao et al. [7] investigated the impact of the narcissistic nature of leadership on employee behaviour, and how the negative behaviour of the leader of the organization leads to job stress among the employees of that organization. Also, the narcissistic behaviour of a leader leads to dissatisfaction among employees and loss of trust towards the organization. Although religion is central to the lives of many people, little attempt has been made to understand its role within work-life balance. Sav [8] in his study researched the encounters of Australian Muslim men and the religion of Islam, which promote a different role for its supporters and includes very much characterized actual demonstrations of worship. The discoveries uphold the recommendation that being religious can be gainful for work-life balance, as opposed to rivaling work and other non-work functions for time and energy. An illustrative work had been performed by Kumari et al. [9] to determine the job satisfaction level of software professionals. Three private software industries, namely HCL Technologies Ltd., IBM India Pvt. Ltd., and Wipro Ltd. had been selected. The respondents of HCL Technologies Limited are found with a higher mean value of 3.44 for working hours as compared to IBM India Pvt. Ltd. and Wipro Ltd. So the respondents of Wipro Limited were found more satisfied in terms of working hours than the other two companies. Secondly, employees of IBM India Pvt. Ltd. got more authority and responsibility than the other two companies. HCL Technologies Ltd. and Wipro Ltd.'s employee's same satisfaction level in terms of appreciation and rewards. The respondents of IBM India Pvt. Ltd., had a mean value of 3.27 whereas for HCL Technologies Limited mean value was 3.3 and Wipro Limited mean value was 3.26. Thus employees of IBM India Pvt. Ltd. enjoy favourable job conditions than other companies. Kumari et al. [10] worked on the occupational health and safety of workers in pharmaceutical industries, Himachal Pradesh, India. The results show that more workers have accepted that workers have been provided with appropriate procedures and instructions before completing the task. Therefore, it can be concluded that management takes it seriously that workers understand the exact course of action before carrying out the task so that it is safer for workers to carry out the operations.

Kumari et al. [11] investigated the job satisfaction level of software employees at IBM India Pvt. Ltd. A well-designed pretested questionnaire was made to study job satisfaction. The sample consisted of 160 professional respondents. Out of these 160 employees, 100 employees were male and 60 were female. Different job dimensions such as working hours, authority and responsibility, welfare facilities appreciation and rewards, career prospects, physical working environment, communication, co-workers, benefits, recognition, job condition, and job security were assessed. It has been observed that the majority of the professionals were satisfied with the arrangement of facilities by the company. A similar investigation had been carried out by Kumari et al. [12] for the software organization. The hypothesis testing was carried out with the linear regression analysis. A positive relationship between career growth opportunities and job satisfaction of employees for HCL Technologies Ltd. was framed. The study showed satisfaction among the professionals in terms of career opportunities offered at their company. Job-related stress results in decreased quality, the quantity of practice, increased costs of health care and decreased job satisfaction leading to organizational inefficiency. Kumari et al. [13] focused on the matter and performed a study. This job stress, its correlations with the various variables were defined and presented in the study. Work stagnation, supervision, administration, ambiguity, promotion policy, role conflict, and income policy were the variables considered. The study indicated that the company maintained a good work culture for its employees. However, the authors found the organizational target as the accountable one which makes the employees stressed in some moments. Significant differences were observed between males and females in terms of work roles such as supervision, administration, and promotion policy.

As the employees of the software industry face a higher level of stress than employees in other private companies, Kumari et al. [14] studied employees of the software industry. Since the software sector is the fastest growing industry, employees are allotted unachievable targets which introduce an increased level of stress. In addition to stress, relaxation techniques practised in the organization, relationship between self-esteem and stress were also parts of the study. 100 randomly chosen employees of HCL Company were considered in the study. The results showed that 98% felt stressed physically and emotionally. 18% of employees said they worked overtime comfortably. The social injustice, organizational culture, and fear of loss of a job are among the reasons behind the stress formed. Kumari et al. [15] surveyed public and private sector firms to examine the taxonomy of job satisfaction. The important factors were incentives and hours of work, management and colleagues, promotion and job security, the difficulty of job like forceful overtime, job content, and prestige. The surveys had been conducted on 500 employees hailing from different sectors in India. Region, sector, and gender-wise analysis highlighted that most of the employees in the Indian industry were not satisfied. The commerce sector and education sector were found comparatively satisfactory for males and females, respectively. Total job satisfaction in the manufacturing sector was found very low. In order of job satisfaction levels, North India has been best followed by South region.

Kumari et al. [16] did a review analysis on job satisfaction and job stress in software companies in India. In this review, most of the articles were based on analysis of the SPSS method and software supported statistical tool. The relationship between job stress and job satisfaction of the software professionals was explored by several researchers. The study is of significance for better understanding about the issues faced by software professionals and helping the companies to improve their performances. The dimensions which can be further explored by future researchers have also been discussed in the study. To determine the factors affecting job satisfaction and know the level of satisfaction, Kumari et al. [17] performed a study on software professionals of IBM India Pvt. Ltd. Gurgaon. It was observed that organizational culture, pay compensation, career growth, social security, and working environment had been important factors that influenced job satisfaction. The questionnaire used in this study was based on a Likert 5-point scale. Descriptive statistics were utilized including mean and standard deviation. The overall job satisfaction of software professionals at IBM India Pvt. Ltd, Gurgaon, was found at the positive level. The finding revealed the relationship among the pay and compensation, career growth, Social Security and working environment, and job satisfaction of software professionals. Kumari et al. [18] investigated the level of job satisfaction variation with the change in role in a software organization. They analysed the staff or officers who give more time to the organization. The SPSS version 20 was used for data analysis. The independent-sample t-test has been used in the study to analyse the job satisfaction level among staff and officers. The findings were supported with the two hypotheses; the officer professionals were more satisfied than the staff in terms of job satisfaction. The appreciation and rewards, communication, co-workers, pay promotion, and supervision are very effective factors. The analysis shows that officers devote more time to the development of the company. Kumari et al. [19] framed the relationship between job stress and job satisfaction among software professionals since job satisfaction is treated as the dependent variable and job stress is treated as an independent variable. The research has been conducted among software professionals in Wipro Ltd. Greater Noida. A simple convenience sampling technique and questionnaire method had been applied for the data collection tool. Descriptive statistics, correlation, and regression analysis are used to analyse data. The Pearson correlation result indicates no significant relationship between job stress and job satisfaction of the company. It was concluded that organizational culture has been the key factor of the employees' job satisfaction.

Long et al. [20] detailed that if a case arises where employee turnover occurs that's not been started by one of the employees then it could be as death or called as dismissal. Long et al. [21] upheld that an employee whose exhibition was beneath normal might be constrained out of an association with decisions like terminating or with boring tedious work hours or tasks, yet if the employee starts the terminations by himself, they were recorded as deliberate. Long et al. [22] expressed that any job or activity is phrased as fulfilling depending upon whether employees are placated, cheerful, and whether their wants and needs are satisfied. Occupation fulfilment was one of the main factors in employees' inspiration, positivity among various employees' spirit in the workplace, and employee objective accomplishment. Solomon et al. [23] featured that if any organization tried to control employee turnover then it could establish an unpredictable and testing task for both the working environment and administrators. Thwala et al. [24] detailed that the activity execution of employees was a significant and key factor in the company SMCFs based in the country Nigeria and this is what alludes to whether a worker carries out his or her responsibility in a proper manner or not. Any work execution comprises of practices that any number of employees accomplish in their work that are applicable towards the fulfilment of the objectives of the company. Kokt and Ramarumo [25] expressed that a difficult work environment could cause expanded degrees of burnout and stress among workers, which could make them withdraw from the company or the industry.

2.1 Research Objectives

- To identify the determinants of employees on the rate of employee's turnover
- To examine the relationship between employees-related factors on employee turnover.

2.2 Significance of the Research

This examination tosses light through important proposals to diminish the weakening level in the company. This investigation can assist the administration with finding the more fragile pieces of the employee, feels towards the association, and likewise helps in changing over those more vulnerable parts into more grounded ones by giving the ideal proposals or arrangements. This investigation has a more extensive range in any sort of company since "weakening" is an overall effect on any company and makes the employees set forth their reasonable troubles and their need factors in the association. This examination can assist the administration with knowing the problems and explanation as to why employees will, in general, change their activities in the job through disappointment factors in the association and likewise assists with recouping by giving the ideal proposals or arrangements.

3 Research Methodology

The research study was based on a descriptive research design that was conducted at Meridian Medicare Limited located in Sloan town and its vicinity. This research study was based on a survey with the help of a structured questionnaire. The questionnaire was framed based on Likert five scales, i.e. strongly disagree, disagree, neutral, agree, and strongly disagree. The sample population was taken 100. The information gathered from various sources was characterized and organized as per the necessity of the investigation. The investigation of the current examination has been done through suitable factual and numerical tools including the percentage investigation, factor investigation, and relapse or regression investigation relying on the targets of

Table 1   KMO and Bartlett's test

KMO and Bartlett's Test		
Kaiser–Meyer–Olkin measure of sampling adequacy		0.718
Bartlett's test of sphericity	Approx. Chi square	399.518
	Degree of freedom	120
	Significance level	0.000

Source The following table shows the output of SPSS 20

the investigation. The data of the present study has been processed using the latest statistical software, i.e. SPSS version 20.

4   Results and Discussion

Factor analysis is a statistical method to describe the variability between observed correlated variables, taking into account a potentially small number of unobserved variables called factors. In the factors analysis, there were following steps which are involved.

4.1   Kaiser–Meyer–Olkin (KMO) and Bartlett's Test

Kaiser–Meyer–Olkin test is used to measure strength relationships among various available variables that have been undertaken. It also determines the sampling competence that has to be greater than 0.5 for satisfactory factor analysis. If any pair of variables has a value less than this, it is considered to drop one of them from the analysis (Table 1).

The KMO test measures the sampling adequacy which falls in the acceptable range with a value of 0.718 Barlett's test of sphericity is significant, thus the hypothesis that the inter-correlation matrix involving these 16 variables is an identity matrix is rejected. Therefore, from the viewpoint of Bartlett's test, factor analysis is feasible.

4.2   Communalities

It shows the amount of the difference in the factors that have been represented by the extracted factor. Extraction Method: Principal Component Analysis (Table 2).

Table 2 Communalities show, 44.9% of the variance in the variables has been accounted for by the extracted factor

	Variables of employee turnover	Initial	Extract Ion
1	Are you satisfied with the training provided for your job to enhance your skills?	1.000	0.808
2	Is your company concerned with the long-term welfare of employees?	1.000	0.685
3	Do you feel you can voice your opinion without fear?	1.000	0.769
4	Does your immediate superior deal with all employees fairly?	1.000	0.653
5	Do you receive timely and accurate communication from the company?	1.000	0.489
6	Does your company maintain the salary level that compares well to other companies in the area?	1.000	0.774
7	Does your organization properly implement the rule and policies laid by the state/central government towards employees?	1.000	0.482
8	Do you receive regular feedback regarding your performance?	1.000	0.717
9	Does your organization have adequate safety and health standards?	1.000	0.837
10	Does your organization has a well-managed and effective grievances system?	1.000	0.449
11	Does your company provide a Provident fund and gratuity fund to the employees?	1.000	0.761
12	Does your employer give you paid holidays, i.e. 26 January, 15 August, 2 October, etc.?	1.000	0.725
13	Does your organization provide job security to the employees?	1.000	0.759
14	Does your organization provide an appropriate leave system for employees?	1.000	0.652
15	Does your organization provide fringe benefits to employees?	1.000	0.805
16	Does your organization tell you to get your family involved in the work achievement reward functions?	1.000	0.700
	Extraction method: principal component analysis		

Source The above mentioned table shows the output of SPSS 20

4.3 Total Variance Explained

Total Variance: There are all the factors extractable from the analysis along with their eigenvalues, the per cent of variance attributable to each other (Table 3).

The first factor accounts for 33.649% of the variance, the second 10.424%, the third 9.524%, the fourth 8.614%, and the fifth 6.958%. All the *r* meaning factors are not significant.

Table 3 Total variance explained

Component	Initial eigenvalues			Extraction sums of squared loadings			Rotation sums of squared loadings		
	Total	% of variance	Cumulative %	Total	% of variance	Cumulative %	Total	% of variance	Cumulative %
1	5.384	33.649	33.649	5.384	33.649	33.649	2.594	16.214	16.214
2	1.668	10.424	44.073	1.668	10.424	44.073	2.324	14.527	30.742
3	1.524	9.524	53.597	1.524	9.524	53.597	2.137	13.359	44.101
4	1.378	8.614	62.211	1.378	8.614	62.211	2.033	12.705	56.805
5	1.113	12.363	69.169	1.113	12.363	69.169	1.928	12.363	69.169
6	0.932	5.825	74.993						
7	0.796	4.977	79.971						
8	0.583	3.643	83.614						
9	0.545	3.404	87.018						
10	0.488	3.052	90.070						
11	0.402	2.515	92.585						
12	0.375	2.343	94.928						
13	0.281	1.759	96.686						
14	0.230	1.438	98.125						
15	0.172	1.074	99.198						
16	0.128	0.802	100.000						

Extraction method: principal component analysis

Source The following table shows the output of SPSS 20

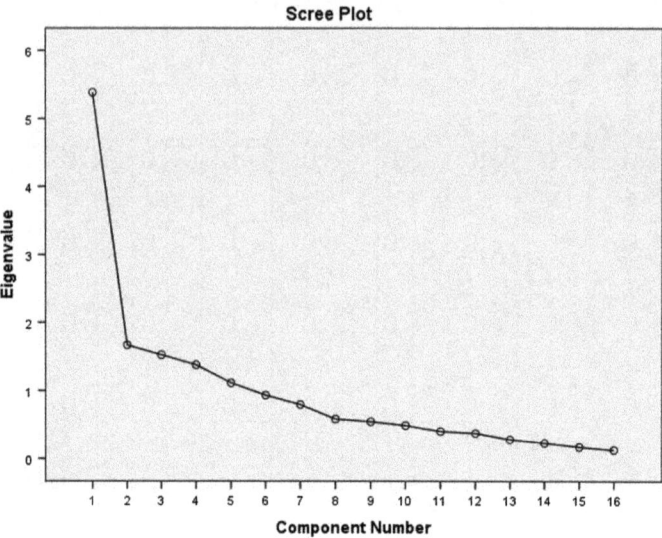

Fig. 1 Scree plot

4.4 Scree Plot

The scree plot is a graphic aid of the eigenvalues against all the factors. It is proposed by Cattell. The graph is used to determine how many factors to retain. It is intended to help in deciding where the "trivial" dimension is begun. In the chart, the curve begins to flatten between 1 and 5. Five factors have been retained (Fig. 1).

4.5 Component Matrix

The loadings of sixteen different variables, which are incorporated with five factors, are shown. The factor contributes more to the variable when the total estimation of the stacking factor is higher in Table 5, there are initial five are the outcomes for the five factors that are separated. The 6th section gives synopsis measurements enumerating how well every variable is clarified by the five segments. The first line of numbers at the base of every segment is the eigenvalues and demonstrates the general significance of each factor in representing the change related to the arrangement of the variable being investigated. The aggregates of squares are for the five components are 5.384, 1.668, 1.524, and 1.378 and 1.113 individually. Factor 1 is representing the most difference and factor five (Table 4).

Table 4 Component matrix

S. No.	Variables of employees turn over	Factors				
		1	2	3	4	5
1	Are you satisfied with the training provided for your job to enhance your skills?	0.685	−0.159	−0.523	−0.175	−0.094
2	Is your company concerned with the long-term welfare of employees?	0.746	−0.086	−0.273	−0.216	0.026
3	Do you feel you can voice your opinion without fear?	0568	−0.534	−0.095	−0.337	0.197
4	Does your immediate superior deal with all employees fairly?	0.495	−0.545	−0.207	−0.026	0.260
5	Do you receive timely and accurate communication from the company?	0.588	−0.028	0.003	0.368	0.081
6	Does your company maintain the salary level that compares well to other companies in the area?	0.581	−0.086	−0.129	0.090	−0.636
7	Does your organization properly implement the rule and policies laid by the state/central government towards employees?	0.462	0.432	−0.008	0.264	−0.111
8	Do you receive regular feedback regarding your performance?	0.676	0.364	0.078	−0.076	−0.341
9	Does your organization have adequate safety and health standards?	0.517	−0.212	0.490	−0.324	−0.424
10	Does your organization have a well-managed and effective grievances system?	0.613	0.149	0.034	−0.220	0.041
11	Does your company provide a Provident fund and gratuity fund to the employees?	0.656	0.084	−0.325	0.459	0.080

(continued)

Table 4 (continued)

S. No.	Variables of employees turn over	Factors				
		1	2	3	4	5
12	Does your employer give you paid holidays, i.e. 26 January, 15 August, 2 October, etc.?	0.419	−0.260	0.084	0.685	0.080
13	Does your organization provide job security to the employees?	0.533	−0.072	0.649	0.196	0.103
14	Does your organization provide an appropriate leave system for employees?	0.654	0.358	0.161	−0.084	0.252
15	Does your organization provide fringe benefits to employees?	0.643	0.119	0.434	−0.216	0.377
16	Does your organization tell you to get your family involved in the work achievement reward functions?	0.250	0.661	−0.333	−0.205	0.218

Extraction method: principal component analysis

[a]5 components extracted
[a] significance value < 0.05 is the conventional threshold for declaring statistical significance
Source The following table shows the output of SPSS 20

4.6   Rotated Component (Factor) Matrix

The VARIMAX pivoted segment investigation factor framework is given in Table 3. The aggregate sum of variance removed is equivalent in the pivoted arrangement as it was in the unrotated one, i.e. 69.169%. Two contrasts are evident in the accompanying table. To start with, the difference has been reallocated so that the factors remain unique. Particularly, in the VARIMAX pivoted factor arrangement, the main factor represents 16.214 of the fluctuation invariance, contrasted with the 33.649% in the unrotated arrangement. As the very second factor represents 14.527% versus 10.424%, the Third factor represents 13.359% versus 9.524%, the Fourth factor represents 12.705% versus 8.614% and Fifth-factor represents 12.363% versus 12.363% in the unrotated arrangement.

In the rotated factor solution, variables 1, 2, 3 and 4 loads significantly on factor 1; variables 9, 13, 14 and 15 loads significantly on factor 2; variables 5, 11 and 12 load significantly factor on 3; variables 14 and 16 loads significantly on factor 4 and variable 6, 8 and 9 loads significantly load on factor 5. Factor 1 has four significant loadings, factor 2 has four significant loadings, factor 3 has three significant loadings,

Table 5 Rotated component (factor) matrix

	Rotated component variables	Component				
		1	2	3	4	5
1	Are you satisfied with the training provided for your job to enhance your skills?	0.717	−0.112	0.190	0.321	0.377
2	Is your company concerned with the long-term welfare of employees?	0.644	0.167	0.166	0.337	0.318
3	Do you feel you can voice your opinion without fear?	0.826	0.269	0.007	−0.090	0.080
4	Does your immediate superior deal with all employees fairly?	0.743	0.108	0.269	−0.123	−0.047
5	Do you receive timely and accurate communication from the company?	0.209	0.234	0.593	0.126	0.151
6	Does your company maintain a salary level that compares well to other companies in the area?	0.218	−0.047	0.282	0.015	0.803
7	Does your organization properly implement the rule and policies laid by the state/central government towards employees?	−0.132	0.160	0.401	0.432	0.303
8	Do you receive regular feedback regarding your performance?	0.051	0.315	0.164	0.426	0.638
9	Does your organization have adequate safety and health standards?	0.201	0.552	−0.141	−0.224	0.649

(continued)

Table 5 (continued)

	Rotated component variables	Component				
		1	2	3	4	5
10	Does your organization has a well-managed and effective grievances system?	0.311	0.375	0.065	0.365	0.273
11	Does your company provide a Provident fund and gratuity fund to the employees?	0.285	−0.008	0.728	0.350	0.167
12	Does your employer give you paid holidays, i.e. 26 January, 15 August, 2 October, etc.?	0.105	0.149	0.805	−0.209	0.032
13	Does your organization provide job security to the employees?	0.000	0.766	0.361	−0.145	0.149
14	Does your organization provide an appropriate leave system for employees?	0.159	0.552	0.203	0.519	0.108
15	Does your organization provide fringe benefits to employees?	0.257	0.812	0.082	0.271	0.004
16	Does your organization tell you to get your family involved in the work achievement reward functions?	0.014	−0.004	−0.054	0.834	−0.032

Extraction method: principal component
Rotation method: varimax with kaiser normalization

[a]Rotation converged in 7 iterations
[a] significance value < 0.05 is the conventional threshold for declaring statistical significance
Source The following table shows the output of SPSS 20

Table 6 Factor loading of employee turnover

S. No.	Extracted factor	Factor loading component	Value
1	Motivational benefits	Training session	0.717
		Long-term welfare	0.644
		Opinion	0.826
		Fair dealing	0.743
2	Compensation	Safety and health	0.552
		Job security	0.766
		Leave system	0.552
		Fringe benefits	0.812
3	Employee welfare	Communication	0.593
		Provident fund and gratuity fund	0.728
		Paid holidays	0.805
4	Appreciation and rewards	Leave system	0.519
		Family member involvement and reward system	0.834
5	Safety and security	Salary level	0.803
		Feedback	0.638
		Safety and health	0.649

Source The following table shows the output of SPSS 20

factor 4 has two significant loadings and factor 5 has three significant loadings. Finally, from the analysis, five summarized factors are obtained (Table 6).

Finally, from the analysis, five summed-up factors are gotten, the information through Component Factor Analysis and utilizing the VARIMAX technique as five variables including their subprocess. Linear regression and correlation applied for analysing the data and results found that motivational benefits, compensation, employee welfare policies, appreciation and reward, and safety and security are significantly related to employee turnover. The results further indicate that all determinants have a significant and strong impact on employee turnover.

4.7 Regression Analysis for Measuring the Impact of Employee Turnover

Linear regression applied for analysing the data and results found that compensation, working environment, and safety and security are significantly related to employee turnover. The results further indicated that all determinants have a significant and strong impact on employee turnover.

Table 7 Multiple linear regression model summary

Model summary					
Model	R	R^2	Adjusted R^2	Std. The error of the estimate	Durbin-Watson
1	0.915[a]	0.837	0.832[b]	0.544	0.455

[a]Predictors: (constant), compensation, working environment, and safety and security
[b]Dependent variable: employee turnover
Source The following table shows the output of SPSS 20

4.8 The Multiple Linear Regression Model Summary

Table 7 gives a model summary that is based on multiple linear regression model summary and overall fit statistics. The estimated R-value speaks to the straightforward correlation and is 0.915 which clearly shows a great correlation relationship between them. The calculated R^2 value demonstrates the amount of the absolute variation in the workers' turnover which is clarified by the free or independent factor to be specific wellbeing and security, gratefulness and prize, pay, and persuasive advantages. For this situation, 83.7% can be clarified, which is huge. The value of Durbin-Waston (d) is 0.455. Therefore, there is no autocorrelation detected in the multiple linear regression data.

4.9 The Multiple Regression ANOVA

The ANOVA Table 8, which reported that the regression equation fitness. Here the data predicts dependent variable was shown in Table 8.

Table 8 shows that the regression model predicts that the dependent variable essentially well. This demonstrates the measurable importance of the regression model that was used. Here, $p < 0.0005$, which is under 0.05, and demonstrates that, by and large, the regression model factually fundamentally predicts the output variable.

Table 8 Regression ANOVA

ANOVA						
Model		Sum of squares	df	Mean square	F	Sig.
1	Regression	145.583	3	48.528	163.942	0.000[b]
	Residual	28.417	96	0.296		
	Total	174.000	99			

[a]Dependent variable: employee turnover
[a] significance value < 0.05 is the conventional threshold for declaring statistical significance
[b]Predictors: (constant), compensation, working environment, safety, and security
Source The following table shows the output of SPSS 20

Accordingly, it is a solid match for the information. The linear regression F-test has the invalid theory that the model clarifies zero difference in the reliant variable, for example R is equivalent to zero ($R = 0$). It is affirmed that the F-test is profoundly critical and the model clarifies a lot of the difference in employee turnover.

4.9.1 The Multiple Regression Coefficients

The Coefficients 11 table provides the necessary information to predict employee turnover from notifying coefficients working environment, safety and security and compensation, as well as determine whether independents variables contribute statistically significantly to the model. The multiple linear regression estimates including the intercept and the significance levels. It was found a non-significant intercept but highly significant with the working environment, safety and security, and compensation. The all variance inflation factor VIF has been found less than 10. In the table, the VIF value of the working environment is observed at 1.897, VIF of safety and security 4.777, and VIF of compensation 4.668. From the VIF value, it is observed that there is not a multi-co-linearity relationship between independent variables.

Model		Unstandardized coefficients		Standardized coefficients			Collinearity statistics	
		B	Std. error	Beta	T	Sig.	Tolerance	
1	(Constant)	−0.421	0.200		−2.103	0.038		
	Working environment	0.204	0.061	0.190	3.337	0.001	0.527	1.897
	Safety and security	0.487	0.102	0.432	4.788	0.000	0.209	4.777
	Compensation	0.481	0.117	0.366	4.108	0.000	0.214	4.668

[a]Dependent variable: employee turnover
Source The following table shows the output of SPSS 20

The graph shows the regularity of residuals with an ordinary P-P plot. The plot gives us an idea that there are no solid deviations between the points and all the points follow the normal line with little to no deviation. This demonstrates the residuals are ordinarily circulated. The dependent variable was derived in Eq. 1

$$Y = a + bX_1 + bX_2 + bX_3 + bX_4 + \varepsilon = r^2 \tag{1}$$

where,
$Y =$ is the dependent variable.
$X =$ is the independent (explanatory) variable.
$a =$ is the intercept.
$b =$ is the slope.

Fig. 2 Normal P-P Plot for regression standardized result

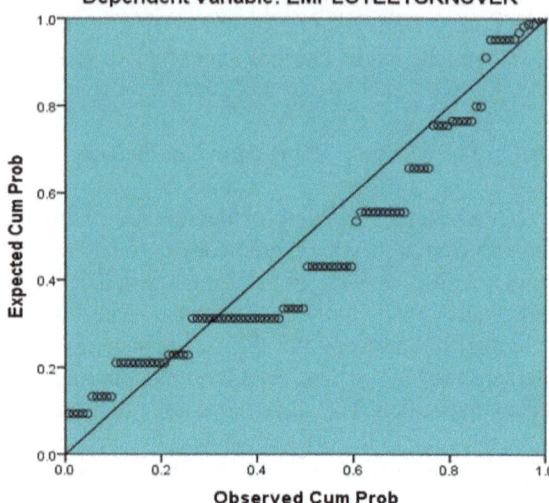

This indicates that there is a strong relationship between, independent variables namely working environment, safety, security, and compensation. A coefficient of 0.421 indicates that the application of a good working environment can reduce employee turnover by a good extent and as much as 0.190. A variable coefficient of 0.204 indicates that a given good working environment at the workplace can help in reducing employee turnover. A coefficient of 0.487 indicates that conductive safety and increasing job security help to reduce an organization's employee turnover by as much as 0.432. A variable coefficient of 0.481 indicates that a good compensative can helps to reduce employee turnover by 0.366.

The regression equation is used to show the relationship between the dependent variables and employee turnover, Fig. 2. Normal P-P Plot for Regression Standardized Result.

$\varepsilon =$ is the residual error.

The values for b_0, b_1 and b_2 are -0.421, 0.204, 0.487 and 0.48 therefore the equation

$$Y = -0.421 + 0.204 + 0.487 + 0.48$$

5 Limitations

- The time duration allocated for data collection was not sufficient.

- All the findings and observations of the study were purely based on the respondent's answers which may be biased.
- The study was restricted to the city of Solan, Himachal Pradesh.
- Sometimes respondents did not respond well to all the questions in the question-naire.

6 Suggestions

The investigation gives significant proposals to diminish the wearing down level in the company. This investigation can assist the administration with finding the more fragile pieces of the employee, feels towards the association, and additionally helps in changing over those more vulnerable parts into more grounded and stronger parts by giving the ideal proposals or arrangements. This examination has a more extensive degree in any sort of association since "wearing down" is a common problem in any organization and it makes the employees set forth their practical troubles and needs in front of the association administrators. This investigation can assist the administration in knowing the various reasons for which their employees will in general change their activity through job disappointment factors in the association and additionally assists with recuperating by giving the ideal recommendations or arrangements.

7 Conclusions

In the light of the discoveries of the examination, it can be very well may be reasoned that employee inspirational advantages were a central point that enormously increases the chances of employees leaving the company thereby increasing the turnover rate of the company. From the investigation we have the main five factors namely are motivational advantages, compensation, employee's government assistance, grate-fulness and prizes, safety, and job security and these were very much influential for employee turnover in the association. The examination additionally discovered that one of the basic boundaries of gathering information is from the leaving employees, post-employment surveys were not consistently done whenever the employees decide to leave the association and this in turns hampered the company's HR management board to foresee or recognize the most important reasons as to why employees decided to leave the association. From the results, it was additionally concluded that if the company pays great compensation, safety, and job security, and likewise gives a favourable environment. At that point, the employee turnover will be very less. The multiple linear regressions also are found that 83.7% of the variation in the employee turnover was explained by selected explanatory variables. It shows that if these variables are taken into consideration by the company it may give the best results.

References

1. Le H, Newman A, Menzies J, Zheng C, Fermelis J (2020) Work-life balance in Asia: a systematic review. Hum Resour Manag Rev 30:100766. https://doi.org/10.1016/j.hrmr.2020.100766
2. Newton C, Bish A, Anibaldi R, Browning V, Thomas D (2018) Stress buffering effects of leader vision. Int J Hum Resour Manag 31:1–27. https://doi.org/10.1080/09585192.2018.1455215
3. Pfister I, Jacob Shaken N, Kälin W, Semmer N (2020) How does appreciation lead to higher job satisfaction? J Manag Psychol. Ahead-of-print. https://doi.org/10.1108/JMP-12-2018-0555
4. Debus M, Unger Dana & König CJ (2020) Insecurity and performance over time: the critical role of job insecurity duration. Career Dev Int 25:325–336. https://doi.org/10.1108/CDI-04-2018-0102
5. Parker SL, Bell K, Gagné M, Carey K, Hilpert T (2019) Collateral damage associated with performance-based pay: the role of stress appraisals. Eur J Work Org Psychol 28(5):691–707. https://doi.org/10.1080/1359432X.2019.1634549
6. Majeed M, Naseer S (2019) Is workplace bullying always perceived as harmful? The cognitive appraisal theory of stress perspective. Asia Pac J Hum Resour. https://doi.org/10.1111/1744-7941.12244
7. Yao Z, Zhang X, Liu Z, Zhang L, Luo J (2019) Narcissistic leadership and voice behavior: the role of job stress, traditionality, and trust in leaders. Chin Manag Stud 14(3):543–563. https://doi.org/10.1108/CMS-11-2018-0747
8. Sav A (2019) The role of religion in the work-life interface. Int J Hum Resour Manag 30(22):3223–3244. https://doi.org/10.1080/09585192.2016.1255905
9. Kumari G, Joshi G, Alam A (2019) A comparative study of job satisfaction level of software professionals: a case study of private sector in India. In: Ray K, Sharma T, Rawat S, Saini R, Bandyopadhyay A (eds) Soft computing: theories and applications. Advances in intelligent systems and computing, vol 742. Springer, Singapore. https://doi.org/10.1007/978-981-13-0589-4_55
10. Kumari G, Khanna S, Bhanawat H, Pandey KM (2019) Occupational health and safety of workers in pharmaceutical industries, Himachal Pradesh. India Int J Innov Technol Explor Eng 8(12):4166–4171. https://doi.org/10.35940/ijitee.L3659.1081219
11. Kumari G, Alam A, JoshiG (2017) Job satisfaction level of software professionals: a case study of IBM India private ltd. Gurgaon. Int J Sci Innov Eng Technol 3. ISBN 978-81-904760-9-6
12. Kumari G, Alam A, JoshiG (2017) The impact of career growth on job satisfaction of software professionals at HCL technologies limited Noida India. Int J Sci Innov Eng Technol 3. ISBN 978-81-904760-9-6
13. Kumari G, Joshi G, Pandey KM (2017) The impact of work role on job stress of software professionals at IBM India Pvt. Ltd., Gurgaon, India. In: 2nd international conference on sustainable computing technique in engineering, science and management (SCESM-2017), 27–28 Jan 2017, pp 109–115
14. Kumari G, Joshi G, Pandey KM (2014) Job stress in software companies: a case study of HCL Bangalore, India. Global J Comput Sci Technol USA 14(7):21–30. ISSN. No. 0975-4172
15. Kumari G, Joshi G, Pandey KM (2014) Analysis of factors affecting job satisfaction of the employees in public and private sector. Int J Trends Econ Manag Technol III(1), 11–19. ISSN. No. 23215518
16. Kumari G, Joshi G, Pandey KM (2015) Job satisfaction and job stress in software companies: a review. Adv Econ Bus Manag (AEBM) 2(7):756–760. Print ISSN: 239-1545, Online ISSN: 2394-1553, April–June 2015
17. Kumari G, Joshi G, Pandey KM (2015) Factors affecting job satisfaction of software professionals at IBM India Pvt., Ltd., Gurgaon, India. Adv Econ Bus Manag (AEBM) 2(12):1202–1207. ISSN: 2394-1545, e-ISSN: 2394-1553
18. Kumari G, Joshi G, Pandey KM (2015) Job satisfaction among software professionals in IBM India Pvt. Ltd., Gurgaon, India: a comparison between officers and staff. Adv Econ Bus Manag (AEBM) 2(7):751–755. ISSN: 2394-1545, Online ISSN: 2394-1553, April–June 2015

19. Kumari G, Joshi G, Pandey KM (2015) Relationship between job stress and job satisfaction in software industries: a case study of Wipro Ltd., Greater Noida India. In: 6th International conference on recent trends in applied physical, chemical sciences, mathematical/statistical, and environmental dynamics, Organized by KrishiSanskriti on 9th Aug 2015 at Jawaharlal Nehru University, pp 83–88. ISBN: 978-81-930585-8-9
20. Long CS, Ajagbe AM, Nor KM, Suleiman ES (2012) The approaches to increase employees' loyalty: A review of employees' turnover models. Aust J Basic Appl Sci 6(10):282–291
21. Long CS, Perumal P, Ajagbe AM (2012) The impact of human resource management practices on employees' turnover intention: a conceptual model. Interdiscip J Contemp Res Bus 4(2):629–641
22. Long CS, Musibau AA (2013) Can employee share option scheme improve firm's performance? A Malaysian case study. Inf Manag Bus Rev 5(3):119
23. Solomon O, Hashim NH, Mehdi ZB, Ajagbe AM (2012) Employee motivation and organizational performance in multinational companies: a study of Cadbury Nigeria Plc. IRACST-Int J Res Manag Technolo (IJRMT) 2(3):303–312
24. Thwala WD, Ajagbe AM, Enegbuma WI, Bilau AA (2012) Sudanese small and medium-sized construction firms: An empirical survey of job turnover. J Basic Appl Sci Res 2(8):7414–7420
25. Kokt D, Ramarumo R (2015) Impact of organizational culture on job stress and burnout in graded accommodation establishments in the Free State province, South Africa. Int J Contemp Hosp Manag 27(6):1198–1213

Chapter 42
A Triumvirate Approach of Blockchain MQTT and Edge Computing Toward Efficient and Secure IoT

Maha A. Abdullah and Omar H. Alhazmi

1 Introduction

Internet of Things (IoT) is one of the emerging technologies that have made a quantum leap in the technical field. However, there were some concerns that may affect the potential benefits of IoT; one of the main concerns is security. The limited resources and memory of IoT devices are what makes this issue a bit challenging. IoT is mostly formed from heterogenous devices with different capabilities, which also needed to be considered. IoT, just like any network, has many protocols that manage it, for instance, Hypertext Transfer Protocol Secure (HTTPS), Extensible Messaging and Presence Protocol (XMPP), Advanced Message Queue Protocol (AMQP), Constrained Application Protocol (CoAP), and Message Queuing Telemetry Transport (MQTT) [1]. However, each of these protocols has certain issues either in security or efficiency and has a domain that is used for. There was much research that has been done in the IoT security field, but a few only have achieved good performance with sufficient security solutions. Moreover, some proposed solutions have merged blockchain technology with the IoT to enhance security. Blockchain has a decentralized, trackable, and secure nature, which will fill the gap for IoT systems leakage. On the other hand, blockchain mostly has a complicated consensus that IoT simple devices may not handle. So, we will need to adapt the blockchain in a way that works in IoT systems. In the next section, we will discuss some of this literature in detail and their proposed solutions.

M. A. Abdullah (✉) · O. H. Alhazmi
Department of Computer Science, Taibah University, Medina, Saudi Arabia
e-mail: tu4160047@taibahu.edu.sa

O. H. Alhazmi
e-mail: Ohhazmi@taibahu.edu.sa

2 Related Work

2.1 Internet of Things (IoT)

Internet of Things (IoT) technology is becoming practical, economical, and more feasible than ever before. IoT is more integrated into devices, as manufacturers embed the technology into all sorts of products and machines. Therefore, billions of devices are connected to the Internet to make the environment of life smarter, connected, automated, profitable, and efficient. Hence, IoT will have variant applications in different fields. Moreover, it is not limited to the applications we see every day. IoT has made a revolution in health care, the economy, industry, and even farming. However, the benefits of IoT application in these fields have unignorable issues due to its reliability and single point of failure due to its centralized nature and security issues.

Most IoT devices are small-sized and have limited memory and limited computation power, which is not enough to handle extra tasks over its main functional tasks. Besides, compared to its benefit, most of the users ignore these issues and are not aware of how security leakage could lead to serious consequences. For example, with its variety of uses, IoT can be used to perform cyber-crime and aid in conventional crimes. Nevertheless, privacy could be a major demand in some sensitive fields like health care. Mirai botnet had exposed the leakage of security in IoT device in 2016 when it controlled many IoT devices, mostly cameras, and performed a DDoS attack. Many websites were affected, like Twitter, GitHub, and Netflix [2]. After the incident, many solutions were proposed to enhance the IoT systems' security; thus, more people realized the importance of device security. One of the recent solutions suggests integrating IoT and blockchain technology to solve the IoT nature problem; we go through some examples in the following section.

2.2 Blockchain

Blockchain concepts as defined in [3] are a secure, public, decentralized, and distributed list of records that are encrypted using hashing, which has put an end to the single point of failure or trusting third-party issues. Nevertheless, the decentralized authority made blockchain powerful; rather, the permanent records almost impossible to be changed or removed. Basically, blockchain is built on the hash function that links the nodes together by carrying the previous node information creating the chain with fixed-size one-way encryption [4]. Furthermore, there are a wide variety of applications of blockchain; according to Lao et al. [5], blockchain was used to deal with many IoT known issues where it can guarantee the level of security without a central server to manage the data transition in a way that goes with the IoT devices nature. Figure 1 illustrates the blockchain decentralized structure and what each node (block) contains.

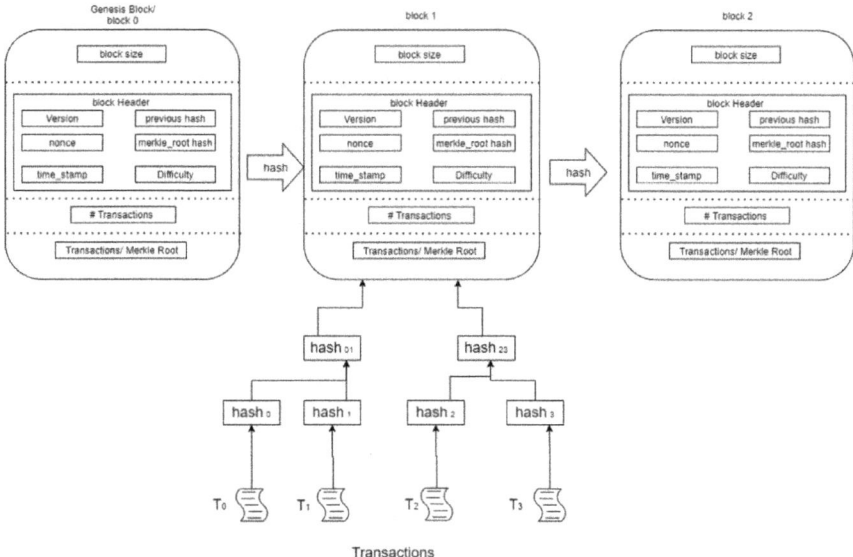

Fig. 1 Sample structure of blockchain [15]

A decentralized infrastructure with efficient data processing, high security, and operations accountability is the main demand for our case. The currently available solutions are based on centralized infrastructure with a high maintenance cost, low interoperability, and data security risks. Table 1 is illustrating the main methodology that was used in the existing solutions.

Sagirlar et al. [6] has proposed a novel hybrid blockchain architecture for IoT. As blockchain is one solution that provides a decentralized platform to IoT, this may become the first step we can take to optimize IoT systems. Moreover, they have used proof of work (PoW) and Byzantine fault tolerant (BFT) consensus protocols in their proposed hybrid-IoT platform. Bitcoin clients and bitcoin simulators are used to test hybrid-IoT architecture, design, and performance analysis.

Another blockchain-based architecture was proposed in [7] for scalable access management in IoT technology. He has proposed decentralized access control systems based on blockchain technology that eliminates the centralized access management systems problems and constraints. This system can be used in isolated administrative domains, provides access control rules available all the time, gives access to multiple managers simultaneously, and allows numerous devices connection through a constrained network. The simulation results showed that the proposed solution is best to access the management system in IoT.

While Dorri et al. [8] provides a case study on smart home tier design with the implementing blockchain for security and privacy of IoT devices. The proposed architecture uses symmetric encryption to achieve security, and the smart home tier can achieve confidentiality, integrity, and availability with its implementation.

Table 1 Previous studies and their contribution

Previous studies	Methodology	Protocols and algorithms	Performance	Notes
[6]	Securing the IoT network with blockchain and divide it to sub-blockchain	Proof of work (PoW) + Byzantine fault tolerance (BFT)	Valid using PoW sub-blockchain design under the sweet-spot 9 guidelines	PoW may not be the best option for IoT devices as it consumes energy to calculate it
[7]	Access management system to secure the IoT network using blockchain	Proof of concept (PoC) prototype for blockchain and CoAP between the IoT devices	Acceptable where it does not involve the IoT devices due to its limitation	The IoT devices are not part of the blockchain so they rely on their management hub; in case of malicious hub, the IoT network will be threaded
[9]	Software-defined networking (SDN) and Blockchain-based cloud architecture for IoT	Proof of service (FoS), matchmaking algorithm by a smart contract and a scheduling algorithm	Very good performance, reducing response time, and detecting real-time attacks	Bringing the computing resources to the edge has showed a big difference in the performance which is the methodology that we use too
[10]	MQTT model and blockchain	MQTT, proof of authority (PoA)	The distributed and interconnected network provides low latency, high throughput performance	There are no experimental results yet, but the proposed model looks desirable

Simulation results show that the proposed architecture provides a safe and secure blockchain environment for the smart home tier.

On the other hand, a software-defined networking (SDN) and blockchain-based cloud architecture for IoT was proposed by Sharma et al. [9]. The proposed distributed cloud architecture is flexible, efficient, scalable, and secure. They have used SDN, blockchain network, and fog computing in its framework. The evaluation results show that it improves performance, reduces induced delay, reduces response time, and provides high security with the ability to detect real-time attacks in IoT networks.

A decentralized infrastructure with efficient data processing, high security, and operations accountability is the main demand for our case. The currently available solutions are based on centralized infrastructure with a high maintenance cost, low interoperability, and data security risks.

2.3 MQTT Protocol and Blockchain

Katende [10] claims that IoT, just like the internet, is controlled by various protocols. Key examples of these protocols include SixFog, MQTT, Cloud of Things (CoT), and Low-power Wide-area Network Technology (LORA) [10]. With the Internet of Things revolving around mist, fog, and cloud computing, one of the most used protocols for the IoT is MQTT, which, once combined with blockchain technology, can be used in securing at the age where fog computing and mist computing operate.

Katende has described MQTT as a protocol that allows for the communication of servers, devices, and subscribers, which is generally what IoT is all about. Clients of the IoT are enabled to receive messages from the servers, which come from the devices. Over time, the use of MQTT has been restricted to low bandwidth, which is the operating bandwidth for the Internet of Things, and this is why it has been used in most of the IoT systems. Figure 2 depicts the structure of MQTT with TCP/IP network.

Fakhri and Mutijarsa [11] claimed that one problem with MQTT is security with the use of plaintext formats for the connect credentials. With security becoming an issue of concern, IoT devices should also focus on enhancing security. According to Li et al. [12], blockchain technology can be one of the measures that IoT systems can turn to for the enhancement of security. Combining the MQTT with blockchain technology would make the IoT systems free from attacks such as ransomware. Thus, MQTT problem will be solved using blockchain by ensuring that message brokers are avoided altogether, and a better system with better operation is enhanced [13].

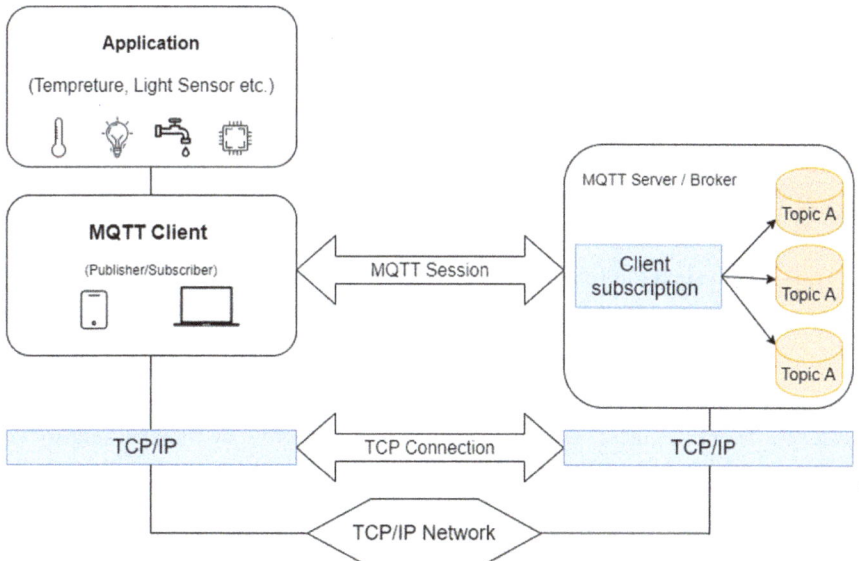

Fig. 2 MQTT architecture [15]

According to Pazhev et al. [14] with blockchain focusing on decentralization, the IoT systems can be decentralized with blockchain, and this will improve the security of these systems. Over time, the MQTT and the blockchain technology have been used independently for the IoT system [11]. However, this can be changed since MQTT has shown some security vulnerabilities but good operations and applicability in IoT, while blockchain has shown useful security features but not as useful in IoT applications as MQTT [15]. This leaves one with the need for balance between the two, MQTT and blockchain technology [16]. This is an area that has not been researched deeply by past researchers and calls for further focus. In this work, we propose an efficient solution that guarantees security by using blockchain and MQTT without affecting the performance by using edge computing for better response. The next section will explain how our system will work.

3 Proposed Solution

As mentioned, IoT is a rich field for research as it is still not totally full blown. Many studies have been done in IoT field to improve the security issue using different approaches. To reach a perfect solution, we need to balance the performance and strong security method because some IoT applications are time-sensitive and need to respond in real time. In our proposed solution, we will use the MQTT protocol and a central edge "fog" server, which will be used as a broker to manage the data flow. Moreover, the IoT network will communicate and publish the data using a secure connection guaranteed by blockchain, as shown in Fig. 3.

As indicated in the figure, the IoT system comprises full nodes and light nodes.

1. Full nodes: They have their own full multipurpose operating system, and with powerful computational and storage power, their power can be utilized to support the system.
2. Light nodes: These are nodes with single-purpose operating systems such as appliances and sensors; such nodes are powerful enough to manage the appliance itself and are not expected to play a role in managing systems.

So, in our system the full nodes here can mine the block and confirm the transactions and have a full list of all the blockchain. While light node will perform the transactions only and store the previous node's information to be connected to the blockchain. Moreover, blockchain ledger is a distributed storage used along with the fog device to store the newly generated data used by nodes to publish and subscribe. However, these valuable collected data will grow by time passing so we have connected our system to the cloud to store the old data that was collected and retrieve it whenever necessary.

In IoT, scalability is a vital requirement; thus, adding new devices dynamically is important for such applications. Blockchain can provide a decentralized building expandable IoT system and add new devices to the environment without burdening

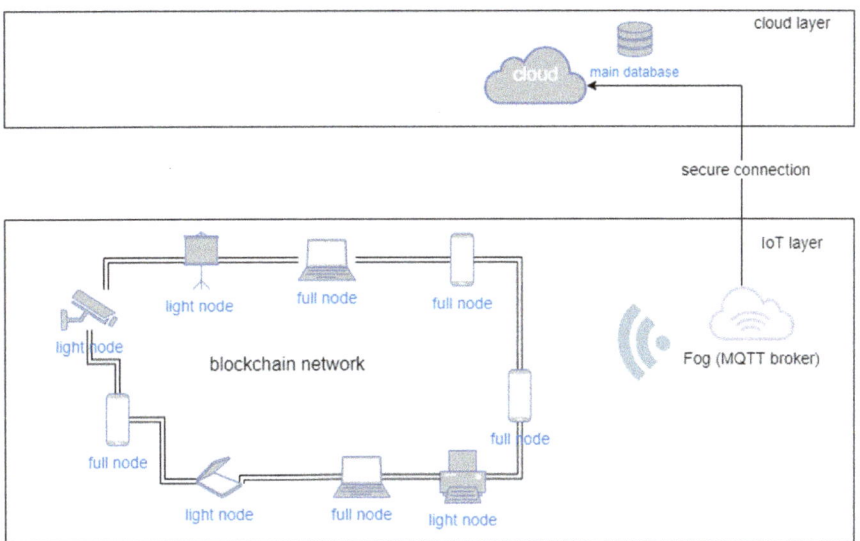

Fig. 3 System architecture

the central authority. We can do it using the full nodes that confirm any transaction, including joining a new device. Figures 4 and 5 show how the process will be done.

To sum up, we proposed a system with circles of trust, where a central computer system will have a blockchain with full nodes authorizing them to have their own circle of trust; thus, they can arrange for new devices to join the network through them. This model will constitute a decentralized and efficient solution with a high security constraint.

4   Discussion

The proposed model presents a distributed authenticated scheme. Many devices create their own group of trusted devices; one of the possible drawbacks could be the ever-growing chain. This chain keeps growing, putting a heavy load even on full nodes as well as time passes. Therefore, to manage this issue, we will have a maximum number of connected devices for each sub-chain connected to the central fog. So, whenever the chain grows to reach this number, the chain will be divided into two sub-chains with at least one full node at each Fig. 6. The maximum number should be determined according to the capability of the network and its constraints. This will create several small domains of trusted devices with the least overhead on the network.

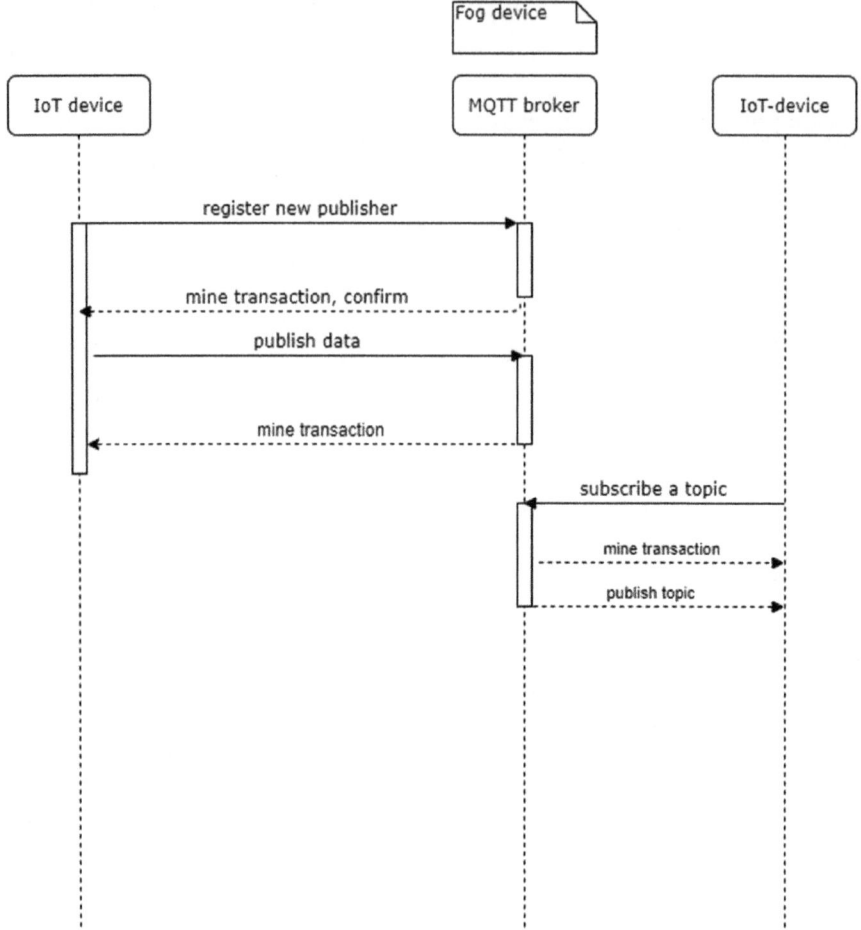

Fig. 4 Sequence diagram for adding new node via MQTT broker

5 Conclusion

IoT technology has been applied to different services and fields. With being ubiquitous and used everywhere, IoT security needed to be guaranteed. In this work, we
propose a new approach that provides a sufficient security level by using blockchain
and enhancement in real-time response by using the MQTT broker at the network's
edge. With this triumvirate of technologies, we believe that this will make a big
difference in the field and be counted as the next generation of IoT. In future, we
will provide some testing results that prove our claims including choose the best
blockchain consensus that achieve the desired results, and we may add more features
for better results.

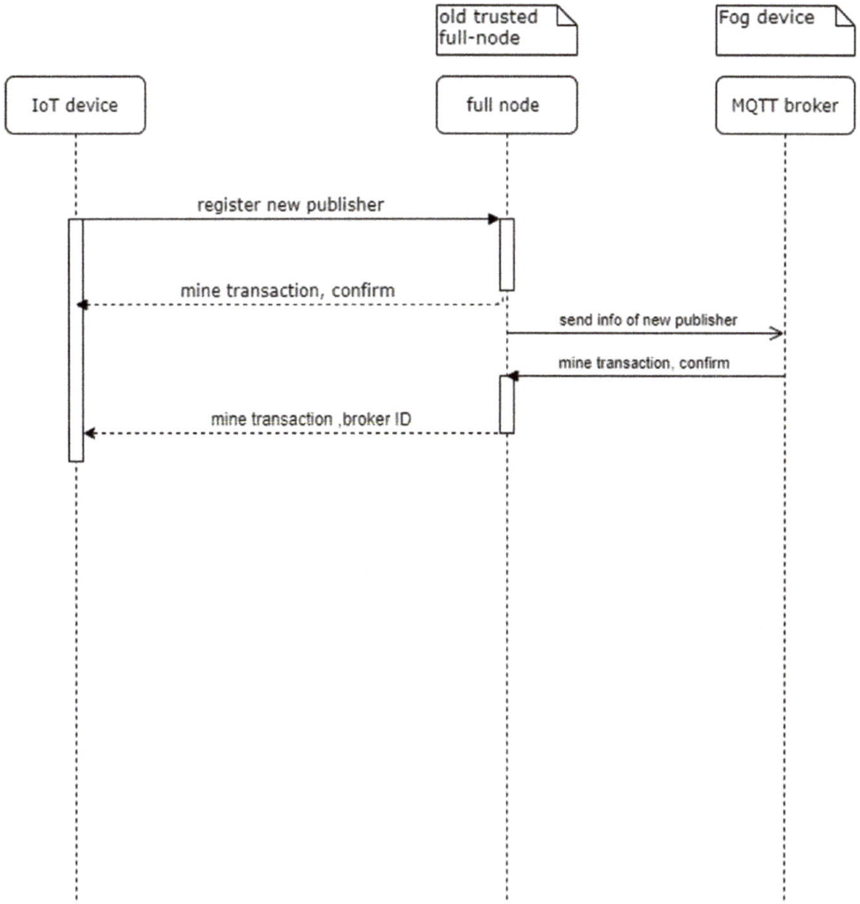

old trusted full-node
Fog device
IoT device
full node
MQTT broker
register new publisher
mine transaction, confirm
send info of new publisher
mine transaction, confirm
mine transaction ,broker ID

Fig. 5 Sequence diagram in case adding new node via a full node

Mini BC1
Mini BC2
Mini BC3
Main Blockchain
Central IoT Database in the Cloud
Fog / Edge Layer MQTT brokers
Mist Layer Full nodes with Mini-Blockchain
Lightweight Nodes

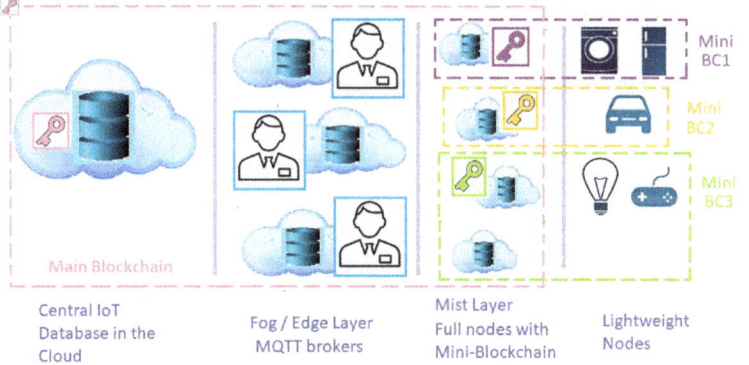

Fig. 6 Chain and sub-chain in our system

References

1. Alhazmi OH, Aloufi KS (2019) Fog-based internet of things: a security scheme. In: 2019 2nd international conference on computer applications & information security (ICCAIS). IEEE, pp 1–6
2. Gupta A (2019) The IoT Hacker's handbook. Apress
3. Khan MA, Salah K (2018) IoT security: review, blockchain solutions, and open challenges. Futur Gener Comput Syst 82:395–411
4. Laurence T (2019) Blockchain for dummies. Wiley
5. Lao L, Li Z, Hou S, Xiao B, Guo S, Yang Y (2020) A survey of IoT applications in blockchain systems: Architecture, consensus, and traffic modeling. ACM Comput Surv (CSUR) 53(1):1–32
6. Sagirlar G, Carminati B, Ferrari E, Sheehan JD, Ragnoli E (2018) Hybrid-iot: hybrid blockchain architecture for internet of things-pow sub-blockchains. In: 2018 IEEE international conference on internet of things (iThings) and IEEE green computing and communications (GreenCom) and IEEE cyber, physical and social computing (CPSCom) and IEEE smart data (SmartData). IEEE, pp 1007–1016
7. Novo O (2018) Blockchain meets IoT: an architecture for scalable access management in IoT. IEEE Internet Things J 5(2):1184–1195
8. Dorri A, Kanhere SS, Jurdak R, Gauravaram, P (2017) Blockchain for IoT security and privacy: the case study of a smart home. In: 2017 IEEE international conference on pervasive computing and communications workshops (PerCom workshops). IEEE, pp 618–623
9. Sharma PK, Chen MY, Park JH (2017) A software defined fog node based distributed blockchain cloud architecture for IoT. IEEE Access 6:115–124
10. Katende MM (2019) Combining MQTT and Blockchain to improve Data Security
11. Fakhri D, Mutijarsa K (2018) Secure IoT communication using blockchain technology. In: 2018 International symposium on electronics and smart devices (ISESD). IEEE, pp 1–6
12. Li X, Jiang P, Chen T, Luo X, Wen Q (2020) A survey on the security of blockchain systems. Futur Gener Comput Syst 107:841–853
13. Ramachandran GS, Wright KL, Krishnamachari B (2018) Trinity: a distributed publish/subscribe broker with blockchain-based immutability. arXiv preprint arXiv:1807.03110
14. Pazhev G, Spasov G, Shopov M, Petrova G (2020) On the use of blockchain technologies in smart home applications. In: IOP conference series: materials science and engineering, vol. 878, No. 1, p 012023. IOP Publishing
15. Satapathy U, Mohanta BK, Panda SS, Sobhanayak S, Jena D (2019) A secure framework for communication in internet of things application using hyperledger based blockchain. In: 2019 10th international conference on computing, communication and networking technologies (ICCCNT). IEEE, pp. 1–7
16. Zhang Y, Kasahara S, Shen Y, Jiang X, Wan J (2018) Smart contract-based access control for the internet of things. IEEE Internet Things J 6(2):1594–1605

Chapter 43
Computer Vision Based Autonomous Fire Detection and IoT Based Fire Response System

Abrar Ahmed Mohammed, Nagur Babu Alapaka, Chaitanya Gudivada, K. P. Bharath, and M. Rajesh Kumar

1 Introduction

This paper aims at providing a solution for erroneous and delayed fire detection and response systems by using cutting-edge image processing technology and the Internet of things. The unreliable fire systems are life-threatening as we are familiar with because of recent incidents around the world. Some of the major disasters are due to fire breakouts which statistically represents 97.1 million fires across the globe between 1993 and 2018 reported by 27–57 countries [1]. Therefore, it is a major responsibility to deploy smart, efficient, scalable, and reliable fire detection and immediate response system in the region of interest. Traditional fire detecting systems using smoke sensors happen to be good for fire detection which alerts humans to manually carry out fire extinguishing process using cylinders, but this process is inefficient because of the response time of sensors to detect the smoke or particle sampling and this can be very critical because more response time causes more spread of fire. Secondly, human involvement is dangerous in many cases without a proper training and guidance of fire experts. Recent innovations through integrating advanced microcontrollers and the Internet of Things can solve this problem.

The research for fast detection of fire summarized that the vision based approach is most efficient to detect fire and has the most probability to detect early fire [2]. Computer vision is more efficient and dynamic to employ instead of smoke sensor. Conventional methods are not as robust and dynamic as video-based image processing methods are thus it is the main reason for the shift to a better performing system [3].

In [4] a fire detection algorithm is discussed which uses CCD camera to capture footage and the counter area is calculated by the polar co-ordinate system, then the

A. A. Mohammed · N. B. Alapaka · C. Gudivada · K. P. Bharath · M. Rajesh Kumar (✉)
School of Electronics Engineering, VIT University, Vellore, India
e-mail: mrajeshkumar@vit.ac.in

S. Kumar et al. (eds.), *Proceedings of International Conference on Communication and Computational Technologies*, Algorithms for Intelligent Systems,
https://doi.org/10.1007/978-981-16-3246-4_43

551

results are placed in a time series, by applying a 2-dimensional Fourier transform a pattern of frequency component distribution is obtained and by entering this data into a neural network flame is detected.

In [5], a machine learning technique involving logistic regression which is robust to color changes, and smoke is proposed. Basically, logistic regression trains a hypothesis function which is sigmoid function and results in a continuous value bounded to the interval [0, 1] in which one can threshold hypothesis $h(x)$ according to $h(x) >$ Th to obtain a crisp value that the corresponding feature vector is a fire feature vector or not, generally the threshold value is 0.5 for experiments.

In [6], an algorithm is developed based on image processing and uses RGB color model to detect the color of fire, and growth of fire is detected by using Sobel edge detection, finally combining both using color-based segmentation on the results fire is detected.

In [7], a fire detection system based on parameters such as color, shape, flame movements was proposed. This approach has been tested on a large database, the MES approach achieved better performance. Also, the database containing fires filmed in different conditions was made publicly available in [8].

In [9], a fire detector that combines color information with a registered background scene is discussed. The color information is determined by the statistical measurement of sample images containing the fire. Histograms of r-g, r-b, g-b are used to extract fire regions over which a possible fire pixel is defined. The foreground objects which are detected are combined with color information to get the output and this is analyzed in consecutive frames to detect the fire.

In [10], an IOT based hardware system that uses both sensors and video processing is proposed. The system monitors both factors and if there is any suspicion of fire then immediately alerts the admin.

In [11], Haar cascade classifier algorithm is used for processing the images from live footage to detect the fire. In Haar cascade algorithm, a large data set with positive and negative samples of fire is used and the footage data is compared with these two samples to detect the fire.

In this paper, a system that can be used as a stand-alone application or as a parallel application-based user's requirements has been proposed. We categorize the idea into two phases named as detection phase and response phase, respectively. The detection phase uses computer vision (OpenCV) technology for fire-detection. Computer vision libraries help in the implementation of statistical digital image processing to every frame of video recorded by CCTV recorders already placed.

The increase in awareness of security and smart homes boosted security camera deployments in the majority of industries, warehouses as well as living homes. By taking advantage of pre-installed cameras the overall project cost can be reduced significantly. The response phase of the micro-controller is implemented using Arduino Uno. The output of the detection phase is sent to the input pin of Arduino as HIGH if the fire is detected and LOW if not. If it detects the fire then fire extinguishers in that region will automatically extinguish the fire. The fire extinguishers are controlled by the Arduino and the fire will be controlled in no time as the response time of Arduino is low. A high-noise alert buzzer also responds to fire detection which acts as an alert system that may have benefits based on the place of usage. Thus, making the system completely autonomous with the benefits of not waiting for the Fire officers to check the burnt place already. This method is very effective, and it is a solution for various fire-detection tasks in real-world problems. Different fire detection algorithms are deeply studied, analyzed, and compared with advantages and disadvantages to give a clear understanding of the efficiencies of the algorithms used in video-based fire detection.

2 Proposed Algorithm

The proposed method uses OpenCV as an image processing library which is very popular and easy to implement a scalable code. The CCTV cameras which are pre-installed in the points of region of monitoring are set to record 24 h round the clock. The coverage of the region should be perfect using the most appropriate points of the region. The CCTV's are collectively accessed through a computer or server of a corporation/home. Therefore, the python application has to be installed on the same computer to give access to live video recordings. To install OpenCV module to python IDE like IDLE install pip application in the scripts folder of python home directory. Now go to the control panel of the pc and select environment variables and provide the path of python home directory path and scripts folder path. This is to ensure python and pip are visible to cmd of the pc. Open command prompt and type pip install OpenCV- python and press enter, this would install the required modules for our use. Now type py in the command line. This should let you run the python interpreter and use import cv2 to check the module is installed or not. The experimentation assumes that an active camera is connected to a computer that has sufficient RAM minimum of 4 GB and sufficient power backup.

2.1 Fire Detection Algorithm

Step-1: Capture the video using VideoCapture() method. The argument takes a whole number indicating which camera is to be monitored.

Step-2: Read the video into frames using the read() method which returns a Boolean datatype of video availability and frames.

Step-3: Resize the frame as required using resize(x,y). The method accepts two arguments x which is the source frame image and y is length and width.

Step-4: Smooth the frame image using GaussianBlur() method which uses the Gaussian kernel. It is used to remove Gaussian noise from the image which helps toward our goal of fire pixel detection. It accepts 2 arguments input resized image and standard deviation of (x,y) direction.

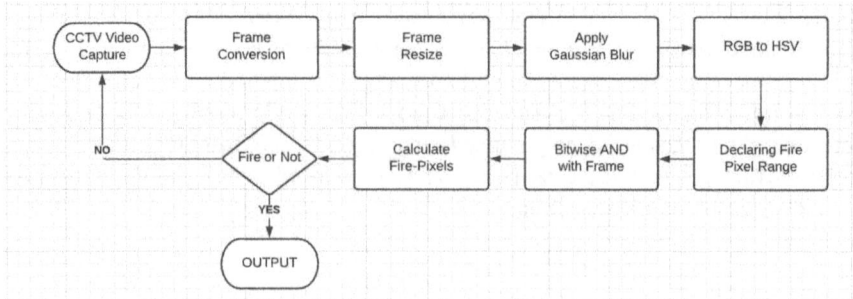

Fig. 1 Flow chart depicting fire detection process

Step-5: The most important step is to convert the resultant RGB image to HSV since it is easy to represent and extract a colored object in this color space. The method cvtColor(img, cv2.COLOR _BGR2HSV) is used to do the same.

Step-6: Use thresholding to object track the fire pixels from HSV using upper and lower pixel color ranges of fire using the built-in method of.inRange().

Step-7: Apply the mask using Bitwise AND operator to earlier resized image and the present object tracking output image to get the fire pixels only if any using bitwise.and() method.

Step-8: Count the pixel density or no. of pixels of fire and add a conditional statement to check if it is a valid number of fire pixels detected to be able to consider as an unintentional and make the flag as HIGH or 1 if the condition is True.

Step-9: Break the loop already if no video input is available to monitor and maintain a key for the forced exit of the program.

Flowchart of the fire detection process. The basic flowchart of the proposed fire detection algorithm is as shown in Fig. 1. Live video is sent through frame converter and frame resizer. The resized frame is smoothed and transformed to HSV color model. Fire pixels are filtered, and output is triggered.

2.2 IoT Based Fire Response System

Our idea is to use python for the frontend and use IoT as a backend system. Thus, it requires both to communicate with each other. The output of the python program when the fire is detected has to be communicated to Arduino to trigger the mechanism of fire extinguishing. Serial communication between python and Arduino is one of the possible solutions. It is achieved by Pyserial module. Import this module and create an object specifying comport number and baud rate. Use this object to create a function to send serial data using the write(b'1') function. Call this function when the

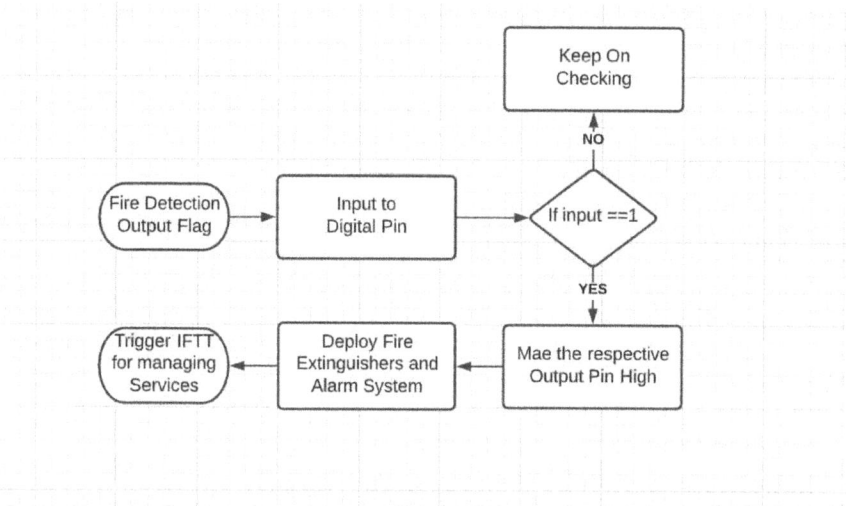

Fig. 2 Flow chart depicting smart fire response system

output flag of python code is high. This sends a byte array of 1 through the com port which has to be connected to Arduino. In Arduino IDE write code to check if any data is being received through the com port. Inside void loop() check if(Serial.available() > 0) if yes then read the data. Check the data is 1 or not. If '1' then make the digital pins HIGH of sprinklers, carbon dioxide diffuser, and Buzzer Alarm system. Repeat this till no data is received from the serial monitor till the fire is extinguished.

Flowchart of IoT based fire response system. The flowchart of the proposed fire response model is as shown in Fig. 2. The python code on front-end sends data to receiving digital pin of Arduino through serial port. It checks for input = 1 and if true output pins are activated to perform suitable fire extinguishing actions.

3 Comparison of Various Fire Detection Algorithms

Color detection which is the oldest fire detection techniques and one of the major image processing based technique. RGB color space is used in this detection technique. This technique is compatible with CCTV and UAVs also. This algorithm is accurate about 90–93% and efficiency is less than 80%. In this technique, some of the defects are background detection variation in lighting and density.

Moving object detection is one of the image processing techniques used for fire detection to remove the disturbance in the background caused by objects in the rest position. Each frame in the video is separated and it checks for the moving object using segmentation. But in this technique, we do not know that the moving object is smoke caused by fire or not. So, this technique is not so efficient and its efficiency is less than 85%.

Flicker detection is one of the techniques used for fire detection in videos. This technique is based on temporal and spatial smoothing. The flicker frequency of the fire is in the range of 1–20 Hz and it is less sensitive to heat radiation and it gives false alarms. So, this technique is also not so efficient.

The flames of the fire are not so steady and they are composed of several varying colors within a small area. The spatial difference analysis technique is focused on this characteristic. By using spatial wavelet analysis, the spatial color variations in pixel values are examined to eliminate ordinary fire-colored objects with a solid flame color. This technique has about 94% accuracy, but in this also some defects are there like in the darker regions this technique is not so effective.

4 Simulation Result of the Proposed System

4.1 Simulation Result of the Fire Detection System

The images of the result of the proposed model simulated in python environment using OpenCV library are below.

Image blurring is an image smoothing technique to remove high-frequency components such as noise. OpenCV provides four major smoothing concepts and we have applied Gaussian blur because it is more efficient in removing Gaussian noise. The blur is the convolution of the original captured frame with a Gaussian kernel. In Fig. 3, we observe that on applying blur the resultant image is smooth and free from noise.

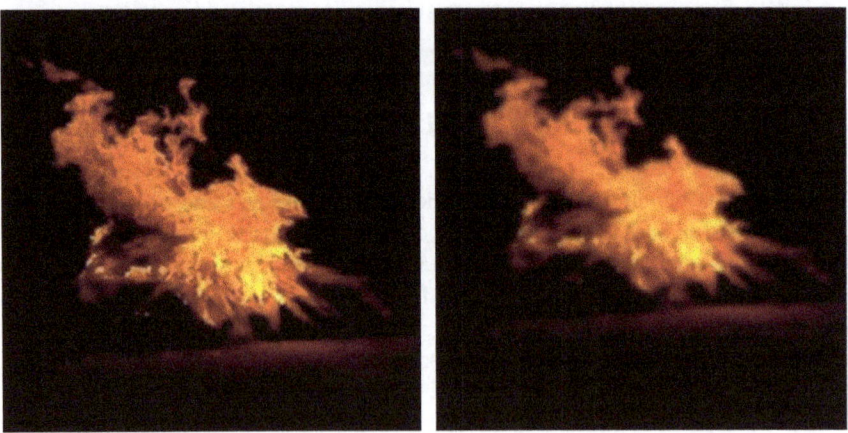

Fig. 3 The original captured frame of fire video and Smoothed frame using Gaussian blur

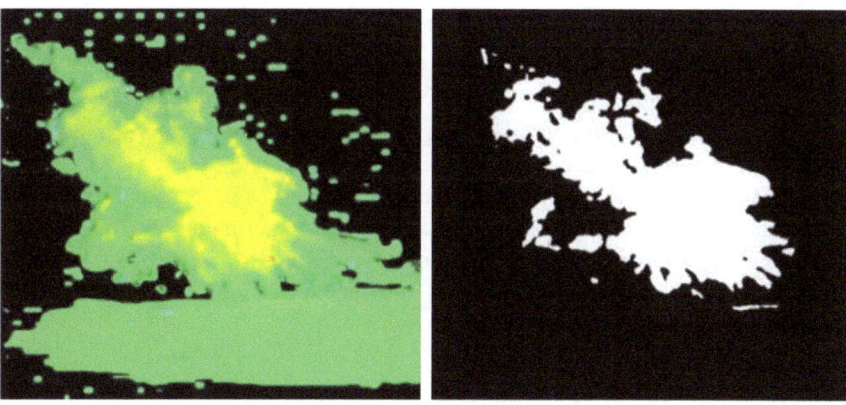

Fig. 4   HSV image of Smoothed frame and Mask of fire pixel range

Changing color-space is very important to achieve object tracking or color tracking. From Fig. 4, the first image shows the HSV image and the second image is a mask of red color alone extracted from the HSV image using a thresholding range of red color in HSV color-space.

The final output image is bitwise AND operation with the original frame and mask. White color in the mask is taken as '1' and AND operation is done resulting in extracting the fire from the frame. The number of pixels of fire is counted from Fig. 5. The algorithm follows the step 8 from Sect. 2.1 as discussed.

Fig. 5   Output image obtained by bitwise AND operation among original frame and mask

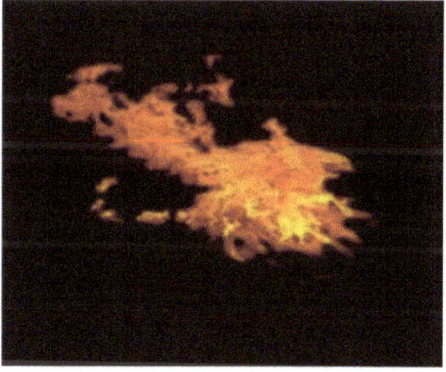

4.2   Simulation of the Fire Response System

We have simulated the fire response system using Arduino UNO using Proteus professional. The Serial COM port3 is used to let python communicate with Arduino serial monitor. Pyserial module has to be installed in python to work.

LED D1 glows when the fire is detected as shown in Fig. 6. This is for result purpose the idea is to install fire extinguishing systems, sprinklers, etc. in the place of LED with appropriate power conversions.

The Arduino displays a "Fire detected" message on the operator's computer for alerting the concerned people as shown in Fig. 7.

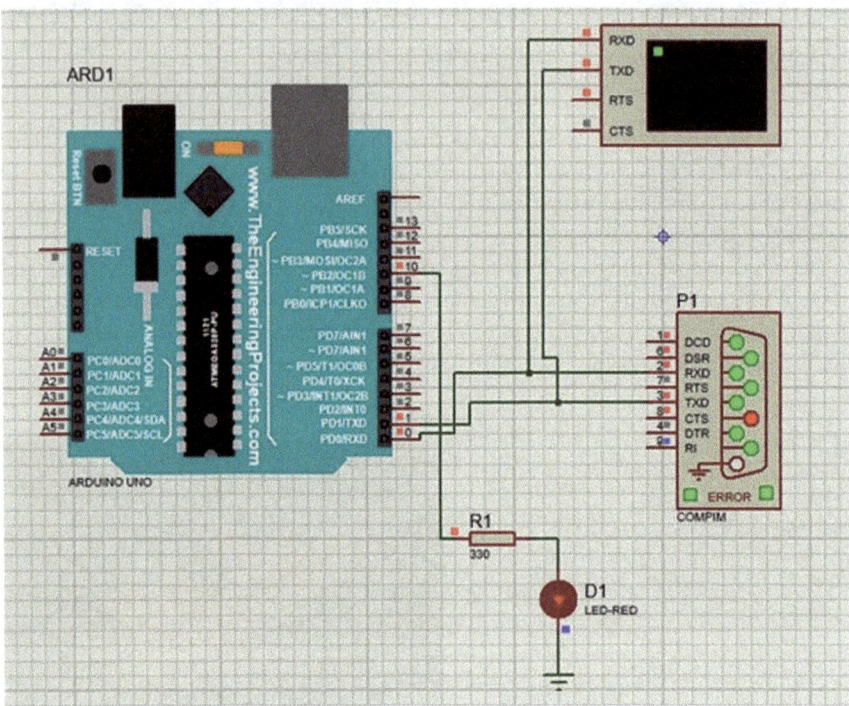

Fig. 6 IoT based Fire response system simulated in proteus environment

```
>>>
RESTART: C:\Users\abrar\AppData\Local\Programs\Python\Python37-32\Firedetection
_dip.py
FIRE DETECTED!!!!!!
FIRE DETECTED!!!!!!
```

Fig. 7 Python program output when the fire is detected

5   Conclusion

The automatic fire detection system is one of the major techniques that can be used by fire centers and firefighters to detect and locate fire occurrences in various environments. Here, our main aim was to stop fire accidents in industries with greater efficiency. So, in this paper, we discussed the latest technology that can help to recognize and reduce the accidents caused by fire. In the detective phase, the fire is detected even when it is low, as soon as the fire is detected by analyzing the frames from CCTV, a quick and fast response is sent to the response phase which is nothing but the IoT part. As we used Arduino UNO in the IoT part it triggers the extinguishers and alarm system in no time to stop the fire which is our main theme. In other techniques rate of false detection is one of the major drawbacks, as we have used the lower and upper frame values we can able to distinguish between a fire caused due to any damage or is it a just simple lighter or match stick fire. Thus, finally with the help of this proposed algorithm fire is detected very effectively and accurately when compared to the remaining methods in use.

Future Scope. The future scope of this project could be implementing a relevant technique to extinguish the fire in the outdoor areas mainly in forests within less time, increasing the efficiency of the algorithm in such a way to find the exact reason for the occurrence of fire parallel with the detection of fire. The other future scope can be if suppose if the algorithm feels that the fire cannot be controlled in any way then send an alert to the manager and closing that particular room so that the spread of fire can be decreased. The proposed model works perfectly well with minimum delay and has proven efficient to detect indoor fires with accuracy. The simulation results show that the proposed system can be added to existing CCTV to make them smart fire detectors.

References

1. Brushlinsky NN, Ahrens M, Sokolov SV, Wagner P (2020) World fire statistics, 2020 No-25. , International Association of Fire and Rescue Services, CTIF. URL: https://www.ctif.org/world-fire-statistics
2. Mahdipour E, Dadkhah C (2014) Automatic fire detection based on soft computing techniques: a review from 2000 to 2010. Artif Intell Rev 42:895–934. https://doi.org/10.1007/s10462-012-9345-z
3. Qureshi WS, Ekpanyapong M, Dailey MN, Rinsurongkawong S, Malenichev A, Krasotkina QuickBlaze O (2015) Early fire detection using a combined video processing approach. Fire Technol 52(5):1293–1317
4. Yamagishi H, Yamaguchi J (2000) A contour fluctuation data processing method for fire flame detection using a color camera. In: 2000 26th annual conference of the ieee industrial electronics society. IECON 2000. 2000 IEEE international conference on industrial electronics, control and instrumentation. 21st century technologies, Nagoya, Japan, 2000, pp 824–829, vol 2. https://doi.org/10.1109/IECON.2000.972229

5. Toulouse T, Rossi L, Celik T et al (2016) Automatic fire pixel detection using image processing: a comparative analysis of rule-based and machine learning-based methods. SIViP 10:647–654. https://doi.org/10.1007/s11760-015-0789-x)
6. Poobalan K, Liew Si (2015) Fire detection algorithm using image processing techniques
7. Foggia P, Saggese A, Vento M (2015) Real-time fire detection for video-surveillance applications using a combination of experts based on color, shape, and motion. IEEE Trans Circuits Syst Video Technol 25(9):1545–1556. https://doi.org/10.1109/TCSVT.2015.2392531
8. https://mivia.unisa.it/  /datasets/
9. Celik T, Demirel H, Ozkaramanli H (2006) Automatic fire detection in video sequences. In: 2006 14th European signal processing conference, Florence, 2006, pp 1-5
10. Imteaj A, Rahman T, Hossain MK, Alam MS, Rahat SA (2017) An IoT based fire alarming and authentication system for workhouse using Raspberry Pi 3. In: 2017 International conference on electrical, computer and communication engineering (ECCE), pp 899–904
11. Pranamurti H, Murti A, Setianingsih C (2019) Fire detection use CCTV with image processing based Raspberry Pi. J Phys Conf Ser 1201:012015. https://doi.org/10.1088/1742-6596/1201/1/012015

Chapter 44
Sentiment Analysis of Text Classification Using RNN Algorithm

Gitashree Borah, Dipika Nimje, G. JananiSri, K. P. Bharath, and M. Rajesh Kumar

1 Introduction

Users of various sites such as Facebook, LinkedIn, and Instagram can share information on a variety of topics ranging from education to entertainment, such as feedback, thoughts, emotions, and opinions. These sites store a huge amount of data in the form of tweets, messages, status updates, links, and so on. Opinion mining looks at how a text feels in relation to a specific source of data. Emotion comprehension is difficult due to slang words, misspellings, short forms, repetitive characters, ethnic language, and new upcoming emoticons. As a result, thinking about the right sentiment of each word is a good habit to get into. One of the most active research fields is sentiment analysis, which is also commonly studied in data mining. Since opinions are central in most human actions and conducts, and sentiment research is used in virtually every business and social sphere. Feeling analysis is very common because of its effectiveness [1, 2]. It is possible to process thousands of documents in order to determine emotions. It has a wide variety of applications because it is a straightforward tool with a high degree of precision:

G. Borah · D. Nimje · G. JananiSri · K. P. Bharath · M. R. Kumar (✉)
School of Electronic Engineering, VIT University, Vellore, India
e-mail: mrajeshkumar@vit.ac.in

G. Borah
e-mail: gitashree.borah2020@vitstudent.ac.in

D. Nimje
e-mail: dipika.nimje2020@vitstudent.ac.in

G. JananiSri
e-mail: jananisri.g2020@vitstudent.ac.in

K. P. Bharath
e-mail: bharathkp25@gmail.com

Purchase of goods or services: We use goods or resources to make the best decision when buying a commodity or utility, which is no longer a difficult task sentiment research allows people to compare competitive products by allowing them to easily evaluate feedback and points of view on any product or service.

Improving standard of product or service: The producers should take the view of the customer, mining, according to the opinion. If the products or services are useful or not, they can develop and upgrade their brand or operations practices.

System Recommendation: A device that can predict a specific item whether should be suggested or not used by assessing and classifying the opinion of the public in compliance with its own expectations.

Decision-making: People's attitudes, opinions, and feelings are crucial to decision-making. If anything is bought, whether books or clothes or technological items, the user first reads the comments of the particular product and these reviews affect the user. The findings of feeling interpretation approaches can be used in research marketing.

Research marketing: Consumer reaction to any goods or services, as well as new government legislation, can be studied using research marketing.

Flame detection: Sentiment analysis makes it easy to keep track of newsgroups, blogs, and social media. In online tweets, posts, forums, and blogs, this method can detect arrogant, insensitive, and overheated language.

The emotional research steps are as follows:

Pre-Processing Phase: To remove noise from the data, it must first be washed.

Extraction of Feature: A token is allocated to each keyword, and the token is now used for analysis.

Step of classification: These keywords are grouped on an RNN basis algorithm [3].

The aim of our study is to access the polarity of emotions such as pleasure, grief, sorrow, dislike, love and text opinions, scores, accessible on these platforms. Through this paper, we develop the classification of feelings at a finely defined level, in which the polarity of a sentence as positive and negative. We've also used techniques-like nultinomial Naive Bayes, support vector machine (SVM), logistic regression, and recurrent neural etworks to classify text (RNN).

2   Literature Survey

Kowsari et al. [1], cover various text feature extractions, existing algorithms and techniques, and methods of evaluation. The Rocchio algorithm, bagging and boosting, logistic regression, Naive Bayes Classifier, k-nearest neighbor, support vector machine, decision tree classifier, random forest, conditional random field, and deep learning are all discussed in this paper.

Yogeshwaran and Yuvaraj [4], text classification using recurrent neural network in Quora. This paper investigates the extraction of sentiment from a well-known Twitter page for microblogging, where users express their opinions. They assume that deep learning algorithms can outperform conventional methods in terms of performance and productivity. This form of research would certainly assist every enterprise in growing its efficiency.

Kaur and Bathla [5], the text classification and its method and a variety of text classification approaches were surveyed. They also looked at and compared various classification systems, such as SVM, ANN, Naive Bayes, and decision trees. They've also looked at performance assessment metrics-like F-measure, G-measure, and precision that have been used in a variety of studies.

Yazdavar et al. [6]. in this article, a new interpretation of emotional analysis with numerical outcomes has been presented in drug reviews. They assessed phrases with quantitative terms to classify them as opinionated or non-opinionated, as well as the polarity shown by the use of fuzzy set theory to classify them. An interview with several doctors from various medical centers was used to create the fuzzy knowledge base. Despite the fact that many studies have been carried out in this field, the numerical (quantitative) knowledge found in the responses cannot be taken into account, leading to the acceptance of a polarity of sentiment.

Murthy [7], he explained that there was no electoral success on Twitter and the various social media networks platforms have been used to boost a candidate's popularity by producing a buzz around them. He found that Twitter had little political success, and that multiple social media networks sites had been used to raise a candidate's popularity by creating a buzz around them.

Amolik et al. [2], in his paper, he used Twitter posts of film reviews and related tweets about those films to create a dataset. Sentiment analysis at the sentence stage is performed on these tweets. It is completed in three steps. The appropriate features are used to build the function vector. Naive Bayes, support vector machine, ensemble classifier, k-means, and artificial neural networks were used to classify tweets into positive, negative, and neutral groups. The SVM is 75% accurate, according to the results.

Sameera et al. [8]. The best features for representing the document vector are defined using a feature selection algorithm in this paper. They tested their hypothesis on AG's news article dataset. Three classifiers are used to build a text classification model: NBM, SVM, and RF.

Munjal et al. developed framework for analysing twits [9] and for opinion dynamics [10, 11]. Recently, Chugh et al. proposed a novel approach for sentiment categorization with enhanced performance.

3 Research Methodology

3.1 RNN (Recurrent Neural Network)

A recurrent neural network (RNN) is a type of neural network in which the previous phase's output is used as an input in the current phase. In traditional neural networks, the inputs and outputs are independent of one another, but in certain situations, such as when predicting the next word of a sentence, the previous words are required, and therefore, the previous words must be remembered. RNN was born, and with the help of a secret layer, it was able to solve the problem. The key and most important feature of RNN is the secret state, which recalls some information about a sequence [4, 12] (Fig. 1).

RNN has a "memory" that remembers all the aspects of the measurement. It uses the same parameters for and input and it executes the same procedure on all inputs or hidden layers to produce the output. Unlike all the other neural networks, the complexity of the parameters is reduced. The deep neural networks architecture is designed to learn through multiple layer connections where each single layer receives connections from previous layer, provides connections to the second layer hidden portion and then to the output layer, i.e., hidden layer. RNNs are a type of neural network in which nodes and links form a directed graph with a sequence between them. It consists of a chain of neural network blocks that transfer data to a successor. Machine learning and data mining are mainly used in the existing method for classifying text documents. In our proposed model for text data classification, we use both deep learning and machine learning algorithms [12].

Step-by-step architecture on how to describe text data (Movie Review) as seen in Fig. 2.

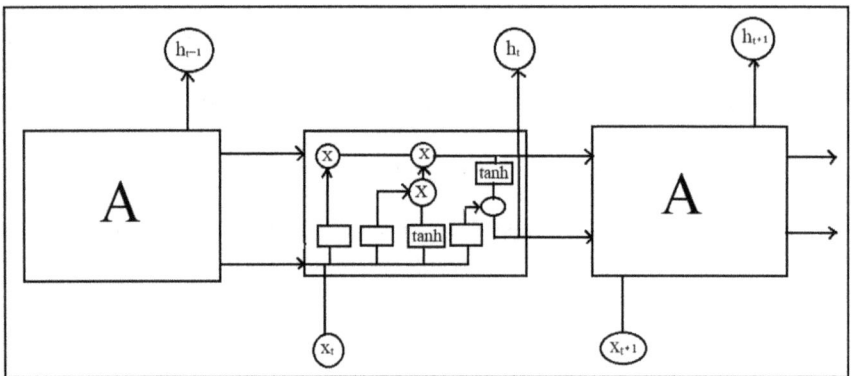

Fig. 1 LSTM block diagram [12]

Fig. 2 Step-by-step architecture on how to describe text data [2]

Data Collection: The initial move is to gather various user data formats that we can test for the results. In our sample, we are using movie reviews data set to evaluate various algorithms.

Pre-processing: Here, a tokenization technique is used to translate a text document into a word document. It is the process of decomposing a text stream into tokens such as words, sentences, symbols, and other essential elements.

Indexing: It is the approach used to reduce and model textual data complexity that is simple to maintain.

Features Selection: It is used to create vector space to improve classifier stability, accuracy, and performance. In terms of appearance and frequency, the core principle of feature selection is to pick subsets from the original paper. These attributes are individual words with their frequency counts. It provides binary weighting to the words, i.e., if the word exists then 0, otherwise 1. Adjectives are often found here because they are important measures of user views. It was used to communicate the views of users, whether positive or evil, like or dislike.

Classification: Classifies textual data automatically into predefined classes. Three retrieval classification and description approaches may be categorized into a document: Supervised, semi-supervised, and unsupervised. And, in modern years, several other methods have also been used.

Output Performance: The last step is to experimentally test the text data instead of analytically, so that we can continue with an accurate judgement on classification [5].

3.2 Working of RNN

RNN converts independent activations into dependent activations by giving all layers the same weights and biases, minimizing the complexity of raising parameters, and memorizing each previous output by feeding each output to the next hidden layer as an input. As a result, all three layers can be combined into a single recurrent layer with the same weights and biases throughout the board.

A recurrent neural network's (RNN) capability is to process an arbitrary length sequence by recursively applying a transition function to its internal hidden state vector of h_t. The current input function f is used to measure the sequence for input activation of the hidden state h_t at time-step t. The x_t symbol and the previous secret h_{t-1} condition [13].

$$h_t = \begin{cases} 0 & t = 0 \\ f(h_{t-1}x_t) & \text{Otherwise} \end{cases} \tag{1}$$

Using the state-to-state transition feature f is popular as the composition of a nonlinearity with an element-wise of both x_t and h_{t-1} effectiveness.

Formula for Current state calculation:

$$h_t = f(h_{t-1}, x_t) \tag{2}$$

where,

h_t—current state

h_{t-1}—previous state

x_t—input state.

Formula for applying feature activation (tanh):

$$h_t = \tanh(W_{hh}h_{t-1} + W_{xh}x_t) \tag{3}$$

where,

W_{hh}—weight at recurrent neuron.

W_{xh}—weight at input neuron.

Formula for Performance calculation:

$$Y_t = \text{Why } h_t \tag{4}$$

where,

Y_t—output

Why—weight at output layer

3.3 Long Short-Term Memory Network

In order to solve the problem of vanishing and collapsing gradients in a deep recurrent neural network, several variants were created. One of the most common is the long short-term memory network (LSTM). A repeating LSTM device, in principle, tries to "remember" all prior knowledge of the network's usage thus far, while "forgetting" irrelevant information. This is done for a number of reasons by integrating different activation function layers known as "gates."

Internal cell state is a vector retained by each recurrent LSTM unit that conceptually represents the knowledge selected by the previous recurrent LSTM unit to be held (Fig. 3).

As mentioned below, a long-short-term memory network comprises of four separate gates for different purposes:

Forget Gate(f): It specifies to what degree the previous data can be forgotten.

Input Gate(i): The scope of the knowledge to be applied to the internal cell state is determined.

Input Modulation Gate(g): It is often treated as a sub-part of the input layer and it is.

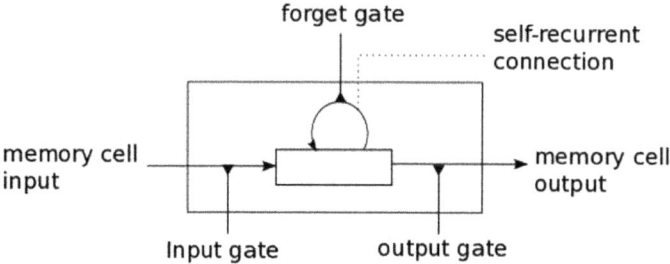

Fig. 3 Memory cell of an LSTM [10]

Not even stated and presumed by many LSTM literatures within the input gate. By applying non-linearity to the information in order to make the data Zero-mean, it's being used to modulate the data that the input gate would write on the inner state cell. Although the acts of this gate are less relevant than others and are often viewed as a term that provides finesse, it is good training to have this gate in the LSTM unit structure [12].

Output Gate(*o*): It decides from the currently internal cell state what output (next Hidden State) to produce.

The fundamental workflow of a long-term memory network is identical to the functionality of a recurrent neural network, including the only exception being that the cell state and the cell state are forwarded on [12] (Fig. 4).

The main difference between RNNs and LSTM architectures is that the LSTM's hidden layer is called a gated unit or gated cell. It is made up of four layers that interact with one another to produce the cell's output, along with the cell's state. After that the next hidden layer is applied to these two objects. Unlike RNNs, which have a single neural net layer of tanh, LSTMs have three logistic sigmoid gates and one tanh layer. Gates have been introduced to limit the amount of data transmitted through the cell. They choose which part of the data will be included in the next cell and which will be discarded. The performance is usually in the 0–1 range, with "0'" indicating "reject all" and "1" indicating "include all" [8].

4   Experimental Result

4.1   *Summary of Data sets*

For our research, we compile data from IMDB movie reviews. We divide our data set into test and training data sets. We use 80% of the training data and 20% of the testing data to determine the learned model's results. The goal of this work is to decide whether the review is positive or negative in terms of the polarity of the review granted (Fig. 5; Tables 1 and 2).

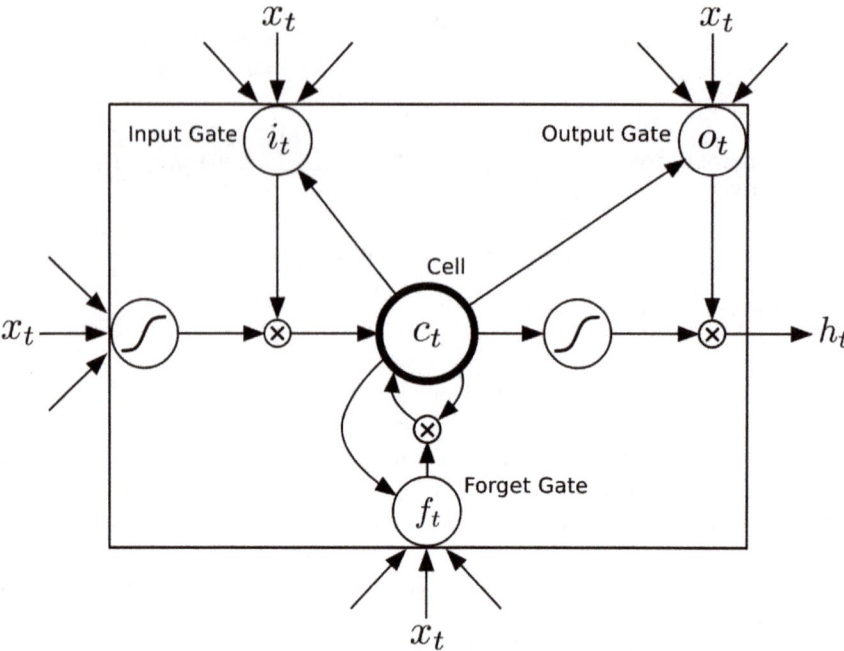

Fig. 4 Cell state of LSTM [13]

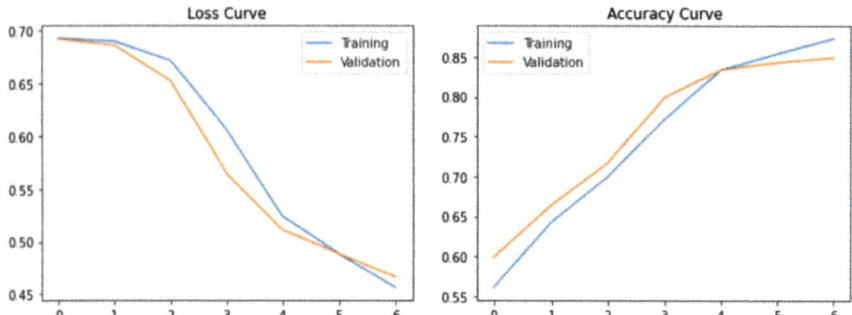

Fig. 5 Loss curve and accuracy curve of training and validation

Table 1 Dataset specifications

Name	IMB dataset
Link	http://www.kaggle.com
Total no. of reviewers used	50,000
No. of reviews trained	40,000
No. of reviews tested	10,000

Table 2 Configuration model for RNN

Configuration of the model	Epochs	Accuracy
Embedding	1	56.09
Layer + LSTM layer + Dense layer	2	64.31
	3	69.98
	4	77.22
	5	83.39
	6	85.41
	7	87.32

Confusion matrix presents a table layout of the different outcomes of prediction and results of a classification problem and helps visualize its outcome. It helps us to identify the correct predictions of a model for different individual classes as well as the errors. The matrices are performance measures which help us find the accuracy of our classifier. Classification models have multiple output categories. Most error measures will tell us the total error in our model, but we cannot use it to find out individual instances of error in our model. Below is the confusion matrix for our model (Fig. 6).

The final accuracies for the algorithms we tested are shown in Table 3.

Fig. 6 Confusion matrix

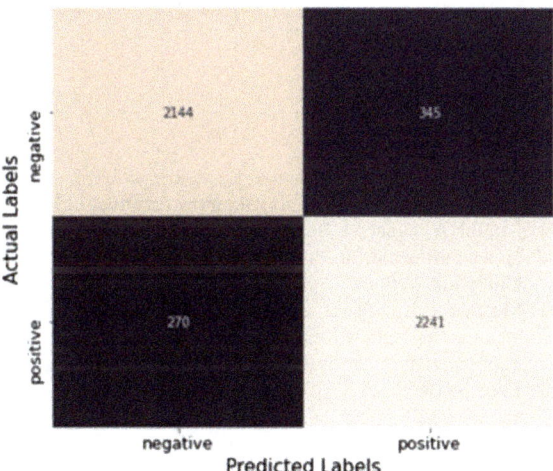

Table 3 Final accuracy of the algorithms used

Algorithms	Accuracy
Support vector machine	58.22
Multinomial naïve bayes	75.17
Logistic regression	75.18
Recurrent neural network	87.70

The algorithms performed in this paper were multinomial Naïve Bayes, SVM, logistic regression and RNN and the best results were given by RNN (LSTM) algorithm using IMDB dataset. The RNN algorithm achieved 87.70% accuracy, in multinomial Naive Bayes we achieved 75.17% accuracy, in SVM we achieved 58.22% accuracy and in logistic regression we got 75.18%.

5 Conclusion

One of the most important applications of data mining is text detection. As a result, we looked into text classification, its process, and some approaches to text classification (algorithms). Different techniques are used to classify the polarity of the input. The algorithms conducted are multinomial Naive Bayes, SVM, logistics regression, and RNN. And by analyzing the output results, we can conclude that RNN algorithm provides the best precision in this textual data classification by evaluating the four algorithms. From this survey, it is concluded that text classification must be evaluated with maximal evaluation matrices that have not been used in several research papers. As well by pre-processing data or using lexicon templates such as Text blob, we can increase the performance of the prediction. We intend to expand this analysis to a greater degree in future. To be considered for multiple embedding models on a vast number of datasets.

References

1. Kowsari K, Jafari Meimandi K, Heidarysafa M, Mendu S, Barnes L, Brown D (2019) Algorithms text classification: a survey data 10(4):150
2. Amolik A, Jivane N, Bhandari M, Venkatesan M (2016) Twitter sentiment analysis of movie reviews using machine learning techniques. School of Computer Science and Engineering, VIT University, Vellore
3. Shaziya H, Kavitha G, Zaheer R (2015) Text categorization of movie reviews for sentiment analysis. Int J Innov Res Sci Eng Technol 4(11)
4. Yogeshwaran D, Yuvaraj N (2019) Text classification using recurrent neural network in Quora
5. Kaur B, Bathla G (2018) Document classification using various classification algorithms: a survey. Int J Fut Revol Comput Sci Commun Eng. 4(2):150–155. ISSN: 2454-4248
6. Yazdavar AH, Ebrahimi M, Salim N (2016) Fuzzy based implicit sentiment analysis on quantitative sentences. J Soft Comput Dec Support Syst 3(4):7–18
7. Murthy D (2015) Twitter and elections: are tweets, predictive, reactive, or a form of buzz? Inf Commun Soc 18(7):816–831. https://doi.org/10.1080/1369118X.2015.1006659
8. Sameera D, Vedavathi K, Prasad KD (2020) A new approach to recurrent neural networks text classification. J Appl Sci Technol Int 29(05):9371–9386
9. Munjal P, Narula M, Kumar S, Banati H (2018) Twitter sentiments based suggestive framework to predict trends. J Stat Manag Syst 21(4):685–693
10. Munjal P, Kumar S, Kumar L, Banati A (2017) Opinion dynamics through natural phenomenon of grain growth and population migration. In: Hybrid intelligence for social networks. Springer, Cham, pp 161–175

11. Munjal P, Kumar L, Kumar S, Banati H (2019) Evidence of Ostwald ripening in opinion driven dynamics of mutually competitive social networks. Phys A 522:182–194
12. https://www.geeksforgeeks.org/introduction-to-recurrent-neural-network/
13. Murthy GSN, Allu SR, Andhavarapu B, Bagadi M, Belusonti M (2020) Text based sentiment analysis using LSTM. Int J Eng Res Technol (IJERT) 9(05). ISSN: 2278-0181, IJERTV9IS050290, May 2020. http://www.ijert.org

Chapter 45
IOT With Blockchain Based Techniques: Review

Bhawana Maurya, Saroj Hiranwal, and Manoj Kumar

1 Introduction

The IoT can associate and communicate billions of things at the same time. It gives different advantages to clients that will change the manner in which those clients collaborate with the innovation [1]. Utilizing an accumulation of low-cost sensors and interconnected items, data can be gathered from our condition that will permit enhancing our method for living [2, 3]. Current IoT frameworks are based on brought together server/consumer show, which requires all gadgets to be associated and confirmed through the server. This model would not have the capacity to give the requirements to extend the IoT framework later on [4, 5]. In this manner, moving the IoT framework into the decentralized way might be the correct choice. One of the famous decentralization stages is blockchain [6].

A blockchain is a disseminated database of records that contains all exchanges that have been executed and shared among contribution parties in the system. This disseminated database is called disseminated record. Every exchange is put away in the appropriated record and should be confirmed by assent of the lion's share of members in the system. All exchanges that have ever constructed are contained in the blockchain [7]. Bitcoin, the decentralized shared computerized cash, is the most prominent model that utilizes blockchain innovation [8]. Incorporating IoT with blockchain will have numerous advantages. The decentralization model of the blockchain will be able to deal with handling of large number of interactions among IoT gadgets [9]. Furthermore, working with the blockchain innovation will

B. Maurya (✉)
Department of I.T, Government Women Engineering College, Ajmer, Rajasthan, India

S. Hiranwal
Department of Computer Engineering, RIET, Jaipur, Rajasthan, India

M. Kumar
Department of Electrical Engineering, Government Polytechnic College, Ajmer, Rajasthan, India

© The Author(s), under exclusive license to Springer Nature Singapore Pte Ltd. 2021   573
S. Kumar et al. (eds.), *Proceedings of International Conference on Communication and Computational Technologies*, Algorithms for Intelligent Systems,
https://doi.org/10.1007/978-981-16-3246-4_45

dispense with the single purpose of disappointment related with the incorporated IoT engineering. In addition, incorporating blockchain with IoT will enable the shared informing document dispersion and self-sufficient coordination between IoT gadgets with no requirement for the brought together server-customer demonstrate [10]. The rest of the paper organization is according to the accompanying: Sect. 2 reviews unique research works in convergence of IoT with blockchain and their benefits and their limitations, and Sect. 3 concludes the review took after by the references.

2 IoT with Block Chain Techniques: A Review

Todays, countless smartplans and items are incorporated with sensors, in this manner empowering them to distinguish real time info from the situation. Nevertheless, this model has altered with the approach of the IoT in which all intellectual things— such as sensors, laptops, smart home gadgets, smart carsalong these lines changing the manner in which we play, live, and work. The gadgets in IoT can be controlled remotely to play out the ideal usefulness. The IOT with Blockchain based systems are talked about as following techniques.

Ouaddah et al. [11] presented security protective access control model based on Blockchain Technology in IoT. The work shows how blockchain, the promising development behind Bitcoin, can be amazingly engaging face those rising challenges. Thus, propose fair access as another decentralized pseudonymous and furthermore protection safeguarding approval the board structure that impacts the consistency of blockchain development to direct get the chance to control in light of a legitimate concern for obliged contraptions.

Cha et al. [12] proposed a plan of a blockchain connected Gateway that safely keeps up client protection inclinations for IoT gadgets in the blockchain organize. Singular security leakage can be anticipated in light of the fact that the entryway viably shields clients touchy information from being gotten to without their assent. A secure computerized signature system is proposed for the motivations behind validation and privacy inclinations. Besides, the blockchain arrange as the basic design of information handling and upkeep to determine security disagreements.

Hassan et al. [13] proposed multithreaded spectrum allocation protocol for IoT based sensor nets. This unique protocol learns and adjusts to the expanding system thickness dependent on the system measurements. It additionally permits every hub inside IoT system to naturally distinguish the neighboring channel properties with the goal that they can swap channels to accomplish most extreme information exchange. This is practiced by persistently removing particular highlights from the system topology. In the wake of separating these highlights, the proposed convention produc-tively chooses the finest channel for an approaching hub, gives the finest network use dependent on now is the ideal time aspects, recognizes and apportions the unused range of neighboring channels via multistage Gaussian outspread premise work and multilayer perceptron-based non-direct support vector machines (SVM) character-ization show. Simulation outcomes exhibit the matchless quality of the presented convention as far as throughput, accuracy, and normal blocking likelihood, and order precision.

Rahman et al. [14] presented a portable and blockchain based screening framework in Internet of Things (IoT).This approach used to get multimodal PDA or tab-based customer association information amid dyslexia testing and offer it through a versatile edge arrange, which uses auto-reviewing calculations to find dyslexia manifestations. Despite calculation based auto-reviewing, the caught portable sight and sound payload is secured in a decentralized vault that can be bestowed to a therapeutic expert for replay and further manual examination purposes. Since the structure is dialect free and subject to Blockchain and a decentralized enormous information vault, dyslexic examples and a monstrous measure of caught interactive media IoT test information can be shared for further clinical research, real examination, and quality affirmation. Regardless, our proposed Blockchain and off-chain-based decentralized and secure dyslexia information stockpiling, the board, and sharing structure will allow security, mystery, and multimodal portrayal of the got test data for compact customers.

Roberto casado-Vara et al. [15] presented a non-direct adaptive closed-loop control framework for enhanced effectiveness in IoT-blockchain management. This approach is novel adaptive closed-loop control framework and quickens looks for model to improve the screen and control effectiveness in IoT systems, especially those which are arranged in blockchain. The non-straight control demonstrate under idea fuses another way to deal with evaluate the perfect number of squares should be at the line of the mineworkers system in order to make the strategy capable utilizing lining theory. Similarly, another system to quicken looks is presented by using hash maps, which makes the checking method snappier, strong, and capable.

Zhang and Wen et al. [16] presented an IoT electric business model utilizing blockchain, extraordinarily intended for the IoT E-business; update numerous components in customary E-business models; understand the exchange of savvy property and paid data on the IoT with assistance of P2P exchange dependent on the blockchain and shrewd contract. They additionally test their plan and influence a far reaching to talk about.

Huh et al. [17] presented an advantageous, cell phone based-finger impression acknowledgment work has been built utilizing integrated automatic log-in platform with incredible security. Here, three elements are used in the platform. To start with, it is conceivable to verify the client in PC, cell phone, and IoT situations through finger impression acknowledgment work. Second, the platform incorporates SDK to create application programming for client validation and IoT administrations. Last is its fortified security utilizing the blockchain theory to get ready adjacent to altering/manufacturing/spilling of a client's unique finger impression data by programmers.

Li et al. [18] proposed a productive and protection safeguarding carpooling plan utilizing blockchain to help and maintain privacy, location matching and data accuracy on IOT. Specifically, we affirm customers in a prohibitively puzzling way. In like manner, embrace private closeness investigation to achieve one-to-numerous vicinity coordinating and grow it to capably set up a mapping between a voyager and a driver.

Blockchain is used to store carpooling records and tree store matrices related to destinations and achieve desired area coordinating using a destination request technique. Finally, we separate the insurance and the security issues and evaluate its execution with respect to computational costs and correspondence overhead.

Viryasitavat et al. [19] proposed blockchain based business process management. Most of BPM frameworks depend vigorously on space specialists or outsiders to manage dependability. In this paper, a mechanized BPM arrangement is examined to choose and make benefits in open business condition, blockchain innovation (BCT) is investigated and recommended to exchange and check the faithfulness of organizations and accomplices, and a BPM structure is produced to delineate how BCT can be incorporated to help quick, dependable, and savvy assessment and exchanging of QOS in the work process piece and the executives.

Singh et al. [20] proposed IoT to make underground genuinely strong organization more shrewd for example Smart-SAGES, additionally by utilizing blockchain a promising improvement had been accomplished for the mining industry as it would help in checking the intrusion and obstruction of attacks like DoS and threat assessment on the SAGES information logger. The SAGES and the information transmission can be made secure.

Liu et al. [21] proposed blockchain and deep reinforcement learning (DRL) to accomplish the most unbelievable extent of amassed information, and guarantee security and steady nature of information sharing. Wide diversion results show that the proposed plan can give higher exactness and more security from various attacks when contrasted with standard database dependent on for various estimations and attacks.

Zhang et al. [22] presented a novel customer security protection framework, privacy guard, hopes to approve clients with full authority of their information on IOT structure. Blockchain with assurance guard framework dependably join new advances of trusted execution environment (TEE). By encoding information will plan and use as impressive arrangements, assurance guard can engage information proprietors to control their information, and can history of their information use. Utilizing far away insistence and TEE, guarantees that information is utilized for the fix intention affirmed by the information proprietor. The method tends to a fundamental and standard security assurances which frequently depend upon cryptography and secure techniques. In view of issue of information utilization control, privacy guard will change into the presumption with the expectation of complimentary market of private data.

Liu et al. [23] presented a blockchain dependent institutionalized independent exchange settlement framework named NormaChain for IoT dependent E-business. Here, the distinctive three-layer sharding blockchain network is designed. Also, by planning a creative decentralized public key searchable encryption scheme, we can reveal unlawful and criminal exchanges and accomplish wrongdoing recognizability. Our new DPEKS scheme cryptographically reliance of a confided in focal expert in the first PEKS conspire and rather grows it to a completely decentralized administration, which conveys the supervision control similarly among all gatherings. All the more critically, by demonstrating NormaChain is secure against chosen cipher text attacks (CCA) and against the taking of the mystery key, we demonstrate that NormaChain keeps a real client's protection from being damaged by banks, chiefs or malignant enemies.

To shield network resources from hacker attack here Li et al. [24] proposed an intrusion sensing technology. IDS is a valuable enhancement to the firewall, which can help the system framework rapidly and enhance the uprightness of the data security foundation. Here, the intrusion detection technology is connected to block chain information security model, and the outcomes demonstrate that the given model has larger location proficiency and adaptation to non-critical failure.

Yu et al. [25] proposed a LR Coin cryptocurrency subject to bitcoin in which the signature figuring utilized for approval. LR coin is appropriate for the conditions where data where information spillage is unavoidable, for example, IoT applications. Our essential objective is proposing a profitable bilinear-based constant spillage solid ECDSA signature. We show the given signature estimation is unforgeable abutting adaptively picked messages attack in the standard bilinear social event model under the relentless spillage setting. Both the hypothetical evaluation and the execution display the practicability of the proposed.

Table 1 explains various IoT with blockchain techniques described in literature along with their benefits and limitations.

3 Conclusion

This paper is a review of the different existing blockchain models with convergence of IoT and reviews the most important works around there. IoT with blockchain has been presented as an upcoming examination territory related to a couple of dug in regions of research, including privacy-preserving access controls, secure mass screening frameworks and decentralized blockchain-based verification frameworks in IoT and so forth. Further this paper talks about the favorable circumstances and issues related to the current assembly of IoT with blockchain models. In future, if work is done on data integrity and scalability then a higher accuracy can be obtained in IoT with blockchain based techniques.

Table 1 Description of advantages and limitations of several IoT with blockchain techniques

Author	Techniques	Benefits	Limitations
Ouaddah et al. [11]	Fair access based Blockchain technology in IoT	It gives the protection safeguarding. It gives a more grounded and straightforward access control	Require emerges for unified entity taking care of approval capacity to scarcely compelled IoT gadgets
Cha et al. [12]	Blockchain connected gateway in IoT	Improving client security and trust in IoT applications while inheritance IoT gadgets are still in use	Significant adjustment gadgets may be required to help the task rationale for security assurance
Hassan et al. [13]	Multithreaded spectrum allocation protocol for IoT	The proposed protocol is incredibly strong and gives high transmission rate	Extend to take the vitality varieties in thought with the ghostly effectiveness of the proposed convention protocol
Rahman et al. [14]	Spatial Blockchain-based safe mass screening structure in IoT based environment	It gives the security, secrecy, and perception of the caught test information for versatile clients	Facing the delay in uploading test modules
Roberto casado-Vara et al. [15]	Adaptive closed-loop control scheme	Speediness model to expand the monitor and control efficacy in IoT networks	Effectiveness of the proposed work is needed to be improved
Zhang and Wen et al. [16]	IoT electric business model: using blockchain innovation	Proposed system running effectively and efficiently Flexible and provides high efficiency	Significant cost reduction is needed
Huh et al. [17]	Automatic log-in platforms	Allow more secure and better assurance of delicate individual data and biometric data	Scalability is difficult
Li et al. [18]	Privacy-preserving carpooling scheme	Privacy, data auditability is good	Unnecessary communication overhead and an increased response delay
Viryasitavat et al. [19]	Business process management (BPM)	Service choices and organizations dependably and expeditiously with straightforward interoperations of dynamic associations	Scalability is difficult Time delays happen

(continued)

Table 1 (continued)

Author	Techniques	Benefits	Limitations
Singh et al. [20]	Self-advancing Goaf edge support (SAGES)'	Can store the information securely	Potential risk and money related misfortune happen
Liu et al. [21]	Blockchain-empowered proficient data collection and secure sharing scheme	Provide higher security, dependability and more grounded protection from some malicious attacks	Higher attack seriousness, the quantity of blocked solicitations will unavoidably increase
Zhang et al. [22]	Novel user privacy protection framework	Data protection Data confidentiality	Data misfortunes happen
Chunchi Liu et al. [23]	Standardized self-governing transaction settlement system based on blockchain	It is secure against the chosen ciphertext attacks (CCA) Accuracy is good	non-supervisability and immense computational overhead
Li et al. [24]	The information security model of block chain on intrusion sensing	Overall, exactness and productivity of intrusion identification is good	Classification issue happen
Yu et al. [25]	Leakage-resilient crypto currency depends on bitcoin	Efficiency is good	No confidentiality No reliability

References

1. Vashi S, Ram J, Modi J, Verma S, Prakash C (2017) Internet of things (IoT): a vision, architectural elements, and security issues. In: International conference on I-SMAC (IoT in social, mobile, analytics and Cloud) (I-SMAC), pp 492–496. IEEE
2. Lai CTA, Jackson PR, Jiang W (2018) Designing service business models for the internet of things: aspects from manufacturing firms. Am J Manag Sci Eng 3(2):7–22
3. Atlam HF, Alenezi A, Walters RJ, Wills GB (2017) An overview of risk estimation techniques in risk based access control for the internet of things. In: Proceedings of the 2nd international conference on internet of things, big data and security (IoTBDS 2017), pp 254–260
4. Domingo MC (2012) An overview of the Internet of Things for people with disabilities. J Netw Comput Appl 35(2):584–596
5. Al-Fuqaha A, Guizani M, Mohammadi M, Aledhari M, Ayyash M (2015) Internet of things: a survey on enabling technologies, protocols, and applications. IEEE Commun Surv Tutor 17(4):2347–2376
6. Karafiloski E, Mishev A (2017) Blockchain solutions for big data challenges: a literature review. In: 17th international conference on smart technologies. IEEE EUROCON 2017, pp 763–768. IEEE
7. Kshetri N (2017) Blockchain's roles in strengthening cybersecurity and protecting privacy. Telecommun Policy 41(10):1027–1038
8. Stanciu A (2017) Blockchain based distributed control system for edge computing. In: 21st international conference on control systems and computer science (CSCS). IEEE, pp 667–671
9. Christidis K, Devetsikiotis M (2016) Blockchains and smart contracts for the internet of things. IEEE Access 4:2292–2303

10. Kumar NM, Mallick PK (2018) Blockchain technology for security issues and challenges in IoT. In: International conference on computational intelligence and data science, ICCIDS, pp 1815–1823
11. Ouaddah A, Elkalam AA, Ouahman AA (2017) Towards a novel privacy-preserving access control model based on blockchain technology in IoT. In: Europe and MENA cooperation advances in information and communication technologies. Springer, Berlin, pp 523–533
12. Cha SC, Chen J-F, Su C, Yeh K-H (2018) A Blockchain connected gateway for BLE-based devices in the Internet of Things. IEEE Access, pp 24639–24649
13. Hassan T, Aslam S, Jang JW (2018) Fully automated multi-resolution channels and multi-threaded spectrum allocation protocol for IoT based sensor nets. IEEE Access 6:22545–22556
14. Rahman MA, Hassanain E, Rashid MM, Barnes SJ, Shamim Hossain M (2018) spatial blockchain-based secure mass screening framework for children with dyslexia. IEEE Access 6:61876–61885
15. Casado-Vara R, Chamoso P, De la Prieta F, Prieto J, Corchado JM (2019) Non-linear adaptive closed-loop control system for improved efficiency in IoT-blockchain management. Inf Fusion
16. Zhang Y, Wen J (2017) The IoT electric business model: using blockchain technology for the internet of things. Peer-to-Peer Netw Appl:983–994
17. Huh J-H, Seo K (2018) Blockchain-based mobile fingerprint verification and automatic log-in platform for future computing. J Supercomput 1–17
18. Li M, Zhu L, Lin X (2019) Efficient and privacy-preserving carpooling using blockchain-assisted vehicular fog computing. IEEE Internet Things J
19. Viryasitavat W, Da Xu L, Bi Z, Sapsomboon A (2018) Blockchain-based business process management (BPM) framework for service composition in industry 4.0. J Intell Manuf:1–12
20. Singh A, Kumar D, Hötzel J (2018) IoT based information and communication system for enhancing underground mines safety and productivity: genesis, taxonomy and open issues. Ad Hoc Netw 115–129.
21. Liu CH, Lin Q, Wen S (2019) Blockchain-enabled data collection and sharing for industrial IoT with deep reinforcement learning. IEEE Trans Ind Inf
22. Zhang N, Li J, Lou W, Thomas Hou Y (2018) PrivacyGuard: enforcing private data usage with blockchain and attested execution. In: Data privacy management, cryptocurrencies and blockchain technology, pp 345–353. Springer, Berlin
23. Liu C, Xiao Y, Javangula V, Hu Q, Wang S, Cheng X (2018) NormaChain: A blockchain-based normalized autonomous transaction settlement system for IoT-based E-commerce. IEEE Internet Things J
24. Li D, Cai Z, Deng L, Yao X, Wang HH (2018) Information security model of block chain based on intrusion sensing in the IoT environment. Cluster Comput 1–18
25. Yu Y, Ding Y, Zhao Y, Li Y, Zhao Y, Du X, Guizani M (2018) LRCoin: leakage-resilient cryptocurrency based on bitcoin for data trading in IoT. IEEE Internet Things J

Chapter 46
Robust Biometric System Using Liveness Detection and Visual Cryptography

Komal and Chander Kant

1 Introduction

Biometric is a secure technique of person authentication over traditional techniques (i.e., password or PIN) in which authentication of the person is based on his/her physiological or behavioral characteristics, therefore eliminating the risk of forgetting the password or PIN [1]. The biometric systems are more vulnerable to various attacks on different levels as compared to the traditional techniques. Figure 1 shows the eight types of attacks in the biometric system. In type 1 attack, imposter may fool the system by using fake characteristics. Type 2 attack is on communication channel between sensor and feature extraction module. In this attack, imposter can acquire the raw data of authenticated user. In type 3 attack, imposter may generate fake feature values and used those instead of original. Type 4 attack is similar to type 2 but this imposter may acquire the feature values of authentic user. In type 5 attack, imposter can generate high matching scores value to get access. In type 6 attack, imposter can modify the information stored in the template. In type 7 attack, imposter can acquire the information which is transmitted over this channel. In type 8 attack, imposter can overwrite the final decision [2].

Among these eight attacks, the most vulnerable points are sensor and template database, and such attacks on sensor level are called presentation attack and could be addressed using liveness detection. Biometric templates store the personal information of the user, thus there is a need to increase the security of template database [3]. There are various security techniques (i.e., steganography, cryptography, watermarking, etc.) exist to protect the template against spoof attacks. In this paper, an approach based on liveness detection and visual cryptography techniques has been used to these issues.

Komal (✉) · C. Kant
DCSA, KU, Kurukshetra, Haryana, India

S. Kumar et al. (eds.), *Proceedings of International Conference on Communication and Computational Technologies*, Algorithms for Intelligent Systems,
https://doi.org/10.1007/978-981-16-3246-4_46

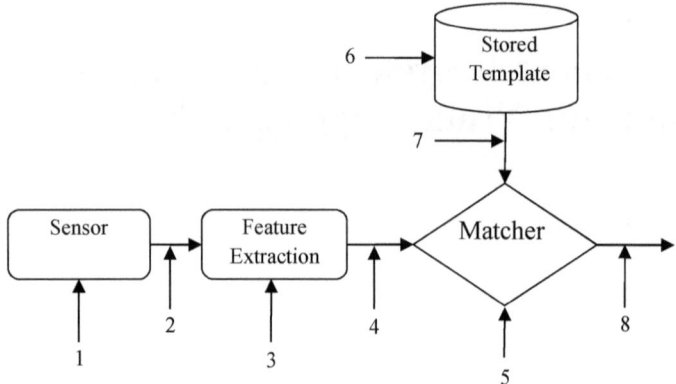

Fig. 1 Attack points in a biometric system

Liveness detection technique analyzes physical qualities and reactions of a person in order to verify if a presented biometric sample of a person at sensor level is a living object or not. This technique is a best way to detect presentation attacks in the biometric system. Liveness detection techniques can be classified into two categories: active liveness detection and passive liveness detection [4].

Active liveness detection is a challenge response method in which the user requires to be involved in the system for liveness check by responding to a challenge generated by the system. Some known examples of this technique are as follows: moving one's head up-down or left-right, recording a short video, blinking, smiling, speaking a series of words or numbers, etc.

Passive liveness detection requires no physical activity by using user's biometric traits. It requires the user do some responsibilities, e.g., enter Captcha, take selfie, etc. Different procedures are feasible for passive liveness, and each differently affects the user experience and on the handling requirements.

In the proposed system, a numerical Captcha-based liveness test is applied before sensor level to detect the bot at the early stage. Further, visual cryptography (VC) technique is used to secure the face template in the system. The basic process of visual cryptography is illustrated in Fig. 2.

Fig. 2 Visual cryptography process

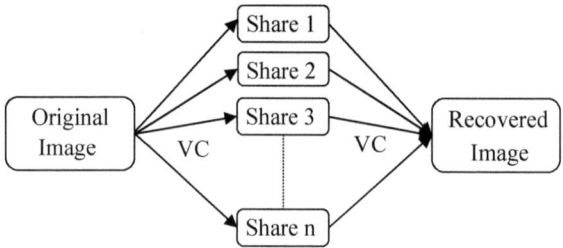

The visual cryptography technique was invented by Naor and Shamir. During the encryption process of the VC, original secrete image is divided into number of shares. These individually do not release any information of the original secrete image, and these can be stored in different places. During decryption process, simple superimposing (OR, XOR, etc.) techniques are used to combine the shares and regenerate the original image. Pixel expansion has been used for generating the shares. Each pixel is divided into number of parts and that depends on the (k, n) VC scheme used. Here, n is number of shares generated from original image and k is number of shares that can expose the original image [5]. In this paper, 2-out-of-2, i.e., (2, 2) VC scheme is used.

This paper is organized as follows: Sect. 2 includes the related works done in the area of liveness detection and visual cryptography techniques. Section 3 describes the algorithms and architecture of the proposed approach. Section 4 shows the experimental results. Last section concludes the paper.

2 Related Work

Kant and Sharma proposed a real-time and non-intrusive method to address the problem of spoofing. This approach is based on skin elasticity and thermal imaging of human face in which non-living objects are identified at very early stage of authentication process. Proposed method is very simple and user friendly [6].

Albakri and Alghowinem studied liveness detection of 3D cameras, where the results show that the performance of the system is higher because of having more flexibility. It was discovered that choosing a wide depth range of 3D camera is significant for anti-spoofing security acknowledgment system, for example, observation cameras utilized in air terminals. Therefore, proposed technique can have a greater accuracy than earlier techniques [7].

Lin et al. proposed a technique using texture features for face anti-spoofing assignment. A contextual patch-based convolution neural network (CP-CNN) is utilized for removing worldwide neighborhood and multilevel deep texture features all the while. At last, weight summation technique is utilized for choice level combination, which assists with summing up the strategy for print attack and replay attack as well as cover attack. Experiments were done on five databases, to show best results of proposed technique [8].

He and Jiang explored the feasibility of Surface Electro MyoGram (sEMG) and muscle activities, as biometric traits. SEMG provides the rate gesture recognition and electrocardiogram (ECG). The proposed system has two identity management modes: identification and verification. Equal error rate (EER) of proposed system is approx. 3.5% in identification mode and 1.1% in verification mode. Experimental results show the effectiveness of the SEMG as a biometric trait [9].

Tian et al. developed an anti-spoofing system using face as a biometric trait. Face features are extracted using computational framework. Convolution neural network (CNN) and SVM together used to classify the face features. This face anti-spoofing method controls the attacks (replay, mask, print, etc.) on the system. Proposed system is used in the real-world scenario [10].

Rohith and Vinay presented two-stage binary image security scheme using (2, 2) VC scheme such that original image cannot be unfold at the first decryption stage. Imposter becomes more relevant to the simple cryptographic algorithms. Performance of the proposed system is compared with three traditional systems. Experimental results show that proposed system has no extra information in the image shares [11].

Vaya et al. reviewed two visual cryptography (VC) schemes out of four schemes. These two schemes are (2, 2) VC and (k, n) VC. Various advantages and disadvantages of these schemes were discussed [12].

Chaudhary et al. proposed an approach to secure the biometric iris template. This approach is based on the (2, 2) VC scheme. Encryption process is simply done by dividing the template image into two shares. One share is stored on user ID card and another stored on database. Decryption process is performed by superimposing both the shares. Performance of the proposed approach is very good with 2.3% equal error rate (EER) [13].

Fathimal and Jansirani proposed a system of examination by e-question paper preparation and secure transmission of current using VC scheme. Institutional seal is used as a key to encrypt the e-question paper. This system makes easy to find out the culprit of question paper leakage. The peak signal to noise ratio may slightly increase with the use of cryptography technique. There was no problem related to pixel expansion [14].

Sanaboina et al. present how the entropy, peak signal–noise ratio (PSNR) values, and mean-square error (MSE) vary with respect to given same image with different sizes in the proposed system [15].

Yan applied a visual cryptography technique to enhance the quality of the text-based Captcha so that it becomes easier for the human beings to recognize the Captcha with naked eyes. Experimental results show the effectiveness of VC-enhanced text-based Captcha (VCETC) system [16].

3   Proposed Work

Based upon the above section, it is clear that biometric systems are vulnerable to various attacks and among these attacks, and an attack upon the sensor and templates in biometric databases is most dangerous. An attack on sensor is known as presentation attack which can be prevented with the help of liveness detection technique. In the proposed approach, numerical Captcha-based technique is used for detecting the bot at the early stage. User has to enter the correct result of the Captcha to get access to the system. On the other hand, visual cryptography technique helps in securing the template database which stores the features set of an individual. If some get access to it, then he/she can steal the personal data and misuse it. Visual cryptography divides the template image into number of shares and stores them in different templates. A

Fig. 3 Samples of
numerical Captcha

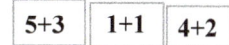

unique ID corresponding to the shares stored in different templates has been provided to the user. These shares individually do not relieve any information about the original image. Thus, the proposed approach based on liveness detection and visual cryptography has been used to resolve the issue of presentation and template attacks on the biometric system.

3.1 Liveness Detection

In this proposed system, a Captcha-based liveness test is applied before sensor level to detect the bots at the early stage. Captcha is an acronym that stands for "Completely Automated Public Turing test to tell Computers and Humans Apart." Such tests are one way of managing bot activity. There are 100 Captcha images, and their corresponding results are stored in the database. Some samples of CAPTCHA images are shown in Fig. 3.

Each time the user entered the system, one of the Captcha images will be selected at random and displayed. Users have to enter the correct result of the Captcha only then he/she will be able to access the system, otherwise he/she declared as bot and must repeat the same procedure. Some samples of Captcha images are shown in Fig. 3.

3.2 Feature Set Extraction of Face

In the proposed system, face image is generally captured with the help of appropriate sensors. Preprocessing of the actual face image consists of segmentation process. Segmentation is basically the process that is used to point out the region of interest (ROI). Local binary pattern (LBP) algorithm is used to point out the feature set of the face image [17]. Euclidean distance in between extracted key points and stored key points in template is marked to calculate matching score (MS). Face feature set extraction process is shown in Fig. 4 [18].

Fig. 4 Feature set extraction process of face

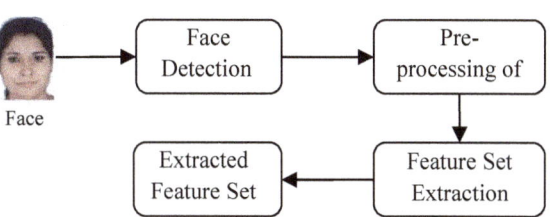

3.3   Visual Cryptography

Visual cryptography is a technique to secure secretes images which perform encryption simply by separating the original image into number of shares. Decryption process performs by superimposing (i.e., OR, XOR etc.) methods. Individually, share does not reveal any information regarding original secret image. Shamir and Naor invented the idea of visual cryptography (VC) [6].

In proposed approach, (2, 2) VC scheme with two sub-pixels is used as represented in Table 1. It divides each pixel of original image into two shares (i.e., Share 1 and Share 2). Representation of different types of shares is shown in Fig. 5. If pixel is black, then there is 50% probability of selection of one of last two columns in Table 1 to encode Share 1 and Share 2. If pixel is white, then there is 50% probability of selection of one of the first two columns in Table 1 to encode Share 1 and Share 2. Each pixel in original image is break-up into two sub-pixels such that every share is composed of 1 white and 1 black pixel. Individual share reveals no information whether a particular pixel is white or black so it is impossible to decrypt the shares. Original image is recovered by simply superimposing the two shares together. The

Table 1   (2, 2) VC scheme

Pixel	White	Black
Probability	50% 50%	50% 50%
Share1	▮◻ ◻▮	▮◻ ◻▮
Share2	▮◻ ◻▮	◻▮ ▮◻
Share1 ⊕ Share2	◻ ◻	▉ ▉

Fig. 5   Representation of shares: **a** vertical, **b** horizontal, **c** diagonal

(a) Vertical Shares   (b) Horizontal Shares

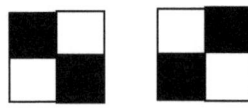

(c) Diagonal Shares

superimposing of two shares (i.e., Share 1 + Share 2) is depicted in last row of Table 1. It will consist of two black sub-pixels if original pixel is black and consist of two white sub-pixels if original pixel is white.

(2, 2) Visual Cryptography Scheme

In this scheme, original image is divided into two shares. Random bit 1 corresponds to black pixels in the original image, if Share 1 has pixel values [1 0] than set the pixel value in Share 2 as [0 1] or vice versa. Random bit 0 corresponds to white pixels in the original image, if Share 1 has pixel values [1 0] than set the pixel value in Share 2 as [1 0] or vice versa. Pseudo-code for the (2, 2) VC scheme is as follows:

```
For i = 1 to size of template
If (random bit == 1)
        If (share 1 = [1 0])
        Share 2 = [0 1]
        Else If (share 1 = [0 1])
        Share 2 = [1 0]
        End If
Else If (random bit == 0)
        If (share 1 = [1 0])
        Share 2 = [1 0]
        Else If (share 1 = [0 1])
        Share 2 = [0 1]
        End If
End If
End of For loop
```

3.4 Proposed Algorithms

Architecture of proposed system is consisting of two phases: Enrollment process and authentication process as shown in Fig. 6. Algorithms for the enrollment process and authentication process are as follows:

Algorithm for Enrollment Process

Begin

1. Capture face image of a user
2. Generate feature set or template of face image. Following steps

   a. Segmentation
   b. ROI
   c. Feature set extraction using LBP algorithm

3. Apply (2, 2) visual cryptography technique on the feature set and convert it into two shares

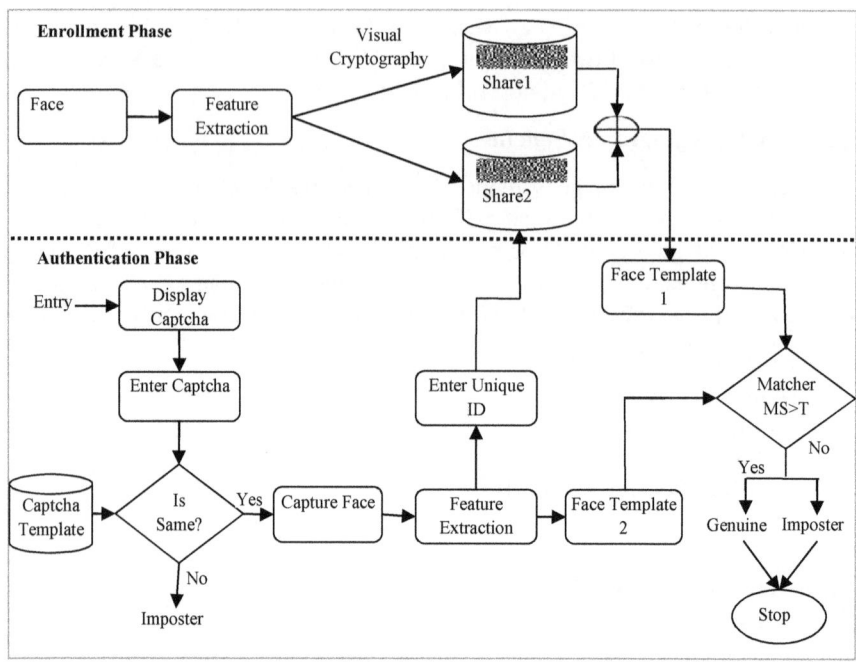

Fig. 6 Architecture of proposed approach

4. Store these shares in different templates
5. Provide unique ID to each user corresponding to their stored shares.

End

Algorithm for Authentication Process

Begin

1. Display Captcha and ask the user to enter Captcha
2. Check entered Captcha with the template database of Captcha
   If (entered Captcha = displayed Captcha) then
      Go to step 3
   Else
      Reject as a bot (non-living)
3. Capture face image of user
4. Generate feature set or template of face image

   a. Segmentation
   b. ROI
   c. Feature set extraction using LBP algorithm

5. Enter unique ID that has been provided at the time of enrollment to extract Share 1 and Share 2.

6. Apply XOR operation on the two extracted shares (Share 1 and Share 2) stored on different templates and generate one template.
7. Compare the current template (obtained in step 4) with enrolled template (obtained in step 6).
8. Find out the matching score (MS) between these two templates.
    If (Matching Score (MS) > Threshold (T)) then
    Accept as a genuine user
    Else
    Reject as an imposter

End

4   Results and Discussion

The performance and effectiveness of proposed approach have been estimated with the help of MATLAB software on available database. At enrollment time, an individual can enroll in biometric system with the face biometric. Visual cryptography technique is applied after feature extraction process to divide the face template into two shares. A unique ID has been provided to the user corresponding to the stored shares in Template 1 and Template 2. At the time of authentication, system first performed liveness test and after passing this test, extract the feature template. User has to enter the Unique ID to extract the Share 1 and Share 2 from template databases and apply XOR operation on both the shares to find out original enrolled face template. This template compared against the face template obtained at the time of authentication. If the matching score (MS) of the templates is greater than threshold, then user is accepted as genuine otherwise reject the user.

In the proposed system, a non-living (bot) user can be detected at the early stage of the system. Figure 7 shows the comparative analysis of the time taken for detection of non-living user in face biometric system and proposed biometric system.

Proposed system also takes less time in case of living user authentication because in proposed system, user has to enter unique ID to extract the shares and then one to

Fig. 7 Time versus biometric systems (face and proposed) graph for non-living user

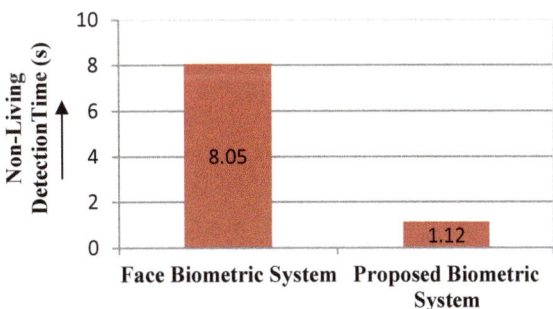

one verification process has been done. On the other hand, in face biometric system, one to many verification process has been used. Figure 8 shows the comparative analysis of the time taken for detection of a living user in face biometric system and proposed system.

In general, the performance of a biometric recognition system is calculated by false acceptance rate (FAR), false rejection rate (FRR) and genuine accept rate (GAR). FAR (%) is the percentage of number of false user accepted against total number of false user presented to the system. FRR (%) is the percentage of number of genuine user rejected with respect to total number of genuine user presented to the system. GAR (%) is calculated as $(1 - \text{FRR} (\%))$.

Figure 9 shows the receiver operating characteristic (ROC) to display the FAR, FRR and GAR of proposed approach at different threshold values. The point at which

Fig. 8 Time versus biometric systems (face and proposed) graph for living user

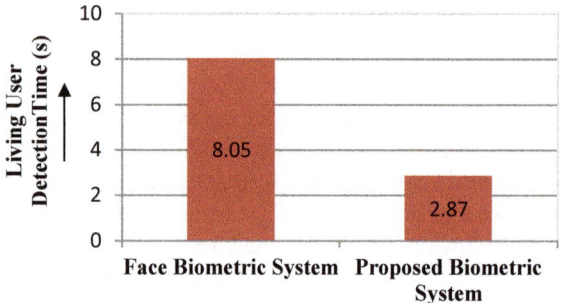

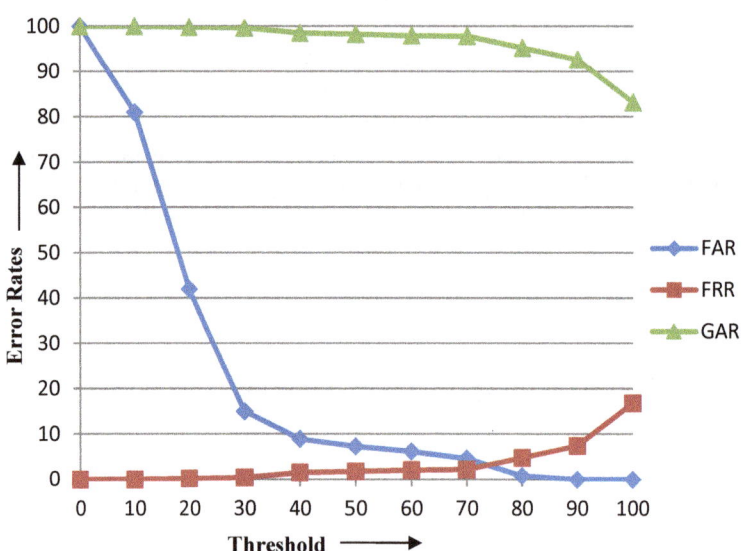

Fig. 9 ROC curves of FAR, FRR, and GAR at different threshold

FAR and FRR cross each other is called equal error rate (EER). The performance of a system is better if value of EER is lower or near to 0%. It is clear from ROC curve that the proposed system shows good recognition performance at EER 2.7.

5 Conclusion

Biometric systems are vulnerable to various types of security attacks, thereby need to protect the system by some security techniques. In this paper, an approach has been proposed to protect the biometric system from presentation attack by non-living objects and also secure the face template using visual cryptography technique. Visual cryptography is security technique to protect biometric templates which divides the template images into number of shares. The individual images in the database have no valuable information about the original image as these are only one part of share. During the authentication process, the original image gets constructed by decryption process by XORing the different shares. Proposed approach protects the system against attacks on sensor level and template database. Experimental results show that this approach is less time consuming and secure approach to protect the system, and it can be combined with other security techniques (i.e., watermarking, steganography, etc.) to increase the performance of the system, but the system might become complex.

References

1. Jain AK, Pankanti, Ross A (2006) Biometric: a tool for information security. IEEE Trans Inf Forensics Secur 1(2):125–144
2. Mwemta J, Kimwele M, Kimani S (2015) A simple review of biometric template protection schemes used in preventing adversary attacks on biometric fingerprint templates. IJCTT 20(1)
3. Schuckers S (2002) Spoofing and anti spoofing measures. Inf Secur Tech Rep 7(4):56–62
4. Babu A, Paul D (2016) A survey on biometric liveness detection using various techniques. Int J Innov Res Comput Sci Commun Eng 4(11):20055–20061
5. Naor M, Shamir A (1995) Visual cryptography. In: Proceedings of the advances in cryptology–eurocrypt'94. Lecture notes in computer science, vol 950, pp 1–120
6. Kant C, Sharma N (2013) Fake face recognition using fusion of thermal imaging and skin elasticity. IJCSC 4(1):65–72
7. Albakri G, Alghowinem S (2019) The effectiveness of depth data in liveness face authentication using 3D sensor cameras. MDPI, Basel
8. Lin B, Li X, Yu Z, Zhao G (2019) Face liveness detection by rPPG features and contextual patch-based CNN. In: Proceedings of the 2019 3rd international conference on biometric engineering and applications, pp 61–68
9. He J, Jiang N (2020) Biometric from surface electromyogram (sEMG): feasibility of user verification and identification based on gesture recognition. Front Bioeng Biotechnol 8
10. Tian Y, Zhang K, Wang L, Sun Z (2020) Face anti-spoofing by learning polarization cues in a real-world scenario. Tianjin Academy for Intelligent Recognition Technologies, pp 1–16
11. Rohith S, Vinay G (2012) A novel two stage binary image security system using (2,2) visual cryptography scheme. Int J Comput Eng Res 2(3):642–646

12. Vaya D, Khandelwal S, Hadpawat T (2017) Visual cryptography: a review. Int J Comput Appl 174(5):40–43
13. Chaudhary S, Nath R, Kant C (2018) Secure iris recognition with visual cryptography. Int J Signal Process Image Process Pattern Recogn 11(4):9–20
14. Fathimal M, Jansirani A (2019) New fool proof examination system through color visual cryptography and signature authentication. Int Arab J Inf Technol 16(1):66–71
15. Sanaboina CS, Odugu SR, Vanamadi G (2019) Secret image sharing using visual cryptography shares with acknowledgement. Int J Innov Technol Explor Eng (IJITEE) 8(11):3474–3481
16. Yan X, Liu F, Yan WQ, Lu Y (2020) Applying visual cryptography to enhance text captchas. MDPI
17. Chaudhary S, Nath R (2015) A new multimodal biometric recognition system integrating iris, face and voice. Int J Adv Res Comput Sci Softw Eng 5(4):145–150
18. Lowe DG (2004) Distinctive image features from LBP. Int J Comput Vis 60(2):91–110

Chapter 47
Environmental Parameters Influencing Perception in the Case of Multimedia Communication

R. V. Shynuⓘ and **R. D. Sambath**ⓘ

1 Introduction

People are important to study along with their environment. The environment is with its unmistakable and elusive angles, associated with the discernment for human perception. Examination in the domain of various types of the environment including nature's impact has been investigated, particularly on human attention, memory, motivation hindrance, fixation, state of mind, and psychological well-being [1]. The influence of environment design has to be investigated along with nature's environment impact over people. Such an environment-based human perception must be investigated so that the productivity of correspondence from various conditions can be perceived. Studies were conducted to comprehend the design and the executives' parts of users' discernment and experience dependent on recreational conditions. The result shows that target-situated experience can be documented in various environmental settings on nature-based context. Examinations demonstrate that the natural components did not bring down members' overall environment encounters. So model-based examination's discoveries are proposed to control future users' conduct research [2]. This information shows the requirement for researching the human perception of an environment. The investigation conducted by Franco et al. [3] states that the presence of nature in visuals just as encountering through windows extemporizes the mind-set and well-being. This shows the pertinence of incorporating nature's environmental characteristics to human conditions for such target-situated experience. Goodey et al. [4] tended to the interest to concentrate over huge scope conditions like built environment assembled conditions concerning discernment. Presentation to nature utilizing controlled laboratory settings exhibited a loosening up encounter

R. V. Shynu (✉) · R. D. Sambath
Department of Architecture, TKM College of Engineering, Kollam, Kerala, India

while contrasting with the pictures of built environment situations [5]. This observational information shows that the presence of common habitats impacts lower attentional commitment and intellectual burden to environmental contrasted and physical conditions. This is fundamentally significant for a designer to accomplish such mental quality through a fabricated or designed environment. Nature plays "therapeutic impacts of associating with the indigenous habitat" as indicated by attention restoration theory [6]. This information on environment impact over human insight connotes the significance of nature as a natural parameter. Exploration uncovers that while considering the impression of the youngsters' age gathering, the guide assumes a critical function in correspondence. This data is essential for a designer so that to comprehend youngsters' discernment to give important learning environment at the beginning phases [7]. This information indicates the importance to comprehend the experience of the user in the media context as well. The experimental investigation recommends that nature introduction-based advantage of protecting the environment for both real and virtual conditions improves prosperity and environmental exchanges [8]. The outcome proposes the chance of ad-libbing the real environment exhibition over the innovation-based environment and expressing the noteworthiness of nature's advantages over human discernment. It is likewise essential to investigate various conditions because of arising mechanical mediations. Since a few environmental conditions are seen dependent on various mental circumstances, it is additionally essential to extemporize the users' psychological state and solace level. Individuals are from real built space settings who were presented to nature gaze for comparable expect nature in his/her environment [9]. This interest is a result of the impact of restorative nature that documents exhausted consideration limit and extemporizes pressure reaction in human autonomic sensory systems. Progressed physiological perusing gadgets like EEG and eye tracking used to assess people's perception and discernment. The investigations investigated with the alpha action identified with unwinding, and beta action concerning consideration completed with these gadgets. The outcomes are captivating that no huge contrasts are found in both real built space green and calm conditions [10]. This outcome prompts the designed environment with affecting variables and contended to examine through controlled environmental settings to distinguish better perception. An enormous number of estimations are accessible in a few examinations utilizing eye tracking and EEG, for a top to the bottom introduction of eye-tracking and EEG measurements allude [11]. The announced zones of engineering innovation, neuroscience, and psychophysiology dependent on eye tracking, EEG, visual, space, real built environment, video, film, cognition, and perception-related points are considered in Table 1 to show the examination interest.

These articles spoke to context demonstrating the boosts mode for leading visual perception. Observational investigation locales assume a critical function in environmental perception-based assignment execution. The examination that completed distinctive natural conditions for the cognizance and discernment-based trials is demonstrated here as boundaries. The information uncovers various boundaries that utilized for the investigation reason. The distinguished boundaries are real environment, semi-immersive, virtual immersive, virtual screen, virtual simulation or

Table 1 Review of selected 44 articles published in journals between 2010 and 2020

Author	Research gap	Key findings
Li et al.	Combined use of EEG and eye tracking	The study shows that the closed-loop personalized virtual art intervention can achieve a higher degree of engagement compared to traditional virtual art intervention [12]
Ahtola et al.	Larger scale (infant and adults)	Research of novel analysis of eye-tracking data during the face distractor competition paradigm explored sustained attention for faces in infants [13]
Wiens et al.	Large samples $n > 150$ (working memory)	Meta-analysis result confirmed the demanding visual tasks can reduce the mismatch negativity (MMN) to auditory distracters [14]
Aspinall et al.	Engagement alertness attention	The new form of high-dimensional correlated component logistic regression analysis reveals the evidence of lower frustration, engagement and arousal, and higher meditation while moving into the green space zone; along with higher engagement while moving out of it [15]
Wang et al.	User group selection based on socio-demography	Statistical analysis demonstrated that the driving duration and rest patterns have enough impacts on the driver's visual behaviors and driving activity [16]
Peterson et al.	Realistic virtual reality	Results show that virtual reality decreased dynamic balance performance and increased physical and cognitive loading while comparing to unaltered viewing at low heights [17]
Chuang et al.	Frequency (theta) band investigation for "neural efficiency"	The study confirmed the degree of temporal stability in attention and arousal regulation during the preparatory period before motor execution through the measurement of Fm θ power [18]

(continued)

Table 1 (continued)

Author	Research gap	Key findings
Li et al.	HMM-based training data set possible patterns	Driving cognitive load in the driver's reaction to the driving demand, which is elevated by the driver's physiological state and traits, as well as the road environment condition [19]
Desvergez et al.	Situation awareness and cognitive load through EEG and eye tracking	Experts visual behavior demonstrated an increased number of total fixation points in comparison with residents performance [20]
Goto et al.	EEG activity related to viewing several products	Results state that ERPs related to single consumer items can predict the behavioral preferences depending upon the type of ERP effects are chosen by the researcher [21]
Tremmel et al.	Virtual reality environment using EEG and EMG	Cognitive workload-based activity generates individual differences in brain activity, which similarly require the development of subject-specific models [22]
Lanata et al.	Eye behavior for imagination performance	Imagery processing affects a magnitude of cognitive, physiological, and behavioral phenomena in many regions such as learning, problem-solving, and consumer experiences [23]
Maggio et al.	VR for rehabilitation, EEG-based approach	EEG and VR-based approach can be a promising aspect to define the most appropriate stimulation protocol, to promote an effective personalization of the rehabilitation task [24]
Yang et al.	Emotions aroused by noise and visual	Perception over landscape plants provides excess noise reduction effects through subjects emotional processing, for psychological noise reduction [25]

(continued)

Table 1 (continued)

Author	Research gap	Key findings
Dimigen et al.	Expectation predictability through EM–EEG relationship	Research shows EEG recordings while normal vision is feasible and useful to consolidate findings for both EEG and eye-tracking studies [11]
Güntekin and Tülay	Beta oscillatory responses	The process of experimental design shows high influences over the perception of international affective picture system (IAPS) pictures [26]
Zheng et al.	Emotion performance	The feature-level fusion and decision-level fusion while integrating the EEG signals and eye-tracking data improve the performance of emotion-based recognition model [27]
Romei et al.	Fluctuations in alpha band	Findings indicate a direct connection between alpha band activity and visual cortex excitability and state that perceptually relevant changes in these measures can increase immediately [28]
Dahal et al.	Behavioral manifestations based on distraction	Results suggest that distracted/non-distracted driving condition can be identified using this approach supporting the practicality of cognition research [29]
Wang et al.	Attention toward two tasks simultaneously	The behavioral activity and the calculated FOAs indicate the immediate switching of human attention/strategies between the different cognitive tasks to optimize their overall performance during the dual-task condition [30]
Nikolaev et al.	Perceptual and cognitive processes	Research proposes the solution for co-registration of EEG and eye movements consists of similarity in eye movement characteristics over the experimental conditions [31]

(continued)

Table 1 (continued)

Author	Research gap	Key findings
Al-Barrak et al.	Impact of urban environments	Neuro place as a better system to classify outdoor places according to the mental condition with a focus on relaxation [32]
Weaver et al.	Limitation over design	The relative salience of task importance and task-irrelevant stimuli can demonstrate situations where an increase in cognitive control is necessary, with individual differences in VWM capacity stating significant variance in the degree of monitoring and control of task-directed eye movement behavior [33]
Maksimenko et al.	Alpha band (8 ± 12 Hz) and beta band (20 ± 30 Hz)	The report may have important applications for monitoring and controlling human alertness in situations in which more attention is required [34]
Olszewska-Guizzo et al.	Alpha waves and beta oscillations	The study demonstrates the different amounts of greenery through a window view can change the brainwave patterns of the viewer [35]
Olszewska-Guizzo et al.	Effects of exposure to environment with varying visual quality	Results are promising in supporting the study of hypothesis and states that exposure to urban green spaces can be connected to mental health outcomes [36]
Valenzi et al.	Alpha asymmetry as a detector for emotions	Research protocol played a key role in delivering high accuracy and, since there is a very high individual difference in the representation of emotions, participants have to be analyzed separately [37]
Campbell and Moran	Perceptual–cognitive judgments	Evidence of expertise differences in visual perceptron task over a simulated green-oriented visual task [38]

(continued)

Table 1 (continued)

Author	Research gap	Key findings
Valtchanov and Ellard	Cognitive and affective responses to environments	Exploring the cognitive and affective responses to environments provides a new direction for the future, suggesting that there is a possibility to have environmental cognitively restorative. Environments are emotionally restorative, and environments can promote both cognitive and emotional restoration [39]
Venkatraman et al.	Biometrics and EEG readings	The significance of neurophysiological measures to complement the traditional approach of measures in improving the prediction of advertising success models [40]
Matukin et al.	Increased sample size	The results state designer-oriented guidelines to advertisers on how to improvise the visual design components of their pamphlets, outdoor advertising, packaging, brochures, and digital advertisements, or how to create web pages or other interactive digital communication regions [41]
Zheng et al.	EEG and eye movements for emotion recognition	Results show the modality fusion with multimodal deep neural networks can improvise the performance compared with a single modality, and the best mean accuracy of 85.11% is gained for four emotions (sad, fear, happy, and neutral) [42]
van Almkerk and Huisman	Involvement of fractal dimensions in restorative effects	The task performed here to offer some tentative results for the involvement of fractals in restorative effects of virtually developed nature, but also has to consider other qualities of nature [43]

(continued)

Table 1 (continued)

Author	Research gap	Key findings
Michael et al.	Sample size to find emotional and cognitive responses	The brain scanning can be used for better understanding the underlying unconscious emotional aspects and cognitive processes that affect consumer thought and activity [44]
Giannakos et al.	Prediction accuracy of users learning performance	Consider the multimodal data that can enable human–computer interaction (HCI) and learning technology for the researchers to examine unscripted, complex tasks in a much more holistic and accurate way [45]
Rogers et al.	Theta as a measure of engagement	The experiment suggests that modulation of frontal theta, developed from a single channel of EEG, expresses the subjective-induced sense based on the EDNA system [46]
Franchak et al.	Behavior and the role of vision	Infants eyes directions are to the relevant areas in the environment to meet changing task demands, continuously switching between different patterns of visual exploration [47]
Berto et al.	Particular cognitive task	Data's of eye movement shows the different kinds of attentions that are engaged for fascinating versus non-fascinating scenes [48]
Valtchanov et al.	Virtual environment featuring a nature or urban	The study demonstrates the computer-generated nature in VR can also promote restorative effects [49]
Codispoti et al.	Top-down template of guides	Affective LPP modulation is not depended on picture color, even under challenging perceptual conditions like picture presentation limited to a brief 24-ms flash [50]

(continued)

Table 1 (continued)

Author	Research gap	Key findings
Chumerin et al.	Classification of the source of fun- hard fun and easy fun	BCI gaming is good in evoking emotions by using more intensive game experience questionnaires and different kinds of elaborate BCI games [51]
Chen and Wu	Research subjects in different academic levels	Voice-based videos, with its relatively low expense, both the lecture capture and picture-in-picture videos may be effective for online learning from the perspectives of improved learning activity and reduced cognitive load [52]
Antle et al.	Tasks that are appropriate to the sample population	The results show the effectiveness of mobile neuro feedback-based interventions to help young kids living in poor condition develop self-regulation skills [53]

interaction, images or static condition or graphics, and real-time render or moving image.

Distinguished environmental influencing or input factors on the exploratory environmental context spoke distinctive subdomains, and they are nature or human, perspective point of vision or peripheral vision, virtual design, real built or designed environment, sound or audio, visual emotion or value or grade, color (black and white or gray) and light or comfort or motivation or reward. Here impacting factor demonstrates the distinguished environmental condition during the test cycle for improving the exhibition and effectiveness for the user.

2 Environmental Parameters Impact on Perception

2.1 Real Environment

The boundary, real environment on the exact investigations is negligibly acquainted concurring with the chose research information. The real environment speaks to human's ongoing prompt environmental factors; it can likewise allude to their temperament of discernment toward the neighborhood environment with no computerized medium or gadgets. Exploratory examination expresses the impact of the environment's current circumstance over consideration limit and therapeutic impact over the focused on autonomic sensory systems in the human body [9]. Since qualities' environmental characteristics have a huge association and impact over the human solace level, it is ideal to think about such boundaries during research examinations. The investigation of Keliher in 1997 states that the children' insight into outside the school in like manner is their openness to the media, particularly TV; it is likewise anticipated that the children elective discernments may regularly result from media and writing. The school encounters on discernment advance a positive perspective on nature and the environment that lead to dynamic nature concern [7]. This information shows the meaning of understanding the idea of the environment in the media for perception study. Also, analysts have given proof to the mental impacts of openness to common settings, utilizing controlled research center trials and field contemplates. The information recommends that the visual impression of indigenous habitats gives low attentional and psychological preparing contrasting and genuine constructed spaces [5]. This outcome shows the significance of considering nature as a boundary since nature affects the human solace level. The outcomes show that individuals presented to the environment's current circumstance feel associated with it and search out comparable normal space in the built environment setting for elevated levels of prosperity [9]. The outcome shows the extension in emulating nature-put together design concept on the virtual environment for improving the prosperity of its user group.

2.2 Virtual Screen

The virtual screen as a boundary demonstrates a compelling use while surveying observational exercises. The virtual screen class-based experimental examinations are intended to gage insight with advanced presentation conditions like LCD monitor, monitor, and touch screen. Research facility settings for artificial conditions are created with such gadgets in assessing members' visual consideration. Results propose the visual consideration increments toward such simulation-based environmental setting [54, 55]. The perception toward an environment utilizing such gear or mechanisms for virtual conditions that created must be dissected. The feeling of "presence" is identified with computer-generated reality research and is as yet an immense domain that must be perceived. Studies anticipated the idea of evoked reality, and proposed models may have critical applications in the potential outcomes of not simply computer-generated reality and investigating the presence [56]. This idea will assist with getting reality and computer-generated reality deductively. The nature of workstation screens regarding resolution, color, and gleam free sound system assumes a huge function in creation feeling of the real world. The execution and design of cave automatic virtual environment (CAVE) through computer-generated experience or logical representation framework in detail as projection innovation applied to augmented reality objectives should be looked after [57]. This innovation will underline the exactness of reality in virtual screens to feel a more sensible environment. Result likewise found that both goal and show size significantly affected undertaking finishing time dependent on a huge, high-goal show delivering the best outcomes to clarify drenching levels [58]. The assignment to convincingly introduce a mimicked world is as yet an issue to explain for augmented reality concerns which gives near the domain of user interface computer-generated reality [59], so it expanded drench level documented through high caliber in computer-generated experience conditions. The outcome recommends that virtual screen as an environment for correspondence would expand the experience more reasonable for the users.

2.3 Semi-immersive

The semi-immersive boundaries survey shows a critical decrease in thinking about such boundaries during the trial systems. Semi-immersive innovation is an arising zone. The productivity of such innovation needs to investigate the discernment and cognizance-based undertaking. Semi-immersive classification shows the examination conducted utilizing 3D virtual conditions like CAVE and augmented reality. The virtual reality (VR)-based innovation ad-libs the association and representation experience of the user gathering. Likewise, the examination shows that there is an adequate extent of occasions to control inside the users' three-dimensional environment with the guide of voice order. Exploration discovering states that the performance and the interaction within the semi-immersive environment are great [60]. Kyriakou and accomplices study exhibited solid examples documented on the group and turned out to be more exact during their intelligence level dependent on VR programming. This VR innovation makes the environmental setting more reasonable. The members saw a more elevated level of authenticity and announced a significant level of essence during the assignment. The observational investigation shows that members adhere to the VR environment rules. Results propose that the member's conduct and emotional measures are impacts by the VR environment through simulation [61]. The standard-based programming shows the capability of a semi-immersive environment where collaboration and presence kept up. The participant connection between real and virtual makes a more consistent and firm insight. The semi-immersive environment results suggest recording members experience more precisely in the objective-driven assignment. This experience brings augmented reality, mixed reality, and human attentional practices which can be all-around mixed inside the semi-immersive space. Thus, the examinations expected that semi-vivid innovation as an incredible arising innovation for engaging human consideration and discernment measure.

2.4 Virtual Immersive

The virtual immersive parameters have thought about not in most of the exact investigations. Here virtual vivid is taken to the setting where head-mounted display

(HMD) and comparable gadgets consider research tests. Immersive virtual environment technology (IVET) can empower the change in perspective inside social brain science examinations. The review expected that IVET inversily effect the study results by compromising the test control authenticity. However, the results show that IVET rules over interactive media designs for socio-mental experimentations [62]. IVET-based examinations can comprehend the standard of conduct on individuals. This innovation additionally gives the occasion to assess virtual collaborations and actual connections. Likewise, customary overviews to address a few inquiries by the investigator can supplant IVET over target populaces. Exploration proposes IVET a compelling instrument to gage human conduct and communication to comprehend social presence [62, 63]. IVE innovation furnishes the members with the experience of being encircled by the computer interaction orchestrated environment [62]. This innovative advancement will be a preferred position to have more biological characters or exhibiting a real environment without bargaining normal space and trial characteristics. The investigation by Deering in 1992 recognized that head-followed sound system shows as a fundamental segment of numerous computer-generated experience frameworks. The precision of the head-following innovation restricts the visual enrollment of the physical and virtual universes. The precision and steadiness of the actual arrangement's adjustment estimations are additionally a noticeable piece of better visual enlistment [64]. So the environment, as well as the detail, has a fundamental part in accomplishing vivid virtual conditions. Virtual immersive has been utilized for the two collaborations and understanding purposes. The investigation shows that consideration commitment can chronicle just as expanded perusing execution in the HMD than in the real environment. Also, the user can focus on various planes in real and virtual space that plays little heed to the screen. Exploratory outcomes express that attentional commitment can assess with the guide of HMD [65]. Since HMD is an arising device, the examination on the conduct and attentional level can analyze for additional investigations.

2.5 Virtual Simulation

Virtual reproductions in the augmented simulation are the cutting-edge innovation for achieving sensible and overwhelming connection with experience. The virtual simulation or collaboration boundary shows that reenactment-based virtual setting for consideration commitment and connections are considered reasonable for observational assessments. Similarly, the boundary virtual simulation alludes to the observational examinations led with personal computer-based reproductions and association with the virtual environment. Reenactment is fundamental in the logical and mechanical cycle for different virtual field tests. Virtual-based recreations broadly acted in the clinical field for exploratory purposes. Result is recognized as a promising innovative strategy for laparoscopic psychomotor abilities. The virtual simulation-based environment advances exact execution and correspondence for target purposes [66]. Studies like continuous human group recreations led to comprehend chain of command made out of virtual people, gatherings, and groups. The novel idea of the research demonstrates the quality of representing virtually simulated crowd more realistically [67]. The reasonable recreated action over a virtual environment may advance the user experience all the more engagingly because of its inclination. The exploration of Courgeon et al. shows joint consideration reenactment utilizing virtual people and eye-following examination that the user just has fractional attention to controlling gaze unexpected showcases. The specialized difficulties incited by investigating the client's focal point of consideration in computer-generated experience are assessed and given a few proposed arrangements. The outcome recommends that reproduced natural designed movement can extemporize consideration and relational abilities [68]. This innovation shows that such boundaries were embraced for the exploratory investigation to answer the few issues recognized in the examination domain, similar to gaze execution and consideration commitment. Virtual people will improve the virtual encounters, preparing help, and in any event, educating and coaching, which will help spare lives by giving substitutes to clinical preparing, careful arranging, and far off telemedicine [69]. Since virtual human is the portrayal of nature quality, understanding the nature calculation for reenacting the example in a virtual environment will be fundamental. This virtual human simulation helps in field tests where social difficulties and financial matters are requested. The outcomes show uninvolved and conversationally inspired visual contrast consideration with the no liveliness setting. Also, virtual human-based practical liveliness impacts gaze conduct. Results recommend that visual consideration toward the virtual human declines while looking toward objective coordinated exercises increments when the members face basic circumstances in between close to home clinical reproductions [70]. Since virtual simulation is as effective as a communicating media, data is prescribed to utilize sensible reenactment to chronicle better human commitment for estimating personal conduct standard on virtual conditions. This information speaks to the user cooperation is productive and alright with such a correspondence medium.

2.6   Images

The boundaries, for example, images, static pictures, and graphical substance of exploratory examinations show the moderate application. Pictures or static or graphics in the boundary allude to the utilization of photos, charts, and texts for assessing consideration and perception levels. Target characterizing objectives accomplished through a comparable cycle with distractors for attentional commitment. The examination explored the jobs of attentional commitment to feature-specific ACS (FACS) and category-specific ACS (CACS). The investigation shows understanding conduct and attentional level dependent on FACS and CACS [71]. The incitement and designed content utilized attentional commitment study and for gaze flicker [72]. The examination by Kahn in 2017 with the guide of pictures for the utilization of visual design to improve participant perception is dissected profoundly, particularly in online stages. The examination expected that users move a greater amount of their task excursion with the aid of web media. The outcomes demonstrate that participants' discernments structure consequently without intellectual intercession, and consideration regarding these environment visual upgrades will be fast. So understanding the principal brain science of perception, attentional, and programmed induction is basic for such a test study [73]. Itti and gathering directed saliency-based visual consideration for fast scene examination models to comprehend visual consideration frameworks. The consequences of organic understanding controlling the engineering demonstrated effective in repeating primate visual frameworks [74]. Such neuronal structural examinations researched with the guide of boundaries for discovering image design-based visual consideration. Distinctive visual guide coordinating planned pictures with qualities structure of nature and artificial images suggested as additional investigation in visual attention and engagement examination.

2.7   Real-Time Render

Real-time render or moving picture boundaries show the emerging application in discernment-based examinations. The classification alludes to the investigations led with recordings and ongoing delivered visuals. Konstantopoulos et al. in 2010 built up the investigation of driver's eye development during the day and night conditions on a controlled virtual environment. The task is dependent on a driving test system with continuous visual rendering method to speak the driving environment. This examination gives a comprehension of the driver's permeability condition and driving experience [75]. This exploration approach shows the unwavering quality of considering ongoing delivered conditions to research eye movement and visual consideration. The idea of a VR-based virtual get together framework shows its interest in mechanical applications and practices. Augmented reality-based devices show their critical incentive for helping the planner in innovative design [76]. Such ongoing delivering conditions were on the ascent and applied in a few virtual environmental stages. Also, the investigation with immersive virtual environment innovation shows a wide scope of possible advantages. The exploration zeroed in an arrangement of how people see and react to regular and constructed conditions. In any case, the innovation used to address long-standing examination inquiries in human discernments and inclinations for fabricated and normal settings stayed restricted because of expanding customer interest for inundation frameworks in the entirety of specialists' scope [77]. The user request permits coordinating both moving pictures and real-time-based render frameworks for breaking down such conditions through the test system. All these examination information offers a scope of possible advantages in the agreement how people see and react to build natural environments on real-time rendering context.

3 Environmental Influencing Factors on Perception

3.1 Perspective Point of Vision

Thinking about perspective and fringe the point of vision, perspective point of vision had a huge impact over perception. This study area required further requests to comprehend its actual property. Findlay and Gilchrist in 2001 examination recognized that psychological cycles are dependent on confidential consideration concerning visual consideration. The eyes bearing in preprocessing data in the event of dynamic vision relies upon the area's visual outskirts [78]. The examination expects human visual commitment may get impacted dependent on the perspective point of visuals in media. Viewpoint vision has a critical part in data handling, and the perspective point of vision will not be confined to the viewpoint levels. However, it is additionally a change in a visual structure. The three-dimensional (3D) structure has a more significant level of data than the two dimensions (2D). The 3D substance on the media pulled in considerably more consideration in the television industry and film. This outcome considered the following key to the future for the user visual experience [79]. The 3D structure advances significantly more data than 2D since 3D is outwardly showing the nature of the viewpoint purpose of vision for higher correspondence. The helpful fields examined while considering fringe vision if there should arise an occurrence of the users' field of vision to go to objects going into the visual field [80]. The exploratory information exhibits that human viewpoint purpose of visuals influences their visual attentional level. The examination by Vater et al. in 2016 fringe state dreams has a critical function in insight. The vision additionally expresses that vision-related viewpoints, for example, spatial vulnerability may be influenced by the visual framework's capacities that influence change in identification rates [81]. This outcome again mirrors the impact of perspective viewpoint on an environment that may get an opportunity in affecting the member's consideration and discernment, especially in multimedia.

3.2 Virtual Design

The design application on the environment impacts attention engagement. Likewise, the virtual design setting has an unmistakable function in attention engagement. For better and advantageous correspondence in a virtual environment, a virtual environment's design assumes a critical job. Presently, a day's computerized reasoning and reproduction innovations are consolidated as a feature of virtual design improvement. The investigations recognized typified specialists' essence in a virtual environment to see others' consideration [82], which is found through the function of gaze and consideration bearing dependent on the virtual environment. Since the design impacts the gaze development, result recognized virtual environment with design specialists could document better consideration. On account of engineering, assembled natural models are proposed for visual discernment. Various discernments distinguished through an assortment of environment condition, particularly a serious level of insight requested on account of an arena or show lobby [83]. The outcome distinguished insight toward the designed environment recognized the impact of discernment impact dependent on the nature of vision. This points out the importance of design language to in-wrinkle the level of discernment in virtual spaces. The examination by Swan in 2001 recognized that design factors influence understudies' learning and fulfillment in virtual space. The chance of collaboration through the environment, clearness of design in the virtual space and arrangement of a dynamic conversation with the members altogether expanded learning and fulfillment [84]. The outcome exhibits that designed virtual space could keep up the fulfillment level of individuals. This outcome predicts that the new gadgets in future advance web-based commerce prospects. This outcome likewise requests to comprehend users' consideration and programmed deduction [73]. The chance of establishing an indigenous habitat through virtual design advances a superior encounter for the members. The visual experience from an elevated position or wide perspectives of an area let loose the mind or illuminate the feelings; additionally, experience can be chronicled through a designed space like constructed conditions [83]. This data opens the opportunities for a designer to build up an environment to upgrade user experience toward their current circumstance. Also, the domain of design-based virtual environment for visual correspondence must be considered for further investigations.

3.3 Real Built Environment

The real built environment speaks two environments, one that will not have any design esteems, and others with the environment created are dependent on design standards. The outcome expected both real and virtual design-based conditions influence the insight and consideration levels of a client in that space. All fabricated zone connection to consideration has considered investigating its impacting level. On a perception test, the listening condition had lower grades than those in the standard listening condition that does not shift while assessing both listening conditions on account of perusing methodology [85]. This outcome shows the need to investigate the natural factors that impact the commitment level and consideration of members. Comparable examination results show that design can give uncommon encounters that invigorate feelings, detects, and scholarly reactions. It likewise can add to the making of excellent conditions. The outcome examines the interruption of consideration because of the impression of the constructed environment configuration factor [86]. While considering the structure arranging angle, result recognized that the structure spatial design decipherability is a fundamental viewpoint for more established individuals [87]. The outcome distinguished that the nature of the constructed environment impacts the users' navigational fulfillment perspectives. This condition coordinates the significance of a design over the space or environment for effective perception and consideration. Human-caused conditions to significantly affect discernment and perception prompt conduct execution. This unmistakable constructed environment chronicled through various design philosophies relying upon their nearby context. The impression of youngsters affected by their nearby environmental factors [88]. Their background and character development are interconnected elements which form them to a model resident. In their persuasive examination, Fusco et al. in 2012 portray how a youngster's perception and early local environment for educational encounters, for example, day by day outdoor environment experience their conduct. The examination recommends the need for expanding indigenous habitat-based commitment in younger students. The result shows the effect of the assembled environment during local environment mode over discernment and visual commitment [88]. Here, nature and the extent of the fabricated environment investigated so how discernment and conduct nature of youngsters comprehended. The information shows the individual contrasts in working through socio-segment and contrasts in psychological and efficient shared traits dependent on their reaction to the built environment [89]. The exploration expresses that the idea of the assembled environment can bring out an enthusiastic reaction. Computational models these days emulate the real conditions for documenting the practical experience for research examination. Users' visual perception, intellectual state, and irregular consideration dependent on the diverted environment created through computational testing. The model investigated likelihood and prescient handling information on psychological displaying, to build up an intellectually conceivable user model for nearby environment situations [90]. All this exploration information expresses the meaning of fabricated environmental relationship toward the insight and discernment angles.

3.4   Sound

It is similarly fundamental to think about sound and visuals for assessing the observers' visual consideration. Sound or audio properties additionally have a conspicuous function in perception and attention. Understandings of nature's sound over people by considering winged creature's voices inspected. Result recognized that particular feathered creature's sound adds to pressure recuperation and consideration rebuilding. Improvements based on rebuilding impact from pressure or decreased excitement and weariness are documented dependent on certain influence. The examination exhibits the member's attention restoration and stress recovery levels influence by the sound [91]. The research discoveries show that keen correspondence association created for an easy attentional center, option in contrast to novel nature. The expanded and the level of audience consideration is tended to by evoked symbolism and consideration on sound fiction dependent on audio effects and shots. Similarly, the outcomes show the consideration of particularly music and portrayal audio cues created in an anecdotal radio dramatization, which expands audience members' psychological symbolism. The exploration recognized the capacity of sound to summon mental pictures and attention on human [92]. The trial directed by

Stiefelhagen et al. exhibits the use of sound to assess focus and attention in the partic
ipant. Contemplated dependent on members' focal point of consideration through
sound and recordings can figure more precise assessment. An expected centering
of human attentional commitment dependent on sound through gaze examination is
accessible [93]. This outcome gives an objective to chronicle better human–computer
interaction collaboration like brilliant gathering rooms through sound quality. All
this information brings up the part of the sound that improves consideration commit
ment. Testing the impact with differing sound improvements for visual consideration
appropriation has significance in cross-disciplinary general media interpretation. The
examination expresses that incited stimuli change the viewer's perception of dynamic
media [94]. The examinations were utilized to comprehend hearable insight in kids
for the improvement of aptitudes related to hearable discernment which shows the
absence of accessible data. Wide scope of psychophysical, electrophysiological, and
acoustical hearable movement discernment led to contemplate insight [95]. The
examination likewise proposes modern specialized methodologies by considering
virtual space incitement strategies coordinating with the kinematic following for
future investigation. This exploration approach may assist with researching method
ically to assess the impact of sound. Results demonstrate that there is a critical
propensity for young ladies to perform with more prominent precision than young
men. Notwithstanding, for youngsters in the upper evaluations, these distinctions can
be ascribed more to inspiration, demeanor, and goal as far as melodic fitness. Result
likewise recognized that young ladies for the most part kept on improving their exhi
bition at each undertaking [96]. So the thought of gender-based sound discernment
is huge for such information investigation. All these examination information show
the significance of sound in the media-based environment since sound thought about
a successful improvement for consideration and core interest.

3.5   Emotion, Value, and Grade

Human perception and attentional levels impact the passionate worth. The inves
tigations distinguished that visual appearance modifies consideration, potentiating
feeling, and uneasiness. People should collaborate with the environment because of
their capacity to identify and organize the preparing of significant data around them.
Result discovered shows emotion and tension associated with particular considera
tion during the analysis. The analyst expresses that upgrading the presence of contrast
can improve consideration execution [97]. Similarly, the examinations dependent
on feeling insight into feeling experience express that enthusiastic pictures cannot
essentially lift passionate responses as a rule yet with feeling help [98]. Hence, the
specialists accepted that feeling enlistment strategy can make a huge commitment to
understanding the neural designs. This information shows the pertinence of consid
ering emotion as an impacting factor over consideration and insight to comprehend
passionate levels dependent on attentional commitment on the environment. The
outcome additionally states negative worth-based pictures showed up more clearly

to the high-dread insight of the members. This information exhibits the consideration completely accomplished through sincerely inspiring pictures [99]. Hur and his partners in 2017 explored working memory and discernment. The outcome shows attentional assets, just as the center relies upon the insight and feeling. The examination demonstrates the conduct impact of feeling on discernment change dependent on high working memory load [100]. This outcome again expresses that human passionate impact has an enormous function in the consideration level. The outcomes demonstrate that willful visual consideration increments with vision performance. The impacts of unbiased pictures-based consideration controls by apparent passionate force were broader than for the negative and positive [101]. This research outcome proposes that consideration is autonomous for more broad mental practice. The ordinary feeling discernment dependent on neural has been additionally tested. Phillips et al. at 2003 recognized the enthusiastic essentialness of a natural upgrade, emotion conduct, and the creation of an emotional state. Feeling conduct permits the age of logically based feeling encounters and practices based on enthusiastic perception [102]. Comparative investigations with the guide of EEG show that without focal consideration, feeling discernment kept up. The investigation determines if feeling insight requires focal attentional assets. It likewise finds that negative feelings catch consideration and get attentional need more successfully than good feelings [103]. All these assessment information recommend that both positive and negative passionate viewpoints can impact the attentional sense. Subsequently, the media-based natural examination for discernment conveys a noticeable part over-enthusiastic worth.

3.6 Color

A few investigations are done by the exploration researcher on color-based consideration exhibitions in a virtual environment. The outcome distinguished that the interior focal point of consideration is comparative with the conditions that advance engine execution. The examination dependent on the constructive outcome of color in the participant environment indicated the utilization of color upgraded engine execution. Color seems to keep people from noticing body-focused signs without express verbal interchanges [104]. Similar examination identified with high contrast tones and shapes to comprehend individuals' view of object completed. The outcomes demonstrate not the profile of an object; however, the objects' colors used for the transaction have a basic condition that improves individuals' impression on object [105]. The outcomes exhibited that color impacts individuals' discernment. All these trial information bring up the impacts of color in the domain of discernment and consideration since color has an influential job. Similarly, it is imperative to consider its impacting level in human–environment consideration and execution. The tones and high contrast have been acquainted with investigating the discernment-based analyses. Studies dependent on highly contrasting subjects by the highly contrasting appearances propose dark subject are predominant in retaining dark contrasted and white subject profile. White countenances are preferable in recollections white appearances over dark appearances. A system for investigating culturally diverse contrasts in consideration dependent on facial highlights was proposed [106]. This outcome shows that color impacts the view of a person toward their current circumstance. Essentially, the highlights that quandary consideration and insight attempted to investigate the capacity to see important items in an incorporated scene dependent on shadings rely upon complex visual cycles [107]. This information shows the impact of color over consideration and discernment. So color as a boundary is similarly essential to contrast with different parameters while exploring the natural discernment and perception undertakings.

3.7 Comfort

The comfort level of an environment is extremely vital for the insight and discernment viewpoints. The outcome distinguished through the experimental examination that light, motivation, and reward are impacting boundaries for accomplishing members' solace level during observational investigations. The exploration by Lourenço et al. in 2019 utilized user comfort on insight and conduct to comprehend the user conduct and diverse utilitarian zones inside the local premises through light and energy designs. The investigation of utilizing light examples in an environment shows more execution on the user. Exploration shows that different variables and rules impact the conduct examples of the user in multiple conditions. Additionally, users' visual comfort boundaries like agreeable brightening level unequivocally affected their presentation [108]. This outcome shows that light contributes a solace level to participants' dependent on its force. The exploration likewise recognized the act of remuneration-based inspiration in the errand commitment measure distinguished in observational investigations. The investigation by Cameron and Pierce in 1994 demonstrates the instance of remuneration-based inspiration errands. Fundamental member inspiration for better execution does not diminish dependent on generally speaking prizes likewise recognized. This information shows substantial prize reflects adversely for doing assignments separately. It was because of the condition when the evacuation of remuneration condition, an insignificant negative impact, was estimated by the time spent on the undertaking performed [109]. A comparative instance of studies exhibits antagonistic social and instructive brain research issues by analysts and neuroscientists analyzed dependent on rewards. The investigation accepted that uplifting feedback, impetuses, and remunerates adversely influence a person's inspiration. The exploration suggests the prize framework could upgrade the member responsibility and viable movement on the errand [110]. Survey information expresses the exhibition upgraded through a prize framework or giving rousing to prosperity and better execution. Likewise, environment solace assumes a noticeable part in the exhibition and proficiency of the user gathering. Similar research outcome shows warm distress in the working spaces which exhibited the impact of representatives' profitability and bearableness. The abstract and non-actual boundaries influence participants' environmental solace insight [111]. This examination expresses that emotional and non-actual boundaries influence client environment comfort and consideration. Isolated ways fundamentally impact the outdoor math and encompassing condition impact participants' comfort and keeping in mind that they focus on the outdoor circumstances on local environment and satisfactory performing space [112]. The solace level can likewise be through a prize framework. Prize-based decrements and increments are in intrinsic motivation on the beneficiary's view. Also, the apportioned prize capacities are fundamentally significant. Prizes can have either sabotaging or upgrading impacts relying upon conditions [113]. Essentially, comfort documented through various viewpoints like the impact of light and inspiration, all these affecting variables recognized in the chosen experimental investigations. The discoveries give data to seeing how actual conditions and virtual environment impact users' view of solace and help design agreeable conditions for execution on facilities.

4   Conclusions

This exploration territory on the human perception of the environment is concentrated on human prosperity and supported the turn of events. Genuine conditions or common settings directly affect human beings. Counterfeit conditions utilizing present-day incitement innovation are a suitable answer for psycho-discernment contemplates. The environment-based consideration examination through various parameters and affecting components has a huge function in the observational works. The exploration considers the diverse environmental setting for examining attentional and perception measure. There is an adequate scope to receive various elements to upgrade the users' performance since recollections are created dependent on environment perception. The designed environment and real nature are accepted to upgrade users' current circumstance discernment and memory for developing effective visual correspondence. Coordinating different parameters and affecting components for environmental discernment would be more compelling at psychological examinations.

References

1. Bratman GN, Hamilton JP, Daily GC (2012) The impacts of nature experience on human cognitive function and mental health. Ann N Y Acad Sci 1249(1):118–136
2. Dorwart CE, Moore RL, Leung Y-F (2009) Visitors' perceptions of a trail environment and effects on experiences: a model for nature-based recreation experiences. Leis Sci 32(1):33–54
3. Franco LS, Shanahan DF, Fuller RA (2017) A review of the benefits of nature experiences: more than meets the eye. Int J Environ Res Public Health 14(8):864
4. Goodey B, Gold JR (1987) Environmental perception: the relationship with urban design. Progr Hum Geogr 11(1):126–133

5. Grassini S et al (2019) Processing of natural scenery is associated with lower attentional and cognitive load compared with urban ones. J Environ Psychol 62:1–11
6. Kaplan S (1995) The restorative benefits of nature: toward an integrative framework. J Environ Psychol 15(3):169–182
7. Keliher V (1997) Children's perceptions of nature. Int Res Geogr Environ Educ 240–243
8. Reddon JR, Durante SB (2018) Nature exposure sufficiency and insufficiency: the benefits of environmental preservation. Med Hypotheses 110:38–41
9. Richardson M et al (2017) Nature: a new paradigm for well-being and ergonomics. Ergonomics 60(2):292–305
10. Neale C et al (2020) The impact of walking in different urban environments on brain activity in older people. Cities Health 4(1):94–106
11. Dimigen O et al (2011) Coregistration of eye movements and EEG in natural reading: analyses and review. J Exp Psychol Gen 140(4):552
12. Li G et al (2020) Closed-loop attention restoration theory for virtual reality-based attentional engagement enhancement. Sensors 20(8):2208
13. Ahtola E et al (2014) Dynamic eye tracking based metrics for infant gaze patterns in the face-distractor competition paradigm. PLoS One 9(5):e97299
14. Wiens S, Szychowska M, Nilsson ME (2016) Visual task demands and the auditory mismatch negativity: an empirical study and a meta-analysis. PLoS One 11(1):e0146567
15. Aspinall P et al (2015) The urban brain: analysing outdoor physical activity with mobile EEG. Br J Sports Med 49(4):272–276
16. Wang Y, Ma C, Li Y (2018) Effects of prolonged tasks and rest patterns on driver's visual behaviors, driving performance, and sleepiness awareness in tunnel environments: a simulator study. Iran J Sci Technol Trans Civ Eng 42(2):143–151
17. Peterson SM, Furuichi E, Ferris DP (2018) Effects of virtual reality high heights exposure during beam-walking on physiological stress and cognitive loading. PLoS One 13(7):e0200306
18. Chuang L-Y, Huang C-J, Hung T-M (2013) The differences in frontal midline theta power between successful and unsuccessful basketball free throws of elite basketball players. Int J Psychophysiol 90(3):321–328
19. Li Y et al (2019) A driver's physiology sensor-based driving risk prediction method for lane-changing process using hidden Markov model. Sensors 19(12):2670
20. Desvergez A et al (2019) An observational study using eye tracking to assess resident and senior anesthetists' situation awareness and visual perception in postpartum hemorrhage high fidelity simulation. PLoS One 14(8):e0221515
21. Goto N et al (2019) Can brain waves really tell if a product will be purchased? Inferring consumer preferences from single-item brain potentials. Front Integr Neurosci 13:19
22. Tremmel C et al (2019) Estimating cognitive workload in an interactive virtual reality environment using EEG. Front Hum Neurosci 13
23. Lanata A et al (2019) Nonlinear analysis of eye-tracking information for motor imagery assessments. Front Neurosci 13
24. Maggio MG et al (2020) Virtual reality based cognitive rehabilitation in minimally conscious state: a case report with EEG findings and systematic literature review. Brain Sci 10(7):414
25. Yang F, Bao ZY, Zhu ZJ (2011) An assessment of psychological noise reduction by landscape plants. Int J Environ Res Public Health 8(4):1032–1048
26. Güntekin B, Tülay E (2014) Event related beta and gamma oscillatory responses during perception of affective pictures. Brain Res 1577:45–56
27. Zheng W-L, Dong B-N, Lu B-L (2014) Multimodal emotion recognition using EEG and eye tracking data. In: 2014 36th annual international conference of the IEEE engineering in medicine and biology society. IEEE
28. Romei V et al (2008) Spontaneous fluctuations in posterior α-band EEG activity reflect variability in excitability of human visual areas. Cerebral Cortex 18(9):2010–2018
29. Dahal N et al (2014) TVAR modeling of EEG to detect audio distraction during simulated driving. J Neural Eng 11(3):036012

30. Wang Y-K, Jung T-P, Lin C-T (2015) EEG-based attention tracking during distracted driving. IEEE Trans Neural Syst Rehabil Eng 23(6):1085–1094

31. Nikolaev AR, Meghanathan RN, van Leeuwen C (2016) Combining EEG and eye movement recording in free viewing: pitfalls and possibilities. Brain Cogn 107:55–83

32. Al-Barrak L, Kanjo E, Younis EMG (2017) NeuroPlace: categorizing urban places according to mental states. PLoS One 12(9):e0183890

33. Weaver MD, Hickey C, Van Zoest W (2017) The impact of salience and visual working memory on the monitoring and control of saccadic behavior: an eye-tracking and EEG study. Psychophysiology 54(4):544–554

34. Maksimenko VA et al (2017) Visual perception affected by motivation and alertness controlled by a noninvasive brain-computer interface. PLoS One 12(12):e0188700

35. Olszewska-Guizzo A et al (2018) Window view and the brain: effects of floor level and green cover on the alpha and beta rhythms in a passive exposure EEG experiment. Int J Environ Res Public Health 15(11):2358

36. Olszewska-Guizzo A et al (2020) Can exposure to certain urban green spaces trigger frontal alpha asymmetry in the brain?—Preliminary findings from a passive task EEG study. Int J Environ Res Public Health 17(2):394

37. Valenzi S et al (2014) Individual classification of emotions using EEG. J Biomed Sci Eng 2014

38. Campbell MJ, Moran AP (2014) There is more to green reading than meets the eye! Exploring the gaze behaviours of expert golfers on a virtual golf putting task. Cogn Process 15(3):363–372

39. Valtchanov D, Ellard CG (2015) Cognitive and affective responses to natural scenes: effects of low level visual properties on preference, cognitive load and eye-movements. J Environ Psychol 43:184–195

40. Venkatraman V et al (2015) Predicting advertising success beyond traditional measures: new insights from neurophysiological methods and market response modeling. J Mark Res 52(4):436–445

41. Deitz GD et al (2016) EEG-based measures versus panel ratings: predicting social media-based behavioral response to super bowl ads. J Adv Res 56(2):217–227

42. Zheng W-L et al (2018) EmotionMeter: a multimodal framework for recognizing human emotions. IEEE Trans Cybern 49(3):1110–1122

43. van Almkerk M, Huisman G (2018) Virtual nature environments based on fractal geometry for optimizing restorative effects. In: Proceedings of the 32nd international BCS human computer interaction conference 32

44. Michael I et al (2019) A study of unconscious emotional and cognitive responses to tourism images using a neuroscience method. J Islam Mark

45. Giannakos MN et al (2019) Multimodal data as a means to understand the learning experience. Int J Inf Manage 48:108–119

46. Rogers JM et al (2020) Single-channel EEG measurement of engagement in virtual rehabilitation: a validation study. Virtual Reality 1–10

47. Franchak JM et al (2011) Head-mounted eye tracking: a new method to describe infant looking. Child Dev 82(6):1738–1750

48. Berto R, Massaccesi S, Pasini M (2008) Do eye movements measured across high and low fascination photographs differ? Addressing Kaplan's fascination hypothesis. J Environ Psychol 28(2):185–191

49. Valtchanov D, Barton KR, Ellard C (2010) Restorative effects of virtual nature settings. Cyberpsychol Behav Soc Netw 13(5):503–512

50. Codispoti M, De Cesarei A, Ferrari V (2012) The influence of color on emotional perception of natural scenes. Psychophysiology 49(1):11–16

51. Chumerin N et al (2012) Steady-state visual evoked potential-based computer gaming on a consumer-grade EEG device. IEEE Trans Comput Intell AI Games 5(2):100–110

52. Chen C-M, Wu C-H (2015) Effects of different video lecture types on sustained attention, emotion, cognitive load, and learning performance. Comput Educ 80:108–121

53. Antle AN et al (2018) East meets west: a mobile brain-computer system that helps children living in poverty learn to self-regulate. Pers Ubiquit Comput 22(4):839–866
54. Couperus JW, Lydic KO (2019) Attentional set and the gradient of visual spatial attention. Neurosci Lett 712:134495
55. Folstein JR, Monfared SS, Maravel T (2017) The effect of category learning on visual attention and visual representation. Psychophysiology 54(12):1855–1871
56. Pillai JS, Schmidt C, Richir S (2013) Achieving presence through evoked reality. Front Psychol 4:86
57. Cruz-Neira C, Sandin DJ, DeFanti TA (1993) Surround-screen projection-based virtual reality: the design and implementation of the CAVE. In: Proceedings of the 20th annual conference on computer graphics and interactive techniques
58. Bowman DA, McMahan RP (2007) Virtual reality: how much immersion is enough? Computer 40(7):36–43
59. Bates J (1992) Virtual reality, art, and entertainment. Presence Teleoperators Virtual Environ 1(1):133–138
60. Gao S, Wan H, Peng Q (2000) An approach to solid modeling in a semi-immersive virtual environment. Comput Graph 24(2):191–202
61. Kyriakou M, Pan X, Chrysanthou Y (2017) Interaction with virtual crowd in immersive and semi-immersive virtual reality systems. Comput Animat Virtual Worlds 28(5):e1729
62. Blascovich J et al (2002) Immersive virtual environment technology as a methodological tool for social psychology. Psychol Inquiry 13(2):103–124
63. Bailenson JN et al (2003) Interpersonal distance in immersive virtual environments. Pers Soc Psychol Bull 29(7):819–883
64. Deering M (1992) High resolution virtual reality. In: Proceedings of the 19th annual conference on computer graphics and interactive techniques
65. Toyama T et al (2015) Attention engagement and cognitive state analysis for augmented reality text display functions. In: Proceedings of the 20th international conference on intelligent user interfaces
66. Grantcharov TP et al (2001) Virtual reality computer simulation. Surg Endosc 15(3):242–244
67. Musse SR, Thalmann D (2001) Hierarchical model for real time simulation of virtual human crowds. IEEE Trans Visual Comput Graph 7(2):152–164
68. Courgeon M et al (2014) Joint attention simulation using eye-tracking and virtual humans. IEEE Trans Affect Comput 5(3):238–250
69. Badler N (1997) Virtual humans for animation, ergonomics, and simulation. In: Proceedings IEEE nonrigid and articulated motion workshop. IEEE
70. Volonte M et al (2018) Empirical evaluation of virtual human conversational and affective animations on visual attention in inter-personal simulations. In: 2018 IEEE conference on virtual reality and 3D user interfaces (VR). IEEE
71. Wu X, Liu X, Fu S (2016) Feature- and category-specific attentional control settings are differently affected by attentional engagement in contingent attentional capture. Biol Psychol 118:8–16
72. Andrews S, Lo S, Xia V (2017) Individual differences in automatic semantic priming. J Exp Psychol Hum Percept Perform 43(5):1025
73. Kahn BE (2017) Using visual design to improve customer perceptions of online assortments. J Retail 93(1):29–42
74. Itti L, Koch C, Niebur E (1998) A model of saliency-based visual attention for rapid scene analysis. IEEE Trans Pattern Anal Mach Intell 20(11):1254–1259
75. Konstantopoulos P, Chapman P, Crundall D (2010) Driver's visual attention as a function of driving experience and visibility. Using a driving simulator to explore drivers' eye movements in day, night and rain driving. Accid Anal Prev 42(3):827–834
76. Jayaram S, Connacher HI, Lyons KW (1997) Virtual assembly using virtual reality techniques. Comput Aided Des 29(8):575–584
77. Smith JW (2015) Immersive virtual environment technology to supplement environmental perception, preference and behavior research: a review with applications. Int J Environ Res Public Health 12(9):11486–11505

78. Findlay JM, Gilchrist ID (2001) Visual attention: the active vision perspective. In: Vision and attention. Springer, New York, NY, pp 83–103
79. Huynh-Thu Q, Barkowsky M, Le Callet P (2011) The importance of visual attention in improving the 3D-TV viewing experience: overview and new perspectives. IEEE Trans Broadcast 57(2):421–431
80. Wolfe B et al (2017) More than the useful field: considering peripheral vision in driving. Appl Ergon 65:316–325
81. Vater C, Kredel R, Hossner E-J (2016) Detecting single-target changes in multiple object tracking: the case of peripheral vision. Atten Percept Psychophys 78(4):1004–1019
82. Peters C (2005) Direction of attention perception for conversation initiation in virtual environments. In: International workshop on intelligent virtual agents. Springer, Berlin, Heidelberg
83. Bittermann MS, Ciftcioglu O (2008) Visual perception model for architectural design. J Des Res 7(1):35–60
84. Swan K (2001) Virtual interaction: design factors affecting student satisfaction and perceived learning in asynchronous online courses. Distance Educ 22(2):306–331
85. Marchand GC et al (2014) The impact of the classroom built environment on student perceptions and learning. J Environ Psychol 40:187–197
86. Peri Bader A (2015) A model for everyday experience of the built environment: the embodied perception of architecture. J Archit 20(2):244–267
87. Tao Y et al (2018) Legibility of floor plans and wayfinding satisfaction of residents in Care and Attention homes in Hong Kong. Australas J Ageing 37(4):E139–E143
88. Fusco C et al (2012) Toward an understanding of children's perceptions of their transport geographies: (non) active school travel and visual representations of the built environment. J Transp Geogr 20(1):62–70
89. Nasar JL (1989) Perception, cognition, and evaluation of urban places. In: Public places and spaces, pp 31–56
90. Pekkanen J et al (2018) A computational model for driver's cognitive state, visual perception and intermittent attention in a distracted car following task. R Soc Open Sci 5(9):180194
91. Ratcliffe E, Gatersleben B, Sowden PT (2013) Bird sounds and their contributions to perceived attention restoration and stress recovery. J Environ Psychol 36:221–228
92. Rodero E (2012) See it on a radio story: sound effects and shots to evoked imagery and attention on audio fiction. Commun Res 39(4):458–479
93. Stiefelhagen R, Yang J, Waibel A (2001) Estimating focus of attention based on gaze and sound. In: Proceedings of the 2001 workshop on perceptive user interfaces
94. Vilaró A et al (2012) How sound is the pear tree story? Testing the effect of varying audio stimuli on visual attention distribution. Perspectives 20(1):55–65
95. Carlile S, Leung J (2016) The perception of auditory motion. Trends Hear 20:233121651664425
96. Pressnitzer D et al (2018) Auditory perception: Laurel and Yanny together at last. Curr Biol 28(13):R739–R741
97. Barbot A, Carrasco M (2018) Emotion and anxiety potentiate the way attention alters visual appearance. Sci Rep 8(1):1–10
98. Baumgartner T, Esslen M, Jäncke L (2006) From emotion perception to emotion experience: emotions evoked by pictures and classical music. Int J Psychophysiol 60(1)
99. Enns JT, Brennan AA, Whitwell RL (2017) Attention in action and perception: unitary or separate mechanisms of selectivity? Progr Brain Res 236:25–52
100. Hur J et al (2017) Emotional influences on perception and working memory. Cogn Emotion 31(6):1294–1302
101. Mrkva K, Westfall J, Van Boven L (2019) Attention drives emotion: voluntary visual attention increases perceived emotional intensity. Psychol Sci 30(6):942–954
102. Phillips ML et al (2003) Neurobiology of emotion perception I: the neural basis of normal emotion perception. Biol Psychiatry 54(5):504–514

103. Shaw K et al (2011) Electrophysiological evidence of emotion perception without central attention. J Cogn Psychol 23(6):695–708
104. De Giorgio A et al (2018) Enhancing motor learning of young soccer players through preventing an internal focus of attention: The effect of shoes colour. PLoS One 13(8):e0200689
105. Piqueras-Fiszman B et al (2012) Is it the plate or is it the food? Assessing the influence of the color (black or white) and shape of the plate on the perception of the food placed on it. Food Qual Prefer 24(1):205–208
106. Ellis HD, Deregowski JB, Shepherd JW (1975) Descriptions of white and black faces by white and black subjects. Int J Psychol 10(2):119–123
107. Treisman A (1998) Feature binding, attention and object perception. Philos Trans R Soc Lond Ser B Biol Sci 353(1373):1295–1306
108. Lourenço P, Pinheiro MD, Heitor T (2019) Light use patterns in Portuguese school buildings: user comfort perception, behaviour and impacts on energy consumption. J Clean Prod 228:990–1010
109. Cameron J, Pierce WD (1994) Reinforcement, reward, and intrinsic motivation: a meta-analysis. Rev Educ Res 64(3):363–423
110. Hidi S (2016) Revisiting the role of rewards in motivation and learning: implications of neuroscientific research. Educ Psychol Rev 28(1):61–93
111. Castaldo VL et al (2018) How subjective and non-physical parameters affect occupants' environmental comfort perception. Energy Build 178:107–129
112. Li Z et al (2012) Investigating bicyclists' perception of comfort on physically separated bicycle paths in Nanjing, China. Transp Res Rec 2317(1):76–84
113. Morgan M (1984) Reward-induced decrements and increments in intrinsic motivation. Rev Educ Res 54(1):5–30

Chapter 48
A Machine Learning Approach to Analyze and Predict the Factors in Education System: Case Study of India

Jeena A. Thankachan and Bama Srinivasan

1 Introduction

Education is the soul of any country for the sake of its development and progress. Prosperity and country building is possible only when a nation has a solid and successful education approach and policy. The literacy rate of India was 12.2% in 1947 which has increased to 74.0% in 2011 census. Although it looks like an accomplishment, still many are there without access to education. The literacy rates in the states of India are dependent on the number of time variants and time-invariant attributes. According to study by KPMG (one of the Big Four accounting organizations) online education market in India is expected to grow to billion from million dollars by 2021 and from million users in 2016 to billion users in 2021. But there are many achievement gaps to be met before achieving the target [1]. Those achievement gaps include availability of digital learning solutions and learning software for teachers to conduct online classes, learning models for effective delivery of education, and other integration of technology methods in the existing Indian education system [2].

In addition to these gaps, Covid-19 pandemic has also drastically affected the education sector as well. Based on the influence of pandemic Covid-19 on education, UNESCO has compiled data globally that lead to the closures of schools and colleges in the quarter-1 of the year 2020 on the scale of localized and national level. The visualization map presented (see Fig. 1), obtained from the dataset [3] analyzed on tableau shows the school closures with note on initiation of distance learning, home-based learning, and online learning for upcoming academic year 2020–21. Initiation

J. A. Thankachan (✉) · B. Srinivasan
Department of Information Science and Technology, Anna University, CEG Campus, Chennai 600025, India

B. Srinivasan
e-mail: bama@auist.net

COVID-19 impact - School closures in 2020 Q1

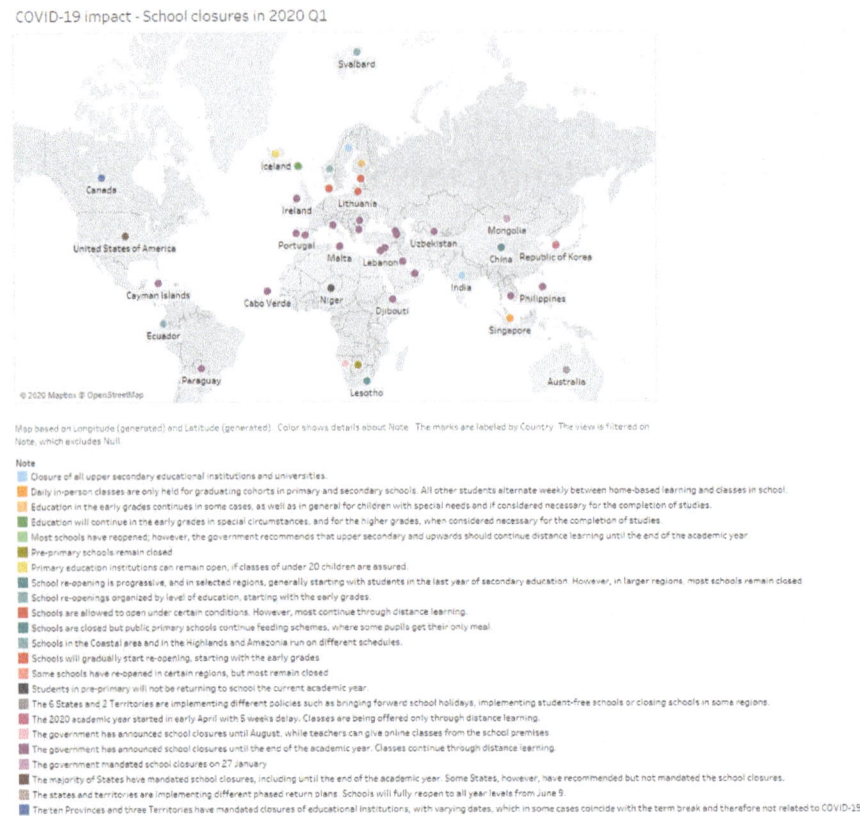

Fig. 1 Covid-19 impact on school closures

of distance learning, home-based learning, and online learning are a challenge for developing and undeveloped nations.

In this paper, challenges of implementing online-based learning in the schools of India had been analyzed and the significant factors contributing to the education system of the nation is determined based on the dataset shared by MHRD (Ministry of Human Resource Development, Govt. of India) [4]. This paper aims to present significant attributes contributing to higher and lower rates of education in India currently. To analyze, this paper first uses PCA to reduce the dimension and then employs three ML algorithms. The main factors that contribute to education are determined with this model.

The study in this paper is structured as follows: Sect. 2 presents the background in the case of research applied on education using machine learning algorithms and Indian education. Section 3 gives a description of the features of Dataset and preprocessing approach. Section 4 describes the dimensionality reduction using Principal

Component Analysis and Sect. 5 provides the details of the analysis using supervised machine learning approaches. Section 6 discusses the implications based on the analysis and Sect. 7 summarizes the work.

2 Background

Use of Artificial Intelligence (AI) tools had been widely accepted and favored in various domains and applications. Similar research and application in the field of education had been more dominant now. While AI may not replace teachers, its synergistic use with other educational technologies along with the teachers would give optimum benefits to the objectives of New Education Policy (NEP)-2019 MHRD, Government of India, so that quality education is available even to rural students [5]. Also, the study in this paper, aids Education for All to meet the learning needs of all children, youth, and adults.

A few studies had been undertaken to improve the overall education system of India by few researchers. But the growth pace to create an impactful change in the educational standards and quality of the system has become sluggish. An overview of some of the recent research trends in Indian education system and machine learning approach in worldwide education is quoted in Sects. 2.1 and 2.2 respectively.

2.1 Indian Education System

A researcher in his theoretical study explains about significance and relevance of Vedic methodologies in the current education system of India. According to him peace, loving society, human beings with good character and ethics are necessary. This can be achieved through Vedic model of imparting education [6]. Few researchers have also presented a critical analysis of ancient and modern education system of India. They have stated that the ultimate goal of the ancient system was self-realization and preservation of noble ideas and culture whereas it does nowhere exist in the current rat race system of education. They have also suggested some of the prime factors to be implemented in order to reinvigorate our learning mechanism [7]. Some have analyzed the challenges faced in Indian education system. Some of the points mentioned were faculty shortage, poor research quality, untrained teachers, irrelevant curriculum, gap in demand and supply. The author concludes that by joint efforts of government, public and private sectors there can be significant change and the educational quality can be highly improved [8].

2.2   Machine Learning Approaches in Education

One of the applications of machine learning is predicting performance of student. It introduces a model that learn the weakness of each student and suggests better methods to improve [9]. Another machine learning model based on assessment has been developed to help students and teachers with feedback on their progress and gap towards the goal [10]. A blended course-based study on finding and solving student's problems in initial stages [11]. An AI-based application on freshmen learning progress is successful or failing [12]. Few studies on applying supervised machine learning algorithm and linear modeling on massive open online courses (MOOCs) and students' perceptions of MOOCs have been performed to find the factors that significantly predict student satisfaction [13]. Recent studies on AI applications in higher education predicts based on an early warning system for the dropouts and to be retained students [14].

From the knowledge gained by above background works, in the proceeding sections, an educational dataset is analyzed using machine learning tools to determine the significant factors contributing towards higher literacy in a developing nation.

3   Dataset and Data Preprocessing

In this paper, a dataset shared by the MHRD is used for the study. This dataset has 680 observations (district wise) with 800 variables for statistical analysis [4]. The dataset is first cleaned and preprocessed with the intension to retain variables that were only significant in contribution to the education system of a nation. The data cleaning is done by checking for missing values if any and by eliminating the repeated values. Nearly 700 attributes are eliminated by replacing them with their sum aggregates into an individual attribute. Further correlation between the attributes obtained by correlation matrix helps to eliminate the attributes with NA values and very less correlation. Thus 800 attributes are reduced to 24 based on its relevance in the future economy of the education system. Hence the preprocessed dataset includes the attributes like State Name, District name, Number of villages, total population, growth rate, sex ratio, overall literacy rate, total schools (government, and private), total rural schools, total enrollment, teachers in government and private, schools approachable by all-weather road, schools with electricity, schools with computers, schools with less enrollment, number of classrooms in schools, teachers in school, professionally qualified government and private teachers and special schools Grade 1–8 (for Blind, Low vision, Hearing Impairment, Speech Impairment, Locomotor Impairment, Mental Retardation, Learning Disability, cerebral palsy, autism and more).

The preprocessed dataset is analyzed for attribute selection using a dimensionality reduction method, Principal Component Analysis. The flow of the study in this paper is represented in the flowchart (Fig. 2).

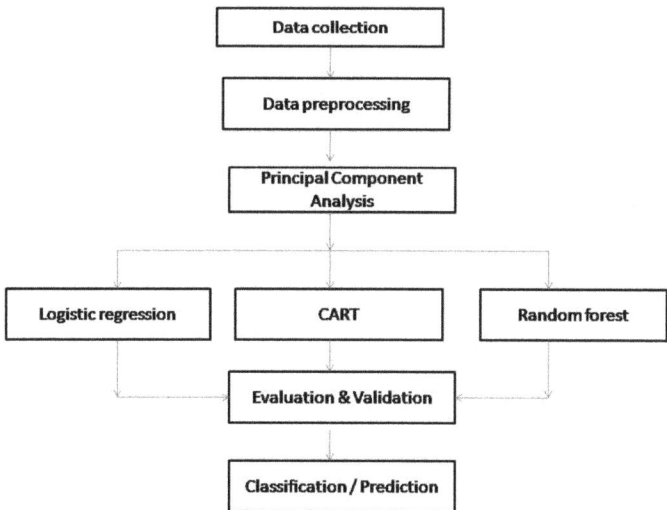

Fig. 2 Flow chart of the analysis

4   Principal Component Analysis

In this paper, the analysis shows that PCA is able to select important attributes from the dataset. The preprocessed and cleaned dataset with 680 observations and 24 variables is analyzed to obtain the principal components. Here the non-numeric attributes like state name and district name are removed. Further, we eliminate the attributes that are constant values and perfectly correlated to itself with standard deviation zero and correlation with other variables has no values. Thus we select 15 attributes that have correlation with each other based on correlation matrix.

The eigenvectors and eigen values of a correlation matrix represent the "core" of a PCA, in order to determine the principal component. In eigenvectors versus eigenvalues graph (Fig. 3) where the first Eigenvector (first principal component)

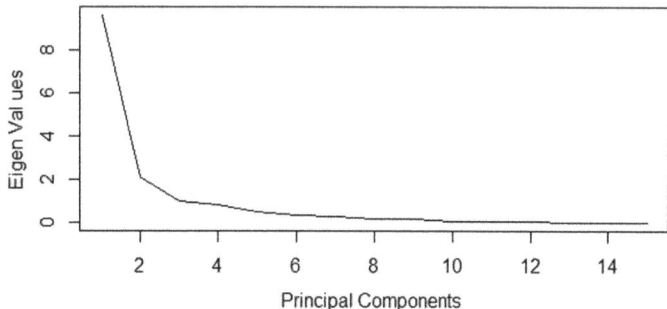

Fig. 3 Principal components versus eigenvalues graph

Table 1 Variance extracted by each component

Var 1	Var 2	Var 3	Var 4	Var 5	Var 6	Var 7	Var 8
64	14	6.4	5	3.3	2.2	1.6	1
Var 9	Var 10	Var 11	Var 12	Var 13	Var 14	Var 15	–
0.8	0	0.2	0.2	0.0	0.0	0.0	–

determines the direction of the new feature space with maximum variance, and the Eigen values determine their magnitude. From the figure, we understand that first component explains 90% of the common variance.

Variance Extracted by first Principal Components is more than the other Principal Components as shown in Table 1. The eigenvectors [2:15] with the least eigenvalues are dropped due to less information retained by them. The large and positive loadings on first principal component indicate that these attributes have a strong effect on that principal component and hence the first principal component is retained. Mathematically the variance and principal components are represented by Eqs. 1 and 2 respectively.

$$\sigma^2 = \frac{\sum_{i=1}^{n}(Xi - \mu)^2}{n-1} \tag{1}$$

where σ^2 is sample variance, Xi is value of ith element, μ is sample mean and n is sample size.

$$PC_i = \sum_{i=1}^{n}\alpha_i X_i \tag{2}$$

where $X_1, X_2 \ldots X_n$ are variables from input feature space and α_i are PCA loadings.

In this paper, the analysis shows that the first principal component with largest variance of 64.01 is modeled as Eq. (3).

$$\begin{aligned} PC_1 = {} & 0.731X_1 + 0.912X_2 + 0.844X_3 + 0.853X_4 \\ & + 0.722X_5 + 0.922X_6 + 0.872X_7 + 0.722X_8 \\ & + 0.606X_9 + 0.957X_{10} + 0.960X_{11} + 0.715X_{12} \\ & + 0.879X_{13} + 0.805X_{14} + 0.012X_{15} \end{aligned} \tag{3}$$

Hence the first Principal Component is labeled after the attributes "VILLAGES", "OVERALL_LI", "SCHTOT", "ENRTOT", "TCHTOTG", "TCHTOTP", "ROAD-TOT", "SELETOT", "SCOMPTOT", "ENR50TOT", "CLSTOT", "TCHTOT", "PTCH", "GRTCH" and "SS" as listed in Table 2.

In the following section, the selected attributes using PCA are analyzed using machine learning tools to determine the significant factors in the education system of a developing nation.

Table 2   Attributes selected using PCA

Variables	Attribute abbreviation	Attribute description
X_1	VILLAGES	No. of villages
X_2	OVERALL_LI	Literacy rate
X_3	SCHTOT	Total (government and private)
X_4	ENRTOT	Elementary_Enrolment_by_School_Category: Total
X_5	TCHTOTG	Teachers_by_School_Category_(Government): Total
X_6	TCHTOTP	Teachers_by_School_Category_(Private): Total
X_7	ROADTOT	Schools_Approachable_by_All_Weather_Road: Total
X_8	SELETOT	Schools_with_Electricity: Total
X_9	SCOMPTOT	Schools_with_Computer: Total
X_{10}	ENR50TOT	Schools_with_Enrolment_<=_50: Total
X_{11}	CLSTOT	Number_of_Classrooms_by_School_Category: Total
X_{12}	TCHTOT	Teachers_by_School_Category: Total
X_{13}	PTCH	Professionally_Qualified_Teachers:_Private: Total Teachers
X_{14}	GRTCH	Professionally_Qualified_Teachers:_Government: Total_Regular_Teachers
X_{15}	SS	Special school

5   Supervised Machine Learning Approach

In this section, three Learning approaches namely Logistic Regression, CART, and Random Forest are applied over the dataset of size 680 observations and 15 attributes to determine the significant attributes. According to UNESCO, literacy rate defines the fraction of population that can read and write belonging to an age group. A high literacy rate suggests the existence of an effective primary education system. To determine the rate of education in India from the given dataset, various attributes have been analyzed and the attribute overall literacy rate of each district has been considered as a target attribute. Overall literacy rate as the target attribute helps to determine the significant factors for classification of districts with high (>75%) and low (<75%) literacy rates.

The dataset considered here for the study contains 680 observations that include 469 observations with negative values (low literacy) and 211 observations with positive values (high literacy). Figure 4 shows the boxplot on the overall literacy rate of the country. The inter-quartile range of overall literacy lies between 66.44 and 80.81, which is a measure of how far apart are the lowest and the highest literacy rates in the districts and states of a developing nation of the case study. There are few outliers below the interquartile range, which reflects few states like districts of Bihar, Telangana, Arunachal Pradesh, etc. with very low literacy rates.

The dimensionally reduced dataset using PCA is analyzed using three machine learning tools: Logistic regression, CART, and Random Forest.

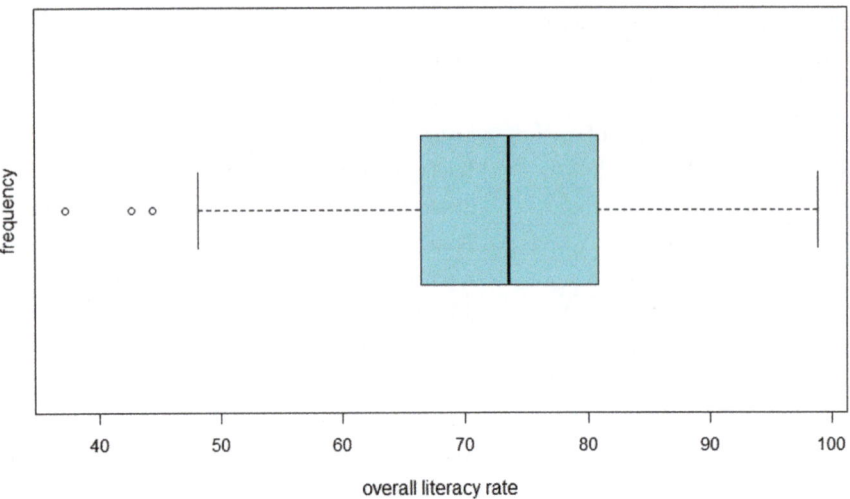

Fig. 4 Boxplot of the target attribute

5.1 Logistic Regression

Logistic regression is a classification algorithm to develop a linear model. In this study, algorithm determines the significant factors on overall literacy rate. In this machine learning model, the dependent variable is categorical. In this, analysis $E(Y|x)$ is the mean value of the dependent variable given the values of the independent variable:

$$E(Y|x) = \beta 0 + x \cdot \beta i \tag{4}$$

where Y is the dependent variable, x is the independent variable, and βi denotes the model parameters.

Formally the model Equation is given by (5):

$$\prod(x) = \frac{e^{(\beta 0 + x \cdot \beta i)}}{1 + e^{(\beta 0 + x \cdot \beta i)}} \tag{5}$$

The transformation of the $\prod(x)$ logistic function is known as the logit transformation, represented in Eq. (6):

$$g(x) = \ln \frac{\prod(x)}{1 - \prod(x)} = \beta 0 + x \cdot \beta i = \beta 0 + x 1 \beta 1 + x 2 \beta 2 + \cdots + x n \beta n \tag{6}$$

where $g(x)$ is called logit function.

$\prod(x)$ in Eq. (5) is the conditional probability that Y is equal to 1 given x, and the quantity $1 - \prod(x)$ gives the conditional probability that Y is equal to zero given

x. To assess the significance of variables in the model, observed value is compared with the predicted value of response variable obtained from models with training and testing dataset. This comparison is done using log likelihood function. Likelihood function for the saturation model and the current model is considered here to obtain the likelihood ratio as follows:

$$D = -2 \ln \frac{\text{likelihood of current model}}{\text{likelihood of saturated model}} \tag{7}$$

where D in Eq. (7) is the statistic variable called the deviance. Deviance determines the goodness of fit of a model.

The target or dependent variable in this study is Overall Literacy Rate, since literacy rates are the test of an educational system. The logistic model used in this study is represented by Eqs. (8) and (9) for binary classification of attributes into high literacy and low literacy respectively:

$$P(\text{high literacy}) = \prod(x) = \frac{e^{g(x)}}{1 + e^{g(x)}} \tag{8}$$

$$P(\text{low literacy}) = 1 - P(\text{high literacy}) = 1 - \prod(x) = \frac{1}{1 + e^{g(x)}} \tag{9}$$

The logit function $g(x)$ derived for this model is given by the Eq. (10):

$$
\begin{aligned}
g(x) = {} & (3.604\text{e} - 01) + \text{VILLAGES}(1.001\text{e} - 04) + \text{SCHTOT}(-5.752\text{e} - 04) \\
& + \text{ENRTOT}(6.026\text{e} - 07) + \text{TCHTOTG}(-4.994\text{e} - 05) \\
& + \text{TCHTOTP}(-2.345\text{e} - 04) + \text{ROADTOT}(2.446\text{e} - 04) \\
& + \text{SELETOT}(3.436\text{e} - 05) + \text{SCOMPTOT}(3.489\text{e} - 05) \\
& + \text{ENR50TOT}(3.213\text{e} - 04) + \text{CLSTOT}(-1.550\text{e} - 05) \\
& + \text{TCHTOT}(9.269\text{e} - 05) + \text{PTCH}(1.852\text{e} - 04) \\
& + \text{GRTCH}(-2.849\text{e} - 05) + \text{SS}(-7.382\text{e} - 06)
\end{aligned}
\tag{10}
$$

The coefficients βi are estimated using logit model. If the significance of respective coefficient is less than 0.05, then this x_i is statistically significant. Hence the input attributes that are statistically significant factors in determining the high literacy of the nation based on the logistic model are given in Table 3.

The optimal cut-off value for the threshold in this case study for logistic regression is 0.604789. At various threshold values, ROC graph is plotted by the true positive rate (TPR) against the false positive rate (FPR) (see Fig. 5). AUC (Area Under the ROC Curve) value for the case study with logistic regression is 84.07 on training dataset, which indicates it to be a good fit model. The logistic regression model on the given dataset gives the concordance 0.831 with accuracy of 0.765 and sensitivity 0.9007. So the higher the concordance, accuracy, and sensitivity, the better is the quality

Table 3 Attributes selected by logistic regression

Attribute abbreviation	Attribute description	*P*-value
SCHTOT	Schools_By_Category: Total (Govt. and Pvt.)	1.01e−08
ENRTOT	Elementary_Enrolment_by_School_Category: Total	0.014374
TCHTOTP	Teachers_by_School_Category_(Private): Total	0.020066
ROADTOT	Schools_Approachable_by_All_Weather_Road: Total	0.000702
ENR50TOT	Schools_with_Enrolment_<=_50: Total	3.01e−06
TCHTOT	Teachers_by_School_Category: Total	0.000338
PTCH	Professionally_Qualified_Teachers:_Private: Total Teachers	0.045881
GRTCH	Professionally_Qualified_Teachers:_Government: Total Teachers	0.049667

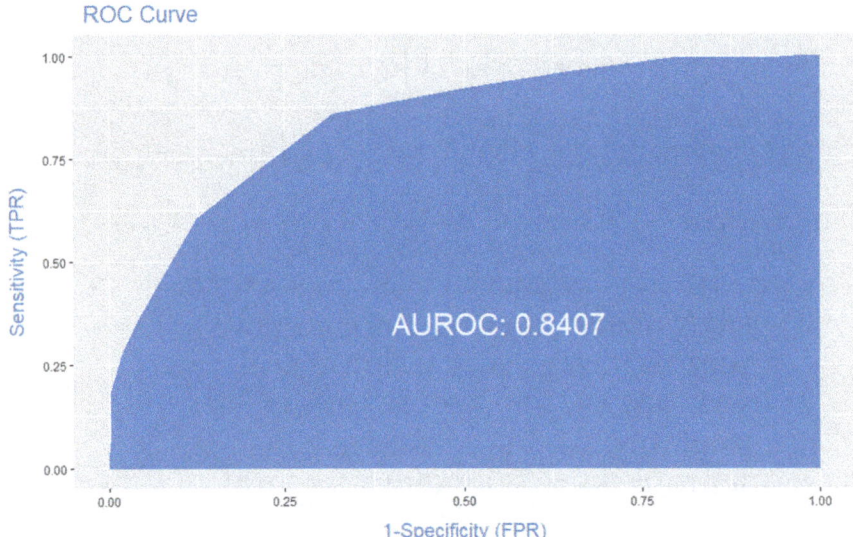

Fig. 5 ROC curve for logistic regression model

of model to classify attributes as significant and insignificant in Indian education system.

Decision Tree. The Decision Tree algorithm used in this study of classification is CART (Classification and Regression Trees). By applying numeric splitting, CART constructs a binary tree built by splitting nodes into two child nodes repeatedly. CART uses the Gini index as a metric. As splitting criterion, we used Gini's impurity index, which is defined as Eq. (11):

$$\text{Gini} = 1 - \sum_{i=1}^{n}\left(p_i^2\right) \qquad (11)$$

where p_i is the probability of an object being classified to a particular class.

The best attribute is selected which has a minimum Gini index. The final result has VILLAGES as the root node called as best predictor. It has root node error of 0.30947 and misclassification rate of 29.47%. The branch nodes and leaf node obtained from splitting are ENRTOT, PTCH, SCOMPTOT, and TCHTOTP. Since the tree is not complex, the pruning produces the same result with better classification model. Overall accuracy of the CART approach is 0.774 with sensitivity 0.797 and hence the best attributes are given in Table 4.

Random Forest. Random Forest is a modeling method using ensemble learning technique. It ensembles a multitude of weak decision trees at training time. Samples in the observation are bootstrapped into different set of bootstrap samples by sampling with replacement. For each bootstrapped sample sets models are developed. The outputs from each model is used for classification based on majority of votes.

Bootstrap aggregation used in random forest is bagging method which uses Out-of-Bag (OOB) error estimate for finding the prediction error on each training sample. In this study Out-of-bag (OOB) error rate is 21.68% which reduces to 3.03% on tuning of random forest (Fig. 6).

Table 4 Attributes selected by decision tree

Attribute abbreviation	Attribute description	Gini
ENRTOT	Elementary_Enrolment_by_School_Category: Total	0.06
SCOMPTOT	Schools_with_Computer: Total	0.04
TCHTOTP	Teachers_by_School_Category_(Private): Total	0.12
PTCH	Professionally_Qualified_Teachers:_Private: Total Teachers	0.60

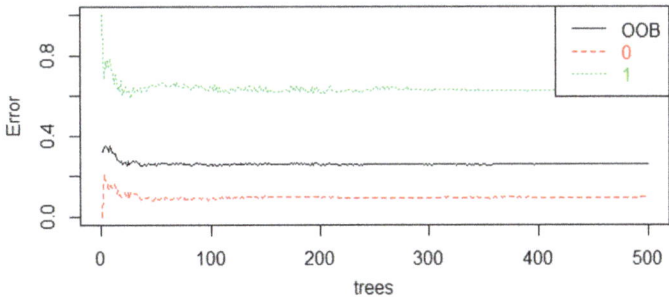

Fig. 6 Error rate versus number of trees

Table 5 List of attributes in importance

Attributes	Mean decrease accuracy	Mean decrease Gini
SCHTOT	20.78	14.36
PTCH	17.66	13.67
VILLAGES	16.71	15.00
TCHTOP	17.53	15.10
ENRTOT	18.08	13.26
CLSTOT	15.78	11.36
SS	16.30	12.79
TCHTOT	15.93	12.33
SCOMPTOT	18.46	15.35
ROADTOT	15.21	11.57
SELETOT	13.49	10.15
GRTCH	11.99	9.37
TCHTOTG	11.75	8.86
ENR50TOT	9.84	11.10

To understand the importance of variables in Random Forest, the following measures are generally used: Mean Decrease in Accuracy and Mean Decrease in impurity. Order of the significance of variables on the basis of and above two measures are given in Table 5.

The attributes importance order (Top 7) on the basis of Mean Decrease Accuracy would be as in Table 6.

The variable importance order (Top 7) on the basis of Mean Decrease Gini would be as in Table 7.

From the two tables, the best subset of significant attributes that contribute to overall literacy of the nation are given in Table 8.

From the results of random forest, we can conclude that SCHTOT (Schools_with_Computer: Total) and SCHTOT (Schools_By_Category: Total) are the significant attributes that contribute to the overall literacy of the nation.

Table 6 Attributes importance order based on mean decrease accuracy

Attributes	Mean decrease accuracy
SCHTOT	20.78
SCOMPTOT	18.46
ENRTOT	18.08
PTCH	17.66
TCHTOP	17.53
VILLAGES	16.71
SS	16.30

Table 7 Attributes importance order based on mean decrease Gini

Attributes	Mean decrease Gini
SCOMPTOT	15.35
TCHTOP	15.10
VILLAGES	15.00
SCHTOT	14.36
PTCH	13.67
ENRTOT	13.26
SS	12.79

Table 8 Best subset of attributes selected by random forest algorithm

Attributes	Attribute description
SCHTOT	Schools_By_Category: Total
SCOMPTOT	Schools_with_Computer: Total
ENRTOT	Elementary_Enrolment_by_School_Category: Total
PTCH	Professionally_Qualified_Teachers:_Private: Total Teachers
TCHTOP	Teachers_by_School_Category_(Private): Total
VILLAGES	Data_Reported_from: No. of villages
SS	Special school

6 Discussions

In this paper, the dataset is evaluated using three machine learning algorithms to determine the significant attributes in the education system of India. The performance of each algorithm is done using the measures overall accuracy, area under curve, sensitivity, precision, and classification error rate.

The *overall accuracy* was used to measure the proportion of the districts in the nation whose literacy level is correctly predicted by each technique. Here all three algorithms are good in accuracy to classify the districts based on the overall literacy rate. The *Area Under Curve (AUC)* is the measure of the accuracy of a classification model's diagnostic ability to classify as high and low literacy factors. *Sensitivity* criterion was used to measure efficiency of each technique in correctly identifying districts final statuses, high or low literacy. Though logistic regression outperforms the other two learning techniques, all the three give a good range of sensitivity above 0.75. Specificity is the measure of classifying as actually low, when it's actually low. *Precision* is the measure of determining firstly the proportion of the districts which were correctly predicted as high in literacy among all high literacy districts including incorrect classification of districts, and secondly the proportion of the districts which were correctly predicted as low in literacy among all low literacy districts that comprise incorrect classifications of districts. Among the three machine

Table 9 Performance of logistic regression on training and test dataset

Measure	Result on training dataset	Result on test dataset	% deviation
Area under curve	0.8407	0.8288	1.19
Concordance	0.8499	0.8312	1.87
Sensitivity	0.5510	0.468	8.3
Specificity	0.9176	0.9007	1.69

Table 10 Performance of CART on training and test dataset

Measure	Result on training dataset	Result on test dataset	% deviation
Area under curve	0.7896	0.7061	8.35
KS	0.4546	0.3325	12.21
Gini	0.3999	0.4037	−0.4
Accuracy	0.7747	0.7463	2.84
Sensitivity	0.4897	0.4531	3.66
Specificity	0.9024	0.8794	2.3
Precision	0.6923	0.6304	6.19

learning algorithms Random forest is giving a high precision of 1. *Classification error rate* is the proportion of observations misclassified over the whole set of observations. *Gini coefficient* is a measure of the dispersion or incorrect classification of a new observation sample.

Performance measures of each machine learning technique—logistic regression, CART, and random forest, is validated in terms of area under curve, accuracy, sensitivity, specificity, precision, and Gini coefficient are presented in Tables 9, 10 and 11 respectively.

It can be observed that the Model Performance values for Training and Testing sets are below the maximum tolerance deviation of ±10%. Hence all three models have performed well and Random forest was the best performing one. Based on the three

Table 11 Performance of random forest on training and test dataset

Measure	Result on training dataset	Result on test dataset	% deviation
Area under curve	1	1	0
KS	1	1	0
Gini	0.6530	0.6306	2.24%
Accuracy	1	1	0
Sensitivity	1	1	0
Specificity	1	1	0
Precision	1	1	0

Table 12   List of significant factors

1.	Total number of schools by category (govt. and private)
2.	Total number of schools with computers
3.	Total elementary enrolment by school category
4.	Schools with enrolment <= 50: total
5.	Professionally qualified teachers: total teachers (govt. and private)
6.	Total teachers by school category (govt. and private)
7.	Total schools approachable by all-weather road
8.	Number of villages
9.	Total number of special schools

ML approaches, the significant factors in Indian education system that contribute towards high literacy rate are listed in Table 12.

7   Conclusion

In this paper, statistical learning and machine learning of school education data is performed by using PCA for dimensionality reduction and the three machine learning techniques logistic regression, CART, and random forest.

From all the three learning techniques, we understand that current Indian education literacy level is dependent on attributes like number of schools in each location, total enrollment of students, number of teachers employed based on qualification and school category, schools approachable by all weather, schools with computers, villages in a district and also the special schools in the nation. Based on these significant attributes determined, it can be concluded that the states with less than 70% literacy rate have suffered and failed during the Covid-19 pandemic to enhance the digital learning of students for all classes and enhance e-learning.

This work is an initial step to analyze the factors currently significant in Indian education system that need to be modified and advanced in order to achieve the goals of National Education Policy 2020 in another 5 years.

References

1. KPMG Home Page (2017) A study by KPMG in India and Google. Online education in India: 2021, May 2017. https://home.kpmg/in/en/home/insights/2017/05/internet-online-education-india.html. Last accessed 27 Jan 2021
2. Luckin R, Holmes W, Griffiths M, Forcier LB (2016) Intelligence unleashed: an argument for AI in education. An open access version is available from UCL discovery. UCL Knowledge Lab, London. ISBN-13: 9780992424886

3. Data World Homepage. https://data.world/liz-friedman/covid-19-impact-on-education. Last accessed 03 July 2020
4. Kaggle Home Page. https://www.kaggle.com/rajanand/education-in-india. Last accessed 20 Jan 2021
5. Ramesh S, Natarajan K (2019) New education policy of India: a comparative study with the education system in the USA. Int J Humanit Soc Sci Invent IJHSSI 8:01–09
6. Rather ZA (2015) Relevance of vedic ideals of education in the modern education system. IOSR J Humanit Soc Sci (IOSRJHSS) 20(1)
7. Dubey KT, Nimje AA (2015) A study of critical comparative analysis of ancient Indian education and present education system. Int J Adv Res Sci Eng 4(1)
8. Kintu MJ, Zhu C, Kagambe E (2017) Blended learning effectiveness: the relationship between student characteristics, design features and outcomes. Int J Educ Technol High Educ 14, article 7
9. Anozie N, Junker BW (2006) Predicting end-of-year accountability assessment scores from monthly student records in an online tutoring system. In: Educational data mining: papers from the AAAI workshop, July 2006. AAAI Press, Menlo Park, CA
10. El-Alfy ESM, Abdel-Aal RE (2008) Construction and analysis of educational tests using abductive machine learning. Comput Educ 51(1):1–16
11. Baepler P, Murdoch CJ (2010) Academic analytics and data mining in higher education. Int J Scholarsh Teach Learn 4(2):17
12. Hew KF, Hu X, Qiao C, Tang Y (2020) What predicts student satisfaction with MOOCs: a gradient boosting trees supervised machine learning and sentiment analysis approach. Comput Educ 145, article 103724
13. Gómez Aguilar DA, García-Peñalvo FJ, Therón R (2014) Analítica Visual en eLearning. Prof Inf 23(3):233–242
14. Singh A (2015) Challenges in Indian education sector. Int J Adv Res Commun Manag 1(3):54–58
15. Lu OHT, Huang AYQ, Huang JCH, Lin AJQ, Ogata H, Yang SJH (2018) Applying learning analytics for the early prediction of students' academic performance in blended learning. Educ Technol Soc 21(2):220–232

Chapter 49
Effectiveness of Connected Components Labelling Approach in Noise Reduction for Image De-fencing

Aditi Awasthi, Deepthi Bhat, Medhini Oak, and N. Kayarvizhy

1 Introduction

Since the integration of image capturing lenses into portable devices like smartphones and digital cameras, the number of images generated has increased by leaps and bounds. Photographers capture images at every opportune moment but they are often hampered by undesirable obstructions like fences, grills, enclosures or reflective surfaces. While they serve as a layer of security in general, from the photographer's perspective, they are a major hindrance as they cannot be removed from the frame by changing the angle of the camera or the plane of focus [1, 2]. The presence of these obstructive structures ruins the aesthetic experience of the image and distracts the viewers from the actual focus of the picture, ruining its visual appeal.

From a machine's perspective, an image is simply an orderly collection of pixels, or rather a signal which can be manipulated mathematically. The detection of fence edges is a non-trivial task—it is difficult to firmly state a specific threshold for how large the intensity change between two neighbouring pixels must be, for us to say that there exists a fence-edge between these pixels [3].

A few manual editing mechanisms are available to photographers to facilitate the removal of occlusions. However, it is seen that the removal of fence-like periodic structures, especially without losing integral parts of the original picture, is particularly tedious as they generally tend to span the entire image. Instead of undertaking the harrowing process of manually editing out such occlusions, we can make use of image processing algorithms, where the fence is automatically detected and filled with minimal manual interference [4, 5]. The image processing-based approach can be implemented by different techniques but the most promising one is based on the frequency domain manipulation of the image [6]. This is because, in spatial

A. Awasthi (✉) · D. Bhat · M. Oak · N. Kayarvizhy
Department of Computer Science and Engineering, B.M.S. College of Engineering, Bengaluru, Karnataka, India

S. Kumar et al. (eds.), *Proceedings of International Conference on Communication and Computational Technologies*, Algorithms for Intelligent Systems, https://doi.org/10.1007/978-981-16-3246-4_49

domain, the colours of an image play an integral role in deciding the effectiveness of the solution. For example, if the background or foreground consists of objects with colours similar to the fence, then the spatial domain-based solutions suffer. Among the frequency domain-based solutions, machine learning techniques for fence detection and noise reduction have proven very effective. However, they suffer from a major drawback—the model has to be trained for every different kind of image and occlusion in order to detect the in-fence and out-of-fence pixels. This results in two issues: One, that the training is time-consuming, which is natural for any machine learning based solution; the other, that the solution is not generic in nature, as the training has to be done all over again for each different type of input image.

In this paper, we demonstrate how this problem can be tackled using the connected components labelling approach to reduce the noise present in the fence mask and help isolate it better after it has been detected using image processing techniques.

We accomplish this task in two steps:

1. obtaining raw fence mask using quasi-periodic texture detection and multi-resolution processing
2. noise-removal using component labelling approach and morphological operations.

2 Related Work

Much of the research in this field is concerned with the detection of regular and near-regular patterns in images. A fence can essentially be interpreted as an overlapping texture in an image. In most previous work, advantage is taken of the fact that the texture of the fence presents itself as a signal which is either periodic or semi-periodic. The smallest repetitive element of a periodic or semi-periodic texture is called a texton or a texel. However, sometimes the fence does not present itself as a periodic signal. Such textures can be classified as random and would have to be detected using different techniques. Thus, the representations of textures vary and many techniques were introduced over time in order to accommodate a variety of fence textures.

Liu et al. [7] first detected fences based on the regularity of the fence texture, by viewing the fence as a deformed lattice. Hettiarachchi et al. [6] proposed a method to detect fences by approaching them as quasi-periodic textures. This algorithm is successful in detecting semi-regular fences in the foreground as well as the background.

Khalid et al. [8] used parallel de-fencing algorithm to detect fence-like occlusions. The algorithm presented in Yang et al. [9] detects fences by using a colour-based classifier from learned samples, obtained from super pixel classification. The lack of fence types, along with simple research objects, caused the poor robustness of these methods, especially when dealing with fences of different shapes and colours.

In recent years, most new methods for fence detection have built upon the foundations already laid by older algorithms. Doubek et al. [10] describes a method for fence retrieval using only repetitive patterns and ignoring the lattice edge. Varalaksh-mamma and Venkateswarlu [11] presents a method which aimed at detecting fences with different shapes, colours and textures. This technique gave reliable results but failed when the image contained a fence whose colour was similar to that of its background. Luo et al. [12] describes a method which detected fences based on image binary morphology. This method proved to be fast and efficient, which can achieve near-real-time conversion even when running on a poorly configured device.

Research has also been conducted on de-fencing of video clips by taking advantage of the fact that consecutive frames when aligned frame-by-frame, can help reconstruct the missing pixels to create a de-fenced video. Mu et al. [7] and Jonna et al. [13] describe methods to remove fences from videos, frame by frame. Xue et al. [14] describes a method which combines the visual information across the image sequence to produce an obstruction-free image. The limitation of these techniques is that the occlusions which are not fronto-parallel to the camera are not removed. Jonna et al. also proposed a method that used convolutional neural networks to detect fence pixels. However, the limitation of all supervised learning algorithms is that their applications are limited by the data set on which they are trained, which, in this case, includes only some fixed-mode mesh fences.

Fence removal from single image has been effectively achieved using machine learning approaches [15, 16]. De-fencing of images captured by a hyperspectral camera and extracting thin occlusion from images has also been attempted [17, 18].

The segmentation techniques mentioned thus far are effective in segmenting coherent objects. However, they are not as effective in segmenting the distributed objects that run over the complete image region.

Connected components labelling method [19] is used to group the pixels which have same value. It produces good results for binary, gray scale as well as colour images. This method has been used to detect specific patterns in images, like the shape of an ear in side-profile photographs [20]. Various algorithms have been suggested to achieve this, like the fast component labelling algorithm in [21] and the age-and-weight-balancing technique in [22].

3 Proposed Methodology

We segregate the fence present in an image by using the mathematical properties of images to detect the quasi-periodic texture of a fence.

3.1 Manipulation in Frequency Domain

We convert the input image from the spatial domain (human understandable format) to the frequency or Fourier domain. This is achieved by applying the Fourier Transform. This transformation facilitates the visualization of the rate at which the pixel values are changing in spatial domain.

For an image of size $M \times N$ pixels, 2-Dimensional Discrete Fourier Transform is given by (1).

$$F(u, v) = \frac{1}{MN} \sum_{x=0}^{M-1} \sum_{y=0}^{N-1} f(x, y) e^{-j2\pi \left(\frac{ux}{M} + \frac{vy}{N} \right)} \tag{1}$$

where,

$u = $ frequency in x direction, $u = 0, \ldots, M - 1$

$v = $ frequency in y direction, $v = 0, \ldots, N - 1$.

In the frequency domain, the value and location are represented by sinusoidal relationships that depend upon the frequency of a pixel occurring within the image. The pixel location is represented by its x and y frequencies and its value is represented by the amplitude. This transformation helps us determine which pixels contain relatively more important information and whether repeating patterns occur.

A Fast Fourier Transform Shift is executed for ease of understanding when we look at the frequency domain image. It rearranges the Fourier transformed image by shifting the zero-frequency component to the centre of the array.

3.2 Quasi-periodic Texture Filtering

The frequency-domain representation of an image can be divided into two major components:

1. High-frequency components, which correspond to edges in an image
2. Low-frequency components, which correspond to smooth regions.

The fences which occur in the image can be categorized as not-so-sharp edges. They are segregated using a high pass filter, given by (2), which allows only high frequencies to go through and blocks low frequencies.

$$H(u, v) = \begin{cases} 0, & D(u, v) \leq D_0 \\ D(u, v), & D(u, v) > D_0 \end{cases} \tag{2}$$

where,

$H(u, v) = $ Signal obtained after filtering out low frequencies

$D_0 = $ threshold for filtering (Fig. 1).

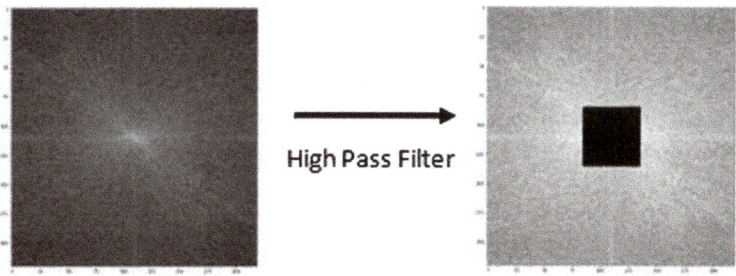

Fig. 1 Action of a high pass filter in the frequency domain

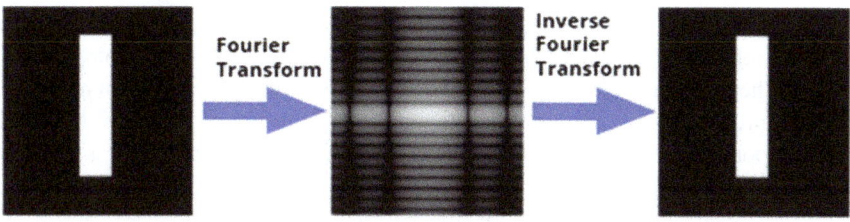

Fig. 2 Conversion from spatial domain to frequency domain and back to spatial domain

The changes thus made can be viewed and better understood after the image is inverse-transformed back to the spatial domain.

For an image of size $M \times N$ pixels, 2-Dimensional Inverse Discrete Fourier Transform is given by (3).

$$f(x, y) = \sum_{u=0}^{M-1} \sum_{v=0}^{N-1} F(u, v) e^{j2\pi \left(\frac{ux}{M} + \frac{vy}{N} \right)} \tag{3}$$

where,

$x = 0, \ldots, M - 1$ and $y = 0, \ldots, N - 1$ (Fig. 2).

3.3 Multi-resolution Processing

A multi-resolution representation provides a simple hierarchical framework to analyze the signal at different resolution levels. Any image can be represented as a multi-resolution pyramid. At each level, the difference (residual) between the image at that level and the predicted image from the next level is stored. We can reconstruct the image by just adding up all the residuals.

Wavelet decomposition of an image produces diagonal (D), vertical (V) and horizontal (H) components, which are employed for feature extraction. We leverage these components to filter out the fence-like components of the image more effectively.

Each stage of the algorithm generates wavelets with sequentially finer representation of signal content. At lower resolutions, it becomes easier to demarcate the fence pixels from the non-fence pixels, which is achieved by thresholding the pixel value of the directional components of a grey-scale image.

3.4 Connected Component Labelling

In order to remove noisy components from the raw fence mask, we try to extract maximal regions of connected mask pixels, also known as connected components. We scan the image and group its pixels into components based on pixel connectivity. We then proceed to discard the insignificant blobs from our mask, since they tend to correspond to noisy areas. This produces a cleaner fence mask with a slightly withered appearance.

3.5 Morphological Closing

We use the Morphological close operation to convert the pixelated fence mask to a smoother fence mask. Closing is equivalent to morphological dilation followed by erosion. It helps retain the original boundary shape and preserves background regions that have a similar shape to the structuring element, or that can completely contain the structuring element, while eliminating all other regions of background pixels.

4 Experimental Results

The algorithm proposed in this article was tested with a number of images with fence occlusions. Some test images were obtained from PSU Near-regular Texture database—http://vivid.cse.psu.edu/index.php?/category/23.

The input image used is shown in Fig. 3, which shows a puma behind a silver-coloured fence. This input image is used against the proposed algorithm. The output of frequency thresholding is fed to the multiresolution processing module. In this module, wavelet decomposition of the image is performed at various levels to extract the horizontal, vertical and diagonal components. A binary 'OR' operation is done to combine these components into a final image. The three resolution levels of the input image and the results obtained by performing the Binary OR operation at each stage can be seen in Figs. 4, 5, 6, 7, 8 and 9.

Fig. 3 Puma behind fence (input image)

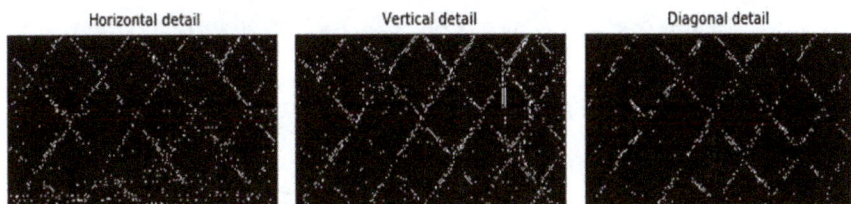

Fig. 4 Components extracted at level 1

Fig. 5 Result of performing
OR operation on level 1

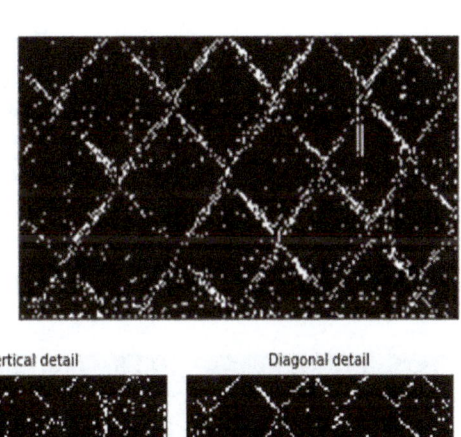

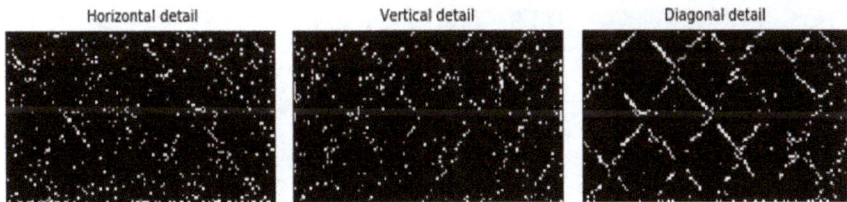

Fig. 6 Components extracted at level 2

Fig. 7 Result of performing
OR operation at level 2

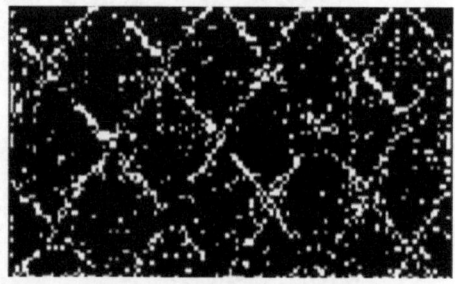

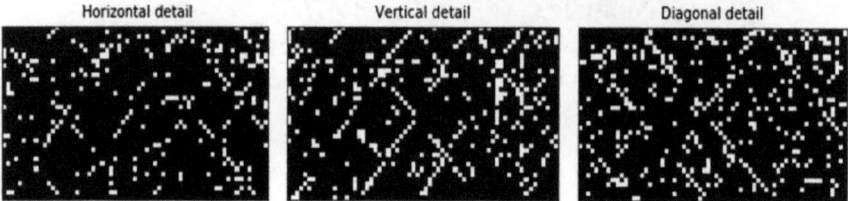

Fig. 8 Components extracted at level 3

Fig. 9 Result of performing
OR operation at level 3

Then, the original image (at level 0) is taken and direct frequency domain thresholding is performed on it to arrive at Fig. 10.

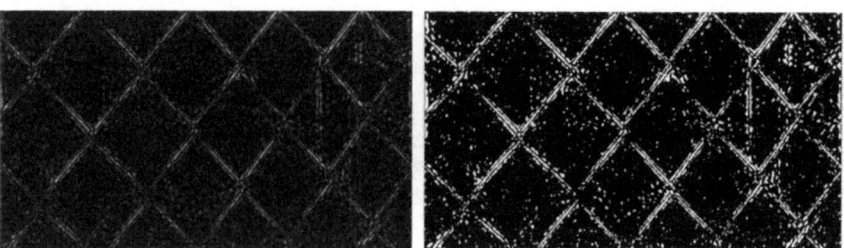

Fig. 10 Thresholding operation on level 0

Fig. 11 Combined image of
4 levels

Fig. 12 Resultant image
after component labelling

The 4 levels of fence masks thus obtained are used to produce a combined image by using adjacent pixel information. This combined image is given in Fig. 11.

The resultant image demarcates the fence components clearly but still contains a lot of noise. To obtain noise reduction, the image is passed over to the component labelling module. Component labelling identifies a threshold and collects only the segments that have at least the required number of pixels. The resultant image with reduced noise obtained using this approach is given in Fig. 12.

Several other images were also used to evaluate the effectiveness of the connected components labelling approach. The images were chosen to test the algorithm across a variety of structural aspects like fence colour, fence shape, its periodicity, etc. The final images with the fence masks detected by our algorithm are shown in Table 1.

We have also compared the results of noise reduction using machine learning techniques with the results produced by our connected components labelling approach. The same input image was used and the results obtained are shown in Table 2.

It can be seen that the noise reduction obtained using connected components approach is comparable to the results obtained using machine learning. While the fences appear as relatively more solid lines when we use the machine learning approach, it should be noted that, in order to achieve this, we need to perform individual evaluation and classification of each pixel into in-fence and out-of-fence groups. This requires a non-generic approach where the model has to be trained based on the domain of images under consideration, which is both complex and time-consuming.

We also see that the machine learning approach falters at certain points—it introduces extra noise in the fence both at the top-right corner and bottom-centre portion,

Table 1 Results of evaluation using other images

Input Image	Resultant fence mask by connected components technique

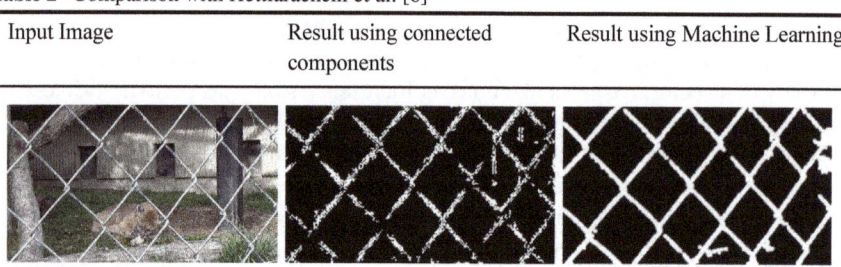

Fig. 13. Input Image 1   **Fig. 14.** Fence Mask 1

Fig. 15. Input Image 2   **Fig. 16.** Fence Mask 2

Table 2 Comparison with Hettiarachchi et al. [6]

Input Image	Result using connected components	Result using Machine Learning

also failing to classify a good chunk of the actual fence in the bottom-right part of the image.

5 Conclusion and Future Work

We have proposed an approach to segregate fence texture from images by frequency domain processing prior to wavelet transformation followed by connected components labelling for noise reduction. This method helps us segregate the fences without any prior training or knowledge of the input image. The results demonstrate that the connected components labelling approach works as effectively as or even better than the machine learning algorithm for noise reduction, providing a compelling alternative that is both faster and generic.

Future work along this line may include inpainting along with fence detection. Frame-by-frame detection of fences in videos containing occlusions can also be attempted. Another approach could be to improve on the accuracy of the fence mask, especially when dealing with low-resolution images, making it continuous and thicker to aid in effective inpainting.

References

1. Jonna S, Nakka KK, Sahay RR (2015) My camera can see through fences: a deep learning approach for image de-fencing. In: 2015 3rd IAPR Asian conference on pattern recognition (ACPR), pp 261–265. https://doi.org/10.1109/ACPR.2015.7486506
2. Jonna S, Satapathy S, Sahay RR (2017) Stereo image de-fencing using smartphones. In: 2017 IEEE international conference on acoustics, speech and signal processing (ICASSP), pp 1792–1796. https://doi.org/10.1109/ICASSP.2017.7952465
3. Park M, Brocklehurst K, Collins RT, Liu Y (2010) Image de-fencing revisited. In: Proceedings of the 10th Asian conference on computer vision—volume part IV. Springer-Verlag, Berlin, Heidelberg, pp 422–434
4. Jonna S, Nakka KK, Sahay RR (2016) Towards an automated image de-fencing algorithm using sparsity
5. Liu Y, Belkina T, Hays J, Lublinerman R (2008) Image de-fencing, pp 1–8. https://doi.org/10.1109/CVPR.2008.4587493
6. Hettiarachchi R, Peters J, Bruce N (2014) Fence-like quasi-periodic texture detection in images. Theory Appl Math Comput Sci 4
7. Mu Y, Liu W, Yan S (2012) Video de-fencing. IEEE Trans Circuits Syst Video Technol 24(7):1111–1121
8. Khalid M, Yousaf M, Murtaza K, Sarwar S (2018) Image de-fencing using histograms of oriented gradients. Signal Image Video Process 12. https://doi.org/10.1007/s11760-018-1266-0
9. Yang J, Wang J, Liu L, Hou C (2015) RIFO: restoring images with fence occlusions. In: 2015 IEEE 17th international workshop on multimedia signal processing (MMSP), pp 1–6. https://doi.org/10.1109/MMSP.2015.7340834
10. Doubek P, Matas J, Perdoch M, Chum O (2010) Image matching and retrieval by repetitive patterns. In: 2010 20th international conference on pattern recognition, pp 3195–3198. https://doi.org/10.1109/ICPR.2010.782
11. Varalakshmamma M, Venkateswarlu T (2019) Detection and restoration of image from multi-color fence occlusions. Pattern Recogn Image Anal 29:546–558. https://doi.org/10.1134/S1054661819030209
12. Luo M-X., Xu W-S, Yu Y-L (2019) Image de-fencing based on binary morphology. DEStech Trans Comput Sci Eng. https://doi.org/10.12783/dtcse/icaic2019/29429

13. Jonna S, Nakka KK, Khasare VS, Sahay RR, Kankanhalli MS (2016) Detection and removal of fence occlusions in an image using a video of the static/dynamic scene. J Opt Soc Am A 33(10):1917–1930. https://doi.org/10.1364/JOSAA.33.001917

14. Xue T, Rubinstein M, Liu C, Freeman WT (2015) A computational approach for obstruction-free photography. ACM Trans Graph (Proc SIGGRAPH) 34(4)

15. Matsui T, Yamaguchi T, Iheara M (2020) Real-world fence removal from a single-image via deep neural network. Electron Imaging 2020(10):26–27. https://doi.org/10.2352/issn.2470-1173.2020.10.ipas-026

16. Matsui T, Ikehara M (2020) Single-image fence removal using deep convolutional neural network. IEEE Access 8:38846–38854. https://doi.org/10.1109/ACCESS.2019.2960087

17. Zhang Q, Yuan Y, Lu X (2016) Image de-fencing with hyperspectral camera. In: 2016 international conference on computer, information and telecommunication systems (CITS), pp 1–5. https://doi.org/10.1109/CITS.2016.7546396

18. Li Y, Wang Y, Piao Y (2017) Extraction of thin occlusions from digital images. In: Lv Y, Le J, Chen H, Wang J, Shao J (eds) Selected papers of the Chinese society for optical engineering conferences held October and November 2016, vol 10255. SPIE, pp 1053–1058. https://doi.org/10.1117/12.2268111

19. He L, Ren X, Gao Q, Zhao X, Yao B, Chao Y (2017) The connected-component labelling problem. Pattern Recogn 70(C):25–43. https://doi.org/10.1016/j.patcog.2017.04.018

20. Prakash S, Jayaraman U, Gupta P (2009) Connected component based technique for automatic ear detection. In: 2009 16th IEEE international conference on image processing (ICIP), pp 2741–2744. https://doi.org/10.1109/ICIP.2009.5414150

21. He L, Chao Y, Suzuki K, Wu K (2009) Fast connected-component labelling. Pattern Recogn 42(9):1977–1987. https://doi.org/10.1016/j.patcog.2008.10.013

22. Dillencourt MB, Samet H, Tamminen M (1992) A general approach to connected-component labelling for arbitrary image representations. J ACM 39(2):253–280. https://doi.org/10.1145/128749.128750

Chapter 50
Modeling Student Confusion Using Fuzzy Logic in e-Learning Environments

Chaitali Samani and Madhu Goyal

1 Introduction

Learners engage themselves with multi-model course content in an online academic setting, including text, videos, and other specific content based on the subject. Learning is a complex process, and learners experience various emotions while learning new knowledge and skills multiple times. Such experiences can be positive with a sense of achievement and motivation to negative emotions like discouragement and Frustration. In classroom environments such as emotional responses, cues like questions, continuous work during class activities, learners' non-verbal expressions, and engagement can be communicated.

In the past few years, a steep advancement in digital educational resources has contributed to the popularity of various Online learning platforms, shortly called MOOCs (Massive Open Online Courses). Such platforms enable learners worldwide to seize an opportunity to learn and improve their knowledge and skills with flexibility. However, such Online Learning Environments often struggle with a high level of attrition rates [1]. Negative emotions like Frustration and anxiety are usually found amongst the learner's final emotional states when they decide to give up on a course, generally feeling disheartened and failing [2].

The classroom environment is less complicated, considering capped class sizes where the teacher-learner ratio can be maintained. In contrast to that MOOCs usually many intakes. MOOCs suffer from another layer of complication due to learners' virtual presence and different learning schedules. Many learners drop-out in the

C. Samani (✉) · M. Goyal
School of Computer Science, Australian Artificial Intelligence Institute, University of Technology Sydney, Sydney, Australia
e-mail: chaitalijayesh.samani@student.uts.edu.au

M. Goyal
e-mail: madhu.goyal-2@uts.edu.au

© The Author(s), under exclusive license to Springer Nature Singapore Pte Ltd. 2021   647
S. Kumar et al. (eds.), *Proceedings of International Conference on Communication and Computational Technologies*, Algorithms for Intelligent Systems,
https://doi.org/10.1007/978-981-16-3246-4_50

first week of the course, but learners who stay longer and drop out gradually also contribute to a high attrition rate [3]. Therefore, it becomes necessary to investigate under which circumstances the learners decide to give up on the course adding up to the drop-out rates for the course. Many researchers believe emotions correlated to the learning process are mainly motivation, Confusion, Frustration, boredom, happiness, and anxiety. It has also been observed that Confusion and motivation correlate positively, while boredom and Frustration present a negative correlation [4].

Confusion bears two folded impacts, both negative and positive. While an optimum amount of Confusion can positively affect overall learning and sense of achievement, contrarily, a learner is more likely to drop out of the course if they experience prolonged Confusion. Prolonged Confusion can lead to Frustration and anxiety, leading them to give up. Evidence suggests that such high drop-outs' primary reason is potentially due to learner's prolonged explicit or implicit Confusion [5].

The rest of this paper is structured as follows: Sect. 2 discusses related works on Confusion and factors affecting Confusion in the learning process. Section 3 describes various factors affecting confusion and Sect. 4 discusses a learning model to postulate a fuzzy rule-based assessment to predict the level of Confusion while attempting a quiz like an assessment in an e-learning environment. Finally, Sect. 5 concludes the paper with provisions for future work.

2 Related Work

Confusion has been studied as a complex epistemic emotion affecting learning and engagement because it is associated with positive and negative outcomes [6, 7]. Affect Dynamics model suggests four epistemic emotions that show a transitional relationship during the learning process [8–10]:

Engagement ↔ Confusion ↔ Frustration ↔ Boredom

This suggests that confusion can create a deep understanding and sense of achievement in learners but can adversely create frustration and boredom. A considerable amount of empirical evidence suggests that Confusion dominates as a key and frequent epistemic emotion during learning for both online [11] and face-to-face learning environments [12, 13]. Hence, for learners to be able to be in the normal flow of their learning process, they will need to manage it productively. It is also a fact that confusion management does not have a one size fits all solution. Depending on the Confusion level, literature provides various methods to help the learner manage Confusion effectively [8].

Various intervention techniques [14–17] are investigated and suggested to prevent the negative transition of Confusion towards Frustration and boredom. Prior studies suggest that Confusion can be beneficial to learning when resolved within certain thresholds, called zone of optimal confusion, both in cognitive and time aspects [18,

19]. It is legitimate to deduce that we must be able to identify the Confusion in time and must also be able to determine the level of Confusion [20] and use such measures to determine how it affects the drop-out [6].

It has also been argued that regardless of confusion detection, helping to predict drop-out, learners who survive with Confusion may still be negatively affecting due to knowledge gaps [21]. However, learners significantly experience negative emotions like Frustration and Boredom before deciding to drop out [2]. It has been studied that persistent confusion leads to low self-efficacy [22], creating adverse attitudes towards learning and finally resulting in drop-out. Thus, helping instructors detect confusion levels in learners by tracking interaction and behavioral patterns in such online learning environments could increase overall retention rates [6].

Confusion results from Cognitive disequilibrium [13]. Cognitive disequilibrium is a state of mind when a learner finds hindered by the normal learning process due to information that creates some uncertainties due to errors or anomalies. The new knowledge or skills is not building properly due to the learners' prior knowledge [9]. This cognitive disequilibrium that generated learning-related affective or emotional states leads to an interest in Confusion and other emotions for many researchers in the field. In short, Confusion can be deemed as an emotional response to Cognitive disequilibrium [20]. As identified earlier that some level of Confusion fosters learning and is also considered as constructive Confusion [23], whilst a prolonged confusion with an inadequate amount of intervention lead learners to negative emotional states like Frustration and boredom [4, 8], and this high level of Confusion is also named as non-constructive Confusion [23].

This leads to the crucial need to detect Confusion timely, foster constructive Confusion, and avoid non-constructive Confusion. In this event where Confusion correlates to both positive and negative extremes, it would be beneficial to review how Confusion is understood in online learning by other researchers.

Researchers have used single and multi-sourced data to determine cognitive-affective states like Confusion, Frustration, and boredom by monitoring learner's postures [24], facial gestures [25], audio and visual expressions [26], eye-tracking strategies [23] as well as tutorial dialogue semantics [27]. However, many such detection models are expensive yet not fully automated and require human expertise to judge such states. Such models using multi-modal information cannot be established as a pervasive approach because it is hard to gain in a large-scale digital environment like MOOCs. Hence, to develop a more scalable solution, we need to focus on non-device-based research approaches studied earlier.

Some early works in identifying Confusion in MOOCs Forum discussions assume that most learners communicate their confusion by posting their concerns on the Forum discussions area or by voting for other similar posts. YouEDU system identified confusion from the Stanford posts dataset using simple bags-of-words and post meta information like post position, votes, etc., to recommend supplementary video snippets [17]. Stanford posts dataset was used again to predict Confusion, urgency, and sentiments using Support Vector Machine (SVM) to achieve 70% or more accuracy to classify Confusion [28]; however, it has been argued to be domain-specific and lacks on being cross-dimensional [21].

Another study claimed to outperform previous classification algorithms that utilized content-based and community-related features for Confusion detection. The study achieved 80% or more accuracy and demonstrated 65% or more accuracy in cross-domain experimentations. Importantly, it established a weak relation of community-related features on Confusion classification and a strong relation to unigrams (bags-of-words) and use of question marks in the posts [29]. Classification of various emotions and confusion were also studied in student-tutor conversation, using lexical analysis [9]. However, all these classification techniques rely heavily on a learner who participates effectively in forum discussions to communicate their Confusion. Confusion can positively lead to deeper understanding. It is also important to determine the Confusion level to see if the learner manages their Confusion under the zone of optimal confusion [2].

An integrated approach was studied where the course interaction patterns were studied together with discussion content using the Confusion Classification model. The study was investigated using two MOOC courses achieving an overall accuracy of 70% or more. The study investigated the most popular click patterns and their prevalence, regardless of learners' participation in Forum discussions [6]. This approach has been argued from the fact that clickstream data is not traced in the event of offline access to the learning material. Self-paced learning on such courses and other constraints can also demonstrate a low accuracy on such a classification model [21].

Apart from various classification models using single or multi-model features, fuzzy logic has also been widely applied in various domains, including educational data mining. These techniques are investigated due to their ability to adapt and deal with vagueness where elements belong to a category rather than exact values resulting in human-like judgments [30]. Fuzzy logic has been studied in e-learning settings to address various needs, features, and contexts.

Fuzzy logic has been investigated to represent a teacher and student assisted evaluation system based on imprecise information and applying a membership function to linguistic labels for fuzzy reasoning [31, 32]. Fuzzy logic has been used to formulate, represent, and analyze individual learners' behavior [33] and group behavior [34, 35]. Early works on Fuzzy logic included an inference of learner's knowledge level and cognitive characteristics using membership functions [36]. Data-driven fuzzy rule induction and inference mechanisms were investigated to evaluate student's academic performance [37].

Fuzzy logic has been studied from various aspects in Fuzzy models' design for learning to represent learners' cognitive and knowledge abilities [38–40]. Despite such vague concepts under investigation, Fuzzy logic has been successful in producing reliable results. However, while much research exists in utilizing Fuzzy logic e-learning systems, little attention has been paid to investigate Fuzzy logic in epistemic emotions like Confusion. Confusion is the central factor in learning, and due to its complex impact on learner's future affective states, it becomes an important factor to explore.

Our work investigates the potentials of using Fuzzy logic techniques to detect confusion levels to address this gap. During the sessions, the learners' clickstream data can be deemed most interactive, even when it is shorter than the time, they spend browsing the course pages. We argue that learners' behavior is most engaging when taking online quizzes as a part of their assessment.

The clickstream behavior of a learner can help detect the underlying Confusion using a Fuzzy logic inference system despite supporting the vagueness, the potential to quantify human-like judgment.

3 Factors Affecting Confusion

The most effective way to determine if someone is confused is usually from the questions they ask their instructors. This learner-instructor communication may be missing on online learning platforms, and even if it is present, it may not be as instant in face-to-face classrooms. Not ignoring that face-to-face also comes with the benefit of body language observations that experienced teachers use in their face-to-face sessions to determine a possible level of Confusion and change their discourse of the class accordingly. Simultaneously, as discussed in the earlier section, multi-modal information used in Confusion detection may not have the potential to become a pervasive approach on online platforms given various constraints.

Hence, such platforms may benefit from a powerful prediction technique like Fuzzy logic by detecting a learner's confusion level when taking the quiz. We propose that attempting the quiz online is the most engaging activity that a learner will do online. Click-stream data generated over that period can provide some meaningful insights.

Yang et al. [6] describe top ten ranked clicked patterns using Exact Pattern Mining method with possible Confusion in two students' group: Students with posts on Forum and students without any posts. They found that both the groups have similar click patterns on MOOC on two courses. That raises us with a hypothesis that if we used the parameters from these highly ranked click patterns and feed to a Fuzzy logic inference system, we would map the learner's confusion level just like an experienced teacher would map in their mind in a face-to-face classroom session.

The proposed Confusion model is based on the above-discussed click patterns. The model comprises factors like Topic browsing, Forum browsing, Productivity, and Performance while the learner is attempting the quiz online. With these parameters, we also propose that the quiz difficulty level may further refine the results in determining the levels of Confusion (Table 1).

4 Fuzzy Rule-Based Confusion Assessment

In this paper, we use the Fuzzy-logic technique to predict the confusion level because the fuzzy logic inference system can be improved, considering the criteria chosen and their levels to incorporate subjectivity brought from various subjects. Also, Confusion

Table 1 Confusion parameters

Parameter	Explanation
Topic browsing	TB(x) = {very low, low, medium, high, very high}. The value of this parameter is pre-determined. This parameter's value is pre-determined and is the number of times a learner visits any learning materials page while attempting the quiz
Forum browsing	FB(x) = {very low, low, medium, high, very high}. The value of this parameter is pre-determined and is the number of times a learner visits the Forum discussion page while attempting the quiz
Productivity	PRO(x) = {very low, low, medium, high, very high}. The value of this parameter describes how long the learner took to complete the quiz. The range is pre-determined
Performance	PER(x) = {very low, low, medium, high, very high}. The value of this determines a learner's raw marks, and level ranges are pre-determined
Level	L(x) = {very low, low, medium, high, very high}. The value of this parameter indicates the level of the question, considering very low to be the easiest questions to very high to be advanced questions. The level is pre-determined at the time of designing the quiz

is an epistemic emotion, and to measure such concepts, we need a model that supports inferring such abstraction, be meaningful, and at the same time be implemented pervasively using some web technologies.

A typical Fuzzy logic system has three components: Fuzzifier, Rule-based inference, and De-fuzzifier. The process involves membership functions, fuzzy logic operators, and If–Then rules that determine various input parameters' values to a fuzzified value from the Universe of Discourse. Then If–Then rules identify a de-fuzzified output value in the output Universe of discourse that is then converted to a crisp value as an output using De-fuzzification [41].

Detecting the clickstream log data along with the performance data can help detect the level of Confusion. Given the two-folded impact on Confusion, it requires great attention to detect Confusion's optimum zone [2]. Hence it may be beneficial to use a comprehensive prediction technique that is easy to apply via a web interface. Such a Fuzzy-logic prediction system can be attached to any quiz assessment to detect the confusion level to help design intervention strategies around it.

Figure 1 suggests the overall proposed Confusion detection in learners using Fuzzy-logic System.

4.1 Membership Functions

Figure 2 shows the Triangular membership function for all parameters in percent as an input with a range of 5 values. The ranges are scaled to percent depending on how they are calculated.

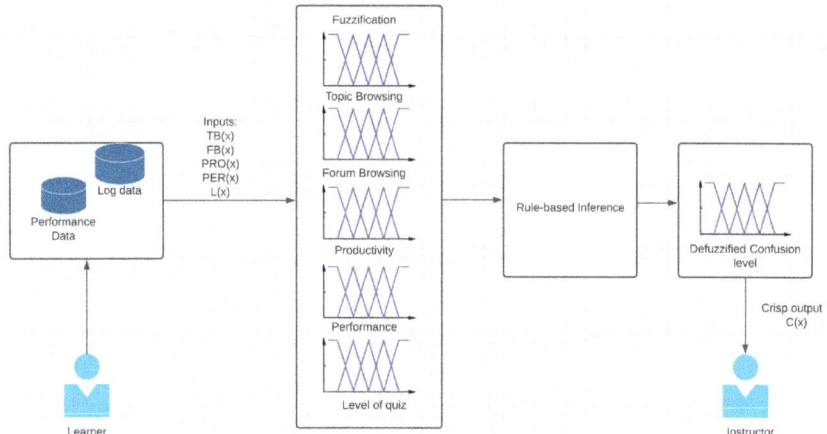

Fig. 1   Confusion detection using fuzzy logic

Fig. 2   Input parameters
triangular function

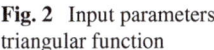
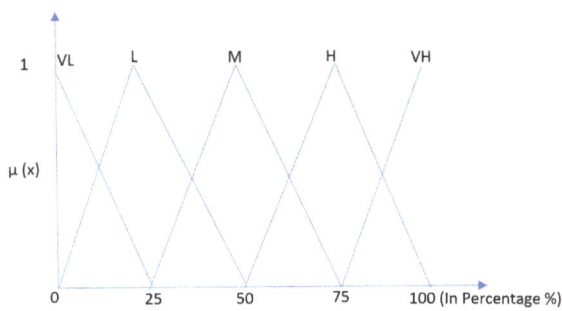

For example, the set $TB(x) = \{$ very low, low, medium, high, very high$\}$ compares the number of times the learner browsed the course pages while attempting the quiz, with the average click-pattern of other learners as a group and determines a percentage.

All input parameters are five linguistic variables: {very low, low, medium, high, very high}.

Furthermore, Fig. 3 shows the Membership functions for all the input parameters.

4.2   Inference Engine

The inference Engine will store all the possible combinations in the form of if–then statements, and the output from the combination will be denoted as $C(x)$ as a value in percentage derived from a possible set of confusion levels: {very low, low, medium, high, very high} using Mamdani Method.

Fig. 3 Membership functions

$$\mu_{vL}(x) = \frac{25 - x}{25} \quad 0 \le x \le 25 \tag{1}$$

$$\mu_L(x) = \frac{x}{25} \quad 0 \le x \le 25$$

$$= \frac{50 - x}{25} \quad 25 \le x \le 50 \tag{2}$$

$$\mu_M(x) = \frac{x - 25}{25} \quad 25 \le x \le 50$$

$$= \frac{75 - x}{25} \quad 50 \le x \le 75 \tag{3}$$

$$\mu_H(x) = \frac{x - 50}{25} \quad 50 \le x \le 75$$

$$= \frac{100 - x}{25} \quad 75 \le x \le 100 \tag{4}$$

$$\mu_{vH}(x) = \frac{x - 75}{25} \quad 75 \le x \le 100 \tag{5}$$

Below are some examples of such rules defined in the Inference engine to determine the Confusion level:

```
[Rule 1]
if (TB(x) == 'high'&& FB(x) == 'high'&& PRO(x) == 'low'&& PER(x) =
= 'medium'&& L(x) == 'low') then:
     C(x) == 'very high'
[Rule 2]
if (TB(x) == 'low'&& FB(x) == 'medium'&& PRO(x) == 'low'&& PER(x)
== 'medium'&& L(x) == 'low') then:
     C(x) == 'high'
[Rule 3]
if (TB(x) == 'low'&& FB(x) == 'low'&& PRO(x) == 'high'&& PER(x) =
= 'medium'&& L(x) == 'low') then:
     C(x) == 'low'
[Rule 4]
if   (TB(x)   ==   'low'&&   FB(x)   ==   'low'&&   PRO(x)   ==
'very high'&& PER(x) == 'very high'&& L(x) == 'low') then:
     C(x) == 'very low'
```

The motivation behind Rule 1 is that if a student must take a lot of assistance with low productivity and performance on low-level questions, the underlying level of Confusion is high and needs to be addressed soon before it frustrates the learners leading them to give up on the course, but as per Rule 4 the student is performing very well on low-level questions without any assistance, then the underlying level of confusion is very low, and the student will likely keep learning on the course.

Confusion in Percent will be calculated based on the simple yet powerful Mamdani Method (centroid) after calculating the maximum of all the input variables' minimum values. The maximum value will then be substituted for the output Membership function. The membership function's aggregate value will be the final output as a percent of the learner's confusion.

$$\mu_R c(x) = \max[\min(\mu_{TB}(x), \mu_{FB}(x), \mu_{PRO}(x), \mu_{PER}(x), \mu_{LEV}(x))] \qquad (6)$$

The crisp value of confusion level is calculated the Mamdani (Centroid) method. In the simplest form, consider that out of five factors. Currently, two factors: Topic Browsing ($TB(x)$) and Performance ($PER(x)$) are known to be 80% and 40%, respectively. With these two factors, if we apply the membership functions, we end up with the following values:

Min(4/5, 2/5) = 2/5 (Topic Browsing HIGH and Performance LOW)
Min(4/5, 3/5) = 3/5 (Topic Browsing HIGH and Performance MEDIUM)
Min(1/5, 2/5) = 1/5 (Topic Browsing VERY HIGH and Performance LOW)
Min(1/5, 3/5) = 1/5 (Topic Browsing VERY HIGH and Performance MEDIUM).

Hence from the above 4 rule inferences, the maximum strength is 3/5. According to our Rule 1, When TB is 'HIGH', and Performance is 'MEDIUM', we are assuming the Confusion level to be 'VERY HIGH.'

Finally, substituting 3/5 to the membership functions of the output variable, we will get two values of Confusion: 90 and 85. Since we are using the Centroid method, our final Confusion level will be:

$$\mu_C(x) = 90 + 852 = 87.5\% \text{ (Very high)} \qquad (7)$$

Hence, the final level of Confusion corresponds to 87.5%, which is very high and may indicate some intervention to avoid drop-out. The other factors have been dis-regarded just to show a simplified version of derivation, but the model with all the factors under consideration will give more confidence in the confusion level calculation.

5 Conclusion

Confusion plays a crucial role in the learning process. Online learning brings much flexibility in learning but at the same time injects a complexity layer to detect Confusion in learners that in a face-to-face academic setting seems to be easy to pick up. The proposed model can be used to detect Confusion levels, when learners are attempting online quizzes. The model could potentially detect Confusion with less computational efforts and more pervasiveness. The model mainly uses the three levels of the Fuzzy-logic Inference system to detect Confusion, helping academics provide timely assistance to learners to preserve the completion rate in an online course. As a part of ongoing work, we plan to experiment and implement the fuzzy neural networks or other machine learning techniques on quiz assessment data obtained from an online platform.

References

1. Jordan K (2014) Initial trends in enrolment and completion of massive open online courses. Int Rev Res Open Dist Learn 15:133–160
2. Arguel A et al (2017) Inside out: detecting learners confusion to improve interactive digital learning environments. J Educ Comput Res 55
3. Wen M, Yang D, Rose C (2014) Sentiment analysis in MOOC discussion forums: what does it tell us? In: Educational data mining 2014. Citeseer
4. Baker RS et al (2010) Better to be frustrated than bored: the incidence, persistence, and impact of learners' cognitive–affective states during interactions with three different computer-based learning environments. Int J Hum Comput Stud 68(4):223–241
5. Conole GG (2013) MOOCs as disruptive technologies: strategies for enhancing the learner experience and quality of MOOCs. Rev Educ Distance 39
6. Yang D, Kraut R, Rose CP (2016) Exploring the effect of student confusion in massive open online courses. J Educ Data Mining 8(1):52–83
7. Lodge JM et al (2018) Understanding difficulties and resulting confusion in learning: an integrative review. Front Educ
8. D'Mello S, Graesser A (2014) Confusion and its dynamics during device comprehension with breakdown scenarios. Acta Physiol (Oxf) 151:106–116
9. D'Mello S, Graesser A (2012) Dynamics of affective states during complex learning. Learn Instr 22(2):145–157
10. D'Mello SK, Graesser A (2012) Language and discourse are powerful signals of student emotions during tutoring. IEEE Trans Learn Technol 5(4):304–317
11. D'Mello S (2013) A selective meta-analysis on the relative incidence of discrete affective states during learning with technology. J Educ Psychol 105(4):1082–1099
12. Lehman B, D'Mello S, Person N (2010) The intricate dance between cognition and emotion during expert tutoring. In: International conference on intelligent tutoring systems. Springer
13. Lehman B et al (2008) What are you feeling? Investigating student affective states during expert human tutoring sessions. In: Intelligent tutoring systems. Springer Berlin Heidelberg, Berlin, Heidelberg
14. Almatrafi O, Johri A, Rangwala H (2018) Needle in a haystack: identifying learner posts that require urgent response in MOOC discussion forums. Comput Educ 118:1–9
15. Chandrasekaran MK et al (2015) Learning instructor intervention from MOOC forums: early results and issues. arXiv preprint arXiv:1504.07206
16. Chaturvedi S, Goldwasser D, Daumé III H (2014) Predicting instructor's intervention in MOOC forums. In: Proceedings of the 52nd annual meeting of the association for computational linguistics, volume 1: long papers
17. Agrawal A et al (2015) YouEDU: addressing confusion in MOOC discussion forums by recommending instructional video clips
18. D'Mello S et al (2014) Confusion can be beneficial for learning. Learn Instr 29:153–170
19. Arguel A, Lane R (2015) Fostering deep understanding in geography by inducing and managing confusion: an online learning approach
20. Lehman B, D'Mello S, Graesser A (2012) Confusion and complex learning during interactions with computer learning environments. Internet High Educ 15(3):184–194
21. Atapattu T et al (2019) An identification of learners' confusion through language and discourse analysis. arXiv preprint arXiv:1903.03286
22. Caprara G et al (2008) Assessing regulatory emotional self-efficacy in three countries. Psychol Assess 20:227–237
23. Pachman M et al (2016) Eye tracking and early detection of confusion in digital learning environments: proof of concept. Australas J Educ Technol 32(6)
24. D'Mello S, Graesser A (2009) Automatic detection of learner's affect from gross body language. Appl Artif Intell 23(2):123–150
25. McDaniel B et al (2007) Facial features for affective state detection in learning environments. In: Proceedings of the annual meeting of the cognitive science society

26. Zeng Z et al (2009) A survey of affect recognition methods: audio, visual, and spontaneous expressions. IEEE Trans Pattern Anal Mach Intell 31(1):39–58
27. D'Mello SK, Dowell N, Graesser AC (2009) Cohesion relationships in tutorial dialogue as predictors of affective states. In: AIED
28. Bakharia A (2016) Towards cross-domain MOOC forum post classification. In: Proceedings of the third (2016) ACM conference on learning @ scale. ACM, Edinburgh, Scotland, pp 253–256
29. Zeng Z, Chaturvedi S, Bhat S (2017) Learner affect through the looking glass: characterization and detection of confusion in online courses. International Educational Data Mining Society
30. Guijarro-Mata-García M, Guijarro M, Fuentes-Fernández R (2015) A comparative study of the use of fuzzy logic in e-learning systems. J Intell Fuzzy Syst 29(3):1241–1249
31. Hawkes LW, Derry SJ, Rundensteiner EA (1990) Individualized tutoring using an intelligent fuzzy temporal relational database. Int J Man Mach Stud 33(4):409–429
32. Hawkes LW, Derry SJ (1996) Advances in local student modeling using informal fuzzy reasoning. Int J Hum Comput Stud 45(6):697–722
33. Beck J, Stern M, Woolf B (1997) Using the student model to control problem difficulty
34. Redondo MA et al (2003) Applying fuzzy logic to analyze collaborative learning experiences in an e-learning environment. USDLA J (United States Distance Learn Assoc) 17:19–28
35. Barros B, Verdejo MF (2000) Analysing student interaction processes in order to improve collaboration. The DEGREE approach. Int J Artif Intell Educ 11(3):221–241
36. Mihalis P, Maria G (1995) An application of fuzzy logic to student modelling. In: IFIP world conference on computers in education. Springer
37. Rasmani KA, Shen Q (2006) Data-driven fuzzy rule generation and its application for student academic performance evaluation. Appl Intell 25(3):305–319
38. Hogo MA (2010) Evaluation of e-learning systems based on fuzzy clustering models and statistical tools. Expert Syst Appl 37(10):6891–6903
39. Di Lascio L, Gisolfi A, Loia V (1998) Uncertainty processing in user-modeling activity. Inf Sci 106(1–2):25–47
40. Xu D, Wang H, Su K (2002) Intelligent student profiling with fuzzy models. In: Proceedings of the 35th annual Hawaii international conference on system sciences. IEEE
41. McNeill FM, Thro E (1994) Fuzzy logic: a practical approach. Academic Press Professional, Inc.

Chapter 51
Numerical Simulation of Damped Welded Profiles Under Variable Amplitude Loading

Imane Amarir, Hamid Mounir, Abdellatif El Marjani, and Zakaria Haji

1 Introduction

Welded structures in engineering can be subjected to constant or variable amplitude loads, especially in the automotive industry [1]. They are then more susceptible to the severe problem of fatigue failure in the long term. For that reason, the fatigue assessment has been extensively analyzed for several decades. On the other hand, there are many correction factors as [2]: chosen material, surface state, geometry size [3], welds process, fluctuating stress, reliability and environment. There are three steps before the failure is expected: Firstly, the occurrence of microscopic cracks and secondly, the crack propagation on all specimen and then all of a sudden, the final rupture [4]. Several methods based on SN or Wohler curve [5] have been proposed by authors [6] as: Eurocode 3, volumetric method and British Standard. In this present paper, damage assessment was built on damped welded structure [7] under variable amplitude loads [5] along with finite element software ANSYS and the strain-life approach [8]. The fatigue life, damage and biaxiality indication stress were discussed to evaluate the fatigue behavior of model. Finally, the Rainflow method [9] and the damage matrix [10] were observed for the cycle counting and the percent of damage.

2 Numerical Method

A three-dimensional numerical method [11] was carried out on a damped welded structure under variable amplitude loads. Thus, the fatigue and static analysis were

I. Amarir (✉) · H. Mounir · A. El Marjani · Z. Haji
Research Team EMISys, Research Centre Engineering 3S, Mohammadia School of Engineers, Mohammed V University, Rabat, Morocco
e-mail: imaneamarir@research.emi.ac.ma

© The Author(s), under exclusive license to Springer Nature Singapore Pte Ltd. 2021   659
S. Kumar et al. (eds.), *Proceedings of International Conference on Communication and Computational Technologies*, Algorithms for Intelligent Systems,
https://doi.org/10.1007/978-981-16-3246-4_51

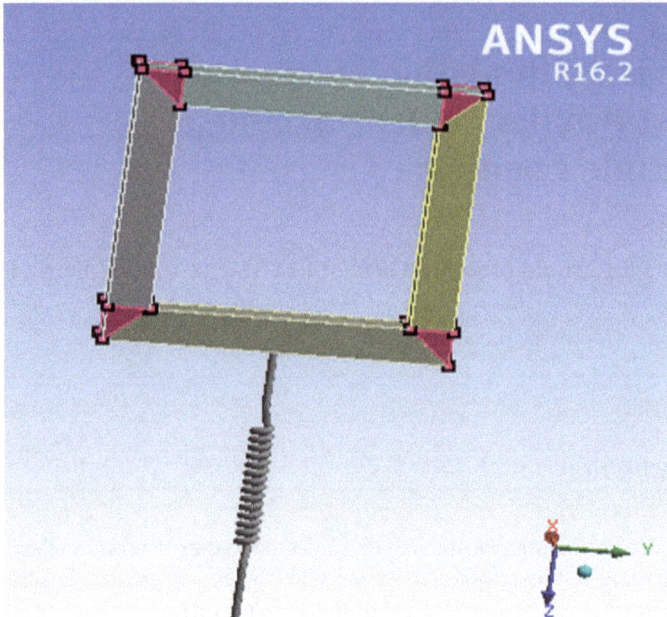

Fig. 1 Geometry of damped welded structure

investigated using one of the most popular finite element software package called
ANSYS [12, 13] which can model the complex design behavior with reasonable
accuracy.

2.1 Geometrical Modeling

The geometry developed in ANSYS in order to define the fatigue life [14] is well
illustrated in Fig. 1, where it is an useful welded schematic in the transportation
domain. Indeed, it is consisted of four rectangular profiles with same dimensions
(length 400 mm) which are welded using the filler metal and the homogeneous weld
process. We have then connected to the lower profile a longitudinal spring with spring
stiffness equal to 2500 N mm^{-1}.

2.2 Wohler Curve of Material

The material applied on our model is the structural steel which is characterized by
the properties and strain-life parameters displayed in Tables 1 and 2.

Table 1 Strain-life parameters of structural steel

Strength coefficient (MPa)	Strength exponent	Ductility coefficient	Ductility exponent	Cyclic strength coefficient (MPa)	Cyclic strain hardening exponent
920	−0.106	0.213	−0.47	1000	0.2

Table 2 Properties of structural steel

Density	$7.85e{-}006 \text{ kg mm}^{-3}$
Thermal conductivity	$6.05e{-}002 \text{ W mm}^{-1} \text{ C}^{-1}$
Resistivity	$1.7e{-}004 \ \Omega \text{ mm}$
Tensile ultimate strength	460 MPa
Tensile yield strength	250 MPa

The Wohler curve [15, 16] indicated the evolution of the alternating stress as a function of the lifetime of the structure. It represented generally a visualization of the behavior of the materials or the whole structure in the fatigue domain through a series of sample tests which consist in subjecting each test specimen to cyclic loading cycles.

Figure 2 shows the S–N curve or so-called Wohler curve for structural steel given by default using the engineering software ANSYS.

The endurance limit refers to 86 MPa which below this stress, the structure will not fail. Furthermore, as the amplitude stress is lower, as the service life of structure is longer.

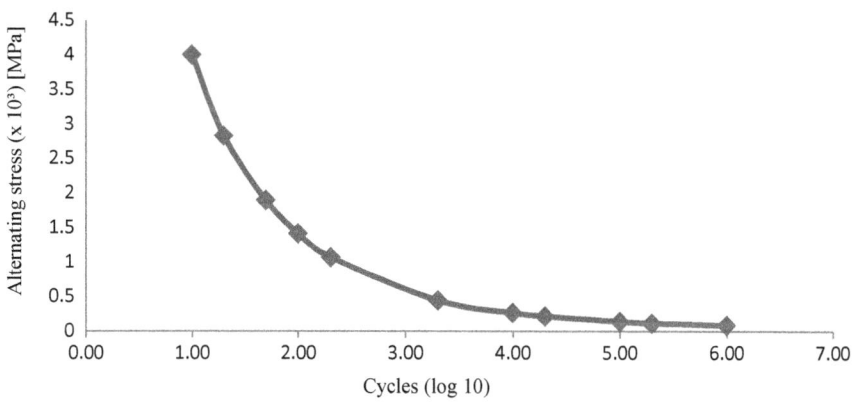

Fig. 2 Wohler curve of steel material

2.3 Meshing and Boundary Conditions

The geometry is modeled in 3D, and then the vertical force applied is 1500 N on the top face of structure; after that, it was necessary to block the movement of structure along the X- and Y-axis because of the use of the longitudinal spring, as shown in Fig. 3.

3 Results and Discussions

We have obtained a set of fatigue resolutions under non-constant amplitude loads from the finite element simulation [17] and the strain-life approach in order to understand the fatigue behavior of our structure used frequently in automotive domain.

Figure 4 is illustrated the different amplitude loads—stress—that we have applied randomly on our welded structure as a function of time (s).

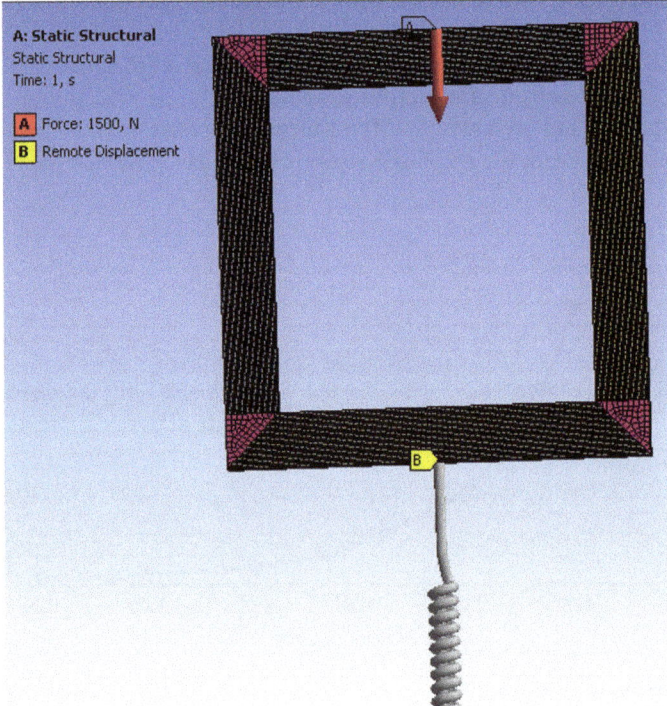

Fig. 3 Mesh and loadings

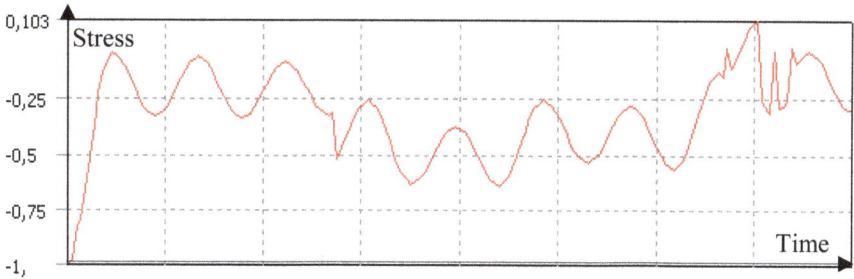

Fig. 4 Variable amplitude loads

3.1 Fatigue Analysis Results

In the Fig. 5, the colored regions are corresponding to the location of critical areas where the occurring of the maximum stress and the minimum number of fatigue life cycles. As a result, it is indicated that the whole structure cannot survive for a long time.

The fatigue damage is an important indicator in the fatigue analysis to understand more the fatigue behavior of the model. In our case of varying stress amplitudes, the total of damage [18] is defined as the damage accumulation at each cycle. From Fig. 6, the damage value is more than 1; consequently, the failure occurs before the design life is achieved.

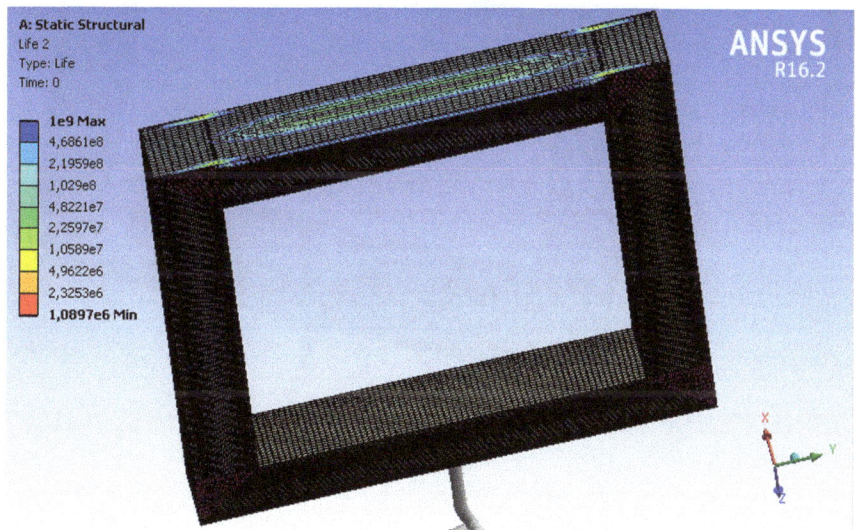

Fig. 5 Fatigue life

Fig. 6 Fatigue damage

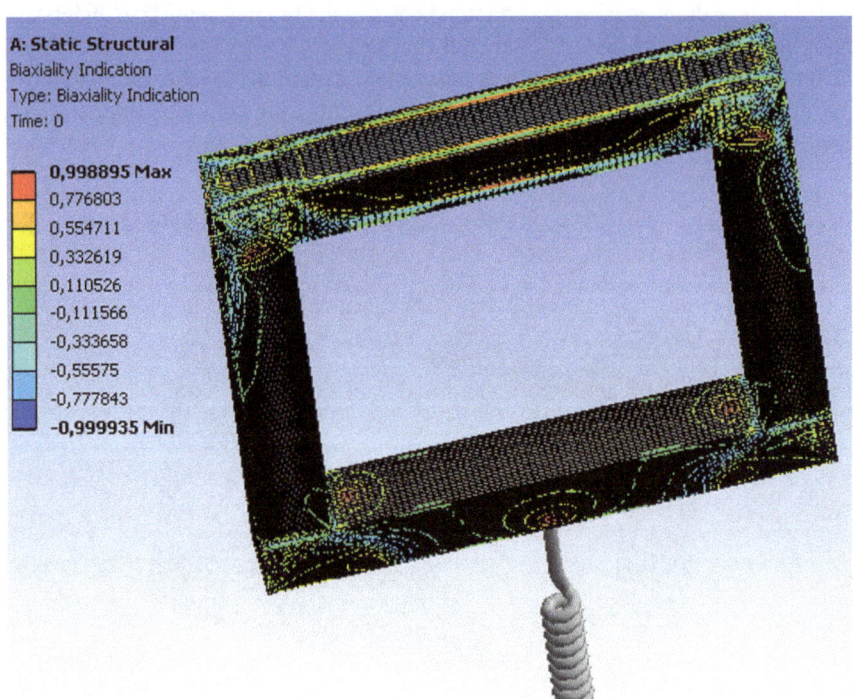

Fig. 7 Biaxiality indication results

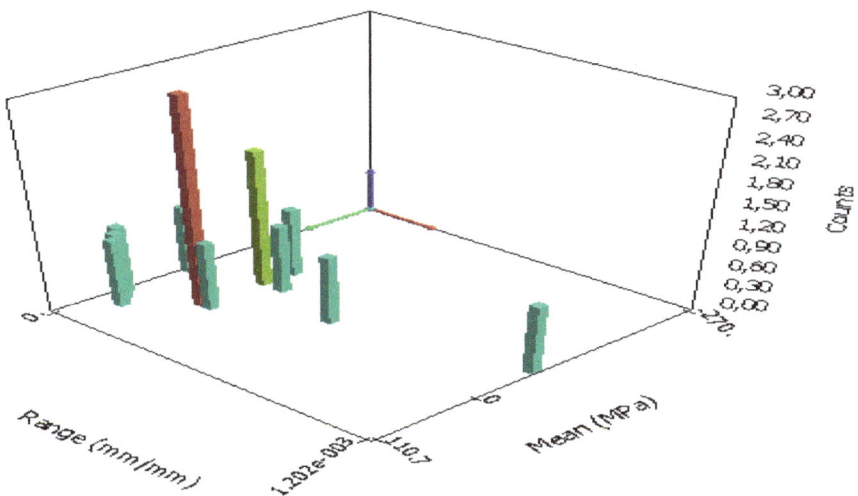

Fig. 8 Rainflow matrix

According to Fig. 7, the biaxiality indication is defined as a qualitative measure of the stress state over the whole structure.

3.2 Rainflow and Damage Matrix

The strain-life approach led to obtain the Rainflow matrix and the damage matrix (3D shape histogram) as shown respectively in Figs. 8 and 9. Indeed, they are plotted at the dangerous area on the model only in case of non-constant amplitude loads where the counting is needed.

Consequently, the Rainflow method is considered as a simplified method to evaluate the alternating and mean stress bins as a function of number of counts, and we have then concluded from Fig. 8 that the most of alternating stresses are fairly low.

On the other hand, the damage matrix is almost similar to the Rainflow matrix. But the difference is that the percentage of damage is plotted as the Z-axis in Fig. 9 instead of the number of counts in Fig. 8.

As a result, the lower stress amplitudes are corresponding to the most number of the counts; by contrast, the higher stress amplitudes occurs at the most of the percent damage.

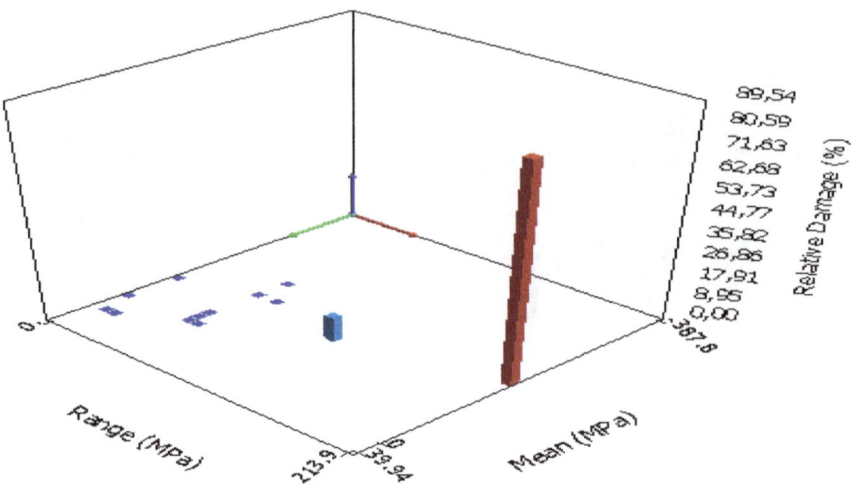

Fig. 9 Damage matrix

4 Conclusion

This article presents the fatigue analysis by finite element simulation on a structure based on damped welded profiles under variable amplitude loads for automotive utilization. The fatigue life, damage and biaxiality indication are then three mains parameters to calculate the lifetime of structure using the simulation software ANSYS and the strain-life approach. Besides, because of the non-constant amplitude loads, it is required to adopt a cycle counting method as Rainflow method and a damage matrix to indicate the percentage of damage at critical areas as a function of the alternating and mean stress bins. To conclude, the results enable designers to evaluate and estimate the fatigue life behavior of this type of model under real-world conditions. In future work, we focus on the experimental results to validate our numerical results.

References

1. Amarir I, Mounir H, El Marjani A (2020) Compilation of fatigue assessment procedures of welded structures for automotive utilization. Mech Based Des Struct Mach 1(1):1–13
2. Jonsson B (2011) Improving weld class systems in assessing the fatigue life of different welded joint designs. In: Fracture and fatigue of welded joints and structures, pp 139–167
3. Amarir I, Mounir H, El Marjani A (2019) Modeling the damping effect of welded rectangular profiles under damped loads for automotive utilization on the durability. In: 5th international conference on optimization and applications (ICOA). IEEE, Kenitra, pp 1–5
4. Yarema SY (1975) Stages of fatigue fracture and their consequences. In: Physicochemical theory of deformation and failure, vol 9, pp 681–686

5. Amarir I, Mounir H, El Marjani A (2018) Effect of temperature and thickness of the weld bead on the durability of welded rectangular profiles under damped loads for electrical vehicles utilization. In: 6th international renewable and sustainable energy conference (IRSEC). IEEE, Rabat, pp 1–6
6. Fuštar B, Lukačević I, Dujmović D (2018) Review of fatigue assessment methods for welded steel structures. Adv Civ Eng 1–16
7. Amarir I, Mounir H, El Marjani A (2018) Numerical modeling of fatigue analysis of welded rectangular profiles under damped loads for automotive utilization. In: 4th international conference on optimization and applications (ICOA). IEEE, Mohammadia, pp 1–5
8. Petinov S (2014) Strain-life approach: application for fatigue design of ship superstructure critical detail. Appl Mech Mater 617:197–202
9. Li Lee Y, Tjhung T (2012) Rainflow cycle counting techniques. In: Metal fatigue analysis handbook, pp 89–113
10. Adumitroaie A, Barbero EJ (2015) Intralaminar damage model for laminates subjected to membrane and flexural deformations. Mech Adv Mater Struct 9:705–716
11. Amarir I, Mounir H, El Marjani A (2019) Damage analysis of welded rectangular profiles for electric vehicles utilization under variable and constant amplitude loads. In: 7th international renewable and sustainable energy conference (IRSEC). IEEE, Rabat, pp 1–5
12. Amarir I, Mounir H, El Marjani A (2019) Modeling the effect of welding process on the durability of welded rectangular profiles under damped loads for automotive utilization. Int J Mech Prod Eng Res Dev (IJMPERD) 1(9):639–648
13. Yuan Y (2018) Fatigue analysis of mobile maintenance platform based on Ansys workbench. MATEC Web Conf 175:03048
14. Alireza R, Arefeh S (2020) Durability of ultra-high performance self-compacting concrete with hybrid fiber. Emerg Mater Res 9(2):1–12
15. Mlikota M, Schmauder S, Božić Ž (2018) Calculation of the Wöhler (S-N) curve using a two-scale model. Int J Fatigue 114:289–297
16. Kłysz S, Lisiecki J, Bąkowski T (2010) Modification of the equation for description of Wöhler's curves. Res Works Air Force Inst Technol 27(1)
17. Rojda O, Berna I, Erkan C (2020) Finite element analysis of shear tests on masonry triplets. Emerg Mater Res 9(1):1–8
18. Jiangchuan L, Hongxia Q, Feifei Z (2020) Reliability analysis of fiber concrete freeze–thaw damage based on the Weibull method. Emerg Mater Res 9(1):1–8

Chapter 52
Application of Machine Learning-Based Pattern Recognition in IoT Devices: Review

Zachary Menter, Wei Zhong Tee, and Rushit Dave

1 Introduction

The Internet of things (IoT) is a term used to describe the network of devices, or things, with embedded software, sensors, transmitters and receivers, and other technology. This network collects, sends, and receives data from other connected devices via the Internet [1]. The emergence of IoT has allowed for a multitude of innovations in many different areas such as home automation, event prediction, and activity recognition, to name a few. Nowadays, many complex calculations and machine learning algorithms that used to require large amounts of processing power can all be run on IoT devices, leading to many exciting and inventive applications [2]. The multitude of embedded sensors in many IoT devices allows for many innovations through pattern recognition in everyday life. These innovations in pattern recognition allow for our mobile and IoT devices to serve new and better functions every day. Because of this, the use of pattern recognition plays a massive role in a large majority of the present work being done in IoT devices. Pattern recognition is the notion of assigning objects to classes. Such patterns that can be recognized include textures, images, speech, biological/physical features, habits, and many other types of patterns. Features of the object are organized in a selected space where an algorithm of technique is used to assign it a class label [3].

Utilizing machine learning algorithms has become popular among IoT devices because it improves IoT-based services such as traffic engineering [4], security [5–8] and security assessment [9], speaker [10] and image [11] recognition, and quality of service optimization [12]. Pattern recognition in IoT devices has also seen much more development and consumer use in smart home [13], automation [14], and cloud-based monitoring and automation [15] systems. Algorithms must be chosen based on

Z. Menter · W. Z. Tee · R. Dave (✉)
University of Wisconsin-Eau Claire, Eau Claire, WI 54701, USA
e-mail: daver@uwec.edu

© The Author(s), under exclusive license to Springer Nature Singapore Pte Ltd. 2021   669
S. Kumar et al. (eds.), *Proceedings of International Conference on Communication and Computational Technologies*, Algorithms for Intelligent Systems,
https://doi.org/10.1007/978-981-16-3246-4_52

their power efficiency and computational cost and their ability to correctly classify features as most IoT-based systems would have constraints with power, memory, or storage [16, 17]. Efficient and intelligent IoT applications have emerged as a result of employing machine learning algorithms into IoT-based devices and services.

The combination of pattern recognition and IoT devices has been deployed in many industries and is often developed on a generic basis [18], allowing for it to be used in many different situations. Various industries have utilized pattern recognition and IoT devices for different causes such as strengthening security and implementing biometric solutions, recognizing vehicle/traffic patterns, predicting complex events, recognizing and classifying human activity, and automating tasks.

2 Background

There is great significance in selecting the best machine learning algorithm for pattern recognition in IoT devices. Given the restraints IoT devices have with memory, storage, or power consumption, an efficient algorithm will allow for fast processing time, low space usage, and low-power consumption. Existing work that is based on human activity recognition (HAR) and biometric security has been developed and surveyed for this article.

Pattern recognition in IoT devices for human activity recognition has seen such an increase in demand and development because the devices and sensors that are capable of recording human motion or the vital signs data have to be light, compact, and wearable [19]. Most commonly, predefined activity models are first used to train classifiers to identify activities performed by humans based on data collected by various wearable sensors [20]. A study carried out by Ketu and Mishra [21] in 2020 published an overall performance evaluation of computing device mastering algorithms for IoT primarily based human activity recognition. Different algorithms consist of different capabilities and performance, and in this case, factors such as the run time, space required, and energy consumed were measured with the aim of selecting the most optimal algorithm for wearable sensors for human activity recognition using a predefined activity recognition dataset. This process included selecting various algorithms and running test cases through them in a virtual simulation that evaluates its accuracy, precision, recall, and F-1 score. The testing concluded that seven of the fourteen algorithms that were selected performed better with higher accuracy; hence, using of one those seven algorithms for human activity recognition under similar circumstances would be more optimal than the others. The seven algorithms include the gradient boosting classifier, random forest, bagging classifier, classification and regression trees, support vector machine, k-nearest neighbor, and the extra trees classifier. Similarly, another common use of machine learning algorithms in IoT devices for pattern recognition is in security systems. One type of security that IoT devices handle is biometrics, which includes recognizing faces or fingerprints [22]. Security is also very important in smart home technology. IoT devices are the main components of every smart home setup. Smart devices generally have wireless access to a user's accounts and home devices. This produces a large need for effective, compact security solutions. These IoT devices are at risk for a number of threats such as information leaks, data mining, denial of service attacks, and various other cyber-attacks [23]. One of the ways these attacks can be detected and prevented is through statistical analysis and machine learning, which can help inspect and detect anomalies in the data being sent over a network [24]. Shi et al. [25] developed a spoof detection model using the support vector machine algorithm. This model was able to correctly identify between legitimate users and spoofed users. In addition, much work has been conducted on applying various machine learning methods to intrusion detection systems. With many new models and developments being made frequently, it is important to review and analyze the work that has been done in order to seek out possible and necessary improvements. Though much has been done in the way of machine learning in human activity recognition and security in IoT devices, each model and method used may have certain advantages and disadvantages. With this study, we seek to explore these advantages and disadvantages to improve the work that is being done in this field.

3 Human Activity Recognition (HAR) in IoT Devices

3.1 Using Smart Computing Inferring In-the-Wild Human Contexts Based on Activity Pattern Recognition

A new approach for activity-aware human context recognition (AAHCR) using both a smartphone and smartwatch together to infer the user's context information based on pattern recognition was introduced in this study. Daily living activities (DLAs) used in the proposed scheme include lying down, running, sitting, standing, and walking.

The methodology of the study is shown in Fig. 1. The data that was used is a publicly available dataset called the extrasensory dataset. It is from a different study that collected data from 60 participants regarding in-the-wild activities. The extrasensory dataset contained binary context labels that corresponded to the selected DLAs.

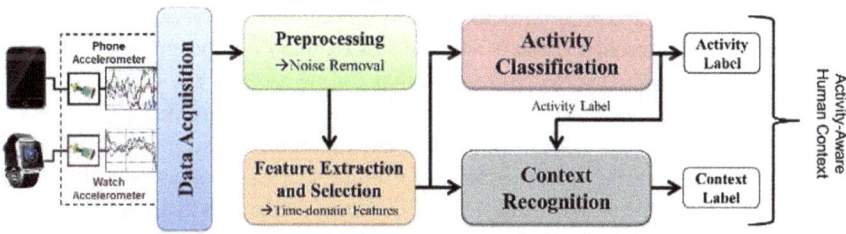

Fig. 1 Proposed methodology for the model that is AAHCR based [26]

The two context labels that each activity consisted of were the behavioral contexts and the phone positions. Such positions include the phone on a table, in a hand, pocket, or bag. The context information that concerns the user's secondary activity is incorporated into the first layer of determining the DLA. The second layer represents the telephone's setting when a particular active work is performed, which the first layer utilizes for position-dependent human context recognition. A low-cost time domain soothing filter was chosen for signal denoising. Signal attributes are determined by the feature extraction process. The separated highlights are then utilized for movement arrangement and the setting acknowledgment. Two kinds of setting acknowledgment tests were directed: position free and position subordinate. In position free, no phone position information is incorporated into the classifiers training. In position reliant, the classifiers are prepared for various telephone positions. The selected pattern recognition put was compared in the model includes decision tree (DT), random forest (RF), k-nearest neighbors (KNN), Bayes net (BN) and multilayer perceptron (MLP-ANN). In both the position-dependent and position-independent tests, the RF classifier showed the best results in the majority of metrics (precision, recall, F-1 score, balance accuracy, kappa, and root mean squared error) when it was paired with the data collected from both the smartphone and watch accelerometer sensors. However, the position-dependent tests offered better results from the metrics compared to that of the position-independent tests. As the RF classifier was determined to be the ideal algorithm for this model, a confusion matrix was then generated to represent the predicted DLAs from a fusion of data from both sensors. Static activities such as lying down, sitting, and standing have a high percentage of correct predictions whereas dynamic activities such as walking or running had a slightly lower percentage of correct predictions. The study concluded that different human behavioral contexts can be used to recognize and predict daily activities. The data provided by extrasensory showed that the best algorithm for pattern recognition in human activities using behavioral contexts was the random forest.

3.2 An IoT Approach for Wearable-Based Human Activity Recognition

A study conducted in 2017 proposed using a remote monitoring component with remote visualization and programmable alarms for human activity recognition (HAR) and validates the approach used. It is made of two main components: a traditional human activity recognition (HAR) system that can be used on any portable or non-cell phone and an application for recognizing and surveilling in a healthcare-related subject. The flow of the model is shown in Fig. 2. The learning phase establishes the relations between the information and exercises. It first collects the data from every one of the sensors the framework is utilizing. Sensors will be subject to the sort of gadget that is being worked for recognition. Time, the sort and length are the factors that need to be recorded in an activity log and the activities should be

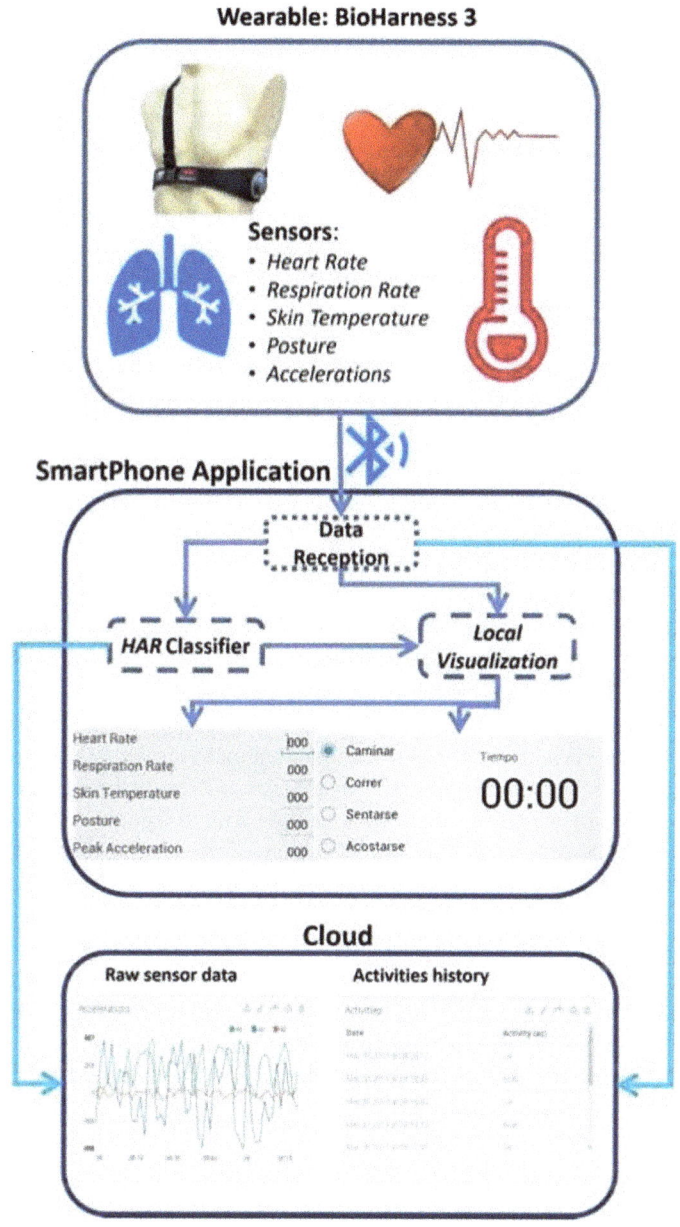

Fig. 2 Flow of the model [27]

carried out in a random order with random durations. The feature extraction step is then based on structural features and statistical features. Structural features often fit in a defined mathematical function, and there is no correlation between the signals in the data. Statistical features extract its features based on the statistical information such as the mean, standard deviation, or correlation. Lastly, the learning phase develops an acknowledgment model that utilizes the information set, activity log to recognize. Following the learning phase, the first step in the recognition phase is the data collection. However, it does not use any prior knowledge on the activities performed so it does not keep an activity log. Similarly, it extracts features in the same fashion as the learning phase using structural and statistical features. The Zephyr Bio harness 3 was chosen as the wearable tracker because of its capability of measuring the required variables and connecting to a smartphone via Bluetooth. A smartphone application is needed to receive the data and handle the transmission and preservation of the wearable's raw data. The wearable's raw data as well as the activities recognized by the HAR classifier is sent to the cloud portion. Ubidots was chosen as the platform. It was able to display the history of known behaviors, as well as heart rate, respiration rate, stance, and acceleration values. Heart rate, respiration rate, stance, three-axis acceleration, peak acceleration, and electrocardiogram magnitude are among the data used. The structure detection algorithm looks for the mathematical function that best fits the structure for groups of data. The classifier's training dataset included 14 weather element samples with no statistical association between them. As a result, each algorithm came up with its own set of rules. A C4.5 algorithm and a Naive Bayes algorithm were the resultant rule sets. The selected classifier was the C4.5 algorithm as it was efficient and used far less space than that of the Naive Bayes algorithm. After proving that it worked with the weather samples, the model was implemented to work with training data to recognize. Following the feature extraction process, the algorithm builds a rule tree for that particular training dataset. There were a total of 13 rules for creating a single recognition model for the various subjects involved during this process. 69 out of 72 tasks were successfully recognized for one of the random test subjects. The system correctly classified test subjects who were lying down or jogging. The few errors made by the system were when it had to classify whether a test subject was sitting or walking. It was then concluded that a human activity recognition system using a smartphone, bioharness, and cloud system could be successfully developed and implemented.

3.3 In Wearable Devices Generalized Activity Recognition Using Accelerometer for IoT Applications

One of the most common implementations of HAR systems includes generalized activity recognition model for wearable devices. A diagram of such a model is shown in Fig. 3. It covers how the automatic detection of different activities works using just

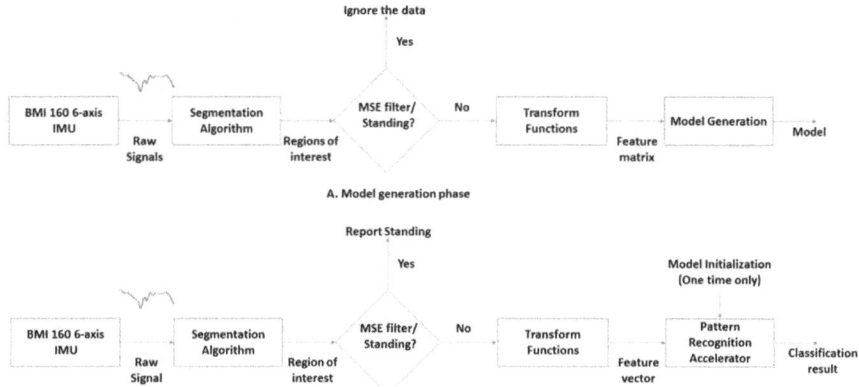

Fig. 3 Proposed flow of the model generation and activity recognition phases [28]

one axis in an accelerometer and the simple features and pattern recognition algorithms that were used that were effective, computationally inexpensive and suitable for wearable devices with constrained resources.

Data was collected in a custom-built device that contained Bosch's BMI 160. The devices were positioned on the sacrum of each test subject and it sampled at a rate of 100 Hz. Data from each subject was collected while they were performing tasks that included walking, running, crawling, ladder climbing, and pronating for 30 s each. The classification model was trained using data obtained from 52 male firefighters and checked on a variety of users at various times. Test data was also obtained from three separate sets from eight subjects, of which there was a mixture of male and female in normal street clothing test subjects. The recognition system only uses data from segmentation, feature extraction, and identification from the accelerometer's Y-axis. To determine whether the test subject was standing, a mean square error (MSE) filter was applied to the segmented results. The feature vectors were clustered using a supervised hierarchical clustering algorithm based on similarity characteristics and labels. For each cluster, a centroid was calculated, and for each centroid, a centroid was calculated and was assigned a label based on majority. The pattern recognition capability was then leveraged by optimizing the number of centroids to use a finite surface around each centroid. To select the final model, multiple iterations with all subject from the training set received new feature vectors, while feature vectors from the test set were left out. In each iteration, a new model is developed, and its performance tested on the test set using metrics such as recall, precision, and accuracy. In the recognition phase, each test vector is assigned a class based on its similarities to patterns. In the event, the test vector does not match any or falls out of the surface of the pattern, and it is assigned an unknown classification (UNK). The decision tree model was only trained using training data that consisted of walking and climbing data in order for the model to be able to distinguish the difference of the two classes. To avoid overfitting, the tree was pruned by adjusting the depth of the tree. The best validation result came from a decision tree with three stages of depth.

The first test consisted of 66 patterns selected based on the datasets that contained 52 subjects. It got an average f1-score of 0.91 and it struggled to identify crawling and pronating the most. The second test was carried out on eight subjects with a mix of males and females and street clothes. The average f1-score was 0.88. The final test also was carried out on eight subjects and the test ended with an average f1-score of 0.91. The model showed accurate results with detecting walking and running but did not see the same success with recognizing crawling.

3.4 On Low-Power Smart Devices Real-Time Motion Pattern Recognition

Another study presented a low-powered smart device known as the Neblina system on modules with hardware variants and expansion modules that targets IoT applications. The accuracy, performance, memory, and power consumption of the Neblina are actively monitored when implementing and testing the proposed motion pattern recognition (MPR) on a fitness activity dataset.

Low-power sensors, on-board memory, and Bluetooth low energy are all part of the Neblina. The base hardware has an efficient 9 axis orientation tracking algorithm, using its magnetic sensors and a framework, shown in Fig. 4, using inertial sensors. The framework contains the shock aware segmentation, feature extraction, and classification. A 3 axis accelerometer and gyroscope is what makes the shock awareness segmentation possible. The sensors have all passed the factory's calibration and filtering techniques. It reads the acceleration that the sensors pick up and compares it to peak constraints to detect specific activities. A segmentation process is based on an overlapping sliding window with the accelerometer and gyroscope sensors. The sensors would have needed to pass factory calibration techniques. The acceleration magnitude is used to determine the minimum peak intensity in the motion segment characterization. The feature extraction uses a time domain histogram instead of more conventional methods. This is because when created with high enough bins, a time domain histogram gives accurate distributions. It is also low in terms of latency

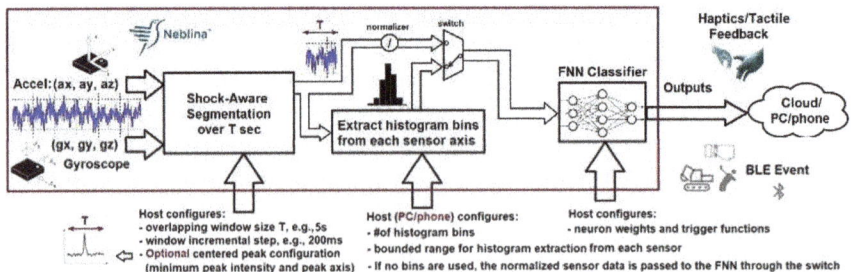

Fig. 4 Flow of model [29]

and computational costs, making it ideal for real-time execution. The classification is performed using a single hidden layer feedforward neural network (FNN), of which the weights and trigger functions are set by the host. It is favorable because the FNN can derive arbitrary nonlinearities from its inputs, and it is also efficient with its latency and memory usage. The FNN is also proven to have better latency and RAM usage compared to the segmentation and histogram extraction and the more conventional KNN classifier. Only 10% of random data points collected were chosen and used to train the classifier. With 60 hidden neurons and 7 epochs, an accuracy of 99.6% was achieved during training. F1-scores for each activity were between 83 and 90% as well for the activities mentioned previously. Furthermore, the Neblina managed to last about 41 h even with the small 100 mAh battery, which proved that this was power efficient.

3.5 For Human Activity Recognition Mobile Healthcare System

The final article that was reviewed presented a healthcare system known as mHealth that is based on IoT and mobile devices. This is part of the m-healthcare system which utilizes mobile devices and wearable body sensors. The MHEALTH dataset was used as it contained ten volunteer's vital sign recordings and body motions for several physical activities. Each subject's left ankle, right wrist, and chest were fitted with sensors. Standing still, sitting/relaxing, lying down, walking, ascending stairs, bending their waist forward, frontal elevation of arms, knees bending, cycling, jogging, biking, and jumping front and back were the twelve activities reported. The human activity identification project used a number of data mining techniques (HAR). The sensor acquired data at rates of 50 Hz. The activities were recorded without any constraints and outside of a laboratory with no controlled variables. The model's predictive ability was then put to the test with ten subjects using eight different algorithms.

Results of the experiment are shown in Table 1. The average accuracy for the random forest and SVM was equal at 99.89%. However, since the random forest is faster than the SVM, it was chosen for HAR. The table below shows the average classification accuracy, F-measure and ROC for each data mining technique that was used 10 times for each subject while testing.

Table 1 Results of average classification accuracy (CA), F-measure (F-M), and area under the ROC curve (AUC) of the different algorithms [30]

	kNN	ANN	SVM	C4.5	CART	Random forest	Rotation forest
Average CA	66.64	99.55	99.89	99.32	99.13	99.89	99.79
Average F-M	0.997	0.996	1	0.99	0.991	0.9989	0.9979
Average AUC	1	1	1	0.998	0.998	1	1

4   Security and Pattern Recognition in IoT Devices

4.1   Activity Pattern Recognition Using Passive Mobile Sensing for Continuous Authentication of Smartphone Users

A study conducted in 2018 used various machine learning classifiers to authenticate smartphone users unobtrusively and continuously by utilizing passive mobile sensing. Common unlocking methods fail to authenticate the user continuously, meaning that if a phone is unlocked, anyone can use it. This proposed method uses machine learning classifiers to detect and recognize physical activity patterns in smartphone users to provide continuous authentication.

In this study, the accelerometer, gyroscope, and magnetometer on the phone were all used. User detection was done by six separate activities: walking, driving, standing, sitting, walking upstairs, and walking downstairs. On all six tasks, the machine was equipped to learn behavioral patterns for various users. Five separate smartphone locations on the body were also used to train the device. These positions were the upper arm, wrist, waist, right thigh, and left thigh. The proposed model is shown in Fig. 5.

The dataset used in this study is a publicly available dataset gathered by a previous study [32] performed on physical activity recognition. Initially, 16 different features that have proven effective in other studies were extracted from the dataset [31]. The output of three different classifiers was compared in this study: Support vector machine (SVM), decision tree (DT), and k-nearest neighbors (KNN). Each of these classifiers was trained separately for each of the ten participants in the study's different behavior patterns.

The performance of the chosen classifiers is shown in Table 2. The SVM classifier on average outperformed both the DT and KNN classifiers with higher accuracy and precision and a lower error rate. One limitation of this system is that it can only identify users based on the activities that the system has been trained for. In the future, more sensors and activities can be added to the system to increase the accuracy and number of situations that the system will work in.

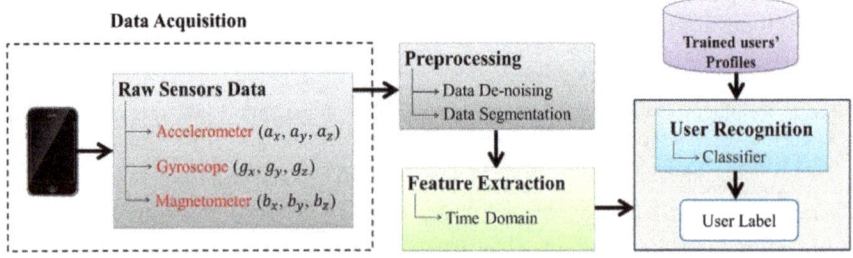

Fig. 5 Flow of the proposed model for user recognition [32]

Table 2 Average performance of the classifiers with smartphone in the waist position [32]

Classifier	Accuracy	Precision	Recall	F-measure	Error rate
DT	0.971	0.972	0.971	0.971	0.029
KNN	0.923	0.924	0.923	0.924	0.078
SVM	0.986	0.986	0.986	0.986	0.014

4.2   Using Smartphone's Built-in Sensors Unobtrusive User Authentication

Common smartphone authentication methods, though simple to perform, are relatively time consuming and obstructive when performed multiple times a day. A study from 2015 sought to solve that problem by introducing an unobtrusive user authentication method based on the micromovements of the user's hands. This method was chosen so that the user can just swipe to unlock their phone and the phone will recognize the user based on their hand movements after unlocking.

The methodology, shown in Fig. 6, includes profiling the smartphone user's hand movements for a short time after unlocking the phone. To collect the required data, the researchers developed an Android app called data collector. They ran the experiment on data collected everywhere between two and ten seconds after unlocking the phone. Seven different statistical values were gathered for different time intervals from all four dimensions of each sensor's data. Therefore, 28 features were extracted from each sensor. Four different machine learning algorithms were chosen for classification: Bayes net (BN), k-nearest neighbor (KNN), multilayer perceptron (MLP), and random forest (RF).

Random forest and multilayer perceptron both yielded the best results with the default parameters, shown in Table 3. MLP performed best on 10 and 6 s of data collection, whereas RF was consistent on all durations. One limitation of this model is that it does not consider the impact of different situations during authentication

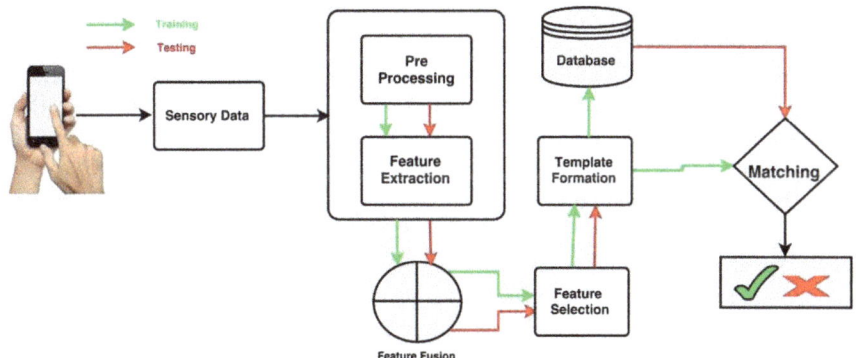

Fig. 6 Flow of the proposed model for user authentication [33]

Table 3 Results of different classifiers for different lengths of data collection [33]

Classifier		2 s	4 s	6 s	8 s	10 s
BN	TAR	0.89	0.89	0.89	0.88	0.89
	EER	0.11	0.11	0.11	0.12	0.12
MLP	TAR	0.93	0.93	0.94	0.94	0.94
	EER	0.07	0.07	0.06	0.06	0.06
KNN	TAR	0.88	0.88	0.89	0.89	0.90
	EER	0.12	0.12	0.11	0.11	0.10
RF	TAR	0.95	0.95	0.95	0.95	0.95
	EER	0.05	0.05	0.05	0.05	0.05

such as running, walking, standing, and sitting. These situations could influence the behavioral patterns being classified as the behavioral patterns of a person standing still may be different than the behavioral patterns of the same person while walking.

4.3 Bot-IoT Attacks Traffic Identification for Internet of Things in Smart City

Another study from 2019 proposed a model and hybrid algorithm for selecting machine learning algorithms for cyber-attack traffic detection. Many organizations have the need for IoT threat detection, but it is not always clear which machine learning algorithm is best suited for the job. This study proposed a framework to solve this problem.

The dataset used in this study was the Bot-IoT dataset. 44 of the most effective features were selected from the dataset. The purpose of the system is to select the best machine learning algorithm for a problem from a set of machine learning algorithms. The set of algorithms that they chose to use were Naïve Bayes, Bayes net, decision tree C4.5, random forest, and random tree. To find the most effective machine learning algorithm, they used a mathematical tool called the bijective soft set. This technique has been effective in multiple other studies related to decision making, so they opted to apply it to this as well. The bijective soft set algorithm, as shown in Fig. 7, calculates soft sets for each machine learning attribute then for each of those soft sets, calculates the correlation AND OR product. After correlation, the algorithm calculates the union operation and the intersection operation to reduce the correlation table to 1 × 1. The proposed algorithm, shown in Fig. 8, is a hybrid machine learning algorithm selection algorithm for anomaly and intrusion detection in IoT networks. The proposed algorithm calculates the performance result values of each of the selected machine learning algorithms. The algorithm then ranks each machine learning algorithm based on threshold values. After this, the algorithm calculates the attributes and runs them through the bijective soft set algorithm.

Algorithm 1: ML algorithm selection based on Bijective soft set:
Input: ML (ML₁, ML₂, ML₃, ... MLₙ)   // set of ML algorithms,
Output: ML algorithm []   // selected effective algorithm

1.   begin
2.   for i = 1 to N   // ML attributes (metrics)
3.   calculate soft set from each [i] for each attributes;
4.   end for
5.   for i = to N;
6.   calculate Correlation both AND OR product n×n;
7.   calculate Union operation either in row or column
            to reduce the correlation table to 1×n or n×1;
8.   calculate Intersection operation either in row or
            column for possible reduction the correlation table
            to 1×1;
9.   end for
10.   Finally, obtain the effective ML algorithm, if the table is
       Reduced to 1×1;
Return Algo;

Fig. 7   Bijective soft set algorithm [34]

The results of the applied machine learning algorithms are shown in Table 4. It can be seen that all of the machine learning algorithms performed very well. However, Naïve Bayes and random tree stood out as the most effective algorithms. This is because, though all were very accurate, Naïve Bayes and random tree both had a very low time taken to build the model (TTBM). The researchers mentioned that though they measured the performance using the five metrics in Table 4, the most important of these metrics is the accuracy and TTBM. This study is a good start with this method, but it is stated that they would like to try it with a larger set of machine learning algorithms and in different scenarios other than anomaly and intrusion detection.

Algorithm 2: ML algorithm selection based on Bijective soft set:
Combined with selected algorithms:

      Input: MLA (M_1, M_2, M_3,…. M_n)   // ML algorithm set,
Output: ML algorithm []   // selected ML algorithm

1.   begin
2.   for i = 1 to M
3.   calculate value [i] for each algorithm;
4.   end for
5.   for i = to N;
6.   calculate ranking (Fi);
7.   if (ranking(F) > δ);
8.   Insert Fi into descending order;
9.   end if
10.   end for
11.   for i = 1 to N   // ML attributes (metrics)
12.   calculate soft set from each [i] for each attributes;
13.   end for
14.   for i = to N;
15.   calculate Correlation both AND OR product n×n;
16.   calculate Union operation either in row or column
            to reduce the correlation table to 1×n or n×1;
17.   calculate Intersection operation either in row or
            column for possible reduction the correlation table
            to 1×1;
18.   end for
19.   Finally, obtain the results of bijective soft set, if the table is
            Reduced to 1×1;
Return Algo;

Fig. 8 Proposed algorithm [34]

Table 4 Results for each of the applied machine learning algorithms [34]

Algorithm	Accuracy	Precision	Recall	TPRate	TTBM
Naïve Bayes	99.79	0.99	0.98	0.99	4.03
Bayes net	99.77	1.00	0.99	0.99	29.26
Decision tree C4.5	99.99	1.00	1.00	1.00	17.1
Random forest	99.99	1.00	1.00	1.00	198.83
Random tree	99.99	1.00	1.00	1.00	4.32

4.4 Based on a Machine Learning Approach Abnormal Behavior Profiling (ABP) of IoT Devices

A study conducted in 2017 sought to build the abnormal behavior profiling of IoT devices using machine learning. Abnormal behavior profiling is important especially in IoT devices because of the wide range of device types and functions. Two different scenarios were tested, one where one piece of data from a sensor was faulty, and one where all of the pieces of data were faulty. This study helped demonstrate how a small modification in sensed data can affect a machine learning algorithms detection accuracy.

A smart building system with heating, ventilation, air conditioning, and a fire alarm was proposed. The temperature, humidity, light level, and voltage will be measured and recorded to a server by the sensors in the building. The proposed threat is of a hacker compromising one of these sensors through a malicious attack. The data used in this study was from the Intel Berkeley Lab. One sensor was chosen, and a profile was built for detecting abnormal behavior using the four attributes of temperature, humidity, light, and voltage.

Two different abnormal datasets were generated as shown in Fig. 9. The first had only one attribute of the sensor modified. The second abnormal dataset had all attributes of the sensor modified. The training set was then fed into both a k-means algorithm and a support vector machine algorithm.

The k-means algorithm, shown in Table 5, worked well at detecting anomalies in the 1-abnormal dataset but performed worse on the 4-abnormal dataset. Conversely, the support vector machine algorithm performed worse on the 1-abnormal dataset

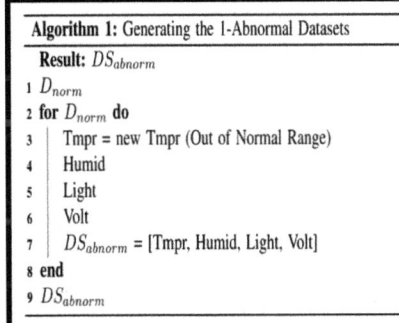

Fig. 9 Algorithms for dataset generation [35]

Table 5 Detection rate using k-means algorithm [35]

Number of clusters	2	4	6	8	10
1-Abnormal	78.3%	88.8%	93.1%	95.0%	97.0%
4-Abnormal	38.4%	63.9%	77.1%	88.1%	92.7%

than it did on the 4-abnormal. They chose to use the k-means algorithm for their abnormal behavior profiling system based on these results because it had the best overall performance on both datasets. This study shows that which machine learning algorithm you select for threat detection is important and may vary based on the situation and system that one is looking to implement.

5 Discussion and Analysis

An analysis of the articles reviewed is shown in Table 6. All the articles that were reviewed under the category of human activity recognition showed that different algorithms suited different methodologies accordingly. The type of data used, features extracted, and power efficiency expectations play crucial roles in choosing the most optimal algorithm. Two articles that were surveyed involved user authentication using either behavior patterns or hand movements. Another article reviewed introduced a system for selecting the most effective machine learning algorithms. The final article for security in IoT devices involved recognizing and profiling IoT devices behaving abnormally.

5.1 Limitations

Some limitations of the models used in human activity recognition models that were reviewed was that the models were only limited to recognize activities based on the training dataset and the feature extraction process. Should there be any new type of activity that the model would want to recognize, the model would have to be retrained with the new data containing the desired activities to re-extract the features again in order to build a working classifier. Furthermore, new parameters would have to be defined for the model generation (feature extraction) which contains the new activity that is being introduced in the model. In the security in IoT devices section, limitations on user authentication include the system not being trained on all activities or situations that may be encountered. This can lead to being locked out of the device when an activity or situation is not recognized. A limitation for the method for the selection of the best machine learning algorithm using the bijective soft set method is that the study was done on a relatively small set of algorithms. Similarly, a limitation for the study done on abnormal behavior profiling of IoT devices is that only two different algorithms were tested, meaning that there may be a superior algorithm that was not tested.

Table 6 Methodology analysis

Title	Methodology	Results	Advantages and disadvantages
Opportunistic sensing for inferring in-the-wild human contexts based on activity pattern recognition using smart computing	A model was built that used behavioral contexts to recognize human activity. A smartphone and watch were used for data collection and RF classifier was used for the algorithm in the model	The RF classifier showed the best results across the metrics used in validating the most optimal algorithm	Data collection was performed by easily accessible devices and results were accurate. The predictive model was limited to predefined contexts and could only predict a limited number of human activities
Wearable-based human activity recognition using an IoT approach	This model uses remote monitoring component with remote visualization and programmable alarms for human activity recognition (HAR)	The C4.5 algorithm was the optimal algorithm as it was efficient and used far less space than that of the Naive Bayes algorithm	While many tasks were successfully recognized, the model struggled with classifying the difference between sitting and walking
Generalized activity recognition using accelerometer in wearable devices for IoT applications	Only one axis from accelerometer data is used to generate a computationally inexpensive recognition model. A supervised hierarchical clustering algorithm was used to cluster the feature vectors	Multiple tests were carried out with different variables and the model had f-1 scores of 0.91, 0.88, and 0.91, respectively	The model showed accurate recognition capability with different environmental variables that were controlled during the tests. However, it struggled to identify crawling and pronating movements
A platform and methodology enabling real-time motion pattern recognition on low-power smart devices	A low-powered smart device known as the Neblina was used in a motion pattern recognition system based on fitness activity	The FNN classifier proved to have better latency and RAM usage compared to the segmentation and histogram extraction and the more conventional KNN classifier	Pattern recognition using the Neblina was accurate and incredibly power efficient, as it managed to operate for 41 h with just a small 100 mAh battery

(continued)

Table 6 (continued)

Title	Methodology	Results	Advantages and disadvantages
IoT-based mobile healthcare system for human activity recognition	MHEALTH data was used to build a recognition model and multiple sensors were placed on subject's body for the human activity recognition	The average accuracy for the random forest and SVM were equal at 99.89%. However, since the random forest is faster than the SVM, it was chosen for HAR	The model that was developed was fast and accurate using a widely utilized dataset
Continuous authentication of smartphone users based on activity pattern recognition using passive mobile sensing	Utilizing a publicly available dataset, user's behavioral patterns were classified and used for recognition and continuous authentication for smartphones	Out of all tested classifiers, SVM performed best on average in all measures	The proposed system allows for continuous user authentication rather than the standard one-time passcode unlock. The system only works with activities that it has been trained on, so any unrecognized activity will reject the user
Please hold on: unobtrusive user authentication using smartphone's built-in sensors	A new authentication method for unobtrusively authenticating smartphone users based on micromovements of the user's hands was introduced. Using collected data, four different classifiers were chosen for the model	The random forest and multilayer perceptron performed the best	This method allows a user to be authenticated unobtrusively, rather than having to input some type of password. The system does not consider the impact of different situations while authenticating
Selection of effective machine learning algorithm and Bot-IoT attacks traffic identification for Internet of things in smart city	Using the bijective soft set algorithm and the Bot-IoT dataset, a system was built to help select the best classifier out of a set of classifiers for the selected dataset	All classifiers performed well, but Naïve Bayes and random tree had the lowest TTBM, thus making them the most effective	The bijective soft set has been used for similar problems and was effective at finding a solution. However, the testing set of classifiers was relatively small

(continued)

Table 6 (continued)

Title	Methodology	Results	Advantages and disadvantages
ProFiOt: abnormal behavior profiling (ABP) of IoT devices based on a machine learning approach	Using data from the Intel Berkeley Lab, a system was built to detect abnormal behavior in IoT devices using the k-means and support vector machine algorithms	Both algorithms performed better than the other in different situations	The system was effective at detecting abnormal behaviors. From the algorithms that were used, there is not a single best algorithm that can be used for every case

6 Conclusion

This article covered the use of pattern recognition in IoT devices for human activity recognition and security models. The literature conducted has proven that there are more optimum algorithms depending on the use case, data collection, and design structure of the model. However, based on the literature reviewed, there are certain algorithms that stand out as the most effective when dealing with IoT devices due to their simplicity, time taken to build models, and accuracy. Different goals and expectations for the model influence the final selection of the optimal algorithm. However, based on the ten articles that were reviewed, the most popular algorithms that were selected in models that consisted of pattern recognition in human activity recognition and security in IoT devices were the k-nearest neighbor, random forest, and support vector machine.

References

1. Atzori L, Iera A, Morabito G (2010) The internet of things: a survey. Comput Netw 54:2787–2805
2. Mahdavinejad M, Rezvan M, Barekatain M, Adibi P, Barnaghi P, Sheth A (2018) Machine learning for internet of things data analysis: a survey. Digit Commun Netw 4:161–175
3. Kucheva L, Whitaker C (2015) Pattern recognition and classification
4. Akbar A, Carrez F, Moessner K, Zoha A (2015) Predicting complex events for pro-active IoT applications, pp 327–332
5. Dean A, Agyeman MO (2018) A study of the advances in IoT security
6. Shelton J et al (2018) Palm print authentication on a cloud platform. In: 2018 international conference on advances in big data, computing and data mining
7. Mason J, Dave R, Chatterjee P, Graham-Allen I, Esterline A, Roy K (2020) An investigation of biometric authentication in the healthcare environment. Array 8:100042. https://doi.org/10.1016/j.array.2020.100042
8. Gunn DJ et al (2019) Touch-based active cloud authentication using traditional machine learning and LSTM on a distributed tensorflow framework. Int J Comput Intell Appl 18:1950022:1–1950022:16
9. Sundaram K, Swarup S (2012) Design of pattern recognition system for static security assessment and classification
10. Kozhirbayev Z, Erol BA, Sharipbay AA (2018) Speaker recognition for robotic control via an IoT device
11. Cui L, Yang S, Chen F (2018) A survey on application of machine learning for Internet of Things. Int J Mach Learn Cyber 9:1399–1417
12. Huskanovicm A, Macan AA, Antolovic Z, Tomas B, Mijac M (2013) Image pattern recognition using mobile devices
13. Fortino G, Giordano A, Guerrieri A, Spezzano G, Vinci A (2015) A data analytics schema for activity recognition in smart home environments
14. Iver R, Sharma A (2019) IoT based home automation system with pattern recognition, vol 8
15. Raghavan S, Tewolde GS (2015) Cloud based low-cost home monitoring and automation system
16. Souza A, Amazonas JR (2015) A novel IoT architecture with pattern recognition mechanism and big data

17. Dave R (2020) Utilizing location data and enhanced modeling based on daily usage pattern for power prediction and consumption reduction in mobile devices: a low power Android application. North Carolina Agricultural and Technical State University
18. Bhamare D, Suryawanshi P (2018) Review on reliable pattern recognition with machine learning techniques
19. Iyer D, Mohammad F, Guo Y (2020) Generalized hand gesture recognition for wearable devices in IoT: application and implementation challenges
20. Kim E, Helal S, Cook D (2009) Human activity recognition and pattern discovery
21. Ketu S, Mishra P (2020) Performance analysis of machine learning algorithms for IoT-based human activity recognition
22. Peixoto A, Vasconcelos F, Guimaraes M, Medeiros A, Rego P, Neto A, Albuquerque V, Filho P (2020) A high-efficiency energy and storage approach for IoT applications of facial recognition, vol 96
23. Abomhara M, Køien G (2015) Cyber security and the internet of things: vulnerabilities. Threats Intruders Attacks 4:65–88
24. Zainab A, Refaat SS, Bouhali O (2020) Ensemble-based spam detection in smart home IoT devices time series data using machine learning techniques
25. Shi C, Liu J, Liu H, Chen Y (2017) Smart user authentication through actuation of daily activities leveraging WiFi-enabled IoT
26. Ehatisham-ul-Haq M, Azam MA (2020) Opportunistic sensing for inferring in-the-wild human contexts based on activity pattern recognition using smart computing. Future Gener Comput Syst 106:375–392
27. Castro D, Coral W, Lopez JL, Rodriguez C, Colorado J (2017) Wearable-based human activity recognition using an IoT approach
28. Al Safadi E, Mohammad F, Iyer D, Smiley BJ, Jain NK (2016) Generalized activity recognition using accelerometer in wearable devices for IoT applications. In: 2016 13th IEEE international conference on advanced video and signal based surveillance
29. Sarbishei O (2019) A platform and methodology enabling real-time motion pattern recognition on low-power smart devices. In: 2019 IEEE 5th world forum on internet of things, pp 269–272
30. Subasi A, Radhwan M, Kurdi R, Khateeb K (2018) IoT based mobile healthcare system for human activity recognition. In: 2018 15th learning and technology conference (L&T), pp 29–34
31. Ehastisham-ul-Haq M, Azam M, Naeem U, Amin Y, Loo J (2018) Continuous authentication of smartphone users based on activity pattern recognition using passive mobile sensing. J Netw Comput Appl 109:24–35
32. Shoaib M, Scholten H, Havinga PJM (2013) Towards physical activity recognition using smartphone sensors. In: 2013 IEEE 10th international conference on ubiquitous intelligence and computing and 2013 IEEE 10th international conference on autonomic and trusted computing. IEEE, Vietri sul Mere, pp 80–87
33. Buriro A, Crispo B, Zhauniarovich Y (2017) Please hold on: unobtrusive user authentication using smartphone's built-in sensors. In: 2017 IEEE international conference on identity, security and behavior analysis (ISBA). IEEE, New Delhi, pp 1–8
34. Shafiq M, Tian Z, Sun Y, Du X, Guizani M (2020) Selection of effective machine learning algorithm and Bot-IoT attacks traffic identification for internet of things in smart city. Future Gener Comput Syst 107:433–442
35. Lee S, Wi S, Seo E, Jung J (2017) ProFiOt: abnormal behavior profiling (ABP) of IoT devices based on a machine learning approach. In: 2017 27th international telecommunication networks and applications conference (ITNAC), pp. 1–6. ResearchGate

Chapter 53
Modified Rectangular Patch Antenna for WLAN and WiMAX Application

Amit Kumar Jain, Tarun Mishra, and Garima Mathur

1 Introduction

Microstrip reception apparatuses have increased more consideration of scientists lately because of their lightweight, little size [1, 2], simple multiplication and integrability with the hardware [3, 4]. In the previous not many decades, the advancement in correspondence with a fast speed spectacularly affected the human life. The development in remote neighborhood (WLAN) symbolizes one of the key interests in the field of correspondence and data [5, 6]. Consequently, the current pattern in rule and business correspondence frameworks has been to grow low profile receiving wires with minimal effort [3] and ostensible weight [7], which can keep up superior over a wide scope of frequencies [8]. This logical pattern has centered to an incredible stretch out for the structuring of microstrip fix reception apparatuses.

Fix receiving wires with basic geometry offer a few points of interest which are not ordinarily shown in other distinctive radio wire setups. The significant focal points controlled by the fix reception apparatuses are their position of safety nature, insignificant weight and modest to create [9]. The strategy utilized for manufacture is the present-day printed circuit board innovation which is exceptionally good with microwave and millimeter-wave incorporated circuits [10] and can fit in with planar and non-planar surfaces [11]. The plans become progressively versatile and adaptable as far as impedance, working recurrence and polarization if the shape and the working method of the fix are chosen [12]. The assortment in structure that is conceivable with microstrip radio wire likely surpasses that of some other sort of receiving wire component [13]. Utilizing the idea of multiband microstrip receiving wire, a multiband rectangular microstrip radio wire is planned and re-enacted right

A. K. Jain (✉) · T. Mishra
Poornima University, Jaipur, India

G. Mathur
Poornima College of Engineering, Jaipur, India

now. There are some products which are accessible to investigate and enhance the framework execution.

WiMAX is a broadband development that approaches fast. Generally speaking, Interoperability for Microwave Access (WiMAX) is an accreditation mark for things that easily finish the reasonableness appraisal with the standard IEEE 802.16. WiMAX is a remote advancement that gives a broadband relationship over noteworthy separations [14].

New planned is proposed for an ordinary microstrip radio wire stacked with three sets of L-formed spaces for triple-band frequency. The principal goal of the paper is to structure the multiband microstrip reception apparatus with the assistance of a solitary component rectangular microstrip receiving wire. The proposed reception apparatus is reproduced utilizing virtual stage IE3D [15] and vector organize analyzer is utilized to estimate the aftereffect of created radio wire [16, 17].

2 Antenna Design and Simulated Result Analysis

Opened receiving wire is moved toward a FR4 substrate with relative permittivity of 4.4 and plane at a stature of 1.59 mm. The ordinary littler scope strip fix receiving wire is modified to achieve proposed radio wire geometry.

2.1 The Proposed Conventional Antenna

The setup of the ordinary printed miniaturized scale strip fix radio wire is appeared in Fig. 1 with $L = 20$ mm, $W = 22$ mm; microstrip line feed is put at proper spot to coordinate with its information impedance 50 Ω.

Re-authorization results obtained for proposed standard microstrip fix radio wire are reflection coefficient $= -24.5$ dB, VSWR $= 1.23$, smith graph impedance $= 54.59 - j3.38$ at resonating repeat 3.4 GHz as showed up in Figs. 2, 3 and 4 exclusively.

Fig. 1 Microstrip patch antenna

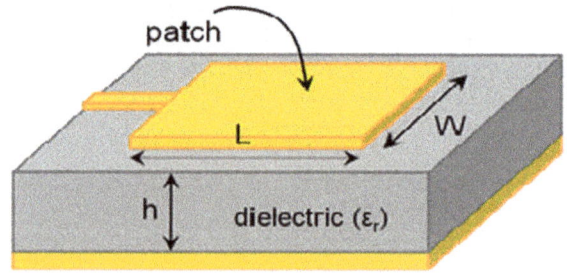

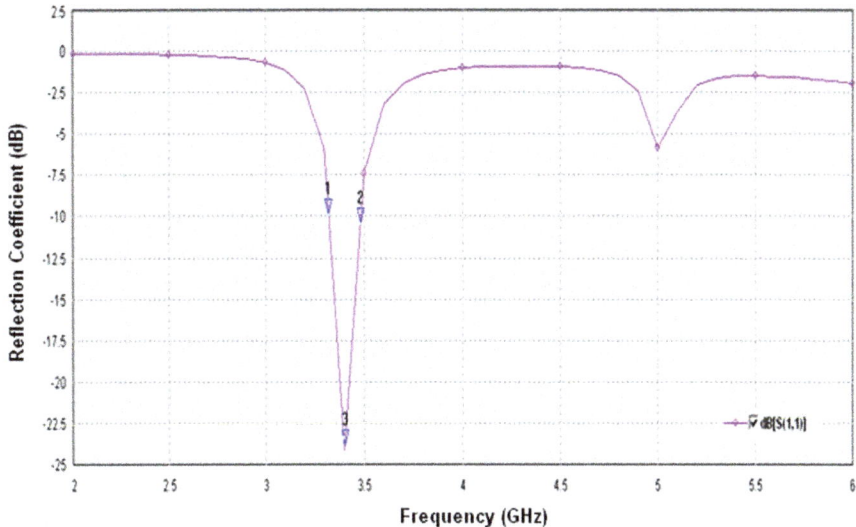

Fig. 2   Varieties in reflection coefficient with recurrence

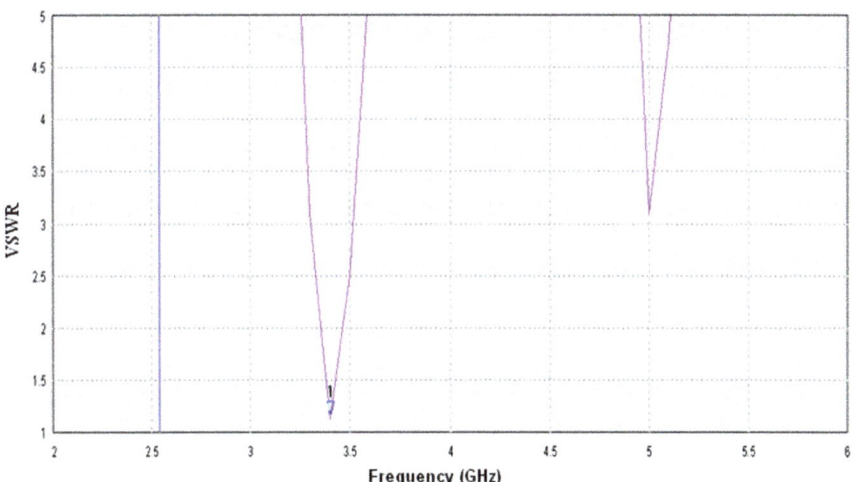

Fig. 3   Variations in VSWR with frequency

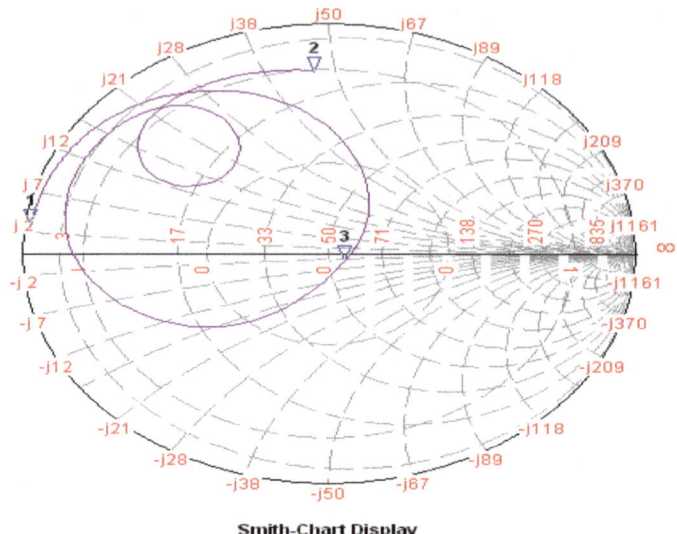

Smith-Chart Display

Fig. 4 Variations in input impedance with frequency

2.2 The Modified Antenna Loaded with Three Pair of L-Shaped Slots

So as to have multiband activity, some transmitting component with an alternate resounding length is presented. Altered radio wire comprises of three sets of L-molded space is carved on the traditional microstrip fix. Proposed adjusted receiving wire is appeared in Fig. 5.

Central parameters for this receiving wire are recorded in Table 1 (Figs. 6, 7 and 8).

3 Proposed Antenna Fabrication and Experimental Result

At long last proposed radio wire stacked by three sets of the L-shape spaces with fix size 20 * 22 mm is created on FR4 substrate utilizing PCB shrewd work. Manufactured proposed receiving wire geometry is appeared in Fig. 9.

Finally, test testing of proposed reception apparatus geometry has been finished by utilizing VNA testing instrument from VNA; three parameters have been estimated for the proposed receiving wire: 1. Reflection Coefficient, 2. VSWR and 3. Smith Chart.

The deliberate reflection coefficient as for recurrence is appeared in Fig. 10. The deliberate two reverberation frequencies are 3.45 and 5.7 GHz. These deliberate qualities are in reasonable concurrence with recreated frequencies. Yet, here one

Fig. 5 Geometry of modified conventional RMSA loaded with three pair of L-shaped slots

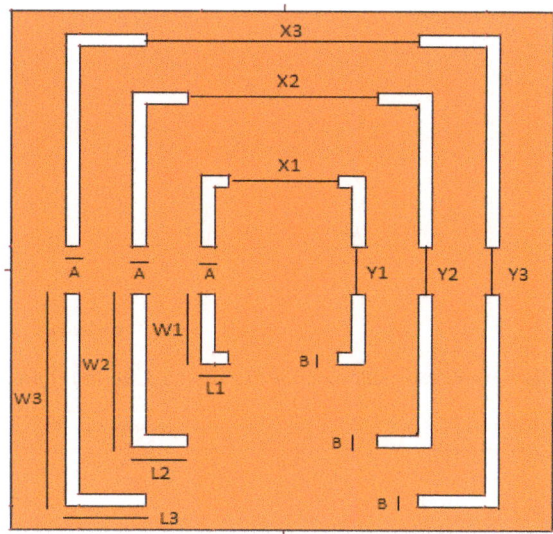

Table 1 Parameters for conventional rectangular patch loaded by three pair of L-shaped slots

Parameter	Value or information
Structure of patch	Rectangular patch
Length in mm	20
Width in mm	22
Dielectric substrate	Glass epoxy FR4
Dielectric constant value	4.4
Value of loss tangent	0.025
Thickness of substrate in mm	1.59
Patch thickness in mm	0.002
Center feed line (mm * mm)	10 * 2
Cut dimension in mm	$L1 = 1, L2 = 2, L3 = 3$ and $W1 = 3, W2 = 6.5, W3 = 9$
Cut slot width in mm	$A = 0.5, B = 0.5$
Cut spacing in mm	$X1 = 4, X2 = 7, X3 = 10$ and $Y1 = 2, Y2 = 2, Y3 = 2$
Frequency range in GHz	2–6.5

recurrence band is missing in estimated result after assess with recreated result, might be because of limited arrangement ground manufacture limit.

It might be seen from chart that the VSWR at all the reverberation recurrence is near wanted 1.0 worth. For recurrence 3.45 and 5.7 GHz, VSWR is 1.04 and 1.4 individually. This shows top notch coordinating between proposed radio wire and information feed line arranges endures (Fig. 11).

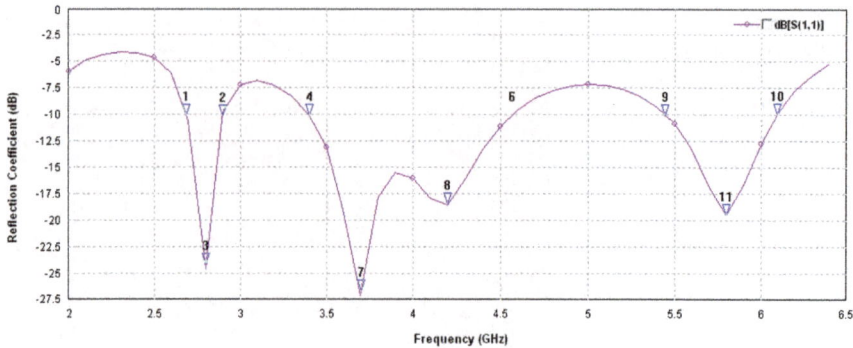

Fig. 6 Varieties in reflection coefficient with 3 sets of L-formed openings with recurrence

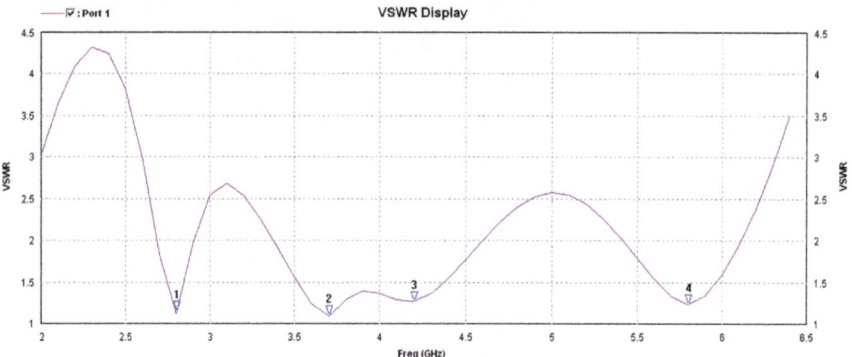

Fig. 7 Variety in VSWR with 3 sets of L-molded openings with recurrence

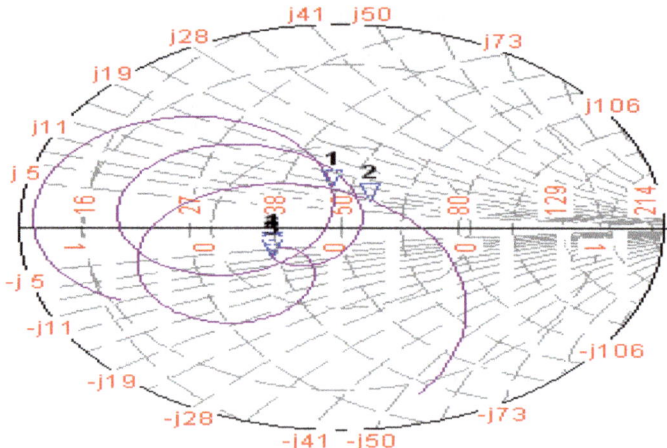

Fig. 8 Variation in input impedance with 3 pair of L-shaped slots with frequency

Fig. 9 Fabricated proposed antenna under test

(a) Patch Top View

(b) Patch Back View

The exploratory aftereffect of info impedance for proposed reception apparatus at full frequencies is appeared in Fig. 12. The deliberate information impedances are $(47.9 - j0.51)$ Ω at recurrence 03.45 GHz, $(53.68 - j17.64)$ Ω at recurrence 05.7 GHz, which are into reasonable contact with input test.

4 Results and Discussion

Changed rectangular receiving wire portrays a dual band conduct of the geometry with great return misfortune and data transmission. Right now, proposed reception apparatus works in three thunderous frequencies allotted in the recurrence groups for WiMAX correspondence frameworks. Despite the fact that little distinction in estimated and mimicked frequencies is acknowledged, that is maybe because of manufacture impediments of fix geometry.

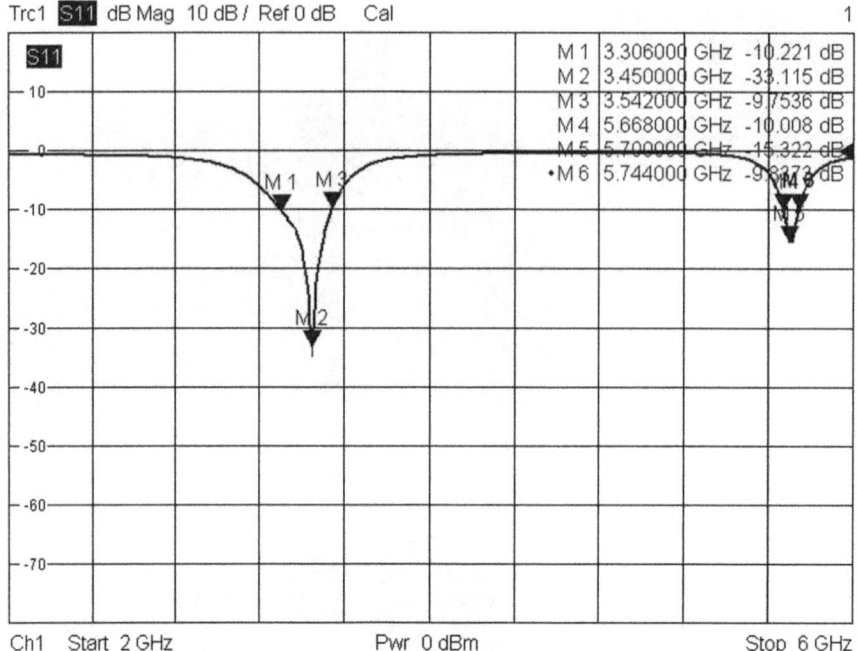

Fig. 10 Measured variations in return loss with respect to frequency

The recreated and trial results got are thought about in Table 2. The organized outcomes further legitimize that reproduced and trial results are almost nearer to one another.

Changes in estimated result are exact moment as contrast with mimicked result, for proposed receiving wire change in first and second full recurrence is 0 and 5.4% as it were.

5 Conclusion

Proposed smaller size opened microstrip fix radio wire structured is recreated and tried for various execution parameters. The primary point of the paper is to improve the arrival misfortune and the transmission capacity of the receiving wire which has been accomplished by cutting the L-molded spaces on base state of radio wire. These planned reception apparatus geometries can be adequately utilized in numerous fields of present-day correspondence frameworks.

Reproduction and estimated aftereffects of proposed reception apparatus plan with three sets of L-molded openings show thunderous frequencies 2.7, 3.65 and 5.7 GHz with accessible impedance band width 07.40, 33.00 and 14.03% separately. Proposed geometry has two frequencies lying in the two endorsed groups for WiMAX

Trc1 S11 SWR 1 U / Ref 1 U   Cal   1

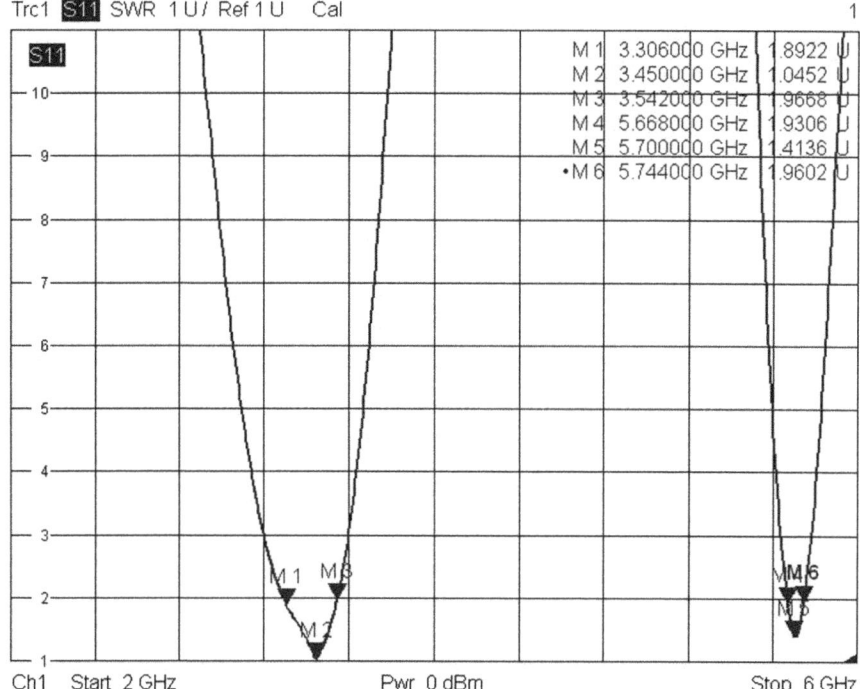

Fig. 11 Variations in VSWR with respect to frequency

correspondence frameworks and according to estimated result radio wire is likewise proper for Wi-Fi correspondence frameworks. It might be additionally reasoned that reception apparatus has diminished size 10.45% and effortlessness for accomplishing better outcomes.

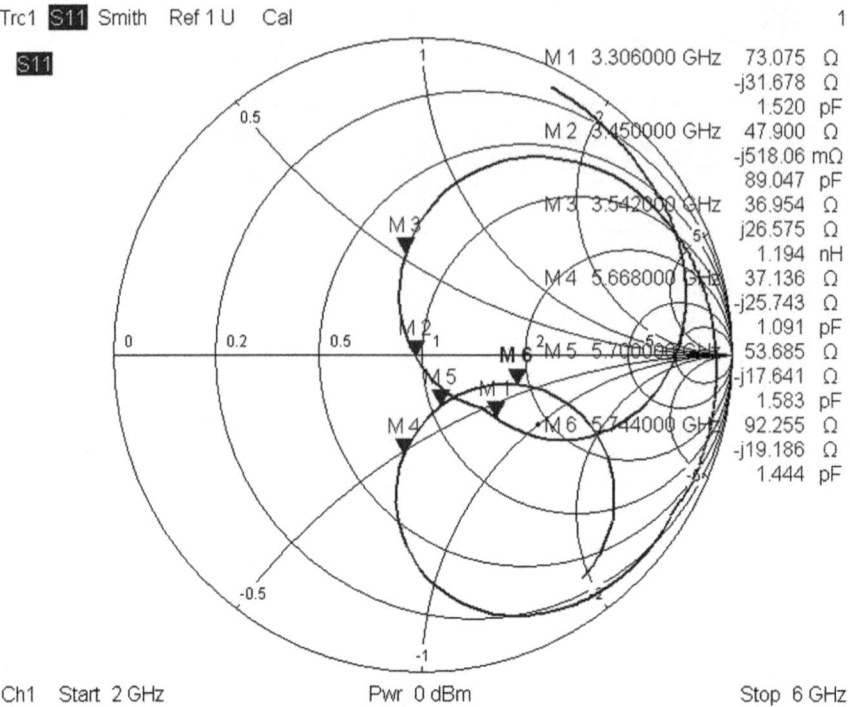

Trc1 S11 Smith Ref 1 U   Cal   1

S11

M 1 3.306000 GHz   73.075 Ω
                    -j31.678 Ω
                     1.520 pF
M 2 3.450000 GHz   47.900 Ω
                    -j518.06 mΩ
                     89.047 pF
M 3 3.542000 GHz   36.954 Ω
                     j26.575 Ω
                     1.194 nH
M 4 5.668000 GHz   37.136 Ω
                    -j25.743 Ω
                     1.091 pF
M 5 5.700000 GHz   53.685 Ω
                    -j17.641 Ω
                     1.583 pF
M 6 5.744000 GHz   92.255 Ω
                    -j19.186 Ω
                     1.444 pF

Ch1   Start 2 GHz   Pwr 0 dBm   Stop 6 GHz

Fig. 12 Input impedance variations with respect to frequency

Table 2 Comparisons between simulated and measured results

Antenna parameters	Simulated results	Measured results
Resonant frequency 1	2.70 GHz	2.58 GHz
Return loss	−25.0	−30.11
VSWR	1.2	1.04
Resonant frequency 2	3.65 Hz	3.45 GHz
Return loss	−27.5	−33.11
VSWR	1.2	1.04
Resonant frequency 3	5.7 GHz	5.7 GHz
Return loss	−19.0	−15.32
VSWR	1.3	1.41

Antenna Parameters: Typical parameters of antennas are gain, bandwidth, radiation pattern, beamwidth, polarization, and impedance

Simulated Results: Results to be generated by virtual performance

Measured Results: Results to be generated by Actual designed device

Resonant Frequency: The resonant frequency can be defined as the natural frequency of an object where it tends to vibrate at a higher amplitude

References

1. Balanis CA (1982) Antenna theory, 2nd edn. Wiley, New York
2. Liu Y, Cai S-T, Xiong X-M, Li W-J, Yang J (2019) A novel wideband circularly polarized modified square-slot antenna with loaded strips. Int J RF Micro Comput Aided Eng
3. Taher N, Zakriti A, Ammar Touhami N, Rahmani F (2020) A tri-band-notched antenna for UWB communication systems. In: 13th international conference interdisciplinary in engineering (INTER-ENG 2019)
4. Gao SC, Li L-W, Leong M-S, Yeo T-S (2003) Dual-polarized slot-coupled planar antenna with wide bandwidth. IEEE Trans Antennas Propag 51(3):441–448
5. Srivastava R, Ayub S, Singh VK, Saini JP (2014) Dual band rectangular and circular slot loaded microstrip antenna for WLAN/GPS/WiMax applications. In: IEEE fourth international conference on communication systems and network technologies, 978-1-4799-3070-8
6. Ozpinar H, Aksimsek S (2020) A novel compact, broadband, high gain millimeter-wave antenna for 5G beam steering applications. IEEE Trans Veh Technol 69(3)
7. Gupta K, Sharma OP (2013) Parametric performance analysis of slotted, stacked and conventional microstrip patch antenna. Int J Enhanc Res Sci Technol Eng 2(3):1–9. ISSN 2319-7463
8. R. Kumar, R. Vijay (2015) A compact multiband frequency agile microstrip slot antenna. In: International conference on advanced computing & communication technologies, IEEE conference publications, vol 5, pp 8–11. ISSN 2327-0632
9. Jain AK, Surana M (2017) Modified rectangular patch antenna loaded with multiple c slots for multiple applications. J Appl Comput Mech 2(3):192–199
10. Nahar T, Sharma OP (2014) A modified multiband bowtie antenna array used for L band applications. Int J Eng Res Technol (IJERT) 3(11):81–85. ISSN 2278-0181
11. Islam MT, Shakib MN, Misran N (2009) Multi-slotted microstrip patch antenna for wireless communication. Progr Electromagn Res Lett 10:11–18
12. Singhal R, Sharma OP (2015) Bandwidth enhancement of hybrid tri-rect slotted microstrip patch antenna. Int J Latest Technol Eng Manag Appl Sci 4(5):48–50. ISSN 2278-2540
13. dos Santos Silveira E, Antreich F, Chagas do Nascimento D (2020) Frequency-reconfigurable SIW microstrip antenna. Int J Electron Commun
14. Yunita T, Usman K, Kurniawan A (2012) Experiment of slotted triangular triple-band antenna for WiMAX/WLAN application in Indonesia. In: TSSA 7th international conference on telecommunication systems, services and applications. IEEE, pp 300–305
15. Surana M, Jain AK (2017) Modified U-slot stacked micro strip patch antenna for ultra-wideband applications in S band, C band & X band. J Appl Comput Mech 3(4):293–301
16. Rajmohan IJ, Hussein MI (2020) A compact multiband planar antenna using modified L-shape resonator slots. Elsevier Ltd
17. Mishra CS, Nayyar A, Kumar S, Mahapatra B, Palai G (2019) FDTD approach to photonic based angular waveguide for wide range of sensing application. Optik 176:56–59

Chapter 54
User Authentication Schemes Using Machine Learning Methods—A Review

Nyle Siddiqui, Laura Pryor, and Rushit Dave

1 Introduction

Within the last few decades, advancements in technology are happening at an increasingly fast rate and devices are rapidly becoming more varied and complex. Concurrently, the amount of sensitive information being stored on personal devices has also seen a dramatic rise, as well as the risk of this information being compromised. There are a multitude of ways to authenticate a single user, with the most common method being static methods, i.e., passwords and personal identification numbers (PINs). Although reliable, once a password or PIN is compromised that account is vulnerable to anyone who has gained access to that sensitive data. To counter this problem, researchers have been investigating the legitimacy of dynamic authentication systems. With the aforementioned shortcomings of static authentication, dynamic authentication offers a wide variety of benefits, such as continuous authentication, increased flexibility, and in some cases, less labor for the user. This survey will be exploring two commonly proposed systems for dynamic authentication and their efficacy; these two areas being physical layer authentication (PLA) and biometric-based authentication methods.

Biometrics are a user's physical or behavioral characteristics those of which can be used in user authentication. Biometrics are categorized into two categories: physiological and behavioral. Physiological biometrics includes scanning physical aspects of the user, for example using a user's palm prints for authentication [1]. On the other

N. Siddiqui · L. Pryor · R. Dave (✉)
University of Wisconsin—Eau Claire, Eau Claire, WI 54701, USA
e-mail: DAVER@uwec.edu

N. Siddiqui
e-mail: SIDDIQUN8701@uwec.edu

L. Pryor
e-mail: PRYORLK8701@uwec.edu

© The Author(s), under exclusive license to Springer Nature Singapore Pte Ltd. 2021 703
S. Kumar et al. (eds.), *Proceedings of International Conference on Communication and Computational Technologies*, Algorithms for Intelligent Systems,
https://doi.org/10.1007/978-981-16-3246-4_54

hand, behavioral biometrics focuses on not a password, or a physical trait of the user, but the user's behaviors in how they interact with the object needing authentication. Many of these biometrics have already been implemented into phones and smart devices for the past couple of years, but they are also being used in other general systems such as the healthcare system to increase their security [2]. However, physiological biometrics have found to be not as cost-effective and not as accurate as researchers hoped for. Physiological biometrics require additional scanners and software to be installed onto the device so that the device can scan the user's physical features [3]. This ultimately ends up increasing the cost to produce the product and also increasing the cost to purchase the product. This poses the question: Is there a better system to fix the problems associated with physiological biometrics? This underperforming authentication scheme has caused researchers to look for a more efficient option, and that efficiency is found using behavioral biometrics in authentication schemes. Behavioral biometrics solves the problem of cost because the biometrics do not require any additional software to be installed into the devices in order to scan behavioral features. Research has been conducted using these behavioral biometrics and different machine learning algorithms in hopes to make our devices more secure and more efficient.

Lightweight dynamic authentication methods alternative to PLA have been proposed in the past, such as using continuous authentication schemes in [4, 5] and using computational cryptographic-based methods in [6–8]. However, in a resource-constrained environment in which many devices and machines are interconnected and communicating, computational cryptography may take up too much time and resources to be an effective authentication method on a large scale. One of PLA's many benefits is its high level of security and significantly less computational cost compared to traditional methods. Additionally, PLA does not require any additional external sensors as opposed to a biometric-based authentication scheme. Examples of these necessary external sensors can be observed in [9], where the authors required two additional sensors to record gyroscopic and directional movements in addition to the resources needed to create the predictive model. The application of machine learning (ML) models to extract patterns from large data sets has been used in conjunction with current PLA methods to better optimize PLA-based systems.

2 Background

Behavioral biometrics encompasses a wide variety of features that can be studied and used for authentication purposes. Behavioral biometrics is an implicit way to check the behavior of the user, which is something that cannot be easily copied. Checking the behavioral trait instead of a physical trait eliminates the problems requiring extra hardware for scanners and the reliance on environmental factors, making the system more efficient and more secure. Behavioral biometrics can include anything from a user's behaviors, such as the users touch patterns, keystroke dynamics, and mouse dynamics [10–12]. Most user authentication schemes using behavioral biometrics follow the same process for their methodology. This includes the steps of: collecting the data, extracting the behavioral biometric features, using machine learning algorithms for classification, and then using those algorithms to create a decision on if the data can be connected to the authorized user. The data is collected either implicitly by an application or scanner on the user's device; the specific behavioral features the authentication system is studying are then extracted from the data collected. From there, machine learning algorithms take in the extracted features and compare them to accurate samples of the user's behaviors. In [13], authors created an application called touchstroke that implicitly monitors the hand micro-movements and touch-stroke patterns of the user. However, it was found in this article that these behavioral features did not perform well while users were in motion, thus lowering the accuracy of the application. This issue with collecting behavioral data while the user was in motion was found to be an issue in other articles as well [5, 14–16]. Also, many studies do not consider how a user's behaviors can change over time, such as the user aging and typing slower, the user not being able to use their dominant hand, and other long-term behavioral changes [14, 17–19]. This survey aims to evaluate multiple studies regarding behavioral-based biometric systems and discuss which features work best and which features need improvements through future research.

Another possible method of dynamic authentication is PLA. PLA is the act of authenticating a user based on the physical attributes of a received signal from said user in a network. The variance of the physical attribute used for authentication can be observed in [4, 20–23]. As opposed to biometric-based authentication, PLA utilizes physical attributes of a received signal to authenticate a user without the need for much additional equipment. With wired and wireless networks' rapidly growing importance due to the increased use of the Internet of things (IoT) devices, as well as the recent developments in 5G networks, the security of these networks is of the utmost importance in ensuring their widespread implementation and reliability. A common example of the use of physical layer authentication in a multiple input multiple output (MIMO) network is the Bob, Alice, and Eve example. This general scheme is most used across the literature. Bob and Alice are trusted users on a MIMO network sending information between themselves. Bob is most commonly receiving a signal from Alice. Eve is a nefarious user, trying to spoof as one of the trusted users by manipulating their own signal's physical attributes. It is then up to Bob to authenticate each signal he receives and determine if it is truly from Alice or a spoofed signal from Eve. PLA methods have been proposed to best mitigate the possibility of accidental authentication. Additionally, PLA can also be used to prevent malicious attacks against a network, such as a denial of service (DOS) attack or man-in-the-middle attacks. As the average social, economic, and personal dependence on wireless networks continues to increase, the investment in the reliability and security of these networks is more than worthwhile. Further optimization of PLA implementations in large-scale networks could lead to efficient, reliable, and computationally cheap authentication methods. Therefore, this survey will aim to review the most recent advancements in this area to identify possible areas of improvement as well as plausibly effective models.

3 Literature Review

3.1 *Biometric-Based Authentication*

Researchers in [24] wrote about their findings from an experiment they conducted testing continuous authentication with behavioral biometrics using machine learning. The purpose of this article was to propose a newly created authentication scheme using graphical user interfaces (GUIs), keystroke dynamics, and mouse dynamics. The data used in this article was collected by the researchers themselves. To collect the data, 31 participants were used, and they were tasked with writing three short articles of 400–500 words that involved a small amount of Internet research. This task of writing the articles allowed researchers to collect data from the participants use of GUIs, their keystroke dynamics, and their mouse dynamics.

The features used for each of the three behavioral biometrics, keystroke dynamics, mouse dynamics, and GUI interactions were selected because of their positive results in previous studies researched by the authors. For the keystroke biometric, two features were used: duration and latency. The duration is the average time each key is being held down and the latency is the average time it takes someone to move from one key to the next. To collect data for mouse dynamics, the researchers calculated any movement that had at least a distance of 30 pixels moved and had to have one of two movement starters and one of three movement enders. The two movement starters were a period of silence, which was any time where there was no movement detected for at least 1 s, and a left button release. The movement enders are again a period of silence, a button press, or a mouse wheel scroll [24]. From there, samples were created using multiple features from the three biometrics that were collected over a 10-min window of 2-min intervals to ensure the system is checking for a genuine user in short intervals, but still gathering enough information to give an accurate decision. This method continues until the user is found to be an impostor to which the computer then locks out the user.

Two different fusion approaches were developed to see which one would work best. The two approaches are outlined in Fig. 1a, combined all features into one sample which then uses feature selection and classification to produce results, and Fig. 1b, had each feature classified individually, then that classifier's confidence probability is sent to the ensemble classifier which would output the final decision. The training data was created by taking two-thirds of all the samples from each

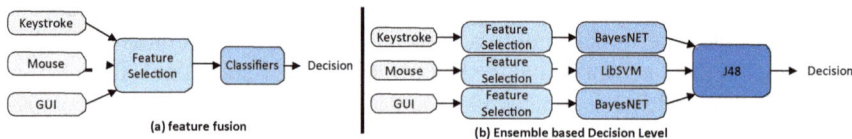

Fig. 1 Architecture of the two fusion approaches: feature fusion and ensemble-based decision level [24]

participant and the other third was used for testing. Different ratios of legitimate users to impostors varied for each user's sample this accounts for the imbalance that would have been created by using too many imposter samples for each user. Therefore, there was an average ratio of 7:6 genuine samples to imposter samples [24]. For the testing data, there were 30 instances of an impostor sample user for each user sample in the last third not used in training. Each of the training and testing sets was then put through each of the fusion models.

For the feature level fusion model, in the identification and authentication portions of the research, the BayesNet classifier performed the best for three out of the four data sets. Table 1 examines the different authentication rates found when using different ratios of legitimate to impostor user instances and Table 2 shows the comparison between the two fusion approaches using false acceptance rate (FAR) and false rejection rate (FRR). Overall, it was proven by these researchers that a fusion of different features does in fact make behavioral biometric authentication more accurate and reliable. It was also proven that multimodal fusion required much less user interaction than individual modality, considering that there were less samples that needed to be taken due to the system testing multiple different dynamics. However, this study again did not use real-life data collection methods, and the participants were all either government employees or students so the scheme may not work as accurately over long periods of time, in certain real-life situations, and with people who are not as knowledgeable with computer systems. Despite some minor limitations, this study is a very good building block for other authentication schemes, especially schemes dealing with keystroke biometrics.

Researchers Attaullah et al. present a new scheme for user authentication using bimodal behavioral biometrics. The purpose of this article is to propose ANSWER-AUTH which is a "behavioral biometric-based authentication scheme" that takes the

Table 1 Authentication rates of different imposter instances [24]

Impostor to legitimate user ratio	FAR	FRR
30:1	3.60	4.17
15:1	3.17	3.88
2:1	2.91	2.32
1:1	2.24	2.10

Table 2 Comparison of feature fusion and ensemble-based decision level fusion authentication rates [24]

Classifier	FAR feature fusion	FAR EDBL fusion	FRR feature fusion	FRR EDBL fusion
Fusion v. Key	3.76 ± 0.48	2.47 ± 0.40	2.51 ± 0.57	2.53 ± 0.37
Fusion v. Mouse	11.67 ± 0.77	2.61 ± 0.01	18.80 ± 1.50	2.51 ± 0.01
Fusion v. GUI	16.29 ± 1.04	2.24 ± 0.45	21.37 ± 2.63	2.10 ± 0.30

way a user unlocks their phone and brings it to their ear and uses that for authentication [14]. However, the authors made this scheme stronger by adding the second element of touchstroke behavior gathered from when the user is unlocking their phone.

The researchers collected over 10,000 samples over three days from a group of 85 participants who unlocked and answered their Android smartphones from three different activities, sitting, standing, and walking; this was all monitored through an application, called Authcollector. The states from which the data is being collected can be seen in Fig. 2. The researchers collected 40 biometric features from each activity for each user, which totaled to be 120 samples for each user and 10,200 samples in total. Figure 3 illustrates an example of how a feature can show extreme differences between two users. From the samples, the researchers extracted a 97-feature long vector to model the phone pickup movement and a 31-feature long vector to model the unlocking movement. Making the final feature vector 128 features long by combining both feature vectors [13]. The researchers then used feature concatenation to provide better and more accurate results. The scheme was tested using one user sample as the genuine sample and compared it with all the other

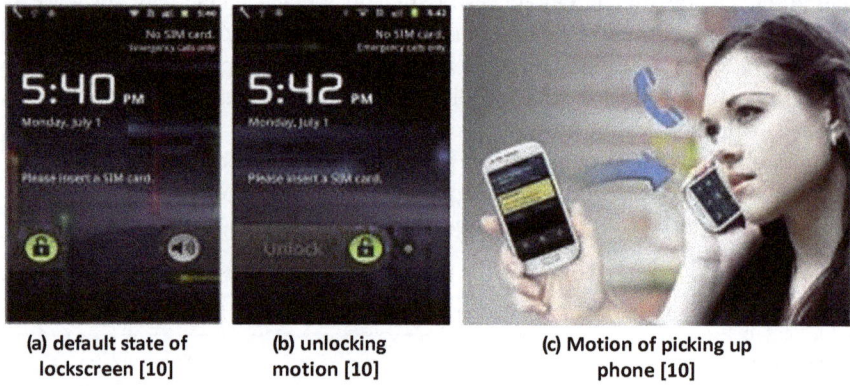

| (a) default state of lockscreen [10] | (b) unlocking motion [10] | (c) Motion of picking up phone [10] |

Fig. 2 Three states being testing in the authentication method [14]

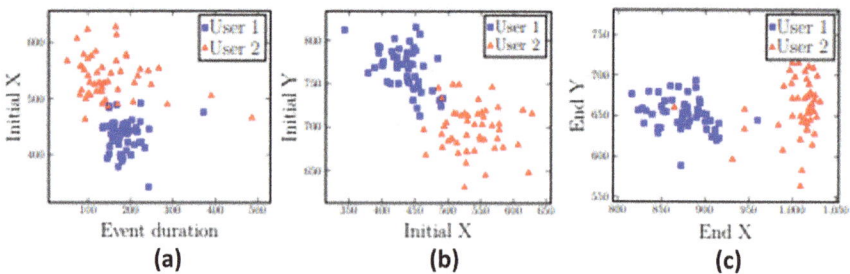

Fig. 3 Examples of the differences between two users's inputs [14]

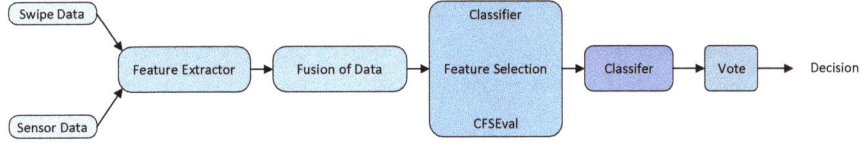

Fig. 4 Outline of the vote classifier and ASC fusion model [14]

users acting as the impostor sample. They then tested authentication with multiple different feature selection algorithms to see which combination would get them the best answers. The classification methods were compared to each other using the following measurements: True acceptance rate (TAR), FAR, true rejection rate (TRR), FRR, and accuracy.

Researchers started with testing only the base classifiers abilities. There were six classifiers chosen for this study: BayesNET, Naive Bayes, support vector machine, k-nearest neighbor, J48 decision tree, and random forest all of which are being used with their default settings. These results were very high, with all classifiers getting high accuracies. However, random forest performed the best out of all classifiers with TARs of 98.87% for sitting, 98.98% for standing, and 96.8% for walking. The researchers then combined the base classifiers with a vote classifier which took each of the base classifiers and had it predict the average probability and returned the average probability distribution for each classifier and the class with the highest probability was chosen. The researchers combined this vote classifier with Weka's AttributeSelectedClassifier (ASC) where the researchers would take each of the six classifiers and add an attribute selection method as seen in Fig. 4. The attribute selection method chosen was CFSEval and the BestFirst (bi-directional) search methods. This method showed slightly higher results in all categories, with random forest once again performing the best with TARs of 99.03%, 99.35%, and 97.0% for sitting, standing, and walking, respectively. The high accuracies in multiple aspects of the study show how secure this authentication system could be in many different varieties and formats. This study, however, did not test multiple different movements such as walking upstairs or laying down. These could affect the security of this system making it more prone to either false positives or false negatives.

In [25], researchers Meng et al. collected and analyzed data to evaluate the performance of their newly created touch gesture-based authentication scheme. The purpose of this article was to propose a new scheme using touch dynamics to create more accurate user authentication. The data set used in this article was created by the researchers themselves. They used a Google/HTC Nexus One phone with a modified Android OS version 2.2. The Android operating system consists of five different layers: the Linux kernel, libraries, android runtime, application framework, and applications. This scheme modifies only the application framework layer which allows them to the input data from the touchscreen. This authentication scheme, outlined in Fig. 5, functions with three major components: the data collection component, the behavior modeling component, and the behavior comparison component. Table 3 gives an example of the touch inputs that were collected. The data was recorded in

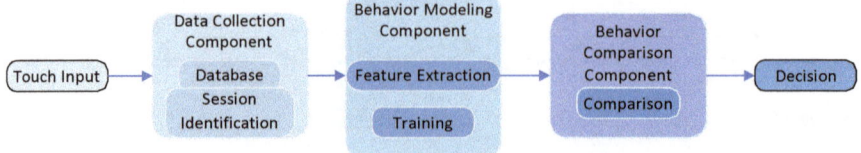

Fig. 5 Architecture of proposed authentication system [25]

Table 3 Example touch input data collected [25]

Input	X-coordinate	Y-coordinate	Time
Press down	475.46866	659.6717	1,770,785
Press move	472.56793	660.3004	1,770,807
Press move	470.2978	660.9292	1,770,814

120 sessions by 20 users, and any session that was less than 10 min was thrown out and re-recorded.

From the data collected by the researchers, they extracted 24 different features of the user's touch dynamics and constructed an authentication signature. The training of the data was split into two different types: initial training and dynamic training. The initial training took the user's first few sessions and created a model of the user's profile and the dynamic training continuously trained the data to include changes to the user's behavior. The results found in this testing period can be found on Fig. 6. The radial basis function network had an average error rate of 7.71% and the back propagation neural network had an average error rate of 11.58%. The other three classifiers had an error rate of 15% or above. However, for all classifiers, the average error rate was about 7.8% which is still too high for there to be any use of this scheme in real-world settings. It was shown that the performance of the classifiers decreases as the variance in features increases, because of this the researchers created an algorithm that combined the best performing classifier, radial basis function network and particle swarm optimization. This brought the average error rate from 7.71 to 2.92% and the FAR down from 7.08 to 2.5% as seen in Fig. 6. The experimental work in this shows promise for the new scheme using touch-based

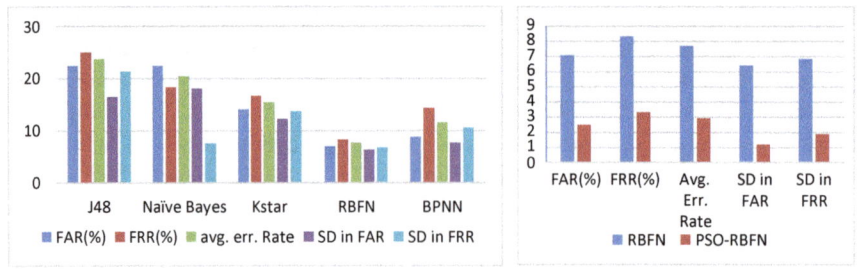

Fig. 6 Evaluation results of classifiers and comparison of PSO-RBFN and RBFN classifiers [25]

dynamics. There are some limitations; however, for example, this system does not distinguish differences between multitouch gestures, which can skew the data since not all gestures are being analyzed correctly. Another problem found in this scheme is that only 20 users were used in this study, so it is hard to conclude if the authentication system is consistent enough for real-world application.

In [13], researchers Akhtar et al. introduced a method of user authentication using touchstroke movements and face patterns. The purpose of this article was to present an "unconstrained and implicit multimodal biometric user authentication system based on user's hand micro-movements, touchstroke, and face patterns" [13]. This authentication scheme combines both behavioral and physiological biometrics in hopes of creating the most secure and accurate system.

In this user authentication scheme, an application would run implicitly in the background of the user's device. This application would implicitly analyze the micro-movements of the user's hand, touchstrokes and take a picture of the user while the user is putting in their 8-digit passcode. This is all analyzed in a very short period of time to ensure that the data can be adequately analyzed, but it would not be able to fall victim to adversary attacks. All three of these biometrics are individually processed and combined using feature concatenation. The resultant feature vector is then put into one of two classifiers: either a random forest or multilayer perceptron. Finally, the classifier checks if the user is an impostor, if so, a notification will be sent to the real owner, and the system will shut down the impostor's access to important data.

To set up their experiment, the researchers created a multimodal data set using the public data sets they obtained and the data set they created themselves. The researchers used the MOBIO face database to collect photographs and videos of 150 subjects in a realistic environment, examples of which are shown in Fig. 7. To collect the data for the movements and touchstrokes, the authors created their own Android application that collected data from multiple sensors already in the smartphone and also collected data from the low-pass and high-pass filters. For training, they used a randomly selected number of either 5 or 10 samples of a genuine user and one impostor as the training set, and then the rest of the samples of genuine users and all imposter users were used for testing purposes. The selected 5 or 10 samples was to eliminate the class imbalance that would be caused by using too many imposter samples. The testing phase was also repeated 94 times for each given user, using a different imposter for each instance. When training with 10 samples, the results were very high. Despite movements, individual modalities were outperformed by a fusion

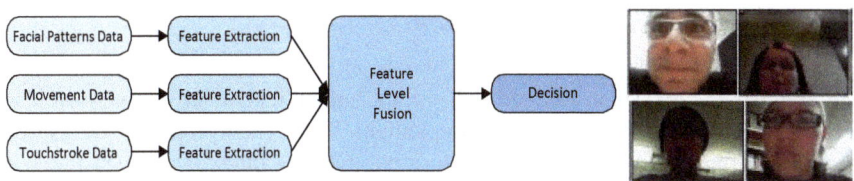

Fig. 7 Architecture of proposed user authentication system and examples from MOBIO face database [13]

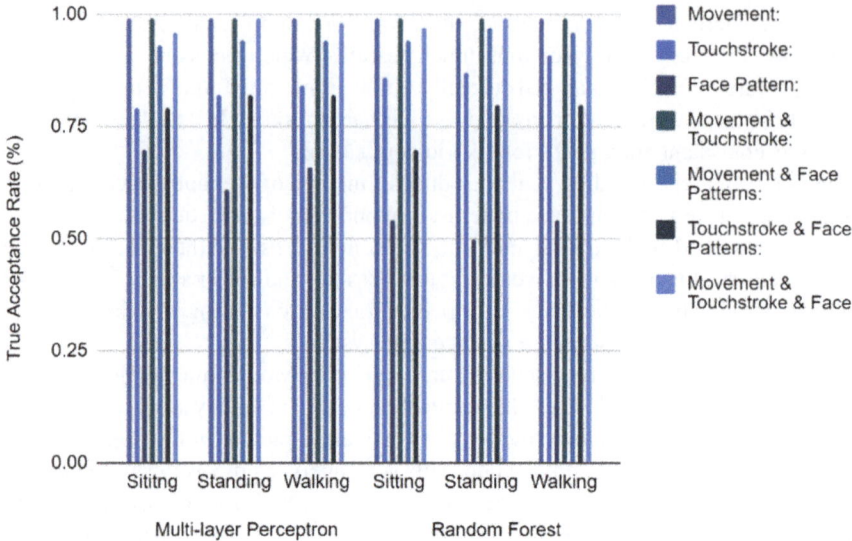

Fig. 8 Comparison of authentication rates for multilayer perceptron and random forest using ten training samples [13]

of all modalities by over 20% for each activity and classifier. Overall random forests classifiers performed the best getting a TAR percentage in the high 90s for almost every category. Each test averaged the TAR for unimodal, bimodal, and trimodal scenarios in three activities: sitting, standing, and walking. These results can be found in Fig. 8.

This fusion of biometrics was proven to have an increase in security against spoofing attacks, as seen by the results. This new authentication method created in this study was able to solve many problems that the authors found in previous authentication systems created. This system was also created generically, so it could be applied to specific applications or systems, or it could be applied to an entire device. The high accuracies in this study proves that having a multimodal authentication system is less likely to fall victim to a spoofing attack. Overall, this study is a great example of how accurate and secure a system using behavioral biometrics can be.

3.2 Physical Layer Authentication

The authors in [26] acknowledged that strong authentication in a large wireless network is important, yet difficult to implement. Thus, they proposed the use of machine learning and channel state information (CSI) matrices to measure the authenticity of a received signal in a MIMO network. They test various machine learning models and one deep learning model, a general adversarial network (GAN) at different signal-to-noise ratios (SNR) to determine which method is most robust, as well as proportionally accurate. Their proposed model would receive a signal from a user on a MIMO network with both the receiver and sender parties equipped with a varying number of receiver and transmitter antennas. They simulate a Bob-Alice-Eve environment. Initial authentication for a user would have already transpired, but after this process was completed, the transmission receiver would save the measurements of these signal's characteristics for future comparison. Thus, this model is used to continuously authenticate a user based on the user's physical signal characteristics. The model evaluates each element in a CSI matrix and calculates its distance from an expected trusted signal.

The authors made use of two separate data sets, meant to test Type I error and Type II error, respectively. The first set consists of signals from six separate transmitters, yet only one among this group should be authenticated. Thus, this data set will reveal how often the model does not authenticate a trusted transmission (Type I error). The second data set consists of five nefarious users attempting to be authenticated by matching the CSI matrix of a single trusted transmitter in the data set. This data set is then testing how often the model is authenticating a user when it should not (Type II error). The main use of machine learning in this article was to assist in outlier and anomaly detection. Thus, the authors concluded that one-class (OC) machine learning problems would be most appropriate for this given experiment. Their best performing algorithm used in this article was the local outlier factor algorithm, or LOF. LOF is an algorithm that evaluates a point based on its proximity to other labeled points assuming that similar data points will be in close proximity to each other. It randomly selects a point in the data as reference, assuming that outliers will be identifiable by the difference in its characteristics compared to this reference point. It is important to note the presence of a GAN as well in this article despite its deep learning-based characteristics. GANs are composed of two sub-models, those two being the generative model (GM) and the discriminative model (DM). In essence, the DM is tasked with the actual classification of data. The GM is tasked with generating data to train the DM on, and more specifically to maximize the error of the DM. If the DM identifies and rejects GM generated data, the GM is then penalized and is required to re-evaluate its parameters to better optimize itself. Concurrently, if the DM is fooled by a fake sample, then the same reevaluation occurs. Tables 4 and 5 show the underlying architecture of the GAN used by the authors of this article.

The results of this article indicated that CSI matrices can be reliably used as a measurement of a signal for physical layer authentication. Additionally, it compared the accuracy of certain ML algorithms, such as the ones discussed above, against the deep-learning GAN model. Regarding both data sets, the most effective algorithm was LOF. At SNR levels of 0 dB, it already had an accuracy rate of 90% and reached 100% accuracy before any other algorithm, including the GAN. Therefore, this article concluded that there are certain ML-based algorithms that have the potential to be quite efficient in PLA-based systems, namely LOF. Additionally, there are viable ML algorithms that can be used for PLA-based systems, yet deep learning models may be an area of interest for future research in order to optimize these types of authentication schemes, especially when sufficiently large data sets are available.

Table 4 DM architecture

Discriminator model

Layer	Output size	Activation function
Input 1 ($x_{1,1}$)	2	–
Input 2 ($x_{1,2}$)	2	–
...	...	...
Input 16 ($x_{4,4}$)	2	–
Concatenated	32	–
Fully connected	64	LeakyReLU
Dropout = 0.2	–	
Fully connected	32	LeakyReLU
Dropout = 0.2	–	–
Output	1	Sigmoid

DM takes a 4 × 4 matrix as input and utilizes dropout layers to prevent overfitting. LeakyReLU layers used an (*alpha* = 0.3)

Table 5 GM architecture

Generator model

Layer	Output size	Activation function
Input	5	–
Fully connected	16	LeakyReLU
Fully connected	32	LeakyReLU
Fully connected	64	Tanh
Output 1	2	Linear
Output 2	2	Linear
...	...	...
Output 16	2	Linear

Authors of [27] similarly use a Bob-Alice-Eve environment (Fig. 9) to propose a PLA-based authentication scheme that improves upon previously proposed methods by using a machine learning algorithm designed to strengthen classification models, namely AdaBoost. This article proposed a classification model based on received physical characteristics of a signal that is boosted by AdaBoost and implicitly sets its own thresholds appropriately. Furthermore, this model was tested using both one and two-dimensional features, the latter being found to strengthen the proposed model's cheating detection rate. A real-world environment was conducted, and the authors concluded that their proposed model was effective in increasing the accuracy rate of proper authentication.

The authors of this article proposed a model that receives multiple frames of data from a signal transmission and creates a decision to authenticate the signal or not by comparing it to previously known authenticated packets. Bob starts by obtaining and saving channel information between himself and Alice for any given frame. After

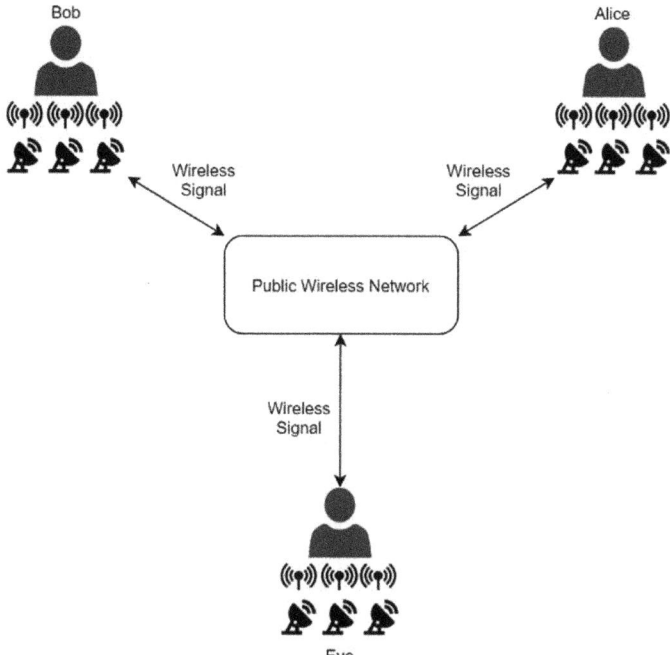

Fig. 9 Bob-Alice-Eve environment

some time, Bob will receive the next data frame and is tasked with estimating whether this unknown channel response was sent by Alice or not by comparing his reference information to this data frame. The channel information is represented in the form of OFDM symbols that represent the physical characteristics of a signal contained in a matrix. Thus, this model is a classification model that is strengthened by an adaptive boosting algorithm, AdaBoost. The authors define two separate test statistics, TA and TB, to test the difference in accuracy when the model utilizes only one statistic to make a decision compared to when they are both used in conjunction. Both test statistics measure the difference between two received transmissions, with a minor difference in considering amplitude and phase information. After the model has been trained, it uses either one or both of the test statistics to create weak classifiers. Once a committee of these classifiers was created, an adaptive boosting algorithm, AdaBoost, was implemented to further strengthen the classifiers through ensemble training. The logical progression of the model is shown in Fig. 10.

After the model has been tested on several frames from an unknown channel response, it can plot the values of the test statistics from a trusted channel against the test statistics calculated for this unknown response. Thus, the model can implicitly decide upon an efficient test threshold by finding the boundary line that separates data points that are "trusted" and "not trusted." In this real-world experiment, the model showed authors that the highest accuracy was achieved with a test threshold

Fig. 10 Authentication
protocol followed by Bob

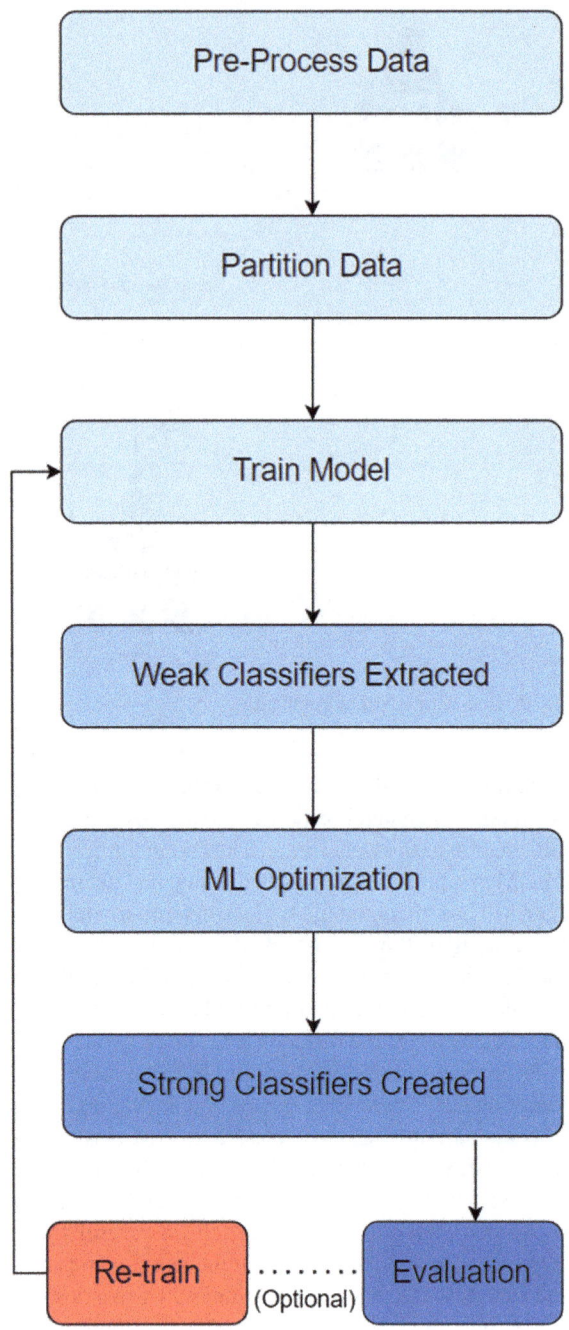

of 0.4. The accuracy rates for when the model was using only TA or TB with no classification boosting for its decision was 79.8% and 65.4%, respectively. However, when the classification boosting was implemented, the accuracy rates for TA rose to 87.1%. When the model was using classification boosting as well as both TA and TB in its decision, the accuracy rate increased to a maximum of 91.3% with only 10% additional computational complexity to incorporate both test statistics into the model. Therefore, this article concluded that in a large-scale wireless network, such as a 5G wireless communication system, PLA-based authentication schemes that use test statistics to train a model for security are both viable and efficient. Furthermore, their performance can be improved using appropriate adaptive classification boosting algorithms and increasing feature dimensionality.

Authors of [28] focused on the performance of different statistical and ML-based decision techniques in a PLA-based system when faced with tailored adversarial attacks. Nearest neighbor (NN) algorithms are used as well as a logarithm of the likelihood ratio test (LLR) to test the efficiency of the model. It is assumed in this article that Eve knows what Bob's decision strategy is and tailor's their attack to maximize the likelihood that an attack is effective and goes undetected. The setup phases for this article were the same as [27], where Bob receives a reference packet from Alice and is then tasked with identifying further incoming packets. A set of N parallel channels between Bob and Alice that carry OFDM transmissions are considered. These transmissions are subject to slow or fast fading, which impacts Bob's authentication rate. The channel impulse response (CIR) of each channel is collected into a vector by Bob. It is also assumed that Eve has access to partial CSI such that Eve can best mimic the statistics of all channels. The researchers evaluate the efficacy of a model off its false alarm (FA) and missed detection (MD) rates. The authors also test an additional protocol which they call the "combined test." This test combines the use of an LLR test as well as a simple comparison between the modulus of a received channel and Bob's initial reference. Note that Eve can forge her own CSI to optimize her chances of success.

In their results, the researchers found that with respect to the sole LLR test, the combined test performs at a lower MD rate as channel variability increases. In general, an OCNN algorithm finds the j-nearest neighbors of a test sample, x, in the target class and the k-nearest neighbors of the first j neighbors and calculates the average distance between each neighbor, respectively. The authors found that when j was fixed to 1, the algorithm was performing at its best potential. Hence, only the results of Eve's attacks on their 1KNN algorithm are shown in Table 6 compared to their LLR and combination test. This article highlighted the most effective methods in PLA-based systems to establish the sender of an unknown signal, detect forged signals and the robustness of PLA in large-scale wireless networks, further reinforcing the validity and plausibility of implementing a PLA scheme in a large network in the real world.

In [29], the authors improve upon their own proposed model from a previous article. This model can be used to authenticate the claimed identity of a wireless device. They test their technique using feature extraction in multiple one and two-dimensional time and frequency domains. 12 cellular communication devices are

Table 6 Average MD and FA probabilities for different test methods for different values of channel variability, denoted by α, and spatial correlation between Alice and Eve, $\rho_{AE} = 0.8$ [28]

α	N	P_{FA}	P_{MD} (1KNN)	P_{MD} (LLR)	P_{MD} (combo)
1	1	0.010	0.206	0.392	0.391
1	2	1.6×10^{-4}	0.110	0.326	0.331
1	3	$<10^{-6}$	0.041	0.319	0.279
1	4	$<10^{-6}$	0.018	0.183	0.171
1	5	$<10^{-6}$	0.012	0.099	0.078
1	6	$<10^{-6}$	0.005	0.052	0.027
0.9	1	0.540	0.206	0.221	0.353
0.9	2	0.799	0.110	0.045	0.127
0.9	3	0.863	0.041	0.016	0.060
0.9	4	0.877	0.018	0.008	0.032
0.9	5	0.817	0.012	0.008	0.035
0.9	6	0.865	0.005	0.003	0.021
0.8	1	0.739	0.206	0.160	0.258
0.8	2	0.928	0.110	0.027	0.067
0.8	3	0.970	0.041	0.007	0.029
0.8	4	0.983	0.018	0.003	0.021
0.8	5	0.982	0.012	0.002	0.017
0.8	6	0.991	0.005	6.1×10^{-4}	9.7×10^{-4}

used to validate and compare the results from different techniques. The improvements made to their model include additional verification/authentication techniques which were not previously present, reviewing the wavelet synchrosqueezed transform (WSST) and its application to PLA, testing different ML algorithms than KNN, and the optimization of the hyperparameters of both WSST and the ML algorithms. The materials used in this article are three phones of four different models (Apple iPhone, Sony Experia, HTC One, and Samsung S5) for a total of 12 phones, an OpenBTS software to monitor communication from each of the phones, and a Universal Software Radio Peripheral (USRP) receiver to collect the signals in space from the transmitting devices. The methodology for the classification of a wireless device starts with the collection of the radio frequency signal in space from all 12 devices. Normalization and synchronization are applied to this data, and it is then fed into 1 of 4 transformations: One-dimensional time domain, one-dimensional Fourier transformation, two-dimensional short time Fourier transform (STFT), or two-dimensional WSST. For the latter 2, hyperparameter optimization occurs before all data is fed into a chosen ML algorithm for authentication or identification. Support vector machines (SVM), KNNs, and decision trees were all used with tenfold methods. Identification or authentication can be based on the entire complex digital output of the RF transmissions, or only on magnitude or phase components. Accuracy rates for authentication

Table 7 Top-five performing algorithms from [29]

ML algorithm	Transformation method	Optimal values	Accuracy
SVM	WSST	$C = 2^7, \Upsilon = 2^8$	0.9236
SVM	STFT	$C = 2^7, \Upsilon = 2^9$	0.8503
KNN	WSST	$K = 3$	0.8388
Decision tree	WSST	$N_s = 12$	0.8225
KNN	1D time domain	$K = 9$	0.7910

methods solely based on magnitude with KNN for different values of K averaged around 80%, while methods solely based on phase components averaged around 75%. Table 7 shows 5 ML algorithm/transformation combination that resulted in the highest accuracy rates.

The authors were able to conclude that their model was viable for both device authentication as well as device identification. As SNR values increased, they observed that all four signal transformations decreased in equal error rate (EER). Additionally, they found that WSST outperformed the other transformation methods at medium to high SNR values. Thus, a PLA model utilizing ML algorithms in combination with WSST was effectively able to both identify and authenticate a device solely based on physical attributes of said device.

4 Discussion and Analysis

In this section, we present the key findings of our survey regarding both biometric-based and PLA-based systems.

4.1 Biometric-Based Authentication

Throughout this survey, there were many different avenues for user authentication using behavioral biometrics discussed. Different behavioral biometrics were compared and contrasted with one another, such as keystroke biometrics, gait patterns, and touchstroke biometrics. While all of them have their advantages, one excelled compared to the rest. Touchstroke biometrics, as seen by the research collected from in these articles, is easily the most secure and effective behavioral biometric to be used for user authentication. Also found was that many researchers used apps to implicitly which, in turn, makes the authentication scheme less costly for both the manufacturers to produce and the users to purchase. Also discussed in this survey was the use of different machine learning algorithms in behavioral biometric-based user authentication. The most commonly used algorithms found in this survey of articles was: BayesNET, random forest, k-nearest neighbor, and support vector machine. While all algorithms performed well in their respective studies, random forest was often the best performing algorithm. This is due to the simplistic nature of the RF algorithm; this algorithm is easy for researchers to picture and understand which makes it easier to use.

Another thing that sticks out about random forest is that it is a very forgiving algorithm, where if there is an error in one of the trees, it does not affect the rest of the model which leads to the higher accuracies that other algorithms are unable to receive.

4.2 Physical Layer Authentication

First and foremost, it appears that supervised learning is the most used from of machine learning in our reviewed PLA-related literature. This may be due in part to the fact that PLA is usually based on interpreting the difference between an incoming signal and a known authentic signal/previously authenticated signal. Supervised learning therefore would require models to be trained on labeled data that allows it to extract patterns of signal authenticity. NNs, SVMs, and decision trees were the most commonly used models in the literature, with these classifying techniques sometimes being strengthened by the addition of ensemble training. Unsupervised algorithms may have an increased authentication rate when tasked with classifying new, never before seen data when compared to supervised learning algorithms, but because of the difficulty in counterfeiting physical layer characteristics of a signal, novel spoofing techniques may not occur frequently enough to justify the use of an unsupervised learning algorithm. Based on the reviewed literature, we conclude that NN algorithms and SVMs are both the most common and efficient ML algorithms for PLA-based authentication systems. This is most likely due to SVM's advantageous performance when only small amounts of data are available and NN's superior speed and performance in lower-dimensional feature spaces when compared to other ML algorithms.

5 Conclusion

Through the review of these surveys, it was found that touchstroke behavior biometrics and the machine learning algorithm of random forest performed better than their behavioral biometric and machine learning counterparts. Touchstroke biometrics, as seen by the research collected from other articles, is easily the most secure and effective behavioral biometric to be used for user authentication. Also, using behavioral biometrics in general makes the authentication scheme less costly for both the manufacturers to produce and the users to purchase. Also discussed in this survey was the use of different machine learning algorithms in behavioral biometric-based user authentication. While all algorithms performed well in their respective studies, random forest was often the best performing algorithm.

Upon review of many articles regarding physical layer authentication-based systems, we conclude that NN algorithms and SVMs outperform other ML algorithms. SVMs and NNs outperform most ML algorithms when there is only a small amount of data available. PLA-based systems tested in controlled settings will not have millions of data points, hence the observed superior performance of these algorithms. However, if implemented into a large-scale wireless network, there may be copious amounts of data in which NNs and SVMs may not perform as well as expected.

6   Limitations and Future Work

There are some limitations that are still causing error in behavioral biometric-based authentication schemes. Many studies, as seen in [25], are using small amounts of participants over a short period of time, so the consistency and validity of the study may not reflect accuracies of the scheme when behavioral changes occur in the user. Also, touchstroke biometrics are found to have issues with collecting data during periods of movement, which can make the system produce more false positives or false negatives. Future work should focus on solving these two issues which would, in turn, increase the accuracy and usability of these schemes.

Despite the efficacy of NNs in the reviewed literature, there are still limitations to these algorithms which leaves room for improvement in the field. For instance, NN algorithms struggle to perform as efficiently as other algorithms when working in high-dimensional spaces, with increasingly large data sets, and are sensitive to outliers. Similarly, SVMs also perform less efficiently than other ML and even deep learning algorithms (as seen in [26]) when larger data sets are available. Additionally, SVMs are quite sensitive to noise. Thus, large data sets and complex dimensional feature spaces are among the most glaring limitations of the algorithms reviewed in this survey. Deep learning has been a heavily researched field recently, and although the main focus of this survey was on machine learning algorithms, the aforementioned limitations are some of the strengths of deep learning algorithms. Therefore, the literature suggests an endeavor to improve and integrate deep learning algorithms into PLA-based systems would be a worthwhile investment. Furthermore, if used in tandem with other ML algorithms, PLA-based systems may be able to reach high enough efficiency and accuracy rates, even better than conventional methods, to reliably be used in large-scale wireless networks.

References

1. Shelton J et al (2018) Palm print authentication on a cloud platform. In: 2018 international conference on advances in big data, computing and data communication systems (icABCD), Durban, pp 1–6. https://doi.org/10.1109/ICABCD.2018.8465479
2. Mason J, Dave R, Chatterjee P, Graham-Allen I, Esterline A, Roy K (2020) An investigation of biometric authentication in the healthcare environment. Array 8:100042. https://doi.org/10.1016/j.array.2020.100042
3. Yousefi N et al (2019) A comprehensive survey on machine learning techniques and user authentication approaches for credit card fraud detection. ArXiv abs/1912.02629
4. Fang H, Wang X, Tomasin S (2019) Machine learning for intelligent authentication in 5G and beyond wireless networks
5. Cheung W, Vhaduri S (2020) Continuous authentication of wearable device users from heart rate, gait, and breathing data. In: 2020 8th IEEE RAS/EMBS international conference for biomedical robotics and biomechatronics (BioRob). https://doi.org/10.1109/biorob49111.2020.9224356
6. Bogdanov A, Knežević M, Leander G, Toz D, Varıcı K, Verbauwhede I (2013) SPONGENT: the design space of lightweight cryptographic hashing. IEEE Trans Comput 62(10):2041–2053
7. Zhang R, Zhu L, Xu C, Yi Y (2015) An efficient and secure RFID batch authentication protocol with group tags ownership transfer. In: IEEE collaboration and internet computing, pp 168–175
8. Ma Z, Liu Y, Wang Z, Ge H, Zhao M (2018) A machine learning-based scheme for the security analysis of authentication and key agreement protocols. Neural Comput Appl. https://doi.org/10.1007/s00521-018-3929-8
9. Wu G, Wang J, Zhang Y, Jiang S (2018) A continuous identity authentication scheme based on physiological and behavioral characteristics. Sensors 18:179
10. Gunn DJ et al (2019) Touch-based active cloud authentication using traditional machine learning and LSTM on a distributed tensorflow framework. Int J Comput Intell Appl 18:1950022:1–1950022:16
11. Krishnamoorthy S et al (2018) Identification of user behavioral biometrics for authentication using keystroke dynamics and machine learning. In: ICBEA'18
12. Shen C et al (2013) User authentication through mouse dynamics. IEEE Trans Inform Forensic Secur 8:16–30
13. Buriro A et al (2015) Touchstroke: smartphone user authentication based on touch-typing biometrics. In: ICIAP workshops
14. Akhtar Z et al (2017) Multimodal smartphone user authentication using touchstroke, phone-movement and face patterns. In: 2017 IEEE global conference on signal and information processing (GlobalSIP), pp 1368–1372
15. Buriro A et al (2016) Hold and sign: a novel behavioral biometrics for smartphone user authentication. In: 2016 IEEE security and privacy workshops (SPW), pp 276–285
16. Mostafa H et al (2019) Behavio2Auth: sensor-based behavior biometric authentication for smartphones. In: ArabWIC 2019
17. Buriro A et al (2019) AnswerAuth: a bimodal behavioral biometric-based user authentication scheme for smartphones. J Inf Secur Appl 44:89–103
18. Meng Y et al (2014) Design of touch dynamics based user authentication with an adaptive mechanism on mobile phones. In: SAC'14
19. Maghsoudi J, Tappert C (2016) A behavioral biometrics user authentication study using motion data from android smartphones. In: 2016 European intelligence and security informatics conference (EISIC), pp 184–187
20. Wu X, Yang Z (2015) Physical-layer authentication for multi-carrier transmission. IEEE Commun Lett 19(1):74–77
21. Hou W, Wang X, Chouinard J, Refaey A (2014) Physical layer authentication for mobile systems with time-varying carrier frequency offsets. IEEE Trans Commun 62(5):1658–1667
22. Wang W, Sun Z, Piao S, Zhu B, Ren K (2016) Wireless physical layer identification: modeling and validation. IEEE Trans Inform Forensic Secur 11(9):2091–2109

23. Liao R, Wen H, Pan F, Song H, Xu A, Jiang Y (2019) A novel physical layer authentication method with convolutional neural network. In: 2019 IEEE international conference on artificial intelligence and computer applications (ICAICA), Dalian, pp 231–235. https://doi.org/10.1109/ICAICA.2019.8873460
24. Bailey KO et al (2014) User identification and authentication using multi-modal behavioral biometrics. Comput Secur 43:77–89
25. Meng Y et al (2012) Touch gestures based biometric authentication scheme for touchscreen mobile phones. In: Inscrypt
26. St. Germain K, Kragh F (2020) Physical-layer authentication using channel state information and machine learning
27. Chen S, Wen H, Wu J, Chen J, Liu W, Hu L, Chen Yi (2018) Physical-layer channel authentication for 5G via machine learning algorithm. Wirel Commun Mob Comput 2018:1–10. https://doi.org/10.1155/2018/6039878
28. Senigagliesi L, Baldi M, Gambi E (2019) Statistical and machine learning-based decision techniques for physical layer authentication. In: 2019 IEEE global communications conference (GLOBECOM), Waikoloa, HI, pp 1–6. https://doi.org/10.1109/GLOBECOM38437.2019.9013609
29. Baldini G, Giuliani R, Steri G (2018) Physical layer authentication and identification of wireless devices using the synchrosqueezing transform. Appl Sci 8:2167

Chapter 55
An Analysis of IoT Cyber Security Driven by Machine Learning

Sam Strecker, Willem Van Haaften, and Rushit Dave

1 Introduction

Based on previous predictions made in 2018, an estimated 50 billion devices are currently connected to the Internet [1]. Most of the devices connected to the Internet are not typical computers, laptops, or smartphones, but instead are devices that are smartwatches, smart refrigerators, thermostats, voice services, security cameras, and much more. These devices have been given various technologies to communicate with the Internet to provide better functionality and efficiency for the end user, and they make up what is considered the Internet of Things. One of the defining characteristics of the Internet of Things is the interconnectivity between its devices. Gateway devices are responsible for collecting data from surrounding sensor devices and transferring it to the Internet. Both the gateway and sensor devices must be secured, and measures need to be in place to prevent infected IoT devices from spreading to other devices. In many cases, the use of machine learning algorithms to secure IoT devices and networks has given promising results and could prove to be extremely beneficial for IoT security.

In the year 2021, there will be an estimated 24 billion IoT devices connected to the Internet [2]. Like most other technologies, the Internet of Things can be exploited by hackers and other deviants in order to weaken networks, deny service to one's system, steal valuable information, or conduct other malicious activities. As the amount of IoT devices have grown, attacks that exploit vulnerable IoT devices have increased exponentially [3]. This makes the security for the Internet of Things extremely important in order to protect the essential systems that they are connected to, in addition to the information that runs through them. Recently, the use of various machine learning techniques in the field of IoT security has been explored with goals like countering man-in-the-middle or selective forwarding attacks [4]. Many of these

S. Strecker (✉) · W. Van Haaften · R. Dave
University of Wisconsin—Eau Claire, Eau Claire, WI 54702, USA

© The Author(s), under exclusive license to Springer Nature Singapore Pte Ltd. 2021
S. Kumar et al. (eds.), *Proceedings of International Conference on Communication and Computational Technologies*, Algorithms for Intelligent Systems,
https://doi.org/10.1007/978-981-16-3246-4_55

techniques prove promising with high prediction accuracies, self-learning models, real-time security, and increased efficiency.

Due to the advancements in technology, we can now utilize machine learning as a practical tool in many programming scenarios, especially in areas like cyber security. Being able to parse through millions of data points, virtually in real time, and progressively self-learn is an incredible advantage when compared to previous methods like active storage scans that were reactionary and costly that only minimized damage instead of preventing it [5]. Machine learning has allowed security specialists to keep up with the ever-changing landscape that makes up cyber security. This study will highlight the uses of machine learning techniques and algorithms in IoT security including intrusion detection, identifying IoT devices, preventative security measures, and how different machine learning techniques compare to one another.

2 Background

The term "The Internet of Things" was created in 1999 by Ashton [6], who used the phrase to describe the use of RFID technology in the supply chain for Procter and Gamble. Nowadays, the term is used in a broader sense, describing every object, or thing, that contains sensors, software, and other hardware which are used to connect to and communicate with the Internet. So far, numerous businesses, homes, cities, and other organizations have deployed individual networks of IoT devices in order to accomplish tasks and goals with greater efficiency such as managing energy consumption, calling the authorities in an emergency, or completing tasks at home with voice commands. One promising usage of an IoT network as part of a city's development can be found in the city of Padova. They utilize IoT devices to monitor street lighting, carbon monoxide levels, noise levels, and more [7]. Although these networks of IoT devices can improve our standard of living, they, like most other technology, can be exploited by hackers and criminals if they are not protected correctly.

The security of IoT devices is one of the most important problems for the Internet of Things that needs to be addressed. New techniques to secure IoT devices are in constant development in order to better protect against cyberattacks on IoT devices and networks, which are equally ever-changing and evolving. However, developing security measures for IoT devices has its own set of challenges. For one, many IoT devices operate under the constraint of low power and, therefore, operate with a low computing power as well [8]. Raza et al. have done research about the use of the constrained application protocol (CoAP), which could be a cost-effective solution to protect the transfer of data for IoT systems that are constrained by computing power making real-time security protection unachievable [9]. In addition, not all IoT devices are of similar structure, which makes vulnerabilities and bugs hard to track [8]. Dorri et al. have done research on the use of blockchaining, a technique that has been used in conjunction with the cryptocurrency, Bitcoin [10]. They have added blockchaining to an IoT system in order to create a security structure that will encompass and protect a household's list of IoT devices, but it would require each household to operate their own private blockchain. Another set of security measures for mobile devices and the cloud environment is the use of biometric authentication methods using machine learning algorithms. The use of biometric authentication methods for mobile devices connected to IoT networks greatly reduces the risk of information being stolen and helps ensure that a person accessing some personal piece of information is the intended user [11–13].

Ultimately, IoT cyber security is still lacking for today's security landscape when compared to other areas with exceptional security like anti-fraud systems for online payments. IoT devices still face challenges in areas of security including malware detection and prevention and object identification. So far, solutions to these areas have been underdeveloped in the field of IoT systems [8]. In addition, the number of botnet attacks using compromised IoT devices is increasing significantly. In September 2016, the website of a computer security consulting firm was hit with 620 Gbps of traffic from an IoT botnet. At the same time, an even bigger distributed denial of service (DDoS) attack, using Mirai malware, peaked at 1.1 Tbps and targeted the webhosting cloud service provider OVH [14]. Attacks like these are only getting more prevalent, and as the number of IoT devices increase, the volumetric data of botnet attacks grow significantly, making it harder for Web infrastructure and website security companies like Cloudflare to mitigate attacks [3]. Due to the growing prevalence of IoT attacks, there is a need to find security measures that will protect IoT devices from being exploited. Thus, this study will explore the potential of using machine learning algorithms to detect several types of cyberattacks in IoT systems and prevent them from taking hold over IoT devices.

3 Literature Review

3.1 'IoT Security Techniques Based on Machine Learning'

Researchers at XMU were concerned about the grave reality that is our current state of Internet of Things security. IoT security is one of the more challenging areas in cyber security to manage. With the rise of machine learning and smart bot attacks, IoT devices must select a protective strategy and identify key attributes in security protocols for compromise in complex networks with and without fog layers [15].

Unsupervised learning has been widely used to improve network security such as authentication, access control, anti-jamming offloading, and malware detection. Support vector machines (SVMs), Naive Bayes, K-nearest neighbor (K-NN), neural networks (NNs), deep NNs (DNNs), and random forest are examples of supervised learning techniques that can be used to classify the network traffic of IoT devices. IoT devices, for example, may use the algorithms mentioned above to detect network intrusion and spoofing attacks, use K-NNs to detect network intrusion and malware, and use NNs to detect network intrusion and DoS attacks. IoT devices can use Naive Bayes for intrusion detection, and random forest can be used to detect malware (Figs. 1 and 2).

The researchers investigated a few different IoT attack models. Multivariate correlation analysis is used to derive geometrical correlations between network traffic features in order to detect DoS attacks. As compared to the triangle area-based nearest neighbors method using the KDD Cup 99 data collection, this model improves detection accuracy by 3.05–95.2% [15]. In a malware detection scheme, an IoT

Fig. 1 Computational comparison, 'IoT security techniques based on machine learning' [15]

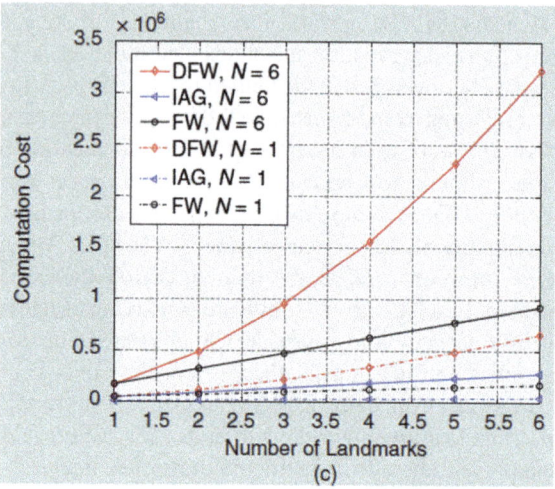

Fig. 2 Network diagram of security model 'IoT security techniques based on machine learning' [15]

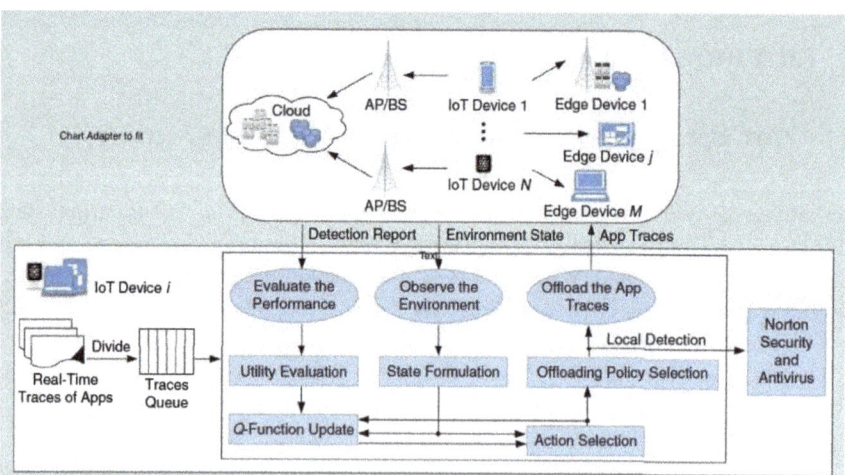

device can apply machine learning to achieve the optimal offloading rate without knowing the source generation and radio bandwidth model of the neighboring IoT devices. The IoT computer measures the detection accuracy gain, detection delay, and energy consumption in order to determine the task performed in a given time frame. In a network of 100 IoT computers, this scheme enhances detection accuracy by 40%, decreases detection latency by 15%, and improves mobile device accuracy by 47% as compared to offloading strategies. XMU researchers discovered a few IoT attack models as well as learning-based IoT protection techniques. IoT authentication, access control, malware detection, and secure offloading are among them [3].

3.2   'Using Machine Learning to Secure IoT Systems'

Faced with solving the issue of detecting anomalous data in IoT systems, Cañedo and Skjellum developed an approach for detecting anomalous data. In this study, Cañedo and Skjellum chose to use an ANN based on its popularity and its usage to monitor the state of IoT devices [16]. The ANN was used to analyze the data that is collected from sensor IoT devices and sent to a gateway IoT device in order to prevent both man-in-the-middle and denial of service attacks. Using a testbed of ten Arduino Uno devices with temperature sensors and Wi-Fi chips to serve as edge devices, each connected to a Raspberry Pi Model 3, which would serve as a gateway device, and a total of 4000 normal temperature recordings were collected. The ANN was trained using half of the collected data, which was chosen at random and then tested using the other half. The training included two input neurons: device ID and temperature. In addition to the half of valid data, ten minutes worth of invalid data was added to the testing set as well. Since the ANN was trained using only valid data, its output neuron would produce a value close to 1.00000 if a value in the testing set was valid, and a value above 1.00000 if a value in the testing set was invalid. This test resulted in a correct prediction of validity over 99% of the time for all the testing dataset. Table 1 is an example of the input and outputs for the ANN with two input

Table 1 'Using machine learning to secure IoT systems' [17]	Device ID	Temperature sensor value	Baseline validity	Prediction
	15	225	Valid	1.00000 (valid)
	16	0	Invalid	1.21264 (invalid)
	8	254	Valid	1.00000 (valid)
	7	75	Invalid	1.21164 (invalid)

Table 2 'Using machine learning to secure IoT systems' [17]

Device ID	Temperature sensor value	Delay (ms)	Baseline validity	Prediction
14	232	1119	Valid	1.00002 (valid)
16	0	951	Invalid	0.00030 (invalid)
8	241	1543	Valid	1.00002 (valid)
7	75	4137	Invalid	0.00031 (invalid)

neurons, where we can see that extraneous temperature values are considered invalid compared to the two valid temperature values.

After the conclusion of the initial results, the ANN was retrained with a third input neuron, the delay between transmissions. However, when the ANN was trained using a dataset filled with only valid data, the ANN had difficulty making correct predictions. Based on these findings, it was determined that the ANN should be trained using a dataset filled with both valid and invalid data. In this time around, the ANN's output neuron would produce a value around 1.00000 for predicting a valid test value, but the output neuron would produce a value around 0.0000 for predicting an invalid test value because of the inclusion of invalid data in the training dataset.

Table 2 is an example of the input and outputs for the ANN that was trained with three input neurons and both valid and invalid data, where we can see that the valid sensor values from the temperature sensors remained above an arbitrary value of 200 once more, and the delay periods were reasonably close to 1000 ms. Thus, these values were correctly predicted as valid by the ANN with an output close to 1.00000. However, the invalid sensor values, which were either well below the same arbitrary value of 200, well above a delay time of 1000 ms, or both, were correctly predicted as invalid by the ANN with an output value nearing 0.0000. Figure 3 shows a visual representation of how the inputs are handled and how the validity is changed by the weights of the plot. This new test resulted in a correct prediction of validity over 99% of the time for all the testing dataset once more. In conclusion, this study developed a method that could successfully detect anomalous data, which shows promise toward preventing man-in-the-middle or denial of service attacks in IoT systems. However, this article has a few limitations, which may have skewed the results to look more promising. First, the second iteration of this experiment, which used three input neurons, was only tested with 360 points of data, which is a relatively small amount of data to test the ANN with. In addition, artificial neural networks generally take a longer time for training compared to other machine learning algorithms [18], which would make them less favorable to other machine learning algorithms such as random forest or Naïve Bayes.

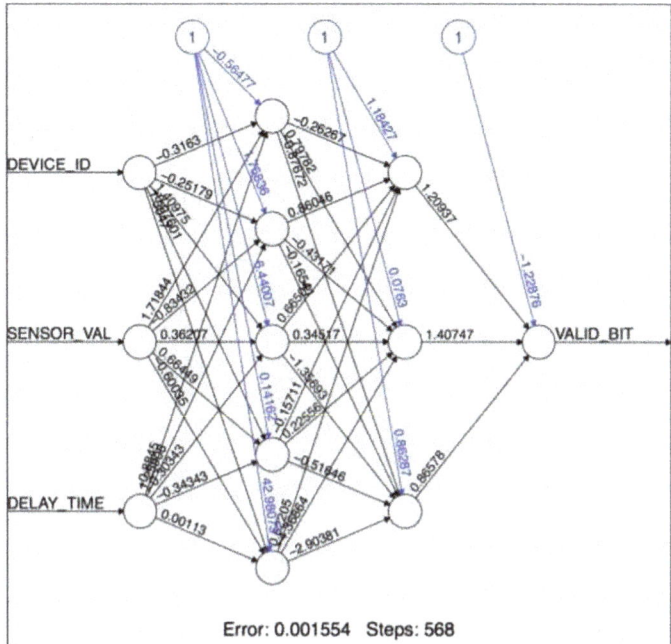

Error: 0.001554 Steps: 568

Fig. 3 ANN plot with three input neurons, 'using machine learning to secure IoT systems' [17]

3.3 'Detection of Unauthorized IoT Devices Using Machine Learning Techniques'

As the usage of IoT devices in the modern workplace expands, so does the potential of an IoT device that is infected with malware, connecting to and infecting an organization's network. This could occur from an attacker who is trying to gain entry to an organization's main systems, or an employee, who has brought their own already compromised IoT device, unaware of its status, and connects said device to their workplace's network. In order to prevent this potential breach of an organization's network, Meidan et al. have conducted a study that aims to limit what types of IoT devices are allowed to connect to an organization's network via a whitelist and use machine learning to determine whether a connecting device should be allowed access to an organization's network. Their study chose to use the random forest supervised machine learning algorithm based on its resistance to bias when using a large number of trees, 500 in this case, and its usage in other studies [19].

Using the testbed of IoT devices shown in Table 3, each device is connected to a central switch, and in turn, a Wi-Fi router, the traffic data of each device is captured over a period of several months by port mirroring it onto a separate, local server in the form of pcap files. These pcap files are run through a feature extractor, which splits the pcap files into the individual TCP sessions, extracts the features of each

Table 3 Detection of unauthorized IoT devices [20]

Device type	Number of models	Number of devices	Number of sessions recorded	Correctly predicted as unknown based on 1 session	Correctly predicted as known based on 1 session
Baby monitor	1	1	51,578	0.96	0.98
Motion sensor	1	2	3834	1	0.98
Smart fridge	1	1	1,018,921	0.97	0.97
Security camera	2	3	14,394	0.98	0.98
Smoke detector	1	1	369	0.97	0.97
Smart socket	1	2	2,808,876	0.98	0.97
Thermostat	1	1	19,015	0.86	0.95
Smart TV	2	2	144,205	0.93	0.96
Smartwatch	3	4	4391	0.81	0.93

session, and constructs a set of feature vectors, which are merged into a dataset. This dataset was split chronologically into three parts: The first part was used to train the classifier using the random forest algorithm, the second part was used to optimize the threshold through which the classifier would identify a device, and the third part was used to test the accuracy of the classifier. This sequence was conducted nine different times, where each time one of the possible device types was left out of the training dataset and was considered as the unknown device type for the testing phase.

Using this method, the classifier achieved an average accuracy of 94% for correctly identifying unknown devices and 97% for correctly identifying known, whitelisted devices based on the results found in Table 3. In order to improve upon the classification accuracies of some devices, the classifier was changed so that it would examine 20 sessions at a time, instead of just one, and select the device that was predicted most often across those 20 sessions. This brought the average accuracies up to 96% for identifying unknown devices and 99% for identifying known devices. In conclusion, this system could prove very beneficial for securing an organization's network when combined with a security information and event management system that can isolate and block connections from unauthorized devices. Although this study proved effective at detecting unauthorized devices, it is limited to identifying devices according to the nine categories it used, such as smartwatch or security camera, and does not differentiate devices by their manufacturer or their model.

3.4 'ProfilloT: A Machine Learning Approach for IoT Device Identification Based on Network Traffic'

Median et al. conducted a similar study to the one above, which shares many aspects with the study seen above, but with the added goal of distinguishing IoT devices from non-IoT devices and a difference in the machine learning algorithms that were used. This study uses a similar testbed to the one seen above, but this time with the addition of several non-IoT device types, including a PC, a laptop, and two types of smartphones.

The data collected in this study was also done in a similar manner, which includes recording the pcap files from a Wi-Fi access point, extracting the features for each session, constructing a dataset of feature vectors for every session, and then chronologically partitioning the dataset into three parts: one for training, one for optimization, and one for testing. However, in this setup, there is no device that would be considered as 'unknown,' thus each session is classified as one device type or another. As shown in Fig. 3, given a multitude of consecutive sessions, S^d, and their feature vectors, the device type, d_i, that the sessions originate from is determined by checking whether the majority of a number of consecutive sessions, s_i^*, have a probability, p_i^s, of belonging to said device, above the optimized threshold, tr^*, of said device. The number of consecutive sessions needed, s_i^*, is determined when the classifier is fed the optimization dataset, such that the device can usually be predicted with a minimum number of consecutive sessions. Thus, the algorithm begins checking the given sessions with the device that needs the lowest number of consecutive sessions and works its way across the rest of the devices in ascending order for said value. The values of s_i^* for each device type can be seen in Table 4 (Fig. 4).

Given the addition of non-IoT device types, the classifier was evaluated based on its ability to distinguish between IoT and non-IoT devices first. In this case, the classification accuracy for PCs and smartphones was extremely close to 100%. Then, the classifier was evaluated on its ability to classify the different IoT devices, in which the classifier achieved an accuracy of 99.281%. The resulting accuracies for each device type can be seen in Table 4. Since the current classifier accuracy did not fully reach 100%, the minimum number of consecutive sessions needed was reevaluated, and the updated classifiers were reevaluated with the testing dataset. It was determined that in order to approach an accuracy of 100%, the minimum number of consecutive sessions needed should be four and a third times higher than what was originally determined. In conclusion, this study demonstrates the ability to accurately identify both IoT and non-IoT device types based on their initial network traffic when connecting to a gateway device.

Table 4 IoT device identification based on network traffic [21]

	Printer	Security camera	Fridge	Motion sensor	Baby monitor	Thermo stat	TV	Smartwatch	Socket
tr*	0.35	0.5	0.2	0.2	0.3	0.2	0.1	0.8	0.25
s*	11	1	3	3	9	45	23	77	1
Acc.	1.00	0.99	1.00	1.00	0.97	1.00	0.98	0.98	1.00

Algorithm 2: IoT Device Classification

procedure CLASSIFYDEVICE(C, S^d)

    Sort C by ascending s_i^*

    for (C_i, tr_i^*, s_i^*) in C **do**

        $a \leftarrow 1$

        $n \leftarrow 0$

        while $a + s_i^* - 1 <= |S^d|$ **do**

            for $sess$ in $\{S^d[a], ..., S^d[a + s_i^* - 1]\}$ **do**

                $p_i^s \leftarrow$ CLASSIFY($C_i, sess$)

                if $p_i^s \geq tr_i^*$ **then**

                    $n \leftarrow n + 1$

                if $n > s_i^*/2$ **then**

                    return d_i

            else

                $a \leftarrow a + 1$

    return 'unknown'

Fig. 4 Algorithm for classifying sessions of IoT devices [21]

3.5 'Machine Learning DDoS Detection for Consumer Internet of Things Devices'

With the goal of finding the best machine learning algorithm for detecting distributed denial of service (DDoS) attacks in IoT network traffic, Doshi, Apthorpe, and Feamster compare the effectiveness of multiple machine learning algorithms for detecting DDoS attacks. This study chose to compare the effectiveness of K-nearest neighbors (K-NN), support vector machine (SVM) with Linear Kernel, decision tree (DT), random forest (RF), and a neural network (NN) at detecting anomalous data and abnormal behavior in IoT systems based on their uses in previous studies on network intrusion detection systems (NIDS) in non-IoT systems.

For the experiment, the testbed included a Raspberry Pi v3, which served as a gateway device, and three IoT devices that would simulate normal traffic: a YI home camera, a Belkin WeMo smart switch, and a Withings blood pressure monitor, the last of which was connected via Bluetooth to an android phone, which was then connected to the gateway device. Ten minutes of data was created through normal interaction with the three devices, resulting in 32,290 packets of normal data. Then, DoS attacks were conducted on a Kali Linux virtual machine to target an Apache Web Server, resulting in 459,565 packets of malicious data, which were merged with the normal packets of data. All of the data was separated by the device from which it originated, after which features of the data, such as packet size or the interval between packets, were extracted. This study chose to use multiple different features from two different categories, stateless and stateful features. The stateless features, packet size, inter-packet interval, and protocol, could all be collected without having to divide

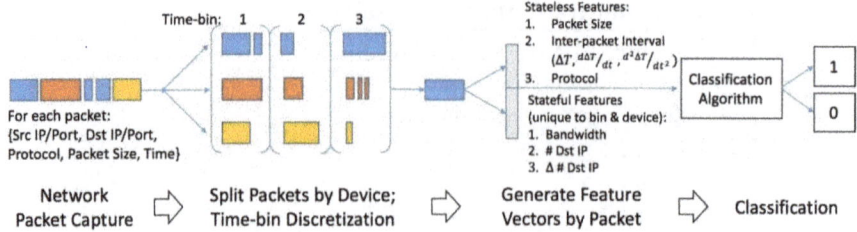

Fig. 5 Analysis of network packet flow features, 'machine learning DDoS detection for consumer Internet of Things devices' [22]

the incoming stream of traffic, which makes stateless features less intensive in terms of performance. However, the stateful features, like bandwidth and the repetition of destination IP addresses, needed the incoming stream of traffic to be divided based on the traffic's device origin and then further divided into time windows in order to be collected. The separation and analysis of the traffic data can be seen in Fig. 5.

Four of the five classifiers were constructed using the Scikit-learn Python library, and the last, the neural network, was created using the Keras library. Then, 85% of the total traffic data was used to train the various machine learning algorithms, and the remaining amount was used to test the resulting classifiers. As a result, it was found that all the algorithms achieved higher scores for the various metrics, as seen in Table 5, except for SVM, which had much lower recall and F1 scores for normal data. The study concluded that in addition to SVM performing the worst, the best performers were DT and K-NN. However, based on the provided table, Table 5, the classifier with the highest performance across all the different metrics seems to be the random forest algorithm.

One of the important realizations of this experiment was the discovery that stateless features were more beneficial to classification than stateful features. The differences in performance for each machine learning algorithm can also be seen in the bottom two columns of Table 5, which demonstrates how the use of only stateless features will diminish each of the classifiers' F1 scores for normal data when identifying normal and malicious data. However, the performance of the random forest classifier when using only stateless features outperforms each of the other classifiers, which reinforces the idea that random forest is the best performing algorithm of the five that were tested.

Table 5 'Machine learning DDoS detection for consumer Internet of Things devices' [22]

	K-NN	SVM	DT	RF	NN
Recall (normal)	0.993	0.870	0.993	0.998	0.989
F1 (normal)	0.995	0.927	0.994	0.998	0.986
Accuracy	0.999	0.991	0.999	0.999	0.999
Only stateless features (F1 normal)	0.967	0.920	0.977	0.981	0.939
Stateless and stateful features (F1 normal)	0.995	0.921	0.995	0.998	0.989

In conclusion, this study found that a machine learning classifier could analyze incoming traffic data using the data's characteristics at a flow-based level to distinguish between normal and attack data in a gateway IoT device. Based on the results for the metrics of evaluation, it appears that the random forest algorithm is the best fit for detecting incoming attacks. However, it should be noted that due to the overwhelming amount of attack data used in this experiment, a classifier which identifies every piece of data as malicious would achieve an accuracy of 93%.

3.6   'Machine Learning-Based IoT Intrusion Detection System: An MQTT Case Study'

The message queuing telemetry transport (MQTT) protocol was investigated by Abertay University researchers. They tested six (ML) techniques to see how effective they were at detecting MQTT-based attacks. There were three abstraction levels of features assessed, namely packet-based, unidirectional flow, and bidirectional flow features. The network consisted of 12 MQTT sensors, a broker, a machine to simulate camera feed, and an attacker. During normal operation, all 12 sensors send randomized messages using the Publish MQTT command (Fig. 6).

The length of the messages is the difference between sensors to simulate different usage scenarios.

To model various use situations, different sensors are used. The messages themselves are created at random. VLC media player, which works with UDP streams, was used to simulate the camera feed. Each of the network emulators dropped packets at a rate of 0.2, 1, and 0.13% to simulate a practical scenario. The background usual activity was left running during the recording of the four attack scenarios. Tiny Core Linux for the sensors, Ubuntu for the camera and camera feed server, and Kali Linux for the malicious users were the operating systems used by the various devices. Each experiment was evaluated using five-fold cross-validation (Table 6).

Overall accuracy was measured using the equation, true positive (TP) represents correctly classified attack instances, true negative (TN) represents correctly classified benign instances, positive (P) represents the number of attack instances, and negative (N) represents the total number of benign instances. Due to their similar characteristics, flow-based features were best suited to differentiate between benign and MQTT-based attacks at the end of the experiment. The weighted average for packet-based features increased from 75.31 to 93.77% and 98.85% for unidirectional and bidirectional flow features, respectively. While the weighted average precision for packet-based features increased from 72.37 to 97.19% and 99.04% for unidirectional and bidirectional flow features, respectively. The k-NN algorithm was used in all of these because it produced the best overall results.

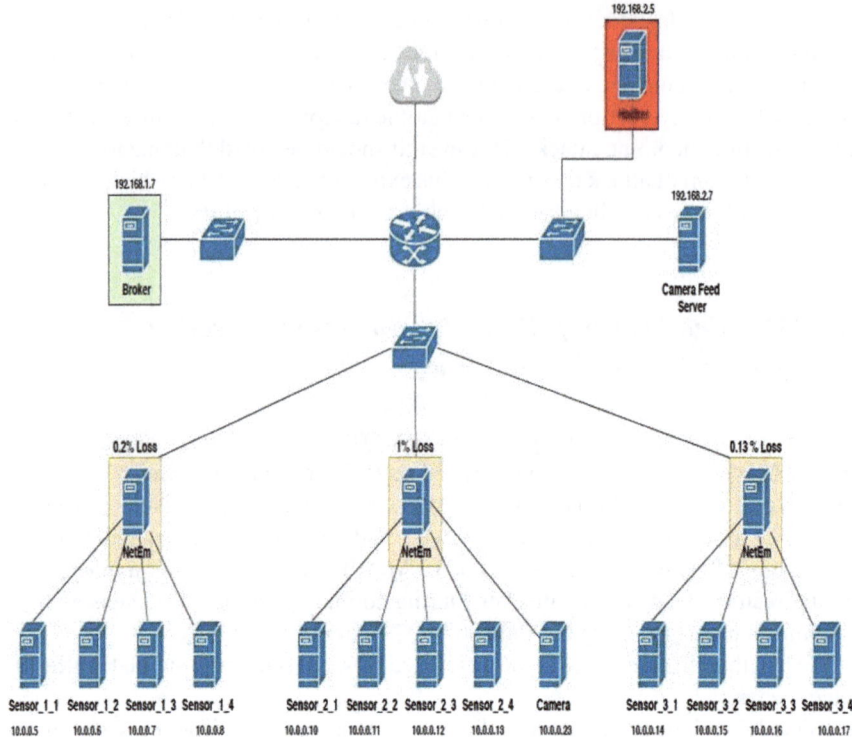

Fig. 6 Cluster node diagram, 'machine learning-based IoT intrusion detection system: an MQTT case study' [23]

Table 6 'Machine learning-based IoT intrusion detection system: an MQTT case study' [23]

File name	Pcap file size	# of benign	# of attack	# of uni-flow benign	# of uni-flow attack	# of uni-flow benign	# of uni-flow attack
Normal	192.5 MB	1,056,230	0	171,836	0	86,008	0
Scan_A	16.2 MB	70,768	40,624	115,600	39,797	5786	19,907
Scan_s UI	41.3 MB	210,819	22,436	34,409	22,436	17,230	22,434
spaarta	3.4 GB	947,177	19,728,942	154,175	28,232	77,202	14,116

3.7 'AD-IoT: Anomaly Detection of IoT CyberAttacks Smart City Using Machine Learning'

Researchers Ibrahim Alrashdi, Ali Alqazzaz, Esam Aloufi, Raed Alharthi, Mohamed Zohdy, and Hua Ming set out to create an anomaly detection system for IoT devices based upon select machine learning algorithms like decision tree, K-nearest neighbor, and random forest. Current intrusion detection systems are not designed for smart devices applications so AD-IoT in the fog layer can help significantly. The AD-IoT system design model consisted of dozens of components involving a large amount of IoT devices connected to distributed fog layers. By watching the fog layer opposed to the local network, or server side, it is closest to the IoT sensor saving alert time and network computing resources. This assumption does rely on networks having a private gateway. The master fog node can intelligently monitor all the communication among the network traffic data. The system is based on ensemble methods, which are improved with the use of random forest and extra tree algorithms. The researchers used the UNSW-NB15 dataset and split it among nodes using ExtraTreeClassifiers to reduce the importance selection to 12 features. The researchers placed the data from the UNSW-NB15 dataset onto a Pandas framework which breaks down the data into more manageable and efficient metrics. This allows them to read, split, convert, and normalize the data efficiently. The number of false positives in current IoT security also needs improvement with demonstrations with a confusion matrix using current technologies. They used a table to break down specific attacks that leverage IoT devices as it requires or works better with botnets. After each model was trained for the specific attack type, different machine learning algorithms like decision tree, K-nearest neighbor, and random forest were used to optimize each step to increase efficiency and lower total time of detection. The binary performance classification was as follows: In the rain forest model, the precision for normal was 99% with a recall rate of 99% while attack mode had a precision of 79% with a recall rate of 97%. They used the UNSW-NB15, KDD99, and NSL-KDD datasets to train their models. In conclusion, the test was highly successful on the datasets they used and should scale with larger datasets or real world (Figs. 7, 8, and 9).

3.8   'Attack and Anomaly Detection in IoT Sensor in IoT Sites Using Machine Learning Approaches'

In addition to DDoS attacks, there are numerous other types of cyberattacks and anomalous data which may be leveraged against IoT devices. Hasan, Islam, Zarif, and Hashem address the need for a safe IoT system that can identify intrusions and recover quickly. To help develop a solution, they proposed the use of a machine learning classifier and aimed to compare the effectiveness of five different machine learning algorithms in intrusion detection, which were logistic regression (LR), SVM, decision tree, RF, and ANN.

For their experiment, they used an open-source dataset from kaggle, which created a virtual IoT environment using distributed smart space orchestration system, in which there were 347,935 instances of normal data and 10,017 instances of anomalous data. Of the anomalous data, there were eight distinct types of attack data: denial of service, data type probing, malicious control, malicious operation, scan,

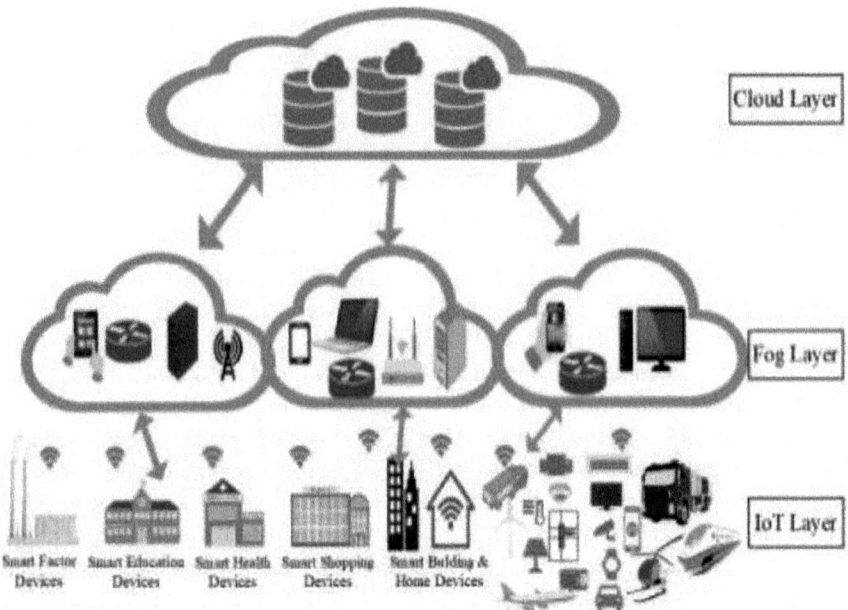

Fig. 7 Network layer diagram [24]

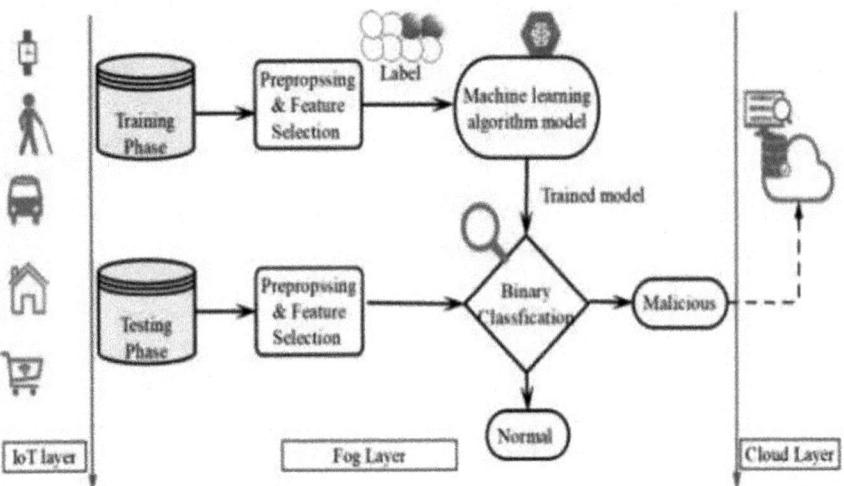

Fig. 8 Fog layer computational flowchart [24]

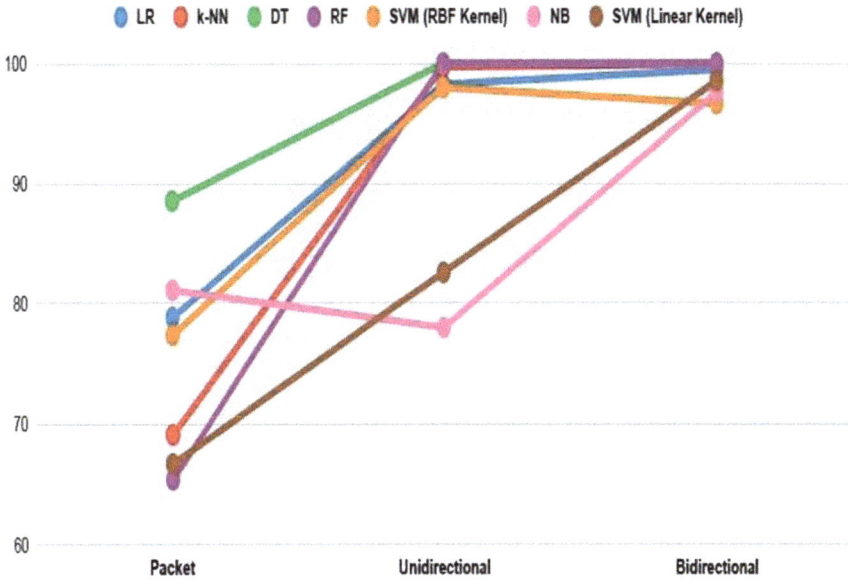

Fig. 9 Comparison graph 'ML algorithm comparison' [24]

spying, and wrong setup. The data from the open-source dataset was then preprocessed, in which missing or unexpected values from the dataset's 'accessed node type' and 'value' columns were given meaningful values. Then, when the feature vectors were being generated, the nominal categorical data from the dataset, which includes features like source address or destination address, was turned into feature vectors using label encoding, which maintains the number of features. After preprocessing the data and developing feature vectors from the data, the set of resulting feature vectors was split into a training set, which consisted of 80% of the data, and a testing set, which held the remaining 20%.

Every machine learning algorithm was given the training dataset to develop their own classifier, and then, each classifier was given the testing dataset and then evaluated based on how the classifiers could differentiate between both normal and attack data and between the eight classes of attack data using several different metrics of evaluation. While the classifiers were given the training and testing datasets, they were cross-validated using five various sample sizes in both cases, which can be seen in Fig. 10. In the two smallest sample sizes of the testing dataset, the accuracies of the DT and ANN classifiers were poor, as it was RF to a lesser extent. However, in the larger sample sizes, the accuracies of DT, ANN, and RF outperformed SVM and LR. SVM and LR were also outperformed by the other algorithms for each fold of cross-validation during training as well. After evaluating the five classifiers, it was found that random forest was the best algorithm compared to the others. RF and DT had the fewest misclassifications, and RF's standard deviation during the testing dataset was slightly better than that of DT and ANN. In conclusion, although RF,

Fig. 10 Five-fold cross-validation with the training set (**a**) and the testing set (**b**), 'attack and anomaly detection in IoT sensors in IoT sites using machine learning approaches' [25]

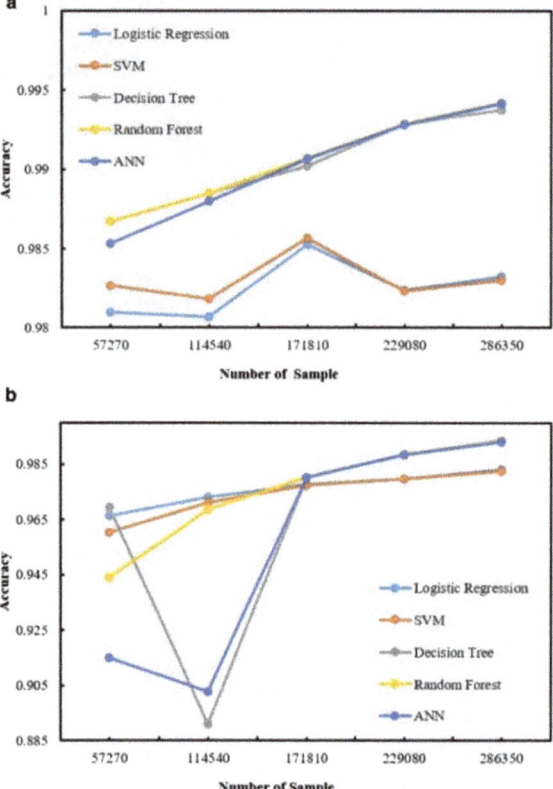

DT, and ANN could all sufficiently classify both normal and attack data given a large enough sample size, it was determined that RF is the best of the chosen algorithms for classifying normal and attack data in an IoT network.

3.9 'Cyber Forensics Framework for Big Data Analytics in IoT Environment Using Machine Learning'

Researchers Gurpal Singh Chhabra and Varinder Pal Singh and Maninder Singh Et created a generalized forensic framework that uses Google's programming model, MapReduce as the backbone for traffic translation, extraction, and analysis of dynamic traffic features. They used open-source tools like Hadoop, Hive, and Mahout and R. Also, comparative analysis of globally accepted machine learning models of P2P malware analysis in mocked real-time is presented. The proposed model was tested using a CAIDA dataset that was run in parallel. According to the model's forensic efficiency parameters, the findings have a sensitivity of 99%. They used the HDFS Hadoop architecture and standard MapReduce Programming to build an environment suite. Hadoop's technology can accommodate unexpected events, and HDFS' multi-cluster architecture is fault tolerant. It produces a copy of every bit of the user's data on a separate cluster computer. As a result, even though a computer crashes, the work performed on the data is unaffected since different copies of each block of data are stored on multiple nodes. Although the entire forensics analysis architecture is divided into four major modules: data collector and knowledge generator, features analytics and extraction module, designing machine learning models, and testing models on different performance matrices, the entire forensics analysis architecture is divided into four major modules. The researchers then used a tool called DunpCap, which had better performance than LINPAC due to its low-level abstraction feature, which was particularly useful when dealing with large quantities of data. Tshark was used in combination with Dumpcap to extract the fields from the traffic/data in comma-separated format, since it offers a range of options for manipulating and cleaning the output in different formats (Fig. 11 and Table 7).

The researchers used the Dumpcap traffic sniffer module on Ubuntu 15X with various ring buffer options to catch successive network traffic and then used Tshark to extract the fields of each packet in three different modes: single, parallel, and remote parallel mode. The researchers used the three nodes they set up earlier for remote parallel mode. With MapReduce shell scripts, the extraction is automatic, and the observation is based on the packet count. The proof traffic collector and information generator begin by sniffing traffic with Dumpcap and then extracting

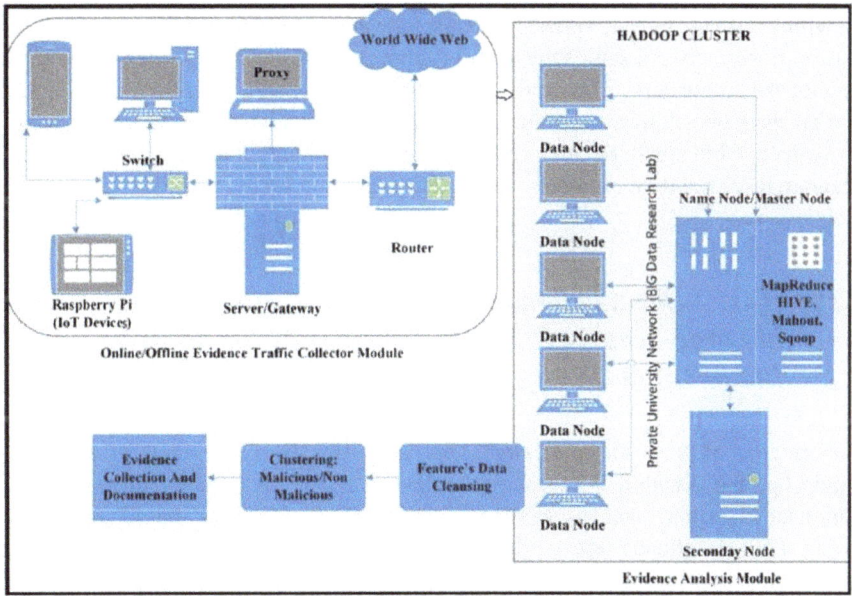

Fig. 11 Hadoop cluster network [26]

Table 7 'Comparative analysis of machine learning models for cyber forensic framework' [26]

Model name	FNR, MR	TR (%)	Sensitivity/recall	Specificity (SPC)	Precision	Type-I error rate	Type-II error rate
Decision tree	0.022	0.992	0.977	0.999	0.999	0.0002	0.022
Ada BOOST	0.011	0.996	0.977	0.999	0.999	0.0002	0.011
Accuracy	1.000	0.997	1.000	1.000	0.971	0.987	1.000
Random forest	0.008	0.997	0.992	0.999	0.99940	0.0003	0.008
SVM	0.217	0.777	0.783	0.771	0.77141	0.229	0.217
Linear model	0.030	0.975	0.970	0.978	0.95575	0.022	0.030
Neural net	0.000	0.667	1.000	0.666	0.00015	0.333	0.000

features with Tshark. In the proposed approach, three approaches were used and compared. The first method is the sequential execution of the script, on a single machine. The second method uses a single machine to run several processes in parallel, and the third method uses a MapReduce script on a Hadoop multi-node cluster. The researchers also tested the performance regression or gains from altering memory size and network bandwidth of the framework. They also tried three different database types MySQL, PostgreSQL, and Sqoop + Hive. The time differences were all within 0.02 ms of each other so the results were inconclusive. The main concern was improper network setups that did not support the constraints needed for the data flow of the framework. Therefore, in the algorithms tested, the most accurate and precise were decision tree and Ada BOOST. In conclusion, the Hadoop framework was effective but smaller-scale networks commonly found in homes resulted in less accurate results and more latency (Fig. 12).

3.10   'Fast Authentication and Progressive Authorization in Large-Scale IoT: How to Leverage AI for Security Enhancement'

Performance of proposed multi-node approach using HDFSResearchers He Fang, Angie Qi, and Xianbin Wang aimed to leverage the functionality of an optimal nonlinear classifier, such as a support vector machine (SVM)-based algorithm. Using a SVM offloads, the work allows you to not use physical keys. Physical keys require more resources and are prone to attacks. Firstly, they used a kernel machine learning-based physical layer authentication scheme to defend against spoofing attacks through tracking the communication link and hardware-related

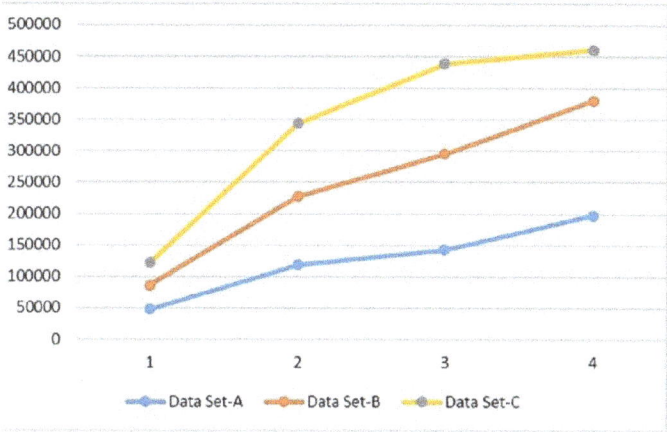

Fig. 12 Dataset comparison 'performance of proposed multi-node' approach using HDFS' [26]

features in time-varying environments. This project created a physical layer authentication scheme that uses machine learning to boost spoofing detection accuracy. The researchers created a watermarking algorithm based on a deep learning long short-term memory framework for dynamic authentication to detect cyberattacks, which allows IoT devices to extract vital information from their generated signal to determine whether they are malicious or not and to dynamically watermark these features into transmitting signals. First, channel probing is used to obtain measurements of selected features. Researchers developed a new quantization technique based on the SVM to derive an ideal nonlinear boundary at the base station to transform these measurements into binary sequences. The optimal nonlinear classifier, unlike the obtained signal strength (RSS)-based quantization technique, eliminates incorrect decisions by reducing measurements near the boundary. Because of the amplified channel reciprocity, the base station then sends the ideal nonlinear boundary to the IoT unit, resulting in extremely similar binary sequences being acquired on both sides. Hash functions could be used to verify that each IoT device and the base station have the same seed, and then, the same PRC could be produced for authentication. Because of the specific and unpredictable features of the communication connection used, the seed created by the base station and IoT device is explicitly hidden from other devices. In this scheme, the AI technique facilitates the security enhancement through training a nonlinear classifier at the base station, which is equipped with high computing and storage capabilities as well as continuous energy supply (Figs. 13 and 14).

In conclusion, by modeling the actions of attackers, game theory can also be used to protect against insider attacks. The developed lightweight authentication scheme and the physical layer key generation scheme were compared. Schemes of authentication scheme for creating physical layer keys authentication scheme with a minimal footprint identifying features continuous and one-time static transmission of the key/seed.

Input: Measurements of features
Output: Authorization level
1: // Fast initial authentication:
2: select one feature for initial authentication, e.g. IP address, channel-related, hardware-related,
   location-related, biometric feature;
3: obtain the initial trust value of that the transmitter is a legitimate device from the evaluation of the
   selected feature for authentication as $q[0]$;
4: // Progressive authentication and authorization:
5: select more features for authentication;
6: while training data, namely for the measurements of features, is available, do
7:      if new measurement (input) is detected as an outlier
8:         decrease the trust value $q[t]$;
9:      else
10:        increase the trust value $q[t]$;
11:     end if
12:     if trust value belongs to set $[a_m, a_{m+1})$, $m = 1, 2, ..., M$
13:        authorize this transmitter with m-level services/resources;
14:     else
15:        terminate the connection with this transmitter;
16:     end if
17: end while

Fig. 13 Authentication model [27]

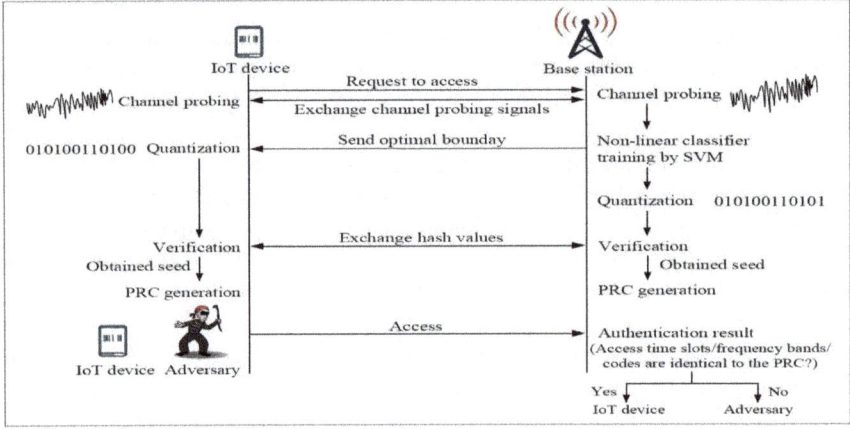

Fig. 14 Attack prevention model 'lightweight SVM authentication scheme' [27]

4   Discussion and Analysis

Overview and Analysis of the Articles under Literature Review

Article	Methodologies	Results	Pros	Cons
[15]	Train six different models based upon precompiled datasets to determine the most accurate algorithm	Detection accuracy of 95.2% with DWF	Lightweight algorithms for less computational power required	Precision was still below 90%
[17]	Collect data from temperature sensors and transfer it to a gateway device. Then, train and test an ANN using said data, first with two input neurons and then three	Correctly predicts validity over 99% of the time for both models with two and three input neurons	Correctly identifies anomalous data for both delay time and value	Conducted using a relatively small dataset
[20]	Collect data from nine different IoT devices running normally and transfer it to a gateway device. Then copy the data and extract a set of feature vectors from the sessions. Then train, optimize, and test a classifier using RF with the data	Correctly identifies a session's device of origin when known 99% of the time and correctly identifies a session as unknown 96% of the time using a window of 20 sessions	Capable of preventing unauthorized device types from entering an organization's network	Accuracies nearing 100% require up to 110 sessions for some device types
[21]	Collect data from multiple IoT and non-IoT devices running normally and transfer it to a gateway device. Then copy the data and extract a set of feature vectors from the sessions. Then train, optimize, and test several classifiers using RF, XGBoost, and GBM with the data	Correctly identifies a session's device of origin 99.281% of the time using the first optimized number of consecutive sessions	Capable of identifying what type of device one is based on its network traffic	Accuracies nearing 100% require a number of consecutive sessions 4.33 times higher than previously determined

(continued)

(continued)

Article	Methodologies	Results	Pros	Cons
[22]	Collect data from three different IoT devices and add data from a simulated DoS attack to the dataset. Then extract feature vectors and train and test the five classifiers with the data	All five of the classifiers reached over 99% accuracies, but RF performed the best both with and without stateful features	Can help prevent DoS attacks by correctly identifying anomalous data from DoS attacks and may work similarly when using only stateless features with RF	The amount of data from the DoS attacks is over 10 times higher than the amount of normal data
[26]	Leveraged a message queuing telemetry transport system to mass collect data from multiple sensors at once	99.04% accuracy for uni-directional and bi-directional features	Results could accurately tell benign versus malicious traffic	Requires large amounts of data throughput
[24]	Utilized fog layer computing to modify intrusion detection systems to be adapted to IoT devices	Random forest had an accuracy of 99% with attack mode at 79% with a recall rate of 97%	Takes pre-existing weathered solutions to save on dev time	Requires complex network to be set up
[25]	Using a simulated dataset, extract feature vectors from said dataset and train and test five classifiers with the data to identify whether its normal or attack, as well as the type of attack	ANN, RF, and DT reached over 99% accuracies, but RF performed the best	Not only identifies normal versus attack data but can also determine what type of attack is occurring	The dataset was obtained through a virtual simulation and may not represent real-world data
[26]	Hadoop cluster-based analysis of data nodes to identify malicious IoT devices	All three database types MySQL, PostgreSQL, Sqop + Hive were within 0.02 ms of one another	Performance throughput increased by 4× with the advanced dataset in the Hadoop cluster	Requires a Hadoop cluster to be properly set up and configured
[27]	Uses a SVM as to not us physical keys. A physical layer authentication scheme using ML to improve spoofing detection	Hash functions can be used to speed up identification, and game theory could be used to increase accuracy	Precise, fast and relatively lightweight because of the SVM	Requires proprietary watermarks in network packets

The article, 'Using Machine Learning to Secure IoT Systems' [17], has put forth a promising system which demonstrates the capability of correctly identifying anomalous data using an ANN. This would provide a solution for both detecting denial of service and man-in-the-middle attacks and allow for an entire IoT network to monitor incoming data to check for compromised IoT edge devices. The article, 'Machine Learning DDoS Detection for Consumer Internet of Things' [22], has also developed a system that can correctly identify anomalous data, but has also chosen to compare the effectiveness of several different machine learning algorithms for detecting anomalous data. This study not only established that the random forest algorithm performed the best of the five algorithms but also established that random forest performed the best when it used only stateless features. The use of only stateless features would make the analysis of the traffic data less cost intensive when using both stateless and stateful features; however, doing so has shown a reduction in the accuracies of each of the five classifiers. 'Attack and Anomaly Detection in IoT Sensors in IoT Sites using Machine Learning Approaches' [25] takes anomalous data detection a step further by not only differentiating between normal and attack data but also by classifying attack data according to what type of cyber-attack it may be. This study concluded that the random forest algorithm performed the best of the five algorithms that were compared, and that the use of this data classification method would not only help detect attack data but would also assist its users by identifying what type of attack is being conducted as well. The article, 'Detection of Unauthorized IoT Devices Using Machine Learning Techniques' [20], has developed a method which can determine what device type an IoT device is when it requests access to some network through the use of network traffic analysis using the machine learning algorithm, random forest. This method was able to identify all nine device types that it used with a near 100% accuracy using a maximum of 110 sessions; thus, this method presents a solution for preventing untrusted devices from connecting to an organization's network if used in conjunction with a security information and event management service. The article, 'ProfilloT: A Machine Learning Approach for IoT Device Identification Based on Network Traffic Analysis' [21], conducts a very similar experiment, but instead, it includes non-IoT devices to identify and does not consider any device type as unknown during classification. This method provides a solution for filtering out non-IoT traffic while also classifying the remaining IoT traffic by device type. In the article 'IoT Security Techniques Based on Machine Learning' [15], they compared six different commonly used algorithms. They concluded, like many others, that random forest is the best choice. The article 'Machine Learning-Based IoT Intrusion Detection System: an MQTT Case Study' [23] uses 12 message queuing telemetry transport (MQTT) sensors in a network of IoT sensors. It acts as a uni, bi, and packet-based flow gate. This allows granular packet inspection that resulted in a final accuracy of malicious intrusions at 99.04%. Like MQTT sensors, 'AD-IoT: Anomaly Detection of IoT CyberAttacks Smart City Using Machine Learning' [24] uses a cloud, fog, and IoT layer to offload network volumetric load from the local network containing the IoT sensors. Using random forest paired with K-nearest neighbor as well as the Panda framework to organize the data, the researchers were able to obtain an accuracy of 97%. While accuracy is one of the largest factors to consider in IoT, cyber security speed is also a large concern. In 'Fast Authentication and Progressive Authorization in Large-Scale IoT: How to Leverage AI for Security Enhancement' the main priority was to develop a framework to support faster authentication. The researchers did so by watermarking authentic network packets which expedited the authentication process significantly. The largest drawback is the use of priority packet monitoring, which can be costly to implement [27]. Finally, 'Cyber forensics framework for big data analytics in IoT environment using machine learning' [26] utilized a Hadoop infrastructure as well but also included a MapReduce program to allow sensitivities of 99% for the incoming data. AdaBoost showed strong results of precision and accuracy at $4\times$ the typical rate.

5 Limitations

First, the classifiers in the article in 3.2 used a relatively small dataset compared to some of the other articles in this study. The classifier with two input neurons trained and tested with 4000 data samples, and the classifier with three input neurons in the article in 3.2 was only tested with 360 pieces of data. While the classifiers did produce high accuracies for the given dataset, they may not hold up against a dataset of a larger size. Since the article in 3.8 utilized a digital dataset, the results of the experiment may differ from the results of a similar experiment conducted using real-world data, especially considering how DoS attacks flood systems with large amounts of data. However, by including the large quantity of data from DoS attacks, it should be noted that a baseline prediction algorithm which classifies every piece of data as anomalous would have a high accuracy as noted in the article in 3.5, which may make the application of the machine learning classifiers less ingenuous. The classification of IoT device types in articles in 3.3 and 3.4 sometimes requires a large number of sessions in order to achieve an accuracy that is close to 100%, and during these sessions, a malicious IoT device might be able to accomplish its goal of attacking the network.

The dataset that was used for most papers including 3.1, 3.5, 3.7, 3.6, and 3.9 was the UNSW-NB15 dataset. Inside the dataset, there are nine different types of attack methods: Fuzzers, Analysis, Backdoors, DoS, Exploits, Generic, Reconnaissance, Shellcode, and Worms [15]. Unfortunately, this dataset was constructed is extremely outdated, as it was constructed in 2015. In addition, the complete lack of botnet attacks all together in the UNSW-NB15 dataset should be enough to turn researchers away, as this is the most popular attack method utilizing IoT devices. Additional attack methods currently being exploited in the real world include traffic analysis, side-channel attacks, replay attacks, man-in-the-middle attacks, and protocol attacks, all of which are not included in the UNSW-NB15 and KDD99, and NSL-KDD datasets. Finally, IoT devices must be safely authenticated using a symmetric key protocol; however, these methods are vulnerable to key-hashing attacks and require a large amount of computing power. For today's IoT applications, the OAuth 2.0 protocol is the most widely used authentication mechanism. But since Oauth 2.0 requires manual authentication from a user, it is vulnerable to cross-site-recovery forgery (CSRF) attacks and which become overwhelmed as the number of devices per user increases. Physical unclonable functions (PUF) have emerged as a solution that takes advantage of manufacturing process variations to create a special and device-specific identity for a physical system [15]. PUF implementations are better than memory-based solutions because they use less resources and take up less space on the die than costly cryptographic ASIC hardware like SHA512 or AES-256.

6 Conclusion

This study has discussed the uses of machine learning algorithms in IoT security by reviewing several articles which conducted various methods and techniques in order to find solutions for some of the problems in IoT security. From the beginning, this study highlighted the enormity of the Internet of Things, as well as its potential to be exploited. Then, it began to focus on the individual cases in which the use of machine learning may benefit IoT security including its use for malware and intrusion detection and the identification of unknown IoT devices. Many of these methods that were featured either concluded that random forest was the best machine learning algorithm for their methods or was using random forest on its own from the get-go. However, others found that by using multiple machine learning algorithms, the accuracy and precision would improve. One algorithm's weaknesses would be complemented by the other algorithm, which produced higher accuracies than a single algorithm could achieve on its own. The use of K-NN and Euclidean gave an accuracy of around 93% and a precision close to 86%, but when K-NN was paired with dynamic time warping (DTW), the accuracy rose to 96% and precision was 91% [5]. In comparison to a single algorithm being utilized like random forest which plummets to an accuracy of 86% and abysmal precision rates in the 82% range.

References

1. Davis G (2018) 2020: life with 50 billion connected devices. In: 2018 IEEE international conference on consumer electronics (ICCE). https://doi.org/10.1109/icce.2018.8326056
2. Bull P, Austin R, Popov E, Sharma M, Watson R (2016) Flow based security for IoT devices using an SDN gateway. In: 2016 IEEE 4th international conference on future internet of things and cloud (FiCloud). https://doi.org/10.1109/ficloud.2016.30
3. Kolias C, Kambourakis G, Stavrou A, Voas J (2017) DDoS in the IoT: Mirai and other botnets. Computer 50(7):80–84. https://doi.org/10.1109/mc.2017.201
4. Mamdouh M, Elrukhsi MAI, Khattab A (2018) Securing the internet of things and wireless sensor networks via machine learning: a survey. In: 2018 international conference on computer and applications (ICCA). https://doi.org/10.1109/comapp.2018.8460440
5. Azmoodeh A, Dehghantanha A, Conti M et al (2018) Detecting crypto-ransomware in IoT networks based on energy consumption footprint. J Ambient Intell Hum Comput 9:1141–1152

6. Ashton K (2009) That 'internet of things' thing. RFID J 22(7):97–114
7. Cenedese A, Zanella A, Vangelista L, Zorzi M (2014) Padova smart city: an urban internet of things experimentation. In: Proceeding of IEEE international symposium on a world of wireless, mobile and multimedia networks 2014. https://doi.org/10.1109/wowmom.2014.691 8931
8. Zhang Z-K, Cho MCY, Wang C-W, Hsu C-W, Chen C-K, Shieh S (2014) IoT security: ongoing challenges and research opportunities. In: 2014 IEEE 7th international conference on service-oriented computing and applications. https://doi.org/10.1109/soca.2014.58
9. Raza S, Shafagh H, Hewage K, Hummen R, Voigt T (2013) Lithe: lightweight secure CoAP for the internet of things. IEEE Sens J 13(10):3711–3720. https://doi.org/10.1109/jsen.2013. 2277656
10. Dorri A, Kanhere SS, Jurdak R (2016) Blockchain in internet of things: challenges and solutions. arXiv preprint arXiv:1608.05187
11. Gunn DJ et al (2019) Touch-based active cloud authentication using traditional machine learning and LSTM on a distributed tensorflow framework. Int J Comput Intell Appl 18:1950022:1–1950022:16
12. Mason J, Dave R, Chatterjee P, Graham-Allen I, Esterline A, Roy K (2020) An investigation of biometric authentication in the healthcare environment. Array 8:100042. https://doi.org/10. 1016/j.array.2020.100042
13. Shelton J et al (2018) Palm print authentication on a cloud platform. In: 2018 international conference on advances in big data, computing and D
14. Kelley T, Furey E (2018) Getting prepared for the next botnet attack: detecting algorithmically generated domains in botnet command and control. In: 2018 29th Irish signals and systems conference (ISSC), Belfast, pp 1–6. https://doi.org/10.1109/ISSC.2018.8585344
15. Xiao L, Wan X, Lu X, Zhang Y, Wu D (2018) IoT security techniques based on machine learning: how do IoT devices use AI to enhance security? IEEE Signal Process Mag 35(5):41–49. https://doi.org/10.1109/MSP.2018.2825478
16. Kotenko I, Saenko I, Skorik F, Bushuev S (2015) Neural network approach to forecast the state of the internet of things elements. In: 2015 XVIII international conference on soft computing and measurements (SCM). https://doi.org/10.1109/scm.2015.7190434
17. Canedo J, Skjellum A (2016) Using machine learning to secure IoT systems. In: 2016 14th annual conference on privacy, security and trust (PST). https://doi.org/10.1109/pst.2016.790 6930
18. Moh M, Raju R (2018) Machine learning techniques for security of internet of things (IoT) and fog computing systems. In: 2018 international conference on high performance computing & simulation (HPCS). https://doi.org/10.1109/hpcs.2018.00116
19. Buczak AL, Guven E (2015) A survey of data mining and machine learning methods for cyber security intrusion detection. IEEE Commun Surv Tutor 18(2):1153–1176
20. Meidan Y et al (2017) Detection of unauthorized IoT devices using machine learning techniques. arXiv preprint arXiv:1709.04647
21. Meidan Y et al (2017) ProfilIoT: a machine learning approach for IoT device identification based on network traffic analysis. In: Proceedings of the symposium on applied computing
22. Doshi R, Apthorpe N, Feamster N (2018) Machine learning DDoS detection for consumer internet of things devices. In: 2018 IEEE security and privacy workshops (SPW). https://doi. org/10.1109/spw.2018.00013
23. Hindy H et al (2020) Machine learning based IoT intrusion detection system: an MQTT case study. arXiv preprint arXiv:2006.15340. https://doi.org/10.1016/j.iot.2019.100059
24. Alrashdi I et al (2019) AD-IoT: Anomaly detection of IoT cyberattacks in smart city using machine learning. In: 2019 IEEE 9th annual computing and communication workshop and conference (CCWC). IEEE
25. Hasan M, Milon Islam M, Islam I, Hashem MMA (2019) Attack and anomaly detection in IoT sensors in IoT sites using machine learning approaches. Internet Things 100059

26. Chhabra GS, Singh VP, Singh M (2020) Cyber forensics framework for big data analytics in IoT environment using machine learning. Multimed Tools Appl 79(23):15881–15900
27. Fang He, Qi A, Wang X (2020) Fast authentication and progressive authorization in large-scale IoT: how to leverage AI for security enhancement. IEEE Netw 34(3):24–29

Chapter 56
A Fast Fault Identification and Classification Scheme for Series Compensated Transmission Lines

Deepika Sharma, Shoyab Ali, and Gaurav Kapoor

1 Introduction

Among the miscellaneous equipment's of power transmission and distribution system, the relaying of transmission lines connecting two power stations through series-compensation is very enforced since the series compensated transmission lines (SCTLs) are principally liable to the happening of faults. Classification of fault in SCTLs is essential to supply proficient and standard power flow.

In [1], a two-ended model-free travelling wave-based fault location scheme for SCTL is presented. In [2], an unbalanced current protection relay based on discrete wavelet transform (DWT) for discriminating between the interior faults and exterior faults is proposed. In [3], a high impedance arc fault (HIAF) discovery technique in the distribution system using empirical mode decomposition (EMD) and artificial neural network (ANN) is introduced. In [4], an online fault detection technique based on WRC-SDT (wavelet noise reduction, Clarke transform, Stockwell transform and decision tree) is proposed for a deep-sea offshore wind farm (9 MW) connected transmission line. In [5], a fault-revealing technique for thyristor-controlled series compensator (TCSC) connected transmission lines throughout the power swing situation is described. The technique is designed using Clarke's transform and teager Kaiser energy operator (TKEO). In [6], a fault location estimation technique for unified power flow controller (UPFC) compensated transmission line using one end voltage and fast discrete orthogonal sparse transform (FDOST) is reported. In [7], a scheme based on travelling wave (TW) and game theory (GT) for TCSC connected transmission lines is presented, and hence, the concept of fault detection/classification

D. Sharma · S. Ali
Department of Electrical Engineering, Vedant College of Engineering and Technology, Bundi, India

G. Kapoor (✉)
Department of Electrical Engineering, Modi Institute of Technology, Kota, India

© The Author(s), under exclusive license to Springer Nature Singapore Pte Ltd. 2021   755
S. Kumar et al. (eds.), *Proceedings of International Conference on Communication and Computational Technologies*, Algorithms for Intelligent Systems,
https://doi.org/10.1007/978-981-16-3246-4_56

has been introduced for TCSC connected transmission line protection. DWT (discrete wavelet transform, Db4 mother wavelet) and mathematical morphological filter (MMF) are used as feature extraction tools. In [8], Hilbert–Huang transform (HHT)-based fault identification and fault classification scheme using average relay energy index (AREI) of the current and voltage signals for power transmission line protection is presented. In [9], travelling wave-aided intrinsic time decomposition-based two-ended fault localization scheme for UPFC (100-MVA and 4-pulse) compensated line is proposed. In [10], supervised relevance vector machine-based dynamic disturbance classification scheme is designed for SCTLs. In [11], a differential protection technique based on the differential phase angle of superimposed current and using decision tree classifier is proposed. In [12], a disturbance detection scheme in SCTL using discrete Fourier transform and sequence space-aided support vector machine (SVM) classifier is presented.

In this paper, a quick differential protection technique is proposed based on DFT-DDWT. The paper is further arranged as. The detailed description of series compensated transmission system (SCTS) connected with differential relay is provided in Sect. 2. The proposed methodology is presented in Sect. 3. The effects of the simulation studies are reported in Sect. 4. A comparative investigation with the other approaches is presented in Sect. 5. The paper is concluded in Sect. 6.

2 Test System of SCTS Connected with Differential Relay

To calculate the implementation of the proposed DFT-DDWT scheme, a distinctive 400-kV series compensated transmission system (SCTS) as displayed in Fig. 1 is implemented for the simulation studies. The simulation studies were carried out using MATLAB/SIMULINK. The model (as demonstrated in Fig. 1) under consideration consists of a 300-km three-phase transmission line with series capacitor banks (SCBs) connected near Bus-A and Bus-B. The SCTL is segregated into two zones of equal length of 150 km both. The first current measurement unit is connected between Bus-A and SCB-1. The second current measurement unit is connected between SCB-2

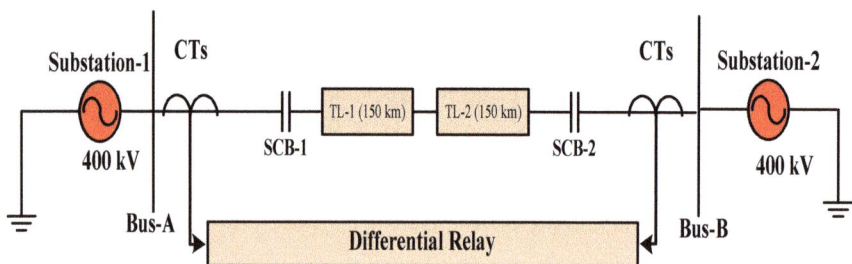

Fig. 1 A two-substation SCTS with a differential relay

and Bus-B. It can be exemplified from Fig. 1 that the SCTL is fed from substation-1 at the sending end and from substation-2 at the receiving end. Substation-1 and substation-2 relate to Thevenin's corresponding of the two power substations.

3 Proposed Scheme for Differential Relaying of SCTL

The DFT-DDWT-based differential relaying scheme (Fig. 2) for SCTL is summarized as follows:

Step-1: Measure three-phase currents (in per-unit) at Bus-A and Bus-B of series compensation-based transmission line.

Step-2: Analyse fault currents using DFT to calculate the fundamental coefficients (FCs) of Bus-A and Bus-B fault currents.

Step-3: Further process the FCs using DDWT and compute the approximate coefficients of DDWT.

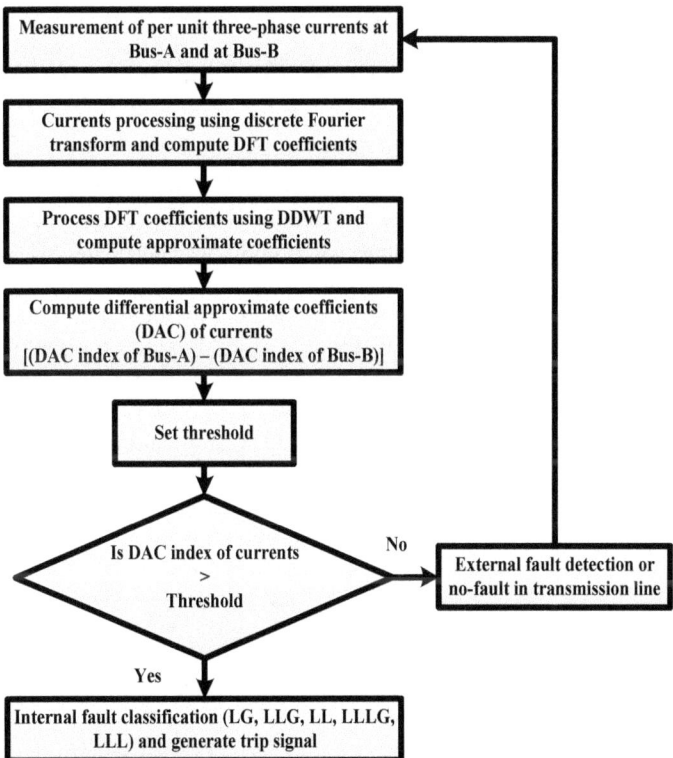

Fig. 2 Proposed differential protection scheme for SCTS

Step-4: Compute the differential approximate coefficients (DACs) of faults currents of Bus-A and Bus-B, i.e. [(DAC index of Bus-A)-(DAC index of Bus-B)].

Step-5: Calculate the amplitude of threshold value (THV) followed by performing various simulation studies.

Step-6: Confirm whether the DAC index of fault currents is greater than the THV. If yes then the differential relay will indicate the classification of internal faults, and it will produce the trip signal. If the DAC index of fault currents is lower than the THV, then the differential relay will indicate the classification of external faults in SCTL.

In this work, the amplitude of THV is preferred as '3' per-unit because in this work the fault currents are measured in per-unit.

4 Simulation Results and Discussions

The functioning of scheme presented in this work is verified on the series compensation-based transmission line model and checked for various fault events.

4.1 Performance During One Capacitor Bank in Operation at a Time

The performance of DFT-DDWT-based design is checked when one unit of capacitors is in operation and the other unit is not in operation, i.e. when one capacitor bank is connected and the other bank is disconnected with SCTL. Hence, the SCTL is set to operate under unsymmetrical fault situation; in precise, a BG fault is operated at a location 150 km from substation-1 on TL-1 of the SCTL with first unit of capacitors in operation and second unit not in operation. Figure 3 depicts the plot of current waveform, and Fig. 4 displays the plot of DACI, achieved in this fault case. The fault is created at 0.1 s, and it is verified that phase B is detected by the protective relay quickly at 0.11 s. The relay took 5 ms for the classification of faulty phase-B. The fault switching time (FST) is 0.1 s for all cases, and the R_{FAULT} and R_{GROUND} values are 12 Ω and 14 Ω, respectively. The DAC indices values of various faults are introduced in Table 1. It is evident that the proposed method can detect all faults accurately.

4.2 Performance During Sudden Load Switching Situation

The behaviour of system is studied under unexpected load switching situation in SCTS by switching a load of 485 MW and 395 MVAr in SCTL on Bus-B. The load

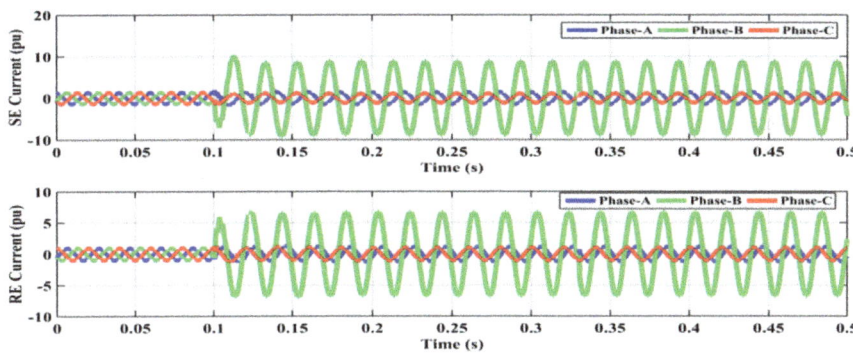

Fig. 3 Current waveforms captured at Bus-A and Bus-B during BG fault at 0.1 s when capacitor unit-1 in operation and capacitor unit-2 not in operation

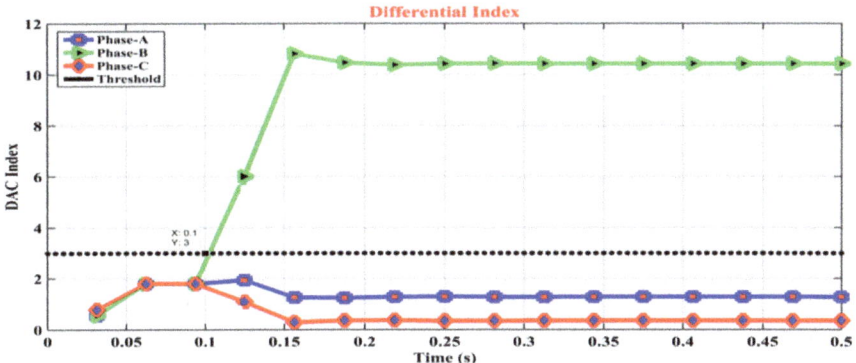

Fig. 4 Performance during BG fault at 0.1 s when capacitor unit-1 in operation and capacitor unit-2 not in operation

Table 1 DAC indices for a single unit of capacitors operation

Fault case	SCB unit	DAC index amplitudes			Fault classification time (ms)		
		DAC-A	DAC-B	DAC-C	FCT-A	FCT-B	FCT-C
BG	Unit-1 ON	1.9415	10.8225	1.8023	–	5	–
ACG	Unit-2 ON	5.1334	1.0555	8.5035	5	–	10
AG	Unit-1 ON	11.1400	1.7978	2.2240	0	–	–
AB	Unit-2 ON	9.4445	3.8737	1.0600	10	20	–
ABCG	Unit-1 ON	19.0117	18.1416	19.3654	5	5	5

is suddenly switched on Bus-B at 0.15 s. The plot of current waveform is presented in Fig. 5. Trip signals initiated by the relay for connected load are given in Fig. 6. During the switching of load in SCTL, the DACI indices of phases A, B and C should be inferior to the limit of threshold. Here, the trip signals of A, B and C are below the threshold limit indicating the faultless situation. The calculated DACI values are presented in Table 2. It can be obviously investigated from Table 2 that the DACI indices of phases A, B and C in all the test cases are lower than the threshold limit. It is clear that the effect of scheme is specific during unexpected load switching situation.

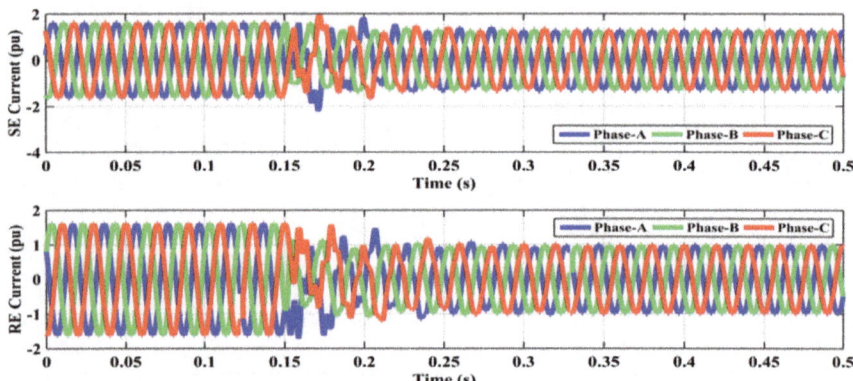

Fig. 5 Current waveforms captured at Bus-A and Bus-B when a load of 485 MW, 395 MVAr is unexpectedly connected to Bus-B at 0.15 s

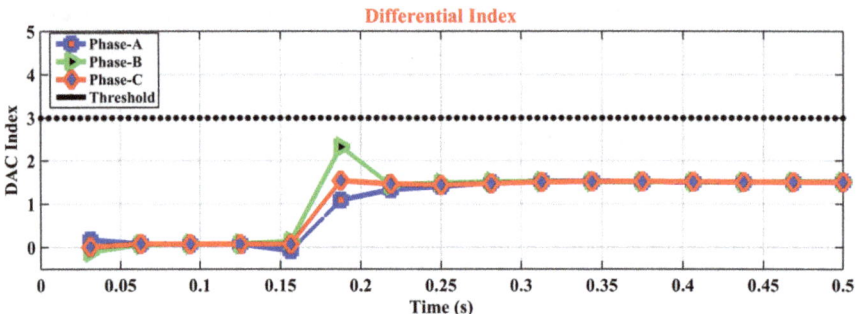

Fig. 6 Performance when a load of 485 MW, 395 MVAr is unexpectedly connected to Bus-B at 0.15 s

Table 2 DAC indices for sudden load switching situation

Active power of Load (MW)	Reactive power of load (MVAr)	FST (s)	DAC index amplitudes			Fault classification time (ms)		
			DAC-A	DAC-B	DAC-C	FCT-A	FCT-B	FCT-C
485	395	0.15	1.5319	2.3272	1.5458	–	–	–
865	690	0.25	1.5381	1.5224	1.5741	–	–	–
1075	875	0.35	1.7008	1.5532	1.6633	–	–	–
580	490	0.15	1.5344	2.1175	1.5294	–	–	–
665	365	0.10	1.6996	1.6063	1.6812	–	–	–

4.3 Performance During Two Different Types of Faults in Two Sections of SCTL

Test system is set to activate when two different faults occur on SCTL in two sections of the SCTL. The plots of current waveform and DACI of the proposed technique for fault (AG fault at 85 km on TL-1 and BG fault at 115 km on TL-2) simulated at 0.25 s on the SCTL are shown in Fig. 7 and Fig. 8, respectively. Here, DACI amplitudes of phase-A and phase-B are beyond the threshold and the DACI amplitude of phase-C is below the threshold representing the incidence of AG and BG faults in two different sections of the SCTL. For all the simulated cases, the R_{FAULT} value is taken as 9 Ω and R_{GROUND} value is taken as 12 Ω, respectively. It is demonstrated that the fault is detected rapidly at the same time, i.e. at 0.25 s, which specifies that the proposed scheme took 0 ms time for detection of AG and BG multiple section faults. Further, the DACI values obtained for different fault cases are presented in Table 3. The operating time of the relay has been computed based on the commencement

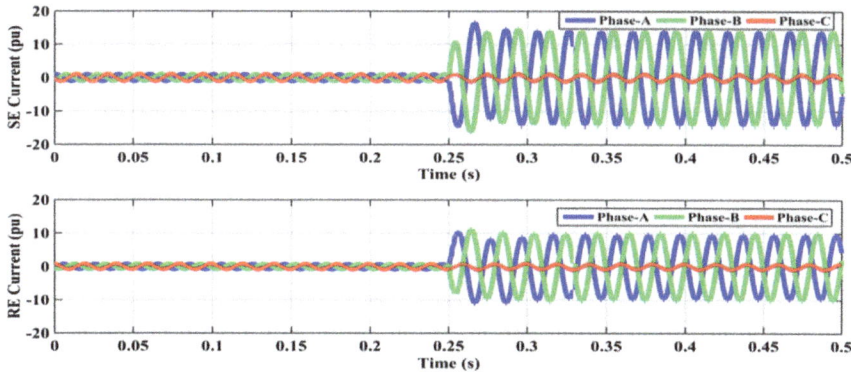

Fig. 7 Current waveforms captured at Bus-A and Bus-B during AG fault at 85 km and BG fault at 115 km switched at 0.25 s

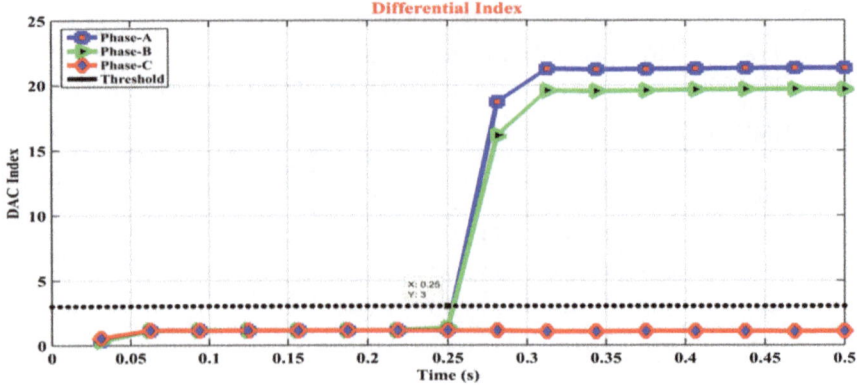

Fig. 8 Performance during AG fault at 85 km and BG fault at 115 km switched at 0.25 s

Table 3 DAC indices for faults in two different sections

Fault case-1 (km)	Fault case-2 (km)	DAC index amplitudes			Fault classification time (ms)		
		DAC-A	DAC-B	DAC-C	FCT-A	FCT-B	FCT-C
AG (85)	BG (115)	21.3384	19.7071	1.1471	0	0	–
CG (130)	AG (70)	5.1080	1.1434	7.8482	15	–	5
BG (120)	CG (80)	1.1463	10.0278	6.8541	–	5	10
AG (92)	CG (108)	17.1567	1.1423	19.2008	5	–	5
CG (125)	BG (75)	1.1469	8.4369	5.3342	–	5	15

of occurrence of fault, and the trip command activated by the protective relay is presented in Table 3. Thus, it is observed that the responsiveness of the protective relay is unaffected by the occurrence of multiple section faults on SCTL.

4.4 Performance During Unexpected Opening of Phase of Circuit Breaker

The consequence of unexpected opening of the circuit breaker phase on the operation of the protective relay is tested by opening the phase of a circuit breaker at a specific time of Bus-B under the occurrence of two-phase with ground (LLG) and two-phase (LL) faults in TL-2 of the SCTS. The SCTL is set to operate under unsymmetrical fault condition; in specific, ABG fault is created at 150 km as of substation-1 in between TL-1 and TL-2 of the SCTS. Now, suddenly, the contact of phase-B of the circuit breaker-2 (CB-2) gets opened at 0.25 s. Figure 9 shows the plot of current waveform and Fig. 10 demonstrates the plot of DACI, achieved in this fault case. The ABG fault is created at 0.15 s, and it is demonstrated that the fault in both phase-A and phase-B are detected at 0.16 s correspondingly. It can be clearly perceived as of Fig. 10 that the protective relay effectively detected the incidence of ABG fault on SCTL even under the unexpected opening of contact of phase-A of circuit breaker CB-2. The protective relay took 10 ms for detection of faulty phases A and B, respectively. The values of fault switching time, R_{FAULT} and R_{GROUND} for all fault cases are 0.15 s, 4 Ω and 7 Ω, respectively. The DACI values obtained for various fault cases are given in Table 4. It is obvious that the protective relay can detect these faults correctly.

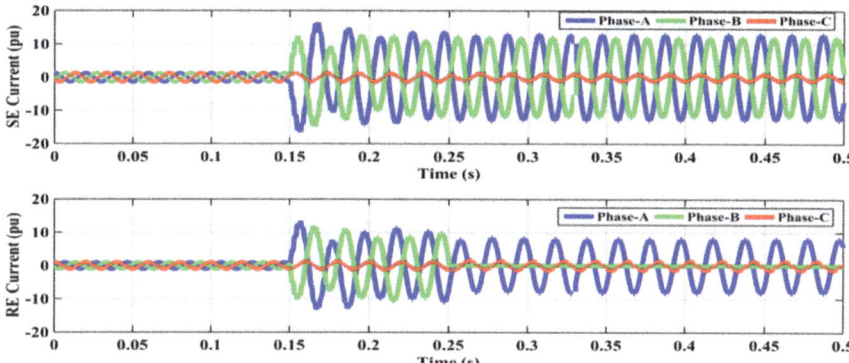

Fig. 9 Current waveforms captured at Bus-A and Bus-B during ABG fault at 0.15 s and sudden opening of phase-B of circuit breaker-2 at 0.25 s

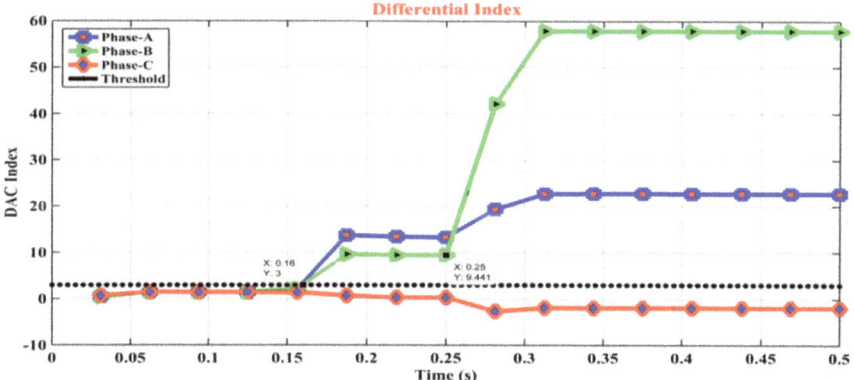

Fig. 10 Performance during ABG fault at 0.15 s and sudden opening of phase-B of circuit breaker-2 at 0.25 s

Table 4 DAC indices for sudden opening of circuit breaker phase

Fault case	FST (s)	Opened phase of CB-2	DAC index amplitudes			Fault classification time (ms)		
			DAC-A	DAC-B	DAC-C	FCT-A	FCT-B	FCT-C
ABG	0.15	B (0.25 s)	22.7687	57.8475	1.5009	10	10	–
ACG	0.15	A (0.2 s)	57.9973	1.4974	22.7515	10	–	10
BCG	0.15	C (0.35 s)	1.5074	22.8130	57.8248	–	10	10
AB	0.15	A (0.30 s)	73.3187	19.1936	2.2079	5	5	–
AC	0.15	C (0.32 s)	19.1578	2.5644	73.2847	10	–	10

5 Comparative Investigation with the Other Approaches

In this section, a comparative exploration has been done with some other approaches of differential protection of transmission line. Table 5 contains the comparative evaluation. The technique shown in [5] is designed with three-phase fault currents of single-circuit transmission line connected with thyristor-controlled series capacitor (TCSC). Clarke's transform and TKEO are utilized for the protection of TCSC line. The travelling wave and game theory-based technique are demonstrated in [7] for the protection of single-circuit transmission line with TCSC, but it necessitates huge calculation inconvenience owing to preparation and analysis event. In [10],

Table 5 Comparative evaluation

Literature							
Parameters for comparison	[5]	[7]	[10]	[11]	[12]	Proposed scheme
Input data	Current	Current and voltage	Current	Current	Current	Current
Scheme	Clarke's transform + TKEO	TW + game theory	SRVM	Decision tree	DFT + SVM	DFT + DDWT
Sudden CB operation	No	No	No	No	No	Yes
Load switching	Yes	No	Yes	Yes	Yes	Yes
Multi-location faults	No	No	No	No	No	Yes
Capacitor bank operation	No	No	No	No	No	Yes

CB circuit breaker, *Yes* considered and *No* not considered

the SRVM-based characteristics of the currents, which are extorted from the relays connected on the terminals of the transmission line, are employed for the protection of transmission line connected with fixed series capacitors (FSCs).

In [11], decision tree-based protection scheme for the transmission line connected with fixed series compensation (FSC) is presented, but the scheme deals with disputes to recognize the faults with fault resistance greater than 300 ohms. In [12], DFT and support vector machine (SVM)-based method are planned for the FSC compensated single-circuit line, but the relay response time is more and the scheme also deals with difficulty to detect the faults with fault resistance greater than 50 ohms. The proposed scheme in this manuscript is authenticated on the three-phase fixed series compensator connected transmission line with unrelated characteristics of power transmission system which are not examined in [5, 7, 10, 11, 12] as described in Table 5. It is noticeable that the method recommends recognition of multi-location faults, sudden opening of circuit breaker contacts, load switching, when one capacitor bank is in operation out of two, and discrimination between external and internal faults as compared with the other schemes.

6 Conclusion

In this work, a differential scheme for series compensation-based transmission system (SCTS) is presented. The DFT and DDWT were used as the fault current processing and decomposition tools correspondingly for the revealing of fault and categorization of faulted phase in transmission system with series compensation. The scheme is supported on the difference of the two-phase fault current of each equivalent phase at both buses of the power system. MATLAB software was used for verifying the performance of the presented scheme for SCTS under various fault circumstances. The simulation effects demonstrated that the protective relay based on DFT and DDWT has the capability to detect the fault and discriminate between external and internal faults. The simulation outcomes substantiate that the protective relay powerfully recognizes the fault and the faulty phase concerned in a fault within a half-cycle time.

References

1. Naidu OD, Pradhan AK (2020) Model Free traveling wave based fault location method for series compensated transmission line. IEEE Access 8:193128–193137
2. Lertwanitrot P, Ngaopitakkul A (2020) Discriminating between capacitor bank faults and external faults for an unbalanced current protection relay Using DWT. IEEE Access 8:180022–180044
3. Lala H, Karmakar S (2020) Detection and experimental validation of high impedance arc fault in distribution system using empirical mode decomposition. IEEE Syst J 14(3):3494–3505

4. Wang XD, Gao X, Liu YM, Wang YW (2020) WRC-SDT based on-line detection method for offshore wind farm transmission line. IEEE Access 8:53547–53560
5. Kumar ML, Sai S, Kumar J, Mahanty RN (2020) Fault detection during power swing in a TCSC-compensated transmission line based on Clark's transform and Teager–Kaiser energy operator. Iran J Sci Technol Trans Electr Eng 1–14
6. Khoramabadi, Hamid RS, Keshavarz A, Dashti R (2020) A novel fault location method for compensated transmission line including UPFC using one-ended voltage and FDOST transform. Int Trans Electri Energy Syst 30(6):–31
7. Khalili M, Namdari F, Rokrok E (2020) Traveling wave-based protection for TCSC connected transmission lines using game theory. Int Trans Electri Energy Syst 30(10):1–25
8. Anand A, Affijulla S (2020) Hilbert-Huang transform based fault identification and classification technique for AC power transmission line protection. Int Trans Electri Energy Syst 30(10):1–15
9. Mishra S, Gupta S, Yadav A (2020) Intrinsic time decomposition based fault location scheme for unified power flow controller compensated transmission line. Int Trans Electri Energy Syst 30(11):1–21
10. Patel U, Chothani N, Bhatt P (2020) Supervised relevance vector machine based dynamic disturbance classifier for series compensated transmission line. Int Trans Electri Energy Syst 1–17
11. Taheri MM, Seyedi H, Nojavan M, Khoshbouy M, Ivatloo BM (2018) High-Speed Decision tree based series-compensated transmission lines protection using differential phase angle of superimposed current. IEEE Trans Power Deliv 33(6):3130–3138
12. Patel UJ, Chothani NG, Bhatt PJ (2018) Sequence-space-aided SVM classifier for disturbance detection in series compensated transmission line. IET Sci Meas Technol 12(8):983–993

Chapter 57
Predicting the Performance of an Electric Submersible Pump Using Recurrent Networks of Long Short-Term Memory (LSTM)

I. V. Karakulov, V. Yu. Stolbov, and A. V. Kluiev

1 Introduction

Equipment failure during operation leads to a reduction in the number of products produced and financial losses. Therefore, it is important to ensure continuous operation of the equipment during extraction of minerals. For this purpose, monitoring systems, routine checks of the equipment condition, and various types of diagnostics are used. Most tools for predicting breakdowns require an expert who must have extensive experience with this type of the equipment. It is also possible to develop an expert system that will store the generalized knowledge of experts [1]. At the same time, it should be taken into account that training experts require a large amount of time and financial costs for their training. There are other ways to predict breakdowns, such as methods based on processing and analyzing existing data. Analysis of historical data allows us to predict the condition of the equipment based on examples and dependencies observed in cases of breakdowns that have already occurred during the operation of the equipment. The best option, in our opinion, is to use the experience of experts and analysis of historical data based on artificial intelligence methods to predict the technical condition of the equipment.

Electric submersible pumps (ESP) are used for mechanized oil production. They are part of a unit for pumping reservoir fluid from oil wells. The depth of the well can reach 7000 m. Since the ESP is located inside the well, maintenance and repair work requires removing it to the surface, finding the fault, repairing or replacing the pump, and loading it back into the well. Technical work can take months, which reduces the profitability of the well. Often, oil wells are located in hard-to-reach places, which also increases the time for equipment repairs. Predicting breakdowns will allow the equipment to work as long as possible and reduce losses when the well is idle.

I. V. Karakulov (✉) · V. Yu. Stolbov · A. V. Kluiev
Perm National Research Polytechnic University, 29 Komsomolsky prospekt, Perm 614990, Russia
e-mail: karakuloviv@yandex.ru

© The Author(s), under exclusive license to Springer Nature Singapore Pte Ltd. 2021   767
S. Kumar et al. (eds.), *Proceedings of International Conference on Communication and Computational Technologies*, Algorithms for Intelligent Systems,
https://doi.org/10.1007/978-981-16-3246-4_57

Each ESP uses monitoring sensors. Sensors take various indicators such as pressure, temperature, vibration, and current when analyzing data obtained using sensors, we can predict the state of the system and diagnose the future failure. For example, as noted in [2], round-the-clock real-time monitoring of the ESP operation is performed at the field under study, which helps to increase the service life of the equipment. Increased service life is achieved by preventing misuse of the pump and tracking excessive loads.

2 Analysis of Solutions

Expert systems based on knowledge and fuzzy logic [1, 3–5] and artificial intelligence (AI) methods and predictive analytics are used for automated assessment of the state of ESP nodes [6–9]. Well-known expert systems use a production model of knowledge representation, which allows to predict the failure of the ESP and determine the reasons for which the failure occurred, by obtaining and comparing trends obtained from sections of the time series of values, measured parameters of the equipment operation, with a certain set of rules that establish acceptable changes in these parameters. The main difficulty in developing such information systems is the formation of an up-to-date knowledge base based on the experience of experts and data on past failures [1].

The use of AI methods should allow for a real-time comprehensive assessment of the state of the platform as a whole and individual equipment in particular. In other words, the created intelligent information system (IIS) should aggregate real-time data on the technical condition of the equipment received from numerous sensors, as well as determine the real need for repair and maintenance work. However, there are only isolated cases of the introduction of AI methods in the oil and gas industry [10, 11]. For example, as notes in [10], the Norwegian oil company Equinor in 2018 announced the creation of a data collection and processing center to improve the management decision-making process. Many large Russian oil and gas companies are also interested in creating such IIS. However, for the development and implementation of IIS in the practice of oil production, it is necessary to work out the applicability of existing AI methods and, in the case of a positive research result, to develop models based on them that, using information received from the equipment in real time, determine its technical state and predict possible breakdowns nodes and failures in the near future. To build such models, based, for example, on neural network technologies, large amounts of data obtained from sensors over a long period of operation are needed, as well as information about the breakdowns that have occurred and the equipment repairs carried out. In addition, in order to identify possible anomalies in the operation of the equipment, it is necessary with the help of experts to accumulate the necessary knowledge, on the basis of which it is possible to form criteria for the onset of various anomalies by multifactorial analysis of data received from sensors for a certain period.

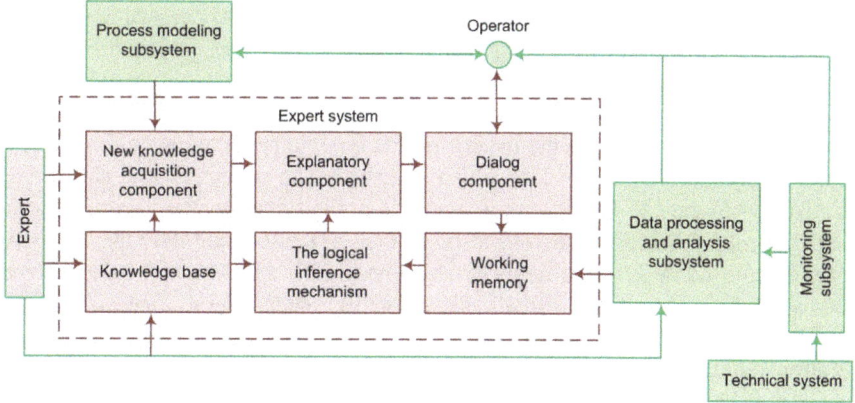

Fig. 1 Diagram of an intelligent information system

Therefore, the purpose of this work is to study the issues of using artificial intelligence methods to assess the technical condition of electric submersible pump units based on deep processing of current data using neural network technologies.

3 IIS Architecture

The module that predicts the technical state of the ESP is built into the IIS architecture (see Fig. 1). IIS work can be divided into three stages. The first stage is monitoring, which receives data from the technical system and stores the received parameter values in the database. At the second stage, the values stored by the monitoring subsystem are processed. The analysis of values is carried out using a data processing and analysis subsystem, an expert system, and a process modeling subsystem. At the third stage, the expert makes the final decision. The expert's decision can be based both on the data obtained from the expert system and the data processing and analysis subsystem, and on personal experience. It should be noted that as anomalies in the operation of the equipment are detected and the necessary data and knowledge are accumulated, the role of the expert system will decrease. This will be possible in the case when trained neural networks will accurately predict possible equipment breakdowns in almost real time without involving experts or an expert system.

4 Architecture of an Artificial Neural Network

According to the author [12], the most popular and widely used classes are autoregressive and neural network forecasting models. Autoregressive methods like ARIMA

[13] and its modifications require the determination of a set of adjustable parameters, the determination procedure of which is not unambiguous. In addition, these methods do not have the necessary degree of flexibility.

Models based on an artificial neural network (ANN) are the most flexible to the nature of time series. Among the models of this type, in applications to forecasting problems, deep LSTM networks are distinguished. The paper [14–17] shows the effectiveness of the LSTM network in predicting oil production in comparison with traditional methods. LSTM proved to be 8–37% more accurate than the ARIMA model when used in different fields in China and India. LSTM was 17–29% more accurate [18], compared to a modification of the decline curve analysis (DCA) method widely used in the oil industry. LSTM networks were also used to analyze vibration and determine the fault tolerance of the equipment in [19–21], where they showed good results.

Therefore, in this paper, the neural network modeling method was chosen as the main method for predicting the state of the ESP. Moreover, the LSTM network was chosen as an artificial neural network (ANN). This network is a modern recurrent ANN that can learn long-term dependencies and preserve the context of historical data for a long time [22, 23].

The following network architecture was used: three hidden LSTM layers, three dropout layers, and one dense layer. Despite the fact that the neural network architecture looks simple, it performed well when predicting time series [24]. A schematic representation of the final architecture is shown in (see Fig. 2).

Fig. 2 Architecture of the neural network

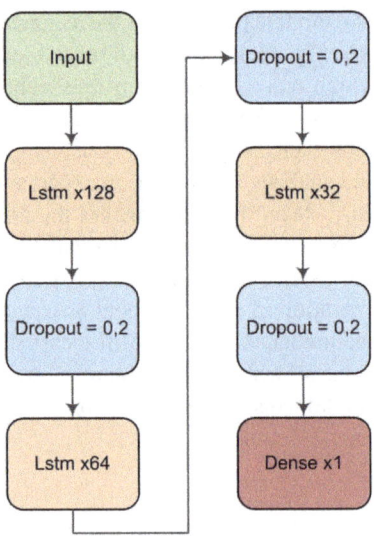

5   Processing of Initial Data

To analyze the state of the ESP, an array of data was obtained from one of the oil production platforms located on the Pechersk sea shelf. Studied data obtained from producing wells, each of which is characterized by the following set of parameters: value (for each phase) of three-phase power supply voltage, V; vibration acceleration along the X-axis, m/s^2, vibration acceleration along the Y-axis, m/s^2. The source of information for all these parameters is the telemetry sensors of the monitoring system, the signals from which are recorded in the form of minute-by-minute values.

As an example of time series analysis, consider the y-axis vibration acceleration data (see Fig. 3) shows a time series, the abscissa of which is the date of data acquisition, and the ordinate is the value of vibration acceleration. Analyzing the initial data, we can conclude that the pump was turned off for a month in 2016, for a month and a half from 2017, and for a week at the end of 2018. The fact of shutdown is also confirmed by data displaying the mains voltage from similar sections of time. It can be noted that after switching on the ESP has a different operating mode, which is due to peaks in vibration values, immediately after the moments of switching off.

Figure 3 shows that the data on the vertical vibration of the pump have a significant spread due to the presence of periods of its stop and a certain number of points with large deviations in the measured values. Therefore, it was decided to remove the places where the phase voltage was zero from the training sample and clear the sample of peaks that are clearly out of normal operation during the preliminary processing of the initial data. One of the data transformations was to reduce the sample size by averaging values over the clock. When getting the values that described the pump operation for the next day, the minute-by-minute measurement required predicting 1440 values, and when averaging the data by hour, it was necessary to find 24 values.

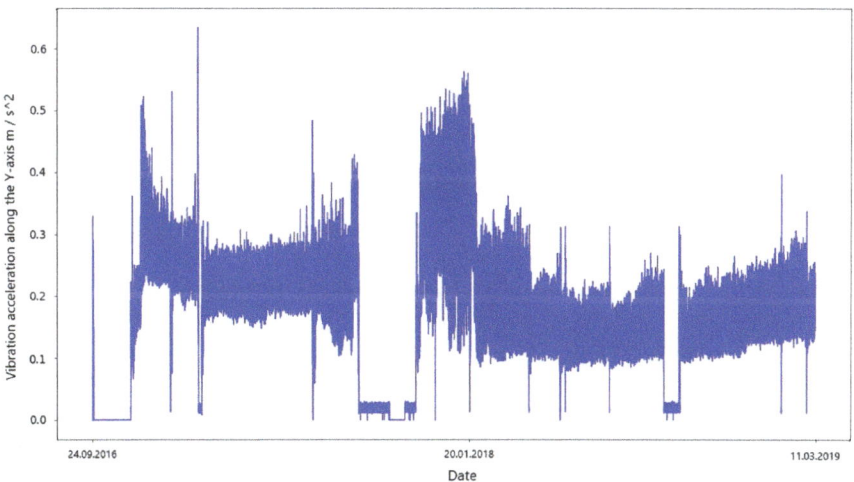

Fig. 3 Y-axis vibration acceleration graph (per minute data)

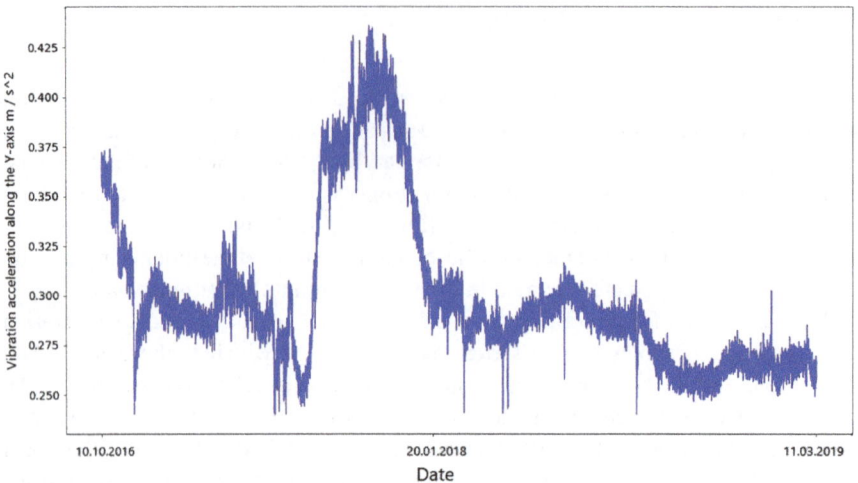

Fig. 4 Processed time series of Y-axis vibration acceleration data (hourly data)

Reducing the number of predicted values will reduce the error for long periods of time. The transformation results are shown in Fig. 4.

Figure 4 shows that due to the hourly averaging of the initial data, it was possible to somewhat smooth out the existing outliers in the data, while preserving all the features of the time series.

6 Neural Network Modeling

Studies on the possibility of forecasting using LSTM neural networks for a short period of time (up to 3 days) were conducted. To train the neural network, we used data of vibration acceleration along the Y-axis in the period from 24.09.2016 for 11.03.2019. The size of the training set was 18,425 elements. The accuracy of the neural network model was estimated using the indicator "Average absolute error in percent," the formula for calculating which has the form:

$$\text{MAPE} = \frac{1}{N} \sum_{t=1}^{N} \frac{\left| Z(t) - \hat{Z}(t) \right|}{Z(t)} * 100\% \tag{1}$$

where N the number of counts of the series, t the discrete time, and $Z(t)$ the values of the original series, $Z(t)$ predicted values.

Predicted time series of pump vibration values for approximately 5 days (from 11.03.2019 to 17.03.2019) (see Fig. 5). For prediction, 12 values from the source data were used in the form of a time window, which was shifted at each step by

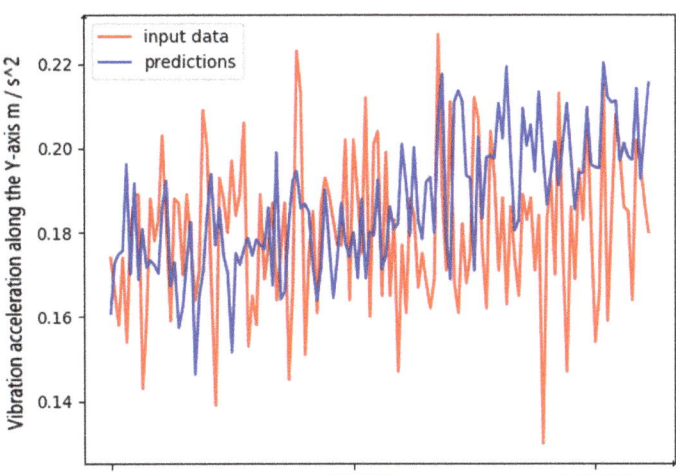

Fig. 5 Forecasting a time series for the next 135 values

one value, and each time the value predicted by the neural network was added. The following 135 hourly pump vibration values were predicted. The percentage of the average error of the trained network over the entire studied time interval (MAPE) was: 11.23%. Figure 5 is marked: input data—real vibration data received from the sensor, and predictions—data predicted by the network. It can be seen that as the prediction interval increases, the error increases and reaches 30% at the end of the interval.

The number of epochs and the number of iterations in one epoch were increased by one and a half times when training the neural network, to improve the accuracy of the forecast. When training on 1500 epochs, the number of iterations in one epoch is 500. For training, a running window of 12 values was also used, the output of the network was equal to one value. Twelve values from the original data were used for prediction, the time window was shifted at each step by one value, and each time the value obtained by the neural network was added. The following 135 values were predicted. The percentage of average error during testing (MAPE) was 10.34%. Figure 6 uses the following notation: input data, predictions-data predicted by the network.

Therefore, additional research to train neural networks and improve the accuracy of the forecast was conducted. To do this, the neural network architecture was changed and the forecast period was reduced to 3 days. Two training options were considered. Training in both cases was performed on an NVIDIA gtx 1080 ti video card, an Intel Core i7 processor, and 16 GB of RAM. For the first option, the training took 4 h (12 s per epoch). For the second option, the training time was 11 h (26 s per epoch).

In the first version, training was performed on 1100 epochs, and the number of iterations in one epoch was 200. We obtained the following error backpropagation

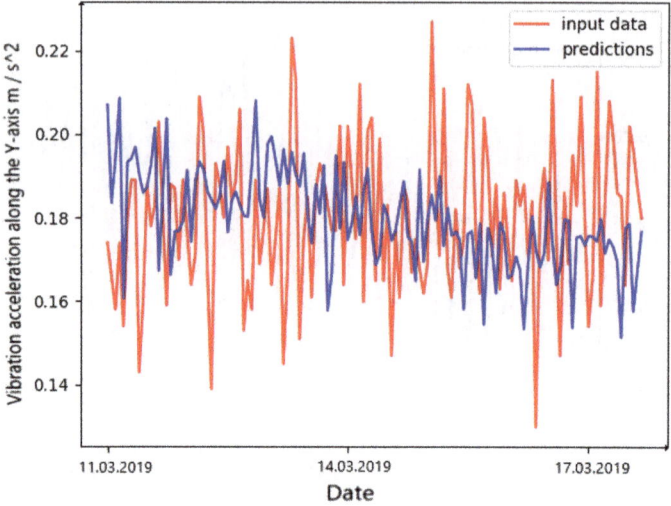

Fig. 6 shows that the accuracy increased over the entire forecast interval, but the maximum deviations of the results could reach 20%, which is hardly acceptable when predicting breakdowns in real time.

function (Fig. 7), where the epoch number is shown on the abscissa axis, and the value of the error function is shown on the ordinate axis. For training, a running window of 12 values was also used, the output of the network was equal to one hourly value

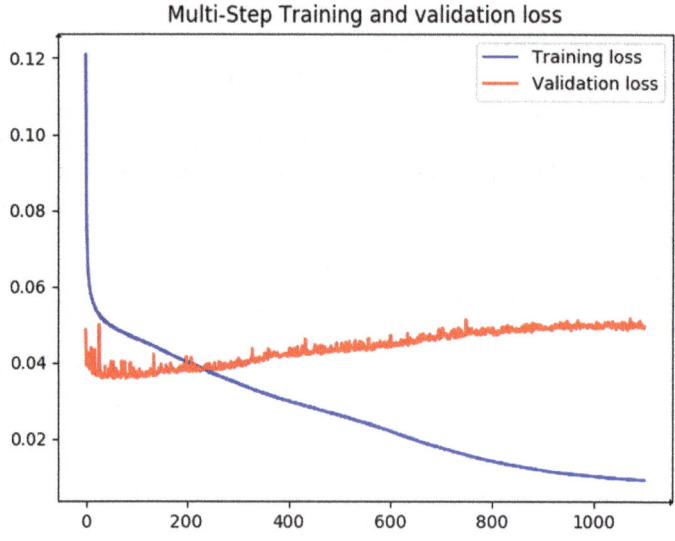

Fig. 7 Error backpropagation function

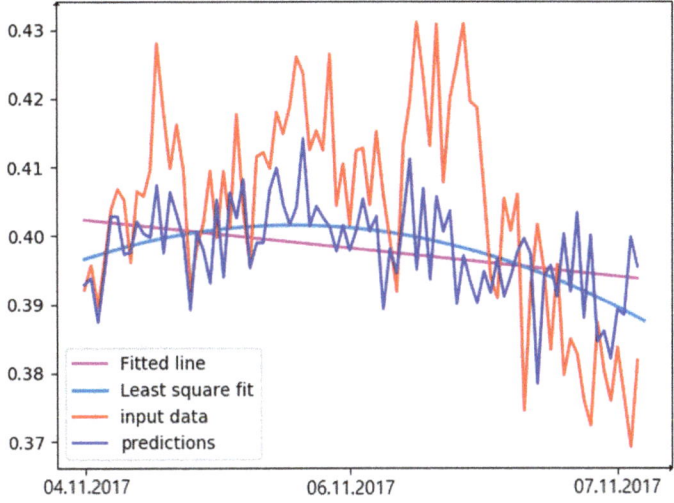

Fig. 8 Forecasting the time series for the next 84 values

of the pump vibration. The figure shows a graph of the error during training in blue and in red during validation.

Time series values were predicted (see Fig. 8). The percentage of average error over the entire time interval (MAPE) was 2.79%. The figure uses the following notation: input data, predictions-data predicted by the network. Figure 8 also shows the predicted trends (linear—purple and nonlinear—green).

Figure 8 shows that the nonlinear trend is predicted quite accurately, but the predicted vibration values at some points are very different from the real ones. Therefore, the second training option with an increased number of iterations up to 500 at each epoch, the number of which was also increased to 1500 was considered. The trained neural network shows good forecast results as shown in Fig. 9. The percentage of the average error over the entire time interval (Test MAPE) was only 2.54%. In the figure: input data, predictions-data received by the network. The figure also shows graphs of linear and nonlinear trends.

Figure 9 shows that the trained neural network almost accurately predicts vibration values in the first 12 h of time (the average deviation of the predicted value from the real value is less than 0.5%), which is a very good result for the neural network.

Further, the error grows and reaches 4% at the end of the forecast interval (see Fig. 10). This result can be considered acceptable in terms of forecast accuracy for the first day, but network training requires a lot of time.

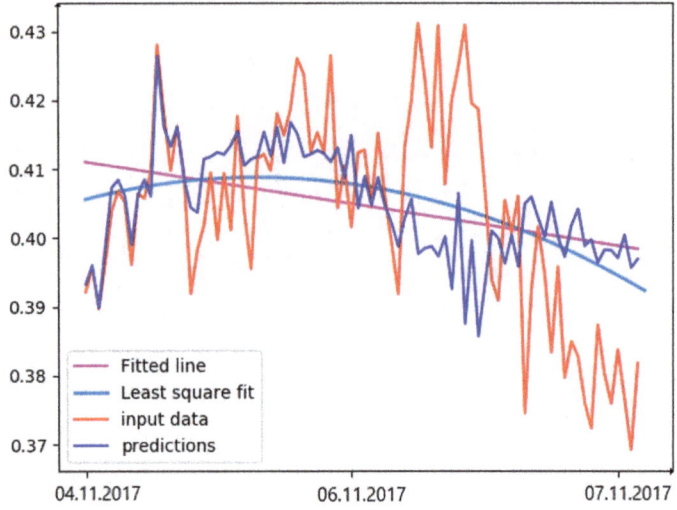

Fig. 9 Forecasting the time series for the third day

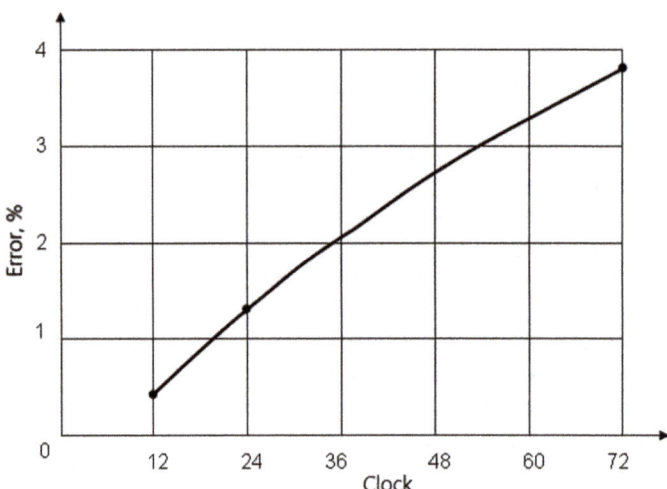

Fig. 10 Change in the average error on the forecast time interval

7 Conclusion

In this work, we tested a method for predicting the vibration of an electric submersible pump using recurrent networks of long short-term memory. Data obtained during the operation of an oil well was used to predict the state of the ESP. Before training the LSTM network, the training data was cleared of peak values that were clearly out of line with the indicators of the main engine operation mode. The values were not considered during periods of time when the voltage was not applied to the unit. When predicting the state of the system for 5 days, the average error was 10.34%. Due to the low prediction accuracy, it was decided to increase the size of the LSTM network and reduce the interval for forecasting. The analysis of the results showed that when predicting the pump operation for the 3rd day, good accuracy is achieved only for the first 12 h, and then the accuracy of the predicted data decreases. At the same time, the neural network captures the trend well within the time series, which indicates the possibility of using neural networks together with an expert system based on the knowledge of experts presented in the form of a set of rules. Determining the trend will allow the expert system to more accurately predict the behavior of the equipment and thereby reduce the cost of its maintenance.

References

1. Istomin DA, Stolbov VY, Platon DN (2020) Expert system for assessment of technical condition of electric centrifugal pump assemblies based on productive presentation of knowledge and fuzzy logic, vol 20(1). Bulletin of the South Ural State University. Ser. Computer Technologies, Automatic Control, Radio Electronics, 2020 pp 133–143 (in Russ.) https://doi.org/10.14529/ctcr200113
2. Camilleri LAP, Macdonald J (2010) How 24/7 real-time surveillance increases ESP run life and uptime. Soc Petrol Eng https://doi.org/10.2118/134702-MS
3. Grassian D, Bahatem M, Scott T, Olsen D. (2017) Application of a fuzzy expert system to analyze and anticipate ESP failure modes. Soc Petrol Eng https://doi.org/10.2118/188305-MS
4. Orchard B (2004) FuzzyCLIPS Version 6.1 d User's Guide. National Research Council of Canada, 82 p
5. Riley G (2001). CLIPS: a tool for building expert systems [online]. GHG Corporation. Available at: http://www.ghg.net/clips/CLIPS.html. Last Accessed 19 May 2019
6. Yan Y, Yan J (2018) Hands-on data science with Anaconda: utilize the right mix of tools to create high-performance data science applications. Packt Publishing-Ebooks Account, 364 p
7. Bruskin SN (2017) Models and tools of predicting analytical research for digital corporation. Vestnik of the Plekhanov Russian Univ Econ 5:135–139 (in Russ.)
8. Okunev AA (2020) Functional data preprocessing application to oil-transfer pumps vibration parameters forecasting. Appl Mathe Control Sci 3:51–72. https://doi.org/10.15593/2499-9873/2020.3.03. (inRussian)
9. Faizullin RV, Hering S (2020) Cointegration analysis method for fault detection based on sensor data. Appl Mathe Control Sci 4:49–6. https://doi.org/10.15593/2499-9873/2020.4.04 (inRussian)
10. Lipatov A (2016) The first complex of predictive analytics in Russia for power and industrial equipment Exposition Oil Gas 4:82–84 (in Russ.)
11. Skobeev A, Maganov D, Roodny V (2020) Umnaya platform a. [Smart platform]. Available at: https://hbr-russia.ru/innovatsii/tekhnologii/803089. Last Accessed 11 June 2020
12. Chuchueva IA (2012) Model' prognozirovaniia vremennykh riadov po vyborke maksimal'nogo podobiia. Dis. kand. tekhn. nauk [Time series prediction model for maximum similarity sample. Cand. Sci. Diss.]. Moscow, 155 p
13. Sahin O, Stewart RA, Faivre G, Ware D, Tomlinson R, Mackey B (2019) Spatial Bayesian Network for predicting sea level rise induced coastal erosion in a small Pacific Island. J Environ Manage 238:41–351 https://doi.org/10.1016/j.jenvman.2019.03.008
14. Bretherton FP, Davis RE, Fandry C (1976) A technique for objective analysis and design of oceanographic experiments applied to MODE-73. In: Deep sea research and oceanographic abstracts, vol 23(7), Elsevier, pp 559–582

15. Tealab A (2018) Time series forecasting using artificial neural networks methodologies: A systematic review. Fut Comput Inf J 3(2):334–340
16. George EPB, Gwilym MJ (1970) Time series analysis forecasting and control. Holden-Day, San Francisco, 553 p
17. Sagheer A, Kotb M (2019) Time series forecasting of petroleum production using deep LSTM recurrent networks Neurocomputing 323:203–213
18. Zollanvari A, Kunanbayev K, Akhavan Bitaghsir SS, Bagheri M (2021) Transformer fault prognosis using deep recurrent neural network over vibration signals. IEEE Trans Instrument Measure 70:1–11, 2021, Art no. 2502011. https://doi.org/10.1109/TIM.2020.3026497
19. ElSaid A, Wild B, Higgins J, Desell T (2016) Using LSTM recurrent neural networks to predict excess vibration events in aircraft engines. In: 2016 IEEE 12th international conference on e-science (e-science). Baltimore, MD, pp 260–269 https://doi.org/10.1109/eScience.2016.7870907
20. ElSaid A, Wild B, Jamiy FE, Higgins J, Desell T (2021). Optimizing LSTM RNNs using ACO to predict turbine engine vibration. In: Proceedings of the genetic and evolutionary computation conference companion (GECCO '17). Association for Computing Machinery, New York, NY, USA, 21–22. https://doi.org/10.1145/3067695.3082045
21. ElSaid A, Desell T, Jamiy FE, Higgins J, Wild B (2018) Optimizing long short-term memory recurrent neural networks using ant colony optimization to predict turbine engine vibration. ArXiv, abs/1710.03753
22. Yasnitsky LN (2016) Intellektual'nyye sistemy [Intelligent Systems]. Moscow, Laboratoriya znaniy Publ., 221 p
23. Sepp H, Jürgen S (1997) Long short-term memory. Neural Computat J 9(8):1735–1780
24. Netbay GV, Oniskiv VD, Stolbov VY, Karimov RR (2020) Management of a local urban heat supply system based on neural network modeling taking into account the weather forecast, vol 20(3). Bulletin of the South Ural State University. Ser. Computer Technologies, Automatic Control, Radio Electronics, pp. 29–38 (in Russ.) https://doi.org/10.14529/ctcr200303

Chapter 58
Energy Efficient Optimized Channel Estimation for Spectrum Sensing in Cognitive Radio Network

M. A. Usha Rani and C. R. Prashanth

1 Introduction

In recent decades, the cognitive radio has gained more attention among the researchers, because it is a promising technology for 5G networks that enables the secondary (unlicensed) users to share the unused spectrum of the primary (licensed) users for enhancing the utilization of spectrum [1, 2]. In addition, the wireless resources are significantly utilized by the secondary users in the overlay cognitive radio systems for achieving high-system performance without disturbing the licensed users [3]. In the wireless communication, the cooperative diversity is an emerging technique that has gained extensive interest, where multiple channels are utilized to communicate the same information symbol [4]. The reliability of the cognitive radio system is significantly improved by combining the cooperative diversity with cognitive radio system [5]. In the overlay cognitive radio systems, the data rate of the unlicensed users is increased in several resource allocation schemes that cause severe interference to the licensed users [6]. In the previous research works, the stochastic resource allocation techniques were also used for multi-channel cognitive radio system, where the sum rate of unlicensed user is increased, while reducing the interference constraints on licensed user [7].

In most of the prior researches, hybrid-cum-improved spectrum access [8], adaptive absolute SCORE algorithm [9], deep neural network [10] are used for resource allocation and spectrum sensing. These algorithms require more number of iterations and also hindering the real time operation. Currently, the use of optimization

M. A. U. Rani (✉) · C. R. Prashanth
Department of Electronics & Telecommunication Engineering, Dr. Ambedkar Institute of
Technology Bengaluru, Bengaluru, India
e-mail: Usharanima.et@drait.edu.in

C. R. Prashanth
e-mail: prashanthcr.et@drait.edu.in

S. Kumar et al. (eds.), *Proceedings of International Conference on Communication
and Computational Technologies*, Algorithms for Intelligent Systems,
https://doi.org/10.1007/978-981-16-3246-4_58

algorithm in cognitive radio systems has gained more popularity in the fields-like smart antennas, multimedia, military network, etc. In this research, ABC optimization algorithm is used to choose the optimal channels that help to improve the spectrum sensing of cognitive radio communication system. Finally, the existing and proposed system performance were analyzed in light of BER value. The major contributions of the research article are presented as follows:

1. ABC algorithm identifies the best channel spectrum to reconstruct efficient spectrum sensing that improves the signal to noise ratio (SNR) of cognitive radio system.
2. With the help of ABC optimization algorithm, the effective relay is identified that helps in transferring the unused spectrum to the secondary users.

Contribution
In the experimental segment, the proposed cognitive radio network is evaluated in terms of bit error rate (BER) by varying the channels (Rayleigh and Rician) and the number of transmitters and receivers such as 4×4 and 2×2 with the channel estimations minimum mean squared error (MMSE), zero forcing (ZF), maximum-ratio combining (MRC) is proposed. From the experimental investigation, the optimized channel estimation achieved reduced error rate compared with others and better BER value.

Motivation
In the proposed system, a new cognitive radio network is proposed. Artificial bee colony (ABC) optimization algorithm is used to detect the occupied and unoccupied bands of secondary users and primary users. Without causing any overlay to the licensed user, assigns the unoccupied bands to the unlicensed users.

Organization of the paper
The paper is organized as follows. Some recent research papers in cognitive radio communication system are surveyed in Sect. 2. Explanation about the proposed cognitive radio system is discussed in Sect. 3. Experimental evaluation of the proposed cognitive radio system is represented in Sect. 4. At last, conclusion is stated in section 5.

2   Literature Survey

Bansal and Rattan [13] presented an optimization methodology for fulfilling the objectives of cognitive radio communication system. The user needs to optimize several objectives of cognitive radio communication system to enhance the quality of service. In this research, modified whale optimization (MWO) algorithm was developed for designing an effective cognitive radio communication system. The modified algorithm utilizes a weight vector on the location of hump-back whales to accomplish exploitation and exploration steps in the search space. From the experimental investigation, the developed optimization method (MWO) attained better performance compared to the existing algorithms in light of throughput and BER estimation. The developed research study was not concentrated on the spectrum usage efficiency that was essential in developing a dynamic cognitive radio communication system.

Hei et al. [14] analyzed about multiple input multiple output (MIMO) cooperative spectrum sensing optimization. By optimizing the dissimilar weights assigned to the received signal of cognitive radios, the probability of the detection was maximized. In this literature, the statistical properties of cooperative spectrum sensing system were assessed for the primary users with single and multiple antennas. In this work, the genetic algorithm was replaced with convex methods to identify the optimal weight vectors without convexity constraints. The experimental consequence shows that the genetic algorithm improved the reliability of cooperative spectrum sensing system related to the convex approaches. In addition, the genetic algorithm was stable and efficient, which exhibits promising result in sensing characteristics. In contrast, the genetic algorithm does not scale well with noise ratio that was considered as one of the major concerns.

Tang et al. [15] presented an improved algorithm for dynamic spectrum allocation in cognitive radio communication system. In this work, improved particle swarm optimization (IPSO) was developed for optimizing the rate and power allocation in cognitive radio system. The non-convex issue was completely solved by combining the mutation searching behavior with PSO. The experimental investigation shows that the developed optimization algorithm delivers high-quality solution to the non-convex optimization issues. The major concern in the developed work was related to whether the cognitive procedure needs to be implemented in a distributed or centralized fashion.

Bhowmik and Malathi [16] developed a new algorithm for spectrum sensing in cognitive radio communication system. In this literature, neural network and Krill-Herd Whale optimizer were utilized for performing spectral sensing in cognitive radio system. Usually in the channel communication, the channel model depends on the transmitter and the input signal. In this work, the developed algorithm determines the unoccupied spectrum that assigns the free spectrum bands to the users, which helps in the reduction of delay. The experimental analysis shows that the developed algorithm attained decent performance in cognitive radio systems in light of detection and false alarm probability. A major concern in the developed algorithm was to model the network layer and medium access control sub layer for cognitive radio system.

In cognitive radio, the secondary and primary networks work simultaneously over the similar spectrum band. Alizadeh et al. [17] identified that the shortest link between the primary receiver and transmitter was not available, so spatial modulation was employed in the secondary network for splitting the transmission space into two spatial domains and amplitude phase modulation (APM). In the developed system, the secondary transmitter re-transmits the primary symbols in AMP domain without causing any interference to the primary receiver. The simulation consequence shows the importance of spatial modulation in cognitive radio and the experiment was validated in light of symbol error probability. For spectrum sensing, delivering a better optimization based solution in cognitive radio was a major concern.

3 Proposed System

In recent times, the rapid growth of wireless communication has created a huge impact on the deployment of new wireless networks in both unlicensed and licensed frequency bands that makes the present spectrum scarcity challenge even worse. In order to deal with the absence of spectrum resources, a new cognitive radio network is proposed in this research paper. In the proposed cognitive radio communication system, ABC optimization algorithm is employed for selecting the optimal channels to enhance the spectrum sensing and also to eliminate the additive white gaussian noise (AWGN) from the original data. The proposed cognitive radio communication system includes following phases; data generation, turbo encoder, modulation quadrature amplitude modulation (QAM) and quadrature phase shift keying (QPSK), pilot insertion, inverse fast fourier transform (IFFT), cyclic prefix, channel (Rician and Rayleigh), removal of cyclic prefix, FFT, channel estimation (MMSE, ZF and MRC), demodulation, turbo decoder and BER optimization using ABC algorithm. The work flow of the proposed cognitive radio communication system is shown in Fig. 1.

3.1 *Input Data and Turbo Encoder and Decoder*

The turbo encoder is accomplished to encode the input signals. The turbo encoder contains two identical recursive systematic convolutional (RSC) encoders, which are partitioned by an inter leaver with the range of $r = 1/2$, where r is represented as code rate. The output of the turbo encoder comprises of systematic outputs from the RSC encoders and the input data. Then, the systematic output from the RSC encoders are eliminated, because it is similar to each other. Thus, the overall code rate becomes $r = 1/3$, as graphically presented in Fig. 2.

Where, C_1 and C_2 are denoted as first RSC encoder outputs and C_3 is indicated as second RSC encoder output. Hence, the turbo decoder receives the inputs from primary and secondary user receiver for decoding the collected signal. The turbo decoder decodes the signal by using maximum likelihood detection (MLD). The turbo decoder comprises of two maximum a posterior (MAP) decoders, which are partitioned by an inner inter leaver. The inter leaver of turbo encoder is identical as the inter leaver of turbo decoder. The packet loss is minimized by using the inter leaver in the turbo decoder, because it controls the erroneous packets during data transmission. The architecture of turbo decoder is graphically indicated in Fig. 3.

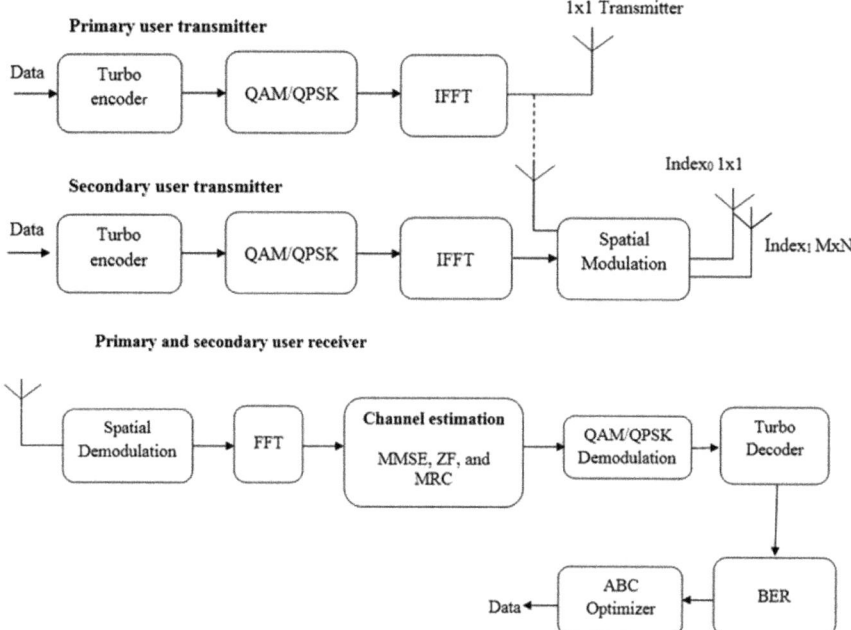

Fig. 1 General work flow of proposed cognitive radio system

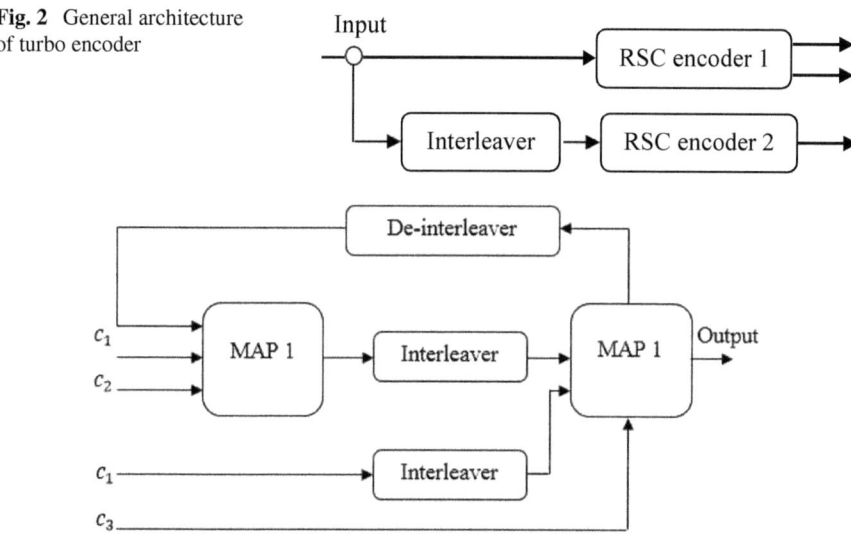

Fig. 2 General architecture of turbo encoder

Fig. 3 Architecture of turbo decoder

3.2   *Modulation*

The modulation is accomplished by using QAM and QPSK modulations. Generally, the combination of high- and low-frequency carrier signal is named as modulation that is mathematically indicated in Eq. (1).

$$f(t) = A \sin(wt + \emptyset) \tag{1}$$

where, A is stated as amplitude or magnitude, θ is denoted as phase angle, and ω is indicated as frequency. These three parameters (amplitude, phase angle and frequency) are altered based on the transmit data. The main aim of the undertaken modulation techniques not only transmit the signals through channels but also help in attaining the function with minimum bandwidth and power efficiency.

A.   IFFT and cyclic prefix modulation

The modulation techniques are selected on the basis of channel requirements. The role of IFFT in cognitive radio communication system is to modulate the sub channels into suitable carriers. The general equation to calculate IFFT/FFT is mathematically denoted in the Eq. (2).

$$X(k) = \sum_{n=0}^{N-1} x(n) e^{-j2\Pi k/N} \tag{2}$$

$$0 \le k \le N - 1$$

where, $x(n)$ is indicated as nth time sample that ranges from 0 to $N-1$, N is represented as number of sample points in the discrete fourier transform (DFT) and $X(k)$ is denoted as DFT frequency output at the kth spectral point that ranges from 0 to N-1. Meanwhile in cognitive ratio communication system, the multipath channel reflections are very high so cyclic prefix is included to reduce the interference between the licensed and unlicensed users.

B.   BER Optimization using ABC Algorithm

The BER value from the channel estimations is optimized by utilizing ABC algorithm. It is a new meta- heuristic algorithm that imitates the behavior of honey bee, where a set of honey bees is named as swarm. The ABC algorithm includes three types of honey bees such as scout bees, onlooker bees and employed bees. Usually, the employed bees search for the food resources in their memory and then reveal the information about the food resources to the onlooker bees. Then, the onlooker bees choose the good food resources that are found by the employed bees. The scout bees are translated for some employed bees that abandon their food resources and search for the new ones [18]. In ABC optimization algorithm, the first part of the swarm comprises of employed bees, and the second part of the swarm consists of onlooker bees. Therefore, the number of onlooker and employed bees is similar to the number of solutions in the swarm. At first, the undertaken optimization algorithm randomly distributes the initial population of swarm size.

Let $X_i = \{x_{i,1}, x_{i,2} \ldots x_{i,D}\}$ denotes the ith solution in the swarm, where D is stated as dimension size. Each employed be X_i creates a new solution V_i in the neighborhood of the current position that is mathematically indicated in Eq. (3).

$$v_{i,j} = x_{i,j} + \emptyset_{i,j} \times (x_{ij} - x_{k,j}) \tag{3}$$

where, $\emptyset_{i,j}$ is indicated as random number that ranges from $[-1, 1]$, X_k is denoted as randomly chosen candidate solution $(i \neq k)$, and j is represented as random dimensional index, which is selected from the set $\{1, 2, \ldots, D\}$. In this algorithm, a greedy selection is applied to generate a new candidate solution V_i. Update the parent value X_i with V_i, if the fitness value of V_i is higher than the parent value X_i, or else keep the same parent value. After completing the searching process, the employed bees reveal the information about the food resources with the onlooker bees. Then, the onlooker bees validate the information of nectar and selects the food resources with a probability associated to its nectar amount. In this algorithm, the probabilistic selection is done by utilizing wheel selection methodology that is denoted in Eq. (4).

$$pi = \frac{fit_i}{\sum_{j-1}^{\text{Swarm size}} \text{fit}_j} \tag{4}$$

where, fiti is stated as fitness function of the ith solution. The higher probability of the food resource ith is selected to better the solution i. Then the food resource is abandoned X_i, if the position is not enhanced over a predetermined number of cycles. At last, the scout bee identifies a new food source by using Eq. (5).

$$x_{i,j} = lb_j + \text{rand}(0, 1) \times (ub_j - lb_j) \tag{5}$$

where, ub_j and lb_j are represented as upper and lower boundaries of the jth dimension, r and $(0, 1)$ is indicated as random number. Pseudo code of ABC algorithm is given below,

Pseudo code of ABC algorithm
Initialize the population
$$X_i = \{x_{i,1}, x_{i,2} \dots x_{i,D}\}$$
Calculate the fitness value of the population and set cycle to one; repeat
For every employed bee
Generate new solution $v_{i,j}$ by using equation   (3)
Evaluate the fitness value
Employ greedy selection procedure fit($v_{i,j}$)
Determine the probability variables p_i for the solution X_i by utilizing equation (4)
End for
For every onlooker be
Choose a solution X_i on the basis of P_i
Generate new solution $V_{i,j}$
Evaluate the fitness value fit($v_{i,j}$)
Employ greedy selection procedure
End for
If there is an abandoned solution for the scout bee;
Replace the solution i with new solution by using equation (5)
End if
Remember the best solution; so far
Cycle = cycle + 1
Output: Best solution is achieved

4 Performance Analysis

The MATLAB (2018a) software was utilized for simulating the proposed cognitive radio communication system. In this paper, the proposed system performance is analyzed with dissimilar channels (Rayleigh and Rician), channel estimations (MMSE, ZF and MRC) and with dissimilar number of transmitters and receivers such as 2 × 2 and 4 × 4 for validating the efficacy of the proposed cognitive radio communication system. In addition, the performance of the proposed communication system is calculating using BER estimation. In BER estimation, the amount of bit errors bits that was mathematically represented in Eq. (6). Similarly, the parameter setting of the proposed cognitive radio communication system is specified in Table 1.

$$BER = \frac{\text{Number of errors}}{\text{Total number of bits transferred}} \qquad (6)$$

Table 1 Simulation Parameters setting of proposed cognitive radio communication system

Number of transmitters	2×2 and 4×4
Number of receivers	2×2 and 4×4
Modulation	QPSK, and QAM
FFT size	64
Fixing the block size for cyclic prefix	16 and 32
Noise environment	AWGN

As shown in Table 1, the simulation parameters have been set for implementing the research work by considering the transmitters, receivers as listed above .The simulation part is explained below.

4.1 Quantitative Analysis

In the proposed cognitive radio communication system, modulation is done by utilizing QAM and QPSK modulations. In this research study, initially 96 random bits are generated for QPSK modulation and 192 random bits are generated for QAM modulation, because it includes four quadrature phases. In this scenario, totally 4800 symbols are generated in the proposed cognitive radio communication system. The output of QPSK and QAM modulations are in the form of 192×1. Then, turbo encoder, IFFT and cyclic prefix are used to avoid the interference between the primary and secondary users. Meanwhile in the receiver end, FFT generates output in the form of 192x4800. In cognitive radio communication network, AWGN is added to the signal transmission by propagating the signal through AWGN fading environment, where the noise ratio ranges from -20 to 20 db.

In this study, the performance of the proposed cognitive radio communication system is analyzed with dissimilar channel estimation techniques such as ZF, MRC, MMSE, and ABC optimization. Table 2 and Fig. 4 represent the BER estimation of proposed cognitive radio system for Rayleigh channel with $T_x = 2$ and $R_x = 2$. From the experimental investigation, the optimized channel estimation achieved better

Table 2 BER analysis of the proposed cognitive radio system for Rayleigh channel with $T_x = 2$ and $R_x = 2$

BER estimation for $T_x = 2$ and $R_x = 2$ (Rayleigh channel)								
Channel estimation	-20	-15	-10	-5	0	5	10	15	20
ZF	0.497	0.476	0.481	0.419	0.339	0.230	0.149	0.103	0.106
MRC	0.491	0.471	0.434	0.375	0.282	0.170	0.080	0.036	0.019
MMSE	0.487	0.465	0.433	0.373	0.282	0.170	0.080	0.033	0.019
ABC optimization	0.486	0.464	0.432	0.369	0.277	0.164	0.074	0.027	0.010

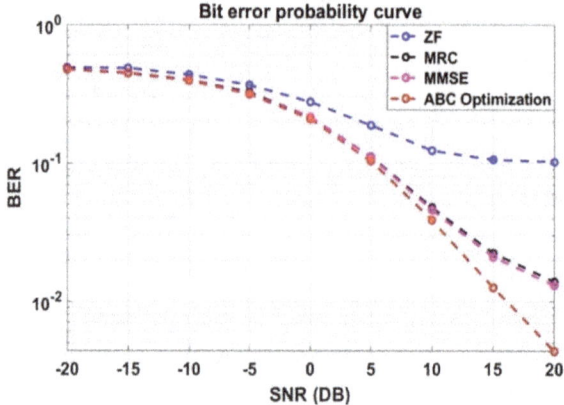

Fig. 4 Graphical comparison of the proposed cognitive radio system for Rayleigh channel with T_x $= 2$ and $R_x = 2$

BER rate compared to other available channel estimations. Hence, the optimization algorithm effectively identifies the unused spectrum of the licensed user and then assigns the free spectrum to the unlicensed user. This process helps in the reduction. Table 2 represents the BER analysis of proposed cognitive radio system for Rayleigh channel with $T_x = 2$ and $R_x = 2$ for SNR values between -20 and $+20$ db. The channel estimation techniques are zero forcing, maximum ratio combining, minimum mean square error and optimization algorithm proposed is rtificial bee colony optimization. The zero forcing channel estimation between SNR values of 0 to -20 db achieved BER rates from 0.339 to 0.497, for SNR values of 0 to $+20$ db achieved BER rates from 0.339 to 0.106. The Maximum ratio combining channel estimation between SNR values of 0db to -20 db achieved BER rates from 0.282 to 0.491 , for SNR values of 0db to $+20$ db achieved BER rates from 0.282 to 0.019. The minimum mean square error channel estimation between SNR values of 0 to -20 db achieved BER rates from 0.282 to 0.487, for SNR values of 0 to $+20$ db achieved BER rates from 0.282 to 0.019. The artificial bee colony optimization between SNR values of 0 to -20 db achieved BER rates from 0.277 to 0.486, for SNR values of 0 to $+20$ db achieved BER rates from 0.277 to 0.010.

The corresponding values graphical representation of the proposed cognitive radio system for Rayleigh channel $T_x = 2$ and $R_x = 2$ for SNR values between -20 db and $+20$ db is plotted shown in Fig. 4. It is observed from plot that bit error rate decreases with increasing SNR. The value of BER is minimum in case of ABC optimization as compared to that of ZF, MRC and MMSE.

4.2 Quantitative Analysis

Similarly, Table 3 and Fig. 5 indicate the BER estimation of the proposed cognitive radio system for Rician channel with $T_x = 2$ and $R_x = 2$. The graphical comparison shows the efficiency of ABC optimization algorithm in channel estimation, which attained low BER compared to other channel estimations (ZF, MRC and MMSE). Table 3 represents the BER analysis of proposed cognitive radio system for Rician channel with $T_x = 2$ and $R_x = 2$ for SNR values between -20 db to $+20$ db. The channel estimation techniques are zero forcing, maximum ratio combining, minimum mean square error and optimization algorithm proposed is artificial bee colony optimization. The zero forcing channel estimation between SNR values of 0 to -20 db achieved BER rates from 0.297 to 0.485, for SNR values of 0 to $+20$ db achieved BER rates from 0.297 to 0.098. The maximum ratio combining channel estimation between SNR values of 0 to -20 db achieved BER rates from 0.261 to 0.485 , for SNR values of 0 to $+20$ db achieved BER rates from 0.261 to 0.011. The minimum mean square error channel estimation between SNR values of 0 to -20 db achieved

Table 3 BER analysis of the proposed cognitive radio system for Rician channel with $T_x = 2$ and $R_x = 2$

BER estimation for $T_x = 2$ and $R_x = 2$ (Rician channel)									
Channel estimation	-20	-15	-10	-5	0	5	10	15	20
ZF	0.485	0.471	0.443	0.383	0.297	0.178	0.111	0.091	0.098
MRC	0.485	0.461	0.424	0.361	0.261	0.136	0.049	0.016	0.011
MMSE	0.485	0.465	0.425	0.359	0.258	0.136	0.048	0.017	0.011
ABC optimiza tion	0.484	0.463	0.422	0.354	0.251	0.129	0.040	0.008	0.002

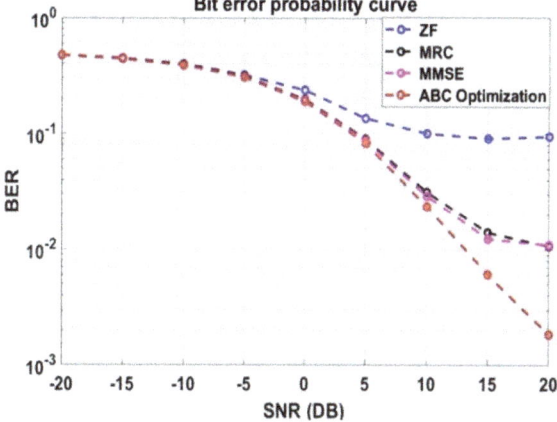

Fig. 5 Graphical comparison of the proposed cognitive radio system for Rician channel with $T_x = 2$ and $R_x = 2$

BER rates from 0.258 to 0.485, for SNR values of 0 to +20 db achieved BER rates from, 0.258 to 0.011. The artificial bee colony Optimization between SNR values of 0 to −20 db achieved BER rates from 0.251 to 0.484, for SNR values of 0db to + 20 db achieved BER rates from 0.251 to 0.002. The corresponding values graphical representation of the proposed cognitive radio system for Rician channel with $T_x = 2$ and $R_x = 2$ for SNR values between -20 to +20 db is plotted shown in figure 5. It is observed from plot that bit error rate decreases with increasing SNR. The value of BER is minimum in case of ABC optimization as compared to that of ZF,MRC and MMSE.

4.3 Quantitative Analysis

Correspondingly, Table 4 and Fig. 6 indicate the BER analysis of proposed cognitive radio system for Rayleigh channel $T_x = 4$ and $R_x = 4$. Figure 4 concluded that the BER of optimized channel estimation is very low compared to individual channel estimations. The proposed optimization algorithm effectively detects the unoccupied

Table 4 BER analysis of the proposed cognitive radio system for Rayleigh channel with $T_x = 4$ and $R_x = 4$

BER estimation for $T_x = 4$ and $R_x = 4$ (Rayleigh channel)									
Channel estimation	−20	−15	−10	−5	0	5	10	15	20
ZF	0.488	0.484	0.434	0.365	0.276	0.188	0.123	0.106	0.102
MRC	0.477	0.448	0.399	0.326	0.215	0.111	0.047	0.022	0.013
MMSE	0.477	0.447	0.398	0.319	0.214	0.110	0.045	0.021	0.013
ABC Optimization	0.474	0.444	0.394	0.315	0.208	0.103	0.038	0.012	0.004

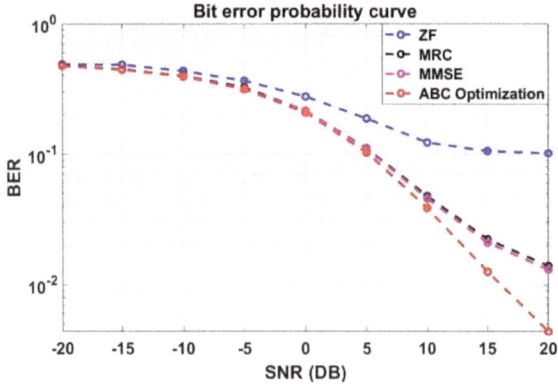

Fig. 6 Graphical comparison of the proposed cognitive radio system for Rayleigh channel with Tx = 4 and Rx = 4

and occupied bands and quickly assigns the unoccupied bands to the secondary users for avoiding overlay and delay in the communication.

Table 4 represents the BER analysis of proposed cognitive radio system for Rayleigh channel $T_x = 4$ and $R_x = 4$ for SNR values between -20 and $+20$ db. The channel estimation techniques are zero forcing, maximum ratio combining, minimum mean square error and optimization algorithm proposed is artificial bee colony optimization. The zero forcing channel estimation between SNR values of 0 to -20 db achieved BER rates from 0.276 to 0.488, for SNR values of 0 to $+20$ db achieved BER rates from 0.276 to 0.102. The maximum ratio combining channel estimation between SNR values of 0 to -20 db achieved BER rates from 0.215 to 0.477, for SNR values of 0 to $+20$ db achieved BER rates from 0.215 to 0.013.

The minimum mean square error channel estimation between SNR values of 0 to -20 db achieved BER rates from 0.214 to 0.477, for SNR values of 0db to $+20$ db achieved BER rates from 0.214 to 0.013. The artificial bee colony optimization between SNR values of 0 to -20 db achieved BER rates from 0.208 to 0.474, for SNR values of 0 to $+20$ db achieved BER rates from 0.208 to 0.004. The corresponding values graphical representation of the proposed cognitive radio system for Rayleigh channel $T_x = 4$ and $R_x = 4$ for SNR values between -20 db and $+20$ db is is plotted shown in Fig. 6. It is observed from plot that bit error rate decreases with increasing SNR. The value of BER is minimum in case of ABC optimization as compared to that of ZF, MRC and MMSE.

4.4   Quantitative Analysis

Table 5 and Fig. 7 indicate the BER performance of the proposed cognitive radio system for Rician channel $T_x = 4$ and $R_x = 4$. Here, the proposed cognitive radio system performance is compared with a few existing works such as MWO [13], genetic algorithm [14], IPSO [15] and neural network and Krill-Herd Whale optimizer [16]. Table 5 represent the BER analysis of proposed cognitive radio system for Rician channel with Tx = 4 and Rx = 4 for SNR values between -20 and $+20$ db.

Table 5 BER analysis of the proposed cognitive radio system for Rician channel with $T_x = 4$ and $R_x = 4$

BER estimation for $T_x = 4$ and $R_x = 4$ (Rician channel)									
Channel estimati on	-20	-15	-10	-5	0	5	10	15	20
ZF	0.476	0.449	0.400	0.319	0.236	0.135	0.100	0.091	0.094
MRC	0.475	0.443	0.395	0.311	0.197	0.090	0.031	0.014	0.010
MMSE	0.476	0.444	0.390	0.308	0.193	0.087	0.029	0.012	0.010
ABC optimiza tion	0.474	0.442	0.388	0.306	0.189	0.082	0.023	0.006	0.001

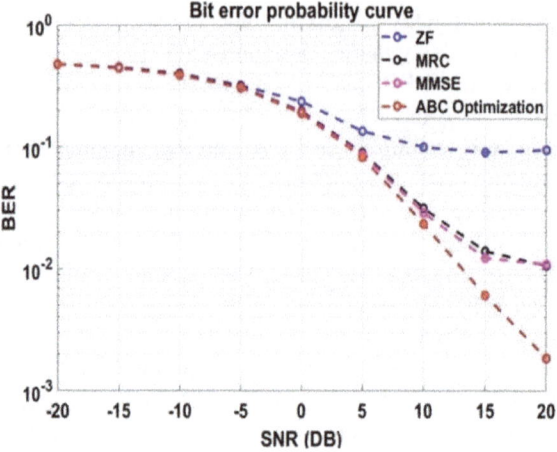

Fig. 7 Graphical comparison of the proposed cognitive radio system for Rician channel with T_x = 4 and R_x = 4

The channel estimation techniques are zero forcing, maximum ratio combining, minimum mean square error and optimization algorithm proposed is artificial bee colony optimization. The zero forcing channel estimation between SNR values of 0 to −20 db achieved BER rates from 0.236 to 0.476, for SNR values of 0 to +20 db achieved BER rates from 0.236 to 0.094. The Maximum ratio combining channel estimation between SNR values of 0 to −20 db achieved BER rates from 0.197 to 0.475, for SNR values of 0 to +20 db achieved BER rates from 0.197 to 0.010. The minimum mean square error channel estimation between SNR values of 0 to −20 db achieved BER rates from 0.193 to 0.476, for SNR values of 0 to +20 db achieved BER rates from 0.193 to 0.010.

The artificial bee colony optimization between SNR values of 0 to −20 db achieved BER rates from 0.189 to 0.474, for SNR values of 0 to +20 db achieved BER rates from 0.189 to 0.001. The corresponding values graphical representation of the proposed cognitive radio system for Rician channel with T_x = 4 and R_x = 4 for SNR values between −20 and +20 db is plotted shown in Fig. 7. It is observed from plot that bit error rate decreases with increasing SNR. The value of BER is minimum in case of ABC optimization as compared to that of ZF, MRC and MMSE.

5   Conclusion

In this research article, an energy efficient ABC algorithm is proposed for achieving cooperative multiplexing, where the primary and secondary systems co- operates each other. It is revealed that the proposed cognitive radio network uses full rate cooperative multiplexing in order to double the data rate. In this research, the ABC optimization algorithm identifies the best channel spectrum for reconstructing efficient spectrum sensing that avoids the overlap and delay during the communication. In addition, the unoccupied bands were effectively detected by ABC algorithm and the optimal tuning of the unoccupied bands was activated. Compared to the prior research works, the proposed cognitive radio network delivered a good performance in spectrum sensing by means of BER. From the experimental study, the proposed cognitive radio system averagely reduced 0.009–0.002 error rate related to the existing works. In future work, a hybrid optimization algorithm can be used to further enhance the spectrum sensing in cognitive radio network.

Acknowledgements Authors would like to thank all the staffs of Department of Electronics and Telecommunication Engineering, Principal, and Dr. Ambedkar Institute of Technology, PVP Welfare Trust for the support.

Conflict of Interest Authors declare no conflict of interest.

References

1. Khan MW, Zeeshan M (2019) QoS-based dynamic channel selection algorithm for cognitive radio based smart grid communication network. AdHoc Netw Elsevier 87:61–75
2. Alam S, Sohail MF, Ghauri SA, Qureshi IM, Aqdas N (2017) Cognitive radio based smart grid communication network. Renewable Sustain Energy Rev Elsevier 72:535–548
3. Bicen AO, Gungor VC, Akan OB (2012) Delay-sensitive and Multimedia communication in cognitive radio sensor networks. Ad Hoc Netw Elsevier 10(5):816–830
4. Rajpoot V, Tripathi VS (2019) A novel proactive handoff scheme with CR receiver based target channel selection for cognitive radio network. Phys Commun 36:1–11
5. Bhattacharya A, Ghosh R, Sinha K, Datta D, Sinha BP (2015) Non contiguous channel allocation for multimedia communication in cognitive radio networks. IEEE Trans Cognitive Commun Netw 1(4):420–434
6. El Shafie A, Khattab T, El-Keyi A (2018) Energy—efficient cooperative cognitive relaying schemes for cognitive radio networks. Phys Commun 30:179–192
7. Dhanasekaran S, Reshma T (2017) Full-rate cooperative spectrum sharing scheme for cognitive radio communications. IEEE Commun Lett 22(1):97–100
8. Thakur P, Kumar A, Pandit S, Singh G, Satashia SN (2016) Advanced frame structures for hybrid spectrum access strategy in cognitive radio communication systems. IEEE Commun Lett 21(2):410–413
9. Nareshkumar S, Bikshalu K (2019) Adaptive absolute SCORE algorithm for spectrum sensing in cognitive radio. Microprocess Microsyst 69:43–53
10. Lee W (2018) Resource allocation for multi—channel underlay cognitive radio network based on deep neural network. IEEE Commun Lett 22(9):1942–1945
11. Elhassan MA, Abd-Elnaby M, El-Dolil SA, El-Samie FEA (2019) Throughput maximization for multimedia communication with cooperative cognitive radio using adaptively controlled sensing time". Multimedia Tools Appl 78(24):34999–35025
12. Bharathi GP, Jeyanthi KMA (2018) An optimization algorithm-based resource allocation for cooperative cognitive radio networks. J Supercomput 76:1–21
13. Bansal S, Rattan M (2019) Design of cognitive radio system and comparison of modified whale optimization algorithm with whale optimization algorithm. Int J Inf Technol 12(4):1–12
14. Hei Y, Li W, Li M, Qiu Z, Fu W (2015) Optimization of multiuser MIMO cooperative spectrum sensing in cognitive radio networks. Cognitive Comput Springer 7(3):359–368

15. Tang M, Xin Y, Long C, Wei X, Liu X (2016) Optimizing power and rate in cognitive radio networks using improved particle swarm optimization with mutation strategy. Wireless Pers Commun 89(4):1027–1043
16. Bhowmik M, Malathi P (2019) Spectrum sensing in cognitive radio using actor—critic neural network with Krill Herd-Whale optimization algorithm. J Wireless Personal Commun 105(1):335–354
17. Alizadeh A, Bahrami HR, Maleki M (2016) Performance analysis of spatial modulation in overlay cognitive radio communications. IEEE Trans Commun 64(8):3220–3232
18. Xu Y, Fan P, Yuan L (2013) A simple and efficient artificial bee colony algorithm. Mathe Probl Eng 9

Chapter 59
An Insight into Handwritten Text Recognition Techniques

Shreya Tiwari, Priyanshi Burad, Netra Radhakrishnan, and Dhananjay Joshi

1 Introduction

Optical Character Recognition (OCR) has been achieving milestones in all fields by reducing labor costs and saving manpower. Nonetheless, HTR, which is a subpart of OCR, is still considered a daunting problem statement, while OCR has been considered a solved problem [1]. Compared to typed text, the high variation in handwriting types across individuals and low consistency of the handwritten text face major challenges in translating it to computer-readable text [2]. HTR could apply to multiple industries such as healthcare, insurance, and banking, or even for organizations aiding the visually impaired. In this paper, we are going to review various techniques which have been used to implement a model for detecting handwritten text from an image. Here we are also reviewing numerous papers about HTR and OCR and based on this we have written the literature review for them.

1.1 Stages of Handwritten text recognition system

These sections describes the main steps used for handwritten text recognition [3, 4] and it is illustrated as shown in Fig 1.

Preprocessing. Preprocessing is the most important stage of any deep learning model; here this stage includes the steps like removal of noise, binarization of input

S. Tiwari (✉) · P. Burad (✉) · N. Radhakrishnan
Department of Computer Engineering,, SVKM's NMIMS (Deemed-to-be-University) Mukesh Patel School of Technology Management & Engineering, Shirpur, India

D. Joshi
Department of Computer Engineering,, SVKM's NMIMS (Deemed-to-be-University) Mukesh Patel School of Technology Management & Engineering, Shirpur, India

© The Author(s), under exclusive license to Springer Nature Singapore Pte Ltd. 2021
S. Kumar et al. (eds.), *Proceedings of International Conference on Communication and Computational Technologies*, Algorithms for Intelligent Systems, https://doi.org/10.1007/978-981-16-3246-4_59

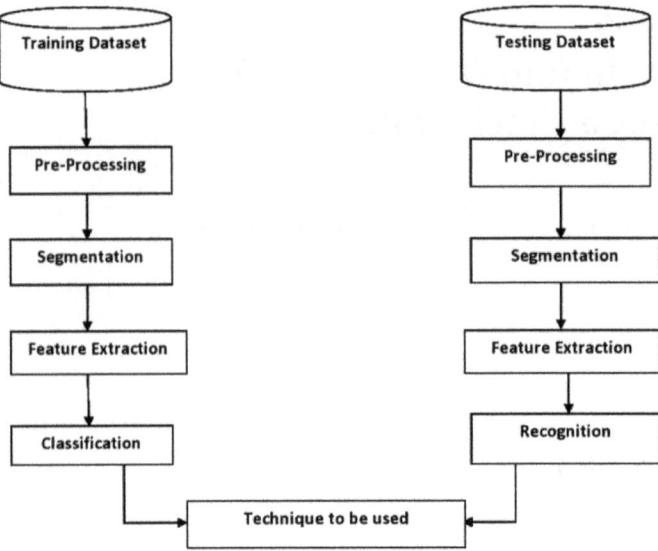

Fig. 1 Shows a representation of the various stages of handwritten text recognition using both training and testing datasets

image [3]. Here, the input character images are first transformed to grayscale images for binarization. This image is in the 0–256 range and it is then further transformed into a binary image that is, either 0 or 1[5]. This conversion decreases the structural complexity and makes it easy for the model to understand the input.

Segmentation. The image segmentation technique is used to extract various information from the image. There are three primarily observed stages of segmentation: first is line segmentation to separate each input line secondly, the words are being separated from the image of that line, the basic objective of word segmentation is to separate an image into words, and lastly, the character segmentation used to separate input image of words into characters [6, 7].

Feature Extraction. Feature extraction is the process in which the most relevant features of the input image have been identified. The feature extraction phase is being implemented after the feature preprocessing phase [5]. If the features extracted are carefully chosen then it is expected that relevant information is being fetched. It is an important phase to focus on because it will give us more insight into the efficiency of the model going to have.

Feature Selection. In feature selection, the most prominent features (attributes) are selected from the dataset [8]. Accurate feature selection results in a substantial reduction in the number of irrelevant (constant, redundant) attribute thereby, reducing the processing time and increasing the accuracy of the model without any loss of information. Feature selection can be done based on a variety of methods namely possible characters and stroke sequence matching, feature matching [4].

Classification. Classification of images recognizes the specified images and generates outputs for the identification of text. Different classification approaches are neural networks, machine learning techniques, generalized algorithms, etc. Classification of the input image is being done based on the most relevant and suitable features and out of that groups are being formed according to the features. The classification of multiple classes applies to those classification activities which have more than two class names.[9, 12] There are 4 main categories of classification namely Binary Classification, Multi-Class Classification, Multi-Label Classification, and Imbalanced Classification. For OCR, Multi-class classification is generally used. It does not include the notion of regular and irregular performance. Instead, examples are listed within a number of recognized groups as belonging to one. A multi-class grouping requires each word in the series of terms to be predicted, where the scale of the language determines the number of possible classes that can be predicted. The most popular algorithms that can be used for multi-class classification include: k-Nearest Neighbors, Decision Trees, Naive Bayes, Random Forest, and Gradient Boosting [10]. Maiwan et al. [11] compared three classification techniques namely Naive Bayes, MLP, and K-Star. They used the NIST handwritten digits and considered only 37 features among the total 256 features. Through this, they concluded that K-Star produced an accuracy of 82.36% which was the highest when compared to NB 67.04%, and MLP 78.35%.

A training set is applied in a dataset to work up a model, whereas a testing set would be used to validate the constructed model. From the training set, all the data points are omitted from the testing set. Since training is a learning phase, after it is done, the model will be able to classify the dataset. Then the testing dataset is applied to the model and this determines the accuracy of the model. If multiple techniques are trailed through this process, the technique to be used can be determined.

1.2   Techniques for Text Recognition

Machine Learning. Machine learning is an AI technology that gives systems the capability to learn and develop from experience automatically without being programmed specifically [13]. Machine learning is highly focused on the development of computer programs that can access and use information to learn on their own. Algorithms in machine learning are classified as either supervised or unsupervised learning. Some of the most used supervised machine learning algorithms are SVM, linear regression, Naive Bayes, decision trees, and logistic regression. Some of the most used unsupervised machine learning algorithms are k-means, Apriori Algorithm, etc.

Neural Network. Neural Network (NN) [14, 15] was originally developed as a highly simplified human nervous system model, they are capable of learning and changing themselves for the changing environment. NNs are mathematical and statistical models that use learning algorithms inspired by the human brain to store knowledge. NNs consists of a collection of connected components known as neurons, the information is passed from one neuron to another neuron. The neurons are aggregated into different layers, each layer performs different transformations on the input data. NN is a magnificent prediction and classification tool. Some basic types of NN are Convolutional Neural Network (CNN), Feedforward Neural Network (FF), Long/Short Term Memory Network (LSTM), and Recurrent Neural Network (RNN).

Deep Learning. Deep Learning [16, 17] is a branch of AI that can mimic the human brain's data processing mechanism and produce identical patterns used by the brain for decision making. It is a division that leverages a collection of nonlinear processing units containing multiple layers for feature transformation and extraction. A Deep Learning model consists of Motherboard, Processors, RAM, PSU. Deep learning requires significant computational power to operate rapidly. The architecture of a DL model includes Convolutional Neural Networks, Recursive Neural Networks, Recurrent Neural Networks, and Unsupervised Pre-trained Networks.

2   Literature Review

2.1   Using CNN and LSTM

Kinthali et al. [18], 2018 proposed a text recognizer system using RNN layers, CNN layers, and CTC layers. The proposed system gave an accuracy of 90.3%. Their solution had 5 CNN layers and 2 RNN layers and a single CTC layer. The IAM Handwriting dataset was used here.

2.2   Using SVM and Multiple Hidden Markov Model

Combining SVM on the MNIST dataset with Multiple Hidden Markov Model MHMM[19] provides a remarkable 99.17% efficiency that surpasses the performance of MHMM alone and KNN classifiers.

2.3   Using RESNET

Jia et al. [20–22], 2018 used a Feature Pyramid Network (FPN) on the top of ResNet50 architecture and then constructed a feature pyramid with three tiers. The ResNEt50 is a 50 layer deep Convolutional Neural Network. In the proposed solution they used a pre-trained ResNet50 model for ImageNet classification. An average F-score of

90.27% was obtained after testing and training on 285 images of handwritten notes and whiteboards. The proposed solution provided substantially better accuracy than the other approaches based on LSTM networks.

2.4 Using CNN and B-LSTM

Hassan et al. [23], 2019 proposed using CNN and B-LSTM for the Urdu text recognition system. Their solution consisted of seven convolutional layers for feature extraction followed by 2 B-LSTM layers for classification. The proposed solution showed an average character recognition rate of 83.69% when experimented on 6000 unique text lines in Urdu language.

Agrawal and Kaur [24] proposed that the accuracy of the text recognition system could be improved by using the Otsu algorithm for image segmentation and the Hough transformation method for skew detection. The proposed solution had an overall accuracy of 93%, their data consisted of text images, where the text is only written in the font style Verdana.

2.5 Separable Multidimensional Using Long Short-Term Memory Recurrent Neural Network (SMDLSTM-RNN)

In HCTR [25], it was proposed that using SMDLSTM-RNN modules provides substantially better precision than the other LSTM-based approaches. The accurate rate that was obtained is 90.02% using the ICDAR-2013 dataset.

2.6 Using CNN and ECOC

Bora et al. [3] proposed a hybridization of the ECOC classifier and CNN architecture. CNN is used to extract functionality where the standard CNN soft-max layer is swapped by the CNN-ECOC classifier. With the help of numerous coding constructs preceded by a linear learner such as the Support Vector Machine SVM, ECOC effectively transforms multi-class classification into binary classification problems. The SVM will effectively prevent overfitting and has a good predictive performance. The SVM also has greater potential to generalize than the Neural Network. There were mainly 4 techniques of CNN paired with ECOC namely LeNet Type 1, LeNet Type 2, AlexNet, and ZfNet. Out of these, the ZfNet with ECOC had the highest accuracy of 97.71 both in training and testing.

2.7   Using CNN and RNN Hybrid Models

Kartik et al. [26] used synthetic data and the normalization and enhancement of domain-specific images, proposing innovative strategies to build a hybrid CNN RNN architecture. A hybrid optimized CNN-RNN system with a considerable emphasis on active teaching is suggested using effective network initialization using pre-training synthetic data, domain-specific data translation and image normalization for slant correction, and distortion for learning significant invariance. With IAM dataset, for unconstrained decoding WER was noted as 12.6 and CER 4.88 whereas for line it was noted that WER was 17.82 and CER 5.7.

2.8   Using Bag of Features and Hidden Markov Model

Rahal et al. [27] presented a Bag of Features Architecture for HTR. This technique was also based on the Deep Sparse Auto-Encoder. A new attribute extraction model was proposed and then HMM was later used to model the sequences. Handwritten text datasets of IFN/ENIT and MNIST were tested for quality evaluation features acquired by our suggested system. The results produced for MNIST dataset are BoF-Deep SAE-HMM (their proposed solution) gave 1.62 CER% and the Training-Testing Configurations are abc-d (%) 1.97, abcd-e (%) 4.93 respectively for the IFN/ENIT datasets.

2.9   Using CNN and RNN CTC layers

Keshav et al. [28] created a model combining CNN, RNN, and CTC layers producing Character error rate: 12.065659% and Word accuracy rate: 71.252174% on the IAM dataset.

2.10   Using BLSTMS

Nguyen et al. [29] have shown the application of Bidirectional LSTMS for online handwritten English text segmentation. Both forward and backward networking improved the segmentation process with a 62% reduction in waiting time and a 50% reduction in processing time than the usual methods.

2.11 Using CNN and SVM

Darmatasiaet al. [30] have used Convolution Neural Networks for feature extraction, and Support Vector Machines (SVM) is being used as a high-end classifier. The authors have found an efficient method in which SVM along with L1 loss function and L2 regularization has outperformed in terms of both computation time and accuracy. By proposed methodology, the recognition rate achieved for numerical characters is 98.85%, on uppercase characters 93.05%, on lowercase characters 83.54%, and 91.37% on a combination of both uppercase characters and numerals.

2.12 Using Faster R-CNN and CNN

Junqinget al. [31] the authors have used two illuminating techniques initially used Faster R-CNN for preprocessing the handwritten character and secondly Convolutional Neural Network for character recognition. The method of accomplishing character recognition by the authors is better than traditional OCR.

2.13 Using K-Means Clustering

Adam et al. [30] The authors have selected the models and features used in scene-text applications, with more scalable and sophisticated feature learning algorithms currently being developed by machine learning researchers. The accuracy achieved by them is 85.5%.

2.14 Using Neural Networks and Deep Learning

Rohan et al. [33] In this paper two important things have been performed for the model first is the pre-trained neural network and the second thing is applying image processing operation on handwritten text image and by doing this they have achieved the accuracy of 94%.

2.15 Using Backpropagation Neural Network

Hussain et al. [34] Here the author has used mainly four various types of Artificial Neural Network Architecture and after comparing and analyzing they showcased that Backpropagation Neural Network outperformed amongst them. They have used the dataset containing handwritten text collected from 20 non-identical writers each of the set consisting of 456 characters of the Mizo language.

Optical character recognition (OCR) is a device used to transform text data into a machine-encoded format. There are various classification methods of handwritten OCR namely ANN, Kernel methods, Statistical methods (parametric and nonparametric), Template matching techniques, structural pattern recognition (graphical and grammar-based methods), and so on. The most common datasets that have been used in research are CEDAR, MNIST (Which is a part of the NIST dataset), UCOM, IFN/ENIT, CENPARMI, HCL2000, IAM. Among these CEDAR and MNIST are used for digits, UCOM is used for the Urdu language, IFN/ENIT is used for the Arabic language, CENPARMI is a collection of Farsi numeric, HCL2000 is a handwritten Chinese character dataset, IAM is a handwritten database of English sentences and words [9] (Table 1).

3 Conclusion

In this paper, we have discussed and evaluated various methods proposed for text recognition systems. Hereafter comparing various research papers based on various parameters we have come across that Convolutional Neural Networks (CNN) works great for feature extraction, also by combining CNN with some other filters or by adding some more layers to CNN, the feature extraction and the features selection process can be greatly improved. Combining CNN with suitable classifiers like LSTM, B-LSTM, ResNest, SVM, etc. gives highly varying results, these results can be improved by improving the preprocessing process on the input image and by using a hybrid model for the text recognition process.

Table 1 Shows the analysis of all the databases and the result of the reviewed papers

Study	Database	Results	Merits
Kinthali et al. [18]	IAM online database (IAM-onDB)	90.3%	The model proved to be both efficient and effective for text recognition
Prasad et al. [19]	MNIST	99.17%	Experiments were conducted in a total of four phases, each phase comprising different features or techniques used
Jia et al. [20]	4893 text-lines	Average F-score: 90.27%	Solutions for improving the robustness of the SegLink model are provided
Hassan et al. [23]	CENPARMI	83.69%	The model was trained on 6000 unique Urdu text lines, the proposed solution highly relies on implicit segmentation of characters
Agrawal et al. [24]	117 documented images	93%	The provided text recognition model had good accuracy on rotated images. It also had good results against scaling
Wu et al. [25]	ICDAR-2013	Maximum Accurate Rate: 90.02%	An improved version of MDLSTM-RNN model was provided which consumed less resources and computational effort
Bora et al. [3]	NIST	ZfNet with ECOC had the highest accuracy of 97.71%	LeNet, AlexNet and ZfNet were compared and then these methods were coupled with ECOC, and further compared in detail to determine accuracy
Dutta et al. [26]	IAM online database (IAM-onDB) and RIMES	For unconstrained decoding WER was noted as 12.6 and CER 4.88 For line, it was noted that WER was 17.82 and CER 5.7	Using synthetic data and domain-specific image normalization and augmentation, the accuracy of the hybrid CNN-RNN was tested

(continued)

Table 1 (continued)

Study	Database	Results	Merits
Rahal et al. [27]	MNIST	BoF-DeepSAE-HMM—1.62 CER% Training-Testing Configurations are abc-d (%) 1.97, abcd-e (%) 4.93 respectively for the IFN/ENIT datasets	An effortless preprocessing of images was done by baseline estimation and slant normalization A dense and steady image representation was noticed by assigning the data to the clusters
Gupta et al. [28]	IAM online database (IAM-onDB)	Character error rate 12.065659% Word accuracy rate 71.252174%	The neural framework consists of many layers that help in improving accuracy and reduce error
Nguyen et al. [29]	IAM online database (IAM-onDB)	86.55%	Reduced waiting and processing time
Darmatasia et al. [30]	NIST SD 19 2nd edition	Numerical characters 98.85% uppercase characters 93.05% lowercase characters 83.54% combination of both uppercase characters and numerals 91.37%	The model is validated using ten folds cross-validation
Junqing et al. [31]	Word dataset, Letter dataset, and EMNIST	97%	The model shows good performance on complex handwriting text
Adam et al. [32]	ICDAR 2003	85.5%	They have applied large-scale algorithms for learning the features automatically from unlabeled data
Rohan et al. [33]	NIST	94%	OpenCV used for Image Processing thus complicated data structures could be used for handling real-time processing
Hussain et al. [34]	20 sets of 456 characters of the Mizo language	Average MSE: 94.5%	Achieved recognition rate of 98%

References

1. https://nanonets.com/blog/handwritten-character-recognition/. Accessed on October 17, 2020
2. Rao Z, Zeng C, Wu M, Wang Z, Zhao N, Liu M, Wan X (2018) Research on handwritten character recognition algorithm based on an extended nonlinear kernel residual network. KSII Trans Internet Inf Syst 12(1)
3. Bora MB, Daimary D, Amitab K, Kandar D (2020) Handwritten character recognition from images using CNN-ECOC. Procedia Comput Sci 167:2403–2409
4. Khedher MZ, Abandah GA, Al-Khawaldeh AM (2005) Optimizing feature selection for recognizing handwritten Arabic characters. WEC (2):81–84
5. https://towardsdatascience.com/pre-processing-in-ocr-fc231c6035a7. Accessed on August 28, 2020
6. Mainkar VV, Katkar JA, Upade AB, Pednekar PR (2020) Handwritten character recognition to obtain editable text. In: 2020 international conference on electronics and sustainable communication systems (ICESC). IEEE, pp 599–602
7. Baldominos A, Saez Y, Isasi P (2019) Hybridizing evolutionary computation and deep neural networks: an approach to handwriting recognition using committees and transfer learning. Complexity
8. Kaushik A, Gupta H, Latwal DS (2016) Impact of feature selection and engineering in the classification of handwritten text. In: 2016 3rd international conference on computing for sustainable global development (INDIACom). IEEE, pp 2598–2601
9. Gavali P, Banu JS (2019) Deep convolutional neural network for image classification on CUDA platform. In: Deep learning and parallel computing environment for bioengineering systems). Academic Press, pp 99–122
10. https://machinelearningmastery.com/types-of-classification-in-machine-learning/. Accessed on January 13, 2021
11. Abdulrazzaq MB, Saeed JN (2019) A comparison of three classification algorithms for handwritten digit recognition. In: 2019 international conference on advanced science and engineering (ICOASE). IEEE, pp 58–63
12. Memon J, Sami M, Khan RA, Uddin M (2020) Handwritten optical character recognition (OCR): a comprehensive systematic literature review (SLR). IEEE Access 8:142642–142668
13. https://expertsystem.com/machine-learning-definition. Accessed on September 8, 2020
14. https://medium.com/towards-artificial-intelligence/main-types-of-neural-networks-and-its-applications-tutorial-734480d7ec8e. Accessed on August 10, 2020
15. https://en.wikipedia.org/wiki/Neural_network. Accessed on August 10, 2020
16. https://www.upgrad.com/blog/deep-learning-vs-neural-networks-difference-between-deep-learning-and-neural-networks/. Accessed on August 12, 2020
17. Chen J, Ran X (2019) Deep learning with edge computing: a review. Proc IEEE 107(8):1655–1674
18. Manchala SY, Kinthali J, Kotha K, Kumar KS, Jayalaxmi J Handwritten text recognition using deep learning with TensorFlow.
19. Prasad BK, Sanyal G (2018) Multiple hidden Markov model post processed with support vector machine to recognize English handwritten numerals. Int J Artif Intell Tools 27(05):1850019
20. Jia W, Zhong Z, Sun L, Huo Q (2018) A CNN-based approach to detecting text from images of whiteboards and handwritten notes. In: 2018 16th international conference on frontiers in handwriting recognition (ICFHR). IEEE, pp 1–6
21. https://iq.opengenus.org/resnet50-architecture/. Accessed on October 1, 2020
22. https://www.mathworks.com/help/deeplearning/ref/resnet50.html#:~:text=ResNet%2D50%20is%20a%20convolutional,%2C%20pencil%2C%20and%20many%20animals. Accessed on August 19, 2020
23. Hassan S, Irfan A, Mirza A, Siddiqi I (2019) Cursive handwritten text recognition using bi-directional LSTMs: a case study on Urdu handwriting. In: 2019 international conference on deep learning and machine learning in emerging applications (Deep-ML). IEEE, pp 67–72

24. Agrawal N, Kaur A (2018) An algorithmic approach for text recognition from printed/typed text images. In: 2018 8th international conference on cloud computing, data science & engineering (Confluence). IEEE, pp 876–879

25. Wu YC, Yin F, Chen Z, Liu CL (2017) Handwritten Chinese text recognition using separable multi-dimensional recurrent neural network. In: 2017 14th IAPR international conference on document analysis and recognition (ICDAR), vol 1. IEEE, pp 79–84

26. Dutta K, Krishnan P, Mathew M, Jawahar CV (2018) Improving CNN-RNN hybrid networks for handwriting recognition. In: 2018 16th international conference on frontiers in handwriting recognition (ICFHR). IEEE, pp 80–85

27. Rahal N, Tounsi M, Hamdani TM, Alimi AM (2019) Handwritten words and digits recognition using deep learning based bag of features framework. In: 2019 international conference on document analysis and recognition (ICDAR). IEEE, pp 701–706

28. Gupta K, Kumar M, Sachdeva N, IMSEC G Character recognition from image using tensorflow and convolutional neural networks

29. Nguyen CT, Nakagawa M (2015) An improved segmentation of online English handwritten text using recurrent neural networks. In: 2015 3rd IAPR Asian conference on pattern recognition (ACPR). IEEE, pp 176–180

30. Fanany MI (2017) Handwriting recognition on form document using convolutional neural network and support vector machines (CNN-SVM). In: 2017 5th international conference on information and communication technology (ICoIC7). IEEE, pp 1-6

31. Yang J, Ren P, Kong X (2019) Handwriting text recognition based on faster R-CNN. In: 2019 Chinese automation congress (CAC). IEEE, pp 2450–2454

32. Coates A, Carpenter B, Case C, Satheesh S, Suresh B, Wang T, Wu DJ, Ng AY (2011) Text detection and character recognition in scene images with unsupervised feature learning. In: 2011 international conference on document analysis and recognition. IEEE, pp 440–445

33. Vaidya R, Trivedi D, Satra S, Pimpale M (2018) Handwritten character recognition using deep-learning. In: 2018 Second international conference on inventive communication and computational technologies (ICICCT). IEEE, pp 772–775

34. Hussain J (2018) A hybrid approach handwritten character recognition for mizo using artificial neural network. In: 2018 international conference on advanced computation and telecommunication (ICACAT). IEEE, pp 1–6

Chapter 60
Optimal Network Reconfiguration of DG Integrated Power Distribution Systems Using Enhanced Flower Pollination Algorithm

S. Dhivya◉ and R. Arul◉

1 Introduction

The distribution networks are built to operate on passive elements that have been mainly designed for transferring power to the destination from the source in a unidirectional way. Hence, the incorporation of the distributed generation (DG) systems is not widely supported by the massive power networks created in the last century. The link between the power systems and photovoltaic (PV)-based DG systems does not react to the changes in the electrical system condition such as a generator having conventional synchronous units. The solar source is having distinct characteristics including power ramps with high slew rate and low inertia with high-velocity response [1]. Multiple studies have addressed the issues related to highly penetrating PV of distributed generation in the advanced technologies of distributed systems and highlighted the effects of variations in voltage [2], differences in frequency and the minimization of technical power losses [3].

In this article, probability-based flower pollination algorithm (PBFPA) is presented with regard to the reduction of technical power losses. The reduction of technical power losses through reconfiguring the distribution network systems shows successful results. For example, a search algorithm is used, which is called ant colony algorithm, and it is effective to address the problems in reconfiguring the optimized network for minimizing the technical power losses [4]. Then, two more methods such as simulated annealing algorithm and genetic algorithm optimization have been compared with this ant colony algorithm. The comparison between these three methods shows that the solution obtained from ant colony algorithm method

S. Dhivya · R. Arul (✉)
School of Electrical Engineering, Vellore Institute of Technology, Chennai 600127, India

S. Dhivya
e-mail: dhivya.s2019@vitstudent.ac.in

provides better results. In order to reduce the disturbances in power quality, a solution was proposed to address the voltage sags, harmonics and reduction of technical power losses through reconfiguring the networks on the basis of differential evolution algorithm. The result obtained from the reconfigured network system presents that the proposed solution is efficient and effective in enhancing the quality indicators of power, while minimizing the technical power losses [5]. An intelligent system is set up to automatically rearrange the distribution network according to the branch adaptation. It is useful in providing solution to real-time issues such as load balance, technical power loss minimization and power quality indicator enhancements. Studies have evaluated this method in a real-time power grid, as well as, determined the promising results associated with the performance indicators of the distributed network systems. A binary group search algorithm method has been used in a study to demonstrate the strategy related to network reconfiguration, having the purpose of minimization of power grid losses [6]. The evaluation of various test scenarios is validated. The reconfiguration of an optimized distributed network system based on the genetic algorithms was explained in a study for improving the reliability of quality indicators of power systems. The cuckoo search algorithm has been used to achieve the improvements in voltage and reduction in technical power losses [7]. The simplicity in the implementation of the algorithm shows the efficiency of the method for solving the problems in reconfiguring the distribution network systems. In another study, the most probable scenario (MPS) method is used for a robust reconfiguration to reduce the active power losses. Furthermore, a concept of receding horizon control is also used in the study, stating that the problems in the reduction of active power losses can be handled effectively. The artificial neural network (ANN) based on a dynamic fuzzy c-means clustering is optimized to solve the reconfiguration problems in the reduction of technical power losses. The ANN method provides various benefits including a simple structure of implementation, high accuracy and short processing time [8].

The operation and control of power saving pose certain challenges to the advanced power grid photovoltaic (PV)-based technology of power generation [9]. Since there is a bidirectional power flow in the utility grid, more intelligent security is needed for unintentional and deliberate islanding. Furthermore, the chance of active power losses is increasing due to the high level of photovoltaic penetration. Hence, it is significant to conduct some reconfiguration studies to determine such active power losses, thereby utilizing the system operation. In this study, the probability-based flower pollination algorithm (PBFPA) acts as an efficient technique for reduction of power losses by processing the reconfiguration of distributed network systems while attaining the global optimization.

2   Literature review

Optimization algorithms have been used for determining the optimal kind of solutions related to the optimization issues by increasing and decreasing a discrete and continuous function related to aspects of inequality and equality criterion. Problems arise with the choice of capacitor capacitance and switching conditions for reactive power compensation [10]. The study highlights the optimization of the particle swarm optimization (PSO) and the degree of minimal power loss in the system [11]. The modified algorithm looks like this: Each particle wants to change its position based on some information, such as the flow rate of the particles, the current position of the particle and the distance between P_{best} and its current position. The velocity of the particles is determined by the sum of the components of inertia, social and individual influence in the group. To determine the optimization of the modified particle group, various frequencies are evaluated. The real value space is the search space, and in this space, the binary PSO used to have set of 1 and 0, and it has chosen standard values [12]. By exploiting the existing structure of power as well as maximal loading, the chances of voltage collapse can be enhancing in the given system of distribution [13].

The use of fireworks algorithm is done for enhancement of the performance of radial distribution network. It is stochastic and robust which is considered to be significant evolutionary algorithms. Its purpose was mainly to reduce the voltage deviation and power loss by assessing the nature of distribution system [14]. A significant aspect is found when new swarm intelligence is deployed with the help of a global optimization procedure for resolving reconfiguration issue. It has also been adaptable under the abnormal conditions. The new method can make sure the power supply, which is being provided to the system's non-faulted areas with less load shedding as well as voltage deviation. It has highly depended upon the standards of IEEE test system.

The reconfiguration of distribution network is the phenomenon which manages the topology of making the modification in open, as well as, closed status of given switches; hence, it is easy to evaluate a radial operating structure which enhances the voltage stability and reduces the loss. With the normal criteria, the distribution power sector would likely minimize the system power losses and make a loading over feeders and transformers [15]. There is a need of enhancing the power quality which has high importance. Sensitive loads would be related to short interruption and less voltage drop when the abnormal faults are found. Supply companies must enunciate the quality of power applied to industries with loads.

Energy sources such as sun-oriented wind are used in recent times. In some cases, energy sources are used to produce power on a limited scope in territories identified with end customers. End customers prefer strength, and the power of abundance is carried over to the lattice. It is known as circulated age and helps limit the expense of age, coal consumption and transmission line power misfortune. The interest of the customers was achieved by the age of the neighborhood and also limits the voltage drop. This investigation led to improved use of the DG to expand the financial development and force security framework [16]. DG power is applied to power source to decide the salient moments of the transmission line's misfortune by solving the reconfiguration issue. The power of the DG is infused into each transport with certain disruption, and reconfiguration problems are planned to evaluate ideal estimates of control factors to decrease transmission line misfortunes [17].

The feeder configuration issues are complex by conflicting with the objectives, which are used to satisfy the normal along with abnormal conditions. Over the past 20 years, a variety of researchers have restored the issues of network reconfiguration with the help of different methods along with voltage profile improvement as well as power loss minimization in power distribution networks [18]. When branch and bound kind of optimization approach is used, the distribution system reconfiguration problem for loss reduction has been solved. The computation time was the most significant issue of the method taken for acquisition of optimal configuration. With the help of two minimum current neighbor chain methods of updating, the reconfiguration problem has been solved. The good is that a new approach for the reconfiguration of distribution network is available with feasible optimal flow as per decomposition approach for load balancing and loss minimization. The differential evolution algorithm is used to improve quality of power issues like voltage sags and harmonics by enhancing the distribution network. Various optimization algorithms such as refined genetic algorithm, plant growth simulation algorithm, fuzzy mutated genetic algorithm, as well as harmonic search algorithm have been used to resolve the reconfiguration issue with several objectives [19].

Particle swarm optimization and discrete artificial bee colony linked with the graph theory have been used to enhance the network of distribution. There are some motivating results in assessing the optimization of the distribution network's problem, but there are some drawbacks as well in constraints like computation time in resolving large-scale system, mutation rate, crossover rate, computing efficiency and convergence property [20]. It is significant to enhance the radial distribution network, when vortex search algorithm is actually used. When radial nature is analyzed, there was the absence of optimal configuration, so there was no mesh loop in the given network. It means that there would be an effect on the network of radial distribution. That is why, it is quite critical to ensure that radial nature of distribution is maintained, which is considered significant during every step of the process [21].

For farmers, the term used for the grasshoppers is pest insects as they are a reason to affect the agricultural crop. It is a matter of fact that a random path can be used by a single grasshopper. This movement can be done in an organized and arranged path being part of the swarm members. The rolling cylinder is the thing taken as for the grasshopper swarm, especially when their search for food source is done to find all kinds of crops to be eaten in the agricultural regions. In a given swarm, there are a range of attraction as well as repletion forces between the grasshoppers. They try to keep a distance from each other on a regular basis. The use of attraction force is found when grasshoppers are at a great distance, rather than being in a comfortable distance, so it means that the development of repulsion force is done when distance of grasshoppers decreases. The agents are indicated by grasshoppers, and best position for the grasshoppers is indicated by the food sources in a given swarm. It means that there are significant steps involved with this proposed algorithm. Randomly created initial valueds of grasshopper positions consider the lower and upper limits of control variables. In this algorithm, j represents number of control variables, and w is number of search agent. The generated grasshopper position indicates switches which can be opened to reconfigure the system [22].

Firefly algorithm is used to enhance the control variables of optimization of voltage stability and power loss of transmission system. This problem has been defined as nonlinear inequality, as well as, the problem of equality constrained optimization with integration of voltage stability limit and real power loss. In the formulation of problem, the control variables for the process are transformer taps as well as unified power flow controllers. The New England 39 bus system has been taken under review to evaluate this new proposed algorithm [23]. The proposed algorithm has been able to provide a range of simulation results. When they are looked for the comparison, their comparison with the coded genetic algorithm is important as it comes with various objectives like limit maximization of voltage stability, as well as minimization of real power loss, as compared to the single purpose of real loss of power minimization. The interior point successive linear programming method is the one, which is said to be a classical optimization method. The firefly algorithm results can be compared with the help of this method, which is effective for minimization of power loss. The proposed algorithm's potential is made sure while doing the management of optimization problems.

Al Samman et al. [24] provide rapid and optimal network reconfiguration with a guided start based on a simplified network approach. Subsequently, this approach has been applied to driven start-ups and population generations and the adjusted population index. The proposed method has been implemented using the firefly method and verified in IEEE-33 and 118 bus test systems. To reduce the unhappiness of strength and improve the voltage profile (VP), Salau et al. [25] presented an ideal strategy to improve DNR problems in constrained organizations. In this case, the editors used the modified selective particle swarm optimization (SPSO) strategy for existing organizations, taking into account various stacking conditions. The main objective of this review was to limit successful power problems and improve the PV of a spread frame using the proposed SPSO strategy. They have adjusted MATLAB R2016b programming and tried to use the IEEE-33 bus radial distribution network (RDN).

The reconfiguration of the ideal organization and the task of the DG using the computation of the modified adaptive progress of the whales, taking into account the flow of probabilistic load, was published by Uniyal et al. [26]. The improvement problem is solved with a adaptive modified whale optimization algorithm (A-MWOA). A procedure, which depends on the calculation of an initial in-depth investigation, was used to verify widespread fears. Nguyen et al. [27] introduced a coyote algorithm (COA), which depends on the public activity of the coyotes, in order to ensure the organizational reconfiguration and the simultaneous localization of general directorates and to reduce the calamities of the real forces. The feasibility of the proposed COA technique was evaluated on two frames with 69 and 119 hubs in two individual situations, including simultaneous reconfiguration and DG localization. Essallah et al. [28] investigated a method to optimize modified particle swarms optimization (MPSOs) to cope with the calamity of limiting dynamic force and to improve the stress profile . The innovation is related to the optimization of binary particle swarms optimization (BPSO) and traditional PSO calculations. In order to evaluate the presentation of the created approach, three different load situations were evaluated during the coordination between DNR and DG. The reconstructions are carried out on two transport test frameworks, namely the specific IEEE-33 bus and the IEEE-69 bus.

Much of the current design uses an actual model-dependent control approach to address the unique problem of DNR. However, this method has two limitations. For starters, model-based calculations without complete and accurate circulation grid boundaries for electrical applications may not be reliable. It should be noted that maintaining precise essential and optional strength limits for dispersion networks that span a large number of hubs is difficult for electrical applications. Second, the computation season for model-based control computations increases dramatically with the number of switches to be controlled, the number of DERs and the length of the working horizon, which is difficult to incrementally use in the reconfiguration network.

The writing models above are based on simplified approaches to monitoring network configuration in different frameworks. However, the optimality of the irregularity required further testing. In addition, an ongoing, concrete test framework should be considered for reconfiguration projects.

3 Problem Formulation

The design of enhanced flower pollination algorithm is applied to the distribution network for the optimal topology to ensure reducing the power losses of technical nature. The loss reduction is involved in the proposed solution by distribution grid's optimal reconfiguration. In the electrical energy systems, the feeder's reconfiguration would be established using opening as well as closing of the switching devices [29]. The changes in switches' position could minimize the line losses and operate the system with high security using contingency methods. The switching must adhere to the radial structure of distribution system. Probability-based flower pollination algorithm (PBFPA) is used for configuration, and hence, the optimal topologies of the process are evaluated for the given systems. The significant aim of reconfiguration of the distribution system is to reduction of losses of technical nature on lines at the time of normal operation.

$$P_{\text{losses}} = \sum_{i=1}^{n} I_i^2 * R_i \tag{1}$$

P_{losses} denotes distribution system's technical losses.

'n' denotes number of lines on the given system.

I_i^2 is i^{th} zone's electrical value.

R_i is the resistance of ith line stretch. In order to resolve this issue, the study proposed probability-based flower pollination algorithm. The reconfiguration problem shown by this system would be managed as a process of combinatorial permutation optimization of 1 and 0.

3.1 Objective Function of the Problem

For optimal reconfiguration, the consideration is done for the radial nature of the network to be significantly constraint. The problem's objective function does come with a formulation for minimizing the real loss of power in the distribution networks.

$$\text{Minimize}(F) = \text{Min}(P_{\text{losses}}) \tag{2}$$

Power flow:

It is essential to prefer load flow method for radial distribution feeders. Primitive impedance-based load flow technique is used to compute the system losses and voltage profile. This system loss is based on size, location and power factor of DG. Non-unity power and unity power factors are the two cases [30]. The size should satisfy the constraint. By using the recursive equations, the distribution network power flow will be computed, and single line diagram can be used to derive it.

Constraints:

As the objective function is mentioned in this study, the constraints for minimizing the power loss are voltage deviation limits and power flow capacity limits given as follows:

(i)   Voltage deviation limits:

$$|V_1 - V_k| \leq \Delta V_{\text{max}} \tag{3}$$

(ii)   Maximum power flow capacity limit of line section between buses k and

$$k = 1 : |S_k| \leq |S_{k.\,\text{max}}| \tag{4}$$

In these given constraints, if anyone is violated, the negligibility would happen for the resultant solution.

4 Proposed Algorithms

Network reconfiguration in distribution systems, taking into account the location and size of the attic, is a complex nonlinear optimization problem. This type of optimization problem cannot be solved with conventional techniques. Therefore, the flower pollination algorithm (FPA) for solving a network reconfiguration problem is being improved while taking into account the location and size of the DG in the delivery system. In this article, the main advantage of the proposed FPA is its ability to achieve an overall solution with the desired speed and accuracy of convergence. In addition, the proposed algorithm can effectively solve the problem of reconfiguring the system taking into account the location and size of the DG, as well as the discrete and continuous nature of the variables being controlled.

Flower pollination algorithm is used for solving the reconfiguration network problem. The characteristics of pollinator behavior, pollination process and flower constancy have some rules such as cross and biotic pollination are terms as global pollination, as it is a fact that pollen-carrying pollinators managing the essence of levy flights [31]. Abiotic and self-pollination are considered to be as one of the local pollinations. The reproduction probability is used to define the essence of flower constancy. The two flowers' behavior is directly proportional to the similar kind of behavior of reproduction probability. Both local and global level pollination are used to assess with the help of switch probability of $p \in [0, 1]$. It happens because there is physical proximity along with other several factors such as local, as well as, wind pollination having the point of friction p in the activities of pollination [32]. According to the given assumptions, it is assumed that each plant comes with a one flower, and only one pollen gamete is produced by a flower. In this problem, it is not necessary to differentiate a pollen, flower, plant or gamete. The given solution for the process is called xi, and this solution is equal to flower gamete or pollen. The multiple flowers do come up with multiple pollen gametes to optimize the problem. The local and global pollination are one of the huge steps in the process of algorithm. The pollinators would be carried by the flower pollens like pollens and insects would go through this long journey because there is a capability of insects to move to long distances. It enunciates reproduction through pollination [33].

4.1 Probability-Based Flower Pollination Algorithm (PBFPA)

In this study, probability-based flower pollination algorithm is presented. By using the characteristics of flower pollination algorithm, this probability with exponential distribution formula is used. Cauchy probability and Mittag–Leffler probability [34] are integrated into flower pollination algorithm.

4.1.1 Cauchy Probability Distribution-Based Flower Pollination Algorithm (CFPA)

The variable X is the real-valued random one, which is actually defined as a heavy right tailed. If there would be slow decay in tail probabilities $P(X > x)$ than any exponential distribution, if

$$\lim_{n \to \infty} \frac{P(X > x)}{e^{-\lambda x}} = \infty \tag{5}$$

The Cauchy distribution is the actual random variable. If it would show its distribution function, then here would be its form;

$$F(x) = \frac{1}{\pi} \frac{\arctan(2(x - \mu))}{\sigma + \frac{1}{2}} \tag{6}$$

where μ is the location parameter and scale parameter is σ

$$U_i^t = X_i^t + \alpha \otimes \times Cauchy(\mu, \sigma) \otimes \left(X_i^t - X_{\text{best}}\right) \tag{7}$$

4.1.2 Mittag–Leffler Probability Distribution-Based Flower Pollination Algorithm (MLFPA)

Mittag–leffler distribution is considered to be a random variable. If it also comes with a distribution function, the following form is shown for it;

$$F_\beta(x) = \sum_{k=1}^{\infty} \frac{(-1)k - 1 \cdot x^{k\beta}}{(1 + k\beta)} \tag{8}$$

where $0 < \beta \leq 1, x > 0$ and $F_\beta(x) = 0$ for $x \leq 0$. For $0 < \beta < 1$, the Mittag–leffler distribution is a heavy-tailed distribution of exponential, and it could minimize when $\beta = 1$.

The convenient expression is generated with the help of Mittag–Leffler random number given by Kozubowski

$$\tau_\beta = -\gamma \ln u \left(\frac{\sin(\beta\pi)}{\tan(\beta\pi\vartheta)} - \cos(\beta\pi)\right)^{\frac{1}{\beta}} \tag{9}$$

where γ is the scale parameter, $u, v\varepsilon(0, 1)$ are uniform random numbers, and τ_β is Mittag–leffler random number.

By using simple random numbers, the randomization will be attained which are derived from normal or uniform distribution,

$$U_i^t = X_i^t + \alpha \, \text{Mittag–Leffler}(\beta, \gamma) \otimes \left(X_i^t - X_{best}\right) \tag{10}$$

where $\text{Mittag–Leffler}(\beta, \gamma)$ represent a random number drawn from Mittag–Leffler distribution.

5 Solution Methodology

In this study, practical 52 bus system [35] is used for improving the optimization of distribution network. MATLAB is used for evaluating the results. By applying Cauchy probability and Mittag–Leffler probability-based FPA in practical 52 bus system, the below solutions have been obtained.

5.1 Practical Indian 52 Bus System

The proposed technique is being evaluated by using the distribution system of Indian radial system. The line and load information in this test system are used from the article [35]. In this test system, 52 buses have been used with three major feeders, and branches of feeding network load are 51. 0.9 lagging is the power factor connected with this load test. The maximum as well as minimum bus voltage limits are 0.9 and 1.05 p.u. in this test system. In this practical distribution system of 52 buses of Indian system, the evaluation is done for both reactive and active losses. The active power injections are noted to be 5071.18 kW at the substation. The primary aim of the study is the reduction of network loss, so that total deviation of load voltage is also reduced. The large number of distributed generated unit used for the system is equivalent to the feeders. The simulation operated for flower pollination algorithm is created in MATLAB environment. The five cases in each scenario (Cauchy probability and Mittag–Leffler probability) will be addressed and calculated. Figure 1 represents the structural units of Indian practical bus-52 system.

6 Simulation Results

It is stated that losses occurred at every line are minimized. The increment is observed in the losses especially when loads are shifted onto feeders. The optimal reconfiguration to minimize the real power flow ensures that the distribution feeders are free from overloading. To determine what kind of performance is shown by the proposed system, the simulated results would be compared between five cases as shown in Tables 1 and 2. In order to avoid the irrelevant incongruities, as well as, the results of other kind of algorithms will be evaluated by the configuration's power flow. To evaluate the comparative performance of CFPA and MLFPA, the reconfiguration issues are solved. The worst and best values over the top solutions are evaluated. It is observed that Mittag–leffler probability has performed better in relation to computation time, and power loss minimization during line, DG and ESS outages. Cauchy probability-based FPA works better in deriving optimal solution during bus outages. In order to evaluate the impacts of parameters of flower pollination algorithm on computation time, convergence behavior and solution quality, this

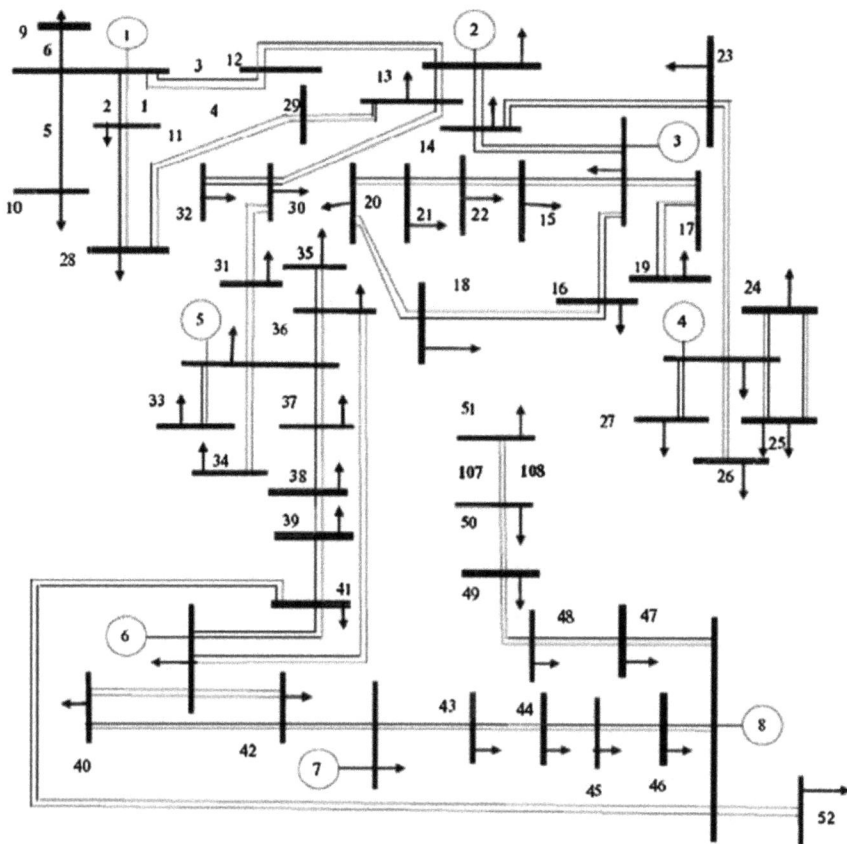

Fig. 1 Practical Indian 52 bus system [35]

Table 1 Optimal switch connections after reconfiguration

Case-1, 3		Case-2, 4, 5	
CFPA	MLFPA	CFPA	MLFPA
5–31	42–14	10–46	9–40
9–33	10–32	47–6	18–31
40–11	36–49	5–48	36–22
6–23	31–51	49–8	11–47
15–51	33–18	16–21	33–34
14–45	11–28	37–11	42–46

Table 2 Comparison of total power loss for all cases

Power loss (kW)		Case-1	Case-2	Case-3	Case-4	Case-5
Before reconfiguration		393.83	418.19	3932.21	419.81	419.81
After	CFPA	387.71	413.20	386.09	417.82	414.82
reconfiguration	MLFPA	388.44	412.98	386.83	414.60	414.60

study is conducted. To evaluate the effects of parameters changes, the five cases are tested. The number of iterations and total power loss has been used to make computation time and converge time for the test system by differentiating the parameters. As the number of location increases, the number of iterations decreases over the same power loss. The higher solution quality has also been attained. The outcome of proposed method outperforms all optimization techniques with respect to power loss minimization. This method solves effectively. Radial behavior is assessed by the proposed method, performance in non-deformed areas with voltage fluctuations and low power loss under exceptional conditions. Restrictions were evaluated using selected range values. The three DGs of 10 kW, 20 kW and 30 kW are randomly located at Bus nos. 12, 24 and 32, respectively. While energy storage system (ESS) of 30kW is located randomly at Bus no. 28, let us assume five cases as follows.

Case 1: Bus outage: We have considered that Bus no. 14 has been faulted for a due course of time. At that point, the reconfiguration has been made using both CFPA and MLFPA. The optimal switch connections are listed using CFPA and MLFPA in Table 1.

Case 2: Line outage: We have assumed that line 17 is cut off. The power loss of 0.58 p.u has been obtained at first line. The line numbers 2, 10, 30 and 45 have more power losses. The total power loss for all cases are enlisted in Table 2.

Case 3: Line and Bus Outage: In this case, both line 17 and bus 14 have been isolated in a similar manner. The power loss of this case is equal to 0.56p.u. The optimal power losses are occurring in this case comparative to all other cases. Its convergence characteristics are shown in Fig. 2.

Case 4: DG outage: The three DGs of 10kW, 20kW and 30kW have been placed in buses 12, 24 and 32 respectively. In this case, the performance of DG has been interrupted for some time. The line nos. 2, 10, 30 and 40 have yielded high power loss.

Case 5: ESS outage: In this case, the ESS has been disconnected from bus 28. The value of power loss is similar to case-4. The MLFPA convergence characteristics have been shown in Fig. 3.

Simulations are performed on MATLAB 2018(A), i3 core processor, 2GHz CPU. Results revealed that MLFPA yields better results during line, DG and ESS outages, whereas CFPA has reached better solution during bus outages and simultaneous instance of line and bus outages. The search capability of constraints could be maximized through independent distribution to reach the global optimal values. Multiple variance search ability could be the great influence of Cauchy probability. Similarly,

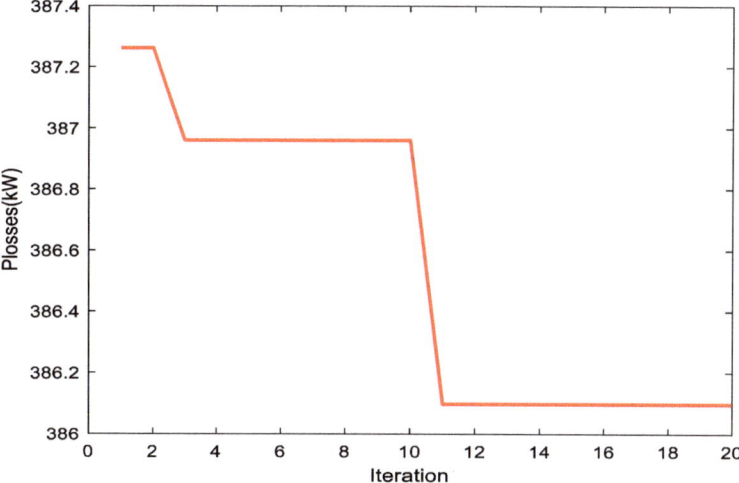

Fig. 2 Plosses versus iterations for case-3 using CFPA

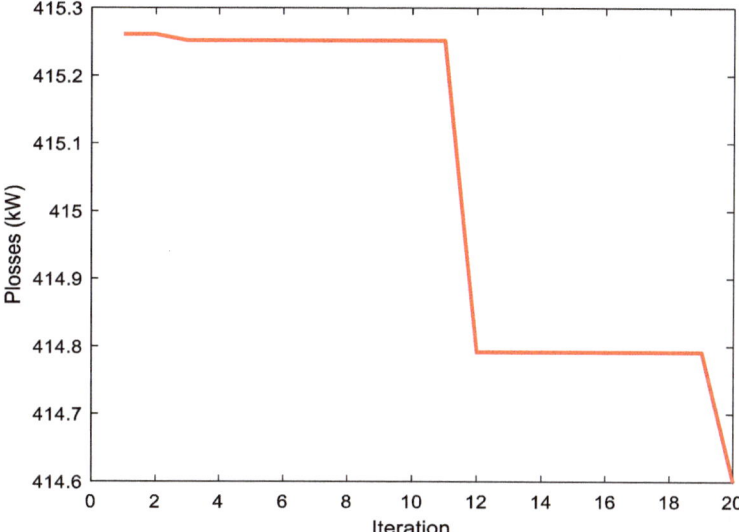

Fig. 3 Power loss versus iterations for case-5 using MLFPA

the finite space in collaborative search of variables could be brought by Mittag–Leffler probability distributions due to its computational tractable feature.

7 Conclusion

The article has focused on developing probability-based flower pollination algorithms for network reconfiguration in practical Indian 52 bus system. As discussed above, there are five different cases related to Cauchy probability and Mittag–Leffler probability-based flower pollination algorithm during network reconfiguration. The results of the five cases are evaluated by the proposed method. The performance of CFPA and MLFPA has been found to be effective than other approaches with respect to solution quality. When there are huge network systems like 52 bus system of Indian radial type, the computational results illustrate that observed advantages of PBFPA are acceptable. The proposed methodology's convergence rate curve is used to develop the ability of resolving reconfiguration issue. The proposed method is easily adapted to large-scale radial distribution networks. The study outcomes revealed that MLFPA yields better results during line, DG and ESS outages. At the same time, CFPA has reached better solution during instance of bus outages and instance of both bus and line outages. It aims to integrate the optimization of primary and secondary distribution networks as future research. In the future, we plan to develop a guide on how to deal with asymptomatic limitations when conducting batch improvement studies. We also plan to improve flexible policy repetition theory by examining various mechanisms of premium collection. Future activities will focus on integrating DGs into DNR to improve the efficiency of clean and orderly distribution systems for renewable energy sources.

References

1. Monteiro RVA, Bonaldo JP, da Silva RF, Bretas AS (2020) Electric distribution network reconfiguration optimized for PV distributed generation and energy storage. Electric Power Syst Res 1–9, https://doi.org/10.1016/j.epsr.2020.106319
2. Lenin K (2018) Real power loss minimization and voltage stability enhancement by hybridization of eagle strategy with particle swarm optimization algorithm. J Instit Eng 14(1):22–34. https://doi.org/10.3126/jie.v14i1.20066
3. Lenin K (2019) Active power loss reduction by particle swarm optimization algorithm. Int J Comput Sci Eng 7(1):904–906 https://doi.org/10.26438/ijcse/v7i1.904906
4. Wang Q, Meng L (2013) Distribution network fault reconfiguration with distributed generation based on ant colony algorithm. Adv Mater Res 732–733:1328–1333. https://doi.org/10.4028/www.scientific.net/amr.732-733.1328
5. Jazebi S, Vahidi B (2012) Reconfiguration of distribution networks to mitigate utilities power quality disturbances. Elect Power Sys Res 91:9–17. https://doi.org/10.1016/j.epsr.2012.04.008
6. Teimourzadeh S, Zare K (2014) Application of binary group search optimization to distribution network reconfiguration. Int J Electr Power Energy Syst 62:461–468. https://doi.org/10.1016/j.ijepes.2014.04.064
7. Nguyen T, Nguyen T, Le B (2020) Optimization of electric distribution network configuration for power loss reduction based on enhanced binary cuckoo search algorithm. Comput Electri Eng 106893. https://doi.org/10.1016/j.compeleceng.2020.106893
8. Monteiro RVA, Guimarães GC, Moura FAM, Albertini MRMC, Silva FB (2016) Long-term sizing of lead–acid batteries in order to reduce technical losses on distribution networks: a

distributed generation approach. Elect Power Sys Res 144:163–174. https://doi.org/10.1016/j.epsr.2016.12.004

9. Degefa MZ, Lehtonen M, Millar RJ, Alahäivälä A, Saarijärvi E (2015) Optimal voltage control strategies for day-ahead active distribution network operation. Elect Power Sys Res 127:41–52. https://doi.org/10.1016/j.epsr.2015.05.018

10. Chittur P, Tant J, Radhakrishna J (2015) Novel methodology for optimal reconfiguration of distribution networks with distributed energy resources. Elect Power Sys Res 127:165–176. https://doi.org/10.1016/j.epsr.2015.05.005

11. Baioletti M, Milani A, Santucci V (2017) Algebraic particle swarm optimization for the permutations search space. In: 2017 IEEE Congress on Evolutionary Computation (CEC). San Sebastian, 1587–1594

12. Pegado R, Ñaupari Z, Molina Y, Castillo C (2019) Radial distribution network reconfiguration for power losses reduction based on improved selective BPSO. Elect Power Sys Res 169:206–213. https://doi.org/10.1016/j.epsr.2018.12.030

13. Napis NF, Fazliana A, Kadir A, Khatib T, Hassan EE (2018) An improved method for reconfiguring and optimizing electrical active distribution network using evolutionary particle swarm optimization. Appl Sci 8:804–822. https://doi.org/10.3390/app8050804

14. Mohamed Imran A, Kowsalya M (2014) A new power system reconfiguration scheme for power loss minimization and voltage profile enhancement using Fireworks Algorithm. Int J Electri Power Energy Syst 62:312–322. https://doi.org/10.1016/j.ijepes.2014.04.034

15. Ying-Yi H, Saw-Yu H (2005) Determination of network configuration considering multiobjective in distribution systems using genetic algorithm. IEEE Trans Power Syst 20(2):1062–1069

16. Mazliham MS, Abu Bakar B, Tahir MJ, Alam M (2018) Distribution system power losses minimization using network reconfiguration. Int J Integr Eng 10(7):1–9

17. Tiwar B, Sharma D (2020) Heuristic approach for power system optimization by comparing the active and reactive power loss in different bus systems. Global J Res Anal 1–5. https://doi.org/10.36106/gjra/4808661

18. Ghatak SR, Sannigrahi S, Acharjee P (2017) Comparative performance analysis of DG and DSTATCOM using improved PSO based on success rate for deregulated environment. IEEE Syst J 12(3):2791–2802

19. Ding F, Loparo K (2016) Feeder reconfiguration for unbalanced distribution systems with distributed generation: a hierarchical decentralized approach. IEEE Trans Power Syst 31(2):1633–1642. https://doi.org/10.1109/tpwrs.2015.2430275

20. Aman MM, Jamson GB, Bakar AHA, Mokhlis H (2014) Optimum network reconfiguration based on maximization of system loadability using continuation power flow theorem. Int J Electr Power Energy Syst 54:123–33

21. Gil-González W, Montoya OD, Rajagopalan A, Grisales-Noreña LF, Hernández JC (2020) Optimal selection and location of fixed-step capacitor banks in distribution networks using a discrete version of the vortex search algorithm. Energies 13:4914

22. Hamour H, Kamel S, Abdel-mawgoud H, Korashy A, Jurado F (2018) Distribution network reconfiguration using grasshopper optimization algorithm for power loss minimization. https://doi.org/10.1109/SEST.2018.8495659

23. Algamal Z (2019) Variable selection in count data regression model based on firefly algorithm. Statist Optimizat Inf Comput 7(2). https://doi.org/10.19139/soic.v7i2.566

24. Al Samman M, Mokhlis H, Mansor NN, Mohamad H, Suyono H, Sapari NM (2020) Fast optimal network reconfiguration with guided initialization based on a simplified network approach. IEEE Access 8:11948–11963

25. Salau AO, Gebru YW, Bitew D (2020) Optimal network reconfiguration for power loss minimization and voltage profile enhancement in distribution systems. Heliyon 6(6):e04233

26. Uniyal A, Sarangi S (2020) Optimal network reconfiguration and DG allocation using adaptive modified whale optimization algorithm considering probabilistic load flow. Electric Power Syst Res 106909

27. Nguyen TT, Nguyen TT, Nguyen NA, Duong TL (2020) A novel method based on coyote algorithm for simultaneous network reconfiguration and distribution generation placement. Ain Shams Eng J

28. Essallah S, Khedher A (2020) Optimization of distribution system operation by network reconfiguration and DG integration using MPSO algorithm. Renewable Energy Focus 34:37–46
29. Oda ES, Abdelsalam AA, Abdel-Wahab MN, El-Saadawi MM (2017) Distributed generations planning using flower pollination algorithm for enhancing distribution system voltage stability. Ain Shams Eng J 8(4):593–603
30. Memarzadeh G, Esmaeili S (2018) Voltage and reactive power control in distribution network considering optimal network configuration and voltage security constraints. ScientiaIranica 1(2):3–12. https://doi.org/10.24200/sci.2018.20565
31. Yang XS (2012) Flower pollination algorithm for global optimization. In: Durand-Lose J, Jonoska N (eds) Unconventional computation and natural computation. UCNC 2012. lecture notes in computer Science, vol 7445. Springer, Berlin, Heidelberg. https://doi.org/10.1007/978-3-642-32894-7_27
32. Nabil E (2016) A modified flower pollination algorithm for global optimization. Expert Syst Appl 57:192–203. https://doi.org/10.1016/j.eswa.2016.03.047
33. Kopciewicz P, Łukasik S (2019) Exploiting flower constancy in flower pollination algorithm: improved biotic flower pollination algorithm and its experimental evaluation. Neural Comput Appl 32(16):11999–12010. https://doi.org/10.1007/s00521-019-04179-9
34. Wei J, Chen Y, Yu Y, Chen Y (2019) Optimal randomness in swarm-based search. Mathematics 7:828. https://doi.org/10.3390/math7090828
35. Sabarinath G, Gowri Manohar T (2019) Application of bird swarm algorithm for allocation of distributed generation in an Indian practical distribution network, I J Intell Syst Appl 7:54–61. https://doi.org/10.5815/ijisa.2019.07.06

Chapter 61
Optimization and Modelling of Storage Conditions in Hydro-cooling of Sapota (*Manilkara zapota*) with Addition of Antimicrobial Agent Using Response Surface Methodology

R. Renu, Kavita Waghray, and P. Dinesh Sankar Reddy

1 Introduction

Fruits should be harvested at the right time for better quality and shelf life. If fruits are harvested during ripened stage, they are more susceptible to mechanical injury, pathogens and physiological disorders which will lead to shorter shelf life [1]. Proper cooling decreases the ripening rate, water loss and decay, hence, preserves the quality and extends shelf-life of the fruit [2]. The ideal storage temperature for sapodillas is either 20 or 0 °C [3]. Both Sapodilla and Sapote Mamey are chilling sensitive fruits [4]. Long distance transportation of Sapota is very difficult due to its less storage life. The short harvesting season limited domestic demand and improper storage facility create glut within the market and consequently loss to the fruit growers [5]. Hydro-cooling is fifteen times faster than other cooling technique, there are differences in time required to cool different products to the target temperature [6]. This is due to differences in products geometry size and the thermal properties [7]. Hydro-cooling can be a substitute to modified atmosphere packaging, or otherwise combination of both can give good results in terms of fruit storage [8]. Synthetic harmful chemicals are used to prevent the postharvest losses, among which some are carcinogenic and cause environmental pollution. An alternative way for synthetic chemicals is use of biological compounds which are nontoxic, specific in their action and safe to environment and for the living beings [9]. Adding $CaCl_2$ to water medium used for

R. Renu (✉) · P. Dinesh Sankar Reddy
JNTUA College of Engineering Anantapur, Ananthapuramu, Andhra Pradesh, India

K. Waghray
University College of Technology, Osmania University, Hyderabad, India

P. Dinesh Sankar Reddy
NIT, Tadepalligudem, Andhra Pradesh, India

© The Author(s), under exclusive license to Springer Nature Singapore Pte Ltd. 2021   823
S. Kumar et al. (eds.), *Proceedings of International Conference on Communication and Computational Technologies*, Algorithms for Intelligent Systems,
https://doi.org/10.1007/978-981-16-3246-4_61

hydro-cooling increased calcium content of fruit tissue and reduced respiration rate, ascorbic acid degradation, and membrane lipid peroxidation for sweet cherry fruit [10]. It is a well-known fact that plant extracts have no negative impact on health and can be managed with low cost [11]. Neem leaves (*Azadirachta indica*) has been used as fungicide and insecticides since centuries [12]. The current study was carried out to evaluate the effect of hydro-cooling with antimicrobial agent (Neem leaf) on physical attributes and on shelf life of Sapota.

2 Materials and Methods

2.1 Hydro-cooling Unit Design

Thermocol box dimensions: Thickness—4 cm, Height—28.5 cm, Length—60 cm, Width—46 cm, Capacity—35 L. In this study, two identical thermocol boxes and overhead showers were used. One of the thermocol box acts as water reservoir to which showers were fixed, and second box kept at the bottom acts as a cooling chamber in which fruit were placed. First box with showers was placed on the supporting stands and cooling chamber placed exactly below the reservoir. After attaining core temperature, the fruits were covered with polyethylene sheet so that water does not seep into the stored fruit.

Antimicrobial Agent: Fresh Neem and Tamarind leaves were plucked, dried, and grinded into powder. The extracts were prepared by adding 100 ml of distilled water to 100 g of each leaf powder separately and kept overnight. This resulted in 100% concentration of each plant extracts [13]. The prepared extract was added into the cooling medium (water) at a rate of 100 per cent of extract per litre of cooling water.

2.2 Cooling Time of Fruit and Temperature Relationship

Hydro-cooling decreases the fruit temperature and maintains relative humidity within the cooling box. The box temperature and humidity was measured using digital humidity–temperature metre, also the core temperature of fruit using a thermocouple throughout the cooling process.

2.3 Percentage Loss in Weight

Sapota fruits (three replicates) were labelled and weighed. Weight of the same labelled fruits were recorded at regular intervals throughout storage period. The average per cent loss in weight was calculated for each treatment, according to the

Table 1 Factors selected for design of experiments

Factor 1	Factor 2	Factor 3
Fruit size (m)	Water medium temperature (°C)	Water flow rate (ml/s)
0.00454	13	0.83
0.00567	18	1.16
0.00662	20	4.16

following equation:

$$\text{Weight loss}(\%) = [(W_0 - W_1)/W_0] \times 100$$

where W_0 is the initial weight and W_1 is the weight measured at sampling date [14].

2.4 Total Soluble Solids

The total soluble solids (Brix) was recorded using handheld refractometer [13].

2.5 Statistical Analysis Using General Factorial Design

Three different set of treatments were applied to study the shelf life extension of Sapota by hydro-cooling. The experiments were performed according to a full Factorial Design using Design Expert (7.0.0). The experiments are 3^3 factorial design, i.e. 27 treatment combinations for each set of treatments. The independent and dependent variables are fruit size(m), cooling medium temperature (°C), cooling medium flow rate (ml/s) and shelf life (Days), respectively.

Total Treatments: 27.

Design: Response Surface Methodology (RSM) (Table 1).

3 Results

3.1 Cooling Time of Fruit and Temperature Relationship

The core temperature of fruit was checked at regular intervals using a thermocouple throughout the cooling process.

Figures 1, 2 and 3 demonstrate the relationship between time and temperature of fruits which were divided into three different fruit sizes at three different water

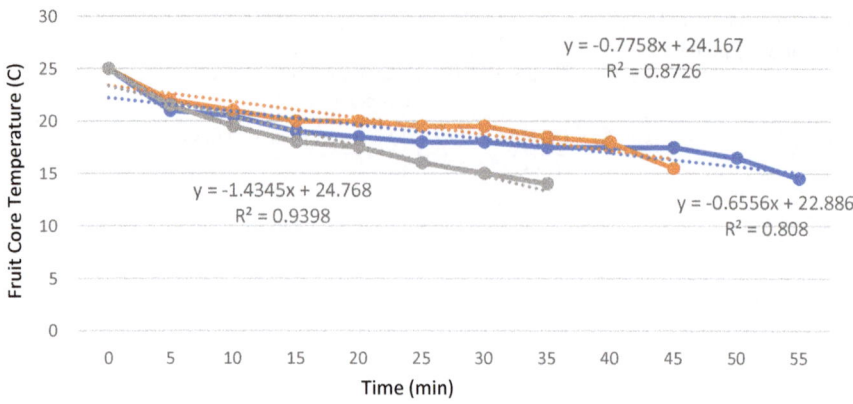

Fig. 1 Time–temperature response of different sized Sapota at medium temperature 20 °C

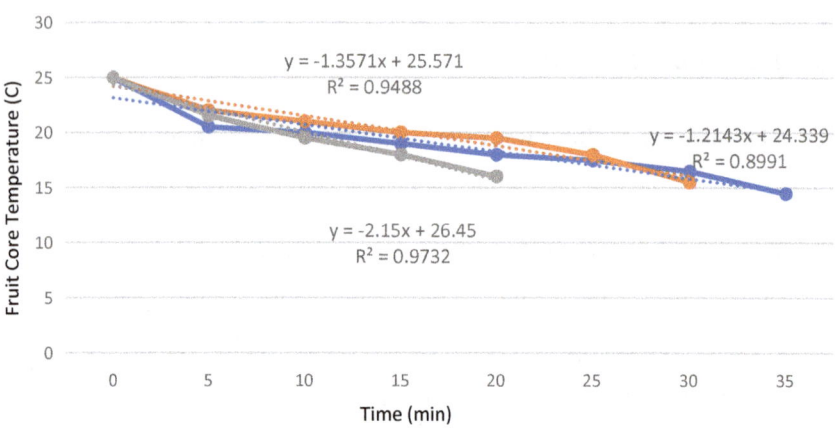

Fig. 2 Time–temperature response of different sized Sapota at medium temperature 18 °C

medium temperature maintained. Linear trend was observed at all the velocities and R^2 value was the highest for small size fruit in all the cases. As described after the cooling time, fruit core temperature was reduced to 10–15 °C. It was observed that Neem leaf extract in the water used for cooling did not affect the cooling trend.

3.2 Percentage Loss in Weight

The percentage loss in weight (PLW) at 13 °C was significantly reduced by hydro-cooling and it was affected by fruit size also; for control sample (without cooling) was as high as 13.0% within 4 days of cooling, whereas for fruit sample subjected

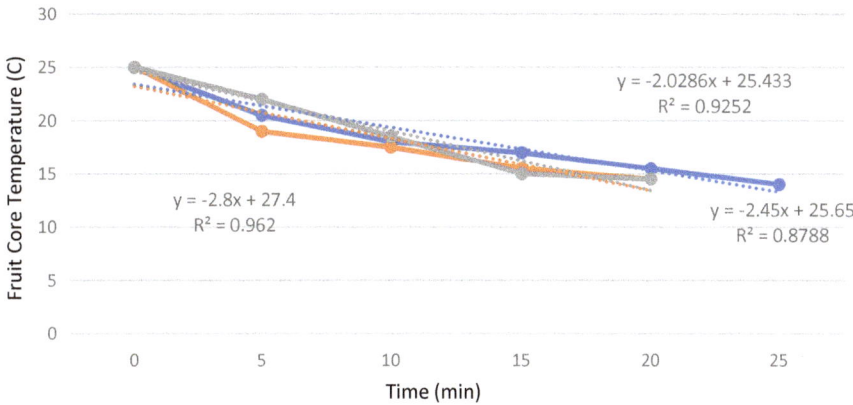

Fig. 3 Time–temperature response of different sized Sapota at medium temperature 13 °C

Table 2 Percentage loss in weight of Sapota stored at different water medium temperature at least water flow rate

Fruit size	Medium temperature (°C)	PLW (%) after 20 days of storage
Large	20	–
	18	0.116a
	13	0.006 bc
Medium	20	–
	18	0.120e
	13	0.051c
Small	20	–
	18	0.083de
	13	0.062 cde

Means followed by the same letters are not significant but different letters are significant at 0.05 level according to LSD method– Indicates that fruit was decayed for the prescribed storage time period

for cooling till 20 days also weight loss did not exceed 1–2%. Physiological loss in weight is due to respiration, transpiration and other biological changes taking place in the fruit. Due to higher water loss, fruits undergo shrivelling stored at high temperature [15]. Hence, as these conditions are restricted due to hydro-cooling there was reduction in percentage loss in weight. The result was on par with observation done by Ravikumar et al. [16] (Table 2).

Table 3 Change in TSS of Sapota at different temperatures at least water flow rate

Fruit size	Medium temperature (°C)	TSS (°B)	
		Before	After 20 days of storage
Large	20	19.0 ± 0.26	–
	18	19.4 ± 0.35	23.5 ± 0.21
	13	19.4 ± 0.31	22.9 ± 0.35
Medium	20	19.5 ± 0.26	–
	18	19.4 ± 0.23	21.0 ± 0.33
	13	18.9 ± 0.13	20.6 ± 0.32
Small	20	19.0 ± 0.29	–
	18	19.0 ± 0.23	22.3 ± 0.12
	13	19.5 ± 0.22	21.5 ± 0.19

– Indicates that fruit was decayed for the prescribed storage time period

The values are indicated as mean ± S.D. of three fruit samples

3.3 Total Soluble Solids

The TSS content of fruits gradually increased throughout the storage period but the values were on par with fresh fruit sample value. Similar observations were obtained by [17]. TSS content increase during ripening process in Sapota fruit can be related to production of more sugars in the fruit due to hydrolysis of starch and slight decline at over ripe stage was due to utilization of sugars during respiration process (Table 3).

3.4 Response Surface Method

Details about the experimental runs, medium temperature, fruit size, and water flow rate are tabulated. As the above-mentioned factors are controlling, the cooling rate which in turn the shelf life of the produce response surface method was applied using software by taking shelf life as the dependent variable (Table 4).

Model Fitting and Analysis of Response

Good storage conditions should be able to maximize the shelf life of the produce. Table 5 presents the summary of the results for fitting a model. A model should be rejected if the result showed significance in the LOF test suggested by Meir et al. [18]. The model F-value of 89.65 suggested that the model is significant. There is only 0.01% chance that a "Model F-Value" this large could occur due to noise. Values of "Prob > F" less than 0.0500 indicate model terms are significant. In this case, A,

Table 4 Shelf life of stored Sapota as a function of medium temperature, fruit size and medium flow rate during forced air cooling

Run	A: Medium temp (°C)	B: Fruit size (m)	C: Flow rate (ml/s)	Shelf life (Days)
1	13	0.00454	0.833	20
2	18	0.00454	1.66	19
3	20	0.00454	4.16	17
4	13	0.00454	0.833	19
5	18	0.00454	1.66	19
6	20	0.00454	4.16	17
7	13	0.00454	0.833	19
8	18	0.00454	1.66	18
9	20	0.00454	4.16	16
10	13	0.00567	0.833	18
11	18	0.00567	1.66	14
12	20	0.00567	4.16	12
13	13	0.00567	0.833	18
14	18	0.00567	1.66	13
15	20	0.00567	4.16	12
16	13	0.00567	0.833	16
17	18	0.00567	1.66	13
18	20	0.00567	4.16	10
19	13	0.00662	0.833	10
20	18	0.00662	1.66	7
21	20	0.00662	4.16	6
22	13	0.00662	0.833	9
23	18	0.00662	1.66	7
24	20	0.00662	4.16	5
25	13	0.00662	0.833	8
26	18	0.00662	1.66	7
27	20	0.00662	4.16	5

B, C, A^2 are significant model terms. Values greater than 0.1000 indicate the model terms are not significant.

The "Pred R-Squared" of 0.9489 is in reasonable agreement with the "Adj R-Squared" of 0.9684. "Adeq Precision" measures the signal to noise ratio. A ratio greater than 4 is desirable. A ratio of 29.856 indicates an adequate signal. This model can be used to navigate the design space.

The suggested model was (**in Terms of Coded Factors**)

$$Shelf\ Life = +15.18 - 5.51 * A - 2.03 * B - 0.63 * C - 0.33$$

Table 5 Results summary of fitting a model in the optimization of storage conditions

Source	Sum of squares	df	Mean square	F Value	p-value Prob > F	Significant/non-significant
Model	650.95	9	72.33	89.65	<0.0001	Significant
A: Fruit size (m)	511.24	1	511.24	633.71	<0.0001	Significant
B: Medium temp (°C)	71.24	1	71.24	88.31	<0.0001	Significant
C: Flow rate (ml/s)	6.91	1	6.91	8.57	0.0094	Significant
AB	1.43	1	1.43	1.77	0.2013	
AC	2.744E−003	1	2.744E−003	3.401E−003	0.9542	
BC	0.19	1	0.19	0.36	0.6337	
Std. dev.	0.90		R-squared		0.9794	
Mean	13.11		Adj R-squared		0.9684	
C.V %	6.85		Pred R-squared		0.9489	
PRESS	33.95		Adeq precision		29.856	

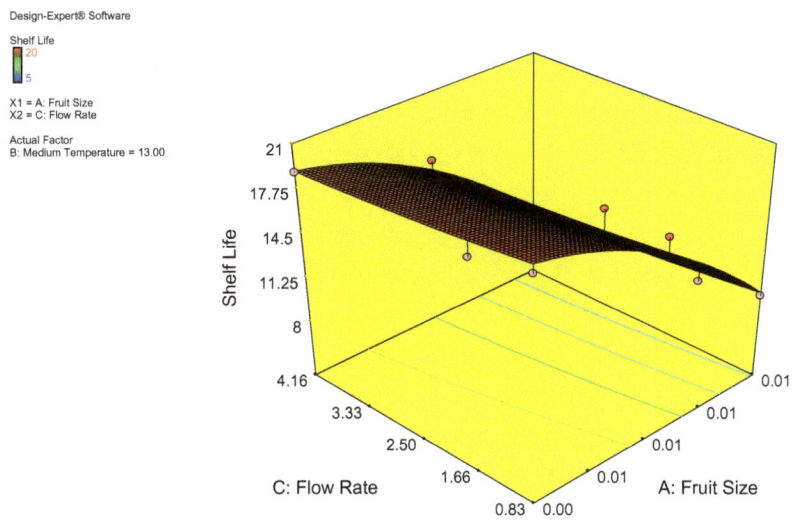

Design-Expert® Software

Shelf Life
20
5

X1 = A: Fruit Size
X2 = C: Flow Rate

Actual Factor
B: Medium Temperature = 13.00

Here 0.00 and 0.01 indicate Fruit size as Least (0.00454) and biggest (0.00662)		

Fig. 4 Relationship between medium temperature, fruit size and water medium flow rate to shelf life for different sized fruits

$$* A * B + 0.015 * A * C + 0.12 * B * C - 1.83 * A^2$$
$$- 0.87 * B^2 + 0.19 * C^2$$

Diagnostic and Optimum Storage Conditions

Figure 4 shows the relationship between the medium temperature, fruit size and water medium flow rate to shelf life for different sized fruits in 3-D plot and Fig. 5 shows the optimization chart predicted for best run out of 27 runs by considering medium temperature, flow rate of water and shelf life for different sized fruits in 3-D plot. From the graphs, it is depicted that best combination of various parameters such as medium temperature, fruit size and water medium flow rate for different sized fruits for hydro-cooling technique is smallest fruit (0.00454 mm), least cooling temperature (13 °C) and least water flow rate (0.83 ml/s) is required to attain longest shelf life of 20 days.

3.5 Microbial Load of Bell Peppers Stored in Three Different Treatments

Microbial load after 20 days of fruit storage was evaluated using total plate count. At lowest temperature (13 °C), microbial spoilage was within acceptable range even at

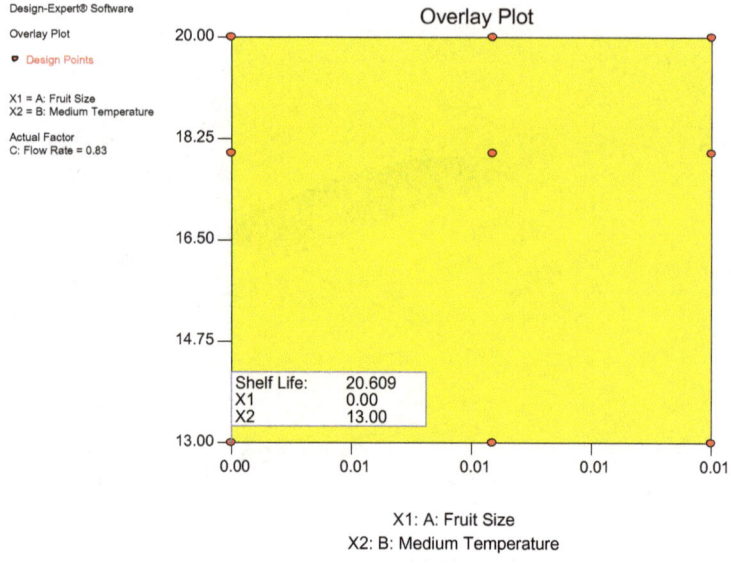

Design-Expert® Software

Overlay Plot

● Design Points

X1 = A: Fruit Size
X2 = B: Medium Temperature

Actual Factor
C: Flow Rate = 0.83

Shelf Life: 20.609
X1   0.00
X2   13.00

X1: A: Fruit Size
X2: B: Medium Temperature

Here 0.00 and 0.01 indicate Fruit size as Least (0.00454) and biggest (0.00662)

Fig. 5 Optimization chart predicted for best run out of 27 runs by considering medium temperature, flow rate of water and shelf life for different sized fruits

Table 6 Microbial load of Sapota before and after 20 days of storage (Cfu)

At 13 °C (cfu/ml)		At 18 °C (cfu/ml)		At 20 °C (cfu/ml)	
Before	After	Before	After	Before	After
3.0×10^{-4}	3.9×10^{-4}	3.1×10^{-4}	4.8×10^{-4}	3.0×10^{-4}	-

– Indicates sample has been observed for Fungal Growth.

the end of 20 days, but at highest temperature (20 oC), fungal growth was observed within 2–5 days of storage. By addition of antimicrobial extract in the cooling water microbial spoilage is prevented during storage (Table 6).

4 Conclusions

In the present study, the shelf life of Sapota was extended up to 20 days when stored under optimum storage conditions in hydro-cooling chamber and with the use of antimicrobial (Neem leaf) agent, whereas control (without cooling and coating) stayed fresh only for 3–5 days. Time taken for cooling smallest to largest sized Sapota fruit was 25 min, 35 min and 55 min, respectively. The most optimum conditions were found to be for smallest Sapota fruit (0.00454 mm), least cooling temperature (13 °C) and least water flow rate (0.83 ml/s) among 27 treatments. As the storage temperature, size of the fruit and water medium flow rate increased shelf-life decreased. Mathematical modelling showed the relation between water medium flow rate, cooling temperature and sample size taken and the response (shelf life) all the parameters were significant. Percentage loss in weight of the fruit sample was found to be maximum for control sample (without cooling) 13% after 4 days of storage, whereas minimum for the cooled sample 0.04–0.123% after 10–20 days of storage period. TSS increased about 5–7% for control, but whereas for cooled Sapota, it was not more than 1–2%. The microbial load of the best treatment combination was also within the permissible limits which as achieved by use of antimicrobial (Neem leaf) agent. The present study reveals that hydro-cooling can be a fastest, less expensive, and simpler way for cooling and extending the shelf life of Sapota. Also, studies can be extended to investigate the change in texture in terms of firmness of fruits subjected for hydro-cooling using antimicrobial agent in the cooling medium.

References

1. Juan JL, Frances J, Montesinos E, Camps F, Bonany J (1999) Effect of harvest date on quality and decay losses after cold storage of "Golden Delicious" apple in Girona, Spain. Acta Hortic 485:195–202
2. Ferreira MD, Brecht JK, Sargent SA, Aracena JJ (1994) Physiological responses of strawberry to film wrapping and precooling methods. Proc Fl State Hort Soc 107:265–269
3. Snowdon ALA (1990) Color atlas of post-harvest diseases of fruits and vegetables: general introduction and fruits, vol 1. CRC Press, Boca Raton
4. Campbell CA (1994) Handling of Florida-grown and imported tropical fruits and vegetables. Horticult Sci 29:975–978. https://doi.org/10.21273/HORTSCI.29.9.975
5. Antala DK, Satasiya RM, Akabari PD, Bhuva JV, Gupta RA, Chauhan PM (2014) Effect of modified atmosphere packaging on shelf life of Sapota fruit. Int J Agric Sci Technol (IJAST) 2(1). https://doi.org/10.14355/ijast.2014.0301.05
6. Manganaris GA, Ilias IF, Vasilakakis M, Mignani I (2007) The effect of hydrocooling on ripening related quality attributes and cell wall physicochemical properties of sweet cherry fruit (*Prunus avium* L.). Int J Refrig 30:1386–1392. https://doi.org/10.1016/j.ijrefrig.2007.04.001
7. Teruel B, Kieckbusch T, Cortez L (2004) Cooling parameters for fruits and vegetables, pp 655–658. https://doi.org/10.1590/S0103-90162004000600014
8. Toivonen PMA (1997) The effects of storage temperature, storage duration, hydro-cooling and micro-perforated wrap on shelf life of broccoli (*Brassica oleracea* L., Italica Group). Postharvest Biol Technol 10:59–65. PII: **SO**925-5214(96)00064-6
9. Abirami LSS, Pushkala R, Srividya N (2013) Antimicrobial activity of selected plant extracts against two important fungal pathogens isolated from papaya fruit. Int J Res Pharmaceut Biomed Sci 1:234–238
10. Wang Y, Xie X, Long LE (2014) The effect of postharvest calcium application in hydro-cooling water on tissue calcium content, biochemical changes, and quality attributes of sweet cherry fruit. Food Chem 160:22–30. https://doi.org/10.1016/j.foodchem.2014.03.0730308-8146/2014
11. Pervin R (2016) Effect of combined botanical extracts on postharvest performances of bitter gourd (*Momordica charantia*). B.Sc. Thesis submitted to Dept. of Food and Nutrition, Khulna City Corporation Women's College (Affiliated to Khulna University), Khulna, Bangladesh

12. Chaturvedi R, Razdan MK, Bhojwani SS (2003) Production of haploids of neem (*Azadirachta indica A. Juss.*) by another culture. Plant Cell Rep 21:531–537. PMid:12789427
13. Shrestha S, Pandey B, Mishra BP (2018) Effects of different plant leaf extracts on postharvest life and quality of mango (*Mangifera indica L.*). Int J Environ Agric Biotechnol (IJEAB) 3(2). https://doi.org/10.22161/ijeab/3.2.14
14. Elansari AM, Mostafa YS (2018) Vertical forced air pre-cooling of orange fruits on bin: Effect of fruit size, air direction, and air velocity. J Saudi Soc Agric Sci. https://doi.org/10.1016/j.jssas.2018.06.006
15. Kader AA (1992) Post-harvest technology of horticultural crops. Div Agric Nat Sci Univ California Spec Publ 3311:56–66
16. Ravikumar M, Desai CS, Raghavendra HR, Pooja N (2018) Effect of pre-cooling in extending the shelf life of banana cv. Grand naine stored under different storage conditions. Int J Chem Stud 6(3):872–878
17. Baloch MK, Bibi F, Jilani MS (2011) Quality and shelf life of mango (*Mangifera indica L.*) fruit: as affected by cooling at harvest time. Sci Horticult 130(2011):642–646. https://doi.org/10.1016/j.scienta.2011.08.022
18. Meir S, Rosenberger I, Aharon Z, Grinberg S, Fallik E (1995) Improvement of the postharvest keeping quality and colour development of bell pepper (cv. 'Maor') by packaging with polyethylene bags at a reduced temperature. Postharvest Biol Technol 5:303–309

Chapter 62
An Intelligent Fault Monitoring System for Railway Neutral Sections

Kennedy Phala, Wesley Doorsamy, and Babu Sena Paul

1 Introduction

Among the existing methods used to detect and isolate the faulty sections on the railway overhead wires, anomaly detection and classification of pantograph failures is an important technique for fault management and resource planning. The current challenge with the neutral section (NS) is the continuous failure of balancing dropper wires, burnt arc runners, worn insulation rods and excessive sparks as a result of unbalanced NS. Failures on the NS itself, poses a challenge because the relevant protection relay will not trip unless there is failure of the catenary wire. Foot patrols and trolley inspections are proving to be labor intensive and inefficient to some extent due to human fatigue and experience. There is no readily available system that can monitor and detect abnormal events on the NS now. The current methods used to inspect the NS is periodical or time-based.

The overhead system, between the NS, is protected from fault currents and insulation failure using protection scheme (voltage and current transformer, distance relays etc.). Although the long sections of catenary wire make it is possible to deploy wired sensors, a wireless system offers attractive alternative due to the scalability and flexibility, as well as physical isolation offered [1]. A system embedded with wireless sensors and cloud computation capabilities is required to discover potential faults autonomously using the Artificial Intelligence sub-fields. Real-time monitoring and fault detection or the rail infrastructure is essential because it enables early detection of problems, allowing for the system to be halted and rectified before the occurrence of

K. Phala
Department of Electrical and Electronic Engineering Technology, University of Johannesburg, Johannesburg, South Africa

W. Doorsamy (✉) · B. S. Paul
Institute for Intelligent Systems, University of Johannesburg, Johannesburg, South Africa
e-mail: wdoorsamy@uj.ac.za

© The Author(s), under exclusive license to Springer Nature Singapore Pte Ltd. 2021
S. Kumar et al. (eds.), *Proceedings of International Conference on Communication and Computational Technologies*, Algorithms for Intelligent Systems,
https://doi.org/10.1007/978-981-16-3246-4_62

catastrophic railway incidents, unplanned downtime and/or costly avoidable damage [2]. All the recorded big data that is unfiltered, unstructured needs to be stored, processed and analyze before any anomaly detection and classification is performed with the use of machine learning algorithms. The prompt increase in research and development in the area of big data, internet-of-things and cloud computing provide a pathway for railway infrastructure to be continuously monitored using intelligent sensors thereby improving network availability and safety, as well as reduce maintenance costs [3]. In fact, there much ongoing research recently toward employing data analytics, IoT sensor and edge computing in rail transportation for monitoring train operating parameters and conditions [4–9].

Rail networks typically consist of multidisciplinary departments—e.g., telecommunications, technical support, signaling, perway track—that need to interact with one another on an operational basis. The interaction of these departments—and the respective operational disciplines—is critical in maintaining safe operation of locomotives. The seamless flow of condition information between the various parties is thus an essential component of successful interaction, and operations. An important aspect of NS anomaly detection and pantograph fault classification is the nature of the track, speed of the locomotives, catenary system and pantograph uplift force. These aspects influence how the NS behaves during operation. Furthermore, maintenance of infrastructure such as overhead wires, track and pantographs are ongoing. Thus, there are subtle operational and infrastructural differences that result in unique operations and practice. This makes automatic fault monitoring via rule-based systems challenging, and thus, intelligent monitoring via data-driven methods are more viable.

The rest of the paper is organized as follows. The next section presents a background of the machine learning techniques that are employed in the presented system. Thereafter, the development of the system including the modeling and testing is presented. The results are then discussed before a summary of the conducted research is given in the conclusion.

2   Background

In this section, the machine learning techniques and algorithms that form part of the fault detection and classification are reviewed. In general, supervised learning requires data comprising training examples, for the purpose of model building [10]. This type of learning is not immediately suitable in the context of the presented application because training data is not readily available. The data need to be split into training and testing where labeled data—in the case of known fault instances—are used. The disadvantage regarding application of this type of learning here is that there is a data imbalance where a comprehensive collection of fault instances are not available. Unsupervised learning does not require training examples for building of a machine learning model, and thus, these are immediately available to be used in the presented system for fault pattern analysis/detection [11]. This type of learning makes it suitable for fault detection/pattern recognition. The core function of machine learning is to provide intelligent methods of interpretation and analysis of data to execute informed decisions on addressing condition monitoring and fault detecting of infrastructure assets. In this case, the feasibility of the system to recognize and suitably distinguish between fault instances, and then utilize this information toward building of a classification system is thus investigated. For the purpose of this investigation, k-means and support vector machines are therefore utilized.

2.1   Clustering

Readily available of data makes k-means suitable to group the dataset based on their similarity. The clustering algorithms involve several types; not all provide models for their clustering to be grouped based on the similarities. K-means algorithm meets the criteria because it is quicker, easy to implement and capable to handle big data. K-means clustering algorithm is a way of grouping data in the cluster, based on their similarity. The algorithm searches for hidden patterns, while objectively classifying the measured raw data from sensors. There are no target variables from this algorithm; it usually self-trained identifying areas lacking the expected outputs or results [12, 13]. The k-means model is constructed, employing the dataset from ThingSpeak™. There are eight fields (AccX, AccY, AccZ, angleX, angleY, angleZ, ambient temperature & object temperature) measured from the pantograph contact-wire interaction. Their fields measured diverse variables from the site, and the data are transmitted to the cloud for computation and diagnostic. K-means uses distance (xy–yi) to determine the similarity or dis-similarity between the dataset of the detected failures. The complexity of the iterations performed on a large dataset, and the speed of convergence render k-mean computationally attractive compared to other algorithms. K-means will be implemented for the objective of grouping the detected faults into clusters.

Algorithm: K-means
Input: M, K where M = set of measured data and K = integer
  Output: K clusters
  Require $M \neq \emptyset$, $K > 0$
  Procedure Generate Cluster (k-fold)
    Initialize K random Centroids
    repeat
    for all Instance i in M **do**
      shortest $\leftarrow 0$
      membership $\leftarrow$ null
        for all Centroid c **do**
          dist $\leftarrow$ Distance(c)
$$d(x_j, y_j) = \left(\sqrt{x_j - y_j}\right)$$
          if dust < shortest **then**
            shortest $\leftarrow$ dist
            membership $\leftarrow$ c
        end if
        end for
      end for
      Recalculate Centroids(c)
    Until convergence
  end procedure

2.2 Support Vector Machines

The SVM is a supervised learning model, based on a classifier model, classifying between two categories by creating a hyperplane in the high-dimensional input space, used for classification. Various SVM kernel functions are identified, such as linear, non-linear, polynomial, Gaussian kernel, and a radial basis function (RBF). SVM kernel performs well, providing continuous data, and that algorithm can learn more from fewer samples. SVM forms part of the experiment was recorded events of the contact-pantograph interaction is classified. The set input data (features x) are required to train the algorithm [14].

$$x = [x_1, x_2, x_3, \ldots, x_n] \tag{2}$$

The label denotes the class

$$c_i = \in \{c_1, c_2, \ldots c_j\} \tag{3}$$

The training set X is defined as a set of N's, containing values such as:

$$X = \{x^t, l^t\} t^N{}_{t=1} \tag{4}$$

The events or non-events labeled are classified using SVM kernel function. SVM performs well on classifying detected faults, provided that input data are trained correctly with no errors.

Algorithm: Training an SVM
Require: x, y and z loaded with trained labelled data $\alpha \Leftarrow 0$ or $\alpha \Leftarrow$ trained SVM
$C \Leftarrow 100$ (training data)
repeat
    for all (x_i, y_i, z_i); (x_j, y_j, z_j); do
        optimize decision boundaries α_i and α_j
    end for
until no changes in α or other resource constraint are met
    Ensure: Retain only the support vectors ($\alpha_i > 0$)

3   Methodology

In the proposed design, two sensors namely accelerometer (ADXL345) and non-contact infrared thermometer sensor (MLX90614) are housed inside an enclosure. These monitor the temperature and motion of the NS to detect any anomaly. GSM/GPRS is used to transmit data to ThingSpeak for cloud computing, thereby enabling real-time monitoring. The circuit is powered by 3 solar panels during the day with charging control and batteries allowing powering of the device at night. The enclosure housing the sensors and circuit are installed on the NS for monitoring and detection. Data analytics are executed online via ThingSpeak. The data processing and analytics are executed online, and in real-time, using Matlab interfaced with the ThingSpeak channels [15, 16]. The monitoring and fault detection system prototype hardware implementation and installation is shown in Fig. 1. It should be noted that the presented monitoring system includes both physical and digital security of the

Fig. 1 Monitoring system hardware implementation with **a** internal circuit, and **b** installation on railway neutral section

sensors and data. The enclosure is shielded and sensors are digitally secured at the point of measurement through to secured communication channels onto ThingSpeak.

3.1 K-means Model

The experimental results recorded into CSV file from the cloud are imported into Matlab where plots and analyzes are performed. Firstly, the results are fitted with varying number of clusters against the error to investigate the monitoring techniques ability to recognize the different conditions in an unsupervised manner. K-means is computed for different values of k ranging from 1 to 10 clusters were for each value of k within-cluster sum of the squared (WSS) error is calculated. Results of the WSS computations are plotted against the number of clusters value (1–10). Figure 2 shows that the optimal number of clusters is at k equals 3 were the graphs begins to flatten (or the elbow). This coincides with the physical ground truth of the conditions or events occurring as shown in Fig. 3.

K-means clustering is initialized using the clusters obtained from elbow method. Each cluster is grouped using the squared Euclidean distance $d(x_j, y_j)$. Clusters are computed closer to the centroids. The 3D scatter shows three clusters of the events recorded through the NS from sensors installed to detect the NS condition.

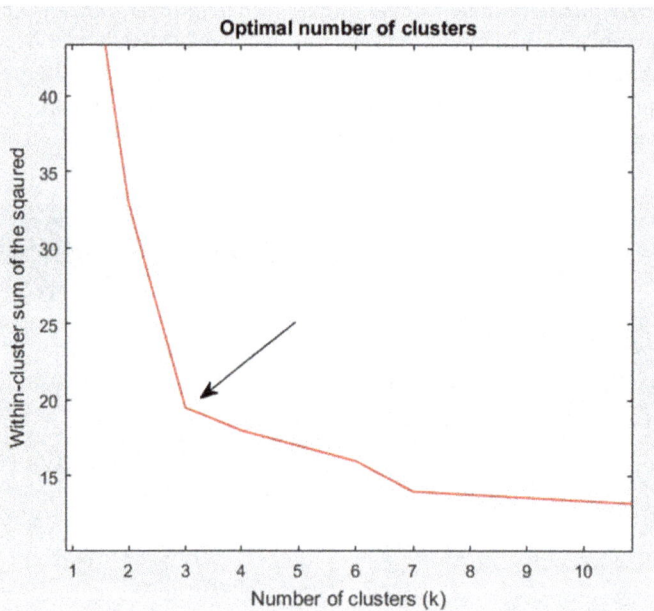

Fig. 2 Elbow method to determine optimal number of clusters via computed WCSS for different number of clusters

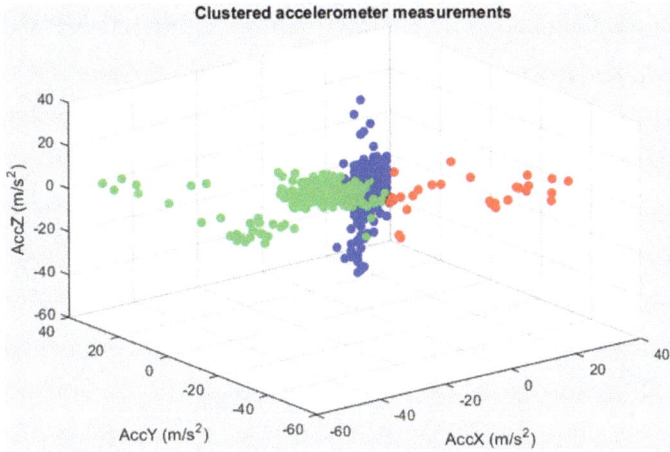

Fig. 3 Clustered accelerometer events as locomotive passes through the NS

Figure 3 shows a 3D scatter plot of clustered accelerometer measurements where the locomotive passes through the NS and are represented by the three different colors shown. The blue color represents the detected pantograph events, green color for contact wire oscillations after the pantograph past the NS and lastly the red color represents the excessive winds experienced during the experiment.

3.2   SVM Linear Kernel Model

The methodology is implemented on the experimental dataset collected from pantograph-contact wire experimental tests. Matlab is used to train and select the best classifier for the locomotive passes were all SVM classifiers are used to train and test the model, with input predictors (acceleration recordings) and response as labels. 80–20 train-test ratio is used to generate and verify the SVM classifier model.

Cross-validation of the training was performed to validate data feature and pattern accuracy, which has been shown to optimize the SVM classification approach [17–20]. Figure 4 shows a scatter plot of two classes represented by different colors, class 1 equals red and class 2 equals green. With the experimental dataset, the linear kernel SVM is used to train and separate the loco pantographs passes from the events recorded with the accuracy of 100%. Class 1 from Fig. 4 indicates the locomotive pantographs recorded, while the locomotive traversing at the NS at different speed intervals (15 km/h, 30 km/h, 45 km/h, and 60 km/h), whereas Class 2 are events detected (contact wire oscillations after pantograph passed the NS and also excessive winds).

Other SVM classifiers were tested and the best classifier model with higher accuracy and lower error is selected as linear kernel SVM as shown on Table 1. The training

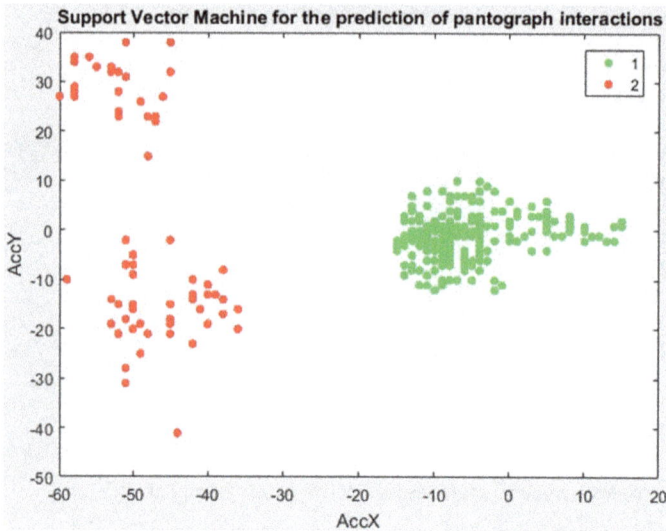

Fig. 4 Clustered accelerometer events were locomotive passing through the NS

Table 1 Results obtained for accuracies trained classifiers

Classifier	Accuracy (%)	Error (%)	Training time (s)
Linear SVM	100	0	19.54
Quadratic SVM	99.4	0.6	35.74
Cubic SVM	99.4	0.6	233.6
Fine Gaussian SVM	90	10	10.38
Medium Gaussian SVM	99.4	0.6	11.65
Coarse Gaussian SVM	89.1	10.9	10.35

data expansion is recommended to improve the SVM classifier for pantographs; large training sets are not readily available to model and simulate faults under measured conditions.

4   Conclusion

This paper presented an intelligent fault monitoring system on the overhead wires to detect failures arising from the NS. The experimental results were achieved through implementation and testing of a hardware prototype that interfaces wirelessly with ThingSpeak™, and provides continuous accelerometer and temperature recordings.

Firstly, an unsupervised approach is used to determine if the system is able to distinguish NS events in an unsupervised manner. The optimal number of clusters and instances correspond with the ground truth. The system was able to group the accelerometer events from the contact-pantograph interactions. Linear kernel SVM classification also proved to be good with the best accuracy to predict locomotive pantograph events. Machine learning is demonstrated to be feasible on the railway overheads where being able to detect and classify events accordingly are essential.

References

1. Aboelela E, Edberg W, Papakonstantinou C, Vokkarane V (2006) Wireless sensor network-based model for secure railway operations, In: IEEE international performance computing and communications conference, pp 1–6. https://doi.org/10.1109/.2006.1629461
2. Efanov D, Osadchy G, Sedykh D, Pristensky D, Barch D (2016) Monitoring system of vibration impacts on the structure of overhead catenary of high-speed railway lines. In: IEEE east-west design & test symposium (EWDTS), pp 1–8. https://doi.org/10.1109/EWDTS.2016.7807691
3. Kaddoumi T, Elhabashy K, Diab M, Watfa MK (2018) The impact of knowledge engineering and cloud computing adoption on business driven IT. Int J Mach Learn Comput 8(5):1–6
4. Saki M, Abolhasan M, Lipman J (2020) A novel approach for big data classification and transportation in rail networks. IEEE Trans Intell Transp Syst 21(3):1239–1249. https://doi.org/10.1109/TITS.2019.2905611
5. N. AlNaimi, and U. Qidwai, U., 2020, "IoT Based on-the-fly Visual Defect Detection in Railway Tracks," In 2020 IEEE International Conference on Informatics, IoT, and Enabling Technologies (ICIoT), pp. 627–631, 2020. doi: https://doi.org/10.1109/ICIoT48696.2020.9089560.
6. Lipare S, Bhavathankar P (2020) Railway emergency detection and response system using IoT. In: 2020 11th international conference on computing, communication and networking technologies (ICCCNT), pp 1–7. https://doi.org/10.1109/ICCCNT49239.2020.9225434
7. Righetti F, Vallati C, Anastasi G, Masetti G, di Giandomenico F (2020) Failure management strategies for IoT-based railways systems. In: 2020 IEEE international conference on smart computing (SMARTCOMP), pp 386–391. https://doi.org/10.1109/SMARTCOMP50058.2020.00082
8. Zhao Y, Yu X, Chen M, Zhang M, Chen Y, Niu X, Sha X, Zhan Z, Li WJ (2020) Continuous monitoring of train parameters using IoT sensor and edge computing. IEEE Sens J. https://doi.org/10.1109/JSEN.2020.3026643
9. Land A, Buus A, Platt A (2020) Data analytics in rail transportation: applications and effects for sustainability. IEEE Eng Manage Rev 48(1):85–91. https://doi.org/10.1109/EMR.2019.2951559
10. Zhang Y, Meratnia N, Havinga P (2010) Outlier detection techniques for wireless sensor networks: a survey. IEEE Commun Surv Tutor 12(2):159–170. https://doi.org/10.1109/SURV.2010.021510.00088
11. Ayodele T (2010) Types of machine learning algorithms. In: New advances in machine learning, pp 19–47. https://doi.org/10.5772/9385
12. Zhang Z, Cluster analysis in data mining. In: K-mean clustering algorithm, http://user.engineering.uiowa.edu/~ie_155/lecture/K-means.pdf
13. Park GY, Kim H, Jeong HW, Youn HY (2013) A novel cluster head selection method based on K-means algorithm for energy efficient wireless sensor network. In: 27th international conference on advanced information networking and applications workshops, Mar 2013, pp 910–915. https://doi.org/10.1109/WAINA.2013.123

14. Patil P, Kulkarni U (2013) SVM based data redundancy elimination for data aggregation in Wireless Sensor Networks. In: International conference on advances in computing, communications and informatics (ICACCI), Aug 2013, pp 1309–1316. https://doi.org/10.1109/ICACCI.2013.6637367
15. Phala KM, Doorsamy W, Paul BS (2019) A study into intelligent neutral section fault monitoring system on the coal line using wireless sensors networks. In: International heavy haul association (heavy haul 4.0—achieving breakthrough performance levels), June 2019, pp 611–615
16. Phala K, Doorsamy W, Paul BS (2019) Detection and clustering of neutral section faults using machine learning techniques for SMART railways. In: 2019 6th international conference on soft computing & machine intelligence (ISCMI), Oct 2019, pp 1–6. https://doi.org/10.1109/ISCMI47871.2019.9004366
17. Shi L, He Y, Li B, Wu Y, Huang Y, Cheng T (2019) Measurement of dynamic tilt angle by compensating gyroscope drift error. IEEE Trans Instrum Meas 1–9
18. Wei X, Hou L, Hao J (2018) Machine fault diagnosis using IIoT, IWSNs, HHT, and SVM. In: IEEE 18th international conference on communication technology (ICCT), pp 1–5
19. Zhan H (2010) Application of rough set and support vector machine in fault diagnosis of power electronic circuits. In: 2nd IEEE international conference on information management and engineering, pp 1–4
20. Liu T, Wang Z (2009) Design of power transformer fault diagnosis model based on support vector machine. In; 2009 international symposium on intelligent ubiquitous computing and education, pp 1–4

Chapter 63
Comparative Analysis of Various Kernel-Based SVM Algorithms for the Classification of Diabetes

Sounak Sinha, Soubhik Chaki, Sukanya Sadhukhan, Priya Das, and Sarita Nanda

1 Introduction

Diabetes is one of the most common and deadliest diseases in this twenty-first century. It is not only a normal disease but also creator of many diseases like heart attack, blindness, kidney diseases, etc. The regular method of identifying this disease is to visit a doctor's chamber and identify the disease [1]. However, this process is very lengthy and every time when a patient visits a doctor's chamber he has to spend a particular amount of money. Diabetes or diabetes mellitus is mainly caused by the inappropriate insulin secretion [2]. This inappropriate insulin secretion results in abnormal blood sugar level. Basically, diabetes is such a potential disease which can cause worldwide healthcare crisis. According to the data given by the organization International Diabetes Federation, 382 million people are already suffering from this diabetes and the number can become 592 million by year 2035 [3]. Machine learning is the scientific way in which the machine or a particular model learns from its own experience.

The main objective of this research has been to develop a system by which diabetic risk of a patient can be predicted with maximum accuracy [4]. This research is based on developing a system based on support vector machine (SVM) using different kernels. Using various kernels with SVM, how much accuracy [5] they are showcasing and how this analysis can be utilized in diabetes prediction are shown in this paper. Our paper is basically divided into different parts explaining introduction first, then some basic theory parts, next experimental setup and result analysis part and last the conclusion part.

S. Sinha · S. Chaki · S. Sadhukhan · P. Das · S. Nanda (✉)
School of Electronics Engineering, Kalinga Institute of Industrial Technology, Bhubaneswar, Odisha 751024, India
e-mail: snandafet@kiit.ac.in

© The Author(s), under exclusive license to Springer Nature Singapore Pte Ltd. 2021   845
S. Kumar et al. (eds.), *Proceedings of International Conference on Communication and Computational Technologies*, Algorithms for Intelligent Systems, https://doi.org/10.1007/978-981-16-3246-4_63

2 Overview of Kernel-Based SVM Algorithm

Support vector machine or SVM is a class of supervised machine learning algorithms that are widely used for classification and regression [6]. A special characteristic of SVM is that it still performs effectively even where the number of dimensions is greater than the number of samples. This property of SVM makes it very efficient even with a small amount of training data, and hence they often outperform neural networks. The idea of SVM algorithm is to find a decision boundary in a multidimensional space to distinctly classify each of the data points into separate classes. The basic difference between logistic regression and SVM is that in SVM two separate hyperplanes are parallelly created on both sides of the hyperplane that separate the data [7]. The separating hyperplane then maximizes the distance between these two planes. In other words, the goal of SVM is to find a hyperplane that has the maximum margins from the closest points of each category [8]. If we consider samples or data points in a D-dimensional space, then the hyperplane will be a (D-1)-dimensional separator. For instance, in a two-dimensional space the hyperplane will be a one-dimensional line separating the data points into different classes.

Suppose we have a set of n number of data points $\{x_1, x_2, x_3, \ldots, x_n\}$ where each point is represented with a feature vector x_i in a D-dimensional feature space, i.e.,

$$x_i \in R^D$$

For classifying more complex data points (nonlinear classification), we use nonlinear functions also called the kernel function to map this feature vector into a higher-dimensional feature space N.

$$\phi : R^D \Rightarrow R^N$$

or

$$\phi(x_i) \in R^N$$

Then, the equation of the separating hyperplane in the new N-dimensional space can be given by

$$w^T \phi(x_i) + b = 0 \tag{1}$$

where w is the weight and an N-dimensional vector, b is the bias and is a scalar and ϕ is a nonlinear function.

In SVM algorithm, we try to create two parallel hyperplanes on each side of the separating hyperplane (which is given by $w^T \phi(x_i) + b = 0$). These parallel hyperplanes can be described with the equations as

$$w^T \phi(x_i) + b = \pm \min y_i [w^T \phi(x_i) + b]; \quad \text{where } i = 1, 2, \ldots, n \tag{2}$$

In Eq. (2), the term min $y_i[w^T \phi(x_i) + b]$ represents the distance of the nearest data points to both of the parallel hyperplanes on both sides [6]. These data points that lie on the two parallel hyperplanes are called support vectors.

For simplification, we normalize this distance of the nearest data points to the parallel hyperplanes to 1. This can be achieved by multiplying a constant term to the weight factor w and the bias b. Therefore, we can rewrite Eq. (2) as

$$w^T \phi(x_i) + b = \pm 1 \tag{3}$$

Figure 1 shows the maximum margin hyperplanes of a trained SVM [9], and Fig. 2 shows the calculation of distance d between two parallel hyperplanes. From Fig. 2, we can find the distance between the two parallel hyperplanes and the goal will be to maximize this distance to get the appropriate decision boundary. This distance d is equal to the projection of the vector $(p-q)$ along the direction of the vector w. From geometry, we can calculate this distance to be 2/|w|. Therefore to find the optimal decision boundary, we need to maximize this distance d or minimize the value |w|.

For the model to classify the data points correctly, the value of the distance from the separating hyperplane should be greater than or equal to one, where the distance will be equal to one for all the data points that are support vectors.

$$y_i[w^T \phi(x_i) + b] \geq 1; \quad \text{where } i = 1, 2, \ldots, n \tag{4}$$

To avoid mis-classification, Eq. (4) should be true at all times; alongside, we need to minimize the value of |w|. The above optimization problem can be solved using Lagrange's multiplier theorem. The Lagrangian function that we are required to minimize can be written as [7]

Fig. 1 Maximum margin hyperplanes of a trained SVM

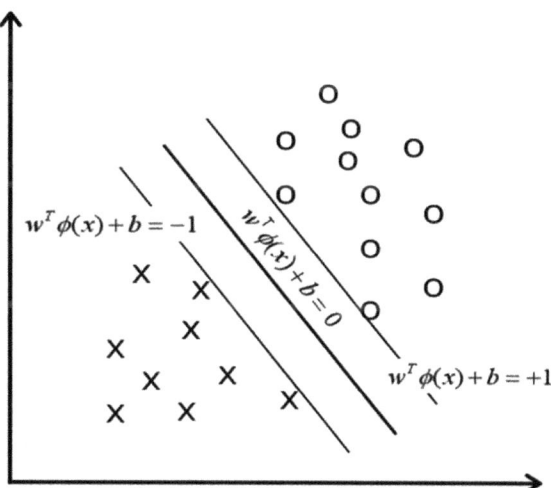

Fig. 2 Calculation of
distance d between two
parallel hyperplanes

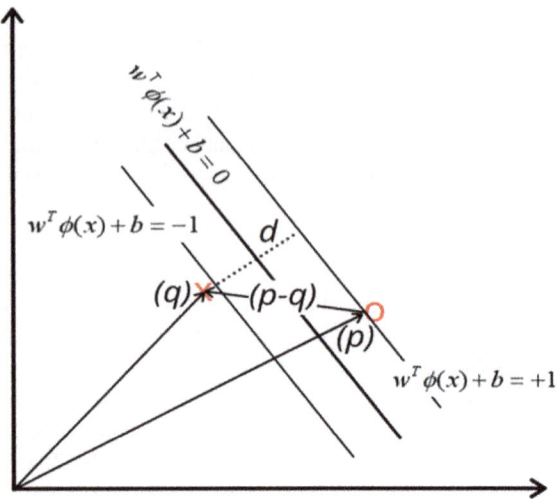

$$L_{(\alpha,w,b)} = \frac{1}{2}|w|^2 - \sum_{i=1}^{n} \alpha_i (y_i[w^T \phi(x_i) + b] - 1) \tag{5}$$

In Eq. (5), α_i is the Lagrange multiplier [10]. In order to perform the optimization for achieving the maximum margin hyperplane, we need to minimize Eq. (5) w.r.t. the weight vector w and bias b (since w and b are independent variables). Therefore, we partially differentiate Eq. (5) w.r.t. w and b and equate it to zero as

$$\frac{\partial L}{\partial w} = 0 \Rightarrow w_0 = \sum_{i=1}^{n} \alpha_i y_i \phi(x_i) \tag{6}$$

and

$$\frac{\partial L}{\partial b} = 0 \Rightarrow \sum_{i=1}^{n} \alpha_i y_i = 0 \tag{7}$$

Substituting the values of Eqs. (6) and (7) in Eq. (5) and simplifying it, we get the following relation

$$L = \sum_{i=1}^{n} \alpha_i - \frac{1}{2} \sum_{i=1}^{n} \sum_{j=1}^{n} \alpha_i \alpha_j y_i y_j \phi(x_i)^T \phi(x_j) \tag{8}$$

From Eq. (8), we have finally inferred that the maximization of the marginal distance d (in Fig. 2) will depend only on the dot product of the support vectors.

Kernelization allows SVM to classify more complex nonlinear data points by mapping them to higher dimensions [11]. In this paper, we have implemented the linear, polynomial [2], RBF, ANOVA RBF, Laplace RBF, sigmoid, Gaussian and Gaussian diagonal functions [12].

3 Experimental Setup

3.1 Briefing About Dataset

The dataset was taken from the Kaggle Web site. The name of the dataset is PIMA Indian Diabetes Database. This dataset is taken from the National Institute of Diabetes and Digestive and Kidney Diseases. The main objective of this dataset is to predict whether a respective person is affected with diabetes or not based on certain diagnostic measures used in this dataset. Several constraints were faced while taking these data from a larger dataset. All patients here are females aged 21 years and of PIMA Indian Heritage. This dataset contains several medical predictor variables and one target variable outcome. Predictor variables include the no. of pregnancies the corresponding female had, her insulin level, her BMI and so on. The main challenge for us was to build a machine learning model which could correctly detect that whether the female person is diagnosed with diabetes or not.

There are 9 columns and 614 data in the dataset. The column names include blood pressure (diastolic blood pressure mm Hg), pregnancies (no. of times pregnant), glucose (plasma glucose concentration in a 2 h oral glucose tolerance dataset), skin thickness (triceps skin fold thickness), BMI (body mass index (weight in kg/height in m^2)), insulin (2 h serum insulin), diabetes pedigree function, age in years and outcome (0 or 1). Two hundred and sixty-eight outcomes of 768 readings are having outcome 1, and other outcomes are 0. Receiving an outcome of 1 signifies that the patient is tested positive for diabetes, whereas outcome 0 signifies the patient is tested negative in diabetes.

3.2 Working on the Dataset

The main objective behind taking the dataset is to predict whether the person is diagnosed with diabetes from early stage such that treatment can be done accordingly. If the disease would be detected early, treatments would be done likely so that it helps in early recovery. So to solve the aim, most commonly used machine learning algorithm Support Vector Machine was taken in which eight kernels were taken in our analysis, namely RBF kernel, ANOVA RBF kernel, Sigmoid kernel, Gaussian kernel, Gaussian diagonal kernel, Laplacian RBF kernel, linear Kernel and polynomial kernel. The entire simulation was done in the software MATLAB.

Initially, codes were written for the eight kernels in the MATLAB language. Then, main code was written inside which the kernels were called. The values for the sigma, gamma and the constants were taken as per trial and error method. Main code was pasted in MATLAB; then one by one, all the kernels were called and the simulation was executed. All the files were kept in a single folder. On executing each single simulation, result was received for all the columns. Two methods were used in the entire process.

In the first method, we trained the model with the dataset and analyzed the results obtained and performance shown by different kernels individually.

In the second method, we used K-fold cross-validation for cross-validating the results obtained from the main method. During cross-validation, we have used the average of all the results obtained from all the iterations.

On using the first process, eight numbers of results were received. For the training ratio, observations were taken for eight kernels. On using the second process, observations were received for eight kernels.

All the results were noted down on a tabular form. The results were noted down and compared based on the six factors, namely accuracy, specificity, sensitivity, recall, precision and F1 score.

Many factors were taken into consideration before predicting the final kernel which is the best for diabetes classification. The factor which played the most important role in finding out the best kernel was accuracy.

Accuracy is defined as the degree to which result of a measurement and calculation conforms to the correct value or standard.

A kernel which would show greatest accuracy would be a viable option to choose for the final results as everyone is aware of the fact that greater the accuracy, greater the performance of the kernel. So, a final kernel which performed the best was chosen which is discussed in the result section.

4   Result Analysis

On observing the result, it was noticed that after the first training process, all the kernels showed accuracy of more than 75%. Gaussian diagonal showed the best performance. It showed accuracy 78.81%, sensitivity 73%, specificity 84.62% and precision 84.35%. Other kernels like Laplace RBF (accuracy 78.57%), RBF (accuracy 78.40%), Gaussian (accuracy 78.57%) and sigmoid (accuracy 78.31%) also performed well. The worst performance was shown by linear and polynomial kernel functions, 76.19%. For testing the performance of all the kernels on previously unseen data, we performed a second K-fold cross-validation with the same dataset.

In the second method, we performed K-fold cross-validation. The performance of all the kernels decreased compared to the first training. The best performance was shown by Gaussian diagonal. It showed an accuracy 73.51%, sensitivity 74.02%, specificity 72.98% and precision 74.03%. Laplace RBF (accuracy 73.06%), linear (accuracy 73.46%) and sigmoid (accuracy 73.24) also preformed well. In this case, ANOVA RBF showed the least accuracy 64.54%. The performance of ANOVA RBF decreased significantly from the previous training. The performance comparison between all the kernels ia shown by the boxplot given in Fig. 4.

Summarizing the model on the basis of classification on the PIDD dataset, Gaussian diagonal, Laplace RBF and sigmoid showed better performance than other kernels. Table 1 shows the results of these kernel functions.

The graphs for the results are shown through Figs. 3 and 4.

Table 1 Result for the variation of different parameters using different kernels

Kernel name	Accuracy (%)	Sensitivity (%)	Specificity (%)	Precision (%)	Recall	F1 score
Linear	76.191	90	63.636	69.231	0.9	0.783
Polynomial with order 3	76.191	80	72.727	72.727	0.8	0.762
RBF	78.404	75	81.808	78.404	0.75	0.769
Gaussian	78.572	70	86.364	82.353	0.7	0.757
Gaussian diagonal	78.814	73	84.629	84.358	0.73	0.757
Laplace RBF	78.572	85	72.727	73.913	0.85	0.791
ANOVA RBF	77.164	69	85.328	82.949	0.69	0.755
Sigmoid	78.313	87.5	69.768	72.917	0.875	0.796

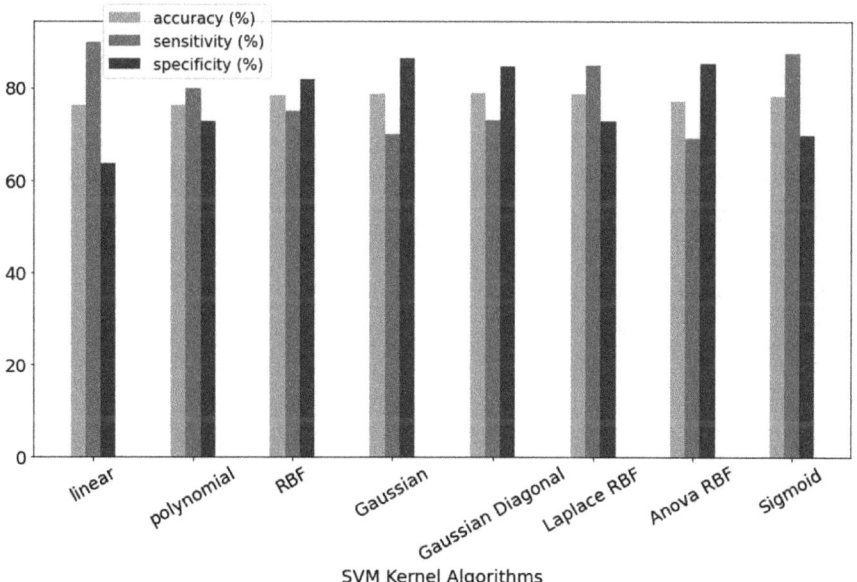

Fig. 3 Accuracy, sensitivity and specificity comparison of different kernels after initial training

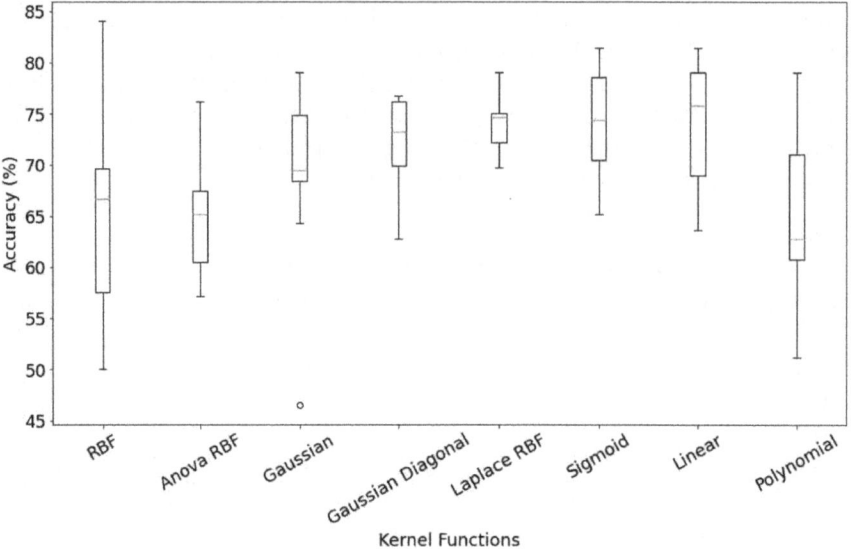

Fig. 4 Accuracy comparison of different kernels after cross-validation

5 Comparison with Related Works

Similar works have been conducted in the field of diabetes classification with various machine learning algorithms. Some of the prominent research works in this field along with their best model performances are listed in Table 2.

While several machine learning techniques have been applied in other research papers for predicting diabetes at early stages, we have performed a detailed comparative analysis of various kernel-based SVM algorithms with the PIMA dataset. As a future research scope, the model can be extended with a larger dataset distributed among all age groups.

Table 2 Comparison based on the accuracy

Authors	Dataset used	Algorithm used	Accuracy (%)
Laabidi et al. [13]	PIMA dataset	RNN	75.9
Kishore et al. [14]		Random forest	74.4
Srinivasa et al. [15]		Naïve Bayes	76.3
Tigga et al. [16]		Random forest	75
Sivakumar et al. [17]	Diabetes dataset and UCI machine learning repository	Naïve Bayes	76.3

6 Conclusion

Through this paper, we have made a comparative analysis between different kernel-based methods of support vector machine (SVM) for the classification of diabetes and we have analyzed and reviewed the performance of sigmoid, Gaussian, Gaussian diagonal, Laplace RBF, RBF, ANOVA RBF, linear and polynomial kernel functions. The performance comparison between them was assessed on the basis of model accuracy, sensitivity, specificity and precision.

We have performed the analysis initially with the training data. Then, we performed a K-fold cross-validation to understand the model's performance on new data. It was observed that out of all the eight kernels Gaussian diagonal, Laplacian RBF and sigmoid showed the best performance with overall accuracy more than 70% for all cases. This inferred that Gaussian diagonal will give a better performance for classification on new data. The future of this research aspects can be extended by pursuing the idea of using these kernel-based SVM algorithms with more larger dataset consisting of all age groups.

References

1. S.D. Patek, et al., Modular closed-loop control of diabetes. IEEE Trans. Biomed. Eng. **59**(11), 2986–2999 (2012). https://doi.org/10.1109/TBME.2012.2192930
2. S. Gupta, H.K. Verma, D. Bharadwaj, Classification of diabetes using Naive Bayes and support vector machine as a technique. In: Sachdeva A, Kumar P, Yadav O, Garg R, Gupta A (eds) Operations management and systems engineering. Lecture Notes on Multidisciplinary Industrial Engineering. Springer, Singapore (2021). https://doi.org/10.1007/978-981-15-6017-0_24
3. IDF Diabetes Atlas 9th edition 2019, https://www.diabetesatlas.org/en/resources/. Accessed 29 Nov 2020
4. Birjais R, Mourya AK, Chauhan R et al (2019) Prediction and diagnosis of future diabetes risk: a machine learning approach. SN Appl. Sci. 1:1112. https://doi.org/10.1007/s42452-019-1117-9
5. Awad M, Khanna R (2015) Support vector machines for classification. In: Efficient learning machines. Apress, Berkley, CA. https://doi.org/10.1007/978-1-4302-5990-9_3
6. Musa AB (2013) Comparative study on classification performance between support vector machine and logistic regression. Int J Mach Learn Cyber 4:13–24. https://doi.org/10.1007/s13042-012-0068-x
7. Muhammad Asraf H, Nooritawati MT, Shah Rizam MSB (2012) A comperative study in kernel-based support vector machine of oil palm leaves nutrient disease. Procedia Eng 41:1353–1359. ISSN 1877-7058. https://doi.org/10.1016/j.proeng.2012.07.321
8. Huang S, Cai N, Pacheco PP, Narrandes S, Wang Y, Xu W (2018) Applications of support vector machine (SVM) learning in cancer genomics. Cancer Genomics Proteom **15**(1), 41–51. https://doi.org/10.21873/cgp.20063
9. Ozer S, Chen CH, Yetik IS (2010) Using K-NN SVMs for performance improvement and comparison to K-highest Lagrange multipliers selection. In: Hancock EE, Wilson RC, Windeatt T, Ulusoy I, Escolano F (eds) Structural, synthetic, and statistical pattern recognition. SSPR/SPR 201. Lecture Notes in Computer Science, vol 6218. Springer, Berlin. https://doi.org/10.1007/978-3-642-14980-1_52

10. Kung S (2014) Support vector machines and variants. In: Kernel methods and machine learning. Cambridge: Cambridge University Press, pp 341–342. https://doi.org/10.1017/CBO978113917 6224.015
11. Sisodia D, Sisodia DS (2018) Prediction of diabetes using classification algorithms, Procedia Comput Sci 132:1578–1585. ISSN 1877-0509. https://doi.org/10.1016/j.procs.2018.05.122
12. Sohail MN, Jiadong R, Uba MM et al (2019) A hybrid forecast cost benefit classification of diabetes mellitus prevalence based on epidemiological study on real-life patient's data. Sci Rep 9:10103. https://doi.org/10.1038/s41598-019-46631-9
13. Laabidi A, Aissaoui M (2020) Performance analysis of Machine learning classifiers for predicting diabetes and prostate cancer. In: 2020 1st international conference on innovative research in applied science, engineering and technology (IRASET), Meknes, Morocco, pp 1–6. https://doi.org/10.1109/IRASET48871.2020.9092255
14. Naveen Kishore G, Rajesh V, Vamsi Akki Reddy A, Sumedh K, Rajesh Sai Reddy T (2020) Prediction of diabetes using machine learning classification algorithms. Int J Sci Technol Res 9(01):1805–1808. ISSN 2277-8616. https://www.ijstr.org/paper-references.php?ref=IJSTR-0120-28234
15. Srinivasa R, Yashashwini, Janakatti S, Venkatesh KB, Yaswanth SP (2020) Prediction of diabetes using machine learning. Int J Adv Sci Technol 29(06):7593–7601. https://sersc.org/journals/index.php/IJAST/article/view/23972
16. Tigga NP, Garg S (2020) Prediction of type 2 diabetes using machine learning classification methods. Procedia Comput Sci 167:706–716. ISSN 1877-0509. https://doi.org/10.1016/j.procs.2020.03.336
17. Sivakumar S, Venkataraman S, Bwatiramba A (2020) Classification algorithm in predicting the diabetes in early stages. J Comput Sci 16(10):1417–1422. https://doi.org/10.3844/jcssp.2020.1417.1422

Chapter 64
Comparison Between Self-organizing Maps and Principal Component Analysis for Assessment of Temporal Variations of Air Pollutants

Loong Chuen Lee and **Hukil Sino**

1 Introduction

Air pollution is one of the major contributing factors to the human health and environment [1]. Since most of the air pollutants are present in gaseous state, for instance, sulfur dioxide (SO_2), nitrogen oxides (NO_2), ozone (O_3) and carbon monoxide (CO), they are readily dissolved and diffused into everywhere on the earth. Hence, the impacts caused by air pollutants can be very huge. Therefore, fast prediction of the air pollutant is one of the important topics of atmospheric and environmental research today. In particular, understanding the spatial-temporal variations of air quality data is of great significant for constructing an accurate air pollutants prediction models.

Studies reporting the spatial-temporal variations of air quality data were numerous; and most of the works employed univariate approaches to explore the spatial and temporal variations [2]. For instance, Bai et al. [3] investigated the temporal variations of pollution characteristics in Shanxi Province using time series plots. Meanwhile, Shen et al. [4] utilized bar plot to illustrate intra-annual variations of concentrations of six air pollutants over a four-year period. Strictly speaking, the benefits of multivariate exploratory tools in investigation of temporal variation of air quality have not yet been fully explored.

Recently, Chang et al. [5] employed a visible two-dimensional topological map of self-organizing maps (SOMs) in extracting the spatio-temporal patterns of long-term regional $PM_{2.5}$ concentration. The maps of SOM have been used to illustrate the spatial distribution of the long-term datasets of 25 monitoring stations and also

L. C. Lee (✉) · H. Sino
Forensic Science Program, CODTIS, Faculty of Health Sciences, Basement 1, PTSL, Universiti Kebangsaan Malaysia, 43600 Bangi, Selangor, Malaysia
e-mail: lc_lee@ukm.edu.my

L. C. Lee
Institut IR4.0 (IIR4.0), Universiti Kebangsaan Malaysia, 43600 Bangi, Selangor, Malaysia

S. Kumar et al. (eds.), *Proceedings of International Conference on Communication and Computational Technologies*, Algorithms for Intelligent Systems,
https://doi.org/10.1007/978-981-16-3246-4_65

the temporal behavior of $PM_{2.5}$ concentrations at various time scales, i.e., yearly, seasonal and hourly. Meanwhile, Neme and Hernandez [6] concluded that SOM is useful to reveal some latent patterns of the pollutant concentrations and evolution of air quality data in Mexico City, from 2003 to 2010.

Principal component analysis (PCA) has been employed for investigation of air pollutant data over the past several decades [7]. It is often employed for exploring the possible pollutant sources and evaluating the correlations among the pollutant concentrations and meteorological variables [8]. Alternatively, PCA can also be used to evaluate the correlation between aerosol organic fractions in particulate matters and their seasonal distribution (i.e., temporal variability) [9].

This paper presents a comparison between SOMs and PCA on investigating the temporal variations of six ambient criteria air pollutants. The air pollutants were particulate matters with an aerodynamic diameter of less than 10 and 2.5 μm (PM_{10}, $PM_{2.5}$), sulfur dioxide (SO_2), nitrogen oxides (NO_2), ground-level ozone (O_3) and carbon monoxide (CO). The main contribution of this paper is to shed light on the potential of SOM and PCA in elucidating temporal variations of gaseous pollutants and particulate matters.

2 Materials and Methods

2.1 Data

Nilai is one of the districts of Negeri Sembilan located at the west coast of Peninsular Malaysia. The continuous air quality monitoring station (CAQM) of Nilai is situated at Taman Semarak (Phase II) (N 2.82050, E 101.81080) and is managed by Alam Sekitar Malaysia Sdn. Bhd. (ASMA) on behalf of the Department of Environment (DOE), Malaysia. The ASMA CAQM station is designed to measure hourly concentrations of ambient air gaseous pollutants (SO_2, NO_2, O_3 and CO) and particulate matters (PM_{10}, $PM_{2.5}$). The Nilai station is surrounded by both residential buildings and manufacturing factories. The six daily air pollutant concentration series were acquired from January 2018 to December 2019.

Typically, air quality data consists of missing values, and thus, various imputation methods are available to resolve the problem [10]. Based on preliminary assessment, the missing data in each of the six data series of 2018 and 2019 were less than 5%. Thus, the mean imputation method was adopted herein to recover the missing values in the imputed data. Mean was first computed for a given column in the data of a year. Then, it was used to replace the missing values in the respective column of the data.

Then, the imputed data were employed to form two types of averaged data. Firstly, the concentrations of a pollutant within a particular month were reduced into a single mean value. As a result, two sets of data matrix with 12 rows and six columns were prepared for year of 2018 and 2019, respectively. Next, for evaluation of hourly variability, we prepared another two sets of data matrix composing of 744 rows and six columns. For instance, the CO was presented with 744 concentration values, i.e., 24 h on 31 days of a year. This was accomplished by averaging the concentrations of CO at a given hour and day across January to December of a year.

2.2 Multivariate Exploratory Tools

The average data obtained were submitted to two unsupervised machine learning approaches, i.e., principal component analysis (PCA) and self-organizing maps (SOM), for exploring the temporal trends of the six pollutants. Statistical analyses were carried out by using the statistical software R (R-project for statistical computing, Ver. 3.6.2). PCA and SOMs were, respectively, executed using `prcomp` and `som` of R package 'kohonen' [11].

Principal component analysis (PCA). PCA aims to reduce the high number of descriptor variables into smaller number of principal components (PCs) [12]. Herein, the descriptor variables refer to the four ambient air gaseous pollutants (SO_2, NO_2, O_3 and CO) and two particulate matters (PM_{10}, $PM_{2.5}$). After PCA, a sample initially presented with six values of the descriptor variables was reduced to a single score value on a given principal component.

The scores value of a sample on a given PC was the linear combination of the four ambient air gaseous pollutants and two particulate matters as defined by Eq. (1):

$$z_j = w_{1,j}x_1 + \cdots + w_{p,j}x_p \tag{1}$$

where z_j refers to the j-th PC; w and x, respectively, denote the loading value and initial concentration value of a given descriptor variable of the sample. The loading value indicates the importance of the six descriptor variables on a particular PC. The variations of the pollutants by 12 months (monthly changes) and 24 h of 365 days, respectively, were illustrated using the scores plot of PCA. Prior to PCA, the data was preprocessed by autoscaling in order to minimize bias caused by different variable scaling and units.

Unsupervised Self-organizing maps (SOMs). SOMs is a class of artificial neural networks and is also known as Kohonen maps. It maps the samples to a two-dimensional array of map units, i.e., neurons, in which very similar samples are mapped to the same unit or to units which are close together in the map [13]. Thus, the map of SOM also known as a topological map since the dissimilarity of the neighboring units is not consistent throughout the map.

Initially, the researcher needs to define size of the SOM map, i.e., number of nodes. Then, each of the nodes is assigned with a weight vector describing the descriptor variables of the air quality data of which are initialized by random values. Herein, the descriptor variables were the six air pollutants. At the same time, all the samples (i.e., 12 months of a year or 24 h of 31 days in a year) are also randomly presented to all the nodes that forming the network.

The node in the map possessing the weight vector most similar to the presented sample is assigned to be the winner. Subsequently, the weight vectors of this node and its closest neighbors in the map are updated in the similar way to become more similar to the presented samples. The amount of change is determined by the learning rate. The weight is updated iteratively until all samples are presented a sufficient number of times to the network.

Basically, the dimension of the topological map is the important parameter to ensure the most meaningful clustering can be obtained from the data. In this work, the number of the neurons was carefully determined that all the nodes must be filled by at least one sample and none of the units appears to be overcrowded.

3 Results and Discussion

3.1 Descriptive Statistics

Table 1 shows the summary statistics for six different pollutant data collected from the Nilai station in 2018 and 2019, respectively. In general, the annual mean PM_{10} and $PM_{2.5}$ concentrations were exceeded that recommended Malaysian Ambient Air Quality Guidelines (MAAQG) standard [14], i.e., 40 and 15 ug/m^3, respectively, except the annual mean concentration of PM_{10} in 2018. In particular, Malaysia has experienced serious haze episodes between August and September of 2019. The Southwest monsoon in the same period of time produced strong wind promoted the movement of suspended particulate PM_{10} and $PM_{2.5}$ from Indonesia to the west coast of Peninsular Malaysia, resulting in severe haze conditions in Nilai, Negeri Sembilan.

Next, it is observed that the annual concentration of NO_2 was much higher than that of CO. Theoretically, both CO and NO_2 are common by-products of the motor

Table 1 Summary statistics of pollutant and meteorological parameters in Nilai station, Negeri Sembilan, Malaysia during January 2018–December 2019

Pollutants	Mean		SD		Median	
	2018	2019	2018	2019	2018	2019
PM_{10} (ug/m^3)	36.42	42.92	20.43	30.20	32.14	35.98
$PM_{2.5}$ (ug/m^3)	26.74	30.18	16.11	25.07	23.90	24.31
SO_2 (ppb)	0.96	1.52	0.54	1.13	0.93	1.15
NO_2 (ppb)	15.00	12.19	6.88	6.16	14.16	11.27
O_3 (ppb)	11.96	8.36	12.71	7.94	7.63	5.98
CO (ppm)	0.62	0.67	0.25	0.30	0.58	0.62

vehicle emissions. However, NO_2 is also one the most common waste released by a factory [15] besides produced by motor vehicle emissions. Since Nilai was surrounded by high number of manufacturing factory, we can easily understand why the concentration of NO_2 was much higher than CO.

Based on Table 1, ground-level ozone (O_3) was recorded higher concentration in 2018 than in 2019. It has been reported in the literature that ground-level ozone is primary produced from NO_x and volatile organic compounds (VOCs). The two compounds are the important precursors for producing ozone through a photochemical reaction of which requires sunlight and heat [16, 17]. In other words, concentration of NO_2 is expected to be positively proportionate with concentration of ozone. Hence, the concentration of NO_2 in 2018 was also found to be higher than that in 2019.

3.2 12-Month Averaged Data (Month-Level Profiles)

The 12-month averaged data were modeled using PCA and SOM, respectively, in order to assess the similarity and dissimilarity among the 12 months of a year by the six air pollutants. Figure 1 shows the PCA results by the year of 2018 and 2019. The biplots were plotted using the first two principal components of which accounting for 85.43%, 80.56% and 74.89% of the total variance, respectively, for the year of 2018, 2019 and both the years. Correspondingly, the temporal variations of the same data in the mapping plot of SOM are presented in Fig. 2.

Biplot of PCA presents the clustering of the 12 months and influence of the six pollutants to the clustering. The SOM results were illustrated via SOM plot and the respective mapping plot. The former plot aimed to reveal the importance of the six pollutants and clustering of the samples (i.e., 12 months) in the nodes was illustrated by the latter plot. The pie size (i.e., radius of a wedge) in the SOM plot reflected concentration magnitude of a particular pollutant (i.e., importance of the descriptor variables in the node). After a few trial-and-error attempts, six nodes (2×3) were decided to be the most optimal dimensions. In general, the interpretations of SOM results were similar to the PCA results.

The biplot for the year of 2018 indicates that August and September were different from the other months in the mean concentrations of PM_{10} and $PM_{2.5}$. The two months have positive values of the first two principal components and therefore registered higher values of PM_{10} and $PM_{2.5}$. Correspondingly, the SOM results also show that only August and September of 2018 showed higher concentrations in the two particulate matters. Additionally, both the biplot of PCA and SOM results reveal that February of 2018 was characterized by SO_2 and O_3.

Meanwhile, August and September were separated from each other and the other ten months in 2019 (Fig. 1b). Generally, the biplot does not allow us to identify the pollutants that responsible for the spatial distribution of the 12 months in 2019. However, map of SOM in 2019 (Fig. 2b) clearly shows that only September is heavily characterized by all the pollutants, except SO_2.

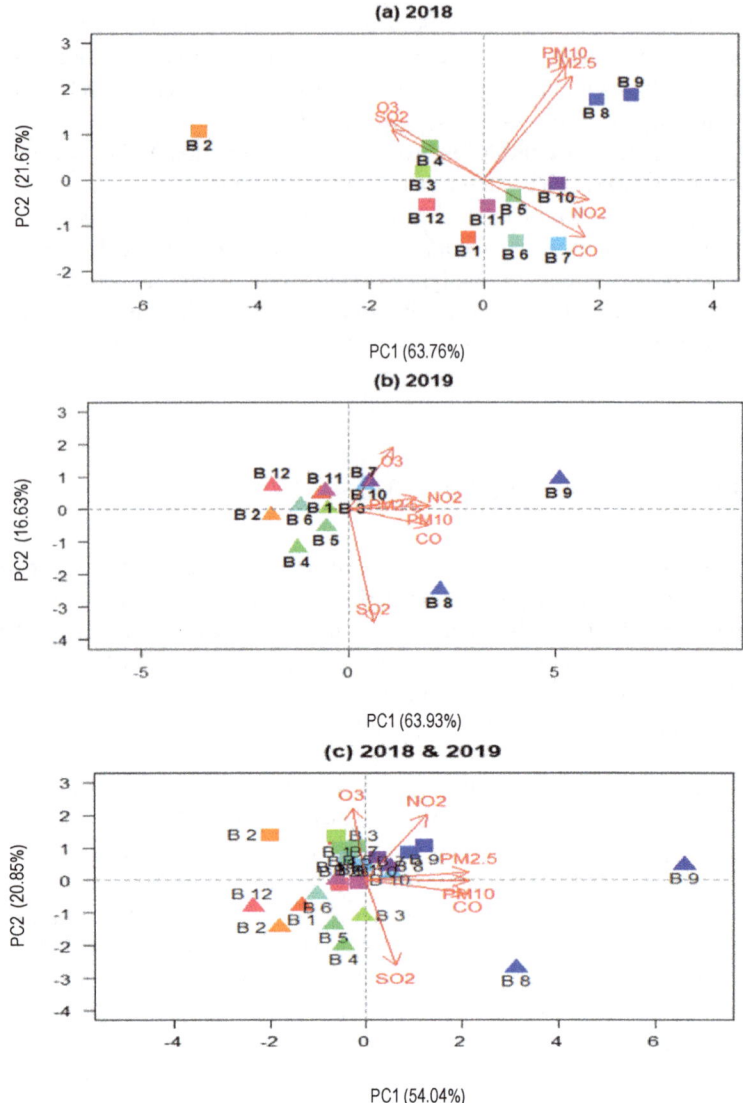

Fig. 1 Biplots of scores and loadings for the first two principal components illustrate temporal variations of month-level profiles of pollutants in **a** 2018; **b** 2019; and **c** combined year of 2018 and 2019. Each point represents the month of January (B1) to December (B12) of a year

Similarly, the contributions of the six pollutants in the 24 months of 2018 and 2019 were hardly determined based on the biplot (Fig. 1c). Nonetheless, SOM results clearly indicate that only August and September of 2019 were most affected by all the pollutants except CO (Fig. 2c). This observation is in accordance with the fact that Malaysia experienced serious haze episodes in September of 2019.

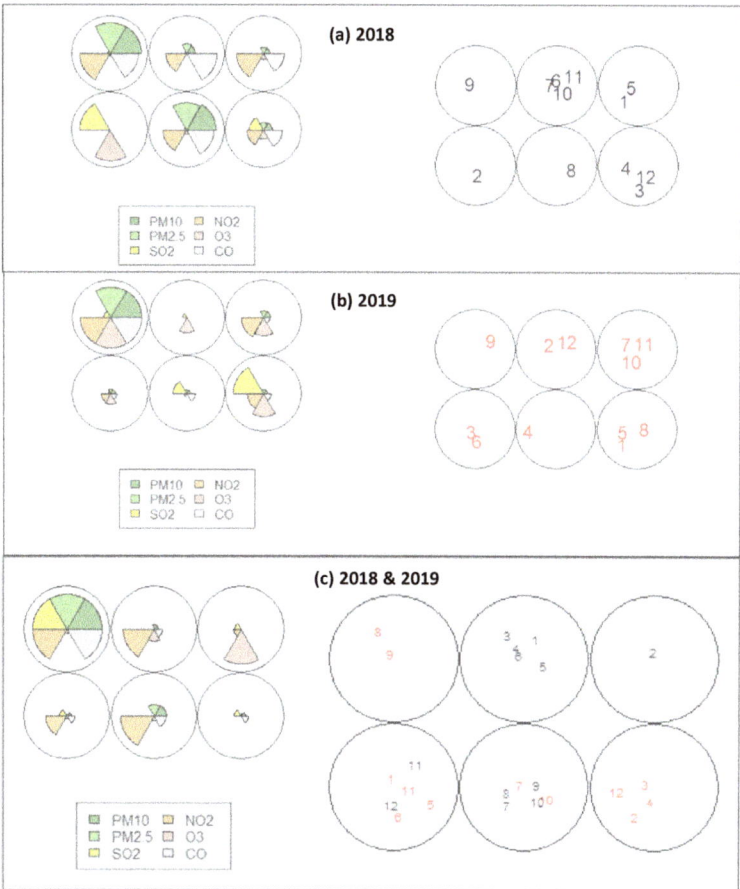

Fig. 2 Temporal variations of month-level profiles of pollutants in **a** 2018; **b** 2019; and **c** combined year of 2018 and 2019 as illustrated by (left) SOM plots and (right) SOM maps. 1 to 12 denote January to December of a year

3.3   24-h Averaged Data (Hour-Level Profiles)

In contrary to most of the previous studies that tended to present only a 24-h data of a year by using a time series plot (e.g., [18, 19]), herein, hourly data were obtained by 31 days of a year. As a result, the 24-h averaged data matrix composes of 744 rows and six columns.

Based on Fig. 3, only the late afternoon periods (blue squares in Fig. 3a, b) were clustered together, of which were positioned positively and negatively on PC1 of 2018 and 2019, respectively. Meanwhile, other time periods were scarcely distributed in the plots without any visible patterns. By referring to the corresponding loadings plots (Fig. 3c, d), it is noted that the ozone was the only pollutant presented the highest

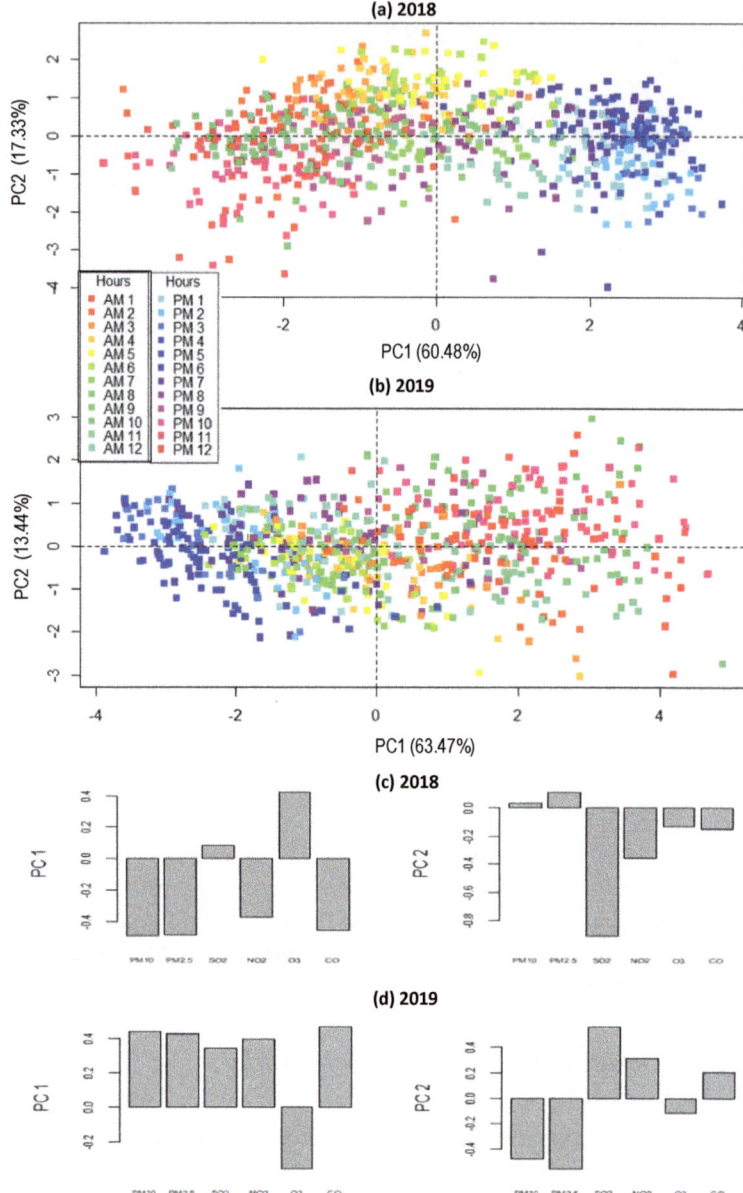

Fig. 3 PCA results of hour-level profiles of pollutants: **a, b** scores plots of 2018 and 2019; and **c, d** the respective loading plots of (left panel) PC1 and (right panel) PC2. Each point in the scores plots represents a particular hour in a day of a year

positive and negative loading values in PC1 of 2018 and 2019, respectively. As a result, it seems sound to state that the temporal pattern was most strongly associated with the ground-level ozone concentrations.

Figure 4 illustrates the respective SOM results of the same data employed to prepare the PCA results as shown in Fig. 3. Again, the late afternoon periods (1300–1800 h) are often clustered in the same nodes and characterized by high ozone levels. Basically, our finding is in agreement with [16] that ozone reached maximal concentrations at noontime. As explained by Nurul Adyani et al. [16], ozone is mainly contributed by photochemical processes on the NO_x and VOC released by

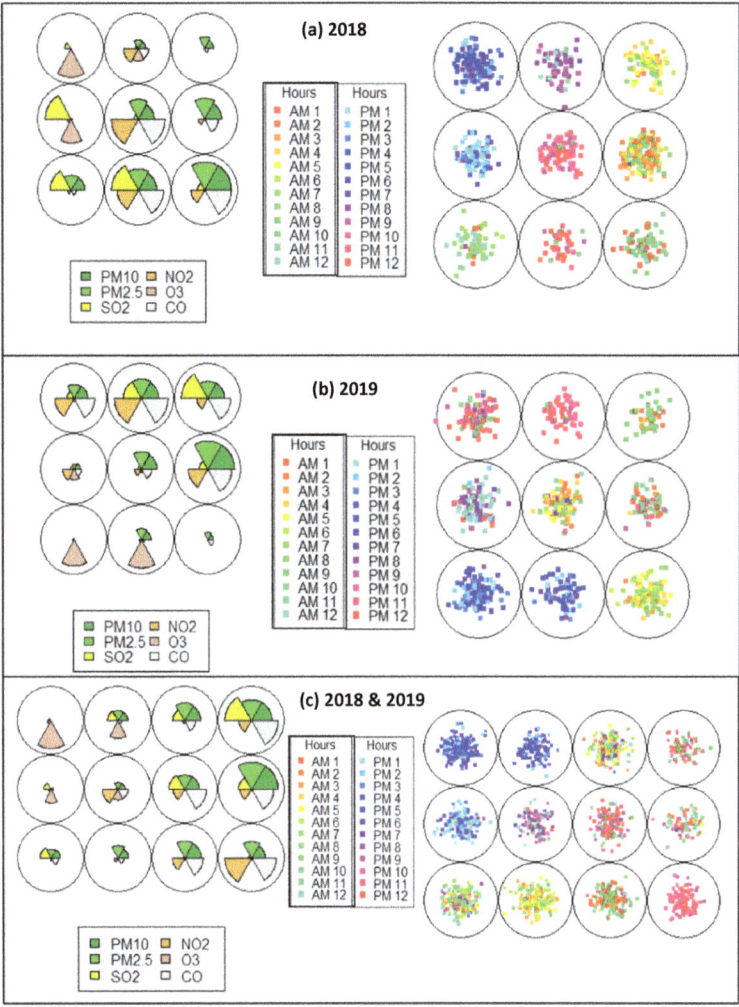

Fig. 4 Temporal variations of hour-level profiles of pollutants in **a** 2018; **b** 2019; and **c** combined year of 2018 and 2019 as illustrated by (left) SOM plots and (right) SOM maps

the motor vehicle. High temperature of the sunlight during late afternoon promotes the formation of ozone through the photochemical processes.

In addition, based on the mapping plot of SOM alone, it is noted that late evening periods (2100–2300 h) were dominated by NO_2 and CO. Meanwhile, early morning (0400–0700 h) is seemed to be the cleanest air since none of the pollutants showing high level of concentration. Essentially, these two patterns are not clearly presented in the scores plots of PCA (Fig. 3).

3.4 General Remarks

Time series plot and bar chart are two favorite illustrative tools employed for studying temporal variations of air quality data. Strictly speaking, time series plot and bar plot are univariate tools while both the PCA and SOM are essentially multivariate illustrative tools.

The principal limitations of univariate approaches are as follows: (a) limited number of data points can be presented in a plot, i.e., normally only annual mean concentrations are presented in the plot, indirectly, this practice reduces the resolution of the air quality data; and (b) simultaneous assessment of all the studied pollutants is infeasible, i.e., multiple plots are required for evaluating temporal variability of various pollutants. In contrary, the principal merit of SOM and PCA is the capability of clustering data by simultaneously considering all the pollutants.

In this work, SOM is shown to be more appropriate than scores plot of PCA for disclosing the inherent clustering pattern of air quality data with high number of samples. The outperformance of SOM than PCA in providing visual clustering result has also been reported by Das et al. [20]. The authors found that SOM was far superior to PCA in revealing the clustering feature for constructing 'retail store personality' in marketing work. Meanwhile, Astel et al. [21] performed the comparison of PCA and SOM by using long-term water quality monitoring data. They found that SOM clustering enabled simultaneous observations of both spatial and temporal changes in water quality. Hence, our future work will consider more number of monitoring stations to enable evaluation of capability of SOM in assessing the spatial-temporal variations of air quality data.

4 Conclusion

This work presented the feasibility of SOM in comparison with PCA in exploring temporal variation in six air pollutants concentrations in Nilai, Malaysia. The temporal trends of pollutants have been studied according to month- and hour-level profiles. Both the multivariate exploratory tools are good in revealing overview of temporal variability in the air quality data. It is worth noting that maps of SOM have revealed novel groupings in the multipollutant data that have gone unnoticed via scores plot of PCA. In conclusion, SOM is outperforming PCA in elucidating the temporal patterns of air quality data.

Acknowledgements The authors acknowledge the Malaysian Department of Environment (DOE) for providing the air quality data. We would like to give our gratitude to Prof. Dr. Mohd Talib Latif from University Kebangsaan Malaysia (UKM) for the collaboration given in doing this research. This research is funded by the CRIM-UKM, GP-2019-K016373.

References

1. Orellano P, Reynoso J, Quaranta N, Bardach A, Ciapponi A (2020) Short-term exposure to particulate matter (PM_{10} and $PM_{2.5}$), nitrogen dioxide (NO_2) and ozone (O_3) and all-cause and cause-specific mortality: systematic review and meta-analysis. Environ Int 142:105876
2. Sanchez-Balseca J, Perez-Foguet A (2020) Spatio-temporal air pollution modelling using a compositional approach. Heliyon 6:e04794
3. Bai X, Tian H, Liu X, Wu B, Liu S, Hao Y, Luo L, Liu W, Zhao S, Lin S, Hao J, Guo Z, Lv Y (2021) Spatial-temporal variation characteristics of air pollution and apportionment of contributions by different sources in Shanxi province of China. Atmos Environ 244:117926
4. Shen F, Zhang L, Jiang L, Tang M, Gai X, Chen M, Ge X (2020) Temporal variations of six ambient criteria air pollutants from 2015 to 2018, their spatial distributions, health risks and relationships with socioeconomic factors during 2018 in China. Environ Int 137:105556
5. Chang F-J, Chang L-C, Kang C-C, Wang Y-S, Huang A (2020) Explore spatio-temporal $PM_{2.5}$ features in northern Taiwan using machine learning techniques. Sci Total Environ 736:139656
6. Neme A, Hernandez L (2011) Visualizing patterns in the air quality in Mexico City with self-organizing maps. In: Laaksonen J, Honkela T (eds) Advances in self-organizing maps. Lecture Notes in Computer Science. Springer, Heidelberg, vol 6731, pp 318–327
7. Smeyers-Verbeke J, Den Hartog JC, Dehker WH, Coomans D, Buydens L, Massart DL (1984) The use of principal component analysis for the investigation of an organic air pollutants data set. Atmos Environ 18:2471–2478
8. Rupakheti D, Yin X, Rupakheti M, Zhang Q, Li P, Rai M, Kang S (2021) Spatio-temporal characteristics of air pollutants over Xinjiang, northwestern China. Environ Pollut 268:115907
9. Padoan S, Zappi A, Adam T, Melucci D, Gambaro A, Formenton G, Popovicheva O, Nguyen D-L, Schnelle-Kreis J, Zimmermann R (2020) Organic molecular markers and source contributions in a polluted municipality of north-east Italy: extended PCA-PMF statistical approach. Environ Res 186:109587
10. Hadeed SJ, O'Rourke MK, Burgess JL, Harris RB, Canales RA (2020) Imputation methods for addressing missing data in short-term monitoring of air pollutants. Sci Total Environ 730:139140
11. Wehrens R, Kruisselbrink J (2019) Supervised and unsupervised self-organizing maps, Ver. 3.0.10
12. Dupont MF, Elbourne A, Cozzolino D, Chapman J, Truong VK, Crawford RJ, Latham K (2020) Chemometrics for environmental monitoring: a review. Anal Methods 12:4597–4620
13. Kohonen T (1997) Self-organizing maps, 2nd edn. Springer, Berlin
14. Department of Environment Malaysia (DOE): Malaysia Environmental Quality Report 2011 Malaysia. Department of Environment, Malaysia (2011)
15. Azmi SZ, Latif MT, Ismail AS, Juneng L (2010) Trend and status of air quality at three different monitoring stations in the Klang Valley, Malaysia. Air Qual Atmos Health 3:53–64
16. Nurul Adyani G, Nor Azam R, Ahmad Shukri Y, Noor Faizah FMDY, Nurulilyana S, Wesam Ahmed AM (2010) Transformation of nitrogen dioxide into ozone and prediction of ozone concentrations using multiple linear regression techniques. Environ Monitor Assessm 165:475–489

17. Chelani AB (2013) Study of Extreme CO, NO_2 and O_3 concentrations at a traffic site in Delhi: statistical persistence analysis and source identification. Aerosol Air Qual Res 13:377–384

18. Khan MF, Latif MT, Juneng L, Amil N, Nadzir MSMN, Hoque HMS (2015) Physicochemical factors and source of particulate matter at residential urban environment in Kuala Lumpur. J Air Waste Manag Assoc 65:958–969

19. Kalbarczyk R, Kalbarczyk E (2020) Meteorological conditions of the winter-time distribution of nitrogen oxides in Poznan: a proposal for a catalog of the pollutants variation. Urban Clim 33:100649

20. Das G, Chattopadhyay M, Gupta S (2016) A comparison of self-organising maps and principal components analysis. Int J Mark Res 58:815–834

21. Astel A, Tsakovski S, Barbieri P, Simeonov V (2007) Comparison of self-organizing maps classification approach with cluster and principal components analysis for large environmental data sets. Water Res 41:4566–4578

Chapter 65
Development of a State Structure Model for a Project-Oriented Organization

Kateryna Kolesnikova⬡, Olga Mezentseva⬡, Oleksii Kolesnikov⬡, and Sergiy Bronin⬡

1 Introduction

The practice of developing management structures usually consists in the fact that generally accepted management structures are adjusted to the tasks of project management [1]. As you know, control objects consist of elements and links between them [2]. Elements are processes that are integrated into a system by information links. The effectiveness of such structures depends on the characteristics of two classes: parametric, reflecting purely the properties of the elements (processes) of the system, and structural, which determines the influence of the connection scheme (topology) on the functioning of the entire system. Usually, most studies consider models and methods belonging to the class of parametric properties [3]. At the same time, improving the structural properties will make it possible to build new management mechanisms—a decision-making system based on models and tools aimed at achieving the project result. The article puts forward and proved the thesis that together with the improvement of the processes that form the basis of the quality management system of a project-managed organization, it is necessary to scientifically substantiate the formation of an improved structure of the QMS.

2 Analysis of Recent Research and Publications

The development of knowledge systems and project management technologies reflects the effectiveness of the methods of using the project approach, which makes it possible to most effectively solve the problem of creating values and achieving the set goals in conditions of limited time, financial, material, human, and other resources

K. Kolesnikova (✉) · O. Mezentseva · O. Kolesnikov · S. Bronin
Taras Shevchenko National University of Kyiv, Volodymyrska Str., 60, Kyiv 01033, Ukraine

[4]. The search for ways to further develop the theoretical foundations of project management should be carried out on the basis of models and methods of analysis of the structural properties of project management systems, reflecting the essential features of the studied systems—projects/programs/project portfolios [5, 6].

Investigation of phenomena and essence, relationships, and patterns in the processes of project/program/portfolio management in the life cycles of projects as managed social or organizational and technical systems, with signs of uniqueness, time and resource constraints, as well as requirements for the quality of products or services, constitutes a substantial aspect of theoretical research in the field of project management. At the same time, the existing approaches to defining and justifying the tactics and strategies for implementing projects are aimed at obtaining certain results and value through research of connections and patterns at the level of organizational management, rather than technologies for manufacturing a project product [4].

The presence of uncertainty in project management in organizational and technical systems is due to a number of features [7, 8]:

- many factors and their interconnection do not allow to single out and study in detail the property of design systems based on the properties of individual elements of systems (due to the emergence of project management systems); therefore, all phenomena occurring in them should be considered in aggregate;
- the absence of deterministic dependencies between the input of design processes and the output due to the presence in the control loop of systems of executors who make decisions or perform the tasks received, depending on their knowledge, skills, and previous experience, which forces us to proceed to a qualitative analysis of such processes; and
- the turbulence of the environment and variability of the nature of processes in time.

Due to these features, such projects are semi-structured systems. Many factors of such systems form a complex system of connections and states, causes and effects that change over time. It is quite difficult to see and understand the logic of the development of events in such multifactor systems. At the same time, in practice, you constantly have to make a decision about what needs to be done (what factors to influence) to improve the state of the project, what will happen to the situation after a certain time, if nothing is done, which of the possible actions will be effective for the achievement of the set goal, etc., [1]. The decision-making procedure can be based on cognitive (cognitive) analysis and modeling of complex processes [9–12].

The development of a method for the analysis of complex semi-structured systems with their further representation in the form of directed graphs was carried out in [13–16]. The authors proposed an analytical method for determining closed loops in complex control systems and showed the suitability of the method for determining inverse and non-return structures in directed graphs. The method also allows you to check the ergodicity of the interaction patterns of project participants [17].

3 Model Development and Use of Modeling Method

The activities of industrial enterprises have always been focused on maintaining a stable range of products, as well as exceeding the planned production volumes. In market conditions, restrictions on the range of products are removed. At the same time, work "for the customer" creates new conditions for working with an open nomenclature, which increases the role of quality management of processes/products. In a competitive environment, the most interested party is the consumer, who takes an active part in the formation of innovative specifications and product characteristics. Development, production of project products in accordance with the requirements of the consumer are the main tasks of a modern enterprise [17]. New concepts of production organization and modern requirements for product quality are the basis of new Ukrainian standards ISO 9001 [18–20], which establish the importance of critical analysis of requirements for product quality (A), product/product support (B), and the formation of a system at enterprises responsibility of performers, distribution of powers of personnel and constant information (C). These aspects of quality management in previous models of the QMS were included in other processes (states) of the enterprise management system [6].

The first task that needs to be solved in the case of improving the existing model of the QMS of a machine tool enterprise can be formulated as follows: What new connections will be formed if new states A, B, and C are included in the system (Fig. 1). Assume that the place of the new states A, B, and C in the general scheme is not accidental but is determined by a certain global goal of the system's functioning. These processes (states) receive certain information at the input, which is the basis for making informed decisions for managing the system.

Process for critical review of product quality requirements (A). The inputs of this process are informational links with states 7 and B (support of product operation processes at the consumer) and with state 2. The result of process A is transferred to states 2 and 18 and is linked to state 8—product design. Thus, it has been established that process (A) is a link in the design management of products/products, taking into account the performance and requirements of the customer (consumer).

Maintenance of the processes of operation of products at the consumer (B). The input of state B receives information from state 7. The results are associated with the process of critical analysis of requirements for product quality (A) and with the main process of the enterprise—state 11.

Formation at the enterprise of a system of responsibility of executors, distribution of powers, and constant information (C). Processes of state C are based on data from state 17 and interaction with state/process 3, which makes it possible to form the competence of personnel (state 4). Thus, it is determined that process C is aimed at increasing the knowledge and competence of personnel.

At the request of the practice at the HC MICRON®, the initiation of the development and implementation of an improved structural model of the project-oriented organization states, based on the new provisions of the ISO 9001 family standards,

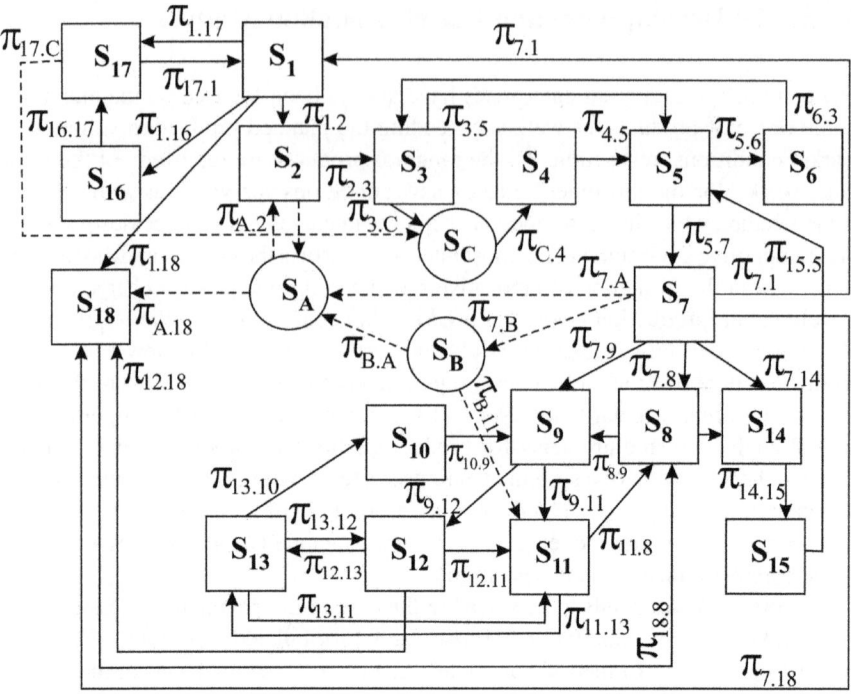

Fig. 1 Marked-oriented graph of the structure QMS of the enterprise HC MICRON® (according to ISO 9001): 1—responsibility of the organization's management; 2—QMS; 3—the process of personnel management; 4—control of personnel competence; 5—product creation management; 6—organization of the professional development process; 7—processes of interaction with customers; 8—product design processes; 9—procurement management; 10—supply quality control; 11—product management processes; 12—control and testing processes; 13—control processes of measuring equipment; 14—document management processes; 15—infrastructure management processes; 16—internal audit; 17—continuous improvement processes; 18—assessment of the fulfillment of customer requirements; A—critical analysis of product requirements; B—product support; and C—a responsibility of executors, authority and constant information

was carried out. The presence of this property is a necessary condition for the system to work.

The practical implementation of the production processes of projects to create products is determined by a number of random factors. Each process in the model corresponds to a certain state. Figure 1 shows the directed graph of the new project-oriented organization structure (new elements are marked as S_A, S_B, S_C).

4 Checking the Ergodicity of the QMS

When developing an improved design-managed organization model, the first step is to confirm that the new process interaction pattern is ergodic. An ergodic graph is understood as a graph consisting entirely of one ergodic class [21–23]. Ergodic graphs are described by a strongly connected graph. This means that in such a system, a transition from any state S_i to any state S_j is possible in a finite number of steps. Since a directed graph is the basis for constructing a Markov chain, all structural properties of a graph are inherited by a Markov chain created on its basis. For an ergodic Markov chain, after a certain time of operation, a stationary regime sets in, in which the probabilities of the states of the system $S \{s_1, s_2, …, s_m\}$ do not depend on time and on the probability distribution at the initial moment of time.

This method is based on the analysis of the graph adjacency matrix and allows determining the properties of complex topological structures. Let us construct the adjacency matrix of the first degree of the improved circuit (Fig. 2).

From state	To state																				
	1	2	3	4	5	6	7	8	9	10	11	12	13	14	15	16	17	18	a	b	c
1	0	1	0	0	0	0	0	0	0	0	0	0	0	0	0	0	1	1	1	0	0
2	0	0	1	0	0	0	0	0	0	0	0	0	0	0	0	0	0	0	0	1	0
3	0	0	0	0	1	0	0	0	0	0	0	0	0	0	0	0	0	0	0	0	1
4	0	0	0	0	1	0	0	0	0	0	0	0	0	0	0	0	0	0	0	0	0
5	0	0	0	0	0	0	1	1	0	0	0	0	0	0	0	0	0	0	0	0	0
6	0	0	1	0	0	0	0	0	0	0	0	0	0	0	0	0	0	0	0	0	0
7	1	0	0	0	0	0	0	0	1	1	0	0	0	1	0	0	0	0	18	1	1
8	0	0	0	0	0	0	0	0	1	0	0	0	0	1	0	0	0	0	0	0	0
9	0	0	0	0	0	0	0	0	0	0	0	1	1	0	0	0	0	0	0	0	0
10	0	0	0	0	0	0	0	0	1	0	0	0	0	0	0	0	0	0	0	0	0
11	0	1	0	0	0	0	0	1	0	0	0	0	1	0	0	0	0	0	0	0	0
12	0	0	0	0	0	0	0	0	0	0	0	1	0	1	0	0	0	1	0	0	0
13	0	0	0	0	0	0	0	0	0	0	1	1	1	0	0	0	0	0	0	0	0
14	0	0	0	0	0	0	0	0	0	0	0	0	0	0	1	0	0	0	0	0	0
15	0	0	0	0	1	0	0	0	0	0	0	0	0	0	0	0	0	0	0	0	0
16	0	0	0	0	0	0	0	0	0	0	0	0	0	0	0	0	1	0	0	0	0
17	1	0	0	0	0	0	0	0	0	0	0	0	0	0	0	0	0	0	0	0	1
18	0	0	0	0	0	0	0	1	0	0	0	0	0	0	0	0	0	0	0	0	0
a	0	1	0	0	0	0	0	0	0	0	0	0	0	0	0	0	1	0	0	0	0
b	0	0	0	0	0	0	0	0	0	0	1	0	0	0	0	0	0	0	1	0	0
c	0	0	0	1	0	0	0	0	0	0	0	0	0	0	0	0	0	0	0	0	0

Fig. 2 Adjacency matrix of the first degree of the improved system of a project-oriented organization

In contrast to the scheme [3], three new states (processes) are included in the new improved structure of the states of the organization (Fig. 1), the machine tool enterprise. New states included in the control system indicated by identifiers a, b, c. Such designations emphasize that the new processes in the management scheme are not arranged in the order in the new management structure.

As the degree of the adjacency matrices grows, the elements of the reach matrix are filled with units in accordance with the sequential displacement of the elements in the direction of the arcs of the directed graph [24]. Here are some intermediate calculations of adjacency matrices and the final result (matrix W^8). The superposition matrix of the second-degree W^2 looks as shown in Fig. 3.

The states of the system 11, 12, and 13 form one closed loop, since the elements of the main diagonal in the obtained superposition matrix of the second stage W^2 are filled with ones.

		1	2	3	4	5	6	7	8	9	10	11	12	13	14	15	16	17	18	a	b	c
From state	1	1	0	0	0	0	0	0	0	0	0	0	0	0	0	0	1	1	0	0	0	0
	2	0	1	0	0	0	0	0	0	0	0	0	0	0	0	0	0	0	0	1	0	0
	3	0	0	0	0	1	1	0	0	0	0	0	0	0	0	0	0	0	0	0	0	0
	4	0	0	0	0	0	0	0	0	0	0	0	0	0	0	0	0	0	0	0	0	0
	5	0	0	1	0	0	1	0	0	0	0	0	0	0	1	0	0	0	0	0	0	0
	6	0	0	1	0	1	0	0	0	0	0	0	0	0	0	0	0	0	0	0	0	0
	7	0	0	0	0	0	0	0	0	0	0	0	0	0	1	0	0	0	0	0	0	0
	8	0	0	0	0	0	0	0	0	1	0	1	1	0	0	0	0	0	0	0	0	0
	9	0	0	0	0	0	0	0	1	0	0	1	0	1	0	0	0	0	1	0	0	0
	10	0	0	0	0	0	0	0	0	0	0	1	1	0	0	0	0	0	0	0	0	0
	11	0	0	0	0	0	0	0	1	1	1	1	1	1	0	0	0	0	0	0	0	0
	12	0	0	0	0	0	0	0	1	0	1	1	1	1	0	0	0	0	0	0	0	0
	13	0	0	0	0	0	0	0	0	1	0	1	1	1	0	0	0	0	0	0	0	0
	14	0	0	0	0	1	0	0	0	0	0	0	0	0	0	0	0	0	0	0	0	0
	15	0	0	0	0	0	0	1	0	0	0	0	0	0	0	0	0	0	0	0	0	0
	16	1	0	0	0	0	0	0	0	0	0	0	0	0	0	0	0	1	0	0	0	0
	17	1	0	0	0	0	0	0	0	0	0	0	0	0	0	0	1	1	0	0	0	0
	18	0	0	0	0	0	0	0	0	1	0	0	0	0	0	0	0	0	0	0	0	0
	a	0	1	0	0	0	0	0	0	0	0	0	0	0	0	0	0	0	0	1	0	0
	b	0	0	0	0	0	0	0	0	0	0	0	0	0	0	0	0	0	0	0	0	0
	c	0	0	0	0	0	0	0	0	0	0	0	0	0	0	0	0	0	0	0	0	0

Fig. 3 Superposition matrix of the second degree

From \ To state	1	2	3	4	5	6	7	8	9	10	11	12	13	14	15	16	17	18	a	b	c
1	1	1	1	1	1	1	1	1	0	0	1	0	0	1	1	1	1	0	1	0	1
2	1	1	1	1	1	1	1	1	1	0	1	1	0	1	1	1	1	1	1	1	1
3	1	1	1	1	1	1	1	1	1	0	1	1	1	1	1	1	1	1	1	1	1
4	1	1	1	1	1	1	1	0	1	0	1	1	1	0	1	1	1	0	1	1	1
5	1	1	1	1	1	1	1	1	1	1	1	1	1	1	1	1	1	1	1	1	1
6	1	1	1	1	1	1	1	1	1	0	1	1	0	1	1	1	1	1	1	1	1
7	1	1	1	1	1	1	1	1	1	0	1	1	1	1	1	1	1	1	1	1	1
8	1	1	1	0	1	1	1	1	1	1	1	1	1	1	1	0	0	1	1	1	1
9	0	1	1	1	1	1	1	1	1	1	1	1	1	1	1	0	0	1	1	0	1
10	0	0	0	0	0	1	0	0	1	1	1	1	1	1	0	0	0	0	1	0	0	0
11	1	1	1	1	1	1	1	1	1	1	1	1	1	1	1	0	0	1	1	1	1
12	0	1	1	1	1	1	1	1	1	1	1	1	1	1	1	0	0	1	1	0	1
13	0	0	1	1	1	0	1	1	1	1	1	1	1	0	1	0	0	1	1	0	0
14	1	1	1	0	1	1	1	1	1	0	1	1	0	1	1	1	1	1	1	1	1
15	1	1	1	1	1	1	1	1	1	0	1	1	1	1	1	0	1	1	1	1	1
16	1	1	1	1	1	1	1	0	0	0	0	0	0	1	0	1	1	0	0	0	1
17	1	1	1	1	1	1	1	0	0	0	0	0	0	1	1	1	1	0	0	0	1
18	0	1	1	0	1	1	1	1	1	1	1	1	1	1	1	0	0	1	1	0	0
a	1	1	1	1	1	1	1	1	1	0	1	1	1	1	1	0	0	1	1	1	1
b	0	1	1	1	1	1	1	1	0	0	1	0	0	1	1	0	0	0	1	0	1
c	1	1	1	1	1	1	1	1	1	0	1	1	0	1	1	1	1	0	1	1	1

Fig. 4 Matrix of superposition of the fifth degree W^5

The fifth power superposition matrix W^5 is shown in Fig. 4. The five contours in the W^5 superposition matrix include almost all states of the system. These closed contours represent the basis for further analysis of the superposition matrices of higher degrees and new contours that include a larger number of states, including contours that intersect. For example, state C, as can be seen from the superposition matrix W^5, simultaneously enters the contour of states 1–7, as well as the contours formed by states 14–15 and 16–17. The presence of units on the main diagonal reflects an essential property of the structure of the system—all elements in closed contours are connected to each other.

The adjacency matrix of the seventh degree W^7 looks as shown on Fig. 5. The two circuits in the W^7 superposition matrix include all processes in the system. These closed contours represent intersecting subsets. In this case, there are processes (states) that are included in two circuits the presence of all units on the main diagonal reflects an essential property of the structure of the system—all elements in closed circuits are interconnected in both directions.

The next superposition matrix W^8 contains one closed loop that includes all states. Any transitions between all elements of the system are possible in this circuit. There

		To state																				
		1	2	3	4	5	6	7	8	9	10	11	12	13	14	15	16	17	18	a	b	c
	1	1	1	1	1	1	1	1	1	1	1	1	1	1	1	1	1	1	1	1	1	1
	2	1	1	1	1	1	1	1	1	1	1	1	1	1	1	1	1	1	1	1	1	1
	3	1	1	1	1	1	1	1	1	1	1	1	1	1	1	1	1	1	1	1	1	1
	4	1	1	1	1	1	1	1	1	1	1	1	1	1	1	1	1	1	1	1	1	1
	5	1	1	1	1	1	1	1	1	1	1	1	1	1	1	1	1	1	1	1	1	1
	6	1	1	1	1	1	1	1	1	1	1	1	1	1	1	1	1	1	1	1	1	1
F	7	1	1	1	1	1	1	1	1	1	1	1	1	1	1	1	1	1	1	1	1	1
r	8	1	1	1	1	1	1	1	1	1	1	1	1	1	1	1	1	1	1	1	1	1
o	9	1	1	1	1	1	1	1	1	1	1	1	1	1	1	1	1	1	1	1	1	1
m	10	1	1	1	1	1	1	1	1	1	1	1	1	1	1	1	0	0	1	1	1	1
s	11	1	1	1	1	1	1	1	1	1	1	1	1	1	1	1	1	1	1	1	1	1
t	12	1	1	1	1	1	1	1	1	1	1	1	1	1	1	1	1	1	1	1	1	1
a	13	1	1	1	1	1	1	1	1	1	1	1	1	1	1	1	1	1	1	1	1	1
t	14	1	1	1	1	1	1	1	1	1	1	1	1	1	1	1	1	1	1	1	1	1
e	15	1	1	1	1	1	1	1	1	1	1	1	1	1	1	1	1	1	1	1	1	1
	16	1	1	1	1	1	1	1	1	1	0	1	1	1	1	1	1	1	1	1	1	1
	17	1	1	1	1	1	1	1	1	1	0	1	1	1	1	1	1	1	1	1	1	1
	18	1	1	1	1	1	1	1	1	1	1	1	1	1	1	1	1	1	1	1	1	1
	a	1	1	1	1	1	1	1	1	1	1	1	1	1	1	1	1	1	1	1	1	1
	b	1	1	1	1	1	1	1	1	1	1	1	1	1	1	1	1	1	1	1	1	1
	c	1	1	1	1	1	1	1	1	1	1	1	1	1	1	1	1	1	1	1	1	1

Fig. 5 Adjacency matrix of the seventh degree W^7

are no logins, as well as the ability to log out. The eight-degree adjacency matrix W^8 looks like shows on Fig. 6.

The carried out structural analysis of the control system demonstrates that the structure of the interaction of the states/processes of the project in the improved management scheme of the machine tool enterprise is ergodic.

5 Development of the Improved Circuit Model

The functioning of an improved quality management system during the manufacturing process of a product at a machine tool enterprise depends on many random processes and factors that cannot be predicted in advance. Among these factors, one can single out as follows: the technical condition of the production equipment, the competence, and motivation of personnel, the level of technological maturity, the microclimate in the team, etc.

		1	2	3	4	5	6	7	8	9	10	11	12	13	14	15	16	17	18	a	b	c
	To state																					
F	1	1	1	1	1	1	1	1	1	1	1	1	1	1	1	1	1	1	1	1	1	1
r	2	1	1	1	1	1	1	1	1	1	1	1	1	1	1	1	1	1	1	1	1	1
o	3	1	1	1	1	1	1	1	1	1	1	1	1	1	1	1	1	1	1	1	1	1
m	4	1	1	1	1	1	1	1	1	1	1	1	1	1	1	1	1	1	1	1	1	1
	5	1	1	1	1	1	1	1	1	1	1	1	1	1	1	1	1	1	1	1	1	1
s	6	1	1	1	1	1	1	1	1	1	1	1	1	1	1	1	1	1	1	1	1	1
t	7	1	1	1	1	1	1	1	1	1	1	1	1	1	1	1	1	1	1	1	1	1
a	8	1	1	1	1	1	1	1	1	1	1	1	1	1	1	1	1	1	1	1	1	1
t	9	1	1	1	1	1	1	1	1	1	1	1	1	1	1	1	1	1	1	1	1	1
e	10	1	1	1	1	1	1	1	1	1	1	1	1	1	1	1	1	1	1	1	1	1
	11	1	1	1	1	1	1	1	1	1	1	1	1	1	1	1	1	1	1	1	1	1
	12	1	1	1	1	1	1	1	1	1	1	1	1	1	1	1	1	1	1	1	1	1
	13	1	1	1	1	1	1	1	1	1	1	1	1	1	1	1	1	1	1	1	1	1
	14	1	1	1	1	1	1	1	1	1	1	1	1	1	1	1	1	1	1	1	1	1
	15	1	1	1	1	1	1	1	1	1	1	1	1	1	1	1	1	1	1	1	1	1
	16	1	1	1	1	1	1	1	1	1	1	1	1	1	1	1	1	1	1	1	1	1
	17	1	1	1	1	1	1	1	1	1	1	1	1	1	1	1	1	1	1	1	1	1
	18	1	1	1	1	1	1	1	1	1	1	1	1	1	1	1	1	1	1	1	1	1
	a	1	1	1	1	1	1	1	1	1	1	1	1	1	1	1	1	1	1	1	1	1
	b	1	1	1	1	1	1	1	1	1	1	1	1	1	1	1	1	1	1	1	1	1
	c	1	1	1	1	1	1	1	1	1	1	1	1	1	1	1	1	1	1	1	1	1

Fig. 6 Eight-degree adjacency matrix W^8

A certain process corresponds to each state of the organization (Fig. 1). The total production time of a product T can be defined as the sum of the duration of the project stay in all processes:

$$T = \sum_{s=1}^{n} t_s \qquad (1)$$

where

t_s is the duration of the s-th project process, $s = 1, 2, \ldots, n$;
n is the number of states.

In each state (process) during the production of a product, the system can be for some time (Fig. 1). The duration of the s-th project process is proportional to the probability of the system being in this state. The ratio of the duration of the s-th project processes to the total production time of the product: $T * PS = t_s/T$ reflects the frequency (probability) that the event will occur. The sum of the probabilities of finding the system in all states:

$$\sum_{s=1}^{n} p_s(t) = \sum_{s=1}^{n} \frac{t_s}{T} = \frac{1}{T} \sum_{s=1}^{n} t_s = 1 \tag{2}$$

The states indicated are a complete group of incompatible events. Possible states\processes of the system will be denoted as the set $S = \{S_1, \ldots, S_n; n = 21\}$ (Fig. 1).

We represent the system using a Markov chain with discrete states and time. The discreteness of time is manifested in the fact that the step number serves as the coordinate of discrete time. At the moments when managerial actions are performed in the system, it changes its state, which leads to a change in probabilities. After an arbitrary step k, with a certain probability, the system S can transition to one of the states: $S = \{S_1, S_2, \ldots, S_n\}$—from the complete group of incompatible events, only one occurs $S_1(k), S_2(k), \ldots, S_n(k)$.

Moreover, according to (2), for each step k, the following condition is satisfied:

$$p_1(k) + p_2(k) + \cdots + p_n(k) = 1 \tag{3}$$

Transitions between the states of the system (Fig. 1) are regulated by job descriptions, although a complete graph can also be considered, in which all states are interconnected. For real structures, some of the transition probabilities will be zero. This is a sign that there is no possibility of one-step transitions between certain states. For system states that have such transitions, at any stage k (at discrete time instants $t_1, t_2, \ldots, t_k$), there are transition probabilities π and $j > 0$ $\{\forall (u, j) \in (1, 2, \ldots, n)\}$ along arcs in one step to other states, as well as the probability of delay in this state. The transition probabilities π and $j > 0$ are determined by experts or based on measurements.

The transition probabilities π and $j > 0$ $\{\forall (u, j) \in (1, 2, \ldots, n)\}$ reflect the features of control in the system. The matrix of transition probabilities in the case of a complete graph, including all possible transitions of the Markov chain with n states (processes), will have the form:

$$\|\pi_{ij}\| = \begin{Vmatrix} \pi_{1.1} & \pi_{1.2} & \pi_{1.3} & \cdots & \pi_{1.n-1} & \pi_{1.n} \\ \pi_{2.1} & \pi_{2.2} & \pi_{2.3} & \cdots & \pi_{2.n-1} & \pi_{2.n} \\ \vdots & \vdots & \vdots & \vdots & \vdots & \vdots \\ \pi_{n-1.1} & \pi_{n-1.2} & \pi_{n-1.3} & \cdots & \pi_{n-1.n-1} & \pi_{n-1.n} \\ \pi_{n.1} & \pi_{n.2} & \pi_{n.3} & \cdots & \pi_{n.n-1} & \pi_{n.n} \end{Vmatrix}$$

Based on the values of the transition probabilities, provided that the initial state of the system is determined, we can find the probability of states $p_1(k), p_2(k), \ldots, p_n(k)$ after an arbitrary k-th step using the formula of total probability:

$$p_i(k) = \sum_{j=1}^{n} \left[p_i(k-1) * \pi_{ji} \right] \big| n = 21; \quad i = 1, 2, \ldots, 21 \tag{4}$$

Based on expression (4), we obtain the general solution for the complete graph of the Markov chain with 21 states:

$$
\begin{Vmatrix} p_1(k+1) \\ p_2(k+1) \\ p_3(k+1) \\ \dots \\ \dots \\ p_{20}(k+1) \\ p_{21}(k+1) \end{Vmatrix}^T = \begin{Vmatrix} p_1(k) \\ p_2(k) \\ p_3(k) \\ \dots \\ \dots \\ p_{20}(k) \\ p_{21}(k) \end{Vmatrix}^T * \begin{Vmatrix} \pi_{1.1} & \pi_{1.2} & \pi_{1.3} & \dots & \pi_{1.20} & \pi_{1.21} \\ \pi_{3.1} & \pi_{3.2} & \pi_{3.3} & \dots & \pi_{3.20} & \pi_{3.21} \\ \pi_{2.1} & \pi_{2.2} & \pi_{2.3} & \dots & \pi_{2.20} & \pi_{2.21} \\ \pi_{3.1} & \pi_{3.2} & \pi_{3.3} & \dots & \pi_{3.20} & \pi_{3.21} \\ \dots & \dots & \dots & \dots & \dots & \dots \\ \dots & \dots & \dots & \dots & \dots & \dots \\ \pi_{20.1} & \pi_{20.2} & \pi_{20.3} & \dots & \pi_{20.20} & \pi_{20.21} \\ \pi_{21.1} & \pi_{21.2} & \pi_{21.3} & \dots & \pi_{21.20} & \pi_{21.21} \end{Vmatrix}
\tag{5}
$$

where

T is the sign of transposition.

In a homogeneous Markov chain, it is assumed that the transition probabilities π and $j > 0$ $\{\forall\, (u, j) \in (1, 2, \dots, n)\}$ are constant. This assumption is acceptable since all operations in the manufacture of the product are performed in accordance with the approved labor intensity standards, technical process regulations, and job descriptions.

The correspondence of the Markov model (5) to the original (Fig. 1) can be justified as follows:

- the topological structures of the original and the directed graph reflecting the Markov chain are similar;
- the actions of the project team at time t_k correspond to steps k of the project, and the system can transition from one state to another in one step;
- during the project, the distribution $\{p_1(k), p_2(k), \dots, p_{21}(k)\}$ of the probabilities of the system states is formed;
- system states make up a complete group of events: $\sum_{i=1}^{21} p_i(k) = 1$;
- the probabilities of transitions π and $j > 0$ $\{\forall\, (u, j) \in (1, 2, \dots, n)\}$ from one state to another depend from the properties of the system; and
- transitions from one state to another make up a complete group of events, one of which must occur: $\sum_{j=1}^{21} \pi_{ji} = 1$, $\{i = 1, 2, \dots, 21\}$.

The correspondence of the properties of the original and the model confirms the conclusion about the correctness of the use Markov's chains for modeling the QMS of the machine tool enterprise HC MICRON®.

Figure 8 shows the results of modeling the states of the system for the initial values of the elements of the matrix of transition probabilities, which are obtained from production regulations of processes and operations (Fig. 7).

	To state																				
	1	2	3	4	5	6	7	8	9	10	11	12	13	14	15	16	17	18	a	b	c
1	0,48	0,4	0	0	0	0	0	0	0	0	0	0	0	0	0	0,05	0,05	0,02	0	0	0
2	0	0,3	0,6	0	0	0	0	0	0	0	0	0	0	0	0	0	0	0	0,1	0	0
3	0	0	0,35	0	0,45	0	0	0	0	0	0	0	0	0	0	0	0	0	0	0	0,2
4	0	0	0	0,60	0,40	0	0	0	0	0	0	0	0	0	0	0	0	0	0	0	0
5	0	0	0	0	0,40	0,20	0,40	0	0	0	0	0	0	0	0	0	0	0	0	0	0
6	0	0	0,20	0	0	0,80	0	0	0	0	0	0	0	0	0	0	0	0	0	0	0
7	0,04	0	0	0	0	0	0,21	0,20	0,1	0	0	0	0	0,07	0	0	0	0,03	0,15	0,2	0
8	0	0	0	0	0	0	0	0,35	0,6	0	0	0	0	0,05	0	0	0	0	0	0	0
9	0	0	0	0	0	0	0	0	0,35	0	0,60	0,05	0	0	0	0	0	0	0	0	0
10	0	0	0	0	0	0	0	0	0,5	0,50	0	0	0	0	0	0	0	0	0	0	0
11	0	0	0	0	0	0	0	0,25	0	0	0,25	0,00	0,50	0	0	0	0	0	0	0	0
12	0	0	0	0	0	0	0	0	0	0	0,25	0,35	0,30	0	0	0	0	0,10	0	0	0
13	0	0	0	0	0	0	0	0	0	0	0,10	0,40	0,30	0,20	0	0	0	0	0	0	0
14	0	0	0	0	0	0	0	0	0	0	0	0	0	0,90	0,10	0	0	0	0	0	0
15	0	0	0	0	0,10	0	0	0	0	0	0	0	0	0	0,90	0	0	0	0	0	0
16	0	0	0	0	0	0	0	0	0	0	0	0	0	0	0	0,90	0,10	0	0	0	0
17	0,60	0	0	0	0	0	0	0	0	0	0	0	0	0	0	0	0,30	0	0	0	0,1
18	0	0	0	0	0	0	0	0,15	0	0	0	0	0	0	0	0	0	0,85	0	0	0
a	0	0,25	0	0	0	0	0	0	0	0	0	0	0	0	0	0	0	0,20	0,55	0	0
b	0	0	0	0	0	0	0	0	0	0	0,15	0	0	0	0	0	0	0	0,15	0,7	0
c	0	0	0	0,25	0	0	0	0	0	0	0	0	0	0	0	0	0	0	0	0	0,75

Fig. 7 Transition probabilities for the case shown on Fig. 1

6 Conclusions

At the stage of initiation and implementation of a new scheme for the organization's QMS, the main processes are the development and refinement of the policy and objectives in the field of quality, administrative management (Fig. 8, curve 1), the creation and implementation of a new scheme (Fig. 8, curve 2), preparation of personnel for work in new conditions (Fig. 8, curves 2, 3, 5, 6), and a critical analysis of product requirements (Fig. 8, curve pA). These processes form the basis for the formation of a project-driven environment in the enterprise. After the tenth step, the probabilities of these processes monotonically decrease to values of 0.1—3% of the project execution time at the 30th step. The probability of finding a project in production is reflected in curve 11. Processes for ensuring the production of a product (Fig. 8, curves 8, 9, 12, 13) are set within the probability of states 0.05–0.10. Assessment of customer satisfaction at the final stage of the implementation of a new scheme is becoming one of the processes to which attention should be paid: p18 (30) > 0.05. As follows from the results obtained using the developed model, the process of forming the conditions of responsibility, distribution of powers, and constant information (C) should be attributed to the main states of the system.

The matrix of transition probabilities corresponds to a certain level of perfection of the enterprise management system. The developed Markov model adequately reflects

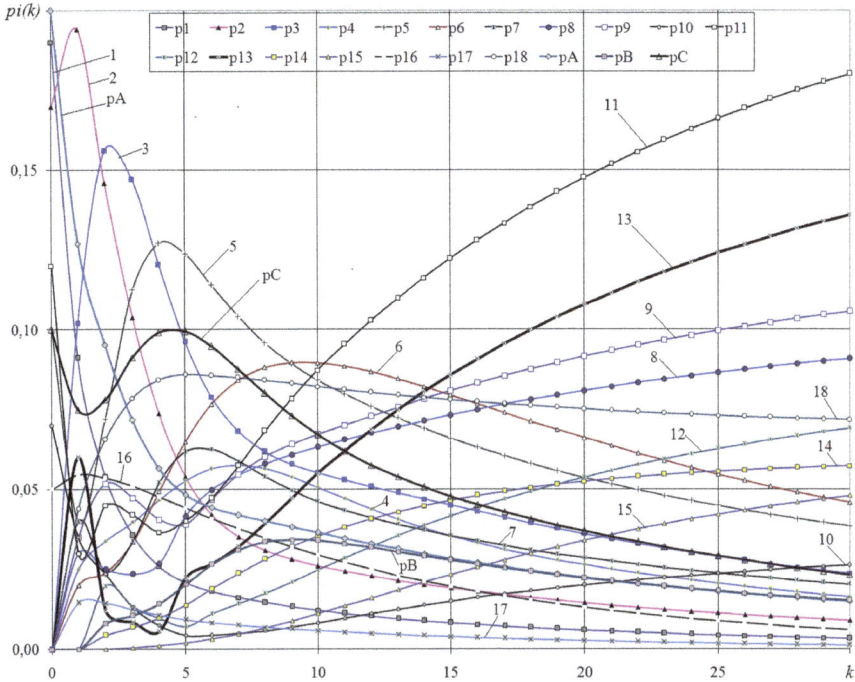

Fig. 8 Changing the probabilities of process states: 1—responsibility of the organization's management; 2—QMS; 3—the process of personnel management; 4—control of personnel competence; 5—product creation management; 6—organization of the professional development process; 7—processes of interaction with customers; 8—product design processes; 9—procurement management; 10—supply quality control; 11—product management processes; 12—control and testing processes; 13—control processes of measuring equipment; 14—document management processes; 15—infrastructure management processes; 16—internal audit 17—continuous improvement processes; 18—assessment of the fulfillment of customer requirements; pA—critical analysis of product requirements; pB—product support; and pC—a responsibility of executors, authority and constant information

the trajectory of development of the project-oriented enterprise in the coordinates of state probabilities.

The work developed a Markov model for an improved structure of the quality management system states, taking into account the recommendations of the new ISO 9001 standard. The hypothesis about the need for scientific substantiation of the new structure of the QMS of a project-managed organization was confirmed. The structure of the QMS has been supplemented with new states: the state of critical analysis of product requirements (S_A), product support throughout the life cycle (S_B), the formation of responsibility, authority, and constant information (S_C).

For the new improved structure of the QMS of the machine tool enterprise, the superposition matrices were calculated up to the eighth degree. Matrix W^8 contains one closed loop that includes all states. Any transitions between system elements are possible in this contour; there are no entrances, as well as the ability to log out. Thus, it has been proved that the structure of the interaction of processes with an improved control scheme of a machine tool enterprise is ergodic.

Acknowledgements The authors are grateful to the management of the HC MICRON®, represented by V. A. Vaisman, for the support and the opportunity for the practical implementation of the elaborated models and methods.

References

1. Turner JR (2007) Project-based management guide. Pr. House of Grebennikov, Moscow, p 552
2. Tutt W (1998) Graph theory. Mir, Moscow, p 424
3. Vaisman VA (2009) Project-managed organizations: models and method of analysis of structural schemes of process control. In: Rybak AI (ed) Science. Notes of the intern. humanizes university, vol 14, pp 4–12
4. Bushuev SD (2005) Development of knowledge systems and project management technologies 2(2):18–24. (Publishing House of Grebennikov, Moscow)
5. Bushuyev SD, Puziychuk AV (2019) Further development of models and methods of project management for value-oriented management. Bull Cherkasy State Technol Univ 2:69–74
6. Georggemünden H, Lehner P, Kock A (2018) The project-oriented organization and its contribution to innovation. Int J Proj Manage 36(1):147–160
7. Bushuyev SD, Bushuiev DA, Yaroshenko RF, Chernova LS (2017) Threats management principles for development programs of high technology industries in turbulent environment. Organ Izarzadzanie 105
8. (2017) A guide to the project management body of knowledge. PMBOK® guide, 6th edn. Project Management Institute Inc.
9. Tolman E (2020) Cognitive maps for rats and humans. http://www.psychology.ru/library/00060.shtml. Last accessed 2020/12/05
10. .Gray W (2018) Cognitive modeling for cognitive engineering. In: Sun R (ed) The Cambridge handbook of computational psychology. Cambridge handbooks in psychology. Cambridge University, Cambridge, pp 565–588
11. Lane HC (2012) Cognitive models of learning. In: Seel NM (ed) Encyclopedia of the sciences of learning. Springer, Boston, MA. https://doi.org/10.1007/978-1-4419-1428-6_241
12. Holt DV, Osman M (2017) Approaches to cognitive modeling in dynamic systems control. Front Psychol 8:2032. https://doi.org/10.3389/fpsyg.2017.02032
13. Sherstiuk O, Kolesnikov O, Lukianov D (2019) Team behaviour model as a tool for determining the project development trajectory. In: IEEE International conference on advanced trends in information theory (ATIT), pp 496–500
14. Lukianov D, Bespanskaya-Paulenko K, Gogunskii V, Kolesnikov O, Moskaliuk A, Dmitrenko K (2017) Development of the Markov model of a project as a system of role communications in a team. East-Eur J Enterp Technol 3(3):21–28
15. Morozov V, Kalnichenko O, Mezentseva O (2020) The method of interaction modeling on basis of deep learning of neural networks in complex IT projects. Int J Comput 19(1):88–96
16. Gogunskii V, Kolesnikov O, Oborska G et al (2017) Representation of project systems using the Markov chain. East-Eur J Enterp Technol 2(3–86):60–65. https://doi.org/10.15587/1729-4061.2017.97883
17. Vaysman VA, Lukianov DV, Kolesnikova KV (2012) The planar graphs closed cycles determination method. Prasi ONPU 1(38):222–227
18. DSTU ISO 9001 (2009) Quality management systems. Requirements (ISO 9001: 2008, IDT). State Standard of Ukraine, Kyiv, p 25
19. DSTU ISO 10002 (2007) Quality management. Customer satisfaction (ISO 10002: 2004, IDT)

20. DSTU ISO 10005 (2007) Quality management systems. Guidelines for quality programs (ISO 10005: 2005, IDT)
21. Griffeath D (1975) Ergodic theorems for graph interactions. Adv Appl Probab 7(1):179–194. https://doi.org/10.2307/1425859
22. Wentzel ES (1988) Operations research: tasks, principles, methodology. Bustard, Moscow, p 388
23. Zheng Y, Pan L, Qian J, Guo H (2020) Fast matching via ergodic Markov chain for super-large graphs. Pattern Recogn 106. https://doi.org/10.1016/j.patcog.2020.107418
24. Mezentseva O, Kolesnikov O, Kolesnikova K (2020) Development of a Markov model of changes in patients' health conditions in medical projects. In: Proceedings of the 3rd international conference on informatics & data-driven medicine, pp 240–251

Chapter 66
Extraction of Pothole Attribute for Road Infra Maintenance Using Kinect

D. Ramesh Reddy, Addanki Pranava, C. D. Naidu, and Prakash Kodali

1 Introduction

Pavements are the most economical means of transport. The construction of roads, whether it be highways or rural roads, has been increasing rapidly. According to the National Highways Authority of India, investments in the construction of roads highly increased. The rate of construction of roads is more than double from the past five years' data according to the Economic Survey 2019, as shown in Fig. 1 [1]. India now contains nearly 5.8 million km of constructed roads. Highways' construction is increased by almost 21.44% compound annual growth with 88% overall highway construction in 2020. The National Highway Authority of India has constructed 3.9K km in the fiscal year 2019–2020. According to the World Economic Forum, India ranks second in road network among 222 countries worldwide, whereas India ranks 46th among 141 countries worldwide.

India currently faces the problem of deterioration of infrastructure due to the roads that are constructed long back, improper construction, and the use of inferior materials for the building. Many factors such as under-construction, pavement holes, sinkholes, street cuts, and uneven roads are responsible for hazardous effects and are the reasons for accidents. Ministry of Road Transport has made a survey report regarding the pavement infrastructure and fatalities caused as shown in Fig. 2 [2].

The pavements are critical factors for road crashes. Pavement holes are one of the chief concerns of the motorist. Rather than depending solely on the authorities to repair the damages, it would be beneficial to the individual motorist to equip the automobile with the sensor to estimate the pavement holes and help the authorities to

D. Ramesh Reddy (✉) · A. Pranava · C. D. Naidu
Department of ECE, VNR VJIET, Hyderabad, India

P. Kodali (✉)
Department of ECE, NIT Warangal, Warangal, India
e-mail: kprakash@nitw.ac.in

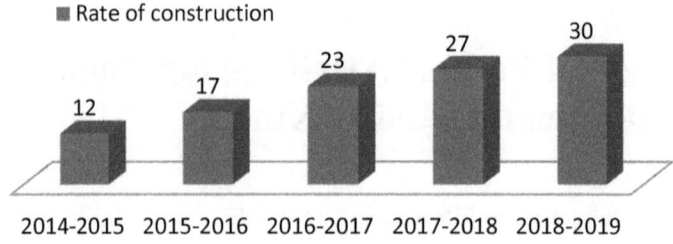

Fig. 1 Economic survey of road construction in India 2019 report

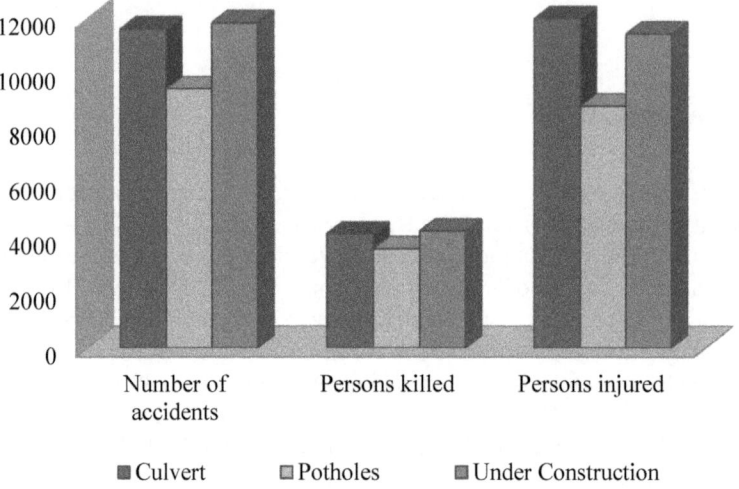

Fig. 2 A survey report by the ministry of road transport due to various road effects from 2015 to 2018

transmit the data. Predictive maintenance is better than preventive care. An accurate estimation of damage has to be made. Prediction of such data assists in analyzing the materials required to reconstruct the damaged infrastructure. This paper offers a cost-effective way of detecting a pothole's length using the Kinect sensor [3]. Hence, the motorist's safety can upgrade by knowing the damage caused due to the pavement holes. The proposed method in this paper uses a three-dimensional Kinect sensor to detect the pothole length.

2 Previous Work

Previous works that involve pavement hole detection consider the hardware devices such as accelerometer, external cameras, or any other vibration sensor. Sattar et al.

[4] made a review report comparing gyroscope, accelerometer, and magnetometer sensors to obtain a three-dimensional rotation view to monitoring the road's surface. This review compares several works done on potholes. Dimple et al. [5] used ADXL 1335 accelerometer with a mobile application named "ROAD MODE" developed that shows details of upcoming holes on the pavement so that the automobile driver can strategize his safety by avoiding the damaged road. This is a conventional way of detecting the holes on the pavement. An Arduino sensor fitted with an accelerometer transmits the coordinates to the cloud server. The server categorizes the severity of the pavement surface based on the obtained coordinates of the Arduino device. Rajmane et al. [6] used Bluetooth, G-Sensor, and an Arduino device. The average and the slope values of the accelerometer coordinates are calculated. This technique helps in analyzing the pavement surface area. The conventional method of using the accelerometer data along with the Arduino increases the hardware complexity. Gavua et al. [7] used an Arduino microcontroller with a light-dependent resistor embedded with the controller to identify the holes on their pathway. The statistics obtained by the controller are then transmitted using a global system for mobile management. The data is forwarded to the authorities and the motorists so that the motorists can also provide their feedback. Vasantha et al. [8] used a microcontroller LPC2148 to warn the authorities regarding the risk detected in a precise location. This system is fitted in the automobile. The system clarifies landslides, gas leakage, accidents, traffic, and even bad weather conditions. LDR sensor, MQ-6, and IR sensor are additionally used. Ramesh Reddy et al. [9] used the depth sensor to spot the defected areas with the non-defected regions and mark the concerned location using the Global Positioning System. Beňo et al. [10] reproduced the three-dimensional view of the static target object or the environment using a grid mapping technique and topological mapping. Lun and Zhao [11] discussed using the triangulation method on the identified items and compared the Kinect version 1 and version 2 sensors' resultant depth frames. Abdullah et al. [12] went through the time of flight technique for evading motor vehicles' collisions. Tanaka and Sogabe [13] distinguished the necessary parts, the target, by measuring the required parameters like X and Y shapes automatically by using the Kinect sensor. Chin et al. [14] used range image sensors to spectate the pavement surfaces and analyze the pavement features. Sungheetha and Rajendran [15] described using the convolutional neural network method for image segmentation techniques. The work portrays the regression and classification techniques applied to the images. Wang et al. [16] normalized accelerometer coordinates to avoid maintaining precise camera angle. Coşkun et al. [17] elaborated usage of vision and depth data on different objects. Sanguino and Gómez [18] used the Hough transformation to classify the three-dimensional objects. Few image processing methods are automated using this method.

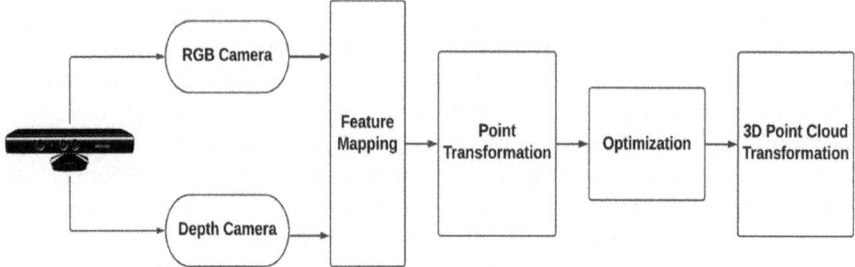

Fig. 3 Three-dimensional reconstruction model using a proposed system

3 Proposed Work

There are two ways of obtaining the parameters of pavement holes on the pavement. One is to grab the frame from the streaming Kinect sensor. The second method uses a standard two-dimensional image to be reconstructed as a three-dimensional frame to enhance the details of the grabbed holes on the pavement. Our paper addresses fragments of both methods. In today's world, the Kinect sensor can be used to measure different parameters in many applications. The Kinect sensor containing the depth and IR camera fit to the motor vehicle in movement helps scrutinize the holes on the pavement. This provides real-time data of the holes whenever the motor vehicle passes through them. The Kinect depth and RGB videos use the computer vision technique of thresholding to seize the video frame as an image frame. As the motor vehicle passes through the pavement, potholes are scrutinized using the depth camera and RGB camera [9]. The two-dimensional image frame can be reconstructed using the reconstruction model, as shown in Fig. 3. The model discussed here can be used on any two-dimensional image frame grabbed by a regular camera and reconstructed into three-dimensional models using the depth sensor.

The Kinect sensor is equipped with the RGB camera, a standard color camera, and the depth camera, which gives a three-dimensional camera view. The speckle points using an IR camera are mapped to the original idea using the color camera's features. The depth image is obtained by mapping the speckle point pattern with the original color frame. This is known as feature mapping. Feature mapping is done internally in the Kinect sensor. The obtained structure is then optimized by removing the unwanted speckles that do not contribute to the original image reconstruction. The whole scene is then constructed in a proper depth view [10].

4 Methodology

Pavement pothole frame from the Kinect depth video involves the following process:

- The real-time motor vehicle streams the video endlessly.

- Computer vision threshold parameter distinguishes the holes present on the road with the normal pavement surface.
- The distinguished surface, which is the detected crack hole, is then grabbed as an image frame, while the motor vehicle is still in motion. Capturing the structure from the streaming video takes only a few milliseconds. The video streaming process does not interrupt the frame grabbing process.
- Further processing is used to perform on the obtained image frames.

The methodology of pavement hole parameter mostly focuses on the post-extraction of the pothole image frame. The image frame of the pothole can use for extracting the essential parameter. Pothole parameters can be found using techniques such as triangulation and time of flight methods.

4.1 Triangulation Method and Time of Flight

The triangulation method uses an infrared (IR) projector that emits IR radiation onto the target, a pothole. The IR ray will reflect from the pothole, which is then captured by a depth sensor and forms a 3D reconstruction of the pothole itself. This method yields the length of the pavement holes and gives a clear insight for comparing physically measured pothole length with the Kinect sensor that measured the length of the potholes [14].

Time of flight method is used for calculating the pothole's Z-coordinate, which is the depth coordinate [15]. Time of flight is a procedure that gives the distance from the sensor to the detected object. This is the fundamental use of the Kinect sensor. For time of flight, sensors define the range from the time it takes light to travel from the sensors to the pothole and return, as shown in Fig. 4. The measurement of distance is entirely dependent on the IR camera. This feature can be enabled in an autonomous automobile to obtain the pavement hole length on the road surface. This feature of Kinect is additionally addressed.

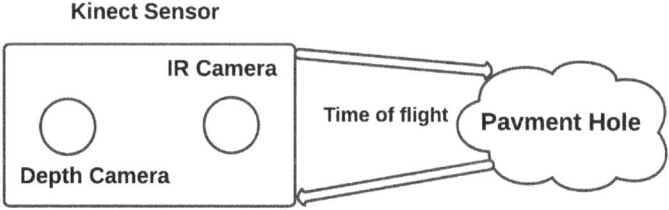

Fig. 4 Time of flight technique of Kinect sensor

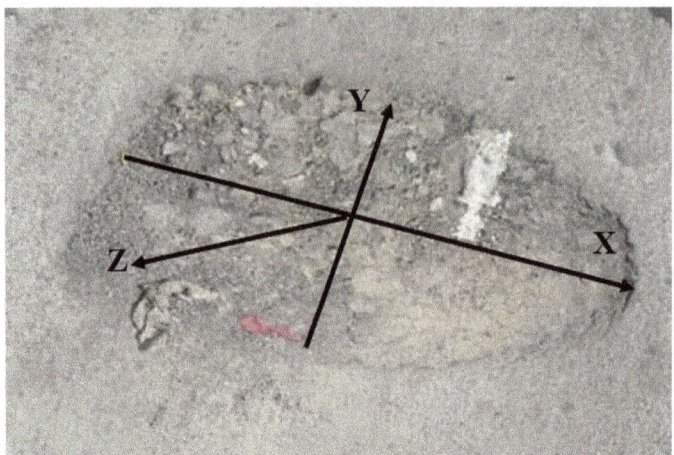

Fig. 5 Method for estimating pothole centroid

```
x,y,z
-0.22424937966362582,0.1611700462701743,-0.39249255932230664
-0.22424937966362582,0.16050597521014062,-0.39249255932230664
-0.22424937966362582,0.15984190415010693,-0.39249255932230664
-0.22424937966362582,0.15917783309007327,-0.39249255932230664
-0.22424937966362582,0.15851376203003958,-0.39249255932230664
-0.22424937966362582,0.1578496909700059,-0.39249255932230664
```

Fig. 6 3D coordinates obtained by Kinect sensor

4.2 Analyzing the Pothole

Kinect generates three coordinates, X, Y, and Z, for the target pothole where X-axis is a horizontal axis, Y-axis is a vertical axis, and Z-axis is the depth axis as shown in Fig. 5. These coordinates help calculate the centroid of the object, i.e., pothole, as shown in Fig. 6. Figure 6 depicts the X-, Y-, and Z-axes, respectively. Z-axis shows the direction of the Kinect sensor.

The pothole length can be obtained by finding the pothole's center using a Kinect sensor [16]. The length is the resultant distance in between the two farther axes. These 3D coordinates are a bit similar to yaw, pitch, and roll.

5 Experimental Results

This proposed system is used to determine the length of the crack in pothole. This parameter can be used in analyzing the holes on the pavement. Preventive methods

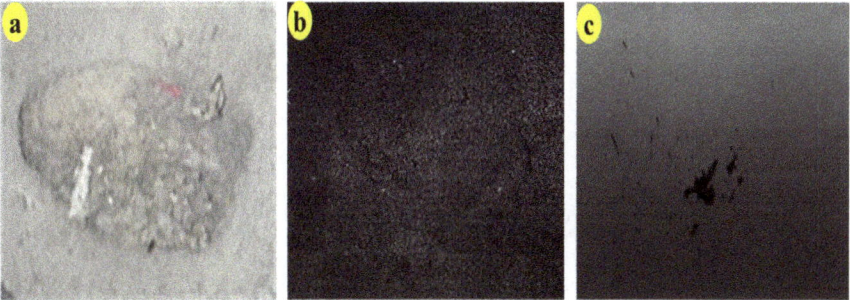

Fig. 7 **a** RGB image on a pothole on the road. **b** Speckle point pattern of the pothole. **c** Speckles mapped with RGB, resulting in a depth image

can be used further to avoid effects caused [17]. Figure 7a shows the color frame grabbed by the RGB camera, Fig. 7b shows the speckle point pattern obtained due to the IR sensor, and Fig. 7c shows the frame grabbed from the depth camera of the Kinect sensor. Figure 8b shows the length of the pavement hole measured in millimeters for the pothole in Fig. 8a.

The proposed system is used to experiment on different potholes, as shown in Fig. 9, with the number of observations in the horizontal axis and the detected hole's length on a pavement in the vertical axis. The hardware arrangement with the sensor fitted to the automobile is shown in Fig. 10.

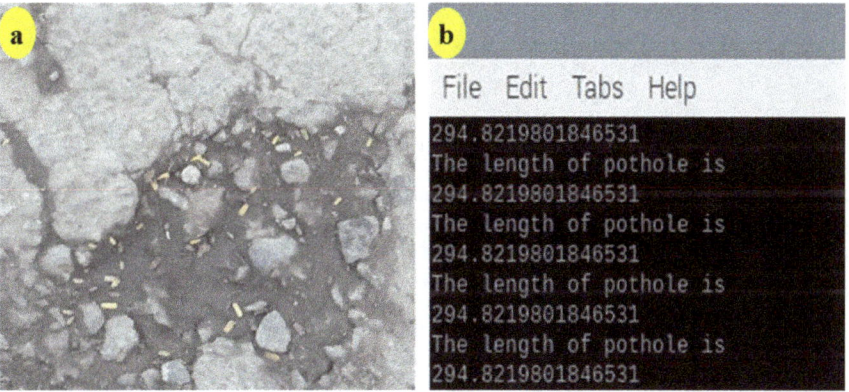

Fig. 8 **a** Image of a pothole on the pavement. **b** Length of pothole estimated is shown on proposed system

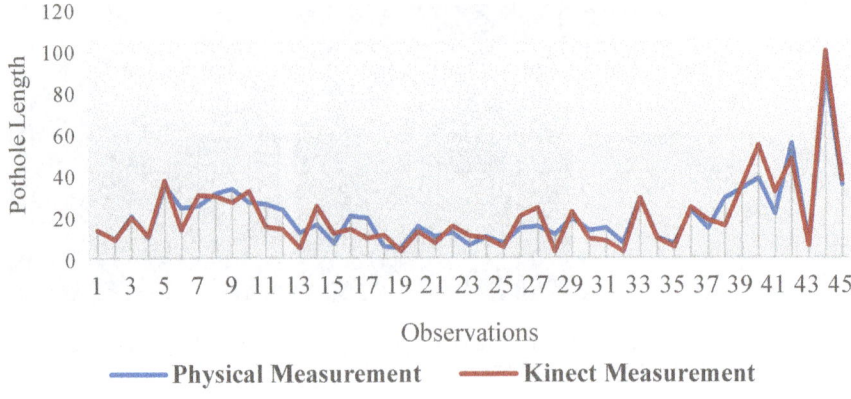

Fig. 9 Comparing original pothole length with sensor measured pothole length

Fig. 10 **a** Installation of a model to a car without changing any parts and **b** hardware setup of the proposed system

6 Conclusion

In this paper, an arrangement of Kinect sensor interfaced with Raspberry Pi has been presented. This paper can be used for analyzing the holes in the pavement. This analysis can help the authorities in rebuilding the road infrastructure. The method addressed is validated by comparing the sensor measured values with the practically measured values, as shown in Fig. 9. The error resulting from comparing the original value with the sensor value is 0.97, making the presented model highly reliable. The method presented in the paper produces the length of the detected hole on the pavement surface. This will help in taking preventive measures for pavement structures. The method can be embedded with the autonomous automobiles with road surface monitoring as one of the features.

References

1. Economic Survey 2019, Highway construction rate in India. https://www.indiatoday.in/diu/story/1562333-2019-07-04
2. Ministry of Road Transport & Highways, Transport Research Wing, Road accidents in India (2018), 2015–2018. https://morth.nic.in/sites/default/files/Road_Accidednt.pdf
3. Motty A, Yogitha A, Nandakumar R (2019) Flag semaphore detection using tensorflow and opencv. Int J Recent Technol Eng (IJRTE) 7(6):2277–3878
4. Sattar S, Li S, Chapman M (2018) Road surface monitoring using smartphone sensors: a review. Sensors 18:3845
5. Dimple S, Monica V, Ashok A, Adarsh C (2016) Monitoring of road irregularities using IOT. Int J Adv Electron Comput Sci (Special issue). ISSN: 2393-2835
6. Rajmane O, Rane V, Bhosale A (2017) Road condition detection using arduino based sensing module and android smartphone. Int J Adv Eng Res Dev Emerg Technol Comput World. ISSN: 2348-4470
7. Gavua E, Okyere-Dankwa S, Agbesi (2017) Pothole detection, reporting and management using internet of things: prospects and challenges 5:2319–6378
8. Vasantha G, Pavithra B, Poornima A, Sriharisudheer G, Sreenivasulu G, Rajagopal R (2018) IoT based smart roads intelligent highways with warning message and diversions according to climate conditions. In: National conference on emerging trends in information, management and engineering sciences (NC'e-TIMES #1.0)—2018, Open access research artcicle
9. Ramesh Reddy D, Kumar P, Naidu CD (2019) Internet of things based pothole detection system using kinect sensor. In: Third international conference on I-SMAC (IoT in social, mobile, analytics and cloud), pp 587–591
10. Beňo P, Duchoň F, Tölgyessy M, Hubinský P, Kajan M (2014) 3D map reconstruction with sensor kinect searching for solution applicable to small mobile robots. In: 23rd International conference on robotics in Alpe-Adria-Danube region, IEEE RAAD 2014—conference proceedings, 10.1109
11. Lun R, Zhao W (2015) A survey of applications and human motion recognition with microsoft kinect. Int J Pattern Recognit Artif Intell 29(05):1555008
12. Abdullah MSH, Zabidi A, Yassin IM, Hassan H (2015) Analysis of microsoft kinect depth perception for distance detection of vehicles. In: IEEE 6th Control and system graduate research colloquium (ICSGRC)
13. Tanaka M, Sogabe A (2017) A measuring system of the legs shape by using the kinect sensor. In: 56th Annual conference of the Society of Instrument and Control Engineers of Japan (SICE). https://doi.org/10.23919/SICE.2017.8105558
14. Chin Y, Ogitsu T, Mizoguchi H (2019) Road surface mapping using kinect sensor for road maintenance. In: International conference on industrial application engineering 2016, pp 273–276
15. Sungheetha A, Rajendran RS (2020) A novel CapsNet based image reconstruction and regression analysis. J Innov Image Process 2:156–164
16. Wang H-W, Chen C-H, Cheng D-Y, Lin C-H, Lo C-C (2015) A real-time pothole detection approach for intelligent transportation system. Math Prob Eng 2015, Article ID 869627
17. Coşkun A, Kara A, Parlaktuna M, Ozkan M, Parlaktuna O (2015) People counting system by using kinect sensor. In: 2015 International symposium on innovations in intelligent systems and applications (INISTA), Madrid, pp 1–7
18. Sanguino TJM, Gómez FP (2015) Improving 3D object detection and classification based on kinect sensor and Hough transform. In: 2015 International symposium on innovations in intelligent systems and applications (INISTA), Madrid, pp 1–8

Chapter 67
Transient Stability Analysis with Optimal Location of FACT Controller Using Hybrid Optimization Technique

P. K. Dhal

1 Introduction

A complicated structure is the power system network. It is important to have power that is reliable, safe, and better regulation. A feasible problem solving strategy is provided by versatile AC transmission systems (FACTS). Private power producers are rising rapidly in many industries in power sector to meet increasing demand. The main objective of the power transmission line is needed to improve reliability, maximum performance, and low cost. As the transfer of power increases with unplanned power flows and greater losses, the system becomes more complex and unsafe. So, the power system security and FACTS devices such as static var compensator can be very efficient. The proper position of the SVC plays a key role in improving the efficiency of the power system without compromising the system's protection. This system provides steady state and transient voltage control for high performance [1, 2]. Several different mathematical methods have been used for the optimal power flow solution.

In recent years, researchers have focused on using intelligent, nature-inspired strategies to tackle the optimization problem of power flow, such as the genetic algorithm (GA), particle swarm optimization (PSO), ant colony optimization (ACO), differential evolution (DE), and harmonic search [3–6]. Ongsakul and Bhasaputra proposed an ideal power flow problem based on hybrid simulated annealing [7]. For optimal positioning and sizing of FACTS controllers, a new multi-objective fuzzy and DE formulation was discussed in [8]. In [9] used a hybrid heuristic technique to solve the loadability of the power system and placed it at SVC. G. I. Rashed presented the DE algorithm to find the optimal location of UPFC in various IEEE bus systems to minimize active and reactive power losses [10]. The loadability of power systems and the optimal positioning of the SVC were presented in [11]. J.S.Huang et al. developed

P. K. Dhal (✉)
Vel Tech Rangarajan Dr. Sagunthala R&D Institute of Science and Technology, Chennai, India

S. Kumar et al. (eds.), *Proceedings of International Conference on Communication and Computational Technologies*, Algorithms for Intelligent Systems,
https://doi.org/10.1007/978-981-16-3246-4_68

an effective static VAR compensator location for improving power system voltage stability under the most demanding operating conditions. In [12] P. K. Roy et al. presented optimal VAR control for improving voltage stability and finding power loss minimization. For a single objective and multi-objective optimization in a fuzzy framework, the conventional approach tackles the issue of the optimum position of the system. So to transform the multi-objective optimization problem into a single fuzzy performance, the fuzzy logic method is applied. It is minimizing the deviation of bus voltage, but lack of power loss calculation in the traditional method. The proposed new approach is therefore used in fifty-seven buses that achieve better results through PSAT software. The remainder of the paper is structured as follows: in Sect. 2, the hybrid approach to solving the problem of optimization is discussed, and in Sect. 3, the results of the simulation and major discussion of the proposed method are given, case 1: optimal power flow without SVC and case 2: optimal power flow with SVC tuned to the BBO-DE algorithm. In Sect. 4, part of the conclusion is eventually outlined in depth.

2 Biogeography-Based Optimization and Differential Evolution Approach to Solving an Optimization Problem

Generally, differential evolution is an evolutionary algorithm focused on the population. It is able to handle no-differential, nonlinear issues. Similarly, the biogeography explains the species are migrated from one island to another. The biological species residences have a high index of habitat suitability. The fundamental optimization algorithm based on biogeography is applied to the problem of optimization. It is a combination of two algorithms, particularly optimization problems which will produce better results. During the crossover process, differential evolution is understood and all control variables are altered together. The individual variable is not individually tuned. The solutions are pushed toward the optimum point very easily, but at a later date, the DE fails to offer better system performance when a fine-tuning operation is needed. However, via migration operations, the BBO approach is very important. In order to achieve the optimum solution, the BBO needs greater processing time. In order to use DE's exploration capacity and BBO's extraction capacity, a single algorithm is used to solve complex optimization problems. It is shown that the operator of hybrid migration balances BBO exploitation and DE searching. The BBO-DE algorithm solution is considered a hybrid technique. The BBO-DE approach proposed is defined as

Step 1: To initialize the search space process population randomly
Step 2: To determine each current population's fitness value
Step 3: Selecting a few elite individuals to stop solutions
Step 4: Assign the number of species based on the habitat suitability index to all the individual habitats
Step 5: Using a hybrid migration operation to change the habitats

Step 6: Measurement of each habitat's mutation rate
Step 7: Change the non-elite individuals in the BBO algorithm's population-based process
Step 8: Go to step number 2, if it fails, and repeat to get the optimum value.

3   Simulation Results and Major Discussion of the Proposed System

The fifty-seven bus systems are taken into account in the IEEE. It requires seven generators, and 42 loads are taken into account as shown in Fig. 1. The machine load is increased by up to 40% points. The weak system buses are evaluated after the iteration phase; i.e., buses 31, 53, 54, and 55 are the weak buses. In contrast to other bus profile, weak buses have been referred to as low voltage profile. Bus 31 has a low voltage profile of 0.818 p.u, bus 53 has 0.803 p.u, bus 54 has 0.762 p.u, and bus 55 has 0.0.737 p.u, according to the fifty-seven bus scheme, when simulation without SVC stopped operating at 2.06 s. The SVC system on bus 54 and bus 55 provides stability because the positive eigenvalue is zero and the system simulation of the system is operating correctly. Bus 31 and bus 53 do not have positive eigenvalues. So, bus 55 is therefore selected to position SVC as shown in Fig. 2 and its results

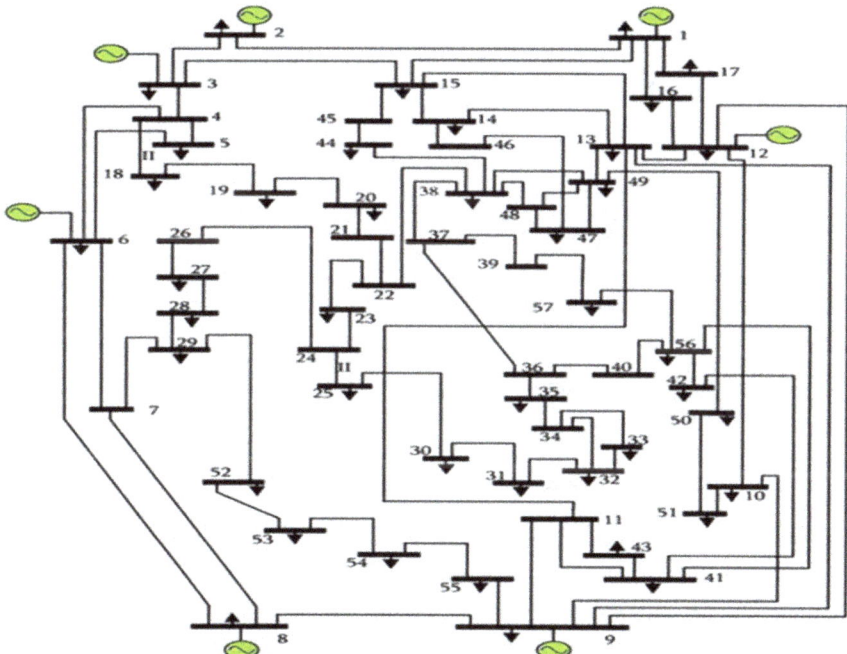

Fig. 1 Simulation diagram of fifty-seven bus system

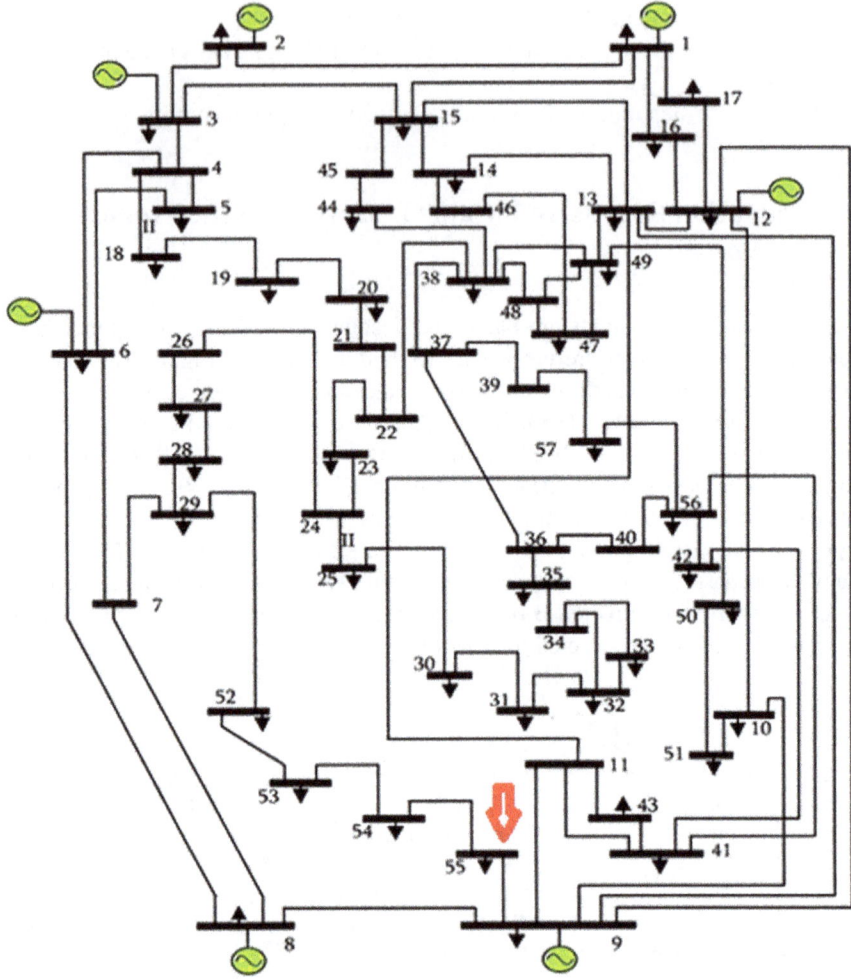

Fig. 2 Simulation diagram of fifty-seven bus system with SVC

are shown in Table 1. The optimal location of SVC is determined by applying the proposed BBO-DE approach.

3.1 Case 1: Optimal Power Flow Without SVC

To show, the effectiveness of the system is performed without SVC. Due to increasing load conditions, a three-phase fault arises in the system. It is observed that there are some weak buses. The weak buses have a low voltage profile without SVC device.

Table 1 Eigenvalue-based parameters in IEEE fifty-seven bus

Sl. No.	Without SVC	SVC used 55 bus location	SVC used 54 bus location	SVC used 53 bus location	SVC used 31 bus location
Dynamic order selection	56	57	57	57	57
Negative eigenvalues	54	55	55	53	52
Positive eigenvalues	0	0	0	2	3

Fig. 3 Voltage outline diagram not considering SVC

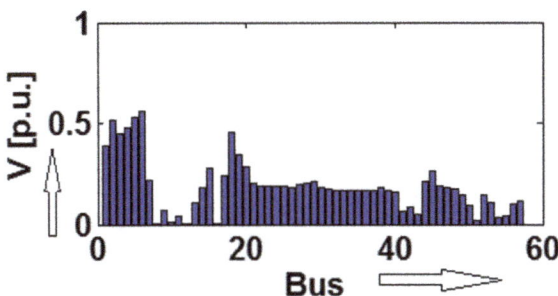

Fig. 4 Real power outline diagram not considering SVC

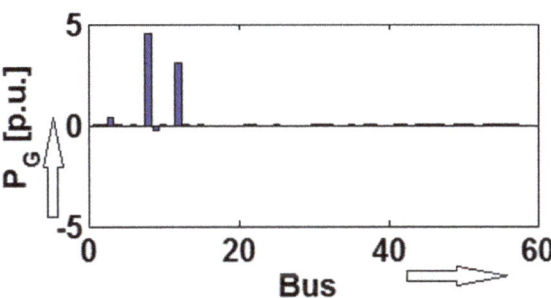

The voltage magnitudes are reduced to below normal value. It is observed that the system gets unbalanced condition. It is shown in Figs. 3, 4, 5, 6, 7, 8, 9, 10, and 11.

3.2   Case 2: Optimal Power Flow with SVC Tuned with BBO-DE Algorithm

The optimal location of the proposed system is considered bus 55. The SVC controller is fixed at that position. All the parameters are improved as shown in Figs. 12, 13, 14, 15, 16, 17, 18, 19, and 20 (Table 2).

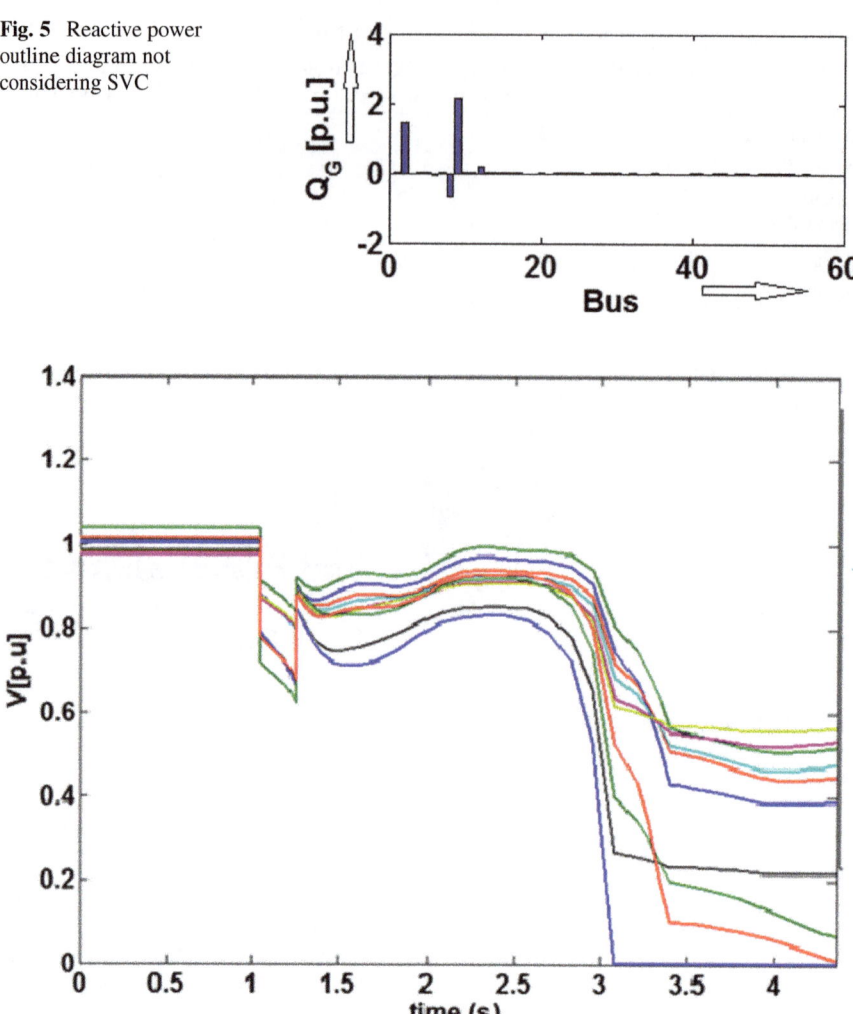

Fig. 5 Reactive power outline diagram not considering SVC

Fig. 6 Voltage waveforms of bus 1–10 diagram not considering SVC

The overall fifty-seven bus system generation and losses are shown in Table 3. The system with SVC has minimized the losses as opposed to the system without SVC in the scheme.

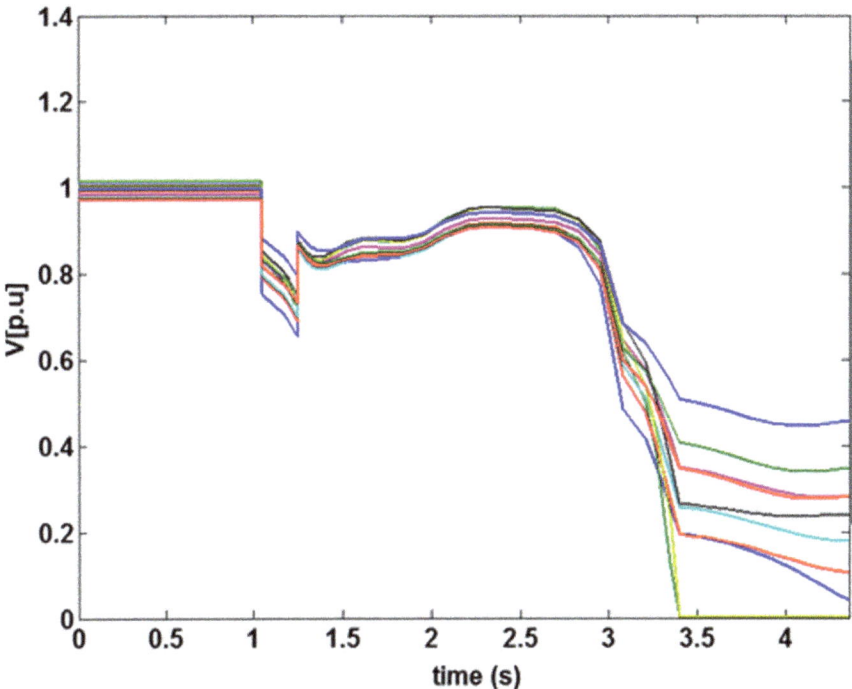

Fig. 7 Voltage waveforms of bus 11–20 diagram not considering SVC

4 Conclusion

In this paper, a hybrid BBO-DE algorithm has been successfully implemented to solve optimal power flow in multi-machine systems. It is important to note that in certain case the shunt compensation at the weakest bus does not result in reactive power compensation for voltage stability condition. The proposed method finds a bus which is strongly tied with many weak buses. This approach has been tested and examined with a selective objective and different constraints to demonstrate its effectiveness. In this study, the BBO-DE algorithm is used to take suitable parameters to stabilize the bus voltages. Through time-domain simulations and comparative results, it has been found that power loss minimization and the augmentation of voltage profile are achieved by suitable tuning of parameters in standard power system networks.

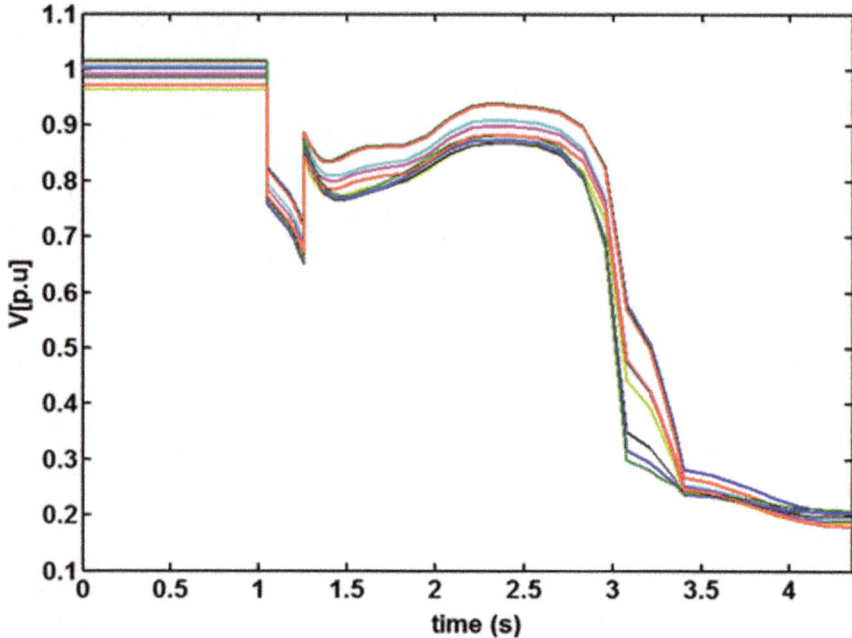

Fig. 8 Voltage waveforms of bus 21–30 diagram not considering SVC

Fig. 9 Voltage waveforms
of bus 31–40 diagram not
considering SVC

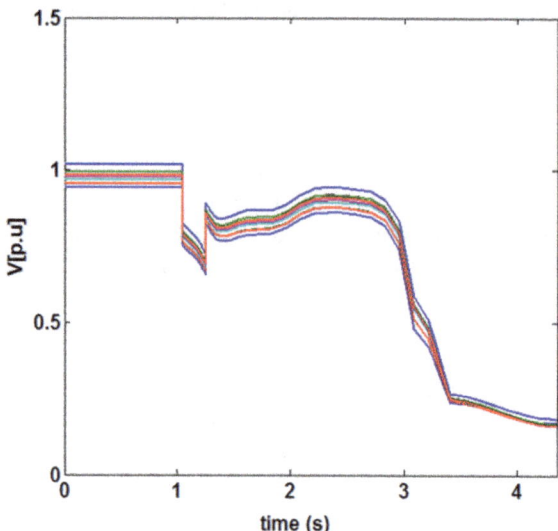

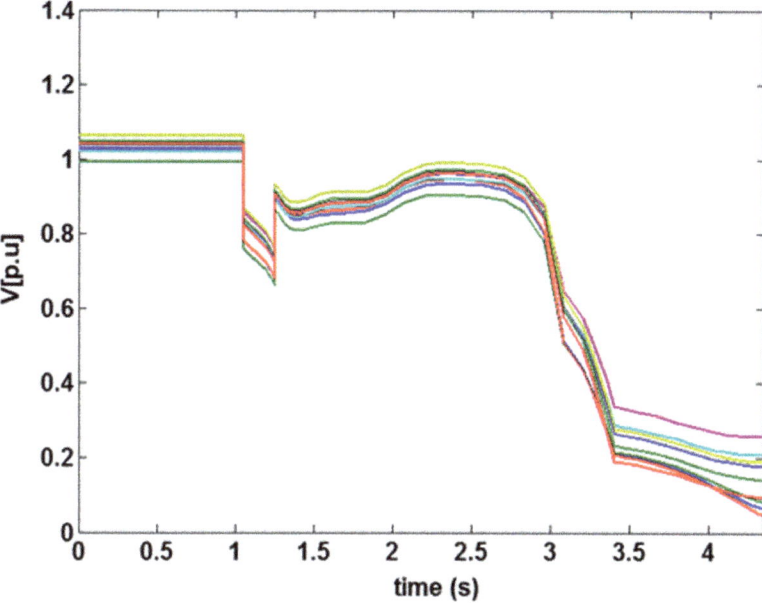

Fig. 10 Voltage waveforms of bus 41–50 diagram not considering SVC

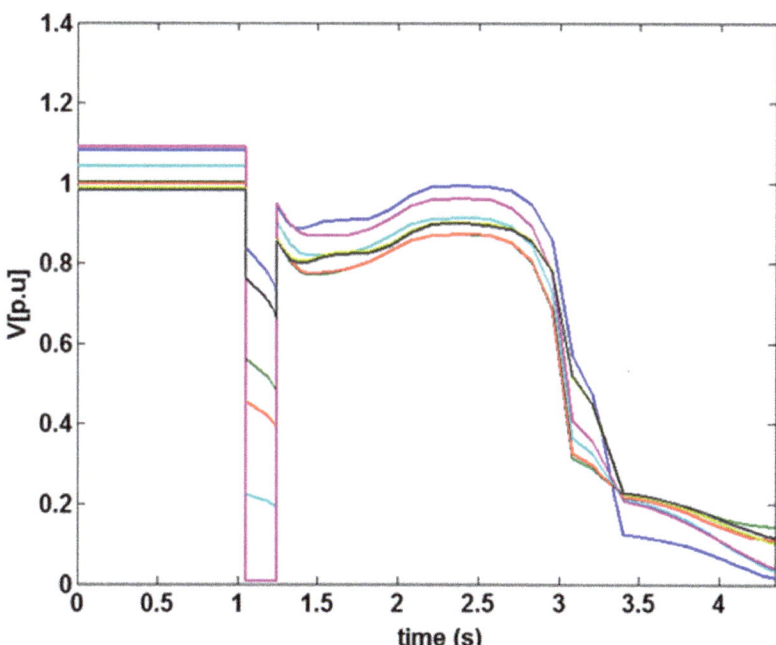

Fig. 11 Voltage waveforms of bus 51–57 diagram not considering SVC

Fig. 12 Voltage outline diagram considering SVC applying with BBO-DE

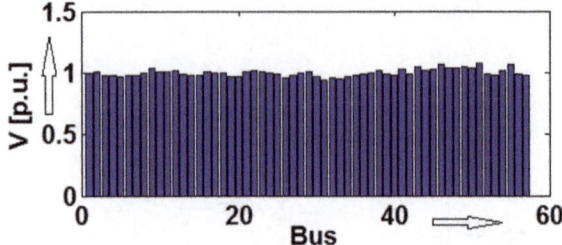

Fig. 13 Real power outline diagram considering SVC applying with BBO-DE

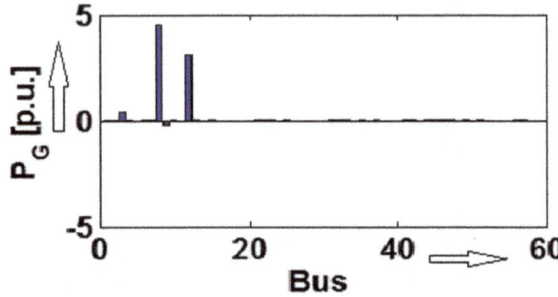

Fig. 14 Reactive power outline diagram considering SVC applying with BBO-DE

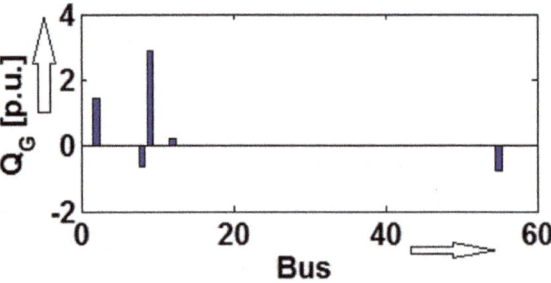

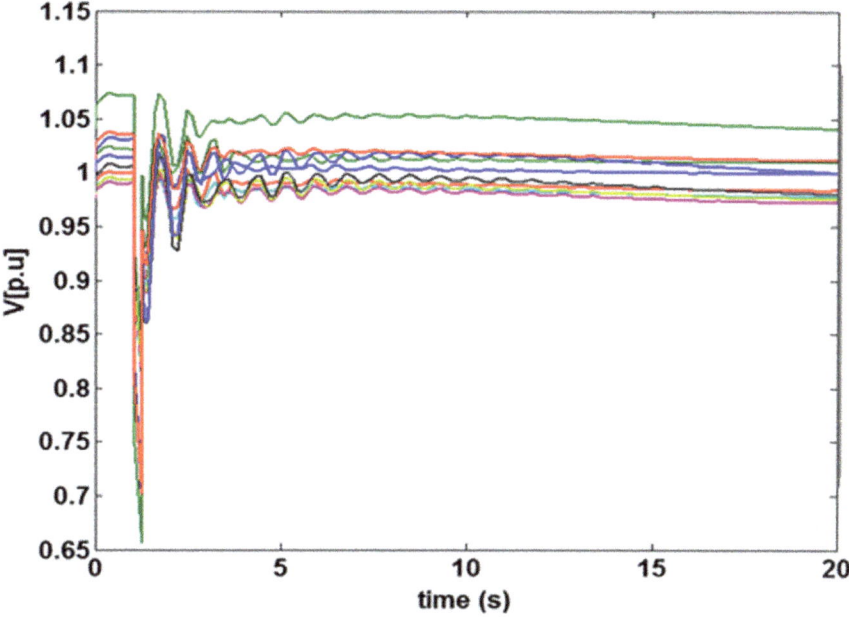

Fig. 15 Voltage waveform of bus 1–10 considering SVC applying with BBO-DE

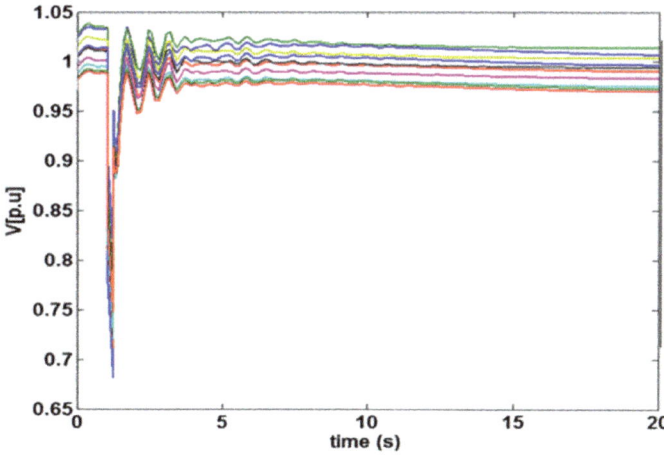

Fig. 16 Voltage waveform of bus 11–20 considering SVC applying with BBO-DE

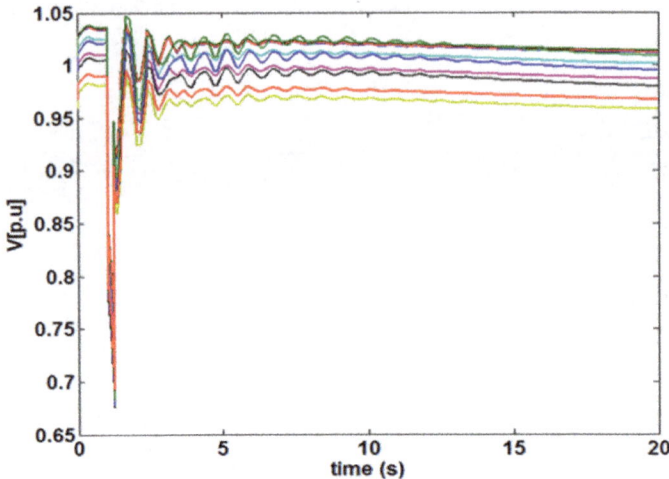

Fig. 17 Voltage waveform of bus 21–30 considering SVC applying with BBO-DE

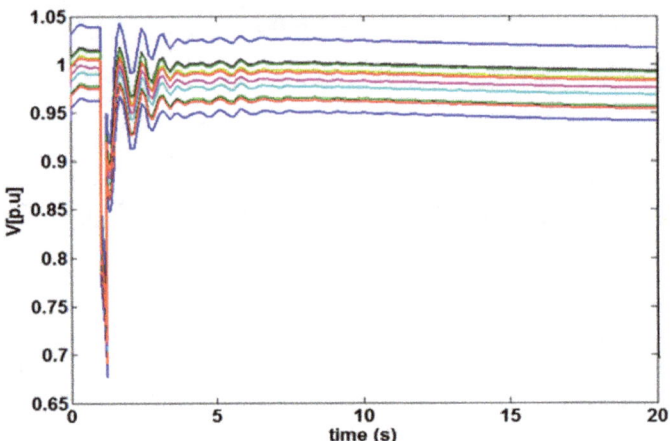

Fig. 18 Voltage waveform of bus 31–40 considering SVC applying with BBO-DE

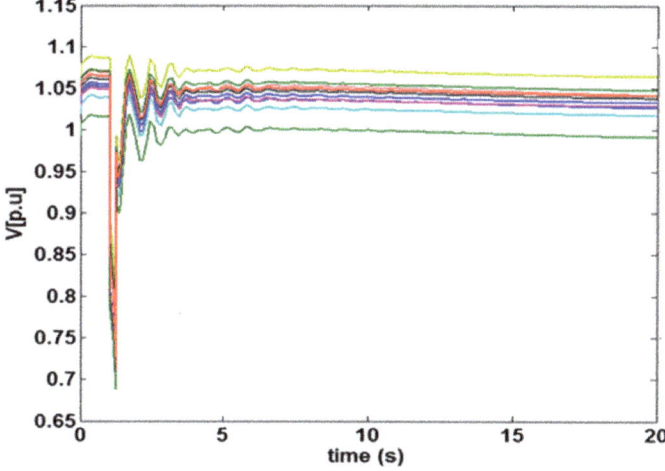

Fig. 19 Voltage waveform of bus 41–50 considering SVC applying with BBO-DE

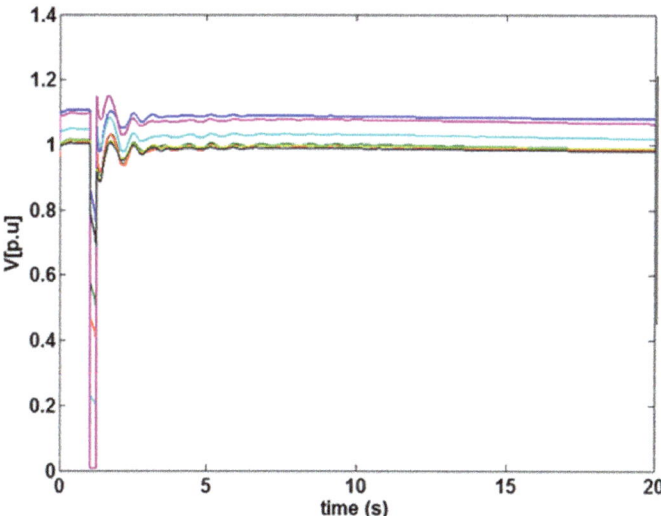

Fig. 20 Voltage waveform of bus 51–57 considering SVC applying with BBO-DE

Table 2 Voltage comparison analysis with and without hybrid algorithm

No. of system buses	Without SVC	Without SVC (increase 40% load)	With SVC (increase 40% load) tuned by BBO-DE algorithm
01	0.9802	0.9791	0.9994
02	1.0000	0.9996	1.0093
03	0.9820	0.9719	0.9833
04	0.9737	0.9637	0.9773
05	0.9710	0.9629	0.9724
06	0.9739	0.9714	0.9758
07	0.9653	0.9529	0.9796
08	0.9950	0.9968	0.9997
09	1.0100	0.9986	1.0407
10	0.9816	0.9694	1.0113
11	1.0029	0.9560	1.0069
12	1.0110	0.9930	1.0140
13	0.9509	0.9472	0.9902
14	0.9358	0.9302	0.9749
15	0.9642	0.9494	0.9831
16	1.0013	0.9780	1.0034
17	0.9749	0.9666	0.9940
18	0.9789	0.9625	0.9966
19	0.9348	0.9180	0.9727
20	0.9620	0.9086	0.9700
21	0.9441	0.9365	1.0117
22	0.9456	0.9386532	1.013987
23	0.9542	0.9361711	1.012458
24	0.9220	0.9171756	1.001365
25	0.9708	0.8759	0.9872
26	0.9124	0.8797	0.9586
27	0.9458	0.9151	0.9800
28	0.9017	0.9388	0.9951
29	1.0061	0.9601	1.0091
30	0.9512	0.8501	0.9677
31	0.9250	0.8181	0.9418
32	0.9391	0.8418	0.9567
33	0.9368	0.8386	0.9544
34	0.9206	0.8741	0.9689
35	0.9275	0.8845	0.9759

(continued)

Table 2   (continued)

No. of system buses	Without SVC	Without SVC (increase 40% load)	With SVC (increase 40% load) tuned by BBO-DE algorithm
36	0.9469	0.8987	0.9854
37	0.9750	0.9100	0.9935
38	1.0159	0.9440	1.0175
39	0.9635	0.9080	0.9921
40	0.9549	0.8964	0.9834
41	1.0174	0.9528	1.0271
42	0.9723	0.8944	0.9918
43	1.0230	0.9825	1.0429
44	1.0165	0.9514	1.0181
45	1.0301	0.9814	1.0289
46	1.0652	1.0065	1.0642
47	1.0393	0.9700	1.0381
48	1.0243	0.9635	1.0331
49	1.0391	0.9820	1.0481
50	1.0321	0.9716	1.0413
51	1.0715	1.0326	1.0811
52	0.9011	0.8508	0.9871
53	0.9237	0.8034	0.9820
54	0.8409	0.7622	1.0195
55	0.9041	0.7373	1.0662
56	0.9080	0.8839	0.9872
57	0.9009	0.8724	0.9800

Table 3   Overall generation power and losses of fifty-seven bus

Total generation	Without SVC	With SVC (BBO-DE)
Total real power in per unit	7.7564	7.7537
Total reactive power in per unit	3.0664	3.1475
Total load		
Total real power in per unit	7.5080	7.5080
Total reactive power in per unit	3.1434	3.1434
Total loss		
Real power in per unit	0.2484	0.2457
Reactive power in per unit	−0.0770	0.0041

References

1. Dutta S, Roy PK (2014) Optimal location of TCSC using hybrid DE/BBO algorithm. In: International conference on non-conventional energy
2. Kazemi BB (2004) Modeling and simulation of SVC and TCSC to study their limits on maximum loadability point. Int J Electr Power Energy Syst 381–388
3. Mirko T, Dragoslav R (2006) An initialization procedure in solving optimal power flow by genetic algorithm. IEEE Trans Power Syst 21:480–487
4. Abou El Ela AA, Abido MA, Spea SR (2010) Optimal power flow using differential evolution algorithm. Electr Power Syst Res 80:878–885
5. Abido MA (2002) Optimal power flow using particle swarm optimization. Electr Power Energy Syst 24:563–571
6. Roa-Sepulveda CA, Pavez-Lazo BJ (2003) A solution to the optimal power flow using simulated annealing. Int J Electr Power Energy Syst 25:47–57
7. Sivasubramani S, Swarup KS (2011) Multi-objective harmony search algorithm for optimal power flow problem. Electr Power Energy Syst 33:745–752
8. Ongsakul W, Bhasaputra P (2002) Optimal power flow with FACTS devices by hybrid TS/SA approach. Electr Power Energy Syst 24:851–857
9. Phadke AR, Fozdar M, Niazi KR (2012) A new multi-objective fuzzy-GA formulation for optimal placement and sizing of shunt FACTS controller. Int J Electr Power Energy Syst 40(1):46–53
10. Rashed GI, Yuanzhang S, Rashed KA, Shaheen H (2012) Optimal location of unified power flow controller by differential evolution algorithm considering transmission loss reduction. In: IEEE International conference on power system technology, Auckland, POWERCON-2012, Oct 30–Nov 2, pp 1–6
11. Huang JS, Jiang ZH, Negnevitsky (2013) Loadability of power systems and optimal SVC placement. Int J Electr Power Energy Syst 45(I):167–174
12. Roy PK, Ghoshal SP, Thakur SS (2012) Optimal VAR control for improvements in voltage profiles and real power loss minimization using biogeography based optimization. Int J Electr Power Energy Syst 43:830–838

Chapter 68
Role of the Fuzzy Cascading and Hierarchical Technique to Reduce the Complexity of Rule Base in the Fuzzy System Development: Case Study of Teaching Faculty Assessment

Vikas J. Magar and Rajivkumar S. Mente

1 Introduction

Fuzzy logic is a mathematical tool that contracts with imprecise, uncertain, unclear, and vague data. This logic was developed by Zadeh in 1965 [1]. Input space is mapped with output space using if-then statements [2]. The set of if-then statements are known as the rule base. Fuzzy inference system (FIS) is a useful tool to get the solution of a complex system with ambiguity and uncertain data. The capability of the system performance is strictly affected by the complexity of the problem [3]. The system having multiple subsystems with multi-input, multi-output for each subsystem improved by the cascaded version of FIS must be implemented rather than developing FIS for each subsystem [4]. The function of the fuzzy expert system mostly depends on the rule base. If the rule base of an expert system is large, then its execution time also increased. Reducing the complexity of an expert system improves the performance. Here, cascading technique is proposed to reduce the rule base complexity of rule base. The proposed method represents the Mamdani-type hierarchical and cascaded fuzzy interface. The expansion of the fuzzy control system is the movement of linking knowledge collection, designing the controller, and defining the rule base. Establishing the rule base is a very critical task in the process of system development [5]. Currently, reducing the overall number of rules and the corresponding computation of these rule bases is the primary issue. With the standard case of the fuzzy system, there is an exponential relationship between fuzzy variables and their rules [6]. If the system has m inputs and n membership functions,

V. J. Magar
Shriram Institute of Information Technology, Paniv, India

R. S. Mente (✉)
Department of Computer Application, PAH Solapur University, Solapur, India

© The Author(s), under exclusive license to Springer Nature Singapore Pte Ltd. 2021   909
S. Kumar et al. (eds.), *Proceedings of International Conference on Communication and Computational Technologies*, Algorithms for Intelligent Systems,
https://doi.org/10.1007/978-981-16-3246-4_69

Fig. 1 General structure of fuzzy system

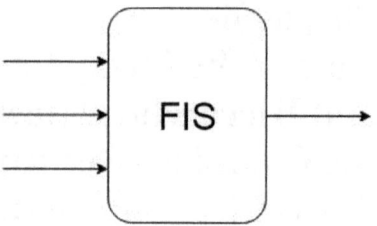

then the total number of rules will be m^n. Suppose if the system has five fuzzy input parameters with five membership functions, then the total number of rules will be $5^5 = 3125$. The number of rules is precisely proportional to m and n. If the rule base is enlarged, then the fuzzy controller becomes complicated to implement. If the complication of the system grows exponentially with the number of variables defined, then the problem of the curse of dimensionality occurs [7]. To avoid the issue of rule explosion and curse of dimensionality, fuzzy cascading and hierarchical model are suitable techniques [8]. The primary advantages of using fuzzy cascading or hierarchical method are the total number of rule boosts linearly as per the number of input variables [9–11].

1.1 General Structure of the Fuzzy System

Figure 1 represents the general structure of the fuzzy inference system. If the fuzzy system has four fuzzy inputs with five membership functions, then the total numbers of rules generated are $5^3 = 225$. If one more variable is included, then the total number of rules will be $5^4 = 625$. Thus, the general fuzzy structure is facing the problem of complexity. To choose the effective rules for a fuzzy system, genetic algorithm is used.

1.2 Hierarchical Structure of Fuzzy System

Figure 2 represents the hierarchical composition of the fuzzy inference system. Hierarchical structure design varies as per the requirement. Here, fuzzy inference system has four inputs. These four inputs are divided into two fuzzy inference systems FIS1 and FIS2. Fuzzy system FIS1 has two parameters. Assume that the whole system has five membership functions. FIS1 has $5^2 = 25$ rules; FIS2 has $5^2 = 25$ rules. The output of FIS1 and FIS2 becomes two inputs for FIS3. FIS3 has $5^2 = 25$ rules. Thus, the total number of rules in the proposed system is FIS1 + FIS2 + FIS3 $= 25 + 25 + 25 = 75$. The whole system covers only 75 rules rather than 625 rules.

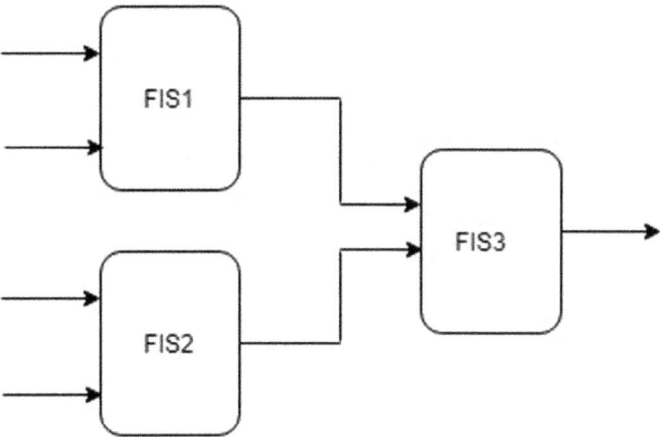

Fig. 2 Hierarchical structure of fuzzy system

1.3   Cascaded Structure of Fuzzy System

Figure 3 represents the cascaded composition of the fuzzy inference system. In the cascaded fuzzy inference system, output of one system becomes the input of another. The combination of new input/inputs and production of the previous fuzzy system is supplied to the next fuzzy system. The whole system is evaluated in cascaded order. Here, the fuzzy system is considered with four variables and having five membership functions. The aggregate number of fuzzy rules will be FIS1 + FIS2 + FIS3 = 5^2 + 5^2 + 5^2 = 75. Thus, using the fuzzy cascading technique whole system covers 75 rules. Thus, it could be found that cascading or hierarchical technique reduces the size of the rule base.

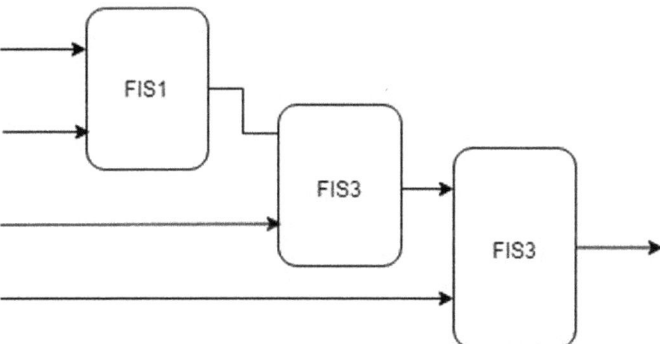

Fig. 3 Cascaded structure of fuzzy system

2 Proposed Method

The case study of teaching staff performance appraisal is carried out to implement the concept of the hierarchical fuzzy system. Feedback about faculty is collected from students through different data collection techniques. A suitable database is developed with appropriate weights. Table 1 represents these factors with the associative weights.

In this research work, five primary factors are carried out to evaluate teaching staff performance. These factors are divided into subsections. Every subsection has assigned some numeric weight as per Table 2. The average of each subsection is forwarded as crisp input to the fuzzy system. In the experimental work, analysis of different fuzzy cascading and hierarchical models is implemented.

2.1 Experimental Work

A. *Algorithm to Generate the Crisp Input*

The essential requirement for the fuzzy expert system is providing input as crisp information. The algorithm mentioned here gives the methodology to produce the crisp input value for the proposed fuzzy expert system. The proposed algorithm provides the step-by-step procedure of evaluating the teaching staff performance using the fuzzy logic.

a. Assign the appropriate weight to every feedback response marked by the students in questionnaire as per the similar entries mentioned in Tables 3 and 4; examine the composed sample data to discover outlier samples by using the below technique.

b. Calculate the sum of weight for all the sample responses.

c. Arrange the calculated sum in ascending order.

d. Find out Quartile 1 (Q_1) and Quartile 3 (Q_3)

e. Calculate the interquartile range (IQR) $= Q_3 - Q_1$

f. Calculate the upper bound (UB) and lower bound (LB) as

$$UB = Q_3 + (1.5 * IQR)$$

Table 1 FIS input parameter and its associated weights

Sr. No.	Attribute	Weight
1	Quality of teaching	350
2	Factors in learning	50
3	Responsibility and punctuality	250
4	Assessment of learning	150
5	Mentoring and counselling	200
Total		1000

Table 2 Weight assigned to each subsection for student feedback appraisal

	Attributes	Evaluation parameter				
		1	2	3	4	5
1	*Quality of teaching*					*350*
1.1	Pace of subject	50	40	30	20	10
1.2	Use of good examples and illustrations	50	40	30	20	10
1.3	Motivation to attend the class	30	24	18	12	6
1.4	Used blackboard efficiently	40	32	24	16	8
1.5	Used audio-visual aids	40	32	24	16	8
1.6	Arranging group discussion, seminar, etc.	40	32	24	16	8
1.7	Stimulated my interest in the subject	50	40	30	20	10
1.8	Audibility and clarity of speech	50	40	30	20	10
2	*Factors in learning*					*50*
2.1	Lecture contributed to my learning	25	20	15	10	5
2.2	Defined learning objectives for each period	25	20	15	10	5
3	*Responsibility and punctuality*					*250*
3.1	Punctuality (arrival in the class and leave from the class)	50	40	30	20	10
3.2	Checking assignment, journal, test result, homework in time	60	48	36	24	12
3.3	Motivate students for extra activities	70	56	42	28	14
3.4	Work dedication	30	24	18	12	6
3.5	Conducting seminars and organizing group discussion	40	32	24	16	8
4	*Assessment of learning*					*150*
4.1	Feedback on assignment was useful	50	40	30	20	10
4.2	Problem sets helped me in learning	50	40	30	20	10
4.3	Answer papers evaluated fairly	50	40	30	20	10
5	*Mentoring/counselling*					*200*
5.1	The teacher was approachable outside the class	50	40	30	20	10
5.2	Teacher is sympathetic to academic problems	70	56	42	28	14
5.3	The teacher is sympathetic to a personal problem	80	64	48	32	16

Table 3 Input parameter values with trapezoidal membership function

Sr. No.	Attribute	Poor	Average	Good	Very good	Excellent
1	Quality of teaching	[0 20 40 70]	[35 80 100 140]	[120 160 180 210]	[200 245 265 280]	[260 300 330 350]
2	Factors in learning	[0 3 6 10]	[7 13 18 20]	[16 21 26 30]	[25 32 36 40]	[35 41 45 50]
3	Responsibility and punctuality	[0 15 35 50]	[40 60 80 100]	[180 110 135 150]	[140 160 180 200]	[180 210 230 250]
4	Assessment of learning	[0 10 20 30]	[25 40 50 60]	[50 65 80 90]	[75 90 105 120]	[110 125 135 150]
5	Mentoring and counselling	[0 10 25 40]	[30 45 60 80]	[70 85 100 120]	[105 125 145 160]	[150 165 180 200]

Table 4 Trapezoidal membership function values for output membership

Linguistic variable	Poor	Average	Good	Very good	Excellent
Trapezoidal values	[0 8 15 20]	[15 22 30 40]	[35 42 50 60]	[50 65 72 80]	[75 82 90 100]

$$LB = Q_1 - (1.5 * IQR)$$

g. To compute outlier samples, compare sample values with upper and lower bound.

   If sample value < lower bound or sample value > upper bound, then

   Outlier (sample) = true,

   Otherwise,
   Outlier (sample) = false.

h. Calculate the percentage of the number of outlier samples.

   If the percentage is negligible/too less, then

   Remove these sample/samples.

   Otherwise,

   The deletion of such samples will be profoundly affected by data losses.

i. Calculate the mean, median, mode, and skewness for every column. Set criteria variable for each section as c_1, c_2, c_3, c_4, and c_5 as zero. Repeat the process until all columns to be finished group by criteria

   If skewness = 0, then

   $C_i = C_i$ + value of mean or median or mode

Otherwise,

If skewness > 0 or skewness < 0

$C_i = C_i +$ value of median

Where the value of 'i' moves from 1 to 5.

j. Provide the input C_1, C_2, C_3, C_4, and C_5 as the crisp input to the fuzzy expert system.

B. *Development of the Fuzzy Inference System*
    The algorithm proposed here generates crisp input for the fuzzy expert system to evaluate the faculty performance. The architecture proposed here contains fuzzification, rule inference mechanism, and defuzzification. To implement the fuzzification, trapezoidal membership functions are used. Its associative parameters are defined with an appropriate crisp input value range.

C. *Fuzzy Model with Input and Output Membership*
    The projected fuzzy inference system is designed using the Mamdani fuzzy model. Five evaluation parameters are quality of teaching, factors in learning, responsibility and punctuality, assessment of learning, and mentoring and counselling used as input for the fuzzy expert system (Fig. 4).

D. *Input Membership Functions*
    Membership function plays a vigorous role in the case of the fuzzy expert system model. In this proposed model, the trapezoidal membership function is used. Here, five trapezoidal membership functions are designed for each section. To design the trapezoidal membership function, 'membership function' editor is used. The membership functions are touched to achieve a better result. The membership function 'quality of teaching' has five linguistic parameters (Fig. 5).

E. *Output Membership Functions*

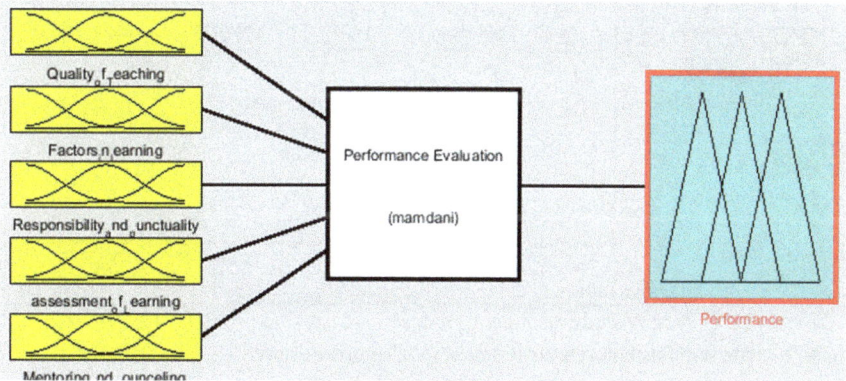

Fig. 4 Fuzzy models to evaluate the faculty performance

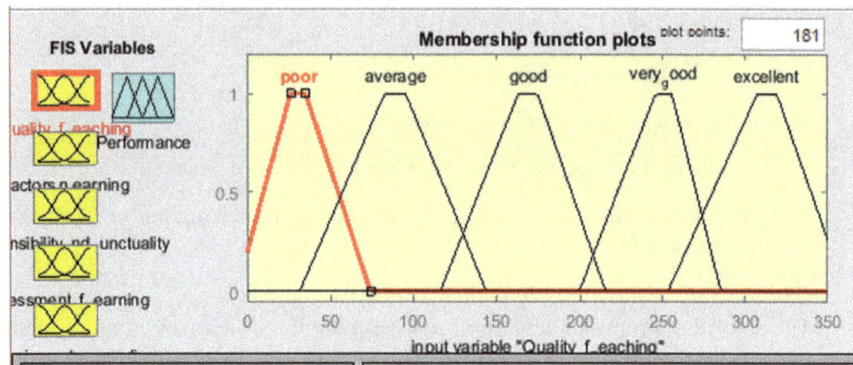

Fig. 5 Membership functions for input 'quality of teaching'

To finalize the performance value using fuzzy, output membership functions with interval 1–100 are segmented into poor, average, good, very good, and excellent with following trapezoidal membership values.

Corresponding output membership function is shown in Fig. 6.
The following are some sample rule bases generated for fuzzy expert system.

1. If (Quality_of_Teaching is poor) and (Factors_in_learning is poor) and (Responsibility_and_punctuality is average) and (assessment_of_Learning is average) and (Mentoring_and_Counceling is good), then (Performance is average) (1).

2. If (Quality_of_Teaching is good) and (Factors_in_learning is poor) and (Responsibility_and_punctuality is average) and (assessment_of_Learning is good) and (Mentoring_and_Counceling is excellent). then (Performance is good) (1) (Fig. 7).

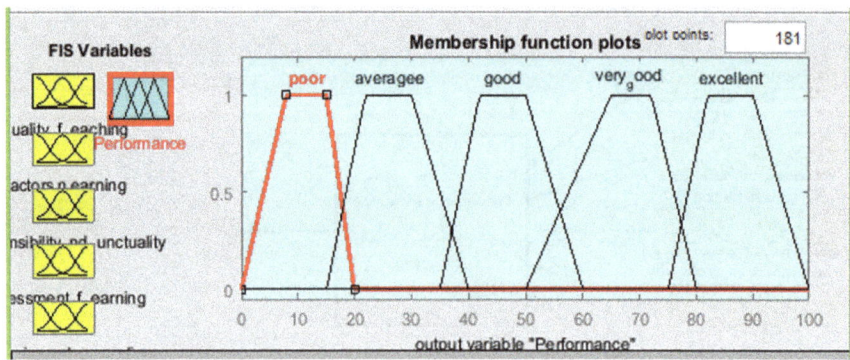

Fig. 6 Membership function values for output variable performance

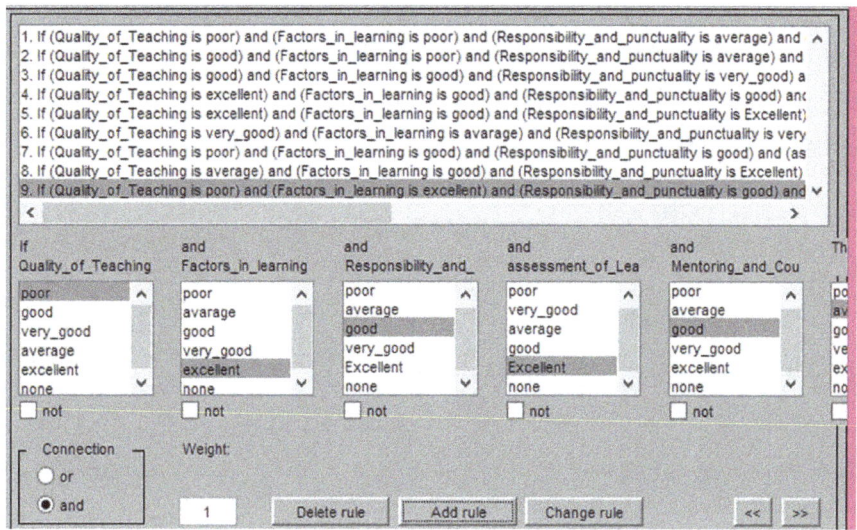

Fig. 7   Rule base editor for fuzzy expert system

3   Experimental Result

Experimental work is carried out on a single-layer fuzzy system, cascaded fuzzy system, and hierarchical fuzzy system. The whole system was implemented on five evaluation parameters with trapezoidal membership functions.

3.1   Experiment on Single-Layer Fuzzy Structure

A single-layer fuzzy system has a simple structure. No middle layer exists between input and output. All the evaluation parameters are directly passed to the fuzzy expert system. The conclusion is directly obtained using fuzzy rule base interfacing. The single-layer system does not accept any input as the output of any other fuzzy system. If the system has an m membership function with n inputs, then the entire number of rules is m^n.

Figure 8 shows a single-layer fuzzy system. This system is complicated because of its huge rule size. In this case, five membership functions are poor, average, good, very good, and excellent. Here, $5^5 = 3125$ rules are generated.

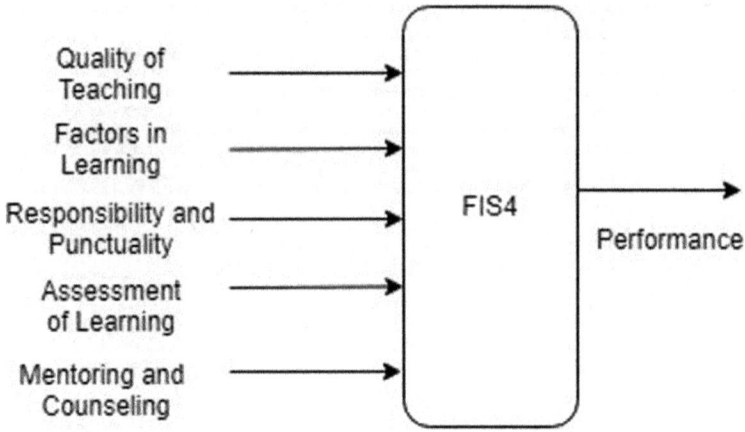

Fig. 8 Single-layer fuzzy system

3.2 Experiment on the Cascaded Fuzzy System

To reduce the complexity of rule base, cascading fuzzy technique is used. Fuzzy cascading is a technique in which the output of one system is passed as input to others. Fuzzy cascading is very useful to reduce the complexity of the rule base (Fig. 9).

In this fuzzy cascading system, the total number of rules is FIS1 + FIS2 + FIS3 + FIS4 = 25 + 25 + 25 + 25 = 100 rules. Thus, fuzzy cascading system reduced the rule base than the general fuzzy system.

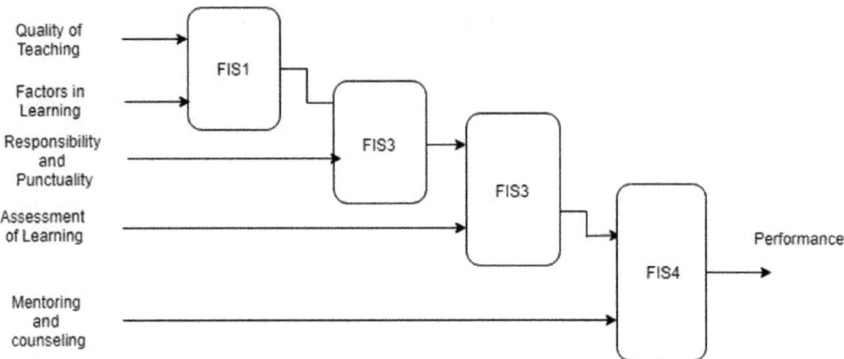

Fig. 9 Cascading structure of fuzzy system

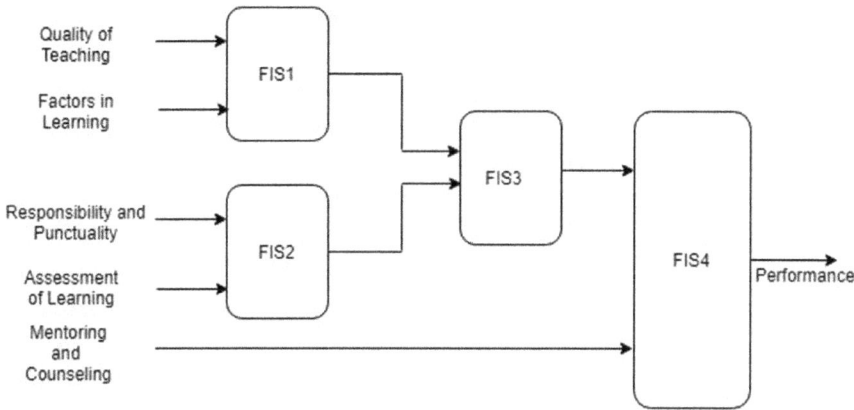

Fig. 10   Hierarchical model for teaching staff performance appraisal system

3.3   *Experiments on Hierarchical Fuzzy System*

Every cascaded system may be known as hierarchical, but every hierarchical system is not known as cascading. A hierarchical system contains the result of two or more fuzzy systems. It is a multilayer system (Fig. 10).

4   Results and Discussion

The proposed research paper works on the same evaluation parameter as having the same fuzzy membership type. Table 5 represents the comparative result between different fuzzy system modules like single-layer, cascaded, and hierarchical fuzzy system models. Results are evaluated using teaching staff feedback records collected from the students using different types of data collection techniques. Implementation was made using MATLAB Fuzzy Toolbox.

It is possible to convert a single-layer fuzzy system into a multilayer fuzzy system by designing an appropriate rule base. The application of the multilayer fuzzy system is reducing the complexity of the rule base. Reduction into the complexity of the rule base decreases the size of an expert system and improves the system performance.

5   Conclusion

It is concluded that multilayer fuzzy system architecture is very suitable to reduce the rule base complexity and size of an expert system. It is found that clustering of these results is found into a similar category. Result values of different fuzzy models have

Table 5 Comparative result of different fuzzy models

No.	Quality of teaching	Factors in learning	Responsibility and punctuality	Assessment of learning	Mentoring and counselling	Single-layer FS	Cascaded FS	Hierarchical fuzzy system	Variation
1	269.80	40.77	203.40	119.20	146.90	73.41	86.93	86.93	81.80
2	268.00	37.14	207.60	114.30	144.70	73.97	86.87	86.87	77.60
3	292.40	42.27	216.20	132.60	171.50	87.01	86.87	86.87	88.70
4	293.40	45.00	224.30	124.30	162.00	87.00	86.98	86.98	86.4
5	294.50	44.67	200.00	131.30	168.9	87.05	86.87	86.87	85.30
6	301.10	40.77	215.10	130.00	158.60	87.09	87.09	87.09	87.20
7	260.00	35.00	188.60	101.30	142.30	69.73	66.38	66.38	71.65
8	271.10	42.27	208.50	118.20	164.20	80.43	86.90	86.90	81.50
9	271.40	40.71	187.10	111.4	160.30	78.49	87.04	87.04	80.00
10	284.40	41.67	201.80	121.70	158.80	85.94	87.08	87.08	93.20
11	293.6	45.88	212.10	131.20	170.40	87.00	86.87	86.87	91.60
12	300.70	40.42	225.20	132.50	182.10	86.93	86.93	86.93	95.60
13	302.20	40.00	230.40	136.70	182.20	86.95	86.93	86.93	96.00

no variation. Thus, it is concluded that fuzzy cascading and hierarchical techniques are useful to minimize the rule base complexity.

References

1. Zadeh LA (1965) Fuzzy sets. Inf Control 8(3):338–353
2. Driankov D, Hellendoorn H (1995) Chaining of fuzzy IF-THEN rules in Mamdani controllers. In: IEEE International conference on fuzzy systems, vol 1. IEEE, Japan, pp 103–108
3. Duan JC, Chung FL (2002) Multilevel fuzzy relational systems: structure and identification. Soft Comput 6(2):71–86
4. Lee ML, Chung HY, Yu FM (2003) Modeling of hierarchical fuzzy systems. Fuzzy Sets Syst 138(2):343–361
5. Ali MA, Lun AK (2019) A cascading fuzzy logic with image processing algorithm-based defect detection for automatic visual inspection of industrial cylindrical object's surface. Int J Adv Manuf Technol 102(1–4):81–94
6. Hu Z, Bodyanskiy YV, Tyshchenko OK, Boiko OO (2016) An evolving cascade system based on a set of neo fuzzy nodes. Int J Intell Syst Appl 9:1–7
7. Rama Sree P, Ramesh SNSVSC (2016) Improving efficiency of fuzzy models for effort estimation by cascading & clustering techniques. Procedia Comput Sci 85:278–285
8. Alsabbah S, Al-Khedher M, Abu K, Mohammad T, Zalata Y (2012) Evaluation of multiregional fuzzy cascade control for pH neutralization process. Int J Res Rev Appl Sci 10
9. Mahapatra SS, Nanda SK, Panigrahy BK (2011) A cascaded fuzzy inference system for Indian river water quality prediction. Adv Eng Softw 42(10):787–796
10. Lendek Z, Babu R, De Schutter B (2008) Stability of cascaded fuzzy systems and observers. IEEE Trans Fuzzy Syst 17(3):641–653
11. Neogi A, Mondal AC, Mandal SK (2011) A cascaded fuzzy inference system for university non-teaching staff performance appraisal. J Inf Process Syst 7(4):595–612

Chapter 69
Social Media Analysis for Sentiment Classification Using Gradient Boosting Machines

Pradeep Kumar and Abdul Wahid

1 Introduction

In recent years, social media has emerged as a personal communication platform to express individual opinions about specific services and products including political, social, legal, and common events of interest among the users. Handling a large number of rapidly growing digital documents has become a tedious task for the automatic categorization of text [1–3]. Prominently text mining (TM), natural language processing (NLP), and machine learning (ML) techniques have been applied extensively to extract useful information from vast unformatted documents [4, 5]. A massive amount of these personnel views and opinions are flooded daily on popular social media including Twitter, Facebook, YouTube, and LinkedIn [6, 7].

Text classification problem deals to handle the huge number of unstructured documents such as web pages, e-mails, social media forum, and postings of other electronic documents. Polysemy and synonymy are other problems with text mining. Polysemy refers to words having multiple meanings. Synonymy refers to the different words having the same or similar meaning. Sentiment analysis (SA) known as opinion mining also or polarity mining deals with computational linguistics, NLP, and other text analytics methods. SA automatically extracts user sentiments from text sources such as words or phrases or complete documents. SA is found as one of the potential research areas in NLP and other diverse fields of data mining [8–10].

Sentiment classification attempts to measure the polarity of a given text document more precisely predicting whether the given reviews and opinions on social media are positive, negative, or neutral. Unstructured text generated from explosive social media such as Facebook, YouTube, and Twitter able to provide substantial clues about services and product reviews leading to better marketing strategies for branding

P. Kumar (✉) · A. Wahid
Department of Computer Science and Information Technology, Maulana Azad National Urdu University, Hyderabad, India

S. Kumar et al. (eds.), *Proceedings of International Conference on Communication and Computational Technologies*, Algorithms for Intelligent Systems,
https://doi.org/10.1007/978-981-16-3246-4_70

the products and maximize the level of customer satisfaction [11–13]. Knowledge extracted from social media can be extremely helpful because a large number of opinions expressed about a specific topic may lead to vital information related to business policy. The impact of all such opinions make them easily understandable by the majority of readers and subsequently set up a trend gradually for recommending some products or services [8, 13].

Major applications of SA include predicting stock market trends, framing policies to promote the product and services, recommender systems, and managing crises. Basic methods for SA include text statistics such as word counts, frequency, and review categories (positive and negative reviews for specific comments or remarks in the given text documents). Therefore, systematic SA of social media can help the stakeholders by providing the extraction of insightful conclusions about the public opinions and variety of topics. Moreover, it poses serious technical challenges also due to noisy, sparse, and multilingual content posted by the users on social media [14–17].

Bootstrap aggregating (bagging) and boosting are popular ensemble techniques. Bagging combines the results of multiple base models to generate improved results. Bootstrapping is a sampling technique with a replacement where subsets of samples are generated from the original dataset. Bagging applies these subsets known as bags to provide the distribution of a complete set. The size of subsets may vary from the original set [18, 19]. Gradient boosting-based classifiers apply different weak learning models like decision trees (DTs) to build up a strong prediction model. Gradient boosting models are capable of handling complex unstructured social media data quite effectively. Among various machine learning methods, gradient boosting machines (GBMs) have shown state-of-the-art results on many standard classification benchmarks.

The primary objective of gradient boosting machines is to minimize the loss function very similar to gradient descent algorithms in a neural network. In an iterative process, new weak learners are added to the model and the weights of the past learners are cemented in place, leaving unchanged samples for the new layers. GBM can be applied to multi-class classification problems and regression problems also.

The rest of this paper is structured as follows: Sect. 2 presents the related work of sentiment analysis and classification; Sect. 3 explains the feature selection and preprocessing of unstructured text; Sect. 4 describes the research methodology applied; Sect. 5 presents the experimental results, and conclusion followed by future directions.

2 Related Work

Many researchers have applied supervised and unsupervised algorithms to perform text classification and sentiment analysis of social media [4–6]. The prior research has focused on sentiment or content classification tasks. Ordinarily, natural text cannot be utilized properly in the analysis without preprocessing. Several techniques of preprocessing have been explored in the literature. Bag of words (BOW) is one of the traditional approaches for sentiment classification to analyze the text features with the help of supervised learning algorithms [7–11]. Using machine learning methods, the words can be filtered from the BOW vector. For example, the appearance of each word in the text can be represented as the feature from such vectors. Other common methods include n-gram, POS tags, and negation-tags are applied to optimize the features. The N-gram and negation-tags are found effective to improvise the precision of the classifier wherein POS tags can be applied for multiple meanings of one word.

Bag of words is a predominantly used method for SA and classification using machine learning algorithms. Pang and Lee extracted features from online reviews about a movie and analyze the user's sentiments with the help of words bag and feature vectors [20]. Bespalov et al. [13] applied high order n-grams for sentiment classification of a text document. They devised an embedding mechanism of n-grams to deal with the curse of dimensionality.

Zou et al. [5] investigated the words bag method using support vector machines (SVM) and naïve Bayes with the help of syntactic features along with part of speech (POS) tags. They constructed word dependencies and syntax trees to establish the grammatical and logistic relationship between words in sentences.

Almatarneh and Gamallo [4] applied supervised learning techniques for class labeled training data based on automatic text classification. A predictive model is designed based on the previous text documents collected from social media and other available repositories. The effective outcomes of these machine learning models depend on several factors such as feature selection, parameter tuning, training of the model, and capability of learning in a dynamic context.

The sentence syntax tree (SST) is another alternative approach applied for designing sentiment classifiers. The sentences are parsed to build a syntax tree to establish the relationship among these words. The sentiment classification model can be built with the help of words polarity, POS features, and syntax. Dave et al. employed machine learning techniques for sentiment classification with the help of top words selected according to their generated points [14]. Almatarneh and Gamallo [4] applied SVM to analyze sentiment classification from the orientation of words point of view.

Pak and Paroubek proposed a sentiment classification model using bag of words for Twitter data as an application of SA in social media. Several researchers and practitioners have applied syntax trees also to establish an internal relationship between the words. Namugera et al. [6] used maximum entropy models for parsing the syntax trees to find the patterns behind the syntax tree. Younis [8] improved the accuracy of parsing the syntax tree using rules and patterns. Nakagawa et al. [9] studied the impact of sentiment dependency on words using CRF with hidden variables. Hotho et al. [10] applied the HMM-based model for the analysis of a sentence's content and sentences. Other approaches for sentiment classification includes lexicon structures for words and their sentiments generating classification rule.

2.1 Social Media Analysis

Social media analytics can be treated as a multiphase activity including capturing, understanding, and presentation. Pre-processing techniques are applied to extract prominent features required for accurate classification and prediction. Preprocessing techniques such as feature extraction, selection, grouping, and evaluating process are applied to classify the text document. Subsequently, machine learning algorithms are applied to train the classifier and then test them to categorize whether the sentiments are positive or negative [12, 13]. Reduction of text dimensionality can be achieved through filtering, lemmatization, and stemming methods. Stop word filtering is the standard filtering approach to remove the words from a dictionary. The main purpose of applying word filtering is to remove words that carry little information such as conjunctions, articles, prepositions, and adjectives bearing no particular statistical relevance [12].

Text documents usually contain various undesirable characters like punctuation marks, special characters, stop words, and digits which may not help to classify the text. Therefore, it has to be preprocessed before applying it to the classifier for effective outcomes. Text cleaning is the first step in any text mining problem where irrelevant details are removed from the document which may not contribute to the vital information of greater interest.

The bag-of-words (BoW) model can be applied to extract the features by considering each word as a feature. Each comment on a specific product or service is treated as a bag of words. BoW creates a dictionary of all the words and their frequency in the text document or dataset used for sentiment classification [15]. Words occurrence is the number of times the word that occurs in the entire corpus. BoW model ignores grammar and word orders and converts each document into numerical vectors. Widely used techniques for vector representation from the text document include word occurrence matrix, word2vec, term frequency (TF), and TF-inverse document frequency (TF-IDF). TF-IDF is an occurrence-based numeric representation of the text document. TF-IDF is a widely used method to transform the text into numerical presentation [13, 14].

Numerically, the TF of a particular word in the text document can be computed as:

$$\text{Term\_Frequency\_word} = \frac{\text{Frequency of word in text document}}{\text{Total words in the document}} \tag{1}$$

Excluding the common words of least importance in the documents that contribute very little to provide the insights are excluded. Therefore, to reduce the impact of minimally used words in the text document, the TF-IDF of a word may be computed as:

$$\text{Inverse\_doc\_Frequency}(w) = \log\left(\frac{\text{Total Number of documents}}{\text{Number of documents containing word } w}\right) \tag{2}$$

3 Research Background

This section examines if certain features alter the probability of unstructured documents for sentiment analysis using GBM, AdaBoost, and XGBM quantitatively.
Bootstrap aggregating (bagging) and boosting are popular ensemble techniques.
Bagging combines the results of multiple base models to get improved results.
Bagging is an ensemble technique for classification works on random subsets of
the original dataset. The final prediction is achieved through voting or by taking
the aggregate of individual predictions. Subsets from the dataset are taken with
replacement.

Bootstrapping is a sampling technique with a replacement where subsets of
samples are generated from the original dataset. Bagging applies these subsets known
as bags to get the distribution of a complete set. The size of subsets may vary from
the original set. Boosting is also widely used as an ensemble technique. The primary
purpose of boosting is to emphasize the samples that are hard to classify accurately.
Boosting builds the multiple models sequentially by assigning equal weights to each
sample initially and then targets misclassified samples in subsequent models. Two
popularly used algorithms are gradient boosting and AdaBoost.

3.1 Gradient Boosting Machines

GBM is an ensemble technique that applies a decision tree as a base classifier. GBM
applies boosted machine learning for e-mails extracted from the spam dataset. GBM
constructs one tree at a time where each new tree helps to rectify errors caused by
the earlier trained tree. Whereas using a random forest classifier the trees does not
correlate with previously constructed trees [21–23]. The gradient descent algorithm
is applied to minimize the error. A set of training samples $X = \{(x_1, y_1), ..., (x_i, y_i)\}$ is taken from the spam datasets and corpus. Where $x_i \in R^n$ and $y_i \in \{+1, -1\}$
denoting the outcomes for ith training sample indicating $+1$ as spam and -1 for
non-spam e-mail. The voted combination of classifiers $F(X)$ can be written as:

$$F(X) = \sum_{t=1}^{T} w_t f_t(x) \tag{3}$$

where $f_t(x): R^n \rightarrow \{+1, -1\}$ are base classifiers, and $w_t \in R$, the weights for each
base classifier in the combined classifiers. A data point (x, y) is classified according
to the sign of $F(X)$ and margin $yF(X)$. A positive value for the margin represents
spam mail, and the negative value corresponds to legitimate mail (non-spam).

3.2   AdaBoost Classifier

AdaBoost classifier applies a sequence of weak learners such as decision trees on modified versions of the text data repeatedly. The AdaBoost method boosts the accuracy of a weak learner by simulating multiple distributions over the training samples. The AdaBoost takes the majority vote of the resulting outcomes. Initially, a set of weights is applied to the training samples and then updated after each round of the training. The weights are updated in such a way that weights of the samples classified incorrectly are increased, whereas the correctly classified samples are assigned lower weights. During the training process updating the weight mechanism focuses the base learner to concentrate on the harder samples. The final prediction of the model is computed on the weighted sum of all classifiers [19, 24].

$$F(X_i) = \text{sign}\left(\sum_{k=1}^{K} \alpha_k f_k(X_i)\right) \tag{4}$$

where K represents the total number of classifiers utilized, $f_k(X_i)$ is the outcome of weak classifier k for corresponding feature X_i, α_k is the weight assigned to classifier k computed as:

$$\alpha_k = \frac{1}{2} \ln\left(\frac{1 - \varepsilon_k}{\varepsilon_k}\right) \tag{5}$$

where ε_k represents the error rate of the classifier, that is, the number of incorrectly classified samples over the training set divided by the total number of the training set, $F(X_i)$ indicates the combination of all the weak classifiers.

3.3   eXtreme Gradient Boosting Machines

eXtreme gradient boosting machines (XGBMs) are an improved gradient boosting classifier. Gradient boosting algorithms can be used for classification and regression problems. GBM classifier performs significantly well on large complex datasets. While classifying social media reviews or customer behavior prediction XGBM demonstrates the importance and impact of DTs boosting as an improved classifier [16, 17, 21, 22]. One of the main reasons behind the success of the XGBM model is its scalability. Distributed and parallel computing provides faster learning enabling quicker model investigation. However, overfitting remains a challenge to overcome that can be controlled through parameter tuning and optimization.

3.4 Optimization of Model Parameters

Gradient Boosting Classifier. This model is generated as an additive model using arbitrary differentiable loss functions for optimizing the classifier accuracy. The main parameters tuned for generalization include loss function, learning rate, number of estimators, number of sub-samples, and error criteria to terminate the learning. Exponential and deviance loss functions are applied to optimize the model. Improved results are obtained while applying exponential loss function in gradient boosting. The non-negative learning rate is adjusted in tune when different estimators are applied between 0 and 1 with a default value of 0.1. A large number of estimators are applied from 100 to 1000 for the higher performance of the model. The number of sub-samples is taken less than or equal to 1. However, a smaller value will lead to the stochastic gradient boosting with low variance and high bias. Different learning rates between 0 and 1 were applied to achieve optimum training and testing the accuracy of the classifier. Mean square error is the function applied to measure the quality of a split. Other error metrics, such as mean absolute error, also may be applied. Moreover, Friedman MSE is generally the best parameter by default leading to a better approximation.

AdaBoost Classifier. Using AdaBoost classifier, we begin with meta-estimators as base estimators where the weights are tuned in such a way that difficult samples are emphasized for classification more accurately. Four vital parameters include the base estimator, the number of estimators, learning rate, and the algorithm applied. Here, two base estimators, LR and DTs, are employed to obtain the generalized results over the social media dataset. The maximum number of estimators varies from 10 to 100 and is applied for model learning. The boosting is terminated earlier also in case the perfect fit is obtained. Different learning rate between 0 and 1 is applied to achieve optimum training and testing accuracy of the classifier. However, the learning rate starts shrinking when a large number of estimators are applied. Two popular algorithms, SAMME and SAMME.R, are explored to get optimal results.

XGBM. eXtreme gradient boosting classifier provides speed enhancement through parallel and distributed computing with the help of cache awareness which makes XGBM quite faster than the original GBM. Additionally, a split finding algorithm is applied to optimize the trees to reduce the overfitting problem that leads to a faster and more accurate classifier over GBM. The GBM generates an additive model in a forward stage-wise manner that allows for the optimization of the differentiable error function. Vital parameters tuned for generalization includes maximum depth of the tree as base learners, number of parallel threads to run the XGB classifier, minimum loss reduction required to make a partition on the leaf node of the tree. The non-negative learning rate between 0 and 1 is applied in tune with different estimators with a default value of 0.1. The number of estimators utilized varies from 100 to 1000 for testing the scalability of the model. Different learning rates between 0 and 1 were applied so that optimum training and testing accuracy of the classifier could be achieved.

4   Proposed Methodology

Sentiment classification can categorize an input text sequence into certain types of scores or ratings such as positive, negative, or neutral. Primary text categorization methods include feature vectors indexed by all the words of a given dictionary to represent text documents. Machine learning algorithms develop multiple models using different datasets collected from the standard repository or corpus, and each model is analogous to an experience.

4.1   *Empirical Data Collection*

Dataset for sentiment classification is collected from Kaggle [25]. Text document for sentiment classification is a sentence extracted from social media that contains review comments about several movies. This dataset contains two field text and sentiments. The text includes the actual review comments about the movie. The sentiment is the response variable containing positive and negative sentiments. The training data contains 7086 sentences categorized as 1 with positive sentiment and 0 with negative sentiment.

4.2   *Experimental Setup*

In this section, we present the experimental set up to illustrate how effectively ensemble techniques (AdaBoost, GBM, XGBM) are applied for sentiment classification. Python machine learning library scikit-learn and natural language tool kit have been used to train and classify the prediction model. Each comment is counted as a record and categorized as positive or negative using machine learning algorithms [26–29].

4.3   *Performance Evaluation Measures*

Accuracy is the most common metric applied for the assessment of a classifier which can be extracted from the confusion matrix referred to as error matrix and classification table [16–18]. The accuracy of the proposed classification model is obtained from combining the number of true positives and true negatives classes divided by the total number of observations that provide the overall accuracy of the predictive models. Training and testing accuracy of the classifier is measured with different learning rates varying between 0 and 1.

4.4   Cross-Validation

Cross-validation is an evaluation method that attempts to generalize the outcomes quantitatively on a dataset. It is conducted irrespective of the training data. Generally, using a round of cross-validation includes splitting the data into two complementary subsets training and testing, performing analysis on training data, and validation analysis using test data. The validation procedure is carried multiple times with different partitions, and the mean value is taken as the results of the model to reduce the scattering.

4.5   Analysis Results

In this section, we compare the performance of individual classifiers applied in our study by evaluating the confusion matrix of each model in terms of accuracy of training and testing with different learning rates of the model. Results achieved from the rigorous experiments conducted on social media data (online movie reviews) are presented in Table 1.

From the dataset two-third of the samples are utilized for training the model and one-third of samples for testing and validation purpose. The total size of the dataset is 50,000 observations divided into two categories as positive sentiments and negative sentiments.

It is observed that the higher value of a word represents greater importance in the text document. However, when corpus size is varied then large size text documents normally have more occurrences of words than smaller sized text. Term frequency normalizes the occurrence of each word within the size of a text document. If any particular term occurs in all the text documents, then the inverse document frequency of that word would be computed as 0. TF-IDF is the product of term frequency and inverse document frequency. After removing, stop words stemming and lemmatization are applied to converts the words into root words. The words which appear in multiple forms having similar meanings such as improve, improved, and improvement needs to be considered as one root word improve only.

5   Discussion and Observations

It is evident from the experimental observations that sentiment analysis for social media using ensemble machine learning techniques (GBM, AdaBoost, XGBM) provides a viable solution for text mining tasks and sentiment analysis to analyze user-generated reviews for specific products and services. From Table 1, it is observed that the XGBM classifier outperforms over three models (GBM, AdaBoost, and bagging

Table 1 Summary of overall accuracy using different models

Classifiers	Learning rate	Accuracy	
		Training	Testing
Ada Boosting	0.050	0.857	0.860
	0.075	0.932	0.936
	0.100	0.963	0.962
	0.250	0.980	0.980
	0.500	0.988	0.986
	0.750	0.990	0.987
	1.000	0.990	0.987
Gradient Boosting	0.050	0.627	0.634
	0.075	0.674	0.689
	0.100	0.675	0.690
	0.250	0.778	0.783
	0.500	0.779	0.784
	0.750	0.780	0.784
	1.000	0.789	0.802
XGradient Boosting	0.050	0.996	0.996
	0.075	0.996	0.996
	0.100	0.996	0.996
	0.250	0.996	0.996
	0.500	0.996	0.996
	0.750	0.996	0.996
	1.000	0.996	0.996
Bagging with KNN	0.050	0.981	0.971
	0.075	0.985	0.974
	0.100	0.989	0.980
	0.250	0.984	0.971
	0.500	0.981	0.968
	0.750	0.988	0.979
	1.000	0.989	0.980

with KNN) applied for sentiment analysis quantitatively. Training and testing accuracy of the XGBM achieved is the maximum (0.996) irrespective of the different learning rate applied (from 0.050 to 1.00 at regular interval of 0.250) performing consistently. However, before making any generalization similar studies needs to be carried out on larger and different scalable datasets with a large number of parameters. Sentiment analysis will provide a competitive edge for the organizations to understand the behavior of their customer for products and services using social media data. This will help them to improve the product branding and maintaining better customer relationships so that revenue could be generated maximum.

SA of unstructured and uncensored modes of delivery will avoid exploiting the public sentiments which have been the common reasons for the downfall and rise of many products or services within organizations across the globe. This will facilitate in improving the business performance and monitoring the products and services from the customer perspective. Therefore, SA can be utilized as a monitoring tool for the assessment of policy decisions or services rendered to the customers or branding their product. Ignorance may lead to dissatisfaction among the customers consequently losing the product or services or downfall in ratings.

6   Conclusion and Future Scope

This paper explored sentiment analysis with automatic extraction and analyzing the reviews and opinions of messages and posts on social media using two novel approaches of machine learning (bagging and boosting). Sentiment analysis could be utilized for monitoring consumer opinions, products, and business intelligence as per local needs and global standards. Machine learning classifiers are found to be potential tools for all stakeholders to monitor and track their branding of products and services from the customer's view particularly in the event of a fluctuating situation of marketing trends. Knowledge extracted from social media can be extremely helpful because a large number of opinions expressed about specific topics or trends may lead to vital information related to business policy.

In this paper, it is demonstrated that ensemble learning techniques can be applied as an effective tool for insight sentiments of social media users. Future work will target reviewing comments, optimal feature selection, and comparing various machine learning algorithms for sentiment classification applied to various benchmark datasets extracted from the social media of a wide range of products and services incorporating authenticity and integrity of digital contents. We also plan to replicate and extend our study for text mining and sentiment analysis more intelligently using advanced machine learning techniques such as deep learning with clustering tweets.

References

1. Isah H, Trundle P, Neagu D (2014) Social media analysis for product safety using text mining and sentiment analysis. In: 14th UK Workshop on computational intelligence (UKCI), Bradford, pp 1–7
2. Mostafa MM (2013) More than words: social networks text mining for consumer brand sentiments. Expert Syst Appl 40(10):4241–4251
3. Weiguo F, Gordon MD (2014) The power of social media analytics. Commun ACM 57(6):74–81
4. Almatarneh S, Gamallo P (2019) Comparing supervised machine learning strategies and linguistic features to search for very negative opinions. Inf 10(1):16
5. Zou H, Tang X, Xie B, Liu B (2015) Sentiment classification using machine learning techniques with syntax features. In: International conference on computational science and computational intelligence (CSCI), Las Vegas, NV, pp 175–179. https://doi.org/10.1109/CSCI.2015.44

6. Namugera F, Wesonga R, Jehopio P (2019) Text mining and determinants of sentiments: Twitter social media usage by traditional media houses in Uganda. Comput Soc Netw 6(3)
7. Balahur A (2013) Sentiment analysis in social media texts. In: Proceedings of the 4th workshop on computational approaches to subjectivity, sentiment and social media analysis, Atlanta Georgia, pp 120–128
8. Younis EMG (2015) Sentiment analysis and text mining for social media microblogs using open-source tools: an empirical study. Int J Comput Appl 112(5):44–48
9. Păvăloaia V-D, Teodor E-M, Fotache D, Danileţ M (2019) Opinion mining on social media data: sentiment analysis of user preferences. Sustainability 11(16):4459
10. Hotho A, Nürnberger A, Paaß G (2005) A brief survey of text mining. LDV Forum GLDV J Comput Linguist Lang Technol 20(1):19–62
11. Ikonomarkis M, Kotsiantis S, Tampakas V (2005) Text classification using machine learning techniques. WSEAS Trans Comput 8(4):966–974
12. Mooney R-J, Nahm U-Y, Mooney R-J (2003) Text mining with information extraction. In: Daelemans W, du Plessis T, Snyman C, Teck L (eds) Multilingualism and electronic language management: proceedings of the 4th international MIDP colloquium. Bloemfontein, Van Schaik, South Africa, pp 141–160
13. Bespalov D, Bing B, Yanjun Q, Shokoufandeh A (2011) Sentiment classification based on supervised latent n-gram analysis. In: Proceedings of the 20th ACM international conference on Information and knowledge management (CIKM'11), New York, NY, USA, pp 375–382
14. Dave K, Lawrence S, Pennock D (2003) Mining the peanut gallery: opinion extraction and semantic classification of product reviews. In: Proceedings of the 12th international conference on the world wide web (WWW-03). ACM Press, New York, pp 519–528
15. Alzamzami F, Hoda M, Saddik A-E (2020) Light gradient boosting machine for general sentiment classification on short texts: a comparative evaluation. IEEE Access 8:101840–101858. https://doi.org/10.1109/ACCESS.2020.2997330
16. Friedman J (2001) Greedy function approximation: a gradient boosting machine. Ann Stat 29(5):1189–1232
17. Hastie T, Tibshirani R, Friedman J (2009) Elements of statistical learning, 2nd edn. Springer, Berlin
18. Ross Q (1993) C4.5: programs for machine learning. Morgan Kaufman Publishers, San Mateo, CA
19. Breiman L (2001) Random forests. Mach Learn 45(1):5–32
20. Pang B, Lee L (2008) Opinion mining and sentiment analysis. Found Trends Inf Retr 2(1–2):1–135
21. Sebastiani F (2002) Machine learning in automated text categorization. ACM Comput Surv 34(1):1–47
22. Breiman L (1999) Pasting small votes for classification in large databases and on-line. Mach Learn 36(1):85–103
23. Louppe G, Geurts P (2012) Ensembles on random patches. In: Machine learning and knowledge discovery in databases, pp 346–361
24. Freund Y, Schapire R (1995) A decision-theoretic generalization of online learning and an application to boosting
25. https://www.kaggle.com/c/si650winter11/data
26. Pedregosa (2011) Machine learning in Python. J Mach Learn Res 12:2825–2830
27. Forman G (2003) An experimental study of feature selection metrics for text categorization. J Mach Learn Res 3:1289–1305
28. Kohavi R (1995) The power of decision tables. In: The eighth European conference on machine learning (ECML-95), Heracleion, Greece, pp 174–189
29. Pak A, Paroubek P (2010) Twitter based system: using Twitter for disambiguating sentiment ambiguous adjectives. In: Proceedings of the 5th international workshop on semantic evaluation, Los Angeles, CA, USA, pp 436–439

Chapter 70
Design of Dual-Band 2 × 2 MIMO Antenna System for 5G Wireless Terminals

Narayan Krishan Vyas⑩, M. Salim, Faleh Lal Lohar⑩, and Rotash Kumar

1 Introduction

Mobile communication technologies and features have evolved rapidly over the last few years and it is growing for the next generation in line with the demands of people's mobile communication or data transmission, which has become one of the most relevant and creative approaches of interest to the area [1]. With great results and wide applications, due to their large sequential bandwidth, all researchers are interested in the region. Subsequently, however, we are indented and need more attention to millimeter wave and sub-millimeter wave bands. There are many types/features of antennas in the world that can be used for applications with 5G mm waves, such as the micro strip antenna and its series. The decrease in radiation quality, however, occurs due to the substantial loss of metallic and surface waves at mm-wave frequencies.

The main patch material is conducting copper and shape of patch is rectangular. The substratum is one of the components of electromagnetic wave channeling. The substrate (ROGGER:5880) characteristics have a significant influence on the efficiency of the antenna. The substrate thickness affects the antenna bandwidth. The ground plane of the antenna was used as the ground. We use simple equations which can used to find the length and patch factor measurements for designing the micro strip antenna and are as follows [2],

N. K. Vyas (✉) · M. Salim
MNIT Jaipur, Jaipur, Rajasthan, India
e-mail: narayankrishanvyas@Rajasthan.gov.in

M. Salim
e-mail: msalim.ece@mnit.ac.in

N. K. Vyas · F. L. Lohar · R. Kumar
Government Engineering College Jhalawar, Jhalawar, Rajasthan, India

© The Author(s), under exclusive license to Springer Nature Singapore Pte Ltd. 2021   935
S. Kumar et al. (eds.), *Proceedings of International Conference on Communication and Computational Technologies*, Algorithms for Intelligent Systems,
https://doi.org/10.1007/978-981-16-3246-4_71

Table 1 Antenna design specifications

S. No.	Parameter	Quality status
1	Frequency	28–38 GHz
2	VSWR	≤ 2
3	S-parameter	< -10 dB
4	Band width	≥ 500 MHz

$$W = c / \left(2 f_0 \sqrt{\varepsilon_r + 1/2} \right)$$

$$\varepsilon_{\text{eff}} = \frac{\varepsilon_r + 1}{2} + \frac{\varepsilon_r - 1}{2} \left[1 + \frac{10h}{W} \right]^{-1/2}$$

$$L = \frac{c}{2 f_0 \sqrt{\varepsilon_{\text{eff}}}} \left[0.824h \left\{ \frac{(\varepsilon_{\text{eff}} + 0.3)(\frac{w}{h} + 0.264)}{(\varepsilon_{\text{eff}} - 0.258)(\frac{w}{h} + 0.8)} \right\} \right]$$

Here

C value is 3×10^8 m/s.

Frequency f_0 (GHz).

Relative permittivity ε_r.

Effective permittivity ε_{eff}.

H is the thickness of substrate (mm).

W is the width of patch antenna (mm).

L is the length of Patch antenna (mm).

Although it is recommended that the ground plane has a size greater than the substrate thickness, In order to obtain the equation as follows for the plane ground scale (Table 1),

$$L_g = L + 12h$$

$$w_g = w + 12h$$

In the last few years, wireless communication technology has been a field of massive change and cutting edge creation. Starting from the first-generation networking systems to the 4G LTE systems of nowadays, a rapid growth and need to migrate voice data, live video streaming, text, GPS data and many more for transmitting and receiving devices [3].

The technology used for millimeter waves contains all the creative features that will make users more demanding in the near future. Here, in order to meet many requirements of modern technology, a new patch antenna concept is proposed [3, 4]. In particular, micro strip patch antennas are very compatible with a number of surfaces, circuits and machines. However, a significant shortcoming of such antennas

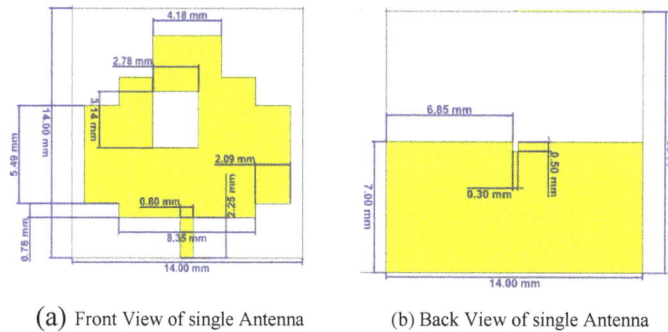

(a) Front View of single Antenna   (b) Back View of single Antenna

Fig. 1 Shows the patch antenna front (**a**) and rear (**b**) view which is initially designed

is their restricted bandwidth. With various techniques used, such as inset feeding [5] or etching various geometric shapes on the ground plane, called defected ground plane structure (DGS) [6].

2 Antenna Array Geometry and Design

For the design of MIMO antenna, first the single antenna is built with patch and cut slots, which form a 2 × 2 MIMO antenna and these antennas are stacked together. There will also be a contrast between the performance characteristics of the two antennas, such as their efficiency [7].

Creating MIMO antenna systems with high integration, miniaturized volume, low interference, is of such great interest but a significant challenge, and low correlation; this is particularly true because of size limitation in terminal devices and placement constraint. In the literature, the most recent research centered on the realization and repetition of self-decoupled compact MIMO antenna sets using polarization or pattern regulation for the sake of N × N MIMO arrays [8] (Fig. 1).

This antenna is made of three filial (branches) opposite the feed line and a rectangle is chopped-off. The proposed antenna is developed for the band of 28–38 GHz. The dimension of patch antenna is given in Table 2.

3 Result of Single Antenna

The substratum circuit is manufactured using ROGERS RT5880 substrate 0.254-mm thickness with tangent loss 0.0009 and relative permittivity 2.2, the cooper thickness is 0.034 mm. The design board with substrate is chosen to have a size of (28 × 28) mm^2. The dimensions used to design for the proposed antenna are given in Table 1.

Table 2 Details of parameters of cut and patch slot of antenna with dimensions in mm

Parameter	Value in (mm)	Parameter	Value in (mm)
Patch length (L)	14	W (Patch width)	14
Feed line width (W_f)	0.80	Feed line length (L_f)	2.25
Substrate height (h)	0.254	Relative permittivity (ε_r)	2.20
Patch cut length (L_P)	3.14	Patch cut width (W_P)	2.78
Material height	0.034		

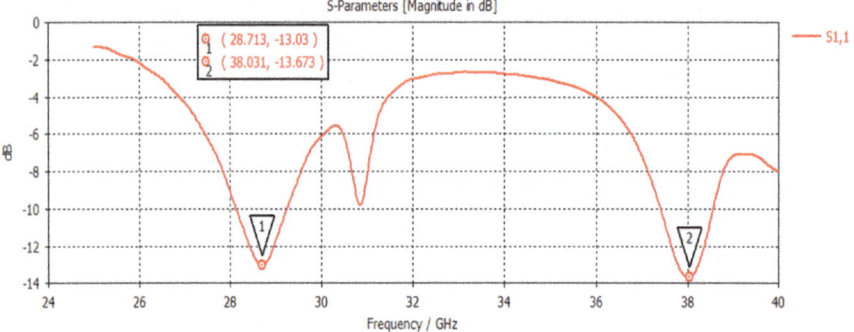

Fig. 2 Shows the S-parameter of single patch antenna

The patch antennas are chosen for mobile unit antennas for MIMO systems. But the main downside is low band-width. This can be achieved by reducing the thickness of the substrate or by lowering the permittivity value. We also suggested that the antenna should be a 3-layer patch antenna. By reducing ground plane areas, we can achieve higher bandwidth. For smaller ground areas, the capacitance between the patch region and the ground plane is lowered and by this the bandwidth increased. The planed four segment 2 × 2 MIMO antenna is shown in Fig. 4 and Figs. 2 and 3 show the graph for S-parameter and surface current at 28 GHz and 38 GHz, respectively, [9, 10] (Figs. 5, 6 and 7).

4 Simulated Results

With considering several parameters and changing the dimensions of cut and patch to achieve acceptable reflection coefficient |S11|, gain and total efficiency required

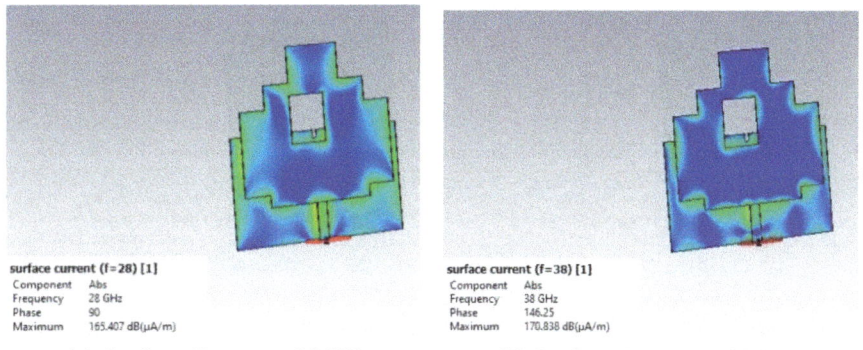

(a) Surface Current at 28GHz   (b) Surface Current at 38GHz

Fig. 3 Shows the surface current for mobile communication frequency at 28/38 GHz

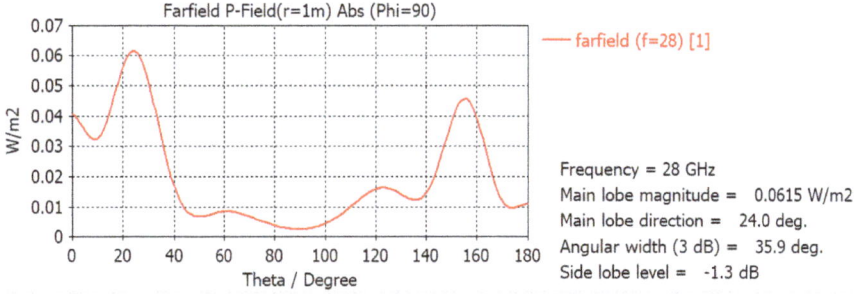

Fig. 4 Shows the power flow for antenna at frequency at 28 GHz

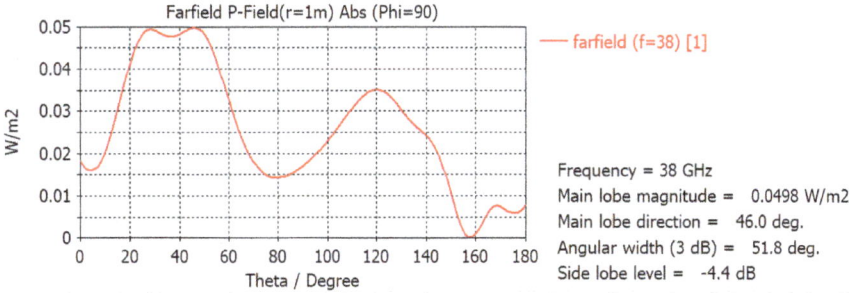

Fig. 5 Shows the power flow for antenna at frequency at 38 GHz

for each frequency. All concept simulations are rendered using computer simulation technology (CST) microwave studio as a commercial software program [3].

Figure 4 shows the 2 × 2 MIMO antennas with different cut of edges on patch and ground in copper conductor. For results, the feed line also have important role for obtaining the desired frequency. 2 × 2 MIMO patch antenna is shown below.

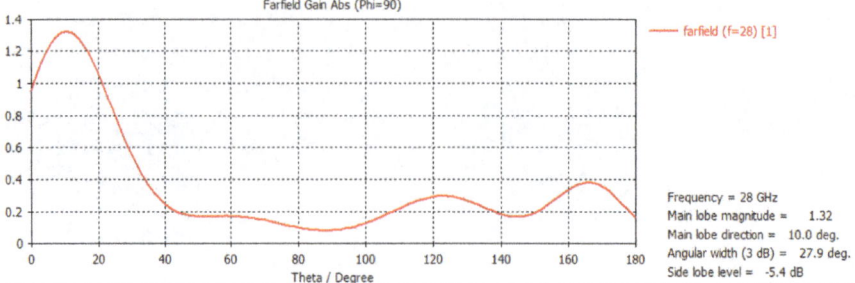

Fig. 6 Shows the gain (IEEE) for antenna at frequency at 28 GHz

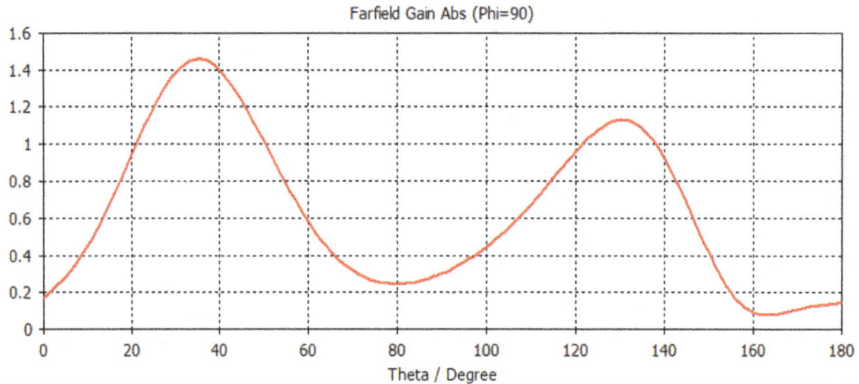

Fig. 7 Shows the gain (IEEE) for antenna at frequency at 38 GHz

Figure 8 displays the configuration and shape of the 2 × 2 MIMO patch antenna, sequentially the effect is obtained when the length of different slots is adjustable for different cuts in lengths. Therefore, among all, the best outcome of the frequency

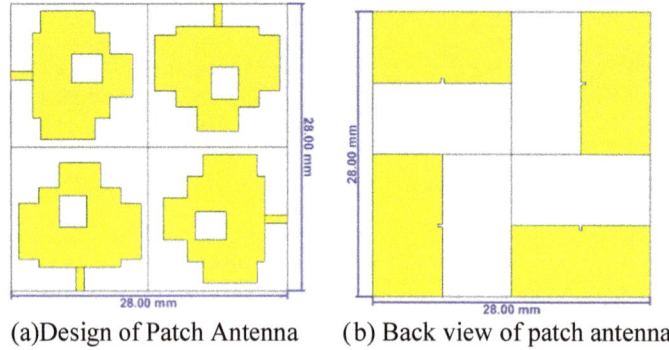

(a)Design of Patch Antenna   (b) Back view of patch antenna

Fig. 8 Shows the final design of 2 × 2 MIMO antenna with patch at front and back side on ground

drop, this is the lowest at the target frequency. In addition, the reflection coefficient, which is −15.83 and −11.87, respectively, for the 28 GHz and 38 GHz frequencies, also shows a strong result. Technically, for good efficiency reflection coefficient must be less than −10 dB, especially in terms of frequency drop and we achieved it for both frequencies with good bandwidth of more than 1 GHz bandwidth. As shown below (Figs. 9, 10, 11 and 12).

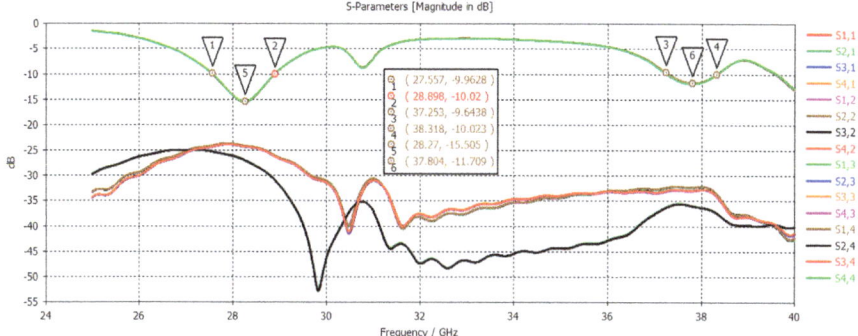

Fig. 9 Shows the S-parameter of 2 × 2 MIMO antenna for resonant frequency

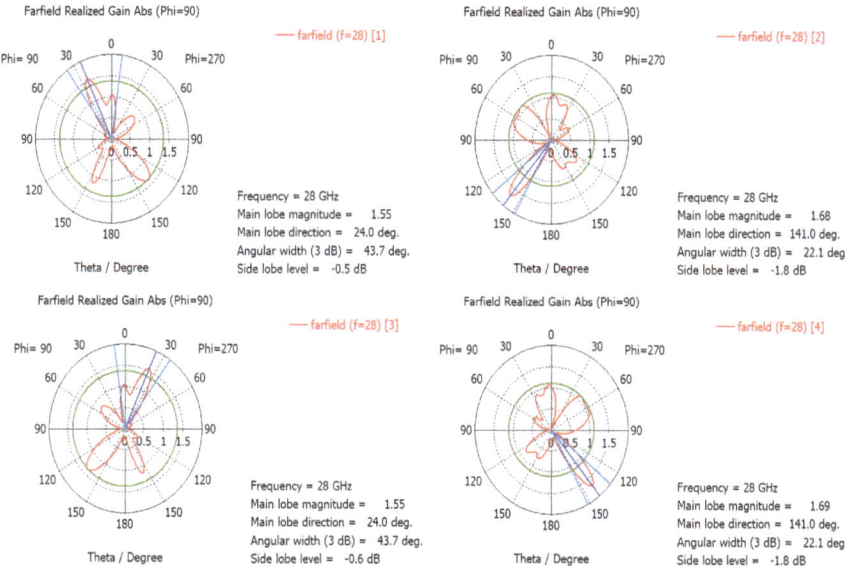

Fig. 10 Shows the farfield of 2 × 2 MIMO antenna at 28 GHz frequency

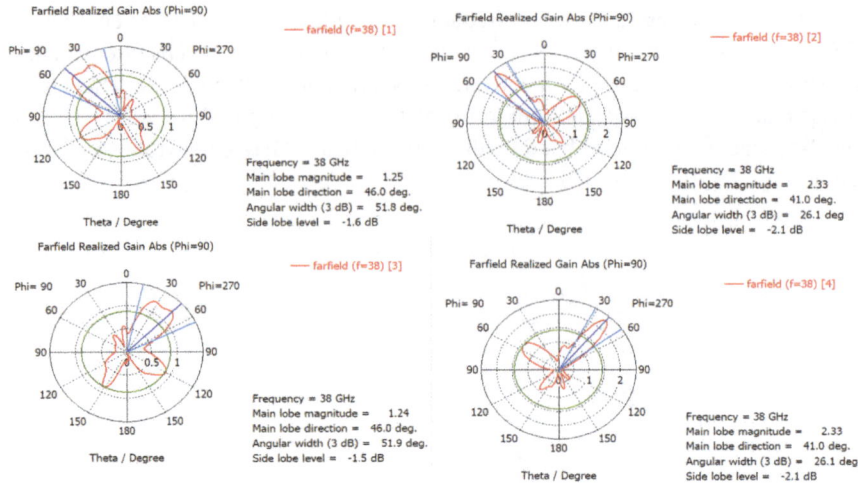

Fig. 11 Shows the farfield of 2 × 2 MIMO antenna at 38 GHz frequency

Fig. 12 Shows IEEE gain of
2 × 2 MIMO antenna at
28 GHz and 38 GHz
frequency

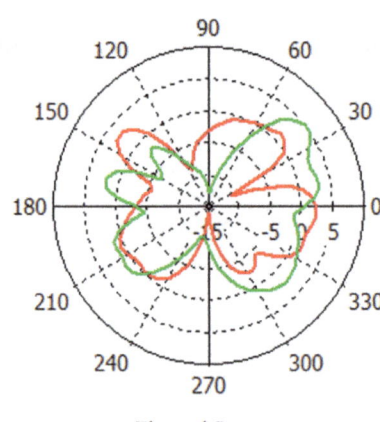

Theta / Degree

5   Conclusion

For 5G applications, four element MIMO antennas are proposed to fulfill the requirements of the reflection coefficient by designing them with a 90° angle to each other
and to achieve a relatively high isolation with a relatively nearby small antenna,
which is suitable for wireless communications today. Proposed antenna designed
with a compact size using substrate Rogers RT5880 to operate for 28–38 GHz. The
findings show that the suggested antenna has good efficiency in terms of radiation
pattern characteristics, reflection coefficient and bandwidth. The final result has a
bandwidth of 1.45 GHz at 28 GHz and 1.7 GHz at 38 GHz, so the results obtained

suggest its promising potential for 5G applications. The design is attractive for 5G mobile terminal MIMO antenna designs.

References

1. Zhang Y, Deng J-Y, Li M-J, Sun O, Guo L-X. A MIMO dielectric resonator antenna with improved isolation for 5G mm-wave applications. IEEE Antennas Wirel Propag Lett. http://doi.org/10.1109/LAWP.2019.2901961
2. Yuwono T, Ismail M. Design of massive MIMO for 5G 28 GHz. 978-1-7281-0108-8/19/$31.00 ©2019 IEEE
3. Daud NN, Jusoh M, Rahim HA, Othman RR, Sapabathy T, Osman MN, Yassin MNM, Kamarudin MR (2017) A dual band antenna design for future millimeter wave wireless communication at 24.25 GHz and 38 GHz. In: 2017 IEEE 13th international colloquium on signal processing & its applications (CSPA 2017), 10–12 Mar 2017, Penang, Malaysia
4. Ojaroudiparchin N, Shen M, Fr G (2015) Multi-layer 5G mobile phone antenna for multi-user MIMO communications. In: Telecommunications forum Telfor (TELFOR). IEEE, pp 559–562
5. Matin MA, Sayeed AI (2010) A design rule for inset-fed rectangular microstrip patch antenna. WSEAS Trans Commun 9(1):63–72. ISSN: 1109-2742
6. Hakanoglu BG, Turkmen M (2017) An inset fed square microstrip patch antenna to improve the return loss characteristics for 5G application. In: 32nd URSI GASS, Montreal, 19–26 Aug 2017
7. Zahid M, Shoaib S, Rizwan M. Design of MIMO antenna system for 5G indoor wireless terminals. Department of Electrical Engineering, HITEC University, Taxila Cantt., Pakistan
8. Piao H, Jin Y, Qu L. Isolated ground-radiation antenna with inherent decoupling effect and its applications in 5G MIMO antenna array. http://doi.org/10.1109/ACCESS.2017
9. Ngo HQ. Massive MIMO: fundamentals and system designs. Linköping studies in science and technology dissertations, No. 1642
10. Ali MMM, Sebak A-R. Design of compact millimeter wave massive MIMO dual-band (28/38 GHz) antenna array for future 5G communication systems. 978-1-4673-8478-0/16/$31.00 ©2016 IEEE

Chapter 71
Histogram of Oriented Gradient-Based Abnormal Weapon Detection and Classification

Jayandrath R. Mangrolia and Ravi K. Sheth

1 Introduction

Automatic video surveillance system can monitor the events, behavior, or objects without the interference of human operators for the purpose of safety and security of a person, process, or group using smart CCTV cameras. Group activities can be monitored through crowd detection and analysis algorithms, while individual activities can be supervised through abnormal activity detection and behavior analysis [1]. An appearance of objects like swords, guns, or knives near public places can be considered as abnormal, and it is inevitable to detect, classify, and recognize those objects immediately in order to send quick notification to concern authority. Input video is converted into frames, followed by background subtraction, feature extraction, and classification. In our proposed approach, firstly, background subtraction is performed by MOG, secondly, HOG and PCA are used to extract important features. Extracted features from training dataset are used to train machine learning techniques in order to create a model, then those models are tested against the features extracted from testing datasets.

J. R. Mangrolia (✉) · R. K. Sheth
School of Information Technology and Cyber Security, Rastriya Raksha University, Ahmedabad, Gujarat, India

R. K. Sheth
e-mail: rks@rru.ac.in

© The Author(s), under exclusive license to Springer Nature Singapore Pte Ltd. 2021   945
S. Kumar et al. (eds.), *Proceedings of International Conference on Communication and Computational Technologies*, Algorithms for Intelligent Systems,
https://doi.org/10.1007/978-981-16-3246-4_72

2 Proposed Algorithm

Our proposed algorithm is efficient for detection of weapons from the video sequences. In the first step, input video is converted into frames, and sample frames are processed to extract the features.

In the second step, background subtraction is performed as a preprocessing on each sample frame, in order to eliminate unwanted objects from the background. Mixture of Gaussian (MOG) is applied for the purpose of background elimination.

In the third step, informative and non-redundant derived values, also referred to as features, are extracted from each background-subtracted sample frames. Histogram of oriented gradients (HOG) is applied for the computation of feature vectors.

In the next stage, eigenvalues and eigenvectors are calculated for the reduction of dimensions using principal component analysis (PCA).

In the last stage, training and testing data set is generated from the final feature vector. SVM and NN are used as a classifier to detect whether the weapon is present in the video or not. Architecture of our proposed algorithm is shown in Fig. 1.

2.1 Background Subtraction

Background subtraction is a procedure that is used to eliminate undesirable objects from the background. Background objects, whose contribution is responsible to mitigate the performance of the system, can be removed using background subtraction. It is also used for the detection of dynamically moving objects from stationary cameras.

Fig. 1 Proposed algorithm

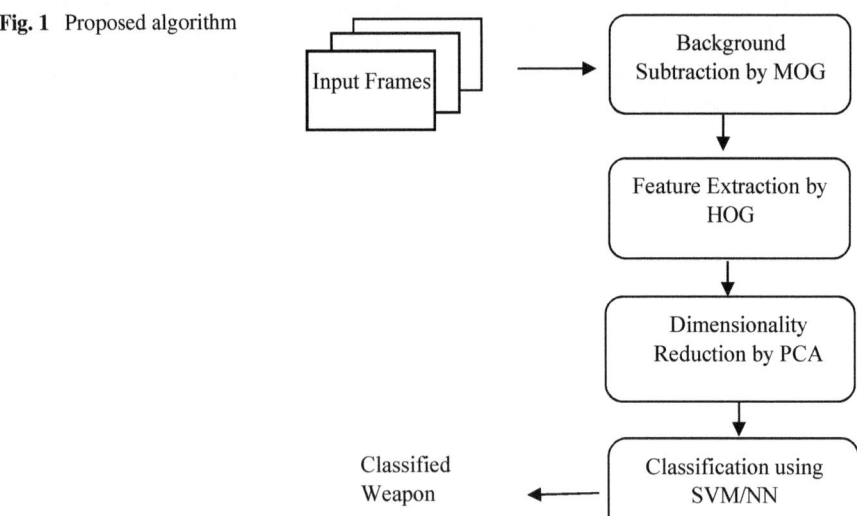

Frame distinction, mean filter, Gaussian average, and background mixture models are various conventional background subtraction techniques. The technique based on MOG (mixture of Gaussian) is used in this article in order to eradicate the unwanted surrounding objects and detect mobile objects. This approach is based on a static context mathematical model, and then, every new video sequence frame is compared with the prototype [2]. The mixture of K Gaussian distributions models each pixel in the scene, where the elementary concept is to subtract the input frame from the reference frame. History modeling, updating, subtraction, and thresholding are the measures involved in the process [2].

MOG is selected because of its ability to track multiple Gaussian distribution simultaneously. It can also maintain the density of each pixel, and it is capable of handling multi model background distributions.

2.2 Histogram of Oriented Gradients

The oriented gradient histogram is a well-known function descriptor that calculates the number of times in the local portions of a picture the gradient orientation occurs [3]. Firstly, the entire image is divided into small regions called cells. Gradient directions and gradient magnitudes are calculated, and bins are defined for each cell. For example, nine bins are defined to represent the nine angles between 0 and 180. At each pixel position, according to the direction of the gradients, magnitude at same pixel position is stored in its corresponding bin. For example, at first pixel position, if the direction is 20 and its corresponding magnitude is 10, then value 10 is stored in the first bin whose angle is 20. Finally, we can obtain a vector of nine values. Next, a block of four adjacent cells is considered to normalize the calculated histograms, so total vectors are 36. Finally, entire window (block of four cells) is moved by shifting each cell. The total number of blocks multiplied by 36 is equal to the size of the final function vector. This vector is informative and help the machine learning technique to take appropriate decision [4].

HOG has many key advantages over other descriptors. HOG feature descriptor is calculated over local cells, so it can be stable in case of geometric and photometric transformations. HOG can be effective in varied backgrounds, lighting conditions, and rotation changes.

2.3 Dimensionality Reduction Using PCA

The principal component analysis (PCA) approach is used to mitigate the size of the data set (feature vector) while improving the interpretability, and at the same time, the loss of information is also minimized [5]. Principal components are extracted in two steps. Firstly, the data covariance matrix of the original data is calculated, and secondly, the eigenvalue decomposition is performed on the covariance matrix. In our work, the final features are calculated by the combination of HOG and PCA.

2.4   Classification Using SVM

Support vector machine (SVM) is a most effective machine learning algorithm that is based on supervised learning paradigm. The algorithm is capable enough to handle both classification and regression problems. The SVM training algorithm generates a model, and that model represents the examples as points in space. These points are plotted in such a way that a simple distance that is as large as possible divides the examples of the various groups [6]. New samples are mapped into the same space from the test data set, and the decision is made for the expected class based on which side of the gap they drop on. Two simple methods exist: (1) maximum (linear) margin classification and (2) nonlinear classification.

2.4.1   Linear (Maximum) Margin Classification

The data set (feature vector) is classified by the hyperplane, which is generated by support vector machine. The selection of the most suitable hyperplane depends on the greatest separation, or margin, between the two data sets. For the hyperplane to be selected, the distance from it to the most adjacent data point on each side is set to be maximum. If such a hyperplane exists, it is referred to as the supreme margin hyperplane. The maximum margin classifier is the classifier that can be elucidated by the linear classifier [6]. For any classification problem, in order to predict the class, for some sample data points, it is highly essential to decide which class a new data point will be in. The data point is interpreted by support vector machines as an n-dimensional vector (a collection of n numbers), and the aim is to identify that with an $n - 1$ dimensional hyperplane whether it is possible to distinguish such points (Fig. 2).

2.4.2   Multiclass SVM

The most significant machine learning approach of multiclass SVM is used for such classification problems where the number of classes are many and predetermined. Support vector machine assigns the labels, where the labels are drawn from a finite set of abundant elements. Any complex multiclass problem can be solved by reducing the single multiclass problem to multiple binary problems where each of the binary problem yields a binary classifier. That binary classifier produces an output function that gives relatively large values for examples from the positive class and relatively small values for examples belonging to the negative class [6].

Fig. 2 Maximum margin
hyperplane [6]

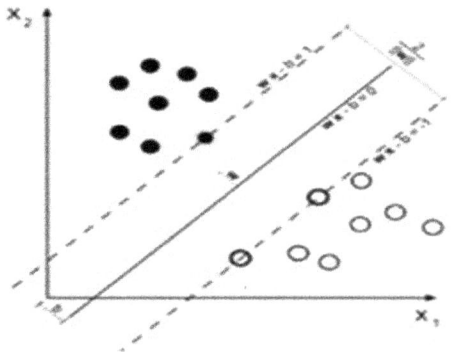

2.5   *Classification Using Neural Network*

One of the popular machine learning algorithms, which is inspired by a human's biological nervous system, is the neural network. It is made up of a finite set of neurons, often called components. Such components work in parallel. The roles of the network define the relations between neurons. The neural network can be trained by changing values of the connections (weights) between elements to perform a precise function [7].

The neural network modification or training is performed in such a way that a specific input contributes to an accurate target output. Based on the contrast of the target output with the computed network output, the network adjustment is carried out and the weights are re-adjusted until the figured output matches the target [7]. In this supervised learning, several input/target pairs are used to train the network. The artificial neural network consists of multiple interconnected artificial neurons, also known as nodes, linked to a network of layers. A typical ANN is illustrated in Fig. 3.

The input value pi of each node and the corresponding weight value wi are added in order to calculate the single value threshold of each node. To formulate the neuron's net input value n, the sum (weighted value), is included with the bias term b. It is essential to shift the sum corresponding to the origin by including the bias, and then, the net input value is entered into the transfer function f which generates the neuron

Fig. 3 Artificial neural
network [7]

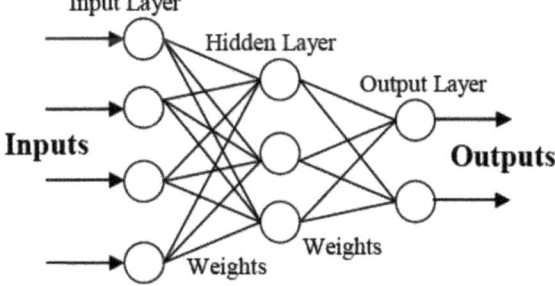

Fig. 4 A neuron [7]

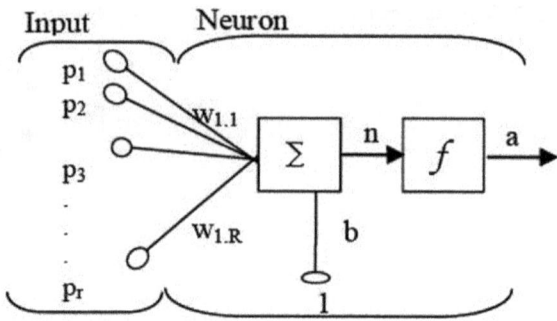

output a [7] (Fig. 4).

$$a = \sum_{i=1}^{r} w_i + p_i + b \tag{1}$$

3 Implementation

Our proposed algorithm is tested to classify two different weapons, knife and gun. Various videos in which a human is moving with such weapons are taken as an input. Each video is processed separately in order to extract frames. From each video, ten frames are selected as a sample for further processing. By considering each frame as a separate image, mixture of Gaussian (MOG) is applied for background subtractions. Input images and its background-subtracted frames are shown in Fig. 5.

Each background-subtracted frame of size (352, 640) is given as an input to calculate HOG feature vector. The size of each cell is kept (10, 10) while window size is (20, 20) in order to compute feature vector. The total possible bins per block is 36 by generating nine bins from each cell, and the size of the final feature vector is 36 × total block number, which is equal to 241,920. Feature vector is resized to (489, 489) and given to principal component analysis in order to reduce the dimensions. PCA compute the eigenvalues and mitigate the dimensions of feature descriptor, so finally, the size of the descriptor is reduced to (0, 489). For all the sample images used for training purpose, the feature descriptor of size 489 vectors is computed and combined in the same array. Total 80 frames from different videos are selected for training, so size of final feature vector is (80, 489).

Fig. 5 Sample and background-subtracted frames

3.1 Feature Classification

Feature vector, computed for training purpose, is given to SVM classifier in order to build a model. Labels are carefully assigned to each feature vector of training data set and linear SVC is implemented, which is based on LIBSVM. Hyperplane is constructed through default parameters in order to correctly classify the testing features. Artificial neural network is constructed through multilayer perceptron classifier, where the size of the hidden layer is set to 100. Target vector is carefully built based on training data set. Feature data set built from 80 frames of different video sequences are given for training, while feature set built from 10 frames is taken as testing for both the classifiers. Classification rates are shown in the confusion matrices (Figs. 6, 7 and 8; Tables 1 and 2).

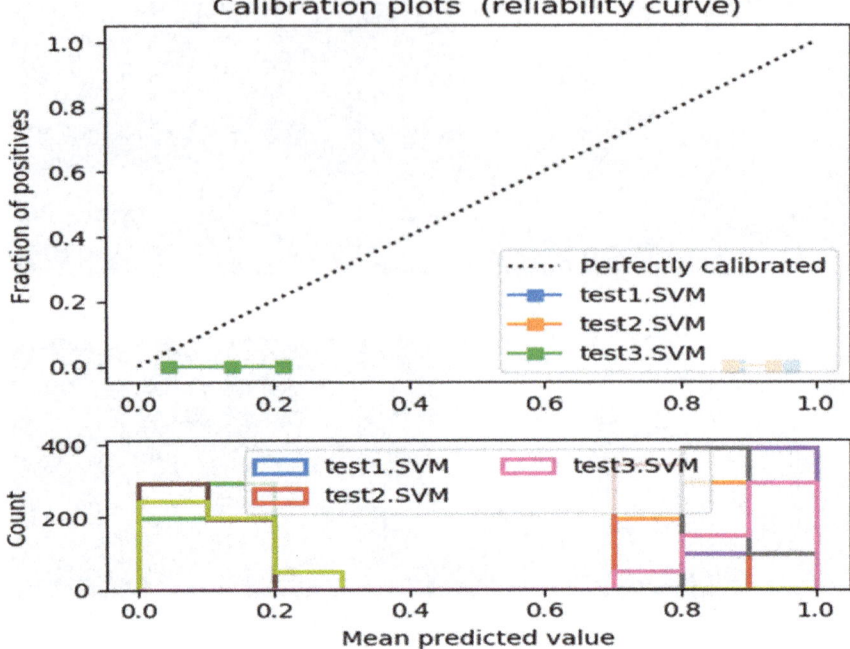

Fig. 6 Calibration curvature of SVM

From the results shown, it can be deduced that support vector machine can provide 100% accuracy in video sequences in which weapons are presented. Efficacy of neural network is less by 10–20% due to misclassification, while video sequences without weapon can be effectively detected by neural network with recognition rate 80%. Conclusively, support vector machine can be the most powerful supervised learning method for videos in which suspicious objects are presented.

4 Conclusion

Security and protection near public places are highly essential in today's era due to fast paced growth of crime and violence. Weapon detection is a part of smart video surveillance that can detect suspicious objects in public places automatically, thence ferocity can be prevented. Our proposed approach, based on the combination of MOG, HOG, and PCA, produces robust feature vector. Histogram of oriented gradient can produce good results, even in case of diverse background and illumination. From the results, it can be deduced that the precision of support vector machine is high as compared to neural network for the videos in which weapons are presented. Support vector machine can achieve 100% effectiveness, while neural network can also produce better results in case of weapon-less video sequences.

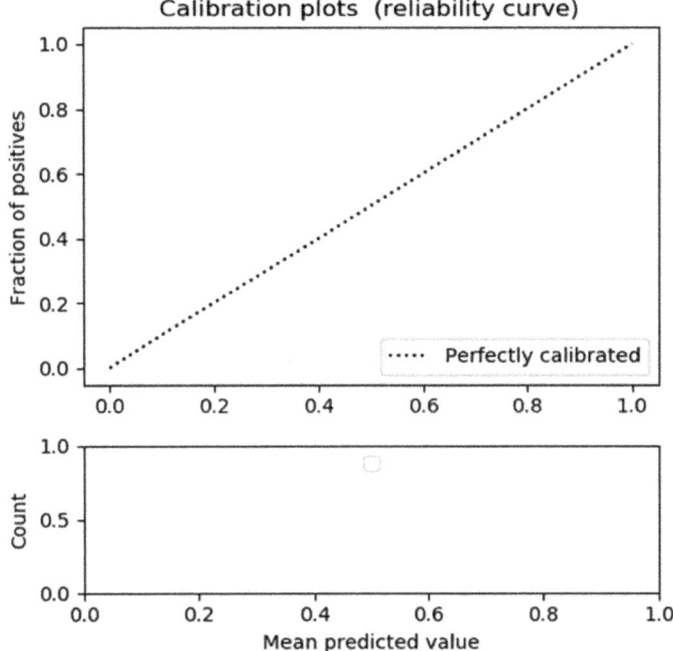

Fig. 7 Calibration curvature of neural network

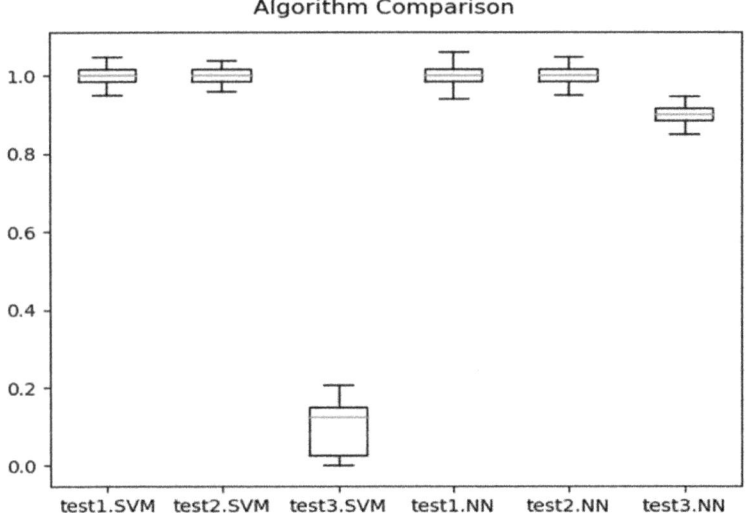

Fig. 8 Comparison of SVM and NN through box and whisker plot

Table 1 Confusion matrix of SVM

Weapon class	Weapon class		
	Class 1 (Gun) (Rate %)	Class 2 (Knife) (Rate %)	Class 3 (Weapon-less) (Rate %)
Class 1 (Gun) (Rate %)	100	0	0
Class 2 (Knife) (Rate %)	0	100	0
Class 3 (Weapon-less) (Rate %)	80	20	0

Table 2 Confusion matrix of neural network

Weapon class	Weapon class		
	Class 1 (Gun) (Rate %)	Class 2 (Knife) (Rate %)	Class 3 (Weapon-less) (Rate %)
Class 1 (Gun) (Rate %)	80	0	20
Class 2 (Knife) (Rate %)	10	90	0
Class 3 (Weapon-less) (Rate %)	10	10	80

References

1. Mangrolia JR, Chauhan NC (2012) Intelligent video surveillance: concept, review and open issue. In: International conference on information, knowledge and research, Mar 2012
2. Chitra M, Kalaisevi Geeta M, Menaka L (2013) Occlusion and abandoned object detection for surveillance application. Int J Comput Appl Technol Res 2(6):708–713
3. Elmir Y, Laouar SA, Hamdaoui L (2019) Deep learning for automatic detection of handgun in video sequences. In: Research on computer sciences, JERI 2019, Saida, Algeria
4. https://www.learnopencv.com/histogram-of-orientedgradients
5. Malagon-Borja L, Fuentes O (2006) Object detection using image reconstruction with PCA. Science Direct
6. Mangrolia JR (2018) Abnormal activity recognition using support vector machine. IJSART 4(1)
7. Mangrolia J (2018) Gabor feature based abnormal event detection. IJIRT 4
8. Tiwari RK, Verma GK (2015) A computer vision based framework for visual gun detection using Harris interest point detector. Procedia Comput Sci 54:703–712
9. Singh A, Sawan S, Hanmandlu M (2009) An abandoned object detection system based on dual background segmentation. In: Advanced video and signal based surveillance, IIT Delhi, India
10. Patole M, Charwad B, Mengade A, Choudhary N, Hondhe VC (2017) Suspicious object detection. Int Res J Eng Technol (IRJET) 4(5)

11. Adalinge D, Patil D, Dongre S, Gokodikar S, Kodmelwar MK (2017) Survey on abandoned object detection using image processing. Int J Innov Res Comput Commun Eng 5(5)
12. Martinez J, Elias Herrero J, Gomez JR, Orrite C (2006) Automatic left luggage detection and tracking using multi-camera UKF. In: IEEE international workshop on PETS, New York, 18 June 2006
13. Kong H, Ponce J (2009) Detecting abandoned objects with a moving camera. Ecole des Ponts ParisTech, France
14. Ke Y, Sukthankar R (2009) Event detection in crowded videos. School of Computer Science, Carnegie Mellon
15. Kausalya K, Chitrakala S (2012) Ideal object detection in video for banking ATM applications. Res J Appl Sci Eng Technol 4(24):5350–5356, Chennai, India
16. Bhondave A, Biradar B, Suryavanshi V, Nakil H (2016) Suspicious object detection using back-tracing technique. Int J Adv Res Comput Commun Eng 5(1)
17. Divya J (2015) Automatic video based surveillance system for abnormal behaviour detection. Int J Sci Res (IJSR) 4(7)

Chapter 72
Prime-Based Encoding Algorithm to Ensure Integrity of Electronic Health Records

K. J. Kavitha and **Priestly B. Shan**

1 Introduction

The growth in the field of communication and Internet has facilitated the healthcare systems to make use of telemedicine, telemetry, etc., to transfer patient details from the remote places for higher diagnosis. The patient details include electronic health records (EHR), scanning images, billing information, etc., and the special care must be taken to ensure safety of these messages stored in database to avoid fraud utilize of the information by illegal money makers, and also the information sent to the expert has not undergone for any kind of tampering so that no wrong diagnosis is made by the corresponding. So, it becomes the responsibility of respective organizations to provide patient's privacy and security to healthcare systems. Patient privacy can be referred to the patient's right to know when, how and to what extent their health information is shared with others, whereas security in healthcare system includes authentication, integrity of the cover object and confidentiality of the information used in the watermark.

In India, EHR system is fully not developed, as there is lack of awareness about the advantages of such systems. The case of 79-year-old Karan Singh who was suffering from lung cancer underwent for 11 years of treatment and needs to carry bulge amount of medical reports each time he visits the hospital but he never heard about system like EHR in any hospital in India. Recently, Kerala government has initiated developing this system by collecting information from nearly 2.58 crore patients

K. J. Kavitha (✉)
JIT, Davangere, India

VTU, Belgaum, India

Sathyabama University, Chennai, India

P. B. Shan
Galgotias University, Noida, India

through eHealth project and is been adopted by 86 government hospitals in the state. The government has to take initial steps towards creating awareness in people and develop such advanced systems in India with proper measures and precautions to ensure safety of patients.

In healthcare systems, medical images (MI) play an important role same as personal health records (PHR) of patient. PHRs help patients to manage their health with the help of important health information stored in electronic form such as vaccination details, laboratory results, consulting dates which helps patients to update and share their records, whereas MIs in healthcare systems have entirely changed the healthcare systems as it allows the practitioners and researchers learn more about diseases by diagnosing and able to come up with a solution to provide better treatment than ever before. MIs also assist doctors to take decisions to treat patients and able to take precautions. With the advances in the technology, MIs are helping doctors to know more about internal problems that would not have done with external examinations and it is a necessary when the patient is passing through online illness as it provides the detailed information, thereby it helps in preventative care.

Overall we may say that patient's EHR is an effective tool of storing information electronically so that it helps in terms of enhancing patient care, improving care coordination and increasing efficiency; moreover, the amount of medical record documentation that is now required is an order of magnitude greater than it was 20 years ago; however, the development in the technology has brought the benefit of storing huge data, but at the same time, it provides the space for providers to commit fraud activities. Security breaches in healthcare department have severe impact on patient's health, and such many security breaches have been reported worldwide such as the medical centres may expose patient records to the public on the Internet having confidence that they were on a server protected with a password, hackers may infiltrate the medical centre's computer system and steal medicine patients' records, accidentally one may send the private correspondence of patients, information to someone else, if a patient's information is disclosed accidentally or unintentionally [1, 2], it may constitute an infringement of privacy, and cause embarrassment, ruin or damage to the individual's career, dismissal from work, loss of health insurance worthiness and financial loss.

Therefore, providing security to such EHR including patient detail and MIs has become an important issue [3] and it is the responsibility of physicians, radiologists and the respective organizations to provide security to the patient details by establishing certain protocols, patient's privacy, etc. Some of the measures taken by many of organizations may be listed as below:

- Staff of a hospital must be trained to maintain patient's security policies.
- Proper storage space and access controls must be provided to the systems stored with patient's details in electronic form.
- Patient should be given the authority to release their EHR to the concerned as and when required obliged by rules and regulations.

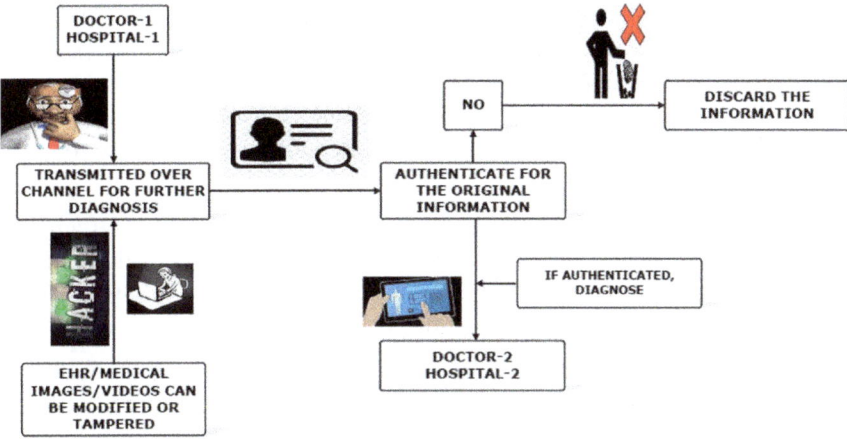

Fig. 1 Scenario for the verification of the EHR

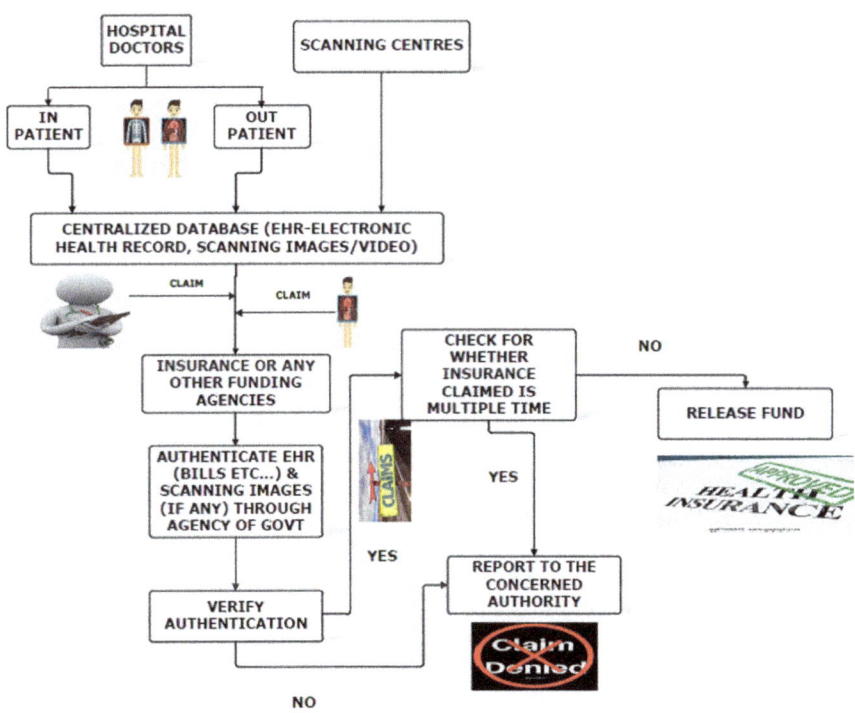

Fig. 2 Scenario of verification of fraud cases by the agency

The following Figs. 1 and 2 show the scenarios of verification of EHR by the receiving end of hospitals and identifying, fraud cases of claiming insurance for multiple times by the respective agencies of government or insurances.

To avoid such fraud cases, the organizations must provide technical protection and administration safeguard methods while at the same time, EHR plus MIs must possess the following attributes [4]:

Confidentiality: It is a form of informational privacy characteristic of certain relationships, such as the physician–patient relationship. Personal information obtained in the course of that relationship should not be revealed to others unless the patient is made aware of this intention and consents to disclosure.

Integrity: As any changes or inaccuracy in data can have an impact on the healthcare process, health information needs to be readily available to the authorized person at the time when it is required.

Security of EHR systems can be implemented by the physical security of the system, providing access only to authorized users, through the implementation of firewall, virtual private networks (VPN), etc. In case of firewall technique, it is able to control the network access to many computers connected to it wherein the main limitation of this type of protection is that the information stored on the disc or in transit is vulnerable; moreover, probability of network breaks down to steal data in terms of packets by the packet sniffers which is very high, whereas VPN provides a secure access to the local area network (LAN) over a shared network and its major limitations is, and the central site must be provided with a permanent Internet connection which allows the remote clients and other site clients to contact at any time, thereby each user is provided with less bandwidth compared to a dedicated line solution.

Although such physical and technical security is adopted by most of the organization, hacking information has not been stopped and steganography, cryptography, hashing, and digital watermarking techniques are been adopted by many researchers to maintain the attributes of medical information. In steganography technique, the information to be protected is hidden in another cover information and cannot be identified with naked eyes, but it requires a lot of overhead to conceal a relatively few bits of information; if the system is revealed once, it becomes practically insignificant. Whereas in cryptography technique, the information to be protected is encrypted using a key in the non-readable format, and this information can be decrypted only if the secret key is known. As this technique protects the information secrecy in unreadable format, such scrawl messages may attract third parties, leading to the destruction of information. Perceptual hashing technique secures the information during transmission to a specific user. Many hashing techniques such as MD5, SHA and CRC-32 are used to generate the hash values to protect information against tampering. The constraints of hashing techniques lies in the role of its implementation and choosing hash function for a specific application is more an art than a science, beneath this its computational cost is very high and it is irreversible and constant. Digital watermarking (DWM) is a method of embedding watermark (WM) information in host information with low degradation to original information. DWM possesses salient features which enable many of technologies to adopt this technique such as in medical, military field as well as in archival-based applications.

1.1 Salient Features of Medical Image Watermarking (MIW)

Some of salient features of DWM techniques are imperceptibility, simplicity, effectiveness, robustness, image fidelity, security, capacity, reversible property and ownership assertion [5], and also, this technique is bounded by many advantages like **saving memory** by combining crucial information as a single avoids detachment-related information from original and thereby helps to escape from the risk of misplace of records and saves bandwidth required for transmission. Although DWM has plenty of advantages, it is not able to avoid the fraud activities by illegal money makers and to avoid such activities, one can combine the positives of steganography, cryptography and DWM techniques as hybrid technique to enhance security and to maintain confidentiality, integrity, authentication, privacy [6], indexing, access control and caption of one's information.

In case of healthcare systems, the patient's information should possess the above features and EHR is embedded in the scanned images by abiding to requirements needed by the system. In medical image management system, most intended requirements are given below [7]:

1. Imperceptibility: the embedded watermark in the image should not be visible to human eye.
2. Robustness: system should be robust enough or able to withstand various attacks possible on images while distribution.
3. Capacity: it is the amount of data payload that can be hidden in the image as a watermark.
4. Authenticity: the data content should only be accessible to legal users after proper authentication. Cryptography is used for this purpose.
5. Reversibility: the image should be recoverable after extracting the watermark.
6. Intactness of ROI: the ROI should not be affected by watermarking process in any way. Damaged ROI can have serious consequences with respect to patient's health.
7. Complexity: less complex algorithm is always profitable with respect to user as well as system by saving execution time.

Before implementing MIW, we may come across various issues that is to be taken care, such as selecting the type of MIW technique whether diagnosis requires either complete or content authentication of information; whether information embedded should be reconstructed or not; information pay load capacity; and protection of WM information from illegal access. And, all these requirements may not be completely implemented by DWM technique alone.

2   Literature Review

The medical centres may expose patient records to the public on the Internet having confidence that they were on a server protected with a password. Hackers may infiltrate the medical centre's computer system and steal medicine patients' records. Accidentally one may send the private correspondence of patients, information to someone else. If a patient's information is disclosed accidentally or unintentionally, it may constitute an infringement of privacy and cause embarrassment, ruin or damage to the individual's career, dismissal from work, loss of health insurance worthiness and financial loss. There is a concern by some researchers that requirements for the patient's consent and anonymity will undermine their research.

Many researchers have contributed their work to safeguard MIs and its related information. Table 1 shows the respective work of few researchers, their contribution and remarks.

Most of MIW involves hybrid techniques to ensure the maintenance of MI attributes, and most of the algorithms are implemented on whole MIs. However, no complete algorithm could be developed to satisfy all requirements of MIs; moreover, more than 90% of the works does not ensure safety and privacy to patient details embedded in MIs. Hence, still there exists scope for MIW techniques for ensuring its safety. Among the existing MIW, hybrid technique is more suitable to make the system more robust against various attacks. As watermarking system alone is not capable of avoiding the fraud being making illegal access to the data, we may combine encoding, encryption and suitable techniques to enhance the security. In this paper, a prime number-based new encoding approach is proposed to ensure safety to PHR while maintaining the features of MIW technique.

3   Proposed Embedding and Prime-Based Encoding Algorithm

In this paper, hybrid medical image watermarking (MIW) is proposed for not only ensuring safety to MI but also to patient information. In this work, embedding is done in low-frequency components and it uses hybrid embedding function defined as Eq. 1:

$$\text{wmi} = (\alpha \times \text{ll}_{ci}) + \left(\frac{\beta}{\alpha} \times \text{ll\_wm}\right) \tag{1}$$

where α = scaling factor $\to 1 \leq \alpha \leq 2$

$$\beta = \text{Payload factor} \to 0.5 \leq \beta \leq 1$$

Table 1 Contribution by authors in MIW domain

References	Contribution	Embedded region	Remarks
[8]	Odd–even embedding algorithm and differential expansion Reversible	Whole image	PSNR = 55 dB, Limiting data payload Suitable for non-reversible schemes and specific images (smooth)
[9]	SHA-1, Advanced classical cipher, DWT JPEG2000	Whole image	Cryptographic security reduces, Lossy algorithm, Suitable for plain text
[10]	Modified LSB, Huffman coding, Average intensity of each block of ROI and LSB of a WM	ROI and RONI	ROI region is distorted, Less susceptible to manipulation attacks Related watermark information is not considered, PSNR = 47–48 dB
[11]	FCT, SVD, Arnold transform, Machine learning (ML)	Whole image (Only retinal scan)	PSNR = 65 dB Time consumption Moderate robustness against attacks
[12]	Number Theoretic Transform (NTT) Diffie–Hellman	Whole image	Vulnerable, Lack of authentication DS cannot be used with Diffie–Hellman Used only for symmetric key
[13]	DWT, DCT, Elliptical CURVE Diffie–Hellman (ECDH)	Whole image	Vulnerable to attacks, Loss of information, Quality metrics-not evaluated, Low data payload
[14]	A novel robust watermarking algorithm for encrypted medical image based on DTCWT-DCT and chaotic map	Whole image	Robust watermarking algorithm for encrypted medical images based on dual-tree complex wavelet transform and discrete cosine transform (DTCWT-DCT) and chaotic map is proposed
[15]	Reversible data hiding (RDH), CDM, IWT, ML algorithms	ROI	Optimum PSNR, Improved robustness Sobel operator used for segmentation

(continued)

Table 1 (continued)

References	Contribution	Embedded region	Remarks
[16]	Medical image(s) watermarking and its optimization using genetic programming	Whole image	In this paper, a genetic programming module has been designed to select the most suitable motion vectors for watermark embedding that reduced the perceptual distortion and improve the watermarking capacity

And prime-based encoding algorithm involving two authentication keys generated for region of interest (ROI); one of the key is used in encoding patient's information and acts as a first layer of security and is defined as:

for i = 1 *to M*/2
$e1(i) = (p(i) + p(N) + K(i)) \times key1(roi)$
$e2(N) = (p(i) - p(N) + K(N)) \times key1(roi)$
$N = N - 1$
end

While another key is used for signature generation, it acts as a second layer of security to the MIW system. In the proposed hybrid of MIW, encryption and coding methods are used to ensure privacy, authenticity; hashing technique ensures the integrity; localization is done to avoid information loss during reconstruction; and thereby the system is made reversible.

The proposed encoding algorithm involves the hybrid functions such as addition and multiplication and larger prime numbers. The significance of larger prime number is that, it is difficult to identify the factors of original prime numbers. It is easy to multiply two larger prime numbers but tracing in reverse is a difficult task, Today's supercomputers also take years together to trace the factors of product of larger prime numbers. The following section explains the implementation details of the proposed system followed by its result.

4 Implementation

The proposed system is mainly implemented using frequency domain transform; integer wavelet transform (IWT), singular value decomposition (SVD). Nowadays, IWT is more popular than DWT, as it overcomes the problem of fractional loss, whereas SVD is a geometric transform and its use in MIW techniques helps in the improvement of robustness [17]. Figure 3 shows the procedure used for embedding watermark information in medical image, and the steps of embedding process are explained below:

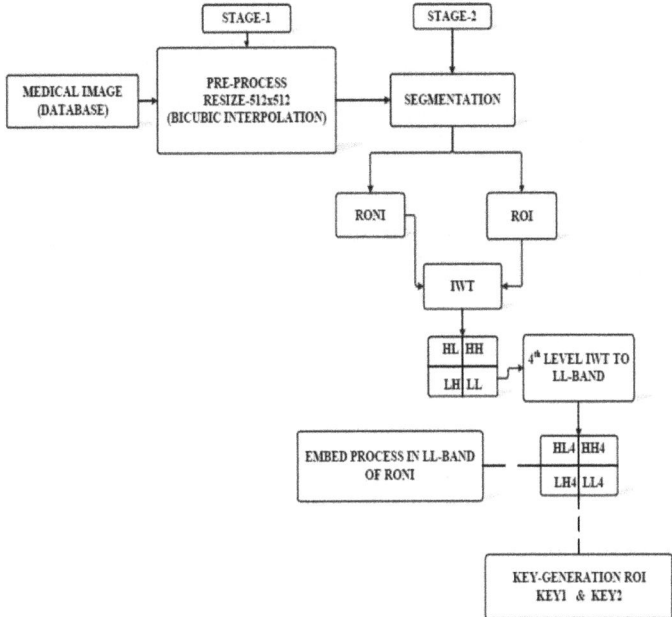

Fig. 3 Block diagram of WM embedding process

4.1 Watermark Embedding Process

In this work, we have used nearly 250 ultra sound (US) and 175 magnetic resonance images (MRI); the MATLAB 2017a, Windows-8 with 3-GB RAM and 500 GB ROM are used for simulation. The medical images (MI) to be secured are read from the database and pre-processed and evaluated for different image sizes. The image is resized to $256 \times 256\ 512 \times 512\ 1024 \times 1024$ using bi-cubic interpolation technique by considering 16-nearest neighbours of a pixel located at (i, j) as shown in Fig. 4 and is implemented using Eq. 2:

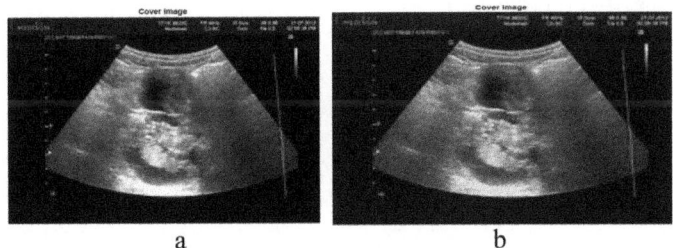

a   b

Fig. 4 **a** Original MI. **b** Resized image using bi-cubic interpolation

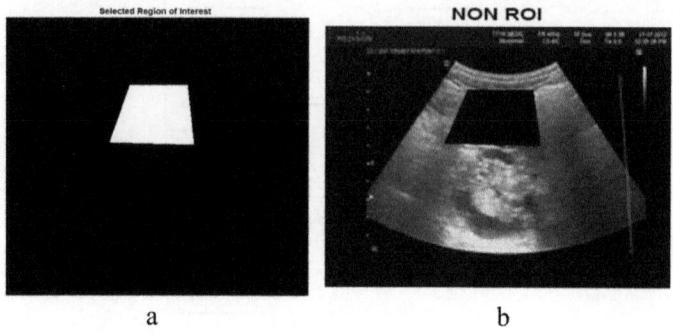

Fig. 5 **a** ROI of MI. **b** RONI of MI

$$p(i, j) = \sum_{m=0}^{3} \sum_{n=0}^{3} a_{mn} i^m j^n \tag{2}$$

The pre-processed image is followed by the next level of pre-processing segmentation stage. Segmentation is one of the main and difficult tasks in medical image processing as a same segmentation technique may not be suitable for all kinds of medical image modalities. However, in medical images, care to be taken to ensure no quantitative loss of information by considering the features such as intensity variation in MI, location in MI, scale variation in MI and orientation in MI during segmentation process. In this paper, we are mainly concentrating only on watermarking process, a simple polygon-based segmentation technique is chosen for simplicity which requires the specification of co-ordinates by the user.

The resized MI is segmented as ROI and RONI polygon-based semi-automatic segmentation technique which needs the specification of co-ordinates either manually or using a simple mouse click to select region of interest and is selected as it is very simple technique to use. The pseudo-code used in this work for segmenting MI is explained below, and segmented images are shown in Fig. 5.

```
---------poly-seg---------------------------
I=read (Pre-processed_image);
[c r]→Specify(co-ordinates)
mask_bin→Poly_seg(I, c,r)
create_buffer→ROI & RONI
if(mask_bin)==1
I→ROI
Else
I→RONI
---------end---------------------------
```

In this work, multiple watermarks are used for ensuring integrity of MIs; three watermarks-related patient are selected and to ensure no tampering in the region of ROI, and key information of ROI is also used in the process of embedding. The

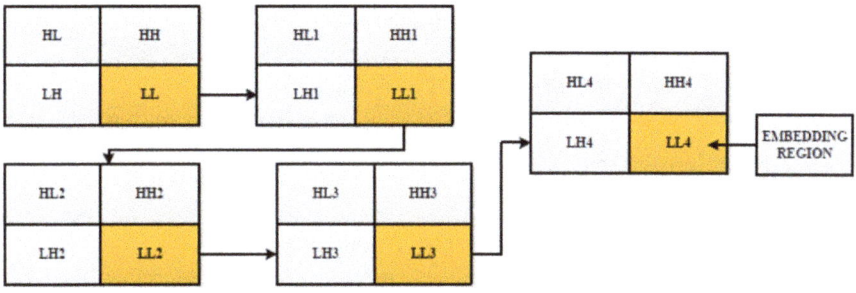

Fig. 6 4th-level decomposition of RONI of MI using IWT

low-frequency region LL of RONI of MI is used for embedding purpose as most of the energy is concentrated in this part, and the magnitude of IWT coefficients is larger and is applied with 4th-level decomposition as shown in Fig. 6.

The 4th-level decomposition is chosen for embedding watermarks in this level ensures robustness and integrity of MI; the lower decomposition level less than two is easily identifiable and vulnerable to various attacks, and also, higher than this 4th level will degrade the quality of images. The middle-frequency components of 4th-level decomposition is also used for embedding purpose, whereas the HH band of MI which embodies edges and texture part of the MIs, embedding in such part may result in information loss. The embedding procedure involves the following procedure:

1. Key information of ROI:

$$key1(roi) = (k \times ll4) + (q \times hl4) = prime\big(sum\big(1d\big(key1_{array}\big)\big)\big)$$

$$key2(roi) = (k \times lh4) + (q \times hh4) = prime\big(sum\big(1d\big(key2_{array}\big)\big)\big)$$

2. Pre-process of watermarks:

$$qr1 = qr\big(rsa\_64\big(proposed_{enc}(wm1)\big)\big)$$

$$qr2 = qr\big(rsa\_64\big(proposed_{enc}(wm2)\big)\big)$$

$$qr3 = qr\big(rsa\_64\big(proposed_{enc}(wm3)\big)\big)$$

$$qrc = cat(qr1, qr2, qr3)$$

'key1' is used to encode watermark information followed by 64-bit RSA encryption, and these encrypted values are encoded into QR-codes and concatenated into a single QR-code.

Embedding of WM in MI using hybrid function:

$$[uw, vw, sw] = svd(qrc);$$

$$512_{bit_{signature}} = (uw, vw, key2)(roi)$$

$$ll_{emb} \rightarrow emb(ll_{mi}\_20th\_bit)$$

$$lh_s \rightarrow \left(\alpha \times lh_s + \left(\frac{\beta}{\alpha} \right) \right) sw$$

$$4th\_level\_iiwt(ll_{emb}, lh_s, hl, hh) \rightarrow LL$$

$$wmi \rightarrow iiwt(LL, LH, HL, HH)$$

The proposed medical image watermarking is evaluated with traditional system and also evaluated against Huffman and arithmetic coding algorithms which is discussed in the next section.

5 Results and Discussion

The proposed system is simulated using MATLAB-2017a with Windows-8 of configuration 500 GB ROM and 3 GB RAM. The performance is evaluated using statistical quality metrics such as PSNR, MSE, NAE, NCC, false positives and negatives.

Figure 7 shows multiple WMs and QR-code generated for encoded-encrypted information and the concatenated QR-code. Figure 8 shows the result of MI, watermarked image and reconstructed images.

Figure 9 shows the result of extraction of QR-code and generating original WM using proposed decoding algorithm and 64-bit RSA decryption.

The results of proposed encoding, Huffman and arithmetic encoding algorithm for the proposed MIW system are shown in Fig. 10. The Huffman and arithmetic coding are lossless coding algorithms. The main limitation of these algorithms is that it requires the knowledge of probability occurrence of symbols which is overcome with proposed prime-based encoding algorithm as prime number plays an important role in securing information.

Table 2 shows the comparison results for all three encoding systems between medical image and watermarked image for CI and WM for different sizes, and Fig. 11 shows performance graph.

Table 3 shows the comparison results for all three encoding systems between medical image and reconstructed MI for different sizes and Fig. 12 the corresponding PSNR.

Fig. 7 Multiple watermarks, respective QR-code and concatenated QR-code

Table 4 shows the comparison results between original WM of different sizes and extracted WM, and Fig. 13 shows the PSNR of extracted watermarks from CIs of different sizes.

The encoding algorithms are compared in terms of compression ratio, compression factor, shrinkage, memory space and bits per pixel (bpp). These parameters are calculated using the following Eqs. 3–6, and Table 5 shows the respective results of all three coding techniques:

$$\text{CR} = \frac{\text{Size of MI after encoding}}{\text{Size of MI before encoding}} \tag{3}$$

$$\text{CF} = \frac{\text{Size of MI before encoding}}{\text{Size of MI after encoding}} \tag{4}$$

$$\text{Shrinkage} = \frac{\text{Size of MI before encoding} - \text{Size of MI after encoding}}{\text{Size of MI before encoding}} \tag{5}$$

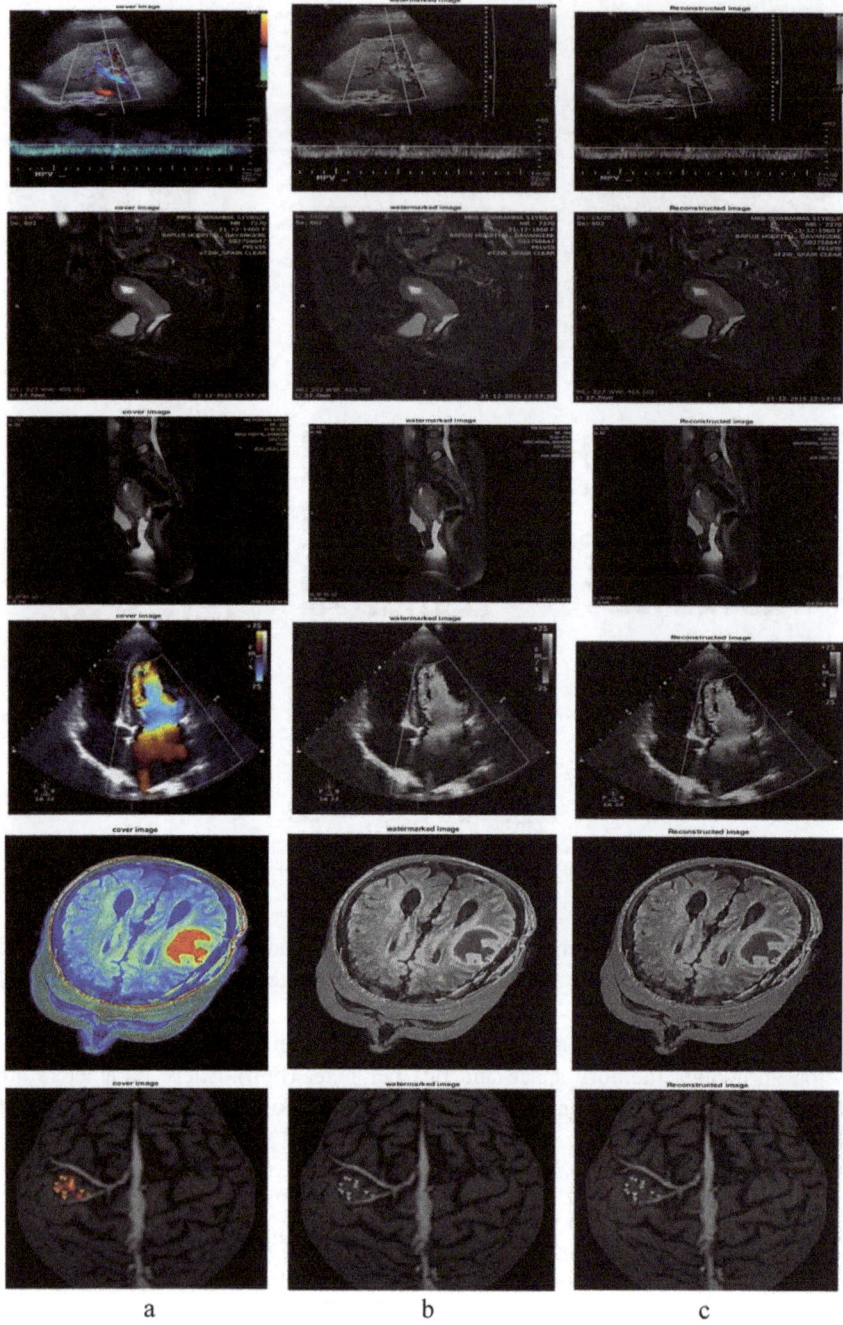

Fig. 8 **a** Original MI. **b** Watermarked image. **c** Reconstructed image

EXTRACTED QR

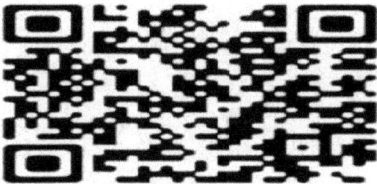

PATIENT DETAIL   PATIENT AADHAAR   HOSPITAL LOGO

text1 =

'Hospital: City centre care

Doctor In charge: Manjunatha K V

Name of the Patient:Ratnamma

Date Of Birth: 19/02/1965

Age: 57 years

Date of Admission: 19/07/2020

Patient ID: CC_IP_77586_022020

Health Issue: KIDNEY_TUMOR

Precaution: Operation'

HOSPITAL LOGO

Patient_ID_Card_Details

Name   : XXXXXX
Father Name   : YYYYYY
Mother Name   : ZZZZZZZ
Year of Admission: 2019
DEPT   : CSE
Register No   : GGGGG
Area of Research : Signal Processing
Papers Published :
        National Journal   : 08
        International Journal : 15
        Conferences   : 20
Patents   : 02
Workshop organized   : 05
Sathyabama University, Chennai
Research scholar signature
Date
Place

Fig. 9 Results of QR-code extraction and reconstruction of WMs

$$bpp = \frac{\text{Size of encoded WM}}{\text{Size of ll of CI}} \qquad (6)$$

MIW is also implemented with DWT using alpha blending embedding and extraction algorithm defined by Eqs. 7 and 8:

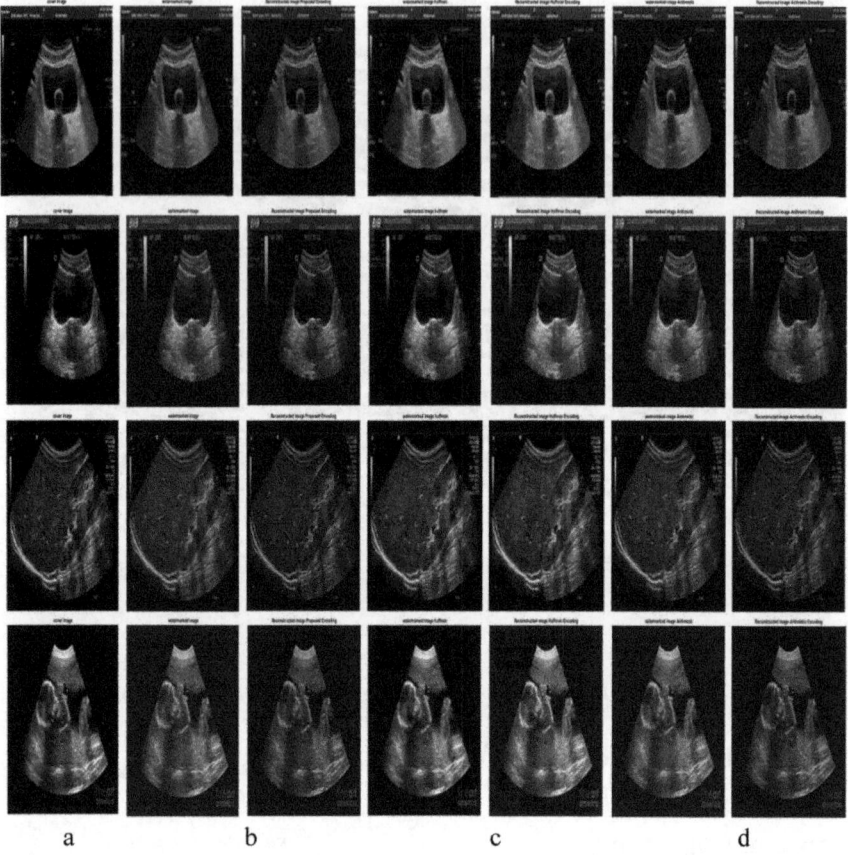

Fig. 10 **a** MI. **b** Proposed encoding. **c** Huffman coding. **d** Arithmetic coding

$$\text{wm}_{\text{im}} = k * \text{ll}_{\text{ci}} + q * \text{ll}_{\text{wm}} \tag{7}$$

$$\text{wm}_{\text{extr}} = k * \text{ll}_{\text{wm}} - q * \text{ll}_{\text{ci}} \tag{8}$$

The results of this method are as shown in Fig. 14 and are evaluated with quality metrics against the proposed system as shown in Table 6 and Fig. 15.

The use of DWT in MIW results in significant information loss and also very sensitive to shifting of input which results in large variations in the filter coefficients and poor directivity; i.e. this wavelet is unable to distinguish in the phase angles. Both MIW systems are subjected to various image attacks and Figs. 16 and 17 shows the results of reconstruction of CI and patient detail after various filter attacks for the proposed system and Table 7 shows, the robustness of both systems against removal of watermark which is evaluated using stirmark bench tool [18] in terms of PSNR.

Table 2 Performance evaluation between MI and WMI

$\beta = 0.5$ to 1 $\alpha = 1$ to 2	PSNR	MSE	NCC	NAE
CI size $= 256 \times 256$ *and WM size* $= 40 \times 40$				
Proposed	77.28	0.001	0.9078	0.0897
Huffman	70.80	0.041	0.899	0.0991
Arithmetic	74.32	0.027	0.901	0.0912
CI size $= 512 \times 512$ *and WM size* $= 40 \times 40$				
Proposed	77.117	0.001	0.9942	0.088
Huffman	70.751	0.0345	0.979	0.099
Arithmetic	75.25	0.02	0.985	0.09
CI size $= 1024 \times 1024$ *and WM size* $= 40 \times 40$				
Proposed	78.003	0.001	0.99	0.0917
Huffman	71.42	0.0271	0.901	0.094
Arithmetic	76.43	0.03	0.908	0.0932

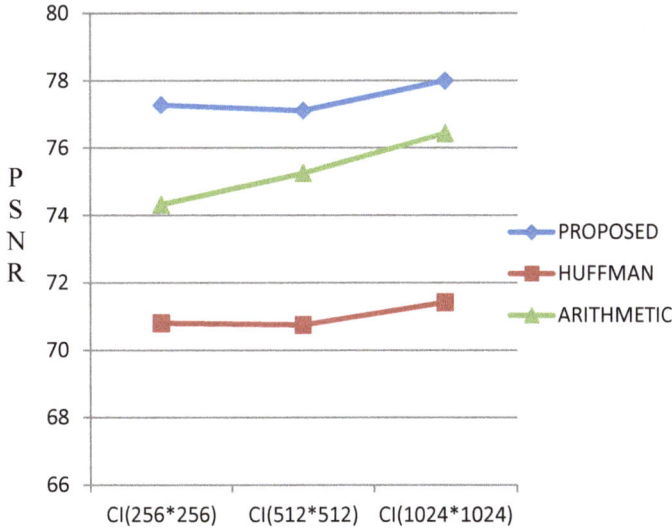

Fig. 11 Performance of proposed, Huffman and arithmetic coding

6 Conclusion and Future Scope

The proposed MIW system results in reduced data payload capacity even for multiple watermarks due to the use of QR-Code and proposed encoding algorithm. The use of IWT and SVD ensures no loss of information and embedding in LL band with hybrid embedding function increase the robustness of the system; moreover, the use

Table 3 Performance evaluation MI and reconstructed MI

$\beta = 0.5$ to 1 $\alpha = 1$ to 2	PSNR	MSE	NCC	NAE
CI size = 256 × 256 and WM size = 40 × 40				
Proposed	76.3	0.001	0.9078	0.0997
Huffman	69.01	0.05	0.809	0.0991
Arithmetic	73.2	0.037	0.89	0.092
CI size = 512 × 512 and WM size = 40 × 40				
Proposed	77.117	0.001	0.9942	0.099
Huffman	69.12	0.043	0.95	0.09
Arithmetic	74.37	0.024	0.96	0.092
CI size = 1024 × 1024 and WM size = 40 × 40				
Proposed	77.003	0.001	0.90	0.997
Huffman	70.38	0.034	0.891	0.092
Arithmetic	75.37	0.0345	0.94	0.099

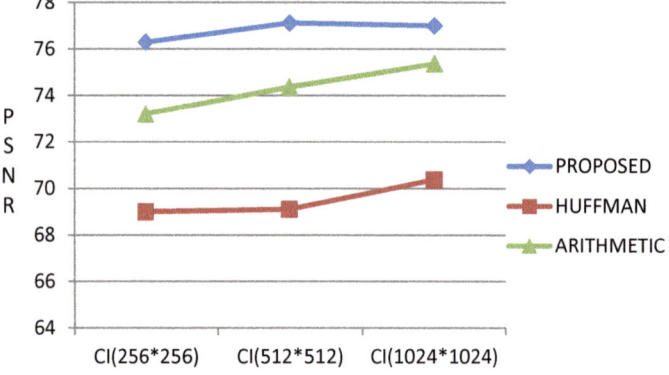

Fig. 12 PSNR of proposed, Huffman and arithmetic coding (MI and reconstructed MI)

of digital signature along with ROI key helps to identify the tampering in MIs. The use of prime-based encoding algorithm not only proves the reduction in size but also ensures the security to the PHR and also results in perceptually good quality images, good compression ratio and less memory space. In this work, ROI is preserved and does not go any modification, instead key is generated using its frequency components and used in signature generation process, thereby it ensures integrity of ROI of MIs. The proposed system is also compared with DWT MIW algorithm, and both the methods are evaluated against various image attacks and are been observed that the encoding-based MIW is more resistant than DWT-alpha blending MIW system The encoding and encryption algorithms may extend for longer bit length to make system more robust, and it is necessary to identify the common and effective segmentation

Table 4 Original WM and extracted WM QR-code

$q = 0.5$ to 1 $k = 1$ to 2	PSNR	MSE	NCC	NAE
CI size = 256 × 256 and WM size = 40 × 40				
WM1	86.93	0.00013	0.999	0.081
WM2	83.2	0.00013	0.999	0.081
WM3	84.02	0.00013	0.999	0.081
CI size = 512 × 512 and WM size = 40 × 40				
WM1	94.99	0.00020	0.999	0.003
WM2	93.71	0.00020	0.999	0.003
WM3	92.3	0.00020	0.999	0.003
CI size = 1024 × 1024 and WM size = 40 × 40				
WM1	100	0.000005	1	0.002
WM2	100	0.000005	1	0.002
WM3	100	0.000005	1	0.002

Fig. 13 PSNR of WMs extracted from CIs of different size

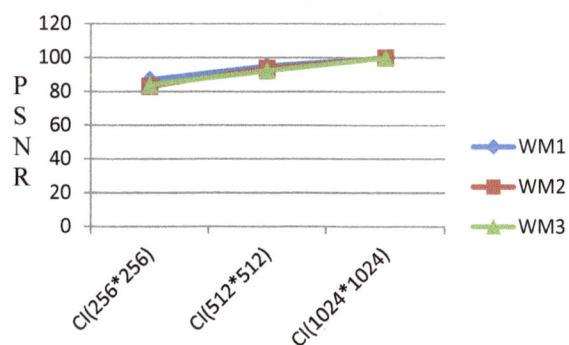

Table 5 Comparison of encoding schemes

Parameters	Huffman	Arithmetic	Proposed
Compression ratio (CR)	97.03	149.12	150.177
Compression factor (deviation) (CF)	0.0103	6.705×10^{-3}	6.658×10^{-3}
Saving percentage (shrinkage)	98.969	99.329	99.335
Memory space	Low	Very low	Very low
Bits per pixel (bpp)	0.4512	0.343262	0.29102

technique for all types of MI. The segmentation technique based on deep learning algorithms may be one of the possible solutions to avoid the loss of information of MI. Further, the use of EHR in India has to be initiated by the governments in all

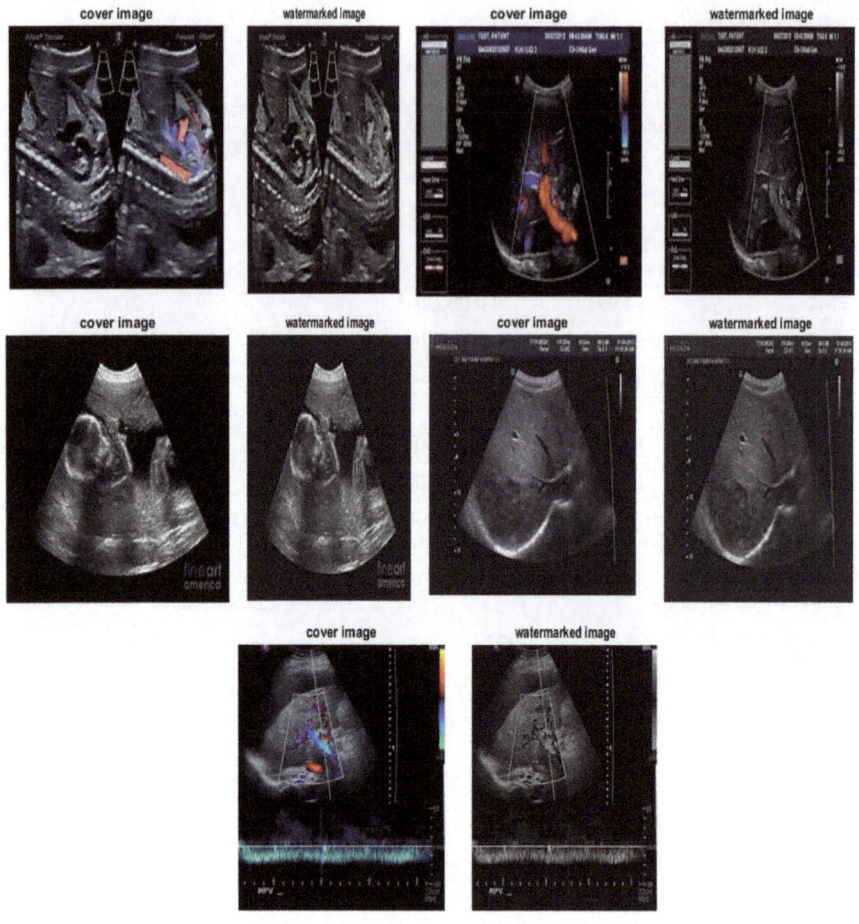

Fig. 14 Results of DWT MIW system using alpha blending algorithm

Table 6 Comparison of DWT and proposed MIW system

Method	CI	PSNR (dB)	SSIM	MSE	NCC
DWT using alpha blending algorithm	Img1	49.98	0.621	0.652	0.8987
	Img2	49.86	0.615	0.671	0.782
	Img3	51.12	0.701	0.502	0.693
	Img4	50.39	0.690	0.594	0.701
	Img5	50.185	0.725	0.623	0.723
Proposed system using hybrid embedding function	Img1	77.33	0.9524	0.0012	0.899
	Img2	76.37	0.9432	0.0015	0.971
	Img3	77.716	0.9563	0.0011	0.987
	Img4	78.04	0.9765	0.00102	0.988
	Img5	76.70	0.9812	0.00139	0.998

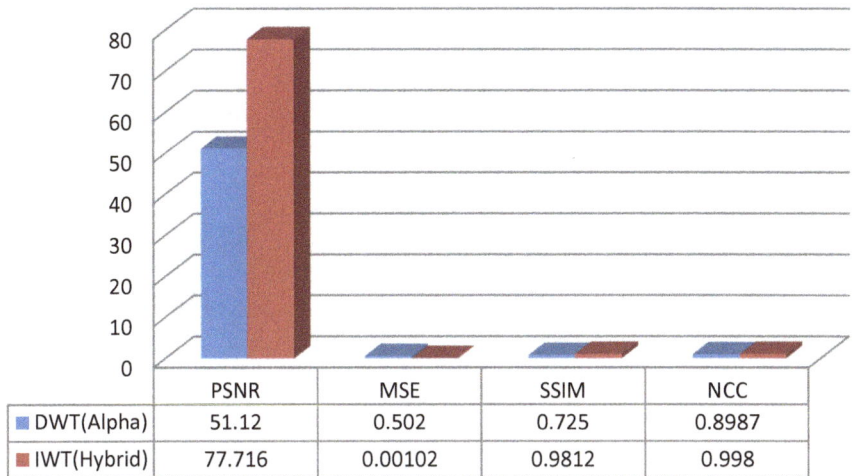

	PSNR	MSE	SSIM	NCC
■ DWT(Alpha)	51.12	0.502	0.725	0.8987
■ IWT(Hybrid)	77.716	0.00102	0.9812	0.998

Fig. 15 Results of DWT and proposed MIW system

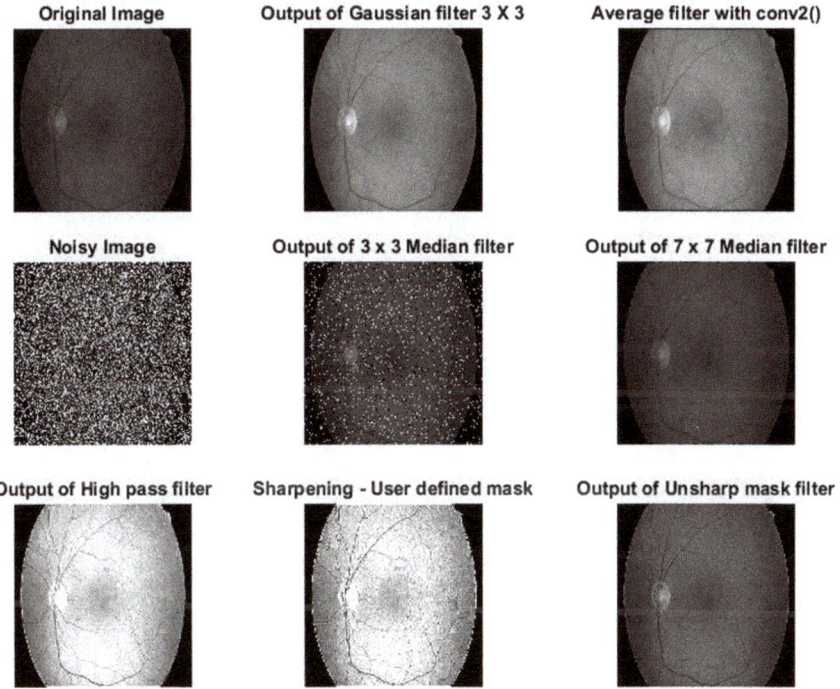

Fig. 16 Reconstruction of CI after filter attacks in proposed MIW system

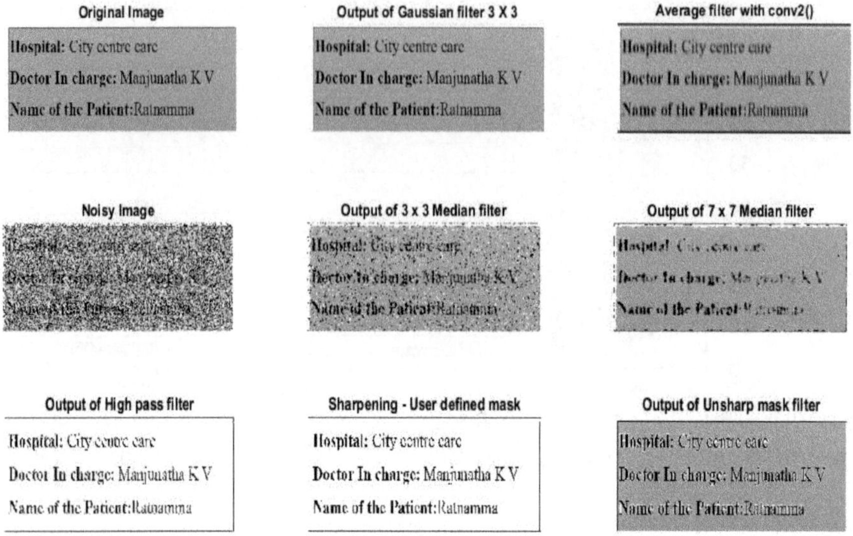

Original Image	Output of Gaussian filter 3 X 3	Average filter with conv2()
Hospital: City centre care Doctor In charge: Manjunatha K V Name of the Patient:Ratnamma	Hospital: City centre care Doctor In charge: Manjunatha K V Name of the Patient:Ratnamma	Hospital: City centre care Doctor In charge: Manjunatha K V Name of the Patient:Ratnamma
Noisy Image	Output of 3 x 3 Median filter	Output of 7 x 7 Median filter
Output of High pass filter	Sharpening - User defined mask	Output of Unsharp mask filter
Hospital: City centre care Doctor In charge: Manjunatha K V Name of the Patient:Ratnamma	Hospital: City centre care Doctor In charge: Manjunatha K V Name of the Patient:Ratnamma	Hospital: City centre care Doctor In charge: Manjunatha K V Name of the Patient:Ratnamma

Fig. 17 Reconstruction of WM after filter attacks in proposed MIW system

Table 7 Robustness evaluation using stirmark bench tool

Types of attacks		IWT-SVD	DWT-SVD
		PSNR (dB)	
Jpeg		74.32	48.34
Filter		Filter size	
Gaussian [0.5]			
	[3 3]	74.7833	43.452
	[6 6]	76.5902	41.012
	[10 10]	77.514	40.25
Median	[3 3]	79.7029	43.46
	[6 6]	78.913	40.39
	[10 10]	79.466	43.76
Convolution	[3 3]	81.93	49.37
	[6 6]	87.97	44.241
	[10 10]	85.997	41.25
Wiener	[3 3]	74.36	48.6
	[6 6]	76.69	47.54
	[10 10]	77.59	49.52
Black and white noise		78.56	50.289
Sharpen		81.109	48.756
Rotation with ± 45°		76.32	39.58
Rotation with ± 90°		78.68	39.362

states for the beneficiary of patients to save their time, money and more than this, early precaution to save anyone's life with proper measures.

References

1. Eswaraiah R, Sreenivasa RE (2014) Medical image watermarking technique for accurate tamper detection in ROI and exact recovery of ROI. Int J Telemed Appl 1–10
2. Seenivasagam V, Velumani R (2013) A QR code based zero-watermarking scheme for authentication of medical images in teleradiology cloud. Comput Math Methods Med 1–16
3. Ahmed M, Hamed T, Obimbo C, Dony R (2013) Improving the security of the medical images. Int J Adv Comput Sci Appl (IJACSA) 4(9):137–146
4. Balamurugan G, Jayarraman KB, Arulalan V (2014) A survey on medical image watermarking techniques. Int J Comput Sci Netw (IJCSN) 3(5):309–317
5. Umaamaheshvari A, Thanuskodi K (2012) Survey of watermarking algorithms for medical images. Int J Eng Trends Technol 3(3):401–410
6. Manivannan M, Suseendran G (2017) A review of watermarking requirements, techniques, documents, human perception and applications for medical images. Int J Innov Res Appl Sci Eng 1(2):58–65
7. Hussain N, Wageeh B, Colin B (2013) A review of medical image watermarking requirements for teleradiology. J Digit Imaging 26(2):326–343
8. Arijit KP, Nilanjan D, Sourav S, Achintya D, Sheli SC (2013) A hybrid reversible watermarking technique for color biomedical images. In: IEEE international conference on computational intelligence and computing research, Enathi, India, Dec 26–28, pp 1–6
9. Umamageswari A, Suresh GR (2014) Novel algorithm for secure medical image communication using ROI based digital lossless watermarking and DS. Int J Appl Eng Res 9(22):12163–12176
10. Adiwijaya, Faoziyah PN, Permana FP, Wirayuda TAB, Wisesty UN (2013) Tamper detection and recovery of medical image watermarking using modified LSB and Huffman compression. In: Second International conference on informatics & applications (ICIA), Lodz, Poland, Sept 23–25, pp 129–132
11. Bilal H, Ramsha A, Bo L, Omar H (2019) An imperceptible medical image watermarking framework for automated diagnosis of retinal pathologies in an eHealth arrangement. IEEE Access 7:69758–69775
12. Abdallah S, Alti A, Laouamer L (2018) Toward a secure and robust medical image watermarking in untrusted environment. In: The international conference on advanced machine learning technologies and applications (AMLTA2018), Advances in intelligent systems and computing, vol 723, pp 693–703
13. Pooja PM, Sreeraj R, Fepslin A, Suthendran K (2018) Combined cryptography and digital watermarking for secure transmission of medical images in EHR systems. Int J Pure Appl Math 118(8):265–269
14. Jing L, Jingbing L, Jieren C, Jixin M, Naveed S, Baoru H, Qiang G, Yang A (2019) A novel robust watermarking algorithm for encrypted medical image based on DTCWT-DCT and chaotic map. Comput Mater Continua CMC 61(2):889–910
15. Bin M, Bing L, Xiao-Yu W, Chun-Peng W, Jian L, Yun-Qing S (2019) Code division multiplexing and machine learning based reversible data hiding scheme for medical image. Secur Commun Netw 2019:1–9
16. Rafi UH, Hani AA (2019) Medical image(s) watermarking and its optimization using genetic programming. (IJACSA) Int J Adv Comput Sci Appl 10(4):163–169

17. Musrrat A, Chang WA, Millie P, Patrick S (2016) A reliable image watermarking scheme based on redistributed image normalization and SVD. Discrete Dyn Nat Soc 1–15
18. Fabien APP, Ross JA, Markus GK (1998) Attacks on copyright marking systems. In: Aucsmith D (ed) Information hiding, Second international workshop, IH'98, Portland, Oregon, USA, 15–17 Apr 1998, Proceedings, LNCS 1525, Springer. ISBN 3-540-65386-4, pp 219–239

Chapter 73
A Compact Multiband CPW Feed Microstrip Fractal Antenna for X-Band and Ku-Band Satellite Communication Applications

E. Kusuma Kumari, Purnima K. Sharma, S. Murugan, and D. Rama Devi

1 Introduction

The discovery of radio waves led to the evolution of wireless communications. Due to the upgradation of technology from time to time, there is a rapid growth in the usage of satellite and radar communications. So, there is a huge demand for ultra-wideband, multiband, miniaturization of antennas all over the world. In the present-day scenario, antenna is the major component in those devices which are used for wireless communications. Antenna has the capability of creating a medium for transmitting or receiving wave signal from one vicinity to other in a free space. As the trend of wireless communications is growing in a good pace, there is not only an increasing probability of using distinct frequency bands in the same device for numerous service applications [1], but also a compact antenna design for the integrated device. It will be an inspiration for those engineers who work on the antennas for developing miniature [2] printed antennas which have the ability for assisting various communication specifications. As today's world needs a cost-effective model with best quality, the demand for planar configuration, low profile, little volume, low cost and ultra-wideband multi-frequency planar antennas has been increased. Compact microstrip patch antennas satisfy all the prerequisites of low profile, low cost and printable circuit technology, so more research work is going on this. IEEE Standard defined the microwave frequency bands X-band with frequency range from 8.0 to 12.0 GHz and Ku-band with frequency range from 12.0 to 18.0 GHz, respectively [3]. Almost all C-band communication satellites use 3.7–4.2 GHz frequency band for their downlinks and 5.925–6.425 GHz frequency band for their uplinks. While considering the applications of X-band, due to its short wavelength, it accommodates high-resolution

E. Kusuma Kumari · P. K. Sharma (✉) · S. Murugan
ECE Department, Sri Vasavi Engineering College, Tadepalligudem, India

D. Rama Devi
ECE Department, MVGR Engineering College, Vizianagaram, India

© The Author(s), under exclusive license to Springer Nature Singapore Pte Ltd. 2021   981
S. Kumar et al. (eds.), *Proceedings of International Conference on Communication and Computational Technologies*, Algorithms for Intelligent Systems,
https://doi.org/10.1007/978-981-16-3246-4_74

imaging, short range tracking, missile guidance, radar airborne intercept [4–6], and majority of the time, it is used for radar contact ranges. Satellite altimetry and high-resolution mapping also use the Ku-band. Within the ranges of 12.87–14.43 GHz, the Ku-band is also used to monitor the satellite.

With regards to this paper, design of a compact multiband microstrip antenna using fractal geometry was presented. Since it resonates in the X-band and Ku-band frequency ranges, the proposed antenna may be ideal for satellite and radar communication systems. As a result, it integrates the X and Ku microwave frequency bands and spans all ranges from 10 to 14 GHz. The designed antenna offers a dual band by using the fractal geometry by introducing reactive loading technique in each iteration by etching the circular slot and rectangular slot on the patch antenna with the microstrip feed line. In the third iteration, the coplanar waveguide feeding technique [7–9] was given to radiating element along with the reactive loading technique, [10] and a triple-band response was reported by the radiating element.

2 Literature Survey

Abol Fazl Azari (Member, IEEE) has worked with just two iterations of fractal geometry implemented antenna structure and implemented a new fractal antenna for ultra-wide and multiband applications, with a gain of 2 dB [11].

Fajar Wahyu Ramadhan has developed a multiband microstrip fractal antenna for radar applications in S-band and C-band. The antenna has a return loss of S11 < 9.54 dB. The antenna is made of Epoxy FR-4 substrate with $r = 4.3$ relative permittivity and $h = 1.6$ mm thickness. The antenna was designed to resonate at three different working frequencies: 2.24 GHz, 3.46 GHz and 4.86 GHz. The maximum bandwidth obtained is 140 MHz [3].

The array is initially planned for service across the UHF, L, S and lower-C bands, according to a paper presented by Markus H. Novak, Member, IEEE, and John L. Volakis, Fellow, IEEE, on Ultra wideband Antennas for Multiband Satellite Communications at UHF Frequencies. This is accomplished by a dual-offset, split unit cell with reduced inter-feed coupling [5].

H. Zahra's paper on Feed Analysis of Tri-Patch Multiband Antenna for Satellite Communication shows us a compact multiband antenna that can work in the S, C, X, Ku and K bands of the electromagnetic spectrum and is ideal for radar and satellite applications. Antenna is single-layer, lightweight and fed with a coaxial probe feed. The location of the feed has been configured, and a report has been prepared [6].

Azzeddin Naghar and others have suggested ultra-wideband and tri-band antennas for satellite applications in the C, X and Ku bands. The use of ultra-wideband and tri-band antennas in the C, X and Ku bands for satellite applications is proposed. The notched-band characteristics are used to achieve the multiband response by generating rejected bands. The radiated patch is etched with two opposite U-slots to achieve this. This property gives you a lot of flexibility when it comes to choosing the frequency bands you like based on the total length and width of the U-slot [4].

Lin Shu, Wu Zhongda, Mao Wandong, Liu Cheng, Cai Runnan, Chen Lijia, Qiu Jinghui and Wang Jinxiang presented multiband triangle fractal nesting printed monopole antenna. Four nesting triangles with different vertex angles and heights make up this antenna. There has been a recording of a multiband of four bands (2.4 GHz, 3.5 GHz, 5.2 GHz, and 5.8 GHz) [12].

A quarter-wavelength or half-wavelength resonant slot cut at a suitable location within the patch edit one more resonant mode close to the patch's resonance frequency, resulting in a dual-band response. Dual- and triple-band responses were obtained with the proper adjustment of the coupling between the slot and the patch's modes. A dual-band rectangular microstrip antenna can be made by cutting a circular slot within the patch. Dual-band rectangular microstrip antennas using these methods are discussed in this article. Furthermore, multiband rectangular microstrip antennas are developed by integrating these dual-band techniques of slots with multiple iterations. Rectangular microstrip antennas with triple and quad frequencies are proposed. The antenna's size and weight must be considered in order for it to be practical. It is preferable to have a lightweight, simple multiband antenna that covers all four wireless frequency bands. Furthermore, being able to monitor the antenna bandwidth through various frequency bands independently is desirable.

3   Antenna Design

3.1   Conventional Microstrip Line Feed Simple Patch Antenna

Initially, the conventional microstrip line feed simple patch antenna was designed on dielectric substrate of FR4 Epoxy with the relative permittivity $\varepsilon_r = 4.4$ and the dimensions of 27.15 mm $\times$ 27.15 mm $\times$ 1.6 mm. The length and width of patch are 20 mm $\times$ 13.8 mm. Microstrip feed line of 50 Ω was connected to the center of the patch. During the simulation, the feed location was adjusted to attain almost entire impedance matching to the 50 Ω source as shown in Fig. 1. To validate the designs, structures are modeled by using HFSS software tool. This structure reports the single band of operation resonating at 6.8 GHz (Table 1).

3.2   Design of Patch Antenna with Circular Slot on Patch (Firstst Iteration of Fractal Geometry)

By observing the improvement in patch antenna performance after implementing the fractal geometry and the improvement in terms of bandwidth and multi-frequency operations without compromising much gain, in this paper, the second antenna was proposed and designed with circular slot on patch as a first iteration. The dimensions of circular slot were shown in Fig. 2. The modeled antenna was simulated, and

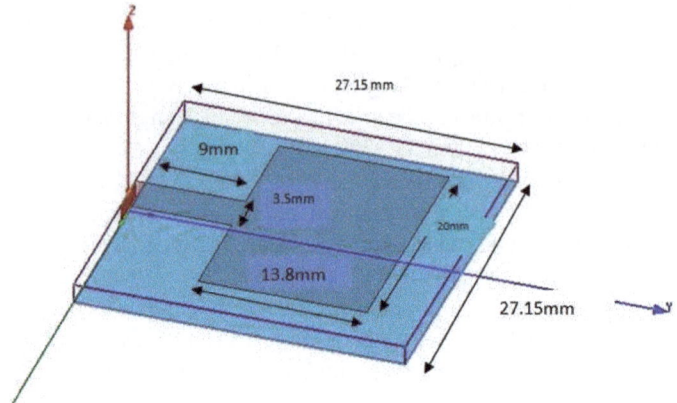

Fig. 1 Dimensions of substrate and patch

Table 1 Antenna design specifications

Antenna parameter	Value
Substrate length	27.15 mm
Substrate width	27.15 mm
Height of substrate	1.6 mm
Dielectric permittivity	4.4
Patch length	20 mm
Patch width	13.8 mm
Feed line length	9 mm
Feed line width	3.5 mm
Radius of circle 1	6 mm
Radius of circle 2	4.3 mm
Radius of circle 3	2.45 mm
Fractal rectangle 1	9 mm × 8 mm
Fractal rectangle 2	7.1 mm × 5 mm

the improved results were observed in terms of bandwidth without losing gain and radiation efficiency. The designed antenna has reported dual-band response by introducing the circular slot in comparison with conventional microstrip line feed patch antenna. The designed antenna was 41% miniaturized compared to conventional patch antenna.

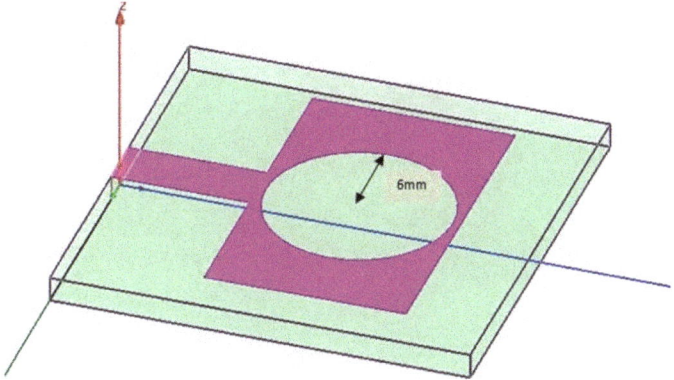

Fig. 2   Design of patch with circular slot

3.3   *Design of Patch Antenna with Second Circular Slot (Second Iteration of Fractal Geometry)*

In this proposed second configuration of antenna as shown in Fig. 3, the fractal geometry was implemented as second iteration in such a way that circular slot was taken on an embedded rectangle. With this second iteration, a better performance analysis was observed through simulation results. After simulation, very good results were reported in terms of dual band, gain, return loss and bandwidth. This antenna offers the dual-band operation, in which the resonating frequencies are in the range of C-band and X-band. The designed antenna reports the 21.4% miniaturization compared to conventional single patch without fractal geometry.

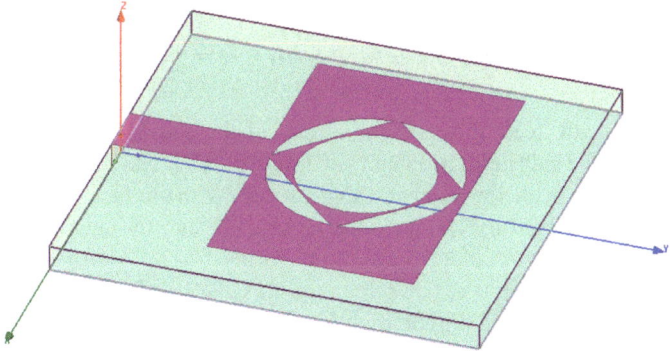

Fig. 3   Design of patch antenna with second circular slot (second iteration of fractal geometry)

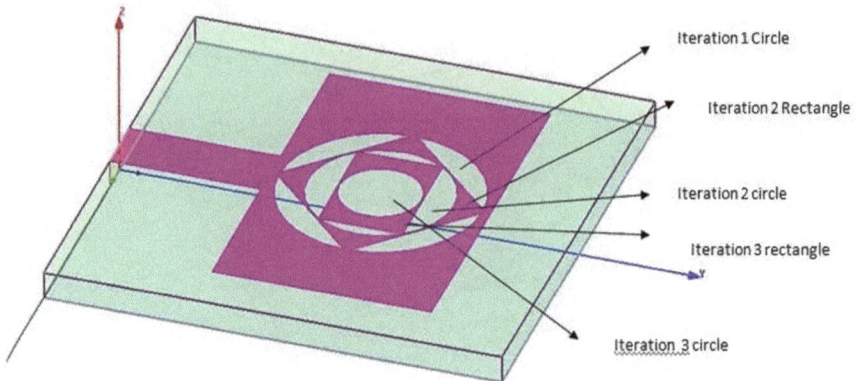

Fig. 4 Design of patch antenna with third circular slot (third iteration of fractal geometry)

3.4 Design of Patch Antenna with Third Circular Slot (Third Iteration of Fractal Geometry)

Figure 4 shows the third configuration of proposed structure in which third iteration of fractal geometry was implemented. A circular shape of 2.45 mm radius slot was done. After simulation, it displays the better performance in terms of multiple resonant frequencies. This antenna offers triple-band operation wherein the resonating frequencies are in the range of X-band and Ku-band. Good performance parameters were obtained without compromising the radiation efficiency and gain. Designed antenna offers good bandwidth range compared to earlier proposed configurations. This structure is 15.58% miniaturized compared to conventional patch antenna.

3.5 Design of Previous Structure with CPW Feeding

Coplanar waveguide feeding technique is one of the promising technologies for antenna structures which helps in enhancing the bandwidth. The enhanced bandwidth of antenna can be suitable for wideband applications. With this fact, the CPW feeding technique was employed for the previous reported design of antenna as shown in Fig. 5.

The fourth configuration of an antenna with CPW feeding was simulated by using HFSS software. After simulation, it was observed that the proposed structure offers quad band of operation which is resonating at four different frequencies. All the resonating frequencies are in the frequency band range of 12–18.2 GHz, i.e., X-band and Ku-band frequency range. This structure shows the better performance compared to previous configurations as such it offers quad band, i.e., single antenna is suitable for four different multi-frequency operations without compromising the gain

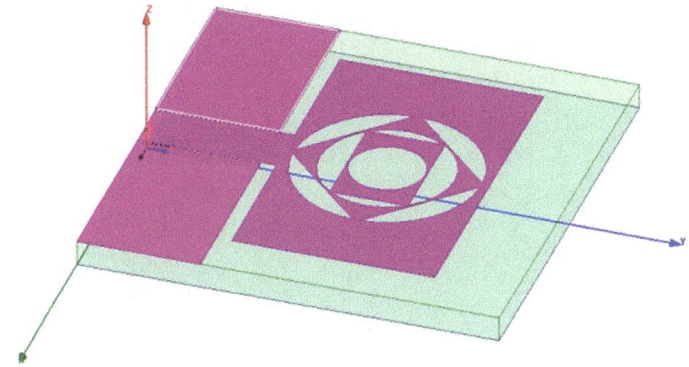

Fig. 5 Third iteration of an antenna with CPW feeding

and radiation efficiency. This fourth configuration of designed antenna was 15.58% miniaturized as compared to original patch antenna without fractal geometry.

4 Results and Discussions

All the above proposed antenna designs using fractal geometry were simulated using HFSS software. The simulated results are in terms of band of operations, radiation efficiency, gain, return loss, bandwidth was observed. Figures 6, 7, 8, 9 and 10 show the simulated results of iteration-wise proposed structures.

The reported results from all the configurations were shown above, and all the performance parameters are comprehensively tabulated and shown in Table 2. From the comparison Table 2, it was clearly understood that the implementation of Fractal geometry along with the reactive loading technique by taking the slots in 3 iterations, the antenna performance was improved in terms of multiband operation without much compromising the percentage of radiation efficiency and gain. Initially, the antenna was fed by microstrip line feeding technique, and results were observed. For the last iteration of antenna structure, CPW feeding technique was connected. Better results were obtained compared to earlier results of designed antenna configurations. Miniaturization of antenna size was also achieved with fractal geometry, and percentage of miniaturization for all antenna configurations is mentioned in comparison Table 2.

5 Conclusion

A simple compact multiband antenna for satellite applications at X-, Ku- and C-bands is proposed in this article. By generating rejected bands, the notched-band traits are

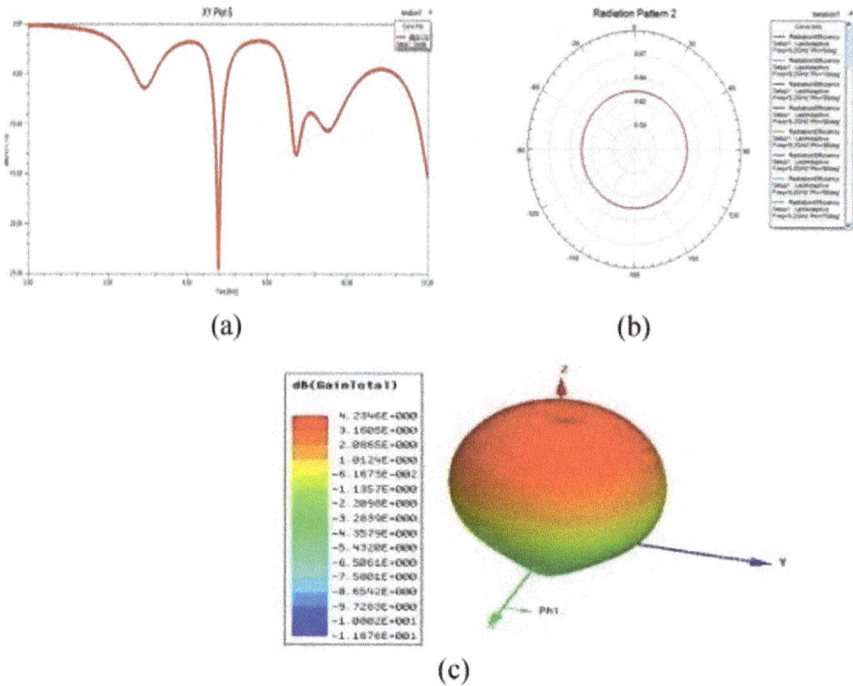

Fig. 6 **a** Return loss, **b** radiation efficiency, **c** 3D polar plot of single patch antenna

using to attain the multiband response. It is obtained by etching repetitive circular slots in the radiated patch in a specific manner. The comparison of performance parameters of all configurations of the structures was done. From the comparison, it was observed that the last iteration structure with CPW feeding shows the better results and offers quad band operation. Quad band operation, i.e., the single antenna structure was resonating at four different frequencies, and these frequencies are in the range of X-band and Ku-band. Fractal geometry and reactive loading technique are used to achieve compactness and multiband operation. The suggested antennas here are ideal for radar and for many wireless applications as well as for satellite applications in the X and Ku bands due to their small size, ease of fabrication and reliable far-field radiation patterns across their operating frequency range.

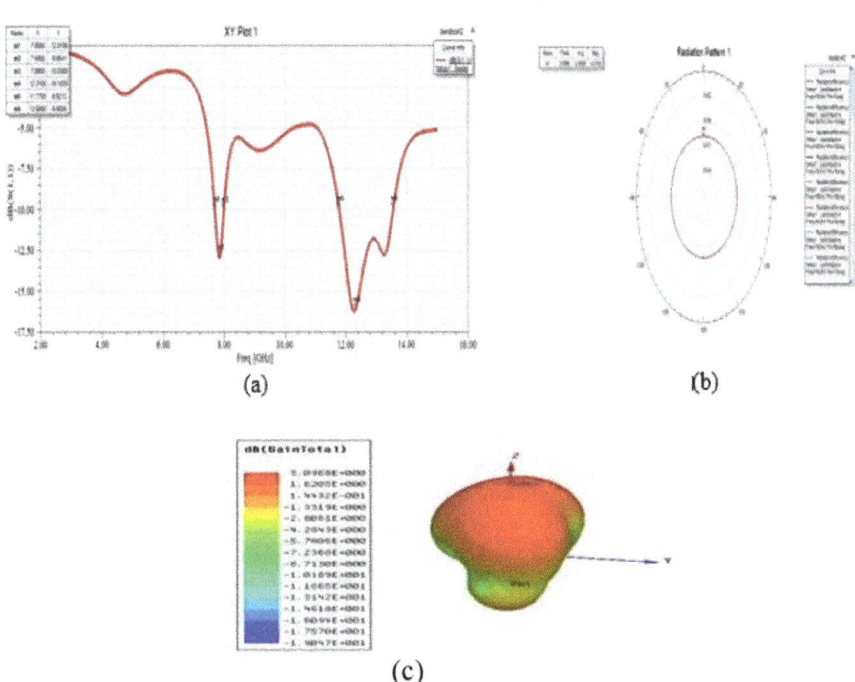

(a)   (b)

(c)

Fig. 7 a Return loss, **b** radiation efficiency, **c** 3D polar plot of gain of first iteration of proposed antenna

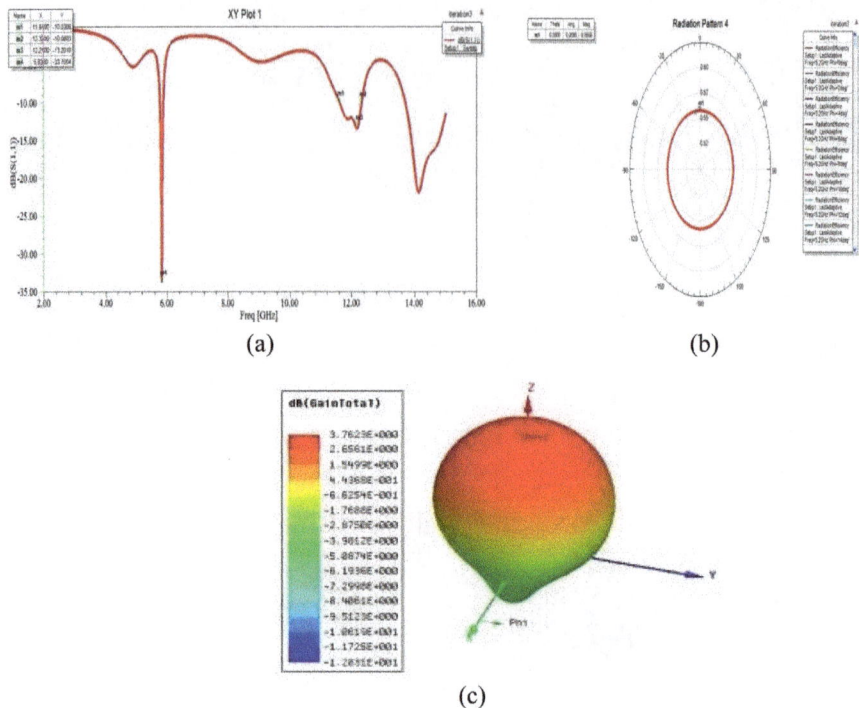

(a)

(b)

(c)

Fig. 8 **a** Return loss, **b** 3D polar plot of gain, **c** radiation efficiency of second iteration proposed antenna

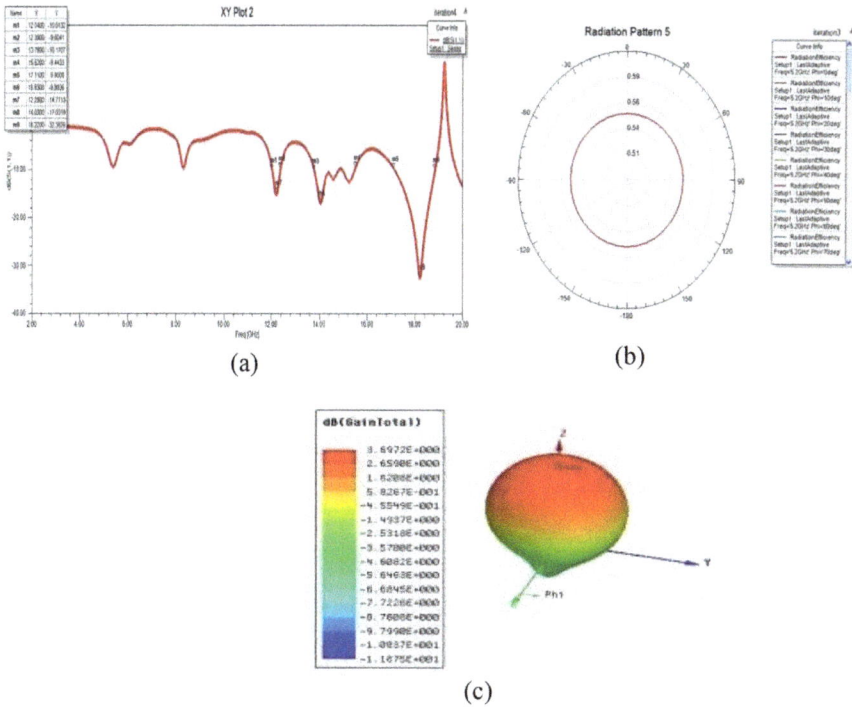

Fig. 9 **a** Return loss, **b** radiation efficiency, **c** 3D polar plot of gain of third iteration of proposed antenna

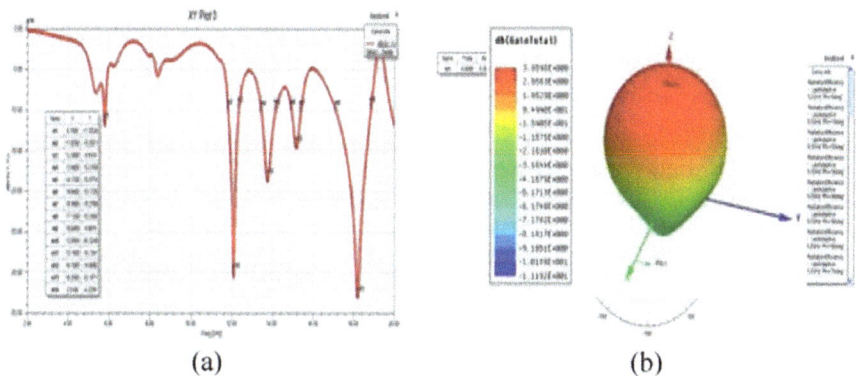

Fig. 10 **a** Return loss, **b** 3D polar plot of gain of third iteration of proposed antenna with CPW feeding

Table 2 Comparison of performance parameters of all proposed structures

S. No.	Performance parameter	Single patch with strip line feeding	Patch with first iteration with strip line feeding	Patch with second iteration with strip line feeding	Patch with third iteration with strip line feeding	Patch with third iteration with CPW feeding
1	Resonating frequency	6.8 GHz	7.85 GHz, 12.3 GHz	6 GHz, 12 GHz	12.25 GHz, 14 GHz, 18.22 GHz	12 GHz, 13.8 GHz, 15.16 GHz, 18.22 GHz
2	Operating bands	Single band	Dual band	Dual band	Triple band	Quad band
3	Radiation efficiency (%)	63	58	55.6	55	58.14
4	Gain (dB)	4.23	3.1	3.76	3.7	3.86
5	% of miniaturization (%)	–	41	21.4	15.58	15.58
6	Maximum return loss (dB)	−25	−16.1	−22	−32	−33
7	Bandwidth	200 MHz	310 MHz, 220 MHz	200 MHz, 400 MHz	350 MHz, 1740 MHz, 1720 MHz	540 MHz, 690 MHz, 440 MHz, 1710 MHz

References

1. Jeemon BK, Shambavi K, Alex ZC (2013) A multi-fractal planar antenna for wireless applications. In: International conference on communication and signal processing, 3–5 Apr 2013, India
2. Zhou Y. A novel, dual-band, miniaturized antenna with fractal-curve patch and TSV-CPW feeder. IEEE. 978-1-5090-4743-7©2016
3. Ramadhan FW, Ismail N, Lindra I (2019) Multi-band microstrip fractal antenna for S-band and C-band radar application. In: Telecommunication and computer engineering (ELTICOM-2019)
4. Naghar A, Aghzout O, Alejos A, Sanchez M, Naghar A (2014) Ultra wideband and tri-band antennas for satellite applications at C-, X-, and Ku bands. IEEE. 978-1-4799-7391-0/14
5. Novak MH, Volakis JL (2015) Ultrawideband antennas for multiband satellite communications at UHF–Ku frequencies. IEEE Trans Antennas Propag 63(4):1334–1341. https://doi.org/10.1109/TAP.2015.2390616
6. Zahra H, Rafique S, Fong W-L (2014) Feed analysis of tri-patch multiband antenna for satellite communication. In: Proceedings of ISAP 2014, Kaohsiung, Taiwan, 2–5 Dec 2014
7. Min K, Lee J-K (2015) Four band patch antenna by CPW feeding for in-building mobile communication. In: 2015 international workshop on antenna technology (iWAT)
8. Shinde PN, Shinde J, Pawar AK (2016) CPW feed electromagnetic Y shaped feeding technique for circular polarization & bandwidth enhancement. In: 2016 international conference on computing communication control and automation (ICCUBEA)
9. Abbak M, Janghi J, Kduman I (2012) Compact slot type CPW-fed ultra-wideband (UWB) antenna. In: 15 international symposium on antenna technology and applied electromagnetics, 2012

10. Anguera J, Puente C, Borja C, Soler J (2007) Dual-freq broadband-stacked microstrip antenna using a reactive loading and a fractal-shaped radiating edge. IEEE Antennas Wirel Propag Letters 6:309–312
11. Azari A (2011) A new fractal antenna for ultra wide- and multi-band applications. In: 2011 17th Asia-Pacific conference on communications (APCC), 2nd–5th Oct 2011 | Sutera Harbour Resort, Kota Kinabalu, Sabah, Malaysia
12. Lin S, Wu Z, Mao W, Liu C, Cai R, Chen L, Qiu J, Wang J (2010) The simulation of the multiband triangle fractal nesting printed monopoleantenna. In: 2010 international conference on microwave and millimeter wave technology

Chapter 74
A System for Generating Alerts for Allergic Outbreaks by Grasses

Antonio Sarasa-Cabezuelo⬵

1 Introduction

Every year between 30 and 40% of the world's population suffers from an allergic disease. The genetic makeup is the main factor that determines that a person presents this pathology. However, there are other factors that influence this situation. In particular, in the case of allergies due to pollen [18], it is also necessary to take into account the increase in temperature or atmospheric pollution that causes an increase in the concentration of pollen found in the atmosphere.

Pollen is a powder of different thicknesses produced by flower stamens, which is responsible for transporting sperm cells to the female reproductive system of other flowers. A single plant is capable of producing thousands upon thousands of pollen grains. Although not all pollinate at the same time, most do so between spring and summer, so the pollen level varies depending on the plant and the time of year. One of the most common and numerous plant families is grasses. It belongs to the order of the monocotyledons, which includes a large number of species used both for human consumption (sugar cane, wheat, rice, corn …) and for animals (pastures, grains …). This family consists of almost 700 genera and about 12,000 species. In this sense, it is estimated that grasses cover 20% of the world's surface.

Sensitivity to allergens found in pollen is called pollinosis [1]. This type of allergy produces asthma, rhinitis and respiratory problems in affected people, worsening their quality of life [12]. Likewise, from the point of view of public health, allergies in general pose an economic problem due to the expense of healthcare resources they produce as well as the sick leave that some of these people need to request.

A. Sarasa-Cabezuelo (✉)
Universidad Complutense de Madrid, Madrid, Spain
e-mail: asarasa@ucm.es

Calle Profesor José García Santesmases, 9, 28040 Madrid, Spain

© The Author(s), under exclusive license to Springer Nature Singapore Pte Ltd. 2021   995
S. Kumar et al. (eds.), *Proceedings of International Conference on Communication and Computational Technologies*, Algorithms for Intelligent Systems,
https://doi.org/10.1007/978-981-16-3246-4_75

This situation can be improved if preventive measures are taken. Since the allergy is associated with the level of pollen in the air, the most immediate preventive measure [19] is to avoid areas near plantations in the months of pollination. However, there are other factors [2] that influence pollen levels that must be taken into account such as air quality, temperature and probability of rain. The requirements to consider these factors are the availability of the information in real time and the availability of the necessary processing resources to analyse them [13]. Currently, both requirements can be satisfied given that the big data phenomenon has led to the development of technologies that allow large amounts of data to be processed quickly and efficiently [11]. In addition, in this same context, a set of initiatives called open data repositories [8] have emerged that aim to provide anyone who requires information on their activities so that they can be exploited. To do this, they offer different ways of accessing information from downloading the data in some data format such as csv, json, xml to the use of a REST API of web services from which queries and requests for information can be done concrete. In this sense, in many cities of the world, there are open data repositories with information on some of the factors that can influence increases in pollen levels in the air of a city. Therefore, it has the processing capacity and the necessary information to be able to implement applications that are capable of predicting outbreaks of allergies [10] due to high pollen concentrations in certain areas of a city [4].

This article presents a web application that implements a service to generate alerts for allergic outbreaks due to the concentration of grass pollen in the province of Madrid. Currently, in the Community of Madrid, there are around 95,959 ha of surface destined to the cultivation of cereals together with other species that grow spontaneously in parks and gardens or on the edge of crops. In this sense, the pollen generated by grasses is considered to be the main cause of pollinosis in the Community of Madrid [17] and in the rest of Spain [18]. To develop the application, a model has been created that allows making predictions by analysing the data retrieved from various open data sources: meteorological information, information on the composition of gases in the atmosphere and the level of pollen in the air. By combining these data, a first prediction is obtained at the beginning of the morning. This prediction then changes throughout the day taking into account the information that the users themselves send to the application [21]. The service offered to the user consists of the automatic sending of notifications about the possible high concentrations of pollen in their area as well as notifications about the allergic outbreaks that are recovered from the internet [22].

The article is structured as follows. Section 2 describes the prediction model that has been developed. Section 3 shows the web application and the implemented functionalities. In Sect. 4, the conclusions and possible lines of future work are presented.

2 Prediction Model

2.1 Background

The allergy outbreak prediction algorithm that has been developed takes into account the following factors [14]: the level of pollen concentration in the air, air quality and weather predictions.

With regard to air quality, it is known that the pollution produced by the particles emitted by the engines of diesel cars, modifies the nature of the pollens, making them more aggressive [15]. According to the Spanish Society of Allergology and Clinical Immunology (SEAIC), the different pollutants end up being deposited in the soil, affecting the natural development of seeds, roots and plants, altering their physiological characteristics [7] (e.g. it affects the pollen calendar of plants and causes them to the pollination period is lengthened) and making the pollens more allergic and potent. In addition, pollution affects the respiratory tract irritating the mucous membranes of the nose, pharynx and lungs [20], which aggravates the symptoms of allergy. This explains [6] why there are more allergies in cities than in the countryside, although the pollen levels are lower. In fact, the increase in the number of patients sensitized to this type of pollen has risen that has coincided with the progressive increase in the number of diesel cars (e.g. in Madrid, around 60% of the vehicles that circulate in Madrid are diesel, while twenty years ago, it was 25%). The algorithm takes into account the concentrations of the following types of pollutants [3]:

- Nitrogen dioxide (NO_2). It is formed as a by-product in high-temperature combustion processes, such as in motor vehicles and power plants. For this reason, it is a frequent pollutant in urban areas. It is the cause of harmful effects on health, especially on the respiratory system.
- Suspended particles < PM10. They are solid particles 10 thousandths of a millimetre in size that due to their size do not pass the body filters, remaining in the nose and throat. The combustion of fossil fuels is one of the main sources of these suspended particles in cities.
- Suspended particles < PM2.5. They are solid particles 2.5 thousandths of a millimetre in size that pass through the body tissues, depositing in the bronchioles, from where they cannot be extracted. A large proportion come from emissions from diesel vehicles.
- Carbon monoxide (CO). It is an intermediate product of the oxidation of a highly toxic hydrocarbon, which can cause death when inhaled at high levels. It is produced by the poor combustion of substances such as gasoline, kerosene, coal, oil, tobacco or wood.
- Ozone concentration (O_3). In the higher layers of the atmosphere, ozone serves as a solar filter and protects against high levels of ultraviolet radiation from the sun. However, at ground level, it can be detrimental. Its exposure may increase

susceptibility to respiratory allergens. Ozone levels are usually lower in polluted urban areas since it reacts with other pollutants such as NO.

- Sulphur dioxide (SO_2). It is a colourless, irritating gas with a pungent odour. He is considered one of those responsible for acid rain. Its exposure carries serious risks to health, as it passes directly into the circulatory system through the respiratory tract. The main source of emissions is the combustion of petroleum products and the burning of coal in power plants and heating plants.
- Nitrogen monoxide (NO). It is a colourless gas and hardly soluble in water. It is one of the pollutants in the atmosphere that is part of acid rain. It is generated, among others, by the explosion engines of cars and expelled into the atmosphere through the exhaust pipes.

The climate also influences pollen allergies [16], on rainy days and cloudy days without wind the symptoms decrease, since the rain cleans the atmosphere and the absence of wind prevents the displacement of pollen [5]. On the contrary, hot, dry and windy weather favours a greater distribution of pollen and, consequently, an increase in symptoms.

2.2 Prediction Model

In order to generate the predictions, the model takes into account three factors that influence the high concentrations of pollen in the air:

- Level of pollen concentration in the air. Pollen is found in the air practically throughout the year; however, the months with the highest incidence are from March to July, with weeks 21, 22 and 23 of the year standing out mainly [9]. In this sense, the pollen level is a seasonal data, and it is possible to define three different alert levels (low, medium and high) to classify each month of the year (Fig. 1). Each of these levels is associated with a numerical value (the low level is associated with the value 0, the medium level 1 and the high level 2), which will gradually increase or decrease depending on the average daily level of grains of pollen per cubic metre of air. This data is retrieved from the website of the Spanish Society of Allergology and Clinical Immunology (SEAIC) for the previous week. Using this data, the average of the week is obtained, and we associate a value to it; if it is less than 200 g/m^3, it is assigned the value 0; if it is between 200 and 1000 g/m^3, the value 1 and for quantities value 2 is assigned. In this way, if the level provided by the pollen calendar is higher than the one assigned taking into account the concentration of pollen in the air, the first is increased by 0.5, while if it is lower, it decreases the level calculated in the same proportion.
- Air quality. These data are retrieved from the air quality page of the Community of Madrid. For each of the localities, 0.1 has been added to the value of the level previously calculated each time any of the pollutants described exceed certain limit values. These limit values are those established by current legislation on air quality (Fig. 2a). Since there are no air quality sensors in all localities, then the

region has been divided into seven zones (Fig. 2b) and the values recorded for each pollutant by the sensors in the same area have been considered. Subsequently, its mean has been extrapolated to all the municipalities in that area.

- Weather forecasts. The meteorological data have been retrieved from the State Meteorological Agency (AEMET) page for each municipality in the next three days in each of the different municipalities of the Community of Madrid:

1. Wind speed (km/h): If the wind speed prediction gives a value higher than 30 km/h, add 0.3 to the calculated alert level since the wind favours the displacement of pollen.
2. Probability of precipitation: If the AEMET predicts a value greater than 30%, it is subtracted from the 0.2 level since the rain reduces the effects by cleaning the air.
3. Minimum relative humidity: With a value greater than 30%, the prediction is decreased by 0.1.
4. Maximum relative humidity: With a value higher than 70%, the calculated level is reduced by 0.1, since having higher humidity reduces allergy symptoms such as nasal congestion and itching.
5. Minimum temperature: Pollen concentrations tend to be higher on warm days. If the minimum temperature is higher than 20 °C, then 0.1 is added to the level.
6. Maximum temperature: If the maximum temperature is higher than 30 °C, then 0.1 is added to the level.

With this calculation, a numerical value is obtained for each municipality and for each of the next three days. This value is rounded and an alert level is assigned (Fig. 3). In this way, the initial prediction that is shown first thing in the morning is calculated.

Throughout the day, the initial prediction may vary depending on the reports that users make through the web. Each time a user makes a report, the alert level for that day is recalculated, and this is done using the following formula:

Fig. 1 Levels associated with the months of the year

MES	NIVEL
Enero	Bajo
Febrero	Bajo
Marzo	Medio
Abril	Alto
Mayo	Alto
Junio	Alto
Julio	Medio
Agosto	Bajo
Septiembre	Bajo
Octubre	Bajo
Noviembre	Bajo
Diciembre	Bajo

a)

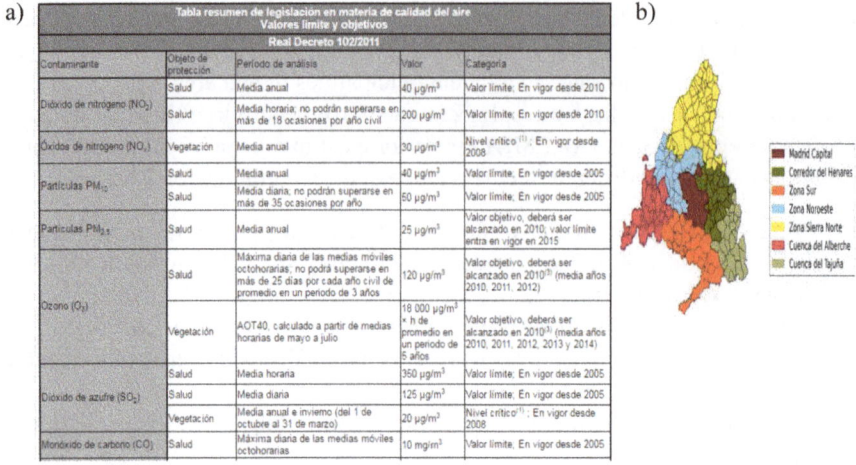

b)

Fig. 2   a Table with limits of pollution levels, **b** division of the territory of Madrid into zones

Fig. 3 Initial prediction
generated by the application

VALOR CALCULADO	NIVEL OBTENIDO
<1	BAJO
>1 y <2	MEDIO
>2	ALTO

$$newLevel = oldLevel + (alert\_level - round(oldLevel)) * 0.45$$

The old level is the existing level so far for that day and that municipality, and the alert_level is the one provided by the user in the report. In this way, if a level immediately higher than the one initially calculated is reported, two reports will be necessary to go to said higher level, but if the reported level is twice higher (initial level: Low, reported level: High), it will be updated at the intermediate level. The same will be done in the event that lower levels are reported. In this way, alert levels are recalculated each time a user makes a report in a municipality.

3   Implemented Web Application

The implemented web application has three main sections (Fig. 4):

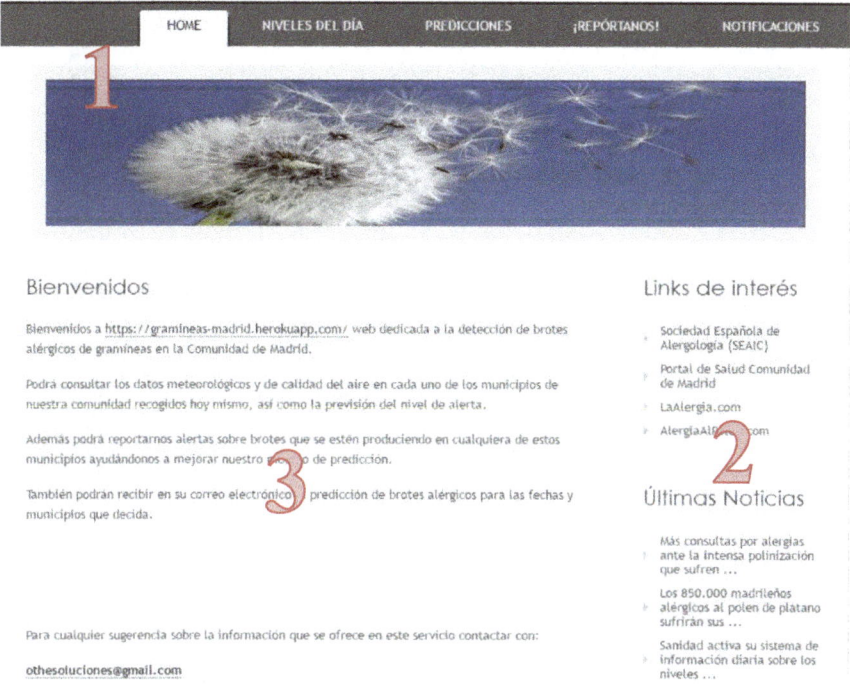

Fig. 4 Main page of the web application

- A header (1) with 5 tabs that allow access to the different contents of the application, which are displayed on a piece of paper located at the bottom of the application.
- A sidebar (2) that contains two sections: Links of interest and latest news. The first contains links to various websites related to pollen allergies: the Spanish Allergy Society, the Madrid Community Health Portal, LaAlergia.com and AlergiaAlPolen.com. And in the second section, ten of the news retrieved daily from the network are loaded using a personalized Google search engine that searches the main newspapers in the area.
- The content panel (3) of the application where the contents are displayed depending on the tab that has been selected in the header.

The content of each of the application's tabs will be described below.

3.1   Home Tab

In this tab, a presentation of the application is shown as content.

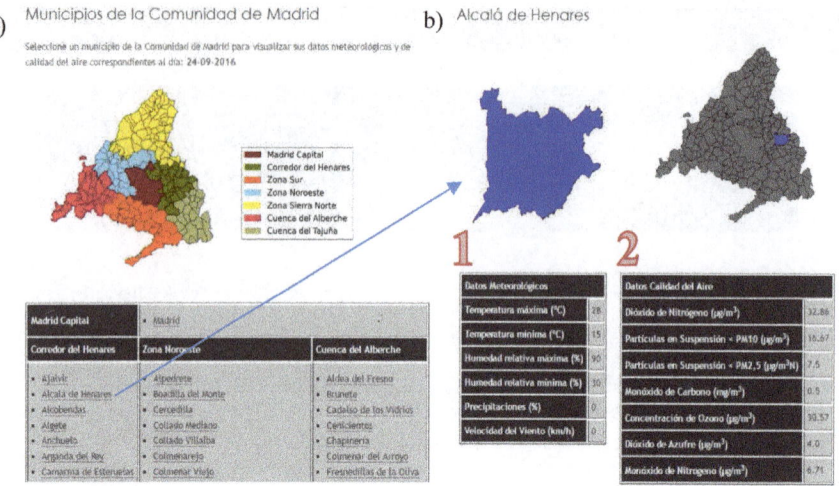

Fig. 5 **a** Geographical areas, **b** specific geographical area

3.2 Levels of the Day Tab

This tab shows a map of the Community of Madrid with seven geographic areas (Fig. 5a): Madrid Capital, Corredor del Henares, Zona Sur, Zona Noroes-te, Zona Sierra Norte, Cuenta del Alberche and Cuenca del Tajuña. For each area, a list of municipalities that compose it is shown, offering for each of them a link to a new screen (Fig. 5b) that draws a graphic with the silhouette of the municipality and next to its location highlighted within the map of the Community of Madrid. In addition, two tables are presented with the meteorological information (1) and air quality of the current day (2).

3.3 Predictions Tab

This tab is divided into three sections (Fig. 6a). The lower section (1) shows two tables with the five municipalities that have a higher alert level for the two days following the current one. The intermediate section (2) shows a table with the same information calculated together with a map of the Community of Madrid, whose municipalities are dynamically coloured according to the alert level. And finally, in the upper section (3), a form is shown with a combo that loads the name of all the municipalities of the Community, and when selecting one of them a screen is displayed (Fig. 6b) with its level of alert for the current day together with two maps, that of the municipality itself and that of its location in the Community, whose colour depends on the value of the alert level calculated for the current day. In addition, the alert level prediction (4) for the next two days also appears, loading equally, but with

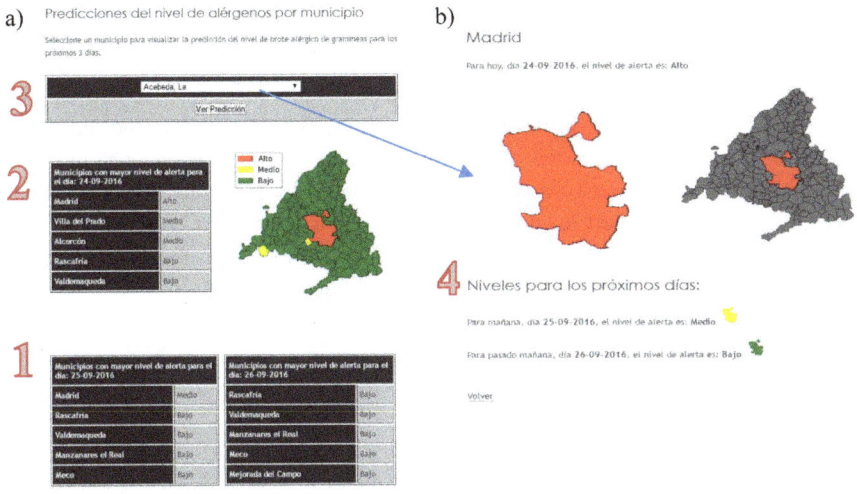

Fig. 6 **a** Predictions tab, **b** specific geographical area

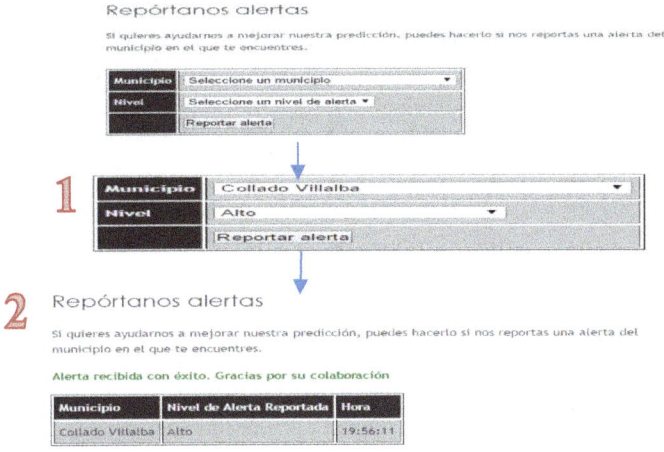

Fig. 7 "Reportamos" tab

a smaller format of the municipality map and with the colour corresponding to its alert level.

3.4 ¡Repórtanos! Tab

This tab shows a form (Fig. 7) that allows the user to enter an alert for a municipality in the Community of Madrid, together with a table with the alerts reported to the system during the day (if none have been reported, the displays a message with this information). In order to it, the user must select from a couple of combos (1) a municipality and an alert level, and click on the "Report Alert" button. Next, the municipality's alert level prediction will be updated and a confirmation message for the alert will be displayed.

3.5 Notifications Tab

This tab shows a form (Fig. 8) that allows a user to receive an email during the chosen period with the alert level of the selected municipality. If the information entered is correct, the subscription is registered in the system and a confirmation message is displayed on the screen (1). Thus, during the indicated period, an email (2) will be sent to the user with the information on the alert level of the municipality of the day, together with that of the two subsequent days and the date on which the subscription created will expire. In addition, if a user subscribes during the same period to the notifications of two or more municipalities, he will receive the information of all the municipalities in a single email.

4 Conclusions and Future Work

In this article, a web application has been presented that implements a service to generate alerts for possible allergic outbreaks due to high levels of pollen from grasses. Although the application is contextualized to the territory of the Community of Madrid, it could nevertheless be adapted to be used in any area.

In order to implement the functionality, a predictive model has been developed that combines information from three open data sources: meteorological information, information about the composition of gases in the atmosphere and levels of pollen in the air. The main functions offered by the application are the generation of alerts for users who subscribe to the alert system, and the possibility of navigating the information generated by the predictive model through different views of maps and tables that show the predictions generated by the system.

Regarding future lines of work, it is proposed:

- Development of a version of the application oriented towards mobile devices. An advantage of this version is that the mobile geolocation information could be used both to offer specific information to the area where the user is, as well as to report information on alert levels in a precise location.

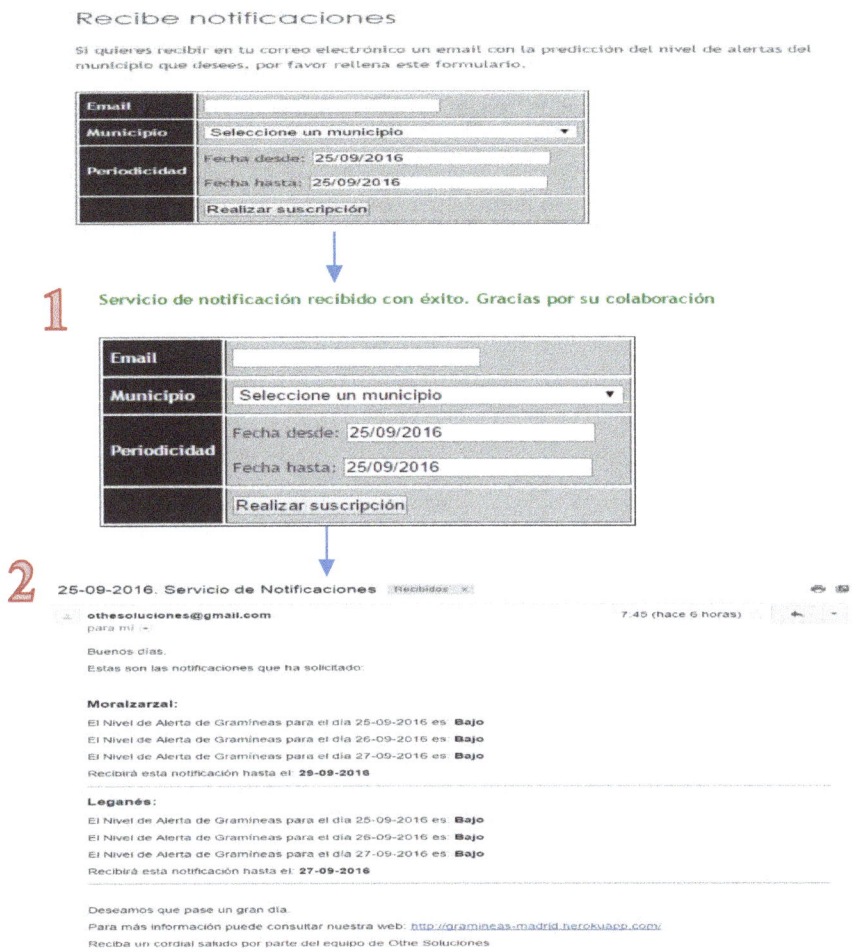

Fig. 8 Notifications tab

- Expansion of the coverage of prediction areas. Currently, the application only allows you to offer information about the Community of Madrid; however, it would be easy to consider other areas. The objective would be to carry out an implementation that is independent of the geographical area in which it is located. What would remain would be the developed prediction model.
- The application is particularized to pollen from grasses; however, there are allergies to other types of pollens such as Urticaceae, Alternaria or Cupresaceas. Information on these types of plants is also available, so it would be easy to extend the functionality of the application so that it is taken into account in the predictive model to obtain results about the pollen levels of these types of plants.

- Analysis of the information of the recovered data. The application retrieves news about allergies on a daily basis; however, this information is not used. The objective is to process the semantic information of the recovered news with the aim of incorporating it into the prediction model and influencing the predictions that are made.
- User management. Currently, the application only allows to subscribe to the notification system; however, no information about users is saved. It would be interesting for users to have an account where they can configure the favourite areas they want to be informed of or store a history of the predictions of the areas they have searched for in the application.

References

1. Aasbjerg K, Backer V, Lund G, Holm J, Nielsen NC, Holse M et al (2014) Immunological comparison of allergen immunotherapy tablet treatment and subcutaneous immunotherapy against grass allergy. Clin Exp Allergy 44(3):417–428
2. Andersson K, Lidholm J (2003) Characteristics and immunobiology of grass pollen allergens. Int Arch Allergy Immunol 130(2):87–107
3. Annesi-Maesano I, Rouve S, Desqueyroux H, Jankovski R, Klossek JM, Thibaudon M et al (2012) Grass pollen counts, air pollution levels and allergic rhinitis severity. Int Arch Allergy Immunol 158(4):397–404
4. Davies JM (2014) Grass pollen allergens globally: the contribution of subtropical grasses to burden of allergic respiratory diseases. Clin Exp Allergy 44(6):790–801
5. D'amato G, Cecchi L (2008) Effects of climate change on environmental factors in respiratory allergic diseases. Clin Exp Allergy 38(8):1264–1274
6. Galán I, Prieto A, Rubio M, Herrero T, Cervigón P, Cantero JL et al (2010) Association between airborne pollen and epidemic asthma in Madrid, Spain: a case-control study. Thorax 65(5):398–402
7. Jutel M, Jaeger L, Suck R, Meyer H, Fiebig H, Cromwell O (2005) Allergen-specific immunotherapy with recombinant grass pollen allergens. J Allergy Clin Immunol 116(3):608–613
8. Kitchin R, Collins S, Frost D (2015) Funding models for open access digital data repositories. Online Inf Rev 39:664–681
9. Marks GB, Colquhoun JR, Girgis ST, Koski MH, Treloar ABA, Hansen P et al (2001) Thunderstorm outflows preceding epidemics of asthma during spring and summer. Thorax 56(6):468–471
10. Mari A, Scala E, Palazzo P, Ridolfi S, Zennaro D, Carabella G (2006) Bioinformatics applied to allergy: allergen databases, from collecting sequence information to data integration. The Allergome platform as a model. Cell Immunol 244(2):97–100
11. Matricardi PM, Dramburg S, Alvarez-Perea A, Antolín-Amérigo D, Apfelbacher C, Atanaskovic-Markovic M et al (2020) The role of mobile health technologies in allergy care: an EAACI position paper. Allergy 75(2):259–272
12. Oh YC, Kim HA, Kang IJ, Cheong JT, Kim SW, Kook MH et al (2009) Evaluation of the relationship between pollen count and the outbreak of allergic diseases. Pediatr Allergy Respir Dis 19(4):354–364
13. Plasek JM, Goss FR, Lai KH, Lau JJ, Seger DL, Blumenthal KG et al (2016) Food entries in a large allergy data repository. J Am Med Inf Assoc 23(e1):e79–e87
14. Puc M (2003) Characterisation of pollen allergens. Ann Agric Environ Med 10(2):143–150

15. Riediker M, Monn C, Koller T, Stahel WA, Wüthrich B (2001) Air pollutants enhance rhinoconjunctivitis symptoms in pollen-allergic individuals. Ann Allergy Asthma Immunol 87(4):311–318
16. Subiza J, Masiello JM, Subiza JL, Jerez M, Hinojosa M, Subiza E (1992) Prediction of annual variations in atmospheric concentrations of grass pollen. A method based on meteorological factors and grain crop estimates. Clin Exp Allergy 22(5):540–546
17. Subiza J, Jerezb M, Jiméneza JA, Narganes MJ, Cabrera M, Varela S, Subiza E (1995) Allergenic pollen and pollinosis in Madrid. J Allergy Clin Immunol 96(1):15–23
18. Subiza J, Feo Brito F, Pola J, Moral A, Fernández J, Jerez M, Ferreiro M (1998) Pólenes alergénicos y polinosis en 12 ciudades españolas. Rev Esp Alergol Inmunol Clin 13(2):45–58
19. Taketomi EA, Sopelete MC, de Sousa Moreira PF, Vieira FDAM (2006) Pollen allergic disease: pollens and its major allergens. Braz J Otorhinolaryngol 72(4):562–567
20. Tang HH, Sly PD, Holt PG, Holt KE, Inouye M (2020) Systems biology and big data in asthma and allergy: recent discoveries and emerging challenges. Eur Respir J 55(1)
21. Taylor PE, Flagan RC, Valenta R, Glovsky MM (2002) Release of allergens as respirable aerosols: a link between grass pollen and asthma. J Allergy Clin Immunol 109(1):51–56
22. Zissler UM, Schmidt-Weber CB (2020) Predicting success of allergen-specific immunotherapy. Front Immunol 11:1826

Chapter 75
Estimation of Road Damage Contents for Earthquake Evacuation Guidance System

Yujiro Mihara, Rin Hirakawa, Hideaki Kawano, Kenichi Nakashi, Yukihiro Fukumoto, and Yoshihisa Nakatoh

1 Introduction

1.1 Current Situation of the Earthquake Disaster

Japan suffers from a variety of natural disasters every year. The number of earthquakes that occur in Japan is particularly high, and the human suffering caused by them is enormous. If we take past major earthquakes as an example, the Hanshin-Awaji earthquake and the Great East Japan Earthquake have caused severe damage, claiming tens of thousands of lives [1]. One of the reasons for the deaths of these victims is the delayed escape of secondary disasters such as fires and tsunamis. In addition, according to a survey conducted by the Cabinet Office after the Great East Japan Earthquake and Tsunami, many said that damage to the roads by themselves and debris on the roads were obstacles [2]. These obstacles have hindered evacuation, and the subsequent tsunami has left many people dead or missing. Therefore, measures are needed to prevent the delay in evacuation.

1.2 Measures Against Earthquake Damage in Japan

Governments and local governments are working on various arrangements and activities to reduce the victims of these disasters. For example, in Japan, we are focusing on earthquake prediction and strengthening of observation and surveying systems. In addition, we establish various laws for reinforcement, improvement of disaster

Y. Mihara · R. Hirakawa · H. Kawano · K. Nakashi · Y. Fukumoto · Y. Nakatoh (✉)
Kyushu Institute of Technology, 1-1 Sensuicho, Tobata, Kitakyushu City, Fukuoka Prefecture, Japan
e-mail: nakatoh@ecs.kyutech.ac.jp

© The Author(s), under exclusive license to Springer Nature Singapore Pte Ltd. 2021   1009
S. Kumar et al. (eds.), *Proceedings of International Conference on Communication and Computational Technologies*, Algorithms for Intelligent Systems,
https://doi.org/10.1007/978-981-16-3246-4_76

prevention facility in local government, we send out information about disaster prevention and crisis management in social networks such as websites or Twitter including evacuation information such as evacuation advice at the time of disaster doing [3]. In addition to hazard maps, we also disclose information on shelters, information on soils, and areas such as hazards predicted in the event of an earthquake. Some local governments are using Twitter to train information sharing among users at the time of evacuation [4]. The national and local governments are taking various measures to deal with the damage caused by the disaster. As part of these measures, there have been many attempts to use social networking sites for evacuation.

1.3 Means of Obtaining Information When the Earthquake Has Occurred

Media and disaster prevention reports can be cited as means of obtaining information in the event of an earthquake. However, in recent years, social network services (SNS) such as Twitter, Facebook, and mixi have attracted attention as a means of obtaining information in the event of an earthquake. They enable get or exchange quickly the information. Also it is excellent in confirming the safety of others. After the Great East Japan Earthquake, many people used SNS to obtain damage information [5]. Further, GPS information can be added to the posting of SNS, and pinpoint information can be transmitted. Research has also been conducted to show the effectiveness of using SNS in disaster evacuation [6]. The results of this study show that the use of social networking sites can be a great way to check on the safety of family and friends in times of disaster. Not only that it is also an excellent tool for exchanging information, such as sharing information about the situation in the corridor.

1.4 Related Study

An example of an evacuation guidance system at the time of a disaster is given [7]. The system uses mobile phones and smart don terminals to guide people along the best route to a predetermined shelter. The system also has a function to report road information with GPS information. The user can evaluate the condition of the passage in three levels: "Passable", "Obstacle", and "Impassable", and then post the information. Then, based on the submissions, the best evacuation route is determined. In this study, we considered that it is necessary to systematically evaluate the passageway situation as a means of understanding the situation of the passageway, considering the possibility of individual differences in the evaluation. In this study, we propose a system that automatically judges the passageway situation and provides evacuation guidance. A method for estimating the damage to the roadway for the necessary treatment portion of the system was also studied.

1.5 Research Purpose

From the above background, in order to minimize the number of victims of the earthquake, it was considered necessary to eliminate the delay in evacuation. For that it is important to choose the shortest route to the shelter. However, in the event of a disaster, there will be various road damages. Therefore, evacuation guidance system is necessary in consideration of those obstacles. However, to identify a safe passage required information on the road. We paid attention to the case of information provision by SNS and the high-real-time performance. Also we proposed a system for providing an optimum evacuation route to a user in linked with SNS. In addition, we have studied a systematic method to grasp the damage situation of the passage. In addition, a systematic method to grasp the damage situation of the passage will be studied.

2 Overview of the Proposed System

In this chapter, the outline of the proposed system in this research and the internal algorithm are described.

2.1 Internal Specification

The purpose of this system is to evacuate from a point outdoors to a predetermined evacuation site. The process of evacuation is shown in Fig. 1. First, when an earthquake occurs, the system obtains information on the earthquake and the necessity of evacuation from the meteorological agency and local government. If there is no need to evacuate in spite of an earthquake, a warning message is displayed. If it is necessary to evacuate, the shortest route to an evacuation site will be shown to the user. The system will then search for posts on social networking sites. If there is information about damage to a pathway in an evacuation route, the system avoids the pathway and suggests the safest and shortest route to the evacuation site. However, it may not always be possible to complete the evacuation by the safe route alone. For example, there are two shortest routes to the evacuation site, both of which are affected by the earthquake. In this case, you can use the safest route. However, in the case of a high emergency, such as during a tsunami, the choice of a detour route may delay the escape. Therefore, using the safe evacuation distance as a threshold, if the route is within the evacuation range, the safe detour is selected, and if it exceeds the evacuation distance, the damage content of the corridor is estimated based on the posted images. This method is discussed in the next section. Then, a relatively safe corridor is determined and selected as the evacuation route. These actions are shown in Fig. 2. Here, the evacuation possible distance is the evacuation possible

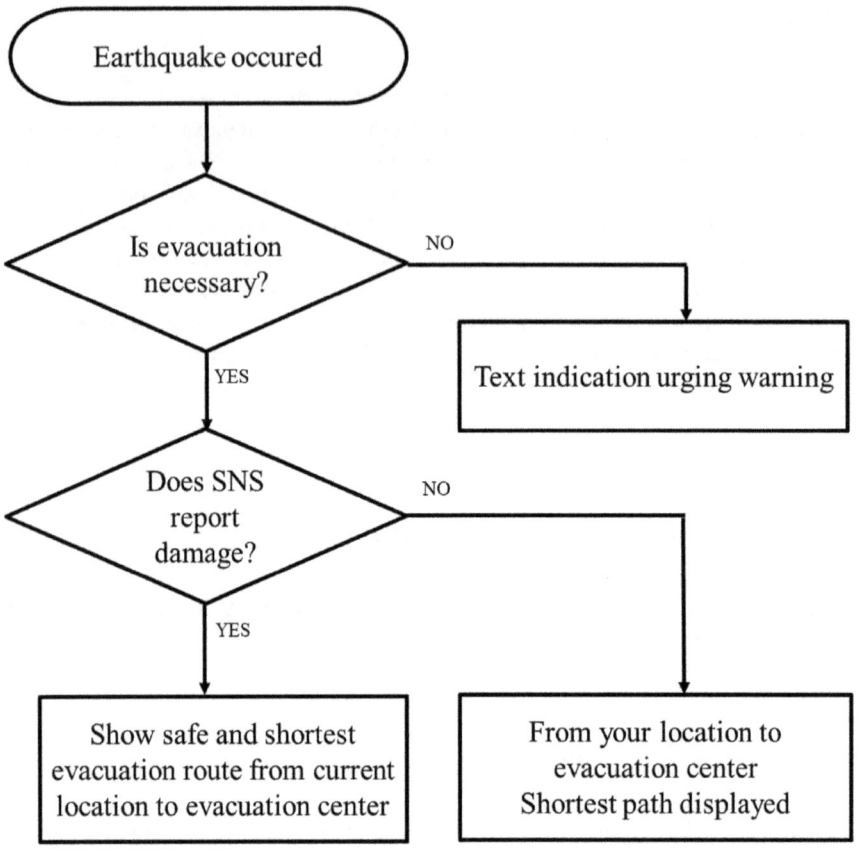

Fig. 1 Route decision flow considering passage damage

distance from the start of evacuation. Based on the arrival time of the tsunami and the general walking speed at the time of evacuation, the starting evacuation distance to the arrival time of the tsunami is calculated. If the evacuation route is chosen, it is determined if it is possible to evacuate based on this distance. If it is not possible to evacuate, the damaged corridor should be determined as the evacuation route, even if it is a little dangerous. The equation is based on the following definition [8], and these contents have been discussed in previous papers [9]

$$\text{Evacuable distance}(L) = V \times (T1 - T2 - T3) \tag{1}$$

where

V Walking speed.
$T1$ Tsunami arrival time.
$T2$ Evacuation start time.
$T3$ Time from evacuation start to current location.

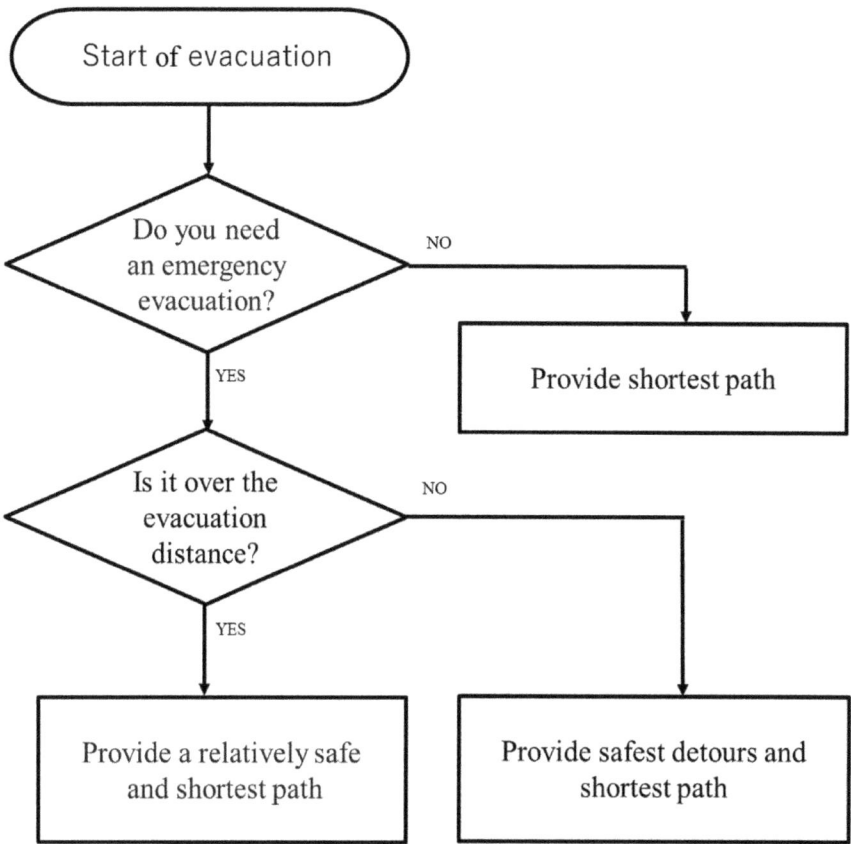

Fig. 2 Route decision flow considering evacuation distance

3 Estimation of Damage to Passage

In this chapter, we examine a method to estimate the damage caused by the disaster based on the corridor images acquired from SNS. They are used to determine whether a passage is appropriate as an escape route. First, the machine learning method and other technologies are described, and the details of the experiments are described.

3.1 Support Vector Machine (SVM)

Support vector machine (SVM) is a supervised learning method for two-class pattern identification. When a plurality of classes exist in the classification of SVM, the classification of the plurality of classes is made possible by using a method of "one-against-all" or "one-against-one". "One-against-all" performs identification in pairs.

So you create as many SVM models as there are classes. On the other hand, "one-against-one" identifies each class one-to-one. Therefore, all classes must be combined [10]. In this study, images were classified by "one-against-one". In this study, the linear function "Liner" was used.

3.2 Convolutional Neural Network (CNN)

A neural network is a mathematical model of neurons, the nerve cells of the human brain. The neural network is divided into three layers: "input layer," "hidden layer," and "output layer," and the nodes in each layer are connected by edges. In the image recognition field, there is a "convolutional layer" that forms a feature map by performing convolutional operations on kernels, especially in the "hidden layer". Then, a method consisting of a "pooling layer" that compresses the data size of the extracted feature map and an "all-combining layer" that performs classification is called a convolutional neural network. In this study, we used CNN for feature extraction and classification.

3.3 Random Forest

Random forest is one of the bagging methods in ensemble learning. In particular, a random forest consists of multiple decision trees, and each decision tree learns independently during the learning process. Each decision tree learns independently during the learning process. In this process, the training data is randomly sampled from the original training data, with some overlap allowed. The output of each decision tree is judged comprehensively to determine the final output [11].

3.4 XGBoost

In this method, multiple weak learning machines are used for learning, and weights are added to the models with low-prediction accuracy to improve the error. By combining these multiple models and making the final prediction as a single model, high accuracy can be achieved. In particular, XGBoost uses decision trees as weak learning machines as well as random forests. The error between the prediction output from the decision tree and the correct label is used as the objective function (loss function), and learning proceeds so that the objective function is minimized [12].

Table 1 Hyperparameters of VGG16

Parameter	Value
Butch size	256
Momentum	0.9
Weight decay	5e−04
Learning rate	1e−05
Epoch	74

3.5 Transfer Learning by VGG 16

The VGG 16 is a CNN model consisting of 16 layers learned in a large-scale image data set "ImageNet". Each parameter used for learning is shown in Table 1 [13]. In this study, we used a learned model of VGG 16 to extract features of training images. The outputs of the convolution layer and the pooling layer of the VGG 16 were used as the input of the classifier by the SVM, CNN, random forest, and XGBoost, as feature values. These feature vectors were trained to classify the images. In doing so, we used grid search to find the optimal parameters.

4 Experiment

The purpose of this research system is to estimate the damage content of the passageway based on SNS posts. The images used for learning were collected from damage reports (picture) posted on Twitter at the time of the actual disaster. It was defined on the basis of those images. Images labeled to those items were learned. Images with multiple defects were excluded. For the training data, we enhanced the data by cropping, rotating, and changing light and dark. These images were used as training images and verification images (Table 2), and ten separate images for each item were prepared, classified as test images, and evaluated. We used four classifiers, SVM, CNN, random forest, and XGBoost, to classify the data. The hyperparameters of four classifiers are shown Table 3. For learning, K-fold cross validation ($K = 5$) was performed, and learning was performed, while replacing training data with validation data, and the average value was calculated.

4.1 Evaluation Method

As a measure of the recognition system, the evaluation was carried out by fitting rate, reproduction rate, F-measure. Each is defined as follows (Formulas 2–4)

$$\text{Precision} = \frac{N1}{N2} \tag{2}$$

Table 2 Categories and number of images

Categories	Number of images	After data augmentation
Collapse of a house	104	728
Road crack	46	322
Collapse of a utility pole	43	301
Road depression	38	266
The inclination of a utility pole	56	273
Shredding of the road	41	392
Collapse of the concrete block	40	287
No damage	40	280

Table 3 Parameters used in the various methods

Model	Hyperparameter	Value
CNN	Optimizer	Adam
	Learning rate	0.0001
	Batch_size	64
SVM	Kernel	Linear
	C	1.0
Random Forest	Criterion	Gini
	N_Estimators	300
	Max_fratures	20
	Min_sumples_split	3.0
	Max_depth	25
	Randomstate	2525
XGBoost	Gamma	0
	Learning_rate	0.3
	Max_depth	3.0
	Subsample	0.8
	Colsample_bytree	1.0
	Colsample_bynode	1.0

$$\text{Recall} = \frac{N1}{N3} \tag{3}$$

$$F - \text{measure} = 2 \times \frac{\text{Recall} \times \text{Precision}}{\text{Recall} + \text{Precision}} \tag{4}$$

where

$N1$ Number of images correctly identified.
$N2$ Number of images identified as correct.
$N3$ Number of images in the correct class.

5 Result

When we compared the percentage of correct answers, CNN and XGBoost were equal, followed by SVM and random forest. Then, when we compared F-measure, which is the harmonic mean of precision and recall, we found that CNN, XGBoost, SVM, and random forest were higher in that order. Therefore, CNN was found to be the best method for damage assessment. Next, we will discuss the details of classification. The results of the classification of each item using each method are shown in Tables 4, 5, 6, and 7. The row is the correct label of the test image, and

Table 4 Classification results for each class (SVM)

Predict class

True class	Collapse of a house	Road crack	Collapse of a utility pole	Road depression	Collapse of the concrete block	The inclination of a utility pole	Shredding of the road	No damage
Collapse of a house	10	0	0	0	0	0	0	0
Road crack	0	6	0	1	0	0	0	3
Collapse of a utility pole	1	0	7	1	0	1	0	0
Road depression	0	2	0	8	0	0	0	0
Collapse of the concrete block	0	2	0	2	6	0	0	0
The inclination of a utility pole	0	0	0	0	0	10	0	0
Shredding of the road	0	0	0	0	0	0	9	1
No damage	0	0	0	0	1	1	2	6

Table 5 Classification results for each class (CNN)

Predict class

True class	Collapse of a house	Road crack	Collapse of a utility pole	Road depression	Collapse of the concrete block	The inclination of a utility pole	Shredding of the road	No damage
Collapse of a house	9	0	1	0	0	0	0	0
Road crack	0	9	0	0	1	0	0	0
Collapse of a utility pole	5	0	4	0	0	0	1	0
Road depression	0	0	0	8	0	2	0	0
Collapse of the concrete block	0	0	0	0	8	1	0	1
The inclination of a utility pole	0	0	0	0	1	8	1	0
Shredding of the road	0	0	0	0	0	0	10	0
No damage	0	0	0	0	2	1	0	7

the column is the prediction of the classification. The respective evaluation values are also shown in Tables 8, 9, 10, and 11. The results show that some categories can be classified with high accuracy by CNN. Overall classification accuracy averages around 79%. The $F1$-measure achieved an accuracy of about 78%. This indicates not only the correct classification but also the low number of omissions. It is also thought that they are able to build a well-balanced model. Here, if attention is paid to a misclassified item, for example, an item of "Road crack" is identified as "Road Depression". In addition, some of the items in "The inclination of a utility pole" are identified as "Collapse of a utility pole". Although these are classified incorrectly, they can be identified as roughly similar categories, and it is inferred that these features can be accurately extracted in the process of learning the passage image. The problem here, however, is the "No damage" item. It is possible that the proposed system in this study will post a passageway image with no damage to estimate the damage of the passageway. In this case, it is suggested that a detour route, which is a detour in spite of a safe passage, may be presented. In the implementation of

Table 6 Classification results for each class (Random Forest)

Predict class

	Collapse of a house	Road crack	Collapse of a utility pole	Road depression	Collapse of the concrete block	The inclination of a utility pole	Shredding of the road	No damage
Collapse of a house	10	0	0	0	0	0	0	0
Road crack	0	8	0	1	0	0	0	1
Collapse of a utility pole	1	0	7	0	0	1	1	0
Road depression	1	3	2	4	0	0	0	0
Collapse of the concrete block	0	1	0	0	8	0	1	0
The inclination of a utility pole	0	0	0	0	0	10	0	0
Shredding of the road	2	0	0	0	0	0	7	1
No damage	0	0	0	0	0	1	2	7

(True class, shown vertically at left margin)

this system, it will be important to extract the passage of "No damage" with high accuracy. Since there is a small number of data in the current learning, it is necessary to examine the method of data augmentation.

6 Conclusion

In this study, we proposed an evacuation guidance system based on pathway information and evacuation distance in order to control human casualties caused by the earthquake. In this study, we proposed an evacuation guidance system based on corridor information and evacuation distance. We classified the damage of the passageway into eight classes, and estimated them using transfer learning. As a result, CNN was the most accurate, with an accuracy of 79%. In future, we will need to expand the training data to improve the accuracy and add more classes to increase the versatility.

Table 7 Classification results for each class (XGBoost)

Predict class

True class	Collapse of a house	Road crack	Collapse of a utility pole	Road depression	Collapse of the concrete block	The inclination of a utility pole	Shredding of the road	No damage
Collapse of a house	10	0	0	0	0	0	0	0
Road crack	0	8	0	0	0	0	0	2
Collapse of a utility pole	0	0	8	0	0	1	1	0
Road depression	0	3	1	4	1	1	0	0
Collapse of the concrete block	0	0	0	0	9	0	1	0
The inclination of a utility pole	0	0	0	0	0	9	1	0
Shredding of the road	1	0	0	0	0	0	8	1
No damage	0	0	0	0	1	0	2	7

Table 8 Evaluation result (SVM)

Categories	Precision	Recall	F-measure	Accuracy
Collapse of a house	1.000	0.909	0.952	
Road crack	0.600	0.600	0.600	
Collapse of a utility pole	0.700	1.000	0.824	
Road depression	0.800	0.667	0.727	
The inclination of a utility pole	0.600	0.857	0.706	
Collapse of the concrete block	1.000	0.833	0.909	
Shredding of the road	0.900	0.818	0.857	
No damage	0.600	0.600	0.600	
Average	0.775	0.786	0.772	
				0.775

Table 9 Evaluation result (CNN)

Categories	Precision	Recall	F-measure	Accuracy
Collapse of a house	0.900	0.643	0.750	
Road crack	0.900	1.000	0.947	
Collapse of a utility pole	0.400	0.800	0.533	
Road depression	0.800	1.000	0.889	
The inclination of a utility pole	0.800	0.667	0.727	
Collapse of the concrete block	0.800	0.667	0.727	
Shredding of the road	1.000	0.833	0.909	
No damage	0.700	0.875	0.778	
Average	0.788	0.811	0.783	
				0.788

Table 10 Evaluation result (Random Forest)

Categories	Precision	Recall	F-measure	Accuracy
Collapse of a house	1.000	0.714	0.833	
Road crack	0.800	0.667	0.727	
Collapse of a utility pole	0.700	0.778	0.737	
Road depression	0.400	0.800	0.533	
The inclination of a utility pole	0.800	1.000	0.889	
Collapse of the concrete block	1.000	0.833	0.909	
Shredding of the road	0.700	0.636	0.667	
No damage	0.700	0.778	0.737	
Average	0.763	0.776	0.754	
				0.763

Table 11 Evaluation result (XGBoost)

Categories	Precision	Recall	F-measure	Accuracy
Collapse of a house	1.000	0.909	0.952	
Road crack	0.800	0.727	0.762	
Collapse of a utility pole	0.800	0.889	0.842	
Road depression	0.400	1.000	0.571	
The inclination of a utility pole	0.900	0.818	0.857	
Collapse of the concrete block	0.900	0.818	0.857	
Shredding of the road	0.800	0.615	0.696	
No damage	0.700	0.700	0.700	
Average	0.780	0.810	0.780	
				0.788

References

1. Appendix to the Cabinet Office's (2018) White paper on disaster prevention. Major natural disasters in our country since 1945
2. Expert Committee on Earthquake and Tsunami Countermeasures Based on the Lessons Learned from the Tohoku District-Off the Pacific Ocean Earthquake of the Cabinet Office. Main issues in future damage estimation based on the Great East Japan earthquake: responses to wide-area disasters caused by trench-type earthquakes, p 22
3. Sugiyama S (2017) Use of SNS in disaster response by local governments. SNS guidebook in disaster response. Information and Communications Technology (IT), Cabinet Secretariat, General Strategy Office, pp 289–295
4. Wakou city Homepage. http://www.city.wako.lg.jp/home/kurashi/bousai/_19053/_19047/_17017/_18386.html. Last accessed 2020/6/27
5. Ito S, Takizawa Y, Obu Y, Yonekura T (2011) Effective indomation sharing in the Great East Japan earthquake. The 26 Japan Society of Social Informatics National Congress
6. Niwa T, Okaya M, Takahashi T (2013) Effect of SNS communication in evacuation simulation. SIG-SAI
7. Ozawa S (2013) Development of a route guidance system for disaster evacuation research report commissioned by the Higashimikawa Regional Disaster Prevention Council
8. Ministry of Land, Infrastructure, Transport and Tourism (2014) Guidelines for building cities resistant to earthquakes and tsunamis
9. Mihara Y, Hirakawa R, Kawano H, Nakashi K, Nakatoh Y (2019) Study on evaluating risks of routes damages at earthquakes for evacuation guidance system, ACIT2019
10. Hsu C-W, Lin C-J (2002) A comparison of methods for multi-class support vector machines. IEEE Trans Neural Netw 13(2):415–425
11. Simonyan K, Zisserman A (2015) Very deep convolutional networks for large-scale image recognition (ICLR)
12. Breiman L (2001) Random forests. Mach Learn 45:5–32
13. Chen T, Guestrin C (2016) XGBoost: a scalable tree boosting system. arXiv:1603.02754v3

Author Index

Lightning Source UK Ltd.
Milton Keynes UK
UKHW051939290822
408042UK00007B/122